QL 391 .P7 S36 1986
Schmidt, Gerald D., 1934-
CRC handbook of tapeworm
 identification

6-13-89

# CRC Handbook of Tapeworm Identification

Author

**Gerald D. Schmidt, Ph.D.**

Department of Biological Sciences
University of Northern Colorado
Greeley, Colorado

CRC Press, Inc.
Boca Raton, Florida

**Library of Congress Cataloging in Publication Data**

Schmidt, Gerald D., 1934-
  Handbook of tapeworm identification.
  Bibliography: p.
  Includes index.
  1. Cestoda—Identification. I. Title.
QL391.P7S36   1986       595.1'21      84-7776
ISBN 0-8493-3280-X

This book represents information obtained from authentic and highly regarded sources. Reprinted material is quoted with permission, and sources are indicated. A wide variety of references are listed. Every reasonable effort has been made to give reliable data and information, but the author and the publisher cannot assume responsibility for the validity of all materials or for the consequences of their use.

All rights reserved. This book, or any parts thereof, may not be reproduced in any form without written consent from the publisher.

Direct all inquiries to CRC Press, Inc., 2000 Corporate Blvd., N.W., Boca Raton, Florida, 33431.

© 1986 by CRC Press, Inc.
Second Printing, 1986

International Standard Book Number 0-8493-3280-X

Library of Congress Card Number 84-7776
Printed in the United States

# PREFACE

When, in 1970, I published my little book, *How To Know The Tapeworms*,[3095] I intended to provide a "...readily available key to the tapeworms, equally useful for the beginning student and the professional parasitologist." The response from those very users was most gratifying, and I am satisfied that my intent was fulfilled. But the book is now out of print, as are Yamaguti's *Systema Helminthum*[4022] and Wardle and McLeod's *The Zoology of Tapeworms*,[3870] the only other keys to tapeworms of the world in the English language. It thus appeared that a new, updated aid to tapeworm identification was needed.

The present volume is intended to fill that void. Based largely on the keys and illustrations in my earlier book, it is expanded considerably by the addition of new descriptions that have appeared in the last 15 years. Further, I have included lists of all species known to me in each genus, together with their synonyms, hosts, and localities. Omissions and errors in such a compilation are inevitable. I made no attempt to obtain every record of cestodes in the world literature, which is nearly impossible, so there will be published host and locality records that are not included here. I am confident, however, that my lists are reasonably complete.

Another improvement over the earlier book is the addition of all synonyms of the taxa which are included. In this I have tried to be thorough, and if the reader cannot find a genus name in this book, as will occasionally happen, it is because I did not know of it, not because I ignored it.

I have attempted to ease the difficulties of tapeworm identification by constructing keys based on adult morphological features of the genera of tapeworms of the world. Every effort was made to keep the keys as simple as possible. Whenever feasible, I have not used cryptic characters, such as muscle bundle arrangement and nerve cord location, that necessitate sectioning or dissection of the specimen.

As before, I am well aware that controversy abounds in the systematics of cestodes and that the system I have chosen will not please all workers in the field. Yet, I feel if one can identify an unknown tapeworm to genus with a minimum of difficulty, he is then in a position to judge for himself the merit of higher categories and can select the scheme that best satisfies him.

In searching the world literature I have drawn upon hundreds of different journals and monographs. Especially useful have been *Helminthological Abstracts* and the *Index-Catalog of Medical and Veterinary Zoology*. I terminated the literature search March 12, 1983, with the realization that many species published before that date had not yet appeared in the abstracting journals. Indeed, a few probably never will.

The reader will notice that throughout the text the names of certain Soviet authors have slightly different spellings. There are differing systems of transliteration from the Cyrillic alphabet in use, and I usually followed the spelling as given in an abstract. Thus, Spassky—Spasskii; Mathevossian—Matevosyan; Yurpalova—Jurpalova; Spasskaya—Spasskaja; etc.

Included here are the names and references of 3,806 species, 601 genera, 66 families, 13 orders, and two subclasses. The Bibliography includes 4,103 references. There are 720 illustrations composing 513 figures.

I wish to thank my friends, Dr. Robert L. Rausch and Dr. John S. Mackiewicz, for their never-ending supplies of knowledge which they were always willing to share. Special thanks are offered to Miss Pauline M. Huber, who typed most of the manuscript and executed many of the drawings.

## THE AUTHOR

**Gerald D. Schmidt, Ph.D.,** is Professor of Zoology and Parasitology at the University of Northern Colorado, Greeley.

Professor Schmidt obtained his B.A. degree in Biology from Colorado State College, Greeley, in 1960, and his M.S. and Ph.D. degrees in Zoology at Colorado State University, Ft. Collins, in 1962 and 1964, respectively.

Professor Schmidt is a member of the American Society of Parasitologists, the American Society of Tropical Medicine and Hygiene, the American Microscopical Society, the Helminthological Society of Washington, the Wildlife Disease Association, and the International Filariasis Association, and is a Fellow of the Royal Society of Tropical Medicine and Hygiene.

Among his awards and honors, he is a member of Lambda Sigma Tau and Sigma Xi. He received the Distinguished Scholar Award and the Harrison Award for Outstanding Teaching from the University of Northern Colorado, and the Henry Baldwin Ward Medal from the American Society of Parasitologists in 1973. He was a NATO Senior Fellow in Science in Australia in 1969, and an OAS Research Fellow in Trinidad-Tobago in 1975.

Professor Schmidt has lectured and conducted research on helminths in the Caribbean, Central America, Mexico, New Zealand, and Oceania, and has consulted on helminth systematics with researchers in most regions of the world. He has published over 135 research papers and three previous books. Currently, he is Secretary-Treasurer of the American Society of Parasitologists.

# TABLE OF CONTENTS

Introduction .................................................................................... 1

General Morphology of Tapeworms ................................................ 2

Techniques of Study ........................................................................ 7

How to Use the Keys ..................................................................... 10

Keys to Subclasses of CESTOIDEA and Orders of Subclass EUCESTODA ........... 11

Order CARYOPHYLLIDEA ......................................................... 17
    Key to the Families in CARYOPHYLLIDEA ............................ 17
    Diagnosis of the Only Genus in BALANOTAENIIDAE ............ 17
    Key to the Genera in LYTOCESTIDAE ................................... 18
    Key to the Genera in CARYOPHYLLAEIDAE ......................... 26
    Key to the Genera in CAPINGENTIDAE .................................. 37

Order SPATHEBOTHRIIDEA ....................................................... 43
    Key to the Families in SPATHEBOTHRIIDEA ......................... 43
    Diagnosis of the Only Genus in SPATHEBOTHRIIDAE ........... 43
    Diagnosis of the Only Genus in CYATHOCEPHALIDAE ......... 44
    Key to the Genera in BOTHRIMONIDAE Fam. N. .................. 45

Order TRYPANORHYNCHA ........................................................ 47
    Key to the Suborders in TRYPANORHYNCHA ....................... 48
    Key to the Families in the Suborder ACYSTIDEA ..................... 48
    Key to the Families in the Suborder CYSTIDEA ...................... 49
        Key to the Genera in DASYRHYNCHIDAE ....................... 52
        Key to the Genera in EUTETRARHYNCHIDAE ................. 55
        Key to the Genera in GILQUINIIDAE ................................ 59
        Key to the Genera in GYMNORHYNCHIDAE ................... 60
        Diagnosis of the Only Genus in HEPATOXYLIDAE ........... 62
        Diagnosis of the Only Genus in HORNELLIELLIDAE ........ 63
        Key to the Genera in LACISTORHYNCHIDAE .................. 63
        Key to the Genera in PARANYBELINIIDAE ...................... 66
        Key to the Genera in OTOBOTHRIIDAE ........................... 67
        Key to the Genera in PTEROBOTHRIIDAE ....................... 69
        Diagnosis of the Only Genus in RHINOPTERICOLIDAE ... 71
        Diagnosis of the Only Genus in SPHYRIOCEPHALIDAE ... 71
        Key to the Genera in TENTACULARIIDAE ....................... 72
        Diagnosis of the Only Genus in MIXODIGMATIDAE ........ 74
        Trypanorhynch Genera of Uncertain Status .......................... 75
        Trypanorhynch Species of Uncertain Status ......................... 76

Order PSEUDOPHYLLIDEA ........................................................ 81
    Key to the Families in PSEUDOPHYLLIDEA ........................... 81
    Key to the Genera in CEPHALOCHLAMYDIDAE .................... 84
    Diagnosis of the Only Genus in HAPLOBOTHRIIDAE ............ 84
    Key to the Genera in DIPHYLLOBOTHRIIDAE ...................... 85
        Diphyllobothriid Species of Uncertain Status ....................... 95

    Key to the Genera in PTYCHOBOTHRIIDAE ........................................ 95
    Key to the Genera in BOTHRIOCEPHALIDAE ..................................... 100
    Key to the Genera in ECHINOPHALLIDAE ........................................ 105
    Family AMPHICOTYLIDAE ............................................................. 106
        Key to the Subfamilies in AMPHICOTYLIDAE ................................. 106
        Key to the Genera in AMPHICOTYLINAE ........................................ 106
        Key to the Genera in ABOTHRIINAE ............................................... 108
        Diagnosis of the Only Genus in BOTHRIOCOTYLINAE ......................... 109
        Key to the Genera in MARSIPOMETRINAE ...................................... 109
    Key to the Genera in PHILOBYTHIIDAE ............................................. 110
    Key to the Genera in TRIAENOPHORIDAE .......................................... 112
    Key to the Genera in PARABOTHRIOCEPHALIDAE ............................... 115

Order LECANICEPHALIDEA ................................................................. 119
    Key to the Families in LECANICEPHALIDEA ........................................ 119
    Diagnosis of the Only Genus in BALANOBOTHRIIDAE ........................... 120
    Diagnosis of the Only Genus in DISCULICEPITIDAE .............................. 120
    Key to the Genera in LECANICEPHALIDAE ........................................ 121
    Diagnosis of the Only Genus in ADELOBOTHRIIDAE ............................. 127

Order APORIDEA ............................................................................. 129
    Key to the Genera in NEMATOPARATAENIIDAE ................................. 129

Order TETRAPHYLLIDEA ................................................................... 131
    Key to the Families in TETRAPHYLLIDEA .......................................... 131
    Diagnosis of the Only Genus in CATHETOCEPHALIDAE ......................... 132
    Key to the Genera in ONCOBOTHRIIDAE ........................................... 132
    Key to the Genera in TRILOCULARIIDAE ........................................... 144
    Key to the Genera in PHYLLOBOTHRIIDAE ........................................ 145
    TETRAPHYLLIDEA of Doubtful or Uncertain Status .............................. 163

Order DIPHYLLIDEA ......................................................................... 165
    Key to the Families in DIPHYLLIDEA ................................................ 165
    Diagnosis of the Only Genus in ECHINOBOTHRIIDAE ........................... 165
    Diagnosis of the Only Genus in DITRACHYBOTHRIDIIDAE ..................... 166

Order LITOBOTHRIDEA ..................................................................... 169
    Diagnosis of the Only Family in LITOBOTHRIDEA ............................... 169
    Key to the Genera in LITOBOTHRIDAE ............................................. 169

Order NIPPOTAENIIDEA .................................................................... 171
    Diagnosis of the Only Family in NIPPOTAENIIDEA .............................. 171
    Key to the Genera in NIPPOTAENIIDAE ............................................ 171

Order PROTEOCEPHALIDEA ............................................................... 173
    Key to the Families in PROTEOCEPHALIDEA ..................................... 173
    Family PROTEOCEPHALIDAE .......................................................... 173
        Key to the Subfamilies in PROTEOCEPHALIDAE ............................. 173
        Key to the Genera in GANGESIINAE ............................................ 173
        Diagnosis of the Only Genus in PROSOBOTHRIINAE ........................ 176
        Diagnosis of the Only Genus in SANDONELLINAE ........................... 177

Key to the Genera in CORALLOBOTHRIINAE ................................. 177
Key to the Genera in ACANTHOTAENIINAE ................................. 180
Key to the Genera in PROTEOCEPHALINAE.................................. 182
Diagnosis of the Only Genus in MARSIPOCEPHALINAE ...................... 190
Family MONTICELLIDAE................................................... 190
Key to the Subfamilies in MONTICELLIDAE ............................... 190
Key to the Genera in ZYGOBOTHRIINAE ................................... 191
Key to the Genera in ENDORCHIINAE ..................................... 193
Diagnosis of the Only Genus in MONTICELLIINAE ......................... 194
Diagnosis of the Only Genus in EPHEDROCEPHALINAE ...................... 195
Diagnosis of the Only Genus in PELTIDOCOTYLINAE ....................... 195
Diagnosis of the Only Genus in RUDOLPHIELLINAE ........................ 196

Order DIOECOTAENIIDEA Ord. N. .......................................... 197
Diagnosis of the Only Family in DIOECOTAENIIDEA ....................... 197
Diagnosis of the Only Genus in DIOECOTAENIIDEA ........................ 197

Order CYCLOPHYLLIDEA ................................................... 199
Key to the Families in CYCLOPHYLLIDEA ................................. 199
Family MESOCESTOIDIDAE................................................. 202
Key to the Subfamilies in MESOCESTOIDIDAE ............................. 202
Diagnosis of the Only Genus in MESOCESTOIDINAE........................ 202
Diagnosis of the Only Genus in MESOGYNINAE ............................ 204
Family TETRABOTHRIIDAE................................................. 204
Key to the Genera in TETRABOTHRIIDAE .................................. 204
Key to the Genera in NEMATOTAENIIDAE................................... 209
Key to the Genera in DIOECOCESTIDAE ................................... 213
Key to the Genera in PROGYNOTAENIIDAE ................................. 217
Key to the Genera in TAENIIDAE ........................................ 221
Key to the Genera in AMABILIIDAE....................................... 227
Key to the Genera in ACOLEIDAE ........................................ 230
Key to the Genera in CATENOTAENIIDAE .................................. 232

Family DAVAINEIDAE..................................................... 237
Key to the Subfamilies in DAVAINEIDAE.................................. 237
Key to the Genera in IDIOGENINAE ...................................... 237
Key to the Genera in OPHRYOCOTYLINAE .................................. 241
Key to the Genera in DAVAINEINAE ...................................... 245
Key to the Subgenera in RAILLIETINA ................................. 252
Species in *Raillietina (Paroniella)* Fuhrmann 1920 ................. 253
Species in *Raillietina (Skrjabinia)* Fuhrmann 1920 ................. 255
Species in *Raillietina (Raillietina)* Fuhrmann 1920................. 256
Species in *Raillietina (Fuhrmannetta)* Stiles et Orleman 1926....... 265
*Raillietina* Species of Unknown Subgeneric Status .................. 266

Family HYMENOLEPIDIDAE ................................................ 267
Key to the Subfamilies in HYMENOLEPIDIDAE.............................. 267
Key to the Genera in FIMBRIARIINAE..................................... 267
Key to the Genera in PSEUDHYMENOLEPIDINAE.............................. 269
Key to the Only Genus in ECHINORHYNCHOTAENIINAE .................... 272
Diagnosis of the Only Genus in DITESTOLEPIDINAE ..................... 272

Key to the Genera in HYMENOLEPIDINAE ................................... 272
Appendix to HYMENOLEPIDINAE from Mammals ........................... 334
Appendix to HYMENOLEPIDINAE from Birds ................................ 335
Addendum: *Cloacotaenia* Wolffhügel 1938 ...................................... 337

Family DILEPIDIDAE ................................................................. 339
Key to the Subfamilies in DILEPIDIDAE ......................................... 339
Key to the Genera in PARUTERININAE ......................................... 339
Key to the Genera in DIPYLIDIINAE ............................................. 352
Key to the Genera in DILEPIDINAE .............................................. 368
Addendum: *Cladotaenia* Cohn 1901 ............................................... 412
Genera in DILEPIDIDAE of Uncertain Status ................................... 413

Family ANOPLOCEPHALIDAE ..................................................... 415
Key to the Subfamilies of ANOPLOCEPHALIDAE ............................. 415
Diagnosis of the Only Genus in TRIPLOTAENIINAE .......................... 415
Key to the Genera in THYSANOSOMATINAE .................................. 416
Key to the Genera in ANOPLOCEPHALINAE .................................. 421
Key to the Genera in LINSTOWIINAE ........................................... 450
Key to the Genera in INERMICAPSIFERINAE .................................. 463

Subclass CESTODARIA ............................................................... 467
Key to the Orders in Subclass CESTODARIA ................................... 467
Order AMPHILINIDEA ............................................................... 467
Key to the Families in AMPHILINIDEA .......................................... 467
Key to the Genera in AMPHILINIDAE ........................................... 468
Key to the Genera in AUSTRAMPHILINIDAE .................................. 471
Order GYROCOTYLIDEA ............................................................ 472
Diagnosis of the Only Family in GYROCOTYLIDEA .......................... 472
Key to the Genera in GYROCOTYLIDAE ....................................... 473

Glossary .................................................................................. 475

Bibliography ............................................................................ 479

Index ..................................................................................... 607

# INTRODUCTION

Tapeworms hold an established and well-recognized place in the hierarchy of the animal kingdom. Their acoelous nature and bilateral symmetry, together with well-organized organ systems, place them in the Phylum Platyhelminthes, along with the Trematoda and Turbellaria. Adaptations to endoparasitism have resulted in complete loss of a digestive system and an increase in reproductive capacity that often staggers the imagination. The unique character of this group of organisms was recognized by Rudolphi, who, in 1809, proposed the class Cestoidea to contain them. The concept still is favored today, although some authorities prefer the name Cestoda.

Nearly every species of vertebrate examined is shown to be host to one or more species of tapeworms. Since there are about 60,000 species of vertebrates, and fewer than 4,000 species of cestodes have been described, it follows that an immense number of species is yet to be found. Unfortunately, comparatively few zoologists are working in the systematics of tapeworms today, partly due to the absence or inaccessability of indentification keys. It is hoped that the present book will partly alleviate this problem.

Tapeworms have long excited in man a sense of bewilderment, and sometimes fear, because they seem to appear spontaneously within a host and, when present, occasionally cause disease. In the last two centuries, even after most scientists accepted the idea that bacteria and other microorganisms were not spontaneously generated, many refused to believe that intestinal worms did not appear *de novo*. How else could they be explained? The pioneering works of Siebold, Küchenmeister, Leuckart, Villot, Braun, and others dispelled superstition forever from scientific thought on these forms and laid the foundation for the modern science of cestodology.

A vast literature has accumulated through the years, describing such aspects as morphology, taxonomy, life cycles, physiology, pathogenesis, and host-relationships of tapeworms. Even so, much remains unknown. In all phases of cestodology, we are only on the threshold. This is indeed a fruitful area for research, and there is much promise that exciting discoveries will be made in the years to come.

## GENERAL MORPHOLOGY OF TAPEWORMS

Although considerable variation of morphology occurs between different orders of tapeworms, there are underlying similarities that unite the orders into the class Cestoda. An understanding of tapeworm anatomy is essential for successful utilization of the keys. The following generalized description is supplemented within the text of this book, especially where specialization has modified the basic pattern.

Tapeworms (Figure 1) usually consist of a chain of segments called *proglottids,* each of which contains one or more sets of reproductive organs. The proglottids are continuously produced near the anterior end of the animal by a process of asexual budding. Each bud moves toward the posterior end as a new one takes its place, and during the process becomes sexually mature. The gravid or senile terminal segments detach or disintegrate. The entire body thus formed is called a *strobila,* and a segmented strobila is said to be *polyzoic.* In some groups the body consists of a single segment, and is then said to be *monozoic.* If each proglottid overlaps the following one, the strobila is said to be *craspedote;* if not, it is called *acraspedote* (Figure 2).

Often between the scolex and the first segments of the strobila there is a smooth, undifferentiated zone called the *neck.* This may be long or short, or absent altogether. The neck, or in its absence the posterior part of the scolex, contains germinal cells that have the potential for budding off the segments, a process called *strobilization.*

At the anterior end is usually found a holdfast organ or *scolex* that is the principal means of locomotion of these animals. Depending on the group, the scolex may be provided with suckers, grooves, hooks, spines, glandular areas, or combinations of these. In some instances the scolex is quite simple, lacking any of these specializations, or it may be absent altogether. In a few species it is normal for the scolex to be lost and replaced in function by a distortion of the anterior end of the strobila. The organ thus produced is called a *pseudoscolex.* A few species are capable of penetrating into the gut wall of the host where the scolex, and often a considerable length of strobila, are encapsulated by host reactions, while the remainder of the strobila dangles into the lumen of the gut.

Since the taxonomy of tapeworms is based primarily upon the anatomy of the organ systems, an understanding of these systems is essential.

**Organ Systems**

*Nervous System*

The nervous system appears to be a modified ladder-type, with a longitudinal cord near each lateral margin and transverse commissures in each segment. The two lateral cords are united in the scolex in a complex arrangement of ganglia and commissures. The nervous system is rarely used as a taxonomic character, although the lateral cords are convenient points of reference for the location of other structures.

*Osmoregulatory System*

As in other groups of Platyhelminthes, the organ of osmoregulation is the *protonephridium,* or *flame cell* (Figure 3). These unicellular glands remove excess fluid from the body parenchyma and discharge it from the body by a series of collecting tubules. The largest of these tubules are called the *osmoregulatory canals* (Figure 4) and are typically of two pairs, one ventrolateral and the other (usually smaller) dorsolateral on each side. These canals may be independent throughout the strobila or may ramify and anastomose in each proglottid. Commonly, a transverse canal near the posterior margin of each segment unites the ventral canals while the dorsal canals remain simple. The dorsal and ventral canals join in the scolex, usually in association with complex branching. Posteriorly, the two pairs of canals unite into an excretory bladder with a single pore. In polyzoic species this bladder is lost with

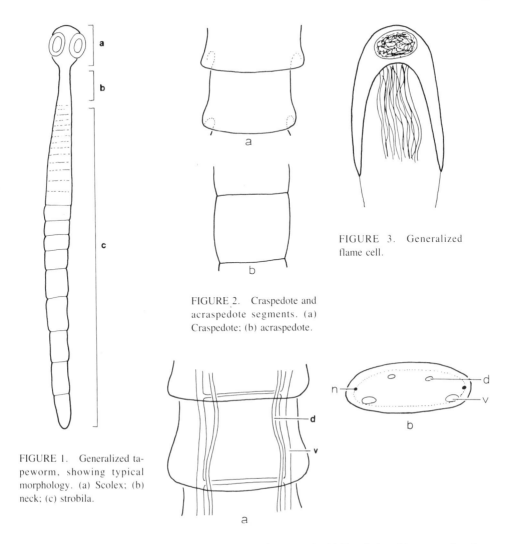

FIGURE 1. Generalized tapeworm, showing typical morphology. (a) Scolex; (b) neck; (c) strobila.

FIGURE 2. Craspedote and acraspedote segments. (a) Craspedote; (b) acraspedote.

FIGURE 3. Generalized flame cell.

FIGURE 4. Osmoregulatory canals. (a) Dorsal view; (b) cross section: d = dorsal vessel; n = lateral nerve cord; v = ventral vessel.

the detachment of the terminal proglottid, and thereafter the canals empty independently at the end of the strobila. In a few instances the major canals also empty through short, lateral ducts.

The major function of the osmoregulatory system seems to be water balance, but some excretion of metabolic wastes also probably occurs. The dorsal canals carry fluid toward the scolex and the ventral canals carry it toward the posterior end. Occasionally, the dorsal canals are absent. The arrangement of major canals is of taxonomic importance.

*Muscular System*

Most tapeworms possess well-defined, longitudinal bundles of muscle fibers and scattered dorsoventral fibers. The scolex is well supplied with muscle fibers, making it extraordinarily motile. In the strobila, the longitudinal muscle bundles often are arranged in a definite layer within the parenchyma, dividing it into a well-defined cortex and medulla (Figure 5). The arrangement of these muscles is of taxonomic importance, but since sectioning of the specimen is usually necessary to observe them, their use is omitted from the keys whenever possible.

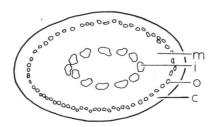

FIGURE 5. Muscle bundle arrangement: c = cortex; i = inner muscle bundle; m = medulla; o = outer muscle bundle.

*Reproductive Systems*

All known tapeworms are *monoecious*, or *hermaphroditic*, with the exception of a few species from birds and two from a stingray, which are *dioecious* or *gonochoristic*. Most commonly, each proglottid contains one complete set each of male and female reproductive organs, although a few species have two complete sets in each segment. A few rare species in birds have one female and two male sets in each proglottid.

As the segment moves toward the rear of the strobila, as described above, the reproductive organs mature and embryonated eggs are formed. Most commonly, the male organs mature first and produce sperm, which are stored until maturation of the ovary. Early maturation of the testes is called *protandry or androgyny*, and is used as a taxonomic character. In fewer species the ovaries mature first, a condition known as *protogyny* or *gynandry*. This too, is used as a taxonomic character.

*Male Reproductive System*

The *male reproductive system* (Figure 6) consists of one to many *testes*, each of which has a fine *vas efferens*. The vasa efferentia unite into a common *vas deferens* which drains the sperm toward the genital pore. The vas deferens may dilate into a spheroid *external seminal vesicle* or it may be highly convoluted, the convolutions functioning in sperm storage, or it may be quite simple. Eventually, the vas deferens leads into a *cirrus pouch*, which is a muscular sheath containing the terminal portion of the male system. Inside the cirrus pouch the vas deferens may form a convoluted ejaculatory duct or swell into an *internal seminal vesicle*. Distally, the duct is modified into a muscular cirrus, the male copulatory organ. The cirrus may be spinous or not and varies considerably in size between species. The cirrus can invaginate into the cirrus pouch and evaginate through the cirrus pore. Often, the male and female genital pores open into a common sunken chamber, the *genital atrium*. This atrium may be simple, or armed with a variety of spines or stylets, or may be glandular or possess accessory pockets. The cirrus pore or the atrial pore may open on the margin or somewhere on a flat surface of the proglottid.

*Female Reproductive System*

The *female reproductive system* (Figure 7) consists of a single *ovary*, which may be large or small, compact or diffuse, and may be located anywhere within the proglottid, depending on the genus. Associated with the ovary are *vitelline cells*, or *vitellaria*, which contribute to eggshell formation and nutrition for the developing embryo. These may be in a single compact vitellarium or scattered as follicles in various patterns. As ova mature they leave the ovary through a single *oviduct* that may have a controlling sphincter, the *ovicapt*. Fertilization usually occurs in the proximal oviduct. Cells from the vitelline glands pass through a common *vitelline duct*, sometimes equipped with a *small vitelline reservoir*, and join with the zygote. Together they pass into a zone of the oviduct surrounded by unicellular

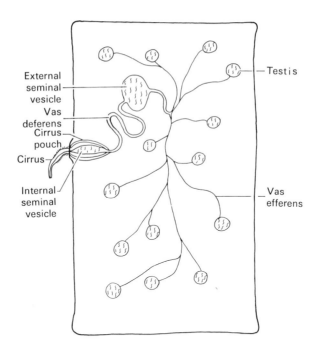

FIGURE 6. Generalized proglottid, showing only male organs.

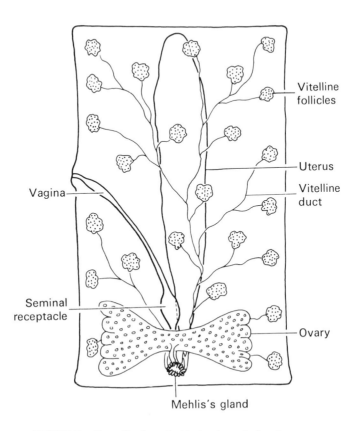

FIGURE 7. Generalized proglottid, showing only female organs.

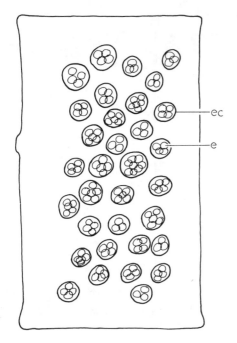 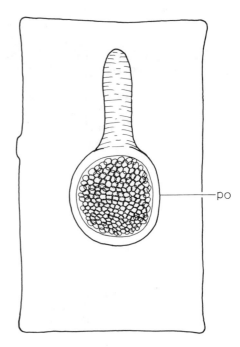

FIGURE 8.  Generalized proglottid with egg capsules: e = Egg; ec = egg capsule.

FIGURE 9.  Generalized proglottid with paruterine organ: po = paruterine organ.

glands called *Mehlis' glands*. The lumen of this zone is known as the *ootype*. The Mehlis' glands secrete a very thin membrane around the zygote and associated vitelline cells. Eggshell formation is then completed from within by the vitelline cells. Leaving the ootype, the embryonating egg passes into the *uterus* where embryonation is completed.

The form of the uterus varies considerably between groups. It may be a simple or convoluted tube, a reticular, lobated or simple sac, or may be replaced by other structures. In some groups the uterus disappears and the eggs, either singly or in groups, are enclosed within hyaline *egg capsules* imbedded within the parenchyma (Figure 8). In other groups one or more fibro-muscular structures, the *paruterine organs,* form attached to the uterus. In this case the eggs pass from the uterus into the paruterine organs, which assume the function of a uterus (Figure 9). The uterus then usually degenerates. Eggs are released from the worm through a preformed *uterine pore* in many groups. In others, the proglottid splits or fragments, releasing the eggs. In many *apolytic* species the gravid proglottids detach from the strobila and are passed from the host, where they crawl about on feces or soil, scattering eggs as they go. In most *anapolytic* species the eggs are first discharged, then the senile segments are released, either singly or in chains.

The female genital pore, the *vaginal pore,* usually opens near the cirrus pore. The vagina may be armed distally with minute spines and may have one or more sphincters along its length. Near the proximal end there is usually a dilation called the *seminal receptacle* that stores sperm received in copulation. From the seminal receptacle a duct continues into the ootype.

A basic understanding of the general anatomy of tapeworms is essential for successful utilization of the keys. Elucidation of anatomical details of a specimen is dependent upon correct techniques of preparation properly used. A brief discussion of techniques is found in the chapter "Techniques of Study" which follows.

## TECHNIQUES OF STUDY

Numerous references are available that expand the following brief account. A very useful manual is Pritchard and Kruse's *The Collection and Preservation of Animal Parasites*.[2684]

## OBTAINING SPECIMENS

Adult tapeworms may be found in almost any species of vertebrate animal, while larval forms are often encountered in a wide variety of vertebrates and invertebrates. Since adults are found most commonly in the intestine and rarely in the coelom or bile or pancreatic ducts, these are the only organs that need be examined. If at all possible, the host should be examined immediately after death, while the tapeworms are still alive. Post-mortem changes, including loss of rostellar hooks, often prevent identification of the specimen. Further, if the worm dies within the gut of its host, it usually is contracted in such a manner that adequate study of it may be impossible. If it is not feasible to examine the host soon after its death, the viscera should be frozen as quickly as possible and kept until needed.

It is more convenient if the intestine is removed before the search begins, either entire in the case of a small animal or in sections if a larger animal is being examined. If the host was shot, a resulting perforation of the gut sometimes results in tapeworms being discovered in anomalous locations.

For examination, place the intestine in a shallow dish or pan of tap water and carefully cut it open with a pair of sharp, small scissors. Care must be taken to avoid cutting worms present in the lumen. Large forms are easily seen, but a dissecting microscope is often required to find small worms. It is imperative that the scolex is not lost, for the classification depends in large part on the study of its characteristics. Remove the worms that are detached. Immersion of the gut in tap water will paralyze the worms, causing them to release their hold on the host mucosa and allowing relaxation of the strobila. Fixation should not be attempted until the worms do not respond to touch. Place all specimens in tap water at room temperature. (Note: Caryophyllidea are best placed directly in steaming hot formalin.)

With large-sized intestines or those with abundant contents, it often is best to remove the obvious worms and then gently scrape the mucosa with a scalpel. Remove the gut and pour the remaining material into a graduated cylinder or other tall container, fill with tap water, and allow the sediment to sink to the bottom. When this has occurred, pour off the supernatant fluid and replace with clean tap water. Repeat this procedure until the supernatant fluid is clear, then pour a small amount of the sediment at a time into a small dish and examine. With this technique it is unlikely that even the smallest worms will be missed.

If the host cannot be examined soon after it was killed, it should be frozen for later investigation. This usually produces inferior tapeworm specimens, but is an alternative. Another technique that often works well is to remove the intestine and drop it directly into boiling water until the contents become very hot, but not cooked. This serves to kill and relax the specimens, after which the lot can be bagged and frozen.

## FIXATION

A fixative should preserve the specimen in a life-like condition, with no brittleness or other unfavorable side effects. Unfortunately, the perfect fixative has not yet been discovered. Two widely used fixatives that are good, inexpensive, easily obtainable, and simple to use are AFA, or a solution of 5% formalin. The former is prepared by mixing 5 parts glacial acetic acid, 10 parts formalin, and 85 parts of 85% ethanol. Either fixative should be poured gently over relaxed and extended, large specimens, while small ones may be dropped directly into it. If the tapeworm is not completely relaxed before fixing, it will contract at this point and become useless for study. The specimens may be stored up to a year in a fixative, or transferred to 70% ethanol with a little glycerine for indefinite storage.

AFA is not recommended for histochemical studies, since it is a coagulative-type fixative. In this case, 4% formalin or solutions of acrolein or gluteraldehyde are suggested.

## STAINING

Several stains are satisfactory for the preparation of tapeworms. Probably the two most commonly used for general studies are *hematoxylin* and *carmine*. Since the former is used in an aqueous solution and the latter is prepared in alcohol, the specimens must be passed through a graded series of concentrations of alcohol to the level of the stain. For example, if the worm was fixed in formalin and one wishes to stain it in a carmine-70% alcohol solution, it should be moved through a series of 30%, 50%, and 70% alcohols before placing in the staining solution. The duration in each concentration depends on the size of the specimen, but 15 min should be sufficient in most cases. Since hematoxylin has a tendency to fade after a few years, its use should be limited to short-term studies.

The two basic methods of staining are *progressive* and *regressive*. In the former the specimen is placed in the stain solution and left there until the correct definition of internal organs is accomplished. At this point, staining is stopped by placing the worm in plain alcohol. Success of this technique depends upon the skill of the worker, for considerable experience is required to differentiate between proper staining and overstaining.

In the *regressive staining technique,* the worms are first overstained, then destained until proper differentiation is accomplished. This method is easier for the beginner and has the advantage of removing stain from the surface layers, thus making the internal organs more visible. For destaining hematoxylin, use 5% aqueous hydrochloric acid and for carmine use 5% acidulated 70% alcohol. Counterstaining is not recommended for either method, for it tends to obscure fine structures.

After staining is completed, dehydrate the specimen by passing it through a graded series of alcohol solutions, from 70% through 100%, for about 15 min each. The specimens are then cleared in preparation for mounting. Several clearing agents are readily available, such as xylene, oil of wintergreen, oil of cloves, terpineol, cedarwood oil, or beechwood creosote. The last mentioned has the distinct advantage of not causing brittleness of the specimen. Clearing takes only a few minutes; by the time the specimen has sunk to the bottom of the dish, maximum clearing has been obtained.

## MOUNTING

Place the specimen from the clearing agent into the mounting medium on a slide. Numerous mountants are available, and most are satisfactory. Canada balsam has the best optical properties for photography but is expensive and tends to become acidic, thus destaining the specimen. (This problem can be avoided by placing a few marble chips in the balsam bottle.) Small specimens can be conveniently arranged on a slide, but large specimens must be cut and arranged in rows, or representative sections only may be selected. If a single specimen is mounted, it may be advantageous to support one side of the coverglass with a small glass chip or piece of capillary tubing. If more than one specimen is placed on a slide, they can be arranged to give even support to the coverglass. Place the coverglass carefully, avoiding the capture of air bubbles in the medium. If an insufficient amount of medium was used, more can easily be added by applying it to the underedge of the coverglass, from which it will flow toward the middle.

The mountant should be hardened before concentrated study of the specimen is attempted. Setting the slide in a safe place at room temperature is satisfactory but slow. The process can be speeded up in a drying oven set at about 56°C. As the medium recedes due to evaporation of the solvent, add more to the edge of the coverglass.

## LABELING AND STORING SLIDES

Each completed slide must be labeled with collecting data or a code number to avoid later confusion. Much mystery exists in the literature and in private collections due to reporting the incorrect host for a parasite.

Stand slide boxes on end so the slides are horizontal to prevent the specimens from gradually drifting to the edge of the slide.

Before the collection becomes very large, it is well worth the time required to set up a method of cataloging the specimens. This will surely prevent much grief in later years when a certain specimen is needed. A file of three-by-five cards, arranged alphabetically with family indices, will serve nicely for several thousand specimens.

## HOW TO USE THE KEYS

The keys in this book are dichotomous; that is, there are two choices at each step. Every attempt was made to avoid ambiguity, but there are often so many exceptions to a general plan within a group that is is possible to make the wrong choice. If the specimen does not key out the first time through, go back to any step where the decision was not clear and run it through the other choice.

One may begin the keys at any level, depending upon how much he knows about the species in question. Thus, if the tapeworm is completely unknown to the worker, he should begin with the key to subclasses, then to orders, then to families, and so on. If he knows the order beforehand, he may begin with the key to families within that order, and proceed.

After keying the specimen to genus, it is important to compare it with the detailed generic description. There are so many undescribed genera that a new form may be in hand and yet key out to a previously described genus.

If it seems certain that the correct generic designation has been determined, then one must begin the laborious task of comparing the specimens with previously described species. The lists of species in this book, while reasonably complete at the time of writing, may have become obsolete by the time the book was published, for new species are being described every month. To bring the list up-to-date, one should make use of the *Zoological Record, Biological Abstracts,* and most especially, *Helminthological Abstracts*. This last journal, dating from 1932 to the present, is probably the most complete and up-to-date of the serial abstracts. Upon comparison, if a specimen matches one of the described species, the work is done. If it matches none of the species, it can only be concluded that a new species has been encountered. It is then the obligation and privilege of the researcher to see that the species is properly described. It must be remembered, however, that variations in morphology within a species often occur. A long series of specimens is of great advantage in determining whether a specimen represents a new species or is only a variation of an already described form. But if only a single specimen differs greatly from all other species, there is no reason why it should not be reported, and even given a new name.

The U.S. Department of Agriculture Parasitological Laboratory, Beltsville, Maryland, has compiled a card file on the literature of parasitology that is probably the most extensive in the world. It is called the *Index-Catalogue of Medical and Veterinary Zoology,* and consists of two main parts, an Author Index and a Host-Parasite Index. These indices are available to persons who wish to visit the Beltsville laboratory for this purpose. The Author Index portion has been printed by the U.S. Government Printing Office, and is supplemented each year. Copies of this index are available to qualified persons and libraries and are extremely useful in systematics work. Each reference cited in this book may be found in the Author Index.

The U.S. National Museum Helminthological Collection is also housed at the U.S. Department of Agriculture Parasitological Laboratory. Here, many type specimens are available on loan to qualified workers.

A glossary describing the terms used in this text is appended at the end. When in doubt as to the proper meaning of any term, the glossary should be consulted for clarification.

The rest of this book is devoted to the taxonomy of tapeworms. I wish you success in your studies of this fascinating group of animals.

# KEYS TO SUBCLASSES OF CESTOIDEA AND ORDERS OF SUBCLASS EUCESTODA

1a. Polyzoic (except orders Caryophyllidea and Spathebothriidea), with one or more sets of reproductive systems per proglottid. Scolex usually present. Shelled embryo with six hooks. Parasites of fishes, amphibians, reptiles, birds, and mammals, one genus maturing in coelom of fresh-water oligochaetes................. ............................................... Eucestoda Southwell 1930. (p.11)

1b. Monozoic, with single set of reproductive organs. No scolex present. Shelled embryo with ten hooks. Parasites of fishes and turtles ........................... ............................................. Cestodaria Monticelli 1891. (p.469)

## KEY TO THE ORDERS IN SUBCLASS EUCESTODA

1a. Strobila with no internal segmentation. One set of hermaphroditic reproductive organs present ........................... Caryophyllidea Beneden *in* Carus 1863.

*Diagnosis:* Body elongate, oval to flat. Scolex (Figure 10) with folds, shallow grooves, loculi, acetabular suckers, or nothing; sometimes with apical sucker, introvert or disc; marked off from body or not. Testes usually medullary (cortical in Balanotaeniidae), anterior to ovary and uterus. Cirrus pouch between ovary and testes. Vitellaria (see Figure 24) lateral or surrounding medulla; cortical, medullary, or both. All gonopores on ventral surface, usually near posterior end. Vagina opening together with sinuous uterus, behind or together with median cirrus pore. Eggs operculate. Parasites of freshwater fishes, except *Archigetes* which matures in coelom of aquatic annelids. (p.17) (For detailed reviews of this order see Mackiewicz.[2040,2045]

1b. Strobila with internal segmentation present. More than one set of reproductive organs present .............................................................. 2

2a. Scolex with no true suckers, bothria, bothridia, or tentacles. No external segmentation........................ Spathebothriidea Wardle et McLeod 1952.

*Diagnosis:* Scolex feebly developed, undifferentiated or with funnel-shaped apical organ or with one or two hollow, cup-like organs (Figure 11). External metamerism absent, internal metamerism present. Genital pores medioventral or dorsal, sometimes alternating. Cirrus pouch medial. Testes in two lateral bands. Ovary dendritic. Vitellaria follicular, lateral or scattered. Uterus coiled, with ventral pore. Parasites of teleost fishes. (p.43)

2b. Scolex with one of the holdfast types listed above (2a). External segmentation usually distinct.............................................................. 3

3a. Scolex with bothridia and four armed proboscides or tentacles................. .................................................. Trypanorhyncha Diesing 1863.

*Diagnosis:* Scolex elongate, with two or four bothridia, and four eversible (rarely atrophied) tentacles armed with hooks (Figure 12). Each tentacle invaginates into internal sheath provided with muscular bulb. Neck present or absent. Strobila apolytic or anapolytic. Genital pores lateral, rarely ventral. Testes numerous. Ovary posterior. Vitellaria follicular, scattered. Uterine pore present or absent. Parasites of elasmobranchs. (p.47)

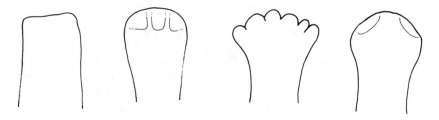

FIGURE 10. Typical scoleces of Caryophyllidea.

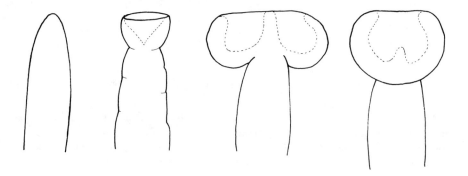

FIGURE 11. Typical scoleces of Spathebothriidea.

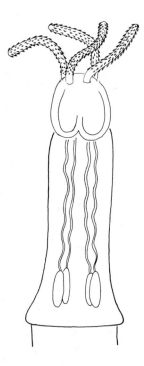

FIGURE 12. Typical scolex of Trypanorhyncha.

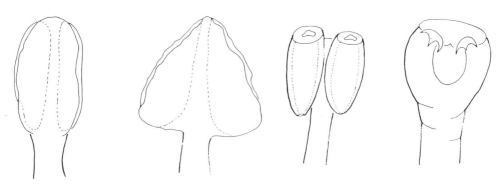

FIGURE 13. Typical scoleces of Pseudophyllidea.

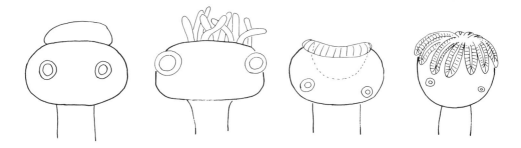

FIGURE 14. Typical scoleces of Lecanicephalidea.

3b. **Scolex with two bothria, tentacles rarely present ...................... Pseudophyllidea Carus 1863 (syn. Diphyllidea Wardle, McLeod et Radinowsky 1974).**

*Diagnosis:* Scolex with two bothria, with or without hooks. (Figure 13). Neck present or absent. Strobila variable. Proglottids anapolytic. Genital pores lateral, dorsal, or ventral. Testes numerous. Ovary posterior. Vitellaria follicular, scattered (except in Philobythiidae). Uterine pore present, dorsal or ventral. Egg usually operculate, containing coracidium. Parasites of fish, amphibians, reptiles, birds, and mammals. (p.81)

3c. **Scolex divided into anterior and posterior regions by horizontal groove, sometimes with small suckers or unarmed tentacles .. Lecanicephalidea Baylis 1920.**

*Diagnosis:* Scolex divided into anterior and posterior regions by horizontal groove (Figure 14). Anterior portion cushion-like, or with unarmed tentacles, capable of being withdrawn into posterior portion, forming a large sucker-like organ. Posterior portion usually with four suckers. Neck present or absent. Testes numerous. Ovary posterior. Vitellaria follicular, lateral or encircling proglottid. Uterine pore usually present. Parasites of elasmobranchs. (p.119)

3d. **Scolex with or without suckers; testes and ovaries without ducts to outside. No external segmentation .............................. Aporidea Fuhrmann 1934.**

*Diagnosis:* Scolex with simple suckers or grooves and armed rostellum (Figure 15). No external metamerism; internal metamerism present or not. Genital ducts and pores, cirrus, ootype, and Mehlis' glands absent. Hermaphroditic, rarely gonochoristic. Vitellarian cells mixed with ovary. Parasites of Anseriformes. (p.129)

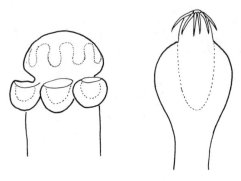

FIGURE 15. Types of scoleces found in Aporidea.

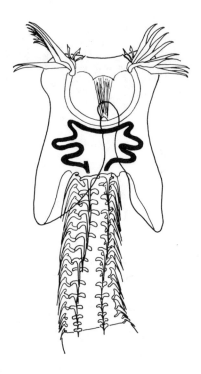

FIGURE 16. A typical scolex of Diphyllidea.

3e. Scolex with bothridia ........................................................... 4
3f. Scolex with suckers, and not divided into anterior and posterior parts. Testes and ovaries with ducts to outside. With external segmentation ............... 6

4a. Scolex with two or four bothridia, no armed rostellum. Genital pores lateral, rarely posterior ................................................................ 5
4b. Scolex with two bothridia, with or without armed rostellum. Genital pores ventral ..................................... **Diphyllidea Beneden** *in* **Carus 1863.**

*Diagnosis:* Scolex with armed or unarmed peduncle (Figure 16). Two spoon-shaped bothridia present, lined with minute spines, sometimes divided by median, longitudinal ridge. Apex of scolex with insignificant apical organ or with large rostellum bearing dorsal and ventral groups of T-shaped hooks. Strobila cylindrical, acraspedote. Genital pores posterior, mid-ventral. Testes numerous, anterior. Ovary posterior. Vitellaria follicular, lateral or surrounding segment. Uterine pore absent. Uterus tubular or saccular. Parasites of elasmobranchs. (p.165)

5a. **Monoecious forms. Vitellaria follicular** .............. **Tetraphyllidea Carus 1863.**

*Diagnosis:* Scolex with highly variable bothridia, sometimes also with hooks, spines, or suckers (Figure 17). Myzorhynchus present or absent. Neck present or absent. Proglottids commonly apolytic, hermaphroditic. Genital pores lateral, rarely posterior. Testes numerous. Ovary posterior. Vitellaria follicular, usually lateral. Uterine pore present or not. Parasites of elasmobranchs. (p.131)

5b. **Dioecious forms. Vitellaria condensed around lobes of ovary** ...................
.................................................**Dioecotaeniidea New Order.**

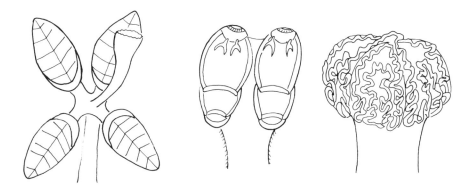

FIGURE 17. A few of the varied types of scoleces in Tetraphyllidea.

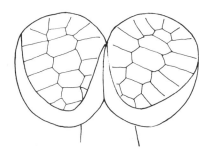

FIGURE 18. A typical scolex of Dioecotaeniidea.

FIGURE 19. A typical scolex of Litobothridea.

*Diagnosis:* Scolex with four large, unarmed bothridia, each subdivided into several loculi (Figure 18). Strobilas dioecious, female larger than male. Female lacking any trace of male organs; ovary bilobed, with seminal receptacle a sac inside one lobe; vagina a median, coiled tube with no pore; vitellaria condensed around posterolateral surfaces of ovarian lobes. Male with very large armed cirrus and internal seminal vesicle. Sperm transfer by hypodermic impregnation; injected cirri break off inside female and come to lie in median, connective tissue sheath, each cirrus containing the sperms from its internal seminal vesicle. Uterus a bilobed sac with ventromedian pore. Parasites of elasmobranchs. (p.197)

| | | |
|---|---|---|
| 6a. | Scolex with one apical sucker only | 7 |
| 6b. | Scolex with four or five suckers | 8 |
| 7a. | Scolex a single sucker, followed by several specialized segments cruciform in cross section. Strobila craspedote.................. Litobothridea Dailey 1969. | |

*Diagnosis:* Scolex a single, well-developed apical sucker (Figure 19). Anterior proglottids modified, cruciform in cross section. Neck absent. Strobila dorsoventrally flattened, with numerous proglottids, each with single set of medullary reproductive organs. Segments laciniated and craspedote; apolytic or anapolytic. Testes numerous, preovarian. Genital pores lateral. Ovary two- or four-lobed, posterior. Vitellaria follicular, encircling medullary parenchyma. Eggs unembryonated. Parasites of elasmobranchs. (p.169)

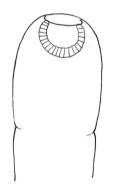

FIGURE 20. A typical scolex of Nippotaeniidea.

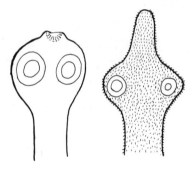

FIGURE 21. Typical scoleces of Proteocephalidea.

FIGURE 22. Typical scoleces of Cyclophyllidea.

7b. **Scolex parenchymatous with single apical sucker. No specialized segments following scolex as described above (7a). Strobila acraspedote ..................... ................................................. Nippotaeniidea Yamaguti 1939.**

*Diagnosis:* Scolex with single sucker at apex, otherwise simple (Figure 20). Neck short or absent. Strobila small. Proglottids each with single set of reproductive organs. Genital pores lateral. Testes anterior. Ovary posterior. Vitelline gland compact, single, between testes and ovary. Osmoregulatory canals reticular. Parasites of teleost fishes. (p.171)

8a. **Scolex with four suckers, sometimes with additional apical sucker or armed rostellum. Vitellaria follicular, usually in lateral margins ........................ ..................................................... Proteocephalidea Mola 1928.**

*Diagnosis:* Scolex with four suckers, occasionally with apical sucker or armed rostellum (Figure 21). Neck usually present. Metamerism usually distinct. Genital pores lateral. Testes numerous. Ovary posterior. Vitellaria follicular, usually lateral. Uterine pore present or absent. Parasites of fishes, amphibians, and reptiles. (p.173)

8b. **Scolex with four suckers. Rostellum present or absent. Vitellaria compact, medial, usually postovarian .............. Cyclophyllidea Beneden *in* Braun 1900.**

*Diagnosis:* Scolex usually with four suckers, rostellum present or not, armed or not (Figure 22). Neck present or absent. Strobila variable, usually with distinct metamerism, hermaphroditic or rarely gonochoristic. Genital pores lateral (ventral in Mesocestoididae). Vitelline gland single, compact, usually posterior to ovary. Uterus variable. Uterine pore absent. Parasites of amphibians, reptiles, birds, and mammals. (p.199)

# ORDER CARYOPHYLLIDEA

In this order, the body is unsegmented and has a single set of male and female repr organs. The scolex is simple, has shallow grooves or loculi, shallow suckers, or is ....cu. It is never armed with hooks. The genital pores are midventral, usually near the posterior end, and may open separately or together within a genital atrium. The testes are numerous, filling most of the median parenchyma anterior to the ovary and uterus. A cirrus pouch is found between the testes and ovary. The ovary is posterior, usually bilobed. The vitellaria are follicular, lateral, sometimes also medial, and often extend behind the ovary. The uterus is a coiled, median tube that opens, often together with the vagina, near the male pore.

All life cycles known involve a freshwater oligochaete annelid intermediate host. The definitive host is infected when it eats the infected oligochaete. Current thought is that adult caryophyllideans are neotenic larvae of a group whose adult forms are extinct. If so, they are equivalent to the plerocercoids of pseudophyllidean cestodes, with which they are closely related. *Archigetes* becomes sexually mature within the coelom of its annelid host, giving it the distinction of being the only tapeworm to mature in an invertebrate.

Caryophyllideans are parasites of freshwater teleost fishes, especially Cypriniformes and Siluriformes, and are found wherever these hosts exist. Major monographs on the order are presented by Mackiewicz.[2040,2045,2046]

## KEY TO THE FAMILIES IN CARYOPHYLLIDEA

1a. Vitellaria and testes entirely cortical ...............................................
    ............................... Balanotaeniidae Mackiewicz et Blair 1978. (p.17)
1b. Testes not cortical ......................................................................... 2

2a. Vitellaria entirely cortical; testes medullary........................................
    ................................................. Lytocestidae Hunter 1927. (p.18)
2b. Vitellaria partly or entirely medullary ........................................... 3

3a. Vitellaria and testes entirely medullary .............................................
    ............................................ Caryophyllaeidae Nybelin 1922. (p.26)
3b. Vitellaria partly medullary, partly cortical; testes medullary ....................
    ............................................... Capingentidae Hunter 1930. (p.37)

## DIAGNOSIS OF THE ONLY GENUS IN BALANOTAENIIDAE

*Balanotaenia* Johnston 1924. (Figure 23)

*Diagnosis:* Scolex fairly well-defined, with frilled, muscular ridges, Neck absent. Two gonopores present, behind ovarian commissure. Cirrus pouch and cirrus well-developed. Internal seminal vesicle small, seminal receptacle well-developed. Ovary basically dumbbell-shaped, occasionally lobated, with commissure arched anteriorly. Vitellaria and testes cortical, not postovarian. Uterus mostly postovarian. Parasites of siluroid fishes.
   Type species: *B. bancrofti* Johnston 1924, in *Tandanus tandanus;* Australia.
   Other species:
      *B. newguinensis* Mackiewicz et Blair 1978, in *Tandanus brevidorsalis;* Papua New Guinea.

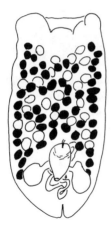
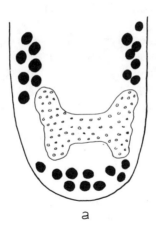
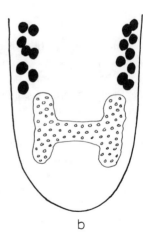

FIGURE 23. *Balanotaenia newguinensis* Mackiewicz et Blair 1978.

FIGURE 24. (a) Postovarian vitellaria present; (b) postovarian vitellaria absent.

## KEY TO THE GENERA OF LYTOCESTIDAE

1a. Postovarian vitellaria present (Figure 24) ........................................ 2
1b. Postovarian vitellaria absent ..................................................... 8

2a. Cirrus and uterovaginal canals open separately ................................. 3
2b. One gonopore present .......................................................... 4

3a. Gonopores in middle third of body ..............................................
    ................*Markevitschia* Kulakowskaya et Achmerov 1965. (Figure 25)

*Diagnosis:* Scolex simple, lacking specializations, narrower than rest of body. Two gonopores, in middle third of body. External seminal vesicle and seminal receptacle absent. Testes numerous, extending from anterior margin of ovary to beginning of neck constriction. Ovary bilobed to H-shaped, medullary. Vitellaria lateral, continuous with postovarian follicles. Uterus not extending anterior to genital atrium. Parasites of cyprinid fish. Russia.

Type species: *M. sagittata* Kulakowskaya et Achmerov 1965, in *Cyprinus carpio;* Russia.

3b. **Gonopores farther posterior** ......................................................
    *Lucknowia* **Gupta 1961 (syn.** *Introvertus* **Satpute et Agarwal 1980). (Figure 26)**

*Diagnosis:* Scolex unspecialized, thinner than rest of body; its tip may serve as an introvert. Two gonopores; cirrus sac and uterovaginal canal open separately at beginning of last seventh of body length. Uterine and vaginal pores common. Ovary a transversely elongated band, both medullary and cortical, overlapping vitelline follicles. Vitellaria mostly lateral, from near front end of body to excretory bladder, postovarian follicles present. Seminal receptacle absent. Uterine coils much convoluted, compactly coiled behind ovarian isthmus, not extending anterior to cirrus pouch. Testes medial to vitelline glands, extending from just behind first vitelline follicles to posterior end of cirrus pouch. Eggs thick-shelled, with polar filament at one end. Parasites of freshwater siluroid fishes. India.

Type species: *L. follilisi* Gupta 1961, in *Heteropneustes fossilis;* India.
Other species:
   *L. raipurensis* (Satpute et Agarwal 1980) comb. n. (syn. *Introvertus raipurensis* Satpute et Agarwal 1980), in *Clarias batrachus;* India.

FIGURE 25. *Markevitschia sagittata* Kulakowskaja et Achmerov 1965.

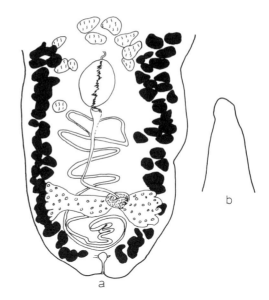

FIGURE 26. *Lucknowia fossilis* Gupta 1961. (a) Posterior end; (b) scolex.

**4a. Ovary indistinctly bilobate** ............... ***Lytocestoides*** **Baylis 1928. (Figure 27)**

*Diagnosis:* Scolex short, conical, with longitudinal grooves but no loculi. Gonoducts separate. Testes extend between scolex and uterus. Ovary indistinctly bilobate, medullary. Vitellaria extensive, postovarian. Uterus not extending anterior to cirrus pouch. Parasites of cyprinoid fish. Africa.
 Type species: *L. tanganyikae* Baylis 1928, in *Alestes* sp.; Tanganyika.
 Other species:
  *L. aurangabadensis* Shinde 1970, in *Barbus collus;* India.

**4b. Ovary shaped like an H or an inverted A ...................................... 5**

**5a. Ovary shaped like an inverted A ...** ***Caryophyllaeides*** **Nybelin 1922. (Figure 28)**

*Diagnosis:* Scolex undifferentiated. Cirrus opening into uterovaginal atrium. Cirrus pouch large, oval. External seminal vesicle absent. Testes median, anterior to uterus. Ovary shaped like an inverted A. Postovarian vitellaria present. Uterus extending anterior to cirrus pouch. Parasites of cyprinid fishes. Scandinavia, Europe, Asia.
 Type species: *C. fennicus* (Schneider 1902) Nybelin 1922, in *Leuciscus* spp., *Tinca tinca, Blicca fjorkna, Abramis brama, Chondrostoma nasus, Stylaria lacustris, Rutilus rutilus, Cyprinus carpio;* Finland, Sweden, Russia, Germany.
 Other species:
  *C. skrjabini* Popoff 1922, in *Leuciscus* spp.; Europe.

FIGURE 27. *Lytocestoides tanganyikae* Baylis 1928.

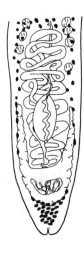

FIGURE 28. *Caryophyllaeides fennicus* (Schneider 1902) Nybelin 1922.

FIGURE 29. *Khawia iowensis* Calentine et Ulmer 1961.

5b. Ovary H-shaped ................................................................ 6

6a. Scolex broad, flat, fimbriate, not separated from body by a well-defined, constricted neck ................................................................
................ **Khawia Hsü 1935 (syn. *Bothrioscolex* Szidat 1937). (Figure 29)**

*Diagnosis:* Scolex lacking loculi, sometimes frilled. One gonopore present. Cirrus pouch well developed. External seminal vesicle absent. Ovary H-shaped, medullary. Vitellaria lateral, with postovarian follicles. Uterus not extending anterior to cirrus pouch. Parasites of cyprinid fishes. Asia, Europe, Japan, U.S. Key to species: Callentine and Ulmer.[431]

Type species: *K. sinensis* Hsü 1935, in *Cyprinus carpio;* China, U.S.
Other species:
  *K. baltica* Szidat 1941, in *Tinca tinca;* Europe.
  *K. dubia* (Szidat 1937) Yamaguti 1959 (syn. *Bothrioscolex dubia* Szidat 1937), in *Carassius carassius;* Germany

FIGURE 30. *Caryoaustralis sprenti* Mackiewicz et Blair 1980.

*K. iowensis* Calentine et Ulmer 1961, in *Cyprinus carpio;* U.S.

*K. japonensis* (Yamaguti 1934) Yamaguti 1959 (syn. *Caryophyllaeus japonensis* Yamaguti 1934; *Bothrioscolex japonensis* [Yamaguti 1934] Szidat 1937), in *Cyprinus carpio;* Japan.

*K. prussica* (Szidat 1937) Yamaguti 1959 (syn. *Bothrioscolex prussica* Szidat 1937), in *Carassius carassius;* Prussia.

*K. rosittensis* (Szidat 1937) Yamaguti 1959 (syn. *Bothrioscolex rosittensis* Szidat 1937), in *Carassius carassius;* Europe.

**6b.   Scolex not fimbriate ............................................................ 7**

**7a.   Scolex bell-shaped, with prominent collar around base and with apical funnel ........................*Caryoaustralus* Mackiewicz et Blair 1980. (Figure 30)**

*Diagnosis:* Scolex bell-shaped, with prominent collar around base, and with apical funnel. Gonopore single, in anterior half of worm. Ovary H-shaped. Uterus not extending anterior to cirrus pouch, mainly postovarian. Vitellaria surrounding testes, extending well into hindbody posterior to ovary. External seminal vesicle absent. Seminal receptacle present. Parasites of Australian freshwater catfish (Plotosidae).

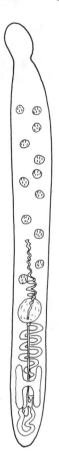

FIGURE 31. *Atractolytocestus huronensis* Anthony 1958.

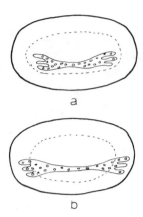

FIGURE 32. (a) Ovarian lobes entirely medullary; (b) ovarian lobes partly cortical.

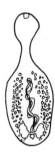

FIGURE 33. *Djombangia penetrans* Bovien 1926.

Type species: *C. sprenti* Mackiewicz et Blair 1980, in *Tandanus ater, T. glencoensis*. Australia.

7b. **Scolex conical, small, narrower than body and separated from it by well-defined, constricted neck** .................... ***Atractolytocestus* Anthony 1958. (Figure 31)**

*Diagnosis:* Scolex simple, may possess an apical introvert. Genital pores opening into common atrium. Cirrus pouch round. Outer seminal vesicle not present. Testes medullary, six to ten in number. Ovary H-shaped, entirely medullary. Vitellaria extensive, both cortical and medullary, postovarian follicles present. Uterus not extending anterior to cirrus pouch. Eggs operculate. Parasites of cyprinid fishes. North America.

Type species: *A. huronensis* Anthony 1958, in *Cyprinus carpio*; U.S.

8a. **Ovarian lobes entirely medullary (Figure 32)** ................................. **9**
8b. **Ovarian lobes partly cortical** ................................................. **11**

9a. **Scolex with terminal sucker** ............... ***Djombangia* Bovien 1926. (Figure 33)**

*Diagnosis:* Scolex nearly spherical, with terminal sucker but no loculi. Well-marked neck present. Gonopores separate, in common atrium near posterior end of body. Cirrus pouch not well developed. External seminal vesicle apparently absent. Testes in lateral medulla,

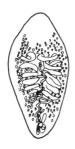

FIGURE 34. *Notolytocestus major* Johnston et Muirhead 1950.

FIGURE 35. *Thalophyllaeus johnstoni* Mackiewicz et Blair 1980.

from neck to ovary. Ovary bilobed, medullary, at posterior end of body. Vitellaria in testicular zone and lateral, but not behind ovary. Uterus median, nearly reaching neck. Eggs spiny. Parasites of siluroid fish. Java, India.

Type species: *D. penetrans* Bovien 1926, in *Clarias batrachus;* Java.
Other species:
  *D. caballeroi* Sahay et Sahay 1977, in *Heteropneustus fossilis;* India.
  *D. indica* Satpute et Agarwal 1974, in *Clarias batrachus;* India.

**9b. Scolex lacking terminal sucker .................................................. 10**

**10a. Scolex undifferentiated; uterus extending far forward from cirrus .............**
**........................ Notolytocestus Johnston et Muirhead 1950. (Figure 34)**

*Diagnosis:* Scolex narrow, undifferentiated. Body short, broad. Cirrus opening into uterovaginal canal. Single gonopore present. Testes mainly lateral because of displacement by uterus. Ovary H-shaped, at posterior end of body. Vitellaria mainly lateral, no follicles postovarian. Uterus mainly medial to testes, extending far forward. Parasites of siluroid fishes (Plotosidae). Australia.

Type species: *N. major* Johnston et Muirhead 1950, in *Tandanus glencoensis, T. ater;* Australia.
Other species:
  *N. minor* Johnston et Muirhead 1950, in *Tandanus tandanus, T. glencoensis, T. ater;* Australia.

**10b. Scolex dome-shaped: uterus not extending anterior to cirrus ....................**
**........................ Thalophyllaeus Mackiewicz et Blair 1980. (Figure 35)**

*Diagnosis:* Scolex dome-shaped, with no loculi. Male and female ducts join to form single gonopore. Ovary H-shaped. Uterus not extending anterior to cirrus, primarily postovarian. Preovarian vitellaria in lateral and median position. Postovarian vitellaria absent. External seminal vesicle absent. Seminal receptacle absent. Parasites of Australian freshwater catfish (Plotosidae).
Type species: *T. johnstoni* Mackiewicz et Blair 1980, in *Tandanus ater, T. glencoensis.*

**11a. Scolex differentiated** .............................................................. **12**
**11b. Scolex undifferentiated** ......................................................... **14**

**12a. Vitellaria surrounding testes** ......... ***Monobothrioides*** **Fuhrmann et Baer 1925.**

*Diagnosis:* Scolex with terminal introvert and longitudinal furrows, but no loculi. Gonopores separate. Cirrus pouch well developed. No external seminal vesicle. Seminal receptacle present. Ovary H-shaped with long, slender lateral cortical processes and median, medullary mass of follicles. No postovarian vitellaria. Uterus not reaching cirrus pouch. Parasites of siluroid fishes. Africa.
Type species: *M. cunningtoni* Fuhrmann et Baer 1925, in *Auchenoglanis orientalis;* Tanganyika.
Other species:
   *M. chalmersius* Mackiewicz et Beverly-Burton 1967, in *Clarias anguillaris;* Zambia.
   *M. tchadensis* Troncy 1978, in *Auchinoglanis biscutatus;* Chad.
   *M. woodlandi* Mackiewicz et Beverly-Burton 1967, in *Clarias mellandi;* Zambia.

**12b. Vitellaria lateral, crescent-shaped in cross section** ............................. **13**

**13a. Ovary inverted A-shaped** .......................... ***Crescentovitus*** **Murhar 1964.**

*Diagnosis:* Scolex well-differentiated, with shallow bothria on dorsal and ventral side, and small terminal introvert. Neck separated from scolex by constriction. Genital pores close together on ventral surface. Vitellaria in lateral cortex, extending to anterior margin of ovary, none postovarian; crescent-shaped in cross section. No external seminal vesicle or seminal receptacle. Ovary inverted A-shaped, arms in lateral cortex and isthmus in medulla. Uterus long. Eight to ten longitudinal osmoregulatory canals with transverse connections. Parasites of siluroid fish. India.
Type species: *C. biloculus* Murhar 1964, in *Heteropneustes fossilis;* India.

**13b. Ovary H-shaped** ......................................... ***Stocksia*** **Woodland 1937.**

*Diagnosis:* Scolex flattened, pointed, with apical cushion and slender longitudinal grooves. Single gonopore present. Cirrus pouch present. External seminal vesicle apparently absent. Testes median, anterior to cirrus pouch. Ovary small, H-shaped, at posterior end of body; lateral lobes in cortex, isthmus in medulla. Vitellaria lateral, crescent-shaped in cross section; postovarian follicles absent. Uterus tightly coiled, anterior to ovary, not extending anterior to cirrus pouch. Parasites of siluroid fish. Africa.
Type species: *S. pujehuni* Woodland 1937, in *Clarias lazera;* Sierra Leone.
Other species:
   *S. lazera* Woodland 1937 — *nomen nudum.*

**14a. Vitellaria lateral.** ......................... ***Bovienia*** **Fuhrmann 1931. (Figure 36)**

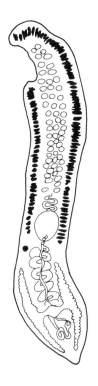

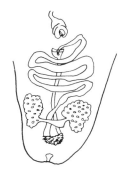

FIGURE 37. *Lytocestus indicus* (Moghe 1925) Moghe 1931.

FIGURE 36. *Bovienia serialis* (Bovien 1926) Fuhrmann 1931.

*Diagnosis:* Scolex small, undifferentiated, not clearly distinguished from neck. Gonopores separate. Cirrus pouch ovoid, anterior of ovary. Cirrus unspined. Seminal receptacle present. Testes median, extending from a point slightly posterior to the posteriormost vitellaria to near cirrus pouch. Ovary H-shaped. Vitellaria lateral, extending from 5.6 to 9.7 mm from anterior end to near ovary. No postovarian follicles. Uterus not extending anterior to male gonopore. Parasites of siluroid fish. Java, Pakistan.

Type species: *B. serialis* (Bovien 1926) Fuhrmann 1931 (syn. *Caryophyllaeus serialis* Bovien 1926), in *Clarias batrachus*; Java.

Other species:
*B. ilishai* Zaidi et Khan 1976, in *Macrura ilisha*; Pakistan.

**14b. Vitellaria surrounding testes ................ *Lytocestus* Cohn 1908. (Figure 37)**

*Diagnosis:* Scolex undifferentiated. Gonopores separate. External seminal vesicle absent. Seminal receptacle absent. Ovary bilobed with lateral lobes cortical. Vitellaria in testicular zone, no postovarian follicles. Uterus not extending anterior to cirrus pouch. Parasites of mormyrid and siluroid fishes. Hong Kong, Burma, Africa, Chad, Singapore, Mollucas.

Type species: *L. adhaerens* Cohn 1908, in *Clarias fuscus*; Hong Kong.
Other species:
*L. birmanicus* Lynsdale 1956, in *Clarias batrachus*; Burma.
*L. filiformis* Woodland 1923, (syn. *L. alestesi* Lynsdale 1956), in *Alestes nurse, Mormyrus caschive*; Sudan.
*L. fossilis* Singh 1975, in *Heteropneustes fossilis*; Nepal.
*L. indicus* (Moghe 1925) Yamaguti 1959, in *Clarias batrachus*; India.
*L. longicollis* Rama 1973, in *Clarias batrachus*; India.

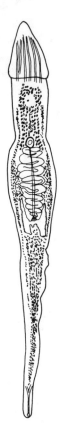

FIGURE 38. *Wenyonia virilis* Woodland 1923.

*L. marcuseni* Troncy 1978, in *Marcusenius harringtoni;* Chad.
*L. parvulus* Furtado 1963, in *Clarias batrachus;* Singapore, Moluccas.

## KEY TO THE GENERA OF CARYOPHYLLAEIDAE

**1a.   Genital pores in anterior half of body...*Wenyonia* Woodland 1923. (Figure 38)**

*Diagnosis:* Scolex undifferentiated or with several longitudinal furrows, constricted off from body or not. Two gonopores present in anterior half of body. External seminal vesicle and seminal receptacle absent. Testes anterior. Ovary H-shaped, medullary. Vitellaria lateral, extending to postovarian zone. Uterus not extending anterior to genital atrium. Parasites of siluroid fishes. Africa.

Type species: *W. virilis* Woodland 1923, in *Synodontis schall;* Nile River.
Other species:
   *W. acuminata* Woodland 1923, in *Synodontis membranaceus;* Nile River.
   *W. kainjii* Ukoli 1972, in *Synodontis nigrita;* Nigeria.
   *W. longicauda* Woodland 1937, in *Synodontis gambiensis, S. frontosus;* Sierra Leone, Chad.
   *W. mcconnelli* Ukoli 1972, in *Synodontis clarias;* Nigeria.
   *W. minuta* Woodland 1923, in *Chrysichthys auratus;* Sudan.
   *W. nilotica* (Kulmatycki 1928), Yamaguti 1959 (syn. *Caryophyllaeus nilotica* Kulmatycki 1928) in *Synodontis schall;* Egypt.

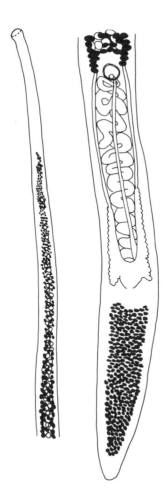

FIGURE 39. *Calentinella etnieri* Mackiewicz 1974.

*W. synodontis* Ukoli 1972, in *Synodontis sorex, S. gambiensis, S. vermiculatus;* Nigeria.
*W. youdeoweii* Ukoki 1972, in *Synodontis gobroni;* Nigeria.

| | | |
|---|---|---|
| 1b. | Genital pores in posterior half of body | 2 |

| | | |
|---|---|---|
| 2a. | Uterus extending anterior of cirrus pouch | 3 |
| 2b. | Uterus entirely behind cirrus pouch | 11 |

| | | |
|---|---|---|
| 3a. | Scolex lacking any loculi or folds | 4 |
| 3b. | Scolex with loculi | 5 |

4a. Neck long, slender.................*Calentinella* Mackiewicz 1974. (Figure 39)

*Diagnosis:* Scolex blunt (cuneiform), without loculi or bothria. Neck and body long, slender. Cirrus joining uterovaginal canal. External seminal vesicle absent. Internal seminal vesicle present. Cirrus pouch weakly developed. Gonopore large, conspicuous. Preovarian vitellaria chiefly lateral, some medial, smaller than testes. Postovarian vitellaria extremely voluminous, its field one and a half to three times ovary length. Ovary with long anterior

FIGURE 40. *Paracaryophyllaeus dubininae* Kulakawskaja 1961.

arms, lacking posterior arms (U-shaped). Seminal receptacle absent. Parasites of North American Catostomidae. Key to single-pored Caryophyllaeidae: Mackiewicz.[2041]

Type species: *C. etnieri* Mackiewicz 1974, in *Erimyzon oblongus;* Tennessee.

**4b.   Neck absent ............... *Paracaryophyllaeus* Kulakowskaya 1961. (Figure 40)**

*Diagnosis:* Scolex slightly widened, without festoons and suckers. Testes in two longitudinal rows. Cirrus pouch rather small. Seminal receptacle present. Ovary H-shaped. Vitellaria begin high and reach to level of ovary. Postovarian vitellaria present. Uterus extending anterior of cirrus pouch. Parasites of cyprinoid fish. Russia.

Type species: *P. dubininae* Kulakowskaya 1961, in *Misgurnus anguillicaudatus;* Russia.

**5a.   Scolex with three pairs of shallow loculi ..... *Hypocaryophyllaeus* Hunter 1927.**

*Diagnosis:* Scolex short, wider than anterior body, with three dorsal and three ventral shallow loculi. Gonopores separate. Cirrus pouch small. External seminal vesicle present. Seminal receptacle apparently absent. Ovary H-shaped, medullary. Postovarian vitellaria present. Uterus extending anterior of cirrus pouch. Parasites of catostomid fishes. North America.

Type species: *H. paratarius* Hunter 1927, in *Carpiodes carpio, C. velifer, Ictiobus caprinella;* U.S.

Other species:
  *H. gilae* Fischthal 1953, in *Gila straria;* Wyoming.

**5b.   Scolex otherwise ................................................................. 6**

**6a.   External seminal vesicle absent .................................................... 7**
**6b.   External seminal vesicle present ................................................... 8**

**7a.   Cirrus opening separately from uterovaginal canal.............................**
**............................................. *Rogersus* Williams 1980. (Figure 41)**

*Diagnosis:* Scolex with a pair of median acetabular suckers and two pairs of lateral loculi. Cirrus opening separately from uterovaginal canal. Uterus not extending anterior of cirrus. Ovary H-shaped. Preovarian vitellaria median and lateral; postovarian vitellaria present or

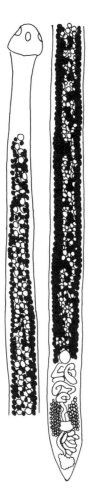

FIGURE 41. *Rogersus rogersi* Williams 1980.

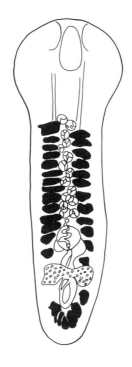

FIGURE 42. *Penarchigetes oklensis* Mackiewicz 1969.

FIGURE 43. *Archigetes iowensis* Calentine 1962.

absent. External seminal vesicle and seminal receptacle absent. Parasites of Catostomidae. Southern U. S.

Type species: *R. rogersi* Williams 1980, in *Moxostoma poecilurum*; U.S.

### 7b.  Cirrus joining uterovaginal canal.. *Penarchigetes* Mackiewicz 1969. (Figure 42)

*Diagnosis:* Scolex with a pair of median bothria, two pairs of lateral loculi, and a small terminal disc. Cirrus joining uterovaginal canal. Ovary H- or dumbbell-shaped. Coils of uterus extending to anterior level of cirrus. Preovarian vitellaria lateral, sometimes continuous with the postovarian vitellaria. External seminal vesicle absent. Parasites of catostomid fishes. North America.

Type species: *P. oklensis* Mackiewicz 1969, in *Minytrema melanops*; U. S.
Other species:
  *P. fessus* Williams 1979, in *Erimyzon sucetta*; U.S.

### 8a.  Cercomer present on adult; ovary dumbbell-shaped; parasites of aquatic annelids (sometimes found in fish) ......................................................
.......... *Archigetes* Leuckart 1878 (syn. *Szidatinus* McCrae 1961). (Figure 43)

*Diagnosis:* Scolex not clearly demarcated from body, with up to six shallow loculi. Cercomer with larval hooks sometimes present at posterior end. Single gonopore present, covered by tegument. External seminal vesicle present. Seminal receptacle present. Ovary dumbbell-shaped. Vitellaria lateral, continuous in region of ovary, with postovarian follicles. Uterus usually not extending anterior of cirrus sac, but occasionally doing so in gravid specimens. Eggs operculate, deposited in a mass between the body and an overlying cuticle. Adults in coelom of tubificid oligochaetes (cercomer present) or intestine of freshwater teleosts (cercomer absent), especially cyprinids. Europe, Asia, Japan, England, Africa, North and South America. Key to species: Kennedy.[1594]

Type species: *A. sieboldi* Leuckart 1878 (syn. *A. appendiculatus* Mrázek 1897; *Biacetabulum sieboldi* Szadat 1937; *B. appendiculatum* Janiszewska 1950), in tubificid oligochaetes; Europe, Asia, North and South America.

Other species:

*A. brachyurus* Mrázek 1908 (syn. *Brachyurus brachyurus* Szidat 1938; *Paraglaridacris silesiacus* Janiszewska 1950; *Glaridacris brachyurus* Yamaguti 1959), in *Limnodrilus* spp., cyprinid fishes; Europe, Russia, Britain.

*A. limnodrili* (Yamaguti 1934) Kennedy 1965 (syn. *Glaridacris limnodrili* Yamaguti 1934; *Brachyurus gobii* Szidat 1938; *Glaridacris gobii* [Szidat 1938] Yamaguti 1959), in *Limnodrilus*, Cyprinidae, Gobitidae; Japan, Russia, Britain.

*A. cryptobothrius* Wisnicwski 1928, in *Limnodrilus;* Europe.

*A. iowensis* Calentine 1962, in *Limnodrilus*, Cyprinidae; U.S.

**8b.  Cercomer absent on adult; ovary not dumbbell-shaped, parasites of freshwater fishes .................................................................... 9**

**9a.  Postovarian vitellaria present ........... *Biacetabulum* Hunter 1927. (Figure 44)**

*Diagnosis:* Scolex with one dorsal and one ventral loculum. One gonopore present; cirrus opening into uterovaginal canal before reaching atrium. External seminal vesicle present. Seminal receptacle present. Ovary H-shaped, medullary. Some vitelline follicles postovarian. Uterus extending anterior of cirrus pouch. Parasites of catostomid fishes. North America.

Type species: *B. infrequens* Hunter 1927, in *Moxostoma anisurum, M. rubreques, Catostomus commersoni, Hypentelium nigricans;* U.S.

Other species:

*B. banghami* Mackiewicz 1968, in *Minytrema melanops, Moxostoma macrolepidotum, M. erythrurum;* U. S.

*B. biloculoides* Mackiewicz et McCrae 1965, in *Catostomus commersoni;* U.S.

*B. hoffmani* Mackiewicz 1972, in *Hypentelium etowanum, Moxostoma erythrurum;* U. S.

*B. macrocephalum* McCrae 1962, in *Catostomus commersoni;* U.S.

*B. oregoni* Williams 1978, in *Catostomus macrocheilus;* Oregon.

**9b.  Postovarian vitellaria absent ..................................................... 10**

**10a.  Scolex with distinct, papilla-like apical sucker .......................................**
**..................................... *Dieffluvium* Williams 1978. (Figure 45)**

*Diagnosis:* Scolex with two shallow loculi, one dorsal, one ventral. A distinct papilla-like apical sucker is present. Neck long, with dorsal and ventral groove just posterior to scolex. One gonopore present, cirrus joins uterovaginal canal. Testes very numerous (972 to 1443), randomly arranged. Cirrus pouch large. External seminal vesicle prominent. Two

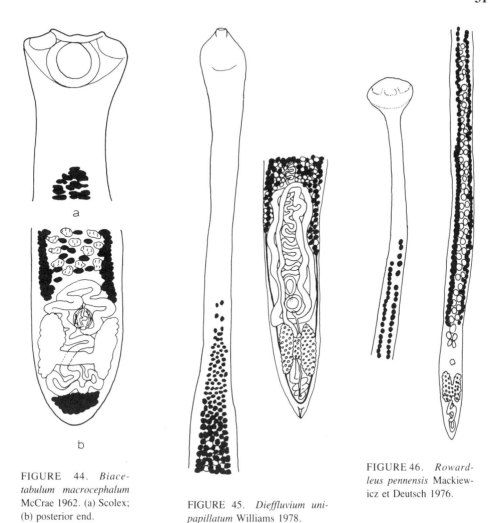

FIGURE 44. *Biacetabulum macrocephalum* McCrae 1962. (a) Scolex; (b) posterior end.

FIGURE 45. *Dieffluvium unipapillatum* Williams 1978.

FIGURE 46. *Rowardleus pennensis* Mackiewicz et Deutsch 1976.

coils of uterus extend well anterior to cirrus pouch and external seminal vesicle, passing into testicular field. Vas deferens conspicuous. Ovary H-shaped. Large seminal receptacle present. Vitellaria numerous, lateral and medial, extending posteriad almost to external seminal vesicle; postovarian follicles absent. Parasites of catostomid fishes. Southeastern U.S.

Type species: *D. unipapillatum* Williams 1978, in *Moxostoma carinatum;* Alabama.

**10b. Scolex blunt, lacking apical sucker ................................................ ........................... *Rowardleus* Mackiewicz et Deutsch 1976. (Figure 46)**

*Diagnosis:* Scolex blunt, with shallow loculi. Neck slender, long. Cirrus and uterovaginal canal open together forming single large gonopore. External seminal vesicle present. Cirrus sac small, round. Preovarian vitellaria in two lateral rows. Postovarian vitellaria absent. Ovary H-shaped. Uterus extending anterior of ovary. Seminal receptacle absent. Parasites of North American Catostomidae.

Type species: *R. pennensi* Mackiewicz et Deutsch 1976, in *Carpiodes cyprinus;* Pennsylvania, U.S.

**11a. Scolex with terminal introvert ........ *Monobothrium* Diesing 1863. (Figure 47)**

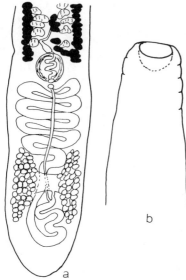
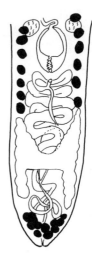

FIGURE 48. *Caryophyllaeus laticeps* (Pallas 1781) Mueller 1787.

FIGURE 47. *Monobothrium ulmeri* Calentine et Mackiewicz 1966. (a) Posterior end; (b) scolex.

*Diagnosis:* Scolex with terminal introvert and six long shallow loculi. Gonopores separate. Cirrus pouch small. External seminal vesicle present or absent. Testes in median field anterior to cirrus pouch. Ovary H-shaped. Vitellaria in testicular zone, some follicles postovarian or not. Uterus not extending anterior of cirrus sac. Parasites of cyprinid and catostomid fishes. Europe, North America. Key to species: Calentine and Mackiewicz.[430]

Type species: *M. wageneri* Nybelin 1922 (syn. *Caryophyllaeus tuba*, renamed), in *Tinca tinca;* Italy.

Other species:

*M. auriculatum* Kulakowskaya 1961, in *Leuciscus danilewskii;* Russia.

*M. fossae* Williams 1974, in *Moxostoma poecilurum;* U.S.

*M. hunteri* Mackiewicz 1963, in *Catostomus commersoni;* U. S.

*M. mackiewiczi* Williams 1974, in *Hypentalium etowanum;* U.S.

*M. ulmeri* Calentine et Mackiewicz 1966, in *Minytrema melanops, Moxostoma erythrurum, Hypentelium nigricans, Erimyzon oblongus;* U.S.

**11b. Scolex lacking terminal introvert** ............................................... 12

**12a. Scolex crenulated, without distinct loculi**.........................................
........................................*Caryophyllaeus* Muller 1787. (Figure 48)

*Diagnosis:* Scolex broad, flattened, lacking loculi, anterior margin frilled. Gonopores separate. External seminal vesicle absent. Cirrus pouch well developed. Seminal receptacle well developed. Ovary H-shaped, medullary. Postovarian vitellaria present. Uterus not extending in front of cirrus pouch. Parasites of cyprinid and catostomid fishes. Europe, Java, Asia, Africa, Japan.

Type species: *C. laticeps* (Pallas 1781) Mueller 1787 (syn. *C. mutabilis* Rudolphi 1802; *C. communis* Schrank 1788; *C. fuhrmanni* Szidat 1937), in *Cyprinus, Carassius, Barbus, Gobio, Rutilus, Tinca, Rhodeus, Abramis, Pelecus, Alburnus, Scardinius, Leuciscus, Chondrostoma, Nemachilus, Cobitis,* etc.; Europe.

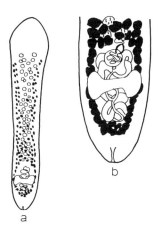

FIGURE 49. *Hunterella nodulosa* Mackiewicz et McCrae 1962. (a) Entire worm; (b) posterior end.

Other species:
   *C. acutus* Bovien 1926, in *Clarias batrachus, Macrones nigriceps;* Java.
   *C. appendiculatus* Ratzel 1868 (syn. *C. mutabilis* [Rudolphi 1802] Sramek 1901), in *Tubifex rivulorum;* Heidelberg.
   *C. armeniacus* Cholodkovsky 1916, in *Lakoeta* sp., *Capoeta* sp.; Russia.
   *C. brachycollis* Janiszewska 1953, in cyprinid fishes; Poland.
   *C. caspicus* Chlopina 1917, in Cyprinidae; Russia.
   *C. chalmersius* Woodland 1924, in *Clarias anguillaris;* Nile River.
   *C. cyprinorum* Zeder 1803, in *Cyprinus* sp.; Europe.
   *C. fimbriceps* Chlopina 1924, in *Cyprinus carpio;* Russia.
   *C. gotoi* Motomura 1927, in *Misgurnus anguillicaudatus;* Japan.
   *C. javanicus* Bovien 1926, in *Clarias batrachus;* Java.
   *C. kashmiriensis* Mehra 1930, in *Schizothorax micropogon;* India.
   *C. microcephalus* Bovien 1926, in *Clarias batrachus, Macrones nigriceps;* Java.
   *C. oxycephalus* Bovien 1926, in *Clarias batrachus;* Java.
   *C. parvus* Zmeev 1936, in *Carassius auratus;* Russia.
   *C. syrdarjensis* Skrjabin 1913, in *Schizothorax intermedius; Aspius aspius, Barbus brachycephalus, Cyprinus carpio, Rutilus rutilus;* Russia.
   *C. tenuicollis* Bovien 1926, in *Clarias batrachus;* Java.
   *C. trisignatus* Molin 1858, in *Gadus merlucius;* Italy.
   *C. truncatus* Siebold et Baird 1853, in *Cyprinus nasus;* Europe.

12b. Scolex not as above ............................................................ 13

13a. Scolex unspecialized, without depressions. Always found in pit in gut wall of host ........................*Hunterella* **Mackiewicz et McCrae 1962. (Figure 49)**

*Diagnosis:* Scolex not set off from body, a simple, unspecialized rounded or conical enlargement without suckers, loculi, or other organs of attachment. Neck absent. Two gonopores; cirrus opens separately, anterior to female gonopore. Testes fill medullary parenchyma from behind scolex to level of cirrus pouch. Cirrus pouch ovoid. External seminal vesicle present. Ovary H-shaped. Coils of uterus not extending anteriorly beyond cirrus pouch. Postovarian vitellaria present. Seminal receptacle present. Egg operculate,

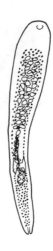

FIGURE 50. *Pliovitellaria wisconsinensis* Fischthal 1951.

unembryonated. Osmoregulatory canals reticular. Parasites of catostomid fishes, forming nodules on gut wall. North America.

Type species: *H. nodulosa* Mackiewicz et McCrae 1962, in *Catostomus* spp.

**13b. Scolex provided with one to three pairs of shallow loculi.................... 14**

**14a. Scolex with one pair of loculi ................................................. 15**
**14b. Scolex with two or three pairs of loculi ...................................... 17**

**15a. Postovarian vitellaria present ......... *Pliovitellaria* Fischthal 1951. (Figure 50)**

*Diagnosis:* Scolex not marked off from rest of body. One dorsal and one ventral loculum present. One gonopore present. Cirrus sac postequatorial. External seminal vesicle present. Testes medullary. Seminal receptacle present. Ovary H-shaped, medullary. Vitellaria medullary, mainly lateral to testes, interrupted lateral to ovary, filling entire medullary space posterior to ovary. Uterus not extending anterior of cirrus sac. Parasites of cyprinid fishes. North America.

Type species: *P. wisconsinensis* Fischthal 1951, in *Notemigonus crysoleucas, Hyborhynchus notatum;* Wisconsin, U.S.

**15b. Postovarian vitellaria absent ..................................................... 16**

**16a. Ovary V-shaped ...................................... *Bialovarium* Fischthal 1953.**

*Diagnosis:* Scolex poorly defined, with one dorsal and one ventral loculum. One gonopore. External seminal vesicle present. Cirrus pouch large. Cirrus joins uterovaginal duct in posterior end of cirrus sac. Seminal receptacle present. Ovary V-shaped, medullary. Vitellaria lateral, no follicles postovarian. Uterus not extending anterior of cirrus pouch. Parasites of cyprinid fishes. North America.

Type species: *B. necomis* Fischthal 1953, in *Necomis biguttatus;* Wisconsin.
Other species:
   *B. giganteum* (Hunter 1927) Fischthal 1953, in *Ictiobus bubulus, I. cyprinella;* Mississippi.
   *B. meridianum* (Hunter 1927) Fischthal 1953 (syn. *Biacetabulum meridianum* Hunter 1927), in *Erimyzon sucetta, E. oblongus;* U.S.

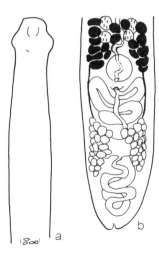

FIGURE 51. *Promonobothrium minytremi* Mackiewicz 1968. (a) Scolex; (b) posterior end.

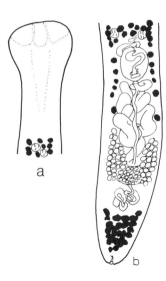

FIGURE 52. *Glaridacris catostomi* Cooper 1920. (a) Scolex; (b) posterior end.

*B. tandani* (Johnston et Muirhead 1950) Fischthal 1953 (syn. *Biacetabulum tandani* Johnston et Muirhead 1950), in *Tandanus tandanus;* Australia.

**16b. Ovary H-shaped ............... *Promonobothrium* Mackiewicz 1968. (Figure 51)**

*Diagnosis:* Scolex with pair of shallow median loculi and two small lateral depressions. Cirrus opening separately from uterovaginal canal. Ovary H-shaped. Coils of uterus not extending anteriorly beyond cirrus pouch. Preovarian vitellaria median and lateral; postovarian vitellaria absent. External seminal vesicle present. Parasites of Catostomidae. North America.
Type species: *P. minytremi* Mackiewicz 1968, in *Minytrema melanops;* Tennessee.

**17a. Scolex with two pairs of shallow loculi and an apical disk ......................
................................................*Paraglaridacris* Janiszewska 1950.**

*Diagnosis:* Scolex with four loculi and a terminal disc, constricted from rest of body. Gonopore single (?). External seminal vesicle present. Testes medullary, in two lateral rows. Seminal receptacle present. Ovary H-shaped. Vitellaria lateral to testes, postovarian follicles present. Uterus not extending anterior of cirrus pouch. Parasites of cyprinid fish. Poland.
Type species: *P. silesiacus* Janiszewska 1950, in *Abramis brama;* Poland.

**17b. Scolex with three pairs of shallow loculi; apical disc present or absent....... 18**

**18a. Two gonopores present........................................................
............ *Glaridacris* Cooper 1920 (syn. *Brachyurus* Szidat 1938). (Figure 52)**

*Diagnosis:* Scolex well defined, with three pairs of loculi. Two gonopores present. Cirrus pouch well developed. Testes median, anterior to cirrus pouch. External seminal vesicle present. Seminal receptacle present. Ovary H-shaped. Vitellaria in testicular zone, some follicles postovarian. Uterus not extending anterior of cirrus pouch. Parasites of catostomid fishes. North America. Key to species: Mackiewicz.[2044]

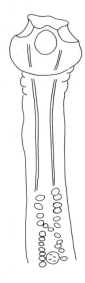

FIGURE 53. *Janiszewskella fortobothria* Mackiewicz et Deutsch 1976.

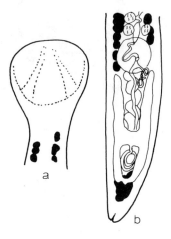

FIGURE 54. *Isoglaridacris bulbocirrus* Mackiewicz 1965. (a) Scolex; (b) posterior end.

Type species: *G. catostomi* Cooper 1920, in *Catostomus commersoni, C. catostomus, Hypentelium nigricans;* U.S.

Other species:

*G. confusa* Hunter 1927, in *Ictiobus bubalus, Dorosoma cepadianum, Catostomus commersoni;* U.S.

*G. gobii* (Szidat 1938) Yamaguti 1959 (syn. *Brachyurus gobii* Szidat (1938), in *Gobio fluviatilis;* Europe.

*G. intermedia* Lyster 1940, in *Catostomus commersoni;* North America.

*G. laruei* Lamont 1921, in *Catostomus commersoni;* Wisconsin, U.S.

*G. oligorchis* Haderlie 1953, in *Catostomus tahoensis;* California, U.S.

*G. terebrans (Linton 1893) Mackiewicz 1974 (syn. Monobothrium terebrans* Linton 1893; *Caryophyllaeus terebrans* [Linton 1893] Hunter 1930), in *Catostomus* spp.; U.S.

*G. vogei* Mackiewicz 1976, in *Catostomus macrocheilus;* U.S.

**18b. One gonopore present ............................................................ 19**

**19a. Scolex with apical disc ...................................................................**
**........................ *Janiszewskella* Mackiewicz et Deutsch 1976. (Figure 53)**

*Diagnosis:* Scolex with dorsal and ventral bothria, four lateral loculi, and apical disc. Neck constricted, short. Ejaculatory duct joins uterovaginal canal. External seminal vesicle present. Cirrus sac large, round. Preovarian vitellaria in two lateral rows. Postovarian vitellaria present. Ovary H-shaped. Uterus not extending anterior of cirrus pouch. Seminal receptacle absent. Parasites of North American Catostomidae.

Type species: *J. fortobothria* Mackiewicz et Deutsch 1976, in *Carpiodes cyprinus;* Pennsylvania, U.S.

**19b. Scolex lacking an apical disc ....... *Isoglaridacris* Mackiewicz 1965. (Figure 54)**

FIGURE 55. *Capingens singularis* Hunter 1927.

*Diagnosis:* Scolex rounded, wider than body, with three dorsal and three ventral shallow loculi. Neck present. Body long, slender. Single gonopore present. Cirrus sac rounded. External seminal vesicle present, inner seminal vesicle absent. Testes begin posterior to first vitelline follicles and extend to cirrus sac. Seminal receptacle weakly developed. Ovary shaped like inverted A; apex of A sometimes not joined. Preovarian vitellaria in lateral rows, not continuous with postovarian vitellaria. Uterus not extending anterior of cirrus pouch. Eggs operculate. Parasites of catostomid fishes. North America.

Type species: *I. bulbocirrus* Mackiewicz 1965, in *Catostomus* spp., *Hypentelium nigricans*; U.S.

Other species:

*I. agminis* Williams et Rogers 1972, in *Erimyzon oblongus, E. sucetta*; U.S.

*I. calentinei* Mackiewicz 1974, in *Catostomus columbianus*; Idaho, U.S.

*I. chetekensis* Williams 1977, in *Moxostoma macrolepidotum*; U.S.

*I. folius* Fredrickson et Ulmer 1965, in *Minytrema melanops, Moxostoma erythrum*; U.S.

*I. jonesi* Mackiewicz 1972, in *Moxostoma duquesnei, M. erythrurum*; U.S.

*I. longus* Fredreckson et Ulmer 1965, in *Moxostoma macrolepidotum*; U.S.

*I. wisconsinensis* Williams 1977, in *Hypentelium nigricans*; U.S.

# KEY TO THE GENERA IN CAPINGENTIDAE

1a. Postovarian median vitellaria present ............................................. 2
1b. Postovarian median vitellaria absent .............................................. 6

2a. Uterine coils extend anterior of cirrus pouch; scolex with two large bothria .... ............................................. *Capingens* **Hunter 1927. (Figure 55)**

*Diagnosis:* Scolex well-defined, one fifth of body length, with a dorsal and a ventral bothrium. Two gonopores present. Cirrus pouch round, between the two lobes of the ovary. External seminal vesicle present. Testes in wide band anterior to ovary. Ovary dumbbell-shaped, medullary, near posterior end, lateral lobes enclosing cirrus sac. Vitellaria extensive, continuous lateral to ovary, some follicles postovarian. Uterus anterior of cirrus pouch. Parasites of catostomid fishes. North America.

Type species: *C. singularis* Hunter 1927, in *Carpiodes carpio, Ictiobus urus*; U.S.

2b. Uterine coils not extending anterior of cirrus pouch; scolex lacking bothria .. 3

3a. Ovary shaped like an inverted A ....... *Adenoscolex* **Fotedar 1958. (Figure 56)**

*Diagnosis:* Scolex smooth, not clearly marked off from rest of body. Neck absent. Two gonopores; cirrus sac and uterovaginal canal open separately at beginning of posterior seventh

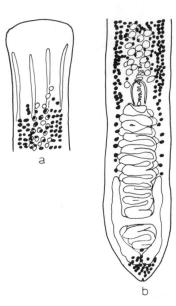

FIGURE 56. *Adenoscolex oreini* Fotedar 1958. (a) Scolex; (b) posterior end.

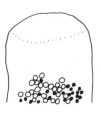

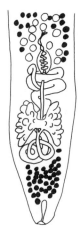

FIGURE 57. *Edlintonia ptychocheila* Mackiewicz 1970.

of body length. External seminal vesicle absent, inner seminal vesicle present. Testes extend from behind scolex to anterior end of cirrus sac. Seminal receptacle present. Ovary entirely medullary, shaped like an inverted A. Postovarian vitellaria present. Uterus never anterior of cirrus sac. Eggs operculated with blunt protuberance near basal end. Parasites of cyprinid fishes. Kashmir.

Type species: *A. oreini* Fotedar 1958, in *Oreinus sinuatus;* Kashmir.

3b.  Ovary not as above............................................................. 4

4a.  Ovary dumbbell-shaped; scolex quite reduced; neck absent......................
..................................................... *Breviscolex* Kulakowskaya 1962.

*Diagnosis:* Scolex smooth, very short, truncated. Neck absent. Genital apertures not described, apparently opening in common atrium. Cirrus large. External seminal vesicle absent. Testes extend from near anterior end of body to level of cirrus sac. Small seminal receptacle present. Ovary butterfly-shaped, variable. Vitellaria begin a little behind anterior testes, mostly lateral, numerous behind ovary. Uterus not extending anterior of ovary. Parasites of cyprinid fishes. Russia.

Type species: *B. orientalis* Kulakowskaya 1962, in *Hemibarbus maculatus, Chilogobio czarskii;* Russia.

4b.  Ovary otherwise; scolex well-developed; neck present........................... 5

5a.  Ovary H-shaped....................... *Edlintonia* Mackiewicz 1970. (Figure 57)

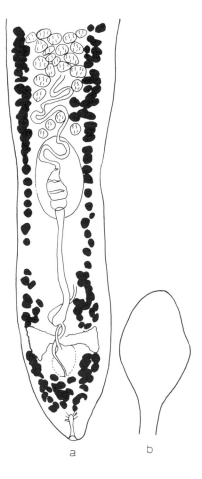

FIGURE 58. *Capengentoides batrachii* Gupta 1961. (a) Posterior end; (b) scolex.

*Diagnosis*: Scolex blunt, lacking loculi or bothria. Cirrus opening separately from uterovaginal canal. Ovary H-shaped. Uterus not extending anteriorly beyond cirrus sac. External seminal vesicle small, narrow, twisted. Vas deferens short, in small area between testes and cirrus sac. Preovarian vitellaria extend to cirrus or uterus in lateral and medial regions. Postovarian vitellaria well developed, not touching ovary. Seminal receptacle poorly developed. Egg operculate, undeveloped when shed. Parasites of North American cyprinid fishes.

Type species: *E. ptychocheila* Mackiewicz 1970, in *Ptychocheilus oregonense*, *Myocheilus caurinus*; British Columbia, U.S.

**5b.    Ovary band-shaped........................................................ *Capin-gentoides* Gupta 1961 (syn. *Pseudocapingentoides* Verma 1971). (Figure 58)**

*Diagnosis*: Scolex smooth, oval, truncated anteriorly and marked off from rest of body. Neck long, narrow. One gonopore; cirrus sac opens into uterovaginal canal at beginning of posterior tenth of body length. Uterine and vaginal pores common. Seminal vesicle present, bell-shaped. Ovary band-shaped or weakly H-shaped. Vitellaria mostly lateral, at level of inner muscular layer; some postovarian follicles present. Uterus never anterior of cirrus pouch. Seminal receptacle absent. Eggs nonoperculate. Parasites of freshwater siluroid fishes. India.

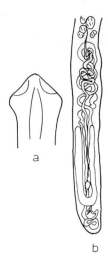

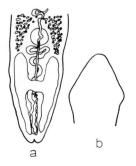

FIGURE 60. *Pseudolytocestus differtus* Hunter 1929. (a) Posterior end; (b) scolex.

FIGURE 59. *Spartoides wardi* Hunter 1927. (a) Scolex; (b) posterior end.

Type species: *C. batrachii* Gupta 1961, in *Clarias batrachus;* India.
Other species:
  *C. indica* (Verma 1971) comb. n. (syn. *Pseudocapingentoides indica* Verma 1971; *Capingentoides heteropneusti* Gupta et Sinha 1980), in *Heteropneustes fossilis;* India.
  *C. moghei* Pandey 1973, in *Channa striatus;* India.
  *C. singhia* Verma 1971, in *Heteropneustes fossilis;* India.

**6a. Ovary U-shaped; uterine coils extending anterior to cirrus pouch ............... ............................................. *Spartoides* Hunter 1929. (Figure 59)**

*Diagnosis:* Scolex well defined, with three pairs of loculi. Neck present. Body slender. Gonopores separate. Outer seminal vesicle present. Testes median, anterior to uterine coils. Ovary U-shaped; lateral lobes partly cortical. Vitellaria mostly medullary, in testicular zone, no postovarian follicles present. Uterus extending between testes and posterior end. Parasites of catostomid fishes. North America.
Type species: *S. wardi* Hunter 1927, in *Carpiodes* spp., *Ictiobus cyprinella;* U.S.

**6b. Ovary not U-shaped; uterine coils not extending anterior to cirrus pouch .... 7**

**7a. Neck absent; ovary H-shaped ........ *Pseudolytocestus* Hunter 1929. (Figure 60)**

*Diagnosis:* Scolex weakly defined, lacking loculi. Gonopores separate. Cirrus pouch well developed. External seminal vesicle present. Testes in broad median field anterior to ovary. Seminal receptacle absent. Ovary H-shaped, mostly medullary, at posterior end of body. Vitellaria in testicular zone; no postovarian follicles. Uterus not extending anterior of cirrus pouch. Parasites of catostomid fishes, North America; siluroid fish, India.
Type species: *P. differtus* Hunter 1929, in *Ictobius bubulus;* U.S.
Other species:
  *P. clariae* Gupta 1961, in *Clarias batrachus;* India.

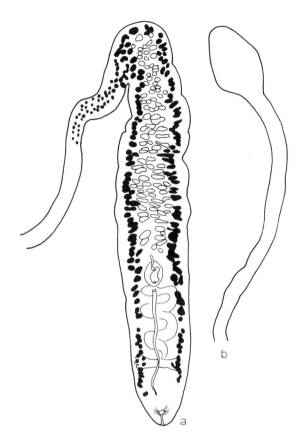

FIGURE 61. *Pseudocaryophyllaeus indica* Gupta 1961. (a) Body; (b) scolex.

**7b. Very long neck present; ovary band-shaped .......................................
....................................... *Pseudocaryophyllaeus* Gupta 1961. (Figure 61)**

*Diagnosis:* Scolex smooth, oval, truncated anteriorly, marked off from rest of body by long slender neck. Cirrus pouch and uterovaginal canal open separately. Postovarian median vitellaria absent. Uterus never extends anterior of cirrus pouch. Seminal receptacle absent. Ovary band-shaped. Testes filling most of medullary parenchyma anterior of cirrus pouch. Seminal vesicle present. Eggs nonoperculated. Osmoregulatory system with four main canals. Parasites of siluroid fishes. India.

Type species: *P. indica* Gupta 1961, in *Clarias batrachus;* India.

## ORDER SPATHEBOTHRIIDEA

This small order is comprised of parasites of teleost fishes, both marine and freshwater. The scolex either is undifferentiated or has a funnel-shaped or hollow, cup-like apical organ. External segmentation is absent, but there are several serially arranged sets of male and female reproductive systems. Genital pores are ventral or alternate dorsal and ventral. The cirrus pouch is median and the testes are in two lateral bands. The ovary is dendritic or bilobed, and the vitellaria are follicular, lateral, or scattered throughout the cortex. The uterus is coiled and median. Life cycles are unknown for any species.

## KEY TO THE FAMILIES IN SPATHEBOTHRIIDEA

1a. **Scolex with no adhesive organ** ................................................................
............................... Spathebothriidae Wardle et McLeod 1952. (p.43)

*Diagnosis:* Scolex and adhesive organs absent. Strobila slender, with pointed ends. External metamerism lacking. Genital pores irregularly alternating on flat surfaces. Testes lateral. Ovary rosettiform. Parasites of marine teleosts.
Type genus: *Spathebothrium* Linton 1922.

1b. **Scolex with one or two adhesive cups, but no true, muscular suckers** ......... 2

2a. **Adhesive cup single, funnel-shaped** ..... Cyathocephalidae Nybelin 1922. (p.44)

*Diagnosis:* Scolex with funnel-shaped apical adhesive organ. Slight constriction separating scolex from strobila. External metamerism absent; internal metamerism present. Testes medullary. Vitellaria cortical. Eggs operculated. Uterovaginal atrium present, with sphincter. Parasites of freshwater teleosts.
Type genus: *Cyathocephalus* Kessler 1868.

2b. **Adhesive cup double, often appearing as one, if contracted** ......................
................................................... Bothrimonidae Fam. N. (p.45)

*Diagnosis:* Scolex with two adhesive cups opening apically, or, when contracted, with a single such cup but still with a ventral septum. External metamerism absent, internal metamerism present. Genital pores ventral or irregularly alternating dorsal and ventral. Testes lateral, medullary. Cirrus opens anterior to vagina. Ovary bilobed. Vitellaria cortical. Uterus convoluted, anterior and opposite to ovary. Uterine pore anterior to vaginal pore, sometimes in common atrium with it. Eggs operculated. Parasites of teleosts.
Type genus: *Bothrimonus* Duvernoy 1842.

## DIAGNOSIS OF THE ONLY GENUS IN SPATHEBOTHRIIDAE

*Spathebothrium* Linton 1922. (Figure 62)

*Diagnosis:* No indication of scolex or adhesive organs present. Strobila flat, pointed at the ends, especially posterior. External metamerism absent. Internal metamerism present, with up to 36 sets of reproductive organs. No longitudinal muscles. Genital pores alternating irregularly from one flat surface to the other, uterine pore lateral to vaginal pore but rarely on opposite surface from it. No uterovaginal atrium. Testes in two lateral bands, continuous throughout strobila. Cirrus pouch small, anterior to vaginal pore. Ovary rosettiform. Vitel-

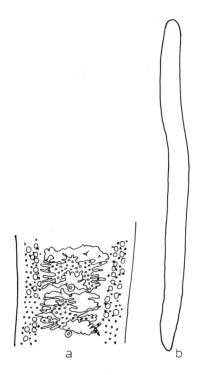

FIGURE 62. *Spathebothrium simplex* Linton 1922. (a) Portion of strobila; (b) entire worm.

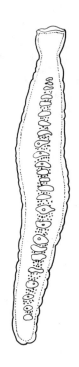

FIGURE 63. *Cyathocephalus truncatus* (Pallas 1781) Kessler 1868.

laria lateral. Uterus convoluted, median. Eggs with very large operculum and filaments at one pole. Parasites of marine teleosts. North Atlantic and North Pacific.

Type species: *S. simplex* Linton 1922, in *Liparis liparis, Crystallias matsushimae;* North Atlantic, Sea of Japan.

## DIAGNOSIS OF THE ONLY GENUS IN CYATHOCEPHALIDAE

*Cyathocephalus* Kessler 1868. (Figure 63)

*Diagnosis:* Scolex with funnel-shaped adhesive organ, separated from strobila by slight constriction. Strobila flat, up to 33 mm long. External metamerism absent, internal metamerism present, with 20 to 45 sets of reproductive organs. Genital pores irregularly alternating from dorsal to ventral surface. Cirrus pouch surrounded by prostatic cells. Cirrus opening anterior to uterovaginal atrium. Testes in lateral medulla. Ovary lobated, with broad isthmus near one flat surface or the other without regard to genital pore. Vitellaria cortical, surrounding body. Uterus convoluted, medial, antiporal, proximal coils surrounded by a great many unicellular glands. Eggs ellipsoidal, with operculum at one end and hook-like knob on the other. Infective embryo lacking hooks (?). Parasites of freshwater teleosts. Circumboreal.

Type species: *C. truncatus* (Pallas 1781) Kessler 1868, in *Esox lucius, Coregonus, Thymallus, Leucichthys, Lota, Perca, Salmo, Salvelinus, Cristovomer, Myoxocephalus*, etc.; Europe, North America, Siberia.

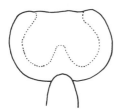

FIGURE 64. *Bothrimonus falax* Lühe 1900.

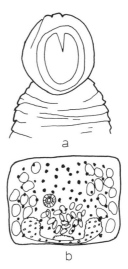

FIGURE 65. *Schizocotyle fluviatilis* Akhmerov 1960. (a) Scolex; (b) proglottid.

## KEY TO THE GENERA IN BOTHRIMONIDAE FAM. N.

**1a.** **Scolex with two apical, rounded adhesive cups, dorsal and ventral; lacking lateral slits. Vitellaria in two lateral, cortical bands............*Bothrimonus* Duvernoy 1842 (syn. *Diplocotyle* Krabbe 1874; *Didymobothrium* Nybelin 1922). (Figure 64)**

*Diagnosis:* Scolex with two forwardly-directed, hollow, sucker-like bothria, one dorsal and one ventral. Depending upon state of contraction, bothria may be expanded or pulled inwards to assume appearance of single, large, apical sucker. Neck absent. External segmentation lacking. Genital pores usually midventral, occasionally middorsal. Male pore anterior to female pore. Cirrus large, muscular. Cirrus pouch ovoid. Up to 20 medullary testes per set of genitalia. Ovary bilobed, subdivided into lobules, crescent-shaped in cross section. Vitellaria follicular, cortical, in two lateral bands. Vagina U-shaped. Uterus enters common pore with vagina. Eggs operculate. Egg with tuft of filaments at one end. Parasites of marine fishes.

Type species: *B. sturionis* Duvernoy 1842 (syn. *Diplocotyle olrikii* Krabbe 1874; *D. coherens* Linstow 1903; *D. nylandica* [Schneider 1902]; *D. rudolphii* Monticelli 1890; *Didymobothrium rudolphii* [Monticelli 1890] Nybelin 1922; *Bothrimonus intermedius* Cooper 1918; *B. nylandicus* Schneider 1902), in *Acipenser, Pseudopleuronectes, Pleuronectes, Solea, Salmo, Apeltes, Salvelinus, Onchorhynchus, Microgadus, Artediellus, Leucichthys, Coregonus,* etc.; North America, Finland, Italy, Greenland, Iceland, Norway, Spitzbergen.

Other species:
*B. fallax* Lühe 1900, in *Acipenser ruthenus;* Rumania (Black Sea).

**1b.** **Scolex with cups as above but each with a lateral slit. Vitellaria completely surround medulla .....................*Schizocotyle* Akhmerov 1960. (Figure 65)**

*Diagnosis:* Scolex short, rounded, with two cup-shaped adhesive organs separated their entire length by septum, and with slit in lateral margin of each. Neck absent. Genital pores ventral. Testes few, medullary, in two lateral fields. Ovary bilobed, medullary, posterior.

Vitellaria cortical, surrounding entire strobila. Uterus rather short, containing large eggs. Parasites of freshwater teleosts. Russia.

Type species: *S. fluviatilis* Akhmerov 1960, in *Pseudespinus leptocephalus;* Russia.

# ORDER TRYPANORHYNCHA

Adults of this group mature only in Elasmobranchii. No complete life cycle is known within the order, but juveniles very commonly are found in molluscs and marine fishes, which presumably are second intermediate or paratenic hosts. The beginning student of trypanorhynchs is likely to be confounded by the seemingly diabolical complexity of hook arrangements on the tentacles. However, no matter how complex the arrangement may be, it is surprisingly constant from one individual to another within the same species. The studies of hook shapes, sizes, and arrangements (oncotaxy) have led to the scheme of identification and classification now in use. It is an arbitrary scheme, but it does work, providing the researcher understands a few basic concepts, each of which has been given a special name. The following brief glossary will be helpful.

1. Antibothridial surface — the surface of a tentacle that is opposite to the closest bothridium.
2. Basal hooks — hooks at the base of the tentacle that are usually different from the metabasal hooks.
3. Bothridial surface — the surface of a tentacle that is toward the closest bothridium.
4. Chainette — a single or double longitudinal row of closely spaced, similar hooks on the external surface of a tentacle (Figure 67).
5. Enigmatic organ — a rounded organ of unknown function located inside the tentacle sheath at or near its junction with the bulb; in life it is reddish orange in color; found only in a few genera.
6. External surface — the surface of a tentacle that is farthest from the scolex.
7. Heteroacanthous armature — hooks arranged in alternating half circles, starting on internal surface and ending on external surface. Chainettes not present.
8. Heteroacanthous atypica — different numbers of hook rows on bothridial and antibothridial surfaces.
9. Heteroacanthous typica — same numbers of hook rows on both sides of tentacle.
10. Heteromorphous armature — hooks change radically in size and/or shape from internal to external surface.
11. Homeoacanthous armature — homeomorphous hooks that are arranged over metabasal tentacle in continuous spirals, or in quincunxes.
12. Homeomorphous armature — hooks are about the same size and shape throughout row.
13. Internal surface — the surface of a tentacle that is closest to scolex.
14. Metabasal hooks — those hooks along length of tentacle except for basal hooks.
15. Ooreceptacle — chamber between lobes of an ovary that lacks an isthmus.
16. Pars bothridialis — portion of scolex equal to length of bothridia (Figure 66).
17. Pars bulbosa — portion of scolex equal to length of bulbs (Figure 66).
18. Pars postbulbosa — portion of scolex posterior to bulbs (Figure 66).
19. Pars vaginalis — portion of scolex from anterior end to level of anterior end of bulbs (Figure 66).
20. Poeciloacanthous armature — metabasal hooks of differing sizes, shapes, and arrangements. Chainette present.
21. Renflement — swelling of base of tentacle; may bear basal hooks.

Proglottids of trypanorhynchs are so similar to those of the Tetraphyllidea and Lecanicephalidea that detached segments may not be identifiable. This is particularly important because mixed infections of these orders are common. Because of the delicate nature of the strobila of most species, great care must be taken to remove them from the spiral intestine of the definitive host without breaking them. A great many species, however, are known only from larvae obtained from paratenic hosts; these still can be identified by their hooks.

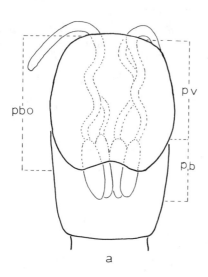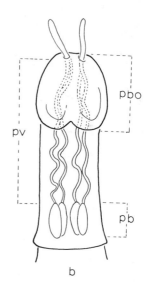

FIGURE 66. (a) Typical scolex of Acystidea; (b) typical scolex of Cystidea. pb — pars bulbosa. pbo — pars bothridialis. pv — pars vaginalis.

## KEY TO THE SUBORDERS IN TRYPANORHYNCHA

1a. **Pars bothridialis extending farther posteriad than pars vaginalis; plerocercus without blastocyst .......................... Acystidea Guiart 1927. (Figure 66a)**
1b. **Pars vaginalis extending farther posteriad than pars bothridialis; plerocercus with blastocyst ............................. Cystidea Guiart 1927. (Figure 66b)**

## KEY TO THE FAMILIES IN THE SUBORDER ACYSTIDEA

1a. **Bothridia each with a pair of ciliated sensory fossettes on the posterior margin. (See also, Otobothriidae in suborder Cystidea) ...................................**
**.............................................Paranybeliniidae Schmidt 1970. (p.68)**

*Diagnosis:* Scolex short and stout, craspedote. Tentacles armed with ascending spirals of similar hooks. Two bothridia present, longer than pars vaginalis. Each bothridium bears a pair of eversible ciliated fossettes on the posterior border. Known only from larvae dredged up in plankton nets. Hosts unknown.
Type genus: *Paranybelinia* Dollfus 1966.

1b. **Bothridia not as above .......................................................... 2**

2a. **Hooks solid; tentacle retractor muscle attached to bottom of bulb; bothridia separate ..................................... Tentaculariidae Poche 1926. (p.72)**

*Diagnosis:* Scolex craspedote. Tentacles short and slender. Retractor muscles attached to base of bulb. Bothridia sessile with separated spiny margins. Hooks solid, in even spirals, irregularly arranged at base of tentacle, similar in size and shape. Bulbs elipsoidal. Proboscis sheaths not twisted or spiral. Proglottids wider than long, anapolytic, acraspedote. Testes medullary, dorsal. Genital pores preequatorial, marginal or submarginal. Cirrus unarmed. Ovary equatorial or more posterior. Parasites of elasmobranchs.
Type genus: *Tentacularia* Bosc 1797.

2b.  Hooks hollow; tentacle retractor muscle not reaching bottom of bulb; bothridia partly or completely fused .......................................................... 3

3a.  Mature proglottid with double set of reproductive organs .......................
...............................................Hepatoxylidae Dollfus 1940. (p.62)

*Diagnosis:* Scolex large, acraspedote. Tentacles short, broad. Hooks hollow, mostly similar, arranged in regularly alternating longitudinal rows. Retractor muscle attached to top of bulb. Bothridia mostly embedded in scolex; lateral and sometimes anterior margins of adjacent bothridia fused. Proglottids wider than long, anapolytic, craspedote. Two sets of reproductive organs in each proglottid. Genital pores marginal, preequatorial. Parasites of sharks.
Type genus: *Hepatoxylon* Bosc 1811.

3b.  Mature proglottid with single set of reproductive organs........................
..............................................Sphyriocephalidae Pintner 1913. (p.71)

*Diagnosis:* Scolex short, wide, craspedote. Tentacles cylindrical. Hooks hollow, similar. Retractor muscle attached to anterior end of bulb. Pars vaginalis and pars bulbosa short. Bothridial margins fused, each bothridium forming deep cavity out of which the tentacle emerges. Proglottids craspedote, anapolytic. Genital pores marginal. One set of reproductive organs per proglottid. Parasites of sharks.
Type genus: *Sphyriocephalus* Pintner 1913.

## KEY TO THE FAMILIES IN THE SUBORDER CYSTIDEA

1a.  Each tentacle with externolateral longitudinal chainette (Figure 67a) or longitudinal band of small hooks (Figure 67b) ....................................... 2
1b.  Tentacles without chainette or longitudinal band of small hooks .............. 6

2a.  Scolex with two bothridia....................................................... 3
2b.  Scolex with four bothridia ..................................................... 5

3a.  Scolex craspedote..........................Dasyrhynchidae Dollfus 1935. (p.52)

*Diagnosis:* Scolex long, craspedote. Tentacles cylindrical, with chainette. Pars vaginalis long; tentacle sheath coiled. Pars bulbosa long; retractor muscles inserted in anterior end of bulb. Two prominent bothridia; each with distinct posterior notch. Proglottids acraspedote, anapolytic, posterior ones longer than wide. Single set of reproductive organs in each proglottid. Genital pores marginal, equatorial or postequatorial. Parasites of sharks.
Type genus: *Dasyrhynchus* Pintner 1928.

3b.  Scolex acraspedote.......................................................... 4

4a.  Chainette confined to external, basal portion of tentacle.........................
............................. Mixodigmatidae Dailey et Vogelbein 1982. (p.74)

*Diagnosis:* Scolex long, with prominent pars postbulbosa. Two bothridia present, each with conspicuous posterior notch. Tentacles long. Basal armature poeciloacanthus with basal, external chainette. Posterior to each chainette are three horn-like spines. Remaining basal armature arranged in longitudinal rows of spiniform hooks, becoming more rosethorn-shaped anteriad. Metabasal armature in ascending spiral half-rows from internal to external surface.

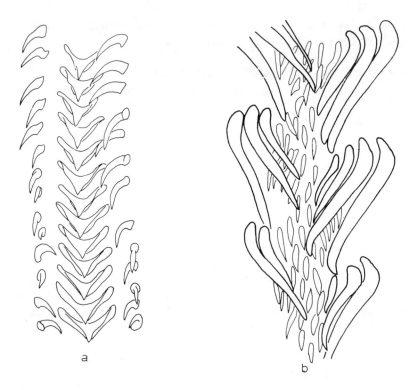

FIGURE 67. (a) Externolateral chainette; (b) externolateral band of hooks.

Pars bulbosa swollen; bulbs much longer than wide. Tentacle sheaths not greatly spiraled. Enigmatic organs present. Strobila anapolytic, acraspedote. Genital pores marginal, preequatorial. Testes medullary, preovarian. Ovary posterior. Cirrus pouch large, dividing testes into two separate fields. Parasites of sharks. Hawaii.

Type species: *M. leptaleum* Dailey et Vogelbein 1982.

**4b. Chainette not confined to external, basal portion of tentacle....................
................................................Lacistorhynchidae Guiart 1927. (p.63)**

*Diagnosis:* Scolex long, spinous, acraspedote. Tentacles with oblique rows of differing hollow hooks and a single or double chainette. Two bothridia present. Tentacle protractor muscle inserted at posterior end of bulb or preequatorial. Pars bulbosa often followed by a swollen region, homologous to appendix of plerocercus. Proglottids acraspedote, apolytic, usually spinous. Genital pores marginal, postequatorial, with muscular pad before and behind or surrounded by sucker. Parasites of elasmobranchs.

Type genus: *Lacistorhynchus* Pintner 1913.

**5a. Bothridia on short stalks, located on apex of scolex ................................
................................................Pterobothriidae Pintner 1931. (p.69)**

*Diagnosis:* Scolex long, slender, acraspedote, pars bulbosa usually swollen. Pars postbulbosa present. Tentacles with hooks of differing types and chainette of V-shaped hooks, or longitudinal band of small hooks. Bothridia pedunculate, mobile, four in number. Tentacle sheaths spiral or not. Proglottids acraspedote, apolytic. Genital pores marginal, postequatorial. Parasites of elasmobranchs.

Type genus: *Pterobothrium* Pintner 1931.

5b. Bothridia without stalks, located on dorsal and ventral surfaces of scolex ......
................................................. Gymnorhynchidae Dollfus 1935. (p.60)

*Diagnosis:* Scolex acraspedote: pars vaginalis longer than pars bothridialis. Pars postbulbosa present. Four sessile bothridia present, with anterior margins fused to scolex and other margins free. Tentacles with ascending spirals of falciform hooks and a double chainette or longitudinal band of very small hooks. Known only from plerocercus with large blastocyst, encysted in marine teleosts. Adult unknown, presumably in Squali.
Type genus: *Gymnorhynchus* Rudolphi 1819.

6a. Lateral or posterior margins of bothridia with two eversible ciliated pits .......
................................................. Otobothriidae Dollfus 1942. (p.67)

*Diagnosis:* Scolex craspedote, rather short. Tentacles shorter than sheaths, armed with oblique rows of large hooks from inner face to middle line of outer face and more numerous rows of smaller hooks over rest of tentacle. Chainette absent. Bothridia wide, shallow, sometimes notched on posterior margin, with two eversible ciliated pits on lateral or posterior margins. Proglottids acraspedote, apolytic. Genital pores marginal, equatorial or postequatorial.
Type genus: *Otobothrium* Linton 1890.

6b. No eversible ciliated pits on bothridia ............................................. 7

7a. Scolex with four bothridia ......................................................... 8
7b. Scolex with two bothridia ......................................................... 9

8a. Tentacles short, with similar hooks, or vestigial or lacking ......................
................................................. Gilquinidae Dollfus 1942. (p.59)

*Diagnosis:* Scolex acraspedote, marked off from the body by a constriction. Tentacles short, with similar hooks evenly distributed in spirals or quincunxes; or the tentacles may be vestigial or lacking, in which case the bulbs are still present. Four bothridia with inner margins fused to scolex. Proglottids acraspedote, apolytic or anapolytic. Genital pores marginal, preequatorial. Parasites of elasmobranchs.
Type genus: *Gilquinia* Guiart 1927.

8b. Tentacles long, slender, with dissimilar, atypical hooks ..........................
................................................. Rhinoptericolidae Carvajal et Campbell 1975. (p.71)

*Diagnosis:* Scolex long, slender, acraspedote, marked off from strobila by constriction. Tentacles long, slender. Armature heteroacanthus, heteromorphus, atypical. Four well-separated, sessile bothridia whose margins are entirely free from scolex. Bulbs long. No enigmatic organ. Strobila acraspedote, apolytic, with many segments. Ovary with four separate lobes lacking connecting isthmus. Median egg chamber (ooreceptacle) present between ovarian lobes. Vitellaria continuous, circumcortical. Nongravid uterus with two posterior diverticulae. Parasites of elasmobranchs.
Type genus: *Rhinoptericola* Carvajal et Campbell 1975.

9a. Testes overreaching osmoregulatory canals; no vitelline follicles behind ovary
...... Eutetrarhynchidae Guiart 1927 (syn. Mustelicolidae Dollfus 1969). (p.55)

*Diagnosis:* Scolex long, acraspedote. Tentacles long, cylindrical. Metabasal hooks on inner surface of different size than those of outer surface, or rarely similar. Basal hooks often greatly dissimilar. Two wide, flattened bothridia, each with or without distinct notch on posterior margin. Proglottids acraspedote, apolytic or anapolytic. Testes numerous, crossing osmoregulatory canals laterally. No postovarian vitellaria. Enigmatic organs present. Parasites of Batoidea.

Type genus: *Eutetrarhynchus* Pintner 1913.

**9b. Testes not overreaching osmoregulatory canals; some vitelline follicles behind ovary .................................... Hornelliellidae Yamaguti 1954. (p.63)**

*Diagnosis:* Scolex rather long, lacking pars postbulbosa; followed by unsegmented neck. Tentacles long and stout, armed with hooks of varying sizes and shapes. Two bothridia present, each with thickened, elevated borders. Proglottids apolytic, acraspedote.

Type genus: *Hornelliella* Yamaguti 1954.

## KEY TO THE GENERA IN DASYRHYNCHIDAE

**1a. Base of scolex expanded............................................................**
**.........*Dasyrhynchus* Pintner 1928 (syn. *Sbesterium* Dollfus 1929). (Figure 68)**

*Diagnosis:* Scolex long, slender, swollen at base, slightly craspedote. Tentacles long, slender, emerging near tip of scolex, armed with one or two chainettes and intercalary rows of small hooks as well as varying sizes and shapes of other hooks. Tentacular sheaths coiled or spiral; bulbs long and slender. Bothridia fused, each fused pair with deep posterior notch. Proglottids acraspedote. Gravid proglottids as long as wide. Genital pores irregularly alternating. Cirrus pouch oval. Cirrus unarmed. Testes very numerous, filling all available space between osmoregulatory canals. Vagina posterior to cirrus pouch, opening into common atrium with cirrus. Ovary rather small, bilobed, near posterior margin of proglottid. Vitellaria encircling proglottid, cortical. Gravid uterus saccular. Parasites of selachians. Atlantic and Pacific oceans. Key to species: Wardle and McLeod.[3870]

Type species: *D. variouncinatus* (Pintner 1913) Pintner 1928 (syn. *Halsiorhynchus variouncinnatus* Pintner 1913; *Tentacularia insignis* Shuler 1938; *Rhynchobothrium insigne* Linton 1924; *Dasyrhynchus insigne* Chandler 1942), in *Carcharias* spp., *Negaprion brevirostris;* Western Atlantic, Gulf of Mexico, New Guinea.

Other species:
- *D. giganteus* (Diesing 1850) Pintner 1928 (syn. *Anthocephalus giganteus* Diesing 1850), in *Chorinemus saliens, Carcharhinus* spp.; Brazil, Hawaii, Florida.
- *D. ingens* (Linton 1921) Pintner 1928 (syn. *Rhynchobothrium ingens* Linton 1921), in *Carcharhinus obscurus, Prionace glauca;* Atlantic and Pacific.
- *D. pacificus* Robinson 1965, in *Sciaena antarctica;* Australia.
- *D. talismani* Dollfus 1935, in *Galeus glaucus;* West Africa.

**1b. Base of scolex not expanded ..................................................... 2**

**2a. Chainette hooks with basal wings............................................*Floriceps* Cuvier 1817 (syn. *Anthocephalus* Rudolphi 1819, in part). (Figure 69)**

*Diagnosis:* Scolex long, slightly craspedote. Tentacles emerging near apex of scolex, each with chainette flanked by satellite hooks. Chainette hooks each biwinged. Pars vaginalis long. Bothridia two, reverse heart-shaped. Tentacle sheaths sinuous. Bulbs much longer

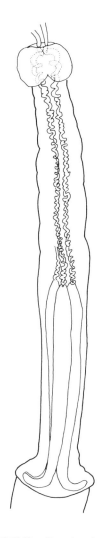

FIGURE 68. *Dasyrhynchus talismani* Dollfus 1935.

FIGURE 69. *Floriceps saccatus* Cuvier 1817.

than wide. Strobila acraspedote. Genital pores irregularly alternating. Cirrus unarmed. Osmoregulatory canals dorsal to cirrus pouch and vagina. Base of cirrus pouch directed anteriormedially in gravid proglottids. Testes filling field between osmoregulatory canals. Vagina opens anterior to cirrus pouch. Seminal receptacle present. Ovary X-shaped in cross section. Vitellaria lateral to muscle bundles, in lateral, interrupted fields. Uterine pore anteriomedian. Parasites of elasmobranchs. Atlantic and Pacific. Key to species Wardle and McCleod.[3870]

Type species: *F. saccatus* Cuvier 1817 (syn. *Anthocephalus elongatus* Rudolphi 1819), in *Mola mola* and other paratenic hosts; Atlantic and Pacific.

Other species:

*F. caballeroi* Cruz-Reyes 1977, in *Negaprion brevirostris*; Mexico.

*F. lichiae* Pintner 1929, in liver of *Scymnus lichia;* Italy.

*F. oxneri* Guiart 1938, in *Coris julis*; Atlantic.

**2b. Chainette hooks without basal wings ..... *Callitetrarhynchus* Pintner 1931 (syn. *Anthocephalus* Rudolphi 1819, in part; *Lintoniella* Yamaguti 1934). (Figure 70)**

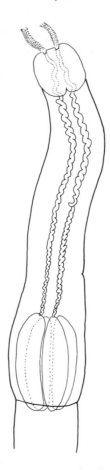

FIGURE 70. *Callitetrarhynchus gracilis* (Rudolphi 1819) Pintner 1931.

*Diagnosis:* Scolex long, very weakly craspedote. Tentacles emerging from anterior margin of bothridia. Pars vaginalis long, glandular; tentacle sheaths sinuous. Bulbs about three times longer than wide. Bothridia two, nearly round, with strong notch in posterior margin. Chainette hooks lacking lateral wings; satellite hooks present. Strobila thin, acraspedote. Genital pore in middle third of margin. Cirrus pouch pyriform. Testes small, partly postovarian. Vagina is behind cirrus pouch. Ovary bilobed, about one fifth from posterior end of proglottid. Vitellaria in lateral cortex. Parasites of selachians. Cosmopolitan.

Type species: *C. gracilis* (Rudolphi 1819) Pintner 1931 (syn. *Anthocephalus gracilis* Rudolphi 1819; *Rhynchobothrium speciosum* Linton 1897; *Callitetrarhynchus speciosum* [Linton 1897] Pintner 1931; *Lintoniella speciosa* [Linton 1897] Yamaguti 1934; *Callitetrarhynchus gracillimum* Pintner 1931; *Tentacularia pseudodera* Shuler 1938), in many teleost paratenic hosts. Adults in *Hypoprion brevirostris, Carcharhinus obscurus, Alutera schoepfii, Carcharhinus leucus;* cosmopolitan.

Other species:

*C. lepidum* (Chandler 1935) Chandler 1942 (syn. *Tentacularia lepida* Chandler 1935), in *Bagre marina, Galeichthys felis;* Texas.

*C. perelica* (Shuler 1938) Yamaguti 1959 (syn. *Tentacularia perelica* Shuler 1938), in *Hypoprion brevirostris;* Florida.

*C. triglae* (Diesing 1863) Yamaguti 1959 (syn. *Tetrarhynchus triglae* Diesing 1863), in *Trigla aspera, Trachinus draco, Chrysophrys aurata;* Europe.

FIGURE 71. *Mecistobothrium myliobati* Heinz et Dailey 1974.

## KEY TO THE GENERA IN EUTETRARHYNCHIDAE

(Syn. Tetrarhynchobothriidae Dollfus 1969, Mustelicolidae Dollfus 1969)

1a. **Bothridia longer than bulbs**.................................................................
    ................................*Mecistobothrium* **Heinz et Dailey 1974. (Figure 71)**

*Diagnosis:* Scolex acraspedote. Bothridia two, sessile, longer than bulbs. Tentacle sheaths not spiral. Tentacle armature heteroacanthus, heteromorphous, forming oblique rows beginning with large rosethorn-shaped hooks on external face and terminating with smaller hooks on the internal face of tentacle. Strobila craspedote. Anterior segments wider than long; terminal segments longer than wide. Testes preovarian, in two longitudinal rows. Genital pores irregularly alternating, in middle third of margin. Cirrus pouch occupying one third proglottid width. Ovary large, filling almost entire posterior fifth of proglottid. Gravid uterus extending to near anterior margin of segment. Parasites of Batoidea. North America.

Type species: *M. myliobati* Heinze et Dailey 1974, in *Myliobatis californica, Urolophus halleri;* California, U.S.

Other species:
  *M. brevispine* (Linton 1890) Campbell et Carvajal 1975 (syn. *Rhynchobothrium brevispine* Linton 1890; *Rhynchobothrium agile* Linton 1897), in *Rhinoptera bonasus;* Chesapeake Bay, U.S.

1b. **Bulbs longer than bothridia**.................................................................. 2

2a. **Basal hooks irregular, except for a circle of similar, rosethorn-shaped hooks at the extreme base**....... *Christianella* **Guiart 1931 (syn. *Armandia* Guiart 1927).**

*Diagnosis:* Scolex acraspedote. Bulbs long, narrow. Two bothridia, each strongly notched on posterior border. Tentacles very long, heteroacanthous, with spiral half turns of similar hooks. Strobila acraspedote, apolytic, consisting of only three to six proglottids. Genital pore in posterior quarter of lateral margin. Cirrus pouch large, reaching median line of proglottid. Cirrus long, unarmed. Testes filling preovarian, intervascular area. Ovary bilobed, posterior. Vitellaria encircling testicular zone. Uterus saccular. Parasites of selachians. Atlantic, Pacific, Mediterranean, Indian Ocean.

Type species: *C. minuta* (Beneden 1849) Guiart 1931 (syn. *Wageneria porrecta* Lühe

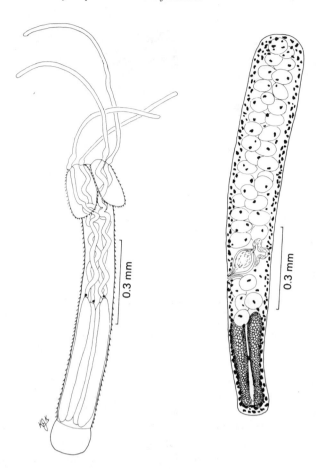

FIGURE 72. *Eutetrarhynchus thalassius* Kovacs et Schmidt 1980. (From Kovacs, K. J. and Schmidt, G. D., *Proc. Helminthol. Soc. Wash.*, 47, 10, 1980. With permission.)

1902, *Rhynchobothrium minutum* Beneden 1849), in *Squatina angelus, Carcharias* sp., *Rhina halavi;* India, Europe.

Other species:

    *C. trygonbrucco* (Wagener 1854) Guiart 1931 (syn. *Tetrarhynchus trygon-brucco* Wagener 1854), in *Trygon brucco, Urobatis halleri;* Mediterranean, California.

2b. **Basal hooks regular or irregular, but lacking basal circle of rosethorn-shaped hooks.................................................................................... 3**

3a. **Prebulbar enigmatic organs present, reddish-orange in life but clearly visible in fixed specimens ................................................................ 4**
3b. **Prebulbar enigmatic organs absent................................................ 5**

4a. **Metabasal hooks very similar in size and shape throughout length of tentacle .. ........................................................ *Eutetrarhynchus* Pintner 1913 (syn. *Renibulbus* Feigenbaum 1975, *Tetrarhynchobothrius* Dollfus 1969). (Figure 72)**

*Diagnosis:* Scolex long, acraspedote, covered with very small spines. Pars vaginalis longer than pars bothridialis and pars bulbosa. Pars postbulbosa small. Tentacle sheaths sinuous, with small red organ of unknown nature (enigmatic organ) at posterior end of each. Bulbs always much longer than wide. Two bothridia, with free posterior and lateral borders. Tentacles emerging from near apex of scolex. Armature heteroacanthus. Strobila anapolytic, craspedote or not, according to species. Genital pores irregularly alternating, postequatorial. Cirrus pouch pyriform, transverse. Testes large, numerous, preovarian. Vagina ventral to cirrus pouch, provided with a distal sphincter. Ovary bilobed, wide. Vitellaria surrounding testicular area. Uterus reaching to near anterior end of proglottid, with lateral branches. Parasites of sharks and rays. Cosmopolitan in warm waters. Key to species: Wardle and McCleod.[3870]

Type species: *E. ruficollis* (Eysenhardt 1829) Pintner 1913 (syn. *Tetrarhynchus ruficollis* Eysenhardt, *Rhynchobothrium ruficolle* Diesing 1887, *Tetrarhynchus corollatus* Rudolphi 1870, *Coenomorphus joyeuxii* Vaullegeard 1893), in *Mustelus* spp., *Acanthias, Raja;* Mediterranean, Atlantic.

Other species:

*E. araya* (Woodland 1934) Yamaguti 1959 (syn. *Tentacularia araya* Woodland 1934, *E. baeri* Lopez-Neyra et Diaz-Ungria 1958), in *Trygon* sp.; Amazon.

*E. caribbensis* Kovacs et Schmidt 1980, in *Urolophus jamaicensis;* Jamaica.

*E. caroyoni* Dollfus 1942, in *Clibanarius misanthropus*, (Crustacea); France.

*E. geraschmidti* Dollfus 1974, in *Urolophus testaceus;* South Australia.

*E. glaber* Dollfus 1969, in *Myliobatis aquila;* Mediterranean.

*E. leucomelanus* (Shipley et Hornell 1906) Dollfus 1942 (syn. *Tetrarhynchus leucomelanus* Shipley et Hornell 1906), in *Dasybatis sephen, Rhynchobatus djeddensis, Dasybatis kuhli;* Sri Lanka.

*E. lineatus* (Linton 1909) Dollfus 1942 (syn. *Rhynchobothrium lineatus* Linton 1909), *Ginglystoma cirratum;* Florida, U.S.

*E. litocephalus* Heinz et Dailey 1974, in *Mustelus californicus, Triakis semifasciata;* Mexico.

*E. macrotrachelus* Heinz et Dailey 1974, in *Mustelus californicus;* California, U.S.

*E. penaeus* (Feigenbaum 1975) comb. n. (syn. *Renibulbus penaeus* Feigenbaum 1975), in *Penaeus brasiliensis* (Crustacea); Florida, U.S.

*E. schmidti* Heinz et Dailey 1974, in *Urolophus halleri, Rhinobatos productus;* California, U.S.

*E. settiense* (Dollfus 1969) comb. n. (syn. *Tetrarhynchobothrium settiense* Dollfus 1969), in *Mustelus mustelus, M. canis, Myliobatis aquila;* Mediterranean.

*E. spinifer* Dollfus 1969, in *Myliobatis aquila;* Mediterranean.

*E. thalassius* Kovacs et Schmidt 1980, in *Urolophus jamaicensis;* Jamaica.

**4b. Innermost hooks in each ascending half row of metabasal armature largest, becoming smaller towards the outer surface ..................................... ..... Parachristianella Dollfus 1946 (syn. Mustelicola Dollfus 1969). (Figure 73)**

*Diagnosis:* Scolex slender, acraspedote; pars bulbosa and pars vaginalis each longer than pars bothridialis. Bulbs much longer than wide. Tentacle sheaths slightly sinuous but not spiral. Bothridia two, rounded, not contiguous apically. Retractors inserted in posterior ends of bulbs. Prebulbar organ distinct. Tentacles moderately long. Armature heteroacanthous, consisting of ascending spiral rows beginning with two large triangular hooks at middle of internal surface, remaining hooks decreasing in size as row ends near middle of external surface. Parasites of Batoidea. France, California.

Type species: *P. trygonis* Dollfus 1946, in *Trygon pastinaca, Urobatis halleri;* France, California.

Other species:

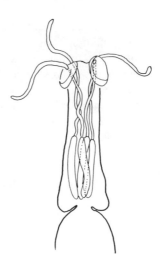

FIGURE 73. *Parachristianella trygonis* Dollfus 1946.

*P. heteromegacanthus* Feigenbaum 1975, in *Penaeus brasiliensis* (Crustacea), Florida, U.S.

*P. monomegacantha* Kruse 1959, in *Rhinobates productus, Dasyatis lata, D. americana, Penaeus duorarum;* California, Hawaii, Florida, Virginia.

5a. **One basal hook on each tentacle much larger than any others on tentacle; remaining hooks similar .................................. *Oncomegas* Dollfus 1929.**

*Diagnosis:* Two flat bothridia, sometimes with faint notch in posterior margin, which is free. Pars bulbosa very long, bulbs much longer than wide. Pars postbulbosa short. Pars vaginalis long; tentacle sheaths highly convoluted. Enigmatic organs absent. Tentacles long, thin. Hooks very numerous, closely spaced. Basal hooks similar to metabasal hooks, except for one hook at base of each tentacle, which is large and stout. Hooks arranged in ascending half-spirals. Proglottids acraspedote. Genital pore about equatorial. Cirrus pouch does not reach midline. Testes in two longitudinal rows in young segments, then filling intervascular parenchyma. Cirrus apparently unarmed. Vas deferens not tightly coiled. Vitellaria surround medulla. Ovary small, at extreme posterior of segment. Vagina posterior to cirrus pouch. Parasites of elasmobranchs. Atlantic, Pacific.

Type species: *O. wageneri* (Linton 1890) Dollfus 1929 (syn. *Rhynchobothrium wageneri* Linton 1890), in *Dasybatis centrura* and several teleost paratenic hosts; also dredged in plankton nets (Dollfus);[809] Europe, Japan, Western Atlantic.

5b. **Not as above. First metabasal hook in each ascending row small, becoming larger near middle of row, then decreasing toward end of row at outer surface ....... ................................................... *Prochristianella* Dollfus 1946.**

*Diagnosis:* Scolex long, slender, acraspedote. Pars bulbosa and pars vaginalis both longer than pars bothridialis. Pars bulbosa swollen. Tentacle sheaths sinuous, not spiral. Prebulbar organs not seen. Bulbs six or seven times longer than wide. Retractors inserted at posterior ends of bulbs. Bothridia two, rounded, not contiguous apically. Tentacles emerging near apex of scolex. Armature heteroacanthus, complex. First hooks in each ascending diagonal row small, enlarging towards middle of row, then decreasing in size toward end of row.

FIGURE 74. *Gilguinia squali* (Fabricius 1794) Guiart 1927.

Acraspedote. Genital pore immediately postequatorial. Testes in two longitudinal rows of 29 to 32 each; no testes postovarian. Ovary occupies the last one seventh of proglottid. Parasites of elasmobranchs. Atlantic, Pacific, Gulf of Mexico.

Type species: *P. pastinaca, Upogebia stelate;* France.

Other species:

*P. aetobatus* Robinson 1958, in *Aetobatus tenuicaudatus;* New Zealand.

*P. fragilis* Heinz et Dailey 1974, in *Rhinobatos productus;* California.

*P. heteracantha* Dailey et Carvajal 1976, in *Rhinobatos planiceps;* Chile.

*P. hispida* (Linton 1890) Campbell et Carvajal 1975 (syn. *Rhynchobothrium hispida* Linton 1890, *P. penaei* Kruse 1959), in *Trygon centrura, Dasyatis say;* Western Atlantic, Gulf of Mexico.

*P. micracantha* Carvajal, Campbell et Cornford 1976, in *Dasyatis lata;* Hawaii.

*P. minima* Heinz et Dailey 1974, in *Urolophus halleri, Platyrhinoidis triseriata;* California.

*P. musteli* Carvajal 1974, in *Mustelus mento;* Chile.

*P. tenuispine* (Linton 1890) Dollfus 1946, in *Trygon centrura;* Massachusetts.

*P. tumidula* (Linton 1890) Campbell et Carvajal 1975 (syn. *Rhynchobothrium tumidula* Linton 1890, *Callitetrarhynchus tumidula* (Linton 1890) Pintner 1931), in *Mustelus canis, M. laevis, Dasyatis lata;* Western Atlantic, Hawaii.

## KEY TO THE GENERA IN GILQUINIIDAE

**1a. Well-developed tentacles present............ *Gilquinia* Guiart 1927. (Figure 74)**

*Diagnosis:* Scolex acraspedote, pars vaginalis longer than pars bothridialis and pars bulbosa. Tentacle sheaths sinuous. Bulbs elliptical. Bothrida four. Tentacles emerge from notches in anterior margin of bothridia. Armature homeoacanthous. Strobila acraspedote. Genital pores irregularly alternating, preequatorial. Vagina ventral to cirrus. Zerney's vesicle well developed. Testes pre- and postovarian. Ovary bilobed, separated from posterior end of proglottid by testes. Vitellaria completely encircling medulla. Uterus extending to near level of genital pore, filling most of proglottid when gravid. Parasites of Squalii. Atlantic, Mediterranean, Pacific.

Type species: *G. squali* (Fabricius 1794) Guiart 1927 (syn. *Taenia squali* Fabricius 1794, *Rhynchobothrium tetrabothrium* Beneden 1849, *Tetrarhynchobothrium affine* Diesing 1854), in *Squalus acanthias, S. sucklii, Mustelus ninnulus;* Atlantic, Pacific, Mediterranean.

Other species:

*G. anteroporus* (Hart 1936) Dollfus 1942 (syn. *Gilquinia tetrabothrium* Wardle 1932, *Tetrarhynchus anteroporus* Hart 1936), in *Squalus suckleyi;* Washington, U.S.

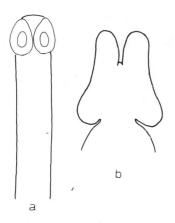

FIGURE 75. *Apororhynchus norvegicum* (Olsson 1868) Nybelin 1918. (a) Scolex; (b) longitudinal section of scolex.

FIGURE 76. *Myrmillorhynchus pearsoni* (Southwell 1929) Bilqees 1980. Circlet of hooks.

*G. nannocephala* (Pintner 1929) Dollfus 1942 (syn. *Tetrarhynchobothrium nannocephala* Pintner 1929); Japan.

**1b.    Tentacles missing, although tentacle sheaths present ............................**
**............................................*Aporhynchus* Nybelin 1918. (Figure 75)**

*Diagnosis:* Scolex acraspedote, long. Four bothridia with free latero-posterior margins. Tentacles not present; tentacle sheaths and bulbs present. Strobila apolytic, acraspedote. Neck present. Genital pores muscular, irregularly alternating, preequatorial. Zerney's vesicle and external seminal vesicle present. Cirrus pouch extends to ventral osmoregulatory canal. Testes numerous, in pre- and postovarian fields. Vagina opens posterior to cirrus. Ovary bilobed, separated from posterior end of proglottid by testes. Vitellaria encircling entire proglottid. Uterine pore present. Parasites of Squalii. North Atlantic, Scandinavia.

Type species: *A. norvegicum* (Olsson 1866) Nybelin 1918 (syn. *Tetrabothrium* Olsson 1866), in *Etmopterus spinax;* Scandinavia.

## KEY TO THE GENERA IN GYMNORHYNCHIDAE

**1a.    Basal hooks not longer than the rest ...............................................**
**................................... *Myrmillorhynchus* Bilqees 1980. (Figure 76)**

*Diagnosis:* Scolex acraspedote. Four sessile bothridia with anterior margins fused to scolex. Pars vaginalis much longer than pars bothridialis and slightly longer or nearly equal to pars bulbosa. Pars bothridialis wider than pars vaginalis and pars bulbosa. Bulbs much longer than wide. Armature heteroacanthous and heteromorphus. Hooks at base of armed portion not longer than on rest of tentacle. Strobila acraspedote. Genital pores unilateral, in posterior one fourth of proglottid. Cirrus pouch spheroid, thin-walled. Testes numerous, prevaginal. Ovary bilobed, posterior. Vitellaria surrounding medulla. Uterus a median tube. Parasites of elasmobranchs; plerocercus in teleosts. India, Pakistan.

Type species: *M. pearsoni* (Southwell 1929) Bilqees 1980 (syn. *Tetrarhynchus pearsoni* Southwell 1929), in *Myrmillo manazo, Cybium guttatum;* Pakistan, India.

**1b.    Basal hooks notably longer than the rest......................................... 2**

FIGURE 77. *Gymnorhynchus gigas* (Cuvier 1817) Rudolphi 1819.

2a. **Tentacles with double chainette; incomplete ring of very large hooks at base of armed portion** ................................................. *Gymnorhynchus* **Rudolphi 1819** (syn. *Anthocephalus* **Rudolphi 1819**, in part; *Floriceps* **Cuvier 1817**, in part; *Vaullegeardia* **Guiart 1927**). (Figure 77)

*Diagnosis:* Scolex acraspedote, pars bothridialis shorter than pars vaginalis and pars bulbosa. Bothridia four, anterior and medial margins fused with scolex, other margins free. Tentacles emerging near apex of scolex. Armature poeciloacanthous, with double chainette; incomplete ring of very large hooks separating armed portion from unarmed portion. Adult anatomy unknown. Plerocercus known from several marine teleosts. Atlantic, Mediterranean, Indian Ocean.

Type species: *G. gigas* (Cuvier 1817) Rudolphi 1819 (syn. *Scolex gigas* Cuvier 1817, *Gymnorhynchus reptans* Rudolphi 1819, *Vaullegeardia moniezi* (Railliet 1899) Guiart 1927), in many marine teleosts; Atlantic.

Other species:

*G. cybiumi* Chincholikar et Shinde 1977, in *Cybium guttatum*; India.

2b. **Tentacles without chainettes; complete ring of very large hooks at base of armed portion** .................................. *Molicola* **Dollfus 1935** (syn. *Anthocephalus* **Rudolphi 1819**, in part; *Gymnorhynchus* **Rudolphi 1819**, in part). (Figure 78)

*Diagnosis:* Scolex acraspedote. Retractors attached to bases of bulbs. Bothridia four, comma-shaped, anterior ends fused to scolex. Tentacles emerging from near apex of scolex. Armature poecilacanthous, chainettes replaced by longitudinal band of small hooks on external surface of tentacle. Complete ring of large hooks separating armed from unarmed portions of tentacles. Strobila acraspedote, apolytic. Genital pores irregularly alternating, preequatorial on lateral margin. Cirrus pouch pyriform, thin-walled. Cirrus long. External seminal vesicle present. Testes numerous, in intervascular medulla. Vagina with distal sphincter, opening into atrium ventral to cirrus. Ovary bilobed, separated from posterior end of proglottid by several testes. Vitellaria surrounding medulla. Uterus elongate, with a ventral, poral outpocketing. Parasites of selachians. Very common, encysted in marine teleosts. Atlantic, Pacific.

Type species: *M. horridus* (Goodsir 1841) Dollfus 1935 (syn. *Gymnorhynchus horridus* Goodsir 1841, *Anthocephalus elongatus* Rudolphi 1819, in part — see Dollfus[785] for other synonyms), in *Isurus glaucus;* Pacific, Atlantic.

Other species:

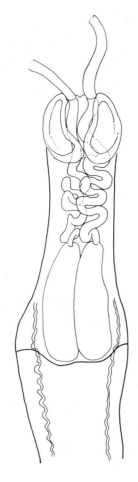

FIGURE 78. *Molicola horridus* (Goodsir 1841) Dollfus 1935.

FIGURE 79. *Hepatoxylon trichiuri* (Holton 1802) Dollfus 1942.

*M. thyrsitae* Robinson 1959 (syn. *Gymnorhynchus (Molicola) thyrsitae* Robinson 1959), in *Thyrsites atun;* New Zealand.

*M. isuri* Robinson 1959 (syn. *Gymnorhynchus (Molicola) isuri* Robinson 1959), in *Isurus glaucus;* New Zealand.

*M. uncinatus* (Linton 1924) Dollfus 1935 (syn. *Rhynchobothrium uncinatus* Linton 1924, *Floriceps uncinatus* [Linton 1924] Yamaguti 1952), in *Vulpecula marina;* Atlantic, Pacific.

## DIAGNOSIS OF THE ONLY GENUS IN HEPATOXYLIDAE

*Hepatoxylon* Bosc 1811 (Figure 79)
(Syn. *Dibothriorhynchus* Blainville in Bremser 1924, *Coenomorphus* Loennberg 1899; *Diplogonimus* Guiart 1931, *Tetrantaris* Templeton 1836)

*Diagnosis:* Scolex wrinkled, acraspedote. Bothridia separated by narrow partition; each bothridium in form of narrow slit. Tentacles short, globular or truncated. Hooks similar, in spiral longitudinal rows, decreasing in size posteriad. Tentacle retractors not inserted on base of sheath. Bulbs ellipsoid to comma-shaped. Proglottids much wider than long. Two sets of genitalia per proglottid, each with a ventral uterine pore. Cirrus unarmed. Testes median to osmoregulatory canals. Ovaries compact, near posterior margin of proglottid. Vitellaria cortical, encircling proglottid. Vagina posterior to cirrus, with outer and inner sphincters. Parasites of selachians. Cosmopolitan.

Type species: *H. squali* (Martin 1797) Bosc 1811 (syn. *H. trichuri* [Holten 1802] Bosc 1811, *Echinorhynchus trichiuri* Holten 1802, *Dibothriorhynchus lepidopteri* Blainville 1828), in *Squalus*, *Isurus*, many marine teleosts. Cosmopolitan.

## DIAGNOSIS OF THE ONLY GENUS IN HORNELLIELLIDAE

### *Hornelliella* Yamaguti 1954

*Diagnosis:* Scolex long, stout. Pars bulbosa and pars vaginalis long, pars postbulbosa absent. Tentacle sheaths provided with muscular rings. Bothridia two, large, lacking posterior margin. Tentacles emerge from anterior margin of bothridia, heteroacanthous. Strobila acraspedote, apolytic. Genital pores irregularly alternating. Genital ducts opening into common muscular pouch, the hermaphroditic organ. Testes numerous, intervascular, none as far posterior as ovary. Ovary four-winged, separated from posterior end of proglottid by transverse band of vitellaria. Vitelline follicles profuse, between transverse and longitudinal muscle sheaths. Uterus ventromedian, extending to anterior end of proglottid; preformed uterine pore present. Eggs nonoperculate. Parasites of elasmobranchs. Mediterranean.

Type species: *H. annandalei* (Hornell 1912) Yamaguti 1954 (syn. *Tetrarhynchus annandalei* Hornell 1912), in *Stegostoma tigrinum*; Sri Lanka, Macassar.
Other species:
*H. palasoorahi* Zeidi et Khan 1976, in *Scoliodon palasoorah*; Pakistan.

## KEY TO THE GENERA IN LACISTORHYNCHIDAE

(Syn. Pseudogrillotiidae Dollfus 1969)

**1a. Bothridia not notched posteriorly**........ *Eulacistorhynchus* **Subhapradha 1957.**

*Diagnosis:* Scolex fairly long, acraspedote. Bulbs long; retractor muscles attached to bases of bulbs. Bothridia two, oval or rounded, lacking posterior notch. Tentacles long, poeciloacanthus; double chainette present. Strobila acraspedote, mature proglottids much longer than wide. Genital pore lateral, in posterior third of proglottid. Cirrus pouch oval, cirrus armed. Testes in single field medial to vitellaria. Ovary compactly lobated, near posterior end of proglottid. Vitellaria lateral. Parasites of sharks. India.

Type species: *E. chiloscyllius* Subhapradha 1957, in *Chiloscyllium griseum*; India.

**1b. Bothridia notched posteriorly** ................................................... 2

**2a. Bulbs very long**............................................ *Grillotia* **Guiart 1927**
(syn. *Heterotetrarhynchus* **Pintner 1929,** *Pintneriella* **Yamaguti 1934). (Figure 80)**

*Diagnosis:* Scolex acraspedote, long, slender, with the tentacles emerging from anterior margins of bothridia. Pars vaginalis long; tentacle sheaths not spiral. Retractor muscles inserted in posterior ends of bulbs. Pars postbulbosa short. Bothridia two, nearly rounded, with small notch in posterior margin. Tentacles poeciloacanthous with hollow hooks of several types. Midline of external surface with longitudinal band of small hooks arranged in sinusoidal pattern, in place of chainette. Midline of internal surface lacking hooks. Strobila acraspedote, apolytic. Genital atrium in posterior half of proglottid margin, irregularly

FIGURE 80. *Grillotia* sp.

alternating. Cirrus pouch ovoid, nearly one fourth width of proglottid. Cirrus unarmed. Testes numerous, some usually postovarian. Vagina ventral to cirrus pouch. Seminal receptacle present. Ovary bilobed, somewhat anterior to posterior end of proglottid. Uterus with lateral branches, extending nearly whole length of proglottid. Vitellaria lateral, mostly medullary. Parasites of elasmobranchs. Cosmopolitan.

Type species: *G. erinaceus* (Beneden 1858) Guiart 1927 (syn. *Tetrarhynchus erinaceus* Beneden 1858, *Rhynchobothrium imparspine* Linton 1890. For other synonyms see Dollfus, 1942), in rays; Atlantic.

Other species: (Note: Dollfus[787,807] proposed the division of this genus into the subgenera *Grillotia, Progrillotia,* and *Paragrillotia.* I have not incorporated this system here; the interested student is referred to Dollfus' papers.)

*G. acanthoscolex* Rees 1944, in *Hexanchus griseus*; Atlantic.

*G. angeli* Dollfus 1969, in *Squatina squatina;* Mediterranean.

*G. bothridiopunctata* Dollfus 1969, in *Caranx trachurus;* Mediterranean.

*G. branchii* Shaharom et Lester 1982, in *Scomberomorus commersoni;* Australia, Malaysia.

*G. dolichocephala* Guiart 1935, in *Centroscymnus coelolepis, Pseudotriacis microdon;* Atlantic.

*G. dollfusi* Carvajal 1971, in *Raja chinensis;* Chile.

*G. institata* (Pintner 1931) (syn. *Heterotetrarhynchus institata* Pintner 1931), in *Heptanchus cinereus, H. griseus, Scymnus lichia;* Mediterranean.

*G. louiseuzeti* Dollfus 1969, in *Dasyatis violacea;* Mediterranean.

*G. megabothridia* (Hart 1936) (syn. *Tentacularia megabothridia* Hart 1936), in *Hexanchus griseus;* Puget Sound.

*G. microthrix* Dollfus 1969, in *Torpedo nobiliana;* Mediterranean.

*G. minor* Guiart 1935, in *Lepidorhinus squamosus;* France.

*G. musculara* (Hart 1936) (syn. *Tentacularia musculara* Hart 1936), Puget Sound.

*G. musculicola* (Yamaguti 1934) Yamaguti 1959 (syn. *Pintneriella musculicola* Yamaguti 1934), in *Pagrosomus unicolor, Epinephelus akaara;* Japan.

*G. pastinacae* Dollfus 1946, in *Trygon pastinaca;* France.

*G. pseudoerinaceus* Dollfus 1969, in *Raja oxyrhynchus;* Mediterranean.

*G. recurvispinus* Dollfus 1969, in *Raja clavata;* Mediterranean.

*G. rowei* Campbell 1977, in *Bathyraja richardsoni;* Western North Atlantic.

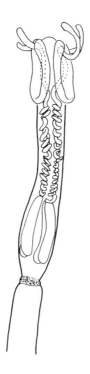

FIGURE 81. *Lacistorhynchus tenius* (Beneden 1858) Pintner 1913.

*G. scolecina* (Rudolphi 1819) Guiart 1927 (syn. *Tetrarhynchus scolecina* Rudolphi 1819, *Tetrarhynchus heptanchi* Vaullegeard 1899), in *Heptanchus cinereus, H. griseus, Torpedo ocellata, Scymnorhinus lichia;* Mediterranean, Atlantic, Pacific.

*G. simmonsi* Dollfus 1969, in *Gingylostoma cirratum;* Florida, U.S.

*G. smarisgora* (Wagener 1854) Dollfus 1946 (syn. *Tetrarhynchus smaris-gora* Wagener 1854, *Tetrarhynchus smaridis-gorae* Diesing 1863), in *Smaris gora, Squatina squatina;* Atlantic.

*G. spinosissima* Dollfus 1969, in *Hexanchus griseus;* Mediterranean.

**2b. Bulbs short ............................................................................ 3**

**3a. Scolex acraspedote................... *Lacistorhynchus* Pintner 1913. (Figure 81)**

*Diagnosis:* Scolex acraspedote, long, pars vaginalis very long. Tentacle sheaths spiral. Pars bulbosa rather short, slightly swollen. Bothridia two, rounded, slightly notched on posterior margin. Tentacles emerging from anterior edge of bothridia. Poeciloacanthous, with simple chainette. Strobila acraspedote, slender. Genital atria irregularly alternating, with muscular pad before and after. Cirrus pouch pyriform. Testes very numerous, filling entire intervascular field, some postovarian. Vagina posterior to cirrus pouch. Ovary four-lobed in cross section, separated from posterior end of proglottid by several testes. Vitellaria lateral. Uterus median, long. Parasites of selachians. Atlantic, Pacific.

Type species: *L. tenuis* (Beneden 1858) Pintner 1913 (syn. *Tetrarhynchus tenuis* Beneden 1858, *Lacistorhynchus planiceps* [Leuckart 1819], *Rhynchobothrium gracile* Diesing 1863, *R. bulbifer* Linton 1889, *R. benedeni* [Crety 1890]. For numerous other synonymies see Dollfus[785]), in *Squalus acanthias, Mustelus canis, Vulpecula marina, Triakis semifasciata, T. maculata;* cosmopolitan.

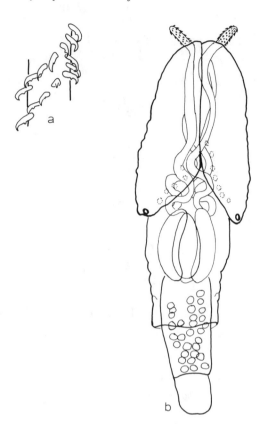

FIGURE 82. *Pseudonybelinia odontacantha* Dollfus 1966. (a) Portion of tentacle with typical hooks; (b) entire juvenile.

**3b. Scolex craspedote .................................. *Pseudogrillotia* Dollfus 1969.**

*Diagnosis:* Scolex long, craspedote. Two bothridia, each with conspicuous posterior notch. Pars vaginalis long, constricted near junction with pars bulbosa. Bulbs longer than wide. Retractor muscles insert in bottom end of bulb. Tentacle sheaths sinuous. Metabasal armature with broad band of small, irregularly arranged hooks on external surface. Basal swelling absent, basal hooks small, thin, with no apparent order. First described as postlarva in muscles of marine fish. Gulf of Mexico, Pacific.

Type species: *P. pleistacantha* Dollfus 1969, in *Pogonias chromia;* Galveston, Texas. Other species:

    *P. basipunctata* Carvajal, Campbell et Cornford 1976, in *Carcharhinus amblyrhyncos;* Hawaii.

## KEY TO THE GENERA IN PARANYBELINIIDAE

**1a. Bulbs of pars bulbosa about three times longer than wide, hooks each with ventral tooth ................................. *Pseudonybelinia* Dollfus 1966. (Figure 82)**

*Diagnosis:* Scolex craspedote. Tentacles inserted near anterior margin of bothridia, armed with hooks in spirals ascending left to right. Hooks similar throughout tentacle, with feeble curve and provided with a tooth-like protuberance on the ventral side. Bulbs three times longer than wide. Posterior margins of bothridia provided with a pair of eversible ciliated

FIGURE 83. *Paranybelinia otobothrioides* Dollfus 1966.

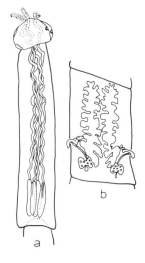

FIGURE 84. *Diplootobothrium springeri* Chandler 1942. (a) Scolex; (b) proglottid.

pits, or fossettes. Adult unknown. Known only from larvae dredged up in plankton nets among nocturnal surface plankton. Atlantic.

Type species: *P. odontacantha* Dollfus 1966, in "plankton"; Cape Verde Islands.

**1b.  Bulbs of pars bulbosa about twice longer than wide, hooks lacking ventral tooth .................................... *Paranybelinia* Dollfus 1966. (Figure 83)**

*Diagnosis:* Scolex craspedote. Tentacles inserted near anterior margins of bothridia, armed with hooks in spirals ascending left to right. Hooks all similar, arcuate, lacking ventral tooth-like projection. Bulbs twice longer than wide. Posterior margins of bothridia provided with a pair of eversible ciliated pits. Adult unknown. Known only from larvae found with nocturnal surface plankton netted in the Atlantic.

Type species: *P. otobothrioides* Dollfus 1966, in "plankton"; Cape Verde Islands.

## KEY TO THE GENERA IN OTOBOTHRIIDAE

**1a.  Two complete sets of reproductive organs per segment ........................ ................................. *Diplootobothrium* Chandler 1942. (Figure 84)**

*Diagnosis:* Scolex long, craspedote. Tentacle sheaths sinuous, longer than tentacles. Bulbs seven or eight times longer than wide. Retractors inserted near anterior ends of bulbs. Bothridia two, each with two lateral ciliated sensory pits. Tentacle armature heteroacanthous, hooks in diagonal rows of nine or ten. Neck present. Strobila acraspedote. Two sets of reproductive organs in each proglottid. Genital pores equatorial in mature segments, postequatorial in gravid ones. Cirrus pouch oval, small. Testes not described. Vagina posterior to cirrus pouch. Ovaries bilobed, midway between median line and lateral margin. Uteri nearly touching. Vitellaria not described. Parasites of Selachii. Gulf of Mexico, Indian Ocean.

FIGURE 85. *Otobothrium* sp.

Type species: *D. springeri* Chandler 1942, in *Platysqualus tudes;* Florida.
Other species:
  *D. tamilnadensis* Reimer 1980, in *Secutor ruconius;* Indian Ocean.

**1b.  One complete set of reproductive organs per segment.......................... 2**

**2a.  Tentacle sheaths strongly convoluted. Plerocercus with short, rounded appendix ............................... *Otobothrium* Linton 1890. (Figure 85)**

*Diagnosis:* Scolex long or short, craspedote. Tentacle sheaths spiral. Retractor muscles inserted at or near anterior end of bulb. Pars bulbosa swollen. Bothridia two, often with posterior notch, each side with a ciliated, eversible sensory pit. Tentacles emerging from anterior margins of bothridia, shorter than their sheaths. Armature heteroacanthous. Strobila acraspedote, apolytic. Genital pores postequatorial. Cirrus pouch small, oval, directed somewhat forward. Cirrus unarmed. Testes numerous, in intervascular field, some postovarian. Ovary bilobed, somewhat removed from posterior margin of proglottid. Vagina and vitellaria not described. Uterus extending to near anterior end of proglottid; no uterine pore. Parasites of selachians. Cosmopolitan. (Dollfus[785] divided this genus into two subgenera, *Otobothrium* and *Pseudotobothrium*. I do not follow that scheme here.)
  Type species: *O. crenacollis* Linton 1890, in *Carcharhinus obscurus, C. platyodon, Scoliodon terraenavae,* larva in numerous marine fishes; Atlantic, Gulf of Mexico.
  Other species:
    *O. balli* Southwell 1929, in *Cybium guttatum, Lethrinus ornatus, Balistes stellatus, Aprion pristipoma;* Sri Lanka, India.
    *O. curtum* (Linton 1909) Dollfus 1942 (syn. *Rhynchobothrium curtum* Linton 1909), in *Galeocerdo tigrinus, Mycteroperca bonaci, Epinephelus striatus;* Caribbean.
    *O. dipsacum* Linton 1897 (syn. *O. insigne* Linton 1905), in *Pomatomus saltatrix, Carcharinus obscurus;* Atlantic, Pacific. Redescribed: Cruz-Reyes.[649]
    *O. ilisha* (Southwell et Prashad 1918) Goldstein 1963 (syn. *Poecilancistum ilisha* (Southwell et Prashad 1918) Dollfus 1930, *Rhynchobothrius ilisha* Southwell et Prashad 1918, *Tentacularia ilisha* [Southwell et Prashad 1918] Southwell 1930), in *Carcharias gangeticus;* India.
    *O. linstowi* Southwell 1912 (syn. *O. magnum* Southwell 1924), in *Pristis cuspidatus, Rhynchobatus djeddensis, Notorhynchus platycephalus;* Sri Lanka, China.
    *O. mugilis* Hiscook 1954, in *Mugil cephalus;* Australia.
    *O. penetrans* Linton 1907, in *Zygaena malleus, Carcharodon lamia, Scoliodon terraenovae, Carcharias limbatus, C. leucus;* Caribbean, Costa Rica.

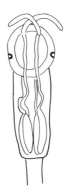

FIGURE 86. *Poecilanstrum* sp.

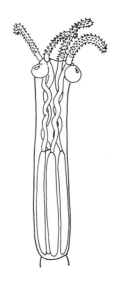

FIGURE 87. *Halysiorhynchus macrocephalus* (Shipley et Hornell 1906) Pintner 1913.

*O. pephrikos* Dollfus 1969, in *Sphyrna zygaena;* Mediterranean.

*Or. pronosomum* (Stossich 1901) Dollfus 1942 (syn. *Rhynchobothrium pronosomum* Stossich 1901), in *Trygon pastinaca;* Mediterranean.

*O. propecysticum* Dollfus 1969, in *Sphyrna zygaena;* Mediterranean.

**2b. Tentacle sheaths scarcely convoluted. Plerocercus with very long, narrow appendix .............................. *Poecilancistrum* Dollfus 1929. (Figure 86)**

*Diagnosis:* Scolex acraspedote, pars vaginalis short, bulbs longer than wide. Pars bulbosa not swollen. Bothridia two, rounded, each with a ciliated, protrusable sensory pit on each lateral margin. Tentacles emerge from anterior margins of bothridia. Armature heteroacanthous. Strobila acraspedote, apolytic. Neck present. Genital pores in postequatorial lateral margin of proglottid. Cirrus pouch oval, containing long, unarmed cirrus. External seminal vesicle present. Testes numerous, in entire proglottid between osmoregulatory canals. Vagina opens posterior to cirrus. Seminal receptacle present. Ovary bilobed. Vitellaria lateral. Uterus reaches near to anterior end of proglottid. Parasites of Squalii. Brazil, India, Texas.

Type species: *P. caryophyllum* (Diesing 1850) Dollfus 1929 (syn. *Rhynchobothrium caryophyllum* Diesing 1850), in *Prionodon leucas, Scoliodon lalandii;* Brazil. Redescribed by Goldstein, 1963.

Other species:

*P. gangeticum* (Shipley et Hornell 1906) Dollfus 1929, in *Carcharias gangeticus;* India.

*P. robustum* (Chandler 1935) Dollfus 1942, in *Cynoscion nebulosus, Pogonias cromis;* Texas.

## KEY TO THE GENERA IN PTEROBOTHRIIDAE

**1a. External surface of tentacles with chainette of V-shaped hooks .................. ...................................... *Halysiorhynchus* Pintner 1913. (Figure 87)**

*Diagnosis:* Scolex long, acraspedote. <u>Tentacle sheaths spiral</u>. Retractors inserted in bases of bulbs. Bothridia four, small, on short stalks arising from apex of scolex. Tentacles with poecilacanthous armature, each external surface bearing a chainette of V-shaped hooks. Strobila acraspedote, apolytic, flattened anteriorly, subcylindrical posteriorly. Genital pores irregularly alternating, near posterior margin of proglottid. Cirrus pouch extending to midline of proglottid. Cirrus unarmed. Testes in two lateral fields anterior to ovary. Ovary bilobed, posterior. Cortical vitellaria encircling entire proglottid. Uterus median, with posterior pore. Parasites of Batoidea. Sri Lanka.

Type species: *H. macrocephalus* (Shipley et Hornell 1906) Pintner 1913 (syn. *Tetrarhynchus macrocephalus* Shipley et Hornell 1906, *T. ruficollis* Shipley et Hornell 1906, *T. shipleyanus* Pintner 1913), in *Trygon walga, Rhynchobatus djeddensis, Dasybatus kuhlii, Pteroplatea micrura;* Sri Lanka, Pakistan.

1b. External surface of tentacles with longitudinal band of small hooks which are not V-shaped .................................................... *Pterobothrium* Diesing 1850 (syn. *Synbothrium* Diesing 1950, *Syndesmobothrium* Diesing 1854).

*Diagnosis:* Scolex long, slender, acraspedote. Tentacle sheaths spiral. Pars bulbosa somewhat swollen. Retractors attached at differing levels, according to species. Bothridia four, each on a short, motile stalk, the lateral pairs of which are connected by a thin membrane. Pars postbulbosa always present. Tentacle hooks poeciloacanthous, each external surface with a band of small hooks. Unsegmented neck absent. Strobila acraspedote, apolytic. Genital pores irregularly alternating, in posterior half of margin. Cirrus pouch thick-walled. Testes numerous, in intervascular field. Vagina ventral to cirrus pouch. Ovary bilobed, near posterior border of proglottid. Vitellaria lateral. Uterine pore anterior. Parasites of Selachii. Cosmopolitan.

Type species: *P. macrourum* (Rudolphi 1819) Diesing 1850, in *Sparus* sp., *Micropogon undulatus, Equula caballa;* Brazil, Uraguay.

Other species:

*P. chaeturichthydis* Yamaguti 1952, in body cavity of *Chaeturichthys hexanemus;* Japan.

*P. crassicollis* Diesing 1850, in *Erythrinus unitaeniatus, Sciaena* sp., *Dorus* sp.; Brazil.

*P. dasybati* Yamaguti 1934, in *Dasybatus akajei;* Japan.

*P. filicollis* (Linton 1889) (syn. *Syndesmobothrium filicolle* Linton 1889, *Synbothrium filicolle* Linton 1889), in *Trygon centrura, Cynoscion regalis, Galeocerdo arcticus, Lobote surinamensis, Mustelus canis, Paralichthys dentatus, P. albiguttus, Pomatomus saltatrix, Pomolobus mediocris, Scomberomorus maculatus, S. cavalla, S. regalis, Brevoortia tyrannus, Mustelus laevis, Carcharhinus milberti, C. obscurus, Lophopsetta maculata, Micropogon undulatus, Pteroplatea maclura, Scolionterraenovae, Dasyatis say, Chorinemus, Clupea ilisha, Cybium guttatum, Harpodon nehereus;* Atlantic, India.

*P. fragile* (Diesing 1850) Diesing 1850 (syn. *Synbothrium fragile, Syndesmobothrium fragile* Diesing 1850), in *Pristis perottetii, Centrophorus* sp.; Brazil.

*P. hawaiiensis* Carvajal, Campbell et Cornford 1976, in *Dasyatis lata;* Hawaii.

*P. heteracanthum* Diesing 1850, in *Micropogon lineatus, Pristipoma coro, Cybium guttatum, Drepane punctata;* Brazil.

*P. hira* Yamaguti 1952, larva in *Ilisha elongata;* Japan.

*P. interruptum* (Rudolphi 1819) Diesing 1850, in *Trichiurus lepturus;* Brazil.

*P. lintoni* (MacCallum 1916) Dollfus 1942 (syn. *Synbothrium lintoni* MacCallum 1916, *Pterobothrium malleum* Linton 1924), in *Dasybatis pastinacus, D. centrura;* Atlantic.

*P. minimum* Linstow 1904, in *Dasybatis;* India.

*P. platycephalum* (Shipley et Hornell 1906) Dollfus 1942 (syn. *Tetrarhynchus platycephalus* Shipley et Hornell 1906), in *Dasybatus walga*, larvae in many marine fishes; Sri Lanka.

FIGURE 89. *Sphyriocephalus tergestinus* Pintner 1913.

FIGURE 88. *Rhinoptericola megacantha* Carvahal et Campbell 1975.

*P. rubromaculatum* (Diesing 1863) Dollfus 1942 (syn. *Rhynchobothrium rubromaculatum* Diesing 1863), in *Trygon pastinaca, T. kuhli;* Sri Lanka.

*P. tangoli* (MacCallum 1921) Dollfus 1942 (syn. *Rhynchobothrium tangoli* MacCallum 1921), in peritoneum of scombriform fish; Borneo.

*P. hemuli* (MacCallum 1921), in *Haemulon plumieri;* New York Aquarium.

## DIAGNOSIS OF THE ONLY GENUS IN RHINOPTERICOLIDAE

*Rhinoptericola* Carvajal et Campbell 1975. (Figure 88)

*Diagnosis:* Scolex acraspedote, delineated from body by constriction. Bothridia oval, arranged in pairs occupying entire width of dorsal and ventral surfaces of scolex. Tentacles as long as pars vaginalis when fully everted. Metabasal armature heteroacanthous, heteromorphous, atypical in arrangement. Basal armature located on slight swelling. Bulbs much longer than wide; retractor muscles inserted at bases. Strobila acraspedote, anterior segments wider than long, posterior segments longer than wide. Genital pore marginal, preequatorial. Testes medullary, preovarian. Cirrus pouch reaches midline of segment. Uterus with two lateral diverticulae at posterior end and reaching anterior margin of segment when nongravid; saccular when gravid. Parasites of elasmobranchs. Western Atlantic.

Type species: *R. megacantha* Carvajal et Campbell 1975, in *Rhinoptera bonasus;* Chesapeake Bay, U.S.

## DIAGNOSIS OF THE ONLY GENUS IN SPHYRIOCEPHALIDAE

*Sphyriocephalus* Pintner 1913 (syn. *Sphyriocephala* Dollfus 1929). (Figure 89)

*Diagnosis:* Scolex thicker than wide. Tentacles emerging from within bothridial cavities. Hooks hollow, metabasal hooks similar, in regularly alternating longitudinal rows. Tentacle sheath and bulb short. Retractor not entering bulb. Bothridia fused, with longitudinal ridge on median side of bottom of deep bothridial cavity. Notch present on posterior margin of each set of fused bothridia. Proglottids wider than long. Genital pores irregularly alternating. Cirrus pouch slender. Cirrus unarmed. Testes in medulla between osmoregulatory canals. Vagina posterior to cirrus pouch, with distal sphincter. Ovary posterior, bilobed. Vitellaria

FIGURE 90. *Tentacularia coryphaenae* Bosc 1797.

encircling proglottid in cortex. Uterus saccular, with submedian pore. Parasites of selachians. Atlantic, Pacific, Mediterranean.

Type species: *S. viridis* (Wagener 1854) Pintner 1913 (syn. *Tetrarhynchus viridis* Wagener 1854), in *Scymnus nicaeensis, Centrophorus granulosus, Scymnorhinus lichia, Isurus glaucus,* etc.; Mediterranean, Japan.

Other species:

*S. alberti* Guiart 1935, in *Centroscymnus coelolepis, Pseudotriacis microdon;* Mediterranean.

*S. dollfusi* Bussieras et Aldrin 1968, in *Thunnus obesus;* West Africa.

*S. richardi* Guiart 1935, in peritoneum of *Synaphobranchus* sp.; Atlantic.

*S. tergestinus* Pintner 1913, in *Alopecias vulpes;* Trieste.

## KEY TO THE GENERA IN TENTACULARIIDAE

**1a. Scolex long; bothridia without free borders..................................................
....................................................*Tentacularia* Bosc 1797. (Figure 90)**

*Diagnosis:* Scolex long, craspedote, subcylindrical. Bothridia separated, without free borders, spinous. Tentacles short, slender, armed with solid hooks in spirals, similar except at base of tentacle. Tentacle sheaths not twisted. Bulbs ellipsoidal. Proglottids acraspedote, anapolytic. Genital pores irregularly alternating, ventromarginal. Cirrus unarmed. Testes numerous, filling dorsal medulla between outer osmoregulatory canals. Seminal receptacle sometimes present. Ovary bilobed in frontal section, X-shaped in cross section. Parasites of elasmobranchs. Cosmopolitan.

Type species: *T. coryphaenae* Bosc 1797 (syn. *Stenobothrium macrobothrium* [Rudolphi 1810], *Tentacularia boscii* Siebold 1850), in *Carcharhinus longimanus,* larvae in many marine fishes; Atlantic, Pacific.

Other species:

*T. bicolor* (Bartels in Nordmann 1832) Dollfus 1942 (syn. *Bothriocephalus bicolor* Bartels in Nordmann 1832), in *Coryphaena equisetis, C. hippurus, Halichelys atra, Scomber pelamys;* Atlantic?

*T. rugosa* (Leuckart 1850) Dollfuss 1942 (syn. *Rhynchobothrius rugosos* Leuckart 1850), in *Squalus carcharias;* Atlantic.

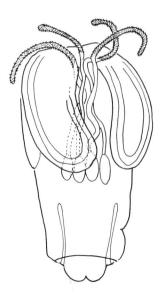

FIGURE 91. *Nybelinia lingualis* (Cuvier 1817) Poche 1926.

1b. Scolex short; bothridia with free borders ........................................
....... *Nybelinia* Poche 1926 (syn. *Acoleorhynchus* Poche 1926, *Aspidorhynchus* Molin 1858 — preoccupied, *Congeria* Guiart 1935, *Pleronybelinia* Sezen et Price 1969, *Rufferia* Guiart 1927, *Stenobothrium* Diesing of Pintner 1913). (Figure 91)

*Diagnosis:* Scolex short, craspedote. Bothridia eparate, with free borders. Tentacles cylindrical, armed with solid, similar hooks in quincunxes. Proglottids craspedote or not, anapolytic. Genital pores irregularly alternating, ventromarginal. Testes medullary. Cirrus pouch long. Cirrus unarmed. Internal seminal vesicle present or not. Ovary X-shaped in cross section. Vitellaria in whole cortex except in region of wings of ovary. Parasites of elasmobranchs. Cosmopolitan. Dollfus[773] divided *Nybelinia* into two subgenera; *Nybelinia* and *Syngenes*. I do not follow that scheme here.

Type species: *N. lingualis* (Cuvier 1817) Poche 1926 (syn. *Tetrarhynchus lingualis* Cuvier 1817, *Aspidorhynchus inflatus* Molin 1858, *Dibothriorhynchus todari* Chiaje 1829, *Tetrarhynchus sepiae* Leuckart 1886, *Tetrabothriorhynchus octopodiae* Diesing 1850, *Tetrarhynchus bisulcatum* Linton 1889, *Rhynchobothrium paleaceum* Rudolphi 1810, *Bothriocephalus tubiceps* Leukart 1886, *Tetrarhynchus megabothrius* Rudolphi 1810), larval stages known from many species of marine fishes; cosmopolitan.

Other species:

*N. africana* Dollfus 1960, in *Galeoides polydactylus, Trigla* sp., *Pagellus* sp., *Mullus barbatus, Serranus cabrilla;* Senegal.

*N. alloiotica* Dollfus 1960, in *Sphyraena guachancho, Coryphaena equisetis;* Senegal.

*N. anantaramanorum* Reimer 1980, in Sparidae (fish); Indian Ocean.

*N. anguillae* Yamaguti 1952, in *Anguilla japonica;* Japan.

*N. anthicosum* Heinz et Dailey 1974, in *Triakis semifasciata, Heterodontias francisci;* California, Mexico.

*N. basimegacantha* Carvajal, Campbell et Cornford 1976, in *Porupeneus multifasciatus;* Hawaii.

*N. bengalensis* Reimer 1980, in *Cynoglossus marcrolepidotus, C.* sp., India.

*N. bisulcata* (Linton 1889) Poche 1926 (syn. *Tetrarhynchus bisulcata* Linton 1889), in *Carcharias obscurus, Carcharhinus milberti, Galeocerdo arcticus, Squalus acanthias, Cestracion zygaena, Xiphias gladius, Sepiella maindroni, Carcharius leucus;* Guatemala, Costa Rica, Japan, Atlantic.

*N. cadenati* Dollfus 1960, in *Epinephelus alexandrinus, Fistularia tabaccaria;* Senegal.

*N. congri* Guiart 1935 (syn. *Gongeria congri* Guiart 1935), in *Synaphobranchus pinnatus;* Atlantic.

*N. dakari* Dollfus 1960, in *Vomer setipennis;* Senegal.

*N. edwinlintoni* Dollfus 1960, in *Sphyrna diplana;* Senegal.

*N. elongata* Shah et Bilqees 1979, in *Pellona elongata;* Pakistan.

*N. equidentata* (Shipley et Hornell 1906) Dollfus 1942 (syn. *Tetrarhynchus equidentata* Shipley et Hornell 1906), in *Trygon walga;* Sri Lanka.

*N. erythraea* Dollfus 1960, in *Cynoglossus sinusarabici;* Senegal.

*N. estigmana* Dollfus 1960, in *Vomer setipennis, Hynnis goreensis, Box boops;* Senegal.

*N. eureia* Dollfus 1960, in *Paraconger notialis, Mustelus canis;* Senegal.

*N. goreensis* Dollfus 1960, in *Sphyrna diplana;* Senegal.

*N. herdmani* (Shipley et Hornell 1906) Dollfus 1942 (syn. *Tetrarhynchus herdmani* Shipley et Hornell 1906), in *Trygon walga;* Sri Lanka.

*N. jayapaulazariahi* Reimer 1980, in *Cynoglossus* sp.; Indian Ocean.

*N. lamonteae* Nigrelli 1938, in *Xiphias gladius;* Nova Scotia.

*N. manazo* Yamaguti 1952, in *Mustelus manazo;* Japan.

*N. narinari* (MacCallum 1917) Dollfus 1930 (syn. *Tetrarhynchus narinari* MacCallum 1917), in *Aetobatis narinari;* U.S.

*N. nipponica* Yamaguti 1952, in *Neobythites macrops, Xystrias grigorijewi, Pseudorhombus pentophthalmus, Argintina kagoshimae;* Japan.

*N. oodes* Dollfus 1960, in *Pristipoma bennetti;* Senegal.

*N. palliata* (Linton 1924) Dollfus 1929 (syn. *Tetrarhynchus palliata* Linton 1924), in *Cestracion zygaena, Lamna ditroois, Zigaena malleus;* Atlantic, Pacific.

*N. perideraea* (Shipley et Hornell 1906) Dollfus 1930 (syn. *Tetrarhynchus* Shipley et Hornell 1906), in *Carcharias gangeticus, Ginglymostoma concolor;* Sri Lanka.

*N. pintneri* Yamaguti 1934, in *Prionace glauca;* Japan.

*N. punctatissima* Dollfus 1960, in *Sphyraena guachancho, Hynnis goreensis, Seriola dumerili;* Senegal.

*N. riseri* Dollfus 1960, in *Raja binoculata;* Senegal.

*N. robusta* (Linton 1890) Dollfus 1930 (syn. *Tetrarhynchus robusta* Linton 1890), in *Trygon centrura, Oxyrhina spallanzanii, Echeneis remora, Trigla lineata;* Atlantic.

*N. rougetcampanae* Dollfus 1960, in *Fiosaccus cutaneus;* Senegal.

*N. senegalensis* Dollfus 1960, in *Hynnis goreensis, Caranx rhonchus;* Senegal.

*N. sphyrnae* Yamaguti 1952, in *Sphyrna zygaena;* Japan.

*N. strongyla* Dollfus 1960, in *Liosaccus cutaneus;* Senegal.

*N. surmenicola* Okada 1929, in *Ommastrephes sloani, Ophiodon elongatus, Theragra* sp.; Pacific.

*N. syngenes* Pintner 1928, in *Sphyrna zygaena;* Japan, Florida.

*N. tenuis* (Linton 1890) Dollfus 1930 (syn. *Tetrarhynchus tenuis* Linton 1890), in *Trygon centrura;* Atlantic.

*N. yamagutii* Dollfus 1960, in *Fiosaccus cutaneus;* Senegal.

## DIAGNOSIS OF THE ONLY GENUS IN MIXODIGMATIDAE

*Mixodigma* Dailey et Vogelbein 1982. (Figure 92)

*Diagnosis:* Pars vaginalis long. Tentacle sheaths sinuous. Pars bulbosa swollen; bulbs

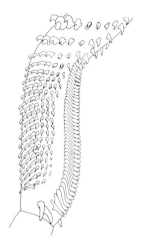

FIGURE 92. *Mixodigma leptaleum* Dailey et Vogelbein 1982. (From Dailey, M. D. and Vogelbein, W., *J. Parasitol.*, 68, 145-149, 1982. With permission.)

long, comprised of six layers of diagonal muscle fibers. Pars postbulbosa prominent. Retractor muscles inserted at base of bulbs. Vitellaria medullary, forming a layer between longitudinal muscles and medullary parenchyma. Cirrus pouch anterior to midsegment. Seminal receptacle and oocapt present. Parasites of deep-water sharks.

Type species: *M. leptaleum* Dailey et Vogelbein 1982, in "Megamouth" shark; Hawaii, U.S.

## TRYPANORHYNCH GENERA OF UNCERTAIN STATUS

1. *Bombycirhynchus* Pintner 1931.
    Type species: *B. sphyraenaicum* Pintner 1931, in *Sphyraena commersoni;* Sri Lanka.
2. *Clujia* Guiart 1935.
    Type species: *C. racovitzai* Guiart 1935, in *Galeus glaucus;* western North Atlantic.
3. *Diesingium* Pintner 1929 (syn. Diesingia Quatrefages 1865, preoccupied; *Diesingiella* Guiart 1931; *Diesingella* Guiart 1931).
    Type species: *D. lomentaceum* (Diesing 1850) Pintner 1929 (syn. *Rhynchobothrium lomentaceum* Diesing 1950), in *Mustelus vulgaris, Galeorhinus laevis;* Atlantic.
    Other species:
        *D. monticelli* (Moniez 1892) Pintner 1929 (syn. *Diesingella monticelli* Moniez 1892] Guiart 1935), in *Lophius piscatorius;* Atlantic.
4. *Microbothriorhynchus* Yamaguti 1952.
    Type species: *M. coelorhynchi* Yamaguti 1952, in *Coelorhynchus* sp.; Japan.
5. *Rhopalothyrax* Guiart 1935.
    Type species: *R. gymnorhynchoides* Guiart 1935, in *Centroscymnus coelolepis;* Azores.
6. *Symbothriorhynchus* Yamaguti 1952.
    Type species: *S. uranoscopi* Yamaguti 1952, in *Uranoscopus oligolepis;* Japan.
7. *Trigonolobum* Dollfus 1929.
    Type species: *T. spinuliferum* (Southwell 1911) Dollfus 1929 (syn. *Tentacularia spinuliferum* Southwell 1911, *Rhynchobothrium laciniatum* Yoshida 1917), in *Rhynchobatus djeddensis;* Sri Lanka.

8.  *Wageneria* Monticelli 1892.
    Type species: *W. proglottis* (Wagener 1854) Monticelli 1892, in *Scymnus nicaeensis*; Mediterranean.
    Other species:
       *W. aculeata* Cohn 1902; host and locality unknown.
       *W. impudens* (Creplin 1846) Cohn 1902, in *Squalus griseus*; locality?

## TRYPANORHYNCH SPECIES OF UNCERTAIN STATUS

The following species are either incompletely described or do not conform to any established modern genus. Most are based on larval forms from fishes or invertebrates.

*Abothrium carchariae* Linstow 1878 ( = *Pierretia carchariae* (Linstow 1878) Guiart 1927), in *Carcharias* sp.
*Anthocephalus rudicornis* Drummond 1839, in *Hippoglossus vulgaris*.
*Bothriocephalus claviger* Leuckart 1819, in *Xiphias gladius*.
*Dibothriorhynchus dinoi* Vannucci Mendes 1945, in Rhizostomata.
*Dibothriorhynchus excisus* Diesing 1854, in *Trigla hirundo*.
*Dibothriorhynchus maccallumi* MacCallum 1921, in *Sphryna tiburo*; Australia.
*Dibothriorhynchus mulli-barbati* Diesing 1854, Diesing 1863, in *Mullus barbatus*.
*Dibothriorhynchus speciosus* MacCallum 1921, in *Mycteroperca venenosa*; N.Y. Aquarium.
*Dibothriorhynchus xiphiae* MacCallum 1921, encysted in *Xiphias gladius*; Woods Hole. Syn. of *D. attenuatus* (Rudolphi 1819) Nigrelli 1938.
*Echinorhynchus quadrirostris* Goeze 1782, in *Salmo salar*. (Acanthocephala?)
*Rhynchobothrium aetobati* MacCallum 1921, in *Aetobatis narinari*; Singapore.
*Rhynchobothrium ambiguum* Diesing 1863, in *Heptanchus, Pristiurus, Raja, Xiphias*.
*Rhynchobothrium binuncum* Linton 1908, in *Dasyatis say*, Florida; *Trygon* sp., Sri Lanka.
*Rhynchobothrium brevispine* Linton 1898, in *Rhinoptera bonasus*.
*Rhynchobothrium exile* Linton 1908, in *Galeocerdo tigrinus*.
*Rhynchobothrium heterospine* Linton 1891, in *Mustelus canis, Squalus acanthias*; Woods Hole.
*Rhynchobothrium hispidum* Linton 1890, adult in *Trygon centrura*, larva in *Tautoga onitis*.
*Rhynchobothrium longicorne* Linton 1891, in *Odontaspis littoralis, Carcharinus taurus*; Woods Hole.
*Rhynchobothrium longispine* Linton 1890; adult in *Trygon centrura*, larva in *Leptocephalus, Microgadus, Paralichthys, Prionotus, Scomber, Scomberomorus, Urophycis*; Woods Hole.
*Rhynchobothrium plicatum* Linton 1905, in *Sphyrna tiburo*.
*Rhynchobothrium rossii* Southwell 1912, in *Dasybatus kuhli, D. walga, Rhynchobatus djeddensis, Stoasodon narinari*.
*Rhynchobothrium simile* Linton 1909, in *Ginglymostoma cirratum*; Florida.
*Rhynchobothrius carangis* MacCallum 1921, in *Caranx hippos*; U.S.
*Rhynchobothrius chironemi* MacCallum 1921, in *Chironemus moadetta*; Singapore.
*Rhynchobothrius commutatus* Diesing 1863, in *Acanthias vulgaris*.
*Rhynchobothrius crassiceps* Diesing 1850, in *Lophius piscatorius*. Transferred by Dollfus (1930) to *Sphyriocephalus* Pintner 1913.
*Rhynchobothrius heteromerum* Diesing 1863 ( = *Tetrarhynchus trygonis-brucconis*) in *Trygon brucco*.
*Rhynchobothrius pilidiatus* Pintner 1927, in *Gadus brandti*.
*Rhynchobothrius smaridium* (Pintner 1893) Stossich 1898, in *Maeno alcedo, M. smaris, M. vulgaris*.
*Rhynchobothrius spiracornutus* Linton 1907, in *Caranx* sp. and *Thynnus* sp. Transferred by Pintner (1931) to *Callotetrarhynchus*, by Southwell (1930) to *Tentacularia*.

*Rhynchobothrius tereticolle* Perrier 1897, in *Mustelus canis*.
*Rhynochobothrius adenoplusius* Pintner 1903, in *Lophius piscatorius*.
*Tentacularia johnstonei* Southwell 1929, in *Dasybatus sephen*.
*Tentacularia macfiei* Southwell 1929, in *Cybium guttatum, Cossyphus axillaris, Chorinemus lysan, C. toloo, Lutjanus argentimaculatus, Serranus stellatus, Balistes* sp.
*Tentacularia michiae* Southwell 1929, in *Rhynchobatus djeddensis, Dasybatis sephen, D. kuhlii*.
*Tentacularia obesa* Southwell 1929, in *Dasybatus sephen*.
*Tentacularia pillersi* Southwell 1929, in *Cossyphus axillaris*.
*Tetrabothriorhynchus merlangi-vulgaris* Diesing 1854, in *Merlangus vulgaris*.
*Tetrabothriorhynchus migratorius* Diesing 1850, for *barbata* Linné 1761.
*Tetrabothriorhynchus scombri* Diesing 1854, in *Scomber scombrus*.
*Tetrarhynchobothrium agile* Linton 1907, adult, in *Rhinoptera bonasus*.
*Tetrarhynchobothrium fluviatile* Linstow 1904, in *Malapterurus electricus*.
*Tetrarhynchus alepocephalus-rostratus* Wagener 1854.
*Tetrarhynchus angusticollis* Carus 1885, in *Raja clavata*.
*Tetrarhynchus aphroditae* Diesing 1854, in *Aphrodite aculeata*.
*Tetrarhynchus appendiculatus* Rudolphi 1809 (= *Echinorhynchus quadrirostris* Goeze renamed, *E. conicus* Zeder), in *Salmo salar*.
*Tetrarhynchus argentinae* Rudolphi 1819, (= *elongatus*), in *Argentina sphyraena*.
*Tetrarhynchus attenuatus* Rudolphi 1819, in *Xiphias gladius, Carcharhinus, Vulpecula, Squalus fernandinus, Thrysites atun, Merluccius capensis*.
*Tetrarhynchus balistes- caprisci* Shipley et Hornell 1904, in *Balistes capriscus*.
*Tetrarhynchus balistidis* Shipley et Hornell 1904, in *Balistes stellatus;* Sri Lanka.
*Tetrarhynchus brevis* Baird 1862, in marine eel.
*Tetrarhynchus carcharias* Welch 1876, in *Carcharias* sp.
*Tetrarhynchus carcharidis* Shipley et Hornell 1906, in *Carcharias melanopterus, Trygon walga;* Sri Lanka.
*Tetrarhynchus cepolae-rubescentis* Diesing 1863, in *Cepola rubescens*.
*Tetrarhynchus ceylonicus* Southwell 1929, in *Ginglymostoma concolor*.
*Tetrarhynchus discophorus* Rudolphi 1819, in *Brama rajai*.
*Tetrarhynchus elongatus* Wagener 1901, larva, in *Mola mola*.
*Tetrarhynchus foveolatus* Rudolphi 1814, in "Hay".
*Tetrarhynchus gadi-aeglefini* Diesing 1863, in *Gadus aeglefinus*.
*Tetrarhynchus gadi-morrhuae* Diesing 1854, in *Gadus morrhua*.
*Tetrarhynchus gracilis* Rudolphi 1819, in *Ammodytes cicerelus;* Mediterranean.
*Tetrarhynchus grossus* Rudolphi 1819, in "Pisce maris laponici".
*Tetrarhynchus gymnorhynchus* Wagener 1854.
*Tetrarhynchus leblondii* Creplin 1851, in *Muraena conger*.
*Tetrarhynchus lepidoleprus-trachyrhynchus* Wagener 1954.
*Tetrarhynchus lichiae vadiginis* Diesing 1863, in *Lichia vadigo*.
*Tetrarhynchus linguatula* Beneden 1853, in *Scymnus glacialis*.
*Tetrarhynchus lintoni* Vaullegeard 1899, in *Trygon centrura*.
*Tetrarhynchus longicollis* van Beneden 1849, in *Mustelus vulgaris, Trygon* spp., *Rhynchobatus djeddensis*.
*Tetrarhynchus lophii-piscatorii* Diesing 1863, in *Lophius piscatorius*.
*Tetrarhynchus lophii-piscatorii (peritonei)* Wagener 1854, in *Lophius piscatorius*.
*Tetrarhynchus lotae* Zschokke 1884, in *Lota, Salmo, Trutta*.
*Tetrarhynchus macroporus* Shipley et Hornell 1906, in *Trygon uarnak*.
*Tetrarhynchus matheri* Southwell 1929, in *Ginglymostoma concolor*.
*Tetrarhynchus megabothrius* Rudolphi 1819, in *Scomber sarda*.

*Tetrarhynchus megacephalus* Rudolphi 1819, in *Squalus stellaris, Etmopterus spinax, Scorpaena porcus*.
*Tetrarhynchus merlangi* Olsson 1869, in *Gadus aeglefinus*, etc.
*Tetrarhynchus minimus* Shipley et Hornell 1904, in *Taeniura melanospilos*.
*Tetrarhynchus minuto-striatus* Baird 1862, in *Brama* sp.
*Tetrarhynchus minutus* van Beneden 1858, adult, in *Squatina angelus;* locality not given.
*Tetrarhynchus morhuae* Rudolphi 1809, syn. *Echinorhynchus quadrirostris, Echinorhynchus quadrirostris gadi-morhuae*, in *Gadus morhua*.
*Tetrarhynchus mulli* Linstow 1878, in *Mullus surmuletus*.
*Tetrarhynchus mulli-rubescentis* Diesing 1863, in *Mullus rubescens*.
*Tetrarhynchus mugil-auratus* Wagener 1854, in *Mugil auratus*.
*Tetrarhynchus notidanus* Risso 1826, in "Monge", *Apes maritimes*.
*Tetrarhynchus papillifer* Poyarkoff 1909, host unknown.
*Tetrarhynchus papillosus* Rudolphi 1809, in *Coryphaena hippuris*.
*Tetrarhynchus phycis-mediterranei* Diesing 1863, in *Phycis mediterraneus*.
*Tetrarhynchus pinnae* Shipley et Hornell 1904, in *Balistes stellatus, B. mitis, Pinna* sp.
*Tetrarhynchus piscium-aliorum* Diesing 1854, based on Beneden 1850, p. 149.
*Tetrarhynchus pleuronectis-limandae* Olsson 1867.
*Tetrarhynchus pleuronectis-maximi* Rudolphi 1819, in *Pleuronectes maximus*.
*Tetrarhynchus quadripapillosus* Baird 1862, in *Alepocephalus* sp.
*Tetrarhynchus rajae-asperae* Diesing 1863, in *Raja aspera*.
*Tetrarhynchus rajae-clavatae* Diesing 1863, in *Raja clavata*.
*Tetrarhynchus rajae-megarhynchae* Diesing 1863, in *Raja megarhynchus*.
*Tetrarhynchus rhynchobatidis* Shipley et Hornell 1906, in *Trygon sephen; Rhynchobatus djeddensis*, Sri Lanka.
*Tetrarhynchus sciaenae-aquilae* Parona in Vaullegeard 1901, in *Sciaena aquila*.
*Tetrarhynchus scomber-gobius* Wagener 1854.
*Tetrarhynchus scomber-pelamys* Wagener 1854.
*Tetrarhynchus scomber-rochei* Wagener 1854.
*Tetrarhynchus scomber-thynnus* Wagener 1854.
*Tetrarhynchus scyllium-canicula* Wagener 1854.
*Tetrarhynchus scymni* Wagener 1854.
*Tetrarhynchus scymnus-micaeensis* Wagener 1854.
*Tetrarhynchus scymni-rostrati* Diesing 1863, in *Scymnus rostratus*.
*Tetrarhynchus scymnus-rostratus* Wagener 1854.
*Tetrarhynchus shipleyi* Southwell 1929, in *Ginglymostoma concolor*.
*Tetrarhynchus smaridis-maenae* Diesing 1863, in *Smaris maena*. Syn. of *smaridium* Pintner 1893.
*Tetrarhynchus solidus* Drummond 1838, in *Salmo*.
*Tetrarhynchus sphyraena-argentei* Vaullegeard 1899.
*Tetrarhynchus strangulatus* Baird 1853, host unknown.
*Tetrarhynchus striatus* Wagener 1854, in *Myliobatis aquila*.
*Tetrarhynchus strumosus* Siebold 1850, in *Brama raji*.
*Tetrarhynchus tenuicaudatus* Leidy 1879, in *Remora*.
*Tetrarhynchus tenuicollis* Rudolphi 1819, in *Pleuronectes pegosa, Lophius piscatorius*.
*Tetrarhynchus thynni* Wagener 1854, in *Thynnus*.
*Tetrarhynchus torpedinis-ocellatae* Diesing 1863, in *Torpedo ocellata*.
*Tetrarhynchus triglae-hirudinis* Diesing 1863, in *Trigla hirundo*.
*Tetrarhynchus triglae-lepidotae* Wagener 1854, in *Trigla hirundo*.
*Tetrarhynchus trygonis-brucconis* Diesing 1863 (syn. of *Rhynchobothrium heteromerum*) , in *Trygon brucco*.

*Tetrarhynchus trygonis-pastinacae* Diesing 1863 (syn. of *Rhynchobothrium rubromaculatum*), in *Trygon pastinaca*.
*Tetrarhynchus uranoscope-scabri* Diesing 1863, in *Uronoscopus scaber*.
*Tetrarhynchus wardii* Garman 1885, in *Chlamydoselachus anguineus*, Japan.
*Tetrarhynque epistocotyle* Leblond 1835, in *Distoma longicolle* in *Muraena conger*.

## ORDER PSEUDOPHYLLIDEA

This is a large order, with species infecting fish, amphibians, reptiles, birds, and mammals. Many species are difficult to study because of their large sizes, numerous and crowded testes and vitellaria, incomplete segmentation, and thick, fleshy strobilae. The scolex has dorsal and ventral depressions called bothria, which are quite agile and serve as holdfast and locomotor organs.

Most cestodologists consider this to be a primitive group of tapeworms, with definite affinities to Caryophyllidea. The life cycle involves a crustacean first intermediate host and a fish (sometimes an amphibian, reptile, bird, or mammal) second intermediate host. In the first intermediate host the oncosphere develops into a procercoid, a simple, elongated form with a posterior appendage, the cercomer. When eaten by the second intermediate host, the procercoid loses its cercomer and enlarges into a plerocercoid. This stage is infective to the definite host.

The classification of this group has slowly been evolving over the past 200 years, and the lower categories are fairly stable. As in most groups of cestodes, there have been revisions of the higher taxa. The most important of these are the papers by Protosova,[2688,2690] to whom the interested reader is referred. I have employed a simple, conservative classification that I believe makes it comparatively easy to identify pseudophyllidean tapeworms.

## KEY TO THE FAMILIES IN PSEUDOPHYLLIDEA

1a. Vagina and cirrus pouch opening medial on flat surface ..................... 2
1b. Vagina and cirrus pouch opening marginal or submarginal, not medial ..... 6

2a. Uterine pore on same surface as those of cirrus pouch and vagina ........... 3
2b. Uterine pore on surface opposite those of cirrus pouch and vagina .......... 5

3a. Cirrus pouch absent ................Cephalochlamydidae Yamaguti 1959. (p.84)

*Diagnosis:* Scolex triangular, with two wide, shallow bothria united at the apex. Neck present. Strobila small, proglottids acraspedote. Genital atrium median, anterior. Testes few, medullary, in two submedian fields anterior to ovary. Ovary bilobed, posterior. Vitellaria follicular, in two lateral cortical fields. Uterus winding broadly, opening posterior to genital atrium. Eggs anoperculate, embryonated. Parasites of amphibians.
  Type genus: *Cephalochlamys* Blanchard 1908.

3b. Cirrus pouch present ....................................................... 4

4a. Adult up to 110 mm long: external metamerism anterior only. Vitellaria medullary. Plerocercoid with four retractable tentacles ...........................
  ........................................... Haplobothriidae Meggitt 1924. (p.84)

*Diagnosis:* Plerocercoid (in liver of freshwater teleosts) with four tentacle-like outgrowths on scolex. Plerocercoid may attain length of 90 mm in intermediate host and show distinct segmentation. Each segment shows secondary segmentation beginning at anterior end.
  Type genus: *Haplobothrium* Cooper 1914.

4b. Adult usually over 110 mm long. External metamerism usually complete. Vitellaria cortical. Plerocercoid without tentacles .......................................
  ........................................... Diphyllobothriidae Lühe 1910. (p.84)

*Diagnosis:* Scolex variable in shape, usually bilaterally compressed, bearing two bothria, sometimes covered with fine spines, rarely with lobated apical cap. Neck conspicuous or not. External metamerism usually distinct, usually acraspedote. Strobila anapolytic. One to fourteen sets of reproductive organs per proglottid. Genital pores ventral, median, independent or together in common atrium. Testes numerous, mostly medullary. External seminal vesicle present. Ovary posterior, bialate, medullary, ventral. Vitellaria follicular, cortical, often in two lateral bands joining in front of genital pores. Uterus spiral, expanded distally, opening ventrally posterior to vaginal pore. Eggs operculate, unembryonated when laid. Parasites of fishes, reptiles, birds, and mammals.

Type genus: *Diphyllobothrium* Cobbold 1858.

5a. **Bothria deep, with inrolled margins. Eggs anoperculate, embryonated ......... ........................................... Ptychobothriidae Lühe 1902. (p.95)**

*Diagnosis:* Scolex bilaterally compressed, rounded apically, with deep bothria having rolled edges. Neck absent. Proglottids acraspedote. Genital atrium opens middorsally. Testes in dorsal medulla, in two lateral fields. Ovary posterior, median, transversely elongated. Vitellaria usually in two lateral, cortical fields. Uterus sinuous, medullary, opening on midventral surface. Eggs anoperculate, embryonated when laid. Parasites of marine and freshwater teleosts and birds.

Type genus: *Ptychobothrium* Lönnberg 1889.

5b. **Bothria deep, shallow, or absent, but lacking inrolled margins. Eggs operculate, unembryonated ....................................................... Bothriocephalidae Blanchard 1849 (syn. Acompsocephalidae Rees 1969). (p.100)**

*Diagnosis:* Scolex variable, usually with apical disk bearing dorsal and ventral notches, and occasionally marginal spines. Bothria deep, shallow, or absent, elongate when present. Neck absent. Metamerism distinct (except *Anatrum*), occasionally with secondary segmentation. Proglottids usually craspedote, with median dorsal and ventral furrows. Genital atrium dorsal, median. Cirrus pouch round, median. Testes medullary in two lateral fields. Seminal receptacle present or not. Ovary posterior, in ventral medulla. Vitellaria usually cortical, some times medullary. Uterus sinuous, with rounded uterine sac. Uterine pore ventral, median. Eggs operculate, unembryonated. Parasites of teleosts.

Type genus: *Bothriocephalus* Rudolphi 1808.

6a. **Cirrus distinctly protrusible, with large spines. Two sets of reproductive organs per segment ......................... Echinophallidae Schumacher 1914. (p.105)**

*Diagnosis:* Scolex on young forms with inconspicuous apical disk and shallow, oval bothria. Adult forms with scolex replaced by pseudoscolex shaped like four-sided truncated pyramid or trapezoid, occasionally funnel-shaped. Neck absent. Proglottids strongly craspedote, with lateral margins swollen. Two sets of reproductive organs per proglottid. Genital pores dorsal, submarginal. Cirrus pouch well-developed. Cirrus with large spines near base. Testes medullary, mostly postequatorial. Ovary lobated, posterior. Vitellaria filling most of cortex. Uterine pore median, ventral. Parasites of marine teleosts.

Type genus: *Echinophallus* Schumacher 1914.

6b. **Cirrus not distinctly protrusible nor spined. One set of reproductive organs per segment ..................................................................... 7**

7a. Vitellarium single, lobate, transversely elongated, posterior to ovary ...........
................................................Philobythiidae Campbell 1977. (p.110)

*Diagnosis:* Scolex unarmed, with bothria; apical disk present or absent. Neck present or absent. Primary and secondary segmentation precedes proglottidization. One set of male and female reproductive systems per segment. Genital pores marginal. Testes medullary. Cirrus unarmed. Ovary compact, posterior. Vitellarium single, lobate, transversely elongated, posterior to ovary. Vagina anterior to cirrus pouch. Uterus triangular, with two lateral branches. Uterine pore median. Eggs single or in multiple clusters in capsules. Parasites of abyssal marine teleosts.

Type genus: *Philobythos* Campbell 1977.

7b. Vitellaria follicular, not compact and posterior to ovary ...................... 8

8a. Eggs anoperculate, embryonated. Scolex unarmed. Uterine pore dorsal or ventral .......................................... Amphicotylidae Ariola 1899. (p.106)

*Diagnosis:* Scolex unarmed, may be replaced by pseudoscolex; bothria distinct. Neck present or absent. Metamerism distinct, strobila serrate or not. Genital pores marginal, usually irregularly alternating. External seminal vesicle absent. Cirrus unarmed. Testes medullary, numerous. Ovary ventral, medullary. Vitellaria variable. Uterus sac-like, with rudimentary opening usually ventral, occasionally dorsal. Eggs anoperculate, embryonated. Parasites of teleosts.

Type genus: *Amphicotyle* Diesing 1863.

8b. Eggs operculate, embryonated or not. Scolex armed or not. Uterine pore ventral .................................................................... 9

9a. Scolex cuboidal, commonly armed, with rounded, shallow bothria, rarely replaced by pseudoscolex. Genital pores marginal. Eggs embryonated or not ....
.......................................... Triaenophoridae Lönnberg 1889. (p.112)

*Diagnosis:* Scolex stout, cuboidal or pyramidal, usually with apical disk, usually armed. May be replaced by pseudoscolex in *Fistulicola*. Bothria shallow, rounded. Neck absent. Metamerism distinct or not. One set of reproductive organs per proglottid. Genital atrium marginal, irregularly alternating. Testes medullary. Ovary usually posterior, bilobed. Vitellaria cortical or medullary. Uterus greatly coiled; uterine sac small. Uterine pore ventral, median, anterior to level of genital atrium. Eggs operculate, embryonated or not. Parasites of teleosts.

Type genus: *Triaenophorus* Rudolphi 1793.

9b. Scolex elongated, with elongated bothria, or replaced by pseudoscolex. Genital pores dorsosubmarginal, rarely marginal. Eggs unembryonated ................
.................................... Parabothriocephalidae Yamaguti 1959. (p.115)

*Diagnosis:* Scolex elongated, with shallow bothria, occasionally replaced by pseudoscolex. Neck absent. Proglottids elongate, craspedote, secondary external metamerism sometimes present. One set of reproductive organs per proglottid. Genital atrium marginal or dorsosubmarginal. Cirrus pouch obliquely transverse. Cirrus armed. Testes numerous, medullary, some follicles postovarian or not. Ovary posterior, medullary. Vitellaria cortical or medullary. Uterus with median sac. Uterine pore ventral, median, anterior. Eggs operculate, embryonated. Parasites of marine teleosts.

Type genus: *Parabothriocephalus* Yamaguti 1934.

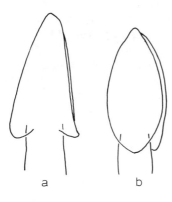

FIGURE 93. *Cephalochlamys namaquensis* (Cohn 1906) Blanchard 1908. (a) Lateral view of scolex; (b) dorsal view of scolex.

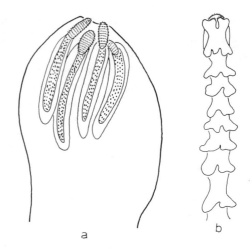

FIGURE 94. *Haplobothrium bistrobilae* Prembati 1969. (a) Primary scolex; (b) pseudoscolex of secondary strobila.

## KEY TO THE GENERA IN CEPHALOCHLAMYDIDAE

**1a. Vitellaria ventral to osmoregulatory canals, extending into postovarian area; testes three to six on each side .......... *Pseudocephalochlamys* Yamaguti 1959.**

*Diagnosis:* Scolex pointed at apex, with fluted bothria. Posterior proglottids broader than long. Testes three to six on each side. Cirrus may form a vesicular dilation before joining vagina. Very short hermaphroditic duct may be present. Genital atrium median, at anterior end of proglottid. Ovary consisting of two imperfectly separated wings. Lateral bands of vitellaria immediately ventral to osmoregulatory canals, extending behind ovary. Uterus winding through testicular area and opening on the right or left of median line behind atrial pore; eggs small. Seminal receptacle conspicuous. Parasites of clawed toads. Africa.

Type species: *P. xenopi* (Ortlepp 1926) Yamaguti 1959 (syn. *Dibothriocephalus xenopi* Orlepp 1926), in *Xenopus laevis;* South Africa.

**1b. Vitellaria lateral to osmoregulatory canals, not extending posterior to ovary; testes 7 to 12 on each side....................................... *Cephalochlamys* Blanchard 1908 (syn. *Chlamydocephalus* Cohn 1906, preoccupied). (Figure 93)**

*Diagnosis:* Scolex with flat bothria. Posterior proglottids longer than wide. Testes 7 to 12 on each side. Cirrus long and thin, opening at anterior corner of longitudinally elongated genital atrium. Ovary dumbbell-shaped, one third width of proglottid, with anterior and posterior lobes strongly developed. Vitellaria lateral to osmoregulatory canals, not extending into postovarian area. Uterus strongly winding in testicular zone, opening in median line just behind atrial pore; eggs fairly large. Parasitic in clawed toads. Africa.

Type species: *C. namaquensis* (Cohn 1906) Blanchard 1908 (syn. *Chlamydocephalus namaquensis* Cohn 1906), in *Xenopus laevis;* Southwest Africa.

## DIAGNOSIS OF THE ONLY GENUS IN HAPLOBOTHRIIDAE

*Haplobothrium* Cooper 1914 (Figure 94)

Primary segments separate and become adult strobilas up to 119 mm long, independent of one another. Adult holdfast with fine spines and shallow dorsal and ventral depressions, presumably bothria. Adult strobila: segmented at anterior end only; one set of genitalia per segment. Genital pores midventral. External seminal vesicle present. Testes in two bands in lateral medulla. Cirrus pouch median, anterior. Cirrus armed. Seminal receptacle present, well-developed. Ovary medullary, posterior. Vitellaria lateral, medullary. Uterus with convoluted proximal duct and large distal sac. One large median and two smaller lateral osmoregulatory canals. Parasites of *Amia calva* (Osteichthyes). North America.

Type species: *H. globuliforme* Cooper 1914, in *Amia calva;* North America.
Other species:
*H. bistrobilae* Premvati 1969, in *Amia calva;* Florida, U.S.

## KEY TO THE GENERA IN DIPHYLLOBOTHRIIDAE

1a. Parasites of reptiles............................................................ 2
1b. Parasites of birds ............................................................... 5
1c. Parasites of mammals ........................................................ 8

2a. Large sucker-like structure at apex of scolex. Bothria rudimentary .............
.............................................. *Scyphocephalus* **Riggenbach 1898.**

*Diagnosis:* Scolex with deep sucker-like apical organ; bothria short, shallow, occupying last fourth of scolex. Neck absent. Proglottids wider than long. Genital atrium midventral, preequatorial. Cirrus pouch occupying entire thickness of medulla. Testes numerous, meeting near anterior and posterior ends of proglottid. Ovary bilobed, posterior. Seminal receptacle present. Vitellaria cortical, surrounding proglottid. Uterus with only two or three loops. Parasites of varanid lizards. Java, Philippines.

Type species: *S. bisulcatus* Riggenbach 1898, in *Varanus salvator;* Java.
Other species:
*S. longus* Sawada et Kugi 1973, in *Varanus salvator;* Southeast Asia.
*S. secundus* Tubangui 1938, in *Varanus salvator;* Philippines.

2b. Scolex not as above, bothria well-developed .................................... 3

3a. Bothria with fused edges, forming two tubes ........................*Bothridium* **Blainville 1824** (syn. *Prodicoelia* **Leblond 1836,** *Solenophorus* **Creplin 1839).**

*Diagnosis:* Bothria fused along margins to form tube-like structures with sphincters at front and rear ends; frilled or slit-like openings present at front end. Proglottids wider than long, proterogynous. Testes in lateral fields. Seminal receptacle present. Genital atrium medioventral, anterior to uterine pore. Cirrus pouch preequatorial, in dorsoventral plane. Ovary bilobed, postequatorial. Vitellaria surrounding inner muscular layer, interrupted in median field. Uterus with thick duct and dorsal sac. Uterine pore ventromedian, posterior to genital pore. Parasites of boid snakes and varanid lizards. Africa, Sri Lanka, Philippines, India, Australia, Asia.

Type species: *B. pithonis* Blainville 1824 (syn. *Prodicoelia ditrema* Leblond 1836, *Solenophorus megalocephalus* Creplin 1839, *S. megacephalus* Diesing 1850, *S. laticeps* Duvernoy 1833, *S. grandis* Creplin 1839, *Dibothrium boe-tigridis* Rudolphi in Creplin 1839, *D. milliapharyngens* Hatch 1891, *Bothridium arcuatum* Baird 1865), in *Python, Morelia, Naja;* Africa, Sri Lanka, India, Philippines.

FIGURE 95. *Duthiersia fimbriata* Diesing 1854.

Other species:
  *B. kugii* Sawada et Kugi 1973, in *Eunectes murinus;* Japan (zoo).
  *B. longicephalum* Sawada et Kugi 1973, in *Python molurus;* Japan (zoo).
  *B. longiorum* Sawada et Kugi 1973, in *Python molurus;* Japan (zoo).
  *B. microdisciformis* Sawada et Kugi 1973, in *Python molurus;* Japan (zoo).
  *B. oboratum* (Molin 1858) Sambon 1907, in *Boa constrictor;* Asia.
  *B. orientalis* Sawada et Kugi 1973, in *Python reticulatus;* Japan (zoo).
  *B. ornatum* Maplestone et Southwell 1923, in *Python spilotes;* Australia.
  *B. oratum* (Diesing 1850) Yamaguti 1959 (syn. *Solenophorus labiatus* Corruccio 1878), in *Constrictor hieroglyphicus, Python sebae;* Africa.
  *B. parvum* Johnston 1913, in *Varanus varius;* Australia.
  *B. sawadai* Sawada et Kugi 1973, in *Epicrates cenchris;* Japan (zoo).

**3b.  Bothria not fused at edges ..................................................... 4**

**4a.  Scolex broad, fan-like; bothria directed somewhat anteriad, sometimes with basal aperture......................................*Duthiersia* Perrier 1873. (Figure 95)**

*Diagnosis:* Scolex bilaterally compressed so that it appears triangular or fan-like. Bothria deep, with frilled or crenulated margins; posterior aperture present in some species. Neck absent. Genital pore median, anterior to uterine pore. Cirrus pouch large, near anterior end of proglottid. Testes numerous, surrounding median genitalia. Vagina with distal sphincter, opening posterior to cirrus pouch. Seminal receptacle present. Ovary bilobed, postequatorial. Vitellaria cortical, in longitudinal lateral bands. Uterus looped. Uterine pore midventral, posterior to cirrus pouch. Parasites of varanid lizards. Africa, Sarawak, Maluccas.
  Type species: *D. expansa* Perrier 1873 (syn. *D. crassa* Woodland 1938, *D. venusta* Woodland 1938), in *Varanus bivittatus;* Moluccas.
  Other species:
    *D. fimbriata* (Diesing 1850) Monticelli et Crety 1891 (syn. *D. elegans* Perrier 1873, *D. latissima* Woodland 1938, *D. robusta* Woodland 1938), in *Varanus* spp.; Africa.
    *D. gomatii* Gupta et Sinha 1980, in *Varanus niloticus;* India.
    *D. sarawakensis* Woodland 1938, in *Varanus salvator;* Sarawak.
    *D. sindensis* Bilqees et Masood 1973, in *Varanus monitor;* Pakistan.

**4b.  Scolex spoon- or finger-shaped; bothria broad and shallow .....................
  .......*Diphyllobothrium* Cobbold 1858. (See p.93 for diagnosis and synonyms.)**

Species in reptiles:
  *D. serpentis* Yamaguti 1935 (syn. *Spirometra serpentis* (Yamaguti 1935) Wardle, McLeod et Stewart 1947), in *Jaja naja;* Taiwan.

FIGURE 96. *Ligula intestinalis* (Linnaeus 1758) Bloch 1782.

5a. Two sets of reproductive organs in each proglottid ..............................
................................................*Digramma* **Cholodkovsky 1915**

*Diagnosis:* Scolex small, triangular, not set off from rest of body. Bothria are short grooves connected apically. Neck absent. Each flat surface with two parallel, longitudinal grooves; ventral surface with a third median groove. Two sets of reproductive organs per proglottid. Genital pores in ventral submedian grooves. Testes in dorsal medulla, interrupted in submedian fields. Ovaries in submedian fields. Vitellaria cortical. Uterus dorsal to vagina, coiled. Parasites of piscivorous birds. Japan, Europe.

Type species: *D. alternans* (Rudolphi 1810) Cholodkovsky 1915 (syn. *Ligula alternans* Rudolphi 1810, *Ligula interrupta* Rudolphi 1810), in many fish-eating birds.

5b. One set of reproductive organs per proglottid ................................. 6

6a. External metamerism only at anterior end; acraspedote .......................
.......................*Ligula* **Bloch 1782** (syn. *Braunia* **Leon 1908**). (Figure 96)

*Diagnosis:* Scolex small, triangular. Bothria each a shallow groove. Neck absent. Strobila with median groove on dorsal and ventral surfaces. External metamerism on anterior portion only; rest with numerous wrinkles. Genital atrium midventral. Testes in single layer in dorsal medulla. Seminal receptacle present. Ovary posterior. Vitellaria cortical. Uterus coiled, in median area. Uterine pore midventral, posterior to genital atrium. Parasites of piscivorous birds. Circumboreal. For a monograph on this and related genera see Dubinina.[831]

Type species: *L. intestinalis* (Linnaeus 1758) Bloch 1782 (syn. *L. cingulum* Pallas 1781, *L. avium* Bloch 1782, *L. piscium* Bloch 1782, *L. abdominalis* [Goeze 1782], *L. gobionis* [Gmelin 1790], *L. leucisci* [Gmelin 1790], *L. alburni* [Gmelin 1790], *L. cobitidis* [Gmelin 1790], *L. simplicissima* Rudolphi 1802, *L. tincae* Zeder 1903, *L. carpionis* Rudolphi 1810, *L. contortrix* Rudolphi 1810, *L. monogramma* Creplin 1839, *L. uniserialis* Rudolphi 1810, *Braunia jassyensis* Leon 1908.) For further synonyms see Dubinina.[831] Adults in many fish-eating birds; plerocercoids common in freshwater fishes; cosmopolitan.

6b. External metamerism over entire strobila...................................... 7

7a. Scolex small, triangular. Bothria shallow, connected by groove over apex......
*Schistocephalus* **Creplin 1829** (syn. *Tortocephalus* **Malhotra, Capoor et Pundir 1980**).

*Diagnosis:* Scolex small, triangular. Bothria are shallow grooves, connected apically. Neck absent. External metamerism complete. Proglottids craspedote, wider than long, greatest width preequatorial. Genital atrium midventral, preequatorial. Cirrus pouch subspherical.

Testes numerous, in single layer in dorsal medulla. External seminal vesicle present. Ovary bilobed, posterior. Vitellaria cortical, continuous, absent in median area. Uterus submedian, regularly alternating. Uterine pore posterior to genital atrium, submedian, regularly alternating. Parasites of piscivorous birds. Circumboreal.

Type species: *S. solidus* (Mueller 1776) Creplin 1829 (syn. *Taenia gasterostei* Fabricius 1780, *Schistocephalus dimorphus* Creplin 1829, *Taenia lanceolata* Bloch 1782, *Schistocephalus rhynchichthydis* Diesing 1863, *Bothriocephalus zschokkei* Fuhrmann 1896, *Tortocephalus songi* Malhotra, Capoor et Pundir 1980), in many fish-eating birds; circumboreal.

Other species:

*S. fahmi* Gagarin, Chertkova et Vshivtsev 1966, in *Lutra lutra;* England.
*S. nemachili* Dubinina 1959, in piscivorous birds; cosmopolitan.
*S. pungitii* Dubinina 1959, in piscivorous birds; cosmopolitan.
*S. thomasi* Garoian 1960, in *Larus argentatus;* U.S.

**7b.    Scolex distinct. Bothria narrow, deep, not connected by apical groove..........
....... *Diphyllobothrium* Cobbold 1858. (See p.93 for diagnosis and synonyms.)**

Species in birds;
*D. arctomarinum* Serdyukov 1969, in *Stercorarius parasiticus;* Russia.
*D. canadensis* Cooper 1921, in *Corvus corax;* Canada.
*D. cordiceps* (Leidy 1872), in fish-eating birds; North America.
*D. dalliae* Rausch 1956, plerocercoid in *Dallia pectoralis;* adult experimentally in *Larus glaucescens* and dog.
*D. dendriticum* (Nitzsch 1824) (syn. *D. strictum* Neveu-Lemaire 1936), in *Larus, Rissa, Sterna,* dog, cat, rat, mouse; Asia, Europe.
*D. ditremum* (Creplin 1825) Lühe 1910, in *Mergus, Colymbus, Urinator, Mergus, Larus, Phalacrocorax;* Europe.
*D. exile* (Linton 1892) Lühe 1910, in *Larus californicus;* Wyoming, U.S.
*D. fissiceps* (Creplin 1829) Lühe 1910, in *Sterna hirundo, Rissa tridactyla;* Europe.
*D. microcordiceps* Szidat et Soria 1957 (syn. *Sparganum microcordiceps* Szidat et Soria 1957, in *Larus marinus;* Argentina.
*D. oblongatum* Thomas 1946, in *Larus* spp.; North America.
*D. osmeri* (Linstow 1878) Lühe 1910, in gulls; locality?
*D. podicipedis* (Diesing 1854) Meggitt 1924, in *Podiceps minor;* Ireland.
*D. vogeli* Kuhlow 1953, in *Larus ridibundus;* Europe.

**8a.    Two sets of reproductive organs in each proglottid............................ 9
8b.    One set of reproductive organs in each proglottid............................ 11
8c.    More than two sets of reproductive organs in each proglottid ...................
.................................. *Hexagonoporus* Gubanov in Delyamure 1955
(syn. *Tetragonoporus* Skriabin 1961, *Polygonoporus* Skriabin 1967). (Figure 97)**

*Diagnosis:* Strobila very long (up to 30 m). Scolex capped with a 12-lobed, fleshy structure. Two short, fleshy bothria present. Neck absent. Strobila immediately behind scolex expanded into "spoon-like" shape with jagged edges. Proglottids much wider than long. Each proglottid with 4 to 14 pairs of reproductive systems, arranged in pairs. Genital pores ventral, anterior, female pore just behind male pore. Testes numerous, scattered throughout the proglottid. Cirrus pouch rounded, cirrus unarmed. External seminal vesicle present. Ovary bilobate, wider than long, posterior. Vitellaria dispersed throughout cortex. Uterus

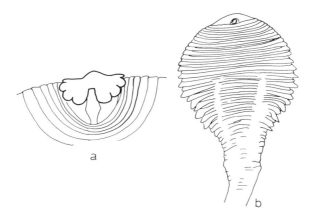

FIGURE 97. *Hexagonoporus physeteris* Gubanov in Delyamure 1955.

FIGURE 98. *Multiductus physeteris* Clarke 1962. Apical view of scolex.

with five to seven loops on each side. Eggs oval with thick shell operculate. Parasites of toothed whales. Russia.

Type species: *H. physeteris* Gubanov in Delyamure 1955 (syn. *Tetragonoporus calyptocephalus* Skriabin 1961, *Polygonoporus giganticus* Skriabin 1967), in *Physeter catodon;* Russia, Antarctica.

**9a. Osmoregulatory canals numerous (35 to 70), anastomosing, confined to medulla** .................................... **Multiductus Clarke 1962. (Figure 98)**

*Diagnosis:* Scolex globular with deep bothria; posterior region overlaps first four proglottids. Neck absent. Strobila long (18 m); proglottids craspedote. Double set of reproductive organs in each segment. Genital atria at same transverse level. Cirrus, vagina, and uterus opening into shallow atrium. Cirrus not described. Cirrus pouch very muscular, ventral to seminal vesicle. Testes numerous, in several layers in medulla. Vagina with strong sphincter near pore, opening behind cirrus. Vitellaria in one median and two lateral fields, between two longitudinal muscle layers. Uterus sac-like. Ovary lacking isthmus; located in ventroposterior region of proglottid. Eggs operculate, with thick shells. Two layers of longitudinal cortical muscle bundles. Excretory system consists of 35 to 70 longitudinal canals, with numerous anastomoses, all in medulla. Parasites of sperm whale. Antarctica.

Type species: *M. physeteris* Clarke 1962, in *Physeter catodon;* Antarctica.

**9b. Osmoregulatory canals not so numerous, cortical** ........................... **10**

**10a. Genital pores at same level** ......................................................
... ***Diplogonoporus* Lönnberg 1872 (syn. *Krabbea* Blanchard 1894). (Figure 99)**

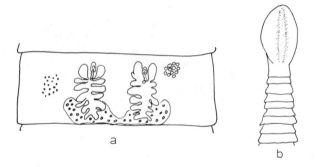

FIGURE 99. *Diplogonoporus tetrapterus* (Siebold 1848) Lönnberg 1892. (a) Proglottid (testes and vitellaria not filled in); (b) scolex.

FIGURE 100. *Baylisia baylisi* Markowski 1952.

*Diagnosis:* Scolex short, bilaterally compressed. Bothria deep, slit-like. Neck absent. Proglottids first arising as primary segments that then subdivide into secondary and tertiary segments. Proglottids short, wide. Two sets of reproductive organs per proglottid. Genital pores ventral, submedian. Testes in median and lateral fields. External seminal vesicle present. Ovary transversely elongated, lobated, posterior. Vitellaria in median and lateral fields. Uterine coils parallel or rosette-shaped. Uterine pores posterior to genital pores. Parasites of baleen whales, seals, sea otter, human. Circumboreal.

Type species: *D. balaenopterae* Lönnberg 1892, in *Balaenoptera borealis;* Scandinavia, Japan.

Other species:

*D. brauni* Leon 1907, in human; Roumania.

*D. fasciatus* (Krabbe 1865) Lönnberg 1892, in *Phoca hispida, Eumetopias jubata;* Iceland, Alaska.

*D. fukuokaensis* Kamo et Miyazaki 1970, in human; Japan.

*D. grandis* (Blanchard 1894) Yamaguti 1959 (syn. *Krabbea grandis* Blanchard 1894), in human; Japan?

*D. peltocephalus* (Monticelli 1893) Ariola 1899, in *Centrolophus ovalis;* Italy.

*D. tetrapterus* (Siebold 1848) (syn. *Bothriocephalus variabilis* Krabbe 1865), in *Phoca vitulina, Callorhinus ursinus* Alaska, North Atlantic?

**10b. One set of genital pores more anterior than the other ............................**
**........................................ Baylisia Markowski 1952. (Figure 100)**

*Diagnosis:* Scolex short. Bothria cup-shaped. Neck absent. External metamerism not corresponding to internal metamerism. Two sets of reproductive organs per proglottid. Genital pores ventral, set into longitudinal furrows, alternating in position so that they are at different levels from one another. Testes in single layer in two lateral fields. External

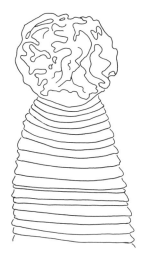

FIGURE 102. *Pyramicocephalus anthrocephalus* (Fabricius 1790) Monticelli 1890.

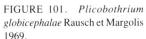

FIGURE 101. *Plicobothrium globicephalae* Rausch et Margolis 1969.

seminal vesicle present. Ovary composed of compact central part giving off branches along the uterine coils; V-shaped in sagittal section. Vitellaria cortical, in continuous sheath interrupted by osmoregulatory canals. Uterus comprising a few horizontal coils in central part of proglottid. Uterine pores posteromedial to genital pores. Parasites of seals. Antarctica.

Type species: *B. baylisi* Markowski 1952, in *Lobodon carcinophagus;* Antarctica.

**11a. Bothria intensely crenulated to form flower-like scolex..................... 12**
**11b. Scolex otherwise ....................................................... 14**

**12a. Entire scolex crenulated. Vitellaria in separate lateral fields ...................**
**........................ *Plicobothrium* Rausch et Margolis 1969. (Figure 101)**

*Diagnosis:* Large cestodes having margins of bothria hypertrophied to form many complex folds. Strobila muscular, with many segments (more than 6000). All segments wider than long. Paired osmoregulatory canals. Genital atrium in ventral midline, with uterine pore opening separately. Testes abundant, forming continuous layer in medullary parenchyma. Testes and vitellaria in separate lateral fields. Parasites of cetaceans; Atlantic.

Type species: *P. globicephalae* Rausch et Margolis 1969, in *Globicephala melaena;* Newfoundland.

**12b. Anterior margins of scolex crenulated. Vitellaria otherwise................. 13**

**13a. Bothria with margins infolded to form cauliflower-like apical organ on scolex. Vitellaria in ventral cortex......................... *Pyramicocephalus* Monticelli 1890 (syn. *Cordicephalus* Wardle, McLeod et Stewart 1947, in part). (Figure 102)**

*Diagnosis:* Bothria with anterior margins folded to form flower-like apical organ on scolex. Strobila wrinkled, external metamerism lacking. Genital pore medioventral. Cirrus pouch well developed. Testes numerous. Ovary posterior, bilobed. Vitellaria in ventral cortex. Uterus median, sinuous. Parasites of seals. Circumboreal.

FIGURE 103. *Baylisiella tecta* (Linstow 1892) Markowski 1952.

FIGURE 104. *Glandicephalus antarcticus* (Baird 1853) Fuhrmann 1920.

Type species: *P. anthocephalus (Fabricius 1780) Monticelli 1890 (*syn. *Alyselminthus lanceolato-lobatus* Zeder 1800, *Taenia anthocephala* Rudolphi 1810, *Anthobothrium tortum* Linstow 1904, *Diphyllobothrium tetrapterus* Siebold 1848, *Bothriocephalus phocaefoetidae* Creplin 1825), in *Phoca barbata, P. hispida, Cystophora cristata, Eumetopias jubatus, Enhydra lutris,* dog, human; Atlantic, Greenland, Alaska.

**13b. Scolex as above. Vitellaria in cortex surrounding entire proglottid .............. ...................................... *Baylisiella* Markowski 1952. (Figure 103)**

*Diagnosis:* Scolex embeds deep into intestinal walls. Bothria powerful, foleaceous at anterior margins. Neck absent. Strobila thick, tapering toward posterior end. Proglottids short, wide. Genital atrium midventral, surrounded by numerous papillae. External seminal vesicle present. Testes medullary, in two or three layers. Ovary posterior, transversely elongate. Vitellaria cortical, surrounding entire proglottid except at genital pore. Uterus comprised of a few irregular transverse coils. Uterine pore posterior to genital pore. Eggs operculate. Osmoregulatory canals cortical, around 100 in number. Parasites of elephant seals. Antarctica.

Type species: *B. tecta* (Linstow 1892) Markowski 1952 (syn. *Cordicephalus tectus* Wardle, McLeod et Stewart 1947), in *Macrorhinus leoninus;* Antarctica.

**14a. Testes intermingled with inner longitudinal muscle bundles ..................... .................................. *Glandicephalus* Fuhrmann 1920. (Figure 104)**

*Diagnosis:* Scolex cylindroid, bluntly pointed apically. Bothria wide apically, lips overlapping posteriorly. Bothrial cavities lined with unicellular glands. Short neck present. Strobila thick, external metamerism well marked. Proglottids shorter than wide, craspedote. Genital atrium midventral. Testes intermingled with inner longitudinal muscle bundles. Ovary ventral, interjected among muscle bundles. Vitellaria cortical, in single layer. Uterus median, with narrow lateral loops surrounded by gland cells. Parasites of seals. Antarctica.

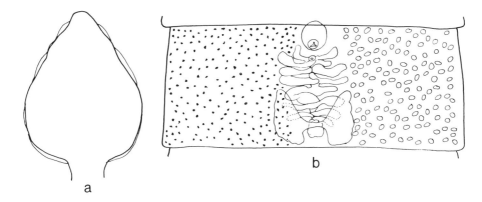

FIGURE 105. *Diphyllobothrium alascense* Rausch et Williamson 1958. (a) Scolex; (b) proglottid.

Type species: *G. antarcticus* (Baird 1853) Fuhrmann 1920 (syn. *Dibothrium antarcticus* Diesing, *Diplogonoporus antarcticus* Zschokke, *Dibothriocephalus antarcticus* Shipley, *Diphyllobothrium antarcticus* Railliet et Henry), in *Ommatophoca rossi;* Antarctica.
Other species:
 *G. perfoliatus* (Railliet et Henry 1912) Markowski 1952 (syn. *Diphyllobothrium clavatum* Railliet et Henry 1912, *D. rufum* Leiper et Atkinson 1914), in *Leptonychotes weddelli;* Antarctica.

**14b. Testes not intermingled with longitudinal muscle bundles.........................
............................. *Diphyllobothrium* Cobbold 1858 (syn. *Adenocephalus* Nybelin 1929, *Cordicephalus* Wardle, McLeod et Stewart 1947, in part, *Diancyrobothrium* Bacigalupo 1945, *Lüheella* Baer 1924, *Lueheella* Schmidt 1970, *Spirometra* Faust, Campbell et Kellogg 1929, *Gatesius* Stiles 1908). (Figure 105)**

*Diagnosis:* Scolex variable in shape. Bothria distinct, narrow, deep, not connected by apical groove. Neck present, sometimes quite short. Strobila often very long, with longitudinal grooves on flat surfaces. Proglottids craspedote, usually wider than long. Genital atrium medioventral, preequatorial, surrounded by small papillae, occasionally. Male and female pores usually opening into common atrium. Testes very numerous, in single layer, continuous across proglottid except in zone of median genitalia. External seminal vesicle present, sometimes partly embedded in wall of cirrus pouch. Ovary bilobed, posterior. Vitellaria very numerous, in single layer, continuous across proglottid except in zone of median genitalia. Uterus usually with loops, which may be parallel, rosette-shaped, or in the form of a simple spiral. Uterine pore posterior to genital pore. Parasites of reptiles, marine and terrestrial mammals, and birds. Cosmopolitan.
Type species: *D. stemmacephalum* Cobbold 1858, in *Delphinus phocaena, D. dussumieri;* Atlantic, Pacific.
Other species:
 *D. alascense* Rausch et Williamson 1958, in dog; Alaska.
 *D. arctomarinum* Serdyukov 1969, in *Stercorarius parasiticus;* Russia.
 *D. atlanticum* Delyamure et Parukhin 1968, in *Arcticephalus pusillus;* South Africa.
 *D. bresslauei* Baer 1927 (syn. *Spirometra bresslauei* (Baer 1927) Wardle, McLeod et Stewart 1947), in *Didelphys aurita;* Brazil.
 *D. cameroni* Rausch 1969, in *Monachus schauinslandi;* Midway Atoll.
 *D. cordatum* (Leuckart 1863) (syn. *Bothriocephalus schistochilos* Germanos 1895, *B. coniceps* Linstow 1905, *B. macrophallus* Linstow 1905, *Dibothriocephalus römeri* Zschokke 1903), in human, dog, *Odobaenus rosmarus, Erignathus barbatus, Phoca groenlandica;* Greenland, Spitzbergen.

*D. dalliae* Rausch 1956, in dog; Alaska.

*D. decipiens* (Diesing 1850) Gedoelst 1911 (syn. *Dibothrium decipiens* Diesing 1850, *Spirometra decipiens* (Diesing 1850) Faust, Campbell et Kellogg 1929), in an unknown cat; Brazil.

*D. didelphydis* (Ariola 1900), in *Didelphys azarae;* Brazil.

*D. elegans* (Krabbe 1865), in *Phoca cristata, P. vitulina, Eumetopias jubata;* Denmark, Greenland.

*D. erinaceieuropaei* (Rudolphi 1819) (syn. *Bothriocephalus decipiens* Railliet 1866, *B. felis* Creplin 1825, *B. maculatus* Leuckart 1848, *B. mansoni* [Cobbold 1882] Blanchard 1888, *B. liguloides* [Diesing 1850], *B. sulcatus* Molin 1858, *Dibothrium serratum* Diesing 1850, *D. mansoni* Ariola 1900, *Diphyllobothrium fausti* Vialli 1931, *Dubium erinacei-europaei* Rudolphi 1819, *Ligula mansoni* Cobbold 1882, *L. pancerii* Polonio 1860, *L. ranarum* Gastaldi 1854, *L. reptans* Diesing 1850, *Sparganum affine* Diesing 1854, *S. ellipticum* Molin 1858, *S. mansoni* [Cobbold 1882] Stiles et Taylor 1902, *S. philippinensis* Tubangui 1924, *S. proliferum* Ijima 1905, *S. reptans* Diesing 1854, *Spirometra decipiens* of Faust, Campbell et Kellogg 1929, *S. erinacei* of Faust, Campbell et Kellogg 1929, *S. houghtoni* Faust, Campbell et Kellogg 1929, *S. okumurai* Faust, Campbell et Kellogg 1929, *S. raillieti* [Ratz 1913] Wardle, McLeod et Stewart 1947, *S. ranarum* Meggitt 1925, *S. reptans* [Diesing 1850] Meggitt 1924, *S. tangalongi* [MacCallum 1921]), in humans and many carnivorous mammals; Europe, Asia, Australia, South Pacific.

*D. fuhrmanni* Hsü 1935, in *Delphinus dussumieri, Neomenis phocaenoides;* South China Sea.

*D. giljacica* Rutkevich 1937 (syn. *Spirometra giljacica* Rutkevich 1937), in human; Russia.

*D. glaciale* Cholodkowsky 1915 (syn. *Adenocephalus pacificus* Nybelin 1929, *A. septentrionalis* Nybelin 1929, *Cordicephalus arctocephalinus* [Johnston 1937]), in *Callorhinus ursinus, Arctocephalus forsteri, A. tasmanicus, A. australis, Neophoca cinerea;* Pacific.

*D. gondo* Yamaguti 1942, in *Globicephalus scammoni;* Japan.

*D. gracilis* Baer 1927 (syn. *Spirometra gracilis* [Baer 1927] Wardle, McLeod et Stewart 1947), in *Felis macrura, F. geoffroyi, F. yagouaroundi;* Brazil, Paraguay.

*D. granaia* Bacigalupo 1948, in dog; Argentina.

*D. hians* (Diesing 1850) (syn. *Bothriocephalus variabilis* Krabbe 1865, *Dibothriocephalus polycalceolus* [Ariola 1896]), in *Phoca monachus, P. hispida, P. barbata, Callorhinus ursinus;* North Atlantic.

*D. janickii* (Furmaga 1953) comb. n., in *Canis lupus, Lynx lynx;* Poland.

*D. krotovi* Delyamure 1955, in *Callorhinus ursinus;* Alaska.

*D. lanceolatum* (Krabbe 1865) (syn. *D. coniceps* [Linstow 1905]), in *Phoca barbata;* Atlantic, Pacific.

*D. laruei* Vergeer 1942, experimental in dogs and cats; North America.

*D. lashleyi* Leiper et Atkinson 1914, in *Leptonychotes weddelli;* Antarctica.

*D. latum* (Linnaeus 1758) Cobbold 1858 (syn. *D. americanus* Hall et Wigdor 1918, *Dibothrium serratum* Diesing 1850, *D. fuscum* (Krabbe 1865), *D. luxi* [Rutkevich 1937], *D. parvum* Stephens 1908, *D. stictus* Talysin 1932, *D. taenioides* Leon 1920, *Diancyrobothrium taenioides* Bacigalupo 1945), in human, dog, cat, many species of piscivorous mammals; circumboreal.

*D. macroovatum* Yurakhno 1973, in *Eschrichtius gibbosus;* Russia.

*D. mansonoides* Mueller 1935 (syn. *Spirometra mansonoides* [Mueller 1935] Wardle, McLeod et Stewart 1947), in *Felis catus, F. geoffroyi, F. pardalis, F. yagouarondi, Dusycyon gymnocercus, Cerdocyon thous;* North and South America.

*D. medium* Fahmy 1954, in *Lutra lutra;* Scotland.

*D. minus* Cholodkovsky 1916, in human, cats, dogs, gulls; Russia.

*D. mobile* (Rennie et Reid 1912), in *Leptonychotes weddelli;* Antarctica.

*D. norvegicum* Vik 1957, in piscivorous birds and mammals; Norway.

*D. pacificum* (Nybelin 1931) Margolis 1956, in *Otaria byronia,* human; Peru.

*D. polyrugosum* Delyamure et Skriabin 1966, in *Orcinus orca;* Antarctica.

*D. ponticum* Delyamure 1971, in *Tursiops truncatus;* Black Sea.

*D. pretoriensis* Baer 1924 (syn. *Spirometra pretoriensis* [Baer 1924] Wardle, McLeod et Stewart 1947), in *Otocyon gegalotis, Lyacon pictus;* Africa.

*D. pterocephalum* Delyamure et Skriabin 1967, in *Cystophora cristata;* Greenland.

*D. quadratum* (Linstow 1892) (syn. *D. resimum* Railliet et Henry 1912, *D. coatsi* Rennie et Reid 1912), in *Ogmorhinus leptonyx;* Antarctica.

*D. scoticum* (Rennie et Reid 1912) (syn. *Dibothriocephalus pygoscelis* Rennie et Reid 1912), in *Hydrurga leptonyx, Otaria byronica;* Antarctica.

*D. subtile* Yamaguti 1942, in *Globicephalus scammoni;* Japan.

*D. theileri* Baer 1925, in *Zebethailurus serval, Felis caffra;* Africa.

*D. trinitatis* Cameron 1936, in *Procyon cancrivora, Nasua nasua;* Trinidad, Paraguay.

*D. urichi* Cameron 1936 (syn. *Spirometra urichi* [Cameron 1936] Wardle, McLeod et Stewart 1947), in *Felis pardalis;* Trinidad.

*D. ursi* Rausch 1954, in *Ursus arctos;* Alaska.

*D. ventropapillatum* Delyamure 1955, in *Hydrurga leptonyx;* Moscow Zoo.

*D. wilsoni* Shipley 1907 (syn. *Dibothriocephalus scotti* Shipley 1907, *D. archeri* Leiper et Atkinson 1914, *D. mobilis* Leiper et Atkinson 1915), in *Ommatophoca rossi, Ogmorhinus leptonyx, Leptonychotes weddelli;* Antarctica.

*D. yonagoensis* Yamane, Kamo, Yazaki, Fukumoto et Maejima 1981, in human; Japan.

## Diphyllobothriid Species of Uncertain Status

*Bothriocephalus columbae* Neveu-Lemaire 1912, in pigeon; France?

*Bothriocephalus longicollis* Parodi et Widakowich 1917, in *Felis yaguarondi;* Argentina.

*Bothriocephalus marginatus* Krefft 1871, in *Macropus* sp.; Australia.

*Bothriocephalus ratticola* Linstow 1904, in *Mus rattus;* Singapore.

*Bothriocephalus similis* Krabbe 1865, in *Canis lagopus, Vulpes vulpes;* Greenland.

*Dibothriocephalus skrjabini* Plotnikow 1932, in human; Russia.

*Dibothrium folium* Diesing 1850, in *Herpestes leucurus;* Senner.

*Diphyllobothrium nenzi* Petrov 1938, in human; Russia.

*Diphyllobothrium tungussicum* Podjapolskaia et Gnedina 1932, in human; Russia.

*Ptychobothrius armatus* Fuhrmann 1902 (syn. *Oncobothriocephalus armatus* [Fuhrmann 1902] Yamaguti 1959), in *Turdus parochus;* Egypt.

## KEY TO THE GENERA IN PTYCHOBOTHRIIDAE

1a. Scolex armed ................................................................ 2
1b. Scolex unarmed............................................................. 4

2a. **Hooks in complete circle on apical disc** ..... ***Circumoncobothrium* Shinde 1968.**

*Diagnosis:* Scolex with two shallow bothria. Apical disc bears a single, continuous circle of spine-like hooks. Neck absent. Proglottids acraspedote, wider than long. Genital atrium middorsal, about equatorial. Testes numerous, mainly in two lateral fields. Cirrus pouch weak. Ovary bilobed, near posterior margin of segment. Vitellaria lateral. Uterus saccate. Parasites of freshwater teleosts. India.

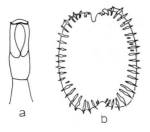

FIGURE 106. *Senga besnardi* Dollfus 1934. (a) Scolex; (b) apical disc.

FIGURE 107. *Polyoncobothrium polypteri* (Leydig 1853) Lühe 1900.

Type species: *C. ophiocephali* Shinde 1968, in *Ophiocephalus leucopunctatus;* India.
Other species:
  *C. bagariusi* Chincholikar et Shinde 1977, in *Bagarius* sp.; India.
  *C. khami* Shinde 1977, in *Ophiocephalus striatus;* India.
  *C. raoii* Shinde et Jadhav 1976, in *Mastacembellus armatus;* India.
  *C. shindei* Shinde et Chincholikar 1977, in *Mastacembellus armatus;* India.

**2b.  Circle of hooks divided into separate fields ................................... 3**

**3a.  Hooks arranged in two semicircles ............ Senga Dollfus 1934. (Figure 106)**

*Diagnosis:* Scolex rectangular, with bilobed apical disc-bearing hooks on the margins. Bothria wide and shallow with thick margins. Neck absent. Strobila short. External metamerism incomplete. Proglottids acraspedote, apolytic or pseudopolytic. Genital atrium middorsal, slightly postequatorial. Testes numerous, mainly in two lateral fields. Ovary posterior, not bilobed. Vitellaria in complete cortical layer around proglottid. Uterine duct extending forward past genital atrium, expanding into uterine sac with ventral, median pore near anterior margin of proglottid. Eggs anoperculate, not embryonated. Parasites of freshwater teleosts. Asia. Key to species: Fernando and Furtado.[918]

  Type species: *S. besnardi* Dollfus 1934, in *Betta splendens* (aquarium), *Ophiocephalus gachua;* India.
  Other species:
    *S. lucknowensis* (Johri 1956), in *Mastacembellus armatus;* India.
    *S. ophiocephalina* (Tseng 1933) Dollfus 1934 (syn. *Anchistrocephalus ophiocephalina* Tseng 1933, *A. polyptera* Southwell 1913, in *Ophiocephalus argus;* China.
    *S. pycnomera* (Woodland 1924) Dollfus 1934 (syn. *Bothriocephalus pycnomera* Woodland 1924), in *Ophiocephalus marulius;* India.
    *S. taunsaensis* Zaidi et Khan 1976, in *Channa gachua;* Pakistan.

**3b.  Hooks arranged into four quadrants ...... Polyoncobothrium Diesing 1854 (syn. Tetracampos Wedl 1861, Oncobothriocephalus Yamaguti 1959.) (Figure 107)**

*Diagnosis:* Scolex nearly rectangular, with apical disc-bearing hooks in four quadrants. Neck absent. Strobila weakly segmented. Genital atrium middorsal, preequatorial. Testes medullary, lateral to ovary. Ovary in posterior medulla. Vitellaria cortical, in two dorsal and two ventral bands. Uterus winding, opening midventrally near anterior margin of proglottid. Eggs anoperculate, unembryonated. Parasites of freshwater fishes. Africa. Genus reviewed: Tadros.[3629]

    Type species: *P. polypteri* (Leydig 1853) Lühe 1900 (syn. *P. septicolle* [Diesing 1854], *P. pseudopolypteri* Meggitt 1930), in *Polypterus bichir, P. endlicheri;* Egypt, Chad, Sudan, Nigeria. Redescribed: Jones.[1440]

    Other species:

        *P. ciliotheca* (Wedl 1861) Dollfus 1935 (syn. *Tetracampos ciliotheca* Wedl 1861), in *Clarias anguillaris:* Egypt.

        *P. clarias* Woodland 1925, in *Clarias anguillaris, C. mossambicus;* Sudan, Uganda.

        *P. cylindraceum* Janicki 1926, in *Clarias anguillaris, C. parvimanus;* Egypt.

        *P. fulgidum* Meggitt 1930, in *Clarias anguillaris;* Egypt.

        *P. gordoni* Woodland 1937, in *Heterobranchus bidorsalis;* Sierra Leone.

        *P. magnum* (Zmeev 1936) Yamaguti 1959, (syn. *Tetracampos magnum* Zmeev 1936), in *Siniperca chuatsi;* Russia.

        *P. parva* (Fernando et Furtado 1963) Blair 1978 (syn. *Senga parva* Fernando et Furtado 1963), in *Channa micropeltes;* Sri Lanka.

        *P. pahangensis* (Furtado et Chau-lan 1971) Blair 1978 (syn. *Senga pahangensis* Furtado et Chau-lan 1971), *Channa micropeltes;* Malaysia.

        *P. scleropagis* Blair 1978, in *Scleropages leichardti;* Australia.

        *P. visakhapatnamensis* (Devi et Rao 1973) Blair 1978 (syn. *Senga visakhapatnamensis* Devi et Rao 1973), in *Ophiocephalus punctatus;* India.

**4a. Vitellaria cortical, surrounding medulla ........................................ 5**
**4b. Vitellaria in lateral fields....................................................... 6**

**5a. Bothrial margins fused except for anterior aperture; bothria connected by apical groove.................................................*Clestobothrium* Lühe 1899.**

*Diagnosis:* Scolex spheroid, with deep bothria. Bothrial margins fused except for anterior aperture; bothria connected by apical groove. Neck absent. Metamerism complete. Margins of strobila serrated. Genital atrium middorsal, equatorial or postequatorial. Testes medullary, in two lateral fields. Ovary transversely elongated, in ventral medulla. Vagina with small seminal receptacle. Vitellaria continuous in cortex, surrounding proglottid. Uterine sac large, opening midventrally. Eggs anoperculate. Parasites of marine teleosts. Mediterranean, Atlantic.

    Type species: *C. crassiceps* (Rudolphi 1819) Lühe 1899, in *Merluccius merluccius, M. bilinearis, M. productus, Fundulus, Hippoglossus, Lophius, Pomatomus sphaeroides, Squalus, Urophycis, Chlorophthalmus;* Europe, North America.

**5b. Bothridial margins not fused; bothria not joined apically .......................
..........................................*Coelobothrium* Dollfus 1970. (Figure 108)**

*Diagnosis:* Scolex egg-shaped, narrowest at apex. Apical disc and spines absent. Bothria very deep, voluminous cavities. Bothrial openings narrow, with nonsalient margins that do not meet at apex of scolex. Neck absent. Immature and mature proglottids wider than long, gravid proglottids longer than wide. External segmentation not clearly seen in older part of strobila. Genital atrium middorsal, anterior. Testes medullary, in two lateral fields. Ovary transversely elongated, in ventral medulla. Vagina with well-developed seminal receptacle. Vitellaria cortical, surrounding medulla. Uterine sac opening midventrally, at anterior border of proglottid. Eggs anoperculate. Parasites of freshwater teleosts. Middle East.

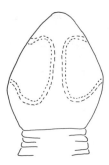

FIGURE 108. *Coelobothrium monodi* Dollfus 1970.

Type species: *C. monodi* Dollfus 1970, in *Varicorhinus damascinus;* Iran.

6a. Vitellaria cortical, not intruding into intermuscular spaces .................. 7
6b. Vitellaria infiltrating between longitudinal muscle bundles ................... 8

7a. Vitellaria dorsolateral; bothria well developed, with margins rolled ............
................................................*Ptychobothrioides* **Yamaguti 1959.**

*Diagnosis:* Scolex longer than broad. Bothria well developed, with margins rolled. Neck present. Genital atrium middorsal. Cirrus pouch containing winding ejaculatory duct. Testes numerous, in two lateral fields of medulla. Seminal receptacle apparently absent. Ovary posterior, transversely elongated, in ventral medulla. Vitellaria in two dorsolateral cortical fields, not intruding into intermuscular space. Uterus sinuous in medulla, with distal sac provided with longitudinal muscle just before opening outside. Uterine pore submedian, anterior to level of genital atrium. Eggs subglobular. Parasites of hawks (may be fish parasite eaten by hawk); Abyssinia.

Type species: *P. spiraliceps* (Volz 1900) Yamaguti 1959 (syn. *Bothriocephalus spiraliceps* Volz 1900), in *Falco concolor;* Abyssinia.

7b. Vitellaria in dorsal and ventral, lateral fields; bothria shallow, not rolled .....
........................................ *Icthybothrium* **Khalil 1971. (Figure 109)**

*Diagnosis:* Scolex short, simple, blunt, lacking armature and apical organ. Bothria two, shallow, open their entire length, not connected at tip of scolex and their margins not inwardly rolled. Strobila acraspedote with incomplete segmentation. Testes in lateral region of medulla parenchyma. Cirrus sac relatively thick-walled, with numerous external gland cells particularly around its proximal end. Genital pore dorsal, median, approximately at the midlevel of proglottid. Cirrus unarmed. Ovary bilobed, near posterior border of proglottid. Uterine duct coiled and sinuous in mature proglottids. Uterine duct distinct, opening ventrally at anterior border of proglottid to one side of median line. Vitelline follicles lateral in cortical parenchyma. Eggs thin-shelled, nonoperculate. Parasites of freshwater teleosts. Africa.

Type species: *I. ichthybori* Khalil 1971, in *Ichthyborus besse;* White Nile.

8a. Uterine pore alternating, on left and right sides of midline .....................
.................................*Ptychobothrium* **Lönnberg 1889. (Figure 110)**

*Diagnosis:* Scolex flattened, unarmed, heart-shaped in lateral view, lacking terminal disc. Bothria well-developed. Neck absent. External metamerism incomplete. Genital atrium dor-

FIGURE 110. *Ptychobothrium belones* (Kujardin 1845) Lönnberg 1889.

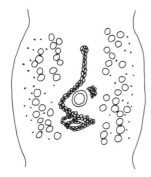

FIGURE 109. *Ichthybothrium ichthybori* Khalil 1971.

FIGURE 111. *Plecatobothrium cypseluri* Cable et Michalelis 1967.

sal. Cirrus pouch thin-walled, lacking prostate cells around proximal end. Testes lateral, in dorsal medulla. Vagina with seminal receptacle. Ovary compact, posterior. Vitellaria intermingled with longitudinal muscle bundles. Uterine sac opening submedian, alternating sides of median line, near anterior margin of proglottid. Eggs oval, thin-shelled. Parasites of marine teleosts. Atlantic, Pacific.

Type species: *P. belones* (Dujardin 1845) Lönnberg 1889 (syn. *Dibothrium restiformis* Linton 1890, *Bothriocephalus restiformis* [Linton 1890] Ariola 1896), in *Belone vulgaris, Tylocurus caribbaeus, T. schismatorhynchus, T. raphidoma, Strongylura notata;* Atlantic, Caribbean, Pacific.

Other species:

*P. clupeoidesii* Chincholikar, Shinde et Deshmukh 1976, in *Chela clupeoides;* India.

**8b.    Uterine pore midventral.................................................... 9**

**9a.    Uterus Y-shaped......... *Plicatobothrium* Cable et Michaelis 1967. (Figure 111)**

*Diagnosis:* Scolex triangular to fan-shaped, lacking apical organ or hooks. Bothria deep, open their entire lengths; not connected apically, lined with minute, hair-like spines. Neck probably absent. Proglottids acraspedote; strobila distinctly segmented only at intervals. Genital pore dorsal, median, posterior to midlevel of proglottid. Uterine pore midventral. Testes in lateral medullary parenchyma. Ovary median, at extreme posterior end of proglottid flanked by arms of Y-shaped uterus of succeeding segment. Vitellaria densest laterally, but encompassing medullary parenchyma except in vicinity of genital pores; follicles between

FIGURE 112. *Alloptychobothrium spilonotopteri* Yamaguti 1968.

FIGURE 113. *Oncodiscus sauridae* Yamaguti 1934. (a) Scolex; (b) hook.

conspicuous layers of longitudinal muscle fibers that are not grouped into distinct bundles. Eggs thin-shelled, anoperculate. Parasites of marine teleosts. Caribbean, Indian Ocean.

Type species: *P. cypseluri* Cable et Michaelis 1967, in *Cypselurus bahiensis;* Curaçao.
Other species:

*P. raoi* (Rao 1959) Khalil 1971 (syn. *Ptychobothrium cypseluri* Rao 1959), in *Cypselurus poecilopterus;* India.

**9b. Uterus a series of median coils** ............................................................
............................*Alloptychobothrium* **Yamaguti 1968. (Figure 112)**

*Diagnosis:* Scolex arrowhead-shaped in lateral view, bilaterally compressed, with bothridial margins strongly crenulated. Neck absent. Strobila may be completely or incompletely segmented. Proglottids wider than long, craspedote or not. Inner longitudinal muscle bundles divided by vitelline layer into two (outer and inner) layers. Testes not numerous, in one layer or two, in submedian medulla immediately dorsal to ventral medullary excretory anastomoses. Cirrovaginal pore opening at bottom of middorsal notch. Ovary multilobulated, arcuate, transversely elongated, median, at posterior end of proglottid. Vitellaria continuous laterally and from segment to segment, but interrupted at notches where uterus and cirrovaginal pore open outside. Uterine coils confined to median field dorsal and anterior to ovary, not forming uterine sac before opening midventrally. Eggs anoperculate. Ventral osmoregulatory canals anastomosing with narrower ventral vessels that run longitudinally, lateral to them. Parasites of flying fish. Pacific Ocean.

Type species: *A. spilonotopteri* Yamaguti 1968, in *Cypselurus spilonopterus;* Hawaii.

## KEY TO THE GENERA IN BOTHRIOCEPHALIDAE

**1a. Holdfast bilaterally flattened, armed with hooks** ...............................
............................................*Oncodiscus* **Yamaguti 1934. (Figure 113)**

*Diagnosis:* Scolex strongly compressed laterally, with median furrow on each side. Bothria with crenulated borders. Apex disk-like, with minute hooks along margin. Neck absent. Metamerism complete. Proglottids craspedote, with median notch on posterior border. Genital atrium middorsal, postequatorial. Cirrus pouch with thin walls, lacking prostatic cells. Testes in lateral medulla. Vagina greatly enlarged, muscular distally. No seminal receptacle. Ovary compact, lobulated, ventral, posterior. Vitellaria very numerous, cortical. Uterine duct coiling forward in median field, expanding into very large uterine sac. Uterine pore midventral, equatorial. Eggs operculate, unembryonated. Parasites of marine teleosts. Japan, India.

FIGURE 114. *Taphrobothrium japonense* Lühe 1899.

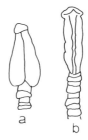

FIGURE 115. *Bothriocephalus cuspidatus* Cooper 1917. (a) Lateral view; (b) dorsal view.

Type species: *O. sauridae* Yamaguti 1934, in *Saurida argyrophanes;* Inland Sea of Japan and East China Sea.
Other species:
    *O. fimbriatus* Subhapradha 1955, in *Saurida tumbil;* India.
    *O. maharashtrae* Jadhav et Shinde 1981, in *Trygon sephen;* India.

**1b.   Holdfast not flattened, unarmed ................................................. 2**

**2a.   Vitellaria medullary .................. *Taphrobothrium* Lühe 1899. (Figure 114)**

*Diagnosis:* Scolex elongate, with prominent apical disk. Bothria long, slender, and shallow. Neck absent. External metamerism incomplete at irregular intervals. Genital atrium middorsal. Cirrus pouch muscular, elliptical. Testes medullary, in two lateral fields. Seminal receptacle absent. Ovary median, transversely elongated, with ends directed dorsolaterally. Vitellaria medullary, scattered among testes. Uterus sigmoid, with terminal sac. Uterine pore ventral, submedian, irregularly alternating. Eggs operculated, embryonated. Parasites of marine teleosts. Japan.
Type species: *T. japonense* Lühe 1899, in *Muraenesox cinereus;* Japan.

**2b.   Vitellaria cortical .......................................................... 3**

**3a.   Scolex with apical disc ............ *Bothriocephalus* Rudolphi 1808. (Figure 115)**

*Diagnosis:* Scolex elongate, with apical disc, bearing indentations on bothrial margins that may be connected by apical groove. Marginal surfaces of scolex concave or convex, may bear longitudinal grooves. Bothria variable. Neck absent. External metamerism complete; secondary pseudometamerism sometimes present on older proglottids. Proglottids acraspedote, anapolytic. Genital atrium dorsomedial. Testes in lateral medulla. Seminal receptacle absent. Ovary posterior, transversely elongate, in ventral medulla. Vitellaria cortical. Uterus and uterine sac median or alternating submedian. Uterine pore median, anterior to genital atrium. Eggs operculate, unembryonated. Parasites of marine and freshwater fishes. Cosmopolitan.
Type species: *B. scorpii* (Mueller 1776) Rudolphi 1808 (syn. *B. punctatus* [Rudolphi 1802], *B. bipunctatus* [Zeder 1800] Lühe 1899), in *Cottus scorpius, Rhombus maximus,* and many other marine fishes of several different families; Atlantic, Pacific, Black Sea, Mediterranean.
Other species:
    *B. abyssmus* Thomas 1952, in *Echiostoma tanneri;* Bermuda.

*B. acheilognathi* Yamaguti 1934 (syn. *B. gowkongensis* Yeh 1955, *B. opsariichthydis* Yamaguti 1934), in *Acheilognathus rhombea, Gnathopogon elongatus, Cyprinus carpio, Ctenopharyngodon idellus, Opsariichthys uncirostris,* other freshwater fishes, mainly cypriniforms.

*B. aegyptiacus* Rysavy et Moravec 1975, in *Barbus bynni;* Egypt.

*B. alessandrinii* Condorelli-Francaviglia 1898, in *Salmo fario;* Italy.

*B. andresi* Porta 1911, in *Eucitharus linguatula;* Italy.

*B. anguillae* (Leeuwenhoek 1722) Rudolphi 1808, in *Anguilla vulgaris, Muraena helena;* locality unknown.

*B. angustatus* Rudolphi 1819 (syn. *B. affinis* Leuckart 1819), in *Scorpaena scrofa;* locality?

*B. angusticeps* Olsson 1868, in *Sebastes norvegicus;* Norway.

*B. apogonis* Yamaguti 1952, in *Apogon lineatus;* Japan.

*B. barbus* Fahmy, Mandour et El-Naffar 1978, in *Barbus bynni;* Egypt.

*B. bengalensis* Devi 1975, in *Caranx plagiotaenia;* Bay of Bengal, India.

*B. bramae* Ariola 1899, in *Brama raji;* Italy.

*B. branchiostegi* Yamaguti 1952, in *Branchiostegus japonicus;* Japan.

*B. breviceps* Guiart 1935, in *Synaphobranchus* sp.; Azores.

*B. brotulae* Yamaguti 1952, in *Brotula multibarbata;* Japan.

*B. callariae* Linstow 1878, in *Gadus morrhua;* locality?

*B. capillicollis* Mégnin 1883, in "carpe de mer"; Norway.

*B. carangis* Yamaguti 1968, in *Caranx helvolus, Carangoides ferdau;* Hawaii.

*B. cepolae* Rudolphi 1819, in *Cepola rubescens;* Europe.

*B. cestus* Leidy 1885, in *Salvelinus* sp.; Canada.

*B. clavibothrium* Ariola 1901, in *Arnoglossus laterna;* Italy.

*B. claviceps* (Goeze 1782) Rudolphi 1810 (syn. *Taenia anguillae* Gmelin 1790), in *Muraena anguilla, Anguilla rostrata, Eupomotis gibbosus, Gasterosteus bispinosus, Chaetobryttus, Esox, Ambloplites, Micropterus, Huro, Lepomis, Percopsis, Stizostedion, Acanthocottus, Apeltes, Cyclopterus, Gladiunculus, Limanda, Pseudopleuronectes;* Europe, North Africa, North America.

*B. cordiceps* (Leidy 1872) Braun 1894, in *Salmo fontinalis;* North America.

*B. cuspidatus* Cooper 1917, in *Stizostedion vitreum, S. canadensis, Amphiodon alosoides, Hiodon tergisus, Esox lucius, Perca flavescens,* other freshwater fishes; North America. Species discussed: Wardle (1932).

*B. ellipticus* Linstow 1880, in *Gadus callarias;* locality unknown.

*B. eriocis* Rudolphi 1819, in *Salmo eriox;* locality?

*B. euryciensis* Schaeffer et Self 1978, in *Eurycea longicauda;* U.S.

*B. fluviatilis* Yamaguti 1952, in *Hymenophysa curta;* Japan.

*B. formosa* Mueller et Van Cleave 1932, in *Percopsis omiscomaycus, Boleosoma nigrum, Percina caprodes, Semotilus atromaculatus, Lepomis gibbosus;* U.S.

*B. ganapattii* Rao 1954, in *Saurida tumbil;* India.

*B. gasterostei* Knoch 1862, in *Gasterosteus pungitius;* locality?

*B. granularis* Rudolphi 1910, in *Cyprinus* sp.; locality?

*B. hirondellei* Guiart 1935, in *Syngnathus pelagica;* Azores.

*B. indicus* Ganapati et Rao 1954, in *Saurida tumbil;* India.

*B. japonicus* Yamaguti 1934, in *Anguilla japonica;* Japan.

*B. kivuensis* Baer et Fain 1958, in *Barbus antianalis;* Africa.

*B. labracis* Dujardin 1845, in *Labrax lupus;* France.

*B. laciniatus* (Linton 1897) Lühe 1899, in *Tarpon atlanticus;* U.S.

*B. lateolabracis* Yamaguti 1952, in *Lateolabrax japonicus;* Japan.

*B. lophii* Rudolphi 1819, in *Lophius piscatorius;* Europe.

*B. luehei* Mola 1912, in *Cottus gobio;* Europe.

*B. manubriformis* (Linton 1889) (syn. *B. histiophorus* Shipley 1909, *B. plicatus* Shipley 1900), in *Tetrapterus* spp., *Histiophorus gladius, H. grayi, Tarpon atlanticus, Makaira audax, Istiompax orientalis;* Atlantic, Pacific, Indian oceans.

*B. minutus* Ariola 1896, in *Syngnathus acus;* Italy.

*B. monorchis* Linstow 1903, in *Mola mola;* Europe.

*B. monticelli* Ariola 1899, in *Trachypterus iris;* Italy.

*B. motellae* Olsson 1893, in *Motella cimbria;* Scandinavia.

*B. musculosus* Baer 1937, in *Cichlosoma biocellata;* aquarium.

*B. neglectus* Loennberg 1893, in *Raniceps niger;* Sweden.

*B. nigropunctatus* Linstow 1901, in *Sebastes norvecious;* Arctic Ocean.

*B. occidentalis* (Linton 1897), in *Sebastodes* sp., *Leptocottus armatus;* North America.

*B. osmeri* Linstow 1878 (syn. *Scolex eperlani* Beneden), in *Osmerus eperlanus;* locality?

*B. palumbi* Monticelli 1889, in *Trigla* sp.; Europe?

*B. parvus* Creplin 1846, in *Ammodytes tobianus;* locality?

*B. platycephalus* Monticelli 1889, in *Beryx decadactyla;* Spain.

*B. prudhoei* Tadros 1967, in *Clarias anguillaris;* Egypt.

*B. phoxini* Molnár 1968, in *Phoxinus phoxinus;* Hungary.

*B. proboscideus* (Batsch 1786) Rudolphi 1810, in *Salmo salar, Trutta trutta, Coregonus oxyrhynchus, Motella mustela;* Europe, Greenland.

*B. rarus* Thomas 1937, in *Triturus viridiscens;* U.S.

*B. rhombi* (Leeuwenhoek) Mola 1928, in *Mullus, Arnoglossus, Rhombus, Rhomboidichthys, Pleuronectes, Solea, Labrus, Trigla;* locality?

*B. salvelini* Lönnberg 1892, in *Salmo alpinus;* Sweden.

*B. sauridae* Ariola 1900, in *Saurida nebulosa;* Zanzibar.

*B. schilbiodes* Cheng et James 1960, in *Schilbiodes insignis;* U.S.

*B. sciaenae* Yamaguti 1934, in *Sciaena schlegeli;* Japan.

*B. speciosus* (Leidy 1858), in *Boleosoma olmstedi;* U.S.

*B. spinachiae* Olsson 1893, in *Gasterosteus spinachia;* Scandinavia.

*B. squaliglauci* Rudolphi 1819, in *Squalus glaucus;* locality?

*B. squalii* Ariola 1900, in *Squalus cavedanus;* Italy.

*B. tetragonus* Ariola 1899, in *Anarrhicas minor;* Europe.

*B. texomensis* Self 1954, in *Hiodon alosoides;* U.S.

*B. tintinnabulus* Guiart 1935, in *Syngnathus phlegon;* Monaco.

*B. trachypteri* Ariola 1896, in *Trachypterus liopterus;* Italy.

*B. trachypteriiris* Ariola 1896, in *Trachypterus iris;* Italy.

*B. trachypteriliopteri* Ariola 1896, in *Trachypterus liopterus;* Italy.

*B. travassosi* Tubangui 1938, in *Anguilla mauritiana;* Philippines.

*B. typhlotritonis* Reeves 1949, in *Typhlotriton spelaeus, T. nereus;* U.S.

*B. vallei* Stossich 1899, in *Mullus barbatus;* Istria.

**3b. Scolex lacking apical disc ..................................................... 4**

**4a. Two smooth bothria present .......... *Penetrocephalus* Rao 1960. (Figure 116)**

*Diagnosis:* Two smooth bothria, like those of *Diphyllobothrium*. No apical disc on scolex. Neck present. Segmentation prominent, proglottids very craspedote. Cirrovaginal pore dorsal, immediately before ovary. Uterine pore median, ventral. Oval cirrus pouch, slightly right or left of middorsal line. Cirrus prominent. Testes in two lateral fields, continuous between segments, about 60 per proglottid. Ovary bilobed, transversely elongated, near posterior margin of proglottid. Vagina opens into atrium. Seminal receptacle present. Vitellaria in

104  CRC Handbook of Tapeworm Identification

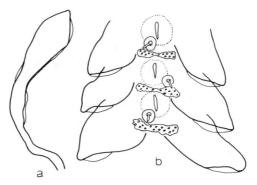

FIGURE 116.  *Penetrocephalus ganapati* Rao 1960. (a) Scolex; (b) portion of strobila.

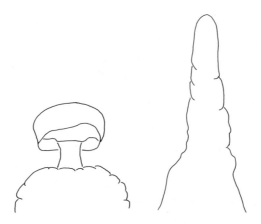

FIGURE 117.  *Anantrum.* Two forms of scolex.

lateral fields. Uterus developing into spherical median sac. Parasites of marine teleosts. India.

Type species: *P. ganapati* Rao 1960, in *Saurida tumbis, S. undosquamis;* India.

**4b.   Bothria absent** ..................................................................
   ***Anantrum* Overstreet 1968 (syn. *Acompsocephalum* Rees 1969).**   **(Figure 117)**

*Diagnosis:* Scolex elongate, clavate, lacking bothria, apical disc or armature, unsegmented neck present. External metamerism lacking, interior metamerism evident. Proglottids wider than long. Testes in lateral medulla. Cirrus pouch oval, at right angles to dorsal surface. Cirrovaginal atrium dorsal, median. Ovary bilobed, in ventral, median, central medulla. Seminal receptacle present. Vitellaria cortical. Uterus with distal sac. Uterine pore ventral, irregularly alternating from one side of median line to the other. Eggs thin-shelled, operculate, unembryonated when laid.

Type species: *A. tortum* (Linton 1905) Overstreet 1968 (syn. *Acompsocephalum tortum* [Linton 1905] Rees 1969, *Dibothrium tortum* Linton 1905), in *Synodus foetens, S. intermedius;* Western Atlantic.

Other species:
   *A. histocephalum* Jensen et Heckmann 1977, in *Synodus lucioceps;* California.

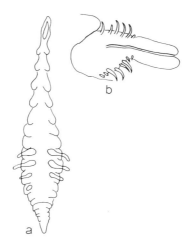

FIGURE 118. *Echinophallus japonicus* Yamaguti 1934. (a) Entire worm; (b) cirrus.

## KEY TO THE GENERA IN ECHINOPHALLIDAE

**1a. Minute spines on posterior margins of proglottids. Holdfast funnel-like ........ .................................................... *Atelemerus* Guiart 1935.**

*Diagnosis:* Scolex funnel-like. Strobila short, external metamerism incomplete. Posterior border of proglottids covered with minute spines. Two sets of reproductive organs per segment. Genital pores dorsal, close to lateral margin. Cirrus pouch cylindrical, equatorial. Cirrus armed. Testes and ovary not described. Uterus in anterior third of proglottid. Uterine pore ventral, anterior. Eggs anoperculate (?). Parasites of marine teleosts. Azores.

Type species: *A. acanthodes* Guiart 1935, in *Centrolophus pompilius;* Azores.

**1b. No spines on posterior margins of proglottids. Holdfast with weak dorsal and ventral sucker-like depressions ................................................... ............*Echinophallus* Schumacher 1914 (syn. *Amphitretus* Blanchard 1894, *Acanthophallus* Lühe 1903, *Paraechinophallus* Protasova 1975). (Figure 118)**

*Diagnosis:* Pseudoscolex present, poorly developed, with shallow bothria. Neck absent. Posterior borders of proglottids unarmed, thickened into lobes. Two sets of reproductive organs per segment. Genital pores dorsal, close to lateral margins. Cirrus pouch large. Cirrus with large spines on base. Testes in two submedian fields median to ovaries, mostly postequatorial. Ovary lobated, posterior. Vitellaria extensive, mostly cortical. Uterus strongly coiled, with weak sac, median to ovary. Uterine pore submedian, near anterior end of proglottid. Eggs operculate. Parasites of marine teleosts. Mediterranean; Japan.

Type species: *E. wageneri* (Monticelli 1890) Schumacher 1914, in *Centrolophus pompilius;* Mediterranean.

Other species:

*E. hyperoglyphe* (Tkachev 1979) comb. n. (syn. *Paraechinophallus hyperoglyphe* Tkachev 1979), in *Hyperoglyphe japonica;* Hawaii.

*E. japonicus* Yamaguti 1934 (syn. *Paraechinophallus japonica* [Yamaguti 1934] Protasova 1975), in *Psenopsis anomela;* Japan.

*E. settii* (Ariola 1895) Schumacher 1914, in *Centrolophus pompilius;* locality?

## FAMILY AMPHICOTYLIDAE

### KEY TO THE SUBFAMILIES IN AMPHICOTYLIDAE

1a. Vitellaria cortical ............................ Amphicotylinae Lühe 1902. (p.106)
1b. Vitellaria medullary .......................................................................... 2

2a. Vitellaria in two lateral fields ................ Abothriinae Nybelin 1922. (p.108)
2b. Vitellaria continuous around proglottid ........................................ 3

3a. Small sucker at rear border of bothrium ..........................................
................................................ Bothriocotylinae Yamaguti 1959. (p.109)
3b. No sucker at rear border of bothrium ..............................................
........................................... Marsipometrinae Cooper 1917. (p.109)

### KEY TO THE GENERA IN AMPHICOTYLINAE

1a. Bothria divided into loculi by transverse grooves ...............................
............................................... *Pseudamphicotyla* Yamaguti 1959.

*Diagnosis:* Scolex with apical disc and elongate surfacial bothria, each of which is divided into loculi by transverse septal grooves. Neck absent. External metamerism distinct anteriorly, indistinct posteriorly. Genital pore marginal, irregularly alternating. Cirrus pouch club-shaped. Cirrus armed. Testes numerous, in single broad field. Vagina muscular. Ovary two-winged, poral. Vitellaria mostly cortical, encircling proglottid. Uterus wide, median, with median terminal sac. Uterine pore ventromedian. Eggs anoperculate. Parasites of marine teleosts. Japan.

Type species: *P. quinquarii* (Yamaguti 1952) Yamaguti 1959 (syn. *Amphicotyle quinquarii* Yamaguti 1952), in *Quinquarius japonicus;* Japan.
Other species:
*P. mamaevi* Tkachev 1978, in *Seriolella tinro;* New Zealand.

1b. Bothria not divided into loculi by transverse grooves ......................... 2

2a. Bothria with posterior accessory sucker. Uterine pore dorsal ...................
.................................................... *Amphicotyle* Diesing 1863.

*Diagnosis:* Scolex with apical disc and rounded bothria, each with posterior partition forming an accessory sucker. External metamerism well marked anteriorly, obscured posteriorly by secondary wrinkles. Proglottids anapolytic. Genital atrium marginal, irregularly alternating. Cirrus pouch large, weakly muscular. Cirrus unarmed. Testes in two lateral fields. Seminal vesicles absent. Ovary asymmetrical, antiporal. Vitellaria cortical, scattered. Uterus simple, with terminal sac. Uterine pore dorsal. Parasites of marine teleosts. Mediterranean.

Type species: *A. heteropleura* (Diesing 1850) Lühe 1902 (syn. *A. typica* Diesing 1863), in *Centrolophus pompilius;* Triest.
Other species:
*A. ceratias* Tkachev 1979, in *Ceratias holboelli;* Antarctica.
*A. kurochkini* Tkachev 1979, in *Sereolella* sp.; Pacific.

2b. Bothria without accessory sucker ................................................. 3

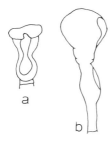

FIGURE 119. *Eubothrium rugosum* (Batsch 1786) Nybelin 1922. (a) Typical scolex; (b) deformed scolex.

FIGURE 120. *Pseudeubothrium xiphiados* Yamaguti 1968.

3a. Neck absent; dorsomedian longitudinal furrows present ........................ ........ *Eubothrium* Nybelin 1922 (syn. *Leuckartia* Moniez 1879). (Figure 119)

*Diagnosis:* Scolex with apical disk and simple bothria, sometimes deformed. Neck absent. Strobila usually with distinct external metamerism; dorsomedial longitudinal furrows present. Testes medial to nerve trunks, in two lateral fields, sometimes almost continuous. Genital atrium marginal, irregularly alternating. Cirrus pouch medium sized. Vagina opens anterior to cirrus. Ovary reniform, median or poral. Vitellaria cortical in two lateral zones. Uterine pore ventral. Parasites of marine and freshwater teleosts. Circumboreal.

Type species: *E. rugosum* (Batsch 1786) Nybelin 1922, in *Gadus mustela, Lota lota, Motella mustela, Haloporphyrus lepidion;* Azores, Europe, Canada, Russia.

Other species:

*E. arcticum* Nybelin 1922, in *Lycodes pallidus;* Greenland.

*E. crassoides* Nybelin 1922, in *Acipenser stellatus;* Russia.

*E. crassus* (Bloch 1779) Nybelin 1922 (syn. *Taenia crassa* Bloch 1779. For other synonyms see Nybelin 1922), in *Lota, Salmo, Salvelinus, Trutta, Coregonus, Oncorhynchus, Thymallus, Esox, Perca, Osmerus, Clupea, Cristiovomer, Stizostedion, Cottus, Gasterosteus, Petromyzon, Acerina, Cyclopterus, Barbus, Squalius, Silurus, Myxocephalus, Leucichthys;* Circumboreal.

*E. fragilis* (Rudolphi 1802) Nybelin 1922, in *Clupea alosa;* Europe.

*E. oncorhynchi* Wardle 1932, in *Oncorhynchus tschwaytscha;* Canada.

*E. parvum* Nybelin 1922, in *Mallotus villosus;* Norway.

*E. salvelini* (Schrank 1790) Nybelin 1922, in *Salvelinus, Oncorhynchus, Salmo, Coregonus, Osmerus, Thymallus, Cristivomer, Ptychocheilus;* Europe, Canada.

3b. Neck present; dorsomedian longitudinal furrows absent ............... *Pseudeubothrium* Yamaguti 1968 (syn. *Pseudeubothrioides* Yamaguti 1968). (Figure 120)

*Diagnosis:* Scolex with apical disc and elongate, simple bothria. Distinct neck present. Proglottids very short, wide, craspedote. No median furrows on flat surfaces of strobila. Testes in dorsal medulla, continuous in median field except in ovarian region. Vas deferens convoluted in dorsal medulla. Cirrus pouch with thick, or thin, muscular wall. Ejaculatory duct convoluted in cirrus pouch. Cirrus bulbous or not, lined with reticular cuticle that appears like spinelets in cross sections. Genital pores lateral, irregularly alternating. Ovary transversely elongated in ventral median medulla. Vitellaria confined to cortex surrounding proglottid, never extending into medulla. Vagina opens posterior to cirrus pouch. Uterine

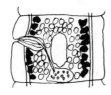

FIGURE 122. *Abothrium gadi* Beneden 1871.

FIGURE 121. *Bathybothrium rectangulum* (Bloch 1782) Lühe 1902.

sac present, muscular; uterine pore ventral, submedian or sublateral. Eggs thick-shelled. Parasites of marine teleosts. Pacific Ocean.

Type species: *P. xiphiados* Yamaguti 1968, in *Xiphias gladias;* Hawaii.
Other species:
   *P. lepidocybii* (Yamaguti 1968) comb. n. (syn. *Pseudeubothrioides lepidocybili* Yamaguti 1968), in *Lepidocybium flavobrunneum;* Hawaii.

## KEY TO THE GENERA IN ABOTHRIINAE

**1a. Scolex normal. Vitellaria only in lateral medulla** .............................................
.........................................*Bathybothrium* **Lühe 1902. (Figure 121)**

*Diagnosis:* Scolex normal, lacking apical disk. Neck? External metamerism indistinct. Median groove present on ventral surface. Genital atrium lateral. Cirrus pouch pyriform, small. Testes numerous, in single field anterior and lateral to uterus. External seminal vesicle present. Ovary thick, posterior, compact. Vitellaria few, in lateral medulla. Uterine sac with lateral outpocketings. Uterine pore ventromedian or slightly aporal. Parasites of cyprinid fishes. Europe.

Type species: *B. rectangulum* (Bloch 1782) Lühe 1902, in *Barbus barbus;* Europe.

**1b. Scolex deformed. Vitellaria not as above.........................................2**

**2a. Vitellaria interspersed with testes.......*Abothrium* Beneden 1871. (Figure 122)**

*Diagnosis:* Scolex deformed. Neck present. Strobila thick. Genital atrium lateral. Cirrus pouch small. Testes numerous, in two lateral fields. Ovary posterior, compact, transversely elongated. Vitellaria intermingled with testes. Uterus sac-like. Uterine pore ventral. Inner longitudinal muscle bundles strongly developed. Eggs thin-shelled. Parasites of marine teleosts. Atlantic, Pacific.

Type species: *A. gadi* Beneden 1871, in *Gadus callarias, G. aeglefinus, G.* spp., *Trachinus, Rhombus, Merluccius, Motella, Melanogrammus, Microgadus, Pollachius, Urophycis, Theragra chalcogramma;* Japan, Atlantic.
Other species:
   *A. acipenserinum* Cholodkovsky 1918, in *Acipenser baeri, A. gueldenstaedtii;* Russia.
   *A. hermaphroductus* Harding 1937, in *Oncorhynchus keta;* Eastern Pacific.
   *A. infundibuliformis* (Rudolphi 1810), in *Salmo salvelinus;* Atlantic.
   *A. longissimum* Cholodkovsky 1918, in salmonids; Russia.
   *A. morrhuae* Cholodkovsky 1918, in *Gadus morrhua, G. callarias;* Russia.

**2b. Vitellaria not interspersed with testes** .............................................
.........................................*Parabothrium* **Nybelin 1922. (Figure 123)**

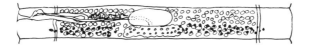

FIGURE 123.  *Parabothrium bulbiferum* Nybelin 1922.

FIGURE 124.  *Bothriocotyle solinosomm* Ariola 1900.

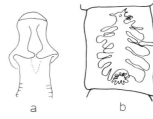

FIGURE 125.  *Marsipometra confusa* Simer 1930. (a) Scolex; (b) gravid proglottid.

*Diagnosis:* Scolex deformed. Genital atrium lateral. Cirrus pouch large, elongated, muscular. Testes numerous, in two fields lateral to ovary and uterus. Vagina with terminal sphincter. Ovary in posterior medulla. Vitellaria posterior, ventral to testes, extending laterally beyond nerve trunks. Uterine sac opening funnel-like in ventromedial region of proglottid. Eggs round, with thin shell. Parasites of marine teleosts (Gadidae). Europe, North America.

Type species: *P. bulbiferum* Nybelin 1922, in *Gadus* spp., *Merluccius merluccius;* North Atlantic, Scandinavia.

## DIAGNOSIS OF THE ONLY GENUS IN BOTHRIOCOTYLINAE

*Bothriocotyle* Ariola 1900 (Figure 124)

*Diagnosis:* Scolex conical, lacking apical disc. Bothria shallow, with small distinct sucker at posterior margin of each. Neck absent. Strobila spiral, concave ventrally, external metamerism distinct, lacking accessory wrinkles. Proglottids broader than long. Genital atrium dorsal, submedian, irregularly alternating. Cirrus pouch large. Cirrus armed. Testes in single broad medullary layer, extending lateral to nerve trunks. Ovary bilobed, in aporal medulla. Vitellaria medullary, ventral. Uterus with long narrow duct and small sac. Uterine pore midventral. Eggs anoperculate, thin-shelled. Parasites of marine teleosts. Mediterranean.

Type species: *B. solinosomum* Ariola 1900, in *Centrolophus pompilius;* Italy.

## KEY TO THE GENERA IN MARSIPOMETRINAE

**1a.  Scolex in shape of truncated pyramid, with well-defined apical disk ............ ........................................ *Marsipometra* Cooper 1917. (Figure 125)**

*Diagnosis:* Scolex in shape of truncated pyramid; apical disk present. Constriction often present between apex and base of scolex. Neck present or absent. External metamerism well marked. Genital atrium marginal, equatorial. Cirrus unarmed. Testes numerous, lateral, may extend anterior to uterus and posterior to ovary. Ovary posterior, may be separated from posterior margin of proglottid by testes or vitellaria. Seminal receptacle present. Vitellaria usually ventral, medullary, lateral to ovary but confluent anterior and posterior to it. Uterine

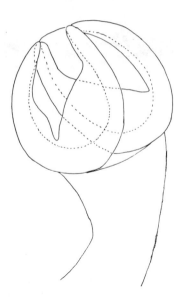

FIGURE 126. *Fissurobothrium uncinum* Roitman 1965.

sac central, with many radial branches. Eggs small, thin-shelled. Parasites of freshwater Chondrostei (Polyodontidae). U. S.

Type species: *M. hastata* (Linton 1897) Cooper 1917, in *Polyodon spathula*; U.S.
Other species:
  *M. confusa* Simer 1930, in *Polyodon spathula*; U.S.
  *M. parva* Simer 1930, in *Polyodon spathula*; U.S.

1b.  **Scolex rounded, lacking apical disk** ...............................................
     ........................... *Fissurobothrium* **Roitman 1965. (Figure 126)**

*Diagnosis:* Scolex lacking apical disk. Bothria deep, with very muscular walls, giving a sucker-like appearance. Neck present. Proglottids wider than long. Genital pores lateral, slightly preequatorial, surrounded by sphincter. Cirrus pouch cylindrical, containing seminal vesicle and unarmed cirrus. Small external seminal vesicle present. Testes numerous, in two lateral fields. Vagina posterior to cirrus pouch. Large seminal receptacle present. Ovary posterior, with two multilobed wings. Vitellaria medullary. Uterus with several large, lateral branches. Eggs anoperculate. Parasites of cyprinid fishes. Russia.

Type species: *F. unicum* Roitman 1965, in *Gobio gobio*; Russia.

## KEY TO THE GENERA IN PHILOBYTHIIDAE

1a.  **Scolex with apical disc; some testes posterior to ovary** ..........................
     ........................... *Philobythos* **Campbell 1977. (Figure 127)**

*Diagnosis:* Scolex unarmed, with bothria and weakly developed apical disc. Strobila distinctly segmented, craspedote, anapolytic. Primary and secondary segmentation precede proglottidization. One set of male and female reproductive systems per segment. Testes medullary, some posterior to ovary. Cirrus pouch small, cirrus unarmed. Vas deferens surrounded by numerous prostate cells. Genital pores marginal, mainly unilateral. Vagina anterior to cirrus pouch. Seminal receptacle present. Ovary compact, median in posterior

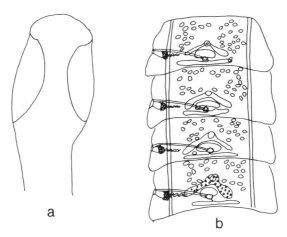

FIGURE 127. *Philobithos atlanticus* Campbell 1977. (a) Scolex and (b) strobila.

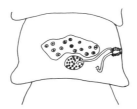

FIGURE 128. *Philobithoides stunkardi* Campbell 1979.

half of segment. Vitellarium lobated, transverse, single, medullary, posterior to ovary. Uterus triangular with preformed, lateral diverticula. Uterine pore ventromedial, preformed. Oncospheres in clusters of three to five per capsule. Parasites of marine teleosts. North Atlantic.

Type species: *P. atlanticus* Campbell 1977, in *Acanthochaenus lutkenii;* Hudson submarine canyon, western North Atlantic.

**1b.  Scolex without apical disk; testes all anterior to ovary ..........................**
**......................................... *Philobythoides* Campbell 1979. (Figure 128)**

*Diagnosis:* Scolex conical. Bothria well formed, apical disc absent. Neck absent. Segments markedly craspedote. Strobila anapolytic. All segments wider than long. Genital pores postequatorial, irregularly alternating. Genital atrium protrusible into papilla-like structure. Cirrus pouch pyriform, delicate. Cirrus unarmed. Testes medullary, in median one fourth of segment. Ovary bilobed, compact. Vitellarium compact, transversely expanded. Vagina anterior to cirrus pouch. Uterine pore median. Uterine duct extends anterior to ovary to join uterus, which is inverted V-shaped. Testes and vitellarium degenerate in gravid segments. Eggs in single capsules. Parasites of marine teleosts. Northwest Atlantic.

Type species: *P. stunkardi* Campbell 1979, in *Alepocephalus agassizi;* Hudson submarine canyon, northwest Atlantic.

FIGURE 129. *Ancistrocephalus microcephalus* (Rudolphi 1819) Monticelli 1890.

FIGURE 130. *Triaenophorus crassus* Forel 1868. (a) Trident hook; (b) scolex.

## KEY TO THE GENERA IN TRIAENOPHORIDAE

1a. Scolex armed ................................................................. 2
1b. Scolex unarmed............................................................. 3

2a. Scolex with many small hooks. Vitellaria in lateral medulla and dorsal cortex
.................................. *Ancistrocephalus* **Monticelli 1890. (Figure 129)**

*Diagnosis:* Scolex with apical disc armed with several alternating rows of small hooks, most numerous on bothrial surfaces. Neck absent. External metamerism distinct, proglottids wider than long. Genital pores marginal, irregularly alternating. Testes in two lateral fields that unite at posterior end of proglottid. Ovary lobulated, posterior. Vitellaria mainly in two lateral fields. Uterus narrow, coiled, with expanded terminal end. Uterine pore opening ventrosubmarginal, irregularly alternating without regard to position of genital pores. Eggs operculate, unembryonated. Parasites of ocean sunfish. Atlantic, Pacific, Mediterranean.
    Type species: *A. microcephalus* (Rudolphi 1819) Monticelli 1890, in *Orthagoriscus mola;* Italy, Japan.
    Other species:
        *A. aluterae* (Linton 1889) Linton 1941, in *Alutera schoepfi;* U.S.

2b. **Scolex with two dorsal and two ventral trident hooks. Vitellaria cortical, diffuse** ........................................................................... *Triaenophorus* **Rudolphi 1793 (syn.** *Tricuspidaria* **Rudolphi 1793). (Figure 130)**

*Diagnosis:* Scolex with apical disc armed with dorsal and ventral pairs of trident-shaped hooks. Bothria rounded, shallow. Neck absent. External metamerism absent; strobila with transverse wrinkles. Genital atrium marginal, irregularly alternating. Cirrus pouch transverse. Testes numerous, in single medullary field. Ovary bilobed, posterior, slightly poral. Vitellaria cortical, diffuse. Uterus weakly coiled, with terminal sac. Uterine pore ventral, slightly poral, anterior to level of genital pores. Eggs operculate. Parasites of teleosts. Circumboreal. Monograph: Kuperman.[1709]
    Type species: *T. lucii* (Mueller 1776) Rudolphi 1793 (syn. *T. nodulosus* [Pallas 1781] Rudolphi 1819, *Taenia tricuspidata intestinalis* Bloch 1779, *Taenia tricuspis* Pallas 1781, *Taenia tricuspidata* Bloch 1782, *Taenia nodosa* Batsch 1786, *Tricuspidaria piscium* Rudolphi 1802, *Triaenophorus nodosus* Dujardin 1845, *T. nodulus* Sramek 1901); in many species of freshwater fishes; circumboreal.

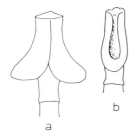

FIGURE 131. *Anoncocephalus chilensis* (Riggenbach 1896) Lühe 1902. (a) Lateral view; (b) dorsal view.

FIGURE 132. *Fistulicola plicatus* (Rudolphi 1819) Lühe 1899.

Other species:
    *T. amurensis* Kuperman 1968, in numerous freshwater fishes; Russia.
    *T. anguillae* Lönnberg 1889, in *Anguilla vulgaris;* Scandinavia.
    *T. crassus* Forel 1868 (syn. *T. robustus* Olsson 1893), in many fishes; Europe, North America.
    *T. meridionalis* Kuperman 1968, in numerous fishes; Russia.
    *T. orientalis* Kuperman 1968, in numerous fishes; Russia.
    *T. stizostedionis* Miller 1945, in *Stizostedion vitreum;* North America.

3a. Vitellaria medullary, ventral to testes .............................................
........................................*Anoncocephalus* Lühe 1902. (Figure 131)

*Diagnosis:* Scolex unarmed, truncated apically, arrowhead-shaped in lateral view, bilaterally compressed. Bothria narrow, deep. Neck absent. External segmentation present. Proglottids wider than long. Genital atrium marginal, postequatorial, irregularly alternating. Cirrus armed. Cirrus pouch pyriform. Testes in two dorsal, lateral fields, united at posterior end of proglottid. Ovary compact, posterior, poral. Vitellaria medullary, ventral to testes. Uterus coiled, narrow, with large, muscular terminal sac. Osmoregulatory canals numerous, anastomosing. Parasites of marine teleosts. Chile.
    Type species: *A. chilensis* (Riggenbach 1896) Lühe 1902 (syn. *Bothriotaenia chilensis* Riggenbach 1896), *Genypterus chilensis;* Chile.

3b. Vitellaria cortical ......................................................... 4

4a. Testes in one continuous field .............. *Fistulicola* Lühe 1899. (Figure 132)

*Diagnosis:* Scolex arrowhead-shaped in lateral view, with unarmed apical disc. Bothria wide, shallow. Scolex sometimes replaced with pseudoscolex. Neck absent. External metamerism evident. Proglottids much wider than long, with leaf-like expansions on lateral margins. Genital pores marginal, irregularly alternating. Testes medullary, in one continuous field, partly exceeding nerve trunks. Ovary poral, ventral. Vitellaria cortical, continuous around proglottid. Uterus wide, strongly coiled, with muscular distal portion. Eggs operculate, embryonated. Parasites of marine teleosts. Atlantic, Pacific, Mediterranean.
    Type species: *F. plicatus* (Rudolphi 1819) Lühe 1899, in *Xiphias gladius;* Atlantic, Pacific, Mediterranean.
    Other species: *F. dalmatinus* (Stossich 1897) Lühe 1902, in *Zeus faber;* Mediterranean.

4b. Testes in two lateral fields ..................................... 5

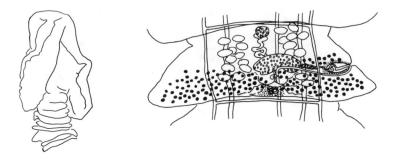

FIGURE 133. *Eubothrioides lamellatum* Yamaguti 1952. Scolex and proglottid.

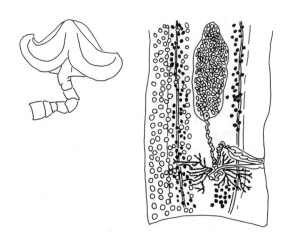

FIGURE 134. *Pistana eurypharyngis* Campbell et Gartner 1982. Scolex and proglottid.

**5a. Segments wider than long .......... *Eubothrioides* Yamaguti 1952. (Figure 133)**

*Diagnosis:* Scolex unarmed, bilaterally compressed, arrowhead-shaped in lateral view, lacking apical disk. Bothria elongate, prominent. Neck absent. Strobila with dorsal and ventral longitudinal grooves. Proglottids broader than long, craspedote. Genital atrium marginal, irregularly alternating. Cirrus pouch small, cortical. Testes few, in two lateral fields medial to nerve trunks. Ovary transversely elongated, posterior, poral. Seminal receptacle absent. Vitellaria cortical, diffuse. Uterus convoluted in median field, with small distal uterine sac. Uterine pore midventral near anterior end of proglottid. Eggs operculate. Parasites of marine teleosts. Japan.

Type species: *E. lamellatum* Yamaguti 1952, in *Zenopsis nebulosa;* Japan.

**5b. Segments longer than wide.... *Pistana* Campbell et Gartner 1982. (Figure 134)**

*Diagnosis:* Scolex unarmed, haustate, with two freely projecting, fleshy bothria. Apical disc lacking. Neck absent. External and internal segmentation present. Segments longer than wide. Poral longitudinal nerve cord ventral to vas deferens and vagina. Genital atrium marginal, irregularly alternating. Cirrus pouch extends into medulla. Cirrus armed. Testes in two lateral bands, medial to nerve trunks. Vagina passes anterior to cirrus pouch. Ovary posterior, submedian, comprised of two multilobed masses. Seminal receptacle absent. Vitellaria occupy entire circumcortical zone. Uterus median, uterine sac, and pore midventral. Eggs operculate, embryonated. Parasites of deep-sea teleosts. Western Atlantic.

FIGURE 135. *Glossobothrium nipponicum* Yamaguti 1952.

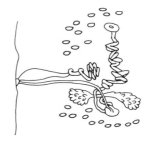

FIGURE 136. *Probothriocephalus muelleri* Campbell 1979.

Type species: *P. eurypharygis* Campbell et Gartner 1982, in *Eurypharynx pelecanoides;* Norfolk and Wilmington submarine canyons, North America.

## KEY TO THE GENERA IN PARABOTHRIOCEPHALIDAE

1a. Genital pores marginal ........................................................ 2
1b. Genital pores dorsal, near lateral margin ...................................... 3

2a. Apical disc present. Bothria with prominant, posterior appendages .............
   ...................................*Glossobothrium* Yamaguti 1952. (Figure 135)

*Diagnosis:* Scolex elongate, with unarmed apical disc. Linguiform, fluted appendage projecting from base of each bothrium. Neck absent. Proglottids short. Genital atrium marginal, irregularly alternating. Cirrus pouch claviform, covered with layer of gland-like cells. Cirrus armed with minute spines. Testes numerous, medullary, medial to osmoregulatory canals with a few postovarian. Vagina forming a fusiform muscular swelling anterior to ovarian isthmus. Ovary bilobed, posterior, slightly poral. Vitellaria diffuse, cortical. Uterus S-shaped, with distal sac. Uterine pore ventral, median, near anterior margin of proglottid. Eggs operculate, unembryonated. Parasites of marine teleosts. Japan.

Type species: *G. nipponicum* Yamaguti 1952, in unknown marine fish related to *Psenopsis anomala;* Japan.

2b. Apical disc and bothrial appendages absent ..................................
   ................................*Probothriocephalus* Campbell 1979. (Figure 136)

*Diagnosis:* Scolex linguiform, bothria shallow, apical disc lacking. Neck present. External segmentation indistinct, internal segmentation lacking. Genital pores marginal, irregularly alternating. Cirrus unarmed. Cirrus pouch small, slender. Testes medullary, in lateral fields, continuous in postovarian space and between segments. Ovary in posterior medulla. Vagina posterior to cirrus pouch. Vitellaria cortical, continuous. Uterine duct and sac medial, invading anterior segment. Uterine pores midventral. Eggs operculate, unembryonated. Parasites of marine teleosts. Northwest Atlantic.

Type species: *P. muelleri* Campbell 1979, in *Alepocephalus agassizi;* Hudson submarine canyon, northwest Atlantic.

3a. Cirrus unarmed. Eggs with conspicuous lateral swelling .......................
   ....................... *Neobothriocephalus* Mateo et Bullock 1966. (Figure 137)

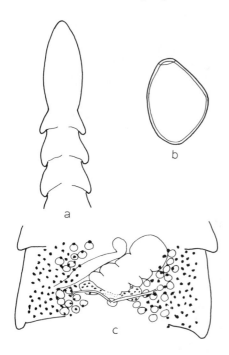

FIGURE 138. *Metabothriocephalus menpachi* Yamaguti 1938. Scolex.

FIGURE 137. *Neobothriocephalus aspinosus* Mateo et Bullock 1966. (a) Scolex; (b) egg; (c) proglottid.

*Diagnosis:* Scolex ovoid, with long, shallow, surficial bothria. Neck absent. Strobila tapered at both ends, most apparently at anterior end. Proglottids slightly craspedote, wider than long except at extreme posterior end. Genital pore dorsal, near lateral margin, irregularly alternating left and right. Genital atrium anterior to vagina. Cirrus pouch large, muscular, with conspicuous swelling at its base; obliquely oriented with basal end at anterior margin of proglottid. Cirrus unarmed. Testes few (25 to 60), medullary, in two lateral fields, continuous between segments. Vagina with prominent sphincter short distance from genital pore. Small seminal receptacle present. Ovary bilobed, submedian, slightly poral, near posterior margin of proglottid. Vitellaria in two wide lateral bands, mostly cortical, but some follicles medullary. Uterus a prominent sac protruding into preceding segment. Eggs thin-shelled, operculate, unembryonated, with prominent lateral swelling. Parasites of marine teleosts. Peru.

Type species: *N. aspinosus* Mateo et Bullock 1966, in *Neptomenus crassus;* Peru, *Seriolella violacea;* Chile.

**3b. Cirrus spined. Egg lacking lateral swelling ..................................... 4**

**4a. Testes in two lateral fields .........................................................**
**............................ *Metabothriocephalus* Yamaguti 1968. (Figure 138)**

*Diagnosis:* Scolex not marked off from neck, with indistinct apical disc followed on each flat surface by a small, oval subapical bothrium surrounded by condensed tissue. Neck present. Strobila comparatively fleshy, with complete segmentation. Proglottids wider than long, with thick posterior margins. Testes medullary, few, in two lateral fields. Cirrus pouch small. Genital atrium very small, preequatorial, dorsomarginal, irregularly alternating. Ovary poral, transversely elongated, not biwinged; posterior. Uterus winding in median and submedian fields; uterine sac opening anterior to cirrus pouch. Vitellaria entirely cortical, diffuse,

FIGURE 139. *Parabothriocephaloides segmentatus* Yamaguti 1934. Scolex.

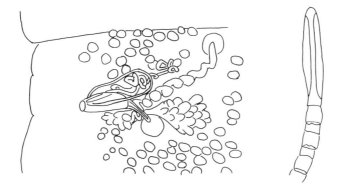

FIGURE 140. *Parabothriocephalus macruri* Campbell, Correia et Haedrich 1982. Scolex and part of proglottid.

continuous between segments. Eggs operculate, unembryonated. Parasites of marine teleosts. Pacific Ocean.

Type species: *M. menpachi* Yamaguti 1968, in *Myripristis* spp.; Hawaii.

**4b. Testes in single, median field ................................................. 5**

**5a. Vitellaria cortical. Scolex lacking, replaced with pseudoscolex with shallow depressions................*Parabothriocephaloides* Yamaguti 1934. (Figure 139)**

*Diagnosis:* Scolex replaced by conical pseudoscolex bearing depressions on surface. Neck absent. Strobila pointed at both ends, broadest near anterior end. Proglottids with swollen posterior margins. Genital atrium equatorial, dorsosubmarginal, irregularly alternating. Cirrus pouch large. Cirrus armed. Testes medullary, between osmoregulatory canals. Vagina with well-developed sphincter, opening behind cirrus. Seminal receptacle present. Ovary posterior, bilobed, slightly poral. Vitellaria diffuse, cortical. Uterus S-shaped, median, with large sac invading preceding segment. Uterine pore midventral, at anterior end of proglottid. Eggs operculated, unembryonated. Parasites of marine teleosts. Japan.

Type species: *P. segmentatus* Yamaguti 1934, in *Psenopsis anomala;* Japan.

**5b. Vitellaria medullary. Scolex present, small .......................................
............................*Parabothriocephalus* Yamaguti 1934. (Figure 140)**

*Diagnosis:* Scolex long, small, lacking apical disk. Neck absent. Strobila slender, filiform near anterior end. Anterior proglottids longer than broad, with salient posterior margins.

Genital atrium dorsosubmarginal, postequatorial. Convoluted vas deferens in front of base of cirrus pouch. Cirrus armed. Testes medullary, medial to nerve trunks. Vagina with narrow proximal duct and expanded distal sac with spinose base, opening posterior to cirrus. Ovary bilobed, posterior, poral. Vitellaria diffuse, cortical. Uterus sigmoid, with median sac invading preceding proglottid. Eggs operculate, unembryonated. Parasites of marine teleosts.

Type species: *P. gracilis* Yamaguti 1934, in *Psenopsis anomala;* Inland Sea of Japan.
Other species:

*P. macrouri* Campbell, Correia et Haedrich 1982, in *Macrourus berglax;* Newfoundland, Canada.

*P. sagitticeps* (Sleggs 1927) Jensen 1976, in *Sebastes paucispinus;* California, U.S.

## ORDER LECANICEPHALIDEA

Members of this order are parasites of the spiral valve intestine of elasmobranchs, especially Batoidea. They are readily recognized by the scolex, which is divided transversely by a horizontal groove. The anterior portion is cushion-like, or has unarmed tentacles that are capable of being withdrawn into the posterior portion, thus forming a large sucker-like organ. The posterior portion usually has four simple suckers. The testes are usually numerous, although as few as four are known. The genital pores are lateral. The ovary is posterior and bilobed, and the vitellaria are follicular, lateral, or encircle the entire proglottid. A median uterine pore is usually present. Complete life cycles are not known, but advanced larvae (metacestodes) are common in marine molluscs.

### KEY TO THE FAMILIES IN LECANICEPHALIDEA

1a. **Paired hooks associated with four suckers on anterior half of scolex ............ ........................................ Balanobothriidae Pintner 1928. (p.120)**

*Diagnosis:* Scolex egg-shaped. Anterior portion with four small suckers; each sucker anterior to two small, double-pronged hooks. Posterior portion of scolex surrounded by membranous collar. Neck present. Genital atrium lateral. Testes in single preovarian field. Vitellaria lateral. Ovary posterior. Parasites of elasmobranchs.
Type genus: *Balanobothrium* Hornell 1912.

1b. **No hooks on scolex ......................................................... 2**

2a. **No suckers on scolex ............. Disculicepitidae Joyeux et Baer 1935. (p.120)**

*Diagnosis:* Scolex lacking hooks or suckers of any kind, divided into anterior, flattened portion of variable shape, and posterior, rounded, corrugated portion. Neck present. Proglottids rectangular to square. Genital pores ventro-submarginal. Testes in irregular clusters, preovarian. Ovary posterior, bilobed. Vitellaria an irregular, postovarian mass (?). Uterus median. Parasites of elasmobranchs.
Type genus: *Disculiceps* Joyeux et Baer 1935.

2b. **Suckers present on anterior or posterior portions of scolex, or both .......... 3**

3a. **Large, cup-like apical sucker on scolex, may be protruded into cushion-like or disc-like structure. Tentacles present or not ...................................... ........................................ Lecanicephalidae Braun 1900. (p.121)**

*Diagnosis:* Scolex with large apical organ that may be retracted to form a sucker-like structure, or protracted to form a cushion-like structure; may be in the form of tentacles or apical disc. Remainder of scolex with four small, rounded suckers. Neck present or absent. Genital atrium lateral. Testes in single preovarian field. Vitellaria lateral or encircling proglottid. Parasites of elasmobranchs.
Type genus: *Lecanicephalum* Linton 1890.

3b. **No suckers or tentacles on anterior half of scolex ............................. ........................................ Adelobothriidae Yamaguti 1959. (p.127)**

*Diagnosis:* Anterior portion of scolex massive, simple; posterior portion membranous,

FIGURE 142. *Disculiceps* sp. Two typical forms of scolex.

FIGURE 141. *Balanobothrium parvum* Southwell 1925.

collar-like, bearing four small, rounded suckers. Proglottids craspedote. Testes in single, large, preovarian field. Vitellaria cortical, encircling entire proglottid. Parasites of elasmobranchs.

Type genus: *Adelobothrium* Shipley 1900.

## DIAGNOSIS OF THE ONLY GENUS IN BALANOBOTHRIIDAE

*Balanobothrium* Hornell 1912. (Figure 141)

*Diagnosis:* Scolex egg-shaped, with membranous collar around base. Anterior portion with four small suckers, each anterior to two small, double-pronged hooks. Entire scolex embedded in gut wall of host. Neck present. Proglottids wider than long. Genital atrium lateral, preequatorial, irregularly alternating. Cirrus pouch entirely poral. Cirrus armed, posterior to vaginal pore. Testes very numerous, in single field anterior to ovary. External seminal vesicle absent. Ovary bilobed, posterior. Vitellaria in two or more lateral rows. Vagina anterior to cirrus pouch. Uterine pore midventral. Uterus median. Parasites of elasmobranchs. Indian Ocean, Pacific.

Type species: *B. tenax* Hornell 1912, in *Stegostoma tigrinum, Trygon walga;* Sri Lanka, Celebes.

Other species:

*B. parvum* Southwell 1925, in *Galeocerdo arcticus, Trygon* sp.; Sri Lanka.

*B. stegostomatis* Yamaguti 1954, in *Stegostoma tigrinum;* Macassar, Celebes.

*B. veravalensis* Jadhav et Shinde 1979, in *Rhinobatus typus;* India.

## DIAGNOSIS OF THE ONLY GENUS IN DISCULICEPITIDAE

(Syn. Discocephalidae Pintner 1928)

*Disculiceps* Joyeux et Baer 1935 (Figure 142)
(Syn. *Discocephalum* Linton 1890).

*Diagnosis:* Scolex with large anterior, flattened, cushion-like portion, variable in shape according to state of contraction. Posterior portion globular, corrugated; both portions lacking hooks or suckers. Neck present. Proglottids acraspedote, rectangular. Genital atrium ventro-submarginal. Cirrus unarmed, opening posterior to vagina. Testes in irregular clusters, preovarian. Ovary bilobed, posterior. Vitellaria an irregular mass, postovarian (?). Vagina opens anterior to cirrus. Seminal receptacle absent. Uterus median with lateral branches.

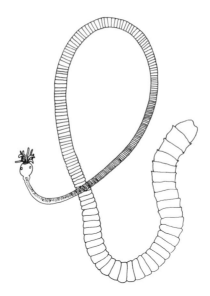

FIGURE 143. *Polypocephalus radiatus* Braun 1878.

Osmoregulatory canals reticular. Parasites of elasmobranchs. North America, Bermuda, France, India.

Type species: *D. pileatus* (Linton 1890) Joyeux et Baer 1935 (syn. *Discocephalum pileatus* Linton 1890), in *Carcharias* spp., *Scoliodon terraenovae, Charodon lamia;* Woods Hole, Bermuda, Bengal, France.

## KEY TO THE GENERA IN LECANICEPHALIDAE

1a. Scolex with unarmed tentacles ................................................. 2
1b. Scolex without tentacles ....................................................... 3

2a. Tentacles originating on anterior half of scolex ...................................
   ... *Polypocephalus* Braun 1878 (syn. *Parataenia* Linton 1889; *Thysanobothrium* Shipley et Hornell 1906; *Anthemobothrium* Shipley et Hornell 1906). (Figure 143)

*Diagnosis:* Scolex divided into two portions. Anterior portion composed of 14 to 16 simple or feather-like retractable tentacles. Posterior portion cushion-like, with four simple suckers. Neck present or not. Proglottids craspedote or not. Genital pores irregularly alternating. Cirrus armed. Testes few, large. Cirrus pouch pyriform. Ovary bilobed, posterior. Vitellaria in bilateral fields, composed of a few large acini. Vagina posterior to cirrus pouch. Small seminal receptacle present. Uterus median. Parasites of elasmobranchs. India, Sri Lanka, North America.

Type species: *P. radiatus* Braun 1878 (syn. *Thysanobothrium uarnakense* Shipley et Hornell 1906; *Parataenia elongata* Southwell 1912); in *Rhinobatus granulosus, Trygon uarnak, T. kuhli, T. sephan, Rhynchobatus djeddensis;* India, Sri Lanka.
Other species:
   *P. affinis* Subhapradha 1951, in *Rhinobatus granulatus;* India.
   *P. alii* Shinde et Jadhav 1981, in *Rhynchobatus djeddensis;* India.
   *P. coronatus* Subhapradha 1951, in *Rhynchobatus djeddensis;* India.
   *P. katpurensis* Shinde et Jadhav 1981, in *Rhynchobatus djeddensis;* India.

FIGURE 144. *Calycobothrium typicum* (Southwell 1911) Southwell 1911.

FIGURE 145. *Staurobothrium aetobatidis* Shipley et Hornell 1905.

*P. lintoni* Subhapradha 1951, in *Rhynchobatus djeddensis;* India.
*P. medusia* (Linton 1889) Yamaguti 1959 (syn. *Parataenia medusia* Linton 1889), in *Trygon centrura, Dasyatis say, D. americana, Rhynchobatus* spp.; India, U.S., Columbia.
*P. pulcher* (Shipley et Hornell 1906) Yamaguti 1959 (syn. *Anthemobothrium pulcher* Shipley et Hornell 1906), in *Trygon sephen;* Sri Lanka.
*P. rhinobatidis* Subhapradha 1951, in *Rhinobatus granulatus;* India.
*P. rhynchobatidis* Subhapradha 1951, in *Rhynchobatus djeddensis;* India.
*P. singhii* Shinde et Jadhav 1981, in *Rhynchobatus djeddensis;* India.
*P. thapari* Shinde et Jadhav 1981, in *Trygon sephen;* India.
*P. vesicularis* Yamaguti 1960, in *Rhinobatus schlegeli;* Japan.
*P. vitellaris* Subhapradha 1951, in *Rhynchobatus djeddensis;* India.

2b.  Tentacles arising on posterior half of scolex ............................*Calycobothrium* Southwell 1911 (syn. *Cyclobothrium* Southwell 1911). (Figure 144)

*Diagnosis:* Scolex divided into two portions. Anterior portion cushion-like, with two to four small suckers. Posterior portion with 14 short, hollow tentacles arranged around the base of the anterior portion like the petals of a flower. Neck present. Proglottids acraspedote, longer than wide. Genital pores marginal, irregularly alternating. Cirrus pouch extending to midline of proglottid. Cirrus armed. Testes numerous, in single intervascular, preovarian field. Small internal seminal vesicle present. Ovary compact, posterior. Vitellaria in bilateral fields. Vagina opens anterior to cirrus. Small seminal receptacle present. Uterus unknown. Parasites of elasmobranchs. Sri Lanka.
Type species: *C. typicum* (Southwell 1911) Southwell 1911 (syn. *Cyclobothrium typicum* Southwell 1911), in *Aetobatis narinari, Brachirus orientalis;* Sri Lanka, Pakistan.

3a.  Anterior portion of scolex reduced to small papilla. Posterior portion with four large, pedunculated suckers......................................................
..........................*Staurobothrium* Shipley et Hornell 1905. (Figure 145)

*Diagnosis:* Anterior portion of scolex represented by a small, unarmed papilla. Posterior portion of scolex with four large, powerful, pedunculated papillae arranged like a Maltese cross. Neck absent. Proglottids craspedote. Genital pores irregularly alternating. Cirrus pouch

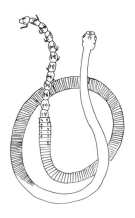

FIGURE 146. *Hexacanalis abruptus* (Southwell 1911) Perrenoud 1931.

pyriform, bent posteriad. Cirrus unarmed. Testes few (24), large, first appearing as central mass, then spreading evenly throughout preovarian field in mature proglottid. Ovary posterior, bilobed. Vitellaria few, large, in lateral fields that join anteriorly. Vagina opens anterior to cirrus. Very large seminal receptacle present. Uterus median. Parasites of elasmobranchs. Sri Lanka.
  Type species: *S. aetobatidis* Shipley et Hornell 1905, in *Aetiobatis narinari;* Sri Lanka.

**3b. Anterior portion of scolex well-developed, may be retractile .................. 4**

**4a. Six osmoregulatory canals present. Scolex rectangular in cross section .........
............................................. *Hexacanalis* Perrenoud 1931. (Figure 146)**

*Diagnosis:* Scolex rectangular in cross section. Anterior region bulbous, nonglandular, retractable. Posterior portion simple, with four suckers. Neck long. Proglottids craspedote. Genital pores irregularly alternating. Cirrus pouch reaches median line. Cirrus unarmed. Testes numerous, anterior to ovary, none anterior to cirrus pouch. Ovary bilobed, about one fourth from posterior end. Vitellaria in bilateral fields, absent anterior to cirrus pouch on poral side. Vagina opens posterior to cirrus pouch. Uterus saccular, median. Osmoregulatory system with six major anastomosing trunks. Parasites of elasmobranchs. Sri Lanka, India.
  Type species: *H. abruptus* (Southwell 1911) Perrenoud 1931 syn. *Cephalobothrium abruptus* Southwell 1911), in *Pteroplatea micrura;* Sri Lanka, India.
  Other species:
    *H. variabilis* (Southwell 1911) Perrenoud 1931 (syn. *Cephalobothrium variabilis* Southwell 1911), in *Pristis cuspidatus, Trygon kuhli;* Sri Lanka.

**4b. Two or four osmoregulatory canals present. Scolex rounded in cross section, divided into anterior pad, disc or sucker; posterior half with four suckers .. 5**

**5a. Anterior pad divided into two semicircular, dorsal and ventral, muscular flaps
............................................. *Flapocephalus* Deshmukh 1979.**

*Diagnosis:* Minute, fragile worms. Scolex divided into two parts by horizontal groove: anterior portion consisting of two semicircular, dorsal and ventral muscular flaps; posterior portion rounded with four simple suckers. Neck absent. Mature segments longer than broad. Genital pores preequatorial, irregularly alternating. Testes in two longitudinal rows in central

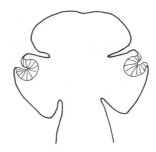

FIGURE 147. *Tetragonicephalum trygonis* Shipley et Hornell 1905.

medulla. Cirrus pouch crosses poral osmoregulatory canals; cirrus not seen. Ovary compact, transverse, near posterior end of segment. Vagina anterior to cirrus pouch. Vitellaria cortical, near lateral margins. Uterus nearly reaches anterior end of segment. Parasites of Batoidea. India.

Type species: *F. trygonis* Deshmukh 1979, in *Trygon sephen;* India.
Other species:
   *F. saurashtri* Shinde et Deshmukh 1979, in *Trygon sephen; India.*

**5b.   Anterior pad not subdivided ................................................... 6**

**6a.   Anterior portion of strobila broadened into a very wide, cobra-like region, then narrows ............................. *Eniochobothrium* Shipley et Hornell 1906.**

*Diagnosis:* Scolex unarmed, with four simple suckers on posterior half. Strobila divided into several regions: first a narrow neck-like region of three or four segments; second, an oval region of about 18 segments that get broader until about the tenth and then narrow again. The segments of this region overlap like a many-caped cloak (giving the genus its name). Third is a very narrow region of about 18 segments, all about the same size; fourth are six or eight mature segments becoming rapidly larger. The last one or two gravid proglottids are about as long as the rest of the strobila. Genital pores alternating. Cirrus sac large, median. Cirrus long, armed. Remaining internal anatomy unknown. Parasites of rays. Indian Ocean.

Type species: *E. gracile* Shipley et Hornell 1906, in *Rhinoptera javanica;* Sri Lanka.
Other species:
   *E. trygonis* Chincholikar et Shinde 1978, in *Trygon sephen;* India.

**6b.   Anterior portion of strobila only slightly swollen, or not at all ............... 7**

**7a.   Testes only in anterior half of segment. Vitellaria mainly posterior to ovary ... ..................... *Tetragonicephalum* Southwell et Hornell 1905. (Figure 147)**

*Diagnosis:* Scolex divided into two parts; anterior portion large, cushion-like, covered with minute spines; posterior portion with four suckers. Neck short. Strobila long, slender. Genital atrium enlarged, with frilled margin, irregularly alternating. Cirrus pouch not fully developed until uterus contains eggs. Cirrus long, unarmed. External seminal vesicle present. Testes few (7 to 12), anterior to level of genital pore, disappearing early. Ovary posterior, present only in same segments as testes. Vitellaria arising postovarian, extending along lateral margins in mature proglottids. Vagina opens posterior to cirrus. Large seminal receptacle present. Uterus median, saccular. Parasites of elasmobranchs. Indian Ocean.

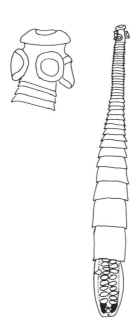

FIGURE 148. *Discobothrium myliobatidis* Dailey et Mudry 1968. Scolex and entire worm.

Type species: *T. trygonis* Shipley et Hornell 1905 (syn. *Tylocephalum trygonis* [Shipley et Hornell 1905], Shipley et Hornell 1906) in *Trygon walga*, *T*. sp.; Sri Lanka, India.

**7b. Testes filling most of segment. Vitellaria lateral along most of length of segment** ................................................................. **8**

**8a. Minute worms, not exceeding 2.5 mm long. Apical disc not invaginable or only slightly so** ....................................................... ***Discobothrium*** **Beneden 1871 (syn. *Hornellobothrium* Shipley et Hornell 1906). (Figure 148)**

*Diagnosis:* Minute worms, usually 1.5 to 2.5 mm total length. Scolex prominent: anterior half in form of prominent disc, slightly invaginable; posterior half with four powerful, round suckers. Neck absent or very short. Proglottids craspedote, wider than long except for terminal ones. Genital pores irregularly alternating. Cirrus pouch not reaching midline of segment. Testes few or many, anterior to ovary. Ovary bilobed, near posterior end. Vitellaria lateral, extending most of length of proglottid. Uterus a median, longitudinal sac. Parasites of rays. Europe, Sri Lanka, Japan, California, Jamaica.

Type species: *D. fallax* Beneden 1871, in *Raja clavata*; Europe.
Other species:

*D. caribbensis* Gardner et Schmidt 1984, in *Urolophus jamaicensis*; Jamaica.

*D. cobraeformis* Shipley et Hornell 1906 (syn. *Hornellobothrium cobraeformis* Shipley et Hornell 1906), in *Aetiobatis narinari*; Sri Lanka.

*D. japonicum* Yamaguti 1934, in *Narke japonica*; Japan.

*D. myliobatidis* Dailey et Mudry 1968, in *Myliobatis californicus*; California, U.S.

*D. quadrisurculi* Khambata et Bal 1954 *(nomen nudum)*, in *Trygon sephen, Rhynchobatus djeddensis, Aetomylaeus maculatus*; India.

*D. redactum* Khambata et Bal 1954 *(nomen nudum)*, in *Trygon sephen, Rhynchobatus djeddensis, Aetomylaeus maculatus*; India.

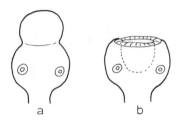

FIGURE 149. *Lecanocephalum* sp. (a) Anterior portion of scolex extruded; (b) anterior portion of scolex withdrawn.

**8b. Size much larger. Anterior cushion invaginable to form large sucker ........... ................................................................ *Lecanicephalum* Linton 1890 (syn. *Aphanobothrium* Seurat 1906, *Kystocephalus* Shipley et Hornell 1906, *Spinocephalum* Deshmukh 1980, *Tylocephalum* Linton 1890). (Figure 149)**

*Diagnosis:* Scolex variable, but clearly divided into two portions. Posterior portion rounded or depressed, with four simple suckers. Anterior portion massive, globular or flattened, with or without cuticular spines, often with glandular cells in its wall, retractable into posterior portion. When retracted, the resulting depression appears as a large apical sucker. Neck usually present. Proglottids craspedote or not. Genital pores marginal, irregularly alternating. Cirrus pouch variable with the species. Cirrus armed or not. Testes usually numerous, in preovarian intervascular field. Ovary two-winged, posterior. Vitellaria bilateral, usually with some postovarian follicles. Vagina opening into atrium posterior to cirrus. Seminal receptacle usually present, often very large. Uterus median, saccular when gravid. One or two pairs of osmoregulatory canals. Parasites of elasmobranchs. Cosmopolitan.

Type species: *L. peltatum* Linton 1890, in *Dasyatis centrura, Pristis cuspidatus, Trygon kuhli, Pteroplates, D. americana;* Columbia, Massachusetts, U.S.

Other species:

*L. aegyptiacus* (Hassan 1982) comb. n. (syn. *Discobothrium aegyptiacus* Hassan 1982), in *Raja circularis;* Egypt.

*L. aetiobatidis* (Shipley et Hornell 1905) comb. n. (syn. *Tetragonocephalum aetiobatidis* Shipley et Hornell 1905, *Tylocephalum aetiobatidis* [Shipley et Hornell 1905] Yamaguti 1959), in *Aetiobatis narinari, Dasybatus walga;* Sri Lanka.

*L. dierama* (Shipley et Hornell 1906) comb. n.(syn. *Tylocephalum dierama* Shipley et Hornell 1906, *Tylocephalum kuhhi* Shipley et Hornell 1906), in *Myliobatis maculata;* Sri Lanka, Pakistan.

*L. madhukarii* (Chincholikar et Shinde 1908) comb. n. (syn. *Tylocephalum madhukarii* Chincholikaret Shinde 1980), in *Trygon sephen;* India.

*L. maharashtrae* Chincholikar et Shinde 1978, in *Trygon sephen;* India.

*L. marsupium* (Linton 1916) comb. n. (syn. *Tylocephalum marsupium* Linton 1916, in *Aetobatus narinari;* Tortugas.

*L. pinguis* (Linton 1890) comb. n. (syn. *Tylocephalum pinguis* Linton 1890), in *Rhinoptera quadriloba, R. bonasus;* Eastern coast, U.S.

*L. rhinobatii* (Deshmukh 1980) comb. n. (syn. *Spinocephalum rhinobatii* Deshmukh 1980), in *Rhinobatus granulatus;* India.

*L. simile* (Pintner 1928) comb. n. (syn. *Tylocephalum simile* Pintner 1928), in *Trygon walga;* locality?

*L. singhii* (Jadhav et Shinde 1981) comb. n. (syn. *Tylocephalum singhii* Jadhav et Shinde 1981), in *Trygon zugei;* India.

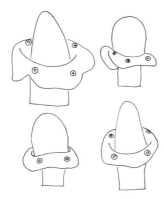

FIGURE 150. *Adelobothrium aetiobatidis* Shipley 1900. Various shapes of scoleces.

*L. squatinae* (Yamaguti 1934) comb. n. (syn. *Tylocephalum squatinae* Yamaguti 1934), in *Squatina japonica;* Japan.

*L. translucens* (Shipley et Hornell 1906) comb. n. (syn. *Kystocephalus translucens* Shipley et Hornell 1906, *Tylocephalum translucens* [Shipley et Hornell 1906] Yamaguti 1959), in *Aetobatis narinari;* Sri Lanka.

*L. yorkei* (Southwell 1925) comb. n. (syn. *Tylocephalum yorkei* Southwell 1925), in *Aetobatis narinari;* India.

**Lecanicephalum Species Unidentifiable from Original Descriptions**

*L. ludificans* (Jameson 1912) comb. n. (syn. *Tylocephalum ludificans* Jameson 1912), encysted in *Margaritifera vulgaris;* Sri Lanka.

*L. margaritiferae* (Seurat 1906) comb. n. (syn. *Tylocephalum margaritiferae* Seurat 1906), encysted in *Margaritifera margaritifera,* adult in *Aetobatis narinari* ?; South Pacific.

*L. minus* (Jameson 1912) comb. n. (syn. *Tylocephalum minus* Jameson 1912) encysted in *Margaritifera vulgaris;* Sri Lanka.

*L. minutum* (Southwell 1925) comb. n. (syn. *Tylocephalum minutum* Southwell 1925), in *Urogymnus* sp.; Sri Lanka.

*L. unionifactor* (Herdmann et Hornell 1903) comb. n. (syn. *Tylocephalum unionifactor* Herdmann et Hornell 1903), encysted in *Margaritifera vulgaris;* Sri Lanka.

## DIAGNOSIS OF THE ONLY GENUS IN ADELOBOTHRIIDAE

*Adelobothrium* Shipley 1900 (Figure 150)

Anterior portion of scolex massive, variably conical. Posterior portion membranous, collar-like, with four small suckers. Neck absent. Proglottids cylindrical, craspedote. Genital pores irregularly alternating. Cirrus pouch oval. Cirrus unarmed. Testes numerous, in entire intervascular, preovarian field. External seminal vesicle present. Ovary bilobed, posterior. Vitellaria cortical, encircling entire proglottid when mature. Vagina opens ventral to cirrus pouch. Seminal receptacle present. Gravid uterus median, sac-like, filling most of medulla. Egg with polar filaments. Parasites of elasmobranchs. Sri Lanka, Loyalty Islands, Tortugas.

Type species: *A. aetiobatidis* Shipley 1900, in *Aetobatis narinari, Rhynchobatus djeddensis;* Loyalty Island, Sri Lanka, Tortugas.

# ORDER APORIDEA

(syn. Aporidea Wardle, McLeod, and Radinovsky, 1974)

This small order contains only the family Nematoparataeniidae, which are parasites of Anseriformes. The scolex has simple suckers or grooves and an armed rostellum. There is no external segmentation, but internal segmentation may occur. Genital ducts and pores, cirrus, ootype, and Mehlis' glands are absent. The testes surround the ovary. The vitellaria develop within the ovary. No life cycle is known, but probably they all use a freshwater invertebrate intermediate host.

## KEY TO THE GENERA IN NEMATOPARATAENIIDAE

**1a. Suckers present on scolex. Huge, glandular rostellum present, with undulating row of hundreds of tiny hooks** ................................................................
.................. *Nematoparataenia* **Maplestone et Southwell 1922. (Figure 151)**

*Diagnosis:* Scolex with four large, forwardly directed suckers and a huge, glandular rostellum armed with an undulating row of about 1000 very small hooks. Short neck present. Strobila cylindrical, up to 10 mm long, with a longitudinal groove on one side. External and internal metamerism lacking. Testes follicular, filling most of medulla, degenerating posteriad. Ovary only in midregion of body, follicular, surrounding testes except in region of lateral groove. Vitellaria follicular, mixed with ovarian cells. Ovarian and vitelline cells in hindbody replaced with eggs. Eggs with two membranes, in clusters in membranous capsules. Two pairs of osmoregulatory canals. Parasites of Anseriformes. Australia, Sweden, Europe, Russia.

Type species: *N. paradoxa* Maplestone et Southwell 1922, in *Chenopis atrata;* Australia.
Other species:
   *N. brabantiae* Cotteleer et Schyns 1961, in "swan"; Belgium.
   *N. minutum* (Endrigkeit 1940) Spasskii et Kornyushin 1975 (syn. *Ophryocotyle minuta* Endrigkeit 1940), in *Cygnus olor;* Nordenburger Sea.
   *N. southwelli* Fuhrmann 1933, in *Cygnus olor;* Sweden.

**1b. Suckers absent. Small rostellum present, with 10 hooks** ...................... 2

**2a. Scolex with four narrow longitudinal grooves** ........................................
........................................ *Apora* **Ginetsinskaya 1944. (Figure 152)**

*Diagnosis:* Scolex with four narrow grooves and rostellum armed with ten hooks. Strobila cylindrical, with longitudinal groove on one side. External and internal segmentation lacking. Testes follicular, medullary, arranged in semicircle in cross section. Vasa efferentia converging toward longitudinal groove but joining together without opening to outside. Ovary follicular, surrounding testes in semicircle. Vitellaria follicular, mixed with ovarian cells. Posteriorly, the ovarian and vitelline cells are replaced with eggs enclosed in membranous capsules. One pair of longitudinal osmoregulatory canals. Parasites under the koilon of anatid birds. Russia.
Type species: *A. dogieli* Ginetsinskaya 1944.

**2b. Scolex without grooves or sucking devices** .........................................
........................................*Gastrotaenia* **Wolffhügel 1938. (Figure 153)**

*Diagnosis:* Scolex rounded, lacking suckers or grooves, with rostellum bearing ten hymenolepid-type hooks. Neck long. Strobila cylindrical, lacking longitudinal groove. External

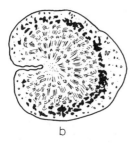

FIGURE 151. *Nematoparataenia southwelli* Fuhrmann 1933. (a) Scolex; (b) cross section of strobila, showing longitudinal groove.

FIGURE 152. *Apora dogieli* Ginetsinskaja 1944. Cross section of strobila.

FIGURE 153. *Gastrotaenia cygni* Wolffhügel 1938.

metamerism lacking, internal metamerism evident. Testes horseshoe-shaped, in intervascular field, with open end directed toward ovary. Ovary follicular, increasing in size to enclose testes. Vitellaria compact, next to ovary. Two pairs of osmoregulatory canals. Some specimens lacking female organs. Gravid specimens not yet reported. Parasitic under the koilon of anatid birds. North and South America, Russia.

Type species: *G. cygni* Wolffhügel 1938.
Other species:

    *G. kazachstanica* Egizbaeva et Nasyrova 1979, in *Arctodiaptomus salinus* (intermediate host), ducks (experimental); Russia.

    *G. paracygni* Czaplinski et Ryzikov 1966, in *Cygnus olor, C. cygnus;* Moscow Zoo.

## ORDER TETRAPHYLLIDEA

Tetraphyllideans are notable for their astonishing variety of scolex forms. Basically there are four bothridia that may be sessile or stalked, crenate, or smooth, or subdivided into loculi or suckers. Many species have hooks, accessory suckers, muscular pads, or any combination of these. Some have an apical myzorhynchus, which is a domed or elongate rostellar organ. The posterior portion of the scolex may be elongated into a pedicel. Commonly, the entire scolex is covered with tiny spines. The strobila may be hyperapolytic, with fertilization occurring after the segment is detached. Proglottid morphology is similar to those of the Trypanorhyncha and Lecanicephalidea, so great care must be taken not to combine segments of more than one species, or even more than one order, thereby creating a composite species!

This order of tapeworms lives in the spiral valve intestine of sharks and rays, where they cling to the mucosa with their extraordinarily mobile bothridia. It is probable that the sizes and shapes of the hooks and adhesive organs correspond to the rugae and villi of the host tissues, thereby contributing to a certain degree of host specificity. When removed alive from the host, these worms must be carefully relaxed in water before fixing, for a contracted scolex may be impossible to identify. However, it is equally useful to study live specimens, for suckers and loculi nearly disappear in some species when they die.

Even though hundreds of species have been described in Tetraphyllidea only about 17% of the known species of elasmobranchs have been examined for cestodes. Apparently, this will be a rich area for research for many years to come.

## KEY TO THE FAMILIES IN TETRAPHYLLIDEA

**1a. Scolex armed** .............................. **Oncobothriidae Braum 1900. (p.132)**

*Diagnosis:* Scolex armed with simple or branched hooks or hook-like horns. Bothridia usually sessile, simple or divided into loculi. Accessory suckers present or not. Myzorhynchus absent. Neck present or absent. Genital pores marginal. Parasites of elasmobranchs.
Type genus: *Oncobothrium* Blainville 1828.

**1b. Scolex unarmed** .................................................................. **2**

**2a. Scolex a single, transverse organ, lacking bothridia, suckers or armature. One to 24 strobilas per scolex** ...................................................... **Cathetocephalidae Dailey et Overstreet 1973. (p.132)**

*Diagnosis:* Scolex a single transverse organ, lacking bothridia, suckers, or armature. Neck present. Strobila slightly craspedote with distinct segmentation. Testes numerous, medullary. Genital pores lateral. Ovary posterior. Vitellaria follicular, cortical, encircling medula. Occasionally multistrobilate with up to 24 strobilae. Parasites of sharks. Caribbean Ocean.
Type genus: *Cathetocephalus* Daily et Overstreet 1973.

**2b. Scolex with four bothridia** ....................................................... **3**

**3a. Each bothridium divided into three to five loculi, arranged in triangle or radially** ............................................................. **...Triloculariidae Yamaguti 1959 (syn. Urogonoporidae Odhner 1904.) (p.144)**

FIGURE 154. *Cathetocephalus thatcheri* Dailey et Overstreet 1973. Scolex with single strobila.

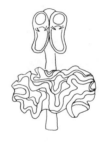

FIGURE 155. *Thysanocephalum crispum* (Linton 1889) Linton 1891.

*Diagnosis:* Scolex unarmed, with four sessile bothridia, each divided into three to five sucker-like loculi arranged in a triangle. Posterior end of bothridia free, anterior margins embedded in scolex. Myzorhynchus absent or rudimentary. Long neck present. Free proglottids spinose anteriorly. Genital pores near median at extreme posterior end of proglottid. Parasites of elasmobranchs.

Type genus: *Trilocularia* Olsson 1867.

3b.  Scolex not as above........................Phyllobothriidae Braun 1900. (p.145)

*Diagnosis:* Scolex unarmed. Bothridia variable, sessile or pedunculated, commonly folded, crumpled or loculate, sometimes simple. Accessory suckers often present. Myzorhynchus sometimes present. Genital pores marginal. Parasites of elasmobranchs.

Type genus: *Phyllobothrium* Beneden 1849.

## DIAGNOSIS OF THE ONLY GENUS IN CATHETOCEPHALIDAE

*Cathetocephalus* Dailey et Overstreet 1973 (Figure 154)

*Diagnosis:* Scolex a single, transverse fleshy organ, lacking bothridia, suckers or armature; anterior surface highly rugose, with fleshy papilliform projections on leading and trailing edges; posterior surface of scolex smooth with a central transverse fold extending entire width. Neck present. Strobila slightly craspedote, apolytic, with distinct segmentation. Longitudinal muscles well developed, forming boundary between cortex and medulla. Testes numerous, medullary. Genital pores lateral, irregularly alternating. Cirrus pouch large. Ovary large, bilobed, posterior. Vitellaria cortical, encircling medulla. Vagina anterior to cirrus pouch. Uterus? Parasites of sharks. Caribbean Ocean.

Type species: *C. thatcheri* Dailey et Overstreet 1973, in *Carcharhinus leucus;* Texas, Mississippi, Florida, Louisiana, Lake Nicaragua, Guatemala, Costa Rica.

## KEY TO THE GENERA IN ONCOBOTHRIIDAE

1a.  **Ruffle-like collar on neck below scolex**........................*Thysanocephalum* **Linton 1889 (syn.** *Myzocephalus* **Shipley et Hornell 1906). (Figure 155)**

*Diagnosis* Scolex small, with four sessile bothridia, each divided by a transverse septum into two loculi. Two stout spines are found on the ends of the septum of each bothridium. Neck present, with fleshy, ruffle-like collar with frilled margins. Cuticle posterior to collar appears scaly. Genital pores irregularly alternating. Cirrus pouch large. Cirrus short and stout, armed apically. Testes very numerous, filling intervitelline field anterior and lateral

FIGURE 156. *Ceratobothrium xanthocephalum* Monticelli 1882.

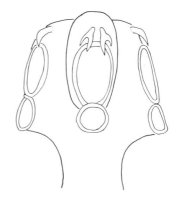

FIGURE 157. *Uncibilocularis trygonis* (Shipley et Hornell 1906) Southwell 1925.

to ovary. Vagina anterior to cirrus pouch. Ovary bilobed, at posterior end of proglottid. Vitellaria in narrow lateral fields. Uterus median, overreaching level of cirrus pouch. Parasites of elasmobranchs. North America, France, Pakistan.

Type species: *T. crispum* (Linton 1889) Linton 1891 (syn. *Phyllobothrium thysanocephalum* Linton 1889, *T. thysanocephalum* [Linton 1889] Braun 1900, *Myzocephalus narinari* Shipley et Hornell 1906), in *Galeocerdo tigrinus, G. arcticus, Sphyrna zygaena*; Atlantic, Indian Oceans.

Other species:

*T. karachii* Zaidi et Khan 1976, in *Trygon bleekeri, Galaecerado rayneri*; Pakistan.

**1b. No such collar ............................................................................ 2**

**2a. Muscular, horn-like appendages on bothridia in place of hooks .................
............................... *Ceratobothrium* Monticelli 1892. (Figure 156)**

*Diagnosis:* Four sessile, undivided bothridia, each directly below large accessory sucker bearing two projecting horn-like appendages consisting mainly of circular muscle fibers. Neck present. Genital pores irregularly alternating. Cirrus pouch thin-walled, long, oblique. Testes numerous, in two lateral intervitelline fields. Vagina anterior to cirrus pouch. Ovary posterior, bilobed in flat view, four-lobed in cross section. Vitellaria in narrow fields lateral to osmoregulatory canals. Uterus median, nearly reaching anterior end of proglottid. Parasites of elasmobranchs. Italy, U.S., Japan.

Type species: *C. xanthocephalum* Monticelli 1892 (syn. *Thysanocephalum riduculum* Linton 1901), in *Lamna cornubica, Isurus dekayi, I. glaucus*; Italy, New England, Japan.

**2b. Bothridia with hooks ..................................................................... 3**

**3a. Bothridia divided by transverse partitions ...................................... 4**
**3b. Bothridia undivided ..................................................................... 11**

**4a. Each bothridium with two loculi................................................... 5**
**4b. Each bothridium with three loculi................................................ 6**

**5a. Hooks slender, forked, with two antler-like prongs...............*Uncibilocularis* Southwell 1927 (syn. *Megalonchos* Baer et Euzet 1962). (Figure 157)**

FIGURE 158.  *Yorkeria parva* Southwell 1927.

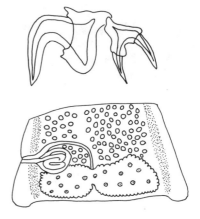

FIGURE 159.  *Acanthobothroides thorsoni* Brooks 1977. Bothridial hooks and mature proglottid.

*Diagnosis:* Four sessile bothridia, each divided into two rounded loculi by a transverse septum. Each bothridium has two dorsal bifid hooks, antler-like in appearance. No accessory suckers. Neck present. Genital pores irregularly alternating. Cirrus pouch small. Cirrus armed at tip. Testes few (15 to 40) in two broad fields medial to osmoregulatory canals. Vagina opens anterior to cirrus, expanded proximally. Ovary posterior, bilobed. Vitellaria in wide lateral bands. Uterus rosette-like, filling proglottid. Minute ventral uterine pore present. Parasites of elasmobranchs. Sri Lanka, India, France, New Zealand.

Type species: *U. trygonis* (Shipley et Hornell 1906) Southwell 1925 (syn. *Prosthecobothrium trygonis* Shipley et Hornell 1906), in *Trygon walga, T. sephen;* Sri Lanka.

Other species:

*U. southwelli* Shinde et Chincholikar 1976, in *Trygon* sp.; India.

*U. veravalensis* Jadhav et Shinde 1981, in *Trygon zugei;* India.

5b.  Hooks simple, not forked. Bothridia paired, at ends of stalks .................... ........................................... *Yorkeria* **Southwell 1927. (Figure 158)**

*Diagnosis:* Scolex with bothridia arranged into two pairs, each pair on the end of a common stalk. Each bothridium divided into a distal small and a proximal large compartment. Two U-shaped hooks on each bothridium, unequal in size, one each end of the septum. Entire scolex spinose. Neck (?). Genital pores irregularly alternating. Cirrus pouch well developed. Cirrus armed. Testes confined anterior to cirrus pouch. Vagina anterior to cirrus pouch, expanded throughout its length. Ovary four-lobed in cross section, about one fourth from posterior end. Vitellaria lateral. Uterus reaching level of cirrus pouch. Parasites of elasmobranchs. Sri Lanka, Australia.

Type species: *Y. parva* Southwell 1927, in *Chiloscyllium indicum;* Sri Lanka.

Other species:

*Y. southwelli* Deshmukh 1979, in *Ginglymostoma concolor;* India.

6a.  Each bothridium with two dissimilar hooks: inner hook with single prong, outer hook with two prongs ............ *Acanthobothroides* **Brooks 1977. (Figure 159)**

*Diagnosis:* Scolex with four sessile bothridia, each with three loculi, apical sucker and pad armed with pair of dissimilar hooks. Inner bothridial hooks with base and single prong, outer hooks bifid with handle. Neck present. Strobila slightly craspedote, with up to 700

FIGURE 160. *Calliobothrium verticillatum* (Rudolphi 1819) Beneden 1850.

proglottids. Genital pores marginal, irregularly alternating. Cirrus pouch large, containing convoluted ejaculatory duct. Cirrus spined. Testes all preovarian, none posterior to cirrus pouch. Vagina anterior to cirrus pouch. Distal vaginal sphincter present. Ovary follicular, with isthmus between lateral halves. Uterus a slightly lobated sac anterior to ovary. Vitellaria marginal. Parasites of Chondrichthyes (Dasyatidae). Caribbean.

Type species: *A. thorsoni* Brooks 1977, in *Himantura schmardae;* near Magdalena, Colombia.

**6b. All bothridial hooks similar ..................................................... 7**

**7a. Each bothridium with two pairs of simple hooks ...............................
.................................... *Calliobothrium* Beneden 1850. (Figure 160)**

*Diagnosis:* Bothridia divided into three loculi by two transverse septa. Two pairs of simple hooks present on muscular pad at anterior end of each bothridium. One to three accessory suckers present on each pad. Neck absent. Genital pores irregularly alternating. Cirrus pouch small, thin-walled. Cirrus armed or not. Testes in intervitelline field anterior to ovary. Vagina opening anterior to cirrus. Seminal receptacle present. Ovary large, bilobed, posterior, occupying entire intervitelline field. Vitellaria lateral. Uterus extending nearly to anterior end of proglottid. Numerous ventral uterine pores forming by rupture of body wall in gravid proglottids. Parasites of elasmobranchs. Europe, Japan, India, North America.

Type species: *C. verticillatum* (Rudolphi 1819) Beneden 1850, in *Mustelus vulgaris, M. schmitti, M. canis, M. laevis, M. equestris, M. plebejus, M. manazo, Raja batis, Squatina angelus, Squalus acanthias, Galeus canis, Hexanchus* sp., *Acanthias* sp.; Europe, North America, England, Mediterranean, Japan, Argentina.

Other species:

*C. eschrichtii* (Beneden 1849) Beneden 1850 (syn. *Onchobothrium elegans* Diesing 1854), in *Mustelus vulgaris, M. hinnulus, M. laevis, M. schmitti, M. canis, Galeus canis, Dasybatus sephen, Cynias manazo;* Europe, New England, India, Japan, Argentina.

*C. leuckartii* (Beneden 1849) Beneden 1850 (syn. *Onchobothrium heteracanthum* Diesing 1854, *Calliobothrium eschrichtii* Beneden of Johntone 1907), in *Mustelus vulgaris, M. hinnulus, M. laevis, Scyllium canicula;* Europe.

*C. nodosum* Yoshida 1917, in *Cynias manazo;* Japan.

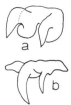

FIGURE 161. *Oncobothrium* spp. Typical hooks.

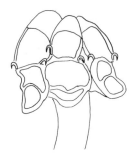

FIGURE 162. *Spiniloculus mavensis* Southwell 1925.

*C. pellucidum* Riser 1955, in *Mustelus californicus;* California, U.S.

7b. Each bothridium with one pair of hooks, simple or forked ................... 8

8a. All hooks simple, may be joined by bar at base ............................... 9
8b. All hooks forked .......................................................... 10

9a. Hooks anterior of front loculi ...... **Oncobothrium** Blainville 1828. (Figure 161)

*Diagnosis:* Bothridia divided into three loculi by two transverse septa and armed with two thorn-shaped hooks. Each pair of hooks may be joined at their bases by a horseshoe-like plate, or they may be separate. Each hook commonly with tubercle- or hair-like process. Neck present. Genital pores irregularly alternating. Cirrus pouch thin-walled. Cirrus armed or not. Vagina anterior to cirrus pouch, commonly greatly dilated near its distal end. Testes in intervitelline field anterior to ovary. Ovary posterior, transversely elongated or U-shaped, with wings sometimes extending anterior to cirrus pouch. Vitellaria lateral. Uterus extending anterior to level of cirrus pouch. Parasites of elasmobranchs. Atlantic, Pacific, Mediterranean.

Type species: *O. pseudouncinatum* (Rudolphi 1819) Beauchamp 1905, in *Raja punctata, R. maculata, R. microcellata, R. batis, R. montagui, Dasyatis centrura, D. say, Torpedo uncinata, T. ocellata, Galeus vulgaris, Trygon pastinaca,* etc.; Atlantic.

Other species:

*O. capoori* Srivastav et Tiwari 1980, in *Torpedo* sp.; India.

*O. convolutum* (Yoshida 1917), in *Cynias manazo;* Japan.

*O. farmeri* (Southwell 1911) Southwell 1930 (syn. *Calliobothrium farmeri* Southwell 1911), in *Trygon kuhli;* Sri Lanka.

*O. ganfini* Mola 1927, in *Scylliorhinus canicula;* Italy.

*O. lintoni* Mola 1927, in *Scylliorhinus stellaris;* Italy.

*O. magnum* Campbell 1977, in *Bathyraja richardsoni;* North Atlantic.

*O. schizacanthium* Lönnberg 1893, host unknown; Java.

9b. Hooks lateroposterior of front loculi .........................................
................................. **Spiniloculus** Southwell 1925. (Figure 162)

*Diagnosis:* Scolex globular, each bothridium divided into three loculi by two transverse septa. Posterolateral corners of anterior loculi each with a single simple hook. Accessory suckers absent. Neck apparently present. Mature proglottids much longer than wide. Genital pores irregularly alternating. Cirrus pouch large. Cirrus armed. Testes numerous, in intervitelline field anterior to ovary. Vagina dilated, anterior to cirrus pouch. Ovary posterior,

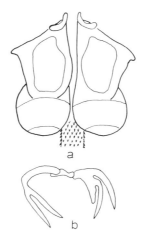

FIGURE 163. *Platybothrium cervinum* Linton 1890, (a) Scolex; (b) hooks.

bilobed. Vitellaria lateral. Uterus median, extending to level of cirrus pouch. Parasites of elasmobranchs. Australia.

Type species: *S. mavensis* Southwell 1925, in *Mustelus* sp.; Moreton Bay, Australia.

**10a. One hook of each pair three-pronged, the other two-pronged ..................
........................................ *Platybothrium* Linton 1890. (Figure 163)**

*Diagnosis:* Bothridia with three loculi, the anterior sometimes being difficult to see. A lateral lobe is sometimes present on the middle loculus, which also has a pair of hooks, one of which is bifid and the other trifid. Hooks may or may not be joined at bases by sclerotized plate. Neck present, spiny. Anterior proglottids spiny. Genital pores irregularly alternating. Testes extending between ovary and anterior end of proglottid, in intervitelline field. Cirrus pouch thin-walled. Vagina anterior to cirrus pouch. Ovary posterior, bilobed. Vitellaria in narrow lateral fields or forming sleeve around reproductive systems. Uterus median, nearly reaching anterior end of proglottid. Parasites of elasmobranchs. North America, Japan, Europe, India.

Type species: *P. cervinum* Linton 1890 (syn. *Phylobothroides spinulifera* Southwell 1912), in *Carcharias obscurus, Prionace glauca;* Newfoundland, New England.

Other species:

*P. auriculatum* Yamaguti 1952, in *Prionace glauca;* Japan.

*P. baeri* Euzet 1952, in *Carcharhinus glaucus;* Italy.

*P. hypoprioni* Potter 1927, in *Hypoprion brevirostris, Carcharhinus leucus;* Florida, Nicaragua, Costa Rica.

*P. musteli* Yamaguti 1952, in *Mustelus manazo;* Japan.

*P. parvum* Linton 1901, in *Carcharhinus milberti, Isurus dekayi, Sphyrna zygaena;* New England.

*P. sardinellae* Hornell et Nayudu 1924, larva in pyloric ceca of *Sardinella longipes;* India.

*P. veravalensis* Deshmukh, Shinde et Jadav 1977, in *Carcharias acutus;* India.

**10b. Both hooks of each pair two-pronged ............................................
........................ *Acanthobothrium* Beneden 1849 (syn. *Acrobothrium* Baer 1948, *Dicranobothrium* Euzet 1953, *Prosthecobothrium* Diesing 1863). (Figure 164)**

**138** *CRC Handbook of Tapeworm Identification*

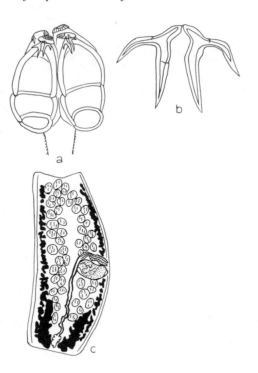

FIGURE 164. *Acanthobothrium floridensis* Goldstein 1964. (a) Scolex; (b) hooks; (c) mature proglottid.

*Diagnosis:* Bothridia trilocular, each with two bifurcate hooks at its anterior end. Small spongy pad or sucker sometimes in front of bothridium. Neck present. Genital pores irregularly alternating. Cirrus pouch well developed. Cirrus usually armed. Testes in entire intervitelline field anterior to ovary. Vagina anterior to cirrus pouch. Seminal receptacle sometimes present. Ovary posterior, variable in shape. Vitellaria lateral. Uterus median, reaching anterior end of proglottid. Parasites of elasmobranchs. Cosmopolitan. Key to species: Goldstein.[1098,1099]

Type species: *A. coronatum* (Rudolphi 1819) Beneden 1849 (syn. *Bothriocephalus bifurcatus* Leuckart 1819, *Calliobothrium corollatum* Monticelli 1887, *Prosthecobothrium urogymni* Hornell 1912, *Taenia dysbiotos* MacCallum 1921), in *Raja* spp., *Scyllium canicula, Squalus stellaris, Squatina angulus, Torpedo marmorata, T. ocellata, Trygon pastinaca, T. kuhli, Mustelus vulgaris, M. laevis, Dasyatis akajei, Leiobatus aquila, Carcharias* spp., etc.; Mediterranean, Atlantic, Pacific, India.

Other species:

*A. aetiobatis* (Shipley 1900), in *Aetiobatis narinari;* Indian Ocean.

*A. amazonensis* Mayes, Brooks et Thorson 1978, in *Potomotrygon circularis;* Brazil.

*A. americanum* Campbell 1969, in *Dasyatis americana;* Virginia, U.S.

*A. annapinkiensis* Carvajal et Goldstein 1971, in *Raja chilensis;* Chile.

*A. australis* Robinson 1965, in *Squalus megalops;* Australia

*A. bajaensis* Appy et Dailey 1973, in *Heterodontus francisci;* California.

*A. benedeni* Lönnberg 1889 (syn. *A. filicolle* var. *benedenii* Beauchamp 1905), in *Raja clavata, Dasyatis violacea, Trygon centrura, Rhinobatus philippi;* Atlantic, Mediterranean, Pacific.

*A. bengalense* Baer et Euzet 1962, in *Trygon sephen;* Indian Ocean.

*A. brachyacanthum* Riser 1955, in *Raja montereyensis;* California.

*A. cartagenensis* Brooks et Mayes 1980, in *Urolophus jamaicensis;* Colombia.

*A. cestracii* Yamaguti 1934 (syn. *A. cestraciontis* Yamaguti 1934), in *Cestracion japonicus;* Japan.

*A. chengi* Cornford 1974, in *Dasyatis lata;* Hawaii.

*A. colombianum* Brooks et Mayes 1980, in *Aetobatis narinari;* Colombia.

*A. confusum* Baer et Euzet 1962, in *Trygon kuhli;* Indian Ocean.

*A. crassicollis* Wedl 1855 (syn. *A. intermedium* Perrenoud 1931, *A. ponticum* Leon-Borcea 1934), in *Trygon pastinaca, Raja* sp.; Europe, Black Sea, New Zealand.

*A. dasybati* Yamaguti 1934, in *Dasybatus akajei, Raja kenojei, Urolophus fuscus;* Japan.

*A. dighaensis* Srivastav et Capoor 1980, in *Trygon marginatus;* India.

*A. dujardinii* Beneden 1849 (syn. *Onchobothrium [Acanthobothrium] papilligerum* Diesing 1854, *Prosthecobothrium dujardinii* [Beneden 1849] Diesing 1854, *Acanthobothrium brevissime* Linton 1909), in *Raja clavata, R. maculata, R. montagui, R. brachyura, R. binoculata, Dasyatis say, Urobatis halleri, Rhinobatis productus;* California, Europe, Black Sea, England, Florida.

*A. edwardsi* Williams 1969, in *Raja fullonica;* England.

*A. electricollum* Brooks et Mayes 1978, in *Narcine brasiliensis;* Colombia.

*A. filicollis* (Zschokke 1887) (syn. *A. paulus* Linton 1924), in *Raja marginata, R. astarias, Dasyatis violacea, Torpedo;* Mediterranean.

*A. floridensis* Goldstein 1964, in *Raja eglanteria;* Gulf of Mexico.

*A. fogeli* Goldstein 1964, in *Gymnura micrura;* Gulf of Mexico.

*A. goldsteini* Appy et Dailey 1973, in *Platyrhinoidis triseriata;* California.

*A. gracile* Yamaguti 1952, in *Narke japonica;* Japan.

*A. grandiceps* Yamaguti 1952, in *Dasybatus zugei, D. akajei;* East China Sea.

*A. hanumantharaoi* Rao 1977, in *Myliobatus nieuhofii;* India.

*A. harpago* (Euzet 1953) (syn. *Dicranobothrium harpago* Euzet 1953), in *Negaprion brevirostris;* Senegal.

*A. herdmani* Southwell 1912 (syn. *S. coronatum* Southwell 1925), in *Trygon kuhli;* Sri Lanka.

*A. heterodonti* Drummond 1937, in *Heterodontus philippi;* Australia.

*A. hispidum* Riser 1955, in *Tetronarce californica;* California.

*A. holorhini* Alexander 1953, in *Holorhinus californicus*; California.

*A. icelandicum* Manger 1972, in *Raja batis;* Iceland.

*A. ijimai* Yoshida 1917 (syn. *Taenia incognita* MacCallum 1921), in *Dasybatus akajei, Raja* spp., *Narcine timleyi, Chiloscyllium* sp.; Sri Lanka, Japan, East China Sea, Irish Sea, North Sea.

*A. karachiense* Bilqees 1980, in *Myrmillo manazo;* Pakistan.

*A. latum* Yamaguti 1952, in *Dasybatus akajei;* Japan.

*A. lineatum* Campbell 1969, in *Dasyatis americana;* Virginia, U.S.

*A. lintoni* Goldstein, Henson et Schlicht 1968, in *Narcine brasiliensis;* Gulf of Mexico.

*A. macracanthum* Southwell 1925, in *Urogymnus* sp.; Sri Lanka.

*A. maculatum* Riser 1955, in *Aetobatis californicus;* California.

*A. mathiasi* Euzet 1959, in *Mustelus mustelus, M. canis;* France.

*A. micracantha* Yamaguti 1952, in *Dasybatus akajei, D. zugei, Pteroplatea micrura;* Japan, East China Sea.

*A. mujibi* Bilqees 1980, in *Myrmillo manazo;* Pakistan.

*A. musculosum* (Baer 1948) (syn. *Acrobothrium musculosum* Baer 1948, *A. intermedium* Perrenoud 1931), in *Dasyatis violacea;* Italy.

*A. olseni* Dailey et Mudry 1968, in *Rhinobatos productus;* California.

*A. parviuncinatum* Young 1954, in *Urobatis halleri, Gymnura marmorata;* California.

*A. parvum* Manger 1972, in *Raja batis;* Iceland.

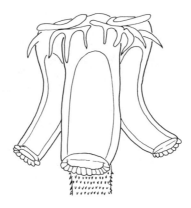

FIGURE 165. *Phoreiobothrium lasium* Linton 1889.

*A. paulum* Linton 1890, in *Trygon centrura, Raja laevis, R. eglanteria, Dasyatis say, Pteroplatea macrura;* New England.
*A. pearsoni* Williams 1962, in *Orectolobus maculatus;* Australia.
*A. psammobati* Carvajal et Goldstein 1969, in *Psammobatis scobina;* Chile.
*A. quadripartitum* Williams 1968, in *Raja naevus;* England.
*A. quinonesi* Mayes, Brooks et Thorson 1978, in *Potomotrygon magdalenae;* Colombia.
*A. rhinobati* Alexander 1953, in *Rhinobatus productus;* California.
*A. robustum* Alexander 1953, in *Rhinobatus productus;* California.
*A. rubrum* Bilqees 1980, in *Myrmillo manazo;* Pakistan.
*A. semivesiculum* Verma 1928, in *Trygon sephen;* India.
*A. septentrionale* Baer et Euzet 1962, in *Raja batis,* Indian Ocean.
*A. terezae* Rego et Dias 1974, in *Paratrygon motora, Elipsurus* sp.; Brazil.
*A. triacis* Yamaguti 1952, in *Triacis scyllium;* Japan.
*A. tripartitum* Williams 1969, in *Raja microocellata;* England.
*A. tsingtaoense* Tseng 1933, in *Dasybatus akajei;* China.
*A. typicum* Olsson 1872, in *Lota vulgaris;* Sweden.
*A. uncinatum* (Rudolphi 1819) Blainville 1828, in *Squalus galeus, Galeus canis, Mustelus hinnulus, Trygon pastinaca, T. kuhli, T. walga, Raja batis, R. punctata, R. clavata, Torpedo oculata;* Europe, Irish Sea, Sri Lanka.
*A. unilaterale* Alexander 1953, in *Holorhinus californicus;* California.
*A. urolophi* Schmidt 1973, in *Urolophus testaceus;* South Australia.
*A. urotrygoni* Brooks et Mayes 1980, in *Urotrygon venezuelae;* Colombia.
*A. wedli* Robinson 1959, in *Raja nasuta;* New Zealand.
*A. woodsholei* Baer 1948, for *A. coronatum* Linton 1901, nec Rudolphi 1819, in *Dasyatis centrura;* Woods Hole, Massachusetts.
*A. zapterycum* Nuñez 1971, in *Zapteryz brevirostris;* Argentina.
*A. zschokkei* Baer 1948, for *Onchobothrium (Calliobothrium) uncinatum* Zschokke 1888, nec Rudolphi 1819, in *Torpedo ocellata;* Italy.

**11a. Bothridia with accessory suckers in front of hooks .......................... 12**
**11b. Accessory suckers absent or surrounding hooks .............................. 13**

**12a. Hooks bifid or trifid ................ *Phoreiobothrium* Linton 1889. (Figure 165)**

*Diagnosis:* Bothridia sessile, shallow, undivided, each with two bifid or trifid hooks on anterior margin. Simple accessory sucker present anterior to each pair of hooks. Spiny neck

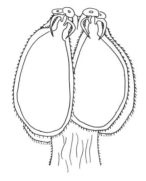
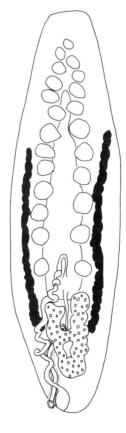

FIGURE 166. *Pomatotrygonocestus magdelensis* Brooks et Thorson 1976. Scolex and mature proglottid.

present. Genital pores apparently irregularly alternating. Cirrus unarmed. Testes in entire intervitelline field anterior to ovary. Vagina anterior to cirrus pouch. Ovary posterior, bilobed. Vitellaria lateral. Uterus nearly reaching anterior end of proglottid. Parasites of elasmobranchs. Atlantic North America.

Type species: *P. lasium* Linton 1889, in *Carcharias obscurus, Carcharhinus commersoni, C. limbatus, C. milberti, Galeocerdo arcticus, Scolidon terraenovae, Vulpecula marina;* New England.

Other species:

*P. exceptum* Linton 1924, in *Cestracion zygaena;* New England.

*P. pectinatum* Linton 1924, in *Cestracion zygaena;* New England.

*P. triloculatum* Linton 1901, in *Carcharhinus obscurus, C. milberti, C. acronotus, C. leucus, Scoliodon terraenovae;* New England, North Carolina, Guatemala, Costa Rica.

**12b. Hooks with single points ........................................................
.................... *Pomatotrygonocestus* Brooks et Thorson 1976. (Figure 166)**

*Diagnosis:* Scolex with four sessile, nonseptate bothridia, each with simple margins, accessory sucker, and pair of simple hooks. Cephalic peduncle absent. Scolex and neck spinose. Genital pores irregularly alternating, near posterior end. Genital atrium absent. Cirrus pouch small, cirrus spinose. Testes in two longitudinal rows anterior to ovary. Vagina anterior to cirrus pouch. Ovary bilobed in cross section, in posterior one third of proglottid; lobes fused posteriorly. Uterus a median, longitudinal, simple sac anterior to ovary. Vitellaria

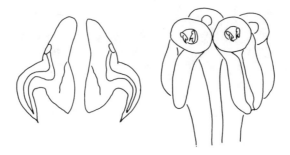

FIGURE 167.   *Pachybothrium hutsoni* (Southwell 1911) Baer et Euzet 1962. Bothridial hooks and scolex.

in two elongate, compact, lateral fields. Parasites of freshwater stingrays (Potamotrygonidae), South America.

Type species: *P. magdalenensis* Brooks et Thorson 1976, in *Pomatotrygon magdalenae;* Magdalena River system, Colombia.

Other species:

*P. amazonensis* Mayes, Brooks et Thorson 1981, in *Pomatotrygon circularis, P. yepezi, P. reticulatus;* Itacuai River, Brazil; Cachiri River, Orinoco River Delta, Venezuela.

*P. orinocoensis* Brooks, Mayes et Thorson 1981, in *Pomatotrygon reticulatus;* Delta of Orinoco River, Venezuela.

*P. travassosi* Rego 1979 (*species inquirendum,* according to Brooks, Mayes et Thorson 1981), in *Pomatotrygon hystrix;* Amazon River, Brazil.

**13a.  Hooks embedded in anterior, muscular pad .........................................**
**........................... *Pachybothrium* Baer et Euzet 1962. (Figure 167)**

*Diagnosis:* Scolex with four elongate, nonseptate bothridia, each with muscular pad at anterior end. One pair of simple, rosethorn-shaped hooks embedded in each pad. Neck present. Strobila acraspedote, with about 100 segments. Genital pores irregularly alternate, slightly preequatorial. Cirrus armed. Testes numerous (50 to 247); none posterior to vagina on poral side. Ovary posterior, bilobed. Vagina anterior to cirrus pouch. Vitellaria lateral. Uterus saccular. Parasites of elasmobranchs.

Type species: *P. hutsoni* (Southwell 1911) Baer et Euzet 1962 (syn. *Phyllobothroides hutsoni* Southwell 1911, *Pedibothrium hutsoni* [Southwell 1911] Southwell 1924), in *Gingylmostoma concolor, Galeocerdo articus, Rhina ancylostoma;* India, Sri Lanka.

**13b.  Anterior, muscular pad absent .................................................... 14**

**14a.  Bothridia embedded in scolex with posterior tips protruding ....................**
**..................................................... *Pinguicollum* Riser 1955.**

*Diagnosis:* Bothridia embedded in scolex except for posterior end which protrudes. Each bothridium with a pair of two-pronged hooks, united at their bases. Accessory suckers absent. Neck long, prominent. Proglottids short, wide. Genital pores irregularly alternating. Cirrus pouch crossing osmoregulatory canals. Testes numerous (110 to 120). Female genitalia not described. Parasites of elasmobranchs. Pacific North America.

Type species: *P. pinguicollum* (Sleggs 1927) Riser 1955 (syn. *Onchobothrium pinguicollum* Sleggs 1927, in *Raja inornata, R. binoculata, R. rhina, R. montereyensis;* West Coast, North America.

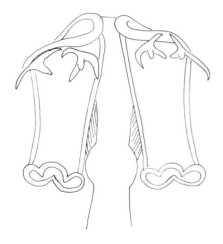

FIGURE 168. *Cylindrophorus triloculatus* Linton 1901.

FIGURE 169. *Pedibothrium globicephalum* Linton 1909. Bothridial hook.

**14b. Bothridia not embedded in scolex ............................................. 15**

**15a. Bothridia cylindrical, hollow ....... *Cylindrophorus* Diesing 1863. (Figure 168)**

*Diagnosis:* Bothridia tubular, hollow, each with a bifid and a trifid hook at the anterior margin. No accessory suckers present. Neck present, spiny. Internal anatomy poorly known. Testes numerous, in intervitelline field anterior to ovary. Vagina anterior to cirrus pouch. Ovary posterior, occupying about one fourth length of proglottid. Vitellaria lateral. Parasites of elasmobranchs. Locality (?).

Type species: *C. typicus* Diesing 1863 (*Tetrabothrium carchariae-rondolettii* Wagner 1854, renamed), in *Carcharias rondoletti;* ? Europe

Other species:

    *C. posteroporus* Riser 1955 (syn. *Platybothrium cervinum* Linton 1890, in part), in *Prionace glauca;* California.

**15b. Bothridia flattened, not hollow ........ *Pedibothrium* Linton 1909. (Figure 169)**

*Diagnosis:* Bothridia flattened, no loculi or accessory suckers, margins smooth. Each bothridium with a pair of double-pronged hooks. Neck distinct. Genital pores irregularly alternating. Cirrus pouch wide. Cirrus armed with very small spines. Testes intervitelline, usually confined anterior to level of cirrus pouch but occasionally reaching antiporal wing of ovary. Vagina anterior to cirrus pouch. Ovary posterior, bilobed. Vitellaria in broad lateral fields, extending behind ovary. Uterus reaching level of cirrus pouch. Parasites of elasmobranchs. North America, India, Sri Lanka.

Type species:

*P. globicephalum* Linton 1909, in *Ginglymostoma cirratum, Pristis cuspidatus;* Florida, India.

Other species:

    *P. brevispine* Linton 1909, in *Ginglymostoma cirratum;* Florida.

    *P. lintoni* Shinde, Jadhav et Deshmukh 1980, in *Stegostoma tigrinum;* India.

    *P. longispine* Linton 1909 (syn. *Phyllobothroides kerkhami* Southwell 1911), in *Ginglymostoma cirratum, Chiloscyllium indicum, Rhynchobatus* sp., *Galeocerdo tigrinus, Rhina ancylostoma;* Florida, India.

    *P. veravalensis* Shinde, Jadhav et Deshmukh 1980, in *Stegostoma tigrinum;* India.

FIGURE 170. *Trilocularia gracilis* Olsson 1867.

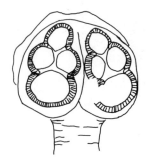

FIGURE 171. *Zyxibothrium kamienae* Hayden et Campbell 1981.

## KEY TO THE GENERA IN TRILOCULARIIDAE

**1a. Each bothridium with three loculi, arranged in triangle ........................ ........................................... *Trilocularia* Olsson 1867. (Figure 170)**

*Diagnosis:* Scolex with four sessile, triangular bothridia, each with three loculi arranged in triangle. No suckers or myzorhynchus. Neck present. Free proglottids with anterior spines. Genital pores and cirrus pouch median near posterior end of proglottid. Testes anterior to ovary. Vagina ventral to cirrus. Ovary posterior, U-shaped with wings median to vitellaria. Vitellaria in lateral medullary fields. Uterus with dorsal duct and ventral elongate uterine sac with 10 to 15 lateral branches on each side. No preformed uterine pore. Parasites of sharks. Scandinavia, Britain, Mediterranean.

Type species: *T. gracilis* Olsson 1867 (syn. *Phyllobothrideum acanthiae-vulgaris* Olsson 1867, *Urogonoporus armatus* Lühe 1901), in *Acanthias vulgaris;* Scandinavia.

**1b. Each bothridium with more than three loculi.................................... 2**

**2a. Each bothridium with four loculi................................................... ........................ *Zyxibothrium* Hayden et Campbell 1981. (Figure 171)**

*Diagnosis:* Scolex with four sessile bothridia, each divided into four loculi by muscular septa. Bothridia fused to scolex anteriorly. Myzorhynchus absent. Cephalic peduncle short. Strobila spineless, slightly craspedote in mature segments. Genital pores marginal, in anterior one-fourth of proglottid, irregularly alternating. Testes preovarian, medullary, in two longitudinal fields lateral to developing uterus. Vagina passes dorsal to cirrus pouch. Cirrus pouch pyriform, thin-walled. Cirrus armed. Ovary posterior, U-shaped, bilobed in cross section. Vitellaria lateral, sometimes confluent posterior to ovary. Uterus an elongate sac filling most of medulla. Embryos lacking shell, enclosed in thin membrane; eggs contained within membranous, intrauterine tube; released by dehiscence of segment. Parasites of elasmobranchs.

Type species: *Z. kamienae* Hayden et Campbell 1981, in *Raja senta;* Gulf of Maine to Scotian Shelf, Western Atlantic.

**2b. Each bothridium with five loculi... *Pentaloculum* Alexander 1963. (Figure 172)**

*Diagnosis:* Bothridia sessile, each divided by permanent, muscular septa into five radially arranged loculi, more or less equal in size. No cephalic peduncle or myzorhynchus. Strobila small. Proglottids slightly craspedote. Genital pores irregularly alternating on posterior third

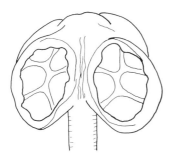

FIGURE 172. *Pentaloculum symmetricum* Alexander 1963.

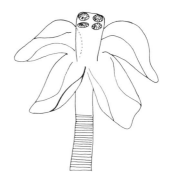

FIGURE 173. *Myzophyllobothrium rubrum* Shipley et Hornell 1906.

of proglottid margin. Cirrus, cirrus pouch, vagina, and uterus not described. Testes fill proglottid anterior to cirrus pouch. Ovary occupying most of space behind cirrus pouch. Vitellaria in narrow, paired bands of small follicles lateral to testes. Parasite of elasmobranchs. New Zealand.

Type species: *P. macrocephalum* Alexander 1963, in *Typhlonarke aysoni;* New Zealand.

## KEY TO THE GENERA IN PHYLLOBOTHRIIDAE

1a. Myzorhynchus with four suckers; may be long and slender or short and dome-shaped ................................................................................................. 2
1b. Myzorhychus without suckers, or absent ........................................ 3

2a. Myzorhynchus prominent, slender ............... *Myzophyllobothrium* **Shipley et Hornell 1906 (syn.** *Rhoptrobothrium* **Shipley et Hornell 1906). (Figure 173)**

*Diagnosis:* Scolex with prominent myzorhynchus bearing four suckers. Bothridia leaf-like, attached at base, with smooth edges. Neck short. Genital pores marginal, equatorial. Cirrus pouch thin-walled. Cirrus well-developed, unarmed. Testes numerous, in entire intervitelline field anterior to ovary. Vagina anterior to cirrus pouch. Ovary posterior, with broad lateral wings. Vitellaria in complete lateral fields. Uterus median, extending to level of cirrus pouch. Parasites of elasmobranchs. Sri Lanka.

Type species: *M. rubrum* Shipley et Hornell 1906 (syn. *Rhoptrobothrium myliobatidis* Shipley et Hornell 1906), in *Aetobatis narinari, Myliobatis maculata;* Sri Lanka, Pakistan.

2b. Myzorhynchus globular; each pair of suckers in contact .......................... ................ *Mixophyllobothrium* **Shinde et Chincholikar 1981. (Figure 174)**

*Diagnosis:* Scolex with anterior, globular "rostellum" bearing four pairs of suckers, each pair in contact. Four bothridia, each thin and convoluted, lacking suckers or hooks. Short neck present. Immature proglottids wider than long; mature proglottids longer than wide, cylindrical. Genital pores equatorial, irregularly alternating. Cirrus pouch massive, reaching midline of segment. Cirrus unarmed. Testes very numerous, preovarian. Ovary posterior, bilobed. Vagina anterior to cirrus pouch. Seminal receptacle present. Vitellaria lateral, in entire length of segment. Uterus and eggs unknown. Parasites of Batoidea.

Type species: *M. okamuri* Shinde et Chincholikar 1981, in *Trygon sephen;* India.

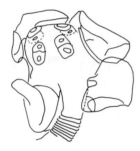

FIGURE 174. *Mixophyllobothrium okamuri* Shinde et Chincholikar 1980.

FIGURE 175. *Pithophorus tetraglobus* (Southwell 1911) Southwell 1925.

FIGURE 176. *Scyphophyllidium giganteum* (Beneden 1858) Woodland 1927.

3a. Bothridia approximately cylindrical, open at both ends ..........................
........................................ *Pithophorus* Southwell 1925. (Figure 175)

*Diagnosis:* Scolex with four globular or cylindrical hollow bothridia; each opens at both anterior and posterior ends. Myzorhynchus present or absent (?). Neck present. Genital pores near equator, irregularly alternating. Cirrus pouch very large, swollen, extending to median line. Cirrus armed at tip in mature proglottids only. Testes numerous in two lateral fields medial to vitellaria. Vagina anterior to cirrus sac. Ovary with two large lobes. Vitellaria lateral. Uterus median, extending nearly entire length of proglottid. Parasite of elasmobranchs. Indian Ocean, Japan.

Type species: *P. tetraglobus* (Southwell 1911) Southwell 1925 (syn. *Orygmatobothrium tetraglobus* Southwell 1911), in *Rhynchobatus djeddensis;* Sri Lanka.
Other species:
*P. pakistanensis* Zaidi et Khan 1976, in *Chilocyllium indicum;* Pakistan.
*P. vulpeculae* Yamaguti 1952, in *Vulpecula marina;* Japan.
*P. yamagutii* Shinde 1978, in *Scolionon* sp.; India.

3b. Bothridia hollow, open at one end only .......................................... 4
3c. Bothridia not as above ...................................................... 8

4a. Bothridia almost sessile, sucker-like ..............................................
................................ *Scyphophyllidium* Woodland 1927. (Figure 176)

*Diagnosis:* Scolex with four globular, sucker-like bothridia, each with irregular, anterior opening. Myzorhynchus and accessory suckers absent. Neck very long. Genital pores in anterior third of proglottid margin, irregularly alternating. All proglottids broader than long. Testes numerous, occupying entire intervitelline field anterior to ovary. Vagina anterior to cirrus pouch. Ovary posterior, bilobed. Vitellaria in two wide lateral fields. Uterus median, somewhat convoluted. Parasite of elasmobranchs. Europe, North America.

Type species: *S. giganteum* (Beneden 1858) Woodland 1927 (syn. *Anthobothrium giganteum* Beneden 1858, *A. elegantissimum* Lönnberg 1889, *A. rugosum* Shipley et Hornell 1906, *A. floraformis* Southwell 1912, *Tetrabothrium maculatum* Olsson 1866), in *Galeus canis, G. vulgaris, Raja batis, R. scabrata, R. radiata, Galeorhinus zyopterus;* Europe, Canada, California, Newfoundland.

Other species:
    *S. arabiansis* Shinde et Chincholikar 1977, in *Trygon* sp.; India.
    *S. pruvoti* (Guiart 1934) (syn. *Diplobothrium pruvoti* Guiart 1934), in *Loligo loligo;* Atlantic.

**4b.** Bothridia not as above ......................................................... 5

**5a.** **Submarginal accessory sucker present on each bothridium; bothridia attached at anterior end**............................. ***Marsupiobothrium*** **Yamaguti 1952.**

*Diagnosis:* Scolex with four sac-like, pyriform bothridia, each with anterior opening provided with sphincter-like muscles and an accessory sucker present at anterior end. Myzorhynchus absent. Neck present. Free proglottids greatly elongated. Genital pore preequatorial. Cirrus pouch and cirrus well developed. Testes numerous. Vagina anterior to cirrus pouch. Ovary with two long slender wings. Vitellaria in lateral fields, some follicles postovarian. Uterus median, extending to level of cirrus pouch. Parasites of selachians. Japan, North America.

Type species: *M. alopias* Yamaguti 1952, in *Alopias vulpinus;* Japan.
Other species:
    *M. forte* (Linton 1924) Yamaguti 1956 (syn. *Orygmatobothrium forte* Linton 1924), in *Cestracion zygaena;* New England.
    *M. rhinobati* Shinde et Deshmukh 1980, in *Rhinobatus granulatus;* India.
    *M. rhynchobati* Shinde et Deshmukh 1980, in *Rhynchobatus djeddensis:* India.

**5b.** **Bothridia without accessory suckers, not attached at anterior end** ............ 6

**6a.** **Bothridia attached along entire length** .............................................
............................................ ***Aocobothrium*** **Mola 1907. (Figure 177)**

*Diagnosis:* Scolex with four sac-like bothridia attached along their entire lengths, each with an enlarged rim around the anterior opening. Myzorhynchus absent. Neck present. Proglottids rounded to oval. Genital pores irregularly alternating. Cirrus pouch at an oblique angle. Testes numerous. Vagina anterior to cirrus pouch. Ovary posterior, bilobed. Vitellaria in lateral bands joined at anterior and posterior ends. Uterus reaching equator of proglottid. Parasite of freshwater teleosts (?). Europe.

Type species: *A. carrucci* Mola 1907.

**6b.** **Bothridia attached at posterior end** ............................................. 7

**7a.** **Vitellaria continuous behind ovary; myzorhynchus absent**......................
........................***Cyatocotyle*** **Mola 1908, for *Cyathocotyle* Muhling 1896.**

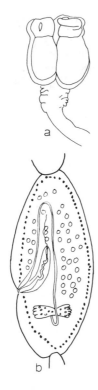

FIGURE 178. *Glyphobothrium zwerneri* Williams et Campbell 1977.

FIGURE 177. *Aocobothrium carrucci* Mola 1907. (a) Scolex; (b) proglottid.

*Diagnosis:* Scolex with four tubular bothridia attached at their posterior ends, each with apical opening. Myzorhynchus absent. Neck present. External segmentation indistinct. Entire strobila with minute spines. Genital pores irregularly alternating. Genital atrium with strong sphincter. Cirrus pouch small. Testes preovarian, numerous. Vagina dorsal and posterior to cirrus pouch. Ovary bilobed, posterior. Vitellaria lateral, joined at anterior and posterior ends of proglottid. Uterus nearly reaching anterior end of proglottid. Parasite of elasmobranchs. Indian Ocean.

Type species: *C. marchesettii* Mola 1908, in *Carcharias limia;* India.

**7b. Vitellaria not continuous behind ovary. Myzorhynchus present ................**
**................................................ *Pseudanthobothrium* Baer 1956.**

*Diagnosis:* Scolex with four cup-shaped bothridia attached at their posterior ends. Myzorhynchus present. Genital pores irregularly alternating, postequatorial. Cirrus pouch small. Cirrus spined. Testes few, large, anterior to level of cirrus pouch. Vagina ventral to cirrus pouch. Ovary posterior, bilobed in surfacial view, X-shaped in cross section. Vitellaria in lateral fields. Parasites of rays. Europe.

Type species: *P. hanseni* Baer 1956 (syn. *Anthobothrium cornucopia* of Heller 1949, after Beneden 1850; *Echeneibothrium minimum* of Rees 1953, after Beneden 1805), in *Raja radiata;* Europe?

**8a. Scolex globular, with four superficial bothridia, each divided into three longitudinal rows of loculi, and separated by shallow longitudinal fissures...........**
**..................... *Glyphobothrium* Williams et Campbell 1977. (Figure 178)**

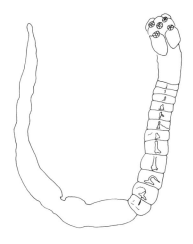

FIGURE 179. *Pelichnibothrium speciosum* Monticelli 1889.

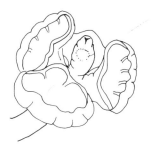

FIGURE 180. *Phormobothrium affine* (Olsson 1867) Alexander 1963.

*Diagnosis:* Bothridial margins fused to scolex. Neck present. Strobila muscular, rounded in cross section; segmentation distinct, craspedote, anapolytic. Testes median, some postovarian. Cirrus armed. Cirrus pouch large, spheroid, about reaching midline. Genital atrium present, preequatorial. Ovary posterior, multilobate. Vitellaria circummedullary, extending full length of segment. Vagina surrounded by thick layer of gland cells, opening anterior to cirrus pouch. Parasites of Batoidea. Western North Atlantic.

Type species: *G. zwerneri* Williams et Campbell 1977, in *Rhinoptera bonasus;* Chesapeake Bay, Virginia, U.S.

8b. Scolex not as above ............................................................. 9

9a. Scolex without myzorhynchus but with apical sucker. Accessory sucker next to anterior margin of each bothridium. ...............................................
................................ *Pelichnibothrium* **Monticelli 1889. (Figure 179)**

*Diagnosis:* Scolex with apical sucker. Myzorhynchus absent. Bothridia sessile, each with a single sucker on the anterior margin. Neck absent. Mature and gravid proglottids oval. Testes medullary, in lateral fields medial to osmoregulatory canals. Genital pores irregularly alternating. Cirrus pouch oblique, thin-walled. Vagina posterior to cirrus pouch. Ovary bilobed, compact. Vitellaria in broad lateral fields. Seminal receptacle present. Uterus median. Parasite of elasmobranchs. Madeira, Japan.

Type species: *P. speciosum* Monticelli 1889 (syn. *P. caudatum* Zschokke et Heitz 1914, *Phyllobothrium salmonis* Fujita 1922), in *Alepidosaurus ferox, Prionace glauca, Lampris regia, Thynnus thynnus;* Atlantic, Japan.

9b. Scolex without apical sucker ................................................... 10

10a. Transverse ridges forming several loculi on adherent surfaces of bothridia .. 11
10b. Bothridia not as above ....................................................... 17

11a. Bothridial ridges incomplete, loculi not completely enclosed .....................
................................ *Phormobothrium* **Alexander 1963. (Figure 180)**

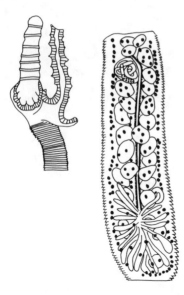

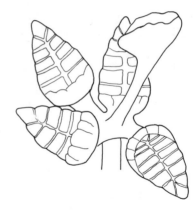

FIGURE 182. *Echineibothrium variable* Beneden 1849.

FIGURE 181. *Duplicibothrium minutum* Williams et Campbell 1978. Scolex and mature proglottid.

*Diagnosis:* Scolex possessing slender, retractile myzorhynchus containing subapical muscular organ. Bothridia pedunculae, basket-shaped, delicate, highly mobile; outer posterior rim and base of bothridium with at least four incomplete longitudinal ridges, anterior rim with two incomplete longitudinal apical ridges forming unenclosed loculus. Cephalic peduncle absent. Genital pores in posterior half of proglottid margin. Testes anterior to cirrus pouch. Vitellaria in lateral bands. Parasite of Rajidae. Cosmopolitan.

Type species: *P. affine* (Olsson 1867) Alexander 1963 (syn. *Echineiobothrium affine* Olsson 1876), in *Raja* spp.; Scandinavia.

**11b. Bothridial ridges complete ..................................................... 12**

**12a. Dorsal and ventral bothridia fused along their lengths into two pairs ...........**
**..................... *Duplicibothrium* Williams et Campbell 1978. (Figure 181)**

*Diagnosis:* Free surfaces of bothridia divided into transverse loculi, except at bases where they form a cup with a few radial loculi. Peduncle well developed. Cuticular annulations present over cephalic peduncle and strobila. Genital pores submarginal, irregularly alternating. Testes medullary, extending into postovarian space. Ovary posterior, with many thin, radiating lobes. Vitellaria circummedullary. Parasites of Batoidea. Western North Atlantic.

Type species: *D. minutum* Williams et Campbell 1978, in *Rhinoptera bonasus;* New England.

**12b. Bothridia not fused into pairs.................................................... 13**

**13a. Myzorhynchus present ..........................................................**
**............... *Echeneibothrium* Beneden 1805 (syn. *Shindeiobothrium* Jadhav, Shinde et Deshmukh 1981, *Tiarabothrium* Shipley et Hornell 1906). (Figure 182)**

*Diagnosis:* Bothridia pedunculate or sessile, with their adhesive surface divided into loculi by transverse, and sometimes also longitudinal ridges. Myzorhynchus present (Figure 183).

FIGURE 183. *Echineiobothrium variable* Benedum 1849. Scolex with myzorhynchus partially evaginated.

Neck present. External segmentation sometimes poorly marked. Genital pores irregularly alternating. Cirrus spined or not. Testes anterior to cirrus pouch. Vagina anterior or ventral to cirrus pouch. Ovary posterior, variable. Vitellaria lateral. Uterus reaching near anterior end of proglottid. Parasites of elasmobranchs. Cosmopolitan. Key to species: Williams.[3928]

Type species: *E. minimum* Beneden 1850 (syn. *E. dubium* Beneden 1858, *Tiarabothrium javanicum* Shipley et Hornell 1906), in *Raja* spp., *Trygon pastinaca*, *Carcharias* sp., *Urobatis halleri*, *Rhinoptera javanica*; Sri Lanka, Europe, Iceland.

Other species:

*E. austrinum* Linton 1924, in "large skate", locality?

*E. bathyphilum* Campbell 1975, in *Raja bathyphila*, Western North Atlantic.

*E. beauchampi* Euzet 1959, in *Raja clavata*; France.

*E. bifidum* Yamaguti 1952, encysted in body cavity of *Trachurus trachurus*; Japan.

*E. bilobatum* Young 1955, in *Urobatis halleri*; California.

*E. burgeri* (Baer 1948), in *Dasyatis centrura*; New England.

*E. ceylonicum* Shipley et Hornell 1906, in *Trygon walga*; Sri Lanka.

*E. demeusiae* Euzet 1959, in *Raja batis*, *R. marginata*, *R. oxyrhynchus*; France.

*E. dolichophorum* Riser 1955, in *Raja rhina*; California.

*E. faxanum* Manger 1972, in *Raja batis*; Iceland.

*E. gracile* Zschokke 1889, in *Raja batis*, *Trygon pastinaca*; locality?

*E. hui* Tseng 1933, in *Dasybatus akajei*; China.

*E. javanicum* Shipley et Hornell 1906, in *Rhinoptera javanica*; Java.

*E. julievansium* Woodland 1927, in *Raja maculata*, *Raja* spp.; Europe.

*E. karbharae* Jadhav, Shinde et Deshmukh 1981 comb. n. (syn. *Shindeiobothrium karbharae* Jadhav, Shinde et Deshmukh 1981), in *Trygon zugei*; India.

*E. macrascum* Riser 1955, in *Raja montereyensis*; California.

*E. meglosoma* Carvajal et Dailey 1975, in *Raja chilensis*; Chile.

*E. minutum* Williams 1966, in *Raja batis*; England.

*E. moucheti* Dollfus 1931, larvae in *Eupagurus cuanensis*, *E. bernhardus*, *E. prideaux*; France.

*E. multiloculatum* Carvajal et Dailey 1975, in *Raja chilensis*; Chile.

*E. myzorhynchum* Hart 1936, in *Raja binoculata*; Puget Sound.

*E. octorchis* Riser 1955, in *Raja montereyensis*; California.

*E. oligotesticulare* Subramanian 1940, in *Rhinobatus granulatus*; India.

*E. palombii* (Baer 1948), in *Dasyatis centrura*; New England.

*E. pollonae* Campbell 1977, in *Bathyraja richardsoni*; Western North Atlantic.

*E. rankini* (Baer 1948), in *Dasyatis centrura*; New England.

*E. rhinobati* Yamaguti 1960, in *Rhinobatis schlegeli*; Japan.

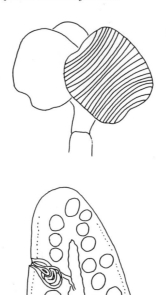

FIGURE 184. *Caulobothrium anacolum* Brooks 1977. Scolex and proglottid.

*E. smitii* Shinde, Deshmukh et Jadhav 1981, in *Trygon sephen;* India.
*E. sobrinum* Campbell 1975, in *Raja erinacea;* Western North Atlantic.
*E. trifidum* Shipley et Hornell 1906, in *Trygon walga;* Sri Lanka.
*E. trygonis* Shipley et Hornell 1906, in *Trygon walga;* Sri Lanka.
*E. tumidulum* (Rudolphi 1819) Diesing 1863 (syn. *Bothriocephalus echeneis* Leuckart 1819), in *Raja* spp.; Ireland, England, France, California.
*E. urobatidium* Young 1955, in *Urobatis halleri;* California.
*E. variabile* Beneden 1850, in *Raja* spp.; Europe, New England, China.
*E. vernetae* Euzet 1956, in *Raja erinacea;* Woods Hole, U.S.
*E. williamsi* Carvajal et Dailey 1975, in *Raja chilensis;* Chile.

**13b. Myzorhynchus absent ........................................................ 14**

**14a. Postovarian testes present.............. *Caulobothrium* Baer 1948. (Figure 184)**

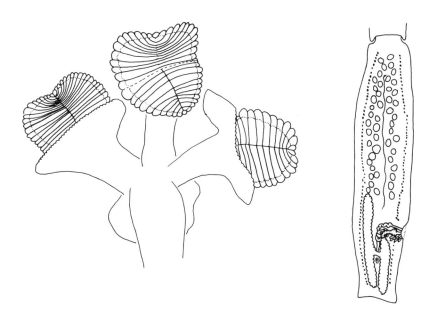

FIGURE 185. *Rhinebothroides glandularis* Brooks, Mayes et Thorson 1981. Scolex and proglottid.

*Diagnosis:* Scolex with four pediculated bothridia, their adherent surfaces divided into loculi by septa. Bothridial margins lobulated or entire in form, hinged into two parts or not. Myzorhynchus absent. Posterior portion of scolex always modified into peduncle. Strobila craspedote or acraspedote, apolytic. Genital pores marginal, irregularly alternating. Genital atrium present or absent. Cirrus pouch well developed, cirrus spined or not. Testes numerous, in median field sometimes posterior to ovary, always present between cirrus pouch and ovary on poral side. Vagina anterior to cirrus pouch. Ovary X-shaped in cross section. Vitellaria in lateral bands that may connect behind ovary, and may encroach into lateral medullary parenchyma. Uterus simple, median, tubular. Parasites of rays (Myliobatidae).

Type species: *C. longicolle* (Linton 1890) Baer 1948 (syn. *Echeneibothrium longicolle* [Linton 1890] Yamaguti 1959, *Crossobothrium laciniatum longicollis* Linton 1890), in *Myliobatis freminvillei, Dasyatis centrura;* New England.

Other species:

*C. anacolum* Brooks 1977, in *Himantura schmardae;* Colombia.

*C. insigna* (Southwell 1911) Baer 1948 (syn. *Rhinobothrium insigna* Southwell 1911), in *Trygon walga;* Sri Lanka.

*C. opisthorchis* Riser 1955 (syn. *Echeneibothrium opisthorchis* (Riser 1955) Yamaguti 1959, in *Aetobatis californicus;* U.S.

*C. multiorchidum* Appy et Dailey 1977, in *Urolophus halleri;* California.

*C. myliobatidis* Carvajal 1977, in *Myliobatis chilensis;* Chile.

*C. tobijei* (Yamaguti 1934) Baer 1948 (syn. *Echeneibothrium tobijei* Yamaguti 1934), in *Myliobatis tobifei;* Japan.

**14b. Postovarian testes absent..................................................... 15**

**15a. Poral lobe of ovary reduced. Genital pore at level of ovary......................**
................ ***Rhinebothroides*** **Mayes, Brooks et Thorson 1981. (Figure 185)**

*Diagnosis:* Scolex with four pedicellated, quadrate bothridia, each with shallow, horizontal loculi. Myzorhynchus absent. Ovary bilobed, with greatly reduced poral lobe in frontal view;

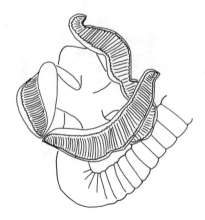

FIGURE 186. *Rhabdotobothrium anterophallum* Campbell 1975.

X-shaped in cross section. Vitelline follicles lateral. Vagina anterior to cirrus pouch, with distal sphincter. Uterus anterior to cirrus pouch, with lateral diverticulae. Testes preovarian. Cirrus pouch at level of ovary, containing internal seminal vesicle. Cirrus spined. Parasites of freshwater stingrays (Pomatotrygonidae). South America. Key to species: Mayes, Brooks et Thorson.[2165]

Type species: *R. moralarai* (Brooks et Thorson 1976) Mayes, Brooks et Thorson 1981 (syn. *Rhinebothrium moralarai* Brooks et Thorson 1976), in *Pomatotrygon magdalenae;* Colombia.

Other species:
*R. circularis* Mayes, Brooks et Thorson 1981, in *Pomatotrygon circularis;* Colombia.
*R. freitasi* (Rego 1979) Brooks, Mayes et Thorson 1981 (syn. *Rhinebothrium freitasi* Rego 1979), in *Pomatotrygon hystrix;* Brazil.
*R. glandularis* Brooks, Mayes et Thorson 1981, in *Pomatotrygon hystrix;* Brazil.
*R. scorzai* (Lopez-Neyra et Diaz-Ungria 1958) Mayes, Brooks et Thorson 1981 (syn. *Rhinebothrium scorzai* Lopez-Neyra et Diaz-Ungria 1958), in *Pomatotrygon hystrix, P. motoro, P. reticulatus, Elipesurus spinicauda;* Venezuela, Brazil.
*R. venezuelensis* Brooks, Mayes et Thorson 1981, in *Pomatotrygon hystrix, P. yepezi;* Venezuela.

**15b. Genital pore anterior to level of ovary ........................................ 16**

**16a. Cephalic peduncle absent ......... *Rhabdotobothrium* Euzet 1953. (Figure 186)**

*Diagnosis:* Scolex with four pedunculated bothridia, each with adherent surfaces subdivided into loculi. Myzorhynchus absent. Scolex nonpedunculated. Short neck present or absent. Strobila craspedote or acraspedote, apolytic. Genital pores lateral, irregularly alternating. Genital atrium present or absent. Cirrus pouch large, cirrus armed. Testes numerous, anterior to ovary, always present between cirrus pouch and ovary. Vagina opens anterior to cirrus pouch. Ovary bilobed in anteroventral view, X-shaped in cross section. Uterus simple, medial, tubular. Vitellaria in lateral bands. Parasites of rays. Mediterranean, Atlantic.

Type species: *R. dollfusi* Euzet 1953 (syn. *Echeneibothrium dollfusi* Euzet 1953, Yamaguti 1959), in *Dasyatis pastinaca;* Mediterranean.
Other species:
*R. anterophallum* Campbell 1975, in *Mobula hypostoma;* New England.

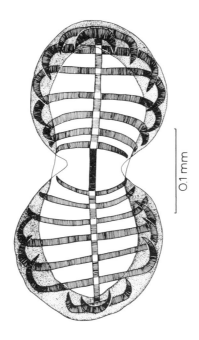

FIGURE 187. *Rhinebothrium* sp. *biorchidum* Huber et Schmidt 1985. Bothridium.

**16b. Cephalic peduncle present .......... *Rhinebothrium* Linton 1890. (Figure 187)**

*Diagnosis:* Scolex with four pedunculated or nonpedunculated bothridia, their adherent surfaces divided by septa into loculi. Bothridial margins lobulated or entire in form, hinged into two parts or not. Myzorhynchus absent. Scolex spined or not. Peduncle and neck present, spined or not. Strobila craspedote or acraspedote, apolytic. Genital pores marginal, irregularly alternating; genital atrium present or absent. Cirrus pouch well developed. Cirrus armed. Testes preovarian, few to numerous, in median field anterior to cirrus pouch; none posterior to cirrus pouch on poral side. Ovary with two, four, six, or eight lobes in dorsoventral view, X-shaped in cross section. Vagina anterior to cirrus pouch. Seminal receptacle present or absent. Vitellaria in lateral bands. Uterus a simple, median tube. Parasites of skates and rays. Cosmopolitan. Table of characteristics of 30 species: Cornford.[621]

Type species: *R. flexile* Linton 1890 (syn. *Echeneibothrium flexile* [Linton 1890] Yamaguti 1959, *E. walga* Shipley et Hornell 1906), in *Trygon centrura, T. walga, T. uarnak, T. kuhli, Dasybatus akajei, Urobatis halleri, Holorhinus californicus;* New England, Sri Lanka, Japan, California.

Other species:

*R. baeri* Euzet 1959, in *Dasyatis violacea;* France.

*R. biorchidum* Huber et Schmidt 1985, in *Urolophus jamacensis;* Jamaica.

*R. chilensis* Euzet et Carvajal 1973, in *Psammobatis lima;* Chile.

*R. corymbum* Campbell 1975, in *Dasyatis americana;* Western North Atlantic.

*R. hawaiiensis* Cornford 1974, in *Dasyatis lata;* Hawaii.

*R. himanturi* Williams 1964, in *Himantura granulata;* Australia.

*R. leiblei* Euzet et Carvajal 1973, in *Psammobatis lima;* Chile.

*R. maccallumi* (Linton 1924) Campbell 1970 (syn. *Echeneibothrium maccallumi* (Linton 1924) Yamaguti 1959), in *Dasyatis centrura;* U.S.

*R. magniphallum* Brooks 1977, in *Himantura schmardae;* Colombia.

*R. paratrygoni* Rego et Dias 1976, in *Elipesurus* sp.; Brazil.

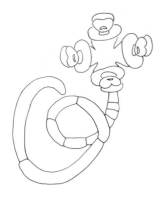
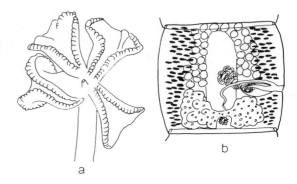

FIGURE 188. *Carpobothrium chiloscyllii* Shipley et Hornell 1906.

FIGURE 189. *Clydonobothrium elegantissimum* (Lönnberg 1889) Euzet 1959. (a) Scolex; (b) proglottid.

*R. rhinobati* Dailey et Carvajal 1976, in *Rhinobatos planiceps;* Chile.
*R. scobinae* Euzet et Carvajal 1973, in *Psammobatis lima;* Chile.
*R. shipleyi* Southwell 1911 (syn. *Echeneibothrium shipleyi* [Southwell 1911]), in *Trygon kuhli, Dasybatis akajei;* Sri Lanka, Japan.

**17a. Each bothridium with two opposed terminal flaps..................................**
**........................... *Carpobothrium* Shipley et Hornell 1906. (Figure 188)**

*Diagnosis:* Scolex with four pedunculate bothridia, each with two opposed flaps bearing minute loculi on their margins. Myzorhynchus absent. Neck short. Mature proglottids much longer than broad. Genital pores irregularly alternating, near middle of lateral margin. Cirrus pouch reaching median line. Cirrus armed. Testes numerous, in whole intervitelline field anterior to ovary. Vagina anterior to cirrus pouch. Ovary bilobed, variable, posterior. Vitellaria in narrow lateral bands. Uterus reaching almost anterior margin of proglottid. Parasite of elasmobranchs. Sri Lanka.

Type species: *C. chiloscyllii* Shipley et Hornell 1906, in *Chiloscyllium indicum, Rhynchobatis djeddensis, Urogymnus asperrimus;* Sri Lanka.

**17b. Bothridia without terminal flaps................................................. 18**

**18a. No accessory suckers ........................................................ 19**
**18b. Accessory suckers present..................................................... 21**

**19a. Myzorhynchus with a subapical, musculo-glandular organ ......................**
**......................................*Clydonobothrium* Euzet 1959. (Figure 189)**

*Diagnosis:* Bothridia thin, pedunculate, highly crumpled, with borders somewhat thickened. Myzorhynchus slender, retractile, with subapical musculo-glandular organ. Cephalic peduncle absent. Proglottids slightly or strongly craspedote. Strobila almost oval in cross section. Apolytic. Genital atrium well developed. Cirrus stout, heavily armed with minute, stout spines. Cirrus pouch well developed. Testes usually arranged two or three abreast. Genital pore irregularly alternating in posterior third of proglottid margin. Vagina thick-walled, anterior to cirrus pouch. Ovary bilobed, near posterior end of proglottid. Vitellaria in two dense lateral fields extending entire length of proglottid. Parasite of Rajidae. Probably cosmopolitan.

Type species: *C. elegantissimum* (Lönnberg 1889) Euzet 1959 (syn. *Anthobothrium*

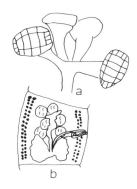

FIGURE 190. *Anthobothrium sexorchidum* Williams 1964. (a) Scolex; (b) proglottid.

*elegantissimum* Lönnberg 1889, *Scyphophyllidium elegantissimum* [Lönnberg 1889] Woodland 1927), in *Galeus* sp., *Raja* sp.; Europe.

Other species:

    *C. leioformum* Alexander 1963, in *Raja nasuta;* New Zealand.

**19b. Myzorhynchus not as above, or absent altogether ............................ 20**

**20a. Ovary two-lobed in cross section. Myzorhynchus present or absent .............**
**.................................................*Anthobothrium* Beneden 1850**
**(syn. *Polipobothrium* Mola 1908, *Spongiobothrium* Linton 1889). (Figure 190)**

*Diagnosis;* Bothridia usually simple, sometimes with loculi. Accessory suckers absent. Myzorhynchus present or absent. Neck present. Genital pores irregularly alternating. Cirrus armed. Testes numerous in single field. Vagina anterior to cirrus pouch. Seminal receptacle usually present. Ovary bilobed, variable. Vitellaria lateral, V-shaped in cross section. Uterus variable between species, reaching level of cirrus pouch or near anterior end of proglottid. Parasite of elasmobranchs. Cosmopolitan.

Type species: *A. cornucopia* Beneden 1850 (syn. *Taenia lilliiformis* MacCallum 1917, *T. rosaeformis* MacCallum 1921), in *Carcharhinus leucus, Carcharias, Galeus canis, Mustelus vulgaris, Galeorhinus, Raja scabrata, Trygon, Lamna, Zygaena;* Europe, Black Sea, Canada, Guatemala.

Other species:

    *A. auriculatum* (Rudolphi 1819) Diesing 1863, in *Squalus, Raja, Galeus, Hexanchus, Lamna, Carcharias, Mustelus, Scyllium, Squatina angelus, Torpedo,* etc.; Atlantic, Mediterranean.

    *A. bifidum* Yamaguti 1952, in *Dasybatus akajei, D. zugei, Pteroplatea micrura;* East China Sea.

    *A. dipsadomorphi* (Shipley 1900), in *Dipsadomorphus irregularis;* India.

    *A. exiguum* Yamaguti 1935, in *Alopias vulpinus, Mustelus manazo;* Japan.

    *A. hickmani* Crowcroft 1947, in *Narcine tasmaniensis;* Tasmania.

    *A. karuatayi* Woodland 1934, in *Glanidium* sp.; Amazon.

    *A. laciniatum* Linton 1890, in *Carcharhinus obscurus, C. leucas, C. longimanus, Isurus, Galeus, Sphyrna, Scoliodon, Raja, Hypoprion, Urobatis halleri;* Brazil, California, Atlantic, Pacific.

    *A. lintoni* (Southwell 1911) (syn. *Spongiobothrium lintoni* Southwell 1911), in *Rhynchobatis djeddensis, Urogymnus asperrimus;* Sri Lanka.

*A. mandube* Woodland 1935, in *Pseudageniosus brevifilis;* Amazon.
*A. minutum* Guiart 1935, in *Galeus glaucus;* Cape Verde Islands.
*A. oligorchidum* Young 1954, in *Urobatis halleri;* California.
*A. panjadi* Shipley et Hornell 1909, in *Myliobatis maculata, Aetobatis narinari;* Sri Lanka.
*A. parvum* Stossich 1895, in *Zygaena malleus, Mustelus manazo;* Europe, Sea of Japan.
*A. piramutab* Woodland 1934, in *Brachyplatystoma vaillanti;* Amazon.
*A. pristis* Woodland 1934, in *Pristis perotteti;* Amazon, Nicaragua, Costa Rica.
*A. pteroplateae* Yamaguti 1952, in *Pteroplatea japonica;* Japan.
*A. quadribothria* (MacCallum 1921) Yamaguti 1959 (syn. *Taenia quadribothria* MacCallum 1921), in *Dasybatus pastinacus;* New England.
*A. rajae* Yamaguti 1952, in *Raja kenojei;* Japan.
*A. sexorchidum* Williams 1964, in *Taeniura lymna;* Australia.
*A. taeniuri* Saoud 1963, in *Taeniura lymna;* Red Sea.
*A. variabile* (Linton 1889) (syn. *Echeneibothrium simplex* Shipley et Hornell 1906, *Polipobothrium vaccarii* Mola 1908), in *Trygon kuhli, T. centrura, Selache maxima;* New England, Sri Lanka, Adriatic Sea.
*A. veravalensis* Shinde, Jadhav et Mohekar 1980, in *Rhynchobatus djeddensis;* India.

**20b. Ovary four-lobed in cross section. Myzorhynchus absent .......................
..................................... *Rhodobothrium* Linton 1889 (syn. *Inermiphyllidium* Riser 1955, *Sphaerobothrium* Euzet 1959, *Proboscidosaccus* Gallien 1949).**

*Diagnosis:* Scolex with four large, subspherical bothridia supported by pedicels. Bothridia trumpet-shaped when relaxed, but adherent surfaces convex and traversed by numerous convolutions forming an irregular pattern when contracted. Bothridial faces round or subtriangular in cross section; margins ruffled. Neither accessory suckers nor rostellum present. Neck present. Segments numerous, apolytic. Genital pores preequatorial, irregularly alternate. Genital atrium well developed. Cirrus pouch elongate, delicate. Cirrus armed. Testes numerous, some may be postovarian. Gravid segments triangular. Ovary tetralobed in cross section. Vitellaria lateral, arching mediad around lateral testes. Uterus median, saccate, may form lateral diverticula. Median uterine pore present. Parasites of elasmobranchs. Atlantic, Pacific.

Type species: *R. pulvinatum* Linton 1889, in *Dasyatis centrura, D. americana;* East Coast of North America.
Other species:
*R. brachyascum* (Riser 1955) Campbell et Carvajal 1979 (syn. *Inermiphyllidium brachyascum* Riser 1955), in *Aetobatus californicus;* U.S.
*R. dollfusi* Euzet 1953 (syn. *Echeneibothrium dollfusi* [Euzet 1953] Yamaguti 1959), in *Dasyatis pastinacea;* France.
*R. enigmaticum* (Gallien 1949 comb. n. (syn. *Proboscidosaccus enigmaticus* Gallien 1949; *Sphaerobothrium lubeti* Euzet 1959, *Rhodobothrium lubeit* [Euzet 1959] Campbell et Carvajal 1979), in *Myliobatis aquila, Mactra solida* (Mollusca); France.
*R. mesodermatum* (Bahamonde et Lopez 1962) Campbell et Carvajal 1979 (syn. *Proboscidosaccus mesodermatis* Bahamonde et Lopez 1962, *Anthobothrium peruanum* Rego, Vicente et Herrera 1968, in *Myliobatis chilensis,* ''clams''; Chile.

**21a. Bothridium with shelf-like anterior thickening................................. 22
21b. Bothridium without shelf-like anterior thickening ............................ 24**

**22a. Vitellaria in ventral medulla from ventral osmoregulatory canal to submedian field ................................. *Gastrolecithus* Yamaguti 1952. (Figure 191)**

FIGURE 191. *Gastrolecithus planus* (Linton 1922) Yamaguti 1952.

FIGURE 192. *Dinobothrium planum* Linton 1922.

*Diagnosis:* Bothridia flat or concave, each with anterior accessory sucker and anterolateral appendage that is split at the tip. Myzorhynchus absent. Neck long. All proglottids broader than long. Genital pores irregularly alternating. Cirrus pouch exceeding osmoregulatory canals. Testes numerous, in two lateral fields anterior and lateral to ovary. Vagina ventral to cirrus pouch, armed in its expanded distal portion. Ovary bilobed, posterior. Vitellaria extending in ventral medulla from ventral osmoregulatory canal to submedian field. Uterus transverse, anterior to ovary. Parasite of basking shark. North America, Europe, Japan.

Type species: *G. planus* (Linton 1922) Yamaguti 1952, in *Cetorhinus maximus;* Atlantic, Pacific.

**22b. Vitellaria in two lateral fields ................................................... 23**

**23a. Ovary four-lobed in cross section... *Dinobothrium* Beneden 1889. (Figure 192)**

*Diagnosis:* Scolex nearly cuboidal, very large (up to 10 mm long), unarmed, lacking myzorhynchus. Bothridia sessile, not folded on margins, each fused at anterior margin to a muscular transverse lobe that is bluntly bifid. Accessory or pseudosuckers may be present on these lobes. Neck absent. Genital pores irregularly alternating. Cirrus armed. Testes numerous, between ovary and anterior end of proglottid in intervitelline field. Vagina anterior to cirrus pouch. Ovary four-lobed, posterior. Vitellaria lateral. Uterus median, with lateral branches. Parasites of elasmobranchs. Europe, Asia, North America.

Type species: *D. septaria* Beneden 1889 (syn. *Diplobothrium simile* Beneden 1889, *D. plicitum* Linton 1922), in *Lamna cornubica, Carcharodon carcharias, Cetorhinus maximus, Todaropsis eblanae, Ommatostrephes illecebrosus;* Belgium, Spain, France, Sweden, China, New England.

**23b. Ovary fan-shaped in cross section ............ *Reesium* Euzet 1955. (Figure 193)**

*Diagnosis:* Scolex with four sessile, spinose bothridia, each with margins entire and fused anteriorly with a horn-like projection bearing an accessory sucker and bifid at the tip. Myzorhynchus absent. Neck spinose. Genital pores regularly alternating. Cirrus pouch exceeds median line. Testes large, 28 to 45 in number, filling intervitelline field anterior to ovary. Vagina posterior to cirrus pouch. Ovary fan-shaped in cross section. Uterus in form of two sacs, anterior and posterior. Vitellaria lateral. Parasites of elasmobranchs. Europe, Japan, Hebrides.

Type species: *R. paciferum* (Sproston 1948) Euzet 1955 (syn. *D. spinosum* Baylis 1950, *D. spinulosum* Yamaguti 1952, *D. humile* Euzet 1952), in *Cetorhinus maximus;* Atlantic, Pacific.

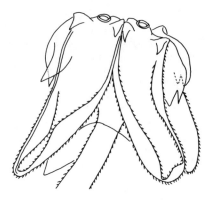

FIGURE 193. *Reesium paciferum* (Sproston 1948) Euzet 1955.

FIGURE 194. *Phyllobothrium kingae* Schmidt 1978. Scolex From Schmidt, G. D., *Proc. Helminthol. Soc. Wash.* 45, 132—134, 1978. With permission.)

24a.  Margins of bothridia greatly crumpled and/or loculated. One or occasionally two accessory suckers present on each bothridium .................... *Phyllobothrium* Beneden 1849 (syn. *Anthocephalum* Linton 1890, *Bilocularia* Obersteiner 1914, *Calyptrobothrium* Monticelli 1893, *Crossobothrium* Linton 1889). (Figure 194)

*Diagnosis:* Scolex without myzorhynchus. Bothridia sessile or pedunculated, with margins usually folded, curled, often with loculi. One or two accessory suckers on each bothridium, often difficult to see in preserved specimens. Neck present or absent. Cirrus pouch thin-walled; cirrus often armed. Genital pores irregularly alternating, rarely unilateral. Testes numerous, in entire intervitelline field anterior to ovary. Vagina usually anterior to cirrus pouch. Seminal receptacle usually present. Ovary posterior, four-lobed in cross section. Vitellaria lateral, V-shaped in cross section. Uterus median, reaching level of cirrus pouch. Elasmobranchs. Pacific, Atlantic, Caribbean and Indian Oceans. Monograph: Williams.[3931]

Type species: *P. lactuca* Beneden 1850 (syn. *P. compactum* Southwell et Prashad 1920, *P. crispatissimum* Monticelli 1899, *P. inchoatum* Leidy 1891), in *Mustelus laevis*, *M. manazo*, *M. vulgaris*, *Galeus canis*, *Galeocerdo arcticus*, *Squatina angelus*, *Lamna spallanzanii*, *Acanthias vulgaris*, *Spinax acanthias*, *Trygon pastinacea*, *Dasybatus kuhli*, *D. walga*, *Raja batis*, *R. clavata*, *Negaprion brevirostris*, etc.; Atlantic, Indian and Pacific Oceans.

Other species:

*P. biacetabulum* Yamaguti 1960, in *Rhinobatus schlegeli;* Japan.

*P. brittanicum* Williams 1968, in *Raja montagui;* England.

*P. brassica* Beneden 1871, in *Spinax acanthias;* Belgium.

*P. centrurum* (Southwell 1925) (syn. *Anthocephalum gracile* Linton 1890, *Anthobothrium gracile* Linton 1890), in *Trygon centrura*, *Hexanchus griseus*, *Dasyatis sabina;* New England, Caribbean.

*P. dagnallium* Southwell 1927, in *Rhynchobatus ancylostomus*, *Chiloscyllium indicum*, *Galeocerdo tigrinus*, *Raja* sp.; Sri Lanka, Canada.

*P. dasybati* Yamaguti 1934, in *Dasybatus akajei*, *Hypoprion brevirostris;* Japan, Florida.

*P. delphini* (Bosc 1802), larvae in cetacean skin. Cosmopolitan (see Testa and Dailey).[3661]

*P. dentatum* (Linstow 1907), in "shark's rectum"; Antarctica.

*P. dohrnii* (Oerley 1885) Zschokke 1889 (syn. *Crossobothrium campanulatum* Klaptocz 1906), in *Heptanchus cinereus*, *Mustelus vulgaris*, *Scymnus lichia*, *Odontaspis littoralis*, *Hexanchus griseus*, *Centrophorus granulosus*, *Cynias manazo;* Japan, Mediterranean, Atlantic.

*P. filiforme* Yamaguti 1852, in *Alopias vulpinus*, *Carcharhinus longimanus;* Japan, Brazil.

*P. floriforme* (Southwell 1912), in *Carcharias* spp.; Sri Lanka.

*P. foliatum* Linton 1890, in *Trygon centrura*, *Carcharhinus obscurus*, *Dasyatus sabrina;* New England, Dry Tortugas.

*P. gracile* Wedl 1855 (syn. *Anthobothrium auriculatum* of Diesing 1863), in *Torpedo marmorata*, *Squatina angelus*, *Heptanchus cinereus*, *H. griseus*, *Rhynchobatus columnae;* Europe, Mediterranean.

*P. ketae* Canavan 1928, in *Oncorhynchus keta;* Alaska.

*P. kingae* Schmidt 1978, in *Urolophus jamaicensis;* Jamaica.

*P. laciniatum* (Linton 1889), in *Odontaspis littoralis*, *Carcharias taurus*, *Squalus sucklii*, *Chimaera monstruosa;* New England, Mediterranean, East China Sea.

*P. leuci* Watson et Thorson 1976, in *Carcharhinus leucas;* Costa Rica, Nicaragua.

*P. loculatum* Yamaguti 1952, in *Heterodontus zebra;* East China Sea.

*P. loliginis* (Leidy 1887) Linton 1897, in *Ommastrephes illecebrosus; O. sagittatus*, etc.; Mediterranean, New England.

*P. magnum* Hart 1936 (syn. *Monorygma macquariae* Johnston 1937), in *Somniosus microcephalus;* Puget Sound, U.S.

*P. marginatum* Yamaguti 1934, in *Squatina japonica;* Japan.

*P. microsomum* Southwell et Helmy 1927, in *Ginglymostoma concolor;* India.

*P. minutum* Shipley et Hornell 1906, in *Carcharias melanopterus;* Sri Lanka.

*P. musteli* (Beneden 1850) (syn. *Anthobothrium musteli* Beneden 1850, *Orygmatobothrium angustum* Linton 1889), in *Mustelus vulgaris*, *Carcharhinus* spp., *Galeocerdo arcticus*, *Triakis scyllium;* Japan, Belgium, Dry Tortugas, New England.

*P. nicaraguensis* Watson et Thorson 1976, in *Carcharhinus leucas;* Costa Rica, Nicaragua.

*P. pammicrum* Shipley et Hornell 1906, in *Carcharias melanopterus*, *Hyplophus sephen*, *Urogymnus asperrimus;* India.

*P. pastinacea* Moktar-Maamouri et Zamali 1981, in *Dasyatis pastinacea;* Tunisia.

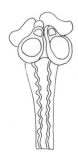

FIGURE 195. *Monorygma perfectum* (Beneden 1853) Diesing 1863.

*P. piriei* Williams 1968, in *Raja naevus;* Scotland.
*P. prionacis* Yamaguti 1934, in *Prionace glauca;* Japan.
*P. pristis* Watson et Thorson 1976, in *Pristis perotteti;* Costa Rica, Nicaragua.
*P. radioductum* Kay 1942, in *Raja binoculata, R. rhina, R. montereyensis, Triakis semifasciata;* West Coast, U.S.
*P. riggii* (Monticelli 1893) (syn. *Calyptrobothrium riggii* Monticelli 1893, *C. minus* Linton 1890, *C. occidentale* Linton 1890, *Bilocularia hyperapolytica* Obersteiner 1914 in part), in *Torpedo occidentalis, Centrophorus granulosus, Scylliorhinus canicula, Raja radiata;* Mediterranean, New England.
*P. rotundum* (Klaptocz 1906), in *Hexanchus griseus;* Trieste.
*P. salmonis* Fujita 1922, larva in *Oncorhynchus* sp.; Japan, British Columbia, Washington, U.S.
*P. serratum* Yamaguti 1952, in *Triacis scyllium;* Japan.
*P. squali* Yamaguti 1952, in *Squalus sucklii;* Japan.
*P. thridax* Beneden 1850 (syn. *P. auricula* Beneden 1858, *P. vagans* Haswell 1902), in *Squatina angelus, Trygon pastinacea, Scymnus glacialis, Laemargus borealis, L. rostratus, Cestracion, Raja, Mustelus;* Belgium, Mediterranean, Atlantic.
*P. triacis* Yamaguti 1952, in *Triacis scyllium, Mustelus* spp, *Hexanchus griseus;* Atlantic, Japan.
*P. tumidum* Linton 1922, in *Carcharodon carcharias, Isurus dekayi, Hemigaleus balfouri, Cynias manazo, Triacis scyllium, Scoliodon terraenovae* Florida, China, Sri Lanka, France, New England.
*P. unilaterale* Southwell 1925, in *Squatina angelus, S. squatina;* France, Italy, England.
*P. williamsi* nom. nov. (for *P. minutum* Williams 1968, preoccupied), in *Raja fullonica;* Scotland.

**24b. Margins of bothridia smooth or slightly sinuous ............................. 25**

**25a. Myzorhynchus present. Single accessory sucker per bothridium ................**
**........................................ Monorygma Diesing 1863. (Figure 195)**

*Diagnosis:* Myzorhynchus present. Bothridia sessile, each hollowed out posteriorly, with thin, smooth margins, and a single anterior accessory sucker. Neck present. Genital pores irregularly alternating. Cirrus pouch thin-walled, cirrus spined. Testes numerous, filling field between osmoregulatory canals anterior to ovary. Vagina opening into genital atrium anterior or ventral to cirrus. Inconspicuous seminal receptacle present. Ovary four-lobed in cross section. Vitellaria lateral. Uterus median, extending to level of cirrus pouch. Eggs spindle-shaped. Parasite of elasmobranchs. North Atlantic, North Pacific, Mediterranean.

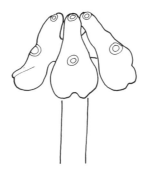

FIGURE 196. *Orygmatobothrium versatile* Diesing 1863.

Type species: *M. perfectum* (Beneden 1853) Diesing 1863 (syn. *M. elegans* Monticelli 1890), in *Laemargus borealis, Scilium catulus, S. stellare, Somniosus brevispinna, S. microcephalus;* Atlantic, Mediterranean, Greenland, Alaska.

Other species:

*M. chlamydoselachi* Lönnberg 1898, in *Chlamydoselachus anguineus;* Sweden.

*M. megacotyla* Yamaguti 1952, in *Cephaloscyllium umbratile;* Japan.

25b. Myzorhynchus absent. Two accessory suckers per bothridium .................. ................................. ***Orygmatobothrium*** **Diesing 1863. (Figure 196)**

*Diagnosis:* Myzorhynchus absent. Each bothridium with an accessory sucker at the anterior end and another near the center. Bothridial margins entire or crumpled. Neck present. Genital pores irregularly alternating. Cirrus pouch thin-walled, somewhat oblique. Cirrus armed or not. Testes numerous, in broad field anterior to ovary. Vagina opening into atrium in front of cirrus. Seminal receptacle present. Ovary four-lobed in cross section, posterior. Vitellaria in broad lateral bands, V-shaped in cross section. Uterus median, extending to level of cirrus pouch. Parasite of elasmobranchs. Europe, Africa, Japan, North America.

Type species: *O. versatile* Diesing 1863 (syn. *O. longicolle* Zschokke 1888, *O. wyatti* Leiper et Atkinson 1914, *O. musteli* [Beneden 1850] of Woodland 1927), in *Galeus canis, Mustelus, Scyllium canicula, Mustelus manazo;* Japan, Mediterranean.

Other species:

*O. crenulatum* Linton 1897, in *Dasyatis centrura;* New England.

*O. musteli* (Beneden 1850), in *Mustelus hinnulus, M. laevis, M. manazo, M. vulgaris, Galeus canis, Scyllium canicula;* Japan, European Atlantic.

*O. paulum* Linton 1897, in *Galeocerdo tigrinus,* New England.

*O. plicatum* Yamaguti 1934, in unidentified skate; Japan.

*O. velamentum* Yoshida 1917, in *Mustelus* spp.; Japan.

*O. zschokkei* Woodland 1927, in *Mustelus laevis, M. vulgaris;* Europe.

## TETRAPHYLLIDEA OF DOUBTFUL OR UNCERTAIN STATUS

1. Order Biporophyllidea Subramanian 1939, based on *Biporophyllaeus madrassensis* Subramanian 1939, is a detached segment of a tetraphyllidean, trypanorhynchan, or lecanicephalidean cestode.
2. Order Anteroporidea and order Lateroporidea were both proposed by Subhapradha 1957, in the same paper, to replace Biporophyllidea (see above).

3. Monoporophyllaeidae Subhapradha 1957 should be suppressed because there is no type genus upon which it is named. It was based on *Anteropora* Subhapradha 1957.
4. *Anteropora indica* Subhapradha 1957 is a detached segment of a tetraphyllidean, trypanorhychan, or lecanicephalidean cestode.
5. *Monoporophyllaeus* Shinde et Chincholikar 1977 was proposed to replace *Anteropora* Subhapradha 1957 (see above). Neither should be recognized as valid until the entire worm is known.
6. *Mastacembellophyllaeus* Shinde et Chincholikar 1977 is based on a detached segment, probably of a pseudophyllidean. Named species, all based on single, detached proglottids are: (type species) *M. nandedensis* Shinde et Chincholikar 1976, in *Mastacembellus armatus;* India. *M. paithanensis* Shinde et Jadhav 1978, in *Pseudeutropius taakree;* India. *M. taakreei* Jadhav et Shinde 1972, in *Pseudeutropius taakree;* India. All these species, along with the genus, should be suppressed until the complete worms are known.
7. *Pleurocercus tandani* Pandey 1973, in *Sciaena* sp., India, is a larval form that cannot be identified with any adult genus.
8. *Pleurocercus puriensis* Pandey 1973, in *Trichurus* sp., India, has the same status as *P. tandani*.
9. *Tritaphros retzii* Lönnberg 1889 probably is valid, but is known from a larval form only.
10. *Spinibiloculus* Deshmukh et Shinde 1980 is insufficiently described to be recognizable.
11. *Yogeshwaria* Chincholikar et Shinde 1976 is a synonym of *Yogeshwaria* Shinde 1966 (same author!). It is impossible to recognize as it is based on a single, very contorted specimen with scolex probably lost. No internal anatomy is described. Therefore *Y. nagabhusheni* Chincholikar et Shinde 1976, in *Trygon* sp. in India is unidentifiable.

## ORDER DIPHYLLIDEA

This order contains two families that probably are not closely related. Each is based on a single genus. Echinobothriidae, with its single genus *Echinobothrium* is comprised of several species from rays in many parts of the world. The genus is easily recognized by its peculiar armature of the scolex and peduncle.

Ditrachybothridiidae consists of a monotypic genus, *Ditrachybothridium* reported in rays. Its scolex and neck are unarmed.

Wardle et al.[3871] declared Diphyllidea to be a *nomen oblitum*, " . . . since it is not presently occupied . . . ", and proposed a new order Diphyllidea to contain Diphyllobothriidae. They came to this conclusion, apparently, because Lühe[2015] did not accept the original designation by Beneden in Carus (1863). They also ignored the fact that Diphyllidea was accepted by Yamaguti,[4022] Schmidt,[3095] and others, to contain *Echinobothrium* spp. Diphyllidea Wardle et al.[3871] then concluded it must be considered a junior synonym of Diphyllidea Beneden in Carus 1863, by preoccupation of the latter. Diphyllobothriidae clearly falls within order Pseudophyllidea, as discussed on p.93.

## KEY TO THE FAMILIES IN DIPHYLLIDEA

**1a. Scolex with dorsal and ventral groups of apical hooks. Cephalic peduncle with longitudinal rows of hooks .............. Echinobothriidae Perrier 1897. (p.165)**

*Diagnosis:* Scolex with well-developed rostellum armed with powerful spines. Peduncle of scolex armed with longitudinal rows of large straight spines. External metamerism distinct, proglottids acraspedote. Genital pores medioventral. Parasites of elasmobranchs.

Type genus: *Echinobothrium* Beneden 1849.

**1b. Scolex and cephalic peduncle lacking hooks......................................**
**..................................... Ditrachybothridiidae Schmidt 1970. (p.166)**

*Diagnosis:* Scolex with short, unarmed peduncle. Apex of scolex blunt, with weakly developed apical organ. External metamerism poorly marked. Vagina opens behind cirrus pouch. Testes in anterior two thirds of proglottid. Ovary bilobed, posterior. Vitellaria diffuse. Uterine pore absent. Parasites of elasmobranchs.

Type genus: *Ditrachybothridium* Rees 1959.

## DIAGNOSIS OF THE ONLY GENUS IN ECHINOBOTHRIIDAE

*Echinobothrium* Beneden 1849 (Figure 197)

*Diagnosis:* Scolex with powerful rostellum armed on dorsal and ventral surfaces with a row of large hooks. Hooks largest in middle of row, decreasing in size toward the ends. Two bothridia present, covered with very small, hair-like spines. Cephalic peduncle armed with longitudinal rows of straight hooks, each with root in shape of transverse bar, or irregular. Neck very short. Strobila small. Posterior proglottids longer than wide. Testes few, medullary, in two longitudinal, preovarian rows. Cirrus median, post-testicular. Cirrus armed. Genital pores median, equatorial or postequatorial, male anterior to female. Ovary bilobed, median, posterior. Vitellaria follicular, in two lateral bands. Uterus median. Uterine pore absent. Eggs may have short filament. Parasites of elasmobranchs. Cosmopolitan.

Type species: *E. typus* Beneden 1849, in *Raja* spp., *Trygon pastinaca;* Atlantic, Mediterranean, Black Sea.

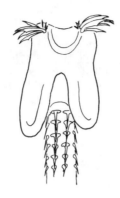

FIGURE 197. *Echinobothrium brachysoma* Pintner 1889.

Other species:
  *E. acanthinophyllum* Rees 1961, in *Raja montagui;* England.
  *E. affine* Diesing 1863 (syn. *E. typus* of Wedl 1855 nec Beneden), in *Raja* spp., *Carcharias, Rhynchobatus;* Triest, France.
  *E. beneden* Ruszkowski 1927, in *Raja* spp.; Europe.
  *E. boisii* Southwell 1911, in *Aetobatis narinari;* Sri Lanka.
  *E. bonasum* Williams et Campbell 1981, in *Rhinoptera bonasus:* Chesapeake Bay, U.S. (also list of species).
  *E. brachysoma* Pintner 1889, in *Raja batis;* Trieste.
  *E. coenoformum* Alexander 1963, in *Raja nasuta;* New Zealand.
  *E. coronatum* Robinson 1959, in *Mustelus lenticulatus;* New Zealand.
  *E. euzeti* Campbell et Carvajal 1980, in *Psammobatis lima;* Chile.
  *E. harfordi* McVicar 1976, in *Raja naevus;* England.
  *E. helmymohamedi* Saoud, Ramaden et Hassan 1982, in *Taeniura lymma;* Red Sea.
  *E. heroniensis* Williams 1964, in *Himantura granulata;* Australia.
  *E. laevicolle* Lespeès 1857, in liver of *Nassa reticulata;* locality?
  *E. longicolle* Southwell 1925, in *Trygon kuhli;* Sri Lanka.
  *E. mathiasi* Euzet 1951, in *Leiobatus aquila;* France.
  *E. musteli* Pintner 1889, in *Mustelus laevis, M. plebejus;* Trieste, France.
  *E. pigmentatum* Nuñez 1971, in *Zapteryx brevirostris;* Argentina.
  *E. raji* Heller 1949, in *Raja scabrata;* Canada.
  *E. reesae* Ramavedi 1969, in *Trygon walga, T. uarnak;* India.
  *E. rhinoptera* Shipley et Hornell 1906, in *Rhinoptera javanica;* Sri Lanka.

## DIAGNOSIS OF THE ONLY GENUS IN DITRACHYBOTHRIDIIDAE

*Ditrachybothridium* Rees 1959 (Figure 198)

*Diagnosis:* Scolex consisting of head and short cephalic peduncle. Two oval, flattened or spoon-shaped bothridia, armed along borders and outer convex face with rows of small spines. Neck unarmed, terminated with a velum. Apex of scolex blunt, with feebly developed apical organ. Proglottids acraspedote. Strobila cylindrical, external metamerism barely marked. Last segments four to five times longer than broad. Genital pores in posterior quarter of proglottid, female behind male. No uterine pore. Cirrus pouch large, half width of segment. Testes numerous (52 to 62) in row one to two deep in center of anterior two thirds of proglottid. Ovary posterior, bilobed. Vitellaria inconspicuous, diffuse, anterior to ovary. Elasmobranchs. Scotland.

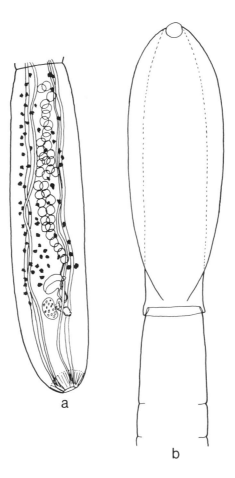

FIGURE 198. *Ditrachybothridium macrocephalum* Rees 1959. (a) Proglottid; (b) scolex.

Type species: *D. macrocephalum* Rees 1959, in *Raja fullonica, R. circularis, Scyliorhinus canaliculus*.

## ORDER LITOBOTHRIDEA

This order contains only the family Litobothridae, with two genera. They are parasites of big-eyed thresher sharks from California and the Soviet Union. No life cycles are known. The scolex is a single, well-developed apical sucker. The anterior proglottids are modified; they are cruciform in cross section. There is no neck. The testes are numerous and preovarian. The cirrus is massive and spined. The genital pores are lateral. The vitelline follicles encircle the proglottid, some extending behind the ovary. The order was established by Dailey.[672]

## DIAGNOSIS OF THE ONLY FAMILY IN LITOBOTHRIDEA

Litobothridae Dailey 1969 (p.169)

*Diagnosis:* Strobila small, metamerism distinct. Scolex a single sucker followed by modified anterior segments. Genital pores lateral, irregularly alternating. Testes numerous, medullary, preovarian. Cirrus pouch well developed. Ovary posterior. Vitellaria follicular, encircling entire proglottid. Osmoregulatory canals medullary. Parasites of Squali.

Type genus: *Litobothrium* Dailey 1969.

## KEY TO THE GENERA IN LITOBOTHRIDAE

**1a. Pseudoscolex with four segments; fourth segment with lateral margins flared outward** .................................. *Litobothrium* **Dailey 1969 (Figure 199)**

*Diagnosis:* Scolex with single apical sucker followed by modified anterior segments that are cruciform in cross section; posterior margins with row of simple spines. Neck absent. Strobila swollen in width behind scolex, decreasing in width before mature segments. Testes numerous, medullary. Cirrus pouch reaches midline. Vas deferens coiled. Ovary median, posterior. Vitellaria encircling proglottid, extending posterior to ovary. Parasites of sharks.

Type species: *L. alopias* Dailey 1969, in *Alopias superciliosus;* U.S.

Other species:

*L. coniformis* Dailey 1969, in *Alopias superciliosus;* U.S.

*L. daileyi* Kurochkin et Slankis 1973, in *Alopias superciliosus;* Russia.

*L. gracile* Dailey 1971, in *Odontaspis ferox;* California, U.S.

**1b. Pseudoscolex with five segments; fourth segment with lateral margins curled inward** ........................ *Renyxa* **Kurochkin et Slankis 1973. (Figure 200)**

*Diagnosis:* Scolex with single apical sucker followed by three modified anterior segments that are cruciform in cross section. Fourth section greatly expanded with velum forming two dorsal and two ventral flaps. First three segments with a row of spines on posterior margins. Fifth segment much constricted but also with four posterior flaps, followed by two still smaller segments. Reproductive proglottids bud off at this point. Segments numerous. Genital pores marginal, preequatorial. Cirrous pouch large, exceeding midline in mature segments. Testes numerous, preovarian. Ovary near posterior margin of proglottid, bilobed. Vitellaria lateral. Parasites of sharks.

Type species: *R. amplifica* Kurochkin et Slankis 1973, in *Alopias superciliosus;* Russia.

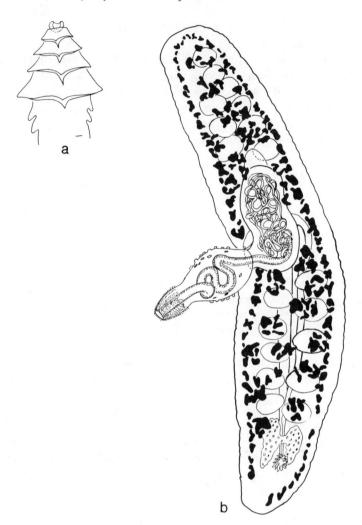

FIGURE 199. *Litobothrium alopias* Dailey 1969. (a) Scolex and (b) proglottid.

FIGURE 200. *Renyxa amplifica* Kurochkin et Slankis 1973.

## ORDER NIPPOTAENIIDEA

These small cestodes are parasites of freshwater fishes in Japan, Russia, and New Zealand. The scolex is rounded, with a single apical sucker, but otherwise simple. The neck, when present, is very short and poorly marked, and contains the genital primordia. The proglottids are nearly cylindrical and contain a set of male and female reproductive organs. The genital pores are lateral, preequatorial, and irregularly alternating. The cirrus is unarmed. The vitellarium is bilobed and directly in front of the ovary. An excellent review of the order was published by Hine.[1248]

## DIAGNOSIS OF THE ONLY FAMILY IN NIPPOTAENIIDEA

Nippotaeniidae Yamaguti 1939 (p.171)

*Diagnosis:* Scolex rounded, with single, large apical sucker and no other attachment devices. Neck very short, poorly marked. Proglottids acraspedote, apolytic. Entire cuticle covered with very minute spines. Single set of reproductive organs per proglottid. Vitellaria bilobed, between ovary and testes. Osmoregulatory canals numerous, anastomosing. Parasites of freshwater teleosts. Japan, Russia, New Zealand. (Family redefined: Hine.[1248])

Type genus: *Nippotaenia* Yamaguti 1939.

## KEY TO THE GENERA IN NIPPOTAENIIDAE

**1a. Proglottids hyperapolytic, maturing separate from strobila. Testes retained in gravid segments .................... *Amurotaenia* Achmerov 1941. (Figure 201)**

*Diagnosis:* Apical sucker very strongly developed, with sphincter at its opening. Anterior proglottids all considerably wider than long, but length increases posteriorly; sides parallel or convex. Proglottids shed when immature; growth and maturity continue after detachment. Testes previtellarian, partially or completely retained in mature and gravid proglottids. Cirrus pouch thin-walled, containing coiled ejaculatory duct. Genital atrium marginal, preequatorial, irregularly alternating. Ovary symmetrical, bilobed, elongated longitudinally, in midregion of segment. Vitellaria preovarian, consisting of two bilateral, spherical to reniform lobes joined transversely. Uterus laterally coiled, or with lateral branches; develops anteriorly in mature proglottids, but extends only partially into testicular region. Eggs with three membranes. Parasites of freshwater teleost fishes. Russia, Japan, New Zealand. (Genus redefined: Hine.[1248])

Type species: *A. percotti* Achmerov 1941, in *Percottus glehni;* Amur River, Russia.
Other species:
 *A. mogurdnae* (Yamaguti et Miyata 1940) Hine 1977, in *Mogurdna obscura;* Japan.
 *A. decidua* Hine 1977, in *Gobiomorphus cotidianus;* New Zealand.

**1b. Proglottids apolytic, not detaching until gravid. Testes not present in gravid segments ............................. *Nippotaenia* Yamaguti 1939. (Figure 202)**

*Diagnosis:* Neck showing genital primordia. Strobila round or oval in cross section. Posterior proglottids longer than wide. Testes in preovarian medulla. Cirrus pouch thin-walled. Cirrus unarmed. Genital atrium marginal, preequatorial, irregularly alternating. Ovary bilobed, in middle third of proglottid. Vitellaria bilobed, directly in front of ovary. Vagina posterior to cirrus pouch. Uterus with close transverse coils, extending nearly entire length of proglottid or confined to posterior half. Testes present in gravid proglottid. Proglottids apolytic. Parasites of freshwater teleosts. Japan, New Zealand.

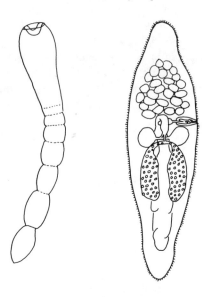

FIGURE 201.  *Amurotaenia decidua* Hine 1977. Complete worm and mature proglottid.

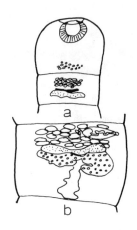

FIGURE 202.  *Nippotaenia chaetogobii* Yamaguti 1939. (a) Scolex; (b) mature proglottid.

Type species: *N. chaenogobii* Yamaguti 1939, in *Chaenobobius annularis, Gobius similis, Mogurnda obscura;* Lake Suwa, Japan.

Other species:

*N. contorta* Hine 1977, in *Retropinna retropinna, Galaxius maculatus;* New Zealand.
*N. fragilis* Hine 1977, in *Retropinna retropinna;* New Zealand.

## ORDER PROTEOCEPHALIDEA

Members of this order are easily recognized as such by the combination of follicular, usually lateral, vitellaria and a scolex with simple suckers. Most are parasites of freshwater fishes, but species are known from sharks, amphibians, and reptiles as well. Taxonomy of the group is not particularly difficult, except that families and subfamilies are separated by the relationships of internal organs to the longitudinal muscle bundles. For example, the vitellaria of Proteocephalidae are medullary, while they are cortical in Monticellidae. A trained microscopist often can discern these relationships in whole mounts, but cross sections may be necessary in some cases. A very important monograph has been written by Freze,[1957] and an important paper by Brooks[381] should be read by anyone interested in this order.

### KEY TO THE FAMILIES IN PROTEOCEPHALIDEA

1a. Vitellaria medullary ...................... Proteocephalidae LaRue 1911 (p.173)
1b. Vitellaria cortical ............................ Monticellidae LaRue 1911 (p.190)

### FAMILY PROTEOCEPHALIDAE

### KEY TO THE SUBFAMILIES OF PROTEOCEPHALIDAE

1a. Armed, nonretractable rostellum present ....... Gangesiinae Mola 1929 (p.173)
1b. Armed rostellum absent ........................................................ 2

2a. Vitellaria surrounding all internal organs, dorsal and ventral as well as lateral. Parasites of elasmobranchs ........................................................
 ........................... Prosobothriinae Yamaguti 1959. (p.176) (Figure 203)
2b. Vitellaria only lateral. Parasites of freshwater or terrestrial hosts ............. 3

3a. Vitellaria postovarian. Apical organ with two lappet-like structures ............
 .............................................. Sandonellinae Khalil 1960. (p.177)

3b. Vitellaria extending all or most of segment length, apical organ lacking lappets ................................................................. 4

4a. Metascolex present, surrounding suckers ..........................................
 ........................................... Corallobothriinae Freze 1965. (p.177)
4b. Metascolex absent ........................................................ 5

5a. Conical, spinose apical organ present, scolex spinose...........................
 ............................................ Acanthotaeniinae Freze 1963. (p.180)
5b. Conical, spinose apical organ absent, scolex spinose or not .................... 6

6a. Testes medullary ............ Proteocephalinae Mola 1929. (p.182) (Figure 204)
6b. Testes in dorsal cortex .........................................................
 ...................... Marsipocephalinae Woodland 1933. (p.190) (Figure 205)

### KEY TO THE GENERA IN GANGESIINAE

1a. Rostellum with several alternating rows of small hooks ........................ 2
1b. Rostellum with not more than two rows of hooks ............................... 3

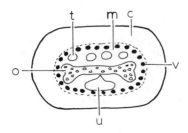

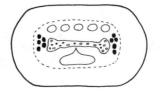

FIGURE 204.   Proteocephalinae.

FIGURE 203.   Prosobothriinae. c — cortex. m — medulla. o — ovary. t — testis. u — uterus. v — vitelline gland.

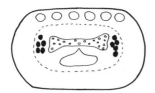

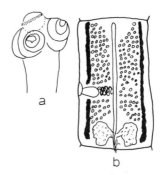

FIGURE 205.   Marsipocephalinae.

FIGURE 206.   *Electrotaenia malopteruri* (Fritsch 1886) Nybelin 1942. (a) Scolex; (b) mature proglottid.

**2a.   No testes in anterior median field. Internal seminal vesicle present .............**
**..........................................*Electrotaenia* Nybelin 1942. (Figure 206)**

*Diagnosis:* Scolex with rostellum armed with several circles of small hooks. Neck present. Acraspedote. Genital atrium deep, muscular. Genital pores irregularly alternating? Cirrus pouch elongated. Inner seminal vesicle present. Testes in two broad lateral fields, median to vitellaria. Vitellaria lateral to osmoregulatory canals. Ovary bilobed, median, at posterior end of proglottid. Uterus median, with lateral branches. Parasites of siluroid fishes. Africa.

Type species: *E. malopteruri* (Fritsch 1886) Nybelin 1942, in *Malopterurus electricus;* Africa.

**2b.   Testes present in anterior median field. No internal seminal vesicle.............**
**..........................................*Silurotaenia* Nyblein 1942. (Figure 207)**

*Diagnosis:* Scolex with rostellum armed with several alternating circles of small hooks. Neck present? Acraspedote. Genital pores irregularly alternating. Genital atrium shallow, not muscular. Cirrus pouch small, rounded. Seminal vesicles absent. Testes fill entire field anterior to ovary. Ovary with two large lateral lobes, at posterior margin of proglottid. Vitellaria lateral to osmoregulatory canals. Uterus median, with lateral branches. Parasites of siluroid fishes. Europe.

Type species: *S. siluri* (Batsch 1786) Nybelin 1942, in *Silurus glanis;* Europe.

**3a.   Vitellaria in short, lateral, preovarian fields. One circle of hooks ..............**
**.................................................*Vermaia* Nybelin 1942. (Figure 208)**

*Diagnosis:* Scolex with rostellum armed with single circle of hooks. Minute spines on margins of suckers, neck, and proglottids. Neck present. Posterior proglottids much longer than wide. Cirrus pouch small, thin-walled. Ducts pass between osmoregulatory canals.

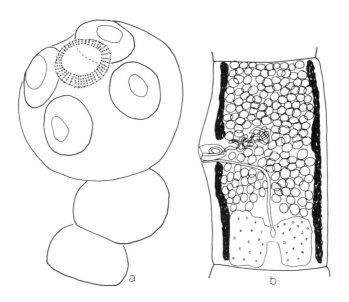

FIGURE 207. *Silurotaenia siluri* (Batsch 1786) Nybelin 1942. (a) Scolex; (b) mature proglottid.

FIGURE 208. *Vermaia pseudotropii* (Verma 1928) Nybelin 1942. (a) Rostellar hooks; (b) proglottid.

Genital pores regularly alternating, in posterior half of proglottid. Seminal vesicles absent. Testes in two lateral fields anterior to ovary. Vitellaria in short, lateral fields between level of genital pore and ovary. Uterus with lateral branches. Eggs embryonated, some developing a neck and bladder (?) while still in uterus. Parasites of siluroid fishes. India.

Type species: *V. pseudotropii* (Verma 1928) Nybelin 1942, in *Pseudotropius gorua;* India.
Other species:
  *V. sarrakowahi* Zaidi et Khan 1976, in *Scoliodon sarrakowah;* Pakistan.

**3b. Vitellaria extending entire lateral fields. One or two circles of hooks ........... ............................................*Gangesia* Woodland 1924. (Figure 209)**

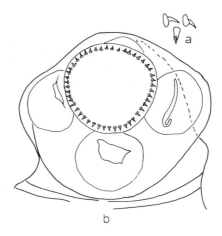

FIGURE 209. *Gangesia polyonchis* Roitman et Frese 1964. (a) Typical hooks; (b) scolex.

*Diagnosis:* Scolex with rostellum armed with one or two circles of hooks. Neck present. Mature proglottids about square. Genital pores irregularly alternating. Cirrus pouch large, cirrus unarmed. Seminal vesicles absent. Testes in continuous field between vitelline fields. Ovary between posterior ends of vitelline fields. Vitellaria lateral, extending entire length of proglottid. Uterus median, with lateral branches. Parasites of siluroid fishes. India, Europe, Russia, Japan, Pakistan. Key to species: Freze.[957]

Type species: *G. bengalensis* (Southwell 1913) Woodland 1924 (syn. *Ophriocotyle bengalensis* Southwell 1913, *Gangesia wallago* Woodland 1924, *G. agraensis* Verma 1928), in *Ophiocephalus striatus, Labeo rohita, Wallago attu;* India, Pakistan.

Other species:

*G. jammuensis* Fotedar et Dhar 1974, in *Wallago atu;* India.

*G. lucknowia* Singh 1948, in *Eutropiichthys vacha;* India, Pakistan.

*G. macrones* Woodland 1924, in *Marones seengha;* India.

*G. mahamdabadensis* Malhotra, Dixit et Capoor 1981, in *Mystis tengra;* India.

*G. oligorchis* Roitman et Freze 1964, in *Pseudobagrus fulvidraco;* Russia.

*G. osculata* (Goeze 1782) Woodland 1924, in *Silurus glanis;* Europe.

*G. parasiluri* Yamaguti 1934, in *Parasilurus asotus, Onchorhynchus gorbuscha, O. keta;* Japan, Russia.

*G. polyonchis* Roitman et Freze 1964, in *Parasilurus asototus;* Russia.

*G. sanehensis* Malhotra, Capoor et Shinde 1980, in *Cirrhina mrigala, Wallago atu;* India.

*G. sindensis* Rehana et Bilquees 1971, in *Wallago atu;* Pakistan.

## DIAGNOSIS OF THE ONLY GENUS IN PROSOBOTHRIINAE

*Prosobothrium* Cohn 1902 (syn. *Lintoniella* Woodland 1927)

*Diagnosis:* Scolex normal, with suckers directed anteriad. Neck present. Genital pores irregularly alternating. Cirrus pouch well developed, containing coiled ejaculatory duct. Cirrus spined. Testes medullary, in broad field anterior to ovary. Vagina anterior or lateral to cirrus pouch. Small seminal receptacle present. Ovary medullary, bilobed, at posterior end of proglottid. Vitellaria medullary, completely surrounding gonads in cross section.

FIGURE 210. *Sandonella sandoni* (Lynsdale 1960) Khalil 1960.

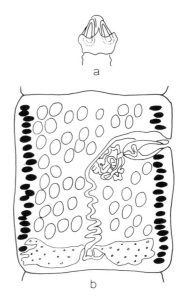

FIGURE 211. *Choanoscolex abscisus* (Riggenbach 1895) LaRue 1911. (a) Scolex; (b) proglottid.

Uterus median, medullary, greatly swollen when gravid. Parasites of sharks. Japan, North America, Europe.

Type species: *P. armigerum* Cohn 1902 (syn. *Cylindrophorus typicus* Diesing, in part; *Ichthyotaenia adherens* Linton 1924), in *Squalus acanthias, Zygaena, Carcharias, Cestracion zygaena, Carcharodon, Prionace;* Atlantic, Mediterranean.

Other species:
 *P. japonicum* Yamaguti 1934, in *Prionace glauca;* Japan.

## DIAGNOSIS OF THE ONLY GENUS IN SANDONELLINAE

*Sandonella* Khalil 1960 (Figure 210)

*Diagnosis:* Scolex club-shaped, longer than broad, with four simple, rounded suckers. Apex of scolex with two muscular lappet-like structures that are notched slightly in the middle edge and attached together posteriorly. Neck present. External segmentation obvious only in early proglottids. All segments wider than long. Genital pores equatorial, irregularly alternating. Testes numerous, anterior and lateral to ovary. Ovary posterior, bilobed. Vagina posterior to cirrus pouch. Vitellaria confined to two compact follicles posterior to ovary. Uterus forms as irregular sac, becoming somewhat bilobed when gravid. Parasites of freshwater teleosts. Africa.

Type species: *S. sandoni* (Linsdale 1960) Khalil 1960 (syn. *Proteocephalus sandoni* Linsdale 1960, in *Heterotis niloticus;* Sudan.

## KEY TO THE GENERA OF CORALLOBOTHRIINAE

**1a. Only base of suckers covered by folds............................................
................................................*Choanoscolex* LaRue 1911. (Figure 211)**

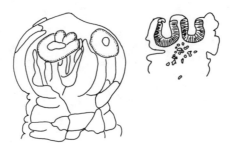

FIGURE 212. *Megathylacoides giganteum* (Essex 1928) Freze 1965. Scolex and section through scolex.

*Diagnosis:* Scolex with swollen base and cone-shaped apex. Base of suckers covered with fold of tissue. Neck present. Terminal proglottids longer than wide. Genital pore in anterior fourth of proglottid margin. Cirrus pouch large, muscular. Cirrus with large lumen. Seminal vesicles absent. Testes in one field between vitellaria. Ovary bilobed, at posterior end of proglottid. Vitellaria in lateral fields, entire length of proglottid. Uterus with lateral branches. Egg with polar swellings. Parasites of siluroid fishes. Paraguay.

Type species: *C. abscisus* (Riggenbach 1895) LaRue 1911, in *Silurus* sp.

1b.  Entire suckers covered by folds.................................................... 2

2a.  Suckers with sphincters ........................................................... 3
2b.  Suckers without sphincters....................................................... 4

3a.  Sphincters complete ............................. *Megathylacus* Woodland 1934.

*Diagnosis:* Scolex large, with four large lobes. Suckers consist of large, thin-walled sacs, each with powerful sphincter at orifice, completely covered by folds. Neck present. Genital pore near middle of proglottid margin. Seminal vesicles absent. Cirrus pouch less than one third proglottid width. Testes in two lateral fields (?). Vagina apparently anterior to cirrus pouch. Ovary transversely elongated, at posterior end of proglottid. Vitellaria lateral to osmoregulatory canals. Uterus with lateral outpocketings. Parasites of siluroid fish. Brazil.

Type species: *M. jandia* Woodland 1934, in *Rhamdia* sp.; Brazil.

3b.  Sphincters incomplete .............................................................
......................*Megathylacoides* Jones, Kerley et Sneed 1956. (Figure 212)

*Diagnosis:* Scolex lacking rostellum or armature. Suckers each with a sphincter that partly surrounds its aperture. Metascolex surrounds suckers. Short neck present. Mature proglottids longer than wide. Genital pores irregularly alternating. Cirrus pouch variable in size between species, not reaching midline of segment. Testes numerous, in one layer. Ovary bilobed, posterior. Vagina anterior or posterior to cirrus pouch. Seminal receptacle indistinct. Vitellaria in lateral bands, not coursing along posterior margin of segment. Uterus with bilateral branches and anterior uterine pore. Parasites of siluroid fishes. North America. Elevated from subgenus by Freze.[957]

Type species: *M. tva* Jones, Kerley et Sneed 1956 (syn. *Corallobothrium (Megathylacoides) tva* Jones, Kerley et Sneed 1956), in *Pilodictis olivaris;* North America.
Other species:

*M. giganteum* (Essex 1928) Freze 1965 (syn. *Corallobothrium giganteum* Essex 1928, *C. [Megathylacoides] giganteum* [Essex 1928] Jones, Kerley et Sneed 1956), in

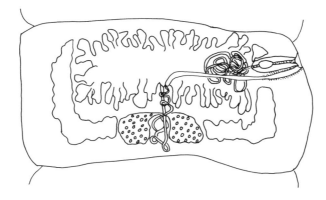

FIGURE 213. *Paraproteocephalus parasiluri* (Zmeev 1936) Chen Yan-hsin 1962. Proglottid.

*Ameiurus melas, A. nebulosus, Ictalurus punctatus, I. lacustris, I. natalis, I. catus, Leptops olivaris;* North America.

*M. intermedium* (Fritts 1956) Befus et Freeman 1973 (syn. *Corallobothrium* Fritts 1959, *Corallotaenia intermedia* [Fritts 1959] Freze 1965), in *Ameiurus nebulosus;* North America.

*M. procerum* (Sneed 1950) Freze 1965 (syn. *Corallobothrium procerum* Sneed 1950, *C. [Megathylacoides] procerum* Jones, Kerley et Sneed 1956), in *Ictalurus furcatus;* North America.

*M. thompsoni* (Sneed 1950) Freze 1965 (syn. *Corallobothrium thompsoni* Sneed 1950, *C. [Megathylacoides] thompsoni* Jones, Kerley et Sneed 1956), in *Ictalurus lacustris;* North America.

**4a.  Postovarian vitellaria not converging..................*Corallotaenia* Freze 1965.**

*Diagnosis:* Scolex with four suckers lacking sphincters. Metascolex weakly developed. Neck short. Acraspedote. Genital pores irregularly alternating. Cirrus pouch crosses osmoregulatory canals. Testes numerous, in single field. Vagina posterior to cirrus pouch. Ovary bilobed, posterior. Vitellaria in lateral fields, not converging at posterior margin. Gravid segments longer than wide. Uterus with lateral branches. Parasites of siluroid fishes. North America.

Type species: *C. parva* (Larsh 1941) Freze 1965 (syn. *Corallobothrium parvus* Larsh 1941), in *Ameiurus nebulosus;* Michigan, U.S.

Other species:

*C. minutia* (Fritts 1959) Befus et Freeman 1973 (syn. *Corallobothrium minutium* Fritts 1959), in *Ameiurus nebulosus;* U.S., Canada.

**4b.  Postovarian vitellaria converging ............................................... 5**

**5a.  Internal seminal vesicle present. Uterine branches mainly anterior and posterior .................*Paraproteocephalus* Chen Yan-hsin 1962. (Figure 213)**

*Diagnosis:* Scolex flattened with four normal suckers and small apical sucker (Figure 214). Metascolex present. Neck absent. Proglottids wider than long, acraspedote. Cirrus pouch well developed, containing internal seminal vesicle. Fourteen to sixteen osmoregulatory canals on each side of strobila. Testes numerous. Ovary posterior, bilobed. Vagina pre- or

FIGURE 214. *Paraproteocephalus parasiluri* (Zmeev 1936) Chen Yan-hsin 1962. Scolex.

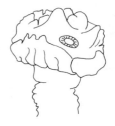

FIGURE 215. *Corallobothrium fimbriatum* Essex 1928.

postcirrus pouch. Vitellaria lateral, converging toward ovary at their posterior ends. Uterus transverse, with anterior and posterior branches. Parasites of siluroid fishes. Russia.

Type species: *P. parasiluri* (Zmeev 1936) Chen Yan-hsin 1962 (syn. *Corallobothrium parasiluri* Zmeev 1936), in *Parasilurus asotus, Silurus soldatovi;* Amur Basin, Russia.

**5b.   Internal seminal vesicle absent. Uterine branches lateral.........................
................................*Corallobothrium* Fritsch 1886. (Figure 215)**

*Diagnosis:* Scolex flat on apical end, with four suckers on flat surface, covered with folds. Neck short, broad. Genital pores irregularly alternating, in anterior half of proglottid margin. Cirrus pouch well developed. Seminal vesicles absent. Testes in single field between osmoregulatory canals. Ovary posterior, variable in shape. Vitellaria lateral, medullary, converging at posterior margin. Uterus with lateral branches. Parasites of siluroid fishes. Africa, U.S.

Type species: *C. solidum* Fritsch 1886, in *Malopterus electricus;* Egypt.
Other species:
   *C. fimbriatum* Essex 1927, in *Ictalurus punctatus, Leptops olivaris, Ameiurus melas;* North America.
   *C. parafimbriatum* Befus et Freeman 1973, in *Ictalurus nebulosus;* Canada.

## KEY TO THE GENERA IN ACANTHOTAENIINAE

**1a.   Eggs contained in capsules..............*Kapsulotaenia* Freze 1963. (Figure 216)**

*Diagnosis:* Scolex and suckers densely covered with spines. Anterior surface of scolex developed into a musculo-glandular piercing organ. Neck present. Mature and gravid proglottids acraspedote, usually longer than wide. Internal musculature very weak. Genital pores alternate irregularly. Testes numerous, in one or two fields. Ovary bilobed, posterior. Vagina usually posterior to cirrus pouch. Vitellaria in two lateral fields. Uterus persistent, with lateral branches. Eggs contained in membranous capsules, each with several eggs. Parasites of varanid lizards. Indonesia, Philippines, New Guinea, Australia.

Type species: *K. sandgroundi* (Carter 1943) Freze 1965 (syn. *Proteocephalus sandgroundi* Carter 1943), in *Varanus komodoensis;* Indonesia.
Other species:
   *K. frezei* Schmidt et Kuntz 1974, in *Varanus salvator;* Philippines.
   *K. saccifera* (Ratz 1900) Freze 1965 (syn. *Ichthyotaenia saccifera* Ratz 1900, *Proteocephalus saccifera* [Ratz 1900] Johnston 1912, *P. saccifera* [Ratz 1900] Baer 1927, *Crepidobothrium saccifera* [Ratz 1900] Meggitt 1927, *Acanthotaenia saccifera* [Ratz 1900] Johnston 1910), in *Varanus* sp.; New Guinea.

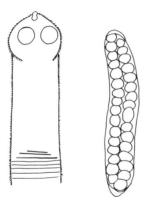

FIGURE 216. *Kapsulotaenia frezei* Schmidt et Kuntz 1974. Scolex and egg capsule.

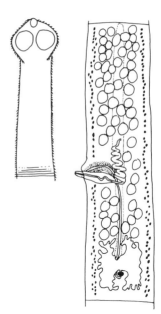

FIGURE 217. *Acanthotaenia daileyi* Schmidt et Kuntz 1974. Scolex and proglottid.

*K. tidswelli* (Johnston 1909) Freze 1965 (syn. *Acanthotaenia tidswelli* Johnston 1909, *Ichthyotaenia [Acanthotaenia] tidswelli* [Johnston 1909] Beddard 1913, *Proteocephalus tidswelli* [Johnston 1909] Johnston 1911, *Acanthotaenia tidswelli* [Johnston 1909] Woodland 1925, *Crepidobothrium tidswelli* [Johnston 1909] Meggitt 1927), in *Varanus varius, V. gouldii, V. belii;* Australia.

*K. varia* (Beddard 1913) Freze 1965 (syn. *Acanthotaenia varia* Beddard 1913), *A. varia* [Beddard 1913] Nybelin 1917, *Ichthyotaenia varia* [Beddard 1913] Woodland 1925, *Proteocephalus varius* [Beddard 1913] Baer 1927, *Crepidobothrium varia* [Beddard 1913] Meggitt 1927), in *Varanus varius, V. gouldi;* Australia.

**1b. Eggs not contained in capsules..................................................**
**.....*Acanthotaenia* Linstow 1903 (syn. *Rostellotaenia* Freze 1963). (Figure 217)**

*Diagnosis:* Apex of scolex cone-shaped or truncated. Suckers normal. Scolex and anterior part of strobila covered with spines. Neck present. External segmentation sometimes indistinct. Genital pores irregularly alternating. Cirrus pouch well developed. Vagina opening anterior or posterior to cirrus. Ovary bilobed, near posterior end of proglottid. Vitellaria lateral, cortical. Parasites of lizards and snakes. Sri Lanka, Celebes, New Guinea, India, Australia, Africa, Komodo, South America, Puerto Rico.

Type species: *A. shipleyi* Linstow 1903, in *Varanus salvator;* Sri Lanka, Celebes.
Other species:

*A. biroi* (Ratz 1900) Johnston 1909 (syn. *Ichthyotaenia biroi* Ratz 1900, *I. biroi* [Ratz 1900] Woodland 1925, *Proteocephalus biroi* [Ratz 1900] Johnston 1911, *Crepidobothrium biroi* [Ratz 1900] Meggitt 1927), in *Varanus* sp., *V. flavescens;* New Guinea, India.

*A. beddardi* (Woodland 1925) comb. n. (syn. *Proteocephalus beddardi* Woodland 1925, *Crepidobothrium beddardi* [Woodland 1925] Meggitt 1927, *Rostellotaenia beddardi* [Woodland 1925] Freze 1965), in *Varanus bengalensis;* India.

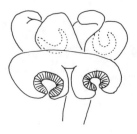

FIGURE 218. *Macrobothiotaenia ficta* (Meggitt 1927) Freze 1965.

*A. daileyi* Schmidt et Kuntz 1974, in *Varanus salvator;* Philippines.

*A. gallardi* (Johnston 1911) Johnston 1913 (syn. *Ophiotaenia gallardi* Freze 1965, *Ichthyotaenia* sp. Johnston 1910, *Proteocephalus gallardi* Johnston 1911, *Crepidobothrium gallardi* [Johnston 1911] Meggitt 1927), in *Pseudechis porphyriacus, P. australis, Notechis scutatus, Denisonia superba;* Australia.

*A. gracilis* (Beddard 1913) Rudin 1917 (syn. *Ichthyotaenia [Acanthotaenia] gracilis* Beddard 1913, *Crepidobothrium gracilis* [Beddard 1913] Meggitt 1927, *Proteocephalus gracilis* [Beddard 1913] Hudson 1934), in *Varanus varius;* Australia.

*A. nilotica* (Beddard 1913) comb. n. (syn. *Rostellotaenia nilotica* Freze 1965, *Ichthyotaenia [Acanthotaenia] nilotica* Beddard 1913, *I. nilotica* [Beddard 1913] Southwell 1916, *I. nilotica* [Beddard 1913] Woodland 1925, *Crepidobothrium nilotica* [Beddard 1913] Meggitt 1927, *Proteocephalus niloticus* [Beddard 1913] Baer 1927, *Acanthotaenia articulata* Rudin 1917, *Crepidobothrium articulata* [Rudin 1917] Meggitt 1927, *Proteocephalus articulatus* [Rudin 1917] Baer 1927, *Acanthotaenia continua* Rudin 1917, *Crepidobothrium continua* [Rudin 1917] Meggitt 1927, *Proteocephalus continuus* [Rudin 1917] Baer 1927) in *Varanus niloticus;* Africa.

*A. overstreeti* Brooks et Schmidt 1978, in *Cyclura stejnegeri;* Puerto Rico.

*A. woodlandi* (Moghe 1926) comb. n. (syn. *Proteocephalus woodlandi* Moghe 1926, *Rostellotaenia woodlandi* [Moghe 1926] Freze 1965), in *Varanus bengalensis, Calotes versicolor;* India.

## KEY TO THE GENERA IN PROTEOCEPHALINAE

1a. Suckers circular, not modified .................................................... 2
1b. Suckers modified .......................................................... 4

2a. Suckers pedunculate ............... *Macrobothriotaenia* Freze 1965. (Figure 218)

*Diagnosis:* Scolex much wider than neck. Suckers each on a stalk. Genital pore slightly postequatorial. Segments longer than wide. Cirrus pouch large, one third width of proglottid. Testes numerous, in two lateral fields. Ovary posterior, transversely elongated. Vitellaria lateral. Vagina posterior to cirrus pouch. Uterus median, with lateral diverticula.
  Type species: *M. ficta* (Meggitt 1927) Freze 1965 (syn. *Crepidobothrium fictum* Meggitt 1927, *Proteocephalus fictus* [Meggitt 1927] Hughes, Baker et Dawson 1941), in *Xenopeltis unicolor, Natrix beddomii, Naja naja;* India.

2b. Suckers sessile ............................................................ 3

3a. Testes in two lateral fields ...... *Ophiotaenia* La Rue 1911 (syn. *Batrachotaenia* Rudin 1917, *Solenotaenia* Beddard 1913, *Testudotaenia* Freze 1965). (Figure 219)

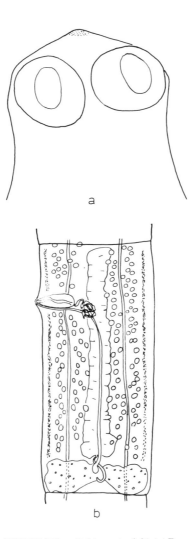

FIGURE 219. *Ophiotaenia dubinini* Freze et Shapiro 1965. (a) Scolex; (b) proglottid.

*Diagnosis:* Scolex unarmed, with typical suckers. Neck present. Genital pores irregularly alternating. Cirrus pouch well developed. Testes in two separate, lateral fields. Vagina anterior, posterior or dorsal to cirrus pouch. Ovary bilobed. H- or M-shaped. Vitellaria lateral. Uterus median, with lateral branches. Parasites of fishes, amphibians, and reptiles. Cosmopolitan.

Type species: *O. perspicua* La Rue 1911, in *Natrix rhombifer, N. sipedon, Thamnophis sirtalis, Natrix taxispilota;* North America, Panama.

Other species:

*O. adiposa* Rudin 1917, in *Bitis arietans;* Africa.

*O. agkistrondontis* Harwood 1933, in *Ancistrodon piscivorus;* Texas, U.S.

*O. alternans* Riser 1942, in *Amphiuma tridactylum;* Louisiana, U.S.

*O. amphiboluri* Nybelin 1917 (syn. *Crepidobothrium adiposa* Meggitt 1927), in *Amphibolurus barbatus, Bitis arietans;* Australia, Africa.

*O. amphiumae* Zeliff 1932, in *Amphiuma tridactylum,* U.S.

*O. andersoni* Jensen, Schmidt et Kuntz 1983, in *Trimerisurus stejnegeri;* Taiwan.

*O. barbouri* Vigueras 1934, in *Tretanorhinus variabilis;* Cuba.

*O. bonariensis* Szidat et Soria 1954, in *Leptodactylus ocellatus;* Argentina.

*O. bufonis* (Vigueras 1942), in *Bufo peltacephalus;* Cuba.

*O. calmetti* (Barrois 1898) LaRue 1911 (syn. *Ichthyotaenia calmettei* Barrois 1898, *Proteocephalus calmettei* [Barrois 1898] Railliet 1899, *Taenia calmettei* [Barrois 1898] Parona 1901, *Crepidobothrium calmettei* [Barrois 1898] Meggitt 1927, *Ophiotaenia calmetti* [Barrois 1898] Wardle et McLeod 1952, *Ichthyotaenia raillieti* Marotil 1898, *Oochoristica racemosa* [Rudolphi 1819] Parona 1901), in *Lachesis lanceolatus, Bothrops atrox;* Martinique, Brazil, Venezuela.

*O. carpatheca* (Sharpilo, Kornyushin et Lisitsina 1979) comb. n. (syn. *Batrachotaenia carpatheca* Sharpilo, Kornyushin et Lisitsina 1979), in *Triturus cristatus;* Russia.

*O. cohospes* Cordero 1946, in *Hydromedusa tectifera;* Uraquay.

*O. congolensis* Southwell et Lake 1939, in *Boodon olivaceus, B. lineatus;* Belgian Congo.

*O. crotophopeltis* Sandground 1929, in *Crotophopeltis torniari, C. hotamboeia;* Tanganyika, Kenya.

*O. cryptobranchi* La Rue 1915, in *Cryptobranchus allegheniensis, Desmognathus* spp., *Plethodon, Pseudotriton, Triturus;* U.S.

*O. dubinini* Freze et Sharpilo 1967, in *Coronella austriaca;* Russia.

*O. elapsideae* Sandground 1929, in *Elapsoidea guenteri;* Tanganyika.

*O. elongata* Fuhrmann 1927, in a snake; Brazil.

*O. europaea* Odening 1963, in snakes; Europe.

*O. faranciae* MacCallum 1921, in *Farancia abacura* (zoo), *Bothrops atrox;* Venezuela.

*O. filarioides* (La Rue 1909) La Rue 1911, in *Amblystoma tigrinum;* U.S.

*O. fima* Meggitt 1927, in *Rhabdophis stolata;* Burma, Calcutta Zoo.

*O. fixa* Meggitt 1927, in *Rhabdophis stolata;* Burma.

*O. flava* Rudin 1917, in *Coluber* sp.; Brazil.

*O. fragilis* (Essex 1929) Wardle et McLeod 1952 (syn. *Proteocephalus fragile* [Essex 1929] Freze 1965, *Crepidobothrium fragile* Essex 1929), in *Ictalurus punctatus, Synodontis schall;* U.S., Egypt.

*O. gabonica* Beddard 1913, in *Bitis gabonica;* Africa.

*O. gracilis* Jones, Cheng et Gillespie 1958, in *Rana catesbiana;* U.S.

*O. grandis* La Rue 1911, in *Ancistrodon piscivorus;* North America.

*O. habanensis* Freze et Rysavy 1976, in *Tropidophis perdalis;* Cuba.

*O. hyalina* Rudin 1917, in a snake; Brazil.

*O. hylae* Johnston 1912, in *Hyla aurea;* Australia.

*O. indica* Johri 1955, in *Naia naia;* India.

*O. japonensis* Yamaguti 1935, in *Elaphe quadrivirgata, Natrix tigrina;* Japan.

*O. jarara* Fuhrmann 1927, in *Bothrops alternatus;* Brazil.

*O. junglensis* (Srivastava et Capoor 1980) comb. n. (syn. *Batrachotaenia junglensis* Srivastava et Capoor 1980), in *Rana tigrina;* India.

*O. kuantanensis* Yeh 1956, in *Naja hannah;* Malaya.

*O. lactea* (Leidy 1855) La Rue 1911, in *Tropidonotus sipedon;* U.S.

*O. loennbergi* (Fuhrmann 1895) La Rue 1911, in *Necturus maculosus;* type locality unknown; also in *Necturus* in U.S.

*O. longmanni* Johnston 1916, in *Aspidites ramsayi;* Australia.

*O. lopesi* Rego 1967, in *Testudo denticulata;* Brazil.

*O. macrobothria* Rudin 1817, in *Elaps corallinus;* Brazil.

*O. magna* Hannum 1925, in *Rana catesbiana* and other species of *Rana, Bufo cognatus, Acris gryllus;* North America.

*O. marenzelleri* (Barrois 1898) La Rue 1911, in *Ancistrodon piscivorus, Pseudaspius cana, Naja nivea, Bitis arietans, B. gabonica;* Africa, U.S.

*O. meggitti* Hilmy 1936, in *Atheris chloroechis, Coronella coronata*; Liberia.

*O. micruricola* (Shoop et Corkum 1982) comb. n. (syn. *Proteocephalus micruricola* Shoop et Corkum 1982), in *Micrurus diastema*; Mexico.

*O. mjobergi* Nybelin 1917, in *Demansia psammophis*; Australia.

*O. moennigi* Fuhrmann 1924, in *Leptodira hotambeia, Natrix piscator, Simotes purpurascens*; Burma.

*O. najae* Beddard 1913, in *Naia tripudians*; India.

*O. nankingensis* Hsü 1935, in *Zoacys dhumnades dhumnades, Homalopsis buccata*; China, Burma.

*O. nattereri* (Parona 1901) La Rue 1911, in *Coluber* sp.; Italy, Mexico?

*O. nigricollis* Mettrick 1963, in *Naja nigricollis*; Southern Rhodesia.

*O. noei* Wolffhügel 1948, in *Calyptocephalus gayi*; Chile.

*O. nybelini* Hilmy 1936, in *Coronella coronata*; Liberia.

*O. olor* (Ingles 1936) Yamaguti 1938 (syn. *Crepidobothrium olor* Ingles 1936, *Batrachotaenia olor* [Ingles 1936] Freze 1965) in *Rana aurora*; California, U.S.

*O. olseni* Dyer et Altig 1977, in *Hyla geographica*; Ecuador.

*O. ophiodex* Mettrick 1960, in *Causus rhombeatus*; Southern Rhodesia.

*O. paraguayensis* Rudin 1917, in *Coluber* sp.; Paraguay.

*O. phillipsi* Burt 1937, in *Trimeresurus trigonocephalus Sri Lanka*.

*O. pigmentata* (Linstow 1907) La Rue 1911, in *Psammodynastes pulverulentus;* Java.

*O. racemosa* (Rudolphi 1819) La Rue 1911, in *Coluber* sp., *Liophis merremii, Salmo irideus, Tropidonotus natrix;* Brazil, Italy, France.

*O. ranae* Yamaguti 1938 (syn. *O. ranarum* Iwata et Matuda 1938), in *Rana nigromaculata;* Japan.

*O. rhabdophidis* Burt 1937, in *Rhabdophis stolata;* Sri Lanka.

*O. russelli* Beddard 1913, in *Vipera russelli;* India.

*O. sanbernardinensis* Rudin 1917, in *Helicops leprieuri;* Paraguay.

*O. saphena* Osler 1931, in *Rana clamitans;* U.S.

*O. schultzei* (Hungerbühler 1910) Dickley 1921 (syn. *Ichthyotaenia schultzei* Hungerbühler 1910, *Crepidobothrium schultzei* [Hungerbühler 1910] Magath 1929, *Batrachotaenia schultzei* [Hungerbühler 1910] Rudin 1917), in *Rana adspersa*; South Africa.

*O. spasskyi* Freze et Sharpilo 1967, in *Viperus berus*; Russia.

*O. striata* Johnston 1914, in *Lialis burtonii*; Australia.

*O. testudo* Magath 1924, in *Amyda spinifera*; U.S.

*O. theileri* Rudin 1917, in *Naja naja, Causus rhombeatus*; Africa.

*O. trimeresuri* (Parona 1898) La Rue 1911, in *Trimeresurus sumatrans*; India.

*O. viperis* (Beddard 1913) Rudin 1917 (syn. *Solenotaenia viperis* Beddard 1913), in *Lechesis alternans*; India.

*O. zschokkei* Rudin 1917, in *Naja naja*; South Africa.

3b. Testes in single field ...........................................................
**Proteocephalus** Weinland 1858 (syn. ***Ichthyotaenia*** Lönnberg 1894). (Figure 220)

*Diagnosis:* Scolex unarmed, with four normal suckers. A fifth apical sucker or apical organ present in some species. Neck present. Genital pores irregularly alternating. Cirrus pouch well developed. Testes in single broad field anterior to ovary. Vagina anterior, posterior, or dorsal to cirrus pouch. Ovary bilobed, transverse, at posterior end of proglottid. Vitellaria lateral. Uterus median, with lateral branches. Parasites of freshwater fishes, amphibians, and reptiles. Cosmopolitan. Key to species: Freze. [957]

Type species: *P. filicollis* (Rudolphi 1802) Weinland 1858, in *Coregonus artedi, C. nigripinnis, C. prognathus, Gasterosteus aculeatus, G. pungitius;* Europe, North America.

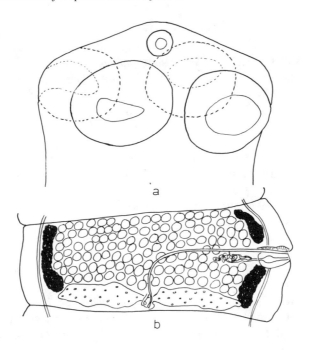

FIGURE 220. *Proteocephalus macrocephalus* (Creplin 1815) Nufer 1905. (a) Scolex; (b) proglottid.

Other species:

P. *aberrans* Brooks 1978 in *Siren lacertina;* U. S.

P. *agonis* Barbieri 1909, in *Alosa finta;* Europe.

P. *albulae* Freze et Kuzakov 1969, in *Coregonus albula;* Russia.

P. *ambloplitis* (Leidy 1887) Benedict 1900 (syn. *Taenia ambloplitis* Leidy 1887), in *Allotis, Ambloplites rupestris, Ameiurus nebulosus, Amia calva, Aphredoderus, Aplites, Aplodinotus, Boleosoma, Catonotus, Centrarchus, Chaenobryttus, Cottus, Cristivomer, Cylindrosteus, Erimyzon, Esox, Eupomotis, Helioperca, Ictalurus, Ictiobus, Labidesthes, Lepibema, Lepisosteus osseus, Lepomis gibbosus, Micropterus dolomieu, M. salmoides, Notropis, Noturus, Opladelus, Perca flavescens, Pomoxis, Salvelinus, Schilbeodes, Sclerotis, Stizostedion, Xenotis;* North America.

P. *amphiumicola* Brooks 1978, in *Amphiuma means;* U.S.

P. *arandasi* Santos et Rolas 1973, in *Liophis miliaris;* Brazil.

P. *arcticus* Cooper 1921, in *Salvelinus marstoni, Salmo clarkii, Onchorhynchus kisutch;* Canada.

P. *atretiumi* Devi 1971, in *Atretium schistosum;* India.

P. *australis* Chandler 1935, in *Lepisosteus osseus;* Texas.

P. *beauchampi* Fuhrmann et Baer 1925, in *Dinotopterus cunningtoni, Chrysichthys brachynema;* North Africa.

P. *belones* (Rudolphi 1810) LaRue 1911, in *Esox belones;* Europe.

P. *bivitellatus* Woodland 1937, in *Tilapia* sp.; Sierra Leone.

P. *buplanensis* Mayes 1976, in *Semotilus atromaculatus;* U.S.

P. *cernuae* Gmelin 1790), LaRue 1911 (syn. *Taenia cernuae* Gmelin 1790), in *Acerina cernua, Esox lucius, Lota lota, Perca fluviatilis, Pygosteus platygaster;* Europe, northern Asia.

P. *chologasteri* Whittaker et Hill 1968, in *Chologaster agassizi:* U.S.

P. *cobraeformis* Haderlie 1953, in *Ptychocheilus grandis;* U.S.

P. *coregoni* Wardle 1932, in *Coregonus atikameg;* Hudson Bay, Canada.

*P. cunningtoni* Fuhrmann et Baer 1925, in *Dinotopterus cunningtoni;* Tanganyika.

*P. cyclops* (Linstow 1877) Nufer 1905, in *Coregonus maraena;* Europe.

*P. dilatatus* (Linton 1889) LaRue 1911, in *Anguilla vulgaris:* U.S.

*P. dinotopteri* Fuhrmann et Baer 1925, in *Dinotopterus cunningtoni;* Tanganyika.

*P. dubius* LaRue 1911, in *Perca fluviatilis;* Lake Geneva.

*P. elongatus* Chandler 1935, in *Lepisosteus osseus;* Texas, U.S.

*P. eperlani* (Rudolphi 1810) Wardle et McLeod 1952, in *Osmerus eperlanus;* Europe.

*P. esocis* (Schneider 1905) LaRue 1911 (syn. *Ichthyotaenia esocis* Schneider 1905), in *Esox lucius;* Asia, Europe.

*P. exiguus* LaRue 1911, in *Coregonus* spp., *Cristovomer, Leucichthys artedi, Micropterus dolomieu, Prosopium;* North America.

*P. fallax* LaRue 1911, in *Coregonus fera;* Lake Lucerne, Switzerland.

*P. fluviatilis* Bangham 1925, in *Micropterus dolomieu;* U.S.

*P. fossatus* (Riggenbach 1895) LaRue 1911 (syn. *Ichthyotaenia fossata* Riggenbach 1895), in *Pimolodes pati;* Paraguay.

*P. glanduliger* Janicki 1928, in *Clarias anguillaris;* Egypt.

*P. gobiorum* Dogiel et Bychowsky 1939, in *Benthophilus macrocephalus, Gobius kessleri, Mesogobius batrachocephalus, Neogobius melanostomus;* Russia.

*P. hanumanthai* Ramadevi 1974, in *Rana cyanophylyctus;* India.

*P. hemisphericus* (Molin 1959) LaRue 1911, in *Anguilla vulgaris;* Europe.

*P. jandia* Woodland 1934, in *Rhamdia* sp., Brazil.

*P. kuyukuyu* Woodland 1935, in *Pseudodoras niger;* Brazil.

*P. largoproglottis* Troncy 1978, in *Synodontis membranaceus;* Chad.

*P. laruei* Faust 1920, in *Coregonus, Leucichthys, Prosopium, Salmo;* North America.

*P. leptosoma* (Lcidy 1888) LaRue 1911, in *Esox reticulatus;* U.S.

*P. longicollis* (Zeder 1800) Nufer 1905, in *Coregonus, Esox, Leuciscus, Lota, Osmerus, Perca, Salmo, Thymallus, Trutta trutta;* Europe.

*P. luciopercae* Wardle 1932, in *Stizostedion canadense, S. vitrius;* Canada.

*P. macdonaghi* (Szidat et Nani 1951) Yamaguti 1959 (syn. *Ichthyotaenia macdonaghi* Szidat et Nani 1951), in *Basilichthys microlepidotus;* Argentina

*P. macrocephalus* (Creplin 1815) Nufer 1905 (syn. *Taenia macrocephala* Creplin 1815), in *Anguilla chrysypa, A. rostrata, A. vulgaris, Osmerus eperlanus?, Stizostedion vitreum?;* Europe, Russia, North Africa, North America.

*P. macrophallus* (Diesing 1850), in *Cichla monoculus;* Brazil.

*P. manjuariphilus* Vigueras 1936, in *Atractosteus tristoechus:* Cuba.

*P. membranacei* Troncy 1978, in *Synodontis membranaceus;* Chad.

*P. microcephalus* Haderlie 1953, in *Micropterus dolomieu;* U.S.

*P. micropteri* (Leidy 1887) LaRue 1911, in *Micropterus nigricans;* U.S.

*P. microscopius* Woodland 1935, in *Cichla ocellaris;* Brazil.

*P. neglectus* LaRue 1911, in *Trutta fario;* Switzerland.

*P. nematosoma* (Leidy 1891), in *Esox lucius;* North America.

*P. niuginii* Schmidt 1975, in *Rana arfarki;* New Guinea.

*P. ocellatus* (Rudolphi 1802), in *Perca fluviatilis;* Europe.

*P. osburni* Bangham 1925, in *Micropterus dolomieu;* U. S.

*P. osculatus* (Goeze 1782) Nybelin 1942 (syn. *Taenia osculatus* Goeze 1782), in *Abramis, Acipenser stellatus, Alburnis, Leuciscus, Siluris glanis;* Europe, Russia.

*P. pamirensis* Dzhalilov et Ashurova 1971, in *Nemachilus stoliczkai;* Russia.

*P. parallacticus* MacLulich 1943, in *Cristivomer namaycush, Salvelinus fontinalis, Salmo fario;* Canada.

*P. parasiluri* Yamaguti 1934, in *Gnathopogon elongatus, Mogurnda obscura, Parasilurus asotus;* Japan.

*P. pearsei* LaRue 1914, in *Ambloplites, Ameirurus, Aplites, Aplodinotus, Centrarchus, Cottus, Huro, Lepibema, Lepomis, Micropterus dolomieu, Notropis, Perca flavescens, Pomoxis,* etc.; U.S.

*P. pentastoma* (Klaptocz 1906), in *Polypterus bichir, P. endicheri;* Nigeria, Sudan. Redescribed; Jones 1980.

*P. percae* (Mueller 1780) Railliet 1899 (syn. *Taenia percae* Mueller 1780), in *Acerina cernua, Coregonus* spp., *Cottus, Esox lucius, Gasterosteus* spp., *Lota lota, Osmerus, Perca fluviatilis, Salmo salvelinus;* Europe.

*P. perplexus* LaRue 1911, in *Amia calva, Lepisosteus platostomus;* U.S.

*P. pinguis* LaRue 1911, in *Esox, Notropis, Perca, Salmo, Salvelinus;* North America.

*P. platystomi* Lynsdale 1959, in *Platystoma* sp.; Amazon River, Brazil.

*P. plecoglossi* Yamaguti 1934 (syn. *P. neglectus* of Kataoka et Momma 1932), in *Plecoglossus altivelis;* Japan.

*P. pollachii* (Rudolphi 1819), in *Merlangus pollachius;* Europe.

*P. pollanicola* Gresson 1952, in *Coregonus pollan;* Ireland.

*P. poulsoni* Whittaker et Zober 1978, in *Amblyopsis spelaea;* U.S.

*P. primaverus* Neiland 1952, in *Salmo clarki;* Washington, U.S.

*P. ptychocheilus* Faust 1920, in *Ptychocheilus oregonensis, Gila straria, Richardsonius balteatus;* U. S.

*P. pugetensis* Hoff et Hoff 1929, in *Gasterosteus cataphractus;* North America.

*P. pusillus* Ward 1910, in *Salmo sebago, Cristivomer namaycush;* U.S.

*P. ritae* Verma 1926, in *Rita rita;* India.

*P. sagitta* (Grimm 1872) LaRue 1911 (syn. *Taenia sagitta* Grimm 1872), in *Cobitis barbatula;* Russia.

*P. salmonidicola* Alexander 1951, in *Salmo gairdnerii;* U.S.

*P. salmonisomus* (Pallas 1813), in *Salmo omul;* Europe.

*P. salmonisumblae* (Zschokke 1884), in *Salmo salvelinus, S. umbla;* Europe.

*P. salvelini* Linton 1897, in *Cristivomer namaycush;* Lake Superior.

*P. simplicissimus* (Leidy 1887), in *Gadus callarias;* locality?

*P. singularis* LaRue 1911, in *Lepisosteus platyrhincus, L. platystomus, L. osseus;* U.S.

*P. skorikowi* (Linstow 1904) LaRue 1911 (syn. *Ichthyotaenia skorikowi* Linstow 1904), in *Acipenser stellatus;* Caspian Sea.

*P. stizostethi* Hunter et Bangham 1933, in *Lepomis marcochirus, Micropterus dolomieu, Percina caprodes, Stizostedion* spp.; North America.

*P. sulcatus* (Klaptocz 1906) LaRue 1911 (syn. *Ichthyotaenia sulcatus* Klaptoca 1906), in *Chrysichthys* sp., *Clarotes laticeps, Polypterus endlichi;* Chad, Sudan.

*P. synodontis* Woodland 1925, in *Synodontis schall;* Nile River.

*P. thymalli* (Annenkova-Khlopina 1923) Gvosdev 1950 (syn. *Ichthyotaenia thymalli* Annenkowa-Khlopina 1923), in *Brachymystax lekok, Thymallus vulgaris;* Russia.

*P. torulosus* (Batsch 1786) Nufer 1905 (syn. *Taenia torulosus* Batsch 1786), in several species of cyprinid fishes; Europe.

*P. trionychinum* (Lönnberg 1894), in *Trionyx ferox;* Florida, U.S.

*P. tumidocollus* Wagner 1953, in *Salmo gairdnerii, Salvelinis fontinalis;* California, U.S.

*P. variabilis* Brooks 1978, in *Natrix cyclopion,* U.S.

*P. vitellaris* Verma 1928, in *Bagarius yarellii;* India.

*P. wickliffi* Hunter et Bangham 1933, in *Leucichthys artedi;* Lake Erie.

**4a.   Suckers uniloculate with indented margin.........................................
................................ *Crepidobothrium* Monticelli 1900. (Figure 221)**

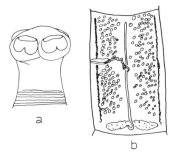

FIGURE 221. *Crepidobothrium garrardi* (Baird 1860) Monticelli 1900. (a) Scolex; (b) proglottid.

FIGURE 222. *Tejidotaenia appendiculatus* (Baylis 1947) Freze 1965.

*Diagnosis:* Scolex large, with four large suckers, each with a ventral notch in the rim. Neck present. Genital pores irregularly alternating. Testes in two large lateral fields. Vagina anterior or posterior to cirrus pouch. Ovary bilobed, near posterior margin of proglottid. Vitellaria marginal. Uterus median, with lateral branches. Parasites of snakes. South America, Mexico.

Type species: *C. gerrardii* (Baird 1860) Monticelli 1900 (syn. *Tetrabothrium gerrardii* Baird 1860, *Tetrabothrium boae* MacCallum 1921), in *Boa constrictor, Eunectes murinus;* South America.

Other species:

*C. brevis* (MacCallum 1921) Yamaguti 1959 (syn. *Tetrabothrius brevis* MacCallum 1921), in *Boa imperator;* Mexico.

*C. dollfusi* Freze 1965 (syn. *C. gerrardi* var. *minus* Dollfus 1932), in unknown boid host; South America.

*C. lachesidis* (MacCallum 1921) Yamaguti 1959 (syn. *Tetrabothrius* MacCallum 1921), in *Lachesis lanceolatus;* South America.

*C. macroacetabula* Kugi et Sawada 1972, in *Eunectes murinus;* Japan (zoo).

**4b. Suckers multiloculate.......................................................... 5**

**5a. Suckers biloculate .........................** *Tejidotaenia* **Freze 1965. (Figure 222)**

*Diagnosis:* Scolex not much wider than neck. Suckers each divided into two unequal cavities. Genital pores irregularly alternating. Cirrus pouch large. Testes in two lateral fields. Ovary posterior, biwinged, large. Vitellaria lateral. Uterus with lateral branches.

Type species: *T. appendiculatus* (Baylis 1947) Freze 1965 (syn. *Proteocephalus appendiculatus* Baylis 1947, *Ophiotaenia appendiculatus* [Baylis 1947] Yamaguti 1959), in *Tupinambis nigropunctatus;* Surinam.

**5b. Suckers tetraloculate ................** *Deblocktaenia* **Odening 1963. (Figure 223)**

*Diagnosis:* Scolex much wider than neck. Suckers transversely elongated, divided by septa into four chambers each. Genital pores preequatorial. Cirrus pouch crosses osmoregulatory canal. Testes in two lateral fields. Ovary posterior, bilobed. Vagina anterior or posterior to cirrus pouch. Uterus with many lateral branches. Parasites of snakes. Madagascar.

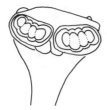

FIGURE 223. *Deblocktaenia ventosaloculata* (Deblock, Rosé et Broussart 1962) Odening 1963.

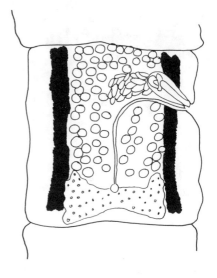

FIGURE 224. *Marsipocephalus rectangulus* Wedl 1861.

Type species: *D. ventosaloculata* (Deblock, Rosé et Broussart 1962) Odening 1963 (syn. *Ophiotaenia ventosaloculata* Deblock, Rosé et Boussart 1962), in *Ichycyphus miniatus;* Madagascar.

## DIAGNOSIS OF THE ONLY GENUS IN MARSIPOCEPHALINAE

*Marsipocephalus* Wedl 1861 (syn. *Lönnbergia* Fuhrmann et Baer 1925) (Figure 224)

*Diagnosis:* Scolex swollen, unarmed, with four simple suckers and no apical organ. Posterior proglottids with median ventral groove. Genital pores irregularly alternating, preequatorial. Cirrus pouch large. Testes in one broad dorsal cortical layer anterior to ovary. Vagina posterior to cirrus pouch. Ovary bilobed, medullary, posterior. Vitellaria in lateral medulla. Uterus in median medulla, with lateral outgrowths. Parasites of siluroid fishes. Africa.

Type species: *M. rectangulus* Wedl 1861, in *Heterobranchus anguillaris;* Egypt.
Other species:
   *M. daveyi* Woodland 1937, in *Heterobranchus bidorsalis;* Sierra Leone.
   *M. heterobranchus* Woodland 1925, in *Heterobranchus bidorsalis;* Nile River.
   *M. tanganyikae* (Fuhrmann et Baer 1925) Janicki 1928 (syn. *Loennbergia tanganyikae* Fuhrmann et Baer 1925), in *Clarias lazera;* Lake Tanganyika.

## FAMILY MONTICELLIDAE

## KEY TO THE SUBFAMILIES OF MONTICELLIDAE

1a. **Testes medullary** ................................................................ 2
1b. **Testes cortical** .................................................................. 3

2a. **Uterus medullary**..........**Zygobothriinae Woodland 1933 (p.191) (Figure 225)**
2b. **Uterus cortical, ovary partly cortical**..............................................
   ............................**Endorchiinae Woodland 1934 (p.193) (Figure 226)**

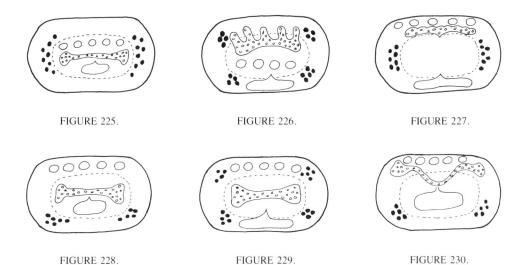

FIGURE 225.  FIGURE 226.  FIGURE 227.

FIGURE 228.  FIGURE 229.  FIGURE 230.

FIGURE 225 to 230.    Diagrammatic cross sections, subfamily Monticellidae. FIGURE 225, Zygobothriinae; FIGURE 226, Endorchiianae; FIGURE 227, Monticellinae; FIGURE 228, Ephedrocephalinae; FIGURE 229, Pelticotylinae; and FIGURE 230, Rudolphiellinae.

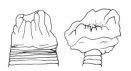

FIGURE 231. *Amphoteromorphus peniculus* Diesing 1850. Two forms of scolex.

3a. Uterus and ovary cortical ........ Monticellinae Mola 1929 (p.194) (Figure 227)
3b. Uterus and/or ovary medullary .................................................. 4

4a. Uterus and ovary medullary .........................................................
 ............................. Ephedrocephalinae Mola 1929 (p.195) (Figure 228)
4b. Ovary medullary, uterus cortical ....................................................
 ............................. Peltidocotylinae Woodland 1934 (p.195) (Figure 229)
4c. Ovary cortical, uterus medullary ....................................................
 ............................. Rudolphiellinae Woodland 1935 (p.195) (Figure 230)

## KEY TO THE GENERA IN ZYGOBOTHRIINAE

(syn. Postgangesiinae Achmerov 1969)

1a. Suckers sunken, entire sucker region surrounded by fleshy collar...............
 ................................. *Amphoteromorphus* Diesing 1850. (Figure 231)

*Diagnosis:* Scolex with suckers sunken, entire sucker region surrounded by fleshy collar. Neck absent. Genital pores irregularly alternating. Cirrus pouch well developed. Cirrus

FIGURE 232. *Zygobothrium megacephalum* Diesing 1850.

FIGURE 233. *Nomimoscolex dorad* (Woodland 1934) Freze 1965.

unarmed. Testes medullary, dorsal to uterus, in broad field anterior to ovary. Vagina anterior to cirrus pouch. Ovary medullary, bilobed, at posterior end of proglottid. Vitellaria lateral, cortical. Uterus medullary, with lateral branches. Parasites of siluroid fishes. South America.

Type species: *A. peniculus* Diesing 1850, in *Brachyplatystoma rousseauxii, Silurus dorada;* Amazon River.

Other species:
   *A. parkarmoo* Woodland 1935, in *Pseudopimelodus zungaro;* Amazon River.
   *A. piraeeba* Woodland 1934, in *Piraeeba* sp., *Brachyplatystoma filamentosum, B. rousseauxii;* Amazon River.
   *A. praeputialis* Rego, Santos, et Silva 1974, in *Cetopsis caecutiens;* Brazil.

**1b.   Suckers not surrounded by fleshy collar .......................................... 2**

**2a.   Genital pores unilateral ............... Zygobothrium Diesing 1850. (Figure 232)**

*Diagnosis:* Scolex with four massive, barrel-shaped suckers, each with prominent irregular hole in one side. Neck absent. Strobila with median groove on each flat surface. Genital pores unilateral. Cirrus pouch well developed. Cirrus unarmed. Testes medullary, dorsal, in broad preovarian field. Vagina posterior to cirrus pouch. Ovary transverse, medullary, at posterior end of proglottid. Vitellaria in lateral cortex. Uterus medullary, with lateral branches. Parasites of siluroid fishes. South America.

Type species: *Z. megacephalus* Diesing 1850, in *Phractocephalus hemiliopteus, Pirarara bicolor;* Amazon River.

**2b.   Genital pores irregularly alternating ............................................. 3**

**3a.   Scolex not spined, with or without apical organ but lacking prominent rostellum............................. Nomimoscolex Woodland 1934. (Figure 233)**

*Diagnosis:* Scolex normal but variable, with or without apical organ. Neck present. Genital pores irregularly alternating. Cirrus pouch large, well developed. Cirrus unarmed. Testes medullary, in broad preovarian field. Vagina anterior or posterior to cirrus pouch. Ovary medullary, bilobed, at posterior end of proglottid. Vitellaria in lateral cortex. Uterus ventral, medullary, with lateral branches and median, ventral pore. Parasites of siluroid fishes. South America.

Type species: *N. piraeeba* Woodland 1934, in *Brachyplatystoma filamentosum;* Amazon River.

Other species:
   *N. alovarius* Brooks et Deardorff 1980, in *Pimelodus clarias;* Colombia.
   *N. kaparari* Woodland 1935, in *Pseudoplatystoma tigrinum;* Amazon River.

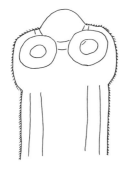

FIGURE 234. *Postgangesia orientale* Achmerov 1969.

FIGURE 235. *Endorchis piraeeba* Woodland 1934.

FIGURE 236. *Myzophorus admonticellia* Woodland 1934.

  *N. lenha* (Woodland 1933) Woodland 1935 (syn. *Proteocephalus lenha* Woodland 1933), in *Platystomatichthys sturio;* Amazon River.
  *N. magna* Rego, Santos, et Silva 1974, in *Pimelodus clarias;* Brazil.
  *N. piracatinga* Woodland 1935, in *Pimelodes pati, Brachyplatystoma filamentosum;* Amazon River.
  *N. sudobim* Woodland 1935, in *Pseudoplatystoma fasciatum, Brachyplatystoma rousseauxii;* Amazon River.

**3b. Scolex spinose, with large apical organ............................................
...................................*Postgangesia* Akhmerov 1969. (Figure 234)**

*Diagnosis:* Scolex not conspicuously set off from long neck. Large, unarmed apical organ present. Suckers simple. Scolex and neck spinose. Genital pores about equatorial. Cirrus pouch pyriform, not exceeding osmoregulatory canals. Vas deferens greatly coiled. Testes numerous(?), medullary, filling complete intervascular, preovarian field. Ovary at posterior end of segment(?), medullary. Vagina posterior of cirrus pouch. Vitellaria lateral, cortical. Uterus(?) medullary a median, longitudinal stem with lateral branches. Parasites of siluroid fishes. Asia.
  Type species: *P. orientale* Akhmerov 1969, in *Silurus soldatovi, Parasilurus asotus;* Amur River, Russia.

## KEY TO THE GENERA IN ENDORCHIINAE

**1a. Suckers with minute spines ............*Endorchis* Woodland 1934. (Figure 235)**

*Diagnosis:* Scolex small, with glandular apical organ. Suckers with triangular or trilocular opening and spinulate margin. Short neck present. Mature proglottids longer than wide. Genital pores irregularly alternating, near anterior end of proglottid. Cirrus pouch small, weakly developed. Cirrus large, unarmed. Testes medullary, in broad field anterior to ovary. Vagina anterior to cirrus pouch. Ovary blocked, mostly medullary, at posterior end of proglottid. Vitellaria lateral in cortex. Parasites of siluroid fishes. South America.
  Type species: *E. piraeeba* Woodland 1934, in *Brachyplatystoma filamentosum;* Brazil.
  Other species:
    *E. mandube* Woodland 1935, in *Pseudogeniosus brevifilis;* Amazon River.

**1b. Suckers not spined ..................*Myzophorus* Woodland 1934. (Figure 236)**

FIGURE 237. *Monticellia rugosa* Woodland 1935.

*Diagnosis:* Scolex unarmed, with apical organ or not. Suckers lacking minute spines on their margins. Neck short or absent. Genital pores irregularly alternating, in anterior half of proglottid margin. Cirrus pouch well developed. Cirrus unarmed. Testes in broad ventral field anterior to ovary. Vagina apparently anterior of cirrus pouch. Ovary bilobed, posterior, with dorsal branches. Vitellaria lateral, often crescentic in cross section. Uterus median, medullary, with lateral outpocketings. Oncosphere lacking hooks (?). Parasites of siluroid fishes. South America.

Type species: *M. admonticellia* Woodland 1934, in *Pirinampus* spp.; Amazon River.
Other species:
   *M. dorad* Woodland 1935, in *Brachyplatystoma rousseauxii*; Amazon River.
   *M. pirara* Woodland 1935, in *Phractocephalus hemiliopterus*; Amazon River.
   *M. sudobim* Woodland 1935, in *Pseudoplatystoma fasciatum*; Amazon River.

## DIAGNOSIS OF THE ONLY GENUS IN MONTICELLIINAE

*Monticellia* LaRue 1911
(syn. *Goezeella* Furhmann 1916, *Spatulifer* Woodland 1934). (Figure 237)

*Diagnosis:* Scolex swollen, unarmed, with simple suckers. Fold of tissue somtimes present around base of scolex. No apical organ. Neck present or absent. Genital pores irregularly alternating, preequatorial. Cirrus pouch with thin walls. Cirrus unarmed. Testes in broad band in dorsal cortex, anterior to ovary. Vagina anterior or posterior to cirrus pouch. Ovary bilobed, mostly cortical, partly medullary. Vitellaria in lateral cortex. Uterus in ventral cortex. Parasites of siluroid fishes. South America.

Type species; *M. coryphicephala* (Monticelli 1891) LaRue, 1911 (syn. *Tetracotylus coryphicephala* Monticelli 1891, *Ichthyotaenia coryphicephala* [Monticelli 1891] Lönnberg 1894, *Taenia coryphicephala* [Monticelli 1891] Monticelli 1900, *Proteocephalus coryphicephala* [Monticelli 1891] Monticelli 1900, *Proteocephalus coryphicephala* [Monticelli 1891] Harwood 1933), in *Silurus* sp.; South America?
Other species:
   *M. diesingii* (Monticelli 1891) LaRue 1911, in *Silurus dargado*; South America?
   *M. lenha* Woodland 1933, in *Platystomatichthys sturio*; Amazon River.
   *M. macrocotyle* (Monticelli 1891) La Rue 1911, in *Silurus megacephalus*; South America?
   *M. megacephala* Woodland 1934, in *Platystomatichthys sturio*; Amazon River.
   *M. piracatinga* Woodland 1935, in *Pimelodus pati*; Amazon River.
   *M. piramutab* (Woodland 1933) Woodland 1935 (syn. *Goezeella piramutab* Woodland 1933), in *Brachyplatystoma vaillanti*; Amazon River.
   *M. rugata* Rego 1975 (syn. *Spatulifer rugata* [Rego 1975] Brooks et Deardorff 1980), in *Calophysus macropterus*; Brazil.
   *M. rugosa* Woodland 1935, in *Pseudoplatystoma fasciatum*; Amazon River.

FIGURE 238. *Ephedrocephalus microecaphalus* Diesing 1850.

FIGURE 239. *Peltidocotyle rugosa* Diesing 1850.

*M. siluri* (Furmann 1916) Woodland 1935 (syn. *Goezeella siluri* Fuhrmann 1916), in Silurid fishes, *Cetopsis caecutiens, Ageneiosus caucanus*; Amazon, Colombia.

*M. spinulifera* Woodland 1935, in *Pseudoplatystoma fasciatum*; Amazon.

*M. surubim* (Woodland 1934) Woodland 1935 (syn. *Spatulifer surubim* Woodland 1934), in "surubim"; Amazon.

## DIAGNOSIS OF THE ONLY GENUS IN EPHEDROCEPHALINAE

*Ephedrocephalus* Diesing 1850. (Figure 238)

*Diagnosis:* Scolex unarmed, with suckers small, sunken; entire sucker region surrounded by fleshy collar. Neck present? Uterine pores present, opening in median ventral groove in strobila. Genital pores irregularly alternating, in anterior half of proglottid margin. Cirrus pouch small. Testes in broad cortical field, dorsal, anterior to ovary. Vagina anterior (or posterior?) to cirrus pouch, with powerful sphincter near distal end. Ovary transverse, entirely medullary. Vitellaria ventral, cortical. Uterus in ventral medulla, with ventral median pore. Eggs longer than wide. Parasites of siluroid fishes. South America.

Type species: *E. microcephalus* Diesing 1850, in *Phractocephalus hemiliopterus*; Brazil.

## DIAGNOSIS OF THE ONLY GENUS IN PELTIDOCOTYLINAE

*Peltidocotyle* Diesing 1850 (syn. *Othinoscolex* Woodland 1933, *Woodlandiella* Freze 1965). (Figure 239)

*Diagnosis*: Scolex dorsoventrally flattened posteriorly, rugose. Each sucker divided into two loculae by distinct transverse muscular septum. Neck absent. Strobila wide behind scolex, tapering toward posterior end. All proglottids broader than long. Genital pores irregularly alternating, at anterior edge of proglottid. Cirrus pouch and cirrus muscular, well developed. Testes in dorsal cortex. Ovary transverse, medullary, at posterior end of proglottid. Vitellaria cortical, lateral. Uterus in ventral cortex, with lateral branches. Parasites of siluroid fishes. Brazil.

Type species: *O. rugosa* Diesing 1850, in *Platystoma tigrinum*; Brazil.

Other species:

*P. lenha* (Woodland 1933) Woodland 1934 (syn. *Othinoscolex lenha* Woodland 1933), in *Platystomatichtys sturio*; Amazon River.

*P. myzofera* (Woodland 1933) Woodland 1934 (syn. *Othinoscolex myzofer* Woodland 1933, *Woodlandiella myzofera* [Woodland 1933] Freze 1965), in *Platystomatichthys sturio*; Amazon River.

## DIAGNOSIS OF THE ONLY GENUS IN RUDOLPHIELLINAE

(syn. Amphilaphorchidinae Woodland 1934)

*Rudolphiella* Fuhrmann 1916 (syn. *Amphilaphorchis* Woodland 1934).

*Diagnosis*: Scolex wrinkled, furrowed, with central prominence bearing suckers. Neck absent (?). Genital pores preequatorial, irregularly alternating. Cirrus pouch with thin walls. Cirrus unarmed. Testes cortical, dorsal, in entire field anterior to, and sometimes lateral to, ovary. Vagina anterior or posterior to cirrus pouch. Ovary mostly cortical. Vitellaria entirely cortical, in lateral fields. Uterus entirely medullary, median, with lateral branches. Eggs with polar elongations. Parasites of siluroid fishes. South America.

Type species: *R. lobosa* (Riggenbach 1895) Fuhrmann 1916 (syn. *Corallobothrium lobosa* Riggenbach 1895, *Ephedrocephalus lobosa* [Riggenbach 1895] Mola 1906), in *Pimelodus pati;* Paraquay.

Other species:

*R. myoides* (Woodland 1934) Woodland 1935 (sun. *Amphilaphorchis myoides* Woodland 1934), in *Pirinampus pirinampus;* Amazon.

*R. piranabu* (Woodland 1934) Woodland 1935 (syn. *Amphilaphorchis piranabu* Woodland 1934), in *Pirinampus pirinampus;* Amazon.

## ORDER DIOECOTAENIIDEA ORD. N.

This order contains only two species, both parasites of the spiral valve intestine of the cow-nosed ray (Batoidea). When I first discovered it (Schmidt[3094]), I was so struck by its unique characteristics that I established a new genus and family for it, and placed it within the Tetraphyllidea. The scolex certainly is tetraphyllidean in nature, but the internal organs are so different from those of any other tapeworm that now I am required to regard these worms as constituting a separate order.

Dioecotaeniids are the only dioeceous cestodes outside the cyclophyllidean family Dioecocestidae, all parasites of charadriiform birds. The female is much larger than the male and has no genital pore; sperm transfer is by hypodermic impregnation. The vagina does not reach the surface of the body, but leads into a seminal receptable located within one lobe of the ovary. The vitellaria are condensed upon the posterolateral surfaces of the ovary. After injection, the cirrus breaks off within the female and comes to lie within a longitudinal, dorsomedian sheath, along with cirri from previous injections. Larval forms have been found in molluscs. A theoretical paper on the evolution of this group was published by Brooks.[382]

## DIAGNOSIS OF THE ONLY FAMILY IN DIOECOTAENIIDEA

Dioecotaeniidae Schmidt 1969 (p.197)

*Diagnosis*: Scolex with four bothridia, each with surficial loculi. Myzorhynchus, hooks, and suckers absent. Neck present. Proglottids acraspedote. Testes arranged in circle in two layers. Male genital pore lateral. Cirrus pouch with large internal seminal vesicle. External seminal vesicle absent. Ovary transversely elongated, containing seminal receptacle in one lobe. Vagina median, convoluted, lacking external pore. Vitelline glands condensed on posterolateral surfaces of lobes of ovary. Uterus a bilobed transverse sac. Uterine pore ventral. Dorsomedian sheath contains injected cirri, continuous from level of first vagina to posterior end. Osmoregulatory canals with six major vessels, lateral ones with ducts to margins of each segment. Parasites of rays (Myliobatidae). North America.

Type genus: *Dioecotaenia* Schmidt 1969.

## DIAGNOSIS OF THE ONLY GENUS IN DIOECOTAENIIDAE

*Dioecotaenia* Schmidt 1969 (Figure 240)

Sexes completely separate, each lacking even accessory genital organs of the other. Sexual dimorphism of size and shape apparent. Scolex with four bothridia on short peduncles, each divided into loculi. External segmentation feeble, proglottids acraspedote. Male genital pore lateral, irregularly alternating. Testes medullary, in two layers, arranged in a circle on both sides of cirrus pouch. Cirrus pouch muscular, large, containing internal seminal vesicle. Cirrus long, armed at base with large hooks; the remainder covered with minute, deciduous spines. Ovary bilobed, medullary. Vagina median, convoluted. Vaginal pore absent. Seminal receptable embedded in substance of one lobe of ovary, irregularly alternating sides. Vitelline glands two, compact, surrounding posterolateral margins of both lobes of ovary. Uterus a bilobed, transverse sac arising anterior to ovary. Uterine pore preformed, medioventral, near posterior margin of proglottid. Sperm transfer by hypodermic impregnation. Medullary, continuous, dorsomedian sheath containing injected cirri present in female. Parasites of cow-nosed ray. Chesapeake Bay.

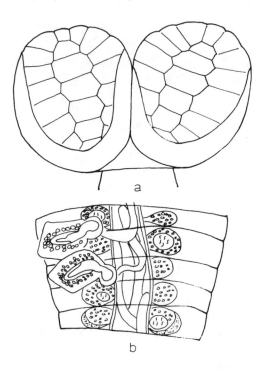

FIGURE 240. *Dioecotaenia cancellata* (Linton 1890) Schmidt 1969. (a) Scolex; (b) mature portion of female strobila, with male proglottids attached.

Type species: *D. cancellata* (Linton 1890) Schmidt 1969 (syn. *Rhinobothrium cancellata* Linton 1890), in *Rhinoptera bonasus*; Atlantic Ocean.
Other species:
  *D. campbelli* Mayes et Brooks 1980, in *Rhinoptera bonasus*; Venezuela.

# ORDER CYCLOPHYLLIDEA

This is the largest order of tapeworms, with more species than all other orders together. Nearly all are parasites of terrestrial or semiaquatic vertebrates, with a notable absence in fishes. They develop in a wide variety of invertebrate intermediate hosts.

A combination of two easily observed characteristics serve to identify a cestode as cyclophyllidean: there is a single, condensed vitellarium (double in Mesocestoididae), and the scolex bears four simple, rounded suckers. Occasional modifications of the suckers are found, but, as in the Proteocephalidea, there seldom can be a doubt as to their basic form.

In recent years there have been attempts to break this large group into several orders, especially by Soviet authors. Also, Wardle et al.[2871] created several new orders, some of which had already been called new orders by other authors. Basically, most of the proposed changes have been the elevation of families to ordinal level and, in some cases, the recognition of infracategories such as tribes, superfamilies, etc. It is not my intent to argue either the phylogenetic significance or the taxonomic consequences of these changes. My goal throughout is to provide the simplest, most accurate method for identifying an unknown tapeworm. However, I have noted major revisions in the various groups, with references for the interested researcher.

## KEY TO THE FAMILIES IN CYCLOPHYLLIDEA

**1a. Genital openings medial.................. Mesocestoididae Perrier 1897 (p.202)**

*Diagnosis:* Scolex with simple, unarmed suckers; lacking rostellum. Single set of reproductive organs per proglottid. Genital pores median, ventral. Uterine pore absent. Uterus present or replaced by paruterine organ. Vitellarium doubled, postovarian. Parasites of birds and mammals.

Type genus: *Mesocestoides* Vaillant 1863.

**1b. Genital openings marginal or rarely submarginal ............................. 2**

**2a. Vitellarium usually anterior to ovary (see also *Arostellina*, Anoplocephalidae). Suckers often with muscular outgrowths.. Tetrabothriidae Linton 1871.(p.204)**

*Diagnosis*: Rostellum lacking. Suckers unarmed, variable from simple to possessing muscular appendages; rarely diminutive or lacking. Neck present or absent. Proglottids craspedote or acraspedote, usually wider than long. Genital atrium unilateral, complex. Cirrus pouch small. Testes numerous. Ovary transversely elongated, with many lobes. Vitellarium anterior, ventral or posterior of ovary. Uterus a transverse tube. Parasites of marine birds and mammals.

Type genus: *Tetrabothrius* Rudolphi 1819.

**2b. Vitellarium not anterior to ovary. Suckers without muscular outgrowths..... 3**

**3a. Strobila cylindrical. Segmentation weak, only in posterior region. One to ten testes per segment ........................ Nematotaeniidae Lühe 1910. (p.209)**

*Diagnosis:* Scolex lacking rostellum and with four simple, unarmed suckers. Strobila cylindrical to oval. External metamerism evident only near posterior end. Genital pores irregularly alternating. Testes numbering one to ten. Ovary compact. One or more paruterine organs containing egg capsules, each with one to many eggs. Parasites of amphibians and reptiles. Considered a new order, Nematotaeniidea, by Wardle et al.[3871]

Type genus: *Nematotaenia* Lühe 1899.

3b. Segmentation well marked throughout most of body .......................... 4

4a. Dioecious forms, either entire worm or regionally................................
................................................ **Dioecocestidae Southwell 1930. (p.213)**

Diagnosis: Dioecious forms, sexes completely separated, or strobilas regionally unisexual. Rostellum present or absent; if present it may be armed or not. Vagina present or absent; if present it may lack external opening. Parasites of birds.

Type genus: *Dioecocestus* Fuhrmann 1900.

4b. Entire strobila monoecious....................................................... 5

5a. Proterogynous forms, testes developed in posterior proglottids only............
.................................................. **Progynotaeniidae Fuhrmann 1936. (p.217)**

*Diagnosis*: Scolex with armed rostellum; may be divided by constriction into anterior portion bearing the rostellum and posterior portion bearing the suckers. Strobila small, with weakly developed musculature. Proglottids proterogynous; hermaphroditic, or with male and female segments regularly alternating. Vagina absent. Parasites of birds.

Type genus: *Progynotaenia* Fuhrmann 1909.

5b. Proterandrous forms, testes developing before ovary.......................... 6

6a. Rostellum, if present, not retractable. Eggshell with radial striations ...........
...............................................**Taeniidae Ludwig 1886. (p.221)**

*Diagnosis*: Rostellum not retractable (absent in *Taeniarhynchus*). With one or two circles of hooks. Strobila usually medium to very large, rarely very small. Gravid proglottids longer than wide. Single set of genitalia per segment. Genital pores irregularly alternating. Testes numerous. Ovary posterior. Gravid uterus median, with lateral branches. Middle eggshell thick, with radial striations. Larval forms bladder worms developing in mammals. Parasites of mammals. An important monograph was published by Abuladze.[5]

Type genus: *Taenia* Linnaeus 1758.

6b. Rostellum retractable or absent. Eggshell lacking radial striations ............ 7

7a. Vaginal pores absent, may be replaced by accessory canal..................... 8
7b. Vaginal pores present .......................................................... 9

8a. Vaginal pore replaced by dorso-ventral accessory canal arising from seminal receptable and opening on one or both flat surfaces............................
................................................**Amabiliidae Ransom 1909. (p.227)**

*Diagnosis*: Scolex with armed rostellum. Strobila small. Proglottids with lateral lappets bearing male genital pores, alternating regularly or irregularly. Single set of male and female genitalia per segment, or in *Amabilia* double male and single female sets. Parasites of birds.

Type genus: *Amabilia* Diamare 1893.

8b.  Vagina swollen as seminal receptable, not replaced by accessory canal .........
........................................Acoleidae Fuhrmann 1906. (p.230)

*Diagnosis*: Scolex with armed rostellum and simple suckers. Proglottids wider than long. Male genital pore regularly alternating. Testes numerous. Vagina represented by a large, transverse seminal receptacle lacking external opening. Uterus sac-like. Parasites of birds.
Type genus: *Acoleus* Fuhrmann 1899.

9a.  Rostellum with many very small T- or hammer-shaped hooks. Suckers often spinous....................................Davaineidae Fuhrmann 1907. (p.237)

*Diagnosis*: Rostellum retractable, with one to several circles of very small, very numerous T- or hammer-shaped hooks. Suckers commonly armed with very small spines that are easily lost from dead specimens. Reproductive systems usually single, but may be double per proglottid. Genital pores lateral. Uterus sac-like or replaced by egg capsules or paruterine organ. Parasites of birds and mammals.
Type genus: *Davainea* Blanchard 1891.

9b.  Rostellar hooks otherwise, or rostellum absent. Suckers rarely spinose ...... 10

10a.  Testes 12 or fewer, usually 1 to 4. Rostellum usually present; if absent, testes number 3 or fewer .......... Hymenolepididae Railliet et Henry 1909. (p.267)

*Diagnosis*: Rostellum present or absent; usually armed when present. Suckers armed or not, rarely in two nearly fused pairs or so small as to be nearly absent. Pseudoscolex sometimes present behind true scolex. Single, rarely double, set of reproductive organs per proglottid. Testes 12 or fewer, usually 1 to 4. Genital pores unilateral, rarely alternating (*Allohymenolepis*). Uterus usually saccular, may be reticular or replaced by egg capsules. Parasites of birds and mammals.
Type genus: *Hymenolepis* Weinland 1958.

10b.  Testes usually more than 12. Rostellum present or absent; if absent, testes numerous .................................................................... 11

11a.  Uterus a median stem with lateral branches, as in Taeniidae...................
........................................ Catenotaeniidae Spasskii 1950. (p.232)

*Diagnosis:* Scolex lacking rostellum. Apical sucker present in young specimens, but disappears as worm matures. Genital pores irregularly alternating. Testes usually postovarian, rarely in two fields lateral to ovary. Ovary lobated, anterior. Vitellaria postovarian, lobated, poral. Uterus with median stem and lateral branches, recalling *Taenia*. Parasites of rodents.
Type genus: *Catenotaenia* Janicki 1904.

11b.  Uterus otherwise......................................................... 12

12a.  Armed rostellum usually present. If absent, gravid proglottids usually longer than broad. Eggs never with pyriform apparatus.................................
........................................ Dilepididae Railliet et Henry 1909. (p.391)

*Diagnosis:* Armed rostellum usually present, occasionally absent or rudimentary and unarmed. Suckers unarmed (except *Cotylorhipis*). Reproductive organs single or double.

Genital pores lateral. Gravid proglottids usually longer than broad. Testes commonly numerous, occasionally few. Gravid uterus usually sac-like, occasionally reticular, or replaced by egg capsules, or with paruterine organ. Parasites of reptiles birds, and mammals.
Type genus: *Dilepis* Weinland 1858.

**12b. Rostellum and scolex armature always lacking. Gravid proglottids usually broader than long. Eggs commonly with pyriform apparatus .......................... .................................. Anoplocephalidae Cholodkovsky 1902. (p.415)**

*Diagnosis*: Scolex lacking rostellum. Suckers unarmed. Neck short or absent. Proglottids numerous, craspedote or acraspedote, usually wider than long. Reproductive systems single or double in each segment. Genital pores marginal. Testes numerous. Uterus saccular, tubular, or reticular, or replaced by egg capsules, or with one to several parauterine organs. Eggs commonly with pyriform apparatus. Parasites of reptiles, birds and mammals.
Type genus: *Anoplocephala* Blanchard 1848.

## FAMILY MESOCESTOIDIDAE

No complete life cycle is known within this family. A unique larval form, the tetrathyridium, is commonly found in mammalian, avian, and reptilian intermediate hosts, and is readily infective to predatory definitive hosts. So far, the first intermediate host has not been found, for eggs are not capable of infecting the vertebrates harboring tetrathyridia. For this reason, together with the paired vitellaria and ventral genital pore, some authorities raise this family to the rank of order Mesocestoidata. The last to do so was Wardle et al.[3871] Concluding that Mesocestoididae has more in common than not with other families in Cyclophyllidea, I retain it in that order. Tschertkowa and Kosupko[3700] created a new family for the monotypic genus *Mesogyna* Voge 1952. I consider this as a subfamily in the present work. Because there is a very great amount of morphological variation within most species, many nominal species may actually be synonyms. An important monograph is presented by Tschertkowa and Kosupko.[3700]

## KEY TO THE SUBFAMILIES IN MESOCESTOIDIDAE

1a. Paruterine organ present................. **Mesocestoidinae Perrier 1897. (p.202)**
1b. Paruterine organ absent.. **Mesogyninae Tschertkowa et Kosupko 1977. (p.204)**

## DIAGNOSIS OF THE ONLY GENUS IN MESOCESTOIDINAE

*Mesocestoides* Vaillant 1863. (Figure 241)

*Diagnosis*: Scolex with four simple suckers. Rostellum absent. Neck present. Proglottids craspedote, apolytic. Cirrus pouch oval, preequatorial. Genital atrium midventral, posterior to equator of cirrus pouch. Seminal vesicle absent. Testes numerous, pre- or postovarian. Ovary bilobed, posterior. Vitellaria in two masses, behind ovary. Vaginal pore anterior or lateral to cirrus. Uterus first developing as a median, sinuous tube: eggs moving into paruterine organ early. Parasites of birds and mammals. Cosmopolitan.
Type species: *M. ambiguus* Vaillant 1863, in *Viverra genetta*; Africa.
Other species:
   *M. alaudae* Stossich 1896, in *Alauda arvensis*; Europe, Asia.
   *M. angustatus* (Rudolphi 1819) Vaillant 1863, in *Meles taxus*; Europe.
   *M. bassarisci* MacCallum 1921, in *Bassariscus astutus*; Mexico.

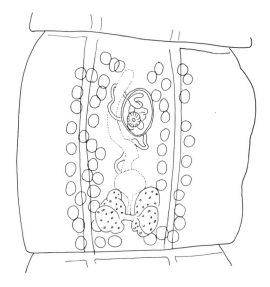

FIGURE 241. *Mesocestoides lineatus* (Goeze 1782) Railliet 1893.

*M. bergini* Tschertkowa et Kosupko 1975, in *Alopex lagopus*; Russia.

*M. caestus* Cameron 1925 (syn. *M. lineatus forma caestus* Witenberg 1934), in *Mellivora capensis, Vulpes vulpes, Canis Familiaris, Felis ocreata;* Africa, Israel.

*M. canislagopodis* (Rudolphi 1810), in *Alopex lagopus, Vulpes vulpes*; Europe.

*M. charadrii* Fuhrmann 1909, in *Erolia minuta*; Africa.

*M. corti* Hoeppli 1925, in *Mus musculus, Procyon lotor, Canis familiaris, Mephitis mephitis*; U.S.

*M. didelphus* Tschertkowa et Kosupko 1977 (syn. *M. variabilis* Mueller 1928 sensu Byrd et Ward 1943), in *Didelphis virginiana*; U.S.

*M. dissimilis* Baer 1933, in *Myonax sanguineus;* Africa.

*M. elongatus* Meggitt 1928, in *Canis lupus*; Egypt.

*M. erschovi* Tschertokowa et Kosupko 1975, in *Vulpes vulpes, V. corsak, Ursus arctos*; Russia.

*M. imbutiformis* (Polonio 1860), in *Anser ferus*; Italy.

*M. jonesi* Ciordia 1955, in *Urocyon cineroargentatus*; U.S.

*M. kirbyi* Chandler 1944, in *Canis latrans, C. lupus, Alopex lagopus, Vulpes vulpes, V. fulva, V. corsak, Procyon lotor, Mephitis mephitis, Gulo gulo;* North America, Russia.

*M. latus* Mueller 1927, in *Mephitis mephitis, Felis ocreata, Didelphis virginiana*; U.S.

*M. lineatus* (Goeze 1782) Railliet 1893 (syn. *Taenia lineata* Goeze 1782, *Taenia literrata* Batsch 1786, *Taenia vulpina* Schrank 1788, *Taenia cataeniformisvulpes* Gmelin 1790, *T. pseudocucumerina* Railliet 1863, *Monoderidum urticulifera* Walter 1866, *Ptychophysa lineata* (Goeze 1782) Hamann 1885, *Halisis lineata* [Goeze 1782] Zeder 1803, *Mesocestoides angustatus* [Rudolphi 1819], *M. literratus* [Batsch 1786] sensu Mueller 1927, *M. variabilis* Mueller 1927, nec Byrd et Ward 1943, *M. tenuis* Meggitt 1931, *M. carnivoricolus* Grundmann 1956, *M. paucitesticulatus* Sawada et Kugi 1973), in many species of carnivora; nearly cosmopolitan.

*M. literratus* (Batsch 1786) (syn. *Taenia literrata* Batsch 1786, *M. canislagopodis* [Rudolphi 1810] sensu Zschokke 1889, *M. lineatus* var. *literrata* Witenberg 1934), in *Vulpes vulpes*; Europe.

*M. longistriatus* Setti 1897, in *Getto selvatico*; Africa.

*M. magellanicus* (Monticelli 1889), host and locality?

*M. manteri* Chandler 1943, in *Lynx lynx;* U.S.

*M. mesorchis* Cameron 1925 (syn. *M. ambiguus* Vaillant 1863 sensu Joyeux et Baer 1932), in *Vulpes ferritatus, Felis ocreata;* Nepal, France.

*M. michaelseni* Lönnberg 1896, in *Canis azarae;* Brazil.

*M. perlatus* (Goeze 1782) Muhling 1898 (syn. *Taenia perlata* Goeze 1782, *T. chrysaeti* Viborg 1795, *T. margaritifera* Creplin 1829, *T. taenius* Creplin 1829, *Halysis perlata* [Goeze 1782] Zeder 1803), in many species of *Falconiformes;* Africa, Asia, Europe, North America.

*M. petrowi* Sadychov 1971 (syn. *M. lineatus* [Goeze 1782] sensu Furmaga, Wisocki 1951, *M. lineatus* [Goeze 1782] sensu Voge 1955), in *Vulpes vulpes, Canis aureus, Felis sylvestris;* Russia, U.S.

*M. zacharovae* Tschertkowa et Kosupko 1975 (syn. *M. lineatus* [Goeze 1782] sensu Mueller 1928, sensu Voge 1955, sensu Dollfus 1965, *M. litteratus* [Batsch 1786] sensu Markowski 1934), in *Vulpes vulpes, Canis aureus, C. lupus, C. familiaris, Felis ocreata;* Russia, Europe.

## DIAGNOSIS OF THE ONLY GENUS IN MESOGYNINAE

### *Mesogyna* Voge 1952

Diagnosis: Scolex with four simple, unarmed, elongate suckers. Rostellum absent. Neck present. Proglottids craspedote, wider than long. Genital pore midventral. Cirrus pouch pyriform, preequatorial. Seminal vesicles absent. Testes in two lateral fields overreaching osmoregulatory canals and extending posterior to level of ovary. Ovary bilobed, near posterior end. Vitellaria bilobed, behind ovary. Vaginal pore not observed. Uterus fills with eggs, not replaced by paruterine organ. Parasites of liver of foxes. U.S.

Type species: *M. hepatica* Voge 1952, in *Vulpes macrotis;* California.

## FAMILY TETRABOTHRIIDAE

This family was raised to ordinal rank by Baer.[145] I can see no justification for excluding this family from Cyclophyllidea. It consists of several species that are common and widespread, mainly in marine birds and mammals. A recent monograph on the family was published by Temirova and Skrjabin.[3645]

## KEY TO THE GENERA IN TETRABOTHRIDAE

**1a. Scolex without suckers, cone-shaped with collar-like base.................................... *Priapocephalus* Nybelin 1922. (Figure 242)**

*Diagnosis:* Scolex conical, basal portion collar-like. Suckers absent. Neck present (?). Proglottids much wider than long, slightly craspedote. Genital atrium not muscular. Cirrus pouch small, pyriform. Cirrus armed. Testes numerous (300 to 400), in two lateral fields. Ovary slightly poral, multilobated. Vitelline gland with many branches, anterioventral to ovary. Vagina lined with hair-like spines, opening ventral to cirrus. Seminal receptacle absent. Uterus sac-like, with dorsal diverticula. Parasites of whales. Antarctica, France, Sweden.

Type species: *P. grandis* Nybelin 1922, in *Balaenoptera borealis, B. musculus, B. physalus, Eublaena glacialis, Physeter catodon;* Antarctica, Atlantic, Pacific.

Other species:

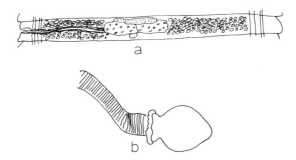

FIGURE 242. *Priapocephalus grandis* Nybelin 1922. (a) Proglottid; (b) scolex.

FIGURE 243. *Strobilocephalus triangularis* (Diesing 1850) Baer 1932.

*P. eschrichtii* Muravijova et Treshchev 1970, in *Eschrichtius gibbosus*; Chukchi Sea.
*P. minor* Nybelin 1928, in *Balaenoptera borealis, B. physalus*; Sweden, France, Antarctica, Russia.

**1b. Scolex with suckers............................................................ 2**

**2a. Suckers small, triangular, sunken into base of swollen scolex...................
........................................ *Strobilocephalus* Baer 1932. (Figure 243)**

*Diagnosis*: Scolex globular, with four small, triangular suckers sunken into its base. Neck present. Proglottids much wider than long. Genital ducts pass between osmoregulatory canals. Cirrus pouch small, muscular. Cirrus armed, opening on papilla within atrium dorsal to vaginal pore. Testes few (24) surrounding female organs, separated by uterus. Genital pores unilateral. Ovary transversely bilobed, extending entire length of proglottid. Vitelline gland anterioventral to ovary. Vagina thick-walled, lined with hair-like spines. Seminal receptacle absent. Uterus a transverse sac extending between dorsal and ventral osmoregulatory canals and possessing a rudimentary middorsal pore. Parasites of Cetacea (dolphins). Atlantic and Pacific.
   Type species: *S. triangularis* (Diesing 1850) Baer 1932 (syn. *Tetrabothrius triangulare* Diesing 1850, *T. triangularis* Fuhrmann 1904, *Prosthecocotyle triangulare* Fuhrmann 1899), in *Delphinus* sp., *Hyperoodon ampullatus, Lagenorhynchus acutus, Mesoplodon bidens, Steno rostratus*; Atlantic and Pacific oceans.

**2b. Scolex not as above............................................................ 3**

**3a. Suckers lacking muscular appendages............................................
........................................ *Anophryocephalus* Baylis 1922. (Figure 244)**

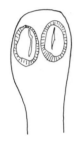

FIGURE 244. *Anophryocephalus anophrys* Baylis 1922.

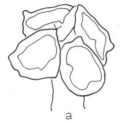 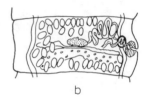

FIGURE 245. *Trigonocotyle monticellii* (Linton 1923) Baer 1932. (a) Scolex; (b) progglottid.

*Diagnosis*: Scolex simple, with four simple suckers lacking appendages, but with slit-like apertures. Neck long. Proglottids wider than long, slightly craspedote. Genital pores unilateral, dextral. Cirrus pouch elongate. Genital atrium with muscular wall ventral to vaginal pore. Testes large, few (30), in dorsal medulla surrounding female organs. Ovary massive, bilobed, extending entire width between osmoregulatory canals. Vitelline gland compact, preovarian. Vagina narrow distally, wide proximally. Uterus in shape of transverse crescent. Parasites of seals. North Atlantic, Russia.

Type species: *A. anophrys* Baylis 1922 (syn. *Tetrabothrius albertinii* Brigenthy 1931), in *Pusa hispida, Cystophora cristata*; Iceland, Norway.
Other species:
  *A. ochotensis* Delyamure et Krotov 1955, in *Eumetopias jubatus*; Okhotsk Sea.
  *A. skrjabini* (Krotov et Delyamure 1955) Murvijova 1968 (syn. *Trigonocotyle skrjabini* Krotov et Delyamure 1955), in *Phoca vitulina, Pusa hispida, Histriophoca fasciata*; Okhotsk Sea.

**3b.   Suckers possessing muscular appendages ........................................ 4**

**4a.   Suckers triangular, with small outgrowth at each corner ........................**
**............................................ Trigonocotyle Baer 1932. (Figure 245)**

*Diagnosis*: Scolex with four large, triangular suckers, each with a small, fleshy appendage at each corner. Neck present. Genital pore unilateral. Genital atrium very muscular. Cirrus pouch pyriform, elongate, muscular. Testes few, large, surrounding female glands. Ovary bilobed, postequatorial. Vitellaria compact, preovarian. Vagina thick-walled. Seminal receptacle absent. Uterus a transverse, intervascular sac. Parasites of toothed whales. Atlantic, Pacific, Mediterranean.

Type species: *T. monticellii* (Linton 1923) Baer 1932 (syn. *Prosthecocotyle monticelli* Linton 1923, *T. lintoni* Guiart 1935, *T. globicephalae* Baer 1954), in *Globicephala melaena, G. macrorhyncha, G. melaena, G. melas, G. scammoni*; Japan, Mediterranean, Atlantic.
Other species:
  *T. spasskyi* Gubanov in Delyamure 1955, in *Orcinus orca*; Okhotsk Sea.

**4b.   Suckers rounded, each with outgrowth on anterior margin .................... 5**

**5a.   Genital atrium not muscular ........................ Chaetophallus Nybelin 1916.**

*Diagnosis*: Scolex rectangular. Suckers rounded, each with single muscular appendage extending laterally from the anterior margin. Neck absent. Proglottids broader than long,

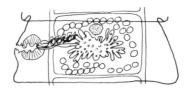

FIGURE 246. *Tetrabothrius macrocephalus* (Rudolphi 1810) Rudolphi 1819.

craspedote. Genital pores unilateral. Cirrus pouch elongate. Genital atrium muscular, suckerlike. Testes entirely postovarian. Ovary multilobate, anterior. Vitellaria median, preovarian. Vagina ventral to cirrus pouch. Seminal receptacle present. Gravid uterus filling proglottid. Parasites of Podicipediformes, Pelecaniformes. Japan, Mexico, Russia.

Type species: *C. robustus* Nybelin 1916, in *Thalassogeron chlororhynchus;* Angola, Africa.

Other species:

*C. umbrella* (Fuhrmann 1898) Nybelin 1916 (syn. *C. musculosus* Szpotanska 1917) in *Diomeda exulans, Phoebetria fuliginosa, Thalassoica antarctica, Macronectes giganteus;* Antarctica.

**5b. Genital atrium muscular** ..........................................................................
.............. ***Tetrabothrius*** **Rudolphi 1819** (syn. ***Amphoterocotyle*** **Diesing 1863,** ***Prosthecocotyle*** **Monticelli 1892,** ***Bothridiotaenia*** **Lönnberg 1896,** ***Oriana*** **Leiper et Atkinson 1914,** ***Porotaenia*** **Szpotanska 1917,** ***Neotetrabothrius*** **Nybelin 1929,** ***Paratetrabothrius*** **Yamaguti 1940,** ***Tetrabothrium*** **Diesing 1856). (Figure 246)**

*Diagnosis:* Scolex rectangular, with four large suckers, each of which has a muscular appendage on its anterior margin. Neck absent. Proglottids wider than long. Genital pores unilateral. Genital atrium muscular. Cirrus pouch rounded, muscular. Testes surrounding female gonads. Ovary multilobated, about equatorial. Vitellaria compact, preovarian. Vagina thick-walled. Seminal receptable present in some species. Uterus saccular or lobated, often with dorsal pore. Parasites of Anseriformes, Charadriiformes, Ciconiiformes, Falconiformes, Pelecaniformes, Podicipediformes, Procellariiformes, Sphenisciformes. Also in whales, seals, dolphins, fox. Cosmopolitan.

Type species: *T. macrocephalus* (Rudolphi 1810) Rudolphi 1819 (syn. *Taenia immerina* Abildgaard 1790, *Tetrabothrius junceus* Baird 1867, *T. arcticus* Linstow 1901, *T. lobatus* Linstow 1905, *T. perfidus* Joyeux et Baer 1934, *T. rostratula* Yamaguti 1940, *Paratetrabothrius orientalis* Yamaguti 1940), in *Brachyhamphus marmoratus, Colymbus auritus, C. cristatus, Gavia stellata, G. arctica, G. adamsi, G. immer, Phalacrocorax aristotelis, Rissa tridactyla, Rostratula bengalensis, Somateria molissima*; Bering Sea, Russia, Atlantic, Pacific.

Other species:

*T. affinis* (Lönnberg 1891) Lönnberg 1892 (syn. *Oriana wilsoni* Leiper et Atkinson 1914, *Tetrabothrius wilsoni* (Leiper et Atkinson 1914) Baylis 1926, in *Balaenoptera musculus, B. borealis, Physeter catodon;* Europe, Africa, New Zealand, Antarctica, Pacific.

*T. argentinum* Szidat 1964, in *Larus maculipennis, L. domicanus;* Argentina.

*T. arsenyevi* Delyamure 1955, in *Balaenoptera borealis*; Antarctica.

*T. baeri* Burt 1976, in *Sula leucogaster;* Sri Lanka.

*T. bairdi* Burt 1978, in *Fregata magnificans;* Jamaica.

*T. bassani* Burt 1978, in *Morus bassanus;* Jamaica.

*T. campanulatus* Fuhrmann 1899 (syn. *Porotaenia brevis* [Szpotanska 1917] Johnston 1935), in *Adamastor cinereus, Diomedea ixulans, Macronectes giganteus, Pagodroma nivea, Procellaris aequinoctialis, Thalassoica antartica*; Arctica, Antarctica.

*T. creani* Leiper et Atkinson 1914 (syn. *T. catherineae* Leiper et Atkinson 1914, *T. aichesoni* Leiper et Atkinson 1914), in *Pterodroma arminjoniana;* Antarctica, Trinidad.

*T. curilensis* Gubanov in Delyamure 1955, in *Physeter catodon;* Pacific Ocean, Antarctica.

*T. cylindraceus* Rudolphi 1819, in *Larus argentatus, L. atricilla, L. canus, L. crassirostris, L. fuscus, L. genei, L. hemprichi, L. hyperboreus, L. marinus, L. melanocephalus, L. pipixcan, L. ridibundus, Puffinus puffinus, Rissa tridactyla, Stercorarius pomarinus, Sterna hirundo, S. maxima, S. sandvicensis, Uria troile, U. aalge, Zema sabini;* Europe, North America, Greenland, Japan.

*T. diomedea* Fuhrmann in Shipley 1900, in *Diomedea chlororhynchus, D. exulans;* Western Pacific.

*T. drygalskii* Szpotanska 1929, in *Sula* sp.; Antarctic.

*T. egregius* Skriabin et Muravijova 1971, in *Balaenoptera physalus;* Balleny Islands, Antarctic Ocean.

*T. erostris* Lönnberg 1896 (syn. *T. lari* Yamaguti 1935), in *Cepphus grylle, C. columba, Fulmarus glacialis, Hydrochelidon, Larus argentatus, L. atricilla, L. canus, L. crassirostris, L. fuscus, L. hyperboreus, L. marinus, L. schistisagus, Pagophila eburnes, Rissa tridactyla, Sterocorarius parasiticus, S. pomarinus, Sterna dougalli, S. hirundo, S. paradisea, Uria aalge;* Europe, North America, Japan, Sri Lanka, Russia.

*T. eudyptidis* Lönnberg 1896 (syn. *Neotetrabothrius eudyptidis* Nybelin 1929, *N. eudyptidis* [Lönnberg 1896] Nybelin 1929), in *Eudyptes cristatus, E. creatopus, Catarrhactes chrysocome, Spheniscus magellanicus;* Antartica.

*T. filiformis* Nybelin 1916, in *Majaquens aequinoctialis;* Antarctica.

*T. forsteri* (Krefft 1871) Fuhrmann 1904 (syn. *Taenia forsteri* Krefft 1871, *Prosthecocotyle forsteri* [Krefft 1871] Monticelli 1892, *P. diplosoma* Guiart 1935, *P. pachysoma* Guiart 1935, *Tetrabothrius delphini* Yamaguti 1942, *T. dalli* Yamaguti 1952), in *Delphinus delphis, D. capensis, Globicephalus melas, Hyperodon ampullatus, Mesoplodon bidens, Phocoenoides dalli, Steno rostratus;* Australia, Mediterranean, Russia.

*T. fuhrmanni* Nybelin 1916, in *Diomedea chlororhynchus, Majaquens aequinoctialis;* Angola, Antarctica.

*T. gracilis* Nybelin 1916, in *Majaquens aequinoctialis, Priocella antarctica, P. glacialoides;* Atlantic.

*T. heteroclitus* Diesing 1850 (syn. *T. auriculatus* Linstow 1880, *T. diomedea* Fuhrmann 1900, *T. intermedius* Fuhrmann 1899, *T. valdiviae* Szpotanska 1917, *T. pseudoporus* Szpotanska 1917, *Porotaenia fragilis* Szpotanska 1917), in *Adamastor cinereus, Daption capensis, Diomedea chlororhynchus, D. exulans, Macronectes giganeteus, Phoebetria palpebrata, Pagodroma nivea, Priocella glacialoides, Procellaria* sp., *Thalassoica antarctica;* North and South America, Russia, Greenland, Australia, Antarctica.

*T. heterosoma* Baird 1853, in *Sula bassana, Fregata aquilla;* Europe, Jamaica.

*T. hoyeri* Szpotanska 1929, in *Sula* sp.; Poland, Antarctica.

*T. innominatus* Baer 1954, in *Steno bredanensis;* Berlin Museum.

*T. jaegerskioldi* Nybelin 1916 (syn. *Tetrabothrius intrepidus* Baylis 1919), in *Alca torda, cepphus grylle, C. carbo, Fratercula arctica, F. cirrhata, Stercorarius parasiticus, Synthliboramphus antiquus, Uria aalge, U. troile;* Arctica, Europe.

*T. joubini* Railliet et Henry 1912, in *Pygoscelis antarctica;* Antarctica.

*T. kowalewskii* Szpotanska 1917), (syn. *Porotanenia kowalewskii* Szpotanska 1917, *Tetrabothrius kowalewskii* [Szpotanska 1917] Johnston 1935, *P. macrocirrosa* Szpotanska 1917), in *Diomedea chlororhynchus, Majaquens aequinoctialis;* Antarctica.

*T. laccocephalus* Spatlich 1909, in *Pagodroma nivea, Priocella glacialoides, Procellaria aequinoctialis, Puffinus creatopus, P. diomedea, P. gravis, P. griseus;* Atlantic.

*T. lutzi* Parona 1901, in *Sphniscus magellanicus, Pygoscelis papua;* Antarctica.

*T. mawsoni* Johnston 1937 (syn. *Tetrabothrius cylindraceus* Leiper et Atkinson 1914 nec Rudolphi 1819), in *Catharacta skua;* Antarctica.

*T. minor* (Lönnberg 1893) Johnston 1937 (syn. *Bothridiotaenia erostris* var *minor* Lonnberg 1893, *Tetrabothrius monticelli* Fuhrmann 1899, *T. strangulatus* Baylis 1914, *T. minutus* Szpotanska 1917), in *Fulmarus glacialis, Diomedea irrorata, Oceanodroma furcata, Thalassoica antarctica;* Arctica, Atlantic, Pacific.

*T. morschtini* Muravijova 1968, in *Larus hyperboreus;* Russia.

*T. mozambiquus* Deblock 1966, in *Phaeton rubircauda;* Mozambique.

*T. nelsoni* Leiper et Atkinson 1914 (syn. *Tetrabothrius glaciloides* Nybelin 1929), in *Diomedea melanophrys, Phoebetria palpebrata*, Antarctica.

*T. pauliani* Joyeux et Baer 1954, in *Pygoscilis papua;* Antarctica.

*T. pelecani* Rudolphi 1819 (syn. *Tetrabothrius priestlyi* Leiper et Atkinson 1914, *T. sulcatus* Linton 1927, *T. fregatae* Szpotanska 1929), in *Fregata aquila, F. magnificens, F. minor, Sula fusca;* Carribbean Ocean, Atlantic Ocean, Europe.

*T. perigrinatoris* Burt 1976, in *Sula leucogaster;* Sri Lanka.

*T. phalacrocoracis* Burt 1977, in *Phalacrocorax aristotelis;* Scotland.

*T. polyorchis* Nybelin 1916, in *Fregata ariel;* Australia, New Caledonia.

*T. procerus* Spätlich 1909 (syn. *Tetrabothrius heteroclitus* Linton 1927 nec Diesing 1850), in *Fulmarus glacialis, Puffinus griseus, P. creatopus, P. diomedea, P. gravis;* Atlantic, Greenland.

*T. prudhoei* Markowski 1955, in *Steno bradanensis, Lagenorhynchus obscurus, L. australis;* Falkland Islands.

*T. reditus* Burt 1978, in *Fregata reditus;* Sri Lanka.

*T. rundi* Nybelin 1928, in *Balaenoptera physalus, Eubalaena glacialis;* Australia, Antarctica, Atlantic.

*T. sarasini* Fuhrmann 1918, in *Sterna paradisea, Thalasseus bergii;* Antarctica, New Caledonia.

*T. schaeferi* Markowski 1955, in *Balaenoptera musculus;* Antarctica.

*T. skoogi* Nybelin 1916, in *Puffinus griseus;* Japan.

*T. sulae* Szpotanska 1929, in *Sula* sp.; Antarctica, Sri Lanka.

*T. torulosus* Linstow 1888 (syn. *T. polaris* Szpotanska 1917, *T. intermedius* var. *exulans* Szpotanska 1917, *T. antarcticus* Fuhrmann 1921, *T. kowalewski* Szpotanska 1925 nec Szpotanska 1917), in *Diomedea exulans, D. nigripes, D. albatrus;* Antarctica, Pacific.

*T. wrighti* Leiper et Atkinson 1914, in *Pygoscelis adeliae, P. papua, Aptenodytes forsteri;* Antarctica.

## KEY TO THE GENERA IN NEMATOTAENIIDAE

**1a. One testis per proglottid. Two paruterine organs united basally, each with three to six eggs..............................*Cylindrotaenia* Jewell 1916. (Figure 247)**

*Diagnosis:* Scolex lacking rostellum. Suckers simple. Neck long. Strobila cylindroid, thinnest near ends. Proglottids acraspedote. External metamerism present only near anterior end. Genital pores irregularly alternating. Cirrus pouch short. Testis single, aporal. Ovary and vitellaria compact, madullary. Vagina posterior to cirrus pouch. Two paruterine organs in each proglottid, dorsal and ventral, each with three to six eggs. Parasites of toads and frogs. Africa, North and South America, India, Sri Lanka.

Type species: *C. americana* Jewell 1916, in *Bufo, Hyla, Rana, Acris, Pseudacris,*

FIGURE 247. *Cylindrotaenia quadrijugosa* Lawler 1939. (a) Scolex; (b) gravid proglottid.

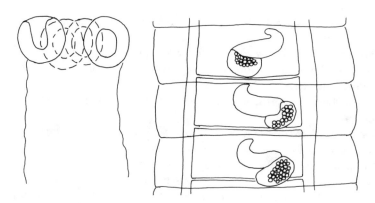

FIGURE 248. *Nematotaenoides ranae* Ulmer et James 1976. Scolex and portion of gravid strobila.

*Leiolopisma, Desmognathus, Leptodactylus, Scaphiopus, Arthroleptis, Plethodon*; North and South America.

Other species:
    *C. philauti* Crusz et Sanmugasunderam 1971, in *Plilautus variabilis*; Sri Lanka.
    *C. quadrijugosa* Lawler 1939, in *Rana pipiens*; Michigan, U.S.
    *C. roonwali* Nama 1972, in *Rana cyanophylctis*; India.

1b.    More than one testis per segment.................................................. 2

2a.    One paruterine organ per segment ................................................ 3
2b.    More than one paruterine organ per segment .................................. 4

3a.    Paruterine organ containing eggs in single compartment..........................
        ............................*Nematotaenoides* Ulmer et James 1976. (Figure 248)

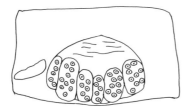

FIGURE 249. *Hexaparuterina mexicana* Palacios et Barroeta 1967. Gravid segment.

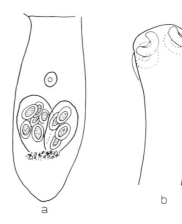

FIGURE 250. *Baerietta diana* (Helfer 1948) Douglas 1958. (a) Gravid segment; (B) scolex.

*Diagnosis*: Scolex arostellate, unarmed, with four simple suckers; not clearly demarcated from strobila. Neck short, cylindrical. External segmentation indistinct anteriorly, pronounced in mature and gravid portions. Mature proglottids broader than long, gravid proglottids longer than broad. Genital pores irregularly alternating. Reproductive organs medullary. Testes three to ten, usually eight. Cirrus pouch well developed, oval, extending to, or slightly into, medulla. Seminal vesicles absent. Ovary globate, midventral, posteroventral to testes. Vitellarium simple, medial to ovary. Uterus tubular, medial. Paruterine organ single, spheroidal to ovoid, containing 20 to 40 eggs when fully developed. Vagina posterior to cirrus pouch.

Type species: *N. ranae* Ulmer et James 1976, in *Rana pipiens;* U.S.

**3b.  Paruterine organ contains eggs in six compartments ..........................
............... *Hexaparuterina* Palacios et Barroeta 1967. (Figure 249)**

*Diagnosis*: Scolex simple, lacing rostellum and hooks. Suckers unarmed. Neck present. External segmentation well marked. Genital ducts ventral to osmoregulatory canals. Cirrus replaced by tuft of long spines. Genital pores irregularly alternating. Cirrus pouch exceeds osmoregulatory canals. Small internal seminal vesicle present. Vas deferens with a few small coils. Testes numerous (13 to 18), posterior to female genitalia, near posterior border of segment. Ovary median, made up of several compact lobules. Vitellarium spherical, postovarian. Vagina posterior to cirrus pouch. Seminal receptacle absent. Uterus first appears as a transverse tube that swells as it is filled with embryos. Then it becomes surrounded by a connective tissue paruterine organ that divides the eggs into six compartments. Parasites of Amphibia.

Type species: *H. mexicana* Palacios et Barroeta 1967, in *Rana montezumae;* Mexico.

**4a.  Two paruterine organs, united basally per proglottid ..........................
............................................. *Baerietta* Hsü, 1935. (Figure 250)**

*Diagnosis*: Scolex lacking rostellum. Suckers simple. Neck long, lacking constriction behind scolex. Strobila covered with fine spines. External metamerism present only on gravid proglottids. Genital pores irregularly alternating. Cirrus pouch cortical and medullary anteriorly, entirely cortical posteriorly. Testes two, in dorsal medulla. Ovary and vitelline gland compact, in ventral medulla posterior to testes. Two paruterine organs, dorsal and ventral, united basally, 1 to 25 eggs in each. Parasites of toads and frogs. Asia, Japan, Africa, North America, New Zealand.

FIGURE 251. *Distoichometra kozloffi* Douglas 1957. Gravid proglottid.

FIGURE 252. *Nematotaenia dispar* (Goeze 1782) Lühe 1899. Gravid proglottid.

Type species: *B. baeri* Hsü 1935, in *Bufo bufo*; China.
Other species:
  *B. allisonae* Schmidt 1980, in *Hoplodactylus maculatus*; New Zealand.
  *B. claviformis* Yamaguti 1954, in *Rana temporaria*; Japan.
  *B. criniae* Hickman 1960, in *Crinia signifera*; Tasmania, Australia.
  *B. desmognathi* Douglas 1957, juvenile in *Desmognathus fusca*; locality?
  *B. diana* (Helfer 1948) Douglas 1956 (syn. *Proteocephalus diana* Helfer 1948), in *Batrachoseps, attenatus*; California, U.S.
  *B. enteraneidis* (Helfer 1948) Yamaguti 1959 (syn. *Proteocephalus enteraneidis* Helfer 1948), in *Anedes lugubris*; California, U.S.
  *B. gerrhonoti* Telford 1965, in *Gerrhonotus multicarinatus*; U.S.
  *B. idahoensis* Waitz et Mehra 1961, in *Plethodon vandydei*; U.S.
  *B. jaegerskioeldi* (Janicki 1928) Hsü 1935 (syn. *Nematotaenia jaegerskioeldi* Janicki 1928), in *Bufo regularis, B. panterinus, Rana madagascariensis*; Egypt, French Somaliland, Rhodesia.
  *B. janicki* (Hilmy 1936) Douglas 1958 (syn. *Nematotaenia janicki* Hilmy 1936), in *Rappia concolor, R. fulvovittata*; Liberia.
  *B. japonica* Yamaguti 1938, in *Hyla arborea, japonica, Rana esculenta, Polypedates schlegeli*; Japan.
  *B. montana* Yamaguti 1954, in *Bufo vulgaris*; Japan.

**4b.  More than two paruterine organs per segment .................................. 5**

**5a.  Paruterine organs numerous, with bases fused .......................................
 ............................................*Distoichometra* Dickey 1921. (Figure 251)**

*Diagnosis*: Rostellum lacking, suckers simple. Neck present. Genital pores irregularly alternating. Strobila cylindroid. Cirrus pouch cortical. Testes two, in dorsal medulla. Ovary and vitelline gland compact, in ventral medulla. Vagina posterior to cirrus pouch. Paruterine organs numerous, with bases fused, each with three to six egg capsules. Parasites of toads. North America.

Type species: *D. bufonis* Dickey 1921, in *Bufo lentiginosus, B.* spp., *Scaphiopus holbrooki*; U.S.A.
Other species:
  *D. kozloffi* Douglas 1958, in *Hyla regilla*; Oregon, U.S.

**5b.  Paruterine organs numerous, bases not fused, scattered throughout parenchyma or arranged in two parallel rows ........*Nematotaenia* Lühe 1899. (Figure 252)**

*Diagnosis*: Scolex simple, lacking rostellum. Neck present. Strobila cylindrical; external metamerism present only near posterior end. Genital pores irregularly alternating. Cirrus pouch intruding into medulla. Testes two, in dorsal medulla. Ovary compact, in ventral medulla. Vitelline gland medullary, dorsal to ovary. Vagina ventral to cirrus pouch. Paruterine organs numerous, with bases not fused; first arranged in two lateral rows, then scattered throughout parenchyma. Parasites of toads, frogs, lizards, salamanders. Europe, Asia, North America, India, Tasmania.

    Type species: *N. dispar* (Goeze 1782) Lühe 1899 (syn. *Taenia dispar* Goeze 1782, *T. dispar salamandrae* Froelich 1789, *T. bufonis* Gmelin 1790, *Halysis obvoluta* Zeder 1803), in *Bufo, Hyla, Rana, Salamandra, Tarentola*; Europe, North America?, India, Morocco.

Other species:

    *N. hylae* Hickman 1960, in *Hyla ewingii, Crinia signifera*; Tasmania.
    *N. kashmirensis* Fotedar 1966, in *Bufo viridis*; India.
    *N. lopezeryrai* Soler 1945, in *Bufo* sp.; Africa.
    *N. mabuiae* Shinde 1968, in *Mabuia carinata*; India.

## KEY TO THE GENERA IN DIOECOCESTIDAE

1a. Completely dioecious, male with double set of reproductive organs ........... 2
1b. Completely or regionally dioecious, male with single set of reproductive organs ................................................................................. 3

2a. Scolex with armed rostellum ..................... *Dioecocestus* Fuhrmann 1900.

*Diagnosis*: Rostellum with single circle of hooks. Scolex with or without suckers. Proglottids wider than long, craspedote. Male strobilas more slender than females, with two sets of reproductive organs per proglottid. Cirrus pouch large, with internal seminal vesicle. Cirrus large, armed. Testes numerous, in two submedian groups. Ovary lobated, transversely elongated, slightly poral. Vitellaria also lobated, posteriodorsal to ovary. Vagina irregularly alternating, ending blindly near cuticle. Seminal receptacle near distal end of vagina. Uterus a strongly lobed sac. Parasites of Ciconiiformes, Podicipediformes. Asia, Australia, Europe, New Guinea, North and South America.

    Type species: *D. paronai* Fuhrman 1900, in *Plegadis guarauna*; Argentina.
    Other species:

        *D. acotylus* Fuhrmann 1904, in *Podiceps dominicus*; Brazil, Jamaica.
        *D. asper* (Mehlis 1831) Fuhrmann 1900 (syn. *Taenia asper* Mehlis 1831, *T. lanceolata* Rudolphi 1805, in part), in *Podiceps cristatus, P. rubricollis, P. griseigena*; Europe.
        *D. fevita* Meggitt 1933, in *Podiceps ruficollis*; India.
        *D. fuhrmanni* Linton 1925, in *Colymbus hollboelli, C. auritus*; Massachusetts, U.S.
        *D. novaeguineae* Fuhrmann 1914, in *Podiceps novaehollandiae, P. capensis*; New Guinea.
        *D. novaehollandiae* (Krefft 1871) Fuhrmann 1900 (syn. *Taenia novaehollandiae* Krefft 1871), in *Podiceps australis*; Australia.

2b. Scolex lacking rostellum or hooks ................................................
............................................ *Neodioecocestus* Siddiqui 1960. (Figure 253)

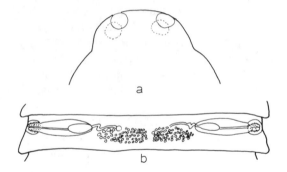

FIGURE 253. *Neodioecocestus cablei* Siddiqui 1960. (a) Scolex; (b) male proglottid.

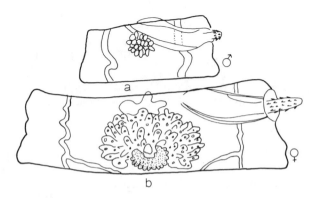

FIGURE 254. *Shipleya inermis* Fuhrmann 1908. (a) Male proglottid; (b) female proglottid.

*Diagnosis*: Dioecious. Scolex lacking rostellum and hooks. Neck very small. Proglottids craspedote, short, and wide. Genital pores bilateral. Two sets of male reproductive organs in each proglottid. Female unknown. Genital atrium well-developed. Genital ducts pass between canals. Cirrus pouch well-developed, exceeding poral canals. Cirrus stout, spined. Internal seminal vesicle present, external lacking. Testes numerous, in two submedian fields, present only in anterior proglottids. Parasites of Podicipediformes. India.

Type species: *N. cablei* Siddiqui 1960, in *Podiceps rufficollis;* India.

3a. Uterus horseshoe-shaped ......................................................... 4
3b. Uterus ring-shaped ................................................................ 5

4a. Scolex lacking rostellum or hooks ..................................................
............................................ **Shipleya** Fuhrmann 1908. (Figure 254)

*Diagnosis*: Scolex large, spheroid, with apical pit and shallow, glandular, longitudinal grooves. Scolex embedded in wall of gut of host. Neck present. Male strobila more slender than female, usually found anterior to female in host intestine. Male genital pores regularly alternating. Cirrus pouch powerful, passing between osmoregulatory canals, and containing internal seminal vesicle. Cirrus large, heavily armed. Testes median, in small, tightly packed cluster. Male has vestigial uterus and seminal receptacle. Female with cirrus pouches and cirri as in male, but lacking testes. Ovary large, median, ventral, strongly lobated. Vitellaria

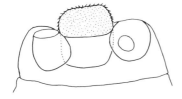

FIGURE 255. *Echinoshipleya semipalmati* Tolkatcheva 1979.

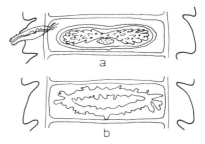

FIGURE 256. *Gyrocoelia perversa* Fuhrmann 1899. (a) Mature proglottid; (b) gravid proglottid.

compact, dumbbell-shaped, posteriodorsal to ovary. Vagina absent. Large seminal receptacle present anterior to ovary. Uterus at first horsehoe-shaped with anterior stem, later filling most of proglottid. Parasites of Charadriiformes. North and South America, Russia.

Type species: *S. inermis* Fuhrmann 1908 (syn. *S. dioica* Spasskii et Gubenov 1959), in *Capella delicata, Gallinago gigantea, Limnodromus griseus;* North and South America, Russia.

**4b. Scolex with massive rostellum covered with minute spines ......................
.................................*Echinoshipleya* Tolkatscheva 1979. (Figure 255)**

*Diagnosis*: Scolex large with massive rostellum covered with minute, equal spines. Short neck present. Male strobila more slender than female. Male genital pores regularly alternating. Cirrus pouch powerful, passing between osmoregulatory canals, with small internal seminal vesicle. Cirrus large, armed at its base with numerous rose-thorn-shaped hooks. Testes in median cluster. Female with cirrus apparatus as in male but lacking testes. Ovary large, transversely elongate or dumbbell-shaped. Vitellarium compact, posterior to ovary. Vagina absent. Uterus horseshoe-shaped with anterior stem, later filling most of proglottid. Parasites of Charadriiformes. Russia.

Type species: *E. semipalmata* Tolkatscheva 1979, in *Limnodromus semipalmatus;* Novosibursk, Russia.

**5a. Rostellum armed. Strobila completely or regionally dioecious ...................
*Gyrocoelia* Fuhrmann 1899 (syn. *Brochocephalus* Linstow 1906). (Figure 256)**

*Diagnosis*: Rostellum armed. Strobilas regionally or individually dioecious. Single set of reproductive organs per proglottid. Male genital pores irregularly alternating. Cirrus pouch large, passing between osmoregulatory canals. Cirrus armed. Testes few, postovarian. Ovary large, median, lobated. Vitelline gland postovarian. Vagina absent. Seminal receptacle present. Uterus ring-shaped, with lateral branches, often opening near posterior border of segment. Parasites of Charadriiformes. Africa, Asia, Australia, Europe, Philippines, Puerto Rico, Sri Lanka, North, Central and South America.

Type species: *G. perversa* Fuhrmann 1899, in *Actophilus africanus, Himantopus himantopus, Hoplopterus spinosus, Limosa rufa, Vanellus* sp.; Africa.

Other species:
*G. albaredai* Lopez-Neyra 1952 (syn. *Infula albaredae* (Lopez-Neyra 1952) Ryzikov et Tolkacheva 1978), in *Himantopus himantopus;* Spain.
*G. australiensis* (Johnston 1910) Johnston 1912, in *Himantopus leucocephalus;* Australia.
Redescribed: Schmidt.[3096]

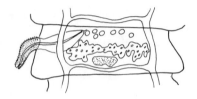

FIGURE 257. *Pseudoshipleya farrani* (Inamdar 1942) Yamaguti 1959.

*G. crassa* (Fuhrmann 1900) (syn. *G. brevis* Fuhrmann 1900, *G. leuce* Fuhrmann 1900), in *Aegialitis collaris, Belanopterus, Charadrius alexandrinus, C. nubicus, C. spinosus, C. suezensis, Hoploxypterus, Tringa, Vanellus;* Egypt, France, Central and South America.

*G. fausti* Tseng 1933, in *Lobivanellus cinereus, Rhynchaea capensis;* China.

*G. fuhrmanni* Rego 1968, in *Charadrius collaris;* Brazil.

*G. kiewietti* Ortlepp 1937, in *Hoplopterus armatus;* South Africa.

*G. milligani* Linton 1927, in *Charadrius nivosus, Crocethia alba, Oxyechus vociferus;* Antigua, North America.

*G. pagollae* Cable et Meyers 1956, in *Pagolla wilsonia;* Puerto Rico.

*G. paradoxa* (Linstow 1906) (syn. *Brochocephalus paradoxa* Linstow 1906), in *Aegialitis mongolica;* Europe, Philippines, Sri Lanka.

**5b. Rostellum unarmed................................................................ 6**

**6a. Completely dioecious. Testes numerous (65 to 75). Vagina present..............**
**............................................................ *Infula* Burt 1939.**

*Diagnosis*: Unarmed rostellum. Strobilas completely dioecious. Proglottids craspedote. Single set of reproductive organs per proglottid. Genital pores usually regularly alternating. Cirrus pouch very large, passing between osmoregulatory canals, and containing seminal vesicle. Cirrus armed. External seminal vesicle absent. Ovary large, biwinged, lobated. Vitellaria lobated, median, postovarian. Vagina opens to exterior, cirrus-like distally. Small seminal receptacle present. Uterus ring-like with marginal branches, opening on dorsal and ventral surfaces in median line at posterior margin of segment. Parasites of Charadriiformes. Africa, Australia, India, Mexico, Sri Lanka.

Type species: *I. burhini* Burt 1939, in *Burhinus oedicnemus indicus, Himantopus himantopus, Lobibyx novaehollandae;* Australia, China, Sri Lanka.

Other species:

*I. macrophallus* Coil 1955, in *Himantopus mexicanus*; Mexico.

**6b. Regionally dioecious. Testes few (about 8). Vagina absent ......................**
**..................................... *Pseudoshipleya* Yamaguti 1959. (Figure 257)**

*Diagnosis*: Scolex with unarmed rostellum. Neck present. Strobilas regionally dioecious. Proglottids craspedote. Single set of reproductive organs per proglottid. Male genital pores alternating irregularly. Cirrus pouch large, ventral to osmoregulatory canals. Cirrus large, armed. Testes about eight in number, anterior. Ovary posttesticular, bialate, lobated. Vitellaria posterior to ovary. Vagina absent. Seminal receptable not observed. Uterus first ring-shaped, later filling most of proglottid. Parasites of Charadriiformes. India.

Type species: *P. farrani* (Inamdar 1942) Yamaguti 1959. (syn. *Shipleyia farrani* Inamdar 1942), in *Himantopus himantopus*; India.

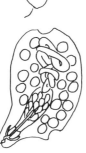

FIGURE 259. *Thomasitaenia nunguae* Ukoli 1965.

FIGURE 258. *Gynandrotaenia stammeri* Fuhrmann 1936. Scolex and mature proglottid.

## KEY TO THE GENERA IN PROGYNOTAENIIDAE

(for Progynotaenidae Fuhrmann 1936)

1a. Male and female proglottids alternating regularly. Scolex may be divided into anterior proscolex with rostellum and posterior metascolex with suckers ..... 2
1b. Mature proglottids hermaphroditic ............................................. 3

2a. Scolex divided into proscolex with armed rostellum and metascolex with armed suckers. Strobila with 14 to 17 proglottids ........................................
.................................*Gynandrotaenia* Fuhrmann 1936. (Figure 258)

*Diagnosis*: Scolex with inflated, flattened, spinous proscolex with retractable rostellum, and a normal metascolex with four armed suckers. Rostellum with a circle of six long-handled hooks. Neck absent. Strobila with large female proglottids alternating regularly with smaller male proglottids. Proterogynous, male systems maturing in 10th or 11th segments from scolex, female in 5th. Male pores irregularly alternating. Vagina absent. Genital atrium deep, protrusible into large cone. Cirrus pouch containing numerous prostatic cells. Cirrus armed. Testes 35 to 45 in number, extending laterally over osmoregulatory canals. External seminal vesicle present. Ovary filling most of medulla. Vitelline gland posterior to ovary. Seminal receptable large. Uterus a median stem, becoming sac-like with lateral pouches when gravid. Eggs with recurved polar processes on middle membrane. Parasites of flamingos. Europe.

Type species: *G. stammeri* Fuhrmann 1936, in *Phoenicopterus roseus*, *P. ruber*, *Phaeniconaias minor*; Europe, Kenya, Russia.

2b. Scolex not divided, rostellum absent, suckers not armed. Strobila with up to 80 proglottids ............................*Thomasitaenia* Ukoli 1965. (Figure 259)

*Diagnosis*: Scolex simple, lacking rostellum or hooks, with four unarmed suckers. Neck absent. Strobila with up to 80 individually dioecious proglottids, male and female regularly

218  CRC Handbook of Tapeworm Identification

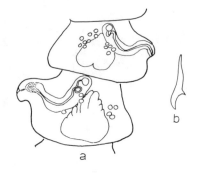

FIGURE 261. *Progynotaenia americana* Webster 1951. (a) Portion of strobila; (b) rostellar hook.

FIGURE 260. *Leptotaenia ischnorhyncha* (Lühe 1898) Cohn 1901.

alternating. Gonads proterogynous. Vagina absent. Male genital pores irregularly alternating. Cirrus pouch well developed. Cirrus large, armed. Testes about 90 in number, overreaching osmoregulatory canals. External seminal vesicle present. Ovary filling most of medulla. Vitelline gland median, postovarian. Seminal receptable present, posterior to uterus. Uterus appearing horseshoe-shaped; when gravid it is a transverse sac, not lobed. Parasites of Charadriiformes. Africa.

Type species: *T. nunguae* Ukoli 1965, in *Himantopus himantopus*; Ghana.

3a. Cirrus pouch regularly alternating. One circle of rostellar hooks.............. 4
3b. Cirrus pouch regularly or irregularly alternating. One or two circles of rostellar hooks............................................................................ 5

4a. Testes poral only ......................... *Leptotaenia* Cohn 1901. (Figure 260)

*Diagnosis*: Rostellum long, slender, with single circle of hooks. Suckers unarmed. Neck lacking. Proglottids few (12 to 15), wider than long. Proterogynous. Vagina absent. Male pores alternating regularly. Cirrus pouch oblique, containing an oval seminal vesicle (?). Cirrus long, armed. Testes poral, 12 to 15 in number, present only in gravid segments. Ovary bilobed, filling most of medulla, present only in first three proglottids. Vitelline gland median, ventral to ovary, extending most of length of segment. Vagina absent. Seminal receptacle longitudinally elongated. Uterus a lobed, transverse sac. Parasites of flamingos. Africa, Cuba, Russia.

Type species: *L. ischnorhyncha* (Lühe 1898) Cohn 1901 (syn. *Taenia ischnorhyncha* Lühe 1898, *Amoebotaenia ischnorhyncha* Cohn 1899), in *Phoenicopterus antiquorum*, *P. ruber*; Africa, Cuba.

Other species:
*L. skrjabini* Shakhtakhtinskaya 1953, in *Phoenicopterus roseus*; Russia.

4b. Testes in two groups, poral and aporal. One circular or zigzag row of rostellar hooks.............................. *Progynotaenia* Fuhrmann 1909. (Figure 261)

*Diagnosis*: Rostellum well developed, sometimes with bilobed apex, with single circle or zigzag row of long-handled hooks. Neck absent. Strobila small, short. Proglottids craspedote. Vagina absent. Male genital pore lateral, regularly alternating. Cirrus pouch long, oblique.

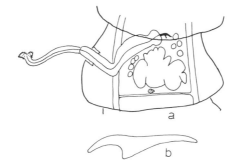

FIGURE 262. *Proterogynotaenia neoarctica* Webster 1951. (a) Proglottid; (b) rostellar hook.

Internal seminal vesicle present. Cirrus armed. Testes developed only in posterior proglottids, in two groups, one on each side of uterus. Ovary bilobed or chevron-shaped, developed only in anterior proglottids. Vitelline gland postovarian, near anterior margin of segment. Seminal receptable between ovary and vitelline gland, or absent. Uterus sac-like or lobated. Parasites of Charadriiformes. Africa, Europe, India, North America.

Type species: *P. jägerskiöldi* Fuhrmann 1909, in *Pulvianus aegypticus*; Egypt.
Other species:

*P. americana* Webster 1951, in *Charadrius melodus, C. semipalmatus, C. vociferus, Crocethia alba*; North America.

*P. evaginata* Fuhrmann 1909 (syn. *P. paucitesticulata* Baczynska 1914), in *Oedicnemus senagalensis, O. oedicnemus*; Africa, Europe.

*P. indica* Johri 1963, in "avocet-like bird", India.

*P. leucura* Singh 1960, in *Chettusia leucura*; India.

*P. longicirrata* Singh 1952, in *Lobipluvia malabarica*; Lucknow.

*P. odhneri* Nybelin 1914 (syn. *P. fuhrmanni* Skrjabin 1914, *P. foetida* Meggitt 1928), in *Aegialitis hiaticula, Charadrius* spp., *Hoploterus spinosus, Oedicnemus oedicnemus, Tringa totanus*; Egypt, Europe.

**5a.  Cirrus pouch doral to osmoregulatory canals........................................
............................................ *Andrepigynotaenia* Davies et Rees 1947.**

*Diagnosis*: Rostellum with two circles of 34 hooks total. Hooks in anterior circle 33 to 39 μm long, those in posterior circle 50 to 55 μm long. Neck absent. Strobila long (over 30 segments), craspedote. Genitalia developing slowly. Vagina absent. Cirrus pouch alternating irregularly. Cirrus pouch dorsal to osmoregularly canals. Cirrus armed at base. Internal seminal vesicle present. Testes only in segments 31 to 35, 50 to 70 in number in two groups, one group on each side of uterus. Poral group less numerous than aporal group, and confined posterior to cirrus pouch. Ovary lobated, anterior. Vitelline gland median, reniform, anteriodorsal to ovary. Seminal receptacle large, anterior to ovary. Uterus sac-like, protruding into preceding segment. Parasites of Charadriiformes. Europe.

Type species: *A. haematopodis* Davies and Rees 1947, in *Haematopus ostralegus*; Wales.

**5b.  Cirrus pouch between osmoregulatory canals.................................... 6**

**6a.  Two circles of rostellar hooks. Cirrus pouch regularly or irregularly alternating
................................. *Proterogynotaenia* Fuhrmann 1911. (Figure 262)**

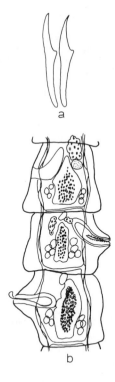

FIGURE 263. *Paraprogynotaenia jimenezi* Rysavy 1966. (a) Rostellar hooks; (b) portion of strobila.

*Diagnosis*: Rostellum well developed, with two circles of hooks. Neck absent. Strobila small, proglottids craspedote. Vagina absent. Cirrus pouch regularly or irregularly alternating. Cirrus pouch passing between osmoregulatory canals, elongate. Cirrus armed. Testes developed only in posterior proglottids, in two fields, on each side of uterus. Ovary developed only in anterior segments, anterior, lobated. Vitelline gland postovarian. Seminal receptacle anterior, between ovary and vitelline gland. Uterus sac-like or lobated. Parasites of Charadriiformes. Aru Islands, Africa, Japan, North America, Russia, Scotland, Taiwan.

Type species: *P. rouxi* Fuhrmann 1911, in *Ochthodromus geoffroyi*; Aru Islands.
Other species:

*P. branchiuterina* Belopol'skaya 1973, in *Squartarola squartarola*; Russia.

*P. dougi* Sandeman 1959, in *Pluvialis apricaria*; Scotland.

*P. flaccida* (Meggitt 1928) Yamaguti 1959 (syn. *Progynotaenia flaccida* Meggitt 1928), in *Recurvirostra avocetta*; Egypt.

*P. neoarctica* Webster 1951, in *Oxyechus vociferous*; Canada.

*P. polytestis* Belopol'skaya 1973, in *Squartarola squartarola*; Russia.

*P. variabilis* Belopol'skaya 1954, in *Squartarola squartarola*; Russia.

**6b.   One circle of rostellar hooks. Cirrus pouch irregularly alternating..............**
**.................................. *Paraprogynotaenia* Rysavy 1966. (Figure 263)**

*Diagnosis*: Rostellum armed with single circle of hooks. Short neck present. Proglottids craspedote, wider than long except last segments. Genital pores irregularly alternating. Cirrus pouch passing between osmoregulatory canals. Testes only in posterior proglottids, in two lateral groups. Internal seminal vesicle present. Cirrus armed. Ovary antiporal. Vitelline

FIGURE 264. *Insinuarotaenia schickobolovi* Spasskii 1948.

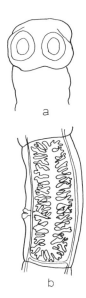

FIGURE 265. *Taeniorhynchus saginatus* (Goeze 1782) Weinland 1858. (a) Scolex; (b) gravid proglottid.

gland postovarian. Vagina absent. Seminal receptacle present. Uterus median, elongated. Parasites of Charadriiformes. Cuba, Taiwan.

Type species: *P. jimenezi* Rysavy 1966, in *Charadrius wilsonia*; Cuba.
Other species:
  *P. charadrii* (Yamaguti 1956) Jensen, Schmidt et Kuntz 1983 (syn. *Proterogynotaenia charadrii* Yamaguti 1956), in *Charadrius alexandrinus, C. dominicus*; Japan, Taiwan.

## KEY TO THE GENERA IN TAENIIDAE

1a. Scolex unarmed.................................................................. 2
1b. Scolex armed .................................................................. 3

2a. Blunt, unarmed apical organ present. Cervical swelling present behind scolex
  .................................... *Insinuarotaenia* Spasskii 1948. (Figure 264)

*Diagnosis*: Scolex with complex muscular apical organ with inner meridional muscle bundles and outer longitudinal and radial muscle fibers. Hooks absent. Four simple suckers present. Constriction separates neck from scolex. Neck expanded proximal to scolex, narrowed distally. Proglottids craspedote. Gravid segments longer than wide. Genital pores irregularly alternating, in anterior third of margin. Testes 200 to 250, surrounding ovary. Ovary posterior, oval, median, separated from posterior margin by one or two rows of testes. Vitelline gland postovarian, bilobed. Uterus with median stem and lateral branches. Parasites of Mustelidae. Russia.

Type species: *I. schikhobolovi* Spassky 1948, in *Meles meles*; Russia.
Other species:
  *I. spasskii* Andreido et Yun 1963, in *Mustela nivalis*; Russia.

2b. No apical organ, rostellum or cervical swelling present ..........................
  .................................... *Taeniarhynchus* Weinland 1858. (Figure 265)

*Diagnosis*: Rostellum absent, four simple suckers present. Strobila large (up to 12 m). Genital pores irregularly alternating. Cirrus pouch pear-shaped. Testes very numerous in single single field, none posterior to vitelline gland. Ovary biwinged, posterior. Vitelline gland postovarian. Vagina opening posterior to cirrus. Uterus a median stem with numerous lateral branches. Larva a cystircercus. Parasites of humans. Cosmopolitan.

Type species: *T. saginatus* (Goeze 1782) Weinland 1858, in *Homo sapiens*. (The many synonyms of this species are listed in Abuladze (1964). Additionally, I regard *T. hominis* (Linstow 1902), *T. confusa* (Ward 1896), and *T. africana* (Linstow 1900), to be synonyms of *T. saginatus*.

3a. One circle of rostellar hooks .................................................................
....................... *Monordotaenia* **Little 1967 (syn.** *Fossor* **Honess 1937).**

*Diagnosis*: Rostellum with a single circle of hooks. Proglottids craspedote. Genital pores irregularly alternating. Cirrus pouch elliptical. Testes numerous, in single field anterior and lateral to ovary. Ovary bialate, posterior. Vitelline gland postovarian, transversely elongated. Seminal receptacle present. Uterus with median stem and lateral branches. Parasites of Mustelidae, Felidae, Canidae; North America, Russia.

Type species: *M taxidiensis* (Skinker 1935) Little 1967 (syn. *Taenia taxidiensis* Skinker 1935, *Fossor angertrudae* Honess 1937), in *Taxidea taxus;* U.S.
Other species:
    *M. alopexi* Obushenkov 1983, in *Alopex lagopus;* Russia.
    *M. honessi* Hendrickson, Grieve et Kingston 1975, in *Canis familaris;* Wyoming, U.S.
    *M. monostephanos* (Linstow 1905) comb. n. (syn. *Taenia monostephanos* Linstow 1905, *Fossor monostephanos* [Linstow 1905] Abuladze 1964), in *Felis lynx;* Russia.

3b. Two circles of rostellar hooks.................................................... 5

4a. Strobila with fewer than six proglottids ..... *Echinococcus* **Rudolphi 1801 (syn.** *Echinococcifer* **Weinland 1858,** *Alveococcus* **Abuladze 1960). (Figure 266)**

*Diagnosis*: Strobila very small, with fewer than six proglottids. Rostellum with a double circle of hooks. Neck absent. Genital pores irregularly alternating. Testes few, anterior and lateral to female organs. Ovary bilobed, posterior. Vitelline gland compact, postovarian. Vagina with terminal sphincter. Seminal receptacle present. Uterus with median stem and short, undivided lateral branches. Gravid uterus only in last proglottid. Larva a hydatid. Parasites of Canidae and Felidae. Cosmopolitan.

Type species: *E. granulosus* (Batsch 1786) Rudolphi 1810. (see Abuladze[5] for many synonyms), in Carnivora, especially Canidae; nearly cosmopolitan.
Other species:
    *E. cameroni* Ortlepp 1934, in *Vulpes vulpes;* England.
    *E. cepanzoi* Szidat 1971, in *Dusicyon gymnocercus;* Argentina.
    *E. felidis* Ortlepp 1937, in *Felis leo;* South Africa.
    *E. lycaontis* Ortlepp 1934, in *Lycaon pictus;* South Africa.
    *E. multilocularis* Leuckart 1863 (syn. *Taenia alveolaris* Klemm 1883, *T. echinococcus*, *Echinococcus sibericensis* Rausch et Schiller 1954, *Alveococcus multilocularis* [Leuckart 1863] Abuladze 1960), in *Alopex lagopus, Canis familiaris, C. latrans, C. lupus, Felis catus, Vulpes corsak, V. fulvus, V. vulpes;* Asia, Europe, North America.
    *E. oligarthra* (Diesing 1863) Lühe 1910 (syn. *Taenia oligarthra* Diesing 1863, *E. cruzi* Brumpt et Joyeux 1924), in *Felis concolor, F. jaguarundi, F. tigrina;* South America.
    *E. pampeanus* Szidat 1967, in *Felis colocolo;* Argentina.

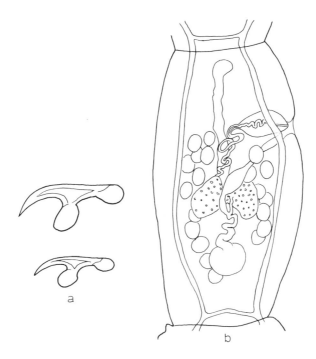

FIGURE 266. *Echinococcus multilocularis* Leuckart 1863. (a) Rostellar hooks; (b) proglottid.

*E. patagonicus* Szidat 1960, in *Dusicyon culpaeus;* Argentina.
*E. vogeli* Rausch et Bernstein 1972, in *Speothos venaticus;* Ecuador.

4b. Strobila with more than six proglottids ............................................
.......................*Taenia* Linnaeus 1758 (syn. *Acanthotrias* Weinland 1858, *Alyselminthus* Zeder 1800, *Cysticercus* Zeder 1800, *Cystotaenia* Leuckart 1863, *Finna* Brera 1809, *Fischiosoma* Brera 1809, *Goeziana* Rudolphi 1810, *Halysis* Zeder 1803, *Hydatigena* Goeze 1782, *Hydatigera* Lamarck 1816, *Hydatis* Maratti 1776, *Hydatula* Abildgaard 1790, *Hygroma* Schrank 1803, *Megocephalos* Goeze 1782, *Multiceps* Goeze 1782, *Neotaenia* Sodero 1886, *Physchiosoma* Brera 1809, *Polycephalus* Zeder 1800, *Reditaenia* Sambon 1924, *Taeniola* Pallas 1760, *Trachelocampylus* Fredault 1847, *Urocystidium* Beddard 1912). (Figure 267)

*Diagnosis:* Rostellum armed with two circles of hooks. Strobila large, with many proglottids. Genital pores irregularly alternating. Cirrus pouch pyriform. Testes numerous, in single field anterior and lateral to female organs. Ovary bialate, posterior. Vitelline gland compact, postovarian. Vagina opening behind cirrus. Uterus with median stem and lateral branches. Larva a cysticercus, coenurus, or strobilocercus, developing in mammals. Parasites of mammals and (rarely) birds. Cosmopolitan. Many species in the following list are poorly known and may not belong to this genus.

Type species: *T. solium* Linnaeus 1758, in humans; nearly cosmopolitan. For several dozen synonyms see Abuladze.[5]

Other species:

*T. acinomyxi* Ortlepp 1938, in *Acinomyx jubatus;* South Africa.
*T. antarctica* Fuhrmann 1922, in *Canis familiaris;* Antarctica.
*T. asiota* Alvey et Kintner 1934, in *Otus asio;* North America.

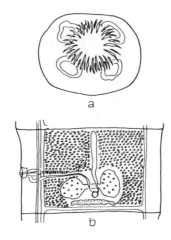

FIGURE 267. (a) *Taenia pisiformis* (Bloch 1780) Gmelin 1790. Apical view of scolex; (b) *T. hydatigena* Pallas 1766. Proglottid.

*T. balaniceps* Hall 1910 (syn. *Hydatigera balaniceps* [Hall 1910] Wardle et McLeod 1952), in *Canis familiaris, Felis rufus;* U.S.

*T. brachyacantha* Baer et Fain 1951, in *Poecilogale albinucha;* Africa.

*T. brauni* Setti 1897 (syn. *Multiceps brouni* (Setti 1897), *T. brachysoma* Setti 1899), in *Canis familiaris, C. mesomelas, Lycaon pictus;* Africa.

*T. bremneri* Stephens 1908, in humans; Nigeria.

*T. brevicollis* Rudolphi 1819, in *Mustela erminea;* Europe.

*T. bubesei* Ortlepp 1938, in *Felis leo, F. tigris;* Africa, Russia.

*T. cervi* Christionsen 1931, in dogs and foxes; Europe.

*T. chamisoni* Linton 1905, larva in porpoise; Atlantic.

*T. clavifer* (Railliet et Mouquet 1919) comb. n. (syn. *Coenurus clavifer* Railliet et Mouquet 1919), in skin of *Myopotamus coypus;* locality?

*T. crassiceps* (Zeder 1800) Rudolphi 1810 (syn. *Alyselminthus crassiceps* Zeder 1800, *Cysticercus longicollis* Rudolphi 1819, *C. talpae* Bendz 1842, *C. taeniacrassicipitis* [Zeder 1800] Joyeux et Baer 1936, *C. multiformis* Hölldobler 1937, *Taenia multiformis* Wardle et McLeod 1952, *T. hyperbores* Linstow 1905), in *Alopex lagopus, Canis familiaris, C. lupus, Felis lynx, Vulpes corsak, V. fulva, V. melanogaster, V. vulpes;* circumboreal.

*T. crassipora* Rudolphi 1819, in *Mephitis, Nasua, Viverra rarica;* Brazil.

*T. crocutae* Mettrick et Beverly-Burton 1961, in *Crocuta crocuta;* Africa.

*T. djeirani* Boev, Sokolova et Tazieva 1964, known in intermediate hosts only (antelopes, gazelles, etc.); Asia.

*T. endothoracica* (Kirschenblatt 1948) Kirschenblatt 1949 (syn. *Coenurus endothoracicus* Kirschenblatt 1948), in *Vulpes vulpes;* Middle East, North Africa, Russia.

*T. erythrea* Setti 1897, in *Canis mesomelas;* Ethiopia.

*T. eunectes* Smith 1908, in *Eunectes murinus;* locality unknown.

*T. foinae* Blanchard 1848 (syn. *T. conocephala* Diesing 1854), in *Mustela foina, M.* sp.; Europe.

*T. gaigeri* (Hall 1916) Baer 1926 (syn. *Multiceps gaigeri* Hall 1916, *Polycephalus gaigeri* [Hall 1916] Sprehn 1932), in *Canis familiaris;* India, Pakistan, Russia, South Africa, Sri Lanka, U.S.

*T. geophiloides* Cobbold 1879, in *Phascolarctos cinereus;* Australia.

*T. glomerata* (Railliet et Henry 1915) Brumpt 1922 (syn. *Multiceps glomeratus* Railliet et Henry 1915, *Coenurus glomeratus* [Railliet et Henry 1915]), larvae in rodents; Africa.

*T. gonyamai* Ortlepp 1938, in *Felis leo;* South Africa.

*T. himalayotaenia* (Malhotra et Capoor 1982) comb. n. (syn. *Hydatigera himalayotaenia* Malhotra et Capoor 1982), in *Felis catus;* India.

*T. hlosei* Ortlepp 1938, in *Acinomyx jubatus;* South Africa.

*T. hyanae* Baer 1926 (syn. *Cysticercus dromedarii* Pellegrini 1945), in *Hyaena brunnea, H. crocuta;* Africa.

*T. hydatigena* Pallas 1766 (for numerous synonyms see Abuladze[5]), in numerous carnivores; cosmopolitan.

*T. hyperborea* Linstow 1905 (syn. *Hydatigera hyperborea* [Linstow 1905] Abuladze 1964), in *Canis lagopus;* Greenland.

*T. ingwei* Ortlepp 1938, in *Felis pardus*; South Africa.

*T. intermedia* Rudolphi 1810 (syn. *Halysis martis* Zeder 1803), in *Martes martes, M. foina, Mustela erminea, M. nivalis, Putorius putorius;* Europe, Scotland.

*T. jakhalsi* Ortlepp 1938, in *Thous mesomelas;* South Africa.

*T. krabbei* Moniez 1879 (syn. *Cysticercus tarandi* Villot 1883, *C. rangifer* Grüner 1910, *Taenia rangifer* (Grüner 1910) Meggitt 1924), in *Alopex lagopus, Canis familiaris, C. latrans, C. lupus, Felis catus, F. lynx, F. rufus;* circumboreal.

*T. krepkogorski* (Sculz et Landa 1934) comb. n. (syn. *Hydatigera krepkogorski* Schultz et Landa 1935), in *Chaus chaus, Felis margarita, F. ocreata, F. silvestris, Vulpes vulpes;* Bahrain, Russia. Redescribed: Bray.[368]

*T. laruei* Hamilton 1940, in *Canis latrans;* North America.

*T. lemuris* (Cobbold 1861) comb. n. (syn. *Coenurus lemuris* Cobbold 1861, *Multiceps lemuris* [Cobbold 1961] Hall 1910), in *Lemur maco;* Madagascar.

*T. lycaontis* Baer et Fain 1955, in *Lycaon pictus;* Africa.

*T. lyncis* Skinker 1935 (syn. *Cysticercus lyncis* [Skinker 1935] Joyeux et Baer 1940, *Hydatigera lyncis* [Skinker 1935] Wardle et McLeod 1952), in *Felis azteca, F. concolor, F. rufus;* North and South America.

*T. macracantha* (Capham 1942) comb. n. (syn. *Multiceps macracantha* Clapham 1942), larvae in white rat; Southern Rhodesia.

*T. macrocystis* (Diesing 1850) Lühe 1910 (syn. *Cysticercus macrocytis* Diesing 1850, *Hydatigera macrocystis* [Diesing 1850] Wardle et McLeod 1952), in *Felis baileyi, F. geoffroyi, F. jaguarundi, F. macroura, F. rufus, F. tigrina, Vulpes vulpes;* North and South America.

*T. melanocephala* Beneden 1861, in *Papio maimon;* locality unknown.

*T. melesi* Petrow et Sadychow 1945, in *Meles meles;* Russia.

*T. michiganensis* Cower 1939, in liver of *Erethizon dorsatum;* U.S.

*T. monostephanos* Linstow 1905, in *Lynx lynx*; Russia.

*T. multiceps* Leske 1780 (see Abuladze[5] for very many synonyms), in *Canis familiaris, C. lupus, C. latrans, C. mesomelas, Alopex lagopus, Nyctereutes procyonoides, Vulpes vulpes;* cosmopolitan.

*T. multiformis* (Hölldobler 1937) (syn. *Cysticercus multiformis* Hölldobler 1937) in skin of fox; Germany.

*T. mustelae* Gmelin 1790, in *Mustela martes, M.* spp., *Putorius nivalis, P. putorius;* Europe, North America.

*T. olngojinei* Dinnik et Sachs 1969, in *Crocuta crocuta;* Tanzania.

*T. omissa* Lühe 1910 (syn. *Echinococcus omissa* [Lühe 1910] Meggitt 1924), in *Felis azteca, F. concolor, F. hipploestes, F. oregonensis, F. tigrina, F. yaguarundi;* North and South America.

*T. opuntioides* Rudolphi 1819, in *Canis lupus;* Europe.

*T. otomys* (Clapham 1942) comb. n. (syn. *Multiceps otomys* Clapham 1942), known only from coenurus larva in *Otomys erroratus;* Africa.

*T. ovata* Molin 1858, in *Alopex lagopus, Vulpes vulpes;* Norway.

*T. ovis* (Cobbold 1869) Ransom 1913 (syn. *Cysticercus ovis* Cobbold 1869, *C. ovipariens* Maddax 1873, *C. oviparus* Leuckart 1886), in *Canis aureus, C. familiaris, Felis concolor, Vulpes vulpes;* cosmopolitan.

*T. packi* (Christenson 1929) Neveu-Lemaire 1936 (syn. *Multiceps packi* Christenson 1929, *Polycephalus packi* [Christenson 1929] Sprehn 1957), in *Canis familiaris, C. latrans, C. lupus, Urocyon cinereoargenteus;* U.S.

*T. paradoxuri* Smith 1908, in *Paradoxurus grayi;* India.

*T. parenchymatosa* Pushmenkov 1945, in *Alopex lagopus, Canis familiaris, C. lupus;* Russia.

*T. parva* Baer 1926 (syn. *Hydatigera parva* [Baer 1926] Wardle et McLeod 1952), in *Genetta ludia, G. tigrina;* Africa.

*T. parviuncinatus* (Kirschenblatt 1939) Kirschenblatt 1948 (syn. *Coenurus parviuncinatus* Kirschenblatt 1939, *Multiceps parviuncinatus* [Kirschenblatt 1939] Abuladze 1964), larvae in *Citellus citellus, Spalax leucodon;* Armenia.

*T. pisiformis* (Bloch 1780) Gmelin 1790 (syn. *Cystotaenia pisiformis* Batsch 1786, *Hydatigena pisiformis* Goeze 1782, *H. utriculenta* Goeze 1782, *Hydatigera laticollis* [Batsch 1786] Wardle et McLeod 1952, *Monostomum leporis* Kuhn 1830, *Taenia caninum solium* Werner 1782, *T. cucurbitina canis* Batsch 1786, *T. laticollis* Rudolphi 1819, *T. novella* Neumann 1896, *T. serrata* Goeze 1782, *T. utricularis;* Hall 1912. *Vermis vesicularis pisiformis* Block 1780, *Vesicaria pisiformis* Schrank 1788. See Abuladze[5] for other synonyms, in many species of Carnivora; Cosmopolitan.

*T. platydera* Gervais 1847, in *Genetta genetta;* North Africa.

*T. polyacantha* Leuckart 1856, in *Vulpes alopex, V. vulpes*; Europe, North America.

*T. polycalcaria* Linstow 1903, in *Felis pardus;* Nepal, Sri Lanka.

*T. polytuberculosus* (Megnin 1880) comb. n. (syn. *Coenurus polytuberculosus* Megnin 1880, *Multiceps polytuberculosus* [Megnin 1880] Hall 1910, *Cysticercus polytuberculosus* [Megnin 1880] Braun 1894, *Coenurus tuberculosus* [Megnin 1880] Kunsemüller 1903); Europe.

*T. pungutchui* Ortlepp 1938, in *Thous mesomelas;* South Africa.

*T. punica* Cholodkovsky 1908, in *Canis familiaris, Casus rhombeatus, Paradoxurus hermaphroditicus;* India, North Africa.

*T. radians* (Joyeux, Richet et Schulmann 1922) comb. n. (syn. *Multiceps radians* Joyeux, Richet et Schulmann 1922), in tissues of white mouse; France.

*T. ramosus* (Railliet et Marullaz 1919) comb. n. (syn. *Multiceps ramosus* Railliet et Marullaz 1919), in skin of *Macacus sinicus;* locality unknown.

*T. regis* Baer 1923, in *Felis leo;* South Africa.

*T. retracta* Linstow 1904, in *Vulpes ferritatus;* Tibet.

*T. rileyi* Loewen 1929 (syn. *Hydatigera rileyi* Abuladze 1964), in *Lynx canadensis;* North America. Redescribed: Rausch.[2803]

*T. saigoni* Le Van Hoa 1964, in *Macacus cynomolgus* (intermediate host); Vietnam.

*T. schavarschi* Matevosyan 1949, in *Larus ichthyaetus;* Russia.

*T. secunda* Olsson 1893, in *Meles taxus;* Scandinavia.

*T. serialis* (Gervais 1847) Baillet 1863 (syn. *Coenurus serialis* Gervais 1847, *C. cuniculi* [Diesing 1863] Cobbold 1864, *C. cerebralis leporiscuniculi* Diesing 1863, *C. lowzowi* Lindemann 1867, *Cystotaenia serialis* [Gervais 1847] Benham 1901), in *Canis aureus, C. familiaris, C. latrans, C. lupus, Felis ocreata, Hyaena hyaena, Nyctereutes procynoides, Urocyon cinereoargenteus, Vulpes fulva, V. vulpes;* cosmopolitan.

*T. sibirica* Dubnitzky 1952 (syn. *T. skrjabini* Romanov 1952), in *Gulo gulo, Martes zibellina;* Russia.

*T. simbae* Dinnik et Sachs 1972, in *Felis leo;* Africa.

*T. skrjabini* (Popov 1937) comb. n. (syn. *Multiceps skrjabini* Popov 1937), in *Canis familiaris, C. lupus;* Russia.

*T. smythi* (Johri 1957) comb. n. (syn *Multiceps smythi* Johri 1957), in *Canis familiaris;* Ireland.

*T. spalacis* (Diesing 1863) comb. n. (syn. *Coenurus spalacis* [Diesing 1863] Joyeux 1923), in tissues of *Spalax capensis, Tachyoryctes splendens;* Ethiopia.

*T. taeniaeformis* Batsch 1796 (for numerous synonyms see Abuladze[5]), in *Canis familiaris, Felis catus, Genetta genetta, Mustela foina, Putorius eriminea,* many other species of carnivores; cosmopolitan.

*T. talicei* Dollfus 1960, in *Ctenomys torquatus* (intermediate host); Uruguay.

*T. tenuicollis* Rudolphi 1819 (syn. *T. mustelae vulgaris* Rudolphi 1810, *T. brevicollis* Rudolphi 1819, *Cysticercus innominatus hypudaci* Leuckart 1857), in *Martes* spp., *Mustela* spp.; Europe, North America, Russia.

*T. triserrata* Meggitt 1928, in *Felis* sp.; Paraguay.

*T. turkmenicus* (Schulz 1931) comb. n. (syn. *Multiceps turkmenicus* Schulz 1931), in intermediate host, *Spermophilopsis leptodactylus;* Russia.

*T. twitchelli* Schwartz 1924 (syn. *Multiceps twitchelli* [Schwartz 1924] Clapham 1942), in *Gulo gulo;* North America.

*T. ursimaritimi* Rudolphi 1810, in *Ursus maritimus;* Europe.

*T. ursini* Linstow 1893, in *Ursus arctos;* Copenhagen Zoo.

## KEY TO THE GENERA IN AMABILIIDAE*

**1a. Rostellum much enlarged, fimbriated, unarmed.................................
............*Diporotaenia* Spasskaya, Spasskii et Borgarenko 1971, (Figure 268)**

*Diagnosis*: Scolex with four simple suckers. Rostellum immensely enlarged, fimbriated, cauliflower-like, unarmed. Neck with collar-like swelling. Proglottids much wider than long. Genital pores regularly alternating. Cirrus pouch slender, cirrus unarmed. External seminal vesicle prominent. Testes, 15 to 16 in single row along breadth of segment. Ovary slightly aporal, with a few lobes. Vitellarium? Seminal receptacle slightly aporal, with accessory duct extending to aporal margin; terminal end with powerful sphincter. Uterus a transverse, lobated sac. Eggs surrounded by simple membranes. Parasites of Colymbiformes. Russia.

Type species: *D. colymbi* Spasskaya, Spasskii et Borgarenko 1971, in *Colymbus ruficollis;* Russia.

**1b. Rostellum otherwise ......................................................... 2**

**2a. Male organs doubled in each proglottid female organs single. Ovary and vitellarium dendritic ...............................................................
................ *Amabilia* Diamare 1893 (syn. *Aphanobothrium* Linstow 1906).**

*Diagnosis*: Rostellum armed. Proglottids craspedote, wider than long. Male reproductive organs doubled in each segment, female organs single. Male genital pores marginal, with large base when cirrus is extruded. Testes in two submedian fields. Cirrus pouch not reaching osmoregulatory canals. Internal seminal vesicle present. Cirrus armed. Ovary and vitellaria

---

\* Review of family: Ryjikov and Tolkatcheva.[296]

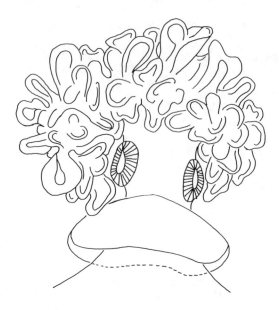

FIGURE 268. *Diporotaenia colymbi* Spasskaja, Spasskii et Borgarenko 1971.

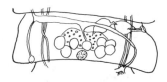

FIGURE 269. *Tatria duodecacantha-* Fuhrmann 1913.

dentritic, median. Seminal receptacle present, with dorso-ventral openings to flat surfaces. Uterus a network of tube-like branches. Eggs fusiform with polar filaments. Parasites of flamingos. Africa, Europe, Sri Lanka.

Type species: *A. lamilligera* (Owen 1832) Diamare 1893 (syn. *Aphanobothrium catenatum* Linstow 1906, *Taenia lamilligera* Owen 1832), in *Phoenicopterus minor, P. roseus, P. ruber*; Europe, Kenya, Sri Lanka.

**2b. Male organs single per proglottid. Female organs not dendritic ............... 3**

**3a. Accessory canal opening from seminal receptacle on one surface or not at all. Testes few.................................*Tatria* Kowalewski 1904. (Figure 269)**

*Diagnosis*: Scolex, rostellum and suckers covered with fine spines. Rostellum long, armed with 8 to 14 large hooks in a single circle, sometimes followed by rostellar spines. Proglottids with large lateral lobes. Single set of reproductive organs per proglottid. Male pores regularly alternating. Cirrus pouch passing between osmoregulatory canals. Cirrus armed. Testes few. Internal and external seminal vesicles present. Ovary compact, sometimes bilobed. Vitellaria median, postovarian. Seminal receptacle with one accessory duct that ends blindly or joins same duct in next proglottid, and another duct on opposite side that sometimes opens to outside. Uterus? Parasites of grebes. Africa, Asia, Europe, Japan, North and South America. Key to species, and development: Rees.[2836]

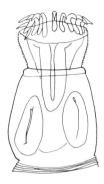

FIGURE 270. *Schistotaenia macrorhyncha* (Rudolphi 1810) Cohn 1900.

Type species: *T. biremis* Kowalewski 1904, in *Podiceps auritus, P. griseigena, P. nigricollis*; Asia, Europe.

Other species:

*T. acanthorhyncha* (Wedl 1855) Kowalewski 1904, in *Colymbus, Fulica, Podiceps*; Europe, Japan.

*T. appendiculata* Fuhrmann 1913, in *Podiceps dominicus*; Brazil.

*T. azerbaijanica* Matevossian et Sailov 1963, in *Podiceps cristatus, P. ruficollis*; Russia.

*T. decacantha* Fuhrmann 1913, in *Podiceps* spp.; Europe, Japan, Siberia.

*T. dubininae* Ryjikov et Tolkatscheva 1981, in *Podiceps* spp.; Russia.

*T. duodecacantha* Olsen 1939, in *Podilymbus podiceps*; U.S.

*T. fimbriata* Borgarenko, Spasskaia et Spasskii 1972, in *Podiceps griseigena*; Russia.

*T. fuhrmanni* Solomon 1932, in *Podiceps* sp.; Africa.

*T. iunii* Korpaczewska et Sulgostowska 1974, in *Podiceps nigricollis*; Russia.

*T. jubilaea* Okorokov et Tkachev 1973, in *Podiceps auritus, P. ruficollis*; Russia.

*T. octacantha* Rees 1973, in *Podiceps ruficollis*; England.

*T. skrjabini* Tretiakova 1948, in *Podiceps ruficollis*; Russia.

**3b. Accessory canals opening from seminal receptacle on both flat surfaces. Testes numerous ................................................................................... 4**

**4a. Genital pores alternating irregularly. Seminal receptacle not continuous from one proglottid to another................ *Schistotaenia* Cohn 1900. (Figure 270)**

*Diagnosis*: Scolex partly spinose. Rostellum strongly developed, with single circle of strong hooks. Suckers sometimes armed. Proglottids wider than long, with marginal extensions. Single set of reproductive organs per segment. Male genital pores marginal, irregularly alternating. Testes dorsal, posterior, in single field. Cirrus pouch weak. Cirrus usually armed. External seminal vesicle present. Ovary lobated but not dendritic; wide. Vitelline gland median, postovarian. Vagina passing between osmoregulatory canals, ending blindly near cuticle. Seminal receptacle median, with dorsal and ventral accessory canals opening on both flat surfaces. Uterus a transverse sac. Eggs spheroid. Osmoregulatory canals in three pairs. Parasites of grebes. Africa, Asia, Europe, North America.

Type species: *S. macrorhyncha* (Rudolphi 1810) Cohn 1900 (syn. *Amabilia macrorhyncha* Cohn 1898, *Drepanidotaenia macrorhyncha* Parona 1899, *Schistotaenia indica* Johri 1959, in the sense of Borgarenko 1972, *Taenia macrorhyncha* Rudolphi 1810), in *Podiceps auritus, P. cristatus, P. griseigena, P. nigricollis, P. ruficollis*; Africa, Europe, Russia.

FIGURE 271. *Pseudoschistotaenia indica* Fotedar et Chisti 1976. Scolex and mature proglottids.

Other species:
  *S. colymba* Schell 1955 (syn. *Tatria antipini* Mathevossian et Okorokov 1959), in *Podiceps auritus, P. cristatus, P. griseigena, P. nigricollis, P. ruficollis*; North America, Russia.
  *S. indica* Johri 1959, in *Podiceps ruficollis*; India.
  *S. macrocirrus* Chandler 1948 (syn. *S. tenuicirrus* Chandler 1948), in *Corvus brachyrhynchos, Podiceps auritus, P. cristatus*; North America.
  *S. mathevossianae* (Okorokov 1956) Ryjikov et Tolkatscheva 1981 (syn. *Tatria erschovi* Mathevossian et Okorokov 1959, *T. jubilaea* Okorokov et Tkatchev 1973, *T. mathevossianae* Okorokov 1956), in *Podiceps auritus, P. griseigena, P. ruficollis*; Russia.
  *S. rufi* Sulgostowska et Korpaczewska 1969, in *Podiceps ruficollis*; Poland.
  *S. scolopendra* (Diesing 1856) Cohn 1900 (syn. *Taenia scolopendra* Diesing 1856, *Tatria scolopendra* [Diesing 1856 Linstow 1908]), in *Podiceps cristatus, P. dominicus*; Antigua, Brazil, Cuba.
  *S. srivastavai* Rausch 1970, in *Podiceps griseigena*; Alaska.

**4b. Genital pores alternating regularly. Seminal receptacle continuous from one proglottid to another .. *Pseudoschistotaenia* Fotedar et Chisti 1976 (Figure 271)**

*Diagnosis*: Small cestodes. Rostellum with single circle of numerous hooks; handle short, guard longer than blade. Suckers aspinose. Neck absent. Proglottids greatly extended laterally. Single set of male and female reproductive systems per proglottid. Genital pores regularly alternating. Testes numerous, medullary, surrounding ovary. External seminal vesicle present, surrounded by gland cells. Cirrus pouch well developed, mainly cortical, expanded distally into a spined knob. Cirrus thin, unarmed. Ovary small, bilobed, median. Vitellarium small, postovarian. Seminal receptacles continuous between segments. Vagina or vaginal accessory canal not seen. Gravid uterus lobated, filling most of proglottid. Parasites of Podicepediformes. India.

Type species: *P. indica* Fotedar et Chisti 1976, in *Podiceps ruficollis*; Kashmir.
Other species:
  *P. pindchii* Fotedar et Chisti 1977, in *Podiceps ruficollis*; India.

## KEY TO THE GENERA IN ACOLEIDAE*

(syn. Diploposthidae Poche 1926)

1a. Vaginal aperture absent ......................................................... 2
1b. Vaginal aperture present ......................................................... 3

---

* Review of family: Ryjikov and Tolkatcheva.[2961]

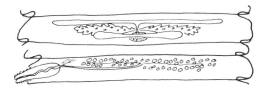

FIGURE 272. *Acoleus vaginatus* (Rudolphi 1819) Fuhrmann 1899.

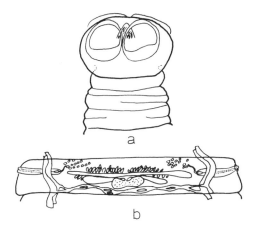

FIGURE 273. *Diplophallus taglei* Olsen 1966. (a) Scolex; (b) proglottid.

**2a. Male and female genitalia single......... *Acoleus* Fuhrmann 1899. (Figure 272)**

*Diagnosis*: Scolex with armed rostellum and simple suckers. Male genital pores regularly alternating. Cirrus pouch large. Cirrus armed. Testes numerous, in a single transverse field. Ovary equatorial, lobated, transversely elongated. Vitellarium compact, median, postovarian. Vagina represented by an enlarged, transverse seminal receptacle lacking external opening. Uterus first appearing as a transverse tube, becoming sac-like when gravid. Eggs with polar thickenings of middle shell. Parasites of Charadriiformes and Rallidae.

Type species: *A. vaginatus* (Rudolphi 1819) Fuhrmann 1899 (syn. *Acoleus armatus* Fuhrmann 1899, *A. hedleyi* Johnston 1910, *A. rugosus* Johnston 1910, *Taenia rugosa* Krefft 1871, *T. vaginata* Rudolphi 1819), in *Belonopterus rayennensis, Himantopus himantopus, H. mexicanus, H. wilsonia, Hoplopterus spinosus, Limosa lapponica, Microsarcops*; Africa, Australia, Europe, North and South America, Russia.

Other species:

*A. longispiculus* (Stossich 1895) Fuhrmann 1899 (syn. *Bothriocephalus longispiculus* Stossich 1895), in *Ortygometra minuta, Porzana parva*; Africa, Europe, India, Turkestan.

*A. meridionalis* (Cholodkovsky 1914) Fuhrmann 1899 (syn. *Bertia meridionalis* Cholodkovsky 1914), in *Himantopus himantopus*; Russia.

**2b. Male genitalia double, female genitalia single..................................... ... *Diplophallus* Fuhrmann 1900 (syn. *Himantocestus* Ukoli 1965). (Figure 273)**

*Diagnosis*: Rostellum with a single circle of hooks, often lost from adults. Proglottids wider than long. Each segment with a double set of male and single set of female reproductive organs. Male genital pores bilateral. Cirrus pouch well developed. Internal and external seminal vesicles present. Cirrus armed. Testes 16 to 100 in number, in two submedian fields. Ovary transversely elongated. Vitelline gland compact, postero-dorsal to ovary. Vaginae absent, represented by transversely elongated seminal receptacle. Parasites of Charadriiformes and chinchilla. Chile, Europe, Peru, Russia, U.S.

> Type species: *D. polymorphus* (Rudolphi 1819) Fuhrmann 1900 (syn. *Himantocestus blanksoni* Ukoli 1965, *Taenia polymorpha* Rudolphi 1819), in *Himantopus himantopus, H. mexicanus, Phalaropus lobatus, Recurvirostra avosetta;* Africa, Europe, Russia.
> Other species:
> *D. andinus* Vote et Read 1953, in *Recurvirostra andina;* Peru.
> *D. coili* Ahern et Schmidt 1976, in *Recurvistra americana;* U.S.
> *D. taglei* Olsen 1966, in *Lagidium peruanum;* Chile.

**3a. Testes arranged in a single median field ............... *Diploposthe* Jacobi 1896.**

*Diagnosis:* Rostellum with single circle of ten hooks. Proglottids wider than long. Double set of male reproductive organs per proglottid. Female organs single, except for vaginae, which are double. Genital pores marginal. Cirrus pouch small. Cirrus large, armed. Testes 3 to 17, in single postovarian field. External seminal vesicle present. Ovary bilobed, median. Vitellaria compact, postovarian. Each vagina opens into atrium ventral to cirrus pouch. Uterus a transverse tube. Parasites of Anseriformes. Africa, Asia, Australia, Europe, North America. Revision: Sulgostowska.[3602]

> Type species: *D. laevis* (Bloch 1782) Jacobi 1896 (syn. *Diploposthe lata* Fuhrmann 1900, *Taenia bifaria* Siebold in Creplin 1846, *T. laevis* Bloch 1782, *T. tuberculata* Krefft 1871), in *Anas, Aythya, Branta, Chaulelasmus, Clangula, Dendrocygna, Fuligula, Netta, Nettion, Nyroca, Oedemia, Querquedela, Spatula*, etc.; Africa, Asia, Australia, Europe, North America.
> Other species:
> *D. skrjabini* Matevosyan 1942 (syn. *D. mathevossiani* Rysavy 1961), in *Netta rufina, Nyroca ferina, N. rufa;* Azerbaidjan, Bohemia.

**3b. Testes arranged in two sub-median fields ... *Jardugia* Southwell et Hilmy 1929.**

*Diagnosis:* Rostellum with single circle of hooks. Proglottids wider than long. Single set of female and double set of male reproductive organs per segment (some segments may contain only a single set of male organs). Genital pores lateral. Genital ducts dorsal to osmoregulatory canals. Cirrus pouch elongate. Cirrus armed. Internal and external seminal receptacles present. Testes few (two to six), preovarian, in two submedian groups. Ovary median, lobated. Vitelline gland lobated, posteromedian to ovary. Two vaginae present, each with proximal seminal receptacle and opening in atrium ventral to cirrus. Uterus a transverse tube. Parasites of herons. Nigeria.

> Type species: *J. paradoxa* Southwell et Hilmy 1929, in *Ardea* sp.; Nigeria.

## KEY TO THE GENERA OF CATENOTAENIIDAE

**1a.** Testes posterior to ovary ....................................................... 2
**1b.** Testes lateral to, anterior to, or surrounding ovary ........................... 5

**2a.** Ovary horseshoe-shaped, symmetrical; vitellarium median ......................
......... *Quentinotaenia* Tenora, Mas-Coma, Murai et Feliu 1980. (Figure 274)

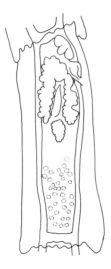

FIGURE 274. *Quentinotaenia mesovitallinica* (Rego 1967) Tenora, Mas-Coma, Murai et Feliu 1980.

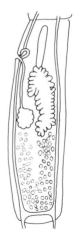

FIGURE 275. *Hemicatenotaenia geosciuri* (Ortlepp 1938) Tenora 1977.

*Diagnosis*: Scolex simple, with four simple suckers. Neck absent. Mature and gravid segments longer than wide, craspedote. Genital pores alternating, in anterior third of proglottid. Testes numerous, posterior to vitelline gland, between osmoregulatory canals. Ovary preequatorial, median, horseshoe-shaped with bend anterior to genital pores. Vitellarium compact, median, posterior to ovary. Parasites of caviomorph rodents. South America.

Type species: *Q. mesovitellinica* (Rego 1967) Tenora, Mas-Coma, Murai et Feliu 1980 (syn. *Catenotaenia mesovitellinica* Rego 1967), in *Galea spixii*; Brazil.

2b. Ovary not horseshoe-shaped, asymmetrical; vitellarium poral ................ 3

3a. Upper margin of ovary not anterior of level of genital pores....................
..................................... *Hemicatenotaenia* **Tenora 1977. (Figure 275)**

*Diagnosis*: Scolex simple, with four simple suckers. Neck present. Proglottids craspedote, mature and gravid segments longer than wide. Genital pores marginal in anterior third of segment, alternating, Cirrus pouch small. Testes posterior to ovary, not crossing osmoregulatory canals. Ovary asymmetrical, median, its anterior margin posterior to level of genital pores. Vitellarium lobated, poral to posterior end of ovary. Uterus a median longitudinal stem with numerous lateral branches. Eggs with or without processes. Parasites of xerine rodents. Africa.

Type species: *H. geosciuri* (Ortlepp 1938) Tenora 1977 (syn. *Catenotaenia geosciuri* Ortlepp 1938), in *Geosciurus capensis, Xerus rutilus, Euxerus erythropus*; Africa.
Other species:
*H. chabaudi* (Dollfus 1953) Tenora 1977 (syn. *Catenotaenia chabaudi* Dollfus 1953), in *Xerus getulus*; Morocco.

3b. Upper margin of ovary anterior to level of genital pores ....................... 4

4a. Field of testes overlapping osmoregulatory canals ................................
.... *Pseudocatenotaenia* **Tenora, Mas-Coma, Murai et Feliu 1980. (Figure 276)**

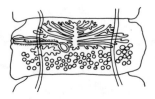

FIGURE 276. *Pseudocatenotaenia matovi* (Genov 1971) Tenora, Mas-Coma, Murai et Feliu 1980.

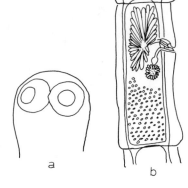

FIGURE 277. *Catenotaenia cricetorum* Kirshenblat 1949. (a) Scolex; (b) proglottid.

*Diagnosis*: Scolex simple, with four simple suckers. Neck present. Strobila acraspedote. Mature segments quadrate or wider than long, gravid segments longer than wide. Genital pores preequatorial. Cirrus pouch not reaching osmoregulatory canals. Testes completely postovarian, overlapping osmoregulatory canals. Ovary asymmetrical, extensive, with lobes anterior to level of genital pores and overlapping osmoregulatory canals. Vitellarium lobated, poral, posterior to ovary, overlapping poral osmoregulatory canals. Gravid uterus with a median stem and numerous lateral branches. Egg membranes with bilateral, auricular processes. Parasites of rodents. Russia, Europe.

Type species: *P. matovi* (Genov 1971) Tenora, Mas-Coma, Murai et Feliu 1980 (syn. *Catenotaenia* Genov 1971), in *Apodemus* spp.; Bulgaria, Spain, Crimea.

**4b. Field of testes not overlapping osmoregulatory canals** .................................
....................................... ***Catenotaenia*** **Janicki 1904. (Figure 277)**

*Diagnosis*: Scolex unarmed, without rostellum. Neck present. Mature and gravid proglottids longer than wide, craspedote. Genital pores preequatorial, irregularly alternating. Cirrus unarmed. Seminal vesicles absent. Testes numerous, in single postovarian field, never crossing osmoregulatory canals. Ovary asymmetrical, strongly lobated, extensive, slightly poral, always extending anterior of genital pore. Vitellarium also lobated, poral to ovary. Uterus taenioid, with median stem and lateral branches. Egg membranes without processes. Parasites of rodents. Europe, Africa, Japan, Russia, North America, Philippines.

Type species: *C. pusilla* (Goeze 1782) Janicki 1904 (syn. *Taenia pusilla* Goeze 1782), in *Mus musculus, M. wageneri, Apodemus sylvaticus, Rattus rattus, R. norvegicus, Clethrionomys* spp., *Microtus agrestis, M. arvalis, Pitymys pinetorium*; Europe, Africa, North America, Japan.

Other species:

*C. afghana* Tenora 1977 (syn. *C. dendritica* [Goeze 1782] sensu Tenora et Kullmann 1970), in *Cricetulus migratorius, Alticola roylei*; Afghanistan.

*C. asiatica* Tenora et Murai 1975, in *Cricetulus barabensis*; Mongolia.

*C. californica* Dowell 1953, in *Dipodomys mohavensis, D. morroensis*; U.S.

*C. compacta* Ortlepp 1962, in *Rattus chrysophilus*; South Africa.

*C. cricetorum* Kirschenblat 1949, in *Mesocricetus auratus, Pollasiomys erythrourus*; Russia.

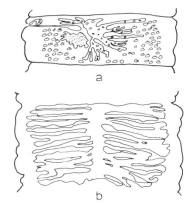

FIGURE 278. *Skrjabinotaenia lobata* (Baer 1925) Spasskii 1951. (a) Mature proglottid; (b) gravid proglottid.

*C. dendritica* (Goeze 1782) Janicki 1904 (syn. *Taenia dendritica* Goeze 1782), in *Sciurus vulgaris, Euxerus erythropus, Crethrionomys rutilus, C. gapperi*; Europe, Africa, North America.

*C. elongata* Ortlepp 1962, in unknown host; South Africa.

*C. indica* Parihar et Nama 1977, in *Tatera indica*; India.

*C. karachiensis* Bilqees 1979, in *Paraechinus micropus*; Pakistan.

*C. kirgizica* Tokobaev 1959, in *Apodemus sylvaticus*; Russia.

*C. kratochvili* Tenora, 1959, in *Apodemus flavicollis*; Czechoslovakia.

*C. kullmanni* Tenora 1977, in *Blanfordimys afghanus, Calomyscus bailwardi*; Afghanistan.

*C. laguri* Smith 1954, in *Lagurus curtatus*; Wyoming, U.S.

*C. linsdalei* McIntosh 1941, in *Thomomys bottae, Perognathus californica, Dipodomys venustus, D. heermanni*; California, U.S.

*C. lucida* Ortlepp 1962, in *Rattus chrysophilus*; South Africa.

*C. peromysci* Smith 1954, in *Peromyscus maniculatus*; U.S.

*C. pigulevski* Uzhakhov 1964, in "Muridae"; Russia.

*C. reggiae* Rausch 1951, in *Marmota caligata*; Alaska.

*C. rhombomidis* Schulz et Landa 1935, in *Rhombomys opimus*; Russia.

*C. ris* Yamaguti 1942, in *Sciurus lis*; Japan.

**5a.  Longitudinal stem of gravid uterus longer than side branches; mature and gravid proglottids longer than wide .... *Skrjabinotaenia* Akhumyan 1946. (Figure 278)**

*Diagnosis*: Scolex unarmed, lacking rostellum, with four small suckers near apex. Neck short or absent. Proglottids craspedote or acraspedote, shorter than wide. Genital pores irregularly alternating. Cirrus pouch small. Cirrus unarmed. Testes posterior to, lateral to, sometimes surrounding ovary; sometimes in two submedian fields. Seminal vesicles absent. Ovary highly lobated, extensive, slightly poral. Vitellaria also lobated, poral to center of ovary. Vagina posterior to cirrus. Seminal receptacle present. Uterus taenioid, with median stem and numerous lateral branches that are shorter than the median stem. Parasites of rodents. Africa, Europe.

Type species: *S. oranensis* (Joyeux et Foley 1930) Akyumyan 1946 (syn. *Catenotaenia oranensis* Joyeux et Foley 1930), in *Meriones shawi*; Algeria.

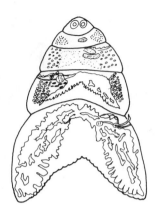

FIGURE 279. *Meggittina gerbilli* (Wertheim 1954) Tenora, Mas-Coma, Murai et Feliu 1980.

Other species:

*S. lobata* (Baer 1925) Spasskii 1951 (syn. *Catenotaenia lobata* Baer 1925, *C. capensis* Ortlepp 1940), in *Apodemus sylvaticus, A. flavicollis, Clethrionomysglareolus, Mastomys erythroleucus, M. coucha, Taterona kempi, Rattus marungensis, Rhabdomys pumilio*; Africa, Europe.

*S. madagascariensis* Quentin et Durette-Desset 1974, in *Brachyuromys betsileoensis*; Madagascar.

*S. media* Quentin 1971, in *Hylomyscus stella*; Central African Republic.

*S. occidentalis* Hunkeler 1972, in *Mastomys erythroleucus, Praomys tullbergi*; Ivory Coast.

*S. pauciproglottis* Quentin 1965, in *Stochomys longicaudatus, Prionomys batesi*; Central African Republic.

*S. psammomi* Mikhail et Fahmy 1968, in *Psammomys obesus*; Egypt.

5b. **Longitudinal stem of gravid uterus much shorter than side branches; mature and gravid proglottids much wider than long.....................................
...... Meggittina Lynsdale 1953 (syn. *Rajotaenia* Wertheim 1954). (Figure 279)**

*Diagnosis*: Scolex very small, with four simple suckers. Neck very wide. Strobila consisting of few very wide, very short proglottids that are acraspedote. Single set of reproductive organs per segment. Genital pores alternating, on anterior third of free border of proglottid. Cirrus pouch curved. Testes numerous (250 to 350) in two fields anterior to ovary, most numerous on aporal side. External and internal seminal vesicles absent. Ovary poral, branched. Vitelline gland medial to ovary. Seminal receptacle present. Uterus a very short median stem with long, lateral branches each with secondary branches. Parasites of rodents. Africa.

Type species: *M. baeri* Lynsdale 1953 (syn. *Catenotaenia baeri* [Lynsdale 1953] Wolfgang 1956), in house and granary rats; Africa.

Other species:

*M. aegyptiaca* (Wolfgang 1956) Tenora, Mas-Coma, Murai et Feliu 1980 (syn. *Catenotaenia aegyptiaca* Wolfgang 1956), in *Meriones* sp., *Gerbilus grbillus, Acomys cahirinus*; Egypt.

*M. cricetomydis* (Hockley 1961) Tenora, Mas-Coma, Murai et Feliu 1980 (syn. *Skrjabinotaenia cricetomydis* Hockley 1961), in *Cricetomys gambianus*; Nigeria.

*M. gerbilli* (Wertheim 1954) Tenora, Mas-Coma, Murai et Feliu 1980 (syn. *Rajotaenia gerbilli* Wertheim 1954, *Skrjabinotaenia gerbilli* [Wertheim 1954] Spasskii 1955), in *Gerbillus pyramidum*; Israel.

# FAMILY DAVAINEIDAE

These tapeworms are parasites of birds and mammals. The intermediate hosts are usually insects or gastropods.

The family is large, and boasts the largest genus of cestodes: *Raillietina*, with about 295 species. There are three subfamilies; Idiogeninae (with a paruterine organ), Ophryocotylinae (with a persistent, sac-like uterus); and Davaineinae (with egg capsules). Some species have been reported from humans, especially in the Orient. Others are common in domestic fowl.

The most readily recognized characteristic of most davaineids is the rostellum. It usually is short, broad, and retractable, and bears circles of very small, very numerous T- or hammer-shaped hooks. The suckers are commonly armed with very small spines that are easily lost from dead specimens. A very important monograph was written by Artyukh.[80] Spasskii[3432] divided the Davaineinae into eight tribes. Wardle et al.[3871] elevated the family to order Davaineidea, and the three subfamilies to level of family. I reject the suggestion on the grounds that it fragments a cohesive order, the Cyclophyllidea.

## KEY TO THE SUBFAMILIES IN DAVAINEIDAE

1a. Paruterine organ present ................... Idiogeninae Fuhrmann 1907. (p.237)
1b. Paruterine organ absent ........................................................ 2

2a. Uterus sac-like, persistent ........... Ophryocotylinae Fuhrmann 1907. (p.241)
2b. Uterus replaced by egg capsules .............. Davaineinae Braun 1900. (p.245)

## KEY TO THE GENERA IN IDIOGENINAE

1a. Rostellum absent ................... *Ascometra* Cholodkovsky 1912. (Figure 280)

*Diagnosis*: Rostellum absent. Suckers each with pair of muscular lappets, unarmed. Neck absent. Proglottids craspedote. Genital pores unilateral. Genital ducts dorsal to osmoregulatory canals. Cirrus pouch crosses osmoregulatory canals. Testes dorsal, numerous, nearly surrounding ovary. Ovary median. Vitelline gland postovarian. Vagina posterior to cirrus pouch. Seminal receptacle present. Uterus sac-like, with anterior paruterine organ. Parasites of Gruiformes (bustards). Africa, Asia.

Type species: *A. vestita* Cholodkovsky 1912 (syn. *Chapmania vestita* [Cholodkovsky 1912] Lopez-Neyra 1954), in *Chlamydotis undulata*, *Houbara macqueeni*, *Lophoitis ruficristata*; Asia Minor, Africa.

Other species:
  *A. baeri* Matevosyan et Movsessian 1970, in ?; Russia?

1b. Rostellum present ............................................................. 2

2a. Rostellum with 10 to 12 circles of hooks ........................................
   ................................... *Sphyroncotaenia* Ransom 1911. (Figure 281)

*Diagnosis*: Rostellum with 10 to 12 circles of small, hammer-shaped hooks. Neck very short. Genital pores unilateral. Cirrus pouch not reaching osmoregulatory canals. Genital ducts ventral to osmoregulatory canals. Testes numerous, medullary. Cirrus armed. Ovary transversely elongated, poral. Vitelline gland dorsal, poral to testes. Vaginal pore posteroventral to cirrus pore. Seminal receptacle proximal. Uterus voluminous, with anterior paruterine organ. Parasites of Otidiformes (bustards). Africa.

FIGURE 280. *Ascometra vestita* Cholodkovsky 1912.

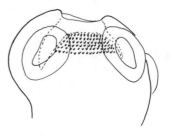

FIGURE 281. *Sphyroncotaenia uncinata* Ransom 1911.

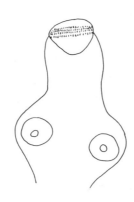

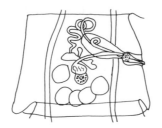

FIGURE 282. *Ersinogenes spinatum* Spassnkaja 1961.

Type species: *S. unicinata* Ransom 1911, in *Neotis caffra*; East Africa.

**2b. Rostellum with fewer than ten circles of hooks ................................. 3**

**3a. Rostellum with five circles of hooks; suckers unarmed ..........................**
**......................................... *Ersinogenes* Spasskaja 1961. (Figure 282)**

*Diagnosis*: Scolex with long rostellum armed with 250 to 300 hooks in 5 circles. Suckers simple, unarmed. Short neck present. Proglottids craspedote. Genital pores equatorial. Cirrus pouch reaches midline of segment. Five testes posterior to ovary. Ovary median, small, bilobed. Vitellarium compact, posterior to ovary. Vagina opens dorsal to cirrus. Seminal receptacle present. Uterus horseshoe-shaped with single, anterior paruterine organ. Parasites of Otidiformes. Russia.

Type species: *E. spinatum* Spasskaja 1961, in *Otis tarda*; Russia.

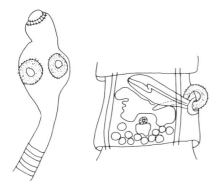

FIGURE 283. *Paraidiogenes mongolica* (Danzan 1964) Movsessian 1970.

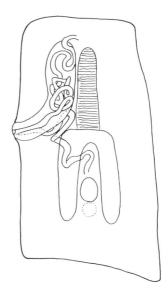

FIGURE 284. *Idiogenes otidis* Krabbe 1868.

**3b. Rostellum with fewer than five circles of hooks, suckers armed or not........ 4**

**4a. Rostellum with three circles of hooks............................................**
**..................................*Paraidiogenes* Movsessian 1970. (Figure 283)**

*Diagnosis*: Rostellum elongate, spiny, with three circles of hammer-shaped hooks. Suckers with about 20 circles of spines. Neck present. Proglottids craspedote; mature wider than long, gravid longer than wide. Cirrus pouch powerful, exceeding midline of segment. Cirrus armed. Testes posterior of ovary. Ovary bilobed, median. Vitellarium compact, postovarian. Uterus horseshoe-shaped, with single, anterior paruterine organ. Parasites of Otidiformes (bustard). Asia.

Type species: *P. mongolica* (Danzan 1964) Movsessian 1970 (syn. *Idiogenes mongolica* Danzan 1964), in *Otis tarda*; Mongolia.

**4b. Rostellum with two circles of hooks ............................................. 5**

**5a. Uterus horseshoe-shaped ..................*Idiogenes* Krabbe 1867. (Figure 284)**

*Diagnosis*: Scolex commonly absent, replaced by pseudoscolex. When present, it has two circles of small, hammer-shaped hooks. Suckers armed. Neck absent. Proglottids craspedote. Genital pores unilateral. Cirrus pouch large. Cirrus armed. Testes in single, posterior field. Ovary dumbbell-shaped. Vitelline gland compact, postovarian. Vagina posterior to cirrus pouch, convoluted. Seminal receptacle proximal. Uterus horseshoe-shaped, posterior to longitudinal paruterine organ. Parasites of Gruiformes, Falconiformes, Passeriformes. Africa, Asia, Europe, North and South America.

Type species: *I. otidis* Krabbe 1868, in *Chlamydotis undulata, Lyssotis melanogaster, Otis cafra, O. tarda, O. tetrax, Trachelotis senegalensis*; Africa, Europe, Russia.
Other species:
*I. bucorvi* Joyeux, Baer et Martin 1936, in *Bucorvus abyssinicus, B. leadbeateri*; Africa.
*I. butasteri* Chatterji 1954, in *Butaster tisa*; India.

FIGURE 285. *Otiditaenia conoides* (Bloch 1782) Beddard 1912.

*I. buteonis* Schultz 1939, in *Buteo swainsoni*; Oklahoma, U.S.

*I. flagellum* (Goeze 1782) Krabbe 1867 (syn. *Taenia flagellum* Goeze 1782, *Idiogenes mastigophora* Krabbe 1879, *Davainea [Chapmania] longicirrosa* Fuhrmann 1906), in *Milvus* spp.; Egypt, Liberia, Tanganyika, Turkestan.

*I. grandiporus* Cholodkovsky 1905, in *Tetrax tetrax*; Russia.

*I. horridus* Fuhrmann 1908, in *Cariama cristata*; Brazil.

*I. horridus* var. *africanus* Hungerbühler 1910, "bird of prey"; Africa.

*I. kolbei* Ortlepp 1938, in *Anthopoides virgojuv, Choriotis kori, Eupodotis senegalensis, Lophotis gindiana, Neotis cafra, Otis tarda*; Africa, Russia.

*I. kori* Ortlepp 1938, in *Lophotis ruficrista gindiana, Otis tarda, O.* sp.; Africa, Russia.

*I. nana* Fuhrmann 1925, in *Chlamydotis undulata, Eupodotis arabus, Otis houbara*; Africa, Russia.

*I. skrjabini* Movsessian 1968, in *Otis tarda*; Mongolia.

*I. travassosi* Ortlepp 1938, in *Milvus migrans*; South Africa.

*I. tuvensis* Spasskaja 1961, in *Chlamydotis undulata*; Russia.

**5b.  Uterus sac-like ............................................................................ 6**

**6a.  Testes mainly behind ovary. Suckers unarmed ......................................**
**................................................. Otiditaenia Beddard 1912 (syn. *Paraschistometra* Woodland 1930, *Schistometra* Cholodkovsky 1912). (Figure 285)**

*Diagnosis*: Rostellum with two circles of small, hammer-shaped hooks. Suckers with two marginal lappets, unarmed. Proglottids wider than long, craspedote. Genital pores irregularly alternating. Cirrus pouch small. Testes numerous, mainly posterior to ovary. Ovary and vitelline gland poral. Uterus a transverse sac, with anterior paruterine organ. Parasites of Gruiformes. Africa, Europe.

Type species: *O. conoideis* (Bloch 1782) Beddard 1912 (syn. *Otiditaenia eupodotidis* Beddard 1912, *Schistometra conoideis* [Bloch 1782] Cholodkovsky 1912, *S. embiensis* Cholodkovsky 1915, *S. togata* Cholodkovsky 1912, *S. wettsteini* Weithofer 1916, *Taenia conoideis* Bloch 1782. *T. cuneata* Batsch 1786), in *Choriotis arabus, C. kori, Lissotis melanogaster, Lophotis ruficristata, Neotis denhami, Otis tarda*; Africa, Europe.

Other species:

*O. korkhaani* (Ortlepp 1938) comb. n., in *Afrotis afroides, Eupodotis barrowi*; South Africa.

FIGURE 286. *Chapmania macrocephala* Fuhrmann 1943.

*O. macqueeni* (Woodland 1930) Baer 1955 (syn. *Paraschistometra macqueeni* [Woodland 1930] Movsessian 1970, *Schistometra macqueeni* Woodland 1930), in *Houbara macqueeni*; Africa.

*O. nigriceps* (Gupta 1976) comb. n., in *Choriotis nigriceps*; India.

**6b. Testes scattered, surrounding ovary. Suckers armed ............................
*Chapmania* Monticelli 1893 (syn. *Capsodavainea* Fuhrmann 1901). (Figure 286)**

*Diagnosis*: Rostellum with two simple or wavy circles of small, hammer-shaped hooks. Suckers armed. Neck present or absent. Proglottids craspedote. Genital pores irregularly alternating. Genital atrium deep. Cirrus pouch reaches osmoregulatory canals. Ovary median. Vitellarium postovarian. Vagina posterior to cirrus pouch. Seminal receptacle proximal. Uterus sac-like, with anterior paruterine organ. Parasites of Gruiformes, Rheiformes, Bucerotiformes. Africa, Europe, South America.

Type species: *C. tauricollis* (Chapman 1876) Monticelli 1893 (syn. *Taenia tauricollis* Chapman 1876, *T. argentina* Zschokke 1888, *Capsodavainea tauricollis* [Chapman 1876] Fuhrmann 1901, *Davainea tauricollis* [Chapman 1876] Fuhrmann 1896), in *Rhea americana*; Argentina, Brazil.

Other species:

*C. brachynrhyncha* (Creplin 1854) Monticelli 1893 (syn. *Davainea brachyrhyncha* [Creplin 1854] Fuhrmann 1908, *Taenia brachyrhyncha* Creplin 1854), in *Dolicholophus cristatus*; Brazil.

*C. macrocephala* Fuhrmann 1943, in *Lissotis melanogaster, Otis Caffra*; Africa.

*C. tapika* (Clerc 1906) Fuhrmann 1908 (syn. *Idiogenes tapika* Clerc 1906, *Otiditaenia tapika* [Clerc 1906] Movsessian 1977), in *Otis arabus, O. tarda, Tetrax tetrax*; Russia.

*C. unilateralis* Skrjabin 1914, in *Bucorax cafer, Bucorvus abyssinicus*; Africa.

## KEY TO THE GENERA IN OPHRYOCOTYLINAE

**1a. Genital pores unilateral ........................................................ 2
1b. Genital pores alternating ....................................................... 4**

**2a. Uterus with median stem and lateral branches ................................
..............................................*Dasyurotaenia* Beddard 1912. (Figure 287)**

*Diagnosis*: Scolex poorly known, apparently with rostellum bearing two circles of hooks. Proglottids very short, wide and muscular. Genital pores unilateral. Testes numerous, mainly

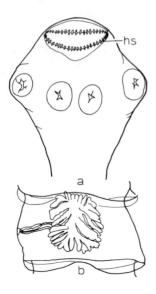

FIGURE 287. *Dasyurotaenia robusta* Beddard 1912. (a) Scolex, hs—hook scars; (b) gravid proglottid.

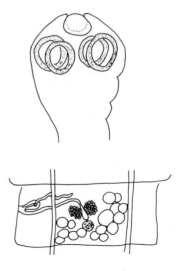

FIGURE 288. *Ophryocotylus dinopii* Srivastav et Capoor 1982. Scolex and proglottid.

lateral, confluent behind vitelline gland. Ovary and vitelline gland transversely elongated, posterior. Large seminal receptacle present. Uterus with median stem and lateral branches. Eggs thin-shelled, middle shell not striated. Parasites of marsupials (Tasmanian devil). Australia.

    Type species: *D. robusta* Beddard 1912, in *Dasyurus ursinus, Sarcophilus satanicus*; Australia.

**2b.**   **Uterus sac-like** ................................................................. 3

**3a.**   **Rostellum with a single circle of hooks** ..........................................
........................ ***Ophryocotylus*** **Srivastav et Capoor 1982. (Figure 288)**

*Diagnosis*: Rostellum with a single circle of hammer-shaped hooks. Suckers armed. Genital pores unilateral. Cirrus pouch not reaching poral osmoregulatory canal. Testes posterolateral to ovary, in continuous field. Ovary multilobate, median. Vitellarium compact, postovarian. Vagina posterior of cirrus pouch. Seminal receptacle present. Uterus sac-like, persistent. Parasites of Piciformes. India.

    Type species: *O. dinopii* Srivastav et Capoor 1982, in *Dinopium benghalense*; Uttar Pradesh, India.

**3b.**   **Rostellum with two circles of hooks** ..............................................
................................. ***Ophriocotyloides*** **Fuhrmann 1920. (Figure 289)**

*Diagnosis*: Rostellar hooks in two simple circles. Suckers armed. Mature proglottids wider than long. Genital pores unilateral. Cirrus pouch variable. Testes mainly lateral to ovary. Ovary median or submedian. Vitelline gland postovarian. Vagina posterior to cirrus pouch. Seminal receptacle small or absent. Uterus sac-like. Parasites of Bucerotiformes, Passeriformes, Piciformes. Africa, Brazil, India.

    Type species: *O. uniuterina* (Fuhrmann 1908) Fuhrmann 1920 (syn. *Davainea uniuterina* Fuhrmann 1908), in *Rupicola rupicola*; Brazil.

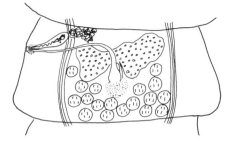

FIGURE 290. *Fernandezia indicus* Singh 1964.

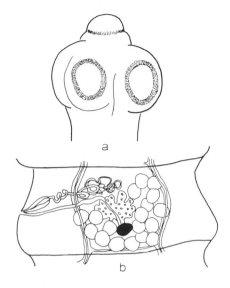

FIGURE 289. *Ophryocotyloides makundi* Singh 1962. (a) Scolex; (b) proglottid.

Other species:

*O. barbeti* Singh 1960, in *Megalaima zeylanica*; India.

*O. bhaleraoi* Inamdar 1944, in *Cinnyris zeylonicus*; India.

*O. corvorum* Gupta et Grewal 1971, in *Corvus splendens*; India.

*O. dasi* Tandan et Singh 1964, in *Megalaima virens*; India.

*O. haemacephala* Singh 1960, in *Megalaima haemacephala*; India.

*O. makundi* Singh 1962, in *Picus xanthopygaeus*; India.

*O. meggitti* Moghe 1933, in *Corvus splendens*; India.

*O. monacanthis* Moghe et Anamdar 1934, in *Dendrocitta rufa*; India.

*O. picusi* Singhe 1962, in *Picus xanthopygaeus*; India.

*O. pinguis* (Fuhrmann 1904) Baer 1927 (syn. *Anoplocephala pinguis* [Fuhrmann 1904] Fuhrmann 1922, *Bertia pinguis* Fuhrmann 1904, *Bertiella pinguis* [Fuhrmann 1904] Douthitt 1915, *Chapmania pinguis* [Fuhrmann 1904] Baer et Fain 1955, *C. unilateralis* Skrjabin 1915, *Otiditaenia pinguis* [Fuhrmann 1904] Baer 1955), in *Bucorvus abissinicus, B. leadbeateri*; Africa.

*O. sharmai* Gupta et Grewal 1971, in *Corvus splendens*; India.

*O. srinagarensis* Malhotra et Capoor 1979, in *Corvus macrarhynchos, C. splendens*; India.

**4a. Suckers unarmed .................. *Fernandezia* Lopez-Neyra 1936. (Figure 290)**

*Diagnosis*: Scolex relatively large. Rostellum armed with two circles of numerous (300 to 1000), small, hammer-shaped hooks. Suckers unarmed. Neck absent. Proglottids wider than long, craspedote. Genital pores irregularly alternating. Cirrus pouch reaches or crosses osmoregulatory canals. Seminal vesicles absent. Testes posterior to ovary. Vitelline gland postovarian. Ovary bilobed, anterior. Vagina postero-ventral to cirrus pouch. Seminal receptacle proximal. Uterus an irregular sac. Parasites of Passeriformes. Africa, Europe, India, Russia.

Type species: *F. goizuetai* Lopez-Neyra 1936, in *Turdus musicus*; Europe, North Africa, Spain.

FIGURE 291. *Ophryocotyle prudhoei* Burt 1962. (a) Scolex; (b) proglottid.

Other species:

*F. indicus* (Singh 1964) Artyukh 1966 (syn. *Ophryocotyle indicus* Singh 1964), in *Turdoides subrufus*; India.

*F. spinosisima* (Linstow 1894) Lopez-Neyra 1936 (syn. *Davainea spinosisima* Linstow 1894), in *Turdus merula*; Africa, Europe, Russia, Israel.

**4b. Suckers armed on anterior margin ........................................ *Ophryocotyle* Friis 1870 (syn. *Burtiella* Spasskii et Kornyushin 1977). (Figure 291)**

*Diagnosis*: Rostellum large, with two undulating circles of small, hammer-shaped hooks. Suckers armed. Genital pores irregularly alternating. Cirrus pouch crosses osmoregulatory canals. Testes in single field posterior to ovary. Genital ducts dorsal to osmoregulatory canals. Ovary median, anterior. Vitelline gland postovarian. Vagina posterior or ventral to cirrus pouch. Seminal receptacle present. Uterus an irregular sac. Parasites of Anseriformes, Bucerotiformes, Charadriiformes, Ciconiiformes. Africa, Europe, Malagasy, North and South America.

Type species: *O. proteus* Friis 1870 (syn. *O. lacazii* Villot 1875), in *Aegialites, Ancylochilus, Calidris* spp., *Charadrius hiaticula, C.* sp., *Erolia alpina, Larus argentatus, L. articilla, L. canus, Limosa fedoa, L. lapponica, Pelidna, Tringa alpina*; Africa, Europe, Scandinavia, South America.

Other species:

*O. alaskensis* Webster 1949, in *Haematopus bachmani*; Alaska.

*O. brasiliensis* Mahon 1957, in *Hoploxypterus cayanus*; South America.

*O. buecki* Joyeux et Baer 1939, in *Lophotibis cristata*; Madagascar.

*O. fuhrmanni* Tendeiro 1953, in *Numenius phaeopus*; Africa, Guinea.

*O. gretellati* Deblock, Rose, Broussart, Capron et Brygoo 1962, in *Aptenodytes patagonica*; Kerquelen Islands, Indian Ocean.

*O. herodiae* Fuhrmann 1909, in *Hagedash hagedash, Theristicus hagedash*; Africa.

*O. insignis* Loennberg 1890, in *Himantopus ostralegus, Mergus serrator, Totanus calidris*: Europe.

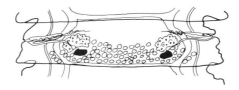

FIGURE 292. *Cotugnia meggitti* Yamaguti 1935.

*O. minutum* Endrigkeit 1940, host unknown; Europe.
*O. prudhoei* Burt 1962, in *Limosa laponica*; Scotland.
*O. turdina* Cholodkowsky 1913, in *Turdus* sp.; Russia.
*O. zeylanica* Linstow 1906 (syn. *Burtiella zeylanica* [Linstow 1906] Spasskii et Kornyushin 1977), in *Lophoceros gingalensis, Neophron pernopterus*; Sri Lanka.

## KEY TO THE GENERA IN DAVAINEINAE

1a. Two sets of reproductive organs per segment .................................. 2
1b. One set of reproductive organs per segment ................................... 3

2a. **Rostellar hooks form four-leaved clover pattern in apical view ..................**
   **......................................................*Abuladzugnia* Spasskii 1973.**

*Diagnosis*: Rostellum armed with small, hammer-shaped hooks, which in apical view form a pattern of a four-leaved clover. Suckers unarmed. Proglottids each with two sets of reproductive organs. Genital ducts dorsal to osmoregulatory canals. Genital pores bilateral. Testes in one medullary field with scattered follicles extending to lateral margins. Ovaries paired, lateral. Vitelline glands postovarian. Vagina posterior to cirrus pouch. One egg in each egg capsule. Parasites of Galliformes. Africa.
   Type species: *A. gutterae* (Ortlepp 1963) Spasskii 1973 (syn. *Cotugnia gutterae* Ortlepp 1963), in *Guttera eduardi*; Africa.

2b. **Rostellar hooks arranged in simple circle in apical view ......................**
   **......................................................*Cotugnia* Diamare 1893**
   **(syn. *Multicotugnia* Movsessian 1969, *Ershovitugnia* Spasskii 1973). (Figure 292)**

*Diagnosis*: Rostellum armed with very small, hammer-shaped hooks. Suckers unarmed. Proglottids each with two sets of reproductive organs. Genital ducts dorsal to osmoregulatory canals. Genital pores bilateral. Testes in one or two median fields. Ovaries paired. Vitelline glands postovarian. Vagina posterior to cirrus pouch. One egg in each egg capsule. Parasites of Anseriformes, Casuariiformes, Columbiformes, Galliformes, Passeriformes, Psittaciformes. Cosmopolitan.
   Type species: *C. digonopora* (Pasquale 1890) Diamare 1893 (syn. *Taenia digonopora* Pasquale 1890), in *Anser* sp., *Columba ferrigineus, C. livia, Gallus gallus, Numida meleagris*; Africa, Burma, India, Indonesia, Philippines.
   Other species:
      *C. akhuminae* Movsesyan 1969, in *Columba livia*; Russia.
      *C. aurangabadensis* Shinde 1969, in *Columba livia*; India.
      *C. bahli* Johri 1934, in *Turtur suratensis*; India.
      *C. bhaleraoi* Mudaliar 1943, in *Gallus gallus*; India.
      *C. brotogerys* Meggitt 1915, in *Brotogerys tirica, Platicercus eximius*; Brazil, Burma, India.

*C. browni* Smith, Fox et White 1908, in *Palaeornis eupatria, P. fasciatus*; Africa, Burma, Sri Lanka.

*C. celebesensis* Yamaguti 1956, in *Geopelia striata*; Celebes.

*C. collini* Fuhrmann 1909 (syn. *Ershovitugnia collini* (Fuhrmann 1909) Spasskii 1973), in *Dromaeus novaehollandiae*; Australia.

*C. columbae* Malviya et Dutt 1969, in *Columba livia*; India.

*C. crassa* Fuhrmann 1909, in *Numida ptilorhyncha, N. rikwae*; Africa.

*C. cuneata* Meggitt 1924, in *Columba livia*; Burma, Egypt, Europe, India.

*C. dayali* Singh 1952, in *Psittacula eupatria*; India.

*C. daynesi* Quentin 1963, in *Gallus gallus*; Madagascar.

*C. dollfusi* Lopez-Neyra 1950, in *Streptopelia turtur*; Spain.

*C. fastigata* Meggitt 1920, in *Anas platyrhyincha, Ptistes coccineopterus*; Burma, India.

*C. fila* Meggitt 1931, in "duck"; Burma, India.

*C. fleari* Meggitt 1927, in "pigeon"; Egypt, India.

*C. fuhrmanni* Baczynska 1914, in *Pavo cristatus, Stigmatopelia cambayensis*; India.

*C. govinda* Johri 1934, in *Milvus govinda*; India.

*C. ilocana* Tubangui et Masilugnan 1934, in *Streptopelia dussumieri*; Philippines.

*C. inaequalis* Fuhrmann 1909, in *Columba livia, Pterocles coronatus, P. orientalis*; Egypt, India.

*C. intermedia* Johri 1934, in *Columba intermedia, Streptopelia turtur*; India, Russia.

*C. januaria* Johri 1934, in *Gallus domesticus*; India.

*C. joyeuxi* Baer 1924 (syn. *Cotugnia* sp. Joyeux 1923), in *Columba punicea, Sphenocercus sphenurus, Turtur senegalensis*; Africa, Burma, India.

*C. longicirrosa* Johri 1939, in *Pavo cristatus*; India.

*C. magna* Burt 1940, in *Columba livia*; Sri Lanka.

*C. margareta* Beddard 1916, in *Caccabis melanocephala, Corvus macrorhynchus, Lophophorus impejanus*; Africa, Arabia, India.

*C. meggitti* Yamaguti 1935 (syn. *C. cuneata* var. *nervosa* Meggitt 1924), in *Columba livia, Streptopelia chinensis, Turtur tranquebaricus*; Africa, Burma, Formosa, India, Vietnam.

*C. meleagridis* Joyeux, Baer et Martin 1936, in *Numida meleagris*; Africa.

*C. noctua* Johri 1934, in *Columba intermedia*; India.

*C. parva* Baer 1925, in *Chalcophaps indica, Columba livia, Corvus macrorhynchus, C.* sp.; Africa, Burma, India.

*C. platycerci* Weerekoon 1944, in *Platycercus icterotis*; Ceylon Zoo (from Australia).

*C. pluriuncinata* Baer 1925, in *Herpestes galea*; Africa.

*C. polyacantha* Fuhrmann 1909, in *Columba eversmanni, C. livia, C. turtur, Stigmatopelia senegalensis, Streptopelia turtur*; Africa, Europe, Russia.

*C. polyacantha* var. *oligorchida* Joyeux, Baer et Martin 1936, in *Stigmatopelia senegalensis*; Africa, Indochina.

*C. polyacantha* var. *paucimusculosa* Meggitt 1927, in "Wandertaube"; Burma, Egypt, Indochina.

*C. polytelidis* Burt 1940, in *Polytelis melanura*; Sri Lanka.

*C. rimandoi* Tubangui et Masiluñgan 1937, in *Columba livia*; Philippines.

*C. seni* Meggitt 1926, in *Brotogerys tirica, Platycerus eximus, Psitttacula manilensis*; Burma, Calcutta Zoo, Sri Lanka.

*C. shindei* nom. nov. (for *C. columbae* Shinde 1969, preoccupied), in *Columba livia*; India.

*C. shohoi* Sawada 1971, in *Acryllium valturinum*; Somalia.

*C. spasskii* Sultanov 1963, in *Pterocles orientalis*; Russia.

*C. srivastavai* Malviya et Dutt 1970, in *Columba livia*; India.

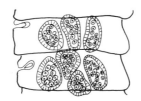

FIGURE 293. *Baerfainia anoplocephaloides* (Baer et Fain 1955) Yamaguti 1959. Gravid proglottids.

*C. streptopeli* Khan et Habibullah 1967, in *Streptopelia decaocto*; Pakistan.

*C. taiwanensis* Yamaguti 1935, in *Columba livia, Pavo cristatus*; Formosa, India, Indochina.

*C. tenuis* Meggitt 1924, for *C. cuneata* var. *tenuis* Meggit 1924, raised to species rank Yamaguti (1935); in *Columba livia*; Burma, Egypt, India.

*C. transvaalensis* Ortlepp 1963, in *Numida meleagridis*; South Africa.

*C. tuliensis* Mettrick 1963, in *Numida meleagris*; Southern Rhodesia.

**3a. Rostellum rudimentary, unarmed ..... *Baerfainia* Yamaguti 1959. (Figure 293)**

*Diagnosis*: Rudimentary, unarmed rostellum present. Suckers armed. Proglottids wider than long. Dorsal osmoregulatory canals absent. Genital pores unilateral. Cirrus pouch extravascular. Internal seminal vesicle present, external absent. Testes few (three to four), one poral, others antiporal. Ovary median. Vitelline gland postovarian. Vagina posterior to cirrus pouch, armed at distal end. Three or four egg capsules per segment, each with 10 to 16 eggs. Parasites of Pholidota. Africa.

Type species: *B. anoplocephaloides* (Baer et Fain 1955) Yamaguti 1959 (syn. *Raillietina (R.) anoplocephaloides* Baer et Fain 1955), in *Manis (Phataginus) tricuspis, M. (Smutsia) gigantea)*; Africa.

**3b. Rostellum well-developed, armed ................................................ 4**

**4a. Strobila very small, of few proglottids. One egg per capsule........... *Davainea* Blanchard 1891 (syn. *Himantaurus* Spasskaja et Spasskii 1971). (Figure 294)**

*Diagnosis*: Scolex globular. Rostellum armed with one or two circles of very small, hammer-shaped hooks. Suckers small. Neck absent. Strobila very small, of few proglottids. Genital pores irregularly alternating or unilateral. Cirrus pouch crosses osmoregulatory canals. Testes few, mainly posterior to ovary. Ovary median or poral. Vitelline gland postovarian. Vagina posterior to cirrus pouch, may have spinous, distal diverticulum. Seminal receptacle present. Egg capsules each with single egg. Parasites of Charadriiformes, Galliformes, Piciformes. Cosmopolitan. Key to species: Artyukh (1966).

Type species: *D. proglottina* (Davaine 1860) Blanchard 1891 (syn. *D. dubius* Meggitt 1916, *D. varians* Sweet 1910, *Taenia proglottina* Davaine 1860), in *Alectoris graeca, Bonas umbellus, Gallus gallus, Lagopus minutus, Perdix perdix*; cosmopolitan.

Other species:

*D. ambajogaiensis* Shinde et Ghare 1977, in *Gallus gallus*; India.

*D. andrei* Fuhrmann 1933, in *Alectoris graeca, Perdix perdix*; Europe.

*D. baeri* Schmilz 1941, in *Gecinus nigrigenis*; Thailand.

*D. chauhani* Rysavy, Barus et Daniel 1975, in *Ithaginis cruentis*; India.

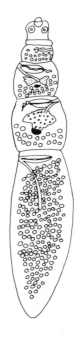

FIGURE 294. *Davainea proglottina* (Davaine 1860) Blanchard 1891.

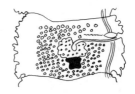

FIGURE 295. *Davaineoides vigintivasus* (Skrjabin 1914) Fuhrmann 1926.

*D. domesticusi* Shinde et Mitra 1980, in *Gallus domesticus*; India.

*D. hewetensis* Dhawan et Capoor 1972, in *Gallus gallus*; India.

*D. himantopodis* Johnston 1911, in *Himantopus himantopus, H. leucocephalus, Microsarcops cinereus*; Australia, India, Russia.

*D. indica* Shinde 1969, in *Gallus gallus*; India.

*D. lagopodis* Rausch 1971, in *Lagopus lagopus*; Alaska, U.S.

*D. meleagridis* Jones 1936, in *Meleagris gallopavo*; U.S.

*D. minuta* Cohn 1901 (syn. *Himantaurus minuta* [Cohn 1901] Spasskaja et Spasskii 1971), in *Charadrius alexandrinus, Himantopus himantopus, H. leucocefalus, Recurvirostra avocetta, Tringa erytropus, T. totanus*; Africa, Europe, Malaysia, Russia.

*D. nana* Fuhrmann 1912, in *Microsarcops cinereus, Numida ptilorhyncha*; Africa, Japan.

*D. paucisegmentata* Fuhrmann 1909, in *Numida meleagris, N. ptilorhyncha*; Africa, Europe.

*D. paucisegmentata* var. *dahomeensis* Joyeux et Baer 1928, in *Numida ptilorhyncha*; France.

*D. tetraoensis* Fuhrmann 1919, in *Bonasa umbellus, Lyrurus tetrix, Tetrao urogalli, T. ruogallis, Tetraster bonasia*; Europe, Russia.

4b. **Strobila medium to large, of many proglottids. One or many eggs per capsule .................................................................. 5**

5a. **Six to 20 osmoregulatory canals ...** ***Davaineoides*** **Fuhrmann 1920. (Figure 295)**

*Diagnosis*: Strobila of numerous proglottids. Genital pores irregularly alternating. Cirrus pouch crosses osmoregulatory canals or not. Testes numerous (90 to 150) mainly surrounding ovary. Ovary bilobed, median. Vitelline gland postovarian. Vagina posterior or dorsal to

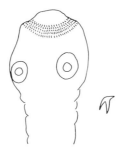

FIGURE 296. *Pentocoronaria rusannae* Matevossian et Movsesyan 1966.

cirrus pouch. Seminal receptacle present. Egg capsules each with single egg. Six to 20 osmoregulatory canals present. Parasites of Galliformes. Brazil.

Type species: *D. vigintivasus* (Skrjabin 1914) Fuhrmann 1926 (syn. *Meggittia vigintivasus* Lopez-Neyra 1931), in *Gallus gallus*; Brazil.

Other species:

*D. polycalceola* (Janicki 1902) Fuhrmann 1920 (syn. *Brumptiella polycalceola* [Janicki 1902] Lopez-Neyra 1931, *Davainea polycalceola* Janicki 1902), in *Acomis muschenbrocki, Arvicanthus abissinicus;* Africa.

5b. Two to four osmoregulatory canals.............................................. 6

6a. Five circles of rostellar hooks....................................................
................ *Pentocoronaria* Matevosyan et Movsesyan 1966. (Figure 296)

*Diagnosis*: Rostellum with five circles of hammer-shaped hooks. Suckers spined. Neck present. Proglottids craspedote, wider than long. Genital pores unilateral, equatorial. Cirrus pouch small. Only ventral osmoregulatory canals present; genital ducts dorsal to poral canal. Testes about 14, posterior and lateral to ovary. Ovary median, central. Vitellarium compact, postovarian. Uterus replaced with numerous egg capsules, each with several eggs. Parasites of Columbiformes. Russia.

Type species: *P. rusannae* Matevosyan et Movsesyan 1966, in *Turtur turtur*; Kirgizia, Russia.

6b. Fewer than five circles of rostellar hooks ...................................... 7

7a. Three circles of rostellar hooks..........................................*Porogynia* Railliet et Henry 1909 (syn. *Polycoelia* Fuhrmann 1907, after King 1849).

*Diagnosis*: Rostellum with three circles of hooks. Suckers armed. Proglottids much wider than long. Genital pores unilateral. Dorsal osmoregulatory canals absent. Testes in single, antiporal field. Ovary and vitelline gland poral; vitelline gland antiporal to ovary. Parasites of Galliformes and Hyracoidea. Africa, Europe.

Type species: *P. paronai* (Moniez 1892) Railliet et Henry 1909 (syn. *Linstowia lata* Fuhrmann 1901, *Polycoelia lata* [Fuhrmann 1901] Fuhrmann 1901, *Taenia paronai* Moniez 1892), in *Francolinsis natalensis, Guttera edouardi, Hyrax* sp., *Numida meleagris, N. ptilorhyncha*; Africa, Europe.

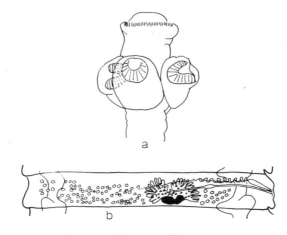

FIGURE 297. *Houttuynia struthionis* (Houtuyn 1772) Fuhrmann 1920. (a) Scolex; (b.) proglottid.

**7b.    Two circles of rostellar hooks................................................. 8**

**8a.    Circles of rostellar hooks followed by several rows of small spines. Suckers unarmed ............................ *Houttuynia* Fuhrmann 1920. (Figure 297)**

*Diagnosis*: Rostellum large, with two circles of hooks followed by several circles of very small spines. Suckers unarmed. Proglottids wider than long, craspedote. Genital pores unilateral. Cirrus pouch well developed. Testes in two lateral fields, antiporal field with more testes than poral field. Ovary bilobed, slightly poral. Vitelline gland postovarian. Vagina posterior to cirrus pouch. Seminal receptacle present. Each egg capsule with several eggs. Parasites of Rheiformes, Struthioniformes. Africa, South America.

    Type species: *H. struthionis* (Houttuyn 1772) Fuhrmann 1920 (syn. *Davainea beddardi* Meggitt 1921, *D. linstowi* Meggitt 1921, *D. struthionis* [Houttuyn 1772] Fuhrmann 1896, *Taenia struthionis* Houttuyn 1773), in *Struthio australis, S. camelus, S. massaicus, S. molybdophanes*; Africa, Palestine.

    Other species:

        *H. streptopelii* Abdel-rahman, Imam et Mohamed 1975, in *Streptopelia senegalensis*; Egpypt.

        *H. struthionis meogaeae* Baer 1928, in *Rhea americana*; South America.

**8b.    No spines behind rostellar hooks. Suckers armed .............................. 9**

**9a.    Genital atrium with massive walls enclosing cirrus pouch and distal vagina .... ................................................. *Dollfusoquenta* Spasskii 1973.**

*Diagnosis*: Scolex very large in proportion to strobila. Suckers powerful. Rostellum with two circles of nearly similar hooks, but circles separated about 20 μm. Blade of hook short, rosethorn-shaped, handle long and slender, guard massive, longer than blade. About 100 hooks total. Neck very short. Osmoregulatory canals ventral to genital ducts. Genital pores unilateral. Genital atrium with massive walls containing many nuclei. Atrial wall surrounding distal vagina and much of cirrus pouch in mature segments; when evaginated it contains entire cirrus pouch and functions as a cirrus pouch itself. Cirrus spinose. Vas deferens convoluted near anterior margin of proglottid. Testes in single, intervascular field posterior to ovary. Vagina apparently dorsal to cirrus pouch. Seminal receptacle present. Ovary

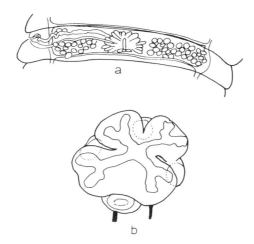

FIGURE 298. *Calostaurus macropus* (Ortlepp 1922) Sanders 1957. (a.) Proglottid; (b) scolex.

multilobate, in anterior half of proglottid. Vitellarium posteriodorsal to ovary. Uterus an irregular sac extending beyond osmoregulatory canals. Parasites of rodents. Africa.

Type species: *D. dollfusi* (Quentin 1964) Spasskii 1973 (syn. *Dilepis dollfusi* Quentin 1964), in *Mastomys* sp.; Central African Republic.

**9b. Genital atrium normal .................................................. 10**

**10a. Two testes per proglottid......................................................
*Diorchirailletina* Yamaguti 1959 (syn. *Manitaurus* Spasskaja et Spasskii 1971).**

*Diagnosis*: Rostellum with two circles of small, hammer-shaped hooks. Genital pores unilateral. Cirrus pouch reaches osmoregulatory canals, genital ducts pass between them. Testes two, antiporal to ovary. Ovary bilobed, median. Vitelline gland postovarian. Each egg capsule with single egg. Parasites of Pholidota. Borneo, Java, Sri Lanka.

Type species: *D. contorta* (Zschokke 1895) Yamaguti 1959 (syn. *Davainea contorta* Zschokke 1895, *Raillietina [R.] contorta* Zschokke 1895] Fuhrmann 1924, *R. [P.] contorta* [Zschokke 1895] Baer et Fain 1955, *Brumptiella contorta* [Zschokke 1895] Lopez-Neyra 1931, *Manitaurus rahmi* [Baer et Fain 1955] Spasskaja et Spasskii 1971, *Raillietina [R.] rahmi]* Baer et Fain 1955), in *Manis (Manis) crassicaudata, M. (Paramanis) javanica, M. (Phataginus) tricuspis*; Africa, Borneo, Java, Sri Lanka, Ivory Coast, Africa.

**10b. More than two testes per proglottid ............................................. 11**

**11a. Rostellar hooks arranged in elborate cross in apical view........................
............................................*Calostaurus* Sandars 1957. (Figure 298)**

*Diagnosis*: Strobila fairly small. Rounded scolex with huge four-lobed or round rostellum armed with two circles of small hooks arranged in an elaborate cross, as viewed apically. Base of rostellum may be armed with numerous spines. Suckers armed with several rows of small hooks. Neck absent. Proglottids craspedote. Genital pores unilateral. Genital atrium small. Genital ducts pass between osmoregulatory canals. Cirrus pouch very small. Cirrus spined. Internal and external seminal vesicles absent. Testes numerous, in two groups lateral

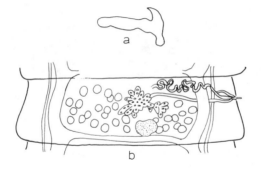

FIGURE 299. *Raillietina bakeri* Chandler 1942. (a) Rostellar hook; (b) proglottid.

to ovary, medial to osmoregulatory canals. Vagina spined distally. Ovary median, composed of several tubular lobules. Vitellaria postovarian. Egg capsules each with a single egg. Parasites of marsupials. Australia. See Beveridge[270] for revision of genus, and Beveridge[276] for table of species characters.

Type species: *C. macropus* (Ortlepp 1922) Sandars 1957 (syn. *Raillietina [Paroniella] macropa* Ortlepp 1922), in *Thylogale wilcoxi, Macropus brunii*; Australia.

Other species:

*C. dorcopsis* Beveridge 1981, in *Dorcopsis veterum*; New Guinea.
*C. mundayi* Beveridge 1975, in *Potorus apicalis*; Tasmania.
*C. oweni* Beveridge 1981, in *Dorcopsis veterum*; New Guinea.
*C. parvus* Beveridge 1981, in *Dorcopsis veterum*; New Guinea.
*C. thylogale* Beveridge 1975, in *Thylogale billardierii*; Tasmania.

11b.  Rostellar hooks arranged in simple circles ..........................................
............................. *Raillietina* Fuhrmann 1920 (syn. *Brumptiella* Lopez-Neyra 1929 in part, *Daoventienia* Spasskii et Spasskaja 1976, *Idiogenoides* Lopez-Neyra 1929 in part, *Johnstonia* Fuhrmann 1920 preoccupied, *Kotlania* Lopez-Neyra 1929 in part, *Kotlanotaurus* Spasskii 1973, *Meggittia* Lopez-Neyra 1929 in part, *Paspalia* Spasskaja et Spasskii 1971, *Roytmania* Spasskii 1973, *Skrjabinotaurus* Spasskii et Jurpalova 1973, *Vadifresia* Spasskii 1973). (Figure 299)

*Diagnosis*: Rostellum with two circles of small, hammer-shaped hooks. Suckers armed or not. Strobila with many proglottids. Genital pores unilateral or irregularly alternating. Cirrus pouch small. Testes numerous. Ovary variable, median or poral. Vitelline gland postovarian. Vagina posterior to cirrus pouch. Seminal receptacle present. Egg capsules with one or several eggs. Parasites of Falconiformes, Anseriformes, Ciconiiformes, Casuariformes, Cuculiformes, Galliformes, Passeriformes, Psittaciformes, Bucerotiformes, Coliiformes, Columbiformes, Piciformes, Pteocliformes, Tinamiformes, Capitoniformes, Caprimulgiformes. Also, Rodentia, Carnivora, Primates, Pholidota, Chiroptera, Lagomorpha. Cosmopolitan.

Type species: *R. tetragona* (Molin 1858) Fuhrmann 1920.

## KEY TO THE SUBGENERA IN *RAILLIETINA*

1a.  One egg per capsule ............................................................. 2
1b.  Several eggs per capsule ........................................................ 3

2a. Genital pores unilateral .......................... *R. (Paroniella)* Fuhrmann 1920 (syn. *Corvinella* Spasskaja et Spasskii 1971, *Metaparonia* Spasskii et Spasskaja 1976, *Nonarmiella* Movsessian 1966, *Numidella* Spasskaja et Spasskii 1971, *Soninotaurus* Spasskii 1973, *Tetraonetta* Spasskaja et Spasskii 1971). (p.253)
2b. Genital pores irregularly alternating ............................................
.............................. *R. (Skrjabinia)* Fuhrmann 1920 (syn. *Armacetabulum* Movsessian 1966, *Daovantienia* Spasskii et Spasskaja 1976, *Daveneolepis* Spasskii 1979, *Delamurella* Spasskii et Spasskaja 1976, *Delmuretta* Spasskii 1977, *Markewitchella* Spasskii et Spasskaja 1972, *Gvosdevinia* Spasskii 1973). (p.255)

3a. Genital pores unilateral .................. *R. (Raillietina)* Fuhrmann 1920 (syn. *Idiogenoides* Lopez-Neyra 1929, *Kotlanotaurus* Spasskii 1973, *Nonarmina* Movsessian 1966, *Paspalia* Spasskaja et Spasskii 1971, *Roytmania* Spasskii 1973, *Skrjabinotaurus* Spasskii et Yurpalova 1973, *Vadifresia* Spasskii 1973). (p.256)
3b. Genital pores irregularly alternating ............................................
.................. *R. (Fuhrmannetta)* Stiles et Orleman 1926 (syn. *Demidovella* Spasskii et Spasskaja 1976, *Mathevossionetta* Movsessian 1966). (p.256)

**Species in *Raillietina (Paroniella)* Fuhrmann 1920**

Type species: *R. (P.) longispina* (Fuhrmann 1909) Fuhrmann 1920 (syn. *Davainea longispina* Fuhrmann 1909, *Brumptiella longispina* [Fuhrmann 1909] Lopez-Neyra 1931), in *Celeus elegans, C. flavescens, Ceophloeus lineatus, Picus* sp.; South America.
Other species:

*R. (P.) acanthovagina* Purvis 1932 (syn. *Nonarmiella acanthovagina* [Purvis 1932] Movsessian 1966), in *Gallus gallus;* Malaysia.

*R. (P.) appendiculata* (Fuhrmann 1909) Fuhrmann 1920 (syn. *Davainea appendiculata* Fuhrmann 1909, *Brumptiella appendiculata* [Fuhrmann 1909] Lopez-Neyra 1931, *Delamurella appendiculata* [Fuhrmann 1909] Spasskii et Spasskaja 1976), in unknown host; New Guinea.

*R. (P.) bargetzii* Mahon 1954, in *Gymnobucco bonopartei;* Africa.

*R. (P.) barmerensis* Mukherjee 1970 (syn. *Corvinella barmerensis* [Mukherjee 1970] Spasskaja et Spasskii 1971), in *Corvus splendens;* India.

*R. (P.) beppuensis* Sawada et Kugi 1976, in *Corvus levaillantii;* Japan.

*R. (P.) blanchardi* (Parona 1898) Fuhrmann 1920 (syn. *Davainea blanchardi* Parona 1898, *Brumptiella blanchardi* [Parona 1898] Lopez-Neyra 1931, *Delamurella blanchardi* [Parona 1898] Spasskii et Spasskaja 1976), in *Rattus siporanus, R. sabanus, R. rajah, Tryonomys swinderianus;* Sarawak, Africa.

*R. (P.) bomensis* Southwell et Lake 1939, in *Melanobucco bidentatus;* Africa.

*R. (P.) bovieni* Baer et Fain 1955 (syn. *Metaparonia bovieni* [Baer et Fain 1955] Spasskii et Spasskaja 1976), in *Manis (Paramanis) javanica;* Java.

*R. (P.) bulbularum* Tubangui et Masiluñan 1937, in *Pycnonotus goiavier;* Philippines.

*R. (P.) capoori* Sravastava et Sawada 1980, in *Francolinus pondicerianus;* India.

*R. (P.) centuri* Rigney 1943, in *Centurus carolinus;* Oklahoma, U.S.

*R. (P.) cirroflexa* Tubangui et Masiluñgan 1937, in *Lichtensteinipicus funebris;* Philippines.

*R. (P.) compacta* (Clerc 1906) Fuhrmann 1920 (syn. *Davainea compacta* Clerc 1906, *Brumptiella compacta* [Clerc 1906] Lopez-Neyra 1931), in *Oriolus galbua, O. auratus, Pyromelana franciscana, Parus caeruleus, Corvus, Diaeum.*

*R. (P.) compacta* (Clerc 1906) var. *polytestis* Spasskii 1946, in *Oriolus oriolus, O. chinensis;* Russia.

*R. (P.) conopophilae* (Johnston 1911) Fuhrmann 1920 (syn. *Davainea conopophilae* Johnston 1911, *Brumptiella conopophilae* [Johnston 1911] Lopez-Neyra 1931), in

*Conopophila albiguris, Entomyza cyanotis, Philemon citreigularis;* Australia, New Guinea.

R. (P.) *coronea* Tubangui et Masiluñgan 1937 (syn. *Corvinella coronea* [Tubangui et Masiluñgan 1937] Spasskaja et Spasskii 1971), in *Corone philippina, Corvus coronoides;* Philippines.

R. (P.) *corvina* (Fuhrmann 1905) Fuhrmann 1920 (syn. *Davainea corvina* Fuhrmann 1905, *D. polycalcarata* Linstow 1906, *Brumptiella corvina* [Fuhrmann 1905] Lopez-Neyra 1931), in *Corvus culminatus, C. macrorhynchus, Macrocorax fuscicapillus, Mino dumonti, Pica rustica;* Thailand, Sri Lanka, Aru Islands, India, Africa.

R. (P.) *cruciata* (Rudolphi 1819) Fuhrmann 1920 (syn. *Taenia cruciata* Rudolphi 1819, *Davainea cruciata* [Rudolphi 1819] Fuhrmann 1909, *Brumptiella cruciata* [Rudolphi 1819] Lopez-Neyra 1931), in *Ceophloeus lineatus, Geginus canus, Brachypterus aurantiacus;* Brazil, Calcutta Zoo.

R. (P.) *culiauana* Tubangui et Masiluñgan 1937, in *Oriolus acrorynchus;* Philippines.

R. (P.) *dendrocopina* Sawada et Kugi 1974, in *Dendrocopos kizuki;* Japan.

R. (P.) *duosyntesticulata* Moghe et Inamdar 1934, in *Xantolaema haematocephala;* India.

R. (P.) *facile* Meggitt 1926 (syn. *Davainea facile* Lopez-Neyra 1936), in *Tragopan satyra;* Burma.

R. (P.) *fecunda* Meggitt 1931, in duck; Burma.

R. (P.) *filiforme* Vigueras 1960, in *Streptopelia turtur;* Cuba.

R. (P.) *fulvia* Meggitt 1933, in *Pterocles orientalis;* Burma.

R. (P.) *huebscheri* Hsü 1935, in *Ixobrychus sinensis;* China.

R. (P.) *japonica* Kugi et Sawada 1972, in *Corvus levaillantii;* Japan.

R. (P.) *karatchvili* Rysavy et Tenora 1979, in *Pica pica;* Afghanistan.

R. (P.) *kashiwarensis* Sawada 1953, in *Gallus gallus;* Japan.

R. (P.) *macassarensis* Yamaguti 1956, in *Gallus gallus;* Celebes.

R. (P.) *magninumida* Jones 1930, in *Meleagris gallopavo, Numida meleagris, Solenophorus pedicularis;* North America.

R. (P.) *minuta* Webster 1947, in *Colinus virginianus;* Texas. U.S.

R. (P.) *molpastina* Moghe et Inamdar 1934, in *Molpastes haemorrhous;* India.

R. (P.) *myzomelae* Yamaguti 1956, in *Myzomela rubrata;* Celebes.

R. (P.) *nedumangadensis* Vijayakumaran et Nadakal 1981, in *Columba livia;* India.

R. (P.) *ngoci* Joyeux et Baer 1937, in *Ducula badia;* Indochina.

R. (P.) *numida* (Fuhrmann 1912) Fuhrmann 1920 (syn. *Davainea numida* Fuhrmann 1912, *Numidella numida* [Fuhrmann 1912] Spasskaja et Spasskii 1971), in *Numida ptilorhyncha, N. meleagris, Guttera* spp.; Africa, U.S., Cuba.

R. (P.) *paradisea* (Fuhrmann 1909), Fuhrmann 1920 (syn. *Davainea paradisea* Fuhrmann 1909, *Brumptiella paradisea* Lopez-Neyra 1931), in *Manucodia chalibeata;* New Guinea.

R. (P.) *parbata* Sharma 1943, in *Gallus gallus;* Nepal.

R. (P.) *perreti* Mahon 1954, in *Pycnonotus barbatus;* Africa.

R. (P.) *pinsonae* Schmelz 1941, in *Gecinus nigrigenis, G. chirrolophus;* Thailand.

R. (P.) *pycnonoti* Yamaguti et Mitunaga 1943, in *Pycnonotus sinensis;* Taiwan.

R. (P.) *retractilis* (Stiles 1895) Fuhrmann 1920 (syn. *Davainea retractilis* Stiles 1895, *Brumptiella retractilis* [Stiles 1895] Lopez-Neyra 1931), in *Lepus zonai, L. melanotis, L. sylvaticus, Sylvilagus arizonae, Rattus rattus;* U.S., Africa.

R. (P.) *reynoldsae* Meggitt 1926, in *Corvus splendins, C. rhipidurus;* India, Burma, Africa, Israel.

R. (P.) *rhynchota* (Ransom 1909) Fuhrmann 1920 (syn. *Davainea rhynchota* Ransom 1909, *Brumptiella rhynchota* [Ransom 1909] Lopez-Neyra, *Soninotaurus rhynchota*

[Ransom 1909] Spasskii 1973), in *Colaptes auratus, Melanerpes erythrocephalus;* U.S.

R. *(P.) siamensis* Schmelz 1941, in *Thereiceryx lineatus, T. phaecostriatus;* Thailand.

R. *(P.) singapurensis* Lee 1966, in *Oriolus chinensis;* Malaysia.

R. *(P.) southwelli* Purvis 1932 (syn. R. *[P.] rangonica* Subramanian 1928), in *Gallus gallus;* Malaya.

R. *(P.) sphecotheridis* (Johnston 1914) Fuhrmann 1920 (syn. *Davainea sphecotheridis* Johnston 1914, *Brumptiella sphecotheridis* [Johnston 1914] Lopez-Neyra 1931), in *Sphecotheris maxillaris, Buchanga stimatops;* Australia, Borneo.

R. *(P.) symonsii* Johri 1939, in *Pavo cristatus;* India.

R. *(P.) tenuiformis* Sawada 1964, in *Gallus gallus;* Sudan.

R. *(P.) tinguiana* Tubangui et Masiluñgan 1937, in *Gallus gallus;* Philippines.

R. *(P.) tragopani* (Southwell 1922) (syn. *Davainea tragopani* Southwell 1922), in *Tragopan* sp.; India.

R. *(P.) urogalli* (Modeer 1790) Fuhrmann 1920 (syn. *Taenia urogalli* Modeer 1790, *T. calva* Baird 1853, *T. microps* Diesing 1850, *Davainea urogalli* [Modeer 1790] Blanchard 1891, *Davainea calva* Shipley 1906, *Davainea urogalli* Shipley 1909, *Meggittia urogalli* [Modeer 1790] Lopez-Neyra 1931, *Tetraonetta urogalli* [Modeer 1790] Spasskaja et Spasskii 1971), in *Perdix perdix, P. robusta, Lagopus lagopus, L. leucurus, L. mutus, L. scoticus, Lyrurus tetrix, Tetrao urogallus, Alectoris rufa, A. graeca, Tragopan satyra, Tetraogallus himalaensis, Amoperdix griseogularis, Alectoris saxalatus;* Europe, Russia.

R. *(P.) woodlandi* Baylis 1934, in *Numida meleagris;* Africa.

**Species in *Raillietina (Skrjabinia)* Fuhrmann 1920**

Type species: R. *(S.) cesticillus* (Molin 1858) Fuhrmann 1920 (syn. *Taenia cesticillus* Molin 1858, *Davainea cesticillus* Blanchard 1891, *Brumptiella cesticillus* [Molin 1858] Lopez-Neyra 1931, *Taenia infundibuliformis* Dujardin 1845 after Goeze 1782, R. *[R.] mutabilis* Rüther 1901), in *Gallus gallus, Meleagris gallopavo, Numida ptilirhyncha, N. meleagris, Lagopus lagopus, L. scoticus, L. lyrurus tetrix, Tetrao urogallus, Tetrastes bonasia, Coturnix coturnis, Perdix perdix, Colinus virginianus, Phasianus colchicus;* Cosmopolitan.

Other species:

R. *(S.) bodkini* Vevers 1923, in *Actitis macularia;* British Guiana.

R. *(S.) boehmi* Pfeiffer 1958, in *Gallus gallus;* Austria.

R. *(S.) bolivari* (Lopez-Neyra 1929) Fuhrmann 1932 (syn. *Meggittia bolivari* Lopez-Neyra 1929), in *Alector rufa, A. graeca;* Spain.

R. *(S.) bonini* (Megnin 1899) Fuhrmann 1932 (syn. *Davainea bonini* Megnin 1899, *D. columbae* Fuhrmann 1909, *Raillietina [Skrjabinia] columbae* Fuhrmann 1920, *Brumptiella bonini* [Megin 1899] Lopez-Neyra 1931, *Markewitchella bonini* [Megnin 1899] Spasskii et Spasskaja 1972), in *Columba* spp., *Palombus torquatus, Streptopelia orientalis;* Europe, Russia, Iran.

R. *(S.) caprimulgi* Burt 1940, in *Caprimulgus asiaticus;* Sri Lanka.

R. *(S.) caucasica* Petrochenko et Kireev 1966, in *Meleagris gallopavo;* Russia.

R. *(S.) centrocerci* Simon 1937, in *Centrocercus urophasianus;* U.S.

R. *(S.) centropi* (Southwell 1922) Fuhrmann 1920 (syn. *Davainea centropi* Southwell 1922, *Brumptiella centropi* [Southwell 1922] Lopez-Neyra 1931, *Daovantienia centropi* [Southwell 1922] Spasskii et Spasskaja 1976), in *Centropus rufipennis;* India.

R. *(S.) circumvallata* (Krabbe 1869) Baer 1925 (syn. *Taenia circumvallata* Krabbe 1869, *Davainea circumvallata* [Krabbe 1869] Blanchard 1891, *Raillietina [Ransomia] circumvallata* [Krabbe 1869] Fuhrmann 1920, R. *[Raillietina] circumvallata* [Krabbe

1896] Fuhrmann 1924, *Meggittia circumvallata* [Krabbe 1869] Lopez-Neyra 1931), in *Coturnix coturnix, Perdix perdix, Alectoris barbata, A. graeca, Caccabis petrosa, C. dactylosoma, C. corunium, Lyrurus tetrix;* Europe, Russia, India, Africa.

R. (S.) *circumvallata cadarachensis* Joyeux et Baer 1938, in *Perdix perdix, Gallus gallus;* Indochina.

R. (S.) *circumvallata siberica* Fediushin 1953, in *Lyrurus tetrix, Perdix perdix;* Russia.

R. (S.) *crepidocotyle* Joyeux et Baer 1935, in *Garrulax chinensis;* Indochina.

R. (S.) *cryptocotyle* Baer 1925 (syn. *Davainea cryptocotyle* Lopez-Neyra 1931), in *Coturnix coturnix;* Africa.

R. (S.) *daviesi* Hughes et Schultz 1942 (syn. *R. [S.] indica* Davies et Evans 1938, preoccupied), in *Alectoris graeca chukar;* India.

R. (S.) *deweti* Ortlepp 1938, in *Numida* sp.; South Africa.

R. (S.) *dhuncheta* Sharma 1943, in *Euplocamus leucomelanus;* Nepal.

R. (S.) *fatalis* Meggitt et Subramanian 1927 (syn. *Brumptiella fatalis* [Meggitt et Subramanian 1927] Lopez-Neyra 1931), in *Rattus norvegicus, Nesocia bengalensis;* India.

R. (S.) *kakia* Johri 1934, in *Corvus splendens;* India.

R. (S.) *lavieri* Joyeux et Baer 1928 (syn. *Davainea lavieri* Lopez-Neyra 1931), in *Centropus* sp.; Africa.

R. (S.) *magnicoronata* (Fuhrmann 1908) Fuhrmann 1920 (syn. *Davainea magnicoronata* Fuhrmann 1909), in *Podager nacunda;* Brazil.

R. (S.) *maroteli* (Neveu-Lemair 1912) Fuhrmann 1920 (syn. *Choanotaenia maroteli* Neveu-Lemair 1912, *Brumptiella maroteli* [Neveu-Lemair 1912] Lopez-Neyra 1931), in *Meleagris gallopavo;* France.

R. (S.) *metacentropi* (Spasskii et Yurpalova 1976) comb. n. (syn. *Daovantienia metacentropi* Spasskii et Yurpalova 1976), *Centropus sinensis;* Vietnam.

R. (S.) *microcotyle* (Skrjabin 1914) Fuhrmann 1920 (syn. *Davainea microcotyle* Skrjabin 1914, *Brumptiella microcotyle* [Skrjabin 1914] Lopez-Neyra 1931), in *Anas platyrhyncha;* Italy.

R. (S.) *oligacantha* (Fuhrmann 1908) Fuhrmann 1920 (syn. *Davainea oligacantha* Fuhrmann 1908), in *Rhynchotus rufescens, Tinamus* sp.; South America.

R. (S.) *petrovi* Tschertkova 1959, in *Gallus gallus;* Europe.

R. (S.) *polyhamata* Sawada et Kugi 1974, in *Numenius phaeophus;* Japan.

R. (S.) *polyuterina* (Fuhrmann 1909) Fuhrmann 1920 (syn. *Davainea polyuterina* Fuhrmann 1909, *Brumptiella polyuterinea* [Fuhrmann 1909] Lopez-Neyra 1931), in *Perdix perdix, Coturnix coturnix, Lyrurus tetrix, Alectoris graeca, Tetrao urogallis;* Africa, India, Asia.

R. (S.) *progenesia* (Movsessian 1968) comb. n. (syn. *Skrjabinia [S.] progenesia* Movsessian 1968), in *Lyrurus tetris, Lagopus lagopus;* Russia.

R. (S.) *pterocleti* Gvosdev 1961 (syn. *Gvosdevinia pterocleti* [Gvosdev 1961] Spasskii 1973), in *Pteroclis orientalis, Syrrhaptes paradoxus;* Russia.

R. (S.) *ransomi* (Williams 1931) Fuhrmann 1932 (syn. *Davainea ransomi* Williams 1931), in *Meleagris gallopavo;* U.S.

R. (S.) *retusa* (Clerc 1903) Fuhrmann 1920 (syn. *Davainea retusa* Clerc 1903, *D. laticanalis* Skrjabin 1914, *Meggittia retusa* [Clerc 1901] Lopez-Neyra 1931), in *Tetrao urogallus, Lyrurus tetrix, Tetrastes bonasia;* Europe, Russia.

R. (S.) *sudanica* Sawada 1964, in *Gallus gallus;* Sudan.

R. (S.) *variabilis* Leigh 1941, in *Tympanuchus americanus;* U.S.

**Species in *Raillietina* (*Raillietina*) Fuhrmann 1920**

Type species: *R. (R.) tetragona* (Molin 1858) Fuhrmann 1924 (syn. *Taenia tetragona* Molin 1858, *T. longicollis* Molin 1858, *Davainea tetragona* [Molin 1858] Blanchard

1891, *Raillietina [Ransomia]j tetragora* [Molin 1858] Fuhrman 1920, *Kotlania Tetragona* [Molin 1858] Lopez-Neyra 1931, *Davainea bothrioplitis* Fillippi 1892, *Raillietina [Raillietina] galli* [Yamaguti 1935] Sawada 1955), in *Gallus gallus, Meleagris gallopavo, Lagopus lagopus, L. mutus, Numida ptilorhyncha, N. meleagris, Guttera eduardi, Pavo cristatus, P. muticus;* cosmopolitan.

Other species:

*R. (R.) afghana* Rysavy et Tenora 1974, in *Blandfordemys afghana;* Afghanistan.

*R. (R.) africana* (Baer 1925) Fuhrmann 1924 (syn. *R. [Ransomia] africana* Baer 1925, *Kotlania africana* [Baer 1925] Lopez-Neyra 1931), in *Herpestes galea;* Africa.

*R. (R.) alagea* (Kotlan 1921) Fuhrmann 1924 (syn. *Davainea alagea* Kotlan 1921, *Idiogenoides alagea* [Kotlan 1921] Lopez-Neyra 1931), in *Cyclopsittacus diophthalmus;* New Guinea.

*R. (R.) allomyodes* (Kotlan 1921) Fuhrmann 1924 (syn. *Davainea allomyodes* Kotlan 1921, *Kotlania allomyodes* [Kotlan 1921] Lopez-Neyra 1931), in *Cyclopsittacus edwardsii;* New Guinea.

*R. (R.) alouattae* Baylis 1947, in *Alouatta macconelli;* South America.

*R. (R.) angusta* Ortlepp 1963, in *Numida meleagris;* South Africa.

*R. (R.) apivori* Makarenko 1963, in *Pernis apivorus;* Russia.

*R. (R.) aruensis* (Fuhrmann 1911) Fuhrmann 1924 (syn. *Davainea aruensis* Fuhrmann 1911, *R. [Ransomia] aruensis* [Fuhrmann 1911] Fuhrmann 1920, *Kotlania aruensis* [Fuhrmann 1909] Lopez-Neyra 1931), in *Trichoglossus cyanogrammus, Lorius lory, Cyclopsittacus edwardsii, C. diophthalmus, Lorius salvodorii, Eos para;* Aru Island, New Guinea.

*R. (R.) asiatica* (Linstow 1901), in human; Russia.

*R. (R.) australis* (Krabbe 1869) Fuhrmann 1924 (syn. *Taenia australis* Krabbe 1869, *Davainea australis* [Krabbe 1869] Blanchard 1891, *R. [Ransomia] australis* [Krabbe 1869] Fuhrmann 1920, *Kotlania australis* [Krabbe 1869] Lopez-Neyra 1931), in *Dromaius novaehollandiae;* Australia.

*R. (R.) baeri* Meggitt et Subramanian 1927 (syn. *Vadifresia baeri* Spasskii 1973, *Kotlania baeri* [Meggitt et Subramanian 1927] Lopez-Neyra 1931), in *Mus coucha, Rattus rattus;* Burma, Africa.

*R. (R.) bakeri* Chandler 1942, in *Sciurus niger, Sigmodon hispidus, Sciurus carolinensis;* U.S.

*R. (R.) bembezi* Mettrick 1962, in *Bubo africanus;* Africa.

*R. (R.) boeti* Joyeux et Baer 1928 (syn. *Kotlania boueti* [Joyeux et Baer 1928] Lopez-Neyra 1931), in *Francolinus bicalcaratus;* Africa.

*R. (R.) buckleyi* Gupta et Grewal 1969, in *Streptopelia senegalensis;* India.

*R. (R.) bumi* Mettrick 1962, in *Bubo africanus;* Africa.

*R. (R.) bycanistis* (Baylis 1919) Fuhrmann 1924 (syn. *Davainea bycanistis* Baylis 1919, *Kotlania bycanistis* [Baylis 1919] Lopez-Neyra 1931), in *Bycanistes subquadratus;* Africa.

*R. (R.) cacatuinae* (Johnston 1911) Fuhrmann 1924 (syn. *Davainea cacatuinae* Johnston 1911, *R. [Ransomia] cacatuinae* [Johnston 1911] Fuhrmann 1920, *Kotlania cacatuinae* [Johnston 1911] Lopez-Neyra 1931), in *Cacatua galeritae;* Australia.

*R. (R.) calcaria* (Fuhrmann 1909) Fuhrmann 1924 (syn. *Davainea calcaria* Fuhrmann 1909, *R. [Ransomia] calcaria* [Fuhrmann 1909] Fuhrmann 1920, *Kotlania calcaria* [Fuhrmann 1909] Lopez-Neyra 1931), in *Corythaeola cristata, Centropus monachus;* Africa.

*R. (R.) calyptomenae* (Baylis 1926) Fuhrmann 1924 (syn. *R. [Ransomia] calyptomenae* [Baylis 1926] Fuhrmann 1924, *Kotlania calyptomenae* [Baylis 1926 ] Lopez-Neyra 1931), in *Calyptomena whichendi;* Sarawak.

*R. (R.) capillaris* (Fuhrmann 1909) Fuhrmann 1924 (syn. *Davainea capillaris* Fuhrmann 1909, *R. [Ransomia] capillaris* [Fuhrmann 1909] Fuhrmann 1920, *Kotlania capillaris* [Fuhrmann 1909] Lopez-Neyra 1931), in *Crypturus* sp.; Brazil.

*R. (R.) carneostrobilata* Vasilev 1967, in turkey, pheasant; Bulgaria.

*R. (R.) carpophagi* Joyeux et Houdemer 1927 (syn. *Kotlania carpophagi* Lopez-Neyra 1931), in *Carpophaga ocnea, Columba livia, Globicera rufigula;* Vietnam, Solomon Islands.

*R. (R.) casuari* (Kotlan 1923) Fuhrmann 1924 (syn. *Davainea casuari* Kotlan 1923, *Raillietnia [Ransomia] casuari* [Kotlan 1923] Fuhrmann 1920, *Kotlania casuari* [Kotlan 1923] Lopez-Neyra 1931, *Kotlanotaurus casuari* [Kotlan 1923] Spasskii 1923), in *Casuarius picticollis;* New Guinea.

*R. (R.) celebensis* (Janicki 1902) Fuhrmann 1924 (syn. *Davainea celebensis* Janicki 1902, *R. [Ransomia] celebensis* [Janicki 1902] Fuhrmann 1920, *R. celebensis* var. *paucicapsulata* Meggitt et Subramanian 1927, *R. formosana* Akashi 1916, *Kotlania celebensis* [Janicki 1902] Lopez-Neyra 1931, *K. formosana* [Akashi 1916] Lopez-Neyra 1931, *Davainea formosana* Akashi 1916), in *Mus maeri, M. norvegicus, Rattus assimilis, Bandicota bengalensis, Epimys meyeri, E. rattus, Nesokia* sp.; Celebes, Philippines, Formosa, China, Indochina, Thailand, Burma, India, Africa, Australia.

*R. (R.) ceylonica* (Baczynska 1914) Fuhrmann 1924 (syn. *Davainea ceylonica* Baczynska 1914, *R. [Ransomia] ceylonica* [Baczynska 1914] Fuhrmann 1920, *Kotlania ceylonica* [Baczynska 1914] Lopez-Neyra 1931), in *Pavo cristatus, Crocopus phoenicopterus, Columba leuconata;* Sri Lanka, India.

*R. (R.) chilmei* Sharma 1943, in *Lerwa nivicola;* Nepal.

*R. (R.) clairae* Schmelz 1941, in *Gecinus canus;* China.

*R. (R.) clavicirrosa* (Fuhrmann 1909) Fuhrmann 1924 (syn. *Davainea clavicirrosa* Fuhrmann 1909, *R. [Ransomia] clavicirrosa* [Fuhrmann 1909] Fuhrmann 1920, *Kotlania clavicirrosa* [Fuhrmann 1909] Lopez-Neyra 1931), in *Francolinus clapertoni, Pteristes lucani, P. leuscocepus holtermulleri;* Africa.

*R. (R.) clerci* Fuhrmann 1920, (syn. *R. [Ransomia] clerci* Fuhrmann 1920, *Davainea crassula* Clerc 1906, after Rudolphi 1819, *Kotlania clerci* [Fuhrmann 1920] Lopez-Neyra 1931), in *Columba* sp.; Russia.

*R. (R.) cohni* (Baczynska 1914) Fuhrmann 1924 (syn. *Davainea cohni* Baczynska 1914, *R. [Ransomia] cohni* [Baczynska 1914] Fuhrmann 1920, *Kotlania cohni* [Baczynska 1914] Lopez-Neyra 1931), in *Gallus gallus, Pteroclidurus exustus, Pterocles arenarius, Numida ptilorhyncha;* Africa, Nepal.

*R. (R.) colinia* Webster 1944, in *Colinus virginianus;* U.S.

*R. (R.) columbiella* Ortlepp 1938, in *Columba livia;* Africa.

*R. (R.) comitata* (Ransom 1909) Fuhrmann 1924 (syn. *Davainea comitata* Ransom 1909, *R. [Ransomia] comitata* [Ransom 1909] Fuhrmann 1920, *Kotlania comitata* [Ransom 1909] Lopez-Neyra 1931), in *Colaptes auritus, Melanerpes erythrocephalus;* North America.

*R. (R.) congolensis* Baer et Fain 1955, in *Pytilia afra;* Africa.

*R. (R.) coreensis* Honda 1939, in *Apodemus agrarius;* Korea.

*R. (R.) coturnixi* Movsessian 1967, in *Coturnix* sp.; Russia.

*R. (R.) cryptacantha* (Fuhrmann 1909) Fuhrmann 1924 (syn. *Davainea cryptacantha* Fuhrmann 1909, *R. [Ransomia] cryptacantha* [Fuhrmann 1909] Fuhrmann 1920, *Kotlania cryptacantha* [Fuhrmann 1909] Lopez-Neyra 1931), in *Columba* sp., *Turtur turtur, T. decipiens, Turturoena sharpei, Stigmatopelia senegalensis;* Egypt, Tunis.

*R. (R.) crypturi* (Fuhrmann 1909) Fuhrmann 1924 (syn. *Davainea crypturi* Fuhrmann 1909, *R. [Ransomia] crypturi* Fuhrmann 1920, *Kotlania crypturi* [Fuhrmann 1909] Lopez-Neyra 1931), in *Crypturus noctivagus;* Brazil.

R. (R.) *cyrtus* (Skrjabin 1914) Fuhrmann 1924 (syn. *Davainea cyrtus* Skrjabin 1914, R. *[Ransomia] cyrtus* [Skrjabin 1914] Fuhrmann 1920, *Kotlania cyrtus* [Skrjabin 1914] Lopez-Neyra 1931), in *Anas* sp.; Paraguay.

R. (R.) *daetensis* Tubangui et Masiluñgan 1937, in *Treron* sp.; Philippines.

R.(R.) *dartevellei* Mahon 1954, in *Gypohierax angolensis;* Africa.

R. (R.) *dattai* Sinha 1960, in *Gallus gallus;* India.

R. (R.) *debilis* (Baylis 1919) Fuhrmann 1924 (syn. *Davainea debilis* Baylis 1919, R. *[Ransomia] debilis* [Baylis 1919] Fuhrmann 1920, *Kotlania debilis* [Baylis 1919] Lopez-Neyra 1931), in *Anastomus lamelligerus;* Africa.

R. (R.) *delalandei* Ortlepp 1938, in *Vinago delalandi;* South Africa.

R. (R.) *demeriensis* Daniels 1895 (syn. *Taenia demerariensis* Daniels 1895, *Raillietina demerariensis* [Daniels 1895] Lopez-Neyra 1931, R. *[R.] brumpti* Dollfus 1930, R. *[R.] davainei* Dollfus 1939, R. *[R.] equatoriensis* Dollfus 1939, R. *[R.] luisaleoni* Dollfus 1939, R. *[R.] leoni* Dollfus 1939, R. *[R.] quitensis* Leon 1935), in human, *Alouatta seniculus;* South America.

R. (R.) *demerariensis* var. *venezolanensis* Lopez-Neyra et Diaz-Ungria 1957, in *Coendou melanurus;* Venezuela.

R. (R.) *douceti* Quentin 1964, in *Turacus,* sp.; Africa.

R. (R.) *echinobothrida* (Megnin 1880) Fuhrmann 1924 (syn. *Taenia echinobothrida* Megnin 1880, *T. botrioplites* Piana 1881, *Davainea parechinobothrida* Magalhães 1898, *Raillietina [Johnstonia] echinobothrida* [Megnin 1880] Fuhrmann 1920, R. *[Fuhrmannetta] echinobothrida* [Megnin 1880] Stiles et Orleman 1926, *Davainea penetrans* Baczynska 1914, *Kotlania echinobothrida* [Megnin 1880] Lopez-Neyra 1931), in *Gallus gallus, Meleagris gallopavo, Perdix perdix, Phasianius colchicus, Gallus bankiva , G. ferrigineus, Numida mitrata, Columba livia;* Cosmopolitan.

R. (R.) *emperus* (Skrjabin 1914) Fuhrmann 1924 (syn. *Davainea emperus* Skrjabin 1914, R. *[Ransomia] emperus* [Skrjabin 1914] Fuhrmann 1920, *Kotlania emperus* [Skrjabin 1914] Lopez-Neyra 1931), in *Buceros seratogynia, Centropus senegalensis;* Egypt, Cameroons.

R. (R.) *erschovi* Movsessian 1965, in *Streptopelia orientalis;* Russia.

R. (R.) *eupodotidis* Dollfus 1957, in *Eupodotis senegalensis*; Africa.

R. (R.) *famosa* (Meggitt 1927) Fuhrmann 1924 (syn. R. *famosa* Meggitt 1927, *Kotlania famosa* [Meggitt 1927] Lopez-Neyra 1931), in *Eclectus pectoralis;* Burma.

R. (R.) *fausti* Schmelz 1941, in *Gecinus canus*; China.

R. (R.) *fischthali* Deardorff, Schmidt et Kuntz, 1976, in *Ducula aenea*; Philippines.

R. (R.) *flabralis* Meggitt 1927 (syn. *Kotlania flabralis* [Meggitt 1927] Lopez-Neyra 1931), in *Dichoceros bicornis;* India, Burma, Vietnam, Sumatra.

R. (R.) *flaccida* Meggitt 1926 (syn. *Kotlania flaccida* [Meggitt 1926] Lopez-Neyra 1931), in *Pterocles orientalis, Passer domesticus;* Burma, Pakistan.

R. (R.) *flaminata* Meggitt 1931, in *Columba punicea, Goura coronata;* Burma.

R. (R.) *fragilis* Meggitt 1931, in *Columba punicea;* Burma.

R. (R.) *frayi* Joyeux et Houdemer 1927 (syn. *Kotlania frayi* [Joyeux et Houdemer 1927] Lopez-Neyra 1931), in *Carpophaga silvatica, Sphenurus siboldi;* Vietnam.

R. (R.) *friedbergeri* (Linstow 1878) Fuhrmann 1924 (syn. *Taenia friedbergeri* Linstow 1878, *Davainea friedbergeri* [Linstow 1878] Blanchard 1891, R. *[Ransomia] friedbergeri* [Linstow 1878] Fuhrmann 1920, *Davainea guivillensis* Megnin 1899, *Kotlania friedbergeri* [Linstow 1878] Lopez-Neyra 1929), in *Pavo cristatus, Meleagris gallopavo, Phasianus colchicus, P. rosenatus;* Europe.

R. (R.) *frontina* (Dujardin 1845) Fuhrmann 1924 (syn. *Taenia frontina* Dujardin 1845, *Davainea frontina* (Dujardin 1845) Blanchard 1891, *Taenia crateriformis* [Rudolphi 1810, in part], R. *[Ransomia] frontina* [Dujardin 1845] Fuhrmann 1920, *Kotlania frontina* [Dujardin 1845] Lopez-Neyra 1931), in *Gecinus viridis, Dendropus major, D. tennirostris, D. medius, Picus martius, P. carelini, Colaptes campestris, Oriolus galbula, Dryocopus martius, Lanius excubitor, campethera nubica, Picus canus;* Indochina, South America, Europe.

*R. (R.) fuhrmanni* (Southwell 1922) Fuhrmann 1924 (syn. *Davainea fuhrmanni* Southwell 1922, *Raillietina fuhrmanni* Meggitt 1926, *Kotlania fuhrmanni* [Southwell 1922] Lopez-Neyra 1931), in *Crocopus phoenicopterus, C. phagrai, Treron curvirostris, Vinago* sp.; India, Burma, Africa, Vietnam.

*R. (R.) fuhrmanni* var. *idiogenoides* Fuhrmann et Baer 1924, in *Streptopelia senegalensis, Vinago delandii, Treron delandii;* Africa.

*R. (R.) fuhrmanni* var. *intermedia* Fuhrmann et Baer 1943, in *Turtur chaleospilos, Oena capensis, Treron calva;* Africa.

*R. (R.) galeritae* (Skrjabin 1914) Fuhrmann 1924 (syn. *Davainea galeritae* Skrjabin 1914, *R. [ Ransomia] galeritae* [Skrjabin 1914] Fuhrmann 1920, *Kotlania galeritae* [Skrjabin 1914] Lopez-Neyra 1931), in *Galerita macrorhyncha; Passer domesticus.* Algeria, Russia, Egypt, India.

*R. (R.) garciai* Whittaker 1973, in *Quiscalus niger;* Puerto Rico.

*R. (R.) garmi* Borgarenko 1981, in *Columba livia;* Russia.

*R. (R.) gendrei* (Joyeux 1923) Fuhrmann 1924 (syn. *R. [Ransomia] gendrei* Joyeux 1923, *Kotlania gendrei* [Joyeux 1923] Lopez-Neyra 1931), in *Vinago calva;* Africa.

*R. (R.) georgiensis* Reid et Nugara 1961, in *Meleagris gallopavo;* U.S.

*R. (R.) gevreyi* Graber 1981, in *Tyto alba;* Africa.

*R. (R.) globirostris* (Fuhrmann 1909) Fuhrmann 1924 (syn. *Davainea globirostris* Fuhrmann 1909, *R. [Ransomia] globirostris* [Fuhrmann 1909] Fuhrmann 1920, *Kotlania globirostris* [Fuhrmann 1909] Lopez-Neyra 1931), in *Perdix perdix, Alectoris barbara, Caccabis petrosa;* Europe, Africa.

*R. (R.) goura* (Fuhrmann 1909) Fuhrmann 1924 (syn. *Davainea goura* Fuhrmann 1909, *R. [Ransomia] goura* [Fuhrmann 1909] Fuhrmann 1920, *Kotlania goura* [Fuhrmann 1909] Lopez-Neyra 1931), in *Goura albertesi, G. coronata;* New Guinea.

*R. (R.) gracilis* (Janicki 1904) Fuhrmann 1924 (syn. *Davainea gracilis* Janicki 1904), in *Mus falvidus, Thryonomys swinderianus;* U.S.

*R. (R.) graeca* Davies et Evans 1938 (syn. *Kotlania graeca* [Davies et Evans 1938] Lopez-Neyra 1931), in *Alectoris graeca, Phasianus colchicus;* India, Russia.

*R. (R.) grobbeni* (Böhm 1925) Fuhrmann 1924 (syn. *R. [Ransomia] grobbeni* Böhm 1925, *Kotlania grobbeni* [Böhm 1925] Lopez-Neyra 1931) in *Gallus gallus;* Austria.

*R. (R.) idiogenoides* Baer 1933, in *Vinago delandii, Treron delandii;* Africa.

*R. (R.) inda* Gupta et Grewal 1971, in *Streptopelia chinensis;* India.

*R. (R.) indica* Meggitt et Subramanian 1927, in *Nesocia bengalensis;* India.

*R. (R.) infrequens* (Kotlan 1923) Fuhrmann 1932, in *Casuarius bennetti picticollis;* New Guinea.

*R. (R.) insignis* (Steudener 1877) Fuhrmann 1924 (syn. *Davainea insignis* [Steudener 1877] Blanchard 1891, *Taenia insignis* Steudener 1877, *R. [Ransomia] insignis* [Steudener 1877] Fuhrmann 1920, *Kotlania insignis* [Steudener 1877] Lopez-Neyra 1931), in *Carpophaga oceanica, Vinago delalandei, Ducula badia, Globicerca oceanica;* Africa, Borneo, Caroline Islands, Java, Malacca, Palau, Sumatra.

*R. (R.) interrupta* (Spasskii et Jurpalova 1973) Nguen Tkhi Ki et Dubinia 1980 (syn. *Skrjabinotaurus interruptus* Spasskii et Jurpalova 1973), in *Streptopelia chinensis, S. orientalis;* Vietnam.

*R. (R.) johri* Ortlepp 1938 (syn. *R. [R.] polychalix* [Kotlan 1920] Johri 1934, after *R. [R] polychalix* Kotlan 1921), in *Psittacula crameri, Treron vernans;* India, Phillipines.

*R. (R.) joyeuxbaeri* Nguen Tkhi Ki et Dubinina 1980, in *Streptopelia tranquebarica;* Vietnam.

*R. (R.) joyeuxi* (Lopez-Neyra 1929) Fuhrmann 1932 (syn. *Kotlania joyeuxi* Lopez-Neyra 1929), in *Columba livia;* Spain.

*R. (R.) kantipura* Sharma 1943, in *Columba livia;* Nepal.

*R. (R.) khalili* Hilmy 1936, in *Turacus persa;* Africa.

*R. (R.) kirghizica* Movsessian 1965, in *Columba livia;* Russia.

*R. (R.) klebergi* Webster 1947, in *Colinus virginianus;* U.S.

*R. (R.) korkei* Joyeux et Houdemer 1927 (syn. *R. [Fuhrmannetta] korkei* Southwell 1930, *Kotlania korkei* Lopez-Neyra 1931), in *Columba* sp., *Alectoris graeca, Streptopelia orientalis;* Russia.

*R. (R.) lateralis* (Bourquin 1906) Fuhrmann 1924 (syn. *Davainea lateralis* Bourquin 1906, *R. [Ransomia] lateralis* [Bourquin 1906] Fuhrmann 1920, *Kotlania lateralis* [Bourquin 1906] Lopez-Neyra 1931), in *Galeopithecus volans;* Sumatra.

*R. (R.) leipoae* Johnston et Clark 1948, in *Leiopoa ocellata;* Australia.

*R. (R.) leptacantha* (Fuhrmann 1909) Fuhrmann 1924 (syn. *Davainea leptacantha* Fuhrmann 1909, *R. [Ransomia] leptacantha* [Fuhrmann 1909] Fuhrmann 1920, *Kotlania leptacantha* [Fuhrmann 1909] Lopez-Neyra 1931), in *Crax alector, C. fasciolata;* Brazil.

*R. (R.) leptosoma* (Diesing 1850) Fuhrmann 1924 (syn. *Taenia leptosoma* Diesing 1850, *T. filiformis* Rudolphi 1819, *Davainea leptosoma* [Diesing 1890] Blanchard 1891, *R. [Ransomia] leptosoma* [Diesing 1850] Fuhrmann 1920, *Kotlania leptosoma* [Diesing 1850] Lopez-Neyra 1931), in *Psittacus erythacus, Ara auricollis, A. macao, A. maracana, A. nobilis, A. severa, Cacathua rosicapilla, Chrysotis purpurea, Canorus guaromba, Pinnus fuscus, Gallerix porphyreolophus;* South and Central America, Mexico, Africa.

*R. (R.) loeweni* Bartel et Hansen 1964 (syn. *Vadifresia loeweni* [Bartel et Hansen 1964] Spasskii 1973), in *Lepus californicus; Sylvilagus auduboni;* U.S. Key to species in North American mammals.

*R. (R.) lutzi* (Parona 1901) Fuhrmann 1924 (syn. *Davainea lutzi* Parona 1901, *R. [Ransomia] lutzi* [Parona 1901] Fuhrmann 1920, *Kotlania lutzi* [Parona 1901] Lopez-Neyra 1931), in *Picus* sp., *Celeus elegans, C. flavescens;* Brazil.

*R. (R.) macracanthos* Paspalewa et Waidowa 1969 (syn. *Paspalia macracanthos* [Paspalewa et Waidowa 1969] Spasskaja et Spasskii 1971), in *Picus viridis;* Russia, Bulgaria.

*R. (R.) macrarhyncha* Nguen Tkhi Ki et Dubinina 1980, in *Treron curvirostra;* Vietnam.

*R. (R.) macrocirrosa* (Fuhrmann 1909) Fuhrmann 1920 (syn. *Davainea macrocirrosa* Fuhrmann 1909, *R. [Ransomia] macrocirrosa* [Fuhrmann 1909] Fuhrmann 1920, *Kotlania macrocirrosa* [Fuhrmann 1909] Lopez-Neyra 1931), in *Turacus buffoni, Centropus senegalensis, C. monachus, Francolinus* sp.; Africa, India.

*R. (R.) macroscolecina* (Fuhrmann 1909) Fuhrmann 1924 (syn. *Davainea macroscolecina* Fuhrmann 1909, *R. [Ransomia] macroscolecina*[Fuhrmann 1909] Fuhrmann 1920, *Kotlania macroscolecina* [Fuhrmann 1909] Lopez-Neyra 1931), in *Lorius garrulus, Dionopsittacus pileatus, Psittacus* sp.; Brazil.

*R. (R.) madagascariensis* (Davaine 1869) Fuhrmann 1924 (syn. *Taenia madagascariensis* Davaine 1869, *Davainea madagascariensis* [Davaine 1869] Blanchard 1891, *R. [Ransomia ] madagascariensis* [Davaine 1869] Fuhrmann 1920, *Kotlania madagascariensis* [Davaine 1869] Lopez-Neyra 1931, *R. [R.] garrisoni* Tubangui 1931, *R. [R.] halli* Vigueras 1943, *R. [R.] siriragi* Chandler et Pradatsundarasar 1957), in human, rats; Madagascar, Formosa, Mauritius, Thailand, South America, Philippines.

*R. (R.) mahonae* Baer et Fain 1955, in *Thryonomys swinderianus;* Africa.

*R. (R.) maplestonei* Southwell 1930, in "macaw" (Calcutta Zoo), *Amazona autumnalis;* Nicaragua.

*R. (R.) mathevossianae* Sultanov 1961, in *Ammoperdix griseogularis;* Russia.

*R. (R.) mehrai* Malviya et Dutt 1971, in *Columba livia;* India.

*R. (R.) michaelseni* Baer 1925 (syn. *Kotlani michaelseni* [Baer 1925] Lopez-Neyra 1931), in *Pterocles variegatus;* Africa.

*R. (R.) micracantha* (Fuhrmann 1909) Fuhrmann 1924 (syn. *Kotlania micracantha* [Fuhrmann 1909] Lopez-Neyra 1931, *Davainea micracantha* Fuhrmann 1909), in *Turtur turtur, Streptopelia senegalensis, Columba livia, Vinago, Stigmatopoelia;* Europe, Africa, Russia.

*R. (R.) microrhyncha* Nguen Tkhi Ki et Dubinina 1980, in *Streptopelia chinensis;* Vietnam.

*R. (R.) microscolecina* (Fuhrmann 1909) Fuhrmann 1924 (syn. *Davainea microscolecina, R. [Ransomia] microscolecina* [Fuhrmann 1909] Fuhrmann 1920, *Kotlania microscolecina* [Fuhrmann 1909] Lopez-Neyra 1931), in *Electus pectoralis, E. rosatus, Cacatua moluccensis, Lorius rotatus;* Aru Islands, Burma.

*R. (R.) multicapsulata* (Baczynska 1914) Fuhrmann 1924 (syn. *Davainea multicapsulata* Baczynska 1914, *R. [Ransomia] multicapsulata* [Baczynska 1914] Fuhrmann 1920, *Kotlania multicapsulata* [Baczynska 1914] Lopez-Neyra 1931), in *Phasianus* sp.; Germany.

*R. (R.) multitesticulata* Perkins 1950, in *Alouatta seniculus;* South America.

*R. (R.) murium* Joyeux et Baer 1936, in *Rattus norvegicus;* Madagascar.

*R. (R.) nagpurensis* Moghe 1925 (syn. *Kotlania nagpurensis* [Moghe 1925] Lopez-Neyra 1931), in *Paloma domestica;* India, Nepal, Australia.

*R. (R.) neyrai* Baer 1955, in *Lophotis ruficristata;* Africa.

*R. (R.) nripendra* Sharma 1943, in *Columba livia;* Nepal.

*R. (R.) oitensis* Sawada et Kugi 1980, in *Accipiter gularis;* Japan.

*R. (R.) oligorchida* (Fuhrmann 1911) Fuhrmann 1924 (syn. *Davainea oligorchida* Fuhrmann 1911, *R. [Ransomia] oligorchida* [Fuhrmann 1911] Fuhrmann 1920, *Idiogenoides oligorchida* [Fuhrmann 1911] Lopez-Neyra 1931), in *Electus pectoralis;* Aru Islands.

*R. (R.) ortleppi* Quentin 1964, in *Vinago waahli;* Africa (Paris Zoo).

*R. (R.) osakensis* Iwata et Tamura 1933, in *Anas platyrhyncha;* Japan.

*R. (R.) palawanensis* Deardorff, Schmidt et Kuntz 1976, in *Chalcophaps indica;* Philippines.

*R. (R.) osipovi* Skrjabin et Popoff 1924, in *Pterocles arenarius;* Armenia.

*R. (R.) parviuncinata* (Meggitt et Saw 1924) Fuhrmann 1924 (syn. *R. [Ransomia] parviuncinata* Meggitt et Saw 1924, *Kotlania parviuncinata* [Meggitt et Saw 1924] Lopez-Neyra 1931), in *Anas platyrhyncha;* Burma, China.

*R. (R.) passeriformicola* Deardorff, Schmidt et Kuntz 1976 in *Gracula religiosa*; Philippines.

*R. (R.) paucitesticulata* (Fuhrmann 1909) Fuhrmann 1924 (syn. *Davainea paucitesticulata* Fuhrmann 1909, *R. [Ransomia] paucitesticulata* [Fuhrmann 1909] Fuhrmann 1920, *Kotlania paucitesticulata* [Fuhrmann 1909] Lopez-Neyra 1931), in *Caloenas nicobarica, Paloma* sp., *Cenopopelia tranquebarica, Streptopelia orientalis, S. chinensis, Geopelia striata;* Celebes, Nicobar Islands, Malaya, New Guinea, Moluccas, Bismark Archipelago, India, Formosa, Indochina, Philippines.

*R. (R.) penelopina* (Fuhrmann 1909) Fuhrmann 1924 (syn. *Davainea penelopina* Fuhrmann 1909, *R. [Ransomia] penelopina* [Fuhrmann 1909] Fuhrmann 1920, *Kotlania penelopina* [Fuhrmann 1909] Lopez-Neyra 1931), in *Penelopa obscura;* Brazil.

*R. (R.) penetrans* (Baczynska 1914) Fuhrmann 1924 (syn. *Davainea penetrans* Baczynska 1914), in *Gallus gallus, Phasianus colchicus, Alectoris graeca, Lyrurus, Tetrao, Lagopus, Tetraogallus;* Africa, Russia.

*R. (R.) penetrans nova* Johri 1934, in Indian Minah; India.

*R. (R.) peradenica* Sawada 1957, in *Gallus gallus;* Sri Lanka.

*R. (R.) permista* Southwell et Lake 1939, in *Campethera permista;* Africa.

*R. (R.) perplexa* Johri 1933, in *Columba intermedia;* India.

*R. (R.) pheidolae* Malviya et Dutt 1971, in *Pheidole rhombinoda* (intermediate host); India.

*R. (R.) pici* Yamaguti 1935, in *Picus awokera;* Japan.

*R. (R.) pintneri* (Klaptocz 1906) Fuhrmann 1924 (syn. *Davainea pintneri* Klaptocz 1906, *R. [Ransomia] pintneri* [Klaptocz 1906] Fuhrmann 1920, *Kotlania pintneri* [Klaptocz 1906] Lopez-Neyra 1931), in *Numida ptylorhincha, N. meleagris, N. mitrata, Guttera adouardi, Herpestes palea;* Africa.

*R. (R.) polychalix* (Kotlan 1921) Fuhrmann 1924 (syn. *Davainea polychalix* Kotlan 1921, *Kotlania polychalix* [Kotlan 1921] Lopez-Neyra 1931), in *Trichoglossus intermedius, Lorius garrulus, Psittacula krameri,* other parrots, *Columba livia;* New Guinea, Australia, India.

*R. (R.) provincialis* (Linstow 1909) Fuhrmann 1924 (syn. *Davainea provincialis* Linstow 1909, *R. [Ransomia] provincialis* [Linstow 1909] Fuhrmann 1920, *Kotlania provincialis* [Linstow 1909] Lopez-Neyra 1931), in *Francolinus adspersus;* Africa.

*R. (R.) pseudocyrtus* Meggitt 1931, in *Anas* sp.; Burma

*R. (R.) psittacea* (Fuhrmann 1911) Fuhrmann 1924 (syn. *Davainea psittacea* Fuhrmann 1911, *R. [Ransomia] psittacea* [Fuhrmann 1911] Fuhrmann 1920, *Kotlania psittacea* [Fuhrmann 1911] Lopez-Neyra 1931), in *Cacathua triton, Eclectus pectoralis;* Aru Islands, New Guinea.

*R. (R.) quadritesticulata* Moghe 1925 (syn. *Kotlania quadritesticulata* [Moghe 1925] Lopez-Neyra 1931), in *Cenopopelia tranquebaria, Goura coronata, Chalcophaps indica;* India.

*R. (R.) quitensis* Leon 1935, in human; Ecuador.

*R. (R.) rybickae* Gupta et Madhu 1981, in *Gallus gallus;* India.

*R. (R.) sartica* (Skrjabin 1914) Fuhrmann 1924 (syn. *Davainea sartica* Skrjabin 1914, *R. [Ransomia] sartica* [Skrjabin 1914] Fuhrmann 1920, *Kotlania sartica* [Skrjabin 1914] Lopez-Neyra 1931), in *Corvus corone, Passer domesticus;* Russia, France.

*R. (R.) sartica* var. *massiliensis* Joyeux et Timon-David 1934, in *Passer domesticus;* France.

*R. (R.) sartica* var. *mediterranea* Joyeux et Gaud 1945, in *Emberiza striolata;* Morocco.

*R. (R.) selfi* Buscher 1975, in *Sylvilagus auduboni;* Oklahoma, U.S.

*R. (R.) senaariensis* (Weithofer 1916) Fuhrmann 1924 (syn. *Davainea senaariensis* Weithofer 1916, *R. [Ransomia] senaariensis* [Weithofer 1916] Fuhrmann 1920, *Kotlania senaariensis* [Weithofer 1916] Lopez-Neyra 1931), in *Columba guinea;* Africa.

*R. (R.) sequens* Tubangui et Masiluñgan 1937, in *Streptopelia dussumieri;* Philippines.

*R. (R.) shantungensis* Winfield et Chang 1936, in *Gallus gallus;* China.

*R. (R.) sigmodontis* Smith 1954, in *Sigmodon hispidus;* U.S.

*R. (R.) sinensis* Hsü 1935 (syn. *R. (R.) sinensis* Winfield et Chang 1936), in "rat", *Columba* sp., *Crocopus phoenicopterus;* China, India.

*R. (R.) singhi* Malviya et Dutt 1971, in *Columba livia;* India.

*R. (R.) skrjabini* Tschertkova 1959, (syn. *R. (R.) volzi* Johri 1939, after Fuhrmann 1905), in peafowl; India.

R. (R.) *somalensis* Sawada 1971, in *Acryllium valturinum;* Somalia.

R. (R.) *spiralis* (Baczynska 1914) Fuhrmann 1924 (syn. *Davainea spiralis* Baczynska 1914, R. *[Ransomia] spiralis* [Baczynska 1914] Fuhrmann 1920, *Kotlania spiralis* [Baczynska 1914] Lopez-Neyra 1931), in *Columba* sp.; New Guinea.

R. (R.) *steinhardti* Baer 1925 (syn. R. *[Ransomia] steinhardti* [Baer 1925] Fuhrmann 1920, *Kotlania steinhardti* [Baer 1925] Lopez-Neyra 1931), in *Numida* sp.; Africa.

R. (R.) *streptopeliae* Gupta et Grewal 1969, in *Streptopelia tranquebaria;* India.

R. (R.) *taiwanensis* Yamaguti 1935, in *Columba livia;* Japan.

R. (R.) *taylori* Baylis 1929, in *Psittacus erythracus;* Africa.

R. (R.) *tetragonoides* (Baer 1925) Fuhrmann 1932 (syn. R. *[Ransomia] tetragonoides* Baer 1925, *Kotlania tetragonoides* [Baer 1925] Lopez-Neyra 1931, R. *[R.] tetragona* var. *cohni* [Baczynska] Lopez-Neyra 1944), in *Numida ptilorhyncha;* Africa.

R. (R.) *tokyoensis* Sawada 1960, in *Columba livia;* Japan.

R. (R.) *torquata* (Meggitt 1924) Fuhrmann 1932 (syn. *Houttuynia torquata* Meggitt 1924, *Kotlania torquata* [Meggitt 1924] Lopez-Neyra 1931), in *Columba livia;* Burma, India, Nepal.

R. (R.) *torquata* var. *rajae* Tubangui et Masiluñgan 1937, in *Columba livia;* Philippines.

R. (R.) *toyohashiensis* Sawada et Chikada 1972, in *Numida galleata;* Japan (zoo).

R. (R.) *trapezoides* (Janicki 1906) Fuhrmann 1924 (syn. *Davainea trapezoides* Janicki 1906, R. *[Ransomia] trapezoides* [Janicki 1906] Fuhrmann 1920, R. *kordofanensis* Meggitt et Subramanian 1927, *Kotlania trapezoides* [Janicki 1906] Lopez-Neyra 1931), in *Mus variegatus, Arvicanthus testicularis, A. niloticus, A. pumilis, A. striatus, Rhabdomys* spp., *Meriones shawi;* Africa.

R. (R.) *trinitatae* (Cameron et Reesal 1951) Baer et Sandars 1956 (syn. R. *[R.] demerariensis* var. *trinitatae* Cameron et Reesal 1951), in *Cunicula paca, Dasyprocta aguti, Proechemys cayennensis;* Trinidad, Venezuela.

R. (R.) *tunetensis* Joyeuxi et Houdemer 1928 (syn. R. *[Ransomia] clerci* [Fuhrmann 1920] Joyeux 1923, *Kotlania tunetensis* [Joyeux et Houdemer 1928] Lopez-Neyra 1931), in *Columba livia, Streptopelia orientalis, S. turtur;* North Africa, India, Russia.

R. (R.) *turaci* Baer 1933, in *Turacus livingstonei;* Africa.

R. (R.) *undulata* (Fuhrmann 1909) Fuhrmann 1924 (syn. *Davainea undulata* Fuhrmann 1909, R. *[Ransomia] undulata* Fuhrmann 1920, *Kotlania undulata* [Fuhrmann 1909] Lopez-Neyra 1931), in *Corythaeola cristata, Gallirex porhyreolophus, Chrisococcyx cupreus;* Africa.

R. (R.) *vietnamense* Nguen Tkhi Ki et Dubinina 1980, in *Streptopelia chinensis;* Vietnam.

R. (R.) *vinagoi* Ortlepp 1938, in *Vinago delalandei;* South Africa.

R. (R.) *vivieni* Joyeux et Baer 1935, in *Coturnix japonica;* Indochina.

R. (R.) *vogeli* Hilmy 1936, in *Vinago calva;* Africa.

R. (R.) *volzi* (Fuhrmann 1905) Fuhrmann 1924 (syn. *Davainea volzi* Fuhrmann 1905, R. *[Ransomia] volzi* [Fuhrmann 1905] Fuhrmann 1920, *Kotlania volzi* [Fuhrmann 1905] Lopez-Neyra 1931), in *Gallus gallus, Columba livia, Pavo cristatus;* Sumatra, India, Russia.

R. (R.) *weissi* (Joyeux 1923) Fuhrmann 1924 (syn. R. *[Ransomia] weissi* Joyeux 1923, *Kotlania weissi* [Joyeux 1923] Lopez-Neyra 1931, *Roytmania weissi* [Joyeux 1923] Spasskii 1973), in *Streptopelia turtur, S. senegalensis, S. orientalis, Columba livia, C. eversmanni;* Africa, Australia, Russia.

R. (R.) *weissi* var. *valiclusa* Joyeux et Baer 1934, in *Columba livia, Streptopelia turtur;* France, Morocco.

R. (R.) *werneri* (Klaptocz 1908) Fuhrmann 1924 (syn. *Davainea werneri* Klaptocz 1908, R. *[Ransomia] werneri* [Klaptocz 1908] Fuhrmann 1920, *Kotlania werneri*

[Klaptocz 1908] Lopez-Neyra 1931), in *Colius leucotis, C. striatus, Galerita cristata;* Africa.

*R. (R.) williamsi* Fuhrmann 1932 (syn. *Raillietina silvestris* Jones 1933), in *Meleagris gallopavo;* U.S.

**Species in *Raillietina* (*Fuhrmannetta*) Stiles et Orleman 1926**

Type species: *R. (F.) crassula* (Rudolphi 1819) Stiles et Orleman 1926 (syn. *Taenia crassula* Rudolphi 1819, *Davainea crassula* Fuhrmann 1909, *D. columbae* [Zeder] Blanchard 1891, *R. [Johnstonia] crassa* Fuhrmann 1920, *Kotlania crassula* [Rudolphi 1891] Lopez-Neyra 1931), in *Columba livia, Turtur turtur, Streptopelia senegalensis, S. orientalis, Caccabis saxatilis, Aplopelia larvata, Coturnix coturnix;* Europe, Africa, North and South America.

Other species:

*R. (F.) bandicotensis* (Olsen et Kuntz 1979) comb. n. (syn. *Fuhrmannetta bandicotensis* Olsen et Kuntz 1979), in *Bandicota indica;* Taiwan.

*R. (F.) birmanica* Meggitt 1926 (syn. *Idiogenes birmanica* [Meggitt 1926] Lopez-Neyra 1929), in *Gallus gallus;* Europe, Burma.

*R. (F.) bucerotidarum* (Joyeux et Baer 1928) Stiles et Orleman 1926 (syn. *Brumptiella bucerotidarum* Joyeux et Baer 1928), in *Melanobucco equatorialis;* Africa.

*R. (F.) caballeroi* Vigueras 1960, in *Oropeleia caniceps;* Cuba.

*R. (F.) elongata* (Fuhrmann 1909) Stiles et Orleman 1926 (syn. *Davainea elongata* Fuhrmann 1909, *R. [Johnstonia] elongata* [Fuhrmann 1909] Fuhrmann 1920, *Kotlania elongata* [Fuhrmann 1909] Lopez-Neyra 1931), in *Tinamus* sp., *Nothura media, Rhynchotus rufescens;* South America.

*R. (F.) globocaudata* (Cohn 1900) Stiles et Orleman 1926 (syn. *Davainea globocaudata* Cohn 1900, *R. [Johnstonia] globocaudata* [Cohn 1900] Fuhrmann 1920, *Kotlania globocaudata* [Cohn 1900] Lopez-Neyra 1931), in *Tetrao urogallus, Tetrastes banasia, Lagopus lagopus;* Europe, Russia.

*R. (F.) hertwigi* (Mola 1907) Stiles et Orleman 1926 (syn. *Davainea hertwigi* Mola 1907, *R. [Johnstonia] hertwigi* [Mola 1907] Fuhrmann 1920, *Kotlania hertwigi* [Mola 1907] Lopez-Neyra 1931), in *Entolboetus fasciatus, Milvus* sp.; Italy, France, Africa.

*R. (F.) laticanalis* (Skrjabin 1914) Stiles et Orleman 1926 (syn. *Davainea laticanalis* Skrjabin 1914, *R. [Johnstonia] laticanalis* [Skrjabin 1914] Fuhrmann 1920, *Kotlania laticanalis* [Skrjabin 1914] Lopez-Neyra 1931), in *Perdix* sp., *Gallus gallus, G. ferugineus;* Brazil.

*R. (F.) leptotrachela* (Hungerbuhler 1910) Stiles et Orleman 1926 (syn. *Davainea leptotrachela* Hungerbuhler 1910, *R. [Johnstonia] leptotrachela* [Hungerbuhler 1910] Fuhrmann 1920, *Kotlania leptotrachela* [Hungerbuhler 1910] Lopez-Neyra 1931, *Demidovella leptotrachela* [Hungerbuhler 1910] Spasskii et Spasskaja 1976), in *Pteroclurus namagua, Turdus semitorquatus;* Africa.

*R. (F.) lophoceri* Ortlepp 1964, in *Lophoceros erythrorhynchus, L. flavirostris;* Africa.

*R. (F.) malakartis* Mahon 1958, in *Coturnix* sp.; Egypt.

*R. (F.) nepalis* Sharma 1943, in *Passer domesticus;* Nepal.

*R. (F.) pluriuncinata* (Crety 1890) Stiles et Orleman 1926 (syn. *Taenia pluriuncinata* Crety 1890, *Davainea pluriuncinata* [Crety 1890] Blanchard 1891, *R. [Johnstonia] pluriuncinata* [Crety 1890] Blanchard 1891, *R. [Johnstonia] pluriuncinata* [Crety 1890] Fuhrmann 1920, *Kotlania pluriuncinata* [Crety 1890] Lopez-Neyra 1931), in *Coturnix communis, Alectoris barbara, Coturnix coturnix;* Europe, Russia.

*R. (F.) pseudoechinobothrida* Meggitt 1926, in *Gallus gallus, Coturnix coturnix;* Burma, Europe, Russia.

R. (F.) *salmoni* (Stiles 1895) Stiles et Orleman 1926 (syn. *Davainea salmoni* Stiles 1895, R. *[Johnstonia] salmoni* [Stiles 1895] Fuhrmann 1920, *Kotlania salmonis* [Stiles 1895] Lopez-Neyra 1931, R. *[Raillietina] stilesiella* Hughes 1941, R. *[F.] stilesiella* Hughes 1941), in *Lepus silvaticus, L. melanotis, Cynomys ludovicianus;* U.S.

R. (F.) *salmoni* (Stiles 1895) Stiles et Orleman 1926 (syn. *Davainea salmoni* Stiles 1895, R. *[Johnstonia] salmoni* [Stiles 1895] Fuhrmann 1920, *Kotlania salmonis* [Stiles 1895] Lopez-Neyra 1931, R. *[Raillietina] stilesiella* Hughes 1941, R. *[F.] stilesiella* Hughes 1941), in *Lepus silvaticus, L. melanotis, Cynomys ludovicianus;* U.S.

R. (F.) *vandenbrandeni* Baylis 1940, in *Psittacus erithacus;* Africa.

## *RAILLIETINA* SPECIES OF UNKNOWN SUBGENERIC STATUS

R. *anatina* (Fuhrmann 1909) Fuhrmann 1920 (syn. *Davainea anatina* Fuhrmann 1909), in *Ana splatyrhyncha;* Italy.

R. *brevicollis* (Froelich 1802) Lopez-Neyra 1931 (syn. *Taenia brevicollis* Froelich 1802, *Davainea brevicollis* [Froelich 1802] Fuhrmann 1920), in *Cucullus canorus;* Europe.

R. *campanulata* (Fuhrmann 1909) Lopez-Neyra 1931 (syn. *Davainea campanulata* Fuhrmann 1909, *Daveneolepis campanulata* [Fuhrmann 1909] Spasskii 1979), in *Perdix* sp.; Brazil.

R. *circumcincta* (Krabbe 1869) Lopez-Neyra 1931 (syn. *Taenia circumcincta* Krabbe 1869, *Davainea circumcincta* [Krabbe 1869] Blanchard 1891, *Kotlania circumcincta* [Krabbe 1869] Lopez-Neyra 1931), in *Garzetta garzetta;* Egypt.

R. *echinata* (Fuhrmann 1909) Baer 1931 (syn. *Davainea chinata* Fuhrmann 1909), in unknown host; New Guinea.

R. *fluxa* Meggitt et Subramanian 1927, in *Rattus norvegicus;* Burma.

R. *funerbis* Meggitt et Subramanian 1927 (syn. *Meggittia celebensis* [Janicki 1902] Lopez-Neyra 1943), in *Rattus norvegicus;* Burma.

R. *globocephala* (Fuhrmann 1920 (syn. *Davainea globocephala* Fuhrmann 1909, *Davainea globocephala* [Fuhrmann 1909] Lopez-Neyra 1931), in *Cassicus affinis;* Brazil.

R. *longicollis* (Molin 1858) Lopez-Neyra 1931 (syn. *Bothriocephalus longicollis* Molin 1858, *Bothriotaenia longicollis* [Molin 1858] Railliet 1893, *Davainea longicollis* [Molin 1858] Stiles 1896), in *Gallus gallus;* Italy.

R. *rangoonica* Subramanian 1928, in *Gallus, Milvus migrans;* India.

R. *rothlisbergeri* Baer 1935, in *Nycticebus tardigradus;* Sumatra.

R. *sphaeroides* (Clerc 1903) Fuhrmann 1920 (syn. *Davainea sphaeroides* Clerc 1903), in *Milvus ater, Buteo vilpinus, B. desertuni, B. menetrieri;* Russia.

R. *tryonomysi* Ortlepp 1938, in *Tryonomys swinderianus;* Africa.

R. *vaganda* (Baylis 1919) Fuhrmann 1920 (syn. *Davainea vaganda* Baylis 1919), in *Haliaetus vocifer;* Uganda, Africa.

# FAMILY HYMENOLEPIDIDAE

This is a large family with many genera and hundreds of species, mostly in wild birds and mammals. There is a tendency toward few testes; most species have 3, but up to 12 may be found. Until recent years most species with three testes have been placed in the genus *Hymenolepis*, which became so large as to be nearly impossible to work with. However the attempts by Spasskii and Spasskaja[3464] and Yamaguti[4022] to break *Hymenolepis* into many new genera have clarified the situation considerably. Further works by various authors have contributed to the classification used here. Wardle et al.[3871] elevated the family to ordinal status, a proposal which I reject.

I include five subfamilies: Fimbriarilinae, with a reticulate uterus and pseudoscolex; Pseudhymenolepidinae, with the uterus replaced with egg capsules; Echinorhynchotaeniinae, with an evaginable, spiny rostellum; Ditestolepidinae, with the uterus continuous between proglottids; and Hymenolepidinae, with the uterus separate in each proglottid.

The family is closely related to Dilepididae and Anoplocephalidae, the three grading together so much that it sometimes is arbitrary to decide which family should receive certain genera. One species, *Vampirolepis nana*, is the most common tapeworm of humans in the world.

## KEY TO THE SUBFAMILIES OF HYMENOLEPIDIDAE

1a. **Pseudoscolex present. Uterus reticulate** ............................................... ........................................**Fimbriariinae Wolffhügel 1899. (p.267)**
1b. **Pseudoscolex absent. Uterus not reticulate** ....................................... 2

2a. **Uterus breaking up into egg capsules**............................................ ..........................**Pseudhymenolepidinae Joyeux et Baer 1935. (p.269)**
2b. **Uterus persistent** ............................................................. 3

3a. **Rostellum long, invaginable, covered with spines** ............................... ................ **Echinorhynchotaeniinae Spasskii et Spasskaja 1975. (p.272)**
3b. **Rostellum retractable, not invaginable** ......................................... 4

4a. **Gravid uterus continous between proglottids. Each pair of suckers almost completely fused** ........................**Ditestolepidinae Yamaguti 1959. (p.272)**
4b. **Uterus separate in each proglottid. All suckers separate** ....................... ........................................**Hymenolepidinae Perrier 1897. (p.272)**

## KEY TO THE GENERA IN FIMBRIARIINAE

1a. **No external segmentation; internal segmentation obsecure. Six osmoregulatory canals** .................................... ***Fimbriaria* Fröhlich 1802. (Figure 300)**

*Diagnosis:* Rostellum with single circle of ten hooks. Scolex very small. Pseudoscolex well developed, lacking genital primordia. External metamerism absent, internal metamerism obscure. Genital pores unilateral. Cirrus pouches crowded. Testes ovoid. Ovary transversely elongated, reticular, nonmetameric. Uterus reticulate, nonmetameric. Osmoregulatory canals six in number. Parasites of Anseriformes. Cosmopolitan.

Type species: *F. fasciolaris* (Pallas 1781) Frölich 1802 (syn. *Taenia malleus* Goeze 1782, *Taenia fasciolaris* Pallas 1781, *T. pediformis* Krefft 1871, *Epision plicatus* Linton 1892, *Notobothrium arcticum* Linstow 1905, *Fimbriaria plana* Linstow 1905), in Anseriformes. Cosmopolitan.

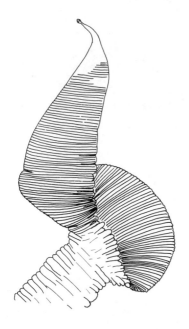

FIGURE 300. *Fimbriaria fasciolaris* (Pallas 1781) Frölich 1802.

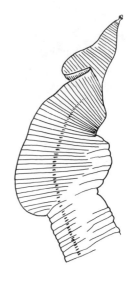

FIGURE 301. *Fimbriarioides intermedia* (Fuhrmann 1913) Fuhrmann 1932.

Other species:
  *F. amurensis* Kotelnikov 1960, in *Anas platyrhynchos;* Russia.
  *F. kubanica* Kotelnikov 1965, in *Anas platyrhynchos;* Russia.

1b. External segmentation present or not, internal segmentation conspicuous. Eight to 11 osmoregulatory canals ................................................... 2

2a. No external segmentation; internal segmentation conspicuous; 9 or 11 osmoregulatory canals ................... ***Fimbriarioides*** Fuhrmann 1932. (Figure 301)

*Diagnosis:* Rostellum with ten hooks in single circle. Pseudoscolex poorly developed, containing genital primordia. External metamerism absent, internal metamerism evident. Genital pores unilateral. Cirrus pouch fusiform. Testes lobated or ovoid. Internal and external seminal vesicles present. Cirrus armed. Ovary reticulate. Vitelline gland median, lobate. Vagina opening ventral to cirrus in large atrium. Uterus reticulate, continuous between segments. Osmoregulatory canals, 9 or 11 in number. Parasites of Anseriformes and Charadriiformes. Europe, North America.

Type species: *F. intermedia* (Fuhrmann 1913) Fuhrmann 1932 (syn. *Fimbriaria intermedia* Fuhrmann 1913), in *Somateria mollissima, S. spectabilis;* Europe, Alaska.
Other species
  *F. haemantopodis* Webster 1943, in *Haemantopus bachmani;* Alaska.
  *F. lintoni* Webster 1943, in *Melanitta perspicillata, M. deglandi, Clangula hyemalis, Fulica americana;* Massachusetts, U.S.
  *F. tadornae* Maksimova 1976, in *Tadorna tadorna;* Russia.

2b. External and internal segmentation conspicuous. Eight osmoregulatory canals ................................................................................. 3

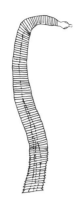

FIGURE 302. *Fimbriariella falciformes* (Linton 1927) Wolffhügel 1936.

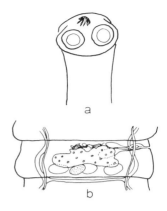

FIGURE 303. *Allohymenolepis mitudori* Yamaguti 1956. (a) Scolex; (b) proglottid.

3a. **Testes lobate. Internal seminal vesicle present** ................................... ........................................*Fimbriariella* **Wolffhügel 1936. (Figure 302)**

*Diagnosis:* Rostellum with ten hooks in single circle. Neck absent. Pseudoscolex poorly developed, containing genital primorida. External and internal metamerism well developed. Proglottids wider than long. Genital pores unilateral. Cirrus pouch fusiform. Cirrus armed at tip. Testes three, lobate. Internal and external seminal vesicles present. Ovary reticulate. Vitelline gland median, lobate. Vaginal pore ventral to cirrus pore. Seminal receptacle present. Uterus first a bilobed sac, becoming reticulate. Osmoregulatory canals eight in number. Parasites of Anseriformes. North America.

Type species: *F. falciformis* (Linton 1927) Wolffhügel 1936 (syn. *Fimbriaria falciformis* Linton 1927, *Fimbriarioides falciformis* [Linton 1927] Fuhrmann 1932), in *Melanitta deglandi;* Massachusetts, U.S.

3b. **Testes rounded. Internal seminal vesicle absent** ................................... ..................................................*Profimbriaria* **Wolffhügel 1936.**

*Diagnosis:* Scolex and pseudoscolex unknown. External and internal metamerism evident. Proglottids wider than long, with lateral fringes. Genital pores unilateral. Cirrus pouch not reaching poral osmoregulatory canal. Cirrus armed. Testes rounded. Internal seminal vesicle absent. Ovary tubular, extending into region of eight osmoregulatory canals. Vitelline gland large, bilobed, dorsal to ovary. Vagina ventral to cirrus pouch, spined at distal end. Seminal receptacle present. Uterus unknown. Parasites of Charadriiformes. Russia.

Type species: *P. multicanalis* (Baczynska 1914) Wolffhügel 1936, in *Scolopax gallinago;* Russia.

## KEY TO THE GENERA IN PSEUDHYMENOLEPIDINAE

1a. **Genital pores irregularly alternating** ................................................... .....................................*Allohymenolepis* **Yamaguti 1956. (Figure 303)**

*Diagnosis:* Rostellum with single circle of ten hooks. Neck present. Proglottids wider than long, craspedote. Genital pores irregularly alternating. Osmoregulatory canals ventral to genital ducts. Cirrus pouch small. Testes three, posterior to ovary. Internal and external seminal vesicles present. Ovary transversely elongate. Vitelline gland postovarian. Vagina

posterior to cirrus pouch. Seminal receptacle large. Uterus first a transverse tube, then saclike, breaking into egg capsules each with one to three eggs. Parasites of Passeriformes (Meliphagidae). Celebes, Philippines.
    Type species: *A. mitudori* Yamaguti 1956, in a melophagid bird, and *Nectorihia jugularis;* Celebes, Philippines.
    Other species:
        *A. palawanensis* Deardorff et Schmidt 1978, in *Nectorinia jugularis;* Philippines.

**1b.**    **Genital pores unilateral** ........................................................... **2**

**2a.**    **Scolex lacking rostellum and armature** .......... ***Sinuterilepis* Sadovskaja 1965.**

*Diagnosis:* Scolex lacking rostellum and armature. Suckers powerfully developed. Neck present. Minute strobila; craspedote. Genital pores unilateral, preequatorial. Cirrus pouch and vagina? Testes three, in a row or triangle. Ovary median, transversely elongated. Vitellarium? Uterus swells into four compartments (capsules?) each with numerous eggs. Parasites of shrews. Russia.
    Type species: *S. spasskyi* Sadovskaja 1965, in *Sorex macropygmaeus;* Russia.
    Other species:
        *S. diglobovary* Sadovskaja 1965, in *Sorex macropygmaeus;* Russia.

**2b.**    **Scolex with rostellum and hooks** .................................................... **3**

**3a.**    **One testis per segment** ........................... ***Globarilepis* Bondarenko 1966.**

*Diagnosis:* Scolex with four simple suckers and rostellum armed with ten aploparaxoid hooks (Figure 342). Neck present. Proglottids numerous, craspedote. Genital pores unilateral. Testes single. Cirrus pouch well developed, cirrus armed. External and internal seminal vesicles present. Vitelline gland postovarian. Seminal receptacle median. Uterus saccular at first, becoming filled with two to several egg capsules, each with four to many eggs. Parasites of Charadriiformes. Russia, North America.
    Type species: *G. spinosus* Bondarenko 1966, in *Capella gallinago, C. stenura, Tringa incana;* Russia, North America.
    Other species:
        *G. mamaevi* Bondarenko 1966, in *Erolia acuminata, Tringa glareola, T. nebularia;* Russia.
        *G. microcirrus* Bondarenko 1966, in *Lymnocryptes gallinula;* Russia.

**3b.**    **Three testes per segment** ......................................................... **4**

**4a.**    **Rostellum with ten hooks in a single circle** ...........................................
        .................................. ***Capiuterilepis* Oschmarin 1962. (Figure 304)**

*Diagnosis:* Scolex rounded, clearly set off from neck. Suckers round, unarmed. Rostellum with ten hooks in single circle. Neck present. Strobila craspedote, anapolytic; all segments wider than long. Genital pores unilateral. Genital ducts dorsal to osmoregulatory canals. Cirrus pouch well developed. Internal and external seminal vesicles present. Testes three, in triangle. Ovary median, lobated. Vitellarium compact, postovarian. Vagina ventral to cirrus pouch. Seminal receptacle prominent, persistent in gravid proglottids. Uterus first appears as a reticulum, then is replaced suddenly with egg capsules, each containing 2 to 15 eggs. Parasites of Passeriformes. Australia, Asia, Europe.

FIGURE 304. *Capiuterilepis australensis* Schmidt 1972. Egg capsule and mature proglottid.

FIGURE 305. *Paradicranotaenia anarmalis* Lopez-Neyra 1943.

Type species: *C. naja* (Dujardin 1845) Oschmarin 1962 (syn. *Taenia naja* Dujardin 1945, *Hymenolepis naja* [Dujardin 1845] Fuhrmann 1906, *Dicranotaenia naja* [Dujardin 1845] Lopez-Neyra 1942. *H.* [Weinlandia] *naja* [Dujardin 1845] Dollfus 1961), in *Certhia, Chloris, Copsychus, Parus, Sitta;* Europe, Russia.

Other species:

    *C. australensis* Schmidt 1972, in *Anthochoera carunculata;* Australia.

    *C. meliphagicola* Schmidt 1972, in *Philemon corniculatus;* Australia.

    *C. pamirensis* Borgarenko 1976, in *Leucosticte brandti;* Russia.

**4b.    Rostellum with more than ten hooks .............................................. 5**

**5a.    Fourteen rostellar hooks. Testes in triangle behind ovary.......................**
**...........................................*Pseudhymenolepis* Joyeux et Baer 1935.**

*Diagnosis:* Rostellum with single circle of 14 hooks. Neck present. External metamerism indistinct. Proglottids apolytic. Genital pores unilateral. Cirrus pouch claviform. Internal seminal vesicle present. Testes three, in triangle with apex directed posteriad, behind ovary. External seminal vesicle present. Ovary horseshoe-shaped with convexity directed anteriad, with compact vitelline gland within the arms. Vaginal pore ventroposterior to cirrus pouch. Seminal receptacle large. Uterus quickly breaking into numerous egg capsules, each containing a single egg. Parasites of shrews. Europe.

    Type species: *P. redonica* Joyeux and Bear 1935, in *Crocidura russula, C. leucodon, Sorex araneus;* Europe.

Other species:

    *P. eburnea* Hunkeler 1970, in *Crocidura theresae, C. poensis, C. jouvenatae;* Ivory Coast, Upper Volta.

    *P. eisenbergi* Crusz et Sanmugusunderam 1971, in *Suncus murinus;* Sri Lanka.

    *P. papillosa* Hunkeler 1970, in *Crocidura flavescens;* Ivory Coast.

**5b.    Single circle of numerous, T-shaped rostellar hooks. One testis poral, two aporal .......................*Paradicranotaenia* Lopez-Neyra 1943. (Figure 305)**

*Diagnosis:* Rostellum short, broad, with numerous T-shaped hooks, each with guard longer than blade. Neck absent. Proglottids shorter than wide, craspedote. Genital pores unilateral. Osmoregulatory canals ventral to genital ducts. Cirrus pouch long. Testes three, one poral, two antiporal. External and internal seminal vesicles present. Ovary bilobed, slightly poral. Vitelline gland compact, postovarian. Vagina posterior to cirrus pouch. Seminal receptacle large, between lobes of ovary. Uterus sac-like, breaking into egg capsules, each with single egg. Parasites of Columbiformes, Galliformes. Europe.

FIGURE 306. *Echinorhynchotaenia tritesticulata* Fuhrmann 1909.

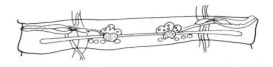

FIGURE 307. *Diplogynia oligorchis* (Maplestone 1922) Baer 1925.

Type species: *P. anormalis* Lopez-Neyra 1943, in *Columba livia;* Spain,
Other species:
    *P. tumens* (Mehlis in Creplin 1846) Yamaguti 1959 (syn. *Paradicranotaenia microps* [Diesing 1850] Lopez-Neyra 1943, *Hymenolepis tetraonis* Wolffhügel 1899), in *Tetrao urogallus, Centrocercus urophasianus, Lagopus scoticus, Lyrurus tetrix, Bonasa umbellus;* Europe, North America.

## DIAGNOSIS OF THE ONLY GENUS IN ECHINORHYNCHOTAENIINAE

*Echinorhynchotaenia* Fuhrmann 1909 (Figure 306)

*Diagnosis:* Rostellum in the form of a long, invaginable proboscis with more than ten hooks at its apex. Handle and blade of same length, guard well developed. Remainder of proboscis covered with minute spines. Suckers unarmed. Strobila craspedote. Proglottids numerous, wider than long. Genital pores unilateral. Genital ducts pass between osmoregulatory canals. Cirrus armed. Internal and external seminal vesicles present. Testes three, posterior and lateral to ovary. Ovary bilobate, median. Vitellarium postovarian. Uterus an irregular sac. Eggs oval, numerous. Parasites of Pelecaniformes, Anseriformes. Africa, Australia, India.

Type species: *E. tritesticulata* Fuhrmann 1909, in *Anhinga rufa;* Africa.
Other species:
    *E. biuncinata* Joyeux et Baer 1943, in "Ephèmeridé"; Morocco.
    *E. lucknowensis* Singh 1956, in *Anhinga melanogaster;* India.
    *E. nana* Maplestone et Southwell 1922, in *Chenopsis atrata;* Australia.

## DIAGNOSIS OF THE ONLY GENUS IN DITESTOLEPIDINAE

*Ditestolepis* Soltys 1952

*Diagnosis:* Rostellum rudimentary, unarmed. Suckers almost completely fused in pairs. Neck present. Proglottids craspedote. Genital pores unilateral. Testes two, one on each side of ovary. Ovary and vitelline gland compact, median. Gravid uterus sac-like, continuous in posterior gravid segments. Parasites of shrews. Europe.

Type species: *D. diaphana* (Cholodkowski 1906) Soltys 1952 (syn. *Hymenolepis diaphana* Cholodkowski 1906), in *Sorex araneus, S. macropygmaeus;* Estonia, Poland.

## KEY TO THE GENERA IN HYMENOLEPIDINAE

1a. Two sets of reproductive organs in each proglottid............................................
................................................ ***Diplogynia* Baer 1925. (Figure 307)**

FIGURE 308. *Acotylepis anacetabula* (Soltys 1954) Yamaguti 1959.

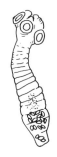

FIGURE 309. *Protogynella blarinae* Jones 1943.

*Diagnosis:* Rostellum with a single circle of hooks. Neck present. Proglottids wider than long. Male and female reproductive systems double. Genital pores bilateral. Cirrus pouch not reaching poral osmoregulatory canals. Cirrus armed. Testes few (usually three in each group), pre- or postovarian. External and internal seminal vesicles present. Osmoregulatory canals ventral to genital ducts. Ovary and vitelline gland compact, poral. Vitelline gland postovarian. Vagina ventral to cirrus. Seminal receptacle present. Uterus a transverse sac. Parasites of Anseriformes, Ciconiiformes. Australia, Java, North America.

Type species: *D. oligorchis* (Maplestone 1922) Baer 1925 (syn. *Diploposthe laevis* of Johnston 1912, *Cotugnia oligorchis* Maplestone 1922), in *Dendrocygna arcuata;* Australia.

Other species:

*D. americana* Olsen 1940, in *Butorides virescens;* Minnesota, U.S.

*D. sandgroundi* (Davis 1944) Davis 1947 (syn. *Cittotaenia sandgroundi* Davis 1944), in *Dendrocygna javanica;* Java.

**1b.    One set of reproductive organs per proglottid ................................... 2**

**2a.    Suckers absent from scolex ........... *Acotylolepis* Yamaguti 1959. (Figure 308)**

*Diagnosis:* Scolex elongate. Suckers lacking. Rostellum with a single circle of minute, U-shaped hooks. Neck absent. Proglottids wider than long, acraspedote. Genital pores unilateral. Cirrus pouch reaching median line of segment. Testes three, one poral, two aporal. Ovary and vitelline glands compact, median. Vagina? Uterus sac-like, forming a single compartment. Parasites of Rodentia. Poland.

Type species: *A. anacetabula* (Soltys 1954) Yamaguti 1959 (syn. *Hymenolepis anacetabula* Soltys 1954), in *Neomys fodiens, N. anomalus;* Poland.

**2b.    Suckers present on scolex (may be vestigial) ..................................... 3**

**3a.    One testis per proglottid ....................................................... 4**
**3b.    More than one testis per proglottid............................................... 7**

**4a.    Rostellum rudimentary, unarmed ....... *Protogynella* Jones 1943. (Figure 309)**

*Diagnosis:* Scolex with four simple suckers and unarmed, rudimentary rostellum. Neck present. Strobila small (0.75 mm long). Proglottids wider than long, about 30 in number. Genital pores unilateral. Cirrus pouch slender. Testes single, lobate or compact. Large external seminal vesicle present. Osmoregulatory canals ventral to genital ducts. Ovary median. Vitelline gland antiporal. Vaginal pore ventral to cirrus pore. Seminal receptacle large, poral to ovary. Gravid uterus saccular. Parasites of shrews. North America.

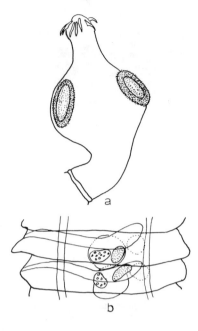

FIGURE 310. *Skrjabinoparaxis tatianae* Krotov 1949. (a) Scolex; (b) proglottids.

Type species: *P. blarinae* Jones 1943, in *Blarina brevicauda, Sorex vagrans, S. bendiri;* U.S.
Other species:
  *P. pauciova* Oswald 1955, in *Blarina brevicauda;* Ohio, U.S.

**4b. Rostellum well developed, armed ................................................. 5**

**5a. Suckers armed. Vitellarium antiporal to ovary .....................................
........................................ Skrjabinoparaxis Krotov 1949. (Figure 310)**

*Diagnosis:* Entire sucker cavity with minute hooks. Rostellum with a single circle of ten hooks. Neck short. Genital pores unilateral. Cirrus pouch slender, nearly reaching antiporal osmoregulatory canals. Testis single, median. External seminal vesicle present. Ovary compact, median. Vitelline gland compact, antiporal to ovary. Gravid uterus sac-like, with few eggs. Parasites of Anseriformes. Russia.
Type species: *S. tatianae* Krotov 1949, in *Anas clypeata;* Russia.
Other species:
  *S. arsenjevi* Oshmarin 1958, in *Nyroca maxilla;* Russia.

**5b. Suckers unarmed. Vittellarium posterior to ovary ............................. 6**

**6a. External seminal vesicle absent. Genital pores opening separately ...............
................................................. Allohaploparaxis Yamaguti 1959.**

*Diagnosis:* Rostellum long, slender, with single circle of hooks bearing handles longer than blades. Suckers unarmed, with posterior margins somewhat reflected, giving scolex arrowhead-like appearance. Neck present. Proglottids wider than long, craspedote. Genital pores lateral, male anterior to and separate from female. Cirrus unarmed. Internal seminal

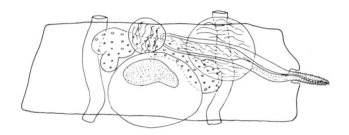

FIGURE 311. *Aploparaksis orientalis* Spasskii et Bobova 1961.

vesicle present, external seminal vesicle absent. Testis single, antiporal. Ovary bilobed. Vitelline gland compact, postovarian. Vagina looped, dilated distally. Seminal receptacle median. Uterus a transverse row of six or more globular pouches eventually filling entire proglottid. Eggshell with bipolar thickenings. Parasites of Anseriformes. Locality not known.

Type species: *A. sagitta* (Rosseter 1906) Yamaguti 1959 (syn. *Drepanidotaenia sagitta* Rosseter 1906), in *Anas platyrhynchos*.

**6b. External seminal vesicle present. Genital pores opening together in a common atrium..................................................................................**
*Aploparaksis* **Clerc 1903 (syn.** *Monorcholepis* **Oshmarin 1961,** *Monotestilepis* **Gvosdev, Maksimova et Kornyushin 1971,** *Haploparaxis* **Clerc 1903,** *Haploparaksis* **Neslobinsky 1911,** *Monorchis* **Clerc 1902 preoccupied). (Figure 311)**

*Diagnosis:* Rostellum prominent, with single circle of hooks with large guards (aploparaxoid Figure 342). Suckers prominent or vestigial. Neck present. Proglottids wider than long. Genital pores unilateral, marginal or submarginal. Cirrus pouch well-developed. Internal and external seminal vesicles present. Testis single. Osmoregulatory canals ventral to genital ducts. Ovary compact or irregularly lobed. Vitelline gland postovarian. Vaginal pore in atrium dorsal or ventral to cirrus pore. Seminal receptacle median. Uterus saccular. Parasites of Charadriiformes, Anseriformes, Passeriformes, Galliformes. Cosmopolitan.

Type species: *A. filum* (Goeze 1782) Clerc 1903 (syn. *Taenia filum* Goeze 1782, *Hymenolepis filum* Railliet 1899, *Drepanidotaenia filum* Cohn 1901, *Monopylidium filum* Parona 1902, *Monorchis filum* Clerc 1902, *Haploparaxis filum* Clerc 1903), in *Scolopax rusticola*, many other Charadriiformes, Anseriformes, Lariformes, and Passeriformes.

Other species:

*A. acanthocirrus* Deblock et Rausch 1968, in *Heteroscelus incanum;* Alaska, U.S.

*A. andrei* Spasskii 1966 in Spasskaja 1966, in *Tringa nebularia;* Russia.

*A. australis* Johnston 1911, in *Gallinago australis;* Australia.

*A. birulai* Linstow 1905, in *Erionetta spectabilis, Somateria fischeri, Aythya marila, Melanitta americana;* Russia.

*A. borealis* Bondarenko et Rausch 1977, in *Stercorarius longicaudus, Anthus cervinus, Calcarius lapponicus;* Alaska, Siberia.

*A. brachyphallos* (Krabbe 1869) Fuhrmann 1932 (syn. *Taenia brachyphallos* Krabbe 1869, *T. filum* var. *brachyphallos* Lönnberg 1890, *Hymenolepis brachyphallos* [Krabbe 1869] Railliet 1899, *Diorchis sepentata* Linstow 1905), in *Tringa, Calidris, Limonites, Charadrius, Lymnocryptus, Erolia, Aegialitis, Arquatella, Gallinago, Pelidna, Numenius;* Greenland, Russia, England, Europe.

*A. bulbocirrus* Deblock et Rausch 1968, in *Ereunetes mauri, E. pusillus;* Alaska, U.S.

*A. bulgarica* Kamburov 1969, in *Scolopax rusticola;* Bulgaria.

*A. caballeroi* Barroeta 1953, in *Larus franklini;* Panama.

A. *chikugoensis* Sawa da et Kifune 1974, in *Podiceps ruficollis;* Japan.

A. *clavata* Spasskaja 1966, in *Limicola falcinellus;* circumboreal.

A. *clavulus* Deblock et Rausch 1968, in *Arenaria melanocephala, Limnodromus scolopaceus;* Alaska, U. S.

A. *clerci* Yamaguti 1935, in *Scolopax rusticola, Pisobia acuminata, Calidris alpina, C. minuta, C. subminuta, Capella gallinago;* Japan, England, U.S.

A. *crassipenis* Deblock et Rausch 1968, in *Erolia alpina;* France.

A. *crassirostris* (Krabbe 1869) Clerc 1903 (syn. *Taenia crassirostris* Krabbe 1869, *Dicranotaenia crassirostris* Stossich 1897, *Hymenolepis crassirostris* Railliet 1899, *Monorchis crassirostris* Clerc 1902), in *Scolopax gallinago, Haematopus ostralegus, Squatarola squatarola, Aegialitis hiaticula, Machetes pugnax, Pisobia damacens, Pelidna alpina, Scolopax rusticola, Lobipes lobatus, Tringa hypoleucos, Galachrysia nuchalis nuchalis, Podiceps ruficollis;* Europe, Africa, Russia.

A. *daviesi* Deblock et Rausch 1968, in *Limnodromus griseus, L. scolopaceus, Arenaria interpres, A. melanocephala;* Alaska, U.S.

A. *diagonalis* Spasskii et Bobova 1961, in *Tringa incana;* Russia.

A. *diminuens* Linstow 1905, in *Crymophilus fulicarius;* Russia, Europe.

A. *dujardini* (Krabbe 1869) Clerc 1903 (syn. *Taenia dujardini* Krabbe 1869, *Monorchis dujardini* Clerc 1902, *Monorcholepis dujardini* [Krabbe 1869] Oshmarin 1961), in *Turdus musicus, Turdus* spp., *Sturnus vulgaris, Asio otis;* Europe, Russia, Japan.

A. *dujardini neoarcticus* Webster 1955, in *Ixoreus naevius;* Alaska, Mexico.

A. *echinovatum* Deblock et Rausch 1968, in *Capella gallinago;* Iran, U.S.

A. *elisae* Skrjabin 1914, in *Fuligula nyroca, Passerella iliaca;* Alaska.

A. *endacantha* Dubinina 1954, in *Anas acuta;* Siberia.

A. *fuligulosa* Solowiow 1911, in *Fuligula cristata;* Europe.

A. *furcigera* (Nitzsch in Rudolphi 1819) Linstow 1905 (syn. *Taenia furcigera* [Nitzsch in Rudolphi 1819] Linstow 1905, *T. rhomboidea* Dujardin 1845, *Dicranotaenia rhomboidea* Railliet 1893, *Dicranotaenia furcigera* Stiles 1896, *Hymenolepis furcigera* Railliet 1899, *Haploparaksis furcigera* Fuhrmann 1926), in *Anas platyrhynchos, A. rubipes, A. acuta, A. discors, Aythya marila, A. collaris, Podiceps ruficollis, Nyroca ferina, Fuligula rufina, Nettion crecca;* Europe, Russia, U.S.A.

A. *galli* Rausch 1951, in *Lagopus* spp.; Alaska, U.S.

A. *groenlandica* (Krabbe 1869) Baer 1956 (syn. *Taenia groenlandica* Krabbe 1869, *Hymenolepis groenlandica* Railliet 1899), in *Clangula hyemalis;* Greenland.

A. *haldemani* Schiller 1951, in *Xema sabini;* Alaska, U.S.

A. *hirsuta* (Krabbe 1882) Clerc 1903 (syn. *Taenia hirsuta* Krabbe 1882, *T. pubescens* Krabbe 1882, *Monorchis hirsuta* Clerc 1902, *Hymenolepis hirsuta* Fuhrmann 1908), in *Scolopax gallinula, Capella media, Limnocryptes minimus, Scolopax rusticola, Tringa ochropus;* Europe, Denmark.

A. *japonensis* Yamaguti 1935, in *Anas platyrhynchos, A. rubipes;* Japan, U.S.

A. *kamayuta* Johri 1935, in *Capella stenura;* Burma.

A. *larina* (Fuhrmann 1921) Spasskaja 1966 (syn. *Haploparaksis larina* Fuhrmann 1921), in *Larus dominicanus, L. melanocephalus;* South Pole, South America, Australia, Europe.

A. *lateralis* Spasskii et Yurpalova 1968, in *Charadrius hiaticula, Phalaropus lobatus;* Russia.

A. *leonovi* Spasskii 1961, in *Calidris minuta, Limicola falcinellus, Numenius madagascariensis;* Russia.

A. *limnocrypti* Bondarenko 1966, in *Lymnocryptes gallinula;* Russia.

A. *moensis* Oshmarin 1963, in *Calidris tenuirostris;* Russia.

A. *moldavica* Spasskaja et Shumilo 1971, in *Gallinago gallinago;* Russia.

*A. murmanica* Baylis 1919, in *Somateria mollissima;* Russian Arctic.

*A. numenii* Kulachkova 1969, in *Numenius arquatus;* Russia.

*A. occidentalis* Prudhoe et Manger 1967, in *Limnodromus griseus;* Texas, U.S.

*A. octacantha* Spasskaja 1950, in *Calidris temmincki;* Russia.

*A. orientalis* Spasskii et Bobova 1961, in *Capella gallinago;* Russia.

*A. parabirulai* Pondarenko 1075, in *Macrarhamphus griseus;* Russia.

*A. parafilum* Gasowska 1932, in *Scolopax rusticola, S. major, Calidris alpina, Capella gallinago, C. media, C. stenura, Tringa ochropus, T. nebularia;* Russia, Europe, Africa, Canada.

*A. penetrans* (Clerc 1902) Clerc 1903 (syn. *Monorchis penetrans* Clerc 1902, *Aploparaksis sobolevi* Oschmarin et Morosov 1948), in *Capella gallinago, C. media, Calidris* sp., *Limnocryptes minima, Scolopax rusticola, Tringa minuta;* Russia, Europe, U.S.

*A. picae* Todd 1967, in *Pica pica;* Montana, U.S.

*A. polystictae* Schiller 1955, in *Polysticta stelleri, Anas acuta, A. formosa, Aythya marila, Clangula hyemalis, Somateria fischeri;* Alaska, Russia.

*A. porzana* (Fuhrmann 1924) Dubinina 1953 (syn. *Hymenolepis porzana* Fuhrmann 1924), in *Porzana porzana, P. pusilla;* Europe, Russia.

*A. pseudofilum* (Clerc 1902) Gasowska 1932 (syn. *Monorchis pseudofilum* Clerc 1902), in *Arenaria interpres, Calidris acuminata, C. alpina, C. minuta, C. ferruginea, C. subminuta, C. ruficollis, Capella gallinago, C. megala, C. stenura, Scolopax rusticola, Tringa glareola, T. hypoleucos, T. nebularia;* Russia, Europe, Japan.

*A. pseudofurcigera* Mathevossian 1946, in *Anas platyrhynchos;* Russia.

*A. pseudosecessivus* Belopolskaja 1967, in *Phalaropus fulicarius;* Russia.

*A. rauschi* Webster 1955, in *Erolia alpina;* Alaska, U.S.

*A. retroversa* Spasskii 1961, in *Calidris temmincki, Macrorhamphus griseus, Tringa glareola, Philomachus pugnax;* Russia.

*A. rissae* Schiller 1951, in *Rissa tridactyla;* Alaska, U.S.

*A. sachalinensis* Krotov 1952, in *Capella solitaria;* Russia.

*A. sanjuanensis* Tubangui et Masiluñgan 1937, in *Gallinago megala, G. gallinago;* Philippines.

*A. schilleri* Webster 1955, in *Erolia alpina, Chalidris melanotos, Arenaria interpres, Gallinago gallinago;* Alaska, Russia.

*A. scolopacis* Yamaguti 1935, in *Scolopax rusticola;* Japan.

*A. secessivus* Gubanov et Mamaev 1959, in *Calidris minuta, Capella gallinago, Numenius borealis, Terekia cinerea, Tringa hypoleucos, T. incana, T. ochropus, T. glareola;* Russia.

*A. sinensis* Tseng 1933, in *Scolopax rusticola;* China.

*A. skrjabini* Spasskii 1945, in *Garrulus glandarius, Anthus* sp.; Russia.

*A. skrjabinissima* Spasskaja 1950, in *Tringa glareola, Capella gallinago;* Russia.

*A. sobolevi* (Tsimbalyuk, Andronova et Kulikov 1968) comb. n. (syn. *Monorcholepis sobolevi* Tsimbalyuk, Andronova et Kulikov 1968), in *Motacilla alba, Anthus hodgsoni, Acanthis flammea, Calcarius lapponicus, Plectrophenax nivalis;* Russia.

*A. spasskii* Bondarenko 1966, in *Capella stenura;* Russia.

*A. stefanski* Czaplinski 1955, in *Anas platyrhynchos;* Poland.

*A. stricta* Spasskii 1961, in *Calidris alpina, Limicola falcinellus;* Russia.

*A. suraishii* Sawada et Kifune 1974, in *Gallinago megala;* Japan.

*A. taimyrensis* Bondarenko 1966, in *Philomachus pugnax;* Russia.

*A. tandani* Singh 1952, in *Tringa hypoleucos;* India.

*A. tinamoui* Olsen 1970, in *Notoprocta perdicaria;* Chile.

*A. turdi* Williamson et Rausch 1965, in *Turdus migratorius, Ixoreus naevius;* Alaska, U.S.

FIGURE 312. *Gopalaia krishnai* Dixit, Capoor et Rengaragu 1980.

*A. uelcal* Spasskii et Yurpalova 1969, in *Numenius arquatus;* Russia.
*A. veitchi* Baylis 1934, in *Querquedula gibberifrons;* Australia.
*A. zemae* Schiller 1951, in *Xema sabini;* Alaska, U.S.

| | | |
|---|---|---|
| 7a. | Two testes per proglottid | 8 |
| 7b. | More than two testes per proglottid | 16 |
| 8a. | Rostellum with single row of many (30 to 100) hooks | 9 |
| 8b. | Rostellum with fewer hooks | 10 |
| 9a. | About 100 rostellar hooks. Testes dorsal and poral to ovary | |
| | *Pseudodiorchis* **Skrjabin et Mathevossian 1948.** | |

*Diagnosis:* Rostellum with large apical sucker and about 100 minute hooks in single circle. Genital pores unilateral, equatorial. Cirrus pouch reaching to median field. Cirrus armed. Testes two, dorsal and poral to ovary. External and internal seminal vesicles absent (?). Ovary bilobed. Vitelline gland anteroventral to ovary. Ovary bilobed. Vitelline gland anteroventral to ovary. Vagina tightly coiled 10 to 16 times. Uterus unknown. Parasites of shrews. North America, Poland.
Type species: *P. reynoldsi* (Jones 1944) Skrjabin and Mathevossian 1948, in *Blarina brevicauda;* U.S.
Other species:
*P. clavatus* Sawada 1967, in *Miniopterus schreiberii;* Japan.

| | | |
|---|---|---|
| 9b. | Rostellar hooks (30 to 35), testes aporal to ovary | |
| | *Gopalaia* **Dixit, Capoor et Rengaragu 1980. (Figure 312)** | |

*Diagnosis:* Scolex with rostellum armed with a single circle of 30 to 35 chelate hooks. Suckers simple. Neck present. Strobila craspedote; segments wider than long. Genital pores unilateral. Genital ducts pass between osmoregulatory canals. Cirrus pouch short. External seminal vesicle present, internal seminal vesicle absent (?). Testes two, aporal. Ovary slightly poral, compact. Vitellarium postovarian. Vagina with dextral loops. Uterus initially bilobed, becoming sac-like. Parasites of bats. India.
Type species: *G. krishnai* Dixit, Capoor et Rengaragu 1980, in *Megaderma lyra;* Nagpur, India.

| | | |
|---|---|---|
| 10a. | Genital pores absent, male and female ducts joined | |
| | *Aporodiorchis* **Yamaguti 1959.** | |

*Diagnosis:* Rostellum with eight hooks in single circle. Proglottids wider than long. Genital pores absent; male and female ducts joining together near lateral margin, on same side throughout strobila. Armed cirrus and internal seminal vesicle present in cirrus pouch. Testes two in number. Ovary lobated, median. Vitellaria horseshoe-shaped, dorsal to ovary. Vagina spinous internally. Seminal receptacle present. Uterus unknown. Parasites of flamingo. Asia.

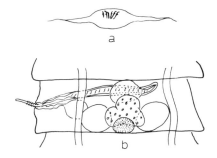

FIGURE 313. *Schillerius longiovum* (Schiller 1953) Yamaguti 1959. (a) Egg; (b) proglottid.

Type species: *A. occlusus* (Linstow 1906) Yamaguti 1959 (syn. *Diorchis occlusus* Linstow 1906), in *Phoenicopterus roseus;* Sri Lanka.

**10b. Genital pores present.......................................................... 11**

**11a. Suckers armed.......................... *Schillerius* Yamaguti 1959. (Figure 313)**

*Diagnosis:* Rostellum with single circle of ten hooks. Suckers armed with small spines. Proglottids wider than long. Genital ducts coursing between osmoregulatory canals. Genital pores unilateral, dextral. Cirrus pouch nearly reaching aporal osmoregulatory canals. Internal and external seminal vesicles present. Cirrus armed at base. Testes two, one on each side of ovary. Ovary median, irregular to trilobate. Vitelline gland compact, postovarian. Vagina ventral and posterior to cirrus pouch. Seminal receptacle present. Uterus first a simple transverse tube, becoming sac-like when gravid. Eggs fusiform, with polar filaments. Parasites of Anseriformes. Alaska, Russia, Japan, Taiwan, Israel.

Type species: *S. longiovum* (Schiller 1953) Yamaguti 1959 (syn. *Diorchis longiovum* (Schiller 1953), in *Anas crecca;* Alaska, U.S.

Other species:

*S. elisae* (Skrjabin 1914) comb. n. (syn. *Aploparaksis elisae* Skrjabin 1914, *Diorchis elisae* [Skrjabin 1914] Spasskii et Frese 1961) in *Anas* spp., *Netta rufina, Aythya ferina, A. fuligula, A. marila, A. nyroca, Tadorna tadorna;* Russia, Europe, Japan

*S. formosensis* (Sugimoto 1934) comb. n. (syn. *Diorchis formosensis* Sugimoto 1934, *D. tshanensis* Krotov 1949), in *Anas platyrhynchos, A. strepera;* Taiwan, Russia.

*S. parvogenitalis* (Mathevossian 1946) Yamaguti 1959 (syn. *Diorchis parvogenitalis* Mathevossian 1946), in *Nyroca ferina, N. fuligula, Anas crecca;* Russia.

*S. ransomi* (Schultz 1940) comb. n. (syn. *Diorchis ransomi* Schultz 1940, *D. accuminata* of Ransom 1909, *D. longibursa* Steelman 1939, *D. moroccana* Dollfus 1975, *D. parvogenitalis* Mathevossian 1945), in *Fulica americana, F. alva, Nyroca ferina;* North America, Russia, Morocco, Israel.

*S. skrjabini* (Udinzew 1937) Yamaguti 1959 (syn. *Diorchis skrjabini* Udinzew 1937), in *Anas crecca, A. platyrhynchos;* Russia.

*S. tringae* (Dubinina 1953) comb. n. (syn. *Diorchis tringae* Dubinina 1953), in *Tringa totanus;* Russia.

**11b. Suckers unarmed ............................................................ 12**

**12a. Ovary poral to testes ......................................................... 13**
**12b. Ovary ventral and/or anterior to testes...................................... 14**

**13a. Rostellar hooks cheliform. Ovary rosette-shaped** .................................
........................................................ *Chelacanthus* **Yamaguti 1959.**

*Diagnosis:* Scolex small. Rostellum with ten small cheliform hooks with rudimentary handle. Testes medial to antiporal excretory canals. Cirrus pouch small, about one fourth as long as proglottid width. Internal seminal vesicle present, external seminal vesicle not reported. Ovary rosette-shaped, in middle third of proglottid near posterior margins. Vitellarium in middle of ovarian lobes near posterior margin of segment. Vagina lined with spines, winding behind cirrus pouch. Seminal receptacle large. Uterus not known. Parasites of aquatic birds. Europe.

Type species: *C. parviceps* (Linstow 1872) Yamaguti 1959 (syn. *Diorchis parviceps* Linstow 1872, *Hymenolepis parviceps* [Linstow 1872] Fuhrmann 1932),in *Mergus serrator;* Europe.

**13b. Rostellar hooks rosethorn-shaped. Ovary not rosette-shaped** .....................
........................................................ *Linstowius* **Yamaguti 1959.**

*Diagnosis:* Scolex rounded. Rostellum spherical in front, with ten rosethorn-shaped hooks. Guard of hooks as long as, and nearly parallel to, blade. Handle comparatively short. Proglottids numerous, broader than long. Genital ducts dorsal to osmoregulatory canals. Testes between ovary and antiporal side. Cirrus pouch cylindrical, reaching to near median line. Cirrus armed. External seminal vesicle not reported. Ovary transversely elongated. Vagina dorsal to cirrus pouch. Seminal receptacle tubular. Uterus intervascular. Parasites of Charadriiformes. Arctic.

Type species: *L. serpentatus* (Linstow 1905) Yamaguti 1959 (syn. *Diorchis serpentatus* Linstow 1905), in *Tringa canutus, Arquatella maritima;* Russia.

**14a. Vagina absent**..............................................*Skorikowia* **Linstow 1905**.

*Diagnosis:* Rostellum enlarged in front, with ten rosethorn-shaped hooks. Guard of hook a little shorter than blade and nearly parallel to it, handle shorter than guard. Proglottids numerous, broader than long. Testes two, symmetrical, dorsal. Cirrus pouch overreaching median line. External and internal seminal vesicles well developed. Cirrus armed. Ovary transversely elongated, ventral. Vitellarium compact, dorsal to ovary, both antiporal. Uterus not reported. Vagina absent (?). Seminal receptacle ventral to cirrus pouch. Parasites of Charadriiformes. Europe.

Type species: *S. clausa* Linstow 1905 (syn. of *Haploparaxis brachyphallos* [Krabbe 1869], according to Fuhrmann 1907, in spite of the presence of two testes), in *Arquatella maritima;* Europe.

**14b. Vagina present**.................................................................... 15

**15a. External seminal vesicle absent** ........................*Jonesius* **Yamaguti 1959.**

*Diagnosis:* Rostellum with ten hooks. Blade of hook short, handle long, guard nodular. Suckers unarmed. Testes two, large, intervascular. Cirrus pouch reaching to median field. Internal seminal vesicle present, external seminal vesicle absent. Cirrus unarmed. Ovary slightly bilobed, ventral, median. Vitelline gland compact, dorsal to ovary. Vagina spined. Seminal receptacle large, median. Uterus an intervascular, transverse sac. Parasites of Ralliformes. North America.

Type species: *J. ralli* (Jones 1944) Yamaguti 1959 (syn. *Diorchis ralli* Jones 1944), in *Rallus elegans;* Virginia, U.S.

FIGURE 314. *Diorchis recurvirostrae* Ahern et Schmidt 1976.

15b. **External seminal vesicle present** ................................................
.... *Diorchis* **Clerc 1903 (syn.** *Diplomonorchis* **Lopez-Neyra 1944). (Figure 314)**

*Diagnosis:* Suckers unarmed. Rostellum with ten hooks in single circle. Neck present. Proglottids wider than long. Genital pores unilateral. Genital ducts dorsal to osmoregulatory canals. Cirrus pouch long or short, cirrus armed or not. Testes two. Internal and external seminal vesicles present. Ovary median or submedian. Vitelline gland compact, posterior, or ventral to ovary. Vagina posterior or ventral to cirrus pouch. Seminal receptacle present. Uterus transversely elongated, with saccular outgrowths. Parasites of Anseriformes, Gruiformes, Columbiformes, Charadriiformes. Cosmopolitan.

Type species: *D. acuminatus* (Clerc 1902) Clerc 1903 (syn. *Drepanidotaenia acuminata* Clerc 1902), in *Anas crecca, A. strepera, Fulica atra, Anas* spp., *Fuligula, Nyroca, Mareca, Nettion,* etc.; Europe.

Other species:

*D. abuladze* Krotov 1949, in *Anas clypeata;* Russia.

*D. aciculasinuatus* (Rosseter 1909) Yamaguti 1959 (syn. *Hymenolepis aciculasinuata* Rosseter 1909), in *Anas platyrhynchos;* England.

*D. alvedea* Johri 1939, in *Streptopelia orientalis;* India.

*D. americanus* Ransom 1909, in *Fulica americana, Gallus gallus, Dendrociatta* sp., *Gallunula chloropus, Fulica atra;* North America, India, Europe.

*D. anivi* Krotov 1953, in *Aythya marila;* Russia.

*D. anomalus* Schmilz 1941, in *Anas* sp.; China.

*D. arsenjevi* (Oshmarin 1958) Spassky 1963, in *Aythya marila;* Russia.

*D. asiatica* Spasskii 1963, in *Anas falcata, A. penelope, A. strepera;* Russia.

*D. balacea* Johri 1960, in *Fulica atra;* India.

*D. brevis* Rybicka 1957, in *Fulica atra, Gallinula chloropus;* Poland.

*D. bulbodes* Mayhew 1929, in *Anas acuta, A. crecca, A. platyrhynchos, A. strepera, Netta rufina, Aythya ferina, A. fuligula, A. marila, A. nyroca, Cidemia fusca, Anas rubepes, Aix sponsa;* North America, Europe, Russia.

*D. chalcophapsi* Johri 1939, in *Chalcophaps indica;* India.

*D. crassicollis* Sugimoto 1934, in *Columba livia, Streptopelia orientalis;* Taiwan, Russia.

*D. danutae* Czaplinski 1956, in *Netta rufina, Aythya ferina;* Poland, Russia.

*D. diorchis* Baer 1962, in *Anas penelope;* Iceland.

*D. donis* Ajinov 1960, in *Anas platyrhynchos;* Russia.

*D. excentricus* Mayhew 1925, in *Erismatura jamaicensis, Aythya affinis, Nyroca fuligula;* U.S., Russia.

*D. flavescens* (Krefft 1871) Johnston 1912 (syn. *Taenia flavescens* Krefft 1871), in *Anas superciliosa, Dendrocygna arcuata, Nettion castaneum, N. crecca, Spatula rhynchotis, Nyroca australis, Arctonetta fischeri;* Australia, New Zealand, China, Alaska.

*D. gigantocirrosa* Singh 1960, in *Fulica atra;* India.

*D. inflata* (Rudolphi 1819) Clerc 1903 (syn. *Taenia inflata* Rudolphi 1819, *Hymenolepis inflata* [Rudolphi 1819] Railliet 1899, *Diplacanthus inflata* [Rudolphi 1819] Cohn 1899), in *Fulica atra, Anas* spp., *Aythya ferina;* Europe, Russia.

*D. jacobii* Fuhrmann 1932, in *Fulica atra*; Europe.

*D. jodhpurensis* Mukharjee 1970, in *Charadrius dubius*; India.

*D. kodonodes* Mayhew 1929, in *Anas discors, A. querquedula*; North America.

*D. lintoni* Johri 1939, in *Marila americana*; U.S.

*D. longicirrosa* Meggitt 1927, in *Fulica atra, Anas crecca*; Egypt, France.

*D. longihamulus* Macko et Rysavy 1968, in *Gallinula chloropus*; Cuba.

*D. magnicirrosus* Moghe et Inamdar 1934, in a "dove"; India.

*D. markewitschi* Pastschenko 1952, in *Anas platyrhynchos*; Russia.

*D. mathevossianae* Krotov 1949, in *Aythya ferina, Clangula clangula*; Russia.

*D. microcirrosa* Mayhew 1929, in *Anas discors, A. querquedula*; North America.

*D. nitidohamulus* Hovorka et Macko 1972, in *Aythya fuligula*; Europe.

*D. nyrocae* Yamaguti 1935, in *Nyroca marila, Tadorna tadorna, Anas* spp.; Japan, Russia.

*D. nyrocoides* Spasskaja 1961, in *Anas crecca*; Russia.

*D. oschmarini* Sudarikov 1950, in *Fulica atra*; Russia.

*D. ovofurcata* Czaplinski 1972, in *Aythya nyroca, A. fuligula*; Poland.

*D. oxyuri* Golovkova 1973, in *Oxyuris leucocephala*; Russia.

*D. pelegicus* Hoberg 1982, in *Aethia pygmaea, A. cristelleta*; Alaska, U.S.

*D. recurvirostrae* Ahern et Schmidt 1976, in *Recurvirostra americana*; U.S.

*D. siamensis* Sawada 1980, in *Anas* sp.; Thailand.

*D. sibiricus* Linstow 1905, in *Erionetta spectabilis, Polysticta stelleri*; Russia, Alaska.

*D. skarbilowitschi* Shachtachtinskaja 1952, in *Colymbus griseigena*; Russia.

*D. sobolevi* Spassakaja 1950, in *Anas platyrhynchos, A. strepera, A. penelope, Anser albifrons, Fulica atra*; Russia.

*D. spasskajae* Spasskii 1963, in *Anas clypeata, A. falcata, Nyroca fuligula*; Russia.

*D. spinata* Mayhew 1929, in *Anas penelope, A. platyrhynchos, A. strepera, A. clypeata*; Russia, Europe, North America.

*D. spiralis* Szpotanska 1931, in *Chenopsis atrata*; Australia.

*D. stefanskii* Czaplinski 1956, in *Anas platyrhynchos, Anser anser, Netta rufina, Aythya nyroca, A. ferina, Oxyura leucocephala, Anas strepera*; Europe, Russia.

*D. tadornae* Kornyushin 1969, in *Tadorna tadorna*; Russia.

*D. tilori* Singh 1952, in *Acridotheres tristis*; India.

*D. turkestanicus* Schultz 1940 (syn. *D. americanus turkestanicus* Skrjabin 1915), in *Gallinula cholropus, Fulica cristata*; Russia, Africa.

*D. tuvensis* Spasskii 1963, in *Anas crecca, Aythya fuligula*; Russia.

*D. vigisi* Krotov 1949, in *Anas crecca*; Russia.

*D. visayana* Tubangui et Masiluñgan 1937, in *Gallinula chloropus*; Philippines.

*D. wigginsi* Schultz 1940 (syn. *D.longae* Schmelz 1941), in *Nyroca vilisneria, Anas crecca, A. querquedula*; U.S.A., China.

16a. Three testes per proglottid ...................................................... 17
16b. More than three testes per proglottid ......................................... 70

17a. Suckers armed (see also, *Drepanidotaenia*) ................................. 18
17b. Suckers unarmed .................................................................. 21

18a. **Rostellum rudimentary, unarmed** ..........................................
................................**Echinolepis Spasskii et Spasskaja 1954. (Figure 315)**

*Diagnosis:* Suckers armed with minute hooks (?). Rostellum rudimentary, unarmed. Neck present. Strobila very thin. Genital pores unilateral, preequatorial. Cirrus pouch reaches

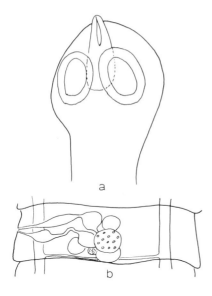

FIGURE 315. *Echinolepis carioca* (Magalhães 1898) Spasskii et Spasskaja 1954. (a) Scolex; (b) proglottid.

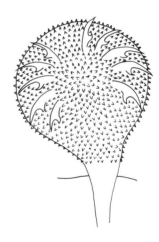

FIGURE 316. *Gonoscolex paradoxus* Saakova 1958. Evaginated gonoscolex.

median line. External and internal seminal vesicles present. Testes three, one poral, two antiporal. Ovary median, bi- or trilobed. Vitelline gland posterodorsal to ovary. Vaginal pore ventroposterior to cirrus pore. Seminal receptacle present. Uterus saccular. Eggs with granular thickenings at poles. Parasites of Galliformes. Cosmopolitan.

Type species: *E. carioca* (Magalhães 1898) Spasskii et Spasskaja 1954 (syn. *Taenia conardi* Zürn 1898, *Hymenolepis pullae* Cholodkovsky 1913, *Davainea carioca* Magalhães 1898, *Hymenolepis carioca* (Magalhães 1898) Ransom 1902, *Dicranotaenia carioca* (Magalhães 1898) Skrjabin et Mathevossian 1945, *Weinlandia rustica* Meggitt 1926, *Hymenolepis rustica* Fuhrmann 1932, *Dicranotaenia rustica* [Meggitt 1926] Skrjabin et Mathevossian 1945), in *Gallus gallus, Meleagris gallopavo, Alectoris graeca, Coturnix coturnix, Bonasa umbellus, Colinus virginianus*. Cosmopolitan.

**18b. Rostellum present, armed .................................................... 19**

**19a. Genital atrium complex, with circles of unequal spines in its depth and with a spiny, dorsal accessory sac capable of being evaginated ........................ ........................................*Gonoscolex* Saakova 1958. (Figure 316)**

*Diagnosis:* Scolex with rostellum bearing ten falciform hooks. Suckers oval, armed with four to six circles of about 40 spines. Neck present. Genital pores unilateral, equatorial. Genital atrium very large, complex; two or three complete circles of unequal spines lie in its depth. In the dorsal wall of the atrium is a huge accessory sac (''gonoscolex'' of Saakova) lined with many small spines and ten large hooks; this organ can be evaginated and does look quite like a rostellum. Cirrus pouch is very large, extending nearly to the aporal side of the segment. Cirrus unarmed and containing a filiform stylet. Vagina in two main parts; distally it comprises two massive lips that nearly fill the atrium; they are separated by a dorsal slit. Proximally, it is saccular with weakly sclerotized lining and muscular walls. Most proximally it is a narrow tube that connects with a voluminous seminal receptacle. Testes three, in a row or slight triangle. Ovary bilobed, compact. Vitellarium ventral to ovary. Uterus and egg unknown. Parasites of Charadriiformes. Europe.

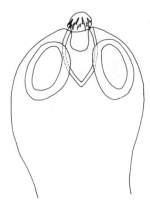

FIGURE 317.  *Pseudoparadilepis ankeli* Brendow 1969.

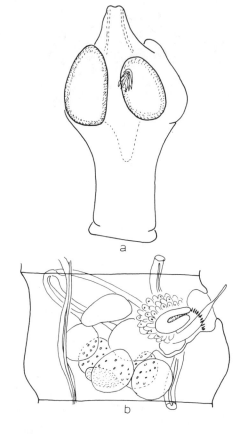

FIGURE 318.  *Echinocotyle longirostris* (Rudolphi 1819) Deblock 1964. (a) Scolex; (b) proglottid.

Type species: *G. paradoxus* Saakova 1958 (syn. *Hymenolepis [Echinocotyle] paradoxus* [Saakova 1958] Deblock et Rosé 1962), in *Limosa limosa;* Roumania.

**19b.  Genital atrium simple or complex, but not as above .......................... 20**

**20a.  Rostellum with two circles of hooks ...............................................**
**................................... *Pseudoparadilepis* Brendow 1969. (Figure 317)**

*Diagnosis:* Scolex with retractable rostellum bearing two circles of hooks. Hooks with spiniform blade, short guard and long handle. Anterior hooks larger than posterior ones. Neck present. Suckers armed with small spines. Proglottids acraspedote, much wider than long. Genital pores unilateral, preequatorial. Genital ducts between osmoregulatory canals. Genital atrium deep. Cirrus pouch slender, nearly reaching midline of segment. Internal and external seminal vesicles present. Testes three, two aporal, one poral. Ovary transversely lobated, in anterior half of proglottid. Vitellarium nearly as large as ovary, in posterior half of segment. Vagina opens dorsal or ventral to cirrus pouch, looping toward a seminal receptacle. Uterus sac-like. Parasites of Insectivora. Germany.

Type species: *P. ankeli* Brendow 1969, in *Sorex minutus;* Germany.

**20b.  Rostellum with one circle of hooks ...............................................**
**..................................... *Echinocotyle* Blanchard 1891. (Figure 318)**

*Diagnosis:* Suckers armed with minute hooks. Rostellum with eight to ten large hooks with reduced guards, in single circle. Neck present. Genital pores unilateral. Cirrus pouch short or long. Cirrus armed. Testes three. External and internal seminal vesicles present. One or two accessory sacs often present in genital atrium. Ovary median or submedian. Vitelline gland compact, postovarian. Vaginal pore posterior or ventral to cirrus pore. Seminal receptacle present. Uterus sac-like. Parasites of Anseriformes, Charadiiformes, Passeriformes; also, Rodentia. Europe, Asia, Taiwan, Africa, North America.

  Type species: *E. rosseteri* Blanchard 1891 (syn. *Taenia lanceolata* Rosseter 1891 after Bloch, *Hymenolepis [Echinocotyle] rosseteri* Clerc 1903), in *Anas platyrhynchos, A. clypeata, A. penelope, A. querquedula, A. crecca, D. discors, Histrionicus histionicus, Tadorna ferruginea;* Europe, India, England, North America, Russia.

Other species:

  *E. anadryensis* Yurpalova et Spasskii 1970, in *Phalaropus lobatus;* Russia.

  *E. anatina* (Krabbe 1869) Blanchard 1891 (syn. *Taenia anatina* Krabbe 1869, *Drepanidotaenia anatina* Railliet 1893, *Dilepis anatina* Cohn 1899, *Hymenolepis anatina* Fuhrmann 1926), in *Anas, Dafila, Nyroca, Fuligula, Anser, Cygnus, Fulica,* etc., Europe, Russia, Formosa, North America.

  *E. birmanica* (Meggitt 1927) Yamaguti 1959 (syn. *Hymenolepis birmanica* Meggitt 1927, *H. neomeggittilis* Hughes 1940), in *Spatula clypeata;* Egypt.

  *E. brachycephala* (Creplin 1829) Spasskaja et Spasskii 1966, in shorebirds, gulls, and ducks; Europe, Asia, Africa.

  *E. clerci* Mathevossian et Krotov 1949, in *Anas crecca, A. clypeata, A. querquedula;* Russia.

  *E. crocethiae* Webster 1947 (syn. *Hymenolepis [Echinocotyle] crocethiae* Webster 1947), in *Crocethia alba, Arenaria interpres;* U.S.

  *E. dolosa* Joyeux et Baer 1928, in *Hypochera ultramarina, Pyromelana franciscana, Spermestes cucullatus, Vidua macroura;* Africa.

  *E. dubininae* (Deblock et Rosé 1962) Spasskaja 1966 (syn. *Echinocotyle nitida* Dubinina 1953 after Clerc 1902, *Hymenolepis [Hymenolepis] nitida* Deblock et Rosé 1962 after Clerc 1902), in *Calidris temmincki, C. minuta, C. maritima, C. ruficollis, Tringa stagnatilis, Larus icthyaetus, L. ridibundus;* Faroe Islands, Russia.

  *E. echinocotyle* (Fuhrmann 1907) Yamaguti 1959 (syn. *Hymenolepis echinocotyle* Fuhrmann 1907), in *Spatula clypeata, Anas platyrhynchos, Cygnus olor, Casarca ferruginea, Philacte canagica;* Europe, India, China, Alaska.

  *E. fimbriata* Spasskii et Yurpalova 1971, in *Tringa glareola, T. ochropus;* Vietnam.

  *E. glareolae* Singh 1952, in *Tringa glareola;* India.

  *E. hypoleuci* Singh 1952, in *Tringa hypoleucos;* India.

  *E. ibanezi* Rego 1973, in *Steganopus tricolor;* Peru.

  *E. kornyushini* Golovkova 1979, in *Chettusia leucura;* Russia.

  *E. linstowi* Daday 1900, in *Diaptomus asiaticus, D. spinosus* (intermedia hosts); Hungary.

  *E. longirostris* (Rudolphi 1819) Deblock 1964 (syn. *Taenia longirostris* Rudolphi 1819, *Nadejdolepis longirostris* [Rudolphi 1819] Spasskii et Spasskaja 1954), in *Clareola austriaca, Philomachus pugnax;* Europe, Russia.

  *E. minutilla* Esch et McDaniel 1965, in *Erolia minutilla;* Oklahoma, U.S.

  *E. minutissima* Singh 1952, in *Anas circia;* India.

  *E. multiglandularis* Baczynska 1914 (syn. *Hymenolepis [Echinocotyle] multiglandularis* Baczynska 1914), in *Larus fuscus, L. ridibundus;* Egypt, Poland.

  *E. nitida* (Krabbe 1869) Clerc 1902 (syn. *Taenia nitida* Krabbe 1869, *Hymenolepis [Echinocotyle] nitida* [Krabbe 1869] Clerc 1902), in *Claidris minuta, C. ferruginea, C. temminckii, C. subminuta, Capella stenura, Tringa glareola, T. hypoleucos, T. nebularia, Limosa lapponica, Pluvialis apricaria;* Europe, Russia, China, India.

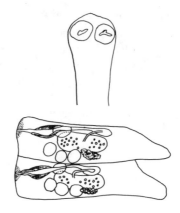

FIGURE 319. *Gvosdevilepis fragmentata* (Gvosdev 1948) Spasskii 1953. Scolex and mature proglottids.

*E. oweni* (Moghe 1933) Yamaguti 1959 (syn. *Drepanidotaenia oweni* Moghe 1933), in *Philomachus pugnax*, ducks; India.

*E. podifufi* Macko 1962, in *Colymbus caspicus;* Czechoslovakia.

*E. polyacantha* Daday 1901, in *Diaptomus asiaticus* (intermediate host); Gobi Desert.

*E. quasioweni* (Dubinina 1954) Yamaguti 1959 (syn. *Hymenolepis quasioweni* Dubinia 1954), in *Tringa glareola, T. stagnatilis, Philomachus pugnax;* Russia.

*E. rauschi* (Schiller 1950) Yamaguti 1959 (syn. *Hymenolepis rauschi* Schiller 1950), in *Oxyura jamaicensis, Anas rubipes;* U.S.

*E. ryjikovi* Jögis 1963, in *Anas clypeata;* Russia.

*E. singhi* Srivastava et Pandey 1980, in *Anas circia;* India.

*E. skrabini* Mathevossian et Krotov 1949, in *Anas clypeata, A. crecca, A. formosa, A. strepera;* Russia.

*E. tenuis* Clerc 1906, in *Totanus* sp., *Pelidna alpina, Philomachus pugnax, Calidris temminckii, Phalaropus lobatus, Tringa glareola, T. stagnalis;* Russia.

*E. ukrainensis* Kornyushin 1969, in *Limicola falcinellus, Calidris testacea;* Russia.

*E. uralensis* Clerc 1902 (syn. *Hymenolepis [Echinocotyle] uralensis* Clerc 1902, *Fuhrmanniella uralensis* Tseng-Shen 1933), in *Tringa hypoleucos, T. totanus, T. erythropos, T. stagnatillis, T. nebularia, T. glareola, Arenaria interpres, Capella gallinago;* Russia, Europe, India, China.

*E. verschureni* (Baer 1959) Macko 1964 (syn. *Hymenolepis verschureni* Baer 1959), in *Hydroprocne tschegrava;* Europe.

| | | |
|---|---|---|
| 21a. | Rostellum absent or rudimentary, unarmed | 22 |
| 21b. | Armed rostellum present | 34 |

| | | |
|---|---|---|
| 22a. | Parasites of mammals | 23 |
| 22b. | Parasites of birds (see also *Cloacotaenia*, p. 337) | 29 |

23a. Testes in a line, all poral to vitelline gland ............................................
........................................... *Gvosdevilepis* Spasskii 1953. (Figure 319)

*Diagnosis:* Scolex lacking rostellum and armature. Suckers unarmed. Neck present. Strobila breaks down into small proglottids, which continue to grow in the intestine of the host independently of the maternal strobila. Proglottids craspedote, asymmetrically broadened on antiporal margin. Testes three, in tranverse row, all poral to vitelline gland. External and internal seminal vesicle present. Genital pores unilateral. Uterus saccular. Parasitic in Lagomorpha. Asia.

   Type species: *G. fragmentata* (Gvosdev 1948) Spasskii 1953 (syn. *Drepanidotaenia fragmentata* Gvosdev 1948), in "rabbits"; Russia.

**23b.  Testes not all poral to vitelline gland .......................................... 24**

**24a.  Proglottids longer than wide. Female organs developing only after segment is detached in intestine of host ....................... *Mathevolepis* Spasskii 1948.**

*Diagnosis:* Entire worm very small (0.6 mm). Scolex unarmed, without rostellum. Proglottids few, elongated. Posterior proglottids containing developing female organs, attaining full maturity after they have been detached in the intestine of the host; anterior segments with male gonads only. Testes three, in a triangle, two being posterior with undifferentiated female gonad between, and the other anterior and antiporal. Genital pores unilateral. Uterus a longitudinal sac, extending past osmoregulatory canals. Parasites of Insectivora. Asia.

   Type species: *M. petrotchenkoi* Spasskii 1948, in "shrew"; Russia.

**24b.  Proglottids wider than long. Female organs not developing as above ....... 25**

**25a.  Gravid uterus becoming a thick-walled capsule................................... ............................................. *Soricina* Spasskii et Spasskaja 1954.**

*Diagnosis:* Strobila small. Suckers unarmed. Scolex lacking armature and rostellum. Proglottids transversely elongated, maturing in groups at the same stage of development. Genital pores unilateral. Testes three, in transverse row. Cirrus pouch claviform, not reaching median line. Seminal vesicles? Ovary and vitellarium in field of median testis. Uterus saccular, forming a thick-walled capsule in gravid proglottids. Parasites of Insectivora. Asia, Europe, North America.

   Type species: *S. soricis* (Baer 1927) Spasskii et Spasskaja 1954 (syn. *Hymenolepis minuta* Baer 1925 after Krabbe 1869), in *Sorex alpinus, S. raddei;* France, Russia.
   Other species:
      *S. bargusinica* Eltyshev 1975, in *Sorex caecutiens, S. araneus;* Russia.
      *S. cirravaginata* Eltyshev 1975, in *Sorex caecutiens, S. araneus;* Russia.
      *S. macyi* (Locker et Rausch 1952) Zarnowski 1956 (syn. *Hymenolepis macyi* Locker et Rausch 1952), in *Sorex vagrans, S. bendirii, S. trowbridgei;* U.S.
      *S. tripartita* Zarnowski 1956, in *Sorex araneus;* Poland.

**25b.  Gravid uterus normal. ..................................................... 26**

**26a.  Testes form a tight triangle in middle of segment, dorsal to ovary and vitellarium. ................................................. *Myotolepis* Spasskii 1954.**

*Diagnosis:* Rostellum rudimentary, unarmed. Suckers simple. Proglottids numerous, transversely elongated. Testes three, close together in median field forming a triangle. Cirrus pouch small. Ovary and vitelline gland compact, ventral to testes. Uterus saccular. Parasites of Chiroptera. Asia.

   Type species: *M. crimensis* (Skarbilovitsch 1946) Spasskii 1954, in *Myotis myotis;* Russia.
   Other species:
      *M. grisea* (Beneden 1873) Spasskii 1954 (syn. *Taenia grisea* Beneden 1873, *Vampirolepis grisea* [Beneden 1873] Spasskii 1954), in *Vespertilio murinus, Eptisicus serotinus, Phinolophus ferrumequinum;* Europe, Japan.

*M. jaisalmerensis* Mukherjee 1970, in *Myotis natteria, M. blythi, Taphozous kachensis;* India.

*M. sawadai* Nama 1974, in *Taphozous perferatus;* India.

26b. Testes not as above .......................................................... 27

27a. Scolex between suckers swollen so as to almost hide suckers ....................
................................ *Cryptocotylepis* Skrjabin et Mathevossian 1948.

*Diagnosis:* Scolex globose, swollen between suckers so as to almost completely hide them. Rostellum vestigial, unarmed. Testes three, in a triangle; two tandem on antiporal side, the other posterior. External seminal vesicle claviform. Genital ducts pass dorsal to osmoregulatory canals. Internal seminal vesicle strongly coiled. Cirrus armed. Genital pores unilateral, anterior. Ovary tripartite, ventral. Vitellarium compact, postovarian. Uterus transversely elongated. Vagina ventral to cirrus pouch; seminal receptacle absent. Parasites of Insectivora. North America.

Type species: *C. anthocephalus* (van Gundy 1935) Skrjabin et Mathevossian 1948 (syn. *Hymenolepis anthocephalus* van Gundy 1935), in *Blarina brevicauda;* U.S. Redescribed: Fasbender (1957).

27b. Scolex not as above .......................................................... 28

28a. Uterus forming an ovoid, median sac, not filling entire proglottid ..............
................................................ *Insectivorolepis* Zarnowski 1956.

Diagnosis: Strobila very small. Scolex relatively large, with rudimentary, unarmed rostellum. Suckers unarmed. Proglottids few, wider than long. Testes three, in transverse row or triangle. Cirrus pouch elongate, thin-walled. External and internal seminal vesicles present. Cirrus armed. Ovary transversely elongated, lobed or not. Vitellarium postovarian. Vagina wide, thin-walled. Seminal receptacle small. Uterus forming ovoid median sac, not filling entire medulla. Parasites of rodents, insectivores and chiropterans. Europe, North America, Japan.

Type species: *I. globosa* (Baer 1931) Zarnowski 1956 (syn. *Hymenolepis globosa* Baer 1931), in *Neomys fodiens;* Switzerland.

Other species:

*I. araii* Sawada 1972, in *Rhinolophus ferrumequinum;* Japan.

*I. globosoides* (Soltys 1954) Zarnowski 1956 (syn. *Dicranotaenia globosoides* Soltys 1954), in *Neomys fodiens, N. anomalus milleri, Sorex minutus;* Poland.

*I. infirma* Zarnowski 1956, in *Sorex araneus, S. minutus;* Poland.

*I. inuzensis* Sawada 1971, in *Rhinolophus ferrumequinum;* Japan.

*I. kenki* (Locker et Rausch 1952) Zarnowski 1956 (syn. *Hymenolepis kenki* Locker et Rausch 1952), in *Sorex* spp.; U.S.

*I. niimiensis* Sawada 1970, in *Rhinolophus ferrumequinum;* Japan.

*I. okomotoi* Sawada 1970, in *Rhinolophus ferrumequinum;* Japan.

*I. osensis* Sawada 1972, in *Rhinolophus ferrumequinum;* Japan.

*I. pulchra* (Voge 1955) Yamaguti 1959 (syn. *Hymenolepis pulchra* Voge 1955), in *Sorex trowbridgei, S. pacificus;* California, U.S.

*I. takashii* Sawada 1967, in *Rhinolophus ferrumequinum;* Japan.

*I. yoshidai* Sawada 1967, in *Rhinolophus ferrumequinum;* Japan.

28b. Uterus a lobated or reticular sac filling entire gravid proglottid ................
.......... *Hymenolepis* Weinland 1858 (syn. *Triorchis* Clerc 1903). (Figure 320)

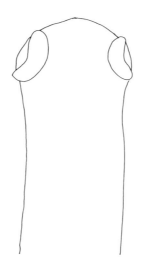

FIGURE 320. *Hymenolepis palmarum* Johri 1956. Scolex.

*Diagnosis:* Rostellum absent or rudimentary, unarmed. Suckers unarmed. Neck usually present. Proglottids usually wider than long, but considerable variation present between (and sometimes within) species in several morphological aspects. Genital pores unilateral. Genital ducts dorsal to osmoregulatory canals. Cirrus pouch usually well developed, containing internal seminal vesicle. Cirrus usually armed. Genital atrium simple. Cirrus stylet absent. Testes three, arranged in row or triangle. External seminal vesicle present. Ovary usually compact, but may be lobated. Vitelline gland postovarian. Vagina posterior or ventral to cirrus, unarmed. Seminal receptacle usually present. Uterus saccular or lobated, filling most of proglottid. Parasites of mammals. Cosmopolitan.

Type species: *H. diminuta* (Rudolphi 1819) Weinland 1858 (syn. *Taenia diminuta* Rudolphi 1819, *T. leptocephala* Creplin 1925, *T. flavopunctata* Weinland 1858, *T. varesina* Parona 1884, *T. minima* Grassi 1886, *Hymenolepis anomala* Splendore 1920, *H. megaloon* Linstow 1901, *H. diminutoides* Cholodkovsky 1912), in *Rattus, Mus, Apodemus, Grammomys, Evotomys, Citellus, Sigmodon, Arvicola, Arvicanthis, Meriones, Microtus, Mesocricetus, Cricetulus, Cercopithecus, Tupaia,* Man. Cosmopolitan.

Other species:

*H. arvicolina* Cholodkovsky 1912, in *Microtus agrestis, Arvicola campestris;* Russia.

*H. cebidarum* Baer 1927, in *Callithrix nigrifrons;* Brazil.

*H. citelli* McLeod 1933, in *Citellus* spp.; North America.

*H. horrida* (Linstow 1901) Lühe 1910 (syn. *Taenia horrida* Linstow 1901, *Oligorchis nonarmatus* Neiland 1952), in *Mus Rattus, Microtus, Arvicola, Tamiasciurus, Thomomys, Lemmus, Clethrionomys;* Europe, North America.

*H. ognewi* Skrjabin 1924 (may be a synonym of *H. diminuta*), in *Rhombomys opimus, Citellus* spp., *Meriones tamaricinus;* Russia.

*H. palmarum* Johri 1956, in *Funambulus palmarum, F. pennanti;* India.

*H. peipingensis* Hsü 1935, in *Talpa* sp.; China.

*H. pennanti* Nama 1974, in *Funambulus pennanti;* India.

*H. relicta* (Zschokke 1887) (syn. *Taenia relicta* Zschokke 1887), in *Mus decumanus, M. musculus, Rattus norvegicus;* Europe.

*H. scalopi* Schultz 1939, in *Scalopus aquaticus;* Oklahoma, U.S.

*H. vogeae* Singh 1956, in *Mus budugia;* India.

FIGURE 322. *Woodlandia phalacrocorax* (Woodland 1929) Yamaguti 1959.

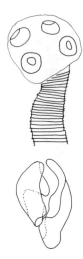

FIGURE 321. *Arhynchotaeniella clausovaginata* (Saakova 1958) comb. n. Scolex and vaginal clamp.

**29a. Testes antiporal to ovary**............*Amphipetrovia* **Spasskii et Spasskaja 1954.**

*Diagnosis:* Scolex unarmed. Testes three, in a transverse row entirely antiporal to ovary. Otherwise very poorly known. Parasites of birds.

Type species: *A. biaculeata* (Fuhrmann 1909) Spasskii et Spasskaja 1954 (syn. *Hymenolepis biaculeata* Fuhrmann 1909), in *Chenolopex aegyptiacus;* West Africa.

**29b. Testes otherwise** ............................................................... 30

**30a. Rostellum completely absent** .................................................... 31
**30b. Rostellum rudimentary or represented by a pit**............................ 32

**31a. Vagina with a distal, sclerotized clamp**................................ *Arhynchotaeniella* **nom. n. (for** *Arhynchotaenia* **Saakova 1958, preoccupied). (Figure 321)**

*Diagnosis:* Scolex large, spheroid, lacking rostellum or armature. Suckers simple, unarmed. Neck absent. Proglottids wider than long, craspedote. Mature and gravid proglottids very imperfectly described. Testes three. Cirrus pouch long, slender. Cirrus armed. Ovary and vitellaria? Vagina with sclerotized, distal clamp, with powerful, associated musculature. Somewhat more proximal is a muscular sphincter. Uterus sac-like. Parasites of Anseriformes. Rumania?

Type species: *A. clausovaginata* (Saakova 1958) comb. n. (syn. *Arhynchotaenia clausovaginata* Saakova 1958), in *Netta rufina;* Danube Delta.

**31b. Vagina lacking a sclerotized clamp** ...............................................
........................................ *Woodlandia* **Yamaguti 1959. (Figure 322)**

*Diagnosis:* Rostellum absent. Suckers unarmed. Proglottids numerous, broader than long. Testes three, one poral, other two antiporal and tandem; all three external to osmoregulatory canals. Cirrus pouch reaching median field, divided into a small distal portion with longitudinal muscles, and a large proximal portion with circular muscles. External seminal vesicle present. Cirrus unarmed. Ovary two-winged. Vitellarium lobed, posteroventral to ovary. Vagina dilated internally. Gravid uterus forming large marginal dilations filled with eggs. Parasites of cormorants. Asia.

Type species: *W. phalacrocorax* (Woodland 1929) Yamaguti 1959 (syn. *Hymenolepis phalacrocorax* Woodland 1929, *Weinlandia phalacrocorax* Woodland 1929), in *Phalacrocorax carbo;* India.

**32a. Rudimentary rostellum completely surrounded by glandular cells. Eggs with very thick outer shell ......................................... *Schmelzia* Yamaguti 1959.**

*Diagnosis:* Strobila very small. Suckers unarmed. Rostellum rudimentary, pyriform, completely encircled by glandular cells. Neck present. Proglottids broader than long. Testes three, one poral, other two antiporal, arranged in transverse row or triangle. Cirrus pouch not very large, with weak musculature. External and internal seminal vesicles present. No accessory sac or stylet. Ovary containing very large germ cells, ventral to testes. Vitellarium median, ventral. Vagina enlarged ventral to cirrus pouch. Seminal receptacle large. Uterus saccular, filling whole parenchyma when gravid, but containing few eggs. Outer egg membrane very thick. Parasitic in pterocliform and cuculiform birds. Asia, Africa.

Type species: *S. linderi* (Schmelz 1941) Yamaguti 1959 (syn. *Hymenolepis linderi* Schmelz 1941), in *Syrrhaptes paradoxus;* China.
Other species:
   *S. rhodesiensis* (Baer 1933) Yamaguti 1959 (syn. *Hymenolepis rhodesiensis* Baer 1933), in *Lybius torquatus* Rhodesia. (Egg membranes not known.)

**32b. Rudimentary rostellum not surrounded by glandular cells. Outer egg membrane not particularly thick...................................................... 33**

**33a. Vitellarium between poral and middle testes......................................... ................................. *Staphylepis* Spasskii et Oschmarin 1954.**

*Diagnosis:* Suckers unarmed. Rostellum rudimentary, unarmed. Proglottids numerous, wider than long. Testes three, in transverse row or forming a triangle. Cirrus pouch usually not reaching median line, but occasionally reaching antiporal osmoregulatory canals. External and internal seminal vesicles well developed. Ovary median. Vitelline gland median, between poral and middle testis, or overlapping the latter. Seminal receptacle rounded. Uterus saccular. Parasites of Galliformes, Columbiformes, Anseriformes, Passeriformes.

Type species: *S. cantaniana* (Polonio 1860) Spasskii et Oschmarin 1954 (syn. *Taenia cantaniana* Polonio 1860, *Davainea oligophora* Magalhães 1898, *D. cantaniana* Raillet et Lucet 1899, *Hymenolepis cantaniana* Ransom 1909, *H. inermis* [Yoshida 1910] Fuhrmann 1932), in *Gallus gallus, Meleagris gallopavo, Pavo cristatus, Phasianus colchicus, Numida meleagris, Coturnix coturnix, Perdix perdix, Colinus virginianus, Tetrastes bonasia, Turnix sustator, Tetrao parvirostris;* cosmopolitan.
Other species:
   *S. cordobensis* (Jordano 1952) Spasskii et Oshmarin 1954 (syn. *Hymenolepis cordobensis* Jordano 1952), in *Columba livia;* Spain.
   *S. infrequens* (Sharma 1943) Yamaguti 1959 (syn. *Hymenolepis infrequens* Sharma 1943), in *Anas platyrhynchos;* Burma.
   *S. inhamata* (Rietschel 1934) Yamaguti 1959 (syn. *Hymenolepis inhamata* Rietschel 1934), in *Eupetomena macroura;* Brazil.
   *S. lamellata* (Woodland 1930) Yamaguti 1959 (syn. *Hymenolepis lamellata* Woodland 1930), in "Australian shield-drake"; London Zoo.

FIGURE 323. *Parafimbriaria websteri* Voge et Read 1954.

*S. tonkinensis* (Joyeux et Baer 1935) Yamaguti 1959 (syn. *Hymenolepis tonkinensis* Joyeux et Baer 1935), in *Surniculus lugubris;* Indochina.

**33b. Vitellarium anterior to middle testis.................................................
........................................*Australiolepis* Spasskii et Spasskaja 1954.**

*Diagnosis:* Suckers unarmed. Rostellum rudimentary, unarmed. Testes three, in transverse row between osmoregulatory canals. Cirrus pouch not reaching median line. Internal and external seminal vesicles present. Ovary and vitelline gland immediately anterior to middle testis. Parasites of Anseriformes, Charadriiformes.

Type species: *A. southwelli* (Szpotanska 1931) Spasskii et Spasskaja 1954 (syn. *Hymenolepis southwelli* Szpotanska 1931, *Sphenacanthus southwelli* [Szpotanska 1931] Lopez-Neyra 1944), in *Chenopsis atrata;* Australia.
Other species:
*A. tushigi* Sawada et Iijima 1964, in *Capella gallinago;* Japan.

**34a. External segmentation lacking, internal segmentation conspicuous ..............
.................................*Parafimbriaria* Voge et Read 1954. (Figure 323)**

*Diagnosis:* Rostellum with single circle of hooks. Neck present. External metamerism absent, internal metamerism evident. Genital pores unilateral. Cirrus pouch elongate. Cirrus armed. Testes three, in antiporal, transverse row. External and internal seminal vesicles present. Ovary lobated, poral to testes. Vitelline gland ventral to ovary. Seminal receptacle present. Uterus in form of irregular sac. Parasites of Podicepediformes. North America.

Type species: *P. websteri* Voge et Read 1954, in *Colymbus nigricollis;* U.S.

**34b. External segmentation evident .................................................. 35**

**35a. Five pairs of osmoregulatory canals ........... *Hymenofimbria* Skrjabin 1914.**

*Diagnosis:* Rostellum with single circle of ten hooks. Proglottids craspedote, much wider than long. Cirrus pouch elongate. Genital pores unilateral. Genital atrium with accessory sac. Testes three, two antiporal, one poral. External and internal seminal vesicles present. Ovary compact, median. Vitelline gland median, postovarian. Uterus sac-like. Ten osmoregulatory canals present. Parasites of Anseriformes (merganser). Russia.

Type species: *H. merganseri* Skrjabin 1914, in *Mergus merganser;* Turkestan, Russia.

**35b. Two pairs of osmoregulatory canals.......................................... 36**

**36a. Rostellum with many small hooks behind apical circle ...................... 37**
**36b. Rostellum with only one (rarely two) circles of hooks ...................... 38**

FIGURE 324. *Vigisolepis barboscolex* Spasskii 1949.

**37a. First circle of hooks large, others smaller, regularly arranged.................. .....................................*Vigisolepis* Mathevossian 1945 (syn. *Echinoproboscilepis* Sadavskaja 1965, *Cucurbilepis* Sadovskaja 1965). (Figure 324)**

*Diagnosis:* Rostellum well developed, with single, wavy circle of 18 to 20 hooks, and numerous smaller hooks posterior to it. Genital pores unilateral. Cirrus pouch reaches median line, with sphincter at base. Testes three, in triangle, one poral, two antiporal. Internal and external seminal vesicles present. Ovary and vitelline gland median. Uterus sac-like. Parasites of shrews. Russia, Europe.

Type species: *V. spinulosa* (Cholodkowsky 1906) Mathevossian 1945 (syn. *Hymenolepis spinulosa* Cholodkowsky 1906, *Echinoproboscilepis kedroviensis* Sadovskaja 1965), in *Sorex araneus, S. minutus, S. vulgaris, S. macropygmaeus;* Russia, Estonia, Poland, Switzerland.

Other species:

*V. barboscolex* Spasskii 1949, in *Sorex araneus, Sorex* sp.; Russia, Poland.

*V. secunda* Sadovskaja 1965, in *Crocidura lasiura;* Russia.

*V. skrjabini* (Sadovskaja 1965) comb. n. (syn. *Cucurbilepis skrjabini* Sadovskaja 1965), in *Sorex* sp.; Russia. (Apparently based on specimen with hooks lost.)

**37b. Hooks all about same size, irregularly arranged................................ ...................................................*Lophurolepis* Spasskii 1973.**

*Diagnosis:* Scolex pyramidal; suckers well developed, unarmed. Rostellum conoid, armed with about 128 tiny hooks apparently not arranged in regular circles. Each hook lacking handle; guard longer than thorn. Neck present. Proglottids slightly craspedote, numerous (500). Genital pores unilateral, in a small genital atrium. Genital ducts dorsal to osmoregulatory canals. Cirrus pouch with thin walls. Internal and external seminal vesicles present. Cirrus spined. Two testes aporal, one poral. Vagina posterior and ventral to cirrus pouch. Seminal receptacle voluminous. Ovary continual between osmoregulatory canals, lobated, mainly anterior. Vitellarium compact, posterior to ovary. Uterus voluminous, a lobated sac, exceeding osmoregulatory canals. Outer egg membrane ornated with numerous bumps. Parasites of rodents. Africa.

Type species: *L. petteri* (Quentin 1964) Spasskii 1973 (syn. *Hymenolepis petteri* Quentin 1964), in *Lophuromys sikapusi;* Boukoko, Central African Republic.

**38a. Scolex with huge hemispherical rostellum armed near its base with a circle of 80 to 90 hooks ..................... *Hilmylepis* Skrjabin et Mathevossian 1942.**

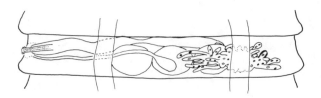

FIGURE 325. *Drepanidotaenia barrowensis* (Schiller 1952) Yamaguti 1959.

*Diagnosis:* Rostellum relatively huge, with a circle of numerous hooks near the base. Suckers situated on a membranous cushion-like organ under the rostellum, with their openings directed anteriorly and slightly laterally. Three testes, one poral and the other two antiporal, one in front of and lateral to the other. A large external and a small internal seminal vesicle present. Cirrus pouch small, cirrus unarmed. Genital ducts dorsal to osmoregulatory canals. Ovary occupying up to one third breadth of proglottid. Vitellarium posteroventral to ovary. Uterus filling entire proglottid. Vagina ventral to cirrus pouch. Seminal receptacle present. Parasites of Insectivora.

Type species: *H. nagatyi* (Hilmy 1936) Skrjabin et Mathevossian 1942 (syn. *Hymenolepis nagatyi* Hilmy 1936), in *Crocidura* sp.; Liberia.

Other species:

*H. kodrensis* Spasskii et Andreiko 1970, in *Sorex araneus*; Russia.

*H. prokopici* Genov 1970, in *Crocidura leucodon*; Bulgaria.

**38b. Scolex not as above.............................................................. 39**

**39a. All three testes poral to ovary ............................................*Drepanidotaenia* Railliet 1892 (syn. *Anserilepis* Spasskii et Talkacheva 1965, *Chimaerolepis* Spasskii et Spasskaja 1972, *Laricanthus* Spasskii 1962). (Figure 325)**

*Diagnosis:* Suckers armed or unarmed. Rostellum with eight to ten hooks in single circle. Neck short. Proglottids much wider than long. Genital pores unilateral. Cirrus pouch well developed. Testes three, in transverse row or triangle, all poral to ovary. External and internal seminal vesicles present. Ovary lobated, antiporal to testes. Vitelline gland posterior or ventral to ovary. Seminal receptacle present. Uterus a transverse sac with saccular outpocketings. Parasites of Anseriformes, Ciconiformes, Podicipediformes, Charadriiformes, Primates. Cosmopolitan.

Type species: *D. lanceolata* (Bloch 1782) Railliet 1892 (syn. *Taenia lanceolata* Bloch 1782, *T. acutissima* Pallas 1781 in part, *T. anserum* Fritsch 1727, *T. anseris* Bloch 1779, *Hymenolepis [Drepanidotaenia] lanceolata* [Bloch 1782] Cohn 1901), in *Anas, Anser, Branta, Cygnus, Chenopsis, Eulabeia, Oxyura, Netta, Aythya, Tadorna, Philacte, Gallus*, occasionally humans and other primates. Cosmopolitan.

Other species:

*D. aporalis* Tscherbovitsch 1945, in *Larus argentatus, L. ichyaetus, L. ridibundus*; Russia.

*D. barrowensis* (Schiller 1952) Yamaguti 1959 (syn. *Hymenolepis barrowensis* Schiller 1952, *Anserilepis* [Schiller 1952] Spasskii et Tolkacheva 1965), in *Anser albifrons, A. fabalis*; Alaska, Russia.

*D. bilateralis* (Linstow 1905) Railliet 1892 (syn. *Hymenolepis bilateralis* Linstow 1905), in *Branta bernicla, Anser anser, Aythya ferina, Cygnus cygnus*; Russia.

*D. elongata* (Fuhrmann 1906) Lopez-Neyra 1942 (syn. *Hymenolepis elongata* Fuhrmann 1906), in *Mylobdophanes coerulescens*; Brazil.

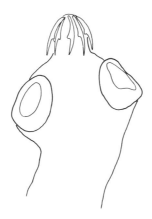

FIGURE 326. *Flamingolepis liguloides* (Gervais 1847) Spasskii et Spasskaja 1954.

*D. lateralis* (Mayhew 1925) Skrjabin et Mathevossian 1945 (syn. *Weinlandia lateralis* Mayhew 1925, *Laricanthus lateralis* [Mayhew 1925] Spasskii 1962), in *Larus glaucescens;* San Juan Island.

*D. lobata* (Szpotanska 1931) Skrjabin et Mathevossian 1945 (syn. *D. lanceolata* var. *lobata* Szpotanska 1931, *D. l. szpotanskaica* Hughes 1940), in *Chenopis atrata;* (?) Poland.

*D. nyrocae* (Yamaguti 1935) Skrjabin et Mathevossian 1945 (syn. *Hymenolepis nyrocae* Yamaguti 1935, *Wardium nyrocae* Ryjikov et Gubanov 1958, *Wardiodes nyrocae* [Yamaguti 1935] Spasskii 1962, *Dicranotaenia [D.] nyrocae* [Yamaguti 1935] Lopez-Neyra 1942), in *Nyroca marila, Anas platyrhynchos, A. querquedula, Cygnus olor, Somateria spectabilis, Philacte canagica, Branta canadensis;* Japan, Russia, Alaska, U.S.

*D. philactes* (Schiller 1951) Yamaguti 1959 (syn. *Hymenolepis philactes* Schiller 1951), in *Philacte canagica,* Alaska.

(Placed in *Parabisaccanthes* by Spasskii and Resnik, 1963).

*D. przewalskii* (Skrjabin 1914) Lopez-Neyra 1942 (syn. *Hymenolepis przewalskii* Skrjabin 1914), in *Anser anser, A. albifrons, A. fabalis, Anas platyrhynchos, A. strepera, A. crecca, A. clypeata, Aythya fuligula, Somateria mollissima, Meleagris gallopavo;* Russia, Eastern Europe.

*D. rapida* (Szpotanska 1931) Yamaguti 1959 (syn. *Hymenolepis rapida* Szpotanska 1931), in *Chenopis atrata;* Australia.

*D. signachiana* Kurashvili 1950, in *Tadorna ferruginea;* Russia.

*D. spinulosa* Dubinina 1954, (syn. *Anatinella spinulosa* [Dubinina 1954] Spasskii 1963), in *Nyroca ferina;* Russia. (Suckers spinose).

*D. watsoni* Prestwood et Reid 1966 (syn. *Chimaerolepis watsoni* [Prestwood et Reid 1966] Spasskii et Spasskaja 1972), in *Meleagris gallopavo;* Arkansas, U.S.

**39b.** Testes not as above .................................................... **40**

**40a.** Uterus reticulate ........ *Flamingolepis* Spasskii et Spasskaja 1954. (Figure 326)

*Diagnosis:* Rostellum with single circle of eight spiniform hooks; stalk of rostellum covered with even circles of small spines. Suckers unarmed. Cirrus pouch not reaching median line. Internal seminal vesicle present. Testes three, lobated, in triangle, one poral and two anti-

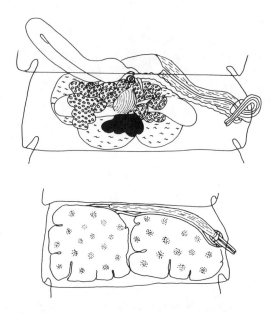

FIGURE 327. *Armadoskrjabinia medici* (Stossich 1890) Spasskii et Spasskaja 1954.

poral. Vas deferens strongly convoluted. Internal seminal vesicle absent. Ovary and vitelline gland median. Uterus reticular. Parasites of flamingos. Europe, Asia, Africa, Cuba.

Type species: *F. liguloides* (Gervais 1847) Spasskii et Spasskja 1954 (syn. *Halysis liguloides* Gervais 1847, *Hymenolepis [Drepanidotaenia] liguloides* Cohn 1901, *Diorchis occlusa* Linstow 1906, *Dicranotaenia liguloides* [Gervais 1847] Skrjabin et Mathevossian 1945, *Sphenacanthus liguloides* [Gervais 1847] Lopez-Neyra 1942, *Aporodiorchis liguloides* [Gervais 1847] Baer 1961), in *Phoenicopterus roseus, P. antiquorum;* France, India, Africa, Sri Lanka.

Other species:

*F. dolguschini* Gvosdev et Maksimova 1968, in *Phoeniconaias roseus;* Russia.

*F. flamingo* (Skrjabin 1914) Spasskii et Spasskaja 1954 (syn. *Hymenolepis flamingo* Skrjabin 1914), in *Phoenicopterus antiquorum, P. roseus;* Russia, France.

*F. megalorchis* (Lühe 1898) Spasskii et Spasskaja 1954 (syn. *Taenia megalorchis* Lühe 1898, *Diplacanthus [Dilepis] megalorchis* Cohn 1899, *Hymenolepis megalorchis* Railliet 1899, *H. [Drepanidotaenia] megalorchis* Cohn 1901, *Weinlandia megalorchis* Mayhew 1925), in *Phoenicopterus roseus;* Europe, Africa, India, Cuba.

*F. tengizi* Gvosdev et Maksimova 1968, in *Phoeniconaias roseus, P. minor;* Russia, Kenya.

**40b. Uterus not reticulate ......................................................... 41**

**41a. Gravid uterus forming two sacs, which may be joined by narrow isthmus... 42**
**41b. Gravid uterus not as above .................................................... 43**

**42a. More than ten rostellar hooks. Cirrus pouch exceeding midline of segment ....**
**..................... *Armadoskrjabinia* Spasskii et Spasskaja 1954. (Figure 327)**

*Diagnosis:* Suckers unarmed. Rostellum with ten or more hooks; handle and blade about equal, guard prominent but shorter than blade. Testes three, one poral, two antiporal. Cirrus pouch muscular, usually very long, may or may not reach to near the antiporal margin of

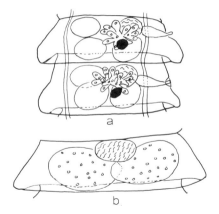

FIGURE 328. *Passerilepis passeris* (Gmelin 1790) Spasskii et Spasskaja 1954. (a) Mature proglottids; (b) gravid proglottid.

proglottid. External and internal seminal vesicle present. Ovary with long digitate lobes, overlapping testes. Vitellarium median. Uterus beginning as a transverse tube, then forming two lobed sacs and occupying whole proglottid. Parasites of aquatic birds.

Type species: *A. medici* (Stossich 1890) Spasskii et Spasskja 1954 (syn. *Taenia medici* Stossich 1890, *Hymenolepis medici* [Stossich 1890] Fuhrmann 1906, *Weinlandia medici* [Stossich 1890] Mayhew 1925, *Dicranotaenia medici* [ Stossich 1890] Skrjabin et Mathevossian 1945, *Echinorhynchotaenia medici* [Stossich 1890] Baer 1959), in *Pelecanus onocrotalus, P. philippensis, P. rufescens;* Europe, Africa, India, Malaya.

Other species:

*A. globulosa* (Szpotanska 1932) Yamaguti 1959 (syn. *Hymenolepis globulosa* Szpotanska 1932), in *Chenopsis atrata;* Australia.

*A. magniuncinata* (Meggitt 1927) Yamaguti 1959 (syn. *Hymenolepis magniuncinata* Meggitt 1927), in *Pelecanus onocrotalus;* Egypt.

*A. rostellata* (Abildgaard 1790) Yamaguti 1959 (syn. *Taenia rostellata* Abildgaard 1790, *T. capitellata* Rudolphi 1810, *Diplacanthus [Dilepis] capitellata* Cohn 1899, *Hymenolepis capitellata* Railliet 1899), in *Colymbus grisegena, Colymbus* spp., *Podiceps hooboelli, Gavis* spp.; Europe, North America.

*A. tubicirrosa* (Baczynska 1914 Yamaguti 1959 (syn. *Hymenolepis tubicirrosa* Baczynska 1914), in *Dicholophus cristatus;* Paraquay.

**42b. Ten rostellar hooks. Cirrus pouch not exceeding midline of segment............ ............................................ *Passerilepis* Spasskii et Spasskaja 1954 (syn. *Mayhewia* Yamaguti 1959, *Satyolepis* Spasskii 1965). (Figure 328)**

*Diagnosis:* Rostellum with single circle of ten hooks. Suckers unarmed. Genital pores unilateral. Cirrus pouch usually not reaching median line. Testes three, in triangle, one poral, two antiporal. Internal and external seminal vesicles present. Ovary and vitelline gland median. Gravid uterus in two separated or nearly separated, sacs. Parasites of Passeriformes, Ciconiformes, Charadriiformes. Europe, Africa, Celebes, Asia, Hawaii, Sri Lanka, Taiwan, South America.

Type species: *P. passeris* (Gmelin 1790) Spasskii et Spasskaja 1954 (syn. *Taenia passeris* Gmelin 1790, *T. fringillarum* Rudolphi 1810, *T. leptodera* Linstow 1879, *Aploparaksis fringillarum* [Rudolphi 1810] Linstow 1904, *Aploparksis linstowi* Kintner 1938, *Hymenolepis fringillarum* Fuhrmann 1908, *Dicranotaenia passeris* [Gmelin 1790] Skrjabin

et Mathevossian 1945), in *Passer, Acanthis, Parus, Aegithalos, Spinus, Fringilla, Sylvia, Dicrurus, Turdus, Coloeus, Corvus, Emberiza, Pyrrhula, Coccothraustes, Riparia, Eurystoma, Alauda, Lanius, Astur, Dryobates, Oenanthe;* Russia, Europe, India, Africa.
Other species:

P. *arciuterus* Yamaguti 1956, in *Ptilotis chrysotis;* Celebes.

P. *brevis* (Fuhrmann 1906) Spasskii et Spasskaja 1954 (syn. *Hymenolepis brevis* Fuhrmann 1906, *Weinlandia brevis* [Fuhrmann 1906] Mayhew 1925), in *Locustella fluviatilis, Copsychus saularis;* Europe.

P. *crenata* (Goeze 1782) Sultanov et Spasskaja 1959 (syn. *Taenia crenata* Goeze 1782, *T. tenuis nodus instructa* Bloch 1782, *T. moculata* Batsch 1786, *T. surpentulus* Schrank 1788, *T. nodosa* Schrank 1788, *T. cornicis* Gmelin 1790, *T. turdorum* Gmelin 1790, *T. punctata* Brugiera 1791, *Taenia corvifrugilegi* Viborg 1795, *Halysis glandarii* Zeder 1803, *T. corvi-cornicis* Rudolphi 1810, *T. angulata* Rudolphi 1810, *T. corvorum* Rudolphi 1819, *Diplacanthus serpentulus* (Schrank 1788) Volz 1899, *Hymenolepis (Drepanidotaenia) serpentulus* [Schrank 1877] Cohn 1901, *Hymenolepis phasianina* Fuhrmann 1907, *H. serpentulus* [Schrank 1788] Fuhrmann 1908, *H. crenata* [Goeze 1782] Kostylew 1915, *Dicranotaenia [D.] serpentulus* [Schrank 1788] Lopez-Neyra 1942, *D. crenata* [Goeze 1782] Skrjabin et Mathevossian 1945, *Hymenolepis serpentulus sturni* Jones 1945, *H. s. turdi* Jones 1945, *Variolepis crenata* [Goeze 1789] Spasskii et Spasskaja 1954, *Mayhewia crenata* [Goeze 1782] Yamaguti 1959, *M. phasianina* [Fuhrmann 1907] Yamaguti 1959, *M. serpentulus* [Schrank 1788] Yamaguti 1959), in *Corvus, Coloeus, Cyanopica, Galerida, Garrulus, Perisoreus, Pica, Nucifraga, Oriolus, Turdus, Fringilla, Anthus, Muscicapa, Eremophila, Emberiza, Calandrella, Passer, Sturnus, Dryobates, Dendrocopus, Picus, Upupa, Coracias, Phasianus, Tetrastes, Tyrurus, Tetrao, Capella, Falco;* Russia, Europe, Japan, North America, Africa.

P. *dahurica* (Linstow 1903) Spasskii et Spasskaja 1954 (syn. *Taenia dahurica* Linstow 1903, *Hymenolepis dahurica* Fuhrmann 1906, *Mayhewia dahurica* [Linstow 1903] Yamaguti 1959), in *Corvus dahuricus, C. levaillanti, Erythrina erythrina, Coracias gallulus;* Russia.

P. *fola* (Meggitt 1933) Spasskii et Spasskaja 1954 (syn. *Hymenolepis fola* Meggitt 1933), in *Rostratula benghalensis;* India.

P. *hemignathi* (Shipley 1898) Spasskii et Spasskaja 1954 (syn. *Drepanidotaenia hemignathi* Shipley 1898, *Hymenolepis hemignathi* [Shipley 1898] Fuhrmann 1932), in *Hemignathus procerus;* Hawaii, U.S.

P. *intermedius* (Clerc 1906) Spasskii et Spasskaja 1954 (syn. *Hymenolepis intermedius* Clerc 1906, *Weinlandia intermedius* Mayhew 1925, *Dicranotaenia [D.] intermedius* [Clerc 1906] Lopez-Neyra 1942, *Mayhewia intermedius* [Clerc 1906] Yamaguti 1959), in *Cuculus canorus, C. intermedius, C. poliocephalus* Russia, Argentina.

P. *japonensis* Sawada et Kugi 1980, in *Corvus macrorhynchus;* Japan.

P. *megacantha* Spasskii, Dang Van Ngy, et Yurpalova 1963, in *Tringa* sp.; Vietnam.

P. *nebraskensis* (Rolan et Leidahl 1969), comb. n. (syn. *Mayhewia nebraskensis* Rolan et Leidahl 1969) in *Columba livia;* Nebraska, U.S.

P. *occidentalis* Spasskii, Dang Van Ngy, et Yurpalova 1963, in *Acridotheres cristatellus, A. grandis;* Vietnam.

P. *oena* (Ortlepp 1938) Spasskii et Spasskaja 1954 (syn. *Hymenolepis oena* Ortlepp 1938), in *Oena capensis;* South Africa.

P. *parina* (Fuhrmann 1907) Spasskii et Spasskaja 1954 (syn. *Hymenolepis parina* Fuhrmann 1907, *Weinlandia parina* Mayhew 1925, *Dicranotaenia [D.] parina* [Fuhrmann 1907] Lopez-Neyra 1942), in *Parus major, P. ater, P. palustris, Aegithalus caudatus, Passer montanus;* Russia.

*P. passerina* (Fuhrmann 1906) Spasskii et Spasskija (syn. *Hymenolepis passerina* Fuhrmann 1906, *Weinlandia passerina* [Fuhrmann 1906] Mayhew 1925), in *Turdus parochus;* Egypt.

*P. pellucida* (Fuhrmann 1906) Spasskii et Spasskaja 1954 (syn. *Hymenolepis pellucida* Fuhrmann 1906, *Weinlandia pellucida* Mayhew 1925, *Dicranotaenia pellucida* Lopez-Neyra 1942), in *Ostinops decumanus, O. viridis, Gymnostinops yurnearium, Spodiopsar cineraceus;* Russia, Brazil.

*P. petrocinclae* (Krabbe 1879) Spasskii et Spasskaja 1954 (syn. *Taenia petrocinclae* Krabbe 1879, *Hymenolepis petrocinclae* Fuhrmann 1908, *Hispaniolepis petrocinclae* [Krabbe 1879] Lopez-Neyra 1942), in *Monticola solitarius;* Russia, Iceland, Europe, Africa.

*P. rysavyi* Spasskii et Spasskaja 1964, in Passeriformes; Europe, Russia.

*P. schmidti* Deardorff et Brooks 1978, in *Cyanocitta cristata;* U.S.

*P. septemsororum* (Burt 1944) Yamaguti 1959 (syn. *Hymenolepis septemsororum* Burt 1944), in *Turdoides griseus;* Ceylon.

*P. spasskii* (Sudarikov 1950) Spasskii et Spasskaja 1954 (syn. *Dicranotaenia spasskii* Sudarikov 1950), in *Coracias garrulus;* Russia.

*P. streptopeliae* (Joyeux et Baer 1935) Sultanov 1963 (syn. *Hymenolepis streptopeliae* Joyeux at Baer 1935, *Drepanidotaenia streptopeliae* [Joyeux et Baer 1935] Lopez-Neyra 1942), in *Streptopelia orientalis, S. capicola, S. senegalensis, Columbia livia;* Russia, Africa.

*P. stylosa* (Rudolphi 1809) Spasskii et Spasskaja 1954 (syn. *Taenia stylosa* Rudolphi 1809, *Diplacanthus stylosa* Volz 1899, *Hymenolepis stylosa* Railliet 1899, *Weinlandia stylosa* Mayhew 1925, *Dicranotaenia stylosa* [Rudolphia 1809] Lopez-Neyra 1942, *Weinlandia corvi* Mayhew 1925, *Hymenolepis corvi* [Mayhew 1925] Fuhrmann 1932, *Dicranotaenia corvi* [Mayhew 1925] Skrjabin et Mathevossian 1945), in *Corvus, Coleus, Balopogon, Garrulus, Pica, Oriolus, Turdus, Phylloscopus, Alauda, Perisoreus, Parus, Acridotheres, Lyrurus;* Russia, North America, Europe, Africa.

*P. taiwanensis* (Yamaguti 1935) Spasskii et Spasskaja 1954 (syn. *Hymenolepis taiwanensis* Yamaguti 1935), in *Ardea purpurea;* Taiwan.

*P. zosteropis* (Fuhrmann 1918) Spasskii et Spasskaja 1954 (syn. *Hymenolepis zosteropis* Fuhrmann 1918, *Weinlandia zosteropis* [Fuhrmann 1918] Mayhew 1925), in *Zosterops minuta, Criniger, Dendrocitta, Garrulax, Melophus, Ploceus;* Loyalty Islands, Burma, India, Himalayas.

43a. Parasites of mammals ........................................................ 44
43b. Parasites of birds ............................................................. 49

44a. Rostellum with two circles of hooks ..................................................
................................*Skrjabinacanthus* Spasskii et Morozov 1959. (Figure 329)

*Diagnosis:* Rostellum stout, with two apical circles of hooks. Hooks slender, with long handle, long curving blade, and well developed guard. Suckers unarmed. Neck present. Proglottids wider than long. Genital pores unilateral. Cirrus pouch long. Cirrus armed. External and internal seminal vesicles present. Testes in triangle, two aporal, one poral. Ovary lobated, extensive. Vitellarium transversely elongated, postovarian. Seminal receptacle present. Uterus? Parasites of shrews; Russia.

Type species: *S. diplocoronatus* Spasskii et Morozov 1959, in *Sorex vir;* Russia.
Other species:
    *S. jacutensis* Spasskii et Morozov 1959, in *Sorex tscherskii,* Sorex sp.; Russia.

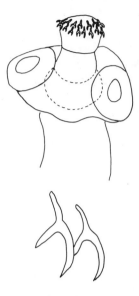

FIGURE 329. *Skrjabinacanthus diplocoronatus* Spasskii et Morozov 1959. Scolex and rostellar hooks.

**44b. Rostellum with one circle of hooks** ............................................. **45**

**45a. Vagina forming an eversible, spinose terminal organ** ...........................
............................................................. *Lockerrauschia* **Yamaguti 1959.**

*Diagnosis:* Strobila very small. Scolex distinctly set off from neck. Rostellum with ten long-handled, cheliform hooks. Proglottids numerous, wider than long. Testes three, in diagonal line, all aporal to median line. Cirrus pouch large, reaching to near aporal margin of proglottid. External and internal seminal vesicles present. Cirrus spinose, with stylet. Ovary ventral to middle testis. Vitellarium? Uterus saccular, filling entire proglottid when fully gravid. Vagina forming an eversible, spinose, terminal organ followed by funnel-shaped portion. Parasites of Insectivora.
    Type species: *L. intricata* (Locker et Rausch 1952) Yamaguti 1959 (syn. *Hymenolepis intricata* Locker et Rausch 1952), in *Sorex vagrans*; Oregon, U.S.

**45b. Vagina not as above** ...................................................................... **46**

**46a. Guard of rostellar hook bifid** .................... *Triodontolepis* **Yamaguti 1959.**

*Diagnosis:* Suckers unarmed. Rostellum acorn-shaped, with large sheath. Rostellar hooks ten or more, Y-shaped in profile; guard bifurcate, nearly as long as or longer than, blade; handle as long as or longer than blade. Proglottids numerous, wider than long. Testes three, one poral, other two antiporal, overlapping each other. Cirrus pouch small. External and internal seminal vesicles present. Ovary and vitellarium median. Uterus forming an oval sac in median field, later occupying whole proglottid. Parasites of rodents, insectivores and Chiroptera.
    Type species: *T. tridontophora* (Soltys 1954) Yamaguti 1969 (syn. *Hymenolepis tridontophora* Soltys 1954), in *Neomys fodiens, Sorex araneus, Crocidura* sp.; Europe.

Other species:
- *T. kurashvilii* Prokopic 1971, in *Neomys fodiens;* Russia.
- *T. miniopteri* (Sandars 1957) Yamaguti 1959 (syn. *Hymenolepis miniopteri* Sandars 1957), in *Miniopterus blepotis;* Australia.
- *T. rysavyi* Prokopic 1972, in *Neomys anomalus;* Czechoslovakia.
- *T. skrjabinini* Spasskii et Andreiko 1968, in *Neomys anomalus, Sorex araneus;* Russia.

**46b. Guard of rostellar hook not bifid............................................. 47**

**47a. Uterus develops very thick, capsule-like wall in detached, gravid segments.....**
***Neoskrjabinolepis* Spasskii 1947 (syn. *Zarnowskiella* Spasskii et Andreiko 1970).**

*Diagnosis:* Suckers unarmed. Rostellum with a circle of small number of hooks. Strobila anapolytic. Genital pores unilateral. Testes three, in transverse row. Cirrus pouch claviform, may or may not overreach osmoregulatory canals. Vitellarium nearly median, between poral and middle testis. In detached gravid proglottids the uterus forms a common egg reservoir, or "syncapsule". Parasites of Insectivora.

Type species: *N. schaldybini* Spasskii 1947, in *Sorex* sp.; Russia.
Other species:
- *N. singularis* (Cholodkovsky 1912) Spasskii 1954 (syn. *Hymenolepis singularis* Cholodkovsky 1912), in *Sorex* sp.; Europe.
- *N. stefanskii* (Zarnowsky 1954) Schaldybin 1964 (syn. *Zarnowskiella stefanskii* [Zarnowsky 1954] Spasskii et Andreiko 1970, *Hymenolepis stefanskii* Zarnowsky 1954, *Vampirolepis stefanskii* [Zarnowsky 1954] Zarnowsky 1955), in *Sorex araneus;* Poland.

**47b. Uterus normal .............................................................. 48**

**48a. Strobila very small, with few segments......................................**
**...*Staphylocystis* Villot 1877 (syn. *Staphylocystoides* Yamaguti 1952, subgenus).**

*Diagnosis:* Strobila small. Rostellum well developed, with rostellar sheath and circle of eight to ten or more hooks. Guard and blade of hook well developed. Proglottids not numerous. Testes three, in a triangle or transverse row; one poral, two antiporal, to vitelline gland, or vice versa. Cirrus pouch may or may not reach median line. External and internal seminal vesicles present. Accessory sac and stylet absent. Ovary compact or slightly lobed, median or submedian. Vitelline gland median or submedian. Uterus horse shoe-shaped, with apex directed forwards, or a simple sac. Parasites of Insectivora and Chiroptera.

Type species: *S. pistillum* (Dujardin 1845) Spasskii 1950 (syn. *Taenia pistillum* Dujardin 1845, *Staphylocystis micracanthus* Villot 1877), in *Sorex, Crocidura;* Europe.
Other species:
- *S. acuta* (Rudolphi 1819) Spasskii 1950 (syn. *Taenia acuta* Rudolphi 1819, *T. obtusa* of Beneden 1873), in *Vesperugo, Eptesicus, Myotis, Nyctalus, Miniopterus, Pipistrellus, Plecotus;* Europe.
- *S. alpestris* (Baer 1932) Spasskii 1950 (syn. *Hymenolepis alpestris* Baer 1932), in *Neomys fodiens;* Switzerland.
- *S. bacillaris* (Goeze 1782) Spasskii 1950 (syn. *Taenia bacillaris* Goeze 1782), in *Scalopus aquaticus, Talpa europaea, Crocidura coerulea, Nyctalus maximus;* Europe, Africa, North America, Japan.
- *S. chrysochloridis* (Janicki 1904) Spasskii 1950 (syn. *Hymenolepis chrysochloridis* Janicki 1904), in *Chrysochloris aurea, C. capensis;* Europe.

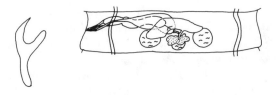

FIGURE 330. *Vampirolepis schmidti* Jensen et Howell 1983; Rostellar hook and mature proglollid.

*S. dodecantha* (Baer 1925) Spasskii 1950 (syn. *Hymenolepis dodecantha* Baer 1925), in *Crocidura occidentalis;* Belgian Congo.

*S. evansi* (Skrjabin et Mathevossian 1942) Yamaguti 1959, in *Oryctolagus cuniculus;* England.

*S. fuelleborni* (Hilmy 1936) Spasskii 1950, in *Crocidura* sp.; Liberia.

*S. furcata* (Stieda 1862) Spasskii 1950, in *Sorex araneus, Neomys fodiens, Crocidura munina, C. caerulea;* Europe, Africa, Asia.

*S. longi* (Oswald 1951) Yamaguti 1959, in *Sorex fumeus, S. bendirii;* U.S.

*S. loossi* (Hilmy 1936) Spasskii 1950, in *Crocidura* sp.; Liberia.

*S. minutissima* (Meggitt 1927) Yamaguti 1959, in *Crocidura murina;* Burma.

*S. murissylvatici* (Rudolphi 1819) Spasskii 1950 (syn. *Taenia murissylvatici* Rudolphi 1819), in *Apodemus sylvaticus;* England, Switzerland.

*S. parvissima* (Voge 1953) Yamaguti 1959, in *Sorex bendirii;* California, U.S.

*S. pauciproglottis* (Neiland 1953) Yamaguti 1959 (syn. *Hymenolepis pauciproglottis* Neiland 1953), in *Sorex vagrans;* Oregon, U.S.

*S. sanchorensis* Nama et Khichi 1975, in *Suncus murinus;* India.

*S. scalaris* (Dujardin 1845) Villot 1877, in *Sorex araneus, Crocidura russula;* Africa, Europe.

*S. sengeri* (Neilrand 1953) Yamaguti 1959 (syn. *Hymenolepis sengeri* Neiland 1953), in *Sorex bendirii;* Oregon, U.S.

*S. serrula* (Oswald 1951) Yamaguti 1959, in *Sorex fumeus;* U.S.

*S. sindensis* Nama 1976, in *Suncus murinus;* India.

*S. solitaria* (Meggitt 1927) Yamaguti 1959 (syn. *Weinlandia solitaria* Meggitt 1927),in *Crocidura murina;* Burma.

*S. sphenomorphus* (Locker et Rausch 1952) Yamaguti 1959 (syn. *Hymenolepis sphenomorphus* Locker et Rausch 1952), in *Sorex vagrans, S. cinereus;* Alaska, U.S.

*S. suncusensis* Olsen et Kuntz 1978, in *Suncus murinus;* Taiwan.

*S. syrdariensis* (Skarbilovitsch 1946) Spasskii 1950, in *Pipistrellus bactrianus;* Russia.

*S. tiara* (Dujardin 1845) Spasskii 1950, in *Sorex araneus, Crocidura* sp.; Europe, Africa.

*S. toxometra* (Baer 1932) Yamaguti 1959 (syn. *Hymenolepis toxometra* Baer 1932), in *Sorex araneus;* Switzerland.

**48b. Strobila medium to large, with many segments.................................**
**.................................................. *Vampirolepis* Spasskii 1954 (syn. *Rodentolepis* Spasskii 1954, *Vampirolepidoides* Yamaguti 1959). (Figure 330)**

*Diagnosis:* Suckers unarmed. Rostellum well-developed, with rostellar sheath and a circle of variable number (up to 50) of Y-shaped, rosethorn- or wrench-shaped hooks. Handle of hook comparatively long, guard broad, nearly as long as, or somewhat longer or shorter than, blade. Proglottids numerous, transversely elongated. Testes three, arrange in transverse row or triangle; middle one may overlap vitelline gland or ovary. Cirrus pouch usually small, may reach median line occasionally. External and internal seminal vesicles present. Gravid uterus extending transversely. Parasites of Chiroptera, Insectivora, Primates, Rodentia, Marsupialia, rarely birds.

Type species: *V. skrjabinariana* (Skarbilovitsch 1946) Spasskii 1954, in *Eptesicus turcomanus;* Russia.

Other species:

*V. akodontis* (Rego 1967) comb. n. (syn. *Rodentolepis akodontis* Rego 1967), in *Akodon arviculoides;* Brazil.

*V. artibei* Zdzitowiecki et Rutkowska 1980, in *Artibeus jamaicensis;* Cuba.

*V. asymmetrica* (Janicki-1904) Spasskii 1954 (syn. *Hymenolepis asymmetrica* Janicki 1904, *H. arvicolae* Galli-Valerio 1930, *Rodentolepis asymmetrica* (Janicki 1904) Spasskii 1954), in *Arvicola arvalis, Microtus nivalis, Evotomys glareolus;* Europe.

*V. australiensis* (Sandars 1957) Yamaguti 1959, in *Rattus assimilis;* Australia.

*V. baeri* Murai 1976, in *Nyctalis noctula;* Hungary.

*V. bahli* (Singh 1958) comb. n. (syn. *Hymenolepis bahli* Singh 1958, *Rodentolepis bahli* [Singh 1958] Yamaguti 1959), in *Crocidura caerulea;* India.

*V. balsaci* (Joyeux et Baer 1934) Spasskii 1954, in *Myotis bechsteini, M. myotis, Eptesicus serotinus, Parabascus lepidotus;* France, Hungary, Czechoslavakia.

*V. bidentatus* Zdzitowiecki et Rutkowska 1980, in *Phyllonycterus poeyi;* Cuba.

*V. blarinae* (Rausch et Kuns 1950) comb. n. (syn. *Hymenolepis blarinae* Rausch et Kuns 1950, *Rodentolepis blarinae* [Rausch et Kuns 1950] Yamaguti 1959), in *Blarina brevicauda;* Wisconsin, U.S.

*V. cercopitheci* (Baer 1934) Yamaguti 1959, in *Cercopithecus nictitans;* W. Africa.

*V. chiropterophila* (Vigueras 1941) Yamaguti 1959, in *Molossus tropidorhynchus;* Cuba.

*V. christensoni* (Macy 1931) Spasskii 1954, in *Myotis lucifugus, M. yumenensis, M. californicus, M. evotis, M. keeni, Chilonycteris parnelli, Eptesicus fuscus;* North America, Cuba.

*V. crassa* (Janicki 1904) comb. n. (syn. *Hymenolepis crassa* Janicki 1904, *Rodentolepis crassa* [Janicki 1904] Spasskii 1954), in *Mus musculus, Rattus norvegicus;* Europe.

*V. criceti* (Janicki 1904) comb. n. (syn. *Hymenolepis criceti* Janicki 1904, *Rodentolepis criceti* [Janicki 1904] Spasskii 1954, *Dicranotaenia criceti* [Janicki 1904] Skrjabin et Mathevossian 1945), *Staphylocystis criceti* [Janicki 1904] Spasskii 1950), in *Cricetus vulgaris;* Europe.

*V. decipiens* (Diesing 1850) Spasskii 1954, in *Chilonycteris rubiginosa, Molossus perotis, Eptesicus fuscus, Tadarida laticaudata, T. minutua;* Brazil, Cuba, Australia.

*V. elongatus* Rego 1962, in *Glossophaga soricina, Phyllostomum hastatus, Molossus rufus;* Brazil.

*V. erinacei* (Gmelin 1790) comb. n. (syn. *Taenia erinacei* Gmelin 1790, *T. compacta* Rudolphi 1810, *T. tripunctata* Braun in Rudolphi 1810, *Hymenolepis steudeneri* Janicki 1904, *Dicranotaenia erinacei* var. *steudeneri* Lopez-Neyra 1942, *Rodentolepis erinecei* [Gmelin 1790] Spasskii 1954), in *Erinaceus* sp.; Europe.

*V. evaginata* (Barker et Andrews 1915) comb. n. (syn. *Hymenolepis evaginata* Barker et Andrews 1915, *Rodentolepis evaginata* [Barker et Andrews 1915] Spasskii 1954), in *Ondatra zibithica, Microtus pennsylvanicus;* U.S. England.

*V. falculata* (Rausch et Kuns 1950) comb. n. (syn. *Hymenolepis falculata* Rausch et Kuns 1950, *Rodentolepis falculata* [Rausch et Kuns 1950] Yamaguti 1959, *Staphylocystis falculata* [Rausch et Kuns 1950] Spasskii 1954), in *Sorex* spp.; U.S.

*V. fujiensis* Sawada 1978, in *Rhinolophus ferrumequinum;* Japan.

*V. gertschi* (Macy 1947) Spasskii 1954, in *Myotis californicus, Eptesicus fuscus, Plecotus townsendi;* U.S.

*V. globirostris* (Baer 1925) comb. n. (syn. *Hymenolepis globirostris* Baer 1925, *Rodentolepis globirostris* [Baer 1925] Spasskii 1954, *Hymenolepis suricattae* Ortlepp 1938), in *Suricatta suricatta*, "rat"; Belgian Congo, South America.

*V. guarany* Rego 1961, in *Molossus crassicaudatus;* Brazil.

*V. hamanni* (Mrazek 1891) Yamaguti 1959, in *Neomys fodiens;* Switzerland.

*V. hattorii* (Sawada 1978) comb. n. (syn. *Rodentolepis hattorii* Sawada 1978), in *Rhinolophus ferrumequinum;* Japan.

*V. hidaensis* Sawada 1967, in *Rhinolophus ferrumequinum, Miniopterus schreibersii;* Japan.

*V. isensis* Sawada 1966, in *Rhinolophus ferrumequinum, R. cornutus;* Japan.

*V. iwatensis* Sawada 1975, in *Rhinolophus cornutus;* Japan.

*V. jacobsoni* (Linstow 1907) comb. n. (syn. *Hymenolepis jacobsoni* Linstow 1907, *Rodentotaenia jacobsoni* [Linstow 1907] Yamaguti 1959, *Staphylocystis jacobsoni* [Linstow 1907] Spasskii 1950), in *Crocidura murina;* Java.

*V. johnsoni* (Schiller 1952) comb. n. (syn. *Hymenolepis johnsoni* Schiller 1952, *Rodentolepis johnsoni* [Schiller 1952] Yamaguti 1959), in *Microtus pennsylvanicus;* Canada.

*V. kerivoulae* (Hubscher 1937) Yamaguti 1959, in *Kerivoula picta;* Java.

*V. khalili* (Hilmy 1936) Spasskii 1954, in *Crocidura* sp.; Liberia.

*V. krishna* (Sharma 1943) comb. n. (syn. *Hymenolepis krishna* Sharma 19543, *Vampirolepidoides krishna* [Sharma 1943] Yamaguti 1959), in *Arvicola torquola;* Nepal.

*V. lineola* (Oswald 1951) comb. n. (syn. *Hymenolepis lineola* Oswald 1951, *Rodentolepis lineola* [Oswald 1951] Yamaguti 1959), in *Sorex fumeus;* U.S.

*V. maclaudi* (Joyeux et Baer 1928) Spasskii 1954, in *Crocidura stampflii;* Africa.

*V. macroscelidarum* (Baer 1926) Yamaguti 1959, in *Macroscelides brachyrhynchus;* locality unknown.

*V. macrotesticulatus* (Sawada 1970) comb. n. (syn. *Rodentolepis macrotesticulatus* Sawada 1970), in *Rhinolophus ferrumequinum;* Japan.

*V. macroti* Zdzitowiecki et Rutkowska 1980, in *Macrotus waterhousei;* Cuba.

*V. magnirostellata* (Baer 1931) Spasskii 1954, in *Neomys fodiens.*

*V. malayensis* (Prudhoe et Manger 1969) comb. n. (syn. *Hymenolepis malayensis* Prudhoe et Manger 1969), in "bats", Malaysia.

*V. manidis* (Baer et Fain 1955) comb. n. (syn. *Hymenolepis manidis* Baer et Fain 1955, *Rodentolepis manidis* [Baer et Fain 1955 ] Yamaguti 1959), in *Manis (Smutsia) gigantea;* Africa.

*V. masaldani* Nama et Khichi 1975, in *Rattus rattus;* India.

*V. meszarosi* (Murai et Tenora 1975) comb. n., in *Atticola roylei;* Mongolia.

*V. microstoma* (Dujardin 1845) comb. n. (syn. *Taenia microstoma* Dujardin 1845, *T. brachydera* Diesing 1854, *T. murisdecumani* Diesing 1863, *Rodentolepis microstoma* [Dujardin 1845] Spasskii 1954), in *Mus musculus, Rattus norvegicus, R. rattus, Meriones shawi, Apodemus, Mastomys, Promomys, Dendromus, Leggada, Microtus;* Europe, Africa.

*V. molus* Srivastav et Capoor 1979, in *Crocidura murianus;* India.

*V. montana* Crusz et Sanmugasanderain 1971, in *Suncus murinus;* Sri Lanka.

*V. multihamatus* Sawada 1967, in *Vespertilio superens;* Japan.

*V. murisvariegati* (Janicki 1904) comb. n. (syn. *Hymenolepis murisvariegata* Janicki 1904, *Rodentolepis murisvariegati* [Janicki 1904] Spasskii 1954), in *Mus variegatus;* Egypt.

*V. nana* (Siebold 1852) Spasskii 1954 (syn. *Taenia nana* Siebold 1852, *Diplacanthus nana* [Siebold 1852] Weinland 1858, *Taenia aegyptiaca* Bilharz in Siebold 1852, *V. fraterna* [Stiles 1906] Spasskii 1954, *Hymenolepis murina* Dujardin 1845, *Lepidotrias nana* [Siebold 1852] Weinland 1858, *Hymenolepis nana fraterna* Stiles 1906, *H. intermedius* Bacigalupo 1927, *H. bacigalupoi* Joyeux et Kobozieff 1928), in Man, *Mus, Rattus, Arvicanthus, Cavia, Eliomys, Apodemus, Cricetus, Micromys, Microtus, Myoxus.* Cosmopolitan.

*V. negevi* (Greenberg 1969) comb. n. (syn. *Rodentolepis negevi* Greenberg 1969), in *Acomys cahirinus*; Israel.

*V. neomidis* (Baer 1931) Spasskii 1954, in *Neomys fodiens;* Switzerland.

*V. novadomensis* Rysavy 1961, in *Myotis mystacinus;* Czechoslavakia.

*V. octocoronata* (Linstow 1879) comb. n. (syn. *Taenia octocoronata* Linstow 1879, *Rodentolepis octocoronata* [Linstow 1879] Yamaguti 1959), in *Myopotamus coypus;* South America.

*V. ogaensis* Sawada 1974, in *Rhinolophus ferrumequinum*; Japan.

*V. olsoni* (Neiland et Senger 1952) comb. n. (syn. *Hymenolepis olsoni* Neiland et Senger 1952, *Rodentolepis olsoni* [Neiland et Senger 1952] Yamaguti 1959), in *Scapanus townsendi;* Washington, U.S.

*V. oregonensis* (Neiland et Senger 1952) comb. n. (syn. *Hymenolepis oregonensis* Neiland et Senger 1952, *Rodentolepis oregonensis* [Neiland et Senger 1952] Yamaguti 1959), in *Ondatra zibethica;* Oregon, U.S.

*V. ozensis* Sawada 1980, in *Plecotus auritus;* Japan.

*V. parva* (Rausch et Kuns 1950) comb. n. (syn. *Hymenolepis parva* Rausch et Kuns 1950, *Rodentolepis parva* [Rausch et Kuns 1950] Yamaguti 1959), in *Sorex* sp.; U.S.

*V. pearsei* (Joyeux et Baer 1930) comb. n. (syn. *Hymenolepis pearsei* Joyeux et Baer 1930, *Rodentolepis pearsei* [Joyeux et Baer 1930] Spasskii 1954), in *Hybomys univittatus;* Africa.

*V. peramelidarum* (Nybelin 1917) Spasskii 1954, in *Perameles macrura, P. nasuta, Thylacis obesulus,* Australia.

*V. petrodromi* (Baer 1933) comb. n. (syn. *Hymenolepis petrodromi* Baer 1933, *Rodentolepis petrodromi* [Baer 1933] Spasskii 1954), in *Petrodromus tetradactylus;* Africa.

*V. pipistrelli* (Lopez-Neyra 1941) Spasskii 1954, in *Pipistrellus pipistrellus;* Spain.

*V. roudabushi* (Macy et Rausch 1946), in *Eptesicus, Nycticeius, Lasiomycteris;* Ohio, U.S.

*V. rysavyi* Tenora et Barus 1960, in *Myotis myotis;* Czechoslavakia.

*V. sandgroundi* (Baer 1933) Yamaguti 1959, in *Pipistrellus nanus;* Africa.

*V. schilleri* (Rausch et Kuns 1950) Yamaguti 1959 (syn. *Staphylocystis schilleri* [Rausch et Kuns 1950] Spasskii 1954, *Hymenolepis schilleri* Rausch et Kuns 1950), in *Sorex cinereus, S. vagrans,* U.S.

*V. schmidti* Jensen et Howell 1983, in *Triaenops persicus;* Tanzania.

*V. sinensis* (Oldham 1929) comb. n. (syn.*Rodentotaenia sinensis* [Oldham 1929 ] Spasskii 1954), in *Cricetulus griseus;* China.

*V. solisoricis* Crusz et Sanmugasanderain 1971, in *Solisorex pearsoni;* Sri Lanka.

*V. somariensis* Malhotra et Capoor 1980, in *Turdoides straitus;* India.

*V. spasskii* Andreiko, Skvortsov et Konovalov 1969, in *Nyctalus noctula;* Russia.

*V. srivastavai* (Rego 1970) comb. n. (syn. *Rodentolepis srivastavai* Rego 1970), in *Zygodontomys pixuna;* Brazil.

*V. straminea* (Goeze 1782) comb. n. (syn. *Taenia straminea* Goez 1782, *Rodentolepis straminea* [Goeze 1782] Spasskii 1954, *Hymenolepis mutata* Neveu-Lemaire 1936), in *Rattus, Mus, Arctomys, Cricetulus, Cricetus;* Russia, Europe.

*V. taruiensis* (Sawada 1967) comb. n. (syn. *Rodentolepis taruiensis* Sawada 1967), in *Rhinolophus cornutus;* Japan.

*V. uncinispinosa* (Joyeux et Baer 1930) comb. n. (syn. *Hymenolepis uncinispinosa*

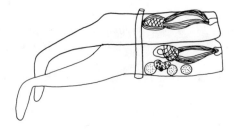

FIGURE 331. *Hispaniolepis villosa* (Bloch 1782) Lopez-Neyra 1942.

Joyeux et Baer 1930, *Rodentolepis uncinispinosa* [Joyeux et Baer 1930] Spasskii 1954), in *Hybomys univittatus, Mastomys erythroleucas;* Nigeria.

*V. virilis* (voge 1955) comb. n. (syn. *Hymenolepis virilis* Voge 1955), in *Sorex trowbridgei;* California, U.S.

*V. wislockii* (Sandground 1938) Yamaguti 1959, in *Solenodon paradoxus;* Santo Domingo.

*V. yoshiyukiae* Sawada 1980, in *Myotis frater;* Japan.

49a. **Antiporal margin of strobila strongly fimbriated** ...............................
.............................................*Hispaniolepis* **Lopez-Neyra 1942 (syn.** ***Hispaniolepidoides*** **Yamaguti 1959,** ***Ortleppolepis*** **Spasskii 1965). (Figure 331)**

*Diagnosis:* Rostellum armed with a single circle of hooks. Suckers simple, unarmed. Neck present. Antiporal margin of proglottid produced into an elongate appendage, becoming longer as the segment matures. Poral margin normal. Testes three, one antiporal, two poral. Cirrus pouch large, with thick muscular wall, commonly reaching median line. Internal and external seminal vesicles present. Ovary median or submedian. Vitellarium postovarian. Vagina may have distal sphincter. Seminal receptacle present. Uterus saccular. Parasites of Gruiformes, Galiformes, Anseriformes; Europe, Africa, Russia.

Type species: *H. villosa* (Bloch 1782) Lopez-Neyra 1942 (syn. *Taenia villosa* Bloch 1782, *T. fimbriata* Batsch 1786, *Hymenolepis villosa* Railliet 1899, *Taenia otidis* Werner 1782, *T. tardae* Goelin 1790, *Hymenolepis cholodkowskyi* Hilmy 1936), in *Otis tarda;* Europe.

Other species:

*H. arcuata* (Kowalewski 1904) Lopez-Neyra 1942 (syn. *Hymenolepis arcuata* Kowalewski 1904, *Weinlandia arcuata* Mayhew 1925, *Microsomacanthus arcuata* [Kowalewski 1904] Spasskii et Spasskaja 1954, *Hispaniolepidoides arcuata* [Kowalewski 1904] Yamaguti 1959), in *Fuligula marila;* Poland.

*H. falsata* (Meggitt 1927) Lopez-Neyra 1942 (syn. *Hymenolepis falsata* Meggitt 1927), in *Otis houbara;* Egypt.

*H. fedtschenkoi* (Solowiow 1911) Lopez-Neyra 1942 (syn. *Hymenolepis fedtschenkoi* Solowiow 1911, *H. gwiletica* Dinnik 1938), in *Tetraogallus himalayensis, T. caucasicus, Lyrurus tetrix, Tetrastes bonasia, Gallus gallus, Megaloperdix nigelli, Numida meleagris;* Russia, Europe, Asia, Africa.

*H. hilmyi* (Skrjabin et Mathevossian 1942) Lopez-Neyra 1942 (syn. *H. tetracis* Hilmy 1936), in *Numida* sp.; Liberia.

*H. kaiseris* (Sharma 1943) Yamaguti 1959 (syn. *Hymenolepis kaiseris* Sharma 1943), in *Caccabis chukar;* Nepal.

*H. multiuncinata* Ortlepp 1963 (syn. *Ortleppolepis multiuncinata* [Ortlepp 1963] Spasskii 1965), in *Guttera edouardi;* Northern Rhodesia.

*H. tetracis* (Cholodkowsky 1906) Spasskii et Spasskaja 1954 (syn. *Hymenolepis tetracis*

Cholodkowsky 1906, *H. dentatus* Clerc 1906, *Weinlandia tetracis* Mayhew 1925), in *Tetrax tetrax;* Europe.

*H. villosoides* (Solowiow 1911) Lopez-Neyra 1942 (syn. *Hymenolepis villosoides* Solowiow 1911), in *Fuligula cristata;* Poland, Switzerland.

49b. Lateral margins of strobila similar ............................................... 50

50a. Poral and most antiporal testes outside osmoregulatory canals, persistent; developing uterus nearly surrounds them............................................
....................................... *Oshmarinolepis* **Spasskii et Spasskaja 1954.**

*Diagnosis:* Suckers unarmed. Rostellum with ten hooks in single circle. Handle and blade of hook nearly equal, guard slightly shorter than blade. Testes three, in transverse row, each outer one outside of excretory stem. Cirrus pouch slender. External and internal seminal vesicles present. Ovary median. Vitellarium postovarian. Uterus extending transversely, later developing outgrowths to embrace to outer testes. Parasites of Ardeiformes, Charadriiformes; Russia, Europe, Africa, Canada, Israel.

Type species: *O. microcephala* (Rudolphi 1819) Spasskii et Spasskaja 1954 (syn. *Taenia microcephala* Rudolphi 1819, *Hymenolepis microcephala* Fuhrmann 1906, *T. leptoptili* Linstow 1901, *T. multiformis* Creplin 1829), in *Ciconia ciconia, Ardea cineria, A. purpurea, Nycticorax nycticorax, Plegadis autumnalis, P. falcinellus, Leptoptilus crumeniferus.*

50b. Testes and uterus not as above .................................................. 51

51a. Genital atrium with accessory sacs .............................................. 52
51b. Genital atrium without accessory sacs (see also *Dicranotaenia coronula,* p. 321) ............................................................................... 60

52a. One accessory sac .............................................................. 53
52b. Two accessory sacs ............................................................. 57

53a. Cirrus stylet and internal seminal vesicle present..................................
....................... *Sobolevicanthus* **Spasskii et Spasskaja 1954. (Figure 332)**

*Diagnosis:* Suckers unarmed. Rostellum with eight or ten long, simple, "skrjabinoid" hooks (Figure 332). Handle of hook usually shorter than blade, guard reduced. Testes three, in triangle or transverse row. Cirrus pouch long, may or may not be spirally coiled, usually overreaching median line; may be longer than segment is wide. Accessory sac usually present in genital atrium. External and internal seminal vesicles present. Cirrus spined, usually with long stylet. Ovary lobed, ventral to testes. Vitellarium postovarian. Uterus saccular. Parasitic in terrestrial and aquatic birds.

Type species: *S. gracilis* (Zeder 1803) Spasskii et Spasskaja 1954 (syn. *Halysis gracilis* [Zeder 1803] *Taenia gracilis* [Zeder 1803] Railliet 1893, *Hymenolepis [Drepanidotaenia] gracilis* Cohn 1901, *H. gracilis* [Zeder 1803] Railliet 1899, *Drepanidotaenia meleagris* Clerc 1902, *Weinlandia meleagris* Mayhew 1925, *Hymenolepis meleagris* Fuhrmann 1932, *Weinlandia gracilis* [Zeder 1803] Meyhew 1925, *Fuhrmanniella gracilis* [Zeder 1803] Tseng-Shen 1932, *Hymenolepis [Weinlandia] gracilis* [Zeder 1803] Neveu-Lemaire 1936, *Sphenacanthus gracilis* [Zeder 1803] Lopez-Neyra 1942, *Dicranotaenia meleagris* [Clerc 1902] Skrjabin et Mathevossian 1945), in *Anas* spp., *Bucephala clangula, Aythya ferina, A. marila, A. nyroca, Netta rufina, Tadorna tadorna, T. ferruginea,*

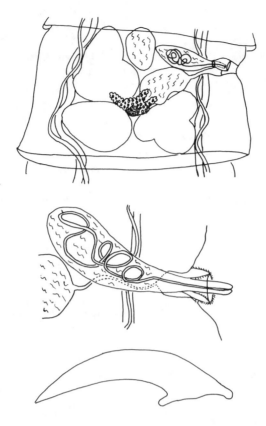

FIGURE 332. *Sobolevicanthus coloradensis* Ahern et Schmidt 1976. Mature proglottid, genital terminalia and skrjabinoid hook.

*Anser anser, Mergus merganser, M. serrator, Oxyura leucocephala, Cygnus cygnus, Phoenicopterus roseus, Fulica atra, Gallus gallus, Meleagris gallopavo;* Russia, E ⇌ pe, North America, Africa, Japan, India.

Other species:

S. *aspirantica* (Zaskind 1959) Maksimova 1963 (syn. *Hymenolepis aspirantica* Zaskind 1959), in *Anser anser, Anas acuta;* Russia.

S. *coloradensis* Ahern et Schmidt 1976, in *Recurvirostra americana;* U.S.

S. *columbae* (Zeder 1800) Spasskii et Spasskaja 1954 (syn. *Taenia columbae* Zeder 1800, *T. turturis* Gmelin 1790, *T. serpentiformis turturis* Fuhrmann 1908, *T. sphenocephala* Rudolphi 1810, *Hymenolepis sphenocephala* Fuhrmann 1908, *Weinlandia sphenocephala* Mayhew 1925, *Hymenolepis columbae* Meggitt 1920), in *Columba livia, Streptopelia turtur, Aplopelia larvata;* Russia, Europe, Africa, India.

S. *cubanus* Rysavy 1966, in *Saurothera merlini;* Cuba.

S. *dafilae* (Polk 1942) Yamaguti 1959 (syn. *Hymenolepis dafilae* Polk 1942, *H. stolli* Schiller 1954 in part), in *Anas crecca, A. platyrhynchos, A. acuta, A. querquedula;* U.S., Russia.

S. *dlouhyi* Czaplinski et Vaucher 1981, in *Amazonetta brasiliensis;* Paraguay.

S. *filumferens* (Brock 1942) Yamaguti 1959 (syn. *Hymenolepis filumferens* Brock 1942), in *Anas discors, A. rubripes, A. crecca;* U.S.

S. *flagellatus* (Fuhrmann 1906) Spasskii et Spasskaja 1954 (syn. *Hymenolepis flagellatus* Fuhrmann 1906), in *Poecilonetta bahamensis;* Brazil, Antigua.

S. *fragilis* (Krabbe 1869) Spasskii et Spasskaja 1954 (syn. *Taenia fragilis* Krabbe 1869,

*Hymenolepis fragilis* Railliet 1899, *Sphenacanthus fragilis* [Krabbe 1869] Lopez-Neyra 1942), in *Anas acuta, A. platyrhynchos, A. strepera, A. crecca, A. querquedula, A. clypeata, Anser anser, A. fabalis, Aythya ferina, A. nyroca, Bucephala clangula, Cygnus olor, Oxyura leucocephala;* Russia, Europe.

*S. gladium* Spasskii et Bobova 1962, in *Aythya marila, Anas crecca, A. formosa, Melanitta deglandi;* Russia.

*S. javanensis* (Davis 1945) Spasskii et Spasskaja 1954 (syn. *Hymenolepis javanensis* Davis 1945), in *Dendrocygna javanica;* Java.

*S. kenaiensis* (Schiller 1952) Czaplinski 1973 (syn. *Hymenolepis kenaiensis* Schiller 1952, *Hymenosphenicanthus kenaiensis* [Schiller 1952] Yamaguti 1959), in *Aythya marilis, Anas formosa, Aythya fuligula, Melanitta migra;* Alaska, Europe, Russia.

*S. krabbeella* (Hughes 1940) Ryjikov 1956 (syn. *Hymenolepis krabbeella* Hughes 1940, *Sobolevicanthus krabbeella* [Hughes 1940] Czaplinski 1956, *Hymenolepis fragilis* [Krabbe 1869] Fuhrmann 1906), in *Anas crecca, A. clypeata, A. formosa, A. platyrhnchos, A. strepera, A. querquedula, Aythya ferina, Bucephala clangula, Aix galericulata;* Russia, Taiwan.

*S. mastigopraeditus* (Polk 1942) Spasskii et Spasskaja 1954 (syn. *Hymenolepis mastigopraeditus* Polk 1942), in *Anas acuta;* U.S.

*S. octacantha* (Krabbe 1869) Spasskii et Spasskaja 1954 (syn. *Taenia octacantha* Krabbe 1869, *Drepanidotaenia octacantha* Clerc 1903, *Hymenolepis octacantha* Railliet 1899, *Weinlandia octacantha* Mayhew 1925, *Sphenacanthus octacantha* [Krabbe 1869] Lopez-Neyra 1942, *Dicranotaenia octacantha* [Krabbe 1869] Skrjabin et Mathevossian 1945), in *Anas* spp., *Aythya ferina, A. fuligula, A. marila, Netta rufina, Bucephala clangula, Melanitta sp., Plectopterus gambensis;* Russia, Europe, Africa.

*S. octacanthoides* (Fuhrmann 1906) Spasskii et Spasskaja 1954 (syn. *Hymenolepis octacanthoides* Fuhrmann 1906, *Dicranotaenia octacanthoides* [Fuhrmann 1906] Skrjabin et Mathevossian 1945, *Sphenacanthus octacanthoides* [Fuhrmann 1906] Lopez-Neyra 1942), in *Larus ridibundus, L. minutus;* Russia, Europe, Africa.

*S. papillatus* (Fuhrmann 1906) Spasskii et Spasskaja 1954 (syn. *Hymenolepis papillatus* Fuhrmann 1906), in *Cairina moschata;* Brazil.

*S. rashidi* Chisthi 1980, in *Anas platyrhynchus;* Kashmir, India.

*S. rugosus* (Clerc 1906) Yamaguti 1959 (syn. *Hymenolepis rugosus* Clerc 1906), in *Columba* sp.; Russia.

*S. serratus birmanicus* Meggitt 1924 (syn. *Hymenolepis joyeuxi* Fuhrmann 1932), in *Columba livia, Streptopelia chinensis, S. orientalis, S. tranquebarica;* Asia, Europe.

*S. stolli* (Brock 1941) Czaplinski 1956 (syn. *Hymenolepis stolli* Brock 1941), in *Anas platyrhynchos, A. acuta, A. formosa;* Russia, U.S.

*S. spasskii* Kornyushin 1969, in *Tadorna tadorna;* Russia.

*S. terraereginae* (Johnston 1911) Yamaguti 1959 (syn. *Hymenolepis terraereginae* Johnston 1911), in *Anseranas semipalmata;* Australia.

*S. wisniewskii* Czaplinski 1956, in *Aythya ferina, Nyroca nyroca;* Poland, Russia.

**53a. Cirrus stylet absent; or if present then internal seminal vesicle is absent .... 54**

**54a. Rostellum with eight hooks ... *Phoenicolepis* Jones et Khalil 1980. (Figure 333)**

*Diagnosis:* Strobila small. Rostellum with eight hooks in single circle, each with rudimentary guard and slightly curved blade. Suckers unarmed, simple. Proglottids numerous, wider than long except when gravid. Numerous pairs of longitudinal muscle bundles. Two pairs of longitudinal osmoregulatory canals ventral to genital ducts. Three testes, one poral, two aporal. External seminal vesicle present, internal seminal vesicle absent. Cirrus pouch

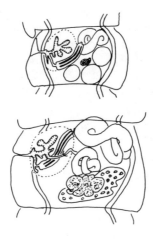

FIGURE 333. *Phoenicolepis nakurensis* Jones et Khalil 1980. Male genitalia (top), female genitalia (bottom).

thin-walled, enclosing robust stylet. Cirrus blunt, covered with minute spines. Accessory sac present, voluminous, eversible, lined with numerous similar spines. Vagina short, opening at apex of small papilla. Seminal receptacle present. Genital atrium deep and convoluted, opening on anterior half of proglottis margin. Uterus sac-like, persistent. Eggs spindle-shaped. Parasites of flamingo; Africa.

Type species: *P. nakurensis* Jones et Khalil 1980, in *Phoeniconaias minor;* Lake Nakuru, Kenya.

**54b. Rostellum with ten hooks ........................................................ 55**

**55a. Accessory sac unarmed, poorly developed ......................................**
**........................................ *Nadejdolepis* Spasskii et Spasskaja 1954.**

*Diagnosis:* Rostellum with a circle of ten nitidoid hooks. Blade of hook long, curved, handle fairly long, slightly curved, guard nearly absent or conical (Figure 345). Testes in transverse row at posterior end of proglottid, one antiporal, two poral, or vice versa. Cirrus pouch crossing median line or not, with thick muscular wall. Cirrus without stylet. Accessory sac present. Ovary two-lobed, compact, Vitellarium postovarian. Uterus saccular. Parasites of Charadriiformes, Passeriformes, Anseriformes.

Type species: *N. nitidulans* (Krabbe 1882) Spasskii et Spasskaja 1954 (syn. *Taenia nitidulans* Krabbe 1882, *Hymenolepis [Echinocotyle] nitidulans* [Krabbe 1882] Fuhrmann 1906, *Echinocotyle nitidulans* [Krabbe 1882] Fuhrmann 1932, *Hymenolepis charadrii* Yamaguti 1935), in *Calidris alpina, Charadrius hiaticula;* Russia, Europe.

Other species:

*N. ansa* Spasskii et Yurpalova 1968, in *Charadriius mongolus;* Russia.

*N. belopolskaiae* (Deblock et Rosé 1962) Spasskaja 1966 (syn. *Hymenolepis belopolskaiae* Deblock et Rosé 1962), in *Crocethia alba;* Russia.

*N. cambrensis* (Davies 1939) Spasskii et Spasskaja 1954 (syn. *Hymenolepis cambrensis* Davies 1939), in *Haematopus ostralegus, Anas platyrhynchos;* England, Russia.

*N. guschanskoi* (Krotov 1952) Spasskii et Spasskaja 1954 (syn. *Dicranotaenia guschanskoi* Krotov 1952, *Hymenolepis guschanskoi* [Krotov 1952] Yamaguti 1959), in *Calidris minuta, Crocethia alba, Calidris alpina, C. ruficollis;* Russia.

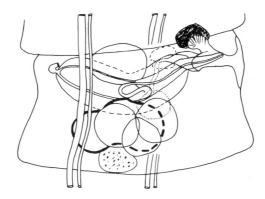

FIGURE 334. *Debloria capetownensis* (Deblock et Rosé 1962) Spasskii 1975.

*N. lauriei* (Davies 1939) Spasskii et Spasskaja 1954 (syn. *Hymenolepis lauriei* Davies 1939), in *Haemantopus ostralegus;* England.

*N. litoralis* (Webster 1947) Spasskaja 1966 (syn. *Hymenolepis [Echinocotyle] litoralis* Webster 1947), in *Crocethia alba, Calidris minuta, C. subminuta, C. ruficollis, C. alpina;* U.S., Russia.

*N. macracanthissima* (Oschmarin 1950) Yamaguti 1959 (syn. *Sphenacanthus macracanthissima* [Oschmarin 1950] Spasskii et Spasskaja 1954, *Hymenolepis macracanthissima* Oschmarin 1950), in *Clangula histeronica;* Russia.

*N. magnisaccis* (Meggitt 1927) Spasskii et Spasskaja 1954 (syn. *Hymenolepis magnisaccis* Meggitt 1927, *Sphenacanthus magnisaccis* [Meggitt 1927] Lopez-Neyra 1942), in *Limonites minuta, Charadrius hiaticula, Erolia maritima;* England, Egypt.

*N. morenoi* Rysavy 1967, in *Arenaria interpres;* Cuba.

*N. musculosa* Belopolskaja 1967, in *Calidris ferruginea;* Russia.

*N. paranitidulans* (Golikova 1959) Spasskii 1962 (syn. *Echinocotyle paranitidulans* Golikova 1959), in *Calidris alpina, Gavia stellata;* Russia.

*N. praeputialis* (Oschmarin 1950) Yamaguti 1959 (syn. *Hymenolepis praeputialis* Oschmarin 1950, *Sobolevicanthus praeputialis* [Oschmarin 1950] Spasskii et Spasskaja 1954), in *Mergus merganser;* Russia.

*N. saguei* Rysavy 1967, in *Crocethia alba;* Cuba.

*N. solowiowi* (Skrjabin 1914) Yamaguti 1959 (syn. *Hymenolepis solowiowi* Skrjabin 1914), in *Fuligula nyroca, Anas acuta, Nyroca ferina;* Russia, Europe.

*N. vallei* (Stossich 1892) Yamaguti 1959 (syn. *Taenia vallei* Stossich 1892, *Hymenolepis vallei* [Stossich 1892] Fuhrmann 1906), in *Tringa minuta;* Italy.

*N. viguerasi* Rysavy 1967, in *Crocethia alba;* Cuba.

**55b. Accessory sac armed with spines .............................................. 56**

**56a. Powerful vaginal sphincter present ........ *Debloria* Spasskii 1975. (Figure 334)**

*Diagnosis:* Rostellum with ten falciform or nitidoid hooks (lacking guard). Suckers unarmed. Short neck present. Strobila craspedote, anapolytic. Genital pores unilateral. Genital ducts dorsal to osmoregulatory canals. Genital atrium deep, with anterior or dorsal accessory sac lined with spines. Cirrus pouch very large, exceeding midline of segment. Testes three, in straight line or slight triangle. Cirrus armed or not. Internal and external seminal vesicles present. Ovary compact, wider than long. Vagina posterior or dorsal to cirrus pouch, with powerful sphincter near distal end. Seminal receptacle present. Uterus saccate. Parasites of Charadriiformes. Europe, Russia.

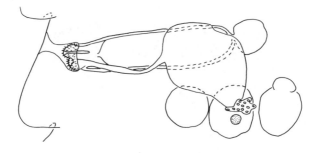

FIGURE 335. *Anatinella tenuirostris* (Rudolphi 1819) comb. n.

Type species: *D. capetownensis* (Deblock et Rosé 1962) Spasskii 1975 (syn. *Hymenolepis [H.] capetownensis* Deblock et Rosé 1962), in *Crocethia alba;* South Africa.
Other species:

*D. crocethia* (Belopolskaia 1953) Spasskii 1975 (syn. *Dicranotaenia crocethia* Belopolskaia 1953, *Nadejdolepis crocethia* [Belopolskaia 1953] Yamaguti 1959, *Hymenolepis [H.] crocethia* Deblock et Rosé 1962, *Echinocotyle crocethia* [Belopolskaia 1953] Spasskaja 1966), in *Crocethia alba;* Russia.

*D. etaplesensis* (Deblock et Rosé 1962) Spasskii 1975 (syn. *Hymenolepis [H.] etaplesensis* Deblock et Rosé 1962), in *Limosa lapponica, Crocethia alba;* France, Russia.

**56b. Vaginal sphincter absent ............................................... *Anatinella* Spasskii et Spasskaja 1954 (syn. *Monosaccanthes* Czaplinski 1967). (Figure 335)**

*Diagnosis:* Scolex with four unarmed suckers and retractable rostellum armed with single circle of ten hooks (handle longer than blade, blade longer than guard). Neck present. Genital pores unilateral. Two pairs of osmoregulatory canals. Genital ducts dorsal to osmoregulatory canals. Three testes in posterior half of proglottid, in transverse row, or with antiporal testis slightly anterior to others. Cirrus pouch well developed; internal and external seminal vesicles present. Base of genital atrium spined; part of it invaginable to form "accessory genital sac," which may protrude into atrium. Cirrus small, armed, lacking stylet. Ovary median, lobated. Vitellarium postovarian, lobated. Uterus forms as transverse sac, fills to become large, irregular sac. Intermediate hosts are freshwater crustaceans. Parasites of Anseriformes, Gaviiformes and Charadriiformes. Europe, Russia, Australia, North America, Africa.

Type species: *A. spinulosa* (Dubinina 1953) Spasskii 1963 (syn. *Drepanidotaenia spinulosa* Dubinina 1953), in *Anas crecca, A. acuta, A. clypeata, A. platyrhynchos, A. querquedula, Netta rufina, Bucephala clangula, Aythya marila;* Russia, Europe.
Other species:

*A. tenuirostris* (Rudolphi 1819) comb. n. (syn. *Taenia tenuirostris* Rudolphi 1819, *Monosaccanthes tenuirostris* [Rudolphi 1819] Czaplinski 1967, *Drepanidotaenia tenuirostris* [Rudolphi 1819] Railliet 1893, *Cysticercus tenuirostris* Rosseter 1897, *Hymenolepis [Drepanidotaenia] tenuirostris* [Rudolphi 1819] Cohn 1901, *Drepanidotaenia tenuirostris* [Rudolphi 1819] Rosseter 1903, *Hymenolepis tritesticulata* Fuhrmann 1907, *H. tenuirostris* [Rudolphi 1819] Lühe 1919, *Dicranotaenia tenuirostis* [Rudolphi 1819] Skrjabin et Mathevossian 1945, *Microsomacanthus tenuirostris* [Rudolphi 1819] Yamaguti 1959, *M. tritesticulatus* [Fuhrmann 1907] Yamaguti 1959), in *Mergus merganser, M. serrator, Mergellus albellus, Anas platyrhynchos;* Europe, Asia.

*A. brachycephala* (Creplin 1829) comb. n. (syn. *Taenia brachycephala* Creplin 1829, *Monosaccanthes brachycephala* [Creplin 1829] Czaplinski 1967, *Hymenolepis brachycephala* [Creplin 1829] Skrjabin et Mathevossian 1945, *H. [Echinocotyle] brachycephala* [Creplin 1829] Deblock 1964, *H. [H.] brachycephala* [Creplin 1829]

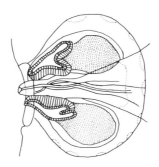

FIGURE 336. *Bisaccanthes bisaccatus* (Fuhrmann 1906) Spasskii et Spasskaja 1954. Genital atrium with associated sacs.

Railliet 1899, *Tschertkovilepis brachycephala* [Creplin 1829] Yamaguti 1959), in Charadriiformes; Europe, Africa.

*A. curiosa* (Szpotanska 1931) comb. n. (syn. *Drepanidotaenia curiosa* Szpotanska 1931, *Monosaccanthes curiosa* [Szpotanska 1931] Czaplinski 1967), in *Chenopsis atrata;* Australia.

57a. Accessory sacs unarmed...................................................... 58
57b. Accessory sacs armed with spines ............................................ 59

**58a. Cirrus without stylet ... *Bisaccanthes* Spasskii et Spasskaja 1954. (Figure 336)**

*Diagnosis* Rostellum with eight hooks; handle and blade equal, guard button-like. Suckers unarmed. Proglottids elongated transversely. Two pairs of osmoregulatory canals. Testes three, one poral, two antiporal, somewhat diagonal. Cirrus pouch very short, not reaching poral canals. Two accessory sacs opening into genital atrium, each with its own envelope, and enclosed in a common sac. Ovary multilobed, median, vitellarium postovarian. Uterus? Parasites of Anseriformes.

Type species: *B. bisaccatus* (Fuhrmann 1906) Spasskii et Spasskaja 1954 (syn. *Hymenolepis bisaccatus* Fuhrmann 1906, *Drepanidotaenia [Drepanidolepis] bisaccata* [Fuhrmann 1906] Lopez-Neyra 1942, *Sobolevicanthus bisaccata* [Fuhrmann 1906] Czaplinski 1956), in *Anas platyrhynchos, A, acuta, A. formosa, A. acuta, A. crecca, A. falcata, A. penelope, A. strepera, Anser fabalis, Cairina moschata, Netta rufina, Nettion brasiliense, Nyroca fuligula, N. nyroca, Casarca ferruginea;* Russia, Europe, Brazil.

**58b. Cirrus with stylet............. *Parabiglandatrium* Gvozdev et Maksimova 1968.**

*Diagnosis:* With four simple suckers and rostellum armed with eight hooks, each with very short guard (skrjabinoid, Figure 338a). Rostellar sac extending well into neck. Neck present. All segments wider than long, craspedote. Genital pores unilateral. Genital atrium deep, with anterior and posterior unarmed accessory sacs. Cirrus pouch large, exceeding osmoregulatory canals, containing internal seminal vesicle. Cirrus thin, armed, containing slender stylet. External seminal vesicle present. Ovary large, multilobate, median. Vitellarium postovarian. Vagina? Uterus an irregular sac, exceeding osmoregulatory canals. Eggs elongate. Parasites of Phaenicopteriformes. Russia.

Type species: *P. phaenicopteri* Gvosdev et Maksimova 1968, in *Phoenicopterus roseus;* Kazakhistan, Russia.

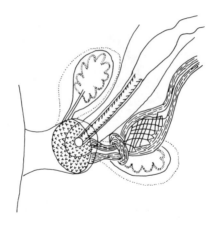

FIGURE 337.  *Biglandatrium biglandatrium* Spasskaja 1961.

**59a. Rostellum with eight hooks ................ *Parabisaccanthes* Maksimova 1963.**

*Diagnosis:* Scolex small. Rostellum with single circle of eight hooks. Rostellar sac extending behind suckers, which are unarmed. Neck very short. First several immature segments broadly expanded to form a cobra-like hood, then narrowing to a typical strobila. Proglottids wider than long, craspedote. Genital pores unilateral, preequatorial; ducts pass between osmoregulatory canals. Genital atrium deep, with anterior and posterior depressions at its depth, each lined with tiny spines. Cirrus pouch very long, exceeding midline of segment. Cirrus slender, armed. Testes three, in shallow triangle. Internal and external seminal vesicles present. Ovary with many short lobes, antiporal. Vitellarium lobated, postovarian. Vagina anterior to cirrus pouch. Seminal receptacle large. Uterus a lobated, transverse sac, often bulging through lateral walls of proglottid. Parasites of Anseriformes (swans); Russia, Europe, Australia, New Zealand

Type species: *P. philactes* (Schiller 1951) Spasskii et Resnik 1963 (syn. *P. cygni* Maksimova 1963, *Hymenolepis cygni* Schiller 1951), in *Cygnus olor, Philacte candica;* Alaska, Russia.

Other species:

*P. bisacculina* (Szpotanska 1931) Maksimova 1963 (syn. *Drepanidotaenia bisacculina* Szpotanska 1931), in *Chenopis atrata, Cygnus* spp.; Australia, Europe.

*P. kazachstanica* Maksimova 1963, in *Cygnus olor;* Russia.

**59b. Rostellum with ten hooks ......... *Biglandatrium* Spasskaja 1961. (Figure 337)**

*Diagnosis:* Scolex with long rostellum armed with ten hooks. Guard of hook shorter than handle and spine. Suckers simple, unarmed. Neck present. Proglottids wider than long, craspedote. Genital pores equatorial, unilateral. Genital atrium deep. In its anterior and posterior depths are a pair of deep glandular pockets, each with a single duct opening into the atrium. The atrium is spined in its fundus, as is the cirrus. Cirrus pouch long, reaching midline of segment. Internal and external seminal vesicles present. Testes three, in triangle, two aporal and one poral. Ovary small, lobated, median. Vitellarium lobated, postovarian. Vagina muscular, with sphincter near distal end, followed immediately by seminal receptacle. Uterus? Parasites of Gaviiformes. Russia.

Type species: *B. biglandatrium* Spasskaja 1961, in *Gavia arctica;* Russia.

**60a. Ovary reticular ....... *Avocettolepis* Spasskii et Kornyushin 1971. (Figure 338b)**

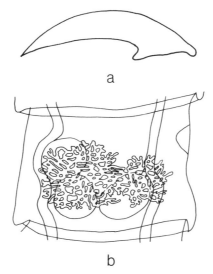

FIGURE 338. *Avocettolepis vaginata* (Baczynska 1914) Spasskii et Kornyushin 1971. (a) Skrjabinoid hook and (b) reticular ovary.

*Diagnosis:* Solex with rostellum armed with eight skrjabinoid hooks (Figure 338a). Neck present. Mature proglottids slightly wider than long, craspedote. Genital pores unilateral, preequatorial. Cirrus pouch well developed, exceeding poral osmoregulatory canals. Internal seminal vesicle present, external seminal vesicle present, but weakly developed. Testes three, in triangle, two aporal, one poral. Ovary reticular, persistent. Vitellarium ventral to posterior branches of ovary. Vagina somewhat coiled, opening posterior to cirrus pouch. Uterus an irregular sac. Parasites of Charadriiformes. Africa.

Type species: *A. vaginata* (Baczynska 1914) Spasskii et Kornyushin 1971 (syn. *Hymenolepis vaginata* Baczynska 1914, *Weinlandia vaginata* [Baczynska 1914] Mayhew 1925, *Sphenacanthus vaginatus* [Baczynska 1914] Lopez-Neyra 1942, *Hymenosphenacanthus vaginata* [Baczynska 1914] Yamaguti 1959, *Hymenolepis innominata* Meggitt 1927, *Microsomacanthus innominata* [Meggitt 1927] Lopez-Neyra 1942), in *Recurvirostra avocetta;* Egypt.

**60b.  Ovary not reticular............................................................. 61**

**61a.  Internal and external seminal vesicles absent ................................. 62**
**61b.  At least one seminal vesicle present ............................................ 63**

**62a.  Cirrus pouch absent, replaced by a complex ejaculatory duct ...................**
**.........................................................................*Cladogynia* Baer 1937.**

*Diagnosis:* Rostellum with single circle of hooks. Suckers unarmed. Neck present. Testes three. External and internal seminal vesicles absent. Cirrus pouch absent, replaced by apparatus formed by a long, strongly muscular proximal ejaculatory duct enclosed in sheath, and a sclerotized dart contained in a pocket which opens on a papilla into the genital atrium. Genital ducts dorsal to osmoregulatory canals. Genital pores unilateral. Ovary strongly branched, extending transversely in ventral medulla across excretory canals. Vitellarium strongly branched, median. Uterus reticular, dorsal to ovary, crossing lateral canals ventrally. Vagina ventral to male apparatus. Seminal receptacle elongate, almost cylindrical. Parasites of Phoenicopteriformes. (Placed in Dilepididae by some authors.)

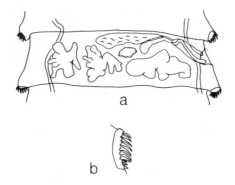

FIGURE 339. *Pararetinometra lateralacantha* Stock et Holmes 1981. (a) Mature proglottid and (b) cluster of lateral spines.

Type species: *C. phoeniconaiadis* (Hudson 1934) Baer 1937 (syn. *Hymenolepis phoeniconaiadis* Hudson 1934), in *Phoeniconaias minor;* Africa.

**62b. Cirrus pouch present................... *Dilepidoides* Spasskii et Spasskaja 1954.**

*Diagnosis:* Rostellum with single circle of numerous hooks. Handle of hook nearly as long as blade, guard prominent. Suckers unarmed. Proglottids not numerous. Testes three, in transverse row, one poral and two antiporal. Cirrus pouch long, slender or claviform. Cirrus with six sizes of spines; with tuft of spines at tip. External seminal vesicle absent. Vas deferens strongly convoluted. Ovary bilobed, median. Vitellarium postovarian. Vagina strongly developed, thick-walled. Seminal reticulum rounded. Parasites of Galliformes. Asia.

Type species: *D. bouchei* (Joyeux 1924) Spasskii et Spasskaja 1954 (syn. *Hymenolepis bouchi* Joyeux 1924), in *Gallus gallus;* Indochina.

**63a. Cirrus with stylet................................................................ 64**
**63b. Cirrus without stylet ........................................................... 65**

**64a. Posterolateral margins of velum of each segment with a row of spines..........**
**.......................... *Pararetinometra* Stock et Holmes 1982. (Figure 339)**

*Diagnosis:* Scolex with retractable rostellum bearing eight hooks with large blade and very short guard. Neck absent. Proglottids craspedote; posterolateral margins of velum with up to 20 crescent-shaped spines. Two pairs of osmoregulatory canals. Testes three, markedly lobated, in transverse row. Cirrus pouch containing stylet. Internal and external seminal vesicles present. Ovary median, strongly lobated. Vitellarium not described. Vagina sclerotized, funnel-shaped at distal end. Uterus a transverse, lobated sac. Mature eggs not described. Parasites of grebes. Canada.

Type species: *P. lateralacantha* Stock et Holmes 1982, in *Podiceps grisegena, P. nigricollis;* Canada.

**64b. Spines absent from margins of velum ................................*Retinometra*  Spasskii 1955 (syn. *Hymenosphenacanthus* Lopez-Neyra 1958, *Hymenosphenacanthoides* Yamaguti 1959, *Sphenacanthus* Lopez-Neyra 1942, *Sphenacanthus [Retinometra]* Spasskii 1955, *Stylopis* Yamaguti 1959). (Figure 340)**

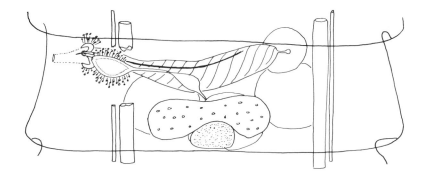

FIGURE 340. *Retinometra deblocki* (Schmidt et Neiland) comb. n.

*Diagnosis:* Rostellum with a circle of eight to ten large, curved hooks. Testes three, in a transverse row or triangle. Cirrus pouch often large. Cirrus with a long narrow stylet. External and internal seminal vesicles present. Accessory sac absent. Vitellarium usually median, postovarian. Ovary usually strongly lobed, median. Gravid uterus occupying whole medulla. Parasites of Anseriformes, Lariformes, Phoenicopteriformes, Galliformes, Columbiformes, Charadriiformes. Europe, Asia, North and South America, Cuba, Japan, Taiwan.

Type species: *R. giranensis* (Sugimoto 1934) Spasskii 1963 (syn. *Hymenolepis giranensis* Sugimoto 1934, *Sphenacanthus giranensis* [Sugimoto 1934] Lopez-Neyra 1942, *Hymenosphenacanthus giranensis* [Sugimoto 1934] Lopez-Neyra 1958, *Sphenacanthus (Retinometra giranensis)* [Sugimoto 1934] Spasskii 1955, *Hymenosphenacanthus giranensis* [Sugimoto 1934] Yamaguti 1959), in *Anas platyrhynchos, A. clypeata, A. querquedula, Eucephala clangula;* Taiwan, Russia.

Other species:

*R. aleuti* (Oschmarin 1950) Spasskii 1963 (syn. *Hymenolepis aleuti* Oschmarin 1950, *Sphenacanthus aleuti* [Oschmarin 1950] Spasskii et Spasskaja 1954, *Hymenosphenacanthus aleuti* [Oschmarin 1950] Yamaguti 1959), in *Larus argentatus;* Russia.

*R. bulbocirrosus* Pfeiffer 1960, in *Cygnus melanocoryphus;* Argentina.

*R. caroli* (Parona 1887) comb. n. (syn. *Taenia caroli* Parona 1887, *Hymenolepis caroli* [Parona 1887] Lopez-Neyra 1942, *Hymenosphenacanthus caroli* [Parona 1887] Yamaguti 1959), in *Phoenicopterus* spp.; Europe, Cuba.

*R. chinensis* Yun 1982, in *Anas platyrhynchus;* China.

*R. cirrostilifer* (Vigueras 1941) comb. n. (syn. *Hymenolepis cirrostilifer* Vigueras 1941, *Hymenosphenacanthus cirrostilifer* [Vigueras 1941] Yamaguti 1959), in *Phoenicopterus ruber;* Cuba.

*R. cyrtoides* (Mayhew 1925) comb. n. (syn. *Weinlandia cyrtoides* Mayhew 1925, *Hymenolepis cyrtoides* [Mayhew 1925] Fuhrmann 1932, *Sphenacanthus cyrtoides* [Mayhew 1925] Lopez-Nayra 1942, *Hymenosphenacanthus cyrtoides* [Mayhew 1925] Yamaguti 1959), in *Erismatura jamaicensis;* U.S.

*R. deblocki* (Schmidt et Neiland 1968) comb. n. (syn. *Hymenolepis [H.] deblocki* Schmidt et Neiland 1968), in *Limnodromus griseus;* Alaska, U.S.

*R. exiguus* (Yoshida 1910) comb. n. (syn. *Hymenolepis exiguus* Yoshida 1910, *Sphenacanthus exiguus* [Yoshida 1910] Lopez-Neyra 1942, *Hymenosphenacanthus exiguus* [Yoshida 1910] Yamaguti 1959), in *Gallus gallus;* Japan, Burma, Hawaii.

*R. fimula* (Meggitt 1933) comb. n. (syn. *Hymenolepis fimula* Meggitt 1933, *Sphenacanthus fimula* [Meggitt 1933] Lopez-Neyra 1942, *Hymenosphenacanthus fimula* [Meggitt 1933] Yamaguti 1959), in *Nyroca ferina;* India.

*R. fista* (Meggitt 1933) comb. n. (syn. *Hymenolepis fista* Meggitt 1933, *Sphenacanthus fista* [Meggitt 1933] Lopez-Neyra 1942, *Hymenosphenacanthus fista* [Meggitt 1933] Yamaguti 1959), in *Nettapus coromandelianus;* India.

R. *fulicatrae* Czaplinska et Czaplinski 1972, in *Fulica atra;* Poland.

R. *guberiana* Czaplinski 1965, in *Cygnus olor;* Poland.

R. *hamulacanthos* (Linton 1927) comb. n. (syn. *Hymenolepis hamulacanthos* Linton 1927, *Sphenacanthus hamulacanthos* [Linton 1927] Lopez-Neyra 1942, *Hymenosphenacanthus hamulacanthos* [Linton 1927] Yamaguti 1959), in *Marila americana;* Massachusetts, U.S.

R. *indica* (Pandey et Tayal 1981) comb. n. (syn. *Stylolepis indica* Pandey et Yayal 1981), in *Uroloncha malabarica;* India.

R. *lari* (Yamaguti 1940) comb. n. (syn. *Sphenacanthus lari* [Yamaguti 1940] Spasskii et Spasskaja 1954, *Hymenolepis lari* Yamaguti 1940, *Hymenosphenacanthus lari* [Yamaguti 1940] Yamaguti 1959), in *Larus argentatus;* Japan.

R. *lintoni* (Lopez-Neyra 1932) Spasskii 1963 (syn. *Hymenolepis macracanthos* Linton 1927 after Linstow 1877, *Sphenacanthus lintoni* [Lopez-Neyra 1932] Lopez-Neyra 1942, *Hymenosphenacanthus lintoni* [Lopez-Neyra 1932] Yamaguti 1959), in *Histrionicus histrionicus, Mergus serrator;* U.S.

R. *longicirrosa* (Fuhrmann 1906) Spasskii 1963 (syn. *Hymenolepis longicirrosa* Fuhrmann 1906, *Hymenosphenacanthus longicirrosa* [Fuhrmann 1906] Yamaguti 1959), in *Anser anser, A. fabalis, A. segetum, Anas crecca; Cygnopsis cygnoides, Aythya ferina, Anser albifrons, A. erythropus, Anas acuta, A. platyrhynchos, A. penelope, Cygnus olor;* Russia, Europe, Asia, India.

R. *longistylosa* (Tseng-Shen 1932) Spasskii 1963 (syn. *Hymenolepis longistylosa* Tseng-Shen 1932, *Sphenacanthus longistylosa* [Tseng-Shen 1932] Spasskii et Spasskaja 1954, *Stylolepis longistylosa* [Tseng-Shen 1932] Yamaguti 1959), in *Anser fabalis;* Russia, China.

R. *longivaginata* (Fuhrmann 1906) Spasskaja 1966 (syn. *Hymenolepis longivaginata* Fuhrmann, 1906, *Hymenosphenacanthus longivaginata* [Fuhrmann 1906] Yamaguti 1959), in *Branta leucopsis, Aix galericulata, Anas platyrhynchos, A. crecca;* Russia, Arctic, Europe.

R. *macracanthos* (Linstow 1877) Spasskii 1963 (syn. *Taenia macracanthos* Linstow 1877, *Hymenolepis macracanthos* [Linstow 1877] Fuhrmann 1906, *H. macracanthoides* Lopez-Neyra 1932, *Sphenacanthus macracanthos* [Linstow 1877] Lopez-Neyra 1942, *Hymenosphenacanthus macracanthos* [Linstow 1877] Yamaguti 1959, *Sphenacanthus macracanthoides* [Lopez-Neyra 1932] Lopez-Neyra 1942), in *Anas platyrhynchos, A. acuta, A. strepera, Mergus merganser, M. serrator, Bucephala clangula, Aythya ferina, A. marila, A. nyroca, A. fuligula, A. leucocephala, Somateria spectabilis, Colymbus cristatus;* Russia, Europe, U.S.

R. *meggitti* (Sharma 1943) comb. n. (syn. *Hymenolepis meggitti* Sharma 1943, *Hymenosphenacanthus meggitt* [Sharma 1943] Yamaguti 1959), in *Anas platyrhynchos;* Burma.

R. *ondatrae* Rider et Macy 1947, comb. n. (syn. *Hymenolepis ondatrae* Rider et Macy 1947, *Hymenosphenacanthoides ondatrae* [Rider et Macy 1947] Yamaguti 1959), in *Ondatra zibethica;* Oregon, U.S.

R. *oshimai* (Sugimoto 1934) Spasskii 1963 (syn. *Hymenolepis oshimai* Sugimoto 1934, *Sphenacanthus oshimai* [Sugimoto 1934] Spasskii et Spasskaja 1954, *Hymenosphenacanthus oshimai* [Sugimoto 1934] Yamaguti 1959), in *Anas platyrhynchos;* Taiwan.

R. *oxyuri* Maksimova in Spasskaja 1966, in *Oxyura leucocephala;* Russia.

R. *pantayi* comb. n. (syn. *Stylolepis meggitti* Pandey et Tayal 1981), in *Streptopelia chinensis;* India.

R. *pauciovatus* (Fuhrmann 1906) comb. n. (syn. *Hymenolepis pauciovatus* Fuhrmann

1906, *Hymenosphenacanthus pauciovatus* [Fuhrmann 1906] Yamaguti 1959), in *Crypturus erythropus;* Brazil.

*R. pittalugai* (Lopez-Neyra 1932) Spasskii 1963 (syn. *Hymenolepis pittalugai* Lopez-Neyra 1932, *Sphenacanthus pittalugai* [Lopez-Neyra 1932] Lopez-Neyra 1942, *Hymenosphenacanthus pittalugai* [Lopez-Neyra 1932] Yamaguti 1959), in *Bucephala clangula, Aythya ferina, Aythya marila, Anas platyrhynchos;* Russia, Spain.

*R. rangoonicus* (Sharma 1943) comb. n. (syn. *Hymenolepis rangoonicus* [Sharma 1943] *Hymenosphenicanthus rangoonicus* [Sharma 1943] Yamaguti 1959), in *Anas platyrhynchos;* Burma.

*R. serrata* (Fuhrmann 1906) Spasskii 1963 (syn. *Hymenolepis serratus* Fuhrmann 1906, *Sobolevicanthus serratus* (Fuhrmann 1906) Yamaguti 1959; in *Turtur turtur, Streptopelia, Columba, Stigmatopelia, Cena*; Europe, Africa, Asia.

*R. skrjabini* (Mathevossian 1945) Spasskii 1963 (syn. *Hymenolepis skrjabini* Mathevossian 1945, *Sphenacanthus skrjabini* [Mathevossian 1945] Spasskii et Spasskaja 1954, *Hymenosphenacanthus skrjabini* [Mathevossian 1945] Yamaguti 1959), in *Aythya ferina, A. fuligula, A. marila, Bucephala clangula, Anas platyrhynchos, A. penelope, A. strepera, Tadorna ferruginea;* Europe, Russia.

*R. tenerrimus* (Linstow 1882) comb. n. (syn. *Taenia tenerrimus* Linstow 1882, *Sphenacanthus tenerrimus* [Linstow 1882] Yamaguti 1959), in *Fuligula cristata, Nyroca marila, Aythya affinis;* Europe, U.S.

*R. venusta* (Rosseter 1897) Spasskaja 1966 (syn. *Taenia venusta* Rosseter 1897, *Drepanidotaenia venusta* Rosseter 1898, *Hymenolepis venusta* [Rosseter 1897] Fuhrmann 1932, *Sphenacanthus venusta* [Rosseter 1897] Lopez-Neyra 1942, *Hymenosphenacanthus venustus* [Rosseter 1897] Yamaguti 1959), in *Anas platyrhynchos, A. acuta, Aythya ferina, Netta rufina;* Russia, Europe, Norway, England.

65a. Tip of rostellum with 8 to 11 conspicuous lobes, each bearing rosethorn-shaped hooks .................................................. **Lobatolepis Yamaguti 1959.**

*Diagnosis:* Scolex spherical, suckers unarmed. Rostellum long, slender, its tip with knoblike enlargement, which has deep marginal lobes, each carrying a hook on the lateral margin. Rostellar hooks 8 to 11, rosethorn-shaped. Handle of hook reduced to knob, guard stout, longer than blade. Proglottids much wider than long throughout strobila. Testes three, in transverse row, one antiporal, two poral to ovary. Cirrus pouch reaching poral osmoregulatory canals. External and internal seminal vesicles present. Ovary transversely elongated, lobed, between and posterior to antiporal testis and two poral testes, slightly antiporal to median line. Vitellarium postovarian. Vagina thin-walled, undulating. Seminal receptacle present. Uterus a transverse sac, lobated. Parasites of grebes.

Type species: *L. lobulata* (Mayhew 1925) Yamaguti 1959 (syn. *Hymenolepis lobulata* Mayhew 1925), in *Podilymbus podiceps;* Michigan, U.S.

65b. Rostellum otherwise ....................................................... 66

66a. Genital atrium very deep, lined with spines at its bottom ........................ ........................ **Echinatrium Spasskii et Yurpalova 1965. (Figure 341)**

*Diagnosis:* Rostellum with ten hooks in single circle. Suckers unarmed. Proglottids wider than long, craspedote. Genital pores unilateral. Cirrus pouch well developed, reaching midline of proglottid. Genital atrium very deep, lined with spines at its bottom. Cirrus spined. Internal and external seminal vesicles well developed. Testes three, one poral, two aporal, in oblique triangle. Ovary lobated, usually with three lobes. Vitellarium simple,

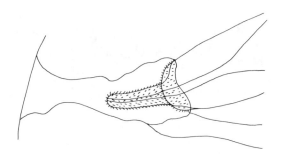

FIGURE 341. *Echinatrium skrjabini* Spasskii et Yurpalova 1965.

FIGURE 342. Aploparaxoid hook.

FIGURE 343. *Dicranotaenia cervotestis* (Ahern et Schmidt 1976) comb. n. Rostellar hook and mature proglottid.

postovarian. Vagina posterior to cirrus pouch. Seminal receptacle prominent. Uterus an irregular, transverse sac. Eggs numerous. Parasites of Anseriformes. Asia.

Type species: *E. skrjabini* Spasskii et Yurpalova 1965, in *Melanitta americana*, *E. deglandi*, *Nyroca marila;* Russia.

Other species:

*E. clanguli* Tolkcheva 1971, in *Clangula hyemalis;* Russia.

*E. filosomum* Spasskii et Yurpalova 1965, in *Nyroca marila;* Russia.

**66b. Genital atrium otherwise.........................................................67**

**67a. Rostellar hooks aploparaxoid (Figure 342), with handle reduced, guard and blade well-developed and about parallel ...................................................**
.......................*Dicranotaenia* **Railliet 1892 (syn.** *Chelacanthus* **Yamaguti 1959,** *Colymbilepis* **Spasskaja 1966,** *Limnolepis* **Spasskii et Spasskaja 1954,** *Wardium* **Mayhew 1925,** *Weinlandia* **Mayhew 1925). (Figure 343)**

*Diagnosis:* Rostellum with single circle of aploparaxoid hooks. Suckers unarmed. Proglottids numerous. Testes three, in triangle or transverse row; one poral, other antiporal. Cirrus pouch usually short, not reaching median line, but occasionally longer. External and internal seminal vesicles present. Ovary lobated. Vitellarium postovarian. Uterus transversely elongated, saccular. Parasites of birds.

Type species: *D. coronula* (Dujardin 1845) Railliet 1892 (syn. *Taenia coronula* Dujardin 1845, *Diplacanthus [Lepidotrias] coronula* [Dujardin 1845] Cohn 1899, *Hymenolepis coronula* [Dujardin 1845] Railliet 1899, *Drepanidotaenia coronula* [Dujardin 1845] Parona 1899, *Hymenolepis [Drepanidotaenia] coronula* [Dujardin 1845] Cohn 1901, *H. megalhystera* Linstow 1905, *Weinlandia coronula* [Dujardin 1845] Mayhew 1925, *Weinlandia introversa* Mayhew 1925, *Weinlandia macrostrobilodes* Mayhew 1925, *Hymenolepis sacciperum* Mayhew 1925, *H. anceps* Linton 1927, *H. introversa* [Mayhew 1925] Fuhrmann 1932, *H. sacciperum* Fuhrmann 1932, *Dicranotaenia deglandi* Skrjabin et Mathevossian 1942, *D. anceps* [Linton 1927] Lopez-Neyra 1942, *Hymenolepis mergi* Yamaguti 1940, *H. apicaris* Sharma 1943, *H. parvisaccata* Shepard 1943, *Dicranotaenia kutassi* Mathevossian 1945, *D. mergi* [Yamaguti 1940] Skrjabin et Mathevossian 1945, *D. coronula micracantha* Skrjabin et Mathevossian 1945, *D. pseudocoronula* Skrjabin et Mathevossian 1945, *Hymenolepis makundi* Singh 1952, *Dicranotaenia makundi* [Singh 1952] Yamaguti 1959, *D. parvisaccata* [Shepard 1943] Yamaguti 1959), in *Anas, Aix, Anser, Branta, Clangula, Dendrocygna, Mergus, Nyroca, Netta, Oidemia, Oxyura, Somateria, Tadorna, Gallus, Fulica;* Russia, Australia, South America, North America, Asia, Japan.

Other species:

*D. aberrata* Ehlert 1969, in *Branta bernica;* Germany.

*D. aequabilis* (Rudolphi 1810) Railliet 1893 (syn. *Taenia aequabilis* Rudolphi 1810, *Hymenolepis aequabilis* [Rudolphi 1810] Railliet 1899, *Drepanidotaenia aequabilis* [Rudolphi 1810] Cohn 1900, *Hymenolepis [Drepanidotaenia] aequabilis* [Rudolphi 1810] Cohn 1901, *Drepanidotaenia musculosa* Clerc 1902, *Hymenolepis musculosa* [Clerc 1902] Fuhrmann 1906, *Dicranotaenia [D.] aequabilis* [Rudolphi 1810] Lopez-Neyra 1942, *D. [D.] musculosa* [Clerc 1902] Lopez-Neyra 1942, *D. musculosa* [Clerc 1902] Yamaguti 1959), in *Cygnus, Anser, Anas, Aythya, Bucephala, Meleagris gallopavo;* Russia, Europe.

*D. alcippina* Srivastava et Capoor 1975, in *Alcippe poioicephala;* India.

*D. amphitricha* (Rudolphi 1819) Lopez-Neyra 1942 (syn. *Taenia amphitricha* Rudolphi 1819, *Hymenolepis amphitricha* Railliet 1899, *H. [Drepanidotaenia] amphitricha* Clerc 1903, *Weinlandia amphitricha* Mayhew 1925, *Wardium amphitricha* Belopolskaja 1970, *Limnolepis amphitricha* [Rudolphi 1819] Spasskii et Spasskaja 1954), in *Tringa hypoleucos, T. totanus, T. glareola, T. ochropus, T. stagnatilis, Calidris temmincki, C. alpina, C. maritima, Scolopax rusticola, Capella gallinago, Aythya marila;* Russia, Europe, North America.

*D. annandalei* (Southwell 1922) Lopez-Neyra 1932 (syn. *Hymenolepis annandalei* Southwell 1922, *Limnolepis annandelei* [Southwell 1922] Spasskii et Spasskaja 1954, *Dicranotaenia annandalei* var. *longosacco* Joyeux et Baer 1939), in *Limosa belgicae;* India.

*D. bernicla* Ehlert 1969, in *Branta bernicla;* Germany.

*D. calumnacantha* (Schmidt 1963) comb. n. (syn. *Hymenolepis calumnacantha* Schmidt 1963, *Wardium calumnacantha* [Schmidt 1963] Bondarenko et Kontrimavichus 1978, *Hymenolepis rybickae* Deblock 1963), in *Capella gallinago;* U.S., Russia, Poland, Syria.

*D. capellae* (Baer 1940) Skrjabin et Mathevossian 1945 (syn. *Hymenolepis capellae* Baer 1940), in *Capella gallinago;* Antigua, U.S.

*D. cervotestis* (Ahern et Schmidt 1976) comb. n. (syn. *Hymenolepis cervotestis* Ahern et Schmidt 1972), in *Recurvirostra americana;* U.S.

*D. chaunense* (Bondarenko et Kontrimavichus 1977) comb. n. (syn. *Wardium chaunense* Bondarenko et Kontrimavichus 1977), in *Gallinago gallinago;* Russia.

*D. chionis* (Fuhrmann 1921) Lopez-Neyra 1942 (syn. *Hymenolepis chionis* Fuhrmann 1921, *Weinlandia chionis* Mayhew 1925), in *Chionis alba;* Antarctic, South America.

*D. cirrosa* (Krabbe 1869) Spasskii 1961 (syn. *Taenia cirrosa* Krabbe 1869, *Monorchis cirrosa* [Krabbe 1869] Clerc 1902, *Aploparaksis cirrosa* [Krabbe 1869] Clerc 1903, *Haploparaksis cirrosa* (Krabbe 1869 Fuhrmann 1932, *Hymenolepis neoarctica* Davies 1938, *Wardium neoarctica* [Davies 1938], Spasskii et Spasskaja 1954, *Dicranotaenia neoarctica* [Davies 1938] Yamaguti 1959), in *Laurus* spp., *Sterna paradisea, S. hirundo, S. macrura, Colymbus caspicus;* Russia, North America, Europe.

*D. clandestina* (Krabbe 1869) Lopez-Neyra 1942 (syn. *Taenia clandestina* Krabbe 1869, *Hymenolepis clandestina* [Krabbe 1869] Railliet 1899), in *Haematopus ostralegus;* Russia, Europe.

*D. clavicirrus* (Yamaguti 1940) Yamaguti 1959 (syn. *Hymenolepis clavicirrus* [Yamaguti 1940] Spasskii et Spasskaja 1954), in *Larus argentatus, L. schistisagus, L. canus, L. marinus;* Japan, Russia, Greenland.

*D. corrariella* Coil 1956, in *Charadrius collaris;* Mexico.

*D. creplini* (Krabbe 1869) Stossich 1898 (syn. *Taenia creplini* Krabbe 1869, *Hymenolepis creplini* [Krabbe 1869] Railliet 1899, *H. pingi* Tseng-Shen 1932, *Dicranotaenia [D.] pingi* [Tseng-Shen 1932] Lopez-Neyra 1942, *D. pingi* [Tseng-Shen 1932] Yamaguti 1959, *Wardium pingi* [Tseng-Shen 1932] Spasskii et Spasskaja 1954), in *Anser albifrons, A. anser, A. fabalis, A. erythropus, A. brachyrhynchus, Branta canadensis, Cygnus cygnus, C. olor, Bucephala clangula;* Russia, Europe, North America, Greenland.

*D. echinorostrae* (Schiller 1957) Yamaguti 1959 (syn. *Hymenolepis echinorostrae* Schiller 1957), in *Aythya affinis;* Alaska, U.S.

*D. ellisoni* (Burt 1944) Yamaguti 1959 (syn. *Hymenolepis ellisoni* Burt 1944), in *Acridotheres tristis;* Sri Lanka.

*D. fallax* (Krabbe 1869) Yamaguti 1959 (syn. *Taenia fallax* Krabbe 1869, *Diplacanthus [Lepidotrias] fallax* [Krabbe 1869] Cohn 1899, *Hymenolepis fallax* [Krabbe 1869] Railliet 1899), in *Somateria mollissima, S. spectabilis; Aythya marila, Fuligula, Mareca, Anas, Clangula;* North Atlantic, Russia, Alaska.

*D. fryei* (Mayhew 1925) Yamaguti 1959 (syn. *Wardium fryei* Mayhew 1925, *Hymenolepis fryei* [Mayhew 1925] Fuhrmann 1932, *H. californicus* Young 1950), in *Larus glaucescens;* North America.

*D. fusa* (Krabbe 1869) comb. n. (syn. *Taenia fusa* Krabbe 1864, *Hymenolepis fusus* [Krabbe 1869] Fuhrmann 1906, *Haploparaxis fusus* [Krabbe 1869] Joyeux et Baer 1928, *Hymenolepis neosouthwelli;* Hughes 1940, *H. pseudofusa* Skrjabin et Mathevossian 1942, *Aploparaksis baeri* Schiller 1951, *Wardium pseudofusum* [Skrjabin et Mathevossian 1942] Spasskii et Spasskaja 1954, *Wardium fusa* [Krabbe 1869] Spasskii 1961), in *Larus, Rhodostethia, Hydroprogne, Rissa, Chilidonias, Sterna;* Europe, Asia, Israel, Sinai, North America, Greenland, Australia.

*D. hamasigi* (Yamaguti 1940) Skrjabin et Mathevossian 1945 (syn. *Hymenolepis hamasigi* Yamaguti 1940, *Limnolepis hamasigi* [Yamaguti 1940] Spasskii et Spasskaja 1954), in *Calidris alpina sakhalina, C. tenuirostris;* Japan, Russia.

*D. himantopodis* (Krabbe 1869) Lopez-Neyra 1942 (syn. *Taenia himantopodis* Krabbe 1869, *Hymenolepis himantopodis* [Krabbe 1869] Fuhrmann 1906, *Dicranotaenia [Dicranolepis] himatopodis* [Krabbe 1869] Lopez-Neyra 1942, *Wardium himantopodis* [Krabbe 1869] Spasskii et Spasskaja 1954), in *Himantopus himantopus, H. mexicanus, Recurvirostra avocetta, Totanus calidris, Haematopus ostralegus;* Africa, Antigua, Europe, China, Russia.

*D. hyodori* Sawada et Kugi 1981, in *Hypsipetes amaurotis;* Japan.

*D. kowalewskii* (Baczynska 1914) Lopez-Neyra (syn. *Hymenolepis kowalewskii* Baczynska 1941, *Wardium kowalewskii* [Baczynska 1914] Spasskii et Spasskaja 1954), in *Fuligula, Nyroca;* Europe.

*D. lagopi* (Bondarenko 1965) comb. n. (syn. *Wardium lagopi* Bondarenko 1965), in *Lagopus lagopus;* Russia.

*D. limicolum* (Spasskii et Dao 1963) comb. n. (syn. *Wardium limicolum* Spasskii et Dao 1963), in *Charadrius alexandrinus;* Vietnam.

*D. manubriatum* (Spasskii et Dao 1963) comb. n. (*Wardium manubriatum* Spasskii et Dao 1963). in *Larus genei;* Vietnam.

*D. mathevossianae* (Kurashvili 1950) comb. n. (syn. *Hymenolepis mathevossianae* Kurashvili 1950, *Wardium mathevossianae* [Kurashvili 1950] Kurashvili 1955), in *Scolopax rusticola;* Russia.

*D. microcirrosa* (Mayhew 1925) Lopez-Neyra 1942 (syn. *Weinlandia microcirrosa* Mayhew 1925, *Hymenolepis microcirrosa* [Mayhew 1925] Fuhrmann 1932), in *Turdus migratorius;* U.S.

*D. multistriata* (Rudolphi 1810) Lopez-Neyra 1942 (syn. *Taenia multistriata* Rudolphi 1810, *Hymenolepis (Drepanidotaenia) multistriata* Cohn 1901, *Hymenolepis multistriata* Skrjabin et Mathevossian 1945, *Dubininolepis multistriata* [Rudolphi 1810] Spasskii et Spasskaja 1954, *Colymbilepis multistriata* [Rudolphi 1810] Spasskaja 1966), in *Colymbus rigricollis, C. griseigena, C. ruficollis;* Russia, Europe, Africa.

*D. ochotensis* (Spasskaja et Spasskii 1973) comb. n. (syn. *Wardium ochotensis* Spasskaja et Spasskii 1973), in *Fluvialis apricarica;* Russia.

*D. pacificum* (Spasskii et Yurpalova 1968) comb. n. (syn. *Wardium pacificum* Spasskii et Yurpalova 1968), in *Larus canis, L. argentatus;* Russia.

*D. paraclavicirrus* (Oschmarin 1963) comb. n. (syn. *Wardium paraclavicirrus* Oschmarin 1963), in *Capella gallinago;* Russia.

*D. paraporale* (Podesta et Holmes 1970) comb. n. (syn. *Wardium paraporale* Podesta et Holmes 1970), in *Aechmorphus occidentalis;* Canada.

*D. pararetracta* (Regel et Bondarenko 1982) comb. n. (syn. *Wardium pararetracta* Regel et Bondarenko 1982), in *Clangula hymalis, Aythya marila;* Russia.

*D. parviceps* (Linstow 1872) comb. n. (syn. *Taenia parviceps* Linstow 1972, *Diorchis parviceps* Linstow 1904, *Chelacanthus parviceps* [Linstow 1872] Yamaguti 1959), in *Mergus serrator;* Europe.

*D. porale* (Meggitt 1927) comb. n. (syn. *Hymenolepis porale* Meggitt 1927, *Sphenacanthus porale* [Meggitt 1927] Lopez-Neyra 1942, *Wardium porale* [Meggitt 1927] Gvosdev 1964, *Glareolepis porale* [Meggitt 1927] Spasskii 1967), in *Glareola pratincola, G. normanni;* Russia.

*D. querquedula* (Fuhrmann 1921) Lopez-Neyra 1942 (syn. *Hymenolepis querquedula* Fuhrmann 1921, *Weinlandia querquedula* Mayhew 1925, *Hymenolepis querquedulae* Gower 1939), in *Anas crecca;* North and South America.

*D. recurvirostrae* (Krabbe 1869) Lopez-Neyra 1942 (syn. *Taenia recurvirostrae* Krabbe 1969, *Drepanidotaenia recurvirostrae* [Krabbe 1969] Cohn 1900, *Hymenolepis recurvirostrae* [Krabbe 1969] Railliet 1899), in *Recurvirostra avocetta, Tringa nebularia;* Africa, Russia.

*D. recurvirostroides* (Meggitt 1927) Yamaguti 1959 (syn. *Hymenolepis recurvirostroides* Meggitt 1927), in *Gallinago* sp.; Egypt.

*D. riggenbachi* (Mola 1913) Yamaguti 1959 (syn. *Hymenolepis riggenbachi* Mola 1913), in *Netta rufina;* Sardinia.

*D. simplex* (Fuhrmann 1906) Lopez-Neyra 1942 (syn. *Hymenolepis simplex* Fuhrmann 1906, *Weinlandia simplex* Mayhew 1925), in *Tadorna tadorna, Anas* spp., *Lophodytes cuculatus;* Europe, China, U.S.

*D. smogorjevskajae* (Kornyushin et Spasskii 1967) comb. n. (syn. *Wardium smogorjevskajae* Kornyushin et Spasskii 1967), in *Tringa totanus;* Russia.

*D. sobolevi* (Bondarenko 1966) comb. n. (syn. *Wardium sobolevi* Bondarenko 1966), in *Charadrius hiaticula;* Russia.

FIGURE 344.    Fraternoid hook.

*D. spasskii* (Schigin 1961) comb. n. (syn. *Wardium spasskii* Schigin 1961), *W. filamentoovatum* Macko 1962), in *Larus minutus, Sterna hirundo;* Russia, Czechoslavakia.

*D. spiculigera* (Nitzsch in Giebel 1857) Sinitzin 1896 (syn. *Taenia spiculigera* Nitzsch in Giebel 1857), in *Fulica atra, Tringa alpina;* Europe.

*D. squatarolae* (Kornyushin 1970) comb. n. (syn. *Wardium squatarolae* Kornyushin 1970), in *squatarola* sp.; Russia.

*D. tenuicollis* (Sawada et Kugi 1975), comb. n. (syn. *Hymenolepis [Weinlandia] tenuicollis* Sawada et Kugi 1975), in *Anas platyrhynchos;* Japan.

*D. tsengi* (Joyeux et Baer 1940) Yamaguti 1959 (syn. *Wardium tsengi* [Joyeux et Baer 1940] Spasskaja 1966), in *Tringa totanus, Himantopus mexicanus, H. himantopus, Recurvirostra avocetta;* Russia.

*D. uragahaensis* (Burt 1944) Yamaguti 1959 (syn. *Hymenolepis uragahaensis* Burt 1944), in *Harpactes fasciatus;* Sri Lanka.

*D. varsoviensis* Sinitzin 1896 (syn. *D. spiculigera* var. *varsoviensis* Sinitzin 1896), in *Tringa alpina;* Poland.

*D. zmorayi* (Macko 1970) comb. n. (syn. *Wardium zmorayi* Macko 1970), in *Anas clypeata;* Czechoslovakia.

67b.  Hooks shaped otherwise........................................................ 68

68a.  **Rostellar hooks fraternoid (Figure 344), with handle well developed, sometimes curved ventrad; guard and blade well developed and about parallel** ............
**Variolepis Spasskii et Spasskaja 1954 (syn. Hybridolepis Spasskii 1959, Troglodytilepis Yamaguti 1956, Decacanthus Yamaguti 1959, Podicipitilepis Yamaguti 1956, Confluaria Ablasov 1953, Dubiniolepis Spasskii et Spasskaja 1954).**

*Diagnosis:* Suckers unarmed. Rostellum with a circle of eight to ten hooks. Guard of hook stout, nearly as long as, and parallel to blade; handle usually rather short, may be curved. Testes three, two antiporal, one poral. Cirrus pouch short, not reaching median line. External and internal seminal vesicles present. Ovary and vitelline gland median. Seminal receptacle present. Uterus saccular. Parasitic mainly in land birds.

Type species: *V. farciminosa* (Goeze 1782) Spasskii et Spasskaja 1954 (syn. *Taenia farciminosa* Goeze 1782, *Hymenolepis farciminosa* [Goeze 1782] Railliet 1899, *Dicranotaenia farciminosa* [Goeze 1782] Lopez-Neyra 1942), in *Garrulus glandarius, Sturnus vulgaris, S. unicolor, Oriolus oriolus, Pica pica, Gracupica nigricollis, Sturnia malabarica, Turdus pilaris, T. merula, T. viscivorus, Corvus levaillanti, C. albus, Acrocephalus agricola, Nuciphaga caryocatactes, Cyanopica cyana, Emberiza cioides, E. aureola, Perisoreus inafustus, Acridotheres tristis, A. albocinctus, Coloeus monedula, Gallus gallus; Aythya ferina;* Europe, Russia, North America, South America, Japan, Asia.

Other species:

*V. arcticus* (Schiller 1955) comb. n. (syn. *Hymenolepis arcticus* Schiller 1955, *Wardium arcticus* [Schiller 1955] Spasskii 1959, *Decacanthus arcticus* [Schiller 1955] Yamaguti 1959), in *Somateria spectabilis, S. mollissima, Arctonetta fischeri;* Alaska, U.S.

*V. asymmetrica* (Fuhrmann 1918) Yamaguti 1959 (syn. *Hymenolepis asymmetrica* Fuhrmann 1918), in *Chalcococcyx plagosus;* New Caledonia.

*V. bilharzi* (Krabbe 1869) Spasskii et Spasskaja 1954 (syn. *Taenia bilharzi* Krabbe 1869), in *Sylvia glactodes, Corvus splendens, Buchanga atra, Tchagra senegala, Luscinia megarhyncha;* Europe, Africa.

*V. brasiliensis* (Fuhrmann 1906) Spasskii et Spasskaja 1954 (syn. *Hymenolepis brasiliensis* Fuhrmann 1906), in *Caprimulgus carolinensis, Nyctiprogne leucophyia;* Brazil.

*V. capillaris* (Rudolphi 1810) comb. n. (syn. *Taenia capillaris* Rudolphi 1810, *Hymenolepis capillaris* [Rudolphi 1810] Railliet 1899, *Dubininolepis capillaris* [Rudolphi 1810] Spasskii et Spasskaja 1954, *Confluaria capillaris* [Rudolphi 1810] Spasskaja 1966), in *Colymbus auritus, C. cristatus, C. glacialis, C. septentrionalis, Gavia arctica;* Europe, Russia.

*V. capillaroides* (Fuhrmann 1906) comb. n. (syn. *Hymenolepis capillaroides* Fuhrmann 1906, *Dicranotaenia capillaroides* [Fuhrmann 1906] Lopez-Neyra 1942, *Dubininolepis capillaroides* [Fuhrmann 1906] Spasskii et Spasskaja 1954, *Confluaria capillaroides* [Fuhrmann 1906] Spasskaja 1966, in *Colymbus* spp.; Asia, Cuba, Central and South America.

*V. caprimulgorum* (Fuhrmann 1906) Spasskii et Spasskaja 1954, (syn. *Hymenolepis caprimulgorum* Fuhrmann 1906), in *Caprimulgus lineatus, Cordeilus rupestris, Podager nacunda;* South America.

*V. columbina* (Fuhrmann 1909) Spasskii et Spasskaja 1954 (syn. *Hymenolepis columbina* Fuhrmann 1909), in *Oena capensis;* Africa.

*V. coronoides* (Tubangui et Masiluñgan 1937) Spasskii et Spasskaja 1954, in *Corone philippina, Corvus coronoides;* Philippines.

*V. fernandensis* (Nybelin 1929) Spasskii et Spasskaja 1954, in *Turdus magellanicus;* Juan Fernandez Island.

*V. furcifera* (Krabbe 1869) comb. n. (syn. *Taenia furcifera* Krabbe 1869, *Diplacanthus [Dilepis] furcifera* [Krabbe 1869] Cohn 1899, *Hymenolepis furcifera* [Krabbe 1869] Szymanski 1905, *Diorchis [Nudorchis] oschmarini* Sudarikov 1905, *Dubininolepis furcifera* [Krabbe 1869] Spasskii et Spasskaja 1954, *Confluaria furcifera* [Krabbe 1869] Spasskaja 1966), in *Colymbus auritus, C. cristatus, C. griseigena, C. caspicus, C. nigricollis, C. ruficollis, Podilymbus podiceps, Fulica atra;* Europe, Russia, Canada, Africa.

*V. globocephala* (Fuhrmann 1918) Spasskii et Spasskaja 1954 (syn. *Hymenolepis globocephala* Fuhrmann 1918), in *Zosterops lateralis;* New Caledonia, Loyalty Island.

*V. hassalli* (Fuhrmann 1924) Spasskii et Spasskaja 1954 (syn. *Hymenolepis hassalli* Fuhrmann 1924), in *Chaloccoccyx pelagicus;* New Caledonia.

*V. hoploporus* (Dollfus 1951) comb. n. (syn. *Hymenolepis [Weinlandia] hoploporus* Dollfus 1951), in *Podiceps cristatus;* Morocco.

*V. hughesi* (Webster 1947) Yamaguti 1959 (syn. *Hymenolepis [H.] hughesi* Webster 1947, *Hybridolepis hughesi* [Webster 1947] Spasskii 1959), in *Charadrius melodus;* Texas, U.S.

*V. japonica* (Yamaguti 1935) comb. n. (syn. *Hymenolepis japonica* Yamaguti 1935, *Dubininolepis japonica* [Yamaguti 1935] Spasskii et Spasskaja 1954, *Confluaria japonica* [Yamaguti 1935] Spasskaja 1966), in *Colymbus ruficollis, C. griseigena;* Japan, Tunisia, Russia.

*V. laticauda* (Yamaguti 1956) comb. n. (syn. *Podicipitilepis laticauda* Yamaguti 1956), in *Podiceps ruficollis;* Japan.

*V. microscolecina* (Fuhrmann 1906) Spasskii et Spasskaja 1954 (syn. *Hymenolepis microscolecina* Fuhrmann 1906, *H. uncinata* Fuhrmann 1906 after Stieda 1862), in *Rupicola rupicola;* Brazil.

FIGURE 345. Nitidoid hook.

*V. orthacantha* (Fuhrmann 1906) comb. n. (syn. *Hymenolepis orthacantha* Fuhrmann 1906, *Sphenacanthus orthacantha* [Fuhrmann 1906] Spasskii et Spasskaja 1954), in *Coscoroba coscoroba*; South America.

*V. planestici* (Mayhew 1925) Spasskii et Spasskaja 1954 (syn. *Hymenolepis planestici* Mayhew 1925, *Weinlandia planestici* Mayhew 1925), in *Turdus migratorius, Acridotheres tristis*; U.S., India.

*V. podicipina* (Szymanski 1905) comb. n. (syn. *Hymenolepis podicipina* Szymanski 1905, *Dicranotaenia podicipina* [Szymanski 1905] Lopez-Neyra 1942, *Dubininolepis podicipina* [Szymanski 1905] Spasskii et Spasskaja 1954, *Confluaria podicipina* [Szymanski 1905] Spasskaja 1966), in *Colymbus auritus, C. caspicus, C. cristatus, C. griseigena, C. nicricollis, C. ruficollis;* Europe, North America, Russia.

*V. pycnonoti* (Tugangui et Masiluñgan 1937) Spasskii et Spasskaja 1954, in *Pycnonotus goiavier;* Philippines.

*V. retracta* (Linstow 1905) Yamaguti 1959 (syn. *Hymenolepis retracta* Linstow 1905, *Wardium retracta* [Linstow 1905] Regel et Bondarenko 1982, *Amphipetrovia retracta* [Linstow 1905] Spasskii et Spasskaja 1954, *A. inflatocirrosa* Oshmarin 1963), in *Erionetta spectabilis; Anas crecca, Clangula hyemalis, Aythya marila;* Russia.

*V. spasskii* (Ablasov 1953) comb. n. (syn. *Confluaria spasskii* Ablasov 1953), in *Aythya fuligula;* Russia.

*V. swiderskii* (Gasowska 1932) comb. n. (syn. *Hymenolepis swiderskii* Gasowska 1932, *Dubininolepis swiderskii* [Gasowska 1932] Spasskii et Spasskaja 1954), in *Urinator arcticus;* Russia.

*V. troglodytis* (Yamaguti 1956) comb. n. (syn. *Troglodytilepis troglodytis* Yamaguti 1956), in *Troglodytes troglodytes;* Japan.

*V. variabilis* (Mayhew 1925) Yamaguti 1959 (syn. *Hymenolepis variabilis* Mayhew 1925, *Dicranotaenia variabilis* [Mayhew 1925] Lopez-Neyra 1942), in *Corvus brachyrhynchus;* Illinois; U.S.

*V. victoriata* (Inamdar 1934) Spasskii et Spasskaja 1954, in *Sturnopaster contra;* India.

*V. woodsholei* (Fuhrmann 1932) comb. n. (syn. *Hymenolepis woodsholei* Fuhrmann 1932, *Dubininolepis woodsholei* [Fuhrmann 1932] Spasskii et Spasskaja 1954), in *Colymbus auritus, C. grisegena;* Massachusetts, U.S.

**68b. Rostellar hooks arcuatioid to nitidoid (Figure 345), with long handle and blade, guard much reduced or absent .................................................. 69**

**69a. Eight rostellar hooks ................... *Octacanthus* Spasskii et Spasskaja 1954.**

*Diagnosis:* Suckers unarmed. Rostellum with a single circle of eight hooks. Handle of hook shorter than blade, with barb-like guard. Proglottids transversely elongated. Testes three, in a triangle, one poral, other two antiporal. Cirrus pouch short or quite long. External and internal seminal vesicles present. Ovary posterior, transversely elongated. Vitellarium postovarian. Uterus saccular. Parasites of birds. Europe, Asia.

Type species: *O. rosenthali* (Mola 1913) Spasskii et Spasskaja 1954 (syn. *Hymenolepis rosenthali* Mola 1913, *Sphenacanthus rosentali* [Mola 1913] Lopez-Neyra 1942), in *Pterocles alchata;* Italy.

FIGURE 346. Arcuatoid hook.

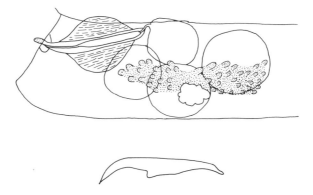

FIGURE 347. *Microsomacanthus compressa* (Linton 1892) Lopez-Neyra 1942. Mature proglottid and rostellar hook.

Other species:
    *O. obvelata* (Krabbe 1879) Spasskii et Spasskaja 1954 (syn. *Taenia obvelata* Krabbe 1897, *Hymenolepis obvelata* [Krabbe 1879] Fuhrmann 1932, *Sphenacanthus obvelata* [Krabbe 1879] Lopez-Neyra 1942), in *Pterocles alchata, P. orientalis, Syrrhaptes paradoxus;* Russia.

**69b. Ten "arcuatoid" (Figure 346) rostellar hooks .................................**
**................................ *Microsomacanthus* Lopez-Neyra 1942 (syn. *Fuhrmanniella* Tseng 1932, *Weinlandia* Mayhew 1925 in part, *Abortilepis* Yamaguti 1959, *Kowalewskius* Yamaguti 1959, *Tschertkovilepis* Spasskii et Spasskaja 1954, *Vigissotaenia* Mathevossian 1968, *Myxolepis* Spasskii 1959, *Hamatolepis* Spasskii 1962, *Anserilepis* Spasskii et Tolkatcheva 1965). (Figure 347)**

*Diagnosis:* Rostellum with a single circle of ten hooks. Handle of hook long, blade long or short, but usually short; Guard very short or absent. Testes three, one poral, two antiporal, or one median and posterior, other two symmetrical and anterior. Cirrus pouch may be strongly muscular, reaching median line or not. External and internal seminal vesicles present, latter occasionally absent. Ovary and vitelline gland median or submedian; ovary usually two- or three-lobed. Uterus sac-like when gravid. Seminal receptacle present. Parsites mainly of aquatic birds. Cosmopolitan.
    Type species: *M. microsoma* (Creplin 1829) Lopez-Neyra 1942 (syn. *Taenia microsoma* Creplin 1829, *Hymenolepis microsoma* Railliet 1899, *Diplacanthus [Dilepis] microsoma* Cohn 1899, *Hymenolepis [Drepanidotaenia] microsoma* Cohn 1901, *Weinlandia microsoma* Mayhew 1925, *Cysticercus limnaei* Villot 1883), in *Bucephala, Clangula, Aythya, Netta, Melanitta, Somateria, Anas, Anser, Larus;* North America, Europe, Asia.
    Other species:
        *M. abortiva* (Linstow 1904) Lopez-Neyra 1942 (syn. *Hymenolepis abortiva* Linstow 1904, *H. volvuta* Linstow 1904, *H. upsilon* Rosseter 1911, *Weinlandia abortiva* Mayhew 1925, *Abortilepis abortiva* [Linstow 1904] Yamaguti 1959), in *Anas acuta, A. platyrhynchos, A. falcata, A. crecca, A. clypeata, Aythya fuligula, A. ferina,*

*Mergus albellus, M. serrator, Melanitta americana, M. deglandi, Anser anser;* Russia, Europe, Egypt, North America.

*M. acus* Spasskii et Jurpalova 1964, in *Melanitta americana;* Russia.

*M. ambiguus* (Clerc 1906) Yamaguti 1959 (syn. *Hymenolepis ambiguus* Clerc 1906), in *Otis tetrax, Lyrurus tetrax;* Africa, Russia.

*M. andrejewoi* (Mathevossian 1945) Kuraschvili 1957 (syn. *Dicranotaenia andrejewoi* Mathevossian 1945, *Sphenacanthus andrejewoi* [Mathevossian 1945] Ablasov 1953, *Myxolepis andrejewoi* [Mathevossian 1945] Spasskaja 1966), in *Melanitta fusca, Anas acuta, A. crecca;* Russia.

*M. baeri* Czaplinski et Vaucher 1977 (syn. *M. fausti* Tseng 1932 *sensu* Spasskii et Spasskaja 1962), in *Aythya fuligula;* Russia, Europe.

*M. barrowensis* (Schiller 1952) comb. n. (syn. *Hymenolepis barrowensis* Schiller 1952, *Drepanidotaenia barrowensis* [Schiller 1952] Yamaguti 1959, *Anserilepis barrowensis* [Schiller 1952] Spasskii et Tolkatcheva 1965, *Vigissotaenia barrowensis* [Schiller 1952] Mathevossian 1986), in *Anser albifrons, A. fabalis, Branta bernicla;* Alaska, Russia.

*M. brevicirrosus* (Fuhrmann 1913) Yamaguti 1959 (syn. *Hymenolepis brevicirrosus* Fuhrmann 1913), in *Celeucides ignota;* New Guinea.

*M. cavoarmata* (Sudarikov 1950) comb. n. (syn. *Hymenolepis cavoarmata* Sudarikov 1950, *Tschertkovilepis cavoarmata* [Sudarikov 1950] Yamaguti 1959), in *Turdus pilaris;* Russia.

*M. childi* (Burt 1940) Yamaguti 1959 (syn. *Hymenolepis childi* Burt 1940), in *Phalacrocorax niger;* Sri Lanka.

*M. compressa* (Linton 1892, Lopez-Neyra 1942 (syn. *Taenia compressa* Linton 1892, *Hymenolepis megalostellis* Solowiow 1911, *H. compressa* [Linton 1892] Skrjabin 1914), in *Anser anser, A. albifrons, Mergus albellus, M. merganser, Anas acuta, A. crecca, A. formosa, A. clypeata, A. querquedula, A. platyrhynchos, A. strepera, A. penelope, Aix galericulata, Aythya marila, A. rufina, A. vallisneria, A. nyroca, A. ferina, A. fuligula, Bucephala clangula, Histrionicus histrionicus, Melanitta americana, Colymbus grieseigena, C. ruficollis;* Russia, Europe, North America.

*M. collaris* (Batsch 1786) Lopez-Neyra 1942 (syn. *Taenia collaris* Batsch 1786, *T. sinuosa* Zeder 1803, *Lepidotrias sinuosa* Weinland 1858, *Drepanidotaenia sinuosa* Railliet 1893, *Hymenolepis sinuosa* 1899, *H. [Drepanidotaenia] sinuosa* Cohn 1901, *Taenia bairdii* Krefft 1871, *Hymenolepis collaris* [Batsch 1786] Fuhrmann 1908, *Weinlandia collaris* Mayhew 1925, *Dicranotaenia collaris* [Batsch 1786] Skrajabin et Mathevossian 1945, *Sobolevicanthus collaris* [Batsch 1786] Ablasov 1953, *Myxolepis collaris* [Batsch 1786] Spasskii 1959), in *Anas, Netta, Bucephala, Mergus, Anser, Melanitta, Gallus;* Asia, North America, Japan, Europe.

*M. conscripta* (Railliet et Henry 1909) comb. n. (syn. *Hymenolepis conscripta* Railliet et Henry 1909, *Tschertkovilepis conscripta* [Railliet et Henry 1909] Yamaguti 1959), in *Anser anser;* Europe.

*M. cormoranti* Ortlep 1938, in *Microcarbo africana, M. melanoleucus;* South Africa, South Australia.

*M. cuneata* (Mayhew 1925) comb. n. (syn. *Hymenolepis cuneata* Mayhew 1925, *Sphenacanthus cuneata* [Mayhew 1925] Lopez-Neyra 1942, *Tschertkovilepis cuneata* [Mayhew 1925] Yamaguti 1959), in wild duck; North America.

*M. diorchis* (Fuhrmann 1913) Lopez-Neyra 1942 (syn. *Weinlandia diorchis* Mayhew 1925, *Hymenolepis diorchis* Fuhrmann 1913), in *Somateria mollissima, Clangula hyemalis;* Russia, Iceland, England. (One of three testes sometimes rudimentary).

*M. ductilis* (Linton 1927) Spasskii et Spasskaja 1954 (syn. *Hymenolepis ductilis* Linton 1927, *Drepanidotaenia [D.] ductilis* Lopez-Neyra 1942), in *Larus argentatus, L.*

marinus, *L. canus, L. schistigagus, L. hyperboreus, L. glaucoides;* U.S., Russia, Greenland.

*M. falcatus* (Meggitt 1927) Lopez-Neyra 1942 (syn. *Hymenolepis falcatus* Meggitt 1927), host and locality not known.

*M. fausti* (Tseng-Shen 1932) Lopez-Neyra 1942 (syn. *Fuhrmanniella fausti* Tseng-Shen 1932, *Dicranotaenia fausti* [Tseng-Shien 1932] Skrjabin et Mathevossian 1945, *Hymenolepis fausti* [Tseng-Shen 1932] Fuhrmann 1932), in *Anas acuta, A. crecca, A. platrhynchos, A. strepera, A. querquedula, Bucephala, Clangula, A. fuligula, A. marila, A. nyroca, A. ferina, Netta rufina, Mergus serrator, M. albellus, Somateria mollissima, Melanitta deglandi;* China, Russia.

*M. filirostris* (Wedl 1855) Lopez-Neyra 1942 (syn. *Taenia filirostris* Wedl 1855, *Hymenolepis filirostris* [Wedl 1855] Fuhrmann 1906), in *Platalea leucoridia, Ibis melanocephala;* India, China, Java.

*M. formosa* (Dubinina 1953) Yamaguti 1959 (syn. *Hymenolepis formosa* Dubinina 1953), in *Aythya fuligula, A. ferina, Bucephala clangula, Anas acuta, A. platyrhynchos;* Russia.

*M. formosoides* Spasskaja et Spassky 1961, in *Bucephala clangula, Melanitta deglandi, M. fusca;* Russia.

*M. francolini* (Joyeux et Baer 1935) Yamaguti 1959 (syn. *Hymenolepis francolini* Joyeux et Baer 1935), in *Francolinus pintadeanus;* Indochina.

*M. fulicicola* (Skrjabin et Mathevossian 1942) Spasskii et Spasskaja 1954 in *Fulica americana;* U.S.

*M. gogonka* Johri 1941, in *Phalacrocorax javanicus;* Burma.

*M. heterospinus* Spasskii et Jurpalova 1964, in *Somateria mollissima;* Russia.

*M. hystrix* Spasskaja et Spasskii 1961, in *Aythya fuligula;* Russia.

*M. jaegerskioeldi* (Fuhrmann 1913) Lopez-Neyra 1942 (syn. *Hymenolepis jaegerskioeldi* Fuhrmann 1913, *Weinlandia jaegerskioeldi* [Fuhrmann 1913] Mayhew 1925), in *Anas crecca, A. platyrhynchos, Clangula hyemalis, Melanitta nigra, M. fusca, Histrionicus histrionicus, Somateria mollissima;* Russia, Alaska, England, Africa.

*M. jamunicus* (Sharma 1943) Yamaguti 1959 (syn. *Hymenolepis jamunicus* Sharma 1943), in *Anas platyrhynchos;* Burma.

*M. kyushuensis* (Sawada et Kugi 1980) comb. n. (syn. *Weinlandia kyushuensis* Sawada et Kugi 1980), in *Phasianus soemmeringii;* Japan.

*M. lari* Belogurov et Kulikov in Spasskaja 1966, in *Larus ridibundus, L. crassirostris;* Russia.

*M. luengoi* (Lopez-Neyra 1942) comb. n. (syn. *Drepanidotaenia [D.] luengoi* [Lopez-Neyra 1942], *Tschertkovilepis luengoi* [Lopez-Neyra 1942] Yamaguti 1959), in *Oidemia americana;* U.S.

*M. magniovatus* (Fuhrmann 1918) Spasskii et Spasskaja 1954 (syn. *Hymenolepis magniovatus* Fuhrmann 1918), in *Myiagra caledonica;* New Caledonia; Loyalty Islands.

*M. mayhewi* (Tseng 1932) Lopez-Neyra 1942 (syn. *Weinlandia mayhewi* Tseng 1932), in *Clangula glaucion; Aix sponsa;* China, U.S.

*M. melanittae* Ryjikov 1962, in *Melanitta deglandi;* Russia.

*M. microskrjabini* Spasskii et Jurpalova 1964, in *Melanitta* sp.; Russia.

*M. mirabilis* Spasskii et Jurpalova 1964, in *Melanitta deglandi, M. fusca;* Russia.

*M. monoposthe* (Dubinina 1954) comb. n. (syn. *Hymenolepis monoposthe* Dubinina 1954, *Tschertkovilepis monoposthe* [Dubinina 1954] Yamaguti 1959), in *Netta rufina, Bucephala clangula;* Russia.

*M. oidemiae* Spasskii et Jurpalova 1964, in *Melanitta americana;* Russia.

*M. pachycephala* (Linstow 1872) Lopez-Neyra 1942 (syn. *Hymenolepis pachycephala* [Linstow 1872] Fuhrmann 1906, *Taenia pachycephala* Linstow 1872), in *Aythya* fuligula, Histrionicus histrionicus, Melanitta deglandi, Anas strepera, A. platyrhynchus, Colymbus grieseigena; Russia, Europe, Japan, North America.

M. paracompressa (Czaplinski 1956) Spasskaja et Spasskii 1961 (syn. Hymenolepis paracompressa Czaplinski 1956), in Anas platyrhynchos, A. crecca, A. falcata, A. querquedula, A. acuta, A. strepera, Bucephala clangula, Histrionicus histrionicus, Aythya fulligula, A. ferina, A. nyroca, Netta rufina, Anser anser; Europe, Russia.

M. paramicrosoma (Gasowska 1932) Yamaguti 1959 (syn. Hymenolepis paramicrosoma Gasowska 1932, Hymenolepis [Drepanidotaenia] microsoma Cohn 1901), in Anas platyrhynchos, A. acuta, A. strepera, A. crecca, A. penelope, Aythya fuligula, A. ferina, A. marila, Netta rufina, Anser anser, Mergus serrator, Gallus gallus; Russia, Europe, North America.

M. parvula (Kowalewski 1904) Spasskii et Spasskaja 1954 (syn. Hymenolepis parvula Kowalewski 1904, Weinlandia parvula [Kowalewski 1904] Mayhew 1925, Dicranotaenia [D.] parvula [Kowalewski 1904] Lopez-Neyra 1942, Kowalewskius parvulus [Kowalewski 1904] Yamaguti 1959, K. yoshidai Yamaguti 1956), in Anas platyrhynchos, A. crecca, A. clypeata, A. poecilorhyncha, A. querquedula, Mergus albellus, Aythya ferina, A. nyroca, Bucephala clangula, Anas rubripes; Russia, Europe, Japan, U.S.

M. pauciannulata (Meggitt 1927) Lopez-Neyra 1942 (syn. Hymenolepis pauciannulata Meggitt 1927, Abortilepis pauciannulata [Meggitt 1927] Yamaguti 1959), in Spatula clypeata, Anas; Egypt.

M. pauciovata (Meggitt 1927) Lopez-Neyra 1942 (syn. Hymenolepis pauciovata Meggitt 1927, H. fidelis Meggitt 1928, H. floreata Meggitt 1930, Abortilepis pauciovata [Meggitt 1927] Yamaguti 1959), in Spatula clypeata; Egypt.

M. pseudorostellatus (Joyeux et Baer 1950) Yamaguti 1959 (syn. Hymenolepis pseudorostellatus Joyeux et Baer 1950, Taenia capitellata Rudolphi 1810 in part), in Gavia immer; Europe.

M. rangdonensis Spasskii, Dang Van Ngy, et Jurpalova 1963, in Anas platyrhynchus; Vietnam.

M. rectacantha (Fuhrmann 1906) Spasskii et Spasskaja 1954 (syn. Hymenolepis rectacantha Fuhrmann 1906), in Charadrius hiaticula, Haematopus ostralegus; Russia, Europe, England.

M. recurvata Spasskaja et Spasskii 1961, in Anas querquedula, Aythya fuligula, A. marila; Russia.

M. setigera (Froelich 1789) comb. n. (syn. Taenia setigera Froelich 1789. Drepanidotaenia setigera [Froelich 1789] Railliet 1893, Tschertkovilepis setigera [Froelich 1789] Spasskii et Spasskaja 1954, Diplacanthus [Dilepis] setigera [Froelich 1789] Cohn 1899, Hymenolepis setigera [Froelich 1789] Railliet 1899, H. anseris Skrjabin et Mathevossian 1942, Vigissotaenia setigera [Froelich 1789] Mathevossian 1968, in Anas, Branta, Cygnus, Aythya, Anser, Somateria, Clangula, Netta, Eulabeia, Gallus; Russia, Europe, India, Japan.

M. shetlandicus Cielecka et Zdzitoweicki 1981, in Larus dominicanus; Antarctica.

M. simulans (Joyeux et Baer 1941) Yamaguti 1959 (syn. Hymenolepis simulans Joyeux et Baer 1941), in Colymbus arcticus; Switzerland.

M. skrjabini Spasskaja 1963, in Histrionicus histrionicus; Russia.

M. sobolevi Spasskii et Jurpalava 1964, in Clangula hyemalis; Russia.

M. spasskii Tolkacheva 1965, in Anas acuta; Russia.

M. spiralicirrata Maksimova 1963, in Aythya fuligula, Anas clypeata, A. acuta, A. crecca, A. platyrhynchos, Tadorna tadorna; Russia.

M. strictophallus Tolkacheva 1971, in Mergus albellus; Russia.

M. styloides (Fuhrmann 1906) Spasskii et Spasskaja 1954 (syn. Hymenolepis styloides

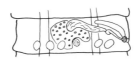

FIGURE 348. *Pentorchis arkteios* Meggitt 1927.

Fuhrmann 1906, *Kowalewskius spinosus* [Linstow 1906] Yamaguti 1959), in *Vanellus aegypticus, Capella gallinago, Rostratula bengalensis;* Egypt, Indochina.

*M. teresoides* (Fuhrmann 1906) Yamaguti 1959 (syn. *Hymenolepis teresoides* Fuhrmann 1906, *Hymenolepis rarus* Skrjabin 1914, *Weinlandia teresoides* [Fuhrmann 1906] Mayhew 1925, *Drepanidotaenia [D.] teresoides* [Fuhrmann 1906] Lopez-Neyra 1942, *Hymenolepis crecca* Singh 1952, *Microsomacanthus rarus* [Skrjabin 1914] Yamaguti 1959, *Hamatolepis teresoides* [Fuhrmann 1906] Spasskii 1962), in *Anas crecca, A. clypeata, A. platyrhynchos, A. strepera, A. penelope, A. acuta, A. falcata, Melanitta fusca, Netta rufina, Aythya nyroca, Chenopsis atrata, Ardea cinera;* Russia, Europe, Australia, Asia.

*M. trichorhynchus* (Yoshida 1910) Spasskii et Spasskaja 1954 (syn. *Hymenolepis trichorhynchus* Yoshida 1910, *Dicranotaenia trichorhynchus* [Yoshida 1910] Skrjabin et Mathevossian 1945), in *Anas platyrhynchos, A. crecca;* Japan, Russia.

*M. trifolium* (Linstow 1905) Spasskii et Spasskaja 1954 (syn. *Hymenolepis trifolium* Linstow 1905, *Weinlandia trifolium* Mayhew 1925), in *Anas platyrhynchos;* Europe.

*M. tuvensis* Spasskaja et Spasskii 1961, in *Bucephala clangula, Aythya fuligula, A. nyroca, Anas falcata;* Russia.

70a. Parasites of mammals ......................................................... 71
70b. Parasites of birds or mammals ................................................ 75

**71a. Five testes in a posterior, straight row .. *Pentorchis* Meggitt 1927. (Figure 348)**

*Diagnosis:* Rostellum unarmed (?). Proglottids wider than long. Genital pores unilateral, provided with sphincter. Cirrus pouch reaches osmoregulatory canal. Testes five, in transverse, postequatorial row. External seminal vesicle absent. Ovary large, median. Vitelline gland compact, postovarian. Vagina posterior to cirrus pouch. Seminal receptacle well-developed. Uterus sac-like. Parasites of Carnivora; Burma.

Type species: *P. arkteios* Meggitt 1927, in *Ursus malayensis;* Burma.

**71b. Testes otherwise ........................................................... 72**

**72a. Four to seven testes, one or two aporal and anterior to others, which are in a transverse row ............. *Paraoligorchis* Wason et Johnson 1977. (Figure 349)**

*Diagnosis:* Rostellum rudimentary, unarmed. Suckers well developed, unarmed. Proglottids numerous, unarmed, broader than long. Genital pores unilateral; genital ducts dorsal to osmoregulatory canals, opening into a common atrium. Testes four to seven, one or two aporal and anterior to others which are in posterior, transverse row. Cirrus pouch exceeding osmoregulatory canals. External and internal seminal vesicles present. Vagina opening posterior to cirrus pore. Seminal receptacle poral. Ovary large, multilobate. Vitellarium postovarian, lobate. Uterus sac-like. Eggs with three membranes, lacking polar specializations. Parasites of Indian rodents.

Type species: *P. taterae* Wason et Johnson 1977, in *Tatera indica;* India.

332  CRC Handbook of Tapeworm Identification

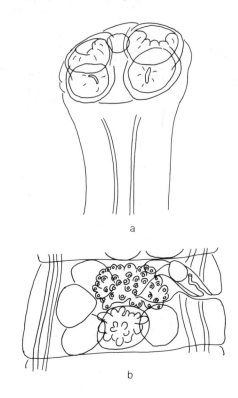

FIGURE 349. *Paraoligorchis taterae* Wason et Johnson 1977. (a) Scolex and (b) mature proglottid.

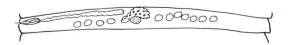

FIGURE 350. *Chitinolepis mjoebergi* Baylis 1926.

**72b. More than seven testes ............................................................. 73**

**73a. Testes (9 to 12) in single, transverse row interrupted by ovary .................**
**............................................. *Chitinolepis* Baylis 1926. (Figure 350)**

*Diagnosis:* Rostellum rudimentary, unarmed. Proglottids much wider than long. Genital pores unilateral. Cirrus pouch elongate, not reaching median line. Testes 9 to 12 in single, transverse row, interrupted by ovary. Internal and external seminal vesicles present. Ovary lobated, median. Vitelline gland compact, postovarian. Seminal receptacle present. Gravid uterus a simple sac as wide as segment. Parasites of Rodentia; Borneo.
Type species: *C. mjoebergi* Baylis 1926, in *Rattus sabanus;* Sarawak.

**73b. Testes otherwise ............................................................. 74**

**74a. Testes (7 to 15) divided by ovary into poral and antiporal groups ...............**
**.................................................... *Hymenandrya* Smith 1954.**

*Diagnosis:* Scolex without rostellum or armature. Proglottids wider than long, craspedote.

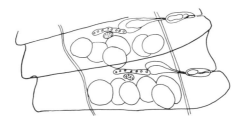

FIGURE 351. *Oligorchis paucitesticulatus* Fuhrmann 1913.

Single set of reproductive organs per segment. Genital pores unilateral. Cirrus pouch crosses poral osmoregulatory canals. External and internal seminal vesicles present. Testes few (7 to 15), divided by ovary into poral and antiporal groups. Ovary lobated, slightly poral. Vitelline gland post-ovarian. Vagina posterior to cirrus pouch. Uterus reticular. Middle egg membrane with polar filament on each end. Parasites of Rodentia. North America.

Type species: *H. thomomyis* Smith 1954, in *Thomomys talpoides;* U.S.
Other species:
   *H. aegyptica* Mikhail et Fahmy 1968, in *Psammomys obesus;* Egypt.

**74b. Testes (8 to 12) grouped laterally and behind ovary............................**
**.....................................................*Pseudoligorchis* Johri 1934.**

*Diagnosis:* Rostellum small, unarmed. Suckers unarmed. Neck absent. Proglottids wider than long. Genital pores unilateral. Cirrus pouch reaching ventral osmoregulatory canal. Testes 8 to 12, surrounding ovary. External and internal seminal vesicles present. Genital ducts passing between osmoregulatory canals. Ovary median. Vitelline gland postovarian. Seminal receptacle large. Uterus an irregular sac. Parasites of Chiroptera. India.

Type species: *P. magnireceptaculatus* Johri 1934, in "bat"; India.

**75a. Armed rostellum present. Genital pores unilateral ..............................**
**........................................*Oligorchis* Fuhrman 1906. (Figure 351)**

*Diagnosis:* Rostellum with single circle of ten or more hooks. Neck present. Proglottids wider than long. Genital pores unilateral, marginal, or dorsal. Cirrus pouch small. Testes three to seven, usually four. Osmoregulatory canals ventral to genital ducts. External and internal seminal vesicles present. Ovary and vitelline gland median. Vaginal pore ventral to cirrus pore. Seminal receptacle present. Uterus large, sac-like. Parasites of Pelicaniformes, Charadriformes, Falconiformes, Passeriformes; Rodentia, Insectivora, Chiroptera. Europe, North and South America, Philippines, Africa, Asia.

Type species: *O. strangulatus* Fuhrman, 1906, in *Elanoides furcatus;* Brazil.
Other species
   *O. brevihamatus* Sawada 1975, in *Rhinolophus ferrumequinum;* Japan.
   *O. cyanocittii* Coil 1955, in *Cyanocita stellera;* Mexico.
   *O. hieraticos* Johri 1934, in *Milvus govinda;* India.
   *O. kwangensis* Southwell et Lake 1939, in *Galachrysia nuchalis;* Africa.
   *O. lahorensis* Khan et Habibulla 1971, in *Corvus splendens;* Pakistan.
   *O. paucitesticulatus* Fuhrmann 1913, in *Aegialitis hiaticula, Vanellus vanellus;* Sweden, France.
   *O. raviensis* Khan et Habibulla 1971, in *Acridotheres ginginianus;* Pakistan.
   *O. toxometra* Joyeux et Baer 1928, in *Gallinago* sp.; West Africa.

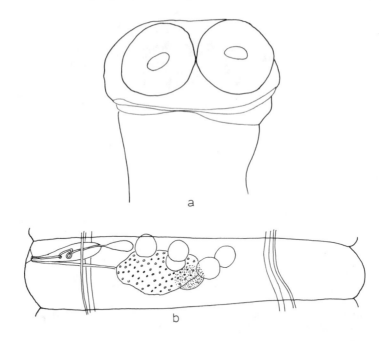

FIGURE 352. *Neoligorchis alternatus* Johri 1960. (a) Scolex; (b) proglottid.

**75b. Rostellum and hooks absent. Genital pores irregularly alternating ..............**
.........................................*Neoligorchis* **Johri 1960. (Figure 352)**

*Diagnosis:* Scolex lacking rostellum or hooks. Neck present. Proglottids acraspedote, external segmentation faintly marked. Genital pores irregularly alternating. Genital atrium small and poorly developed. Genital ducts pass between osmoregulatory canals. Cirrus pouch crosses canals; poral to ovary. Cirrus unarmed. External seminal vesicle present. Testes five or six, dorsal, anterior and lateral to ovary or aporal side; medial to osmoregulatory canals. Vagina narrow, posterior to cirrus pouch. Seminal receptacle absent. Ovary transversely elongated, not lobed, slightly poral. Vitelline gland ventral and aporal to ovary. Uterus an irregularly lobed sac occupying entire segment. Eggs spindleshaped, outer membrane tapering at both poles. Charadriiformes. India.

Type species: *N. alternatus* Johri 1960, in *Rostratula bengalensis;* India.

## APPENDIX TO HYMENOLEPIDINAE FROM MAMMALS

The following species from mammals are placed in *Hymenolepis sensu lato* because of inaccessibility or inadequacy of published descriptions, or inadequate opportunity for me to study them.

*H. aduncihami* Hunkeler 1972, in *Crocidura bottegi, C. poensis;* Ivory Coast.

*H. angusta* Prudhoe et Manger 1969, in bats; Malaysia.

*H. arvicolae* (Blanchard 1891) (Syn. *Taenia inermis* Linstow 1898), in *Arvicola campestris, Microtus agrestis;* Europe

*H. asketus* Brooks et Mayes 1977, in *Blarina brevicauda;* Nebraska, U.S.

*H. bakamovi* Hunkeler 1972, in *Crocidura flavescens, C. jouvenatae, C. poensis, C. lamottei;* Ivory Coast.

*H. barrosii* (Moniez 1880), in *Talpa europaea;* Europe.

*H. bellieri* Hunkeler 1972, in *Crocidura poensis, C. bottegi, C. flavescens, C. jouvenatae, C. theresae;* Ivory Coast.

*H. bennetti* Freeman 1960, in *Napaeozapus insignis;* Canada.

*H. biliaris* (Villot 1877) Mas-Coma et Jourdane 1977, in shrews; Europe.

*H. brustae* Vaucher 1971, in *Crocidura suaveolens;* Europe.

*H. capensis* Janicki 1904, in *Chrysochloris aurea;* Europe.

*H. chalinolobi* Andrews et Daniel 1974, in *Chalinolobus tuberculatus;* New Zealand.

*H. contracta* Janicki 1904, in *Mus musculus, Rattus rattus, R. norvegicus;* Europe.

*H. crociduri* Mikhail et Fahmy 1977, in *Crocidura olivieri;* Egypt.

*H. fodientis* Voucher 1971, in *Neomys* sp.; Europe.

*H. gilloni* Hunkeler 1972, in *Crocidura poensis, C. bottegi, C. flavescens;* Ivory Coast.

*H. hipposideri* Prudhoe et Manger 1969, in *Hipposideros* spp.; Malaysia.

*H. iriei* Sawada 1972, in *Rhinolophus ferrumequinum;* Japan.

*H. lamtoensis* Hunkeler 1972, in *Crocidura flavescens, C. jouvenatae;* Ivory Coast.

*H. lasionycteridis* Rausch 1975, in *Lasionycteris noctivagans, Myotis californicus, M. yumanensis, M. evotis, M. volans, M. lucifugus, Eptesicus fuscus;* U.S.

*H. mathevossianae* Akhumian 1946, in *Mesocricetus branti;* Armenia, Russia.

*H. micracantha* Collins 1972, in *Tatera brantsi;* South Africa.

*H. minimedius* Johri 1959, in *Pteropus medius;* India.

*H. moniezi* Parona 1893, in *Pteropus medius;* India.

*H. mopoyemi* Hunkeler 1972, in *Crocidura theresae, C. flavescens, C. jouventae, C. poensis;* Ivory Coast.

*H. mujibi* Bilquees et Malik 1974, in *Suncus marinus;* Pakistan.

*H. multihami* Hunkeler 1972, in *Steatomys* sp., *Dasymys incomtus, Uranomys ruddi;* Upper Volta.

*H. murinae* Voucher 1971, in shrew; Burma.

*H. nishidai* Sawada 1982, in *Rhinolophus ferrumequinum;* Japan.

*H. odaensis* Sawada 1968, in *Rhinolophus ferrumequinum, Miniopterus schreibersi;* Japan.

*H. olivieri* Mikhail et Fahmy 1977, in *Crocidura olivieri;* Egypt.

*H. parvus* Sawada 1967, in *Rhinolophus ferrumequinum;* Japan.

*H. peromysci* Tinkle 1972, in *Peromyscus maniculatus, P. boylei;* U.S.

*H. phyllostomi* Voucher 1982, in *Phyllostomus hastatus;* Peru.

*H. pitymi* Yarinski 1952, in *Pitymys pinetorum;* Tennessee, U.S.

*H. pribilofensis* Olsen 1969, in *Sorex pribilofensis;* Alaska, U.S.

*H. procera* Janicki 1904, in *Myotis terrestris, Meriones shawi, Arvicola amphibius;* Switzerland, Russia, Tunisea.

*H. pseudofurcata* Voucher 1971, in shrew; Burma.

*H. rashomonensis* Sawada 1972, in *Rhonolophus ferrumequinum, R. cornutus;* Japan.

*H. steatomidis* Hunkeler 1972, in *Steatomys* sp., *Dasymys incomtus, Uranomys ruddi;* Upper Volta.

*H. subrostellata* Sawada 1970, in *Rhinolophus ferrumequinum, R. cornutus;* Japan.

*H. sunci* Crusz et Sanmugasundaram 1971, in *Suncus murinus;* Sri Lanka.

*H. sunci* Vaucher et Tenora 1971, in *Suncus marinus;* Afghanistan.

*H. suslika* Shaldybin 1965, in *Citellus suslica;* Russia.

*H. tateri* Collins 1972, in *Tatera brantsi;* South America.

*H. tsuzurasensis* Sawada 1972, in *Rhinolophus ferrumequinum;* Japan.

*H. uranomidis* Hunkeler 1972, in *Uranomys ruddi, Lemniscomys striatus, Mastomys erythroleucus, Arvicanthus niloticus, Dasymys incomtus;* Upper Volta.

*H. voucheri* Hunkeler 1972, in *Crocidura flavescens, C. theresae, C. poensis, C. jouvenatae, C. lamottei, C. odorata;* Upper Volta.

## APPENDIX TO HYMENOLEPIDINAE FROM BIRDS

The following species from birds are placed in *Hymenolepis sensu lato* because of inaccessibility or inadequacy of published descriptions, or inadequate opportunity for me to study them.

*H. acirrosa* Fuhrmann 1943, in *Upupa africana;* Africa. Scolex unknown.

*H. acridotheridis* (Parona 1980), in *Acridotheres albocinctus;* Burma.

*H. alaskensis* Deblock et Rausch 1967, in *Haemantopus bachmani;* Alaska, U.S.

*H. almiquii* Perez-Vegueras 1960, in *Atopogale cubana;* Cuba.

*H. angularostris* Sugimoto 1934, in *Anas platyrhynchos;* Taiwan. Hooks unknown.

*H. anseris* Skrjabin et Mathevossian 1942, for *H.* sp. of Micacic et Erlich 1940, in domestic goose; Yugoslavia.

*H. aploparaksioides* Deblock 1964, in *Haemantopus ostralegus;* France.

*H. arenariae* Cabot 1969, in *Arenaria interpres;* Ireland.

*H. arguei* Pomeroy et Burt 1964, in *Larus argentatus;* Canada.

*H. armata* Fuhrmann 1906, in *Columba gymnophthalma;* Brazil. No scolex described.

*H. baschkiriensis* (Clerc 1902), in *Larus canus;* Russia.

*H. belopolskaiae* Deblock et Rosé 1962, in *Crocethia alba;* Russia.

*H. breviannulata* Fuhrmann 1906, in *Molybdophanes coerulescens;* Brazil.

*H. cameroni* Singh 1960, in *Anas poecilorhyncha;* India.

*H. chenopsis* Palmer 1981, in *Cygnus stratus;* Australia.

*H. ciconia* Johri 1960, in *Ciconia ciconia;* India.

*H. coraciae* (Rudolphi 1819) Parona 1902, in *Coracias garrula;* Europe.

*H. cryptacantha* (Krabbe 1869), in *Glareola pratinicola;* Europe.

*H. dubia* Daday 1901 (larval form), in *Diaptomus allucandi;* definitive host unknown.

*H. dusmeti* (Lopez-Neyra 1942) for *Hymenolepis* sp. of Linton 1927 (syn. *Drepanidotaenia [D.] dusmeti* Lopez-Neyra 1942), in *Larus delawarensis;* U.S.

*H. ellisi* Johnston et Clark 1948, in *Pelecanus conspicillatus;* Australia.

*H. exilis* (Dujardin 1845), in *Gallus gallus;* Europe, Burma.

*H. fanatica* Meggitt 1927, in *Phoenicopterus* sp.; Cairo, Egypt.

*H. finta* Meggitt 1933, in *Ketupa ceylonensis;* India.

*H. flaminata* Meggitt 1930, in *Otis houbara;* Egypt.

*H. fona* Meggitt 1933, in *Cygnus olor;* India.

*H. foveata* Meggitt 1933, in *Anas poecilorhyncha;* India.

*H. fructifera* Meggitt 1927, in *Spatula clypeata;* Egypt.

*H. fruticosa* Meggitt 1927, in *Spatula clypeata;* Egypt,

*H. fungosa* Meggitt 1928, in *Otis houbara;* India.

*H. furcouterina* Davis 1945, in *Anhinga melanogaster;* Celebes. No scolex known.

*H. futilis* Meggitt 1927, in *Platalea leucorodis;* Egypt.

*H. gallinae* Fuhrmann in Parona 1900, in *Gallus gallus;* Argentina.

*H. graeca* Johri 1960, in *Alectoris graeca;* India.

*H. guadeloupensis* Graber et Euzeby 1976, in *Anas platyrhynchus;* Guadeloupe.

*H. guschanskoi* (Krotov 1952), in *Calidris minuta, Crocethia alba;* Russia.

*H. ibidis* Johnston 1911, in *Platibis flavipes;* Australia. No hooks available.

*H. jaenschi* Johnston et Clark 1948, in *Pelecanus conspicillatus;* Australia.

*H. jasuta* Johri 1960, in *Coturnix coturnix;* India.

*H. jerratta* Johri 1960, in *Erolia minuta;* India.

*H. lali* Singh 1960, in *Sterna aurantia;* India.

*H. limnodromi* Prudhoe et Manger 1967, in *Limnodromus griseus;* U.S.

*H. linea* (Goeze 1782) in *Perdix, Caccabis, Coturnix, Gallus, Alectoris, Alauda, Anser, Lagopus;* Europe.

*H. lintonella* Fuhrmann 1932, in *Colymbus hollboelli;* Massachusetts, U.S.

*H. liphallos* (Krabbe 1869) Railliet 1899, in *Cygnus atratus;* Europe.

*H. lobata* Fuhrmann 1906, in *Poecilonetta bahamensis;* Brazil. No scolex.

*H. longiovata* Johri 1962, in *Erolia minuta;* India.

*H. longocylindrocirrus* Deblock et Rosé 1964, in *Charadrius hiaticula;* France.

*H. mahonae* Burt 1969, in *Aegithina tiphia;* Borneo.

*H. malaccensis* Lee 1966, in *Dendrocygna javanica;* Malaysia.

*H. mandabbi* Beverly-Burton 1960, in *Aythya fuligula;* England (?)

*H. meggitteella* Lopez-Neyra 1942, for *H. uliginosa* (Krabbe) of Meggitt 1927, in *Otis houbara.*

*H. meleagris* (Clerc 1902) in *Meleagris gallapavo, Gallus gallus;* Europe, Russia.

*H. mesacantha* (Daday 1900), larval form in *Diaptomus asiaticus;* Gobi Desert.

*H. micrancristrota* (Wedl 1855) in *Cygnus atratus, C. cygnus, Oidemia nigra;* Europe.

*H. minor* Ransom 1909, in *Lobipes lobatus, Phalaropus fulicarius;* Greenland.

*H. parvirostellata* (Linstow 1901) in *Eurystomas afer;* Africa.

*H. patersoni* Deblock et Rosé 1962, in *Crocethia alba;* South Africa.

*H. pigmentata* (Linstow 1872), in *Anas marila;* Europe.

*H. pocilifera* (Linstow 1879), in *Fulica atra;* Germany.

*H. poralis* Meggitt 1927, in *Glareola pratinicola;* Egypt, Russia. No scolex.

*H. pseudoinflata* Skrjabin et Mathevossian 1942, in *Fulica atra.*

*H. pseudosetigera* Baer 1962, in *Anas crecca*; Ireland.

*H. pusilla* Podesta et Holmes 1970, in *Aythya affinis, A. valisineria;* Canada.

*H. ratzi* (Daday 1900) larval form in *Diaptomus asiaticus;* Gobi Desert.

*H. robertsi* Baylis 1934, in *Anas gibberfrons;* Australia.

*H. rudolphica* Hughes 1940, in *Glareola pratinicola, Lymnocryptes minima;* Europe.

*H. septaria* Linstow 1906, in *Upupa ceylonensis;* India.

*H. shen-tsengi* Hughes 1940, in *Rostratula capensis;* China, Philippines.

*H. skrjabinissima* Krotov 1952 in *Calidris subminuta;* Russia.

*H. smythi* Singh 1960, in *Nettapus coromandelianus;* India.

*H. sphaerophora* (Rudolphi 1810), in *Scolopax arquata, Capella gallinago, Numenius tenuirostris, Scolopax rusticola;* Europe.

*H. spinocirrosa* Podesta et Holmes 1970, in *Aythya affinis, S. valisineria, A. americana;* Canada.

*H. stellorae* Deblock, Biquet et Capron 1960, in *Larus ridibundus;* France.

*H. tanakpuris* Johri 1960, in *Alectoris graeca;* India.

*H. tichodroma* Fuhrmann 1908, in *Tichodroma muraria;* Europe.

*H. trichosoma* (Linstow 1882), in *Fuligula ferina;* locality not known.

*H. trombidacantha* Podesta et Holmes 1970, in *Aythya affinis, A. valisineria, A. americana;* Canada.

*H. uliginosa* (Krabbe 1882) (syn. *H. fuliginosa* Mayhew 1925), in *Numenius phaeopus, Gruiformes, Charadriiformes;* Europe, Egypt.

*H. upuparum* Joyeux et Baer 1955, in *Upupa epops;* France.

*H. vilocirrus* Deblock et Rausch 1967, in *Limosa haemastica;* Alaska, U.S.

*H. vistulae* Czaplinski 1960, in *Mergus merganser;* Czechoslovakia.

*H. wardlei* Dubey et Pande 1964, in *Anas peocilorhyncha;* (?) India.

## ADDENDUM

The following genus was inadvertently left out of the preceding key.

### *Cloacotaenia* Wolffhügel 1938
### (syn. *Orlovilepis* Spasskii et Spasskaja 1954). (Figure 353)

*Diagnosis:* Scolex very large, relative to size of strobila. Rostellum rudimentary, unarmed,

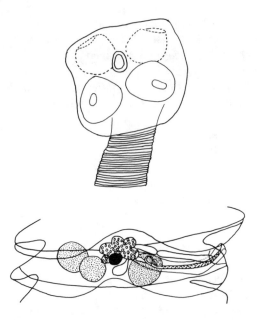

FIGURE 353. *Cloacotaenia megalops* (Nitzsch in Creplin 1829) Wolffhügel 1938.

with small central pit. Neck present. Proglottids much wider than long, craspedote, with long velum. Testes three, one poral, two antiporal. Internal and external seminal vesicles present. Cirrus pouch elongate, crossing excretory canals dorsally. Cirrus armed. Genital pores dextral. Ovary and vitellarium median, pretesticular. Uterus a simple sac, not reaching lateral canals. Vagina opening ventral to cirrus. Seminal receptacle present. Parasites of the cloaca of anseriform birds, commonly observed hanging from the anus.

Type species: *C. megalops* (Nitzsch in Creplin 1829) Wolffhügel 1938 (syn. *Orlovilepis megalops* [Nitzsch in Creplin 1829] Spasskii et Spasskaja 1954, *Taenia cylindrica* Krefft 1871, *T. anatis-marilae* Creplin 1825, *Hymenolepis megalops* Parona 1899), in many species of anseriform birds; cosmopolitan.

# FAMILY DILEPIDIDAE

This is a dominant family of tapeworms, especially in birds. It has three subfamilies: Paruterininae (with paruterine organs), considered by some authors to represent a separate order; Dilepidinae (with a permanent uterus); and Dipylidinae (with the uterus replaced by egg capsules). Classification of this family is in a constant state of flux, and thus is somewhat confused. Dilepididae is closely related to Hymenolepididae and Anoplocephalidae.

Dilepidids are parasites of reptiles (rarely), birds, and mammals. Most parasitize wild birds and mammals and thus are of little importance to human welfare. An exception is *Dipylidium caninum*, a cosmopolitan parasite of domestic dogs and cats that has been found many times in humans, especially children. Its medical consequences are negligible.

Several major monographs have been written on this group in recent years. Among these are Mathevossian,[2141,2146] and Spasskaja and Spasskii.[3385] Wardle et al.[3871] elevated Dilepididae to order Dilepididea, while raising the subfamilies to familial status.

## KEY TO THE SUBFAMILIES OF DILEPIDIDAE

1a. Paruterine organs present............... Paruterininae Fuhrmann 1907. (p.339)
1b. Paruterine organs absent...................................................... 2

2a. Uterus replaced by egg capsules containing one or more eggs ...................
................................................Dipylidiinae Stiles 1896. (p.352)
2b. Uterus reticular, ring-shaped or saccular .........................................
..............................................Dilepidinae Fuhrmann 1907. (p.368)

## KEY TO THE GENERA IN PARUTERININAE

1a. **Rostellum with four circles of hooks; those in first circle triangular........*Neyraia* Joyeux et Timon-David 1934 (syn. *Biuterinoides* Ortlepp 1940). (Figure 354)**

*Diagnosis:* Rostellum with four circles of hooks of varying sizes. Neck long. Proglottids wider than long. Genital ducts pass between osmoregulatory canals. Genital pores alternating irregularly. Cirrus pouch nearly reaches median line. Cirrus unarmed. Testes few (seven to ten), intervascular, surrounding ovary or lateral to it. Ovary bilobed, median, posterior. Vitelline gland postovarian. Vaginal pore posterior to cirrus pore. Seminal receptacle small. Paruterine organ single, large, with median constriction. Uterine sacs two, near posterior end of segment. Parasites of Upupiformes. Europe, Africa.

Type species: *N. intricata* (Krabbe 1878) Joyeux et Timon-David 1934 (syn. *Taenia intricata* Krabbe 1882, *Biuterina lobata* Fuhrmann 1908, *Biuterinoides upupai* Ortlepp 1940, *Neyraia upupai* [Ortlepp 1940] Yamaguti 1959), in *Upupa epops, U. africana, Phoeniculus somaliensis;* Europe, India, Egypt, South Africa, Russia.

Other species:

*N. aegypti* (Omran, El-Naffar et Mandour 1981) comb. n. (syn. *Biuterinoides aegypti* Omran, El-Naffar et Mandour 1981), in *Upupa epops;* Egypt.

*N. hookensis* Chibichenko 1974, in *Upupa epops;* Russia.

*N. krabbei* Kalyankar et Palladwar 1977, in *Upupa epops;* India.

*N. moghei* Shinde 1972, in *Upupa epops;* India.

*N. parva* Mahon 1958, in *Upupa epops;* Africa.

*N. sultanpurensis* Srivastav 1980, in *Upupa epops;* India.

1b. **Rostellum otherwise ......................................................... 2**

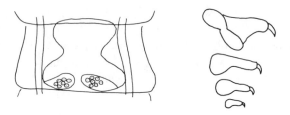

FIGURE 354.. *Neyraia intricata* (Krabbe 1878) Joyeux et Timon-David 1934. Gravid proglottid and rostellar hooks.

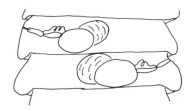

FIGURE 355. *Zosteropicola clelandi* Johnston 1912. Gravid proglottids.

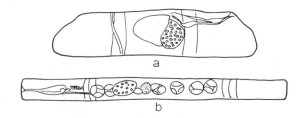

FIGURE 356. *Anoncotaenia mexicana* Voge et Davis 1953. (a) Gravid proglottid; (b) mature proglottid.

**2a. Rostellum with one circle of hooks .. *Zosteropicola* Johnston 1912 (Figure 355)**

*Diagnosis:* Rostellum with a single circle of hooks. Proglottids craspedote. Genital ducts pass ventral to osmoregulatory canals. Genital pores regularly alternating (?). Cirrus pouch extravascular. Testes few, posterior. Ovary median, bilobed. Vitelline gland postovarian. Vaginal pore posterolateral to cirrus pore. Uterus rounded, somewhat poral. Paruterine organ anterior to uterus. Parasites of Passeriformes (Zosteropidae). Australia.

Type species: *Z. clelandi* Johnston 1912, in *Zosterops coecrulescens;* Australia.

**2b. Rostellum otherwise ......................................................... 3**

**3a. Rostellum and hooks absent ...................................... 4**
**3b. Rostellum present, hooks present or absent .................................. 12**

**4a. Mature oncospheres vermiform............................ *Anoncotaenia* Cohn 1900 (syn. *Anurina* Fuhrmann 1909, *Amerina* Fuhrmann 1901). (Figure 356)**

*Diagnosis:* Rostellum absent. Scolex unarmed. Proglottids wider than long. Strobila cylindroid. Genital pores irregularly alternating. Genital ducts ventral to osmoregulatory canals.

Cirrus pouch short. Testes few, anterior. Ovary and vitelline gland compact. Vagina ventral or posterior to cirrus pouch. Seminal receptacle present. Uterus small, ovoid. Paruterine organ lateral or anterior to uterus. Mature oncospheres vermiform. Parasites of Passeriforms. Europe, Asia; North, Central, and South America; Australia, Oceanica, Japan.

Type species: *A. globata* (Linstow 1879) Cohn 1900 (syn. *Taenia globata* Linstow 1879, *T. rudolphiana* Linstow 1879, *T. breviceps* Linstow 1879, *Anonchotaenia clava* Cohn 1900, *Taenia alaudae* Cerruti 1901, *Amerina inermis* Fuhrmann 1901, *Halisis loxiae* Zeder 1873, *Taenia clavata* Marchi 1870), in a wide variety of passeriform birds. Cosmopolitan.

Other species:

*A. arhyncha* Fuhrmann 1918, in *Alauda arvensis, Anthus spinoleta, A. hodgsoni, Emberiza rustica, E. sachalinensis, E. spodecephala, E. variabilis, Muscicapa narcissina, Parus major, Zosterops griseotincta;* New Caledonia, New Hebrides, Australia, Russia, Japan, Asia.

*A. brasiliensis* Fuhrmann 1908, in *Cassicus affinis, C. haemorrhous, Himatione sangvinea, Loxops virens, Vestiaria coccinea,* South America.

*A. castellanii* Fuhrmann et Baer 1943, in *Eurocephalus ruppeli;* Africa.

*A. chauhani* Mukherjee 1967, in *Turdoides sommervilli;* India. (syn. *A. gaugi*?)

*A. dendrocitta* (Woodland 1929) Fuhrmann 1932 (syn. *Rhabdometra dendrocitta* Woodland 1929), in *Dendrocitta rufa, D. vagabunda;* India, Indochina.

*A. gaugi* Singh 1952, in *Turdoides sommervillei;* India.

*A. indica* Singh 1964, in *Muscicapa sundara;* India.

*A. longiovata* (Fuhrmann 1901) Fuhrmann 1932 (syn. *Amerina longiovata* Fuhrmann 1901), in *Curacus aterrimus, Icterus cayennensis, Loxops* sp., *Pipilo aberti, Plegadis guarauna, Notiopsar curaeus;* South America.

*A. macrocephala* Fuhrmann, 1908, in *Hirundella* sp., *Hirundo* sp., *Progne chalybea, Progne purpurea, Progne tapara;* Central and South America.

*A. magniuterina* Rysavy 1957, in *Parus atricapillus, Passer domesticus;* Czechoslovakia.

*A. megaparuterina* (Capoor et Srivastava 1966) Matevossian 1969 (syn. *Mogheia megaparuterina* Capoor et Srivastava 1966), in ''bird''; India.

*A. mexicana* Voge et Davis 1953, in *Carpodacus mexicanus, Pipilo erythrophthalmus;* Mexico.

*A. oriolina* Cholodkowsky 1906, in *Oriolus galbula;* Russia.

*A. piriformis* Fuhrmann 1929, in *Pachycephala morariensis, P. caledonica;* New Caledonia.

*A. quiscali* Rausch et Morgan 1947, in *Quiscalus versicolor, Agelaius phoeniceus;* U.S.

*A. sbesteriometra* Joyeux et Baer 1935, in *Motacilla cinerea;* Indochina.

*A. transcaucasica* Bauer 1941 (syn. *Orthoskrjabinia transcaucasica* [Bauer 1941] Matevossian 1969), in *Aegithabulus caudatus, Parus ater, Sitta europea;* Russia.

*A. trochili* Fuhrmann, 1908, in *Dupetomena macrura;* Brazil.

*A. zanthopygiae* Yamaguti 1956, in *Zanthopygia narcissima;* Japan.

**4b.    Mature oncospheres not vermiform ............................................. 5**

**5a.    Genital pores open irregularly on dorsal and ventral surfaces, sublateral or medial.........................*Anomaloporus* Voge et Davis 1953. (Figure 357)**

*Diagnosis:* Scolex unarmed, lacking rostellum. Neck long. Proglottids acraspedote, wider than long except some gravid ones, which are longer than wide. Genital pores irregular on either side, lateral, sublateral or medial. Atrium weakly developed. Cirrus not described.

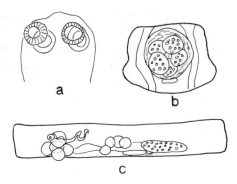

FIGURE 357. *Anomaloporus hesperiphonae* Voge et Davis 1953. (a) Scolex; (b) gravid proglottid; (c) mature proglottid.

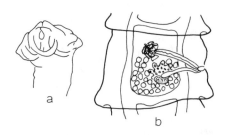

FIGURE 358. *Octopetalum longicirrosum* Baer 1925. (a) Scolex; (b) proglottid.

Inner and outer seminal vesicles absent. Testes number 7 to 14, in one or two groups, poral to ovary. Vagina posterior to cirrus pouch. Ovary compact, on aporal side. Vitellarium anterior, posterior or lateral to ovary. Paruterine organ anterior of uterus. Eggs further enclosed in fibrous capsules, or not. Parasites of Passeriformes, Apodiformes. Mexico.

Type species: *A. hesperiphonae* Voge et Davis 1953, in *Hesperiphona abeilli;* Mexico.
Other species:
  *A. lambi* Voge et Davis 1953, in *Streptoprocne semicollaris;* Mexico.

**5b.    Genital pores lateral............................................................. 6**

**6a.    Each sucker almost covered by two lobe-like flaps ...............................
........................................*Octopetalum* Baylis 1914. (Figure 358).**

*Diagnosis:* Rostellum absent, scolex unarmed. Suckers each with a pair of lappets, separated by a median cleft. Proglottids craspedote, longer than wide when gravid. Genital pores irregularly alternating. Genital ducts dorsal to osmoregulatory canal. Dorsal osmoregulatory canal not described. Cirrus pouch crossing osmoregulatory canal or not. Testes numerous in intervascular medulla. Ovary slightly poral. Vitelline gland posterior or lateral to ovary. Vaginal pore posterior to cirrus. Seminal receptacle present. Uterus sac-like posterior. Paruterine organ anterior to uterus. Parasites of Galliformes. Africa, France.

Type species: *O. gutterae* Baylis 1914, in *Guttera edouardi, Numida ptilorhyncha, N. meleagris;* Africa, France.
Other species:
  *O. numida* (Fuhrmann 1909) Baylis 1914 (syn. *Rhabdometra numida* Fuhrmann 1909,

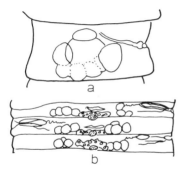

FIGURE 359. *Multiuterina skrjabini* Mathevossian 1948. (a) Gravid proglottid; (b) mature proglottid.

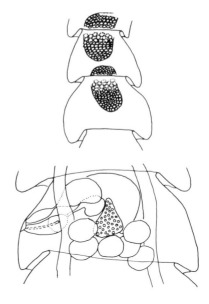

FIGURE 360. *Lallum magniparuterina* Johri 1960. Gravid and mature proglottids.

*O. longicirrosum* Baer 1925, *Unciunia sudanea* Woodland 1928), in *Numida meleagris, N. ptilorhyncha, N. maxima, N. mitrata;* Africa.

6b.  Suckers without flaps ......................................................... 7

7a.  Testes number five to seven..................................................... 8
7b.  Testes number 20 or more ..................................................... 9

8a.  Uterus a multiseptate sac. Genital pores alternating irregularly ................
    .................................. **Multiuterina** Mathevossian 1948. (Figure 359)

*Diagnosis:* Rostellum absent. Scolex unarmed. Gravid proglottids wider than long, craspedote. Genital pores irregularly alternating, equatorial. Genital ducts passing between osmoregulatory canals. Cirrus pouch well developed. Internal seminal vesicle present. Testes few (six to seven), in two lateral groups. Ovary median. Vitelline gland postovarian. Vagina posterior to cirrus pouch. Gravid uterus a multiseptate sac. Paruterine organ rounded, anterior of uterus. Parasites of Passeriformes. Russia.

Type species: *M. skrjabini* Matevossian 1948, in *Oriolus oriolus;* Russia.
Other species:
   *M. dubininae* Matevossian 1969, in *Coccothraustes coccothraustes, Dryobates leucotos, D. major, Emberiza elegans, Eophona personata, Fringilla montifringilla;* Russia.
   *M. junlanae* Matevossian 1969, in *Apodemus agrarius;* Russia.
   *M. spasskyi* Rysavy 1965, in *Dendrocopus major, Oriolus oriolus;* Czechoslovakia.

8b.  Uterus not multiseptate. Genital pores unilateral.................................
    ................................................*Lallum* **Johri 1960. (Figure 360)**

*Diagnosis:* Scolex spheroid, no rostellum, suckers unarmed. Short neck present. Proglottids craspedote. Genital pores unilateral, in small atrium at middle of lateral margin.

FIGURE 361. *Metroliasthes lucida* Ransom 1900. Gravid proglottid.

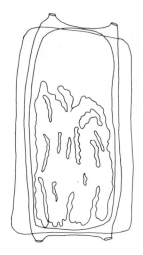

FIGURE 362. *Dictyuterina cholodkowskii* (Skrjabin 1914) Spasskii in Spasskaja et Spasskii 1971. Gravid proglottid.

Genital ducts pass dorsal to osmoregulatory canals. Cirrus pouch reaches or slightly crosses poral osmoregulatory canals. Internal seminal vesicle absent, external seminal vesicle present. Five to seven testes, postovarian, intervascular. Ovary not lobed, in anterior half of proglottid, sometimes extending into posterior part of preceding segment. Vitellarium triangular, dorsal and posterior to ovary. Vagina anterior to cirrus pouch. Large seminal receptacle present. Uterus a transverse sac. Paruterine organ posterior to uterus, extending into following proglottid. Parasites of Anseriforms. India.

Type species: *L. magniparuterina* Johri 1960, in *Nettion crecca;* India.

**9a.    Gravid uterus a double sac ........... *Metroliasthes* Ransom 1900. (Figure 361)**

*Diagnosis:* Rostellum absent. Scolex unarmed. Proglottids craspedote, may be longer than wide. Genital pores irregularly alternating. Genital ducts pass between osmoregulatory canals. Cirrus pouch crossing osmoregulatory canals or not. Cirrus armed. Testes in two lateral groups. Ovary median. Vitelline gland posterior to ovary. Vagina posterior to cirrus pouch. Gravid uterus in form of two sacs, side by side. Paruterine organ anterior to uterine sacs. Parasites of Galliformes. Cosmopolitan.

Type species: *M. lucida* Ransom 1900, in *Alectoris graeca, Coturnix coturnix, C. rufa, Gallus bankiva, G. gallus, Guttera eduardi, Meleagris gallopavo, Numida meleagris N. ptilorhyncha, Perdix perdix;* North and South America, Europe, Africa, India, Australia, Russia.

Other species:
  *M. coturnix* Sawada et Funabashi 1972, in *Coturnix coturnix;* Japan.
  *M. fulvida* Meggitt 1933, in *Oriolus chinensis;* Calcutta Zoo.

**9b.    Gravid uterus otherwise....................................................... 10**

**10a.    Uterus reticular, with paruterine organ at anterior end..........................
            ............... *Dictyuterina* Spasskii in Spasskaja et Spasskii 1971. (Figure 362)**

*Diagnosis:* Scolex with rostellum armed with two circles of hooks. Suckers powerful,

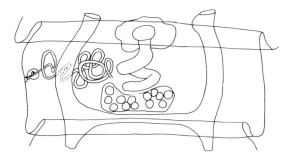

FIGURE 363. *Lyruterina nigropunctata* (Crety 1890) Spasskaja et Spasskii 1971. Gravid proglottid.

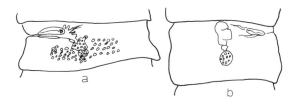

FIGURE 364. *Rhabdometra nullicolis* Ransom 1909. (a) Mature proglottid; (b) gravid proglottid.

unarmed. Neck present. Gravid proglottids longer than wide; acraspedote. Genital pores marginal, irregularly alternating. Testes numerous, posterior and lateral to ovary. Cirrus pouch well developed. Ovary and vitellarium median. Vagina posterior to cirrus pouch. Uterus reticular, with paruterine organ at anterior end. Seminal receptacle present. Parasites of Passeriformes. Russia.

Type species: *D. cholodkowskii* (Skrjabin 1914) Spasskii in Spasskaja et Spasskii 1971 (syn. *Paruterina cholodkowskii* Skrjabin 1914, *Deltokeras cholodkowskii* [Skrjabin 1914] Mathevossian 1965), in *Otomela romanowi, Lanius cristatus:* Russia.

**10b.  Uterus not reticular ......................................................... 11**

**11a.  Paruterine organ sinuous ........................................................**
**........................................*Lyruterina* Spasskaja et Spasskii 1971. (Figure 363)**

*Diagnosis:* Scolex lacking rostellum or armature. Suckers simple. Short neck present. All segments wider than long, craspedote. Genital pores equatorial, irregularly alternating. Genital ducts pass between osmoregulatory canals. Cirrus pouch small, not reaching ventral osmoregulatory canal. Vas deferens coiled. Testes numerous, posterior and lateral to uterus. Ovary small, medial, lobated. Vitellarium compact, postovarian. Uterus a small, transverse sac near posterior margin of segment, with sinuous paruterine organ extending to anterior margin of proglottid. Parasites of Galliformes. Russia, Europe.

Type species: *L. nigropunctata* (Crety 1890) Spasskaja et Spasskii 1971 (syn. *Taenia nigropunctata* Crety 1890, *Rhabdometra nigropunctata* [Crety 1890] Cholodkowsky 1906), in *Coturnix, Caccibis, Lyrurus, Tetrastes, Alectoris, Perdix;* Europe, Russia.

**11b.  Paruterine organ not sinuous .....................................................**
**...................................*Rhabdometra* Cholodkowsky 1906. (Figure 364)**

*Diagnosis:* Rostellum absent. Scolex unarmed. Proglottids craspedote, wider than long. Genital pores irregularly alternating. Genital ducts pass between osmoregulatory canals. Cirrus pouch claviform, may cross osmoregulatory canals. Testes in single, posterior field. Ovary median. Vitelline gland postovarian. Vagina posterior to cirrus pouch. Seminal receptacle present. Uterus a single, median sac. Paruterine organ anterior to uterus. Parasites of Galliformes. Africa, Russia, North America, Europe.

Type species: *R. tomica* Cholodkowsky 1906, in *Francolinus pictus, Lagopus lagopus, Lyrurus tetrix, Tetrao urogalloides, T. parvirostris, Tetraster bonasia;* Russia, India.

Other species:

*R. alpinensis* Olsen, Haskins et Braun 1978, in *Lagopus leucurus;* U.S.

*R. cylindrica* Beddard 1914, in *Caccabis melanocephala;* Africa.

*R. dogieli* Gvosdev 1954, in *Tetraogallus himalayensis, Alectoris graeca;* Russia.

*R. lygodaptrion* Beverly-Burton et Thomas 1980, in *Lagopus lagopus;* Canada.

*R. nigromaculata* Dubinina 1950, in *Phasianus colchicus, Alectoris graeca, Coturnix coturnix;* Russia.

*R. nullicollis* Ransom 1909, in *Centrocercus urophasianus, Pedioecetes phasianellus;* U.S.

*R. odiosa* (Leidy 1887) Jones 1929 (syn. *Taenia odiosa* Leidy 1887), in *Colineus virginianus, Lophortyx californica, Oreotyx picta, Podiocetes phasianellus, Tympanuchus pallidicincutus;* U.S.

*R. pteroclesi* Sultanov 1963, in *Pterocles orientalis;* Russia.

*R. setosa* Fediushin 1953, in *Lyrurus tetrix;* Russia.

**12a. Rostellum rudimentary, lacking hooks .......................................... 13**
**12b. Rostellum with a double circle of hooks .......................................... 14**

**13a. Gravid uterus tree-like or fungiform. Paruterine organ on a short stalk ........**
**..................................... Dendrometra Jordano et Diaz-Ungria 1956.**

*Diagnosis:* Scolex unarmed, rostellum rudimentary. Neck poorly defined. Proglottids acraspedote. Genital pores irregularly alternating. Genital ducts pass between osmoregulatory canals. Testes scattered randomly throughout proglottid, mainly lateral, and posterior to ovary. Cirrus pouch well developed, in front of vagina. Cirrus not described. Seminal receptacle? Ovary median, irregularly shaped. Vitelline gland compact, postovarian. Uterus first a simple transversely arched sac, then shaped like an inverted mushroom, with paruterine organ anterior and attached to uterus by slender stalk. Eggs contained in several capsules. Parasites of Pelecaniformes (frigate bird). South America.

Type species: *D. ginesi* Jordano et Diaz-Ungria 1956, in *Fregata magnificens;* Venezuela.

**13b. Gravid uterus not as above ................................................. *Ortho-*
***skrjabinia* Spasskii 1947 (syn. *Skrjabinerina* Mathevossian 1948). (Figure 365)**

*Diagnosis:* Rostellum rudimentary, unarmed. Proglottides craspedote, wider than long. Genital pores irregularly alternating. Genital ducts ventral to osmoregulatory canals. Cirrus pouch may cross osmoregulatory canals. Internal seminal vesicle present. Testes few (9 to 12) in two lateral fields. Ovary compact, median. Vitelline gland postovarian. Uterus first simple, then dividing into several lobes in transverse row. Paruterine organ anterior to uterus. Oncosphere elongate. Parasites of Passeriformes. Europe, Africa, Russia, North American.

Type species: *O. bobica* (Clerc 1903) Spasskii 1947 (syn. *Amerina inermis* Clerc 1912, after Fuhrmann 1906, *Anonchotaenia bobica* Clerc 1903, *Skrjabinerina bobica* (Clerc 1903) Mathevossian 1948), in *Chrysomitris spinus, Lullula arborea, Parus ater, P.*

FIGURE 365. *Orthoskrjabinia bobica* (Clerc 1903) Spasskii 1947. Gravid proglottid.

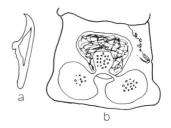

FIGURE 366. *Biuterina campanulata* (Rudolphi 1819) Fuhrmann 1908. (a) Rostellar hook; (b) gravid proglottid.

atricapillus, *P. major, Sitta europea, S.* sp., *Tshagra senegala, Fringilla coelebsi, Passer domesticus, Picoides tridactylus, Parus palustra;* Europe, Russia, Africa.

Other species:

*O. conica* (Fuhrmann 1908) Spasskii 1947 (syn. *Anonchotaenia conica* Fuhrmann 1908, *Skrjabinerina conica* [Fuhrmann 1908] Mathevossian 1947), in *Dryobates major, D. minor, Jungipicus nanus, Nuctifraga caryocatactes, Pericrocotus roseus, Picus canus, Parus major;* Europe, Russia.

*O. oshmarini* (Spasskii 1946) Mathevossian 1948 (syn. *Anonchotaenia oshmarini* Spasskii 1946, *Skrjabinerina oshmarini* [Spasskii 1946] Mathevossian 1948), in *Lanius minor;* Russia.

*O. rostellata* (Rodgers 1941) Mathevossian 1948 (syn. *Anonchotaenia rostellata* Rodgers 1941, *Skrjabinerina rostellata* [Rodgers 1941] Mathevossian 1948), in *Cardinalis cardinalis, Fringilla montifringilla;* U.S., Russia.

**14a. Uterus breaking down into egg capsules, surrounded by paruterine tissues.....
.......................................................*Deltokeras* Meggitt 1927.**

*Diagnosis:* Rostellum armed with two circles of triangular hooks. Proglottids craspedote. Genital pores unilateral or irregularly alternating. Cirrus pouch may cross osmoregulatory canals. Testes numerous, lateral and posterior to ovary. Ovary median. Vitelline gland postovarian. Vagina posterior to cirrus pouch. Seminal receptacle present. Uterus breaking into egg capsules surrounded by paruterine tissue. Parasites of Passeriformes. Asia, Africa, Europe, South America, Oceanica.

Type species: *D. ornitheios* Meggitt 1927, in *Urocissa occipitalis;* Burma.

Other species:

*D. campylometra* Joyeux et Baer 1928, in *Penthetriopsis macrura, Pyromelana franciscana;* Africa.

*D. delachauxi* Hsü 1935, in *Lanius schach;* China.

*D. granatensis* Lopez-Neyra 1943, in *Lanius colurio;* Spain.

*D. multilobatus* Losen 1939, in *Selenchides melanoleucus;* New York Zoo.

*D. synallaxis* Mahon 1957, in *Synallaxis rutilans;* Brazil.

**14b. Uterus not breaking into egg capsules.........................................15**

**15a. Uterus divided into two symmetrical sacs behind spherical paruterine organ. Hooks triangular .....................................................*Biuterina*
Fuhrmann 1902 (syn. *Triaenorhina* Spasskii et Shumilo 1965). (Figure 366)**

*Diagnosis:* Rostellum with two circles of triangular hooks Proglottids craspedote. Genital pores irregularly alternating. Genital ducts pass between osmoregulatory canals. Cirrus pouch small. Testes numerous, anterior, lateral, sometimes posterior to ovary. Ovary bilobed, posterior, median. Vitelline gland postovarian. Vagina ventral or posterior to cirrus pouch. Uterus first a single sac, then dividing into two connected sacs. Paruterine organ anterior to uterus. Parasites of Upupiformes, Coraciiformes, Caprimulgiformes, Passeriformes. Africa, Asia, Europe, New Guinea, South and Central America.

Type species: *B. clavulus* (Linstow 1888) Fuhrmann 1908 (syn. *Taenia clavulus* Linstow 1888, *Davainea clavulus* [Linstow 1888] Blanchard 1891, *Biuterina paradisea* Fuhrmann 1902), in *Manucodia chalybeata, Paradisea raggiana, Ptilorchis alberti;* Australia, New Guinea.

Other species:

*B. africana* Joyeux et Baer 1928, in *Pomatorhynchus senegalis, P. australe, P. anchietae;* Africa.

*B. campanulata* (Rudolphi 1819) Fuhrmann 1908 (syn. *Taenia campanulata* Rudolphi 1819), in *Muscicapa audax, M. columbina, Taenioptera vilata, Thamnophilus sulfuratus;* Brazil.

*B. clerci* Spasskii 1946, in *Emberiza citrinella;* Russia.

*B. colluriones* Mathevossian 1950, in *Lanius collurio;* Russia.

*B. coracii* Chiriac 1963, in *Coracias garrulus;* Rumania.

*B. cylindrica* Fuhrmann 1908, in *Astimastillas falhensteini, Tachyphonus critatus, T. melaleucus;* Brazil, Panama.

*B. dicruri* Singh 1964, in *Dicrurus leucophaeus;* India.

*B. distincta* Fuhrmann 1908, in *Graculus* sp.; Brazil.

*B. dunganica* Skrjabin 1914, in *Oriolus galbula;* Rusisa.

*B. fallax* Meggitt 1928, in *Merops apiaster;* Africa.

*B. fuhrmanni* Schmelz 1941, in *Emberiza aureola;* China.

*B. garrulae* (Mathevossian 1950) Mathevossin 1964 (syn. *Paruterina garrulae* Mathevossian 1950), in *Coracias garrulus;* Russia.

*B. globosa* Fuhrmann 1908, in *Tityra semifasciata;* Brazil, Mexico.

*B. meggitti* (Johri 1931) Mathevossian 1964 (syn. *Paruterina meggitti* Johri 1931), in *Pophoceros birostris, L. erythrorhynchus;* India, Africa.

*B. meropina* (Krabbe 1869) Fuhrmann 1908 (syn. *Taenia meropina* Krabbe 1869), in *Lullula arborea, Merops superciliosis;* Egypt, Madagascar, Europe.

*B. meropina macrancristrota* Fuhrmann 1908, in *Dryocopus angolensis, Melitophagus albifrons, Merops albicollis, M. apiaster, M. nubicoides;* Africa, India, Russia.

*B. mertoni* Fuhrmann 1911, in *Paradisea apoda;* Aru Island, New Guinea.

*B. morgani* (Rausch et Schiller 1949) Mathevossian 1965 (syn. *Paruterina morgani* Rausch et Schiller 1949), in *Salpinctes obsoleutus;* U.S.

*B. motacillabrasiliensis* (Rudolphi 1819) Fuhrmann 1932 (syn. *Taenia motacillabrasiliensis* Rudolphi 1819), in *Motacilla* sp.; Brazil.

*B. motacillacayanae* (Rudolphi 1819) Fuhrmann 1908 (syn. *Taenia motacillacayanae* Rudolphi 1819), in *Dachnis cayana;* Brazil.

*B. passerina* Fuhrmann 1908, in *Alauda arvensis, Emberiza citrinella, E. schoeniclus, Galerida cristata, Lanius spenocerus, Lullula arborea, Motacilla flava, Sturnus vulgaris;* Europe, Russia.

*B. pentamyzos* (Mettrick 1960) Mathevossian 1964 (syn. *Paruterina pentamyzos* Mettrick 1960), in *Prinops plumata;* Africa.

*B. planirostris* (Krabbe 1879) Fuhrmann 1908 (syn. *Taenia planirostris* Krabbe 1879), in *Alauda* sp.; Turkestan.

*B. rectangula* Fuhrmann 1908 (syn. *Triaenorhina rectangula* [Fuhrmann 1908] Spasskii et Shumilo 1965), in *Coracias galbula, C. garrulus, C. caudata;* Europe, Africa, Russia.

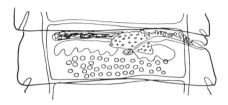

FIGURE 367. *Culcitella crassa* Fuhrmann 1906.

*B. reynoldsi* (Daly 1958) Mathevossian 1964 (syn. *Paruterina reynoldsi* Daly 1958), in *Corvus brachyrhynchos;* U.S.

*B. singhi* Mathevossian 1965 (syn. *B. meggitti* Singh 1959, preoccupied), in *Dicrurus caerulescens;* India.

*B. sobolevi* Sudarikov 1950, in *Saxicola rubetra;* Russia.

*B. trapezoides* Fuhrmann 1908, in *Acrocephalus phragmitis, Caprimulgus* sp., *Emberiza* sp., *Molothrus pecoris*; Brazil, Russia.

*B. triangula* (Krabbe 1869) Fuhrmann 1908 (syn. *Taenia triangulus* Krabbe 1869), in *Anthus trivialis, Motocilla alba, Turdus pilaris, Saxicola rubetra, Sylvia hortensis, Acrocephalus shoenbaenus, Phyllocephalus fuscatus, Erithacus rubecula, Luscinia mesarhynchos, Phoenicurus phoenicurus;* Europe, Russia.

*B. trigonacantha* Fuhrmann 1908, in *Synallaxis phryganophila;* Brazil.

*B. ugandae* Baylis 1919, in *Chalcomitra gutturalis;* Africa.

*B. uzbekiensis* Mathevossian 1964 (syn. *Paruterina rectangula* [Fuhrmann 1908] Sultanov et Spasskaja 1959), in *Caracias garrulus;* Russia.

*B. zambiensis* (Mettrick 1960) Mathevossian 1964 (syn. *Paruterina zambiensis* Mettrick 1960), in *Campephaga phoenica;* Africa.

**15b.** Uterus not as above ......................................................... 16

**16a.** Uterus a transversely elongated sac with short branches, behind paruterine organ.................................. ***Culcitella*** **Fuhrmann 1906. (Figure 367)**

*Diagnosis:* Rostellum with two circles of long-handled hooks. Proglottids craspedote. Genital pores unilateral or irregularly alternating. Genital ducts pass between osmoregulatory canals. Testes numerous, posterior and lateral to ovary. Vagina posterior or ventral to cirrus pouch. Seminal receptacle present. Ovary median or poral. Vitelline gland postovarian. Uterus a transversely elongated sac with short branches. Paruterine organ anterior to uterus. Parasites of Accipitriformes. Africa, Central and South America.

Type species: *C. rapacicola* Fuhrmann 1906, in *Astrunia nitida, Geranospizias caerulescens, Ictinia palumbea;* South America.

Other species:

*C. bresslaui* Fuhrmann 1927, in *Rupornis leucorrhoea;* Brazil.

*C. crassa* Fuhrmann 1906, in *Spizaetus ornatus;* Central and South America.

*C. fuhrmanni* (Southwell 1925) Baer 1933 (syn. *Lateriporus fuhrmanni* Southwell 1925), in *Aquila rapax, Circaetus cinereus, Falco biarmicus;* Africa.

**16b.** Uterus not as above ......................................................... 17

**17a.** Testes dorsal to ovary and vitellarium, five in number............................ ..................................... ***Notopentorchis*** **Burt 1938. (Figure 368)**

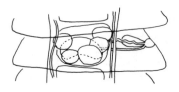

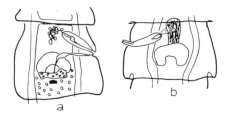

FIGURE 369. *Paruterina angustata* Fuhrmann 1906. (a) Mature proglottid; (b) gravid proglottid.

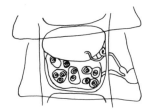

FIGURE 368. *Notopentorchis micropus* Singh 1952.

*Diagnosis:* Rostellum with two circles of triangular hooks. Proglottids craspedote. Genital pores irregularly alternating. Genital ducts ventral to osmoregulatory canals. Cirrus pouch extravascular. Testes five. Ovary compact, median. Vitelline gland postovarian. Vagina posterior to cirrus pouch. Seminal receptacle present. Uterus first double, then a single sac. Paruterine organ anterior to uterus. Parasites of Apodiformes. Sri Lanka, India.

Type species: *N. collocaliae* Burt 1938, in *Collocalia unicolor;* Sri Lanka.
Other species:
   *N. micropus* Singh 1952, in *Micropus affinis;* India.

**17b. Testes posterior, sometimes also partly lateral or anterior to ovary and vitellarium, more than five in number ................................................. 18**

**18a. Testes posterior and lateral, sometimes partly anterior to ovary. Uterus horseshoe-shaped ...................... *Paruterina* Fuhrmann 1906. (Figure 369)**

*Diagnosis:* Rostellum with two circles of hooks. Proglottids craspedote. Genital pores irregularly alternating or unilateral. Genital ducts pass between osmoregulatory canals. Cirrus pouch may cross osmoregulatory canals. Testes mainly posterior and lateral to ovary. Ovary median. Vitelline gland postovarian. Vagina posterior to cirrus pouch. Seminal receptacle present. Uterus horseshoe-shaped or transversely elongate. Paruterine organ anterior to uterus. Parasites of Accipitriformes, Bucerotiformes, Cuculiformes, Cypseliformes, Passeriformes, Strigiformes. Europe, Asia, Java, North and South America, Africa.

Type species: *P. candelabraria* (Goeze 1782) Fuhrmann 1906 (syn. *Taenia candelabraria* Goeze 1782), in *Aegolius acadica, Asio accepitrinus, A. flammeus, A. otus, Bubo bubo, B. virginianus, Nictala tengmalmi, Nyctea scandiaca, Otus bakkamoena, Scops scops, Strix aluco, S. uralensis, Surnia ulula, Speotyto syrnium,* etc.; circumboreal.

Other species:
   *P. angustata* Fuhrmann 1906, in *Scops brasilianus;* South America.
   *P. bovieni* Hübscher 1937, in *Macropteryx longipennis;* Java.
   *P. bucerotina* Fuhrmann 1909, in *Lophoceros nasutus;* Africa.
   *P. chlorurae* Rausch et Schiller 1949, in *Chloroura chloroura;* U.S.
   *P. daouensis* Joyeux, Baer et Martin 1936, in *Bucorvus abyssinicus;* Africa.
   *P. garrulae* Mathevossian 1950, in *Coracias garrulus;* Russia.
   *P. guineensis* Joyeux et Baer 1925, in *Coccystes cafer;* Africa.

FIGURE 370. *Sphaeruterina punctata* Johnston 1914. Mature proglottids.

*P. iduncula* Spasskii 1946, in *Apus apus, Apus pacificus;* Russia.

*P. isoniciphora* Dollfus 1958, in *Apus pallidus;* Morocco.

*P. javanica* Hübscher 1937, in *Macropteryx longipennis;* Java.

*P. kirghisica* Mathevossian 1950, in *Syliva curruca;* Russia.

*P. otidis* Baczynska 1914, in *Asio accepitrinus;* Europe.

*P. parallelepideda* (Rudolphi 1810) Fuhrmann 1908 (syn. *Taenia parallelepipeda* Rudolphi 1810), in *Lanius excubitor L. minor, L. collurio, Vanga rufa;* Europe, Africa.

*P. podocesi* Danzan 1964, in *Podoces hendersoni;* Mongolia.

*P. quelea* Mettrick 1963, in *Quelea quelea;* Central Africa.

*P. rauschi* Freeman 1957, in *Strix varia, Bubo virginianus, Argolius acadica;* North America, Israel.

*P. septotesticulata* Moghe et Inamdar 1934, in *Coracius indica;* India.

*P. similis* (Ransom 1909) Linton 1927 (syn. *Rhabdometra similis* Ransom 1909), in *Coccyzus americanus, Dryobates villosus;* U.S.

*P. skrjabinini* Mathevossian 1950, in *Coracias garrulus;* Russia.

*P. southwelli* Hilmy 1936, in *Lophoceros semifasciatus, Tockus alboterminatus;* Africa.

*P. vesiculigera* (Krabbe 1882) Fuhrmann 1926 (syn. *Taenia vesiculigera* Krabbe 1882, *Anomotaenia vesiculigera* [Krabbe 1882] Fuhrmann 1908), in *Apus apus, A. malba, A. pacificus, Hirundapus caudacutus;* Europe, Russia.

**18b. Testes posterior to ovary. Uterus spherical or irregular.........................**
**........................................ *Sphaeruterina* Johnston 1914. (Figure 370)**

*Diagnosis:* Rostellum with two circles of hooks. Neck absent. Proglottids craspedote. Genital pores alternating irregularly. Genital ducts pass between osmoregulatory canals. Cirrus pouch extravascular. Cirrus short. Testes few, postovarian. Ovary bilobed, slightly poral. Vitelline gland? Vagina posterior to cirrus pouch. Seminal receptacle present. Uterus rounded. Paruterine organ anterior to uterus, with apical dilation. Parasites of Jaccamariformes, Passeriformes. New Caledonia, Europe, South America, Australia.

Type species: *S. punctata* Johnston 1914 (syn. *Biuterina punctata* Fuhrmann 1918), in *Pachycephala rufiventris, P. xantherythraea;* New Caledonia, Australia.

Other species:

*S. fuhrmanii* (Baczynska 1914) Fuhrmann 1932 (syn. *Paruterina fuhrmanni* Baczynska 1914), in *Bucco* sp.; Brazil.

*S. longiceps* (Rudolphi 1819) Fuhrmann 1932 (syn. *Taenia longiceps* Rudolphi 1819, *Biuterina longiceps* Fuhrmann 1908), in *Cassicus affinis, Oriolus cristatus, Ostinops decumanus, Cairina moschata;* North and South America.

*S. purpurata* (Dujardin 1845), *Dilepis purpurata* (Dujardin 1845) Fuhrmann 1908, *Paruterina purpurata* [Dujardin 1845] Joyeux et Timon-David 1934), in *Lanius sen-*

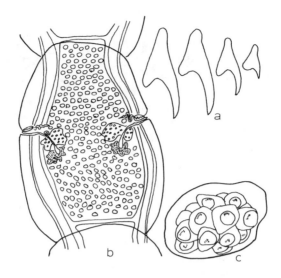

FIGURE 371. *Dipylidium otcocyonis* Joyeux, Baer et Martin 1936. (a) Rostellar hooks; (b) proglottid; (c) egg capsule.

*ator, Melizophilus undatus, Sylvia cinerea, S. communis, S. melanocephala, S. sylvia;* Europe.

## KEY TO THE GENERA IN DIPYLIDIINAE

1a. Two sets of reproductive organs in each proglottid ............................. 2
1b. One set of reproductive organs in each proglottid ............................. 5

2a. Rostellum with a single circle of hooks ...... *Diskrjabiniella* Mathevossian 1954

*Diagnosis:* Rostellum with a single circle of hooks. Two sets of reproductive organs per proglottid. Genital pores bilateral. Testes numerous. Ovary compact. Each egg capsule with a single egg. Parasites of Accipitriiformes (vulture), Columbiformes; Africa.
  Type species: *D. avicola* (Fuhrmann 1906) Mathevossian 1954 (syn. *Dipylidium avicola* Fuhrmann 1906, *Progynopylidium avicola* [Fuhrmann 1906], *Diplopylidium avicola* [Fuhrmann 1906] Fuhrmann 1932), in *Gyps kolbi;* Africa.
  Other species:
    *D. columbae* (Fuhrmann 1908) Mathevossian 1954 (syn. *Dipylidium columbae* Fuhrmann 1908, *Progynopylidium columbae* [Fuhrmann 1908] Lopez-Neyra 1929, *Diplopylidium columbae* [Fuhrmann 1908] Fuhrmann 1932), in *Columba* sp.; Egypt.

2b. Rostellum with several circles of hooks.......................................... 3

3a. Each egg capsule with several eggs ................................................
................................. *Dipylidium* Leuckart 1863 (syn. *Alyselminthus* Bloch 1782, *Halisis* Zeder 1800 in part, *Microtaenia* Sedwick 1884). (Figure 371)

*Diagnosis:* Rostellum with several circles of rosethorn-shaped hooks. Mature and gravid proglottids longer than wide, acraspedote, constricted at intersegments. Each proglottid with two sets of reproductive organs. Testes numerous, in entire intervascular field. Genital pores postequatorial. Ovary bilobed. Vitelline gland postovarian. Vagina ventral or posterior to

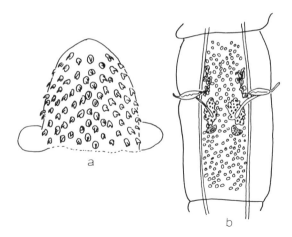

FIGURE 372. *Joyeuxiella rossicum* (Skrjabin 1923) Fuhrmann 1935. (a) Rostellum; (b) proglottid.

cirrus pouch. Uterus first reticular, then breaking into egg capsules each with several eggs. Parasites of Carnivora, rarely in reptiles or humans. Cosmopolitan.

Type species: *D. caninum* (Linnaeus 1758) Leuckart 1863 (for a very long list of synonyms see Mathevossian[2141]), in *Canis* spp., *Hyaena striata*, *Felis* spp., *Zilethalurus viverina*, *Vulpes* spp., *Nictereutes procyonoides*, *Alopex lagopus*, *A. beringensis*; cosmopolitan.
Other species:

*D. buencaminoi* Tubangui 1925, in *Canis familiaris*; Philippines.

*D. dongolense* Beddard 1913, in *Genetta dongolana*; Africa.

*D. genettae* (Gervais 1847) Diamare 1893, in *Viverra genetta*; France.

*D. otocyonis* Joyeux, Baer et Martin 1936, in *Otocyon megalotis*; Africa.

**3b.   Each egg capsule with a single egg .............................................. 4**

**4a.   Cirrus anterior to vagina ................................................*Joyeuxiella* Fuhrmann 1935 (syn. *Joyeuxia* Lopez-Neyra 1927, preoccupied). (Figure 372)**

*Diagnosis:* Rostellum with several circles of rosethorn-shaped hooks. Mature proglottids usually wider than long. Two sets of reproductive organs per proglottid. Genital pores bilateral, preequatorial. Cirrus pouch crosses osmoregulatory canals. Testes numerous, filling intervascular space. Ovary lobated. Vitelline gland postovarian. Vagina posterior to cirrus pouch. Uterus breaks into egg capsules, each with a single egg. Parasites of Carnivora. Europe, Africa, Palestine, Asia.

Type species: *J. pascualei* (Diamare 1893) (syn. *Diplidium pasqualei* Diamare 1893, *D. chyseri* Ratz 1897, *D. fuhrmanni* Baer 1924, *Joyeuxia chyseri* [Ratz 1897], *J. fuhrmanni* [Baer 1924], *J. aegyptica* Meggitt 1927, *J. pasqualaeiformis* Lopez-Neyra 1927), in *Canis familiaris*, *C. lupus*, *Felis catus*, *F. caffra*, *F. sylvatica*, *Zibethailurus serval*, *Genetta rubiginosa*; Europe, Africa, Palestine.
Other species:

*J. echinorhynchoides* (Sonsino 1889) Fuhrmann 1935 (syn. *Taenia echinorhynchoides* Sonsino 1889), in *Canis aureus*, *C. rostrata*, *C. nilotica*, *Megalotis cerdo*, *Vulpes vulpes*, *Fennecus zerda*; Africa, Russia, Palestine.

*J. rossicum* (Skrjabin 1923) Fuhrmann 1935 (syn. *Dipylidium rossicum* Skrjabin 1923), in *Felis catus*, *Canis familiaris*; Russia.

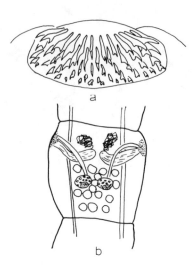

FIGURE 373. *Diplopylidium noelleri* (Skrjabin 1924). (a) Rostellum; (b) mature proglottid.

**4b. Cirrus posterior to vagina** .......................................... ***Diplopylidium* Beddard 1913 (syn. *Progynopylidium* Skrjabin 1924). (Figure 373)**

*Diagnosis:* Rostellum with several circles of hooks. Each proglottid with double sets of reproductive organs. Genital pores bilateral, preequatorial. Testes mainly posterior to level of ovaries. Cirrus pouch large, nearly reaching median line of segment. Ovary lobated. Vitelline gland postovarian. Vagina open anterior to cirrus. Egg capsules each with single egg. Parasites of Carnivora. Europe, Africa, Asia, Palestine, Cyprus.

Type species: *D. genettae* Beddard 1913, in *Genetta dongolana;* London Zoo (Africa?). Other species:

*D. acanthotretra* (Parona 1887) Beddard 1913 (syn. *Cysticercus acanthotetra* Parona 1886, *Diplopylidium trinchesii* [Diamare 1892], *D. triseriale* [Lühe 1898], *D. guinquecoronatum* [Lopez-Neyra et Munoz-Medina 1921], *D. fabulosum* Meggitt 1927, *Dipylidium trinchesii* Diamare 1892, *D. triseriale* Lühe 1898, *D. quinquecoronatum* Lopez-Neyra et Munoz-Medina 1921, *Progynopylidium fabulosum* [Meggitt 1927]), in *Felis catus, Vivera civetta, Genetta afra, Otocyonis megalotis* (many amphibians and reptiles known as transport hosts); Europe, Middle East.

*D. monoophorum* (Lühe 1898) Beddard 1913 (syn. *Dipylidium monoophorum* Lühe 1898, *Progynopylidium monophorum* [Lühe 1898] Skrjabin 1924), in *Genetta afra, Vivera civetta;* Africa.

*D. noelleri* (Skrjabin 1924) (syn. *Progynopylidium nölleri* Skrjabin 1924, *Dipylidium trinchesei* Lopez-Neyra 1927, *Progynopylidium monoophoroides* Lopez-Neyra 1928), in *Canis familiaris, Felis catus, Vulpes rilatica;* China, India, Palestine, Cyprus, Russia, Spain, Egypt, Turkey.

*D. skrjabini* Popov 1935, in *Felis catus* (experimental); Russia.

*D. zschokkei* (Hüngerbühler 1910) Beddard 1913 (syn. *Dipylidium zschokkei* Hüngerbühler 1910, *Progynopylidium zschokkei* [Hüngerbühler 1910] Skrjabin 1924), in *Ichneumia leucrura, Cynictis penicillata;* Africa.

**5a. Rostellum unarmed (hooks may have been lost)** .............................. **6**
**5b. Rostellum armed** ........................................................... **7**

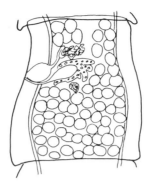

FIGURE 374. *Eugonodaeum bybralis* Johri 1951.

**6a.** Genital pores alternating regularly or irregularly .............................. 8
............... *Pseudochoanotaenia* **Burt 1938 (syn.** *Ptilotolepis* **Spasskii 1969).**

*Diagnosis:* Rostellum retractable, definitely unarmed. Suckers large, unarmed. Proglottids craspedote, wider than long except when gravid. Genital pores alternating regularly or irregularly. Cirrus pouch crosses osmoregulatory canals. Seminal vesicles absent. Testes few (10 to 23), postequatorial. Cirrus armed with very small spines. Ovary bilobed, nearly reaching osmoregulatory canals. Vitelline gland postovarian. Vagina posterior to cirrus pouch. Seminal receptacle proximal. Uterus first reticular, later breaking into egg capsules each with single egg. Parasites of Apodiformes, Passeriformes. Sri Lanka, Australia.

Type species: *P. collocaliae* Burt 1938, in *Collocalia unicolor;* Sri Lanka.
Other species:

    *P. meliphagidarum* (Johnston 1911) Mathevossian 1953 (syn. *Choanotaenia meliphagidarum* Johnston 1911, *Ptilotolepis meliphagidarum* [Johnston 1911] Spasskii 1969), in *Meliphaga leucotis, Philidronyris novaehollandiae,* other Meliphagidae; Australia. (Validity confirmed: Schmidt[3096]).

**6b.** Genital pores unilateral ............. *Eugonodaeum* **Beddard 1913. (Figure 374)**

*Diagnosis:* Rostellum unarmed (hooks may have been lost). Proglottids craspedote. Genital pores unilateral. Cirrus pouch large. Cirrus armed. Testes mainly posterior. Genital ducts pass between poral osmoregulatory canals. Ovary slightly poral. Vitelline gland postovarian. Vagina posterior to cirrus pouch. Egg capsules each with single egg. Parasites of Charadriiformes, Accipitriformes. Asia, South America.

Type species: *E. oedicnemi* Beddard 1913, in *Burhinus bistriatus, Oedicnemus oedicnemus;* South America.
Other species:

    *E. burmanense* Johri 1951, in *Charadrius placidus;* Burma.
    *E. bybralis* Johri 1951, in *Aquila vindhiana;* Burma.
    *E. pycnonoti* Khan et Habibullah 1967, in *Pycnonotus jocosus;* Pakistan.

**7a.** Genital pores unilateral ......................................................... 8
**7b.** Genital pores alternating ...................................................... 14

**8a.** Ovary poral ...................................................................... 9
**8b.** Ovary median ................................................................. 10

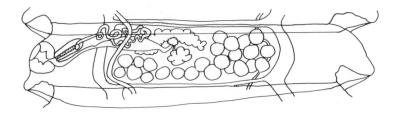

FIGURE 375. *Nasutaenia nasuta* (Fuhrmann 1908) Baer et Bona 1960.

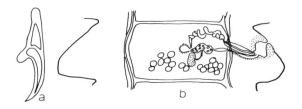

FIGURE 376. *Malika kalawewaensis* Burt 1940. (a) Rostellar hook; (b) mature proglottid.

9a. Wide osmoregulatory canal ventral on poral side, dorsal on aporal side
...................................*Nasutaenia* **Baer et Bona 1960. (Figure 375)**

*Diagnosis:* Rostellum armed (number of hooks unknown). Hook with long handle and short guard. Proglottids wider than long. Osmoregulatory canals reversed on right and left sides: wide vessel ventral on poral side, dorsal on aporal side; narrow vessels similarly reversed. Genital pores unilateral, sinistral. Genital ducts pass between osmoregulatory canals. Atrium deep, surrounded by very powerful sphincter. Cirrus pouch not reaching poral osmoregulatory canals. Seminal vesicles absent. Vas deferens coiled, near cirrus pouch. Testes numerous, posterior and lateral to ovary, intervascular. Ovary slightly poral, with compact lobes. Vitellarium compact, postovarian. Vagina posterior to cirrus pouch; seminal receptacle conspicuous, persistent. Uterus at first reticular, then breaking into capsules each with a single egg. Parasites of Ardeiformes. Brazil.

Type species: *N. nasuta* (Fuhrmann 1908) Baer et Bona 1960 (syn. *Dilepis nasuta* Fuhrmann 1908), in *Theristicus caudatus;* Brazil.

9b. Osmoregulatory canals normal .......................................................
...... *Malika* **Woodland 1929 (syn. *Parachoanotaenia* Rego 1967). (Figure 376)**

*Diagnosis:* Rostellum with double circle of hooks. Proglottids craspedote, most wider than long. Genital ducts pass between osmoregulatory canals. Genital pores unilateral. Genital atrium large. Cirrus pouch long. Cirrus unarmed. Testes mainly postovarian. Ovary poral, anterior. Vitelline gland median or postovarian. Vagina posterior to cirrus pouch. Seminal receptacle small. Uterus first with lateral branches, later breaking into egg capsules, each with several eggs. Parasites of Charadriiformes, Passeriformes. Sri Lanka, Asia, Brazil.

Type species: *M. oedicnemus* Woodland 1929, in *Burhinus oedicnemus;* India.
Other species:
*M. chauhani* Pandey et Tayal 1981, in *Burhinus oedicnemus;* India.
*M. daviesi* Mathevossian 1963 (syn. *Dilepis undula* [Schrank 1788] Davies 1935), in *Corvus frugilegus, Sturnus vulgaris, Turdus merula, T. musicus;* England.

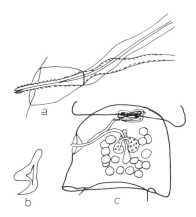

FIGURE 377. *Aleurotaenia planicipitis* Cameron 1928. (a) Genital atrium; (b) rostellar hook; (c) proglottid.

*M. himantopodis* Burt 1940, in *Himantopus himantopus;* Sri Lanka.
*M. kalawewaensis* Burt 1940, in *Oedicnemus indicus;* Sri Lanka.
*M. numida* Woodland 1929, in *Numida ptilorhyncha;* Sudan.
*M. pittae* Inamdar 1933, in *Pitta brachyura;* India.
*M. skrjabini* Krotov 1953, in *Limosa limosa;* Russia.
*M. woodlandi* Pandey et Tayal 1981, in *Burhinus oedicnemus;* India.
*M. zeylanica* Burt 1940, in *Burhinus oedicnemus;* Sri Lanka.

**10a. Testes posterior, sometimes also lateral to ovary .............................. 11**
**10b. Testes surrounding ovary ..................................................... 13**

**11a. Genital ducts ventral to osmoregulatory canals ...................................**
**................................... *Aleurotaenia* Cameron 1928. (Figure 377)**

*Diagnosis:* Rostellum with a single circle of rosethorn-shaped hooks. Neck long. Proglottids craspedote. Genital ducts pass ventral to osmoregulatory canals. Genital pores preequatorial, unilateral. Cirrus armed. Testes few, posterior and lateral to ovary. Vas deferens convoluted in median, anterior part of segment. Ovary bilobed, median. Vitelline gland compact, between lobes of ovary. Vagina ventral to cirrus, with distal dilation. Uterus first a bilobed sac, later breaking into egg capsules, each usually with a single egg. Parasites of Carnivora. Trinidad.

Type species: *A. planicipitis* Cameron 1928, in *Felis planiceps;* Trinidad.

**11b. Genital ducts dorsal to osmoregulatory canals ................................. 12**

**12a. Rostellum with two circles of hooks .......... *Neodilepis* Baugh et Saxena 1974.**

*Diagnosis:* Rostellum with two circles of hooks. Suckers unarmed. Proglottids wider than long. Testes usually four, posterior and lateral to female genital complex. Cirrus pouch long, extending far past poral osmoregulatory canals. Internal seminal vesicle present, external seminal vesicle absent. Cirrus armed. Genital pores unilateral, in anterior third of margin. Genital ducts dorsal to osmoregulatory canals. Ovary bilobed, median. Vitellarium postovarian. Vagina posterior to cirrus pouch. Seminal receptacle present. Uterus initially bilobed, becoming branched, then breaks into egg capsules each with 1 to 20 eggs. Parasites of Pelecaniformes (shags). Australia.

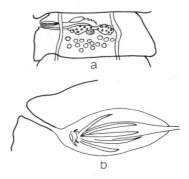

FIGURE 378.  *Spiniglans microsoma* Southwell 1922. (a) Proglottid; (b) cirrus pouch.

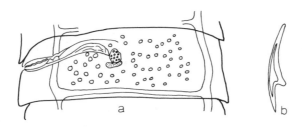

FIGURE 379.  *Seminluncinus dacelonis* Johnston 1909. (a) Proglottid; (b) rostellar hook.

Type species: *N. maxima* (Goss 1941) Baugh et Saxena 1974 (syn. *Dilepis maxima* Goss 1941, *Paradilepis maxima* [Goss 1941] Mathevossian 1959), in *Microcarbo melanoleucus:* Australia.

**12b.  Rostellum with one circle of hooks. ...............................................
.......................................... Spiniglans Yamaguti 1959. (Figure 378)**

*Diagnosis:* Rostellum with single circle of hooks. Suckers large, prominent. Neck absent. Strobila small, of few proglottids. Segments craspedote, wider than long. Genital ducts dorsal to osmoregulatory canals. Genital pores unilateral (?). Cirrus pouch extravascular. Cirrus with short spines at right angles on tip and with very long subapical spines. Testes few (16 to 20), posterior to ovary. External seminal vesicle present. Ovary bialate, median, preequatorial. Vitelline gland postovarian. Vagina posterior to cirrus pouch. Seminal receptacle present. Uterus first appears as a preovarian, transverse sac, then breaks into egg capsules each with one egg. Parasites of Passeriformes, Galliformes. India.

Type species: *S. microsoma* (Southwell 1922) Yamaguti 1959 (syn. *Choanotaenia microsoma* Southwell 1922, *Prochoanotaenia microsoma*[Southwell 1922] Meggitt 1924), in *Ploceus atrigula, Melophus melanicterus;* Calcutta Zoo.

Other species:
  *S. southwelli* Pandey et Tayal 1981, in *Coturnix coturnix;* India.

**13a.  Single circle of rostellar hooks....... Similuncinus Johnston 1909. (Figure 379)**

*Diagnosis:* Rostellum with a single circle of hooks. Neck present or absent. Proglottids craspedote, wider than long. Genital pores unilateral. Cirrus pouch extravascular or crossing

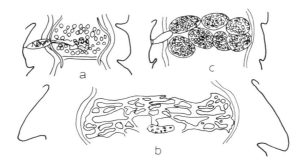

FIGURE 380. *Capsulata edonensis* Sandeman 1959. (a) Mature proglottid; (b) early uterus; (c) gravid proglottid.

ventral to osmoregulatory canals, along with vagina. Cirrus unarmed. Testes surrounding ovary, fewer anteriorly. Ovary rather compact, median or slightly poral. Vitelline gland postovarian. Vagina posterior to cirrus pouch. Seminal receptacle present. Uterus first branched, then breaking into egg capsules. Parasites of Coraciiformes and Charadriiformes. Australia, Asia.

Type species: *S. dacelonis* Johnston 1909, in *Dacelo gigas;* Australia.
Other species:

*S. leonovi* Tsimbalyuk, Andronova et Kulikov 1968, in *Tringa glareola, Phalaropus lobatus;* Russia.
*S. pavlovskyi* Krotov 1953, in *Pinicola enucleator;* Russia.
*S. totaniochropodis* Inamdar 1934, in *Totanus ochropus;* India.

**13b. Double circles of rostellar hooks....... *Capsulata* Sandeman 1959. (Figure 380)**

*Diagnosis:* Rostellum armed with a double circle of hooks. Suckers face slightly forward. Neck very short. Segmentation begins in diffuse area near posterior end, with external segmentation appearing last at posterior end. Maturation of genitalia precedes from posterior to anterior; segments then lost from posterior to anterior until only 40 to 50 remain. Mature proglottids acraspedote. Genital pores unilateral. Genital ducts pass between osmoregulatory canals. Genital atrium small. Cirrus pouch exceeds poral canals, and contains coiled ejaculatory duct. Cirrus short, armed. Inner and outer seminal vesicles absent. Testes (30 to 45) completely surrounding female organs. Vagina opens ventral to cirrus. Seminal receptacle present. Ovary median, of four to nine fan-like lobes. Vitelline gland postovarian. Uterus first reticular, then replaced by a few egg capsules each containing many round eggs. Inner longitudinal muscle bundles numerous (50 to 60); outer bundles few, small. Parasites of Charadriiformes. Scotland.

Type species: *C. edonensis* Sandeman 1959, in *Limosa lapponica;* Scotland.

**14a. Testes in two groups, one anterior and one posterior to ovary...................
.................................... *Kowalewskiella* Baczynska 1914. (Figure 381)**

*Diagnosis:* Rostellum with a single circle of hooks. Neck present. Mature proglottids longer than wide. Genital ducts pass between osmoregulatory canals. Genital pores irregularly alternating. Cirrus pouch crosses osmoregulatory canals. Cirrus armed. Testes in two groups, one anterior and one posterior to ovary. Ovary bilobed, median or poral. Vitelline gland compact, postovarian. Vagina posterior to cirrus pouch. Seminal receptacle present. Uterus first sac-like, then breaking into capsules, each with one egg. Parasites of Charadiiformes. Russia, Sri Lanka, North America.

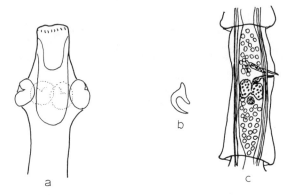

FIGURE 381. *Kowalewskiella cingulifera* (Krabbe 1869) Lopez-Neyra 1952. (a) Scolex; (b) rostellar hook; (c) mature proglottid.

Type species: *K. longiannulata* Baczynska 1914, in *Tringa stagnatilis, Rhyacophilus glareola;* Russia.

Other species:

*K. cingulifera* (Krabbe 1869) Lopez-Neyra 1952 (syn. *Taenia cingulifera* Krabbe 1869, *T. marchii* Parona 1887, *Monopylidium cingulifera* [Krabbe 1869] Clerc 1903, *Choanotaenia cingulifera* [Krabbe 1869] Fuhrmann 1932, *C. hypoleucia* Singh 1952), in *Arenaria, Calidris, Capella, Charadrius, Crocethia, Glottis, Limonites, Numenius, Philomachus, Totanus, Tringa;* Europe, Asia, Guadeloupe.

*K. glareolae* (Burt 1940) Lopez-Neyra 1952 (syn. *Choanotaenia glareolae* Burt 1940), in *Tringa glareola;* Sri Lanka.

*K. lobipluviae* (Burt 1940) Mathevossian 1963 (syn. *Choanotaenia lobipluviae* Burt 1940, *Onderstepoortia lobipluviae* [Burt 1940] Yamaguti 1959), in *Lobipluvia malabarica;* Sri Lanka.

*K. stagnatilidis* (Burt 1940) Lopez-Neyra 1952 (syn. *Choanotaenia stagnatilidis* Burt 1940), in *Tringa stagnatilis;* Sri Lanka.

*K. susanae* Burt 1969, in *Tringa glariola;* Borneo.

*K. totani* Self et Janovy 1965, in *Totanus flavipes;* Kansas, U.S.

**14b.  Testes not as above ........................................................... 15**

**15a.  Testes encircling ovary except on pore side ..... *Onderstepoortia* Ortlepp 1938.**

*Diagnosis:* Rostellum with a single circle of *Taenia*-like hooks. Proglottids craspedote. Genital ducts pass between osmoregulatory canals. Testes numerous, surrounding ovary except on poral side. Ovary crescentic, slightly poral. Vitelline gland postovarian. Vagina posterior to cirrus pouch. Seminal receptacle dorsal to ovary. Uterus replaced with egg capsules, each with one egg. Parasites of Charadriiformes. Africa, Ceylon.

Type species: *O. taeniaformis* Ortlepp 1938 (syn. *Kowalewskiella taeniaeformis* [Ortlepp 1938] Mathevossian 1963, *Choanotaenia burhini* Burt 1940), in *Burhinops capensis, Burhinus oedicnemus;* Africa, Sri Lanka.

Other species:

*O.coronati* Mettrick 1961, in *Stephanibyx coronatus;* Africa.

**15b.  No testes anterior to ovary ....................................................... 16**

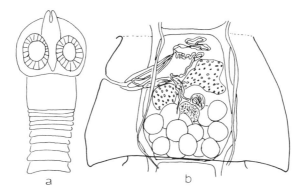

FIGURE 382. *Choanotaenia uncinata* Fuhrmann 1918. (a) Scolex; (b) proglottid.

16a. One circle of rostellar hooks ................................................. 17
16b. Two circles of rostellar hooks ................................................ 18

17a. Genital ducts pass between osmoregulatory canals ................ *Choanotaenia* Railliet 1896 (syn. *Icterotaenia* Railliet et Henry 1909, *Choanofuhrmannia* Lopez-Neyra 1935, *Macracanthus* Moghe 1925, *Megalacanthus* Moghe 1925, *Molluscotaenia* Spasskii et Andreiko 1971 in part, *Prochoanotaenia* Meggitt 1924, *Spreotaenia* Spasskii 1969, *Sobolevitaenia* Spasskaja et Makarenko 1965). (Figure 382)

*Diagnosis:* Rostellum with a single circle of hooks. Suckers rarely armed. Proglottids craspedote. Genital ducts pass between osmoregulatory canals. Genital pores irregularly alternating. Cirrus pouch extravascular or not. Cirrus armed or not. Testes numerous, mainly posterior to ovary. Ovary bilobed or compact, usually median. Vitelline gland compact, postovarian. Vagina posterior to cirrus pouch. Seminal receptacle present. Uterus sac-like, lobed, or reticular, breaking into egg capsules, each with one egg. Parasites of Accipitriformes, Coraciiformes, Charadriiformes, Galliformes, Passeriformes, Piciformes, Gruiformes, Anseriformes, Strigiformes, Upupiformes. Also, in Insectivora, Rodentia. Cosmopolitan.

Type species: *C. infundibulum* (Bloch 1779) Railliet 1896 (syn. *Taenia infundibulum* Bloch 1779, *T. infundibuliformis* Goeze 1782, *Choanotaenia infundibuliformis* Lucet et Marotil 1904, *Drepanidotaenia infundibulum* [Bloch 1779] Stossich 1895, *Taenia lagenicollis* Megnin 1898, *Monopylidium infundibulum* [Bloch 1779] Clerc 1903, *Choanotaenia polyorchis* [Klaptocz 1908] Baer 1925, *C. dutti* Mukherjee 1964, *C. fuhrmanni* Skrjabin 1914), in *Gallus gallus, Caccabis petrosa, C. rufa, Cincus cinereus, Coturnix coturnix, Phasianus colchicus, Francolinus pintdeanus, Meleagris gallopavo, Numida meleagris, Pavo cristatus, Perdix perdix,* etc., *Macaca mulatta* (captive); cosmopolitan.

Other species:

*C. abassenae* Joyeux, Baer et Martin 1936, (syn. *Spreotaenia abassenae* [Joyeux, Baer et Martin 1936] Spasskii 1967), in *Spreo superbus;* Africa.

*C. acridotheresi* Saxena 1972, in *Acridotheres tristis, A. ginginiesis;* India.

*C. angolensis* Mettrick 1960, in *Pitta angolensis;* Africa.

*C. anthusi* Spasskaja 1958 (syn. *Sobolevitaenia anthusi* [Spasskaja 1958] Spasskaja et Makarenko 1965), in *Anthus trivialis, A. richardi;* Russia.

*C. aurantia* Singh 1956, in *Sterna aurantia;* India.

*C. baicalensis* (Eltyshev 1975) comb. n. (syn. *Molluscotaenia baicalensis* Eltyshev 1975), in *Sorex araneus;* Russia.

*C. bhattacharai* Chatterji 1954, in *Anas querguedula;* India.

*C. birostrata* Belgurov et Zueva 1968, in *Gallinago gallinago, G. stenura;* Russia.

*C. burti* Lopez-Neyra 1952, in *Tringa glareola, T. totanus;* Sri Lanka.

*C. cayennensis* (Fuhrmann 1907) Fuhrmann 1932, in *Belonopterus cayennensis;* Brazil.

*C. cayennensis* var. *africana* Joyeux, Gendre et Baer 1928, in *Gallinago* sp.; French West Africa.

*C. centuri* Rysavy 1966, in *Centurus superciliaris;* Cuba.

*C. chionis* Fuhrmann 1921 (syn. *Icterotaenia chionis* Baer 1925, *Paricterotaenia chionis* [Fuhrmann 1921] Fuhrmann 1932), in *Crocethia alba;* South America.

*C. cholodkowskyi* Krotov 1953, in *Alauda arvensis;* Russia.

*C. cirrospinosa* (Patwardhan 1935) Mathevossian 1963 (syn. *Paricterotaenia cirrospinosa* Patwardhan 1935), in "snipe"; India.

*C. coronata* (Creplin 1829) Mathevossian 1963 (syn. *Taenia coronata* Creplin 1829, *Paricterotaenia coronata* [Creplin 1829] Fuhrmann 1932), in *Aegialitis nivosa, Charadius alexandrinus, Numenius arquata, Burhinus oedicnemus, B. senegalensis, Scolopax rusticola;* Europe, Asia, Africa, Israel, Sri Lanka.

*C. corvi* Joyeux, Baer et Martin 1937, in *Corvus rhipidurus;* Africa.

*C. croaxum* Mukherjee 1970, in *Corvus splendens;* India.

*C. decacantha* (Fuhrmann 1913) Lopez-Neyra 1951 (syn. *Icterotaenia decacantha* Baer 1925, *Paricterotaenia decacantha* [Fuhrmann 1913] Fuhrmann 1932, *Choanotaenia joyeuxi* Tseng 1932), in *Gallinago* sp., *Pelidna alpina, Scolopax rusticola, Tringa ochropus, T. maritima;* Africa, India, North America.

*C. dispar* Burt 1940, in *Lobipluvia melabarica;* Sri Lanka.

*C. dogieli* Krotov 1954 (syn. *Anomotaenia dogieli* [Krotov 1954] Mathevossian 1963), in *Calidris tenuirostris;* Russia.

*C. estavarensis* Euzet et Jourdane 1968, in *Neomys fodiens;* France.

*C. fotedori* Chisti 1973, in *Acridotheres tristis;* India.

*C. gondwana* Inamdar 1934, in *Passer domesticus;* India.

*C. himalayana* Fotedar et Chishiti 1977, in *Passer domesticus, Gallus gallus;* India.

*C. hypoleucia* Singh 1952, in *Tringa hypoleucos;* India.

*C. ibanezi* Rego et Vicente 1968, in *Numenius phaeophus;* Peru.

*C. kapurdiensis* Mukherjee 1970, in *Cursorius cursor;* India.

*C. larimarina* Elce 1962, in *Larus marinus;* Wales.

*C. littorie* De Muro 1934, in *Athene noctura;* Italy.

*C. macrocephala* Sawada et Kugi 1976, in *Scolopax rusticola;* Japan

*C. magnihamata* Burt 1940, in *Burhinus oedicnemus;* Sri Lanka.

*C. mancocapaci* Ibanez-Herrera 1966, in *Crotophaga sulcirostris;* Peru.

*C. marchali* (Mola 1907) Lühe 1910 (syn. *Taenia marchali* Mola 1907), in *Gallinula chloropus;* Italy.

*C. nebraskensis* Hansen 1950 (syn. *Rodentotaenia nebraskensis* [Hansen 1950] Mathevossian 1963), in *Microtus ochrogaster, Sciurus rufiventris;* U.S.

*C. nilotica* (Krabbe 1869) Railliet 1896 (syn. *Taenia nilotica* Krabbe 1896), in *Cursorius isabellinus, C. yallicus;* North Africa.

*C. numenii* Owen 1946, in *Numenius americanus;* U.S.

*C. orientalis* (Spasskii et Konovalov 1969) comb. n. (syn. *Sobolevitaenia orientalis* Spasskii et Konovalov 1969), in *Motacilla alba, M. flava, Calcarius lapponicus;* Russia.

*C. orioli* Joyeux et Baer 1955, in *Oriolus oriolus;* France, Switzerland.

*C. paranumenii* Clark 1952, in *Numenius americanus;* U.S.

*C. parina* (Dujardin 1895) Cohn 1899 (syn. *Taenia parina* Dujardin 1845, *Drepanidotaenia parina* [Dujardin 1845] Stossich 1898, *Icterotaenia parina* [Dujardin 1845]

Baer 1925, *Paricterotaenia parina* [Dujardin 1845] Fuhrmann 1932, *Anomotaenia parina* [Dujardin 1845] Lopez-Neyra 1952), in *Parus coeruleus, P. major, P. cinereus, Passer montanus, P. domesticus, Sturnus vulgaris, Eremophila alpestris, Coloenus monedula, Aegithalos caudatus;* Europe, Russia.

*C. passerina* (Fuhrmann 1907) Fuhrmann 1932 (syn. *Anomotaenia passerina* [Fuhrmann 1907] Lopez-Neyra 1952, *Monopylidium passerina* Fuhrmann 1907), in *Fringilla ruficeps, Passer domesticus, P. montanus, P. hispaniolensis, Parus major, P. caeruleus;* Europe, Russia, Africa.

*C. perisorei* Spasskaja 1957 (syn. *Anomotaenia perisorei* [Spasskaja 1957] Mathevossian 1963), in *Perisoreus infaustus;* Russia.

*C. peromysci* (Erickson 1938) Hansen 1950 (syn. *Prochoanotaenia peromysci* Erickson 1938), in *Peromyscus maniculatus;* U.S.

*C. picusi* Singh 1964, in *Picus squamatus;* India.

*C. platycephala* (Rudolphi 1810) Fuhrmann 1932 (syn. *Taenia platycephala* Rudolphi 1810), in *Motacilla luscinia, Galerita, Alauda, Anthus, Locustella, Saxicola, Ruticilla, Phoenicurus, Sylvia, Pratincola;* Europe, Africa, Asia.

*C. porosa* (Rudolphi 1810) Cohn 1899 (syn. *Taenia porosa* Rudolphi 1810, *Drepanidotaenia porosa* [Rudolphi 1810] Stossich 1898, *Parachoanotaenia* [Rudolphi 1810] Lühe 1910, *Icterotaenia porosa* [Rudolphi 1810] Baer 1925, *Paricterotaenia porosa* [Rudolphi 1810], Fuhrmann 1932, *P. gongyla* [Cohn 1900] Baer 1925), in *Larus, Rissa, Sterna*; Asia, Europe, Israel, North America.

*C. prunellae* Yamaguti 1953, in *Prunella collaris;* Japan.

*C. ridibundum* Deblock, Capron et Rosé 1960, in *Larus ridibundus;* France.

*C. rostellata* (Fuhrmann 1908) Baylis 1925 (syn. *Monpylidium rostellata* Fuhrmann 1908), in *Himantopus mexicanus;* Brazil.

*C. rostrata* (Fuhrmann 1918) Baylis 1925 (syn. *Monopylidium rostrata* Fuhrmann 1918, *Anomotaenia rostrata* [Fuhrmann 1918] Mathevossian 1963), in *Halcyon sancta;* New Caledonia.

*C. sciuricola* Harwood et Cooke 1949 (syn. *Rodentotaenia sciuricola* [Harwood et Cooke 1949] Mathevossian 1963), in *Sciurus niger;* U.S.

*C. scolopacina* Lopez-Neyra 1944 (syn. *C. joyeuxibaeri* Lopez-Neyra 1952, *C. cayennensis* var. *scolopacis* Joyeux et Baer 1939), in *Scolopax rusticola;* France.

*C. shohoi* sawada 1964, in *Gallus gallus;* Sudan.

*C. similis* (Spasskii et Konovalov 1969) comb. n. (syn. *Sobolevitaenia similis* Spasskii et Konovalov 1969), in *Anthus corvinus;* Russia.

*C. skrjabini* Ivanitskii 1940 (syn. *Anomotaenia skrjabini* [Ivaniskii 1940] Mathevossian 1963), in *Charadrius hiaticula;* Russia.

*C. slesvicensis* (Krabbe 1882) Clerc 1903 (syn. *Taenia slesvicensis* Krabbe 1882, *Icterotaenia slesvicensis* [Krabbe 1882] Baer 1925, *Paricterotaenia slesvicensis* [Krabbe 1882] Fuhrmann 1932, *Choanotaenia trigienciensis* Joyeux et Baer 1939), in *Scolopax rusticola;* Europe, Russia.

*C. sobolevi* (Spasskaja et Makarenko 1965) comb. n. (syn. *Sobolevitaenia sobolevi* Spasskaja et Makarenko 1965), in *Anthus richardi;* Russia.

*C. sonoti* Mukhurjee 1964, in *Acridotheres tristis;* India.

*C. soricina* (Cholodkovsky 1900) (syn. *Amoebotaenia subterranea* of Meggitt 1924, *Monopylidium scutigerum* of Baer 1928), in *Sorex araneus;* Europe.

*C. spasskii* (Mamaev et Okhotina 1968) comb. n. (syn. *Prochoanotaenia spasskii* Mamaev et Okhotina 1968), in *Mogera robusta;* Russia.

*C. speotytonis* Rausch 1948, in *Speotyto cunicularia;* Colorado, U.S.

*C. spermophili* (McLeod 1933) Hansen 1950 (syn. *Prochoanotaenia spermophili* McLeod 1933, *Rodentotaenia spermophili* [McLeod 1933] Mathevossian 1963), in *Citellus tridecemlineatus, C. richardsoni, Peromyscus maniculatus;* Canada, U.S.

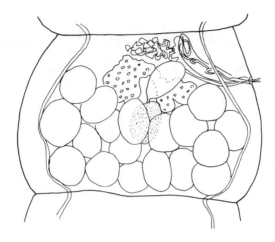

FIGURE 383. *Irvitaenia mukteswarensis* Singh 1962.

*C. srivastavai* Mukherjee 1967, in *Turdoides somervillei;* India.

*C. stercorarium* (Baylis 1919) Fuhrmann 1932 (syn. *Monopylidium stercorarium* Baylis 1919, *M. [Macracanthus] stercorarium* Moghe 1925), in *Stercorarius pomatorhinus, Pratercula arctica, Larus canus;* Arctica.

*C. sternina* (Krabbe 1869) Cohn 1899 (syn. *Taenia sternina* Krabbe 1869, *Icterotaenia sternina* [Krabbe 1869] Baer 1925, *Paricterotaenia sternina* [Krabbe 1869] Fuhrmann 1932), in *Larus canus, L. ridibundus, Sterna fluviatilis, S. macrura, S. hirundo;* Iceland, Greenland, Russia.

*C. strigium* Joyeux et Timon-David 1934, in *Otus scops, Stercorarius pomarinus, Athene noctua;* Europe, North Africa.

*C. sylvarum* Oliger 1950, in *Tetrastes bonasia;* Russia.

*C. tandani* Singh 1964, in *Myiophoneus caeruleus;* India.

*C. taylori* Johnston 1912, in *Malurus cyanochlamys;* Australia.

*C. tetrastes* Belopolskaya 1963, in *Tetrastes bonasia;* Russia.

*C. thraciensis* Kamburov 1969, in *Scolopax rusticola;* Bulgaria.

*C. tringae* Joyeux, Baer et Martin 1937, in *Tringa* sp.; Africa.

*C. uncinata* Fuhrmann 1918 (syn. *Paricterotaenia uncinata* [Fuhrmann 1918] Fuhrmann 1932, *Icterotaenia unicoronata* Baer 1925, *Paricterotaenia unicoronata* [Baer 1925] Fuhrmann 1932), in *Collocalia leucopygia, C. unicolor;* New Caledonia.

*C. unicoronata* (Fuhrmann 1908) Fuhrmann 1932 (syn. *Monopylidium unicoronata* Fuhrmann 1908, *Anomotaenia unicoronata* [Fuhrmann 1908] Clerc 1911, *Choanofuhrmannia unicoronata* [Fuhrmann 1908] Lopez-Neyra 1935, *Choanotaenia corvi* Joyeux, Baer et Martin 1937), in *Turdus merula, T. pilaris, T. viscivorus, Corvus rhipidurus;* Europe, Russia.

*C. upupae* Fuhrmann 1943, in *Upupa africana;* Africa.

**17b. Genital ducts dorsal to osmoregulatory canals ....................................**
**.............................................. *Ivritaenia* Singh 1962. (Figure 383)**

*Diagnosis:* Scolex large, well-developed. Rostellum armed with single circle of hooks. Neck short. Proglottids acraspedote, broader than long, except gravid ones that are much longer than broad. Genital pores regularly alternating. Genital atrium shallow. Genital ducts dorsal to osmoregulatory canals. Cirrus pouch well developed, extending one third across proglottid, proximal end, curving anteriad. Cirrus small, unarmed, surrounded by hair-like

FIGURE 384. *Imparmargo baileyi* Davidson, Doster et Prestwood 1974.

processes that protrude from genital pore. Inner and outer seminal vesicles absent. Testes 15 to 18 in number, in one field posterior and lateral to ovary. Vagina posterior to cirrus pouch. Seminal receptacle present. Ovary distinctly bilobed, slightly poral in anterior half of proglottid. Vitellarium compact, oval, posterior to ovarian isthmus. Uterus sac-like, replaced by egg capsules each with one to four eggs. Parasites of Piciformes. India.

Type species: *I. mukteswarensis* Singh 1962, in *Dendrocopos auriceps;* India.

**18a. Genital ducts pass between osmoregulatory canals .......................... 19**
**18b. Genital ducts pass dorsal to osmoregulatory canals .......................... 20**

**19a. Genital pores regularly alternating. Small tuft of delicate spines protruding from genital atrum ................................................................**
**................ *Imparmargo* Davidson, Doster et Prestwood 1974. (Figure 384)**

*Diagnosis:* Small cestodes, with few proglottids. Rostellum with two circles of hooks. Suckers powerful, simple. Neck absent. Immature and mature segments craspedote, wider than long, longer on poral than aporal side. Gravid proglottid single, longer than wide. Genital pores preequatorial, regularly alternating. Genital ducts passing between osmoregulatory canals. Cirrus exceeds poral canal. Genital atrium with a small tuft of delicate spines protruding. Cirrus unarmed. Testes in single field posterior to ovary. Ovary compact, median. Vitellarium compact, postovarian, slightly poral. Vagina posterior to cirrus pouch. Seminal receptacle prominent. Uterus replaced with egg capsules, each with single egg. Parasites of Galliformes. U.S.

Type species: *I. baileyi* Davidson, Doster et Prestwood 1974, in *Meleagris gallipavo;* Virginia, U.S.

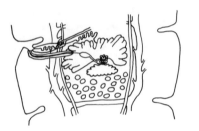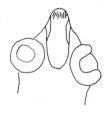

FIGURE 385. *Monopylidium musculosa* (Fuhrmann 1896) Fuhrmann 1899.

**19b. Genital pores irregularly alternating. No spines in genital atrum................**
**................................................. *Monopylidium* Fuhrmann 1899**
**(syn. *Rodentotaenia* Mathevossian 1953 in part, *Viscoia* Mola 1929). (Figure 385)**

*Diagnosis:* Rostellum with two circles of hooks. Suckers unarmed. Proglottids craspedote. Genital ducts pass between osmoregulatory canals. Genital pores irregularly alternating. Cirrus pouch extravascular or not. Cirrus armed or not. Testes numerous, mainly posterior to ovary. Ovary bilobed or compact, vitellarium postovarian. Vagina posterior to cirrus pouch. Seminal receptacle present. Uterus sac-like, lobed, or reticular, breaking into egg capsules, each with one egg. Parasites of birds and mammals. Cosmopolitan.

Type species: *M. muscolosa* (Fuhrmann 1896) Fuhrmann 1899 (syn. *Davainea musculosa* Fuhrmann 1896, *Choanotaenia musculosa* [Fuhrmann 1896] Fuhrmann 1932, *Anomotaenia musculosa* [Fuhrmann 1896] Mathevossian 1863), in *Oriolus oriolus, Sturnus vulgaris, S. varius, Sylvia hortenesis;* Europe, Russia.

Other species:

*M. arcticum* Baylis 1919 (syn. *Choanotaenia arcticum* [Baylis 1919] Fuhrmann 1932), in *Arquatella maritima, Tringa ochropus, Plilomachus pugnax;* Arctic, Volga Delta, Africa.

*M. bondarevae* (Shaekenov 1978) comb. n. (syn. *Rodentotaenia bondarevae* Shaikenov 1978), in *Apodemus sylvaticus, Meriones erythrourus, Sicista subtilius;* Russia.

*M. crateriformis* (Goeze 1782) Fuhrmann 1899 (syn. *Taenia crateriformis* Goeze 1782, *Choanotaenia crateriformis* [Goeze 1782] Fuhrmann 1932), in *Dryobates leucotos, D. major, D. medius, D. minor, Dryocopus martius, Picus canus, P. viridis, Jynx torquilla;* Europe, Russia.

*M. columbae* (Borgarenko 1976) comb. n. (syn. *Icterotaenia columbae* Borgarenko 1926), in *Columba livia;* Russia.

*M. exigua* (Dujardin 1845) Spasskaja et Shumilo 1973 (syn. *Taenia exigua* Dujardin 1845, *Choanotaenia exigua* [Dujardin 1845] Baylis 1948), in *Passer domesticus, Fringilla coelebs, Troglodytes troglodytes;* England, Europe.

*M. fieldingi* Maplestone et Southwell 1923 (syn. *Choanotaenia fieldingi* [Maplestone et Southwell 1923] Fuhrmann 1932, *Anomotaenia fieldingi* [Maplestone et Southwell 1923] Mathevossian 1963), in *Craticus destructor;* Australia.

*M. filamentosum* (Goeze 1782) Baer 1932 (syn. *Taenia filamentosa* Goeze 1782, *T. blanchardi* Mola 1907, *Hymenolepis filamentosa* [Goeze 1782] Meggitt 1927, *Multitesticulata aegyptica* Meggitt 1927, *Viscoia blanchardi* Mola 1929, *Choanotaenia filamentosum* [Goeze 1782] Joyeux et Baer 1936, *Rodentotaenia filamentosum* [Goeze 1782] Mathevossian 1963), in *Talpa europaea;* Europe.

*M. galbulae* (Gmelin 1790) Skrjabin 1914 (syn. *Taenia galbulae* Gmelin 1790, *Choanotaenia galbulae* [Gmelin 1790] Cohn 1899, *Icterotaenia galbulae* [Gmelin 1790] Railliet et Lucet 1909, *Parachoanotaenia galbulae* [Gmelin 1790] Lühe 1910, *An-*

FIGURE 386. *Panuwa lobivanelli* Burt 1940.

*omotaenia galbulae* [Gmelin 1790] Fuhrmann 1932), in *Coleus monedula, Corvus cornix, C. splendens, Oriolus oriolus, O. galbula;* Europe, Russia.

*M. hepatica* Baer 1932 (syn. *Choanotaenia hepatica* [Baer 1932] Joyeux et Baer 1936, *Rodentotaenia hepatica* [Baer 1932] Mathevossian 1963), in *Sorex araneus;* Europe.

*M. iola* (Lincicome 1939) comb. n. (syn. *Choanotaenia iola* Lincicome 1939), in *Turdus migratorius, Sturnus vulgaris;* U.S., Israel.

*M. manipurensis* (Patwardhan 1935) Lopez-Neyra 1952 (syn. *Choanotaenia manipurensis* Patwardhan 1935, *Anomotaenia manipurensis* [Patwardhan 1935] Mathevossian 1963), in "snipe"; India.

*M. merionidis* (Shaikenov 1978) comb. n. (syn. *Rodentotaenia merionidis* Shaikenov 1978), in *Meriones erythrourus;* Russia.

*M. moldavica* Shumilo et Spasskaja 1975, in *Turdus philomelos, T. merula;* Russia.

*M. ratticola* (Sandars 1957) comb. n. (syn. *Choanotaenia ratticola* Sandars 1957, *Rodentotaenia ratticola* [Sandars 1927] Mathevossian 1963), in *Rattus assimilis;* Australia.

*M. southwelli* (Fuhrmann 1932) comb. n. (syn. *Choanotaenia southwelli* Fuhrmann 1932, *Anomotaenia southwelli* [Fuhrmann 1932] Mathevossian 1963), in *Lobivanellus lobatus;* Australia.

*M. spinosocapite* (Joyeux et Baer 1955) comb. n. (syn. *Choanotaenia spinosocapite* Joyeux et Baer 1955, *Sobolevitaenia spinosocapite* [Joyeux et Baer 1955] Galkin 1979, *Anomotaenia spinosocapite* [Joyeux et Baer 1955] Mathevossian 1963), in *Garrulus glandarius, Pica pica, Sturnus vulgaris, Turdus merula;* Europe, Russia.

*M. timuri* Borgarenko 1976, in *Calandrella cinerea;* Russia.

*M. tsengi* Sandeman 1959, in *Microsarcops cinereus, Rhynchea upensis;* China.

**20a. Egg capsules each with one egg..................................................**
**......*Panuwa* Burt 1940 (syn. *Cholodkovskia* Mathevossian 1953). (Figure 386)**

*Diagnosis:* Rostellum with two circles of hooks. Proglottids wider than long, craspedote. Genital ducts dorsal to osmoregulatory canals. Genital pores alternating irregularly. Testes numerous, posterior to ovary. Ovary bilobed, equatorial, slightly poral. Vitelline gland postovarian. Vagina posterior to cirrus pouch. Seminal receptacle present. Egg capsules each with one egg. Parasites of Charadriiformes. Sri Lanka, India, Vietnam.

Type species: *P. lobivanelli* Burt 1940, in *Lobivanellus indicus;* Sri Lanka.
Other species:

*P. caballeroi* Singh et Singh 1960, in *Chettusia leucura;* India.

*P. metaskrjabini* Spasskii et Yurpalova 1968, in *Lobivanellus indicus;* Vietnam.

*P. stylicirrosa* Singh 1964, in *Picus xanthopygeus;* India.

*P. vogeae* Singh et Singh 1960, in *Chettusia leucura;* India.

**20b. Egg capsules each with several eggs, with long filaments extending from packet..............................................*Cinclotaenia* Macy 1973. (Figure 387)**

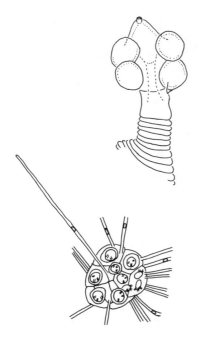

FIGURE 387. *Cinclotaenia filamentosa* Macy 1973. Scolex and egg cluster.

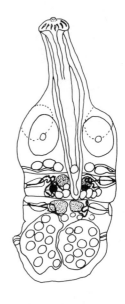

FIGURE 388. *Mirandula parva* Sandars 1965. Entire worm.

*Diagnosis:* Rostellum with two circles of hooks. Suckers large, simple, unarmed. Short neck present. Proglottids craspedote, wider than long. Genital ducts dorsal to osmoregulatory canals. Cirrus pouch just reaches ventral osmoregulatory canals. Seminal vesicles absent. Cirrus unarmed. Testes numerous, mainly postovarian. Ovary wide, with thin lobes, in anterior half of segment. Vitellarium transversely elongated, postovarian. Vagina posterior to cirrus pouch. Seminal receptacle present. Uterus saccate, lobed. Several eggs in clusters (capsules?), with long filaments extending from packet. Parasites of passeriform birds (Cinclidae). U.S.

Type species: *C. filamentosa* Macy 1973, in *Cinclus mexicanus;* Oregon, U.S.

## KEY TO THE GENERA IN DILEPIDINAE

**1a. Two sets of reproductive organs per segment** .......................................
.............................................*Mirandula* **Sandars 1956. (Figure 388)**

*Diagnosis:* Very small worm with two sets of reproductive organs per segment. Scolex with four weakly developed, unarmed suckers and an extremely large, retractile rostellum bearing a double circle of hooks. Rostellar sac thick-walled, extending backward into middle of mature segment or even into penultimate segment. Neck absent. Strobila consisting of only a few segments. Genital ducts dorsal to osmoregulatory canals. Testes usually in two groups of four, lateral to median line. Cirrus pouch reaches poral canal. No external or internal seminal vesicles. Cirrus armed with small spines. Genital pores preequatorial. Ovaries and vitelline glands compact, anterior to testes. Uterus at first a transverse sac, then slightly bilobed. Eggs relatively large. Parasites of marsupials.

Type species: *M. parva* Sandars 1956, in *Perameles nasuta;* South Queensland, Australia.

**1b. One set of reproductive organs per segment** ...................................... 2

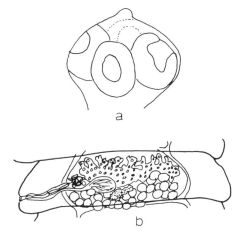

FIGURE 389. *Arctotaenia tetrabothrioides* (Lönnberg 1890) Baer 1956. (a) Scolex; (b) proglottid.

2a. Genital pores unilateral ........................................................... 3
2b. Genital pores alternating ...................................................... 22

3a. Rostellum lacking ....................... **Arctotaenia Baer 1956. (Figure 389)**

*Diagnosis:* Rostellum absent, scolex unarmed. Proglottids craspedote, wider than long. Genital pores unilateral. Genital ducts pass between osmoregulatory canals. Cirrus pouch crosses osmoregulatory canals. Testes 25 to 30, in single, postovarian field. Ovary anterior, lobated, extending width of medulla. Vitelline gland postovarian, slightly poral. Vagina posterior to cirrus pouch. Uterus first a single sac, becoming lobated when gravid. Parasites of Charadriiformes. Russia, Norway, Greenland.

Type species: *A. tetrabothrioides* (Lönnberg 1890) Baer 1956 (syn. *Taenia tetrabothrioides* Lönnberg 1890), in *Tringa alpina, Erolia maritima;* Norway, Greenland.
Other species:
   *A. recapta* (Clerc 1906) Baer 1956 (syn. *Dilepis recapta* Clerc 1906, *Lateriporus recapta* [Clerc 1906] Mathevossian 1963), in *Erolia minuta, Tringa stagnalis, Erolia maritima;* Greenland, Russia.

3b. Rostellum present .............................................................. 4

4a. Genital pores submarginal ...................................................... 5
4b. Genital pores marginal .......................................................... 6

5a. One circle of rostellar hooks, rostellum slightly bifurcated at tip ................
    ................................... **Trichocephaloidis Sinitzin 1896. (Figure 390)**

*Diagnosis:* Rostellum slightly bifurcated on the tip, with a single circle of hooks. Proglottids craspedote, nearly cylindrical. Genital pores unilateral sublateral. Genital ducts dorsal to osmoregulatory canals. Cirrus pouch large, crossing osmoregulatory canals. Testes few, posterior. Ovary median. Vitelline gland postovarian. Vagina dorsal to cirrus pouch, dilated at distal end. Uterus sac-like. Parasites of Charadriiformes. Russia, Japan, North America, Europe, Caribbean.

Type species: *T. megalocephala* (Krabbe 1869) Sinitzin 1896 (syn. *Taenia megalocephala*

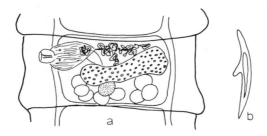

FIGURE 390. *Trichocephaloidis megalocephala* (Krabbe 1869) Sinitzen 1896. (a) Proglottid; (b) rostellar hook.

Krabbe 1869, *Trichocephaloides inermis* Sinitzin 1896, *Dilepis hamasigi* Yamaguti 1940, *Lateriporus hamasigi* [Yamaguti 1940] Belopolskaja 1953), in *Calidris maritima, C. arenaria, C. subminuta, C. alpina, C. temminckii, C. minuta, Larus canus, Tringa calidris, T. totanus, Philomachus pugnax, Limicola falcinellus, Ancylochilus subarquatus;* Russia, North America, Europe.

Other species:

*T. beauporti* Graber et Euzeby 1976, in *Tringa flaviceps, Microplama himantopus, Gallinago gallinago, Squatarola squatarola, Quiscalus lugubris;* Guadaloupe.

*T. birostrata* Clerc 1906, in *Calidris minuta, C. testacea;* Russia.

*T. charadrii* Lavroff 1908, in *Charadrius hiaticula;* Russia.

*T. temminckii* Belopolskaja 1958, in *Calidris temminckii;* Russia.

**5b.  Two circles of rostellar hooks, rostellum not bifurcated...... *Vogea* Johri 1959.**

*Diagnosis:* Scolex with well-developed rostellum armed with double circle of large triangular hooks. Suckers unarmed. Neck present. Genital pores unilateral and submarginal. Genital atrium absent. Cirrus opens on circular muscular pad. Vagina opens on groove surrounding pad. Genital ducts dorsal to vessels. Internal and external seminal vesicles absent. Testes numerous (44 to 51), anterior and posterior to ovary. Ovary transversely elongated, in anterior half of segment, median to osmoregulatory vessels. Vitellaria compact, postovarian. Seminal receptacle present. Uterus a simple transverse sac. Parasites of Passiformes. India.

Type species: *V. vestibularis* Johri 1959, in *Argyra malcolmi;* India.

**6a.  Genital atrium very large, deep and muscular................................... 7**
**6b.  Genital atrium not as above .................................................... 9**

**7a.  Genital atrium complex, with spines, bristles or diverticulae ....................**
**............................ *Neogryporhynchus* Baer et Bona 1960. (Figure 391)**

*Diagnosis:* Rostellum with two circles of 20 powerful hooks, each with long handle. Suckers simple. Neck present. Genital ducts pass between osmoregulatory canals. Genital pores unilateral, dextral or sinistral. Genital atrium very large, complex, with spines, bristles of diverticulae. Cirrus pouch massive, containing coiled ejaculatory duct. Vas deferens an extensive, coiled duct. Testes four, posterior or dorsal to ovary. Cirrus armed, often with a tuft of spines emerging from its tip. Ovary with a few massive lobes. Vitellarium compact, postovarian. Vagina dorsal to cirrus pouch on left side, ventral to cirrus pouch on right side. Uterus horseshoe-shaped, with convexity at anterior end, somewhat lobated. Parasites of Ardeiformes. Circumboreal.

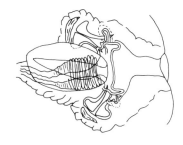

FIGURE 391. *Neogryporhynchus cheilancristrotus* (Wedl 1855) Baer et Bona 1960. Genital atrium.

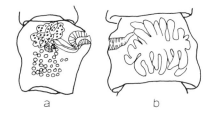

FIGURE 392. *Valipora parvispine* Linton 1927. (a) Mature proglottid; (b) gravid proglottid.

Type species: *N. cheilancristrotus* (Wedl 1855) Baer et Bona 1960 (syn. *Taenia cheilancristrota* Wedl 1855, *Dilepis macropeos* Clerc 1906, *Acanthocirrus cheilancristrota* [Wedl 1855] Fuhrmann 1907, *Gryporhynchus cheilancristrotus* [Wedl 1855] Ransom 1909, *Gryporhynchus tetrorchis* Hill 1941, *Gryporhynchus nycticoracis* Yamaguti 1956, *Gryporhynchus pusillus* Spasskaya, Spasskii et Evakina 1974), in *Botaurus stellarii, Ardea purpurea, A. cinerea, A. herodias, Nycticorax nycticorax;* Europe, Russia, Japan, North America.

Other species:

  *N. lasiopeius* Baer et Bona 1960, in *Ardea purpurea;* France.

**7b.  Genital atrium muscular but not as above ..................................... 8**

**8a.  Entire genital atrium muscular ........................................... *Valipora* Linton 1927 (syn. *Neovalipora* Baer 1962, *Baerbonaia* Deblock 1966, *Diagonaliporus* Krotov 1951, *Platyscolyx* Spasskaja et Spasskii 1971). (Figure 392)**

*Diagnosis:* Rostellum with a single circle of hooks. Proglottids craspedote, wider than long. Genital pores unilateral. Genital ducts pass between osmoregulatory canals. Cirrus pouch elongate. Cirrus long, unarmed. Genital atrium with very muscular walls, sometimes eversible. Testes mainly posterior, none anterior, to ovary. Ovary medium, anterior. Vitelline gland postovarian. Vaginal pore posterior to cirrus pore. Seminal receptacle present. Uterus a lobated sac. Parasites of Ardeiformes, Gaviiformes, Charadriiformes, Anseriformes. Europe, Asia, North America, Madagascar. Key to species in Ardeiformes: Deblock.[701]

Type species: *V. mutabilis* Linton 1927, in *Nycticorax nycticorax?, Columba livia;* U.S., Europe, India.

Other species:

  *V. amethiensis* Srivastava et Capoor 1975, in *Bubulcus ibis;* India.

  *V. baeribonae* (Deblock 1966) comb. n. (syn. *Baerbonia baeribonae* Deblock 1966), in *Egretta dimorpha;* Ile de Europa (Mozambique Channel), Africa.

  *V. glomovaginata* Baer et Bona 1960, in *Ardea purpurea;* France.

  *V. pachipora* Baer et Bona 1960, in *Egretta intermedia;* Java.

  *V. parvispine* Linton 1927 (syn. *Neovalipora parvispine* [Linton 1927] Baer 1962, *Platyscolex parvispine* [Linton 1927] Spasskaja et Spasskii 1971), in *Gavia immer;* U.S., Europe, Iceland.

  *V. parvitaeniunca* Baer et Bona 1960, in *Egretta sacra;* Australia.

  *V. portei* (Deblock, Rosé, Broussart, Capron et Brygoo 1962) comb. n. (syn. *Diagonaliporus portei* Deblock, Rosé, Broussart, Capron et Brygoo 1962), in *Capella macrodactyla;* Madagascar.

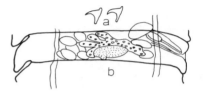

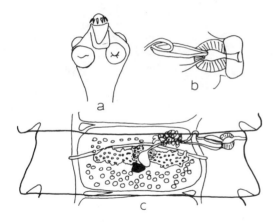

FIGURE 394.  *Pseudandrya monardi* Fuhrmann 1943. (a) Rostellar hooks; (b) proglottid.

FIGURE 393.  *Mashonalepsis dafyddi* Beverly-Burton 1960. (a) Scolex; (b) genital atrium; (c) proglottid.

*V. schikhobalovae* (Krotov 1951) Yamaguti 1959 (syn. *Diagonaliporus schikhobalovae* Krotov 1951), in *Scolopax rusticola;* Russia.

*V. skrjabini* (Krotov 1951) Yamaguti 1951 (syn. *Diagonaliporus skrjabini* Krotov 1951), in *Capella solitaria;* Russia.

*V. spasskyi* (Krotov 1951) Yamaguti 1959 (syn. *Diagonaliporus spasskyi* Krotov 1951), in *Clangula hyemalis;* Russia.

*V. sultanpurensis* Capoor, Srivastava et Chauhan 1975, in *Nycticorax nycticorax;* India.

**8b.   Genital atrium muscular only at proximal end ......................................
................................ *Mashonalepis* Beverly-Burton 1960. (Figure 393)**

*Diagnosis:* Scolex with well-developed rostellum bearing single circle of hooks. Suckers weak, unarmed. Neck present. Genital pores unilateral. Genital atrium with powerful, interior sphincter. Genital ducts dorsal to vessels. Cirrus pouch weak, not reaching poral vessel. Cirrus unarmed. No internal or external seminal vesicle. Seminal receptacle present. Testes numerous (42 to 65), surrounding ovary except anterior poral side. Ovary transversely elongated. Vitellaria compact, postovarian. Gravid uterus a deeply lobed transverse sac. Only one pair of osmoregulatory canals. Parasites of Ciconiiformes (heron). Africa.

Type species: *M. dafyddi* Beverly-Burton 1960, in *Ardea cinerea;* Southern Rhodesia.
Other species:
  *M. ardeius* Mettrick 1967, in *Nycticorax nycticorax;* Zambia.

**9a.   One circle of rostellar hooks ..................................................... 10
9b.   Two circles of rostellar hooks ................................................... 11**

**10a.  Testes mainly antiporal. Uterus reticular ...........................................
..................................... *Pseudandrya* Fuhrmann 1943. (Figure 394)**

*Diagnosis:* Rostellum with a single circle of hooks. Proglottids craspedote, wider than long. Genital pores unilateral. Genital ducts dorsal to osmoregulatory canals. Cirrus pouch crosses osmoregulatory canals. External and internal seminal vesicles present. Testes about ten, mainly antiporal. Ovary large, lobated, slightly poral. Vitelline gland lobated, postovarian. Vaginal pore posterior to cirrus pore. Seminal receptacle very large. Uterus reticular. Parasites of Carnivora. Africa.

Type species: *P. monardi* Fuhrmann 1943, in *Paracynictis selousi;* Angola.

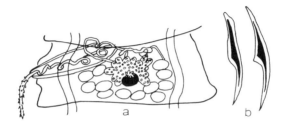

FIGURE 395. *Lateriporus clerci* (Johnston 1912) Fuhrmann 1972. (a) Proglottid; (b) rostellar hooks.

Other species:

P. *mkuzii* Ortlepp 1963, in *Ichneumia albicauda;* South Africa.

**10b. Testes mainly postovarian, uterus saclike .................................. *Lateriporus* Fuhrmann 1907 (syn. *Fuhrmanacanthus* Spasskii (1966). (Figure 395)**

*Diagnosis:* Rostellum with a single circle of hooks. Proglottids craspedote, wider than long. Genital pores unilateral. Genital ducts pass dorsal to osmoregulatory canals. Cirrus pouch crosses osmoregulatory canals. Cirrus armed. Accessory sac may be present. Testes mainly postovarian. Ovary median. Vitelline gland postovarian. Vagina ventral to cirrus pouch. Seminal receptacle present. Uterus sac-like. Parasites of Anseriformes, Charadriiformes, Passeriformes. Greenland, Asia, Europe, Africa, North and South America, Scotland.

Type species: *L. teres* (Krabbe 1869) Fuhrmann 1907 (syn. *Taenia teres* Krabbe 1869, *Hymenolepis teres* [Krabbe 1869] Railliet 1899), in *Harelda glacialis, Somateria mollissima, S. spectabilis, Nyroca marila, Larus melanocephalus, Clangula hyemalis, Arctonetta fischeri;* Alaska, Greenland, Russia.

Other species:

*L. aecophylus* Oschmarin 1950, in *Clangula histrionica;* Russia.

*L. australis* Jones et Williams 1967, in *Chionis alba;* Scotland.

*L. biuterinus* Fuhrmann 1908, in *Nethium brasiliense, Dendrocygna autumnalis, Sarcidiornis carunculata, Chenoplex jubatus, Carina moschata, Cedemia fusca;* Brazil, Europe.

*L. clerci* (Johnston 1912) Fuhrmann 1932 (syn. *Taenia cylindrica* Clerc 1902), in *Larus canus, L. argentatus, L. melanocephalus, L. minutus, L. ridibundus, Sterna fluviatilis, Anas acuta, A. strepera, Hydrochelidon* sp.; Europe, Russia.

*L. destitutus* (Loennberg 1889) Fuhrmann 1908 (syn. *Taenia destitutus* Loennberg 1889), in *Tadorna tadorna;* Europe.

*L. eranui* Sailov 1962, in *Ardea cinerea;* Russia.

*L. geographicus* Cooper 1921, in *Somateria mollissima, Clangula hyemalis;* Canadian Arctic.

*L. gnedini* Sailov 1962, in *Larus argentatus, L. ridibundus, Hydroprogne esdiergrawa;* Russia.

*L. karajasicus* Kuraschvili 1957, in *Ardea cinerea;* Russia.

*L. mahdiaensis* Joyeux 1923, in *Ardea purpura;* Algeria, Tunis.

*L. mathevossianae* Ryjikov et Gubanov 1962, in *Melanitta deglandi;* Russia.

*L. merops* Woodland 1928, in *Merops apoaster;* Sudan, Egypt.

*L. propteres* Fuhrmann 1907, in *Nettion brasiliense;* Brazil.

*L. skrjabini* Mathevossian 1946, in *Nyroca marila, Bucephala clangula, Aythya fuligula, A. fernia;* Russia.

*L. solitariae* Borgarenko 1975, in *Gallinago solitaria;* Russia.

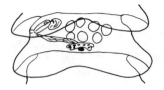

FIGURE 396. *Proorchida lobata* Fuhrmann 1908.

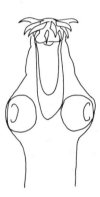 

FIGURE 397. *Amirthalingamia macracantha* (Joyeux et Baer 1935) Bray 1974. Proboscis and apical view of hooks.

*L. spinosus* Fuhrmann, 1908, in *Canchroma cochlearia, Ardea purpurea;* Brazil.

**11a. Testes completely anterior to ovary ..*Proorchida* Fuhrmann 1908. (Figure 396)**

*Diagnosis:* Rostellum with two circles of hooks. Proglottids craspedote, wider than long. Genital pores unilateral. Genital ducts dorsal to osmoregulatory canals. Testes about seven, anterior to ovary. External seminal vesicle present. Ovary median, posterior. Vitelline gland postovarian. Vagina posterior to cirrus pouch. Seminal receptacle present. Gravid uterus an irregular sac. Parasites of Ciconiiformes. South America.

Type species: *P. lobata* Fuhrmann 1908, in *Cancroma cochlearis;* Brazil, Guiana.

**11b. Testes lateral and/or posterior to ovary, or surrounding it.................... 12**

**12a. Anterior circle of hooks bilaterally symmetrical in the following sequence of sizes: small, large, small, small, large, small, large, small, small, large ...............
.......................................... *Amirthalingamia* Bray 1974. (Figure 397)**

*Diagnosis:* Scolex with four simple suckers and a rostellum bearing 20 large hooks in two circles. Anterior circle consists of ten hooks of two sizes arranged in a bilaterally symmetrical pattern in the sequence: small, large, small, small, large, small, large, small, small, large. Hooks in posterior circle smaller, equal. Genital pores unilateral. Genital ducts pass between osmoregulatory canals. External seminal vesicle absent. Ejaculatory duct convoluted, swollen, functioning as internal seminal vesicle. Cirrus armed. Testes numerous, in two lateral fields. Ovary median, bilobed. Uterus a lobed sac. Parasites of cormorants; larvae in freshwater fish. Africa.

Type species: *A. macracantha* (Joyeux et Baer 1935) Bray 1974 (syn. *Paradilepis ma-*

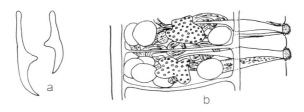

FIGURE 398.. *Paradilepis simoni* Rausch 1949. (a) Rostellar hooks; (b) proglottids.

*cracantha* Joyeux et Baer 1935), in *Phalacrocorax carbo, P. africanus* (larvae in *Tilapia nilotica);* Mali, Sudan.

**12b. Both circles of hooks radially symmetrical ..................................... 13**

**13a. Testes only lateral to ovary, three antiporal and one poral ......................
...................................................*Paradilepis* Hsü 1935 (syn.
*Meggittiella* Lopez-Neyra 1942, *Skrjabinolepis* Mathevossian 1945). (Figure 398)**

*Diagnosis:* Rostellum with two circles of hooks. Proglottids craspedote. Genital pores unilateral. Genital ducts dorsal to osmoregulatory canals. Cirrus armed. Testes four, three antiporal, one poral. Ovary about median. Vitelline gland about postovarian. Vagina ventral to cirrus pouch. Seminal receptacle present. Uterus sac-like. Parasites of Pelecaniformes, Accipitriformes, Ciconiiformes, Galliformes. Asia, Europe, Africa, Australia, New Guinea, Ceylon, North America, Cuba. Key to species: Freeman.[950]

Type species: *P. scolecina* (Rudolphi 1819) Hsü 1935 (syn. *Taenia scolecina* Rudolphi 1819, *Dilepis scolecina* [Rudolphi 1819] Fuhrmann 1908, *Paradilepis duboisi* Hsü 1935, *P. brevis* Burt 1940), in *Phalacrocorax carbo, P. capillatus, P. fusicollis, P. pygmaeus, P. africanus;* Europe, Africa, Asia, Australia, Russia.

Other species:

*P. burmanensis* (Johri 1941) Freeman 1954 (syn. *Oligorchis burmanensis* Johri 1941), in *Phalacrocorax jayanicus;* Burma.

*P. caballeroi* Rysavy et Macko 1971, in *Phalacrocorax auritus;* Cuba.

*P. delachauxi* (Fuhrmann 1909) Hsü 1935 (syn. *Oligorchis delachauxi* Fuhrmann 1909, *Dilepis scolecina* of Joyeux et Baer 1928, *Paradilepis depidocolpos* Burt 1936), in *Phalacrocorax africanus, P. niger;* Africa, Sri Lanka.

*P. diminuta* Huey et Dronen 1981, in *Ajaia ajaja;* Texas, U.S.

*P. ficticia* (Meggitt 1927) Hsü 1935 (syn. *Skrjabinolepis ficticia* [Meggitt 1927] Mathevossian 1945, *Hymenolepis ficticia* Meggitt 1927, *Meggittiella ficticia* [Meggitt 1927] Spasskii 1952), in pelican; Burma.

*P. kempi* (Southwell 1921) Hsü 1935 (syn. *Dilepis kempi* Southwell 1921, *Meggittella kempi* [Southwell 1921] Lopez-Neyra 1942, *Hymenolepis kempi* [Southwell 1921] Mayhew 1925), in *Phalacrocorax niger, P. pygamaeus;* India.

*P. lloydi* (Southwell 1926) (syn. *Meggittiella lloydi* [Southwell 1926] Spasskii 1952), in stork; Africa.

*P. longivaginosus* (Mayhew 1925) Freeman 1954 (syn. *Oligorchis longivaginosus* Mayhew 1925), in *Pelecanus erythrorhynchus;* U.S.

*P. macracantha* Joyeux et Baer 1936, in *Phalacrocorax africanus;* Africa.

*P. maliki* Khalil 1961, in *Threskiornis ethiopicus;* Sudan.

*P. minima* (Gross 1941) Freeman 1954 (syn. *Dilepis minima* Goss 1941), in *Phalacrocorax varius, P. ater, P. sulcirostrus, Microcarbo melanoleucus;* Australia.

*P. multihamata* (Meggitt 1927) Hsü 1935 (syn. *Hymenolepis multihamata* Meggitt 1927, *Meggittiella multihamata* [Meggitt 1927] Lopez-Neyra 1942, *Skrjabinolepis multihamata* [Meggitt 1927] Mathevossian 1945), in *Milvus aegypticus;* Egypt.

*P. patriciae* Baer et Bona 1960, in *Platibis flavipes;* Australia.

*P. phalacrocoracis* Ukoli 1967, in *Phalacrocorax africanus;* Ghana.

*P. rugovaginosus* Freeman 1954, in *Pandion haliaetus;* Canada.

*P. simoni* Rausch 1949, in *Pandion haliaetus;* Wyoming, U.S.

*P. urceus* (Wedl 1855) Hsü 1935 (syn. *Taenia urceus* Wedl 1855, *Hymenolepis urceus* [Wedl 1855] Meggitt 1927), in *Plegadis falcinellus, Platalea leucorodia, Ibis ibis, Ciconia* sp., *Harpoprion caerulescens, Pelecanus onocrotalus, Milvus cinereus, M. migrans;* Africa, India, Russia.

*P. variacanthos* (Southwell et Lake 1939) (syn. *Skrjabinolepis variacanthos* [Southwell et Lake 1939] Mathevossian 1945, *Meggittiella variacanthos* [Southwell et Lake 1939] Spasskii 1952, *Hymenolepis variacanthos* Southwell et Lake 1939), in *Ibis ibis;* Belgian Congo.

*P. yorkei* (Kotlan 1923) Freeman 1954 (syn. *Dilepis yorkei* Kotlan 1923, *Oligorchis yorkei* [Kotlan 1932] Mayhew 1925), in *Megapodius brunneiventris;* New Guinea.

13b. Testes not as above............................................................... 14

14a. Genital ducts ventral to osmoregulatory canals...... **Metadilepis** Spasskii 1949.

*Diagnosis:* Rostellum sucker-like, without sac. Hooks in two circles. Neck present. Proglottids craspedote, wider than long. Genital pores unilateral. Genital ducts ventral to osmoregulatory canals. Cirrus pouch containing convoluted ejaculatory duct. Testes in two lateral groups. Ovary median. Vitelline gland postovarian. Vagina with spinous distal end. Uterus sac-like. Parasites of Caprimulgiformes. North and South America.

Type species: *M. globacantha* (Fuhrmann 1913) Spasskii 1949 (syn. *Dilepis globacantha* Fuhrmann 1913), in *Caprimulgus europaeus, C. ruficollis;* Europe.

Other species:

*M. caprimulgorum* (Fuhrmann 1908) Spasskii 1949 (syn. *Dilepis caprimulgorum* Fuhrmann 1908), in *Cordeiles virginianus, Hydropsalis climacocereus;* North and South America.

14b. Genital ducts dorsal to or in between osmoregulatory canals ................. 15

15a. Ventral osmoregulatory canals normal in position on poral side, dorsal to true dorsal canal on antiporal side.................................................... 16

15b. Osmoregulatory canals normal .................................................. 17

16a. Genital ducts pass between osmoregulatory canals .............................
.............................................. **Ophiovalipora** Hsü 1935. (Figure 399)

*Diagnosis:* Rostellum with two circles of hooks. Neck short. Proglottids wider than long. Ventral osmoregulatory canal normal in position on poral side, dorsal to true dorsal canal on antiporal side. Genital ducts pass between osmoregulatory canals. Genital pores unilateral. Genital atrium large. Cirrus pouch large, crosses osmoregulatory canals. Vas deferens convoluted near proximal end of cirrus pouch. Testes surrounding ovary except on poral side. Ovary bilobed. Vitelline gland postovarian. Vagina opens ventral to cirrus. Seminal receptacle absent (?). Gravid uterus an irregular sac. Parasites of snakes, herons. China, Celebes, North and Central America.

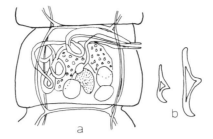

FIGURE 399. *Ophiovalipora gorsakii* Yamaguti 1956. (a) Proglottid; (b) rostellar hooks.

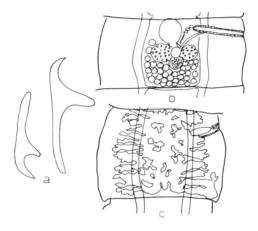

FIGURE 400. *Dendrouterina botauri* Rausch 1948. (a) Rostellar hooks; (b) mature proglottid; (c) gravid proglottid.

Type species: *O. houdemeri* Hsü 1935, in *Elaphe carinata;* China.
Other species:

*O. gorsakii* (Yamaguti 1956) Mathevossian 1963 (syn. *Parvitaenia gorsakii* Yamaguti 1956), in *Gorsakius goisagi;* Celebes.

*O. lintoni* (Olsen 1937) Coil 1950 (syn. *Dilepis unilateralis* Linton 1927, *Dendrouterina lintoni* Olsen 1937), in *Butorides virescens;* U.S.

*O. micracantha* Yamaguti 1954, in *Varanus salvator;* Celebes.

*O. minuta* Coil 1950, in *Butorides virescens;* Indiana, U.S.

*O. nycticoracis* (Olsen 1937) Coil 1950 (syn. *Dendrouterina nycticoracis* Olsen 1937), in *Nycticorax nycticorax;* Minnesota, U.S

**16b. Genital ducts dorsal to osmoregulatory canals .....................................**
**..................................... *Dendrouterina* Fuhrmann 1912. (Figure 400)**

*Diagnosis:* Rostellum with two circles of hooks. Ventral osmoregulatory canal normal in position of poral side, dorsal to true dorsal canal on antiporal side. Proglottids craspedote. Genital pores unilateral. Genital ducts dorsal to osmoregulatory canals. Cirrus pouch crosses osmoregulatory canals. Cirrus armed. Testes mainly posterior and lateral to ovary, a few may be anterior. Ovary median. Vitelline gland postovarian. Vagina posterior to cirrus pouch. Seminal receptacle present. Gravid uterus O- or U-shaped, with many branches. Parasites of herons. Africa, North and South America, India, Australia.

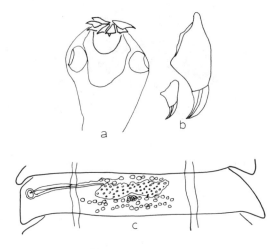

FIGURE 401. *Cyclorchida vestibularis* (Johri 1959) Mathevossian 1963. (a) Scolex; (b) rostellar hooks; (c) proglottid.

Type species: *D. herodiae* Fuhrmann 1912, in *Herodias garzetta;* Africa.
Other species:
  *D. botauri* Rausch 1948, in *Botaurus lentiginosus, B. stellaris, Ixobrychus exilis;* U.S.
  *D. egrettae* Spasskii et Yurpalova 1967, in *Egretta garzetta;* Vietnam.
  *D. fovea* Meggitt 1933, in *Dendrocitta rufa;* Calcutta Zoo.
  *D. mackoi* Mathevossian 1963 (syn. *D. karajasicus* Kuraschvili 1957 of Macko 1959), in *Ardea cinerea;* Czechoslovakia.
  *D. pilherodiae* Mahon 1956, in *Pilherodias pileatus;* Brazil.

17a.  Testes surrounding ovary ............................................................
  ...................................... **Cyclorchida Fuhrmann 1907. (Figure 401)**

*Diagnosis:* Rostellum with two circles of hooks, each hook with stout handle and small blade. Proglottids craspedote, wider than long. Genital pores unilateral. Genital ducts pass between osmoregulatory canals. Cirrus pouch extravascular. Testes surrounding ovary. Ovary lobated, transversely elongated. Vitelline gland postovarian. Vaginal pore dorsal to cirrus pore. Uterus an irregular, transverse sac. Parasites of Ciconiiformes, Passeriformes, Carnivora (civit). Africa, Asia.
  Type species: *C. omalancristrota* (Wedl 1855) Fuhrmann 1907 (syn. *Taenia omalancristrota* Wedl 1855), in *Platalea leucorodia, P. regia, Plegadis falcinellus;* Africa, Australia, Russia.
  Other species:
    *C. crassivesicula* Vevers 1923, in *Paradoxurus hermaphroditicus;* Malaya.
    *C. foteria* Meggitt 1933, in *Geocichla citrina;* Calcutta Zoo.
    *C. fuhrmanni* Hilmy 1936, in *Podica senegalensis;* Liberia.

17b.  Testes mainly posterior to ovary................................................. 18

18a.  Genital atrium with two or four spines in special pockets........... *Gryporhynchus* **Nordmann 1932 (syn. *Acanthocirrus* Fuhrmann 1907). (Figure 402)**

*Diagnosis:* Rostellum with two circles of rostellar hooks. Proglottids craspedote, wider than long. Genital pores unilateral. Genital atrium large, with side pockets containing two

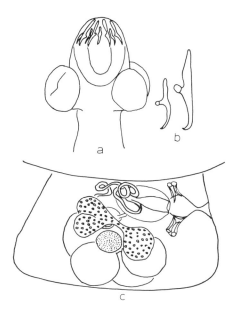

FIGURE 402. *Gryporhynchus tetrorchis* Hill 1941. (a) Scolex; (b) rostellar hooks; (c) proglottid.

or more spines. Genital ducts pass between osmoregulatory canals. Cirrus pouch large, reaching median line or farther. Testes few, mainly posterior to ovary. Ovary bilobed, median. Vitelline gland postovarian. Vaginal pore posterior to cirrus pore. Seminal receptacle present. Uterus sac-like or U-shaped. Parasites of Ciconiiformes. Asia. Europe, Greenland, Africa, Japan, North America, Russia.

Type species: *G. pusillum* Nordmann 1932 (syn. *Taenia macropeos* Wedl 1855, *Dilepis macropeos* [Wedl 1955] Clerc 1906), in *Ardeola grayi, Nycticorax nycticorax, Ardea purpurea, A. cinerea, Botaurus stellaris, Haliaeetus albicilla;* Europe, Asia, India, Russia.

Other species:

*G. macrorostratum* (Fuhrmann 1907) Furhmann 1932 (syn. *Acanthocirrus macrorostratum* Fuhrmann 1907), in *Anthus pratensis;* Egypt.

*G. nycticoracis* Yamaguti 1956, in *Nycticorax nycticorax;* Japan.

*G. retirostris* (Krabbe 1869) Belopolskaja 1953 (syn. *Taenia retirostris* Krabbe 1869, *Dilepis retirostris* [Krabbe 1869] Zschokke 1903, *Acanthocirrus retirostris* [Krabbe 1869] Baer 1956), in *Anthus pratensis, Arenaria interpres, Calidris alpina, C. maritima, Tringa macrura, Eremophila alpestris;* Russia, Europe, Greenland, Africa.

*G. tetrorchis* Hill 1941, in *Ardea herodias;* Oklahoma, U.S.

**18b. Genital atrium lacking such spines ............................................. 19**

**19a. Genital ducts dorsal to osmoregulatory canals....................................**
**............*Dilepis* Weinland 1858 (syn. *Ascodilepis* Guidal 1960, *Anomolepis* Spasskii, Yurpalova et Kornyushin 1868, *Birovilepis* Spasskii 1975, *Brasiliolepis* Spasskii 1965, *Ershovilepis* Mathevossian 1963, *Fuhrmanolepis* Spasskii et Yurpalova 1967 *Spasskytaenia* Borgarenko 1981 in part). (Figure 403)**

*Diagnosis:* Rostellum with two circles of hooks. Proglottids wider than long, craspedote. Genital pores unilateral. Genital ducts dorsal to osmoregulatory canals. Cirrus pouch long.

# 380  CRC Handbook of Tapeworm Identification

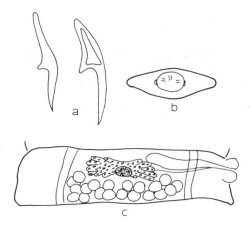

FIGURE 403. *Dilepis glareola* Dubinina 1953. (a) Rostellar hooks; (b) egg; (c) proglottid.

Testes numerous, mainly posterior to ovary. Ovary medium. Vitelline gland posterior to ovary. Vagina variable. Seminal receptacle present. Uterus sac-like. Parasites of Passeriformes, Accipitriformes, Ciconiiformes, Caprimulgiformes, Charadriiformes, Cypseliformes, Galliformes, Gruiformes, Pelecaniformes, Psittaciformes, Procellariiformes. Also, Insectivora, Primates. Cosmopolitan.

Type species: *D. undula* (Schrank, 1788) Weinland 1858 (syn. *Taenia undula* Schrank 1788, *T. angulata* Dujardin 1845, *Dilepis angulata* [Dujardin 1845] Clerc 1900, *Hymenolepis undulata* Parona 1899, *Dilepis undulata* Volz 1900, *Drepanidotaenia undulata* Rosseter 1906, *Southwellia ransomi* Chapin 1926, *Hymenolepis pyramidalis* Sinitzin 1896, *Dilepis turdi* Yamaguti 1935), in *Anthus* spp., *Fringilla coelobs, Coleus monedula, Corvus* spp., *Galerida cristata, Garrulus glandarius, Monticola* spp., *Nucifraga caryocatactes, Oriolus* spp., *Passer* spp., *Pica pica, Spodiopsar cineraceus, Sturnus vulgaris, Troglodytes troglodytes, Merula atrigularis, Turdus* spp., *Saiurus* sp., *Numenius borealis, Trypanocorax, Acridotheres*, etc.; Europe, Asia, North America, Russia, Israel.

Other species:

*D. ardeae* (Rausch 1955) (syn. *Cyclustra ardeae* Rausch 1955), in *Ardea herodias*, U.S.

*D. ardeolae* Singh 1952, in *Ardeola grayi*; India.

*D. attenuata* (Dujardin 1845) Fuhrmann 1908 (syn. *Taenia attenuata* Dujardin 1845), in *Anthus pratensis, Anorthura troglodytes, Fringilla coelebs, Passer domestica, P. montanus;* Europe.

*D. australiensis* Johnston 1911, in *Himantopus leucocephalus;* Australia.

*D. averini* (Spasskii et Yurpalova 1967) comb. n. (syn. *Fuhrmanolepis averini* Spasskii et Yurpalova 1967, *Anomolepis averini* [Spasskii et Yurpalova 1967] Spasskii, Yurpalova et Kornyushin 1968), in *Phalaropus lobatus;* Russia.

*D. bicoronata* Fuhrmann 1908 (syn. *Brasiliolepis bicoronata* [Fuhrmann 1908] Spasskii 1865), in *Harpiprion cayennensis, Tantalus* sp., *Mesembrinibis cayennensis;* Brazil.

*D. brachyarthra* Cholodkovsky 1906, in *Corvus cornix, C. corone, Turdus* spp., *Corvus frageligus, Pica pica;* Europe, Russia.

*D. campylancristrota* (Wedl 1855) Fuhrmann 1908 (syn. *Taenia campylancristrota* Wedl 1955), in *Ardea cinerea, Ardeola grayi, Herodias garzetta;* India, Burma.

*D. capellae* Yamaguti 1935 (syn. *Anomolepis capellae* [Yamaguti 1935] Spasskii, Yurpalova et Kornyushin 1968), in *Capella solitaria, C. stenura, Rostratula benghalensis;* Japan, Burma.

*D. caprimulgina* Neslobinsky 1911, in *Caprimulgus europaeus;* Russia.

*D. crassirostrata* Fuhrmann 1908, in *Tigrisoma brasiliense;* Brazil.

*D. cypselina* Neslobinsky 1911, in *Cypselus apus, Dendrocitta leucogaster;* Russia, India.

*D. distincta* (Loennberg 1889) Fuhrmann 1908 (syn. *Taenia distincta* Loennberg 1889), ih *Larus canus;* Scandinavia.

*D. dollfusi* Quentin 1964, in *Mastomys* sp.; Central African Republic.

*D. fovea* (Meggitt 1933) Yamaguti 1959 (syn. *Dendrouterina fovea* Meggitt 1933), in *Dendrocitta rufa;* Calcutta Zoo.

*D. fuhrmanni* Railliett et Henry 1909 (for *D. unilateralis* Fuhrmann 1908), in *Hoploxypterus cayanus;* Brazil.

*D. glareola* Dubinina 1953 (syn. *Anomolepis glareola* [Dubinina 1953] Spasskii, Yurpalova et Kornyushin 1968), in *Tringa glareola;* Russia.

*D. harvathi* Kotlan 1923, in *Megapodius brunneiventris;* New Guinea.

*D. hilli* Polk 1941, in *Florida caerulea;* Oklahoma, U.S.

*D. hoplites* (Linstow 1903) Fuhrmann 1908 (syn. *Taenia hoplites* Linstow 1903), in *Ardea* sp.; Russia.

*D. irregularis* Southwell et Lake 1939, in *Rostratula bengalensis;* Africa.

*D. jacobsoni* Baylis 1929, in *Eucichla cyanura;* Java.

*D. javanica* Baylis 1929, in *Eucichla cyanura;* Java.

*D. lebasquei* Joyeux et Baer 1935, in *Francolinus pintadeanus;* Indochina.

*D. leptophallus* Kotlan 1923, in *Megapodius brunneiventris;* New Guinea.

*D. limosa* Fuhrmann 1907, in *Limosa limosa, Numenius phaeopus, Tringa ochropus;* Egypt, Europe, Russia.

*D. macrocephala* Fuhrmann 1908, in *Psophia crepitans;* Brazil.

*D. macrosphincter* Fuhrmann 1909, in *Ardeola ralloides, Ardea purpurea;* Africa, Russia, India.

*D. megacirrosa* Ortlepp 1940, in *Chrysochloris asiatica;* South Africa.

*D. megalorhyncha* (Krabbe 1869) Baer 1956 (syn. *Taenia megalorhyncha* Krabbe 1869, *Drepanidotaenia megalorhyncha* [Krabbe 1869] Lopez-Neyra 1942, *Hymenolepis megalorhyncha* Railliet 1899), in *Erolia maritima;* Greenland.

*D. modigliani* (Parona 1898) Fuhrmann 1908 (syn. *Hymenolepis modigliani* Parona 1898), in *Corone enca, C. levaillanti, C. macrorhynchus;* Europe.

*D. monedulae* Neslobinsky 1911, in *Coloeus monedula, Anthus pratensis:* Russia.

*D. nymphoides* Clerc 1903 (*Taenia paradoxa* Krabbe 1869, renamed), in *Erolia minuta, E. subminuta, Limonites damacensis, Scolopax rusticola;* Russia.

*D. ochropodis* Neslobinsky 1911 (syn. *Anomolepis ochropodis* [Neslobinsky 1911] Belopolskaya 1978, *Spasskytaenia ochropodis* [Neslobinsky 1911] Borgarenko 1981), in *Totanus ochropus;* Russia.

*D. odhneri* Fuhrmann 1909, in *Oedicnemus senegalensis, Burhinus capensis;* Africa.

*D. oligorchida* Fuhrmann 1906, in *Busarellus nigricollis;* Brazil.

*D. orientalis* Yamaguti 1956, in *Turdus aureus;* Japan.

*D. papillifera* Fuhrmann 1908, in *Florida caerulea;* Brazil.

*D. pifanoi* Diaz-Ungria et Jordano 1958, in *Nyctibius griseus;* Venezuela.

*D. rostratulae* Sawada et Kifune 1974, in *Rostratula bengalensis;* Japan.

*D. sedowi* Skrjabin 1926, in *Puffinus anglorum, Larus maximus, Thalasseus maximus;* Antarctica.

*D. sobolevi* Spassky 1946, in *Luscinia luscinia;* Russia.

*D. sphaerocephala* (Rudolphi 1819), in *Chrysochloris aurea;* Europe.

*D. transfuga* (Krabbe 1869) Fuhrmann 1908 (syn. *Taenia transfuga* Krabbe 1869, *Ascodilepis transfuga* [Krabbe 1869] Guidal 1960), in *Platalea ajaja;* South America.

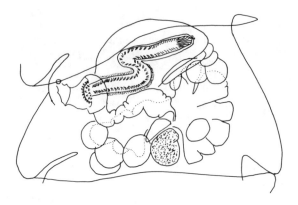

FIGURE 404. *Clelandia parva* Johnston 1909.

*D. trichocephalus* Linstow 1905, in *Cercopithecus pyrrhorotus;* West Africa.
*D. tringae* Cholodkowsky 1912, in *Tringa platyrhyncha, Limicola falcinellus;* Russia.
*D. turdi* Yamaguti 1935, in *Turdus aureus;* Japan.
*D. unilateralis* (Rudolphi 1819) Clerc 1906 (syn. *Taenia unilateralis* Rudolphi 1819), in *Herodia egretta, Ardea cinerea, Ardeola grayi, Butorides virescens, Garzetta garzeeta;* Europe, South and North America.

**19b.    Genital ducts pass between osmoregulatory canals ........................... 20**

**20a.    Testes, ovary and uterus extending laterally beyond osmoregulatory canals. Cirrus pouch extending to aporal osmoregulatory canals .. *Megacirrus* Beck 1951.**

*Diagnosis:* Rostellum with two circles of hooks. Neck absent. Proglottids craspedote, wider than long. Genital pores unilateral. Genital ducts pass between osmoregulatory canals. Genital atrium small. Cirrus pouch reaching aporal osmoregulatory canal. Cirrus unarmed. Testes numerous, posterior, exceeding osmoregulatory canals. Ovary bilobed, median, exceeding osmoregulatory canals laterally. Vitelline gland postovarian. Vagina dorsal to cirrus pouch. Seminal receptacle present. Gravid uterus an irregular sac, extending laterally between osmoregulatory canals. Parasites of megapodiformes. South Pacific.
   Type species: *M. megapodii* Beck 1951, in *Megapodius laperouse;* Palau Islands.

**20b.    Testes, ovary and uterus intervascular. Cirrus pouch large but not extending to aporal osmoregulatory canals ................................................. 21**

**21a.    Cirrus stout, heavily armed .............. *Clelandia* Johnston 1909. (Figure 404)**

*Diagnosis:* Rostellum with 20 hooks in two circles. Neck short. Proglottids craspedote. Genital pores preequatorial, unilateral. Genital ducts pass between osmoregulatory canals. Cirrus stout, heavily armed. Cirrus pouch massive, extending into antiporal half of proglottid. Testes 11 to 13, posterior and antiporal to ovary. Ovary bilobed, each lobe somewhat divided. Vitellarium compact, postovarian. Uterus U-shaped, becoming an irregular sac when gravid. Parasites of Ciconiiformes. Australia.
   Type species: *C. parva* Johnston 1909, in *Ephippiorrhynchus asiaticus;* Jervis Bay, Australia.

**21b.    Cirrus unarmed .......................... *Metabelia* Mettrick 1963. (Figure 405)**

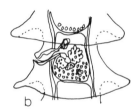 

FIGURE 405. *Metabilia aetodex* Mettrick 1963. (a) Scolex; (b) proglottid.

FIGURE 406. *Cotylorhipis furnarii* (Del Pont 1906) Blanchard 1909.

*Diagnosis:* Medium-sized worms. Rostellum armed with two circles of hooks. Suckers unarmed. Neck present. Proglottids craspedote. Genital pores unilateral, on middle third of lateral margin. Atrium small. Genital ducts pass between osmoregulatory canals. Cirrus unarmed. Internal and external seminal vesicles absent. Testes 25 to 34, postovarian, postvitelline. Vagina posterior to cirrus pouch. Seminal receptacle present. Ovary median, lobed. Vitelline gland compact, postovarian. Uterus filling segment, not extending between canals. Parasites of Falconiformes. Africa.

Type species: *M. aetodex* Mettrick 1963, in *Aquila rapax;* Southern Rhodesia.

**22a.  Rostellum lacking ............................................................ 23**
**22b.  Rostellum present, armed or not ............................................... 25**

**23a.  Suckers membranous, with marginal hooks arranged like sticks of a fan ....... ...................................... *Cotylorhipis* Blanchard 1909. (Figure 406)**

*Diagnosis:* Rostellum absent. Suckers membranous, with long, marginal hooks arranged like sticks of a fan. Proglottids craspedote; gravid segments longer than wide. Genital pores irregularly alternating. Cirrus armed. Other internal organs not described. Parasites of Passeriformes. Argentina.

Type species: *C. furnarii* (del Pont 1906) Blanchard 1909 (syn. *Taenia furnarii* del Pont 1906), in *Furnaria rufus*.

**23b.  Suckers unarmed ............................................................ 24**

**24a.  Uterus with a median stem and numerous lateral branches..................... ...................................... *Anoplotaenia* Beddard 1911. (Figure 407)**

*Diagnosis:* Scolex large, unarmed, arostellate, with four large, powerful suckers, each with middorsal, circular pit. Genital pores irregularly alternating. Cirrus pouch spherical. Testes numerous, encircling ovary and vitelline gland. Ovary bilobed, median. Vitelline gland compact, postovarian, separated from posterior margin of proglottid by numerous testes. Genital atrium with very powerful musculature, pierced laterally by vagina. Uterus with median stem and numerous lateral branches. Parasites of marsupials (Tasmanian devil). Australia.

Type species: *A. dasyuri* Beddard 1911, in *Dasyurus ausinus, Sarcophilus satanicus;* Australia.

**24b.  Uterus a deeply lobed, transverse sac .............*Ethiopotaenia* Mettrick 1961.**

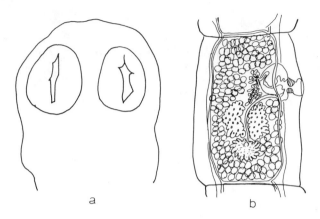

FIGURE 407. *Anoplotaenia dasyuri* Beddard 1911. (a) Scolex; (b) proglottid.

*Diagnosis:* Scolex unarmed, no rostellum. Suckers unarmed. Neck present. Genital pores alternating irregularly. Genital atrium small, on anterior third of margin. Genital ducts passing between osmoregulatory vessels. Cirrus pouch extends beyond poral osmoregulatory canals. Internal seminal vesicle present. External seminal vesicle absent. Testes 20 to 27, in two lateral groups, sometimes joining posteriorly. Vagina posterior to cirrus. Seminal receptacle present. Ovary lobed, small, median, in anterior half of segment. Vitellaria postovarian. Uterus a deeply lobed transverse sac. Parasites of Passeriformes. Africa.

Type species: *E. trachyphonoides* Mettrick 1969, in *Trachyphonus vaillantii;* Southern Rhodesia.

25a. Rostellum unarmed............................................................26
25b. Rostellum armed ............................................................28

26a. Scolex massive. Genital atrium with thick, muscular walls .....................
       ........................................ **Platyscolex** Spasskaja 1962. (Figure 408)

*Diagnosis:* Scolex robust, with powerful, unarmed suckers. Rostellum rudimentary, unarmed. Neck present. Strobila long, craspedote. Genital pores alternating, in anterior one fourth of margin. Genital atrium large, thick-walled. Genital ducts pass dorsal to osmoregulatory canals. Cirrus pouch slender, exceeding osmoregulatory canals. Vas deferens coiled, not voluminous. Testes numerous, postovarian, intervesicular. Ovary follicular, in anterior half of segment, exceeding osmoregulatory canals. Vitellarium lobated, median, postovarian. Vagina opens posterior to cirrus. Seminal receptacle present. Uterus a lobated sac. Eggs numerous. Parasites of Anseriformes. Scandinavia, Asia.

Type species: *P. ciliata* (Fuhrmann 1913) Spasskaja 1962 (syn. *Anomotaenia ciliata* Fuhrmann 1913), in Anseriformes; Sweden, Siberia.
Other species:
   *P. notabilicirrus* Leonov et Belogurov 1970, in *Clangula hyemalis;* Wrangel Island (Arctic Ocean).

26b. **Scolex not massive. Genital atrium normal** ..................................27

27a. **Genital pores alternating regularly** ..... **Neoangularia** Singh 1952. (Figure 409)

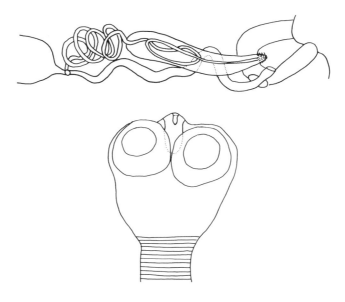

FIGURE 408. *Platyscolex ciliata* (Fuhrmann 1913) Spasskaja 1962. Genital terminalia and scolex.

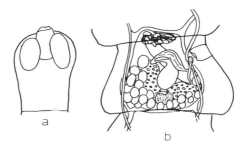

FIGURE 409. *Neoangularia ababili* Singh 1952. (a) Scolex; (b) proglottid.

*Diagnosis:* Rostellum present but unarmed. Proglottids wider than long, craspedote. Genital pores regularly alternating. Testes posterior and lateral to ovary. Vas deferens convoluted near anterior margin of proglottid. Cirrus pouch very large, oblique. Tip of cirrus armed. Ovary bilobed, posterior to cirrus pouch. Vitelline gland postovarian. Vaginal pore anterior to cirrus pore. Vagina swollen, seminal receptacle present. Uterus a lobed sac. Parasites of Apodiformes (swifts). India.

Type species: *N. ababili* Singh 1952, in *Micropus affinis;* India.

**27b. Genital pores alternating irregularly (see also *Cladotaenia*, p. 412)..............
... *Unciunia* Skrjabin 1914 (syn. *Emberizotaenia* Spasskaja 1970). (Figure 410)**

*Diagnosis:* Rostellum present but unarmed. Proglottids craspedote. Genital pores alternating irregularly. Genital atrium deep, preequatorial. Testes mainly postovarian. Cirrus pouch crosses osmoregulatory canals. Cirrus usually armed. Ovary median. Vitelline gland postovarian. Vagina posterior to cirrus pouch. Seminal receptacle present. Uterus a compartmented sac. Parasites of Falconiformes, Charadriiformes, Anseriformes. Africa, India, South America.

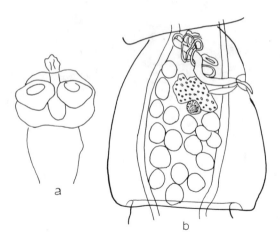

FIGURE 410. *Unciunia burmanensis* (Johri 1951) Mathevossian 1963. (a) Scolex; (b) proglottid.

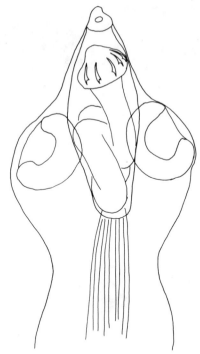

FIGURE 411. *Taeniarhynchaena micropalamae* Burt 1983.

Type species: *U. trichocirrosa* Skrjabin 1941, in *Polyborus* sp.; Paraguay.
Other species:
    *U. acapillicirrosa* Moghe 1933, in *Anas platyrhynchos;* India.
    *U. falconis* Lin 1976, in *Falco tinnunculus;* China.
    *U. hypsipetis* Lin 1976, in *Hypsipetes flavatus;* China.
    *U. polyorchis* (Klaptocz 1908) Spasskaja et Spasskii 1971 (syn. *Monopylidium infundibuliformis* var. *polyorchis* Klaptoz 1908, *Choanotaenia polyorchis* [Klaptoz 1908] Baer 1925), in *Milvus aegyptius, M. korshun, M. govinda, M. migrans;* Africa.

28a. Genital pores alternating regularly ............................................. 29
28b. Genital pores alternating irregularly ........................................... 37

29a. One circle of rostellar hooks ..................................................... 30
29b. Two circles of rostellar hooks .................................................... 32

30a. **Rostellum elongate, capable of being coiled within rostellar sac** .................
    ..................................... ***Taeniarhynchaena* Burt 1983. (Figure 411)**

*Diagnosis:* Scolex spheroidal, with four simple suckers. Rostellum elongate, capable of being coiled within rostellar sac. Apex of rostellum bifid, with single row of five hooks each side. Neck present. Strobila small, acraspedote, about 2.5 mm long. Genital pores regularly alternating, in anterior third of proglottid. Genital ducts dorsal to osmoregulatory canals. Cirrus pouch nearly reaching midline of proglottid. Cirrus unarmed. Vas deferens a large, coiled mass. Testes 14, postovarian. Ovary lobated, median. Vitellarium postovarian, ovoid. Vagina posterior to cirrus pouch; seminal receptacle present. Gravid uterus reticular, persistent. Parasites of Charadriformes. North America.

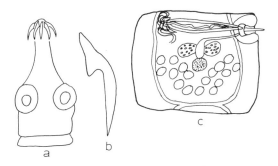

FIGURE 412. *Bakererpes fragilis* Rausch 1947. (a) Scolex; (b) rostellar hook; (c) proglottid.

Type species: *T. micropalamae* Burt 1982, in *Micropalama himantopus;* Hudson Bay, Canada.

**30b. Rostellum not capable of being coiled within rostellar sac** .................... 31

**31a. Genital ducts passing between osmoregulatory canals. Testes in a postovarian cluster**..................................... *Bakererpes* **Rausch 1947. (Figure 412)**

*Diagnosis:* Rostellum with one circle of hooks. Strobila small, with few proglottids. Segments wider than long, strongly convex on poral side. Genital atrium large, with muscular walls lined with small spines. Genital ducts pass between osmoregulatory canals. Cirrus pouch very large, at least reaching median line of segment. Cirrus armed. Testes posterior to ovary. Ovary median. Vitellarium postovarian. Vagina posterior to cirrus pouch. Seminal receptacle present. Uterus a large sac. Parasites of Charadriiformes. North America.

Type species: *B. fragilis* Rausch 1947, in *Chordeiles minor;* Ohio, U.S.
Other species:
    *B. addisi* Webster 1948, in *Chordeiles minor;* Texas, U.S.

**31b. Genital ducts dorsal to osmoregulatory canals (between them in *A. setosa* Burt 1940). Testes in single, posterior row...** *Amoebotaenia* **Cohn 1900. (Figure 413)**

*Diagnosis:* Rostellum armed with a single circle of hooks. Strobila small, of few segments. Proglottids craspedote or not. Genital pores alternating regularly. Atrium may have long bristles. Genital ducts dorsal to osmoregulatory canals (between them in *A. setosa*). Cirrus pouch extervascular. Testes few (6 to 20), in single transverse row posterior to ovary. Ovary median, usually transversely elongated. Vitelline gland postovarian. Vagina usually posterior to cirrus pouch. Seminal receptacle present. Uterus an irregular sac. Parasites of Galliformes, Charadriiformes, Piciformes. Cosmopolitan.

Type species: *A. cuneata* (Linstow 1872) Cohn 1900 (syn. *Taenia cuneata* Linstow 1872, *T. sphenoides* Railliet 1892, *Amoebotaenia sphenoides* [Railliet 1892] Meggitt 1914, *Dicranotaenia sphenoides* Railliet 1896), in *Gallus gallus, Perdix perdix;* Cosmopolitan.
Other species: (some of these probably belong elsewhere)
    *A. awogera* Yamaguti 1956, in *Picus awokera;* Japan.
    *A. brevicollis* Fuhrmann 1907, in *Charadrius nubicus, Hoplopterus spinosus;* Africa.
    *A. cohni* Kalyankar et Palladwar 1977, in *Gallus gallus;* India.
    *A. domesticus* Shinde, Ghare et Suryawanshi 1980, in *Gallus gallus;* India.
    *A. indiana* Shinde 1972, in *Gallus gallus;* India.
    *A. kharati* Kalyankar et Palladwar 1977, in *Gallus gallus;* India.

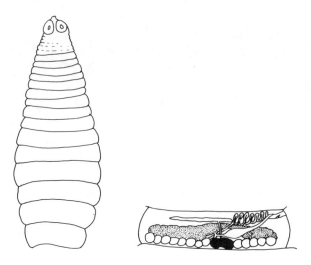

FIGURE 413. *Amoebotaenia cuneata* (Linstow 1872) Cohn 1900. Entire worm and mature proglottid.

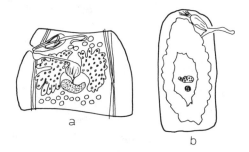

FIGURE 414. *Cyclustera capito* (Rudolphi 1819) Fuhrmann 1901. (a) Mature proglottid; (b) gravid proglottid.

A. *longirostellata* Sawada et Kugi 1976, in *Scolopax rusticola;* Japan.
A. *longisacculus* Yamaguti 1956, in *Gallus gallus;* Celebes.
A. *madrasiensis* Dixit et Capoor 1981, in *Passer domesticus;* India.
A. *maharashtri* Shinde 1972, in *Gallus gallus;* India.
A. *megascolecis* Shinde 1972, in *Gallus gallus;* India.
A. *oligorchis* Yamaguti 1935, in *Gallus gallus;* Japan.
A. *oophorae* Belopolskaja 1970, in *Pluvialis apricaria;* Russia.
A. *pekinensis* Tseng 1932, in *Charadrius veredus;* China.
A. *puthurensii* Pillai et Peter 1971 (nomen nudum), in *Gallus gallus;* India.
A. *setosa* Burt 1940, in *Lobipluvia malabarica;* Sri Lanka.
A. *spinosa* Yamaguti 1956, in *Gallus gallus;* Celebes.
A. *trapezoides* Pillai et Peter 1971 (nomen nudum), in *Gallus gallus;* India.
A. *vanelli* Fuhrmann 1907, in *Vanellus dongolanus;* Egypt.
A. *yamasigi* Yamaguti 1956, in *Scolopax rusticola;* Japan.

**32a. Gravid uterus ring-like, surrounding ovary............................................**
**........................................... *Cyclustera* Fuhrmann 1901. (Figure 414)**

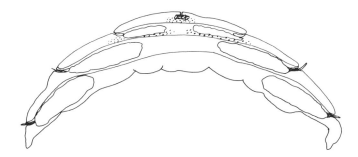

FIGURE 415. *Eurycestus falciformis* Burt 1979. Entire worm.

*Diagnosis:* Rostellum with two circles of hooks. Genital pores regularly alternating. Genital atrium muscular. Genital ducts pass between osmoregulatory canals. Cirrus pouch preequatorial, crossing osmoregulatory canals. Testes surrounding ovary. Ovary lobated, median. Vitelline gland compact, postovarian. Vagina posterior to cirrus pouch. Seminal receptacle present. Gravid uterus ring-like, surrounding remnants of ovary and vitelline gland. Parasites of Ciconiiformes (herons). Europe, Madagascar, North and South America.

Type species: *C. capito* (Rudolphi 1819) Fuhrmann 1901 (syn. *Taenia capito* Rudolphi 1819), in *Ajaja ajaja, Platalea leucorodia, Pseudotantalus ibis;* Brazil, Mexico, Africa, Russia, Madagascar.

Other species:

*C. fuhrmanni* Clerc 1906, in *Botaurus stellaris;* Europe.

**32b. Gravid uterus saccular, sometimes strongly lobated ........................... 33**

**33a. Gravid proglottids about 150 times as wide as long ............................**
**........................................... *Eurycestus* Clark 1954. (Figure 415)**

*Diagnosis:* Scolex well-developed, deeply imbedded in the host intestinal mucosa. Rostellum long, slender, with double circle of 14 to 16 hooks. Suckers well-developed, with tiny hooks, at least on their anterior margins. Scolex usually lost when one tries to detach specimen from host; also is commonly absent from mature specimens. Proglottids very wide and very short, with five to eight acraspedote proglottids. Gravid proglottids about 150 times as wide as long. Genital pores preequatorial, becoming located on anterior margin of proglottid, regularly alternating. Genital atrium deep, evaginable with a short hermaphroditic canal. One set of male and female reproductive organs per segment. Cirrus pouch powerfully developed, cirrus large, armed. Testes poral, in narrow field, overlapping poral wing of ovary. Ovary median, with wide, narrow wings, mainly aporal to testes. Vitellarium compact, postovarian. Vagina posterior to cirrus pouch, convoluted. Seminal receptacle small. Uterus transverse, with sac-like outgrowths. Osmoregulatory canals reticular, lacking longitudinal stems. Eggs with polar elongations of outer membrane. Parasites of Charadiiformes (avocet). North America, France.

Type species: *E. avoceti* Clark, 1954, in *Recurvirostra aermicana;* U.S., France.
Other species:

*E. falciformis* Burt 1979, in *Recurvirostra americana;* North Dakota, U.S.

*E. latissimus* Burt 1979, in *Recurvirostra americana;* North Dakota, U.S.

**33b. Gravid proglottids not nearly as wide relative to length ....................... 34**

**34a. Gravid uterus strongly lobated ....................... *Liga* Weinland 1857 (syn. *Fuhrmannia* Parona 1901, *Rallitaenia* Spasskii et Spasskaja 1974). (Figure 416)**

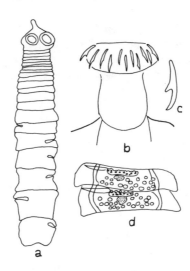

FIGURE 416. *Liga brevis* (Linstow 1884) Mathevossian 1963. (a) Entire worm; (b) rostellum; (c) rostellar hook; (d) mature proglottid.

*Diagnosis:* Rostellum with two circles of hooks. Suckers armed or not. Strobila small, of few segments. Proglottids craspedote. Genital pores alternating regularly. Genital ducts dorsal to osmoregulatory canals. Cirrus pouch crosses osmoregulatory canals or not. Testes mainly posterior to ovary. Ovary median, anterior. Vitelline gland postovarian. Vagina posterior to cirrus pouch. Seminal receptacle present. Gravid uterus an irregular sac. External eggshell may have polar knobs. Parasites of Piciformes, Charadriiformes, Passeriformes, Gruiformes; also, *Sorex*. North and South America, Europe, Russia, Africa, China.

Type species: *L. punctata* Weinland 1856) Weinland 1857 (syn. *Taenia punctata* Weinland 1856, *Fuhrmannia brasiliensis* Parona 1901), in *Picus auratus, Picus* sp.; U.S.A., Brazil.

Other species:

*L. alternans* (Cohn 1900) Ransom 1909 (syn. *Taenia alternans* Cohn 1900), in *Tringa totanus;* Europe.

*L. brevis* (Linstow 1884) Mathevossian 1963 (syn. *Taenia brevis* Linstow 1884, *Amoebotaenia brevis* [Linstow 1884] Fuhrmann 1908), in *Charadrius pluvialis, Recurvirostra avocetta, Charadrius hiaticola, C. aegypticus, C. mongolus, Pluvialis apricarius, Squatarola squatarola, Anas acuta;* Europe, Russia.

*L. dubinini* Bauer 1941, in *Turdus merula;* Russia.

*L. facilis* (Meggitt 1927) Szpotanska 1932 (syn. *Anomotaenia trivialis* Meggitt 1927, *A. facilis* Meggitt 1927), in *Oedicnemus crepitans;* Egypt.

*L. gallinulae* (Beneden 1858) Dollfus 1934 (syn. *Taenia gallinulae* Beneden 1858, *Rallitaenia gallinulae* [Beneden 1858] Spasskii et Spasskaja 1975), in *Gallinula chlorops, Fulica americana;* U.S., Eurasia.

*L. leonovi* Belogurov et Zueva 1968, in *Charadius mongolus;* Russia.

*L. leucoranica* Sailov 1962, in *Larus minutus;* Russia.

*L. porzanae* (Ablasov, Spasskaja et Karubekova 1976) comb. n. (syn. *Rallitaenia porzanae* Ablasov, Spasskaja et Karubekova 1976), in *Porzana pusilla;* Kirgizia, Russia.

*L. shentsengi* Sandeman 1959, in *Charadriius veredus;* China.

**34b. Gravid uterus not strongly lobated ............................................. 35**

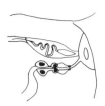

FIGURE 417. *Neoliga swifti* (Shinde 1969) Kayton et Kritsky 1983. Genital terminalia.

**35a. Testes in two groups, anterior and posterior to ovary.... *Thaparea* Johri 1953.**

*Diagnosis:* Rostellum with two circles of large and small hooks. Proglottids craspedote, wider than long. Genital pores alternating regularly. Genital ducts dorsal to osmoregulatory canals. Cirrus pouch large, oblique, nearly reaching anterior margin of segment. Testes in two groups, anterior and posterior to ovary. Internal seminal vesicle present. Ovary medium. Vitelline gland postovarian. Genital atrium shallow. Seminal receptacle present. Gravid uterus sac-like, extending laterally across osmoregulatory canals. Parasites of Charadriiformes. India.

Type species: *T. magnivesicula* Johri 1953, in *Capella gallinago;* India.

**35b. Testes posterior, lateral and dorsal to ovary ................................... 36**

**36a. Genital ducts dorsal to osmoregulatory canals ...................................
 .......... *Neoliga* Singh 1952 (syn. *Mehdiangularia* Shinde 1969). (Figure 417)**

*Diagnosis:* Rostellum with two circles of similar hooks. Neck and anterior proglottids spinose. Segments craspedote, wider than long. Genital pores regularly alternating. Genital ducts dorsal to osmoregulatory canals. Cirrus pouch large, oblique, enclosing convoluted ejaculatory duct. Cirrus slender, partly armed. Testes lateral, posterior and dorsal to ovary. Ovary median, bilobed. Vitelline gland postovarian, lobated. Vagina swollen, with sphincter, opening anterior to cirrus. Seminal receptacle present. Uterus large, sac-like. Parasites of Apodiformes (swifts). India.

Type species: *N. diplacantha* Singh 1952, in *Apus affinis;* India.
Other species:
  *N. affinis* (Shinde 1968) Kayton et Kritsky 1984 (syn. *Sureshia affinis* Shinde 1968), in *Apus affinis;* India.
  *N. alii* (Shinde 1968) Kayton et Kritsky 1984 (syn. *Sureshia alii* Shinde 1968), in *Apus affinis;* India.
  *N. depressa* (Siebold 1836) Spasskii et Spasskaja 1959 (syn. *Taenia depressa* Siebold 1836, *Anomotaenia depressa* [Siebold 1836] Fuhrmann 1908, *Amoebotaenia frigida* Meggitt 1927, *Liga frigida* [Meggitt 1927] Fuhrmann 1932, *Liga depressa* [Siebold 1836] Joyeux et Baer 1936), in *Apus apus, A. melba, A. pacificus, Hirundapus caudacutaus, Delichon urbica, Hirundo rustica;* Europe, Africa, Central America.
  *N. orientalis* Spasskaja et Spasskii 1971, in *Apus pacificus, A. apus;* Russia.
  *N. singhi* Shinde, Jadhav et Kadam 1981, in *Apus affinis;* India.
  *N. swifti* (Shinde 1969) Kayton et Kritsky 1984 (syn. *Mehdiangularia swifti* Shinde 1969), in *Apus affinis;* India.

**36b. Genital ducts pass between osmoregulatory canals ...............................
 ............................................. *Chettusiana* Singh 1959. (Figure 418)**

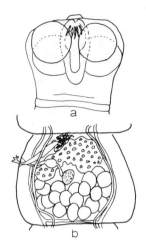

FIGURE 418. *Chettusiana indiana* Singh 1959. (a) Scolex; (b) proglottid.

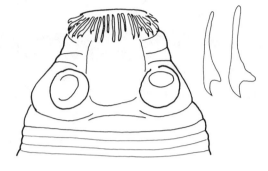

FIGURE 419. *Vitta magniuncinata* Burt 1938. Scolex.

*Diagnosis:* Scolex globular, sharply demarcated from neck. Suckers large, unarmed. Rostellum bears a double circle of hooks. Neck short. Proglottids somewhat craspedote. Genital pores regularly alternating. Genital atrium small, with hair-like processes protruding from it. Genital ducts passing between osmoregulatory canals. Cirrus pouch extends to near anterior middle of proglottid. Cirrus not described. Internal and external seminal vesicles absent. Vagina posterior to cirrus pouch. Seminal receptacle present. Ovary bilobed, aporal lobe larger; in anterior half of proglottid. Vitellaria compact, irregular, postovarian. Uterus sac-like, sometimes extending past osmoregulatory canals. Eggs rounded to oval. Parasites of Charadriiformes. India.

Type species: *C. indiana* Singh 1959, in *Chettusia leucura;* India.

37a. One circle of rostellar hooks .................................................. 38
37b. Two circles of rostellar hooks ................................................. 44

38a. Circle of hooks wavy or zigzag ............................................... 39
38b. Circle of hooks regular ......................................................... 41

39a. Hooks in circle alternating one anterior, two posterior, etc. ...................
................................................................. *Vitta* Burt 1938. (Figure 419)

*Diagnosis:* Scolex broad basally. Suckers powerful, unarmed. Rostellum large, with two circles of hooks, alternating one anterior hook between each pair of posterior hooks, thus:

ıı'ıı'ıı'ıı

Neck absent. Proglottids wider than long, craspedote. Genital pores irregularly alternate. Genital ducts dorsal to osmoregulatory canals. Genital atrium deep. Cirrus unarmed. Testes numerous, posterior, dorsal and lateral to ovary. Ovary with numerous fan-like follicles. Vitellarium postovarian. Seminal receptacle present. Uterus an irregular sac. Outer egg membrane with a tapering process at each pole. Parasites of swallows. Sri Lanka, Europe, Asia, North America.

Type species: *V. magniuncinata* Burt 1938, in *Hirundo rustica;* Sri Lanka.
Other species:

FIGURE 420. *Angularella beema* (Clerc 1906) Strand 1928. Rostellum.

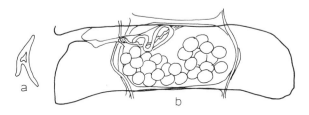

FIGURE 421. *Pseudangularella thompsoni* Burt 1938. (a) Rostellar hook; (b) proglottid.

*V. alexandri* Kornyushin 1966, in *Glariola nordmanni;* Russia.
*V. minutiuncinata* Burt 1938, in *Hirundo rustica;* Sri Lanka.

**39b.  Hooks in simple wavy circle ................................................. 40**

**40a.  Cirrus pouch small, mostly or entirely lateral to osmoregulatory canals ........ *Angularella* Strand 1928 (syn. *Angularia* Clerc 1906, preoccupied). (Figure 420)**

*Diagnosis:* Rostellum armed with single, zigzag row of hooks. Proglottids wider than long, craspedote. Genital pores alternating irregularly. Genital ducts dorsal to osmoregulatory canals. Cirrus pouch small, mainly extravascular. Testes numerous, mainly postovarian. Ovary mainly dorsal to cirrus pouch. Seminal receptacle present. Gravid uterus a lobated sac. Parasites of Passeriformes, Apodiformes. Europe, Asia, Sri Lanka, Taiwan, North and Central America.
   Type species: *A. beema* (Clerc 1906) Strand 1928 (syn. *Angularia beema* Clerc 1906), in *Clivicola riparia, Hirundo rustica, Delichon urbica, Riparia riparia, Iridoprocne albilinea;* Russia, Europe, North and Central America.
Other species:
   *A. audubonensis* Stamper et Schmidt 1984, in *Petrochelidon pyrrhonota;* Colorado, U.S.
   *A. hirundina* (Fuhrmann 1907) Spasskaja et Spasskii 1971 (syn. *Vitta hirundina* [Fuhrmann 1907] Spasskii 1968, *Anomotaenia hirundina* Fuhrmann 1907), in *Delichon urbica, Riparia riparia;* Russia.
   *A. ripariae* Yamaguti 1940, in *Riparia paludicola;* Formosa.
   *A. swifti* Singh 1952, in *Micropus affinis;* Lucknow, India.
   *A. taiwanensis* Yamaguti 1940, in *Hirundo daurica;* Formosa.
   *A. urbica* Spasskaja et Spasskii 1971, in *Riparia riparia, Delichon urbica;* Russia.

**40b.  Cirrus pouch large, medial to osmoregulatory canals ............................ ........................................ *Pseudangularia* Burt 1938. (Figure 421)**

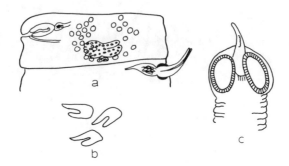

FIGURE 422. *Laterorchites bilateralis* (Fuhrmann 1908) Fuhrmann 1932. (a) Proglottid; (b) rostellar hooks; (c) scolex.

*Diagnosis:* Rostellum massive, armed with single, zigzag circle of hooks. Proglottids craspedote. Genital pores alternating irregularly. Genital atrium very deep. Cirrus pouch large, entirely medial to osmoregulatory canals. Cirrus armed basally. Testes numerous, mainly postovarian. Internal and external seminal vesicles present. Ovary large, bialate. Vitellaria postovarian. Distal portion of vagina surrounded by glandular cells, separated from seminal receptacle by powerful sphincter. Uterus a lobated sac. Parasites of Apodiformes. Sri Lanka, Morocco.

Type species: *P. thompsoni* Burt 1938, in *Collocalia unicolor, Apus pallidus;* Sri Lanka, Morocco.

Other species:

*P. triplacantha* Burt 1938, in *Collocalia unicolor;* Sri Lanka.

**41a. Ovary posterior. Testes in two lateral fields .......................................**
**................................. *Laterorchites* Fuhrmann 1932. (Figure 422)**

*Diagnosis:* Rostellum long, slender, with a single circle of hooks. Genital pores alternating irregularly. Proglottids craspedote, wider than long. Testes in two lateral groups. External seminal vesicle present. Cirrus armed. Ovary compact, posterior. Vitelline gland postovarian. Vagina posterior to cirrus pouch. Uterus sac-like. Parasites of Podicipediformes, Facloniformes. Central America, India.

Type species: *L. bilateralis* (Fuhrmann 1908) Fuhrmann 1932 (syn. *Choanotaenia bilateralis* Fuhrmann 1908), in *Podiceps dominicus;* Central America.

Other species:

*L. rajasthanensis* Mukherjee 1970, in *Falco jugger;* India.

**41b. Ovary about central ........................................................... 42**

**42a. Uterus reticular............................................... *Krimi* Burt 1944.**

*Diagnosis:* Rostellum with single circle of hooks. Strobila small, of few segments. Proglottids craspedote. Genital pores irregularly alternating. Genital ducts pass between osmoregulatory canals. Cirrus pouch may cross osmoregulatory canals. Testes numerous, posterior. Vas deferens convoluted near proximal end of cirrus pouch. Ovary compact, anterior. Vitelline gland postovarian. Vagina posterior to cirrus pouch. Seminal receptacle present. Gravid uterus reticular. Parasites of Piciformes, Rallidae. Ceylon, India, Europe.

Type species: *K. chrysocolaptis* Burt 1944, in *Chrysocolaptes guttacristatus;* Sri Lanka.
Other species:

*K. rallida* Macko 1966, in *Rallus aquaticus;* Czechoslovakia.

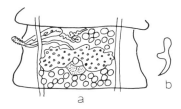

FIGURE 423. *Tubanguiella buzzardia* (Tubangui et Masiluñgan 1937) Yamaguti 1959.

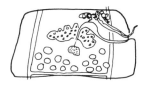

FIGURE 424. *Polycercus pauciannulata* (Fuhrmann 1908) comb. n.

42b. Uterus saccular ................................................................. 43

43a. Testes in two groups, anterior and posterior to ovary............................
................................................. ***Tubanguiella* Yamaguti 1959. (Figure 423)**

*Diagnosis:* Rostellum with single circle of hooks. Proglottids acraspedote, wider than long. Genital pores alternating irregularly. Genital ducts pass between osmoregulatory canals. Cirrus pouch crosses osmoregulatory canals. Testes numerous, in anterior and posterior groups. Ovary bilobed, large. Vitelline gland postovarian. Vagina posterior to cirrus pouch. Seminal receptacle present. Uterus a transversely elongated, lobated sac. Parasites of Ardeiformes, Accipitriformes (buzzard). Philippines, Pakistan.

Type species: *T. buzzardia* (Tubangui et Masiluñgan 1937) Yamaguti 1959 (syn. *Kowalewskiella buzzardia* Tubangui et Masiluñgan 1937), in *Butaster indicus;* Philippines.
Other species:
   *T. ardeola* Bilqees et Jehan 1978, in *Ardeola bacchus;* Pakistan.

43b. **Testes posterior to ovary..................... *Polycercus* Villot 1883 (syn. *Paricterotaenia* Fuhrmann 1932, *Sacciuterina* Mathevossian 1963, *Stenovaria* Spasskii et Borgarenko 1973, *Schmidneila* Spasskii et Spasskaja 1973). (Figure 424)**

*Diagnosis:* Rostellum with single circle of hooks. Proglottids craspedote. Genital pores alternating irregularly. Genital ducts pass between osmoregulatory canals. Cirrus pouch usually crosses osmoregulatory canals. Testes numerous, posterior to ovary. Ovary median, lobated, preequatorial. Vitelline gland postovarian. Vagina posterior to cirrus pouch. Seminal receptacle present. Uterus sac-like. Parasites of Columbiformes, Charadriiformes, Caprimulgiformes, Passeriformes, Accipitriformes. Cosmopolitan.

Type species: *P. lumbrici* Villot 1883 (syn. *Taenia paradoxa* Rudolphi 1802, *Polycercus paradoxa* [Rudolphi 1802] Spasskaja et Spasskii 1970, *Drepanidotaenia paradoxa* [Rudolphi 1802] Parona 1899, *Choanotaenia paradoxa* [Rudolphi 1802] Cohn 1899, *Parachoanotaenia paradoxa* [Rudolphi 1982] Lühe 1910, *Icterotaenia paradoxa* [Rudolphi 1802] Railliet et Henry 1909, *Paricterotaenia paradoxa* [Rudolphi 1802] Fuhrmann 1932, *Sacciuterina paradoxa* [Rudolphi 1802] Mathevossian 1963, *S. paradoxa* var. *gasowskae* Mathevossian 1963, *Taenia choatica* Geibel 1866, *Amoebotaenia lumbrici* [Villot 1883] Joyeux et Baer 1939, *Paricterotaenia lumbrici* [Villot 1883] Sandeman 1959), in *Haematopus, Charadrius, Scolopax, Numenius, Aegialites, Tringa, Capella, Lymnocryptes,* etc.; Asia.
Other species:
   *P. aegyptiaca* (Krabbe 1869) comb. n. (syn. *Taenia aegyptiaca* Krabbe 1869, *Choanotaenia aegyptiaca* [Krabbe 1869] Clerc 1903, *Icterotaenia aegyptiaca* [Krabbe 1869] Baer 1925, *Paricterotaenia aegyptiaca* [Krabbe 1869] Fuhrmann 1963), in *Cursorius*

*isabellinus, Gallinago gallinago, G. media, Scolopax rusticola;* Egypt, Russia, Germany.

P. *albani* (Mettrick 1958) comb. n. (syn. *Paricterotaenia albani* Mettrick 1958, *Sacciuterina albani* [Mettrick 1958] Mathevossian 1963), in *Sturnus vulgaris;* England.

P. *apterygis* (Benham 1900) comb. n. (syn. *Drepanidotaenia apterygis* Benham 1900, *Choanotaenia apterygis* [Benham 1900] Fuhrmann 1908, *Icterotaenia apterygis* [Benham 1900] Baer 1925, *Paricterotaenia apterygis* [Benham 1900] Fuhrmann 1932), in *Apteryx bulleri;* New Zealand.

P. *arguata* (Clerc 1906) comb. n. (syn. *Choanotaenia arguata* Clerc 1906, *Icterotaenia arguata* [Clerc 1906] Baer 1925, *Paricterotaenia arguata* [Clerc 1906] Fuhrmann 1932, *Sacciuterina arguata* [Clerc 1906] Mathevossian 1963), in *Numenius arquatus, Columba palumbus;* China, Russia.

P. *asymmetrica* (Fuhrmann 1908) comb. n. (syn. *Choanotaenia asymmetrica* Fuhrmann 1908, *Icterotaenia asymmetrica* [Fuhrmann 1908] Baer 1925, *Paricterotaenia asymmetrica* [Fuhrmann 1908] Fuhrmann 1932), in *Caprimulgus* sp.; Brazil.

P. *baczynskae* (Mathevossian 1963) Spasskaya et Spasskii 1970 (syn. *Amoebotaenia baczynskae* Mathevossian 1963, *A. brevicollis* Fuhrmann 1907 according to Baczynska 1914), shore birds; Russia.

P. *barbara* (Meggitt 1926) comb. n. (syn. *Choanotaenia barbara* Meggitt 1926, *Paricterotaenia barbara* [Meggitt 1926] Fuhrmann 1932), in *Passer montanus, Garrulax leucolophus, Prunella collaris;* Burma, Japan.

P. *burti* (Sandeman 1959) Spasskaja et Spasskii 1970 (syn. *Paricterotaenia burti* Sandeman 1959), in *Lymnocryptes minimus, Numenius arquatus;* England.

P. *campanulata* (Fuhrmann 1908) comb. n. (syn. *Choanotaenia campanulata* Fuhrmann 1908, *Icterotaenia campanulata* [Fuhrmann 1908] Baer 1925, *Paricterotaenia campanulata* [Fuhrmann 1908] Fuhrmann 1932), in *Opisthocomus hoazin;* Brazil.

P. *chlamyderae* (Krefft 1871) comb. n. (syn. *Taenia chlamyderae* Krefft 1871, *Choanotaenia chlamyderae* [Krefft 1871] Johnston 1912, *Icterotaenia chlamyderae* [Krefft 1871] Baer 1925, *Paricterotaenia chlamyderae* [Krefft 1871] Fuhrmann 1932) in *Chlamydera maculata;* Australia.

P. *clerci* (Spasskii et Spasskaja 1965) Spasskaja et Spasskii 1970, in *Scolopax rusticola;* Russia.

P. *crassitestata* (Fuhrmann 1908) comb. n. (syn. *Choanotaenia crassitestata* Fuhrmann 1908, *Icterotaenia crassitestata* [Fuhrmann 1908] Baer 1925, *Paricterotaenia crassitestata* [Fuhrmann 1908] Fuhrmann 1932, *Sacciuterina crassitestata* [Fuhrmann 1908] Mathevossian 1963), in *Pteroglossus inscriptus;* Brazil.

P. *delachauxi* (Baer 1925) Fuhrmann 1932 (syn. *Icterotaenia delachauxi* Baer 1925, *Burhinotaenia delachauxi* [Baer 1925] Spasskaja et Spasskii 1965), in *Burhinops capensis, Oedicnemus oedicnemus, O. crepitans;* Africa, Spain.

P. *delachauxi* var. *mesacantha* Lopez-Neyra 1935, in *Oedicnemus oedicnemus;* Granada.

P. *dobrogica* (Chiriac 1963) comb. n. (syn. *Paricterotaenia dobrogica* Chiriac 1963), in *Coracias garrulus;* Rumania.

P. *dodecacantha* (Krabbe 1869) comb. n. (syn. *Taenia dodecacantha* Krabbe 1869, *Icterotaenia dodecacantha* [Krabbe 1869] Baer 1925, *Paricterotaenia dodecacantha* [Krabbe 1869] Fuhrmann 1932), in *Larus* spp.; Europe, Russia.

P. *dubininae* (Sandeman 1959) comb. n. (syn. *Paricterotaenia dubininae* Sandeman 1959), in *Capella gallinago;* locality unknown.

P. *embryo* (Krabbe 1869) comb. n. (syn. *Taenia embryo* Krabbe, 1869, *Choanotaenia embryo* [Krabbe 1869] Clerc 1911, *Icterotaenia embryo* [Krabbe 1869] Baer 1925, *Paricterotaenia embryo* [Krabbe 1869] Fuhrmann 1932), in *Scolopax gallinago, S. major, Hoplopterus spinosus, Capella media, C. gallinago, Limnocryptes;* Europe, Russia, Africa.

*P. falsificata* (Meggitt 1927) comb. n. (syn. *Choanotaenia falsificata* Meggitt 1927, *Paricterotaenia falsificata* [Meggitt 1927] Fuhrmann 1932, *Stenovaria falsificata* [Meggitt 1927] Spasskii et Borgarenko 1973), in *Oedicnemus crepitans;* Egypt.

*P. fuhrmannoides* (Skrjabin 1914) comb. n., nom. n. (syn. *Choanotaenia fuhrmanni* Skrjabin 1914, *Icterotaenia fuhrmanni* [Skrjabin 1914] Baer 1925, *Paricterotaenia fuhrmanni* [Skrjabin 1914] Furhmann 1932), in *Circus cinereus, Milvus migrans;* Russia, Ethiopia.

*P. fuhrmanni* (Tseng 1932) Spasskaja et Spasskii 1970 (syn. *Amoebotaenia fuhrmanni* Tseng 1932, *Paricterotaenia fuhrmanni* [Tseng 1932] Sandeman 1959), in *Capella gallinago;* Palaearctic, Nearctic.

*P. gongyla* (Cohn 1900) comb. n. (syn. *Choanotaenia gongyla* Cohn 1900, *Icterotaenia gongyla* [Cohn 1900] Baer 1925, *Paricterotaenia gongyla* [Cohn 1900] Furhmann 1932), in *Larus ridibundus, L. fuscus;* North Africa, France.

*P. innominata* (Meggitt 1927) comb. n. (syn. *Choanotaenia innominata* Meggitt 1927, *Paricterotaenia innominata* [Meggitt 1927] Furhmann 1932), in "finch"; Burma.

*P. intermedia* (Fuhrmann 1908) comb. n. (syn. *Choanotaenia intermedia* Fuhrmann 1908, *Icterotaenia intermedia* [Fuhrmann 1908] Baer 1925, *Paricterotaenia intermedia* [Fuhrmann 1908] Fuhrmann 1932, *Sacciuterina intermedia* [Fuhrmann 1908] Mathevossian 1963), in *Gallinago gigantea, G. undulata, Scolopax* sp.; Brazil.

*P. inversa* (Rudolphi 1819) comb. n. (syn. *Taenia inversa* Rudolphi 1819, *Choanotaenia inversa* [Rudolphi 1819] Furhmann 1908, *Parachoanotaenia inversa* [Rudolphi 1819] Lühe 1910, *Icterotaenia inversa* [Rudolphi 1819] Railliet et Henry 1909, *Taenia oligotoma* Rudolphi 1819, *T. gennarii* Parona 1887, *Drepanidotaenia gennarii* [Parona 1887] Parona 1900, *Paricterotaenia inversa* [Rudolphi 1819] Furhmann 1932), in *Sterna nigra, S. paradisea;* Europe.

*P. laevigata* (Rudolphi 1819) comb. n. (syn. *Taenia laevigata* Rudolphi 1819, *Choanotaenia laevigata* [Rudolph 1819] Clerc 1906, *Monopylidium laevigata* [Rudolphi 1819] Furhmann 1909, *Parachoanotaenia laevigata* [Rudolphi 1819] Lühe 1910, *Icterotaenia laevigata* [Rudolphi 1819] Railliet et Henry 1909, *Paricterotaenia laevigata* [Rudolphi 1819] Furhmann 1932), in *Charadrius* spp., *Numenius arguatus, Burhinus capensis;* Africa, Europe, Russia.

*P. macracantha* (Fuhrmann 1908) comb. n. (syn. *Choanotaenia macracantha* Fuhrmann 1908, *Icterotaenia macracantha* [Fuhrmann 1908] Baer 1925, *Paricterotaenia macracantha* [Fuhrmann 1908] Fuhrmann 1932, *Sacciuterina macracantha* [Fuhrmann 1908] Mathevossian 1963, *Hamatofuhrmannia macracantha* [Fuhrmann 1908] Spasskii 1969, *Parachoanotaenia macracantha* [Fuhrmann 1908] Rego 1968), in *Myothera* sp., *Helodromus ochropus, Phalaropus lobatus, Holopterus, Microsarcops, Rostratula, Rhynchaea capensis, Lobivanellus cinereus;* China, Brazil, Sri Lanka, Europe, Egypt.

*P. magnicirrosa* (Meggitt 1926) comb. n. (syn. *Choanotaenia magnicirrosa* Meggitt 1926, *Paricterotaenia magnicirrosa* [Meggitt 1926] Furhmann 1932), in *Acridotheres tristis*; Burma.

*P. mariae* (Mettrick 1958) comb. n. (syn. *Paricterotaenia mariae* Mettrick 1958, *Sacciuterina mariae* [Mettrick 1958] Mathevossian 1963), in *Erithacus rubecula melophilus;* England.

*P. mathevossiani* (Schmidt et Neiland 1971) comb. n. (syn. *Sacciuterina mathevossiani* Schmidt et Neiland 1971, *Schmidneila mathevossiani* [Schmidt et Neiland 1971] Spasskii et Spasskaia 1973), in *Rhamphocaenus rufiventris;* Nicaragua.

*P. megacantha* (Rudolphi 1819) comb. n. (syn. *Taenia megacantha* Rudolphi 1819, *Choanotaenia megacantha* [Rudolphi 1819] Furhmann 1907, *Icterotaenia megacantha* [Rudolphi 1819] Baer 1925, *Paricterotaenia megacantha* [Rudolphi 1819] Furhmann

1932) in *Caprimulgus* spp., *Nyctibius jamaicensis, Lurocalis semitorquatus;* South America, Africa, Europe.

*P. megistacantha* (Fuhrmann 1909) comb. n. (syn. *Choanotaenia megistacantha* Fuhrmann 1909, *Icterotaenia megistacantha* [Fuhrmann 1909] Baer 1925, *Paricterotaenia megistacantha* [Fuhrmann 1909] Fuhrmann 1932, *Burhinotaenia megistacantha* [Fuhrmann 1909] Spasskaja et Spasskii 1965), in *Oedicnemus senegalensis;* Africa.

*P. milvi* (Singh 1952) comb. n. (syn. *Paricterotaenia milvi* Singh 1952, *Sacciuterina milvi* [Singh 1952] Mathevossian 1963), in *Milvus migrans, M. korschun;* India, Russia.

*P. olgae* (Krotov 1953) comb. n. (syn. *Paricterotaenia olgae* Krotov 1953, *Sacciuterina olgae* [Krotov 1953] Mathevossian 1963), in *Scolopax rusticola;* Russia.

*P. parvirostris* (Krabbe 1869) comb. n. (syn. *Taenia parvirostris* Krabbe 1869, *Choanotaenia parvirostris* [Krabbe 1869] Cohn 1899, *Parachoanotaenia parvirostris* [Krabbe 1869] Lühe 1910, *Icterotaenia parvirostris* [Krabbe 1869] Baer 1925, *Paricterotaenia parvirostris* Baer 1925, *Sacciuterina parvirostris* [Krabbe 1869] Mathevossian 1963), in *Hirundo urbica, H. rustica;* Europe.

*P. passerellae* (Cooper 1921) comb. n. (syn. *Choanotaenia passerellae* Cooper 1921, *Icterotaenia passerellae* (Cooper 1921) Baer 1925, *Paricterotaenia passerellae* [Cooper 1921] Fuhrmann 1932), in *Passerella iliaca;* Alaska, U.S.

*P. pauciannulata* (Fuhrmann 1908) comb. n. (syn. *Choanotaenia pauciannulata* Fuhrmann 1908, *Icterotaenia pauciannulata* [Fuhrmann 1908] Baer 1925, *Paricterotaenia pauciannulata* [Fuhrmann 1908] Fuhrmann 1932, *Sacciuterina pauciannulata* [Fuhrmann 1908] Mathevossian 1963), in *Caprimulgus europaeus, Podager nacunda, Chordeiles minor;* U.S., Brazil, Egypt, Europe.

*P. producta* (Krabbe 1869) comb. n. (syn. *Taenia producta* Krabbe 1869, *Choanotaenia producta* [Krabbe 1869] Cohn 1899, *Icterotaenia producta* [Krabbe 1869] Cohn 1899, *Icterotaenia producta* [Krabbe 1869] Baer 1925, *Paricterotaenia producta* [Krabbe 1869] Fuhrmann 1932), in *Picus viridis;* Europe.

*P. ransomi* (Linton 1927) comb. n. (syn. *Choanotaenia ransomi* Linton 1937, *Paricterotaenia ransomi* [Linton 1927] Fuhrmann 1932), in *Gavia immer, Sterna hirundo, Rissa tridactyla, Larus* spp., U.S., Europe, Russia.

*P. rhynchopis* (Fuhrmann 1908) comb. n. (syn. *Choanotaenia rhynchopis* Fuhrmann 1908, *Icterotaenia rhynchopis* [Fuhrmann 1908] Baer 1925, *Paricterotaenia rhynchopis* [Fuhrmann 1908] Fuhrmann 1932), in *Rhynchops intercedens;* Brazil.

*P. rotunda* (Clerc 1913) comb. n. (syn. *Choanotaenia rotunda* Clerc 1913, *Icterotaenia rotunda* [Clerc 1913] Baer 1925, *Paricterotaenia rotunda* [Clerc 1913] Fuhrmann 1932, *Sacciuterina rotunda* [Clerc 1913] Mathevossian 1963), in *Capella gallinago, C. stenura;* Russia, Europe.

*P. stellifera* (Krabbe 1869) comb. n. (syn. *Taenia stellifera* Krabbe 1869, *Choanotaenia stellifera* [Krabbe 1869] Cohn 1899, *Icterotaenia stellifera* [Krabbe 1869] Baer 1925, *Paricterotaenia stellifera* [Krabbe 1869] Fuhrmann 1932, *Sacciuterina stellifera* [Krabbe 1869] Mathevossian 1963), in *Capella gallinago, Scolopax rusticola, Tringoides hypoleucus;* Europe, Russia.

*P. sujerensis* (Mukherjee 1970) comb. n. (syn. *Paricterotaenia sujerensis* Mukherjee 1970), in *Oenanthe deserti;* India.

*P. turdi* (Spasskaja 1957) Spasskaja et Spasskii 1970 (syn. *Paricterotaenia turdi* Spasskaja 1957, *Choanotaenia turdi* [Spasskaja 1957] Mathevossian 1963), in *Turdus* spp.; Russia.

*P. zonifera* (Johnston 1912) comb. n. (syn. *Paricterotaenia zonifera* Johnston 1912), in *Zonifer tricolor;* Australia.

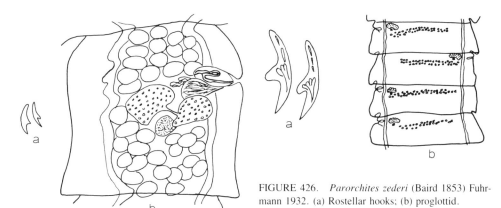

FIGURE 426. *Parorchites zederi* (Baird 1853) Fuhrmann 1932. (a) Rostellar hooks; (b) proglottid.

FIGURE 425. *Bancroftiella forna* Meggitt 1933. (a) Rostellar hooks; (b) proglottid.

**44a. Testes in two fields, anterior and posterior to ovary .............................
........................................ *Bancroftiella* Johnston 1911. (Figure 425)**

*Diagnosis:* Rostellum with two circles of hooks. Proglottids craspedote. Genital pores alternating irregularly. Genital ducts pass between osmoregulatory canals. Cirrus pouch elongate, preequatorial. Testes numerous, in two groups, one anterior and one posterior to ovary. Ovary median, bilobed. Vitellarium postovarian. Vagina posterior to cirrus pouch. Seminal receptacle present. Uterus an irregular sac. Parasites of Ciconiiformes (herons), Charadriiformes, Passeriformes. Also, kangaroo. Australia, Celebes, Sumatra, Moluccas, Japan, India, Vietnam.

Type species: *B. tenuis* Johnston 1911, in *Macropus ualabatus;* Australia.
Other species:

*B. ardeae* Johnston 1911, in *Nycticorax caledonicus;* Australia, Celebes.

*B. forna* Meggitt 1933, in *Tringa hypoleucos;* Calcutta Zoo.

*B. glandularis* (Fuhrmann 1905) Johnston 1911 (syn. *Anomotaenia glandularis* Fuhrmann 1905), in *Herodias timoriensis, Notophoyx novaehollandiae;* Sumatra, Australia, Celebes, Moluccas.

*B. sudarikovi* Spasskii et Yurpalova 1970, in *Egretta garzetta;* Vietnam.

*B. toratugumi* Yamaguti 1956, in *Turdus aureus;* Japan.

**44b. Testes posterior and/or lateral, an occasional few anterior to ovary .......... 45**

**45a. Ovary poral .................................................................. 46**
**45b. Ovary medial ................................................................ 49**

**46a. Testes in a transverse band .......... *Parorchites* Fuhrmann 1932. (Figure 426)**

*Diagnosis:* Rostellum with two circles of hooks. Neck swollen. Scolex imbedded in gut wall. Proglottids craspedote, wider than long. Genital pores alternating irregularly. Genital ducts dorsal to osmoregulatory canals. Cirrus pouch small, extravascular. Testes numerous, in a continuous band posterior to ovary. Ovary compact, anterior, poral. Vitelline gland postovarian. Uterus sac-like. Parasites of penguins. Antarctica.

Type species: *P. zederi* (Baird 1853) Fuhrmann 1932 (syn. *Taenia zederi* Baird 1853, *Anomotaenia zederi* [Baird 1853] Clausen 1915), in *Aptenodytes forsteri, Aptenodytes* sp., *Pygocelis antarctica, P. papua;* Antarctica.

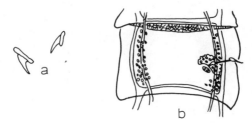

FIGURE 427. *Laterotaenia nattereri* Fuhrmann 1906. (a) Rostellar hooks; (b) proglottid.

**46b. Testes not in a transverse band** ................................................. **47**

**47a. Testes in a single field posterior to and partly lateral to ovary** .................
.................................................... ***Yogeshwaria* Shinde 1968.**

*Diagnosis:* Scolex with four simple suckers and rostellum armed with 22 hooks in two circles. Rostellar sac extends beyond posterior margin of suckers. Neck absent. Mature and gravid proglottids wider than long. Genital pores irregularly alternating, preequatorial. Cirrus pouch reaches osmoregulatory canals. Vas deferens convoluted, anterior to level of cirrus sac. Testes numerous, mainly posterior to ovary, a few at same level. Ovary bilobed, simple, somewhat poral. Vitelline gland posterior to ovary. Vagina posterior to cirrus pouch; seminal receptacle present. Gravid uterus a simple sac. Gravid proglottid with prominent transverse band of muscle (artifact due to contraction?). Eggs simple. Parasites of Charadriiformes. Asia.

Type species: *Y. malabarica* Shinde 1968, in *Vanellus malabaricus;* India.

**47b. Testes in two lateral fields** ...................................................... **48**

**48a. Testes extending most of length of proglottid** ......................................
........................................ ***Laterotaenia* Fuhrmann 1906. (Figure 427)**

*Diagnosis:* Rostellum small, with two circles of hooks. Proglottids craspedote, wider than long. Genital pores alternating irregularly. Genital ducts pass between osmoregulatory canals. Cirrus pouch small, extravascular. Testes numerous, in two longitudinal, lateral groups, both anterior and posterior to level of ovary. Ovary poral, equatorial. Vitelline gland posterior to ovary. Vagina posterior to cirrus pouch. Seminal receptacle present. Uterus sac-like. Parasites of Falconiformes. Brazil.

Type species: *L. nattereri* Fuhrmann 1906, in *Gypagus papa;* Brazil.

**48b. Testes postovarian** ................... ***Parvirostrum* Fuhrmann 1908. (Figure 428)**

*Diagnosis:* Rostellum small, with two circles of hooks. Proglottids wider than long, acraspedote. Genital pores alternating irregularly. Cirrus pouch small, extravascular. Testes few, in two lateral groups, not extending length of segment. Ovary small, bilobed, somewhat poral. Vitelline gland postovarian. Vagina posterior to cirrus pouch. Seminal receptacle small. Uterus sac-like. Parasites of Falconiformes, Passeriformes. Brazil, India.

Type species: *P. reticulatum* Fuhrmann 1908, in *Picolaptes fuscicapillus, Dendrornis elegans, D. rostripallens:* Brazil.

Other species:
*P. magnisomum* Southwell 1930, in a vulture; India.

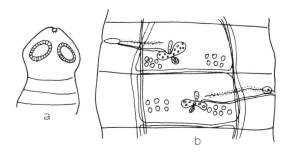

FIGURE 428. *Parvirostrum reticulatum* Fuhrmann 1900. (a) Scolex; (b) proglottids.

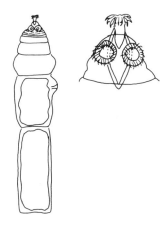

FIGURE 429. *Nototaenia fileri* Jones et Williams 1967. Entire worm, and scolex.

**49a. Suckers armed with a single row of hooks ........................................**
**........................... *Nototaenia* Jones et Williams 1967. (Figure 429)**

*Diagnosis:* Minute cestodes with few segments. Rostellum with two circles of hooks. Sucker rims armed with a single row of hooks (not spines), similar in shape to those on the rostellum, but smaller. Genital pores preequatorial, irregularly alternating. Cirrus pouch powerful, reaching midline of proglottid. Cirrus armed. Genital ducts pass dorsal of osmoregulatory canals. Testes numerous, dorsal to and surrounding ovary. Ovary bilobed, median. Vitellarium posterior to ovary. Vagina voluminous, armed with spines. Seminal receptacle large. Uterus saccate. Parasites of Charadriiformes. Antarctica.

Type species: *N. fileri* Jones et Williams 1967, in *Chionis alba;* South Orkney Island, Antarctica.

**49b. Suckers unarmed ....................................................... 50**

**50a. Vagina with a sclerotized clamp ..... *Sureshia* Ali et Shinde 1967. (Figure 430)**

*Diagnosis:* Rostellum with two circles of similar hooks. Suckers unarmed. Neck short. Strobila spined or not. Genital pores irregularly alternate. Genital ducts pass dorsal to osmoregulatory canals. Cirrus pouch long, may greatly exceed poral osmoregulatory canals.

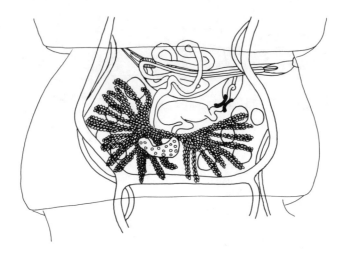

FIGURE 430.  *Sureshia trychopeus* Kayton et Kritsky 1984.

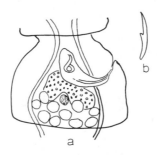

FIGURE 431.  *Chitinorecta agnosta* Meggitt 1927. (a) Proglottid; (b) rostellar hook.

Testes numerous, dorsal to ovary. Ovary with two highly branched lobes and narrow isthmus. Vitellarium compact, postovarian. Vagina with sclerotized clamp. Seminal receptacle preovarian. Uterus a lobated sac. Parasites of Apodiformes (swifts). Asia, North America.

Type species: *S. micropusia* Ali et Shinde 1967, in *Micropus affinis;* India.
Other species:
  *S. aurangabadensis* Shinde et Ghare 1980, in *Micropus affinis;* India.
  *S. depressa* (Siebold 1836) Kayton et Kritsky 1984 (syn. *Taenia depressa* Siebold 1836, *Anomotaenia depressa* [Siebold 1836] Fuhrmann 1908), in *Cypselus apus, Hirundo rustica, Chelidonaria urbica, Hirundapus caudacutus;* Europe, Africa.
  *S. elloraii* Shinde et Ghare 1980, in *Micropus affinis;* India.
  *S. macropterygis* (Huebscher 1937) Kayton et Kritsky 1984 (syn. *Anomotaenia macropterygis* Huebscher 1937, *Parvitaenia macropterygis* [Huebscher 1937] Yamaguti 1959), in *Macropteryx longipennis;* Java.
  *S. shindei* Shinde et Ghare 1980, in *Micropus affinis;* India.
  *S. singhi* Shinde et Ghare 1980, in *Micropus affinis;* India.
  *S. trychopeus* Kayton et Kritsky 1984, in *Aeronautes saxtalis;* Colorado, U.S.

50b.  Vagina without a sclerotized clamp............................................. 51

51a.  Testes and ovary partly cortical ........*Chitinorecta* Meggitt 1927. (Figure 431)

FIGURE 432. *Lapwingia adelaidae* Schmidt 1972. Entire worm, showing uterus.

*Diagnosis:* Rostellum with two circles of hooks. Strobila short, of few segments. Proglottids craspedote. Genital pores alternating irregularly. Genital ducts pass between osmoregulatroy canals. Cirrus pouch crosses median line of segment. Vas deferens convoluted at anterior margin of proglottid. Testes few, partly cortical, partly medullary. Ovary large, mainly medullary, but with ventral, cortical branches. Vitellarium postovarian. Vaginal pore dorsal to cirrus pore. Uterus lobated. Parasites of Charadriiformes. Egypt.

Type species: *C. agnosta* Meggitt 1927, in *Hoplopterus spinosus;* Egypt

Other species:

*C. metaskrjabini* Spasskaja 1973, in *Charadrius mongolus;* Russia.

51b.  Testes and ovary entirely medullary............................................52

52a.  Uterus reticular.......................................................................53
52b.  Uterus saccular or horseshoe-shaped ........................................54

53a.  Genital atrium spinous..................... **Lapwingia** Singh 1952. (Figure 432)

*Diagnosis:* Rostellum with two circles of hooks. Gravid proglottids longer than wide. Genital pores irregularly alternating. Genital atrium spinous. Genital ducts pass between osmoregulatory canals. Cirrus pouch oblique, crossing osmoregulatory canals. Vas deferens convoluted near proximal end of cirrus pouch. Testes mainly posterior to ovary. Ovary bilobed preequatorial. Vitelline gland postovarian. Vagina posterior to cirrus pouch. Seminal receptacle present. Uterus reticular, extensive. Parasites of Charadriiformes. India, Australia.

Type species: *L. reticulosa* Singh 1952, in *Lobipluvia malabarica;* India.

Other species:

*L. adelaidae* Schmidt 1972, in *Lobibyx novaehollandae;* Australia.

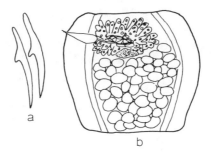

FIGURE 433. *Anomotaenia nymphaea* (Schrank 1790) Fuhrmann 1908. (a) Rostellar hooks; (b) proglottid.

*L. malabarica* Shinde 1972, in *Vanellus malabaricus;* India.
*L. singhi* Shinde 1972, in *Vanellus indicus;* India.
*L. yogeshwarii* Shinde 1972, in *Vanellus indicus;* India.

**53b. Genital atrium not spinous.............................. *Anomotaenia* Cohn 1900 (syn. *Alcataenia* Spasskaja 1971, *Borgarenkolepis* Spasskaja et Spasskii 1977, *Cracticotaenia* Spasskii 1966, *Diplochetos* Linstow 1906, *Gruitaenia* Spasskii, Borarenko et Spasskaja 1971, *Laritaenia* Spasskaja et Spasskii 1971, *Rissotaenia* Spasskaja et Kolotilova 1972, *Spasskytaenia* Oshmarin 1956 in part, *Molluscotaenia* Spasskii et Andreiko 1971 in part, *Dictymetra* Clark 1952). (Figure 433)**

*Diagnosis:* Rostellum with two circles of hooks. Proglottids craspedote, wider than long or longer than wide. Genital pores irregularly alternating. Genital ducts pass between osmoregulatory canals. Cirrus pouch crosses osmoregulatory canals. Testes numerous, mainly posterior to ovary. Ovary median. Vitelline gland postovarian. Vagina posterior to cirrus pouch. Seminal receptacle present. Uterus reticular. Parasites of Charadriiformes, Falconiformes, Cuculiformes, Passeriformes, Ciconiformes, Trogoniformes, Piciformes, Anseriformes, Galliformes, Gruiformes, Strigiformes, Apterygiformes, Cypseliformes. Also, in *Sorex.* Cosmopolitan.

Type species: *A. microrhyncha* (Krabbe 1869) Cohn 1900 (syn. *Taenia microrhyncha* Krabbe 1869), in *Capella gallinago, Charadrius apricarius, C. curonicus, C. dubius, C. hiaticola, C. minor, C. pluvialis, Erolia minuta, Philamachus pugnax, Tringa glareola, T. stagnatilis;* Europe, Asia.

Other species:

*A. acollum* Fuhrmann 1907, in *Crotophaga ani, Cuculus varius;* Brazil, India, Ceylon, Indochina.

*A. acrocephala* Fotedar et Chisti 1977, in *Acrocephalus stentorus;* India.

*A. alata* Spasskii et Konovalov 1971, in *Philomachus pugnax, Calidris minuta;* Russia.

*A. anadyrensis* Spasskii et Konovalov 1967, in *Charadrius hiaticula;* Russia.

*A. angolensis* (Mettrick 1960) Mathevossian 1963 (syn. *Choanotaenia angolensis* Mettrick 1960), in *Pitta angolensis;* Africa.

*A. anthusi* (Spasskaja 1958) Mathevossian 1963 (syn. *Choanotaenia anthusi* Spasskaja 1958), in *Anthus trivialis;* Russia.

*A. asymmetrica* Johnston 1913, in *Herodias timoriensis;* Australia.

*A. bacilligera* (Krabbe 1869) Fuhrmann 1908 (syn. *Taenia bacilligera* Krabbe 1869), in *Scolopax rusticola, Capella, Limnocryptes;* Europe, Brazil, Africa.

*A. baeri* (Spasskii 1968) comb. n. (syn. *Dichoanotaenia baeri* Spasskii 1968, *Anomotaenia larina* Baer 1956), in *Lariformes;* Europe.

*A. belopolskaja* (Spasskaja et Spasskii 1973) comb. n. (syn. *Dictymetra belopolskaja* Spasskaja et Spasskii 1973), in *Eromophila alpestris;* Russia.

*A. binzui* Yamaguti 1956, in *Anthus hodgesoni;* Japan.

*A. borealis* (Krabbe 1869) Fuhrmann 1908 (syn. *Taenia borealis* Krabbe 1869), in *Emberiza nivalis, E. citrinella, Motacilla alba, M. flava, Enneoctonus collurio, Muscicapa, Phylloscopus, Acrocephallus, Turdus, Cinclus cinclus;* Europe, Asia, Africa.

*A. brasiliensis* Fuhrmann 1907, in *Trogon surucura;* Brazil, Uruguay, Paraguay.

*A. brevis* (Clerc 1902) Fuhrmann 1908 (syn. *Choanotaenia brevis* Clerc 1902), in *Dendrocopus major, Perisoreus infaustus;* Russia, Europe, Africa.

*A. caenodex* Mettrick et Beverly-Burton 1962, in *Turdus viscivorus;* England.

*A. caledonica* Fuhrmann 1918 (syn. *Choanotaenia caledonica* [Fuhrmann 1918] Lopez-Neyra 1935), in *Zosterops minuta, Z. xarthochroa;* New Caledonia, Loyalty Islands.

*A. chandleri* (Moghe 1925) Mathevossian 1963 (syn. *Monopylidium chandleri* Moghe 1925, *Choanotaenia chandleri* [Moghe 1926] Fuhrmann 1932), in *Sarcogrammus indicus;* India.

*A. charadrii* Gvosdev 1964, in *Charadrius asiaticus;* Russia.

*A. cingulata* (Linstow 1905) Fuhrmann 1908 (syn. *Dilepis cingulata* Linstow 1905), in *Pelidna alpina;* Europe.

*A. citrus* (Krabbe 1869) Fuhrmann 1908 (syn. *Taenia citrus* Krabbe 1869, *Choanotaenia citrus* [Krabbe 1869] Clerc 1903, *Monopylidium cayennensis* Fuhrmann 1907, *Dichoanotaenia citrus* [Krabbe 1869] Lopez-Neyra 1944), in *Belonopterus cayennensis, Calidris alpina, C. minuta, Charadius asiaticus, C. dubius, Capella gallinago, C. stenura, C. media, Limnocryptes gallinula, Philomachus pugnax, Scolopax rusticola, Terekia cinerea, Tringa hypoleucos, T. ochropus, T. glareola, T. incana;* Europe, China, Brazil, Russia.

*A. clavigera* (Krabbe 1869) Cohn 1900 (syn. *Taenia clavigera* Krabbe 1869, *Choanotaenia clavigera* [Krabbe 1869] Clerc 1903, *Dichoanotaenia clavigera* [Krabbe 1869] Lopez-Neyra 1944), in *Arenaria interpres, Calidris alpina, C. canutus, C. maritima, C. minuta, Limonites damacensis, Limosa limosa, Scolopax rusticola, Vanellus vanellus;* Greenland, Europe.

*A. cyanthiformoides* Fuhrmann 1908, in *Cypseloides senex;* Brazil.

*A. dehiscens* (Krabbe 1879) Fuhrmann 1908 (syn. *Taenia dehiscens* Krabbe 1879), in *Cinclus aquaticus, Passer domesticus;* Turkestan, Switzerland, Egypt.

*A. discoidea* (Beneden 1868) Fuhrmann 1908 (syn. *Taenia discoidea* Beneden 1868, *T. multiformis* Krabbe nec Creplin 1869, *Choanotaenia discoidea* [Beneden 1868] Joyeux et Baer 1943, *Dichoanotaenia* discoidea [Beneden 1868] Lopez-Neyra 1947), in *Ciconia alba;* Europe, Africa.

*A. dominicanus* (Railliet et Henry 1912) comb. n. (syn. *Icterotaenia dominicanus* Railliet et Henry 1912, *Paricterotaenia dominicanus* [Railliet et Henry 1912] Fuhrmann 1932, *Anomotaenia antarctica* Fuhrmann 1925, *Pseudanomotaenia antarctica* [Fuhrmann 1925] Mathevossian 1963, *Anomotaenia micracantha dominicana* [Railliet et Henry 1912] Fuhrmann 1920, *Rissotaenia dominicanus* [Railliet et Henry 1912] Spasskaja et Kolotilova 1972), in *Larus, Rissa, Belonopterus, Uria*, etc.; North Atlantic, Russia, North and South America.

*A. dubia* Meggitt 1927, in *Cerchneis tinnunculus;* Egypt.

*A. dubininae* Mathevossian 1963 (syn. *Choanotaenia* sp. Dubinina 1953), in *Cuculus canorus;* Russia.

*A. ericetorum* (Krabbe 1869) Cohn 1901 (syn. *Taenia ericetorum* Krabbe 1869), in *Charadrius pluvialis, C. apricarius;* Europe.

*A. eroliae* Yurpalova et Spasskii 1971, in *Calidris ferruginea;* Russia.

*A. eudromii* Spasskaja 1954, in *Eudromius* sp.; Russia.

A. *fortunata* Meggitt 1927, in *Cerchneis tinnunculus, Coturnix coturnix;* Egypt, Europe.

A. *ganii* (Spasskaja et Shumilo 1971) comb. n. (syn. *Dictymetra ganii* Spasskaja et Shumilo 1971), in *Actitis hypoleucus;* Moldavia, Russia.

A. *globulus* (Wedl 1855) Fuhrmann 1908 (syn. *Taenia globulus* Wedl 1855, *Choanotaenia globulus* [Wedl 1855] Clerc 1902, *Dichoanotaenia globulus* [Wedl 1855] Lopez-Neyra 1944), in *Calidris minuta, Charadrius domincus, Limnocryptus gallinula, Philomachus pugnax, Squatorola squatorola, Tringa glareola, T. hypoleucos, T. incana, T. ochropus;* Europe, Africa, Russia.

A. *guiarti* (Tseng 1932) Mathevossian 1963 (syn. *Monopylidium guiarti* Tseng 1932, *Choanotaenia guiarti* [Tseng 1932] Fuhrmann 1932), in *Charadrius dubius, Philomachus pugnax, Aegilatitis minor, A. curonica, Glareola pratincola;* China, Russia, Sri Lanka.

A. *heimi* Quentin 1964, in *Lophuromys sikapsu;* Central African Republic.

A. *heterocoronata* Fuhrmann 1918 (syn. *Choanotaenia heterocoronata* [Fuhrmann 1918] Lopez-Neyra 1935), in *Acridotheres tristis;* New Caledonia, Loyalty Islands.

A. *hilmylum* Mathevossian 1963 (syn. *Choanotaenia infundibulum* of Hilmy 1936), in *Milvus migrans;* Liberia

A. *hoeppli* Tseng 1933, in *Gypaetus barbatus;* China.

A. *hydrochelidonis* Dubinina 1953 (syn. *Choanotaenia hydrochelidonis* Dubinina 1953, *Laritaenia hydrochelidonis* [Dubinina 1953] Spasskaja et Spasskii 1971), in *Hydrochelidon leucoptera, Clidonias nigra, Larus minutus, L. ridibundus;* Russia.

A. *iwanizkyi* nom. n. (syn. *Choanotaenia skrjabinia* Iwanitzky 1940, *Anomotaenia skrjabini* [Iwanizky 1940] Mathevossian, preoccupied), in *Charadius hiaticula;* Russia.

A. *jurii* Spasskii et Konovalow 1967, in *Calidris temminckii, Phalaropus lobatus, Tringa glareola;* Russia.

A. *kashmirensis* Fotedar et Chisti 1973, in *Sturnus vulgaris;* India.

A. *konoresniki* nom. n. (syn. *Dichoanotaenia minuta* Konovalov et Resnik 1968), in *Charadrius domincus;* Russia.

A. *larina* (Krabbe 1869) Cohn 1901 (syn. *Taenia larina* Krabbe 1869, *Pseudanomotaenia larina* [Krabbe 1869] Mathevossian 1963, *Rissotaenia larina* [Krabbe 1869] Spasskaja et Kolotilova 1972), in *Larus, Rissa, Sterna, Stercorarius, Cepphus,* etc.; Russia, Greenland, Iceland, Alaska.

A. *latissima* (Spasskii, Borgarenko et Spasskaja 1971) comb. n. (syn. *Gruitaenia latissima* Spasskii, Borgarenko et Spasskaja 1971), in *Grus grus;* Russia.

A. *leuckarti* (Krabbe 1869) Fuhrmann 1908 (syn. *Taenia leuckarti* Krabbe 1869), in *Ardea cinerea;* Europe.

A. *luehei* Gasoska 1932, in *Numenius arguatus;* Europe.

A. *macracantha* (Fuhrmann 1907) Mathevossian 1963 (syn. *Monopylidium macracantha* Fuhrmann 1907, *Choanotaenia macracantha* [Fuhrmann 1907] Fuhrmann 1932), in *Charadrius alexandrinus, Hoplopterus spinosus, Lobivanellus cinereus, Rostratula capensis, Tringa hypoleucos, T. ochropus;* Egypt, China, India, Russia.

A. *macracanthoides* Fuhrmann 1907, in *Vanellus* sp.; Egypt.

A. *megascolecina* Ukoli 1967, in *Egretta garzetta;* Ghana.

A. *microphallos* (Krabbe 1869) Fuhrmann 1908 (syn. *Taenia microphallos* Krabbe 1869, *Choanotaenia microphallos* [Krabbe 1869] Clerc 1903, *Dichoanotaenia microphallos* [Krabbe 1869] Lopez-Neyra 1934), in *Charadrius morinellus, Calidris minuta, Limonites danacensis, Vanellus vanellus, Tringa glareola;* Europe, Russia.

A. *minuta* (Benham 1900) Fuhrmann 1908 (syn. *Drepanidotaenia minuta* Benham 1900), in *Apteryx manetlli;* New Zealand.

A. *mollis* (Volz 1900) Fuhrmann 1906 (syn. *Taenia mollis* Volz 1900), in *Falco minor, Cerchneis tinnuculus, Gyps fulvus, Milvus migrans, Pseudogyps africanus, Hypotriorchis cerchneis, Erythropus;* Egypt, Europe, Russia.

*A. multifilamenta* (Bondarenko et Kontrimavichus 1980) comb. n. (syn. *Dichoanotaenia multifilamenta* Bondarenko et Kontrimavichus 1980), in *Arenaria melanocephala;* Alaska, U.S.

*A. murudensis* Baylis 1926, in *Garrulax schistochlamys;* Sarawak.

*A. mutabilis* (Rudolphi 1819) Fuhrmann 1907 (syn. *Taenia mutabilis* Rudolphi 1819), in *Crotophaga minor, C. major, Guira guira;* South and Central America.

*A. nymphaea* (Schrank 1790) Fuhrmann 1908 (syn. *Taenia nymphaea* Schrank 1790, *Choanotaenia nymphaea* [Schrank 1790] Clerc 1910, *Dichoanotaenia nymphaea* [Schrank 1790] Lopez-Neyra 1944), in *Bartramia longicauda, Glareola pratincola, Hoplopterus spinosus, Numenius arguatus, N. phaeopus, Tringa hypoleucos, Cursorius cursor;* Europe, Asia.

*A. occidentalis* (Belpolskaya 1977) comb. n. (syn. *Dichoanotaenia occidentalis* Belopolskaya 1977), in *Charadrius dubius;* Russia.

*A. oligorhyncha* Singh 1960, in *Lobivanellus indicus;* India.

*A. orientalis* Spasskii et Konovalow 1971, in *Tringa hypoleucos;* Russia.

*A. ovifusa* (Spasskii et Konovalov 1967) comb. n. (syn. *Dichoanotaenia ovifusa* Spasskii et Konovalov 1967), in *Terekia cineria;* Russia.

*A. ovolaciniata* (Linstow 1877) Fuhrmann 1908 (syn. *Taenia ovolaciniata* Linstow 1877), in *Hirundo rustica;* Europe.

*A. papilla* (Wedl 1855) Fuhrmann 1908 (syn. *Taenia papilla* Wedl 1855, *Drepanidotaenia papilla* [Wedl 1855] Stossich 1897), in *Ardea purpurea;* Europe.

*A. paucitesticulata* Fuhrmann 1908, in *Cypseloides senex;* Brazil.

*A. platyrhyncha* (Krabbe 1869) Cohn 1900 (syn. *Taenia platyrhycha* Krabbe 1869, *Dichoanotaenia platyrhyncha* [Krabbe 1869] Lopez-Neyra 1944, *Spasskytaenia platyrhyncha* [Krabbe 1869] Oshmarin 1956), in *Calidris minuta, C. temmincki, Charadrius apricarius, Limonites damacensis, Tringa calidris, T. glareola, T. ochropus, T. stagnatilis, T. totanus, Philomachus pugnax, Helodromus odaopus;* Europe, Russia.

*A. plegadis* (Dubinina et Dubinin 1940) Mathevossian 1963 (syn. *Choanotaenia plegadis* Dubinina et Dubinin 1940), in *Plegadis falcinellus;* Russia.

*A. porata* Macko 1968, in *Porzana parva;* Czechoslovakia.

*A. praecox* (Krabbe 1882) Spasskaja 1957 (syn. *Taenia praecox* Krabbe 1882), in *Anthus pratensis, Emberiza citrinella, E. pussilla, E. schoeniclus, Fringilla montifringilla, Luscinia svecica, Ruticilla erythrogastra;* Russia.

*A. procirrosa* Fuhrmann 1909, in *Francolinus clappertoni;* Africa.

*A. pseudomicrorhyncha* Belopolskaja 1977, in *Calidris testacea;* Russia.

*A. quadrata* (Rudolphi 1819) Fuhrmann 1908 (syn. *Taenia quadrata* Rudolphi 1819), in *Muscicapa atricapilla, M. collaris;* Europe, Asia, Africa.

*A. reticulata* Spasskii et Konovalov 1967, in *Phalaropus reticulata;* Russia.

*A. reutensis* Spasskaja et Shulimo 1971, in *Gallinago gallinago;* Russia.

*A. rhinocheti* Johnston 1911, in *Rhinochetus jubatus;* New Caledonia, Loyalty Islands.

*A. ricci* (Fuhrmann et Baer 1943) Mathevossian 1963 (syn. *Choanotaenia ricci* Fuhrmann et Baer 1943, *Dictymetra ricci* [Fuhrmann et Baer 1943] Clark 1952), in *Sphenorhynchus abdimii;* Europe.

*A. riparia* Dubinina 1954, in *Riparia riparia;* Russia.

*A. rosickyi* Rysavy 1962, in *Fringilla coelebs;* Czechoslovakia.

*A. rostrata* (Fuhrmann 1918) Mathevossian 1963 (syn. *Monopylidium rostratum* Fuhrmann 1918, *Choanotaenia rostrata* [Fuhrmann 1918] Baer 1925), in *Halcyon sancta;* New Caledonia.

*A. secunda* (Fuhrmann 1907) Mathevossian 1963 (syn. *Monopylidium secunda* Fuhrmann 1907, *Choanotaenia secunda* [Fuhrmann 1907] Fuhrmann 1932), in *Belonopterus cayennensis*; Brazil.

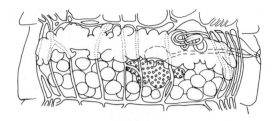

FIGURE 434. *Rauschitaenia ancora* (Mamaev 1959) Bondarenko et Tomilovskaja 1979.

*A. sinensis* (Joyeux et Baer 1935) Lopez-Neyra 1952 (syn. *Choanotaenia sinensis* Joyeux et Baer 1935), in *Parisomius dalhousiae;* Indochina.

*A. skrjabini* Ginetsinskaja et Naumov 1958, in *Arenaria interpres;* Russia.

*A. spasskajae* nom. n. (syn. *A. skrjabini* Spasskaja 1958, preoccupied), in *Emberiza leucocephala;* Russia.

*A. steatomidis* Hunkeler 1972, in *Steatomys* sp.; Ivory Coast.

*A. stentorea* (Froehlich 1802) Fuhrmann 1908 (syn. *Taenia stentorea* Froehlich 1802, *T. variabilis* Rudolphi 1810, *Anomotaenia variabilis* Cohn 1900, *Choanotaenia variabilis* Clerc 1902, *Dichoanotaenia stentorea* [Froehlich 1802] Lopez-Neyra 1944), in *Ancylochilus subarquatus, Calidris alpina, Capella gallinago, Philohela minor, Tringa glareola, Squatorola helvetica, Tringa calidris, Tringoides hypoleucus, Vanellus vanellus, V. gregarius;* Russia, China, Europe, North America.

*A. subterranea* Cholodkovsky 1906, in *Sorex* sp.; Russia, Hungary.

*A. telescopica* Barker et Andrews 1915 Syn. *Rodentotaenia telescopia* Barker et Andrews 1915), in *Fiber zibethicas;* U.S.

*A. tordae* (Fabricius 1780) Fuhrmann 1908 (syn. *Taenia tordae* Fabricius 1780), in *Alca pica, Uria troile, U. lomvia, U. aalge, Cepphus grylle, Cerorchinea monocerata;* Russia, Greenland, Moneron Islands.

*A. tugarinovi* (Dubinina 1950) Mathevossian 1963 (syn. *Choanotaenia tugarinovi* Dubinina 1950), in *Saxicola torquata;* Russia.

*A. tundra* (Spasskii et Konovalov 1907 (syn. *Dichoanotaenia tundra* Spasskii et Konovalov 1967), in *Macrorhamphus griseus;* Russia.

*A. variabilis* (Rudolphi 1802) Cohn 1900 (syn. *Taenia variabilis* Rudolphi 1802), in *Vanellus vanellus, Philohela minor, Capella gallinago;* Europe, U.S.

*A. volvulus* (Linstow 1906) Fuhrmann 1908 (syn. *Diplochetos volvulus* Linstow 1906, *Choanotaenia dispar* Burt 1940, *Dictymetra volvulus* [Linstow 1906] Spasskaja et Spassky 1973), in *Sarciophorus malabaricus, Pluvialis apricaria;* India, Sri Lanka, Russia.

**54a. Osmoregulatory canals reticular, branching in each segment ....................
................ *Rauschitaenia* Bondarenko et Tomilovskaja 1979. (Figure 434)**

*Diagnosis:* Scolex globular, with four simple suckers. Rostellum with two circles of hooks. Neck present. Proglottids craspedote, wider than long. Genital pores irregularly alternating. Osmoregulatory canals reticular, with many branches in each segment. Cirrus pouch reaches about one fourth width of segment. Cirrus spined. Testes numerous, postovarian. Vas deferens coiled. Ovary anterior, nearly as wide as segment. Vitellarium lobated, postovarian. Seminal receptacle present. Uterus an irregular sac. Parasites of Charadriiformes. Russia.

Type species: *R. ancora* (Mamaev 1959) Bondarenko et Tomilovskaja 1979 (syn. *Anomotaenia ancora* Mamaev 1959), in *Gallinago gallinago;* Russia.

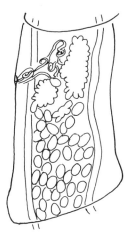

FIGURE 435. *Pseudanomotaenia constricta* (Molin 1858) Mathevossian 1963.

**54b. Osmoregulatory canals not reticular** .......................................... 55

**55a. Genital atrium with a tuft of long bristles**............ *Spinilepis* **Oshmarin 1972.**

*Diagnosis:* Scolex well-developed, with two circles of unequal hooks, each hook with small guard. Suckers unarmed. Neck present. Proglottids craspedote, wider than long when mature. Genital pores irregularly alternating. Genital ducts pass between osmoregulatory canals. Testes numerous (17), posterior to female system. Cirrus pouch weakly developed; ejaculatory duct long, coiled; cirrus with tuft of long setae. Vas deferens extensive. Vagina posterior to cirrus pouch. Ovary highly lobated. Vitellarium posterior to ovary. Uterus a single sac. Eggs with small polar knobs. Parasites of galliform birds. Russia.
Type species: *S. turnicis* Oshmarin 1972, in *Turnix tanki;* Primorye, Russia.

**55b. Genital atrium otherwise**..................................................... 58

**56a. Uterus saccular** ............ *Pseudanomotaenia* **Mathevossian 1963. (Figure 435)**

*Diagnosis*: Rostellum with two circles of hooks. Proglottids craspedote, wider than long or longer than wide. Genital pores irregularly alternating. Genital ducts pass between osmoregulatory canals. Cirrus pouch crosses osmoregulatory canals. Testes numerous, mainly posterior to ovary. Ovary median. Vitelline gland postovarian. Vagina posterior to cirrus pouch. Seminal receptacle present. Uterus an irregular sac. Parasites of Falconiformes, Alciformes, Charadriiformes, Gruiformes, Lariformes, Passeriformes. Cosmopolitan.
  Type species: *P. constricta* (Molin 1858) Mathevossian 1963 (syn. *Taenia constricta* Molin 1858, *T. coronina* Krabbe 1869, *T. affinis* Krabbe 1869, *T. puncta* Linstow 1872, *Anomotaenia puncta* Cohn 1901, *Choanotaenia constricta* [Molin 1858] Clerc 1903, *Anomotaenia constricta* [Molin 1858] Fuhrmann 1908), in *Accipiter nisus, Coloeus monedula, Corvus brachyrhynchus, C. corax, C. cornix, C. corone, C. frugilegus, C. ossifragus, Monedula furrium, Pica pica, Phylloscopus trochilus, Turdus iliacus, T. merula, T. musicus, T. ruficollis, T. pillaris, T. viscivorus;* Europe, Asia, South America, Russia.
  Other species:

*P. accipitris* (Johnston 1911) Mathevossian 1963 (syn. *Anomotaenia accipitris* Johnston 1911), in *Accipiter cirrhocephalus;* Australia.

*P. antarctica* (Fuhrmann 1921) Mathevossian 1963 (syn. *Anomotaenia antarctica* Fuhrmann 1921), in *Larus dominiacanus;* Antarctica.

*P. arkita* (Mathevossian 1950) Mathevossian 1963 (syn. *Anomotaenia arkita* (Mathevossian 1950), in *Corvus corone;* Russia.

*P. armillaris* (Rudolphi 1810) Mathevossian 1963 (syn. *Taenia alcae* Fabricius 1780, *T. armillaris* Rudolphi 1810, *T. socialis* Krabbe 1869, *Anomotaenia armillaris* [Rudolphi 1810] Baer 1956, *Alcataenia armillaris* [Rudolphi 1810] Spasskaja 1971), in *Alca torda, Cerorhinca monocerata, Uria aalge, U. lomvia;* Greenland, Iceland, Russia.

*P. campylacantha* (Krabbe 1869) Mathevossian 1963 (syn. *Taenia campylacantha* Krabbe 1869, *Alcataenia campylacantha* [Krabbe 1869] Spasskaja 1971, *Anomotaenia campylacantha* [Krabbe 1869] Fuhrmann 1908), in *Cepphus grylle, Uria brunnichi, Herodias garzetta, Ardeola grayi;* Greenland, Arctic region, Russia, Calcutta Zoo.

*P. chelidonariae* (Spasskaja 1957) Mathevossian 1963 (syn. *Anomotaenia chelidonariae* Spasskaja 1957), in *Chelidonaria urbica, Delichon urbica;* Europe, Russia.

*P. cyathiformis* (Froelich 1791) Mathevossian 1963 (syn.) *Taenia cyathiformis* Froelich 1791, *Dichoanotaenia cyathiformis* [Froelich 1791] Lopez-Neyra 1944, *Anomotaenia cyathiformis* [Froelich 1791] Lopez-Neyra 1952), in *Hirundo rustica, Apus apus, A. melba, Chacturia zonaria;* Europe, Russia.

*P. cyathiformoides* (Fuhrmann 1908) Mathevossian 1963 (syn. *Anomotaenia cyathiformoides* Fuhrmann 1908), in *Cypseloides senex;* Brazil.

*P. filovata* (Clark 1952) Mathevossian 1963 (syn. *Anomotaenia filovata* Clark 1952), in *Charadrius vociferus;* U.S.

*P. gallinagilis* (Davies 1938) Mathevossian 1963 (syn. *Anomotaenia gallinagilis* Davies 1938), in *Capella gallinago;* Wales.

*P. isacantha* (Fuhrmann 1908) Mathevossian 1963 (syn. *Anomotaenia isacantha* Fuhrmann 1908), in *Emberiza* sp.; Brazil.

*P. meinertzhageni* (Baer 1956) Mathevossian 1963 (syn. *Anomotaenia meinertzhageni* Baer 1956, *Alcataenia meinertzhageni* [Baer 1956] Spasskaja 1971), in *Cepphus grylle, Uria lomvia;* Greenland.

*P. micracantha* (Krabbe 1869) Mathevossian 1963 (syn. *Taenia micracantha* Krabbe 1869, *Anomotaenia micracantha* [Krabbe 1869] Fuhrmann 1932), in *Larus canus, L. fuscus, L. glaucus, L. glaucoides, L. marinus, L. ridibundus, Pagophila eburnea, Larus argentatus, Rissa rissa, R. tridactyla, Hydrochelidon* sp., *Sterna hirundo;* Iceland, Greenland, New Zealand, Alaska, Russia.

*P. otidis* (Skrjabin 1914) Mathevossian 1963 (syn. *Anomotaenia otidis* Skrjabin 1914), in *Otis tetrax, O. tarda;* Russia.

*P. parachelidonariae* Jaron 1967, in *Delichon urbica, Hirundo rustica;* Russia.

*P. paramicrorhyncha* (Dubinina 1953) Mathevossian 1963 (syn. *Anomotaenia paramicrorhyncha* Dubinina 1953), in *Charadrius dubius, C. hiaticula, Calidris testacea, Capella stenura, Phylomachus pugnax, Tringa stagnatilis, T. totanus, T. hypoleucus, T. glareola, T. nebularia;* Russia.

*P. passerum* (Joyeux et Timon-David 1934) Mathevossian 1963 (syn. *Anomotaenia passerum* Joyeux et Timon-David 1934), in *Turdus merula, Prunella atrogularis, Hirundo rustica, Emberiza leucocephala;* Europe, Russia.

*P. penicellata* (Fuhrmann 1908) Mathevossian 1963 (syn. *Anomotaenia penicillata* Fuhrmann 1908), in *Gymnostinops yuracarium;* Brazil.

*P. prinopsia* (Mettrick 1959) Mathevossian 1963 (syn. *Anomotaenia prinopsia* Mettrick 1959), in *Prinops plumata;* Africa.

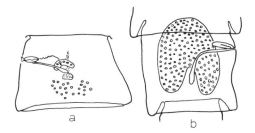

FIGURE 436. *Proparuterina aruensis* Fuhrmann 1911. (a) Mature proglottid; (b) gravid proglottid.

*P. pyriformis* (Wedl 1855) Mathevossian 1963 (syn. *Taenia pyriformis* Wedl 1855, *Anomotaenia pyriformis* [Wedl 1855] Fuhrmann 1908), in *Crex crex*; Europe, Russia.

*P. rowettiae* (Liang-Sheng 1957) Mathevossian 1963 (syn. *Anomotaenia rowettiae* Liang-Sheng 1957), in *Rowettia goughensis*; London Zoo.

*P. rustica* (Neslobinsky 1911) Mathevossian 1963 (syn. *Anomotaenia rustica* Neslobinsky 1911), in *Hirundo rustica, Chelidonaria urbica, Riparia riparia*; Europe, Russia, Morocco, U.S.

*P. skrjabiniana* (Spasskaja 1958) Mathevossian 1963 (syn. *Anomotaenia skrjabiniana* Spasskaja 1958), in *Emberiza leucocephala*; Russia.

*P. tarnogradskii* (Dinnick 1927) Mathevossian 1963 (syn. *Anomotaenia tarnogradskii* Dinnick 1927), in *Cinclus cinclus*; Russia.

*P. trapezoides* (Fuhrmann 1906) Mathevossian 1963 (syn. *Anomotaenia trapezoides* Fuhrmann 1906), in *Urubutinga zonura, Milvus ater*; Brazil, Russia.

*P. trigonocephala* (Krabbe 1869) Mathevossian 1963 (syn. *Taenia trigonocephala* Krabbe 1869, *Anomotaenia trigonocephala* [Krabbe 1869] Fuhrmann 1932), in *Turdus ruficollis, T. ericetorum, Motacilla flava, M. alba, M. citreola, Oenantha oenantha, Luscinia svesica, Emberiza aureola, E. schoeniclus*; Greenland, Sweden, Europe, Russia.

*P. undulatoides* (Fuhrmann 1908) Mathevossian 1963 (syn. *Anomotaenia undulatoides* Fuhrmann 1908), in *Atticora fasciata*; Brazil.

*P. verulamii* (Mettrick 1958) Mathevossian 1963 (syn. *Anomotaenia verulamii* Mettrick 1958), in *Turdus cricetorum cricetorum*; England.

**56b. Uterus horseshoe-shaped ..................................................... 57**

**57a. Testes posterior of ovary .......... *Proparuterina* Fuhrmann 1911. (Figure 436)**

*Diagnosis:* Rostellum with two circles of hooks. Proglottids craspedote. Genital pores alternating irregularly. Genital ducts pass between osmoregulatory canals. Cirrus pouch small. Testes in a posterior, median group. Ovary small, median. Vitelline gland postovarian. Vagina posterior to cirrus pouch. Seminal receptacle present. Gravid uterus in form of inverted-U. Parasites of Caprimulgiformes. Australasia.

Type species: *P. aruensis* Fuhrmann 1911, in *Podargus papuensis*; Australasia.
Other species:
*P. lali* Baugh et Saxena 1976, in *Passer domesticus*; India.

**57b. Testes mainly posterior, but a few dorsal and anterior of ovary .................**
**............................................. *Parvitaenia* Burt 1940. (Figure 437)**

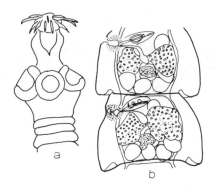

FIGURE 437. *Parvitaenia ardeolae* Burt 1940. (a) Scolex; (b) proglottids.

*Diagnosis:* Rostellum with two circles of hooks. Gravid proglottids longer than wide. Segments craspedote. Genital ducts pass between osmoregulatory canals. Genital pores alternate irregularly. Genital atrium deep. Cirrus armed or not. Cirrus pouch large, mostly or entirely intervascular. Testes few, mainly, posterior to ovary, but a few dorsal and anterior to ovary. Ovary bilobed, large. Vitelline gland postovarian. Vagina posterior to cirrus pouch. Seminal receptacle present. Uterus first bilobed, then horseshoe-shaped. Parasites of Ciconiiformes, Falconiformes. Ceylon, Celebes, Japan, Java, Mexico, South America, Philippines.

Type species: *P. ardeolae* Burt 1940, in *Ardeola grayi;* Sri Lanka.

Other species:

*P. ambigua* Baer et Bona 1960, in *Cochlearius cochlearius;* Brazil.

*P. buckleyi* Saxena 1970, in *Milvus migrans;* India.

*P. caribaenis* Rysavy et Macko 1973 (syn. *P. heardi* Schmidt et Courtney 1973), in *Ardea herodias;* Cuba, U.S.

*P. clavipera* Baer et Bona 1960, in *Ardea novaehollandiae;* Australia.

*P. cochlearii* Coil 1955, in *Cochlearius cochlearius;* Mexico.

*P. echinatia* Mettrick 1967, in *Nycticorax nycticorax;* Zambia.

*P. heckmanni* Jensen, Schmidt et Kuntz 1983; in *Spilornis cheela;* Philippines.

*P. ibisae* Schmidt et Bush 1972 (syn. *P. eudocimi* Rysavy et Macko 1973), in *Eudocimus albus, Rhynchops nigra, Phalacrocorax auritus, P. mexicanus, Pelecanus occidentalis, Casmerodius albus;* Cuba, Florida, U.S.

*P. macrophallica* Baer et Bona 1960, in *Cochlearius cochlearius;* Brazil.

*P. microphallica* Baer et Bona 1960, in *Cochlearius cochlearius;* Brazil.

*P. nycticoracis* (Yamaguti 1935) Yamaguti 1959 (syn. *Anomotaenia nycticoracis* Yamaguti 1935), in *Nycticorax nycticorax;* Japan.

*P. pseudocyclorchida* Baer et Bona 1960, in *Ardea novaehollandiae;* Australia.

*P. samfyia* Mettrick 1967, in *Ardea purpurea;* Zambia.

*P. yamagutii* Gaikwad et Shinde 1980, in *Milvus migrans;* India.

## ADDENDUM

The following genus was inadvertently omitted from the preceding key:

*Cladotaenia* Cohn 1901 (syn. *Paracladotaenia* Yamaguti 1935)

*Diagnosis:* Rostellum small, with two circles of hooks. Proglottids craspedote. Genital pores irregularly alternating. Cirrus pouch not reaching osmoregulatory canals. Testes nu-

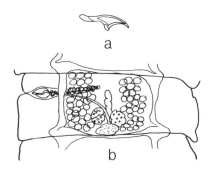

FIGURE 437A. *Cladotaenia globifera* (Batsch 1786) Cohn 1901. (a) Rostellar hook; (b) proglottid.

merous, in two submedian fields, occasionally meeting behind vitelline gland. Ovary two-winged, posterior. Vitelline gland postovarian. Internal seminal vesicle present or not. Small, proximal seminal receptacle present. Uterus with median stem and lateral branches, usually not intruding into anterior quarter of proglottid, but doing so in a few species. Parasites of Falconiformes, *Mustela*. Europe, Asia, Africa, Taiwan, North America.

Type species: *C. cylindracea* (Bloch 1782) Cohn 1901 (syn. *Taenia cylindracea* Bloch 1782), in *Falco* spp., *Circus* spp. etc.; Europe, Rusia.

Other species:

*C. accipitris* Yamaguti 1935, in *Accipiter virgatus;* Taiwan.

*C. aquilastur* Mettrick 1963, in *Hieraaetus dubius;* Southern Rhodesia.

*C. armigera* (Volz 1900), in *Falco nudicus;* Egypt.

*C. banghami* Crozier 1946, in *Haliaetus leucocephalus;* Ohio, U.S.

*C. cathartis* Hwang 1961, in *Cathartes aura;* U.S.

*C. circi* Yamaguti 1935, in *Circus aeruginosus;* Taiwan, Siberia, Europe, Africa, North America (?).

*C. falcoris* (Sawada et Kugi 1980) comb. n. (syn. *Paracladotaenia falcoris* Sawada et Kugi 1980), in *Falco peregrinus;* Japan.

*C. fania* Meggitt 1933, in *Hieraetus pennatus, Choriotis kori;* Calcutta Zoo.

*C. freani* Ortlepp 1938, in black eagle hawk; South Africa.

*C. feuta* Meggitt 1933, in *Circus assimilis, Gypaetus barbatus;* Calcutta Zoo, Indochina.

*C. foxi* McIntosh 1940, in *Falco peregrinus;* U.S.

*C. globifera* (Batsch 1786), in *Falco, Circus, Asio, Astur, Buteo, Milvus, Pernis,* etc.; Europe, Africa, North America.

*C. oklahomensis* Schmidt 1940, in *Buteo jamaicensis;* Oklahoma, U.S.

*C. mustelis* (Sawada et Kugi 1979) comb. n. (syn. *Paracladotaenia mustelis* Sawada et Kugi 1979), in *Mustela siberica;* Japan.

*C. secunda* Meggitt 1928, in unknown host; Egypt.

*C. spasskyi* Kobyshev 1971, in *Aquila rapax;* Russia.

*C. vulturi* Ortlepp 1938, in vulture; Africa.

## GENERA IN DILEPIDIDAE OF UNCERTAIN STATUS

*Copesoma* Sinitzin 1896.

Type species: *C. papillosum* Sinitzin 1896, in *Limonites damacensis;* Russia.

*Tetracisdicotyla* Fuhrmann 1907.

Type species: *T. macrosclecina* Fuhrmann 1907, in *Butorides virescens;* Brazil.

# FAMILY ANOPLOCEPHALIDAE

This is a very large family of tapeworms, most of which are found in mammals. One large genus, *Oochoristica*, is commonly found in reptiles. The scolex always lacks a rostellum and armature. The proglottids are usually numerous and wider than long. The reproductive systems are single or double in each segment. The uterus is variable according to the subfamily. The eggs commonly have an outer shell elongated and crossed at one pole, the *pyriform apparatus* (Figure 438). Insects and mites serve as intermediate hosts.

The family has five subfamilies; Triplotaeniinae, with two strobilas per scolex; Thysanosomatinae, with one or more paruterine organs; Anoplocephalinae, with tubular, saccular, lobed, or reticular uterus; Linstowiinae, with each egg capsule having a single egg; and Inermicapsiferinae, with each egg capsule having several eggs. These subfamilies are considered as families by some authors, and Wardle et al.[3871] created the order Anoplocephalidea for the single family Anoplocephalidae. This appears to be unnecessary to me.

The monograph by Spasskii[3402] is indispensible for anyone working with this family. He considers the group to be a suborder, following Skrjabin.[3274]

## KEY TO THE SUBFAMILIES IN ANOPLOCEPHALIDAE

1a. Strobila divided into two in neck region ..............................................
  ................................................ Triplotaeniinae Beveridge 1976. (p.415)
1b. Stroblia single ................................................................ 2

2a. Uterus with one or more paruterine organs........................................
  .................................... Thysanosomatinae Skrjabin 1933 (p.416)
2b. Uterus without paruterine organs .................................................. 3

3a. Gravid uterus tubular, saccular, lobed or reticular, never breaking into egg capsules ..............................................................................
  ... Anoplocephalinae Blanchard 1891 (syn. Monieziinae Spasskii 1951). (p.421)
3b. Gravid uterus breaking into egg capsules ........................................ 4

4a. Egg capsules each with a single egg ...............................................
  .................................... Linstowiinae Fuhrmann 1907. (p.450)
4b. Egg capsules each with several eggs ..............................................
  .................................... Inermicapsiferinae Lopez-Neyra 1943. (p.463)

## DIAGNOSIS OF THE ONLY GENUS IN TRIPLOTAENIINAE

*Triplotaenia* Boas 1902 (Figure 439)

*Diagnosis:* Suckers arranged in dorsal and ventral pairs; tissue posterior to each pair drawn out into pointed process. Strobilas arising on each side of scolex posterior to level of suckers. Anterior portion of each strobila very slender, becoming wider posteriorly, with one margin thickened and the other thin and fringed. Internal and external metamerism reduced. Genital pores unilateral. Testes few, unilateral. Cirrus pouches four or five per testis. Cirrus unarmed. Ovary and vitellaria compact, medial to testis. Seminal receptacle absent. Uterus poral, a transversely elongated sac. Eggs with pyriform apparatus. Parasites of marsupials. Australia.

Type species: *T. mirabilis* Boas 1902, (syn. *Cittotaenia mirabilis* (Boas 1902) Lopez-Neyra 1954) in *Petrogale penicillata;* Australia.
Other species:

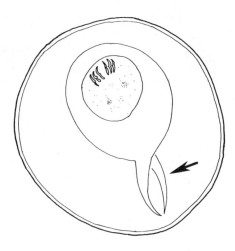

FIGURE 438. Diagrammatic anoplocephalid egg, showing pyriform apparatus (arrow).

FIGURE 439. *Triplotaenia mirabilis* Boas 1902. (a) Scolex with bases of strobilas; (b) region of one strobila.

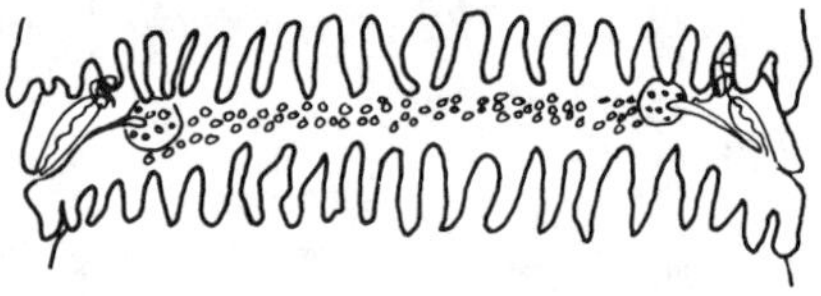

FIGURE 440. *Thysanosoma actinioides* Diesing 1834.

*T. fibriata* Beveridge 1976, in *Macropus giganteus, M. fuliginosus, Petrogale* sp.; Australia.

*T. undosa* Beveridge 1976, in *Wallabia bicolor, Macropus eugenii, M. fuliginosus, M. giganteus*; Australia.

## KEY TO THE GENERA IN THYSANOSOMATINAE

**1a. Posterior margin of proglottid fringed** ............................................. ..................................... *Thysanosoma* Diesing 1835. (Figure 440)

*Diagnosis:* Posterior margin of each proglottid with fringe-like outgrowths. Two sets of reproductive organs per proglottid. Genital pores bilateral. Cirrus pouch small, cylindrical. Cirrus slender, unarmed (?). Testes numerous, between two ovaries. Seminal vesicles absent. Ovary and vitellaria mixed into a compound germovitellarium, which is poral. Vagina posterior to cirrus pouch. Seminal receptacle small. Uterus a sinuous tube that breaks into many paruterine organs, each with several eggs. Eggs lacking pyriform apparatus. Parasites of ruminents. North and South America.

Type species: *T. actinioides* Diesing 1835 (syn. *Taenia fimbriata* Diesing 1850), in *Cervus dichotomus, Bos taurus, Blastocercus paludosus, Mazama nana, M. rufus, Capreolus capra, Ovis aries, Taurotragus oryz, Antilocapra americana, Odocoileus* spp., etc.; North and South America, Africa (?), India.

FIGURE 441. *Wyominia tetoni* Scott 1941.

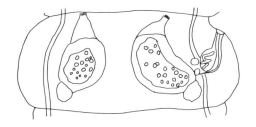

FIGURE 442. *Aliezia indica* Shinde 1969. Gravid proglottid.

Other species:
  *T. misrai* Nama 1974, in goat; India.

**1b.  Posterior margin of proglottid not fringed** ........................................ **2**

**2a.  Reproductive organs doubled in each proglottid** ............................... **3**
**2b.  Reproductive organs single in each proglottid** .................................. **4**

**3a.  Testes in single row in posterior two thirds of proglottid** .........................
  ................................................. ***Wyominia* Scott 1942. (Figure 441)**

*Diagnosis:* Suckers pedunculated. Posterior margins of proglottids not fringed. Double set of reproductive organs in each segment. Genital ducts passing between osmoregulatory canals. Genital ducts bilateral; male pore anterior, lateral; female pore posterior, somewhat dorsal. Cirrus pouch narrow, reaching ventral osmoregulatory canal. Cirrus unarmed. Testes 50 to 62, in single row in posterior two thirds of proglottid. Seminal vesicles absent. Ovary and vitelline gland poral. Seminal receptacle large, between ovary and vitelline gland. Uterus first a transverse tube anterior to testes, then breaking into paruterine organs each with about six eggs. Parasites of ruminants. North America.
  Type species: *W. tetoni* Scott 1941, in bile ducts, gall bladder and small intestine of *Ovis canadensis;* Wyoming, U.S.

**3b.  Testes in two clusters near lateral margins** ..........................................
  ................................................. ***Aliezia* Shinde 1969. (Figure 442)**

*Diagnosis:* Scolex unarmed, with four simple suckers. Neck present. Proglottids wider than long. Two rows of 10 to 12 interproglottid glands present near anteriolateral margins of each proglottid. Two sets of reproductive systems in each segment. Four to six testes in a cluster near each side, extending over osmoregulatory canals. Cirrus pouch near anterior margin in mature segments. Cirrus unarmed. Vas deferens short, coiled. Ovary median to testes. Vitelline glands not described. A single paruterine organ develops on each side of proglottid. Vagina opens posterior to cirrus. Parasites of sheep. India.
  Type species: *A. indica* Shinde 1969, in *Ovis bharal;* India.

**4a.  Two paruterine organs per proglottid** ....... ***Stilesia* Railliet 1893. (Figure 443)**

*Diagnosis:* Proglottids wider than long, craspedote. Genital pores unilateral, irregularly alternating. Genital atrium deep. Cirrus pouch small. Testes few, in two lateral fields. Seminal vesicles absent. Ovary and vitellaria united into poral germovitellarium. Vaginal

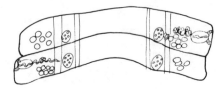

FIGURE 443.   *Stilesia vittata* Railliet 1896.

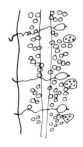 

FIGURE 444.   *Avitellina arctica* Kolmakov 1938.

pore dorsoposterior to cirrus pore. Uterus first a bilobed sac with slender isthmus, then replaced by two paruterine organs. Parasites of ruminants. Asia, Africa, Europe.

Type species: *S. globipunctata* (Rivolta 1874) Railliet 1893 (syn. *Taenia globipunctata* Rivolta 1874, *T. ovipunctata* Rivolta 1874, *Stilesia ovipunctata* [Rivolta 1874] Railliet 1893), in *Ovis aries, Capra hircus, Kobus ellipsiprimnus, Hippotragus equinus, Camelus bactrianus, Bos bovis;* Asia, Africa, West Europe.

Other species:

*S. hepatica* Wolffhuegel 1903 (syn. *S. sjöstedti* Fuhrmann 1909), in *Bos taurus, Capra hircus, Ovis aries, Bubulus caffer, Aepiceros melampus, Cephalophus monticola, Hippotragus equinus, Tragelophus scriptus, Tetraceros quadricornis, Cobus defassa, Camelus dromedarius;* Asia, Africa.

*S. okapi* Leiper 1936, in *Okapia johnstoni;* Africa.

*S. vittata* Railliet 1896, in *Camelus dromedarius, C. bactrianus, Ovis aries, Capra hircus;* Africa, Europe, Asia. Redescribed: Martinez-Gomes and Hernandez-Rodriques (1973).

4b.   One or several paruterine organs per proglottid................................ 5

5a.   One paruterine organ per proglottid ............................................ 6
5b.   Several paruterine organs per proglottid ....................................... 8

6a.   Ovary and vitelline gland united into single germovitellarium ....................
........................................................................... *Avitellina* Gough 1911 (syn. *Hexastichorchis* Blei 1922, *Anootypus* Woodland 1928). (Figure 444)

*Diagnosis:* Strobila large, narrow, proglottids wider than long. External metamerism poorly marked or absent in anterior half of strobila. Genital pores irregularly alternating. Osmoregulatory canals ventral to genital ducts. Cirrus pouch small. Seminal vesicles absent. Testes in two lateral groups, each subdivided by osmoregulatory canals. Ovary and vitellaria united into poral germovitellarium. Seminal receptacle present. Paruterine organ single, containing fibrous capsules each with several eggs. Parasites of ruminants. Europe, Asia, Africa, Philippines, North America. Monograph: Raina[2736]

Type species: *A. centripunctata* (Rivolta 1874) Gough 1911 (syn. *A. laciniosa* Blei 1922, *Hexastichorchis pinterneri* Blei 1922, *Taenia centripunctata* Rivolta 1824, *Stilesia centripunctata* [Rivolta 1874] Railliet 1893, *Avitellina lahorea* Woodland 1927, *A. sudanea* Woodland 1927, *A. southwelli* Nagaty 1929, *A. woodlandi* Bhalerao 1936), in *Bostaurus, Bubalus bufellus, Capra hircus, Ovis aries, Camelus* sp., *Aepiceros melappus, Cephalophus grimmia, Hippotragus equinus, Oreotragus oreotragus, Pediotragus sharpei,*

*P. horstocki, Taurotragus oryx, Gazella granti, Ourebia ourebi, Silvicapra;* Europe, Asia, Africa.

Other species:

*A. aegyptiaca* Nagaty 1929, in *Cephalophus* sp., *Camelus dromedarius;* Rhodesia, Egypt.

*A. arctica* Kolmakov 1938 (syn. *Avitellina arctica* Kolmakov 1938), in *Rangifer tarandus, Capreolus pygargus;* Russia.

*A. bigemina* Amin 1939; in ovines; India.

*A. bubalinae* Tubangui 1930, in *Bubalus bubalis;* Philippines.

*A. chalmersi* Woodland 1927, in *Ovis aries;* North America; India.

*A. edifontaineus* (Woodland 1928) Southwell 1929 (syn. *Anootypus edifontaineus* Woodland 1928), in *Taurotragus oryx;* East Africa.

*A. goughi* Woodland 1927 (syn. *A. centripunctata* of Gough 1911 in part), in *Bos taurus, Ovis aries, Capra hircus;* South Africa, India.

*A. monardi* Fuhrmann 1932 (syn. *Anootypus monardi* [Fuhrmann 1932] Fuhrmann 1933), in *Taurotragus oryx;* Angola, West Africa.

*A. nagatyi* Ezzat 1945, in *Ammotragus lervia, Ovis aries, Camelus dromedarius;* Egypt.

*A. ricardi* (Woodland 1928) Southwell 1928 (syn. *Anootypus ricardi* Woodland 1928), in *Kobus* sp., East Africa (?).

*A. sandgroundi* Woodland 1935, in *Hippotragus equinus;* Belgian Congo.

*A. spirillometra* Amin 1939, in ovines; India.

*A. tatia* Bhalerao 1936, in *Capra hircus;* India.

**6b. Ovary and vitelline gland separate, although ovary may surround vitelline gland ................................................................................. 7**

**7a. Proglottids narrower than scolex. External seminal vesicle absent............... ........... Mogheia Lopez-Neyra 1944 (syn. *Baeria* Moghe 1933, preoccupied).**

*Diagnosis:* Scolex large, neck absent. Proglottids wider than long, narrower than scolex. Genital pores irregularly alternating. Genital ducts passing between osmoregulatory canals. Cirrus pouch small. Testes few, antiporal. Seminal vesicles absent. Ovary slightly poral. Vitelline gland posteromedial to ovary. Vagina posterior to cirrus. Seminal receptacle small. Uterus persistent as a spherical sac filling nearly entire length of segment, with single spherical paruterine organ. Eggs without pyriform apparatus. Parasites of Passeriformes. India.

Type species: *M. orbiuterina* (Moghe 1933), in *Turdoides somervillei;* India.

Other species:

*M. bayamegaparuterina* Capoor 1967, in *Ploceus philippinus;* India.

*M. megaparuterina* Capoor et Srivastava 1966, in "bird"; India.

**7b. Proglottids wider than scolex. External seminal vesicle present. Testes both poral and antiporal to ovary ... *Columbia* Srivastava and Capoor 1965. (Figure 445)**

*Diagnosis:* Scolex simple, narrower than strobila. Proglottids wider than long. Genital pores irregularly alternating. Cirrus pouch not reaching osmoregulatory canals. Testes in poral and antiporal groups. Internal and external (?) seminal vesicles present. Ovary bilobed, poral. Vitelline gland compact, transversely elongated, postovarian. Vagina irregularly dorsal and ventral to cirrus pouch. Uterus a transverse, lobulated sac. Paruterine organ single. Parasites of Columbiformes. India.

Type species: *C. allahabadi* Srivastava and Capoor 1965, in *Columba livia;* India.

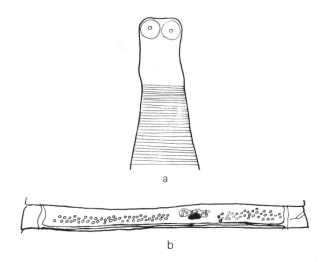

FIGURE 445. *Columbia allahabadi* Srivastava et Capoor 1965. (a) Scolex and (b) mature proglottid.

FIGURE 446. *Thysaniezia ovilla* (Rivolta 1878) Skrjabin 1926.

8a. About 300 paruterine organs per proglotid ..........................................
......... *Thysaniezia* Skrjabin 1926 (syn. *Helictometra* Baer 1927). (Figure 446)

*Diagnosis:* Proglottids wider than long. Genital pores irregularly alternating. Cirrus pouch oblique, small. Cirrus armed. Testes numerous in two lateral, extravascular fields. Genital ducts passing between osmoregulatory canals. Vas deferens convoluted anterior to cirrus pouch, lateral to poral osmoregulatory canals. Ovary lobated, poral. Vitelline gland compact, postovarian. Vagina posterior to cirrus pouch. Seminal receptacle present. Uterus first a transverse tube, replaced by numerous (300) paruterine organs. Parasites of ruminants. Europe, Asia, Africa, Australia, Argentina.

Type species: *T. ovilla* (Rivolta 1878) Skrjabin 1926 (syn. *Taenia ovilla* Rivolta 1878, *T. giardi* Moniez 1879, *T. aculeata* Perroncito 1882, *T. brandti* Cholodkovsky 1894, *Moniezia ovilla* [Rivolta 1878] Moniez 1891, *M. ovilla* var. *macilenta* Moniez 1891, *Thysanosoma giardi* [Moniez 1879] Stiles et Hassal 1893, *Thysanosoma ovilla* [Rivolta 1878] Railliet 1895, *Helictometra giardi* [Moniez 1879] Baer 1927, *Thysaniezia ovilla* [Rivolta 1878] Skrjabin 1926), in *Bos taurus, Ovis aries, Aepyceros melampus, Taurotragus oryx, Tragelaphus scriptus, Bubalis caama, Sus scrofa* (?), *Galumna obvious, Scheloribates laevigatus;* Europe, Asia, Africa, Australia, Argentina.

Other species:

*T. aspinosa* Nama 1974, in goat; India.

*T. cannochaeti* (Fuhrmann 1943) Spasskii 1951 (syn. *Helictometra cannochaeti* Fuhrmann 1943), in *Cannochaetus taurinus;* Angola, Africa.

*T. himalayai* Fotedar et Bambroo 1978, in *Ovis aries;* Kashmir, India.

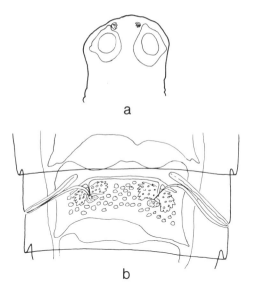

FIGURE 447. *Moniezioides rouxi* Fuhrmann 1918. Scolex and proglottid.

**8b.  Eight to twelve paruterine organs per proglottid ........ *Ascotaenia* Baer 1927.**

*Diagnosis:* Proglottids wider than long. External metamerism indistinct in anterior half of proglottid. Genital pores irregularly alternating. Cirrus pouch small. Testes few, in two fields separated by ovary. Osmoregulatory canals ventral to genital ducts. Ovary poral. Vitelline gland rudimentary. Uterus first a transverse tube, then replaced by 8 to 12 paruterine organs, each with several eggs. Parasites of ruminants. Russia.

Type species: *A. pygargi* (Cholodkovsky 1902) Baer 1927 (syn. *Thysanosoma pygargi* Cholodkovsky 1902, *Avitellina pygargi* Cholodkovsky 1902) Spasskii 1951, in *Capreolus pygargus*; Siberia.

## KEY TO THE GENERA OF ANOPLOCEPHALINAE

| | | |
|---|---|---|
| 1a. | Parasites of birds | 2 |
| 1b. | Parasites of mammals | 17 |
| 2a. | Two sets of reproductive organs in each proglottid | 3 |
| 2b. | One set of reproductive organs in each proglottid | 6 |
| 3a. | Suckers each with two muscular projections | 4 |
| 3b. | Suckers without muscular projections | 5 |
| 4a. | Four glands present anterior to each sucker. Uterus simple ............................................................ *Moniezioides* Fuhrmann 1918. (Figure 447) | |

*Diagnosis:* Scolex with four suckers, each with two protuberances. Four groups of small glands present between suckers. Neck absent. Proglottids wider than long, craspedote. Two sets of reproductive organs in each segment, except for uterus, which is single. Genital pores lateral. Cirrus pouch slender, crossing osmoregulatory canals. Testes in single, dorsal, medullary field, median and posterior to ovaries. Ovaries bialate, submedian. Vitelline glands compact, submedian. Vaginas posterior to cirrus pouches. Seminal receptacles present.

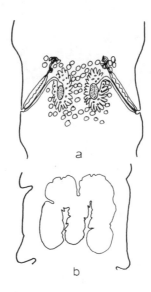

FIGURE 448. *Paronia coryllidis* Burt 1939. (a) Mature proglottid; (b) gravid proglottid.

Uterus first a single, transverse tube, assuming two swellings when gravid. Parasites of parrots. South Pacific.

Type species: *M. rouxi* Fuhrmann 1918, in *Trichoglossus haematodes;* New Caledonia.

**4b.    Glands not present anterior to suckers. Uterus with posterior outpocketings ...
..................................................................*Coelodela* Shipley 1900.**

*Diagnosis:* Scolex with four suckers, each with two lateral, muscular projections. Proglottids wider than long, craspedote. Each segment with double sets of reproductive organs except uterus, which is single. Genital pores lateral. Cirrus pouch bulbous. Genital atrium deep. External and internal seminal vesicles present. Testes numerous in single anterior field. Ovaries lobated, in lateral medulla. Vitelline glands postovarian. Vagina opens anterior to cirrus. Seminal receptacle present. Uterus a transverse tube with posterior outpocketings. Parasites of Columbiformes. New Guinea.

Type species: *C. kuvaria* Shipley 1900 (syn. *Cittotaenia kuvaria* [Shipley 1900] Fuhrmann 1901, *C. columbae* Skrjabin 1915), in *Carpophaga vanwicki, Leucotreron jambu, Columba* sp.; Malayan Archipelago, Bismarck Archipelago.

**5a.    Gravid uterus horseshoe-shaped, without anterior and posterior diverticulae...
............................................ *Paronia* Diamare 1900. (Figure 448)**

*Diagnosis:* Suckers simple. Neck absent. Proglottids wider than long. Two sets of reproductive organs in each segment. Genital ducts dorsal to osmoregulatory canals. Ventral canals commonly with valves at origins of transverse anastomoses. Cirrus pouch elongate, crossing osmoregulatory canals. Testes numerous, in dorsal intervascular field. Internal seminal vesicle present. External seminal vesicle absent. Ovaries multilobate, submedian. Vitelline glands compact, interovarian. Vagina ventral, dorsal or posterior to cirrus pouch. Seminal receptacle present. Uteri horseshoe-shaped, sometimes with outgrowths or joined by dorsal branches. Parasites of Passeriformes. Piciformes. Columbiformes, Rhamphastiformes, Psittaciformes. Australia, New Guinea, Africa, South America, Ceylon, Thailand, India, Taiwan, Sumatra.

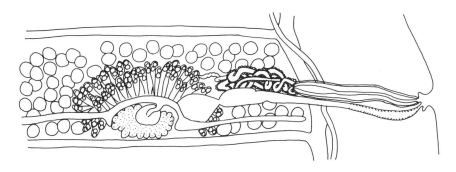

FIGURE 449. *Stringopotaenia psittacea* (Fuhrmann 1904) Beveridge 1978.

Type species: *P. trichoglossi* (Linstow 1888) Spasskii 1951 (syn. *Taenia trichoglossi* Linstow 1888, *Paronia carrinoi* Diamare 1900, *Moniezia carrinoi* [Diamare 1900] Fuhrmann 1901, *Moniezia trichoglossi* [Linstow 1888] Kotlan 1921), in *Cyclopsittacus suavissimus, Lorius erythrothorax, Trichoglossus nigrigularis, T. novaehollandae, T. swainsoni;* Australia, New Guinea.

Other species:

*P. africana* (Joyeux et Baer 1927) Spasskii 1951 (syn. *Cittotaenia africana* Joyeux and Baer 1927), in *Bucorax* sp., *Bycanistes buccinator;* Africa.

*P. ambigua* (Fuhrmann 1902) Fuhrmann 1918 (syn. *Moniezia ambigua* Fuhrmann 1902), in *Amazona amazonica;* Brazil.

*P. beauforti* (Janicki 1906) Fuhrmann 1918 (syn. *Moniezia beauforti* Janicki 1906), in *Cyclopsittacus diophthalmus;* New Guinea.

*P. biuterina* Burt 1939, in *Coryllis beryllinus;* Sri Lanka.

*P. bocki* Schmilz 1941, in *Megalaima virens, Cyanops ramsayi;* Thailand.

*P. calcauterina* Burt 1939, in *Molpastes haemorrhous;* Sri Lanka.

*P. columbae* (Fuhrmann 1902) Fuhrmann 1918 (syn. *Paronia carrinoi* Diamare 1900 in part, *Moniezia columbae* Fuhrmann 1902), in *Columba* sp., *Ptilonopus* sp.; Sumatra, India.

*P. coryllidis* Burt 1939, in *Coryllis beryllinus;* Sri Lanka.

*P. galli* Nama 1978, in *Gallus gallus;* India.

*P. pycnonoti* Yamaguti 1935, in *Pycnonotus sinensis;* Taiwan.

*P. variabilis* (Fuhrmann 1904) Fuhrmann 1908, (syn. *Moniezia variabilis* Fuhrmann 1904), in *Rhamphastos culminatus, R. dicolorus, R. erythrorhynchus, R. monilis, R. toco;* South America.

*P. zavattarii* Fuhrmann et Baer 1943, in *Colius striatus;* Abyssinia.

**5b. Gravid uterus not horseshoe-shaped, with anterior and posterior diverticulae.. ................................. *Stringopotaenia* Beveridge 1978. (Figure 449)**

*Diagnosis:* Scolex rounded, unarmed. Suckers unarmed. Proglottids numerous (more than 100 in gravid strobila), craspedote, much wider than long. Longitudinal osmoregulatory canals paired. Genital ducts dorsal to osmoregulatory canals. Vagina ventral to cirrus sac on right side of strobila, dorsal to it on left side. Internal seminal vesicle present, external seminal vesicle absent. Testes numerous throughout most of medulla, not crossing osmoregulatory canals, but overlying and lateral to ovary and cirrus pouch. Vaginal pore posterior to cirrus pore in genital atrium. Seminal receptacle present. Ovaries in lateral quarters of segment. Developing uterus in middle of proglottid, with two U-shaped loops passing over vitellaria. Gravid uterus similar in shape with anterior and posterior diverticulae. Eggs with pyriform apparatus. Parasites of Psittacidae. New Zealand.

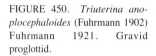

FIGURE 450. *Triuterina anoplocephaloides* (Fuhrmann 1902) Fuhrmann 1921. Gravid proglottid.

FIGURE 451. *Aporina alba* Fuhrmann 1902.

Type species: *S. psittacea* (Fuhrmann 1904) Beveridge 1978, in *Stringops habroptilus;* New Zealand.

6a.   Gravid uterus a three-lobed sac .....   **Triuterina Fuhrmann 1921. (Figure 450)**

*Diagnosis:* Scolex with globular scolex lacking a rostellum. Four oval, unarmed suckers and a neck present. Proglottids somewhat wider than long. Single set of reproductive organs per segment. Genital pores irregularly alternating. Genital ducts passing between osmoregulatory canals. Cirrus pouch crosses ventral osmoregulatory canals, very muscular, with strong sphincter at base. Testes numerous, occupying most of medulla. Ovary lobated, poral. Vitelline gland postovarian. Vagina ventral to cirrus, with distal sphincter. Seminal receptacle present. Uterus a three-lobed sac, two lobes lateral, one lobe anterior. Parasites of parrots. Africa.

Type species: *T. anoplocephaloides* (Fuhrmann 1902) Fuhrmann 1921 (syn. *Taenia anoplocephaloides* Fuhrmann 1902), in *Psittacus erythracus;* Africa. Redescribed: Jones.[1441]
Other species:
  *T. uteriloba* Dollfus 1975, in *Polcephalus guliemmi;* Africa.

6b.   Gravid uterus a transverse sac or tube, or reticular............................ 7

7a.   Genital atrium atrophied, genital pores absent in mature proglottids ......... 8
7b.   Genital pores normal........................................................ 9

8a.   Genital ducts dorsal to osmoregulatory canals, gravid uterus overreaches osmoregulatory canals.................... **Aporina Fuhrmann 1902. (Figure 451).**

*Diagnosis:* Single set of reproductive organs per segment. Genital pores irregularly alternating in immature proglottids, atrophied in mature and gravid ones. Genital ducts passing dorsal to osmoregulatory canals. Cirrus pouch small, poorly developed. Testes numerous, surrounding ovary. Ovary large, fan-shaped, median. Vitelline gland postovarian, somewhat poral. Vagina posterior to cirrus pouch. Uterus a transverse sac that overreaches osmoregulatory canals laterally, then bends forward. Parasites of Psittaciformes. Brazil, India
Type species: *A. alba* Fuhrmann 1902, in *Pyrrhura* sp.; Brazil.
Other species:
  *A. chauhani* Ghosh 1975, in *Columba livia;* India.
  *A. nakayamai* Sawada et Kifune 1974, in *Columba livia;* Japan.

8b.   Genital ducts pass between osmoregulatory canals; gravid uterus contained between osmoregulatory canals................ **Neoaporina Saxena et Baugh 1973.**

*Diagnosis:* Scolex with an unarmed apical cone. Suckers simple. Proglottids wider than long. Genital pores absent. Genital ducts pass between osmoregulatory canals. Cirrus pouch present, ventral to vagina. Vas deferens and vagina do not cross each other. Testes evenly

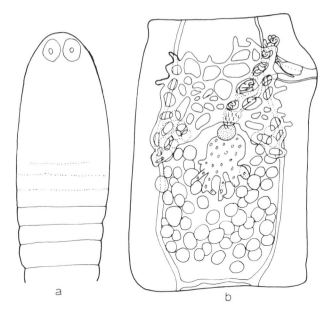

FIGURE 452. *Arostellina reticulata* Neiland 1955. (a) Scolex; (b) proglottid.

distributed in intervascular field. Ovary slender, transversely elongated. Vitellarium median, postovarian. Uterus a transverse tube with anterior and posterior diverticula; confined to intervascular field. Parasites of Anseriformes. Arctic.

Type species: *N. borealis* (Linstow 1905) Saxena et Bough 1973 (syn. *Aporina borealis* Linstow 1905, *Paricterotaenia borealis* [Linstow 1905] Fuhrmann 1932, *Choanotaenia borealis* [Linstow 1905] Fuhrmann 1908, *Icterotaenia borealis* [Linstow 1905] Baer 1925), in *Harelda glacialis, Clangula, Tadorna, Motacilla, Emberiza;* Arctic, France, Russia, Vietnam.

**9a.   Vitelline gland anterior to ovary........ *Arostellina* Neiland 1955. (Figure 452).**

*Diagnosis:* Scolex small, not demarcated from wider neck. Most proglottids longer than wide, acraspedote. One set of reproductive organs per proglottid. Genital ducts dorsal to osmoregulatory canals. Genital pores irregularly alternating. Testes posterior and lateral to ovary. Seminal vesicles absent. Cirrus pouch small, not reaching osmoregulatory canals. Ovary lobulated, median. Vitelline gland compact, preovarian. Vagina opening posterior to cirrus. Seminal receptacle present. Uterus reticular, arising anterior to ovary, occupying entire proglottid when gravid. Parasites of hummingbird. South and Central America.

Type species: *A. reticulata* Neiland 1955, in *Phaeochroa curieri;* Nicaragua, Brazil.

**9b.   Vitellarium posterior to ovary ................................................. 10**

**10a.  Testes in continuous, transverse band anterior to ovary ........................
........................................... *Taufikia* Woodland 1928. (Figure 453)**

*Diagnosis:* Scolex with four large suckers. Proglottids wider than long, craspedote. Single set of reproductive organs per proglottid. Dorsal osmoregulatory canals absent. Genital pores irregularly alternating. Cirrus pouch not reaching poral osmoregulatory canal. Testes few, in single, preovarian field. Internal seminal vesicle present, external absent. Ovary bilobed,

FIGURE 453. *Taufikia edmondi* Woodland 1928.

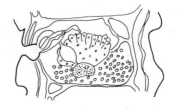

FIGURE 454. *Monoecocestus anoplocephaloides* (Douthitt 1915) Harkema 1936.

slightly poral. Vitelline gland postovarian. Vagina posterior to cirrus pouch. Uterus saccular. Parasites of Accipitriformes. Africa.

Type species: *T. edmondi* Woodland 1928 in *Gyps rueppeli, Torgos tracheliotus;* Africa.
Other species:
   *T. ghoshi* Capoor 1967, in *Neophron percnopterus;* India.
   *T. iranica* Dollfus 1963, in *Aegypius monachus;* Iran.
   *T. magnisomum* (Southwell 1930) Spasskii et Spasskaja 1974, in vultures; India.

**10b.   Testes not as above ........................................................... 11**

**11a.   Uterus first appears reticular, later becomes saccular ................ *Monoecocestus* Beddard 1914 (syn. *Schizotaenia* Janicki 1904, preoccupied). (Figure 454)**

*Diagnosis:* Neck absent. Scolex simple. Proglottids wider than long, craspedote. Single set of reproductive organs per segment. Genital pores lateral, regularly or irregularly alternating. Cirrus pouch well developed. Cirrus armed. Testes numerous, in posterior medulla. External and internal seminal vesicles present. Ovary medial or poral. Vitellarium postovarian. Vaginal pore anterior to cirrus pore, sometimes absent. Seminal receptacle present. Uterus first appears reticular, later becomes saccular. Eggs with pyriform apparatus. Parasites of Struthioniformes, Rodentia, Artiodactyla; North, Central, and South America.

Type species: *M. decrescens* (Diesing 1856) Fuhrmann 1932 (syn. *Taenia decrescens* Diesing 1856), in *Dicotyles albirostris, D. torquatus;* Brazil.
Other species:
   *M. americana* (Stiles 1895) Spasskii 1951 (syn. *Andrya americana* Stiles 1895, *Bertia americana* [Stiles 1895] Stiles 1896, *Bertiella americana* [Stiles 1895] Stiles at Hassall 1902, *Schizotaenia americana* [Stiles 1895] Janicki 1904), in *Erethizon dorsatum, E. epixanthum, Ondatra zibethica;* North America.
   *M. anoplocephaloides* (Douthitt 1915) Harkema 1936 (syn. *Schizotaenia anoplocephaloides* Douthitt 1915), in *Geomys breviceps;* South America.
   *M. diplomys* Nobel et Tesh 1974, in *Diplomys darlingi;* Panama.
   *M. erethizontis* Beddard 1914, in *Erethizon dorsatum;* North America.
   *M. giganticus* Buhler 1970, in *Erethizon dorsatum;* Colorado, U.S.
   *M. gundlachi* Vigueras 1943, in *Capromys pilorides;* South America.
   *M. hagmanni* (Janicki 1904) Spasskii 1951 (syn. *Schizotaenia hagmanni* Janicki 1904), in *Hydrochoerus capybara;* South America.
   *M. hydrochoeri* (Baylis 1928) Spasskii 1951 (syn. *Schizotaenia hydrochoeri* Baylis 1928), in *Hydrochoerus capybara;* Paraguay.
   *M. mackiewiczi* Schmidt et Martin 1978, in *Phyllotis* sp.; Paraguay.
   *M. macraobursatum* Rego 1961, in *Hydrochoeris hydrochoeris;* Brazil.
   *M. minor* Rego 1960, in *Cavia aperia;* Brazil.

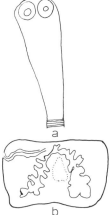
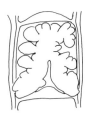

FIGURE 456. *Biporouterina psittaculae* Burt 1973. Gravid uterus.

FIGURE 455. *Hemiparonia cacatuae* (Maplestone 1922) Baer 1925. (a) Scolex; (b) gravid proglottid.

*M. myopotami* Sutton 1973, in *Myocastor coypus;* Argentina.
*M. parcitesticulatus* Rego 1960, in *Cavia porcellus, C. aperia;* Brazil.
*M. rheiphilus* Voge et Read 1953, in *Pterocnemia pennata;* Peru, Chile.
*M. sigmodontis* (Chandler et Suttles 1922) Spasskii 1951 (syn. *Schizotaenia sigmodontis* Chandler et Suttles 1922), in *Sigmodon hispidus;* U.S.
*M. thomasi* Rausch et Maser 1977, in *Glaucomys sabrinus;* Oregon, U.S.
*M. torresi* Olsen 1976, in *Ctenomys maulinus;* Chile.
*M. variabilis* (Douthitt 1915) Freeman 1949 (syn. *Schizotaenia variabilis* Douthitt 1915), in *Erethizon dorsatum, E. epixanthum;* North America.

**11b. Uterus always a sac or transverse tube........................................ 12**

**12a. Gravid uterus horseshoe-shaped, with outgrowths...............................
........................................ Hemiparonia Baer 1925. (Figure 455)**

*Diagnosis:* Scolex long, not set off from neck. Proglottids not much wider than long, craspedote. Single set of reproductive organs per segment. Genital pores unilateral. Cirrus pouch long. Testes in single field posterior and antiporal to ovary. Internal and external seminal vesicles absent. Genital ducts dorsal to osmoregulatory canals. Ovary median. Vitelline gland postovarian. Vagina ventral to cirrus pouch. Seminal receptacle present. Uterus horseshoe-shaped, with convexity directed anteriorly.

Type species: *H. cacatuae* (Maplestone 1922) Baer 1925 (syn. *Schizotaenia cacatuae* Maplestone 1922), in *Cacatua galerita;* Australia.
Other species;
*H. bancrofti* (Johnston 1912) Schmidt 1972 (syn. *Dilepis bancrofti* Johnston 1912), in *Barnardius barnardi, Platycercus eximius;* Australia. Redescribed; Schmidt.[3096]
*H. merotomocheta* Woodland 1930, in *Cacatua leadbeateri;* Australia.

**12b. Gravid uterus not horseshoe-shaped............................................ 13**

**13a. Gravid uterus crosses osmoregulatory canals to open through lateral uterine pores ........................................ Biporouterina Burt 1973. (Figure 456)**

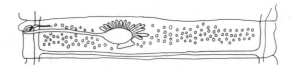

FIGURE 457. *Killigrewia delafondi* (Railliet 1892) Meggitt 1927.

*Diagnosis:* Scolex with four simple suckers. Neck present. Proglottids wider than long when immature, longer than wide when gravid. Genital ducts dorsal to osmoregulatory canals. Genital pores preequatorial, irregularly alternating. Testes numerous, in single field between osmoregulatory canals. External and internal seminal vesicles present. Cirrus pouch well developed, emptying into nearly spherical genital atrium. Cirrus unarmed. Ovary large, fan-like, filling most of anterior half of proglottid. Vitelline gland compact, concave anteriorly, posterior to ovary. Vagina thick-walled and surrounded by gland cells distally, thin-walled proximally, separated from large, spheroid seminal receptacle by distinct constriction. Uterus arises as inverted letter V with one or more anterior diverticula, increasing in size as more diverticula arise laterally; it is bounded by longitudinal osmoregulatory canals except posterolaterally where it crosses these vessels in later gravid segments to open to the exterior through uterine pores. Embryophore with pointed polar processes. Parasites of Psittaciformes. Sri Lanka.

Type species: *B. psittulae* Burt 1973, in *Psittacula calthropae;* Sri Lanka.

**13b. Gravid uterus lacking uterine pores ............................................. 14**

**14a. Testes mainly in two groups, poral and aporal to ovary .........................**
**................................................. *Killigrewia* Meggitt 1927 (syn.**
***Nepalesia* Sharma 1943, *Pseudoaporina* Saxena et Baugh 1973). (Figure 457)**

*Diagnosis:* Proglottids wider than long. Single set of reproductive organs per segment. Genital pores irregularly alternating, persisting throughout strobila. Osmoregulatory canals ventral to genital ducts. Cirrus pouch small. Testes in poral and antiporal groups. Ovary slightly poral. Vitelline gland postovarian. External and internal seminal vesicles present. Seminal receptacle present. Uterus a transverse sac without growths. Parasites of Columbiformes. Asia, Taiwan, Europe, Africa, Australia, North and South America.

Type species: *K. frivola* Meggitt 1927, in *Columba livia;* Egypt, India.

Other species:

*K. delafondi* (Railliet 1892) Meggitt 1927 (syn. *Aporina delafondi* [Railliet 1892] Baer 1927, *Taenia sphenocephala* Rudolphi 1810, *T. delafondi* Railliet 1892, *Bertia delafondi* [Railliet 1892] Fuhrmann 1901, *Bertiella delafondi* [Railliet 1892] Railliet et Henry 1909, *Killigrewia pamelae* Meggitt 1927, *K. streptopeliae* Yamaguti 1935), in *Columba livia, C. intermedia, C. oenass, Turturoena sharpi, Crossophthalmus gymnophthalmus, Ectopistes migratorius, Platycercus pennanti, Turtur auritus, Streptopelia turtur, S. chinensis, S. dissumieri, S. orientalis, S. semitorcuata, Turtur senegalansis, Zenaidura macroura;* Europe, Asia, Africa, North America.

*K. fuhrmanni* (Skrjabin 1914) Meggitt 1927 (syn. *Aporina fuhrmanni* Skrjabin 1914), in *Ornis;* Eastern Bolivia.

*K. oenopopeliae* Yamaguti 1935 (syn. *Aporina oenopopeliae* [Yamaguti 1935] Spasskii 1951), in *Oenoplopeliae tranquebarica;* Taiwan.

**14b. Testes mainly anterior and lateral to ovary .................................... 15**

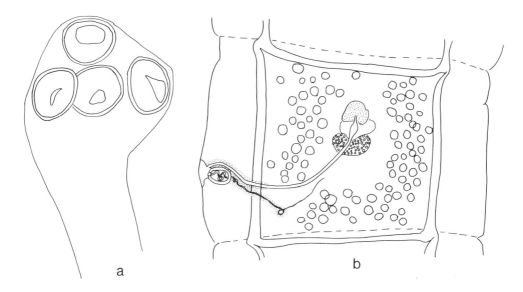

FIGURE 458. *Neophronia lucknowensis* Saxena 1967. Scolex and proglottid.

**15a. Ovary bilobed. Seminal receptacle absent.................... *Gidhaia* Johri 1934.**

*Diagnosis*: Scolex unknown. Dorsal osmoregulatory canals absent. Genital pores irregularly alternating. Reproductive organs single. Genital ducts dorsal to osmoregulatory canal. Cirrus pouch not reaching osmoregulatory canal. Testes few, mainly anterior and lateral to ovary. Ovary bilobed, nearly median. Vitelline gland postovarian. Vagina posterior to cirrus pouch. Seminal receptacle absent. Uterus a transverse sac with lateral ends subdivided. Parasites of vulture. India.

Type species: *G. indica* Johri, 1934, in *Gyps indicus;* India.
Other species:
   *G. kolhapurensis* Kadam, Shinde et Jadhav 1981, in *Torgos calvus;* India.

**15b. Ovary otherwise. Seminal receptacle present ................................. 16**

**16a. Ovary compact, median................... *Neophronia* Saxena 1967. (Figure 458)**

*Diagnosis:* Scolex unarmed, lacking rostellum. Suckers simple. Neck present. Mature proglottids wider than long, acraspedote. Gravid proglottids longer than wide. Genital pores alternate irregularly, in anterior one third of proglottid. Genital atrium deep. Cirrus unarmed. Cirrus pouch spheroid, not reaching osmoregulatory canals. Interior and exterior seminal vesicles present in gravid segments. Vas deferens not highly convoluted. Testes numerous, surrounding ovary, most numerous on aporal side. Ovary compact, median. Vitellarium postovarian. Vagina posterior to cirrus pouch; small seminal receptacle present. Gravid uterus median, with lateral lobes. Parasites of vultures. India.

Type species: *N. lucknowensis* Saxena 1967, in *Neophron percnopterus;* India.
Other species:
   *N. irregularis* Saxena 1968, in *Neophron percnopterus;* India.
   *N. luteus* Saxena 1968, in *Neophron percnopterus;* India.
   *N. melanotus* Saxena 1968, in *Neophron percnopterus;* India.
   *N. percnopteri* (Singh 1952) Saxena 1969 (syn. *Aporina percnopteri* Singh 1952), in *Neophron percnopterus;* India.

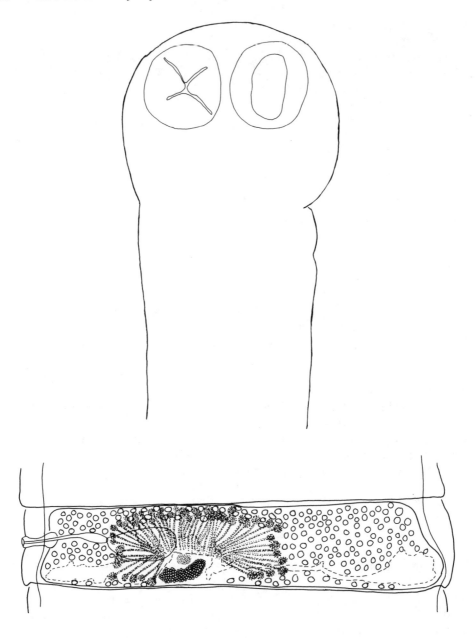

FIGURE 459. *Pulluterina nestoris* Smithers 1954. Scolex and proglottid.

**16b. Ovary fan-shaped, nearly length of proglottid** .......................................
.......................................... ***Pulluterina* Smithers 1954. (Figure 459)**

*Diagnosis:* Scolex rounded, unarmed, with four unarmed suckers deeply sunken into the surrounding tissue. A longitudinal furrow exits between the two dorsal and the two ventral suckers. Neck long. Proglottids wider than long. Single set of reproductive systems per segment. Dorsal osmoregulatory canals apparently absent. Genital pores alternating irregularly. Cirrus pore dorsal to osmoregulatory canal. Testes numerous, mainly lateral to ovary, but several anterior and posterior to it. Internal seminal vesicles present. Ovary fan-shaped, slightly poral, nearly length of proglottid. Vitellarium postovarian. Vagina ventral to cirrus

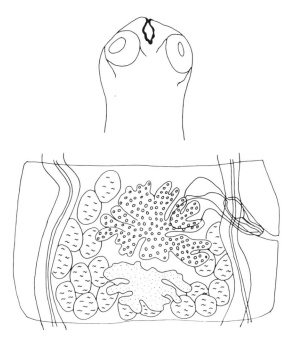

FIGURE 460. *Sudarikovina taterae* Hunkeler 1972. Scolex and proglottid.

pouch. Seminal receptacle present. Uterus first a posterior, transverse tube that arches over vitellarium; first fills laterally, then fills entire proglottid as simple sac. Uterine pores sometimes present on posterolateral margins of gravid proglottids. Eggs with polar points on middle shell. Parasites of parrots. New Zealand.

Type species: *P. nestoris* Smithers 1954, in *Nestor notabilis;* New Zealand. Redescribed: Weeks.[3891]

| | | |
|---|---|---|
| 17a. | One set of reproductive organs in each proglottid | 18 |
| 17b. | Two sets of reproductive organs in each proglottid | 32 |
| 18a. | Uterus reticular | 19 |
| 18b. | Uterus a transverse tube or sac | 25 |

**19a. Scolex with small apical organ........ *Sudarikovina* Spasskii 1951. (Figure 460)**

*Diagnosis:* Scolex unarmed, with small apical organ. Neck present. Proglottids craspedote, somewhat wider than long. One set of reproductive systems per proglottid. Genital pores unilateral. Two sets of simple osmoregulatory canals present. Genital atrium small. Genital ducts dorsal to osmoregulatory canals. External and internal seminal vesicles present. Testes in semicircular field posterior and lateral to ovary. Vaginal pore dorsal or anterior to cirrus pore. Seminal receptacle well-developed. Uterus reticular, persistent. Eggs lacking pyriform apparatus. Parasites of African rodents.

Type species: *S. monodi* (Joyeux et Baer 1930) Spasskii 1951 (syn. *Andrya monodi* Joyeux et Baer 1930, *Aprostatandrya [Sudarikovina] monodi* [Joyeux et Baer 1930] Spasskii 1951), in *Xerus (Euxerus) erythropus;* Nigeria.

Other species:

*S. africana* (Baer 1933) Spasskii 1951 (syn. *Andrya africana* Baer 1933, *Apro-*

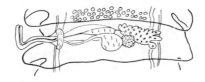

FIGURE 461.  *Parabertiella campanulata* Nybelin 1917.

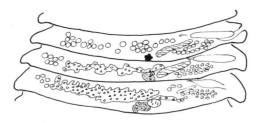

FIGURE 462.  *Anoplocephala spatula* (Linstow 1901) Janicki 1910.

statandrya *[Sudarikovina] africana* [Baer 1933] Spasskii 1951), in *Tateria lobengulae;* Southern Rhodesia.

*S. taterae* Hunkeler 1972, in *Tateria kempi, T. guineae, Taterillus gracilus;* Ivory Coast.

19b. Scolex lacking apical organ .................................................... 21

20a. Vagina anterior of cirrus pouch, or absent ..........................................
.................................... **Monoecocestus** Beddard 1914. (See p.426).
20b. Vagina posterior of cirrus pouch ............................................... 21

21a. Ovary and vitellarium antiporal ...... **Parabertiella** Nybelin 1917. (Figure 461)

*Diagnosis:* Small cestodes (23 to 30 mm). Scolex four-lobed. Proglottids craspedote, wider than long. One set of male and female reproductive organs per segment. Genital pores alternate regularly. Two pairs of lateral osmoregulatory canals. Genital ducts pass dorsal of osmoregulatory canals. Genital atrium with strong sphincter. Internal seminal vesicle and spherical seminal receptacle present. Cirrus armed. External seminal vesicle absent. Testes in single group anterior to ovary. Ovary and vitellarium aporal. Uterus a transverse tube with anterior and posterior pouches. Egg with pyriform apparatus. Vagina atrophied after insemination; insemination occurring before male reproductive system reaches maturity. Parasites of Australian marsupials.

Type species: *P. campanulata* Nybelin 1917, in *Hemibelideus lemuroides;* Australia. Redescribed: Beveridge (1976).

21b. Ovary median or nearly median, or poral ..................................... 22

22a. Ovary extending entire width of medulla. Genital pores unilateral ..............
... ***Anoplocephala*** Blanchard 1848 (syn. *Plagiotaenia* Peters 1871). (Figure 462)

*Diagnosis:* Proglottids wider than long, craspedote. Single set of reproductive organs per segment. Genital pores unilateral. Genital ducts dorsal to osmoregulatory canals. Scolex may have a pair of lappets on each side. Cirrus pouch well developed. Testes numerous, medullary. Ovary multilobate, slightly poral, filling most of medullary width. Vitelline gland postovarian. Vagina ventroanterior to cirrus pouch. Seminal receptacle present. Uterus a transverse sac. Eggs with pyriform apparatus. Parasites of Hyracoidea, Perissodactyla, gorilla. Cosmopolitan.

Type species: *A. perfoliata* (Goeze 1782) Blanchard 1848 (syn. *Taenia perfoliata* Goeze 1782, *T. equina* Pallas 1781, *T. equina perfoliata* Goeze 1782, *T. quadrilobata* Mueller 1789, *T. quadriloba* Gmelin 1790, *Alyselminthus lobatus* Zeder 1800, *Halysis perfoliata* [Goeze 1782] Zeder 1803, *Taenia perfoliata megnini* Cobbold 1879, *T. incruis* Huber 1896, *Alyselminthus perfoliatus* [Goeze 1782] Blainville 1828), in horse, donkey, mule, zebra; cosmopolitan.

Other species:

*A. diminuta* Sandground 1933, in *Rhinoceros sondaicus;* Malaya.

*A. gigantea* (Peters 1856) Blanchard 1891 (syn. *Taenia gigantea* Peters 1856, *Plagiotaenia gigantea* [Peters 1856] Peters 1871, *Schizotaenia gigantea* [Peters 1856] Douthitt 1915, *Anoplocephala vulgaris* Southwell 1921, *Plagiotaenia vulgaris* [Southwell 1921] Stunkard 1926, *P. longa* Stunkard 1926), in *Diceros bicornis, Ceratotherium simum;* Africa.

*A. gorillae* Nybelin 1927, in *Gorilla beringei;* Central Africa.

*A. latissima* Deiner 1912 (syn. *Taenia magna* Murie 1870, *Schizotaenia latissima* [Deiner 1912] Douthitt 1915, *Plagiotaenia latissima* [Deiner 1912] Stunkard 1926), in *Rhinoceros unicornis, Dicerorhynus sumatrensis;* India, Malayan Archipelago.

*A. magna* (Abildgaard 1789) Sprengel 1905 (syn. *Taenia magna* Abildgaard 1789, *T. equi* Mueller 1780 in part, *T. equina* Pallas 1781 in part, *T. plicata* Zeder 1800, *Alyselminthus plicatus* Zeder 1800, *Halysis plicata* ]Zeder 1800] Zeder 1803, *Taenia megalocephala* Cobbold 1874, *T. zebrae* Rudolphi 1808, *Anoplocephala zebrae* Railliet 1891, *A. restricta* Railliet 1893, *A. plicata* var. *peduculata* Railliet 1893, *A. plicata* var. *strangulata* Railliet 1893, *A. plicata* var. *restricta* Railliet 1893, *A. plicata* var. *servei* Bounhiol 1912), in horse, mule, donkey, zebra; cosmopolitan.

*A. manubriata* Railliet, Henry et Bouche 1914, in *Elephas indicus;* India.

*A. opatula* Allen et Lawrence 1936, in *Heterohyrax syricus hindei;* Kenya.

*A. rhodesiensis* Yorke et Southwell 1921 (syn. *T. zebrae* Collin 1891, *Anoplocephala zebrae* Fuhrmann 1909, *A. perfoliata* var. *zebrae* Baer 1923), in zebras, horse, monkey; Africa.

*A. spatula* (Linstow 1901) Janicki 1910 (syn. *Taenia [Anoplocephala] spatula* Linstow 1901), in *Heterohyrax brucei, H. mossambrica, Procavia capensis;* Africa.

**22b. Ovary not extending entire width of medulla. Genital pores unilateral or alternating** ............................................................. **23**

**23a. Testes occupying entire width of medulla** .............................. ***Pseudanoplocephala* Baylis 1927 (syn. *Hsuolepis* Yang, Zhai et Chen 1957). (Figure 463)**

*Diagnosis:* Scolex small. Strobila up to 1.5 m. Neck narrow. Proglottids wider than long, craspedote. Genital pores unilateral. Dorsal osmoregulatory canals absent. Cirrus pouch elongate. Internal and external seminal vesicles present. Testes in two groups, one on each side of ovary. Ovary median. Vitelline gland postovarian. Vagina ventral to cirrus pouch. Seminal receptacle present. Uterus a transversely elongate sac with numerous outgrowths. Eggs lacking pyriform apparatus. Parasites of wild and domestic swine. Ceylon, China, Japan.

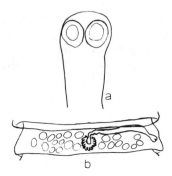
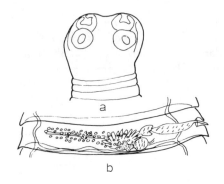

FIGURE 463. *Pseudanoplocephala crawfordi* Baylis 1927. (a) Scolex; (b) proglottid.

FIGURE 464. *Paranoplocephala mamillana* (Mehlis 1831) Baer 1927. (a) Scolex; (b) proglottid.

Type species: *P. crawfordi* Baylis 1927 (syn. *Hsuolepis shengi* Yang, Zhai et Chen 1957, *H. shensiensis* Lang et Cheng 1963), in swine; Ceylon, China.

Other species:

*P. nipponensis* Hatsushika, Shimizu, Kawakami, et Sawada 1978, in *Sus scrota;* Japan.

**23b. Testes mainly antiporal to ovary..................................................... 24**

**24a. Uterus first appears as a tube, becomes reticulate at the ends, then is totally reticulate ..........................................................................*Paranoplocephala* Lühe 1910 (syn. *Aprostatandrya* Kirschenblat 1938). (Figure 464)**

*Diagnosis:* Cestodes of median size. Segments usually wider than long, with relative length increasing posteriad. Scolex with prominent, motile suckers. Osmoregulatory canals simple, of two lateral pairs. Single set of reproductive systems per segment. Genital pores unilateral or irregularly alternating. Genital ducts dorsal to osmoregulatory canals. Vaginal pore posterior to cirrus pore. Internal and external seminal vesicles present. Testes numerous, aporal or aporal and anterior to ovary. Ovary lobed, poral or median. Vitellarium postovarian. Seminal receptacle present. Uterus at first a transverse tube, becoming reticulate at the ends, then totally reticulate, with anterior and posterior diverticula. Egg with pyriform apparatus. Parasites of Rodentia. Europe, Asia, North and South America. Genus redefined; Rausch.[2801]

Type species: *P. omphalodes* (Hermann 1783) Lühe 1910 (syn. *Taenia omphalodes* Hermann 1783, *Halysis omphalodes* [Hermann 1783] Zeder 1803, *Anoplocephala blanchardi* Monieze 1891, *Taenia blanchardi* [Moniez 1891] Braun 1894, *Anoplocephala omphalodes* [Hermann 1783] Janicki 1904, *Bertiella omphalodes* [Hermann 1783] Meggitt 1921, *Anoplocephala campestris* Cholodkovsky 1912, *Paranoplocephala blanchardi* [Moniez 1901] Baer 1927), in *Arvicola, Chionomys, Clethrionomys, Microtus, Pitymys, Lagidium;* Europe, North America, Russia.

Other species:

*P. gracilis* Tenora et Murai 1980, in *Microtus agrestis;* Czechoslovakia.

*P. mascomai* Murai, Tenora et Rocamora 1980, in *Microtus cabrerae;* Spain.

*P. petauristae* (Sawada et Kugi 1979) comb. n. (syn. *Aprostatandrya [A.] petauristae,* Sawada et Kugi 1979), in *Petaurista leucogemys;* Japan.

**24b. Uterus fully reticulate when first formed.... *Andrya* Railliet 1893. (Figure 465)**

*Diagnosis:* Scolex spheroid. Neck short. Proglottids wider than long, craspedote. Single

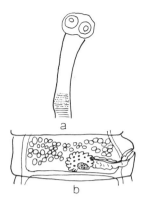

FIGURE 465. *Andrya cuniculi* (Blanchard 1891) Railliet 1893. (a) Scolex; (b) proglottid.

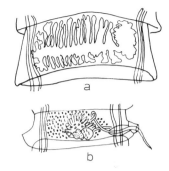

FIGURE 466. *Flabelloskrjabinia tapirus* (Chin 1938) Spasskii 1951. (a) Gravid proglottid; (b) mature proglottid.

set of reproductive organs per segment. Genital pores unilateral or irregularly alternating. Genital ducts dorsal to osmoregulatory canals. Testes numerous, mainly antiporal and anterior to ovary. Ovary multilobate, poral or median. Vitellarium compact, postovarian. External and internal seminal vesicles present. Vaginal pore posterior to cirrus pouch. Seminal receptacle present, large. Uterus fully reticulate when formed. Eggs with pyriform apparatus. Parasites of Rodentia, Lagomorpha. Europe, Africa, North America, Russia.

Type species: *A. rhopalocephala* (Riehm 1881) Railliet 1893 (syn. *Taenia rhopalocephala* Riehm 1881, *Alyselminthus pectinatus* [Goeze 1782] Zeder 1900, *Anoplocephala rhopalocephala* [Riehm 1881] Blanchard 1891, *Andrya pectinata* [Goeze 1782] Railliet 1893), in *Lepus timidus, L. europaeus, L. tibetanus, Oryctolagus cuniculus;* Europe, Africa, Asia.

Other species:

*A. arctica* Rausch 1952, in *Dicrostonyx groenlandicus, Lemmus trimucronatus, Clethrionomys rutilus, Microtus miurus;* Alaska, U.S.

*A. bialowiezensis* Soltys 1949, in *Clethrionomys glareolus, Microtus arvalis;* Poland.

*A. cuniculi* (Blanchard 1891) Railliet 1893, (syn. *Taenia rhopalocephala* Riehm 1881, *Anoplocephala cuniculi* Blanchard 1891, *Moniezia cuniculi* [Blanchard 1891] Blanchard 1893), in *Lepus timidus, L. europaeus, Oryctolagus cuniculus;* Europe, U.S.

*A. dasymidis* Hunkeler 1972, in *Dasymys incomtus, Mylomys lowei;* Ivory Coast.

*A. gundii* (Joyeux 1923) Spasskii 1951 (syn. *Andrya primordialis* var. *grundii* [Joyeux 1923]), in *Stenodactylus grundii;* Africa.

*A. primordialis* Douthitt 1915, in *Sciurus hudsonicus, Tamiasciurus, Phenacomys, Clethrionomys, Microtus, Dicrostonyx, Lemmus;* North America.

25a. Genital pores unilateral (rarely alternating in *Anoplocephaloides*) ............ 26
25b. Genital pores alternating ................................................... 27

26a. Parasites of tapirs ............... ***Flabelloskrjabinia* Spasskii 1951. (Figure 466)**

*Diagnosis:* Neck absent. Proglottids wider than long, craspedote. Single set of reproductive organs per segment. Genital pores unilateral. Cirrus pouch crosses osmoregulatory canals. Cirrus long. Testes numerous, mainly antiporal. Ovary fan-shaped, median. Vitelline gland postovarian. Seminal receptacle present. Uterus an irregular, transverse sac anterior to vitellaria and seminal receptacle. Eggs with pyriform apparatus. Parasites of tapirs. Philippines (?), Thailand, Brazil.

Type species: *F. tapirus* (Chin 1938) Spasskii 1951 (syn. *Anoplocephala tapirus* Chin 1938), in *Tapirus* sp.; Philippine Islands (?).

Other species:

*F. globiceps* (Diesing 1856) Spasskii 1951 (syn. *Anoplocephala globiceps* [Diesing 1856] Lühe 1895, *Taenia globiceps* Diesing 1856, *Paranoplocephala globiceps* [Diesing 1856] Baer 1927 in part), in *Tapiris terrestris;* Brazil.

*F. indicata* (Sawada et Papasarathorn 1966) Rausch 1976 (syn. *Paranoplocephala indicata* Sawada et Papasarathorn 1966), in *Tapirus indicus;* Thailand.

### 26b. Parasites of rodents, lagomorphs and perissodactyls .............................
.... *Anoplocephaloides* Baer 1923 (syn. *Galligoides* Tenora et Mas-Coma 1978).

*Diagnosis:* Size of strobila variable, from small to medium, sometimes long. All segments wider than long, with relative length increasing posteriad. Scolex usually wider than neck. Osmoregulatory canals simple, in two bilateral pairs. Genital ducts dorsal to osmoregulatory canals. Genital pores usually unilateral, rarely irregularly alternating. Vaginal pore ventral to cirrus pore. Internal and external seminal vesicles present. Testes numerous, with majority aporal to ovary. Ovary labate, usually poral. Uterus first a transverse tube, developing anterior and posterior diverticula, filling gravid segments. Egg with pyriform apparatus. Parasites of rodents, lagomorphs and perissodactyls. North America, Africa, Japan, Asia, Central America.

Type species: *A. infrequens* (Douthitt 1915) Baer 1923 (syn. *Anoplocephala infrequens* Douthitt 1915, *Paranoplocephala infrequens* [Douthitt 1915] Baer 1927), in *Geomys bursarius, Thamomys talpoides;* North America.

Other species:

*A. acanthocirrosa* (Baer 1924) Rausch 1976 (syn. *Paranoplocephala acanthocirrosa* Baer 1924, *P. a. kivuensis* Baer 1959, *P. otomyos* Collins 1972), in *Otomys bisulcatus, O. kempi, O.* sp.; Africa.

*A. arfaai* (Mobedi et. Ghadirian 1977) comb. n. (syn. *Schizorchis arfaai* Mobedi et Ghadirian 1977, *Galligoides arfaai* [Mobedi et Ghadirian 1977] Tenora et Mas-Coma 1978), in *Apodemus sylvaticus;* Iran, Spain.

*A. baeri* Rausch 1976, in *Apodemus argenteus;* Japan.

*A. blanchardi* (Moniez 1891) Baer 1924 (syn. *Anoplocephala blanchardi* Moniez 1891, *A. campestris* Cholodkovsky 1912, *Taenia blanchardi* [Moniez 1891] Braun 1894, *Paranoplocephala blanchardi* [Moniez 1891] Baer 1927), in *Microtus, Arvicola, Evotomys, Apodemus;* Europe, Japan.

*A. dentata* (Galli-Valerio 1905) Spasskii 1951 (syn. *Anoplocephala dentata* Galli-Valerio 1905, *Paranoplocephala brevis* Kirschenblat 1938), in *Arvicola* spp., *Microtus* spp., *Clethrionomys rutilus;* Eurasia, Korea.

*A. floresbarroetae* Rausch 1976, in *Sylvilagus brasiliensis* (bile duct); Costa Rica.

*A. isomydis* (Setti 1892) Rausch 1976 (syn. *Taenia isomydis* Setti 1892, *Paranoplocephala isomydis* [Setti 1892] Baer 1949), in *Isomys abyssinicus, Otomys tropicalis, Oenomys hypoxanthus;* Africa.

*A. kontrimavichusi* Rausch 1976, in *Synaptomys borealis;* Alaska.

*A. lemmi* (Rausch 1952) Rausch 1976 (syn. *Paranoplocephala lemmi* Rausch 1952), in *Lemmus sibericus;* Alaska, Canada, Siberia.

*A. mamillana* (Mehlis 1831) Rausch 1976 (syn. *Taenia mamillana* Mehlis 1831, *Anoplocephala mamillana* [Mehlis 1831] Blanchard 1891, *Paranoplocephala mamillana* [Mehlis 1831] Baer 1927), in *Equus caballus, E. burchelli;* cosmopolitan.

*A. neofibrinus* (Rausch 1952) Rausch 1976 (syn. *Paranoplocephala neofibrinus* Rausch 1952), in *Neofiber alleni;* southeastern U.S.

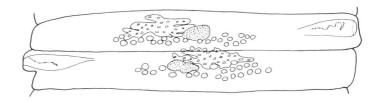

FIGURE 467. *Lentiella machadoi* Rego 1964.

*A. romerolagi* Kamiya, Suzuki et Villa-R. 1979, in *Romerolagus diazi;* Mexico.

*A. ryjikovi* (Spasskii 1950) Rausch 1976 (syn. *Paranoplocephala ryjikovi* Spasskii 1950), in *Marmota baibacina, M.* sp.; middle Asia.

*A. transversaria* (Krabbe 1879) Baer 1927 (syn. *Taenia transversaria* Krabbe 1879, *Anoplocephala transversaria* [Krabbe 1879] Blanchard 1891, *Paranoplocephala transversaria* [Krabbe 1879] Baer 1927), in *Marmota* spp.; Eurasia.

*A. troeschi* (Rausch 1946) Rausch 1976 (syn. *Paranoplocephala troeschi* Rausch 1946), in *Microtus* spp.; North America.

*A. variabilis* (Douthitt 1915) Rausch 1976 (syn. *Anoplocephala variabilis* Douthitt 1915, *Paranocephala variabilis* [Douthitt 1915] Hansen 1947), in *Thomomys talpoides, Geomys bursarius, Microtus* spp., *Lemmus sibericus;* North America.

*A. wigginsi* (Rausch 1954) Rausch 1976 (syn. *Paranoplocephala wigginsi* Rausch 1954), in *Citellus parryi;* Alaska, U.S.

*A. wimerosa* (Moniez 1880) Rausch 1976 (syn. *Taenia wimerosa* Moniez 1880, *Andrya wimerosa* [Moniez 1880] Railliet 1891, *Anoplocephala wimerosa* [Moniez 1880] Blanchard 1891, *Paranoplocephala wimerosa* [Moniez 1880] Baer 1927), in *Lepus timidus, Oryctolagus cuniculus;* Eurasia.

**27a. Genital pores alternating regularly** ..................... ***Perutaenia*** **Parra 1953.**

*Diagnosis:* Neck absent. Proglottids wider than long, craspedote. Single set of reproductive organs per proglottid. Genital pores regularly alternating. Genital ducts dorsal to osmoregulatory canals. Cirrus pouch crossing osmoregulatory canals. Cirrus armed. Testes few (15 to 20), dorsal, mainly preovarian. Internal and external seminal vesicles present. Ovary median, bilobed. Vitelline gland postovarian. Vagina ventroanterior to cirrus pouch. Seminal receptacle (?). Uterus an irregular, transverse sac. Eggs with pyriform apparatus. Parasites of Rodentia. Peru.

Type species: *P. threlkeldi* (Parra 1952) Parra 1953 (syn. *Paranoplocephala threlkeldi* Parra 1952), in *Lagidium peruanum;* Peru.

**27b. Genital pores irregularly alternating** ........................................... **28**

**28a. Vitelline gland poral to ovary** ................ ***Lentiella*** **Rego 1964. (Figure 467)**

*Diagnosis:* Scolex small, simple. Neck absent. Proglottids few (24 to 28), wider than long. Single set of reproductive organs per segment. Genital pores irregularly alternating. Cirrus pouch well developed, containing armed cirrus. Testes few, in a single field posterior to ovary and vitellarium. Ovary lobated, slightly aporal. Vitellarium compact, mainly poral to ovary. Vagina ventral to cirrus pouch. Seminal receptacle small. Uterus an irregular, transverse sac, mainly preequatorial. Eggs with pyriform apparatus. Osmoregulatory system not described. Parasites of rodents. South America.

Type species: *L. machadoi* Rego 1964, in *Proechimys gayennensis;* Brazil.

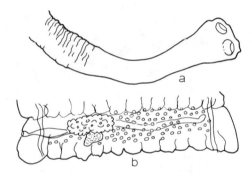

FIGURE 468. *Bertiella hamadryadis* Pierantoni 1928. (a) Scolex; (b) proglottid.

**28b. Vitelline gland posterior to ovary**.............................................. **29**

**29a. Ovary poral**.................................................... ***Bertiella*** **Stiles et Hassall 1902 (syn.** ***Bertia*** **Blanchard 1891,** ***Prototaenia*** **Baer 1927). (Figure 468)**

*Diagnosis:* Scolex quadrangular to rounded; suckers sessile or pedunculate. Proglottids craspedote, usually wider than long. One pair of reproductive systems per segment. Genital pores alternating irregularly. Two pairs of longitudinal osmoregulatory systems. Genital ducts dorsal to osmoregulatory canals. Internal seminal vesicle present. Vas deferens strongly developed, functioning as external seminal vesicle. Testes in single band or two groups anterior to uterus. Ovary and vitellarium central or poral. Seminal receptacle often present. Uterus single, sac-like, dorsally exceeding osmoregulatory canals. Pyriform apparatus usually present. Parasites of Primates (including humans), Rodentia, Marsupialia. Australia, Asia, Europe, Africa, Oceania, North and South America, Japan. Key to Australian species: Beveridge.[271]

Type species: *B. studeri* (Blanchard 1891) Stiles et Hassall 1902 (syn. *Bertia studeri* Blanchard 1891, *Bertia satyri* Blanchard 1891, *Taenia [Bertia] conferta* Meyner 1895, *Taenia studeri* [Blanchard 1891] Braun 1896, *T. satyri* [Blanchard 1891] Braun 1896, *Bertiella satyri* [Blanchard 1891] Stiles et Hassall 1902, *Bertiella conferta* [Meyner 1895] Stiles et Hassall 1902, *Bertia polyorchis* Linstow 1905, *Bertiella cercopitheci* Beddard 1911), in human, *Simia satyrus, Anthropopithecus troglodytes, Hylobates hoolock, Cercopithecus* spp., *Macacus rhesus, Hamadryas hamadryas, Cynomolgus cynomolgus, C. irus, C. sinicus;* Africa, South Asia, Philippines, Mauritius, Puerto Rico, Cuba.

Other species:

*B. aberrata* Nybelin 1917 (syn. *Prototaenia aberrata* [Nybelin 1917] Baer 1927), in *Pseudocheirus herbertensis, P. peregrinus;* Australia.

*B. anapolytica* Baylis 1934, in *Rattus rattus;* Sumatra.

*B. boholensis* Spasskii 1951 (syn. *B. elongata* [Bourquin 1905] Chu 1931), in *Galeopithecus volans;* Philippines.

*B. congolensis* Baer et Fain 1951, in *Colobus polycomus;* Belgian Congo.

*B. cynocephali* Spasskii 1951 (syn. *B. studeri* [Blanchard 1891 of Chu, 1931]), in *Galeopithecus volans;* Philippines.

*B. edulis* (Zschokke 1899) Stiles et Hassall 1902 (syn. *Bertia edulis* Zschokke 1899, *Prototaenia edulis* [Zschokke 1899] Baer 1927), in *Phalanger ursinus;* Celebes.

*B. elongata* (Bourquin 1905) Stiles et Hassall 1902 (syn. *Bertia elongata* Bourquin 1905, *Prototaenia elongata* [Bourquin 1905] Baer 1927), in *Galeopithecus temmincki;* Sumatra, Java.

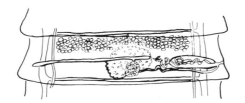

FIGURE 469. *Indotaenia indica* Singh 1962.

*B. fallax* Meggitt 1927, in *Cebus capuchinus;* Egypt.

*B. finlayi* Mazza, Parodi et Fiora 1932, in *Lagidium tucumanus;* Argentina.

*B. foederata* Beveridge 1976, in *Pseudocheirus peregrinus;* Australia.

*B. forcipata* Linstow 1904, in *Lagidium peruanum;* Peru.

*B. hamadryadis* (Pierantoni 1928) (syn. *Bertia hamadryadis* Pierantoni 1928), in *Hamadryas hamadryas;* Italy (zoo).

*B. kapul* Baylis 1934, in *Phalanger maculatus;* Admiralty Islands.

*B. lemuriformis* Deblock et Capron 1959, in *Lichanotus laniger;* Madagascar.

*B. lopezneyrai* Flores Borroeta 1955, in *Nasua narica;* Guatemala.

*B. mawsonae* Beveridge 1976, in *Schoinobates volans;* Australia.

*B. mucronata* (Meyner 1895) Stiles et Hassall 1902 (syn. *Taenia [Bertia] mucronata* Meyner 1895, *Bertia mucronata* [Meyner 1895] Stiles 1896), in *Alouatta nigra, Cebus fatuellus, Calicebus nigrifrons;* Guiana, Paraguay.

*B. musasabi* Yamaguti 1942, in *Petaurista leucogenys;* Japan.

*B. obesa* (Zschokke 1898) Stiles et Hassall 1902 (syn. *Taenia obesa* Zschokke 1898, *Bertia obesa* [Zschokke 1898] Zschokke 1899, *Prototaenia obesa* [Zschokke 1898] Baer 1927), in *Phascolarctos cinereus;* Australia.

*B. okabei* Sawada et Kifune 1974, in *Macaca iris;* Asia.

*B. pellucida* Nybelin 1917 (syn. *Prototaenia pellucida* [Nybelin 1917] Baer 1927), in *Hemibelideus lemuroides;* Australia.

*B. petaurina* Beveridge 1976, in *Schoinobates volans;* Australia.

*B. plastica* (Sluiter 1896) Stiles et Hassall 1902 (syn. *Taenia plastica* Sluiter 1896 *Bertia plastica* [Sluiter 1896] Stiles 1896, *Prototaenia plastica* [Sluiter 1896] Baer 1927), in *Galeopithecus temmincki;* India, Sumatra, Java.

*B. pseudochiri* Nybelin 1917 (syn. *Prototaenia pseudochiri* [Nybelin 1917] Baer 1927), in *Pseudocheirus herbetensis;* Australia.

*B. rigida* (Janicki 1906) Spasskii 1951 (syn. *Bertia rigida* Janicki 1906, *Prototaenia rigida* [Janicki 1906] Baer 1927), in *Phalangista* sp.; New Guinea.

*B. sarasinorum* (Zschokke 1899) Stiles et Hassall 1902 (syn. *Bertia sarasinorum* Zschokke 1899, *Prototaenia sarasinorum* [Zschokke 1899] Baer 1927), in *Phalanger ursinus;* Celebes.

*B. trichosuri* Khalil 1970, in *Trichosurus vulpecula, T. caninus;* Australia.

*B. undulata* Nybelin 1917 (syn. *Prototaenia undulata* [Nybelin 1917] Baer 1927), in *Hemibelideus lemuroides;* Australia.

**29b. Ovary median ............................................................ 30**

**30a. Testes in single group ..................... *Indotaenia* Singh 1962. (Figure 469)**

*Diagnosis:* Scolex unarmed. Suckers weakly developed. Neck short. Proglottids slightly craspedote, wider than long. Genital pores irregularly alternating. Genital atrium shallow. Genital ducts pass dorsal to osmoregulatory canals. Cirrus pouch cylindrical. Cirrus unarmed.

FIGURE 470. *Schizorchis altaica* Gvosdev 1951.

Internal and external seminal vesicles absent. Testes numerous (103 to 121), in a single group anterior to all other genitalia. Vagina opens posterior to cirrus. Seminal receptacle absent. Ovary median, fan-shaped, near rear of proglottid. Vitellaria compact, overlapping posterior third of ovary. Uterus a transverse sac with shallow diverticula, exceeding osmoregulatory canals. Pyriform apparatus on eggs well developed. Parasites of rodents. India.

Type species: *I. indica* Singh 1962, in *Petaurista inornatus;* India.

**30b.   Testes in two groups** ............................................................. **31**

**31a.   Short neck present. Ovary fan-shaped** ...............................................
............................................*Schizorchis* **Hansen 1948. (Figure 470)**

*Diagnosis:* Scolex unarmed, small. Short neck present. Proglottids wider than long, craspedote. Single set of reproductive organs per segment. Genital pores irregularly alternating. Cirrus pouch crosses poral osmoregulatory canals. Cirrus unarmed. Internal seminal vesicle present, external seminal vesicle absent. Testes numerous, in two lateral groups in posterior half of proglottid. Ovary slightly poral or median. Vitelline gland postovarian. Vaginal pore posterior or ventral to cirrus. Seminal receptacle visible only in gravid proglottids. Uterus an irregular, transverse sac. Eggs with pyriform apparatus. Parasites of lagomorphs. North America, Asia.

Type species: *S. ochotonae* Hansen 1948, in *Ochotona princeps;* U.S.
Other species:
 *S. altaica* Gvozdev 1951, in *Ochotona alpina;* Russia.
 *S. caballeroi* Rausch 1960, in *Ochotona collaris;* Alaska, U.S.
 *S. changtuensis* Wu 1965, in *Lepus oiostolus;* Tibet.
 *S. esarsi* Lovekar, Seth et Deshmukh 1972, in *Mus musculus;* India. (This species probably belongs in Linstowiinae).
 *S. tibetana* Wu 1965, in *Ochotona daurica;* Tibet.
 *S. yamashitai* Rausch 1963, in *Ochotona hyperborea;* Japan.

**31b.   Neck absent. Ovary bilobed.** .........................................................
........................*Schizorchoides* **Bienek et Grundmann 1973. (Figure 471)**

*Diagnosis:* Scolex unarmed, lacking rostellum. Suckers relatively large. Neck absent. Proglottids craspedote. Single set of reproductive systems per segment. Genital pores irregularly alternating. Genital atrium present. Cirrus unarmed. External seminal vesicle present. Testes numerous, in two lateral groups in posterior half of segment. Ovary bilobed, median. Vitellarium postovarian. Vaginal pore and vagina posterior to cirrus pore and cirrus sac. Uterus longitudinal, sac-like, filling most of proglottid when gravid. Egg membranes lacking pyriform apparatus. Parasites of rodents. North America.

Type species: *S. dipodomi* Bienek et Grundmann 1973, in *Dipodomys merriami;* Utah, U.S.

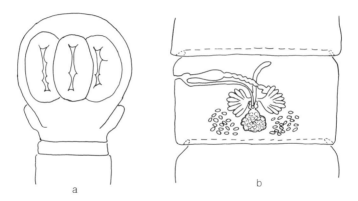

FIGURE 471. *Schizorchoides dipodomi* Bienek et Grundmann 1973. Scolex and proglottid.

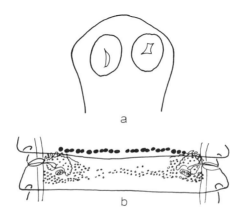

FIGURE 472. *Moniezia expansa* (Rudolphi 1810) Blanchard 1891. (a) Scolex; (b) proglottid.

32a. Uterus reticular..................................................................33
32b. Uterus a transverse tube or sac..............................................36

33a. Interproglottid glands present .......... *Moniezia* Blanchard 1891. (Figure 472)

*Diagnosis:* Scolex simple. Strobila very long. Proglottids wider than long, craspedote. Interproglottid glands present. Genital pores marginal, equatorial. Genital ducts dorsal to osmoregulatory canals. Cirrus pouch reaches poral osmoregulatory canal. Internal seminal vesicle present. Cirrus armed with minute spines. Testes in single, intervascular field, mainly posterior to and between ovaries. Ovary fan-shaped, poral, Vitelline gland compact, postovarian. Vagina posteroventral to cirrus pouch on one side and posterdorsal to cirrus pouch on the other. Seminal receptacle present. Uterus first reticular, then filling entire proglottid. Eggs with pyriform apparatus. Parasites of Perissodactyla, Artiodactyla, Primates (including human). Cosmopolitan.

Type species: *M. expansa* (Rudolphi 1805) Blanchard 1891 (syn. *Taenia ovina* Goeze 1782, *T. vasi* Bloch 1782, *T. capreoli* Viborg 1795, *Halysis ovina* [Goeze 1782] Zeder 1803, *Taenia expansa* Rudolphi 1810, *Alyselminthus expansus* [Rudolphi 1810] Blainville 1828, *Taenia denticulata* Mayer 1837, *Moniezia oblongiceps* Stiles et Hassal 1893, *M. trigonophora* Stiles et Hassall 1893, *M. rangiferina* Kolmakov 1938, *M. minima*

Marotel 1912, *M. nullicollis* Moniez 1891), in *Bos taurus, Capra hircus, Ovis aries, Bubalus bufellus, Camelus bactrianus, C. dromedarius, Rangifer tarandus, Antilocapra americana, Antilope cervicapra, Blastocercus campestris, Capra pyrenaica, Capreolus capreolus, Cephalopus monticola, Cervus elaphus, Gazella dorcas, Ibex ibex, Mazama nana, M. rufus, Kobus ellipsiprymnus, Ovibos moschatus, Rangifer caribou, Rupicapra rupicapra, R. tragus, Tetraceros quadricornis, Cariacus, Coassus, Giraffa, Alces, Aepyceros, Odocoileus, Neomorhaedus, Silvicapra, Hippopotamus,* man, rabbit, dog; cosmopolitan.

Other species:

*M. ambigua* Fuhrmann 1902, in *Chrysotis amazonica;* South America.

*M. amphibia* Linstow 1902, in *Hippopotamus amphibius;* South Africa.

*M. autumnalia* Kuznetsov 1967, in sheep, cattle; Russia.

*M. baeri* Skrjabin 1931, in *Rangifer tarandus;* Russia.

*M. benedeni* (Moniez 1879) Blanchard 1891 (syn. *Alyselminthus denticulatus* [Rudolphi 1810] Blainville 1928, *Taenia denticulata* Rudolphi 1810, *T. benedeni* Moniez 1879, *T. alba* Perroncito 1879, *Moniezia alba* [Perroncito 1879] Blanchard 1891, *M. alba* var. *dubia* Moniez 1891, *M. denticulata* [Rudolphi 1810] Blanchard 1891, *M. neumanni* [Moniez 1891] Blanchard 1891, *M. planissima* Stiles et Hassal 1893, *M. triangularis* Marotel 1912, *M. alba* var. *nova* Sauter 1917, *M. alba* var. *longicollis* Sauter 1917, *M. conjugens* Sauter 1917, *M. crassicollis* Sauter 1917, *M. crassicollis* var. *nova* Sauter 1917, *M. latifrons* Sauter 1917, *M. parva* Sauter 1917, *M. planissima* var. *lobata* Sauter 1917, *M. pellucida* Blei 1920, *M. translucida* Jenkins 1923), in *Ovis, Bos, Capra, Bubalus, Odocoileus, Alces, Bison, Aepyceros, Taurotragus, Anoa, Alcelaphus, Redunca, Adenota, Hippotragus, Dictotyles, Capreolus, Lama,* etc.; cosmopolitan.

*M. bipapillosa* (Leidy 1875) Johnston 1909 (syn. *Taenia bipapillosa* Leidy 1875), in *Phascolomys mitchelli;* Australia.

*M. caprae* (Rudolphi 1810) Stiles et Hassall 1893 (syn. *Taenia caprae* Rudolphi 1810), in *Capra hircus;* Europe.

*M. crucigera* (Nitzsch in Geibel 1866) Railliet 1893 (syn. *Taenia crucigera* Nitzsch in Geibel 1866), in *Capreolus capreolus;* Europe.

*M. kuznetsovi* Butylin 1974, in sheep and goat; Russia.

*M. mettami* Baylis 1934, in *Phacochoerus aethiopicus;* Uganda.

*M. monardi* Fuhrmann 1931, in *Redunca amadirum;* Angola.

*M. pallida* Monnig 1926, in *Equus caballus;* South Africa.

*M. rugosa* (Diesing 1850) Luhe 1895 (syn. *Taenia rugosa* Diesing 1850), in *Brachyteles arachnoides, Cebus fatuellus;* South America.

*M. rupicaprae* Galli-Valerio 1929, in *Rupicapra rupicapra;* Europe.

*M. skrjabini* Bator 1971, in sheep and goat; Mongolia.

*M. trigonophora* Stiles et Hassall 1893, in *Bos taurus, Orvis aries, Antilopa cervicapra, Tetraceros quadricornis, Cephalophus grimmia, Pediotragus sharpei, P. horstocki;* cosmopolitan.

*M. vogti* (Moniez 1879) Stiles et Hassal 1896 (syn. *Taenia vogti* Moniez 1879, *Anoplocephala vogti* [Moniez 1879] Moniez 1891), in *Ovis aries;* France.

**33b. Interproglottid glands absent ................................................. 34**

**34a. Reticular uterus arching anteriorly over genital glands ........................**
**........................................... *Fuhrmannella* Baer 1925. (Figure 473)**

*Diagnosis:* Scolex not described. Proglottids wider than long, craspedote. Double set of reproductive organs per segment. Genital pores marginal. No interproglottid glands. Genital

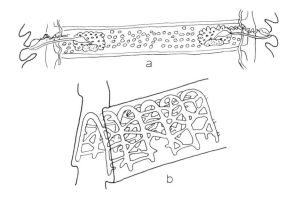

FIGURE 473. *Fuhrmanella transvaalensis* Baer 1925. (a) Mature proglottid; (b) uterus.

ducts pass dorsal to osmoregulatory canals. Cirrus pouch not reaching osmoregulatory canals. Cirrus unarmed. Testes numerous, in single field between ovaries and occasionally extending lateral to ovaries. Ovaries compact, poral. Vitelline gland dorsal to ovary. Vaginal pore ventral to cirrus pore. Seminal receptacle present. Uterus reticular, arching anteriorly over genital glands. Eggs with pyriform apparatus. Parasites of rodents. Africa.

Type species: *F. transvaalensis* Baer 1925, in *Thryonomys swinderianus;* South Africa.

**34b. Uterus otherwise............................................................ 35**

**35a. Internal and external seminal vesicle absent ........... Cittotaenia Riehm 1881.**

*Diagnosis:* Strobila ribbon-like, of moderate size. Scolex small, unarmed. Suckers unarmed. Proglottids numerous (more than 100 in gravid strobilae), craspedote, wider than long. Longitudinal osmoregulatory canals paired, with or without accessory longitudinal vessels and numerous anastomosing supplementary vessels connected to them. Reproductive systems paired. Genital ducts pass dorsally to longitudinal osmoregulatory canals. Vagina ventral to cirrus sac. Internal and external seminal vesicles absent. Testes numerous, in single band or two groups. Seminal receptacle present. Ovaries in lateral quarters of medulla. Developing uterus a single, transverse tube, very slightly reticulated, not crossing longitudinal osmoregulatory canals. Gravid uterus developing anterior and posterior diverticula, finally becoming sac-like. Pyriform apparatus present. Parasites of Lagomorpha and Rodentia. Genus redefined: Beveridge.[274]

Type species: *C. denticulata* (Rudolphi 1804) Stiles at Hassall 1896 (syn. *Taenia denticulata* Rudolphi 1904, *Alyselminthus denticulata* [Rudolphi 1804] Blainville 1928, *Taenia goezei* Baird 1853, *Cittotaenia latissima* Riehm 1881, *Dipylidium latissimum* [Riehm 1881] Riehm 1881, *Taenia latissima* [Riehm 1881] Neumann 1888, *Moniezia goezei* [Baird 1853] Blanchard 1891, *M. denticulata* [Rudolphi 1804] Blanchard 1891, *M. latissima* [Riehm 1881] Blanchard 1891, *Ctenotaenia goezei* [Baird 1853] Railliet 1893, *C. denticulata* [Rudolphi 1804] Stiles et Hassall 1896), in *Oryctolagus cuniculis, Lepus timidus, L. europaeus;* Europe, Morocco.

Other species:

*C. krishnai* Nama 1974, in domestic goat (?); India. (Probably not in this genus).

*C. viscaciae* (Spasskii 1951) Beveridge 1978 (syn. *Mosgovoyia viscaciae* Spasskii 1951), in *Viscacia viscacia;* Chile.

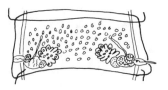

FIGURE 474. *Diandrya composita* Darrah 1930.

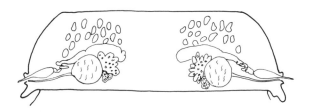

FIGURE 475. *Progamotaenia diaphana* (Zschokke 1907) Baer 1927.

35b. Internal and external seminal vesicles present ...................................
............................................ *Diandrya* **Darrah 1930. (Figure 474)**

*Diagnosis:* Strobila large, ribbon-like, with numerous segments. All segments craspedote, wider than long. Scolex unarmed. Two pairs of simple, lateral osmoregulatory canals present. Two sets of reproductive systems per segment. Genital pores bilateral, postequatorial. Genital ducts dorsal to osmoregulatory canals. Vaginal pore posterior to cirrus pore. Seminal receptacle present. External and internal seminal vesicles present. Testes numerous, filling most of field between and anterior to ovaries. Ovaries and vitellaria posterior, in lateral thirds of segment. Uterus reticular, originating bilaterally, then fusing as one. Gravid segments filled with eggs, which lack pyriform apparatus. Parasites of North America rodents.
Type species: *D. composita* Darrah 1930, in *Marmota flaviventris, M. caligata, M. broweri, M. olympus;* North America.
Other species:
   *D. vancouverensis* Mace et Shepard 1981, in *Marmota vancouverensis;* Vancouver Island, British Columbia, Canada. (This may be a synonym of *D. composita*).

36a. Testes divided into two groups in front of ovaries ............................ 37
36b. Testes in a continuous band or behind ovaries ................................. 38

37a. Uterus doubled ..................................................................
   ... *Progamotaenia* **Nybelin 1917 (syn. *Hepatotaenia* Nybelin 1917). (Figure 475)**

*Diagnosis:* Scolex rounded, four-lobed. Proglottids craspedote, usually wider than long, some species with fringed velar margin. Reproductive systems duplicated. Genital pores in posterior half of lateral margin. Two main pairs of osmoregulatory canals; accessory canals present in some species. Genital ducts dorsal to osmoregulatory canals. Internal and external seminal vesicles present. Testes in two groups anterior to ovary and medial to osmoregulatory canals. Vaginal pore posterior to cirrus pore. Seminal receptacle very large, persistent in gravid proglottids. Uterus doubled, simple or with anterior and posterior diverticulae. Pyriform apparatus, when present, with numerous reflexed filaments. Parasites of marsupials. Australia, New Guinea. Key to species: Beveridge.[271]

Type species: *P. bancrofti* (Johnston 1912) Nybelin 1917 (syn. *Cittotaenia bancrofti* Johnston 1912), in *Onychogalea frenata, O. unguifera, Wallabia bicolor, Setonix brachyurus*. Australia.

Other species:

*P. aepyprymni* Beveridge 1976, in *Aepyprymnus rufescens;* Queensland, Australia.

*P. diaphana* (Zschokki 1907) Baer 1927 (syn. *Moniezia diaphana* Zschokke 1907, *Cittotaenia diaphana* [Zschokke 1907] Douthitt 1915, *Hepatotaenia diaphana* [Zschokke 1907] Nybelin 1917), in *Vombatus ursinus, Lasiorhinus latifrons;* Australia.

*P. festiva* (Rudolphi 1819) Baer 1927 *(Taenia festiva* Rudolphi 1819, *Moniezia festiva* [Rudolphi 1819] Blanchard 1891, *Hepatotaenia festiva* [Rudolphi 1819] Nybelin 1917, *H. fellicola* Nybelin 1917, *Cittotaenia festiva* [Rudolphi 1819] Theiler 1924, *Progamotaenia* sp. Sandars 1957), in *Macropus giganteus, M. robustus, M. parryi, M. agilis, Wallabia bicolor, Petrogale penicellata, Megaleia rufa, Setonix brachyurus, Lagorchestes conspicillatus, Vombatus ursinus;* Australia.

*P. gynandrolinearis* Beveridge et Thompson 1979, in *Lagorchestes conspicillus;* Australia.

*P. lagorchestis* (Lewis 1914) Nybelin 1917 (syn. *Cittotaenia lagorchestis* Lewis 1914), in *Lagorchestis conspicillatus, Thylogale Stigmatica, T. billardierii, T. thetis, Macropus agilis, M. rufogriseus;* Australia.

*P. macropodis* Beveridge 1976, in *Macropus giganteus, M. fuligenosus, M. rufogriseus, M. eugenii, Wallabia bicolor;* Australia.

*P. ruficola* Beveridge 1978, in *Macropus rufus;* New South Wales, Australia.

*P. thylogale* Beveridge et Thompson 1979, in *Lagorchestes conspicillatus;* Australia.

*P. johnsoni* Beveridge 1980, in *Lagorchestes conspicillatus;* Australia.

*P. spearei* Beveridge 1980, in *Thylogale stigmatica;* Australia.

**37b. Uterus single.........................................*Adelataenia* Beveridge 1976.**

*Diagnosis:* Medium to large worms. Scolex large, with suckers on apices of prominent lobes. Neck short or absent. Proglottids much wider than long, craspedote. Velum fringed with 12 to 30 long, finger-like lobes that may extend completely over following proglottid. Two sets of male and female reproductive systems per proglottid. Genital pores postequatorial. Two sets of lateral osmoregulatory canals; accessory canals may also occur. Genital atria small. Genital ducts dorsal to osmoregulatory canals. Internal and external seminal vesicles present. Testes in two groups, anterior to ovary and developing uterus, medial to osmoregulatory canals. Vaginal pore posterior to cirrus pore. Seminal receptacle well developed, near inner margins of osmoregulatory canals, persistent in gravid segments. Vagina atrophies after insemination. Uterus single with anterior and posterior diverticula when gravid. Pyriform apparatus with several fine filaments. Parasites of marsupials. Australia, New Guinea.

Type species: *A. villosa* (Lewis 1914) Beveridge 1976 (syn. *Cittotaenia villosa* Lewis 1914, *Triplotaenia villosa* [Lewis 1914] Spasskii 1951, *Progamotaenia villosa* [Lewis 1914] Beveridge 1976), in *Lagorchestes conspiculatus;* Australia.

Other species:

*A. zschokkei* (Janicki 1906) Beveridge 1976 (syn. *Cittotaenia zschokkei* Janicki 1906, *Progamotaenia zschokkei* Nybelin 1917), in *Macropus agilis, Petrogale penicillata, Thylogale stigmatica, Lagorchestes conspicillatus;* New Guinea, Australia.

**38a. Scolex globular, with prominent apical glands, found imbedded deeply into host tissue ................. *Ectopocephalium* Rausch et Ohbayashi 1974. (Figure 476)**

FIGURE 476. *Ectopocephalium abei* Rausch et Ohbayashi 1974. Scolex.

FIGURE 477. *Fuhrmannodes talboti* Schmidt 1975. Entire worm.

*Diagnosis:* Cestodes of small size. Anterior portion of strobila modified, imbedded deeply into host tissue, causing conspicuous nodules on the outside of the intestine. Scolex globular, with four suckers and prominent apical glands. Two pairs of single osmoregulatory canals. Two sets of male and female reproductive systems. Ovaries lobulated, in lateral quarters of segment. Vitellarium weakly lobed, ventral and posteriomedial to ovary. Genital pores bilateral, about equatorial. Genital atrium shallow. Vagina posterior to cirrus sac, opening posterior to cirrus pore. Genital ducts dorsal to osmoregulatory canals. Testes numerous; in single field between osmoregulatory canals; both dorsal and ventral to female organs. Uterus first a transverse tube that exceeds osmoregulatory canals, developing extensive anterior and posterior diverticulae when gravid. Egg with pyriform apparatus. Parasites of lagomorphs. Asia.

Type species: *E. abei* Rausch et Ohbayashi 1974, in *Ochotona roylei, O. macrotis;* Nepal.

**38b. Scolex not as above, not imbedded in wall of host intestine ................... 39**

**39a. Testes in continuous band anterior of ovaries ................................. 40**
**39b. Testes otherwise ........................................................... 41**

**40a. Uterus double. Strobila strongly proterogynous ....................... *Fuhrmannodes* Strand 1942 (syn. *Baeriella* Fuhrmann 1932, preoccupied). (Figure 477)**

*Diagnosis:* Scolex large, with prominent suckers. Neck short. Proglottids wider than long, craspedote, with lobed velum. Two sets of highly proterogynous reproductive organs per segment. Genital pores bilateral. Cirrus pouch small. Testes numerous, in continuous, preequatorial band. Internal and external seminal vesicles present. Ovary compact, transversely elongated. Vitelline gland postovarian. Seminal receptacle present. Uterus double, lobated, filling entire medulla when gravid. Eggs with pyriform apparatus. Parasites of marsupials. Australia, New Guinea.

Type species: *F. proterogyna* (Fuhrmann 1932) Strand 1942 (syn. *Baeriella proterogyna* Fuhrmann 1932, *Progamotaenia proterogyna* [Fuhrmann 1932] Lopez-Neyra 1954), in *Macropus rufus, M. agilis* (?). Geneva Zoo, Australia.

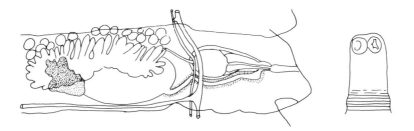

FIGURE 478. *Wallabicestus ewersi* Schmidt 1975. Half of proglottid, and scolex.

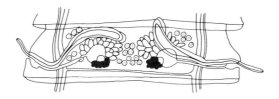

FIGURE 479. *Phascolotaenia comani* Beveridge 1976.

Other species:
  *F. talboti* Schmidt 1975, in *Macropus agilis;* New Guinea.

**40b. Uterus single. Proterogyny not conspicuous.........................................
................................... Wallabicestus Schmidt 1975. (Figure 478)**

*Diagnosis:* Medium-sized worms, dorsoventrally flattened. Scolex unarmed. Neck absent. Proglottids craspedote, wider than long. Osmoregulatory canals in two pairs, simple. Two sets of reproductive systems per segment; (primary segments may have only one set). Genital pores bilateral. Genital atria variable. Genital ducts dorsal to osmoregulatory canals. Testes medullary in single band extending across proglottid between osmoregulatory canals, anterior to ovaries. Cirrus armed with very small spines. Internal seminal vesicle present, external seminal vesicle present but variably enlarged. Ovaries in lateral thirds of segment, with many slender lobes. Vitelline gland compact, posterodorsal to ovary. Vagina posterior to cirrus pouch, lined with delicate spines or cilia, surrounded by small unicellular glands. Seminal receptacle very large, persistent in gravid segments. Uterus single, simple, extending across posterior half of entire proglottid, passing posterior to vagina and nearly reaching surface of lateral margin of segment. Eggs with pyriform apparatus. Parasites of marsupials. New Guinea, Australia.

  Type species: *W. ewersi* Schmidt 1975 (syn. *Progamotaenia ewersi* [Schmidt 1975] Beveridge 1976), in *Macropus agilis, M. rufogriseus, M. fuliginosus, M. giganteus, M. parryi, M. eugenii, M. robustus, Wallabia bicolor;* New Guinea, Australia.
  Other species:
    *W. effigia* (Beveridge 1976) comb. n., in *Macropus fuliginosus;* Australia.

**41a. Testes in two lateral fields as well as in median ................................
................................... Phascolotaenia Beveridge 1976. (Figure 479)**

*Diagnosis:* Small size (30 to 40 mm), scolex and suckers unarmed. Neck present. Proglottids craspedote, unfringed, wider than long. Two pairs of male and female reproductive systems per segment. Double genital pores in posterior half of lateral margin of proglottid.

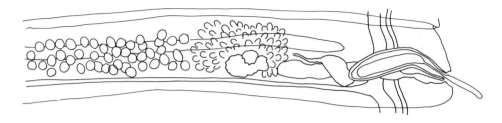

FIGURE 480. *Ctenotaenia marmotae* (Froehlich 1802) Railliet 1893.

Two pairs of lateral osmoregulatory canals. Genital ducts dorsal to osmoregulatory canals. Cirrus pouch large. Internal and external seminal vesicles present. Testes in lateral band posterior to developing uterus, extending to osmoregulatory canals; two small lateral groups of testes anterior to cirrus pouches. Female systems poral. Vagina posterior and ventral to cirrus pouch. Seminal receptacle large. Uterus single, with anterior and posterior diverticula. Pyriform apparatus present. Parasites of Australian Vombatidae.

Type species: *P. comani* Beveridge 1976, in *Vombatus ursinus;* Australia.

**41b. Testes in single field ............................................................ 42**

**42a. Testes in single band between ovaries .. Ctenotaenia Railliet 1893. (Figure 480)**

*Diagnosis:* Strobila broad, ribbon-like, of moderate size. Scolex small, unarmed. Suckers unarmed. Proglottids numerous (over 100 in gravid strobilae), craspedote, much wider than long. Osmoregulatory canals paired. Genital ducts dorsal to osmoregulatory canals. Reproductive systems paired. Female pore posterior to cirrus pore in genital atrium. Cirrus sac dorsal to vagina. Internal and external seminal vesicles present. Testes numerous, medial to ovaries, in single band. Seminal receptacle present. Ovaries in lateral quarters of medulla. Single transverse uterus in each segment, not crossing longitudinal osmoregulatory canals. Gravid uterus with anterior and posterior pockets. Egg membrane with pyriform apparatus. Parasites of rodents. Europe, Asia.

Type species: *C. marmotae* (Froehlich 1802) Railliet 1893 (syn. *Taenia marmotae* Froehlich 1802, *Moniezia marmotae* [Froelich 1802] Blanchard 1891), in *Marmota* spp., *Citellus* spp.; Europe. Redescribed: Beveridge.[274]

**42b. Testes otherwise ................................................................ 43**

**43a. Testes between, posterior to and sometimes anterior to ovaries............... 44**
**43b. Testes only posterior to ovaries .................................................. 45**

**44a. Proximal end of cirrus pouch directed strongly anteriad. Parasites of reindeer
 ........................................... Eranuides Semenova 1972. (Figure 481)**

*Diagnosis:* Scolex unknown. All proglottids wider than long. Two sets of reproductive systems per segment. Genital pores slightly preequatorial. Genital atrium muscular. Posterior end of cirrus pouch directed anteriad. Vas deferens convoluted. Testes numerous, posterior to and filling space between ovaries. Ovary a cluster of follicles. Vitellarium lobated, postovarian. Vagina posterior to cirrus pouch. Gravid uterus an irregular sac with diverticula, extending far past osmoregulatory canals. Parasites of reindeer. Russia.

Type species: *E. mathevossianae* Semenova 1972, in *Rangifer rangifer;* Taimyr Penninsula.

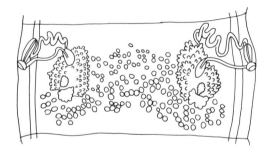

FIGURE 481. *Eranuides mathevossianae* Semenova 1972.

FIGURE 482. *Paramoniezia johnstoni* Beveridge 1976.

**44b. Cirrus pouch not as above. Parasites of Australian wombats ..................
...................... *Paramoniezia* Maplestone et Southwell 1923. (Figure 482)**

*Diagnosis:* Anoplocephalinae of moderate size. Scolex and suckers unarmed. Proglottids craspedote, transversely elongated. Two sets of male and female reproductive systems present. Genital ducts dorsal to osmoregulatory canals. Genital pores bilateral. Testes numerous, distributed throughout medulla. Internal seminal vesicle present. Vagina posterior and ventral to cirrus sac. Seminal receptacle elongate. Uterus single, with anterior and posterior diverticula. Eggs with pyriform apparatus. Parasites of Vombatidae, Australia.

Type species: *P. suis* Maplestone et Southwell 1923, in a "bush pig" (wombat: Marsupialia: probably *Lasiorhinus barnardi);* Australia.

Other species:

*P. johnstoni* Beveridge 1976, in *Vombatus ursinus;* Australia.

**45a. Uteri develop as longitudinal grooves extending through field of testes..........
....................................... *Diuterinotaenia* Gvosdev 1961. (Figure 483)**

*Diagnosis:* Scolex lacking apical organ or armature. Suckers simple. Neck present. Strobila acraspedote; proglottids wider than long. Two sets of reproductive systems per segment. Genital pores slightly postequatorial. Cirrus pouch usually not reaching poral osmoregulatory canals. Testes numerous in continuous postovarian field. Genital ducts dorsal to osmoregulatory canals. Ovaries lobated, in anterior half of segment. Vitelline glands posterior to ovaries. Vagina posterior to cirrus pouch. Uteri first form as longitudinal slits lateral to ovaries, and extend ventral into testicular field. Gravid uteri longitudinally oriented, lobated sacs. Parasites of lagomorphs. Asia.

Type species: *D. spasskyi* Gvosdev 1961, in *Ochotona pusilla;* Russia.

**45b. Uteri develop as transverse grooves ........................................... 46**

**46a. Developing uterus near anterior margin of segment, crossing osmoregulatory canals ventrally .................... *Pseudocittotaenia* Tenora 1976. (Figure 484)**

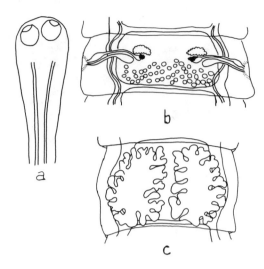

FIGURE 483. *Diuterinotaenia spasskyi* Gvosdev 1961. Scolex, mature, and gravid proglottids.

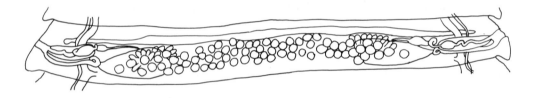

FIGURE 484. *Pseudocittotaenia praecoquis* (Stiles 1895) Beveridge 1978.

*Diagnosis:* Strobila small, over 50 proglottids in gravid strobilae. Scolex rounded, unarmed. Suckers unarmed. Proglottids wider than long. Longitudinal osmoregulatory canals paired. Genital ducts dorsal to osmoregulatory canals. Reproductive systems paired. Vaginal pore posterior to cirrus pore in genital atrium. Vagina ventral to cirrus sac. Internal seminal vesicle present, external seminal vesicle present or absent. Testes numerous, posterior to developing uterus, either restricted between ovaries or extending laterally to them. Seminal receptacle prominent. Ovaries in lateral quarters of proglottid. Uterus a single, transverse tube near anterior edge of segment, crossing longitudinal osmoregulatory canals laterally, terminating anterior to cirrus pouches. Gravid uterus sac-like, with anterior and posterior diverticula. Pyriform apparatus present. Parasite of Rodentia. North America.

Type species: *P. praecoquis* (Stiles 1895) Beveridge 1978, Syn. *Cittotaenia praecoquis* [Stiles 1895] Stiles et Hassall 1896, *Ctenotaenia praecoquis* Stiles 1895), in *Geomys bursarius, Thomomys talpoides, T. tenellus;* Iowa, Utah, Wyoming, U.S.

Other species:

*P. glandularis* Beveridge 1978, in *Thomomys talpoides;* Utah, Wyoming, U.S.

**46b. Developing uterus about equatorial, crossing osmoregulatory canals dorsally or not at all .................................................................................................**
 **.... *Mosgovoyia* Spasskii 1951 (syn. *Neoctenotaenia* Tenora 1976). (Figure 485)**

*Diagnosis*: Strobila broad, ribbon-like, of moderate size. Scolex small, unarmed. Proglottids numerous (over 100 in gravid strobilae), craspedote, much wider than long. Lateral

FIGURE 485. *Mosgovoyia pectinata* (Goeze 1782) Spasskii 1951.

osmoregulatory canals paired. Genital ducts dorsal to osmoregulatory canals. Reproductive systems paired. Vaginal pore posterior to cirrus pore in genital atrium. Cirrus pouch dorsal to vagina. Internal seminal vesicle present, external seminal vesicle absent. Testes numerous, entirely posterior to developing uterus, in single band or two groups, either restricted between ovaries or extending laterally beyond them. Distal vagina surrounded by glandular cells. Seminal receptacle present. Ovaries in lateral quarters of medulla. One or two transverse uteri per segment. Gravid uterus sac-like, with or without anterior and posterior diverticula. Pyriform apparatus present on egg membrane. Parasites of Lagomorpha and rarely Rodentia. Europe, Africa, Asia, North America, Norway, (?) South America. Genus redefined: Beveridge.[274]

Type species: *M. pectinata* (Goeze 1782) Spasskii 1951 (syn. *Taenia pectinata* Goeze 1782, *Alyselminthus pectinatus* [Goeze 1782] Zeder 1800, *Taenia leporina* Rudolphi 1810, *Halysis pectinata* [Goeze 1782] Zeder 1810, *Dipylidium pectinatum* [Goeze 1782] Riehm 1881, *Moniezia pectinata* [Goeze 1782] Blanchard 1891, *Ctenotaenia pectinata* [Goeze 1782] Railliet 1893, *Cittotaenia bursaria* Linstow 1906, *Cittotaenia pectinata* [Goeze 1782] Stiles et Hassall 1896), in *Oryctolagus cuniculus*, *Lepus* spp., *Sylvilagus* spp., *Sciurus caroliensis*, *Citellus undulatus;* continents listed above.

Other species:

*M. ctenoides* (Railliet 1890) Beveridge 1978 (syn. *Dipylidium leuckarti* Riehm 1881, *Taenia leuckarti* [Riehm 1881] Neumann 1888, *T. ctenoides* Railliet 1890, *Moniezia leuckarti* [Riehm 1881] Blanchard 1891, *Ctenotaenia leuckarti* [Riehm 1881] Railliet 1893, *C. leuckarti* [Riehm 1881] Stiles et Hassall 1896, *C. ctenoides* [Railliet 1890] Spasskii 1951, *Cittotaenia ctenoides* [Railliet 1890] Stiles 1896, *C. ctenoides* [Riehm 1881] Arnold 1938), in *Oryctolagus cuniculus*, *Lepus* spp., *sylvilagus* spp. Europe, North America.

*M. indica* Nama 1980, in *Lepus* sp.; India.

*M. oitana* Sawada et Kugi 1974, in *Lepus brachyurus;* Japan.

*M. variabilis* (Stiles 1895) Beveridge 1978 (syn. *Taenia pectinata* [Goeze 1782] Curtice 1892, *Cittotaenia variabilis* [Stiles 1895] Stiles et Hassall 1896, *C. variabilis angusta* Stiles 1896, *C. variabilis imbricata* Stiles 1896, *Ctenotaenia variabilis* Stiles 1895), in *Sylvilagus* spp.; North America.

## KEY TO THE GENERA IN LINSTOWIINAE

1a. **Parasites of reptiles**............................................................. 2
1b. **Parasites of birds or mammals** ................................................ 5

2a. **Two sets of reproductive organs per segment**......................................
.............................. *Panceriella* **Stunkard 1969 (syn.** *Pancerina* **Fuhrmann 1899 preoccupied,** *Panceria* **Sonsino 1895 preoccupied). (Figure 486)**

*Diagnosis:* Scolex simple, neck absent. Gravid proglottids may be longer than wide. Two sets of reproductive organs per segment. Genital pores bilateral. Cirrus pouch well developed. Seminal vesicles absent. Testes in two lateral groups, each surrounding ovary except on

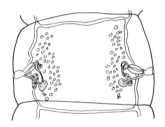

FIGURE 486. *Panceriella varani* (Stossich 1895) Stunkard 1969.

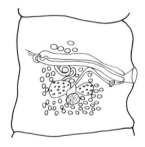

FIGURE 487. *Seminoviella amphisbaenae* (Rudolphi 1819) Spasskii 1951.

poral side. Genital ducts passing between osmoregulatory canals. Ovaries compact or bilobed, poral. Vitelline glands postovarian. Two uteri formed, each breaking into egg capsules each with one egg. Parasites of varanid lizards. Africa, Palestine.

Type species: *P. varani* (Stossich 1895) Stunkard 1969 (syn. *Panceria arenaria* Sonsino 1895, *Pancerina varanii* [Stossich 1895] Fuhrmann 1899, *Taenia varani* Stossich 1895) in *Varanus griseus, V. aranarius;* Egypt, Algeria, Palestine.

2b.   One set of reproductive organs per segment ..................................... 3

3a.   Gravid proglottids two to six times as long as broad. Anterior two-fifths of mature proglottid free of reproductive organs ................. *Diochetos* Harwood 1932.

*Diagnosis:* Scolex small, neck present. Strobila of few proglottids. Mature and gravid segments two to six times as long as wide. Genital pores irregularly alternating, preequatorial. Osmoregulatory canals two, poral one ventral to genital ducts. Testes numerous, tending to be arranged in two lateral groups. Ovary and postovarian vitelline gland small, median. Anterior two fifths of mature proglottids free of reproductive organs. Uterus breaking into sparse egg capsules, each with single egg. Parasites of phrynosomid lizards. North America.

Type species: *D. phrynosomatis* Harwood 1932 (syn. *Oochoristica phrynosomatis* [Harwood 1932] Loewen 1940), in *Phrynosoma cornutum;* North America.
Other species:
   *D. parvovaria* (Steelman 1939) Yamaguti 1959 (syn. *Oochoristica parvovaria* Steelman 1939), in *Phrynosoma cornutum;* Texas, U.S.

3b.   Gravid proglottids not over twice as long as broad. Anterior two fifths of mature proglottids not free of reproductive organs ..................................... 4

4a.   Vaginal sphincter present. Cirrus pouch very long ...............................
        ........................................ *Semenoviella* Spasskii 1951. (Figure 487)

*Diagnosis:* Proglottids longer than wide, acraspedote. Genital pores irregularly alternating. Genital ducts passing between osmoregulatory canals. Cirrus pouch very large, passing median line of segment. Seminal vesicles absent. Testes encircling ovary. Ovary bilobed, median. Vitelline gland postovarian. Vagina posterior to cirrus pouch, with distal sphincter. Uterus breaks into egg capsules, each with single egg. Parasites of worm lizards. South America.

Type species: *S. amphisbaenae* (Rudolphi 1819) Spassky 1951 (syn. *Taenia amphisbaenae*

FIGURE 488. *Oochoristica indica* Mizra 1945. (a) Scolex; (b) proglottid.

Rudolphi 1819, *Oochoristica amphisbaenae* [Rudolphi 1819] Lühe 1898), in *Amphisbaena alba;* South America.

4b. Vaginal sphincter absent. Cirrus pouch not particularly long.................
......*Oochoristica* Lühe 1898 (syn. *Skrjabinochora* Spasskii 1948). (Figure 488)

*Diagnosis:* Proglottids acraspedote, gravid segments usually longer than wide. Genital pores irregularly alternating. Genital atrium usually muscular. Genital ducts passing between or dorsal to osmoregulatory canals. Testes few or many. Cirrus pouch small. Seminal vesicles absent. Ovary bilobed, median or slightly poral. Vitelline gland postovarian. Vagina posterior to cirrus pouch, lacking sphincter. Seminal receptacle present. Uterus rapidly breaking into egg capsules, each containing single egg. Parasites of lizards, snakes, and turtles. Africa, Europe, North and South America, Asia, Ceylon, Java, Australia, Macassar, Celebes, Philippines, New Zealand.

Type species: *O. tuberculata* (Rudolphi 1819) Lühe 1898 (syn. *Taenia tuberculata* Rudolphi 1819, *T. rotundata* Molin 1859, *T. pseudopodis* Krabbe 1879, *Oochoristica rotundata* [Molin 1859] Parona 1900, *O. pseudopodis* [Krabbe 1879] Zschokke 1905), in *Acanthodactylus paradalis, Chalicides ocellatus, Eumeces schneideri, Lacerta agilis, L. lepida, L. muralis, L. ocellata, L. viridis, Ophisaurus apus, O. pallasi, Uromastix acanthinurus, Varanus griseus;* Europe, North Africa, Russia.

Other species:

*O. ameivae* (Beddard 1914) Baer 1924 (syn. *Linstowia ameivae* Beddard 1914, *O. brasiliensis* Fuhrmann 1927, *O. fuhrmanni* Hughes 1940), in *Ameiva surinamensis, A. dorsalis, Psammophis sibilans;* South America, Jamaica, South Africa.

*O. americana* Harwood 1932, in *Farancia abacura;* Texas, U.S.

*O. anniellae* Stunkard et Lynch 1944, in *Anniella pulchra;* California, U.S.

*O. anolis* Harwood 1932, in *Anolis caroliensis;* Texas, U.S.

*O. aulicus* Johri 1961, in *Lycodon aulicus;* India.

*O. australiensis* Spasskii 1951 (syn. *O. trachysauri* [MacCallum 1921] Johnston 1932), in *Trachysaurus rugosus;* Australia.

*O. bailea* Singal 1961, in *Hemidactylus flaviveridis;* India.

*O. bivitellolobata* Loewen 1940, in *Cnemidophorus sexlineatus;* U.S.

*O. brachysoma* Depouy et Kechemir 1973, in *Sphenops boulengeri;* Algeria.

*O. bresslaui* Fuhrmann 1927, in *Tropidurus hispidus;* Brazil.

*O. calotes* Nama et Khichi 1974, in *Calotes versicolor;* India.

*O. celebesensis* Yamaguti 1954, in *Mabuia* sp.; Celebes.

*O. chabaudi* Dollfus 1954, in *Chalcides mionecton;* North Africa.

*O. chavenoni* Deblock, Rosé, Broussert, Capron et Brygoo 1962, in *Chamaeleo verrucosus;* Madagascar.

*O. chinensis* Jensen, Schmidt et Kuntz 1983, in *Japalura swinhonis;* Taiwan.

*O. courdurieri* Deblock, Rosé, Broussart, Capron et Brygoo 1962, in *Chamaeleo paradalis;* Madagascar.

*O. crassiceps* Baylis 1920 (syn. *O. sigmoides* Moghe 1926, *O. fusca* Meggitt 1927), in *Psammophis subtaeniatus, Calotes versicolor;* East Africa.

*O. crotalicola* Alexander et Alexander 1957, in *Crotalus viridus;* U.S.

*O. cryptobothrium* (Linstow 1906) LaRue 1914 (syn. *Ichthyotaenia cryptobothrium* Linstow 1906, *Oochoristica rostellata* [Zschokke 1905] Baer 1927 in part), in *Chrysopelea ornata;* Sri Lanka.

*O. danielae* Deblock, Rosé, Broussart, Capron et Brygoo 1962, in *Chamaeleo verrucosus;* Madagascar.

*O. darensis* Dollfus 1957, in *Uromastix acanthinurus;* Morocco.

*O. elaphis* Harwood 1932, in *Elaphe obsoleta;* Texas, U.S.

*O. elongata* Dupouy et Kechemir 1973, in *Agama mutabilis;* Algeria.

*O. eumecis* Harwood 1932, in *Eumeces fasciatus;* North America.

*O. excelsa* Tubangui et Masiluñgan 1936, in *Mabuia multifasciata;* Philippines.

*O. fibrata* Meggitt 1927, in *Boiga cyaneus, B. multimaculata, Pityophis sayi;* Burma.

*O. gallica* Dollfus 1954, in *Psammodromus hispanicus, P. algirus;* Europe.

*O. gallica* var. *pleionorcheis* Dollfus 1954, in *Lacerta lepida;* Europe.

*O. gracewileyae* Loewen 1940, in *Crotalus atrox;* North America.

*O. hainanensis* Hsü 1935, in undetermined lizard; China.

*O. hemidactyli* Johri 1955, in *Hemidactylus flaviviridis;* India.

*O. indica* Misra 1945, in *Calotes versicolor;* India.

*O. insulaemargaritae* Lopez-Neyra et Diaz-Ungria 1957, in *Ameiva ameiva;* Margarita Island, Venezuela.

*O. javaensis* Kennedy, Killick et Beverly-Burton, in *Gehyra mutilata, Cosymbotus platyurus, Hemidactylus frenatus;* Java.

*O. jodhpurnesis* Nama 1977, in *Hemidactylus flaviviridis;* India.

*O. junkea* Johri 1950, in *Gecko gecko;* India.

*O. khalili* Hamid 1932, in *Psammophis shokari;* Egypt.

*O. langrangei* Joyeux et Houdemer 1927, in *Liolepis belliana;* Indochina.

*O. longicirrata* Dupouy et Kechemir 1973, in *Scincus scincus;* Algeria.

*O. lygosomae* Burt 1933, in *Lygosoma punctatum;* Sri Lanka.

*O. lygosomatis* Skinker 1935 (syn. *O. parva* Baylis 1929, *O. parva* [Sandground 1926] Meggitt 1934, *O. parva* Stunkard 1938, *O. baylisi* Baer 1935), in *Lygosoma chalcides;* Java.

*O. mandapemensis* Johri 1958, in *Calotes versicolor;* India.

*O. microscolex* Della Santa 1956, in *Hemidactylus coctaei;* India.

*O. natricis* Harwood 1932, in *Natrix rhombifer, Python sebae, Naja melancoleuca, Sepedon haemacheta, Disphelidus typus;* North America, Africa.

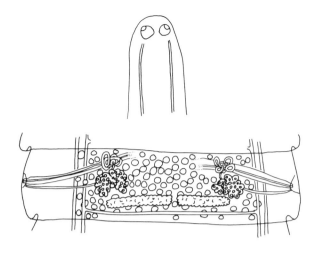

FIGURE 489. *Echidnotaenia tachyglossi* (Johnston 1913) Beveridge 1980. Scolex and proglottid.

*O. novaezealandae* Schmidt et Allison 1985, in *Leiomopisma nigriplantare;* New Zealand.

*O. osheroffi* Meggitt 1934, in *Pituphis sayi, Crotalis viridis;* U.S.

*O. ophia* Capoor, Srivastava et Chauhan 1976, in unknown snake; India.

*O. parvula* (Stunkard 1938) Stunkard 1938 (syn. *O. parva* Stunkard 1938), in *Coleonyx elegans;* Yucatan, Mexico.

*O. porrogenitalis* Dupouy et Kechemir 1973, in *Stenodactylus petriei;* Algeria.

*O. pseudocotylea* Dollfus 1957, in *Eumeces algeriensis;* North Africa.

*O. rostellata* Zschokke 1905, in *Zamenis gemonensis, Z. viridiflavus, Coluber jugularis, Cerastes cornutus, Chrysopelia ornata, Malpolon monspessulana, Lycodon* sp.; Italy, France, Algeria, Somalie.

*O. salensis* Dollfus 1954, in *Malpolon monspessulana;* Algeria.

*O. scelophori* Voge et Fox 1950, in *Scelophorus occidentalis;* U.S.

*O. sobolevi* (Spasskii 1948) Spasskii 1951 (syn. *Skrjabinchora sobolevi* Spasskki 1948).

*O. tandani* Singh 1957, in *Lycodon aulicus;* India.

*O. thapari* Johri 1934, in *Calotes* sp.; India.

*O. theileri* Fuhrmann 1924, in *Agama hispida, Chamaeleo tempeli;* South Africa.

*O. trachysauri* (MacCallum 1921) Baer 1924 (syn. *Taenia trachysauri* MacCallum 1921), in *Trachysaurus rugosus;* New York Zoo (Australia).

*O. travassosi* Rego et Ibanez 1965, in *Leiocephalus* sp.; Peru.

*O. truncata* (Krabbe 1879) Zschokke 1905 (syn. *T. truncata* Krabbe 1879, *Oochoristica agamae* Baylis 1919, *O. africana* Malan 1939, *O. a.* var. *ookispensis* Malan 1939), in *Agama, Hemidactylus, Gerrosaurus, Cerastes, Dendropus, Eryx, Psammophis, Chameleon;* Africa, Southern Asia.

*O. varani* Nama et Khichi 1972, in *Varanus monitor;* India.

*O. vacuolata* Hickman 1954, in *Egernia whitii;* Tasmania, Australia.

*O. vanzolinii* Rego et Oliveira-Rodrigues 1965, in *Dryadophis bifossatus, Hemidactylus mabouia;* Brazil.

*O. whitentoni* Stellman 1939, in *Terrapene triunguis, Ctenosaura pectinata;* U.S., Mexico.

*O. zonuri* Baylis 1919, in *Zonurus tropidosternus, Gerrosaurus* sp.; Africa, Europe.

5a. **Reproductive systems doubled ..... *Echidnotaenia* Beveridge 1980. (Figure 489)**

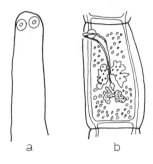
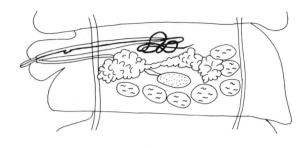

FIGURE 491. *Cleberia oligorchis* Rego 1967.

FIGURE 490. *Cycloskrjabinia taborensis* (Loewen 1934) Spasskii 1951. (a) Scolex; (b) proglottid.

*Diagnosis:* Strobila 46 to 100 mm long. Scolex and suckers unarmed. Proglottids craspedote, extended transversely. Longitudinal osmoregulatory canals paired. Transverse canal connects left and right ventral canals at posterior margin of each proglottid. Genital ducts cross osmoregulatory canals ventrally. Genitalia paired. Cirrus sac opens to genital atrium anterior to vagina. Internal and external seminal vesicles absent. Testes numerous, scattered throughout medulla. Seminal receptacle present. Ovaries situated in lateral quarters of proglottid medulla. Vitellaria greatly extended aporally, meeting in midline. Mehlis' gland spherical. Uterus absent. Eggs embedded singly in egg capsules throughout medullary parenchyma. Parasites of monotremes. Australia.

Type species: *E. tachyglossi* (Johnston 1913) Beveridge 1980 (syn. *Cittotaenia tachyglossi* Johnston 1913), in *Tachyglossus aculeatus;* Queensland, Australia.

**5b.    Reproductive systems single..................................................... 6**

**6a.    Testes separated into anterior and posterior groups.............................**
**.................................... *Cycloskrjabinia* Spasskii 1951. (Figure 490)**

*Diagnosis:* Proglottids longer than wide. Genital pores irregularly alternating. Cirrus pouch spherical. Testes in preovarian and postovarian fields. Seminal vesicles absent. Ovary and vitelline gland bilobed, median. Vagina posterior to cirrus pouch. Uterus not described; eggs scattered in parenchyma. Parasites of Chiroptera. North America, Czechoslovakia.

Type species: *C. taborensis* (Loewen 1934) Spasskii 1951 (syn. *Oochoristica taborensis* Loewen 1934), in *Lasiurus borealis;* U.S., *Myotis myotis;* Czechoslovakia.

**6b.    Testes not separated into anterior and posterior groups....................... 7**

**7a.    Genital pores unilateral ....................................................... 8**
**7b.    Genital pores alternating...................................................... 11**

**8a.    Testes few (4 to 9, usually 7).................. *Cleberia* Rego 1967. (Figure 491)**

*Diagnosis:* Strobila small, scolex unknown. Proglottids craspedote, wider than long. Genital pores unilateral. Cirrus armed with small spines. Testes few, posterior and lateral to ovary. Vagina posterior to cirrus pouch. Seminal receptacle present. Genital atrium in large papilla-like extension. Cirrus pouch reaching median of segment, slender, Vas deferens convoluted. Ovary median, bilobed, transversely elongated. Vitellarium compact, posterior to ovary. Uterus first a transverse tube, then replaced with egg capsules, each with a single

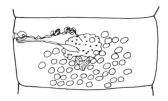

FIGURE 492. *Oschmarenia oklahomensis* (Perry 1934) Spasskii 1951.

egg. Pyriform apparatus not seen. Parasites of South American rodents. (This species may belong in Dipylidiinae.)

Type species: *C. oligorchis* Rego 1967, in *Agouti paca;* Brazil.

**8b. Testes numerous** ................................................................. 9

**9a. No testes anterior to ovary** ............ ***Oschmarenia* Spasskii 1951. (Figure 492)**

*Diagnosis:* Proglottids craspedote, wider than long. Genital pores unilateral. Cirrus pouch small. Seminal vesicles absent. Testes in single or double field behind or lateral to ovary. Ovary and vitelline gland median or submedian. Vagina posterior to cirrus pouch. Uterus? Eggs scattered singly in parenchyma. Parasites of Marsupialia, Mustelidae, Canidae; Burma, Africa, North America.

Type species: *O. incognita* (Meggitt 1927) Spasskii 1951 (syn. *Thysanotaenia incognita* Meggitt 1927), in *Macropus ruficollis;* Burma (?).

Other species:

*O. genettae* (Ortlepp 1937) Spasskii 1951 (syn. *Anoplocephala genettae* Ortlepp 1937), in *Genetta rubiginosa;* Africa.

*O. mephitis* (Skinker 1935) Spasskii 1951 (syn. *Oochoristica mephitis* Skinker 1935), in *Mephitis elongata, M. nigrans, M. varians, Urocyon cinereoargenteus;* U.S.

*O. okalahomensis* (Peery 1939) Spasskii 1951 (syn. *Oochoristica oklahomensis* Peery 1939), in *Spilogale interrupta, S. putorius, Mephitis mephitis;* North America.

*O. pedunculata* (Chandler 1952) Yamaguti 1959 (syn. *Oochoristica pedunculata* Chandler 1952), in *Mephitis* sp.; Minnesota, U.S.

*O. wallacei* (Chandler 1952) Yamaguti 1959 (syn. *Oochoristica wallacei* Chandler 1952), in *Spilogale interrupta;* Minnesota, U.S.

**9b. A few or many testes anterior to ovary** ....................................... 10

**10a. Ovary extending between dorsal and ventral osmoregulatory canals** ............
........................ ***Multicapsiferina* Fuhrmann 1921 (syn. *Zschokkea* Fuhrmann 1902, *Zschokkeella* Ransom 1909, *Linstowia* Zschokke of Fuhrmann 1901).**

*Diagnosis:* Genital pores unilateral. Genital ducts passing between osmoregulatory canals. Cirrus pouch small. Testes numerous, lateral, posterior, and anterior to ovary. Ovary poral, between poral osmoregulatory canals. Vitelline gland lateral or posterior to ovary. Vagina posterior to cirrus pouch. Seminal receptacle long. Uterus breaking into egg capsules, each with single egg. Parasites of Galliformes. Africa.

Type species: *M. linstowi* (Parona 1885) Fuhrmann 1921 (syn. *Taenia linstowi* Parona 1885, *Hymenolepis linstowi* [Parona 1885] Parona 1900, *Linstowia linstowi* [Parona 1885] Fuhrmann 1901, *Zschokkea linstowi* [Parona 1885] Fuhrmann 1902, *Zschokkeella*

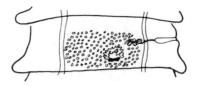

FIGURE 493. *Sobolevina otidis* (Meggitt 1927) Spasskii 1951.

FIGURE 494. *Tupaiataenia quentini* Schmidt et File 1977. Anterior edge of sucker, and mature proglottid.

*linstowi* [Parona 1885] Ransom 1909, *Inermicapsifer linstowi* [Parona 1885] Baer 1925), in *Numida ptilorhyncha;* West Africa.

**10b. Ovary medial to osmoregulatory canals .........................................
................................... Sobolevina Spasskii 1951. (Figure 493)**

*Diagnosis:* Proglottids craspedote, wider than long. Genital pores unilateral. Cirrus pouch small. Seminal vesicles absent. Testes numerous, in single field surrounding ovary. Median to osmoregulatory canals. Ovary and postovarian vitelline gland slightly poral. Uterus not described. Eggs lie singly in parenchyma. Parasites of Otidiformes. Egypt.

Type species: *S. otidis* (Meggitt 1927) Spasskii 1951) (syn. *Inermicapsifer otidis* Meggitt 1927), in *Otis houbara;* Egypt.

**11a. Anterior margins of suckers split into two opposable lappets ....................
................................... Tupaiataenia Schmidt et File 1977. (Figure 494)**

*Diagnosis:* Scolex lacking rostellum, hooks or apical organ. Suckers round, powerful, unarmed; anterior margin of each sucker split and produced into two apposable, muscular lappets. Opening into suckers is a longitudinal slit. Neck present, wider than scolex. Proglottids numerous, craspedote; immature segments wider than long, mature segments about as wide as long, gravid segments longer than wide. Osmoregulatory canals reticulated throughout proglottid. Muscle bundles feebly developed. One set of reproductive organs per segment. Genital pores irregularly alternating. Genital atrium well developed, lacking sphincter muscles, smaller than cirrus pouch. Genital ducts dorsal to osmoregulatory canals. Testes mainly postovarian, some lateral to ovary, between outer limits of osmoregulatory canals. Cirrus pouch ovoid, crossing outermost limits of poral osmoregulatory canals. Cirrus unarmed. External seminal vesicle absent, proximal ejaculatory duct slightly swollen to form internal seminal vesicle. Ovary median, with bilateral wings subdivided into slender lobes.

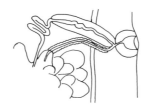

FIGURE 495. *Markewitschitaenia rodentinum* (Joyeux 1927) Sharpilo et Kornyushin 1975. Genital terminalia.

Vitelline gland posterior to ovary. Vagina posterior to cirrus pouch, distal two thirds surrounded by large unicellular glands. Distal end of vagina entering, with cirrus, into common hermaphroditic canal that then enters genital atrium. Seminal receptacle small but discrete in mature segments, becoming immensely swollen and persisting in gravid segments. Uterus develops as a thin-walled, reticular sac, which quickly fills and disappears, leaving developing eggs scattered throughout the medullary parenchyma. Oncospheres each surrounded by single, thick membrane. Parasites of tree shrews. Southeast Asia.

Type species: *T. quentini* Schmidt et File 1977, in *Tupaia glis;* Thailand.

**11b. Suckers simple................................................................ 12**

**12a. Cirrus absent. Genital ducts join to form common hermaphroditic duct opening into genital atrium.............................................................**
**..................*Markewitschitaenia* Sharpilo et Kornyushin 1975. (Figure 495)**

*Diagnosis:* Scolex lacking rostellum, hooks or apical organ. Suckers round, simple, unarmed. Neck present, wider than scolex. Proglottids numerous, acraspedote. Mature and gravid segments longer than wide. Osmoregulatory canals reticulated throughout proglottid. One set of reproductive systems per segment. Genital pores preequatorial, irregularly alternating. Genital atrium well developed. Genital ducts dorsal to osmoregulatory canals. Testes postovarian, between outer limits of osmoregulatory canals. Cirrus pouch elongate, crossing outermost limit of poral osmoregulatory canals. Functional cirrus absent. Seminal vesicles absent. Ejaculatory duct swollen proximally. Ovary median with bilateral wings subdivided into slender lobes. Vitellarium posterior to ovary, lobated. Vagina posterior to cirrus pouch. Distal end of vagina entering, with male duct, into common hermaphroditic canal that then enters genital atrium. Seminal receptacle absent. Uterus breaks into numerous egg capsules, each with single egg. Parasites of rodents (snake accidental host); Africa, Russia.

Type species: *M. rodentinum* (Joyeux 1927) Sharpilo et Kornyushin 1975 (syn. *Mathevotaenia rodentinum* [Joyeux 1927] Spasskii 1951, *Oochoristica erinacei* var. *rodentinum* Joyeux 1927), in *Meriones shawi, Mus musculus, Elaphe dione;* Algeria, Russia.

**12b. Cirrus present, hermaphroditic canal absent .................................. 13**

**13a. Testes anterior to ovary.......................................................**
**.......................*Sinaiotaenia* Wertheim et Greenberg 1971. (Figure 496)**

*Diagnosis:* Scolex lacking rostellum, armature, or apical organ. Suckers simple. Neck present, wider than scolex. Immature proglottids wider than long, mature proglottids somewhat longer than wide, gravid proglottids very long, up to ten times longer than wide.

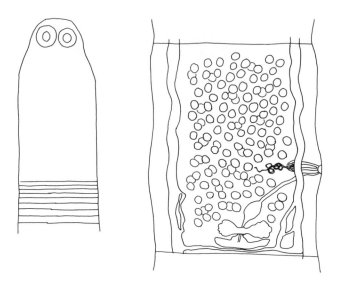

FIGURE 496. *Sinaiotaenia witenbergi* Wertheim et Greenberg 1971. Scolex and mature proglottid.

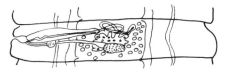

FIGURE 497. *Linstowia semoni* (Zschokke 1846) Zschokke 1899.

Osmoregulatory canals simple. Genital atrium shallow, simple. Genital pores irregularly alternating, in middle third of segment. Genital ducts pass between osmoregulatory canals. Cirrus pouch pyriform, exceeding poral osmoregulatory canals. Seminal vesicles absent. Testes numerous, medullary, anterior to ovary. Ovary small, bilobed, near posterior margin of segment; vitellarium transversely elongated, postovarian. Vagina posterior to cirrus pouch. Seminal receptacle prominent. Eggs embedded singly in medullary parenchyma. Parasites of gerbilline rodents. Israel, Sinai.

Type species: *S. witenbergi* Wertheim et Greenberg 1971, in *Gerbillus dasyurus, G. gerbillus, Meriones calurus, M. crassus;* Israel, Egypt.

**13b. Testes posterior and lateral to ovary ............................................. 14**

**14a. Genital ducts pass ventral of osmoregulatory canals.............................**
**............................................ Linstowia Zschokke 1899. (Figure 497)**

*Diagnosis:* Proglottids numerous, craspedote, wider than long. Genital pores irregularly alternating. Cirrus pouch long or short. Seminal vesicles absent. Testes numerous in continuous, extensive field. Genital ducts passing ventral to osmoregulatory canals. Ovary and postovarian vitelline gland median. Vagina posterior or ventral to cirrus pouch. Seminal receptacle present. Uterus first a transverse tube, then breaking into egg capsules, each with one egg. Pyriform apparatus absent. Parasites of monotremes and marsupials. Australia, Brazil.

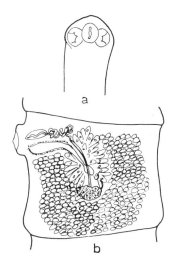
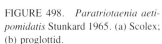

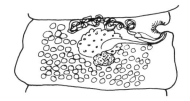

FIGURE 499. *Atriotaenia incisa* (Railliet 1899) Spasskii 1951.

FIGURE 498. *Paratriotaenia aetipomidatis* Stunkard 1965. (a) Scolex; (b) proglottid.

Type species: *L. echidnae* (Thompson 1893) Zschokke 1899 (syn. *Taenia echidnae* Thompson 1893), in *Tachyglossus aculeata, Perameles obesula;* Australia.

Other species:

*L. iheringi* Zschokke 1904 (syn. *L. brasiliensis* Janicki 1904, *Oochoristica iheringi* [Zschokke 1904] Baer 1924, *O. brasiliensis* [Janicki 1904] Baer 1924), in *Didelphys tristriata, Marmosa elegans, Metachirus opossum, Metachirops quica, Peramys americana;* Brazil.

*L. semoni* Zschokke (1896) Zschokke 1899 (syn. *Taenia semoni* Zschokke 1896, *Linstowia semoni* var. *acanthocirrus* Neblin 1917), in *Perameles obesula, P. nasuta, P. macrura;* Australia.

*L. tasmanica* Spasskii 1951 (for *L. echidnae* [Thompson 1893] of Kerr 1935), in *Tachyglossus setosa;* Tasmania, Australia.

**14b. Genital ducts pass between or dorsal to osmoregulatory canals ............... 15**

**15a. Vaginal sphincter present .......... *Paratriotaenia* Stunkard 1965. (Figure 498)**

*Diagnosis:* Scolex simple, neck present. Proglottids about as long as wide, slightly craspedote. Genital pores irregularly alternating. Genital atrium large, protrusible. Cirrus pouch small. Seminal vesicles absent. Testes numerous, in continuous field lateral and posterior to ovary. Lobated ovary and postovarian vitelline gland median. Vagina posterior to cirrus pouch, with powerful distal sphincter. Seminal receptacle present. Uterus not observed. Eggs single in parenchyma; pyriform apparatus absent. Parasites of Primates. South America.

Type species: *P. oedipomidatis* Stunkard 1965, in *Oedipomidas oedipus.*

**15b. Vaginal sphincter absent ..................................................... 16**

**16a. Genital atrium forms large, sucker-like organ ......................................**
**.................................... *Atriotaenia* Sandground 1926. (Figure 499)**

*Diagnosis:* Proglottids wider than long, craspedote. Genital pores irregularly alternating. Genital atrium may form large, sucker-like organ. Genital ducts passing dorsal to osmore-

FIGURE 500. *Mathevotaenia symmetrica* (Baylis 1927) Akhumian 1946.

gulatory canals that often are reticular. Testes mainly posterior and lateral to ovary, occasionally a few anterior. Seminal vesicles absent. Ovary and vitelline gland median. Vagina posterior to cirrus pouch, lacking sphincter. Seminal receptacle present. Uterus not observed. Eggs scattered singly in parenchyma. Parasites of mustelid and procyonid carnivores. Europe, Asia, North and South America.

Type species: *A. sandgroundi* (Baer 1935) Spasskii 1951 (syn. *Oochoristica sandgroundi* Baer 1935, *O. parva* [Sandground 1926] Meggitt 1934, *Atriotaenia parva* Sandground 1926), in *Nasua nasua;* Brazil.

Other species:

*A. baltazardi* Quentin 1967, in *Galea spizii;* Brazil.

*A. incisa* (Railliet 1899) Spasskii 1951 (syn. *Oochoristica incisa* Railliet 1899, *O. figurata* Meggitt 1927), in *Meles meles, Cricidura caerulea* (shrew); Europe, Asia.

*A. procyonis* (Chandler 1942) Spasskii 1951 (syn. *Oochoristica procyonis* Chandler 1942), in *Procyon lotor;* U.S.

### 16b. Genital atrium simple ............ *Mathevotaenia* Akhumian 1946. (Figure 500)

*Diagnosis:* Proglottids craspedote, wider than long. Genital pores irregularly alternating. Genital ducts passing between or dorsal to osmoregulatory canals. Genital atrium is not sucker-like. Cirrus present. Testes numerous, lateral, and posterior to ovary. Seminal vesicles absent. Ovary and postovarian vitellarium median. Vagina posterior to cirrus pouch. Seminal receptacle absent or present. Uterus replaced by thin egg capsules, each with single egg. Parasites of rodents, marsupials, insectivores, viverrids, mustelids, lemurs, monkeys, bird(?), edentates. Europe, Asia, Japan, North and South America, Africa, Panama, Tasmania.

Type species: *M. symmetrica* (Baylis 1927) Akhumian 1946 (syn. *Catenotaenia symmetrica* Baylis 1927, *Oochoristica ratti* Yamaguti et Miyata 1937, *O. symmetrica* [Baylis 1927] Meggitt 1934, *O. lemuris* [Beddard 1916] Meggitt 1934), in *Mus musculus, Rattus rattus, R. norvegicus, Cricetulus migratorius, Rattus alexandrinus, Nycticebus tardigradus;* Armenia, India, Japan, U.S.

Other species:

*M. aegyptica* Mikhail et Fahmy 1968, in *Jaculus orientalis;* Egypt.

*M. aethechini* Dollfus 1954, in *Aethechinus algirus;* Morocco.

*M. antrozoi* (Voge 1954) Yamaguit 1959 (syn. *Oochoristica antrozoi* Voge 1954, *Atriotaenia antrozoi* [Voge 1954] Saxena et Baugh 1978), in *Antrozous pallidus;* California, U.S.

*M. bivittata* (Janicki 1904) Yamaguti (1959 (syn. *Oochoristica bivittata* Janicki 1904, *Linstowi [Opossumia] bivitta* [Janicki 1904] Spasskii 1951), in *Caluromys langier, C. philander, Didelphys aurita, D. marsupialis, D. paraguayensis, Marmosa elegans, M. murina, Metachirus medicaudatus, M. opposum;* South and Central America.

*M. brasiliensis* Kugi et Sawada 1970, in *Saimiri sciureus;* Brazil.

*M. cruzsilvai* Mendonça 1981, in *Macaca irus;* Lisbon Zoo.

*M. cubana* Zdzitowiecki et Rutkowska 1980, in *Phyllonycteris poeyi, Erophylla sezekorni;* Cuba.

*M. deserti* (Millemann 1955) Yamaguti 1959 (syn. *Oochoristica deserti* Millemann 1955, in *Dipodomys merriami, Citellus leucurus;* U.S.

*M. didelphidis* (Rudolphi 1819) Spasskii 1951 (syn. *Taenia didelphydis* Rudolphi 1819, *Oochoristica didelphydis* [Rudolphi 1819] Janicki 1904, *O. murina* Zschokke 1904), in *Marmosa elegans, M. murina;* South America.

*M. dipi* (Parona 1900) Yamaguti 1959 (syn. *Taenia dipi* Parona 1900, *Skrjabinia dipi* [Parona 1900] Lopez-Neyra 1954), in *Dipus aegyptius, Jaculus jaculus;* North Africa.

*M. erinacei* (Meggitt 1920) Spasskii 1951 (syn. *Oochoristica erinacei* Meggitt 1920, *O. amphisbeteta* Meggitt 1924, *Mathevotaenia amphisbeteta* [Meggitt 1924] Spasskii 1951), in *Herpestes auropunctatus, H. albopunctatus, Erinaceus algirus, Elephantulus, Meriones, Mus;* Africa, Europe, India, Burma.

*M. herpestis* (Kofend 1917) Spasskii 1951 (syn. *Oochoristica herpestis* Kofend 1917), in *Galerella sanguineus, Atelerix spiculus, A. spinifex;* Sudan.

*M. hardoiensis* Johri 1961, in *Herpestes javanicus;* India.

*M. ichneumontis* (Baer 1924) Spasskii 1951 (syn. *Oochoristica ichneumontis* Baer 1924), in *Galerella gracilis, Myonax cauni, Putorius* sp.; Africa.

*M. immatura* Rego 1963, in *Glossophaga soricina;* Brazil.

*M. kerivoulae* (Prudhoe et Manger 1969) comb. n. (syn. *Oochoristica kirivoulae* Prudhoe et Manger 1969), in bats; Malaysia.

*M. marmosae* (Beddard 1914) Spasskii 1951 (syn. *Oochoristica marmosae* Beddard 1914, *Linstowia [Paralinstowia] iheringi* [Zschokke 1904] Baer 1927 in part), in *Marmosa elegans;* Brazil.

*M. megastoma* (Diesing 1850) Spasskii 1951 (syn. *Taenia megastroma* Diesing 1850, *Oochoristica megastoma* [Diesing 1850] Zschokke 1905, *Bertiella fallax* Meggitt 1927), in *Cebus, Alouata, Brachyteles, Callithrix, Hapale, Cebus, Senicebus;* South America.

*M. nyctophili* Hickman 1954, in *Nyctophilus geoffroyi;* Tasmania, Australia.

*M. ornithis* Saxena et Baugh 1978, in *Passer domesticus*(?); India.

*M. paraechinis* Nama 1975, in *Paraechinus micropus;* India

*M. paraguayae* Schmidt et Martin 1978, in *Euphractus sexcinctus;* Paraguay.

*M. parva* (Janicki 1904) Spasskii 1951 (syn. *Davaineia parva* Janicki 1904, *Oochoristica parva* [Janicki 1904] Baer 1935); in *Erinaceus* sp.

*M. pennsylvania* (Chandler et Melvin 1951) Yamaguti 1959 (syn. *Oochoristica pennsylvanica* Chandler et Melvin 1951), in *Blarina brevicauda;* U.S.

*M. sanchovensis* Nama et Khichi 1973, in *Herpestes* sp.; India.

*M. skrjabini* Spasskii 1949, in *Erinaceus auritus;* Central Asia.

*M. surinamensis* (Cohn 1902) Spasskii 1951 (syn. *Taenia surinamensis* Cohn 1902, *T. acephala* Creplin et Cohn 1902, *Oochoristica surinamensis* [Cohn 1902] Cohn 1903), in *Dasypus novemcinctus, Dasypus* sp., *Didelphys aurita, D. marsupialis, Priodontes gigas;* South America.

*M. tetragonocephala* (Bremser in Diesing 1856) Spasskii 1951 (syn. *Taenia tetragonocephala* Bremser in Diesing 1856, *T. tetragonacephalus* [Bremser 1856] Macalistes 1874, *Oochoristica wageneri* Janicki 1904), in *Myrmecophaga tridactyla, Tamandua tetradactyla;* South America.

## KEY TO THE GENERA IN INERMICAPSIFERINAE

1a. **Ovary and vitelline gland median or nearly median** ............................ 2
1b. **Ovary and vitelline gland definitely poral** ........................................ 3

2a. **Genital pores unilateral** ............. *Thysanotaenia* Beddard 1911. (Figure 501)

**464** CRC Handbook of Tapeworm Identification

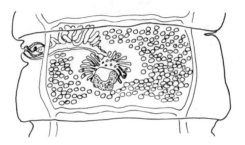

FIGURE 501. *Thysanotaenia lemuris* Beddard 1911.

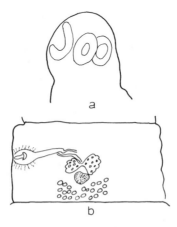

FIGURE 502. *Megacapsula leiperi* Wahid 1961. (a) Scolex; (b) proglottid.

*Diagnosis:* Proglottids craspedote, wider than long. Genital pores unilateral, preequatorial. Cirrus pouch pyriform. Cirrus armed. Osmoregulatory canals ventral to genital ducts. Testes numerous (100 to 190), lateral and posterior to ovary; sometimes anterior to ovary on antiporal side. Vas deferens wide, convoluted. Ovary about median, bialate. Vitelline gland compact, postovarian. Vagina opening behind cirrus. Seminal receptacle proximal. Uterus first a transverse tube, rapidly breaking down into egg capsules, each with several eggs. Parasites of lemur, man. Malagasy, Cuba, South America.

Type species: *T. lemuris* Beddard 1911, in *Lemur varius, L. macao;* Madagascar.
Other species:
 *T. cubensis* (Kouri 1938) Spasskii 1951 (syn. *Raillietina cubensis* Kouri 1938, *Inermicapsifer cubensis* [Kouri 1938] Kouri 1943), in Cuba, Chile.

**2b. Genital pores irregularly alternating ... *Megacapsula* Wahid 1961. (Figure 502)**

*Diagnosis:* Scolex not separated from neck by constriction. Neck short. Proglottids acraspedote, broader than long. Genital pores irregularly alternating. Genital atrium small. Genital ducts (?). Cirrus pouch robust, reaching poral osmoregulatory canals. Cirrus not described. Inner and outer seminal vesicles absent. Testes (26 to 30) in small postovarian field. Vagina posterior to cirrus pouch. Seminal receptacle present. Ovary median or very slightly poral, bilobed. Vitelline gland median, postovarian. Uterus first a transverse sac, then replaced by egg capsules each containing several eggs. Parasites of lizards. Africa.

Type species: *M. leiperi* Wahid 1961, in *Agama cyanogaster;* Africa.

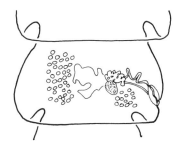

FIGURE 503. *Inermicapsifer settii* Janicki 1910.

3a. No testes poral to vitelline gland....................*Metacapsifer* Spasskii 1951.

*Diagnosis:* Proglottids wider than long, except for gravid ones that may be as long as wide; craspedote. Genital pores unilateral. Testes few (6 to 8) in antiporal half of segment. Cirrus pouch small. Ovary and postovarian vitelline gland poral. Uterus breaks down into egg capsules, each with several eggs. Parasites of Rodentia. Africa.

Type species: *M. aberratus* (Baer 1924) Spasskii 1951 (syn. *Inermicapsifer abberatus* Baer 1924), in *Mus moggi;* South Africa.

3b. Testes present poral to vitelline gland............................................. 4

4a. Testes absent anterior of ovary on poral side......................................
............................ *Inermicapsifer* Janicki 1910 (syn. *Arhynchotaenia* Pagenstecher 1877 preoccupied, *Hyracotaenia* Beddard 1912). (Figure 503)

*Diagnosis:* Proglottids craspedote, wider than long, except for gravid ones. Genital pores unilateral. Cirrus pouch small. Genital ducts passing between osmoregulatory canals. Testes numerous, in single or double field but absent anterior to vagina on poral side. Ovary and postovarian vitelline gland poral. Vagina posterior to cirrus pouch. Seminal receptacle small. Uterus breaks down into egg capsules each containing several eggs. Parasites of Rodentia, Hyracoidea, Pholidota, and accidentally in man. Africa, Syria, Brazil, Sinai.

Type species: *I. hyracis* (Rudolphi 1808) Janicki 1910 (syn. *Taenia hyracis* Rudolphi 1810, *Arhynchotaenia critica* Pagenstecher 1878, *Anoplocephala hyracis* [Rudolphi 1810] [Moniez 1891], *Taenia [Anoplocephala] ragazzi* Setti 1891, *Anoplocephala hyracis* var. *hepatica* Nassonow 1897, *Inermicapsifer capensis* Beddard 1912, *Hyracotaenia hyracis* Beddard 1912, *Zschokkeella hyracis* [Beddard] Beddard 1912, *Z. capensis* [Beddard] Beddard 1912, *Inermicapsifer criticus* [Pagenstecher 1878] Bischoff 1913), in *Procavia abissinicus, P. capensis, P. syricus;* Africa.

Other species:

*I. abyssinicus* Bischoff 1912 (syn. *I. interpositus* Janicki of Baer 1927), in *Procavia* sp.; East Africa.

*I. angolensis* Voelker 1960, in *Procavia capensis;* Angola.

*I. apospasmation* Bischoff 1912 (syn. *I. parvulus* Bischoff 1912), in *Procavia burcei, Procavia* sp.; Africa.

*I. arvicanthidis* (Kofend 1917), in *Arvicanthis testicularis, Mus moggi, M. rufinus, Golunda campanae, Otomys irroratus,* human, *Rattus* sp., *Steatomys pratensis, Hybomys univittatus, Mastomys erythroleucos, Taterona kempi;* Africa.

*I. congolensis* Mahon 1954, in *Cricetomys gambianus;* Belgian Congo.

*I. gondokorensis* (Klaptocz 1906) Bischoff 1913, in *Procavia slatini;* Africa.

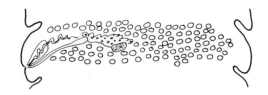

FIGURE 504. *Pericapsifer pagenstecheri* (Setti 1897) Spasskii 1951.

*I. guinensis* (Graham 1908) Baer 1924 (syn. *Davainea guinensis* Graham 1908, *Thysanosoma gambianum* Beddard 1911, *Thysanotaenia gambianum* Beddard 1911, *Zschokkeela gambianum* Beddard 1912, *Inermicapsifer zanzibarensis* Meggitt 1921, *Zschokkeella guinensis* [Graham 1908] Southwell et Maplestone 1921), in *Cricetomys gambianum, Dasymys nudipes, Mus damarensis, Rattus rattus, R. norvegicus, Mastomys coucha, Rhabdomys pumilio, Pelomys flater;* Africa.

*I. interpositus* Janicki 1910, in *Procavia capensis, P. syriaca;* Africa, Syria.

*I. isometra* Voelker 1960, in *Procavia capensis;* Angola.

*I. leporis* Ortlepp 1938, in *Lepus capensis;* South Africa.

*I. lopas* Bischoff 1912, in *Heterohyrax brucei, Procavia* sp.; Africa.

*I. muricola* (Baylis 1915) Spasskii 1951 (syn. *Zschokkeella muricola* Baylis 1915), in "mouse" (*Rattus rattus?*).

*I. prionodes* Bischoff 1912 (syn. *I. p.* var. *intermedia* Bischoff 1912), in *Procavia brucei;* Africa.

*I. remotus* (Linstow 1905), in *Cercopithecus pyrrhonotus;* Brazil. (Identity doubtful).

*I. rhodesiensis* Mettrick 1959, in *Manis temminckii;* Southern Rhodesia.

*I. setti* Janicki 1910, in *Procavia capensis;* Africa.

*I. sinaitica* Spasskii 1951 (syn. *I. interpositus* var. *sinaitica* Bischoff 1912), in *Procavia syriaca;* Asia Minor.

**4b.  Testes present anterior to ovary on poral side .....................................
............................................ *Pericapsifer* Spasskii 1951. (Figure 504)**

*Diagnosis:* Proglottids wider than long, craspedote. Genital pores unilateral. Cirrus pouch small. Testes numerous, distributed throughout medulla, both before and behind vagina. Ovary and postovarian vitelline gland poral. Vagina posterior to cirrus pouch. Seminal receptacle present. Uterus forming egg capsules, each with several eggs. Parasites of Hyracoidea. Africa.

Type species: *P. pagenstecheri* (Setti 1897) Spassky 1951 (syn. *Hydrocotaenia procaviae* Beddard 1912, *Inermicapsifer paronae* Bischoff 1912, *Anoplocephala pagenstecheri* Setti 1897, *Inermicapsifer pagenstecheri* [Setti 1897] Janicki 1910), in *Procavia capensis, Heterohyrax brucei;* Africa.

Other species:

*P. norhalli* (Baer 1924) Spasskii 1951 (syn. *Inermicapsifer norhalli* Baer 1924), in *Procavia capensis, Procavia* sp.; Africa.

*P. tanganyikae* (Baer 1933) Spasskii 1951 (syn. *Inermicapsifer tanganyikae* Baer 1933), in *Procavia landemanni;* Tanganyika, Africa.

## SUBCLASS CESTODARIA

Members of this small but significant group are parasites of primitive fishes and of Australian turtles. Cestodarians have a single set of male and female reproductive organs. No scolex is present, and there is no indication of segmentation. Some species are quite large for monozoic tapeworms. No complete life cycle is known. The first larval stage (the lycophora) hatching from the egg has ten hooks on its anterior end, compared with six on members of the subclass Eucestoda. They may be quite common, infecting nearly every fish of certain species examined. Often only two are present in a host, suggesting an exclusion of competing individuals.

These worms are of no known medical or economic importance. Their phylogenetic position remains problematical, although several lucid studies have elevated them to levels of separate classes, detached from the Cestoidea. Thus the Amphilinidea and Austramphilinidae are regarded as forming the class Amphilinida by Dubinina,[835] while the Gyrocotylidea is aligned with the Class Monogenea by Llewellyn.[1905] I am of the opinion that both groups, while not closely related to one another, still form a cohesive and definitely tapeworm assemblage apart from the Eucestoda. Therefore, I retain them as a subclass with the class Cestoidea.

## KEY TO THE ORDERS IN SUBCLASS CESTODARIA

1a. Genital pores at or near posterior end. Uterine pore anterior .................. ................................................... **Amphilinidea Poche 1922.**

*Diagnosis:* Body flattened, elongated, with indistinct hold-fast mechanism at anterior end. Genital pores near posterior end. Testes usually in two lateral, preovarian fields. Ovary posterior. Vitelline glands lateral. Uterus N-shaped or looped. Uterine pore near anterior end. Parasites of fishes and turtles (p.467)

1b. Genital and uterine pores in anterior fourth of body........................... ................................................... **Gyrocotylidea Poche 1926.**

*Diagnosis:* Body flattened, elongated, with indistinct holdfast mechanism at anterior end. Posterior end forming a crenulated rosette or a long, slender cylinder. Male pore anterior, ventral. Testes anterior, in two lateral fields. Ovary posterior. Vaginal pore anterior, dorsal. Vitellaria follicular, lateral. Uterine pore anterior, ventral. Parasites of marine fishes. (p.472)

## ORDER AMPHILINIDEA

## KEY TO THE FAMILIES IN AMPHILINIDEA

1a. Uterus N-shaped, with terminal ascending limb lateral ....................... ............................................. **Amphilinidae Claus 1879. (p.468)**

*Diagnosis:* Small proboscis-like or sucker-like apical organ present at anterior end. Body flat, elongate, with rounded margins. Testes numerous, in two lateral fields or scattered throughout body. Cirrus and vaginal pores at or near posterior end of body, opening separately or in common atrium. Ovary posterior. Vitellaria follicular, in lateral fields. Uterus slender, N-shaped, with descending limb median. Uterine pore anterior. Osmoregulatory canals two, with common posterior pore. Parasites of coelomic cavity of Teleostei and Chondrostei.

Type genus: *Amphilina* Wagener 1858.

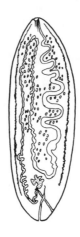

FIGURE 505. *Amphilina foliacea* (Rudolphi 1819) Wagener 1858.

**1b.    Uterus looped, with terminal ascending limb median............................
........................................Austramphilinidae Johnston 1931. (p.471)**

*Diagnosis:* Rostellum-like organ with sheath at anterior end. Body flat, slender. Testes numerous, mostly lateral. Cirrus and vaginal pores at or near posterior end of body, opening separately. Ovary posterior. Vitellaria lateral. Seminal receptacle large, median. Uterus with terminal ascending limb median. Uterine pore anterior. Parasites of Chelonia and fishes.
Type genus: *Austramphilina* Johnston 1931.

## KEY TO THE FAMILIES IN AMPHILINIDEA

**1a.    Vaginal pore marginal .................. *Amphilina* Wagener 1958. (Figure 505)**

*Diagnosis:* Body leaf-shaped, flattened. Anterior end bluntly pointed, with sucker-like depression. Posterior end rounded or slightly concave. Testes numerous, scattered throughout body. Cirrus long, slender, armed with hooks at the end, opening at posterior end of body. Ovary lobated, irregular, in posterior third of body. Vitellaria lateral. Vagina crossing vas deferens, opening on margin of body near posterior end. Accessory seminal receptacle absent. Descending and first ascending limbs of uterus mostly on same side of body. Uterine pore near anterior end. Parasites of sturgeons. Europe, Japan, North America.
   Type species: *A. foliacea* (Rudolphi 1819) Wagener 1859 (syn. *Monostomum foliaceum* Rudolphi 1819, *A. neritina* Selensky 1874), in Acipenseridae; Europe, Russia.
   Other species:
      *A. japonica* Goto et Ishii 1936 (syn. *A. bipunctata* Riser 1948), in Acipenseridae; Japan, North America.

**1b.    Vaginal pore medial............................................................. 2**

**2a.    Seminal receptacle absent...................................................... 3**
**2b.    Seminal receptacle present..................................................... 4**

**3a.    Cirrus opening at posterior end of body .........................................
........................................... *Gephyrolina* Poche 1926. (Figure 506)**

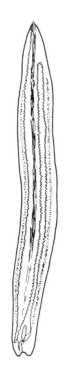

FIGURE 506. *Gephrolina paragonopora* Woodland 1923.

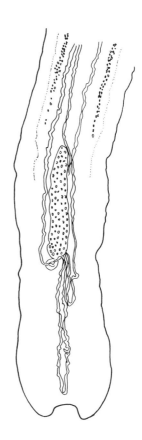

FIGURE 507. *Hunteroides mystei* Johri 1959. Posterior end.

*Diagnosis:* Anterior end bluntly pointed, with protrusible proboscis. Posterior end rounded except for notches at genital pores. Body elongate, flattened. Testes in lateral fields. Cirrus short, unarmed, opening at posterior end. Ovary elongate, not lobated, near posterior end of body. Vitellaria lateral. Vagina not crossing vas deferens, opening dorsal to cirrus at posterior end of body. Accessory seminal receptacle absent. Descending limb of uterus mostly on same side of body as first ascending limb. Uterine pore anterior, beside proboscis. Eggs with short polar processes. Parasites of siluroid fishes. India.

Type species: *G. paragonopora* (Woodland 1923) Poche 1926 (syn. *Amphilina paragonopora* Woodland 1923), in *Macrones aor, M. seenghala, Bagarius yarreli;* India.

**3b.   Cirrus opening at level of posterior end of ovary ..................................
................................................ Hunteroides Johri 1959. (Figure 507)**

*Diagnosis:* Scolex well-developed, highly muscular, with constriction separating it from rest of body. Body large (60 mm) mostly covered with glandular cells. Column of muscles continue from scolex deep into body, bifurcating most of the way. Vitellaria in two narrow lateral fields, extending posteriorly to ovary, apparently in cortical parenchyma. Ovary an elongate mass, medullary, in longitudinal axis of body. Vaginal opening slightly anterior to cirrus opening. Descending limb of uterus median. Uterine pore apparently anterior. Vagina makes a long posterior loop almost to posterior margin of body. Cirrus sac oval, posterior to ovary, slightly left of median line. Testes extend from 6 mm behind scolex to just in front of ovary, lateral to uterine loops, medial to vitellaria. Parasites of siluroid fish. India.

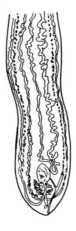

FIGURE 508. *Gigantolina magna* Southwell 1915. Posterior end.

Type species: *H. mystei* Johri, 1959, in *Mystus seenghala;* India. (Considered a synonym of *Gephyrolina paragonopora* by Dubinina (1982)).

**4a. Vaginal pore single** ........................ ***Gigantolina* Poche 1922. (Figure 508)**

*Diagnosis:* Both ends blunt. Body very long (38 cm), flattened. Cuticle with reticular sculpturing. Testes very numerous, in lateral fields. Cirrus short, opening at posterior end of body. Ovary bilobed, each lobe dendritic. Vitellaria lateral. Vagina opens dorsally, anterior to cirrus, with distal sphincter. Large seminal receptacle present. Descending limb of uterus median, ascending limbs lateral. Uterine pore ventral, near anterior end. Eggs lacking polar processes. Parasites of marine teleosts. Sri Lanka. (Dubinina[835] considers this genus to represent a distinct family, Gigantolinidae Dubininina 1982).

Type species: *G. magna* (Southwell 1915) Poche 1922 (syn. *Amphilina magna* Southwell 1915), *Diagramma crassispinum;* Sri Lanka.

**4b. Vaginal pore double** .................................................................. 5

**5a. Seminal receptacle about one third as long as body** ..............................
.................................................... ***Schizochoerus* Poche 1922.**

*Diagnosis:* Body long, narrow, flattened. Ends bluntly pointed. Retractile proboscis on anterior end. Testes numerous, lateral. Cirrus opens at posterior end of body. Ovary round, near posterior end. Vitellaria lateral. Vagina with double opening anterior to cirrus; sphincter absent. Seminal receptacle about one third body length. Descending limb of uterus on same side of body as last ascending limb. Uterine pore near anterior end. Eggs lacking polar processes. Parasites of teleosts. Brazil. (Poche[2645] and Dubinina[835] consider this genus to represent a distinct family, Schizochoeridae Poche 1922).

Type species: *S. liguloides* (Diesing 1850) Poche 1922 (syn. *Monostomum liguloideum* Diesing 1850, *Amphilina liguloidea* Monticelli 1892), in *Arapaima gigas;* Brazil.

**5b. Seminal receptacle about one sixth as long as body** .............................
............................................ ***Nesolecithus* Poche 1922. (Figure 509)**

*Diagnosis:* Body leaf-shaped, flattened, with rather pointed ends. Proboscis with short

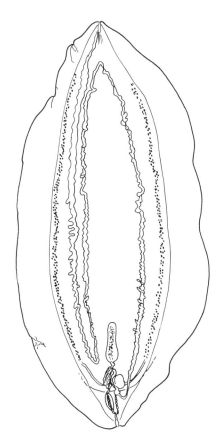

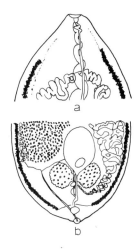

FIGURE 510. *Gyrometra albotaenia* Yamaguti 1954. (a) Anterior end; (b) posterior end.

FIGURE 509. *Nesolecithus africanus* Dönges et Harder 1966.

retractor muscle. Testes numerous, lateral. Cirrus opens at posterior end of body. Ovary round, near posterior end of body. Vitellaria lateral. Vagina with double opening slightly anterior to cirrus pore. Seminal receptacle one twelfth to one sixth length of body. Uterus with descending limb on same side of body as ascending limb. Uterine pore at anterior end of body. Parasites of teleosts. Brazil, Africa.

Type species: *N. janicki* Poche 1922 (syn. *Amphilina liguloidea* Janicki 1908), in *Arapaima gigas;* Brazil.

Other species:

*N. africanus* Dönges et Harder 1966, in *Gymnarchus niloticus;* Nigeria.

## KEY TO THE GENERA IN AUSTRAMPHILINIDAE

1a. **Vaginal pore separated from male pore** ............................................
............................................ **Gyrometra Yamaguti 1954. (Figure 510)**

*Diagnosis:* Body long, ribbon-like. Anterior end conical, with apical pit; posterior end rounded, with button-like papilla. Testes, small, numerous, in two narrow lateral bands beginning at different levels shortly behind anterior end and ending at level of ovary or anterior to it. Male pore on posterior papilla. Ovary bilobed, near posterior end. Vaginal pore middorsal, slightly anterior to male pore. Vagina with distal sphincter. Very large seminal receptacle anterior to ovary, with small accessory receptacle ventral near its posterior end. Vitellaria in two lateral narrow bands from near anterior end to level of vaginal pore.

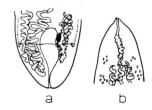

FIGURE 511. *Austramphilina elongata* Johnston 1931. (a) Posterior end; (b) anterior end.

Terminal ascending limb of uterus median, laterally coiled, opening into apical pit along with frontal glands. Parasites of coelom of pristopomid fishes. Macassar, Celebes, Indian Ocean.

Type species: *G. albotaenia* Yamaguti 1954, in *Diagramma* sp., Celebes. (Dubinina[835] considers this species to be a synonym of *Gigantolina magna).*

Other species:

    *G. kunduchi* Khalil 1977, in *Plectorhynchus pictus;* Indian Ocean. *(Dubinina*[835] considers this species to be a synonym of *Gigantolina magna.)*

**1b.    Vaginal pore joined with male pore ................................................**
**........................................................................ *Austramphilina***
**Johnston 1931 (syn. *Kosterina* Ihle et Ihle-Landenberg 1932). (Figure 511).**

*Diagnosis:* Body long, thin, pointed on anterior end, with apical pit; rounded at posterior end. Tegument pitted. Testes numerous, lateral, extending from near where median limb of uterus crosses other limbs to just anterior to level of ovary. Male genital pore at posterior end of body. Ovary compact, not bilobed, near posterior end of body at base of seminal receptacle. Vitellaria in narrow lateral bands. Vagina opening together with male pore. Distal end of vagina ciliated, with diverticulum halfway along its length. Seminal receptacle very large, with two small accessory receptacles. Terminal ascending limb of uterus median, with lateral loops, opening at base of apical pit. Parasites of coelom of freshwater tortoise. Australia.

Type species: *A. elongata* Johnston 1931. (syn. *Kosterina kuiperi* Ihle et Ihle-Landenberg 1932), in *Chelodina longicollis;* Australia.

# ORDER GYROCOTYLIDEA

## DIAGNOSIS OF THE ONLY FAMILY IN GYROCOTYLIDEA

### Gyrocotylidae Benham 1901

*Diagnosis:* Anterior end provided with holdfast mechanism. Posterior end with cylindrical or funnel-like opening that may be frilled. Body elongate, may have crenulated margins. Testes lateral, anterior, cirrus pore ventral, submedian, near anterior end. Ovary follicular, posterior to uterus. Vitellaria follicular, lateral. Uterine pore preequatorial, ventral. Eggs operculate, unembryonated. Parasites of Chondrichthyes, mainly Holocephali.

Type genus: *Gyrocotyle* Diesing 1850.

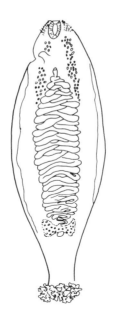

FIGURE 512. *Gyrocotyle rugosa* Diesing 1850.

## KEY TO THE GENERA IN GYROCOTYLIDAE

**1a. Body margins crenulated, posterior end with ruffle ............................
....... Gyrocotyle Diesing 1850 (syn. *Crobylophorus* Kroyer 1852). (Figure 512)**

*Diagnosis:* Body elongate, flattened, with spinous cuticle and crinkled margins. Anterior end with cone containing sucker-like cavity. Posterior end with rosette-like adhesive organ, variably crenulated according to state of contraction and species differences. Testes mainly lateral, in anterior part of body. Male genital pore anterior, submedial, ventral. Ovary midequatorial. Vitellaria lateral, anterior to ovary, confluent posterior to ovary. Vaginal pore anterior, submedial, dorsal. Vagina long, mainly median. Seminal receptacle present. Uterus preovarian, median with lateral loops, large uterine sac present. Uterine pore midventral, posterior to male pore. Parasites of Holocephali, Selachii. Cosmopolitan.

Type species: *G. rugosa* Diesing 1850, in *Callorhynchus antarcticum, C. milii;* nearly cosmopolitan.

Other species:

*G. fimbriata* Watson 1911, in *Hydrolagus colliei, Chimaera monstrosa;* North and South Pacific.

*G. maxima* Macdonagh 1927, in *Mustelus asterias;* Argentina.

*G. meandrica* Mendivil-Nerrera 1946, in *Callorhynchus callorhynchus;* Uruguay.

*G. medusarum* Linstow 1903, in *Phyllorhiza rosacea;* Pacific.

*G. nigrosetosa* Haswell 1902, in *Chimaera ogilbyi;* Australia.

*G. plana* Linton 1924, in *Callohynchus antarcticum;* South Africa.

*G. urna* (Grube et Wagener 1852) Olsson 1886 (syn. *Amphiptyches urna* Grube et Wagener 1852, *Crobylophorus chimaerae* Kroyer 1852), in *Chimaera monstrosa;* Italy.

*G. urna* var. *magnispinosa* Lynch 1946, and *G. u.* var. *parvispinosa* Lynch 1946, in *Chimaera monstrosa, C. ogilbyi, Hydrolagus colliei;* Atlantic and Pacific.

FIGURE 513. *Gyrocotyloides nybelini* Fuhrmann 1931.

**1b. Body margins not crenulated, posterior end a long, cylindrical tube ............
................................ *Gryocotyloides* Fuhrmann 1931. (Figure 513).**

*Diagnosis:* Body elongate, flattened, with smooth margins. Anterior end broad, with deep, weak sucker; posterior end attenuated, cylindrical, containing a tube with powerful distal sphincter and opening dorsally at its anterior end. Cuticle lacking spines. Testes numerous, lateral, preovarian. Ventral cirrus pore opens into preequatorial, submedian atrium anterior to level of vaginal pore. Ovary median near posterior end of broad portion of body. Vitellaria? Vaginal pore dorsal, behind level of cirrus pore. Uterus nearly straight, median, preovarian, with large distal sac. Uterine pore midventral, behind level of vaginal pore. Eggs operculate, unembryonated when laid. Two excretory pores near anterior end of body. Parasites of Holocephali.

Type species: *G. nybelini* Fuhrmann, 1931, in *Chimaera monstrosa;* locality unknown.

# GLOSSARY OF TERMS USED IN THIS BOOK

Accessory sac — a sac in the wall of a proglottid, opening into the genital atrium.

Acetabulum — a muscular sucker.

Acini — saccular glands comprised of a few cells grouped together, possessing a common duct.

Acoelous — lacking a body cavity (coelom).

Acraspedote — condition when an anterior proglottid does not overlap the next posterior one.

Alveolar hydatid — the larval form of *Echinoccocus multilocularis*, characterized by being subdivided into many compartments containing protoscolesces and by an exogenous, infiltrative type of growth.

Anapolytic — condition when terminal proglottids become gravid before detaching.

Androgyny — type of maturation when the male organs develop before the female organs do.

Anoperculate — absence of an operculum on an egg.

Apical disk — a flattened knob on the apex of a scolex.

Apolytic — condition when the terminal proglottids detach before becoming gravid.

Atrium — a sunken chamber into which open various reproductive ducts.

Bilateral symmetry — type of symmetry when each lateral half is the mirror image of the other.

Blastocyst — a bladder at the posterior end of a plerocercus, into which the rest of the body can withdraw.

Bothridium — a muscular holdfast organ, of various shapes, on the scoleces of the Tetraphyllidea, Diphyllidea, and Trypanorhyncha.

Bothrium — a dorsal or ventral, longitudinal groove in the scolex of pseudophyllideans.

Brood capsule — a cyst inside a hydatid, containing protoscoleces.

Cephalic peduncle — an elongated portion of a scolex posterior to the suckers, bothria or bothridia.

Cercomer — a knob-like appendage, often bearing the hooks of the oncosphere, on a larval tapeworm.

Chainette — an externo-lateral, longitudinal row of hooks on the tentacles of some Trypanorhyncha.

Cirrus — the male copulatory organ.

Cirrus pore — the opening through which the cirrus is extruded.

Cirrus pouch — a muscular organ containing the introverted cirrus and the ejaculatory duct or internal seminal vesicle.

Coenurus — a larval form of *Taenia* with a few to several scoleces but no brood capsules or daughter cysts.

Coracidium — a ciliated oncosphere, found in the Pseudophyllidea and Trypanorhyncha.

Cortex — region exterior to the outer longitudinal muscle bundles.

Craspedote — condition when an anterior proglottid overlaps the next posterior one.

Cysticercoid — a larval form of tapeworm developing from an oncosphere, with a solid body, lacking a bladder.

Cysticercus — a larval form of some *Taenia* developing from an oncosphere, possessing a fluid-filled bladder and a single scolex.

Daughter cyst — a bladder containing fluid and protoscolices that has formed by exogenous budding of the germinal epithelium of a unilocular hydated.

Decacanth — the ten-hooked larva that hatches from the egg of a cestodarian; also called lycophora.

Definitive host — the host in which a parasite attains sexual maturity.

Dioecious — condition where male and female gonads are found in separate individuals.
'Egg capsule — a structure containing one or more eggs of a tapeworm, in the absence of a uterus.
Ejaculatory duct — that portion of the vas deferens enclosed within the cirrus pouch.
Endogenous budding — inward proliferation of germinal epithelium of a hydatid, resulting in a brood cyst or daughter cyst.
Exogenous budding — proliferation of germinal epithelium of a hydated outward through the cyst wall, often resulting in a daughter cyst.
External seminal vesicle — a dilation of the vas deferens before it reaches the cirrus pouch, which stores sperm cells.
Extravascular — lateral to the osmoregulatory canals.
Fimbriate — with finger-like projections.
Flame cell — an excretory cell with a tuft of cilia extending into a fine efferent tubule.
Fossette — a ciliated, sensory pit.
Genital atrium — a sunken, pit-like structure into which the genital ducts open.
Gonochoristic — dioecious, having the sexes separated into different individuals.
Gonoduct — a duct of a reproductive system.
Gonopore — genital pore, exterior oriface of a reproductive tract.
Gravid — filled with eggs.
Gynandry — condition when the female reproductive system matures earlier than the male system.
Heteracanthus — hook arrangement on a trypanorhynchan tentacle in which there is no chainette and the hooks are in alternating oblique rows extending from the internal to the external surfaces. The hooks may be similar in size and shape (homeomorphous) or dissimilar (heteromorphous).
Hexacanth — the larva that hatches from the egg of the Eucestoda, bearing six hooks.
Homeoacanthus — hook arrangement on a trypanorhynchan tentacle in which the hooks are approximately alike in size and shape and are arranged in spirals or quincunxes.
Hydatid — a complex cysticercus in *Echinococcus* that normally contains a great number of protoscolesces, and buds daughter and brood cysts by endogenous or exogenous proliferation of germinal epithelium.
Hydated sand — free protoscolesces in a hydated cyst.
Hypodermic impregnation — sperm transfer by forcefully piercing of the body wall by the cirrus.
Intermediate host — a host in which development of a parasite occurs, but sexual maturity is not accomplished.
Internal seminal vesicle — a dilation of the ejaculatory duct within the cirrus pouch, that serves to store sperm cells.
Intervascular — medial to osmoregulatory canals.
Introvert — an organ capable of being withdrawn into the body or scolex.
Koilon — the horny, acellular lining of a bird's gizzard.
Loculum — a shallow cavity.
Lycophora — the ten-hooked larva that emerges from the egg of a cestodarian.
Medulla — interior to the outer longitudinal muscle bundles.
Mehlis' glands — unicellular glands surrounding the ootype that contribute a thin membrane around the zygote and its accessory materials.
Metascolex — the posterior portion of a divided scolex.
Monoecious — hermaphroditic, containing gonads of both sexes within a single individual.
Monozoic — not strobilated, the entire body consisting of a single unit.
Multilocular hydated — the specialized cysticercus of *Echinococcus multilocularis*, in which much exogenous budding occurs resulting in infiltration of host tissues.

Myzorhynchus — a slender, muscular stalk arising from the apex of the scolex in some Tetraphyllidea.

Neck — the unsegmented zone between the scolex and strobila of many tapeworms.

Neoteny — the occurrence of sexual maturity in an otherwise juvenile form.

Oncosphere — a hexacanth, the six hooked larva that emerges from the egg of a eucestodan.

Ootype — the area where the oviduct, vitelline duct, and uterus join.

Operculum — a lid-like covering of the opening of an egg shell.

Osmoregulatory canals — the main canals of the excretory system, usually longitudinal with transverse anastomoses but occasionally reticular.

Osmoregulatory system — the excretory system of tapeworms, mainly used in water balance.

Ovary — the female gonad, the origin of ova.

Oviduct — a tube extending from the ovary to the ootype.

Ovicapt — a muscular sphincter on the oviduct.

Parasitologist — a harmless drudge who peers into dark places.

Paratenic host — a host in which a larval stage of a parasite may successfully maintain itself but cannot further develop; often serves as an ecological bridge between intermediate and definitive hosts.

Parenchyma — a loosely organized mass of cells and fibers.

Pars bothridialis — region of scolex of Trypanorhyncha from apex to posterior margins of bothridia.

Pars bulbosa — region of scolex of Trypanorhyncha occupied by the tentacle bulbs.

Pars postbulbosa — any extensive region of the scolex of a trypanorhynchan posterior to the pars bulbosa.

Pars vaginalis — region of scolex of Trypanorhyncha traversed by tentacle sheaths.

Paruterine organ — a fibrous appendage to the uterus that receives the eggs and retains them in a sac at one end, while the uterus sometimes degenerates.

Plerocercoid — the third-stage larva of pseudophyllidean and proteocephalan cestodes, characterized by a solid body.

Plerocercus — the third-stage larva of some Trypanorhyncha, characterized by a posterior bladder (blastocyst) into which the rest of the body can withdraw.

Poeciloacanthus — hook arrangement on a trypanorhynchan tentacle in which a chainette or longitudinal band of small hooks is present on the externo-lateral surface.

Polyzoic — condition when the body is formed of two or more proglottids.

Prebulbar organ — a small organ, often red, of unknown function located anterior to a tentacle bulb in certain trypanorhynchans; also called inenigmatic organ.

Procercoid — the second-stage larva of several orders of tapeworms; usually bears the hexacanth hooks near the posterior end.

Proglottid — a tapeworm segment.

Proscolex — the anterior portion of a divided scolex.

Protandry — condition in a monoecious cestode when the male reproductive system matures first.

Protogyny — condition in a monoecious cestode when the female reproductive system matures first.

Protonephridium — see flame cell.

Protoscolex — the scolex of a larval tapeworm; it may have the same dimensions and armature of the adult.

Pseudoscolex — a holdfast made of distorted anterior proglottids, in the absence of the true scolex.

Quincunx — an arrangement of five objects such that four form a rectangle with the fifth in its center.

Scolex — the holdfast and locomotor organ of a tapeworm, usually considered to be at the anterior end.

Seminal receptacle — a dilation of the vagina that receives and stores sperm from the male system.

Sparganum — a plerocercoid whose identity is unknown.

Strobila — the body of a tapeworm.

Strobilization — the process of budding new proglottids, usually in the neck region. (But see *Haplobothrium*).

Strobilocercoid — a cysticercoid that shows strobilization.

Strobilocercus — the cysticercus of *Taenia taeniaeformis* that bears an immature strobila anterior to the bladder.

Testis — the male gonad, the origin of sperm.

Tetrathyridium — the cysticercoid of *Mesocestoides* that has a solid body and a scolex not surrounded by special membranes.

Unilocular hydatid — a specialized cysticercus found in most species of *Echinococcus*, bearing a great many protoscolesces internally and budding daughter cysts endogenously and rarely exogenously.

Uterine pore — the opening through which eggs escape from the uterus; may be preformed or appear spontaneously upon disintegration of the gravid proglottid.

Uterovaginal canal — a canal into which both uterus and vagina open.

Uterus — the organ receiving eggs from the oviduct; it may hold them until their release or be replaced by egg capsules or paruterine organs.

Vagina — the duct leading from the exterior to the oviduct, through which the sperm travel to reach the ova.

Vaginal pore — the external opening of the vagina.

Vas deferens — the duct receiving sperm from the vasa efferentia and through which they move toward the cirrus pouch.

Vas efferens — a delicate canal extending from a testis to the vas deferens.

Velum — the overlapping portion of a craspedote proglottid.

Vitellarium — the organ that provides cells and materials used for eggshell formation.

Vitelline duct — the canal through which vitelline cells and products move from the vitellarium to the ootype.

Vitelline reservoir — an enlarged portion of vitelline duct that stores vitelline products until used.

Zerney's vesicle — a saccular diverticulum arising from the point where the seminal vesicle joins the cirrus pouch.

# REFERENCES

1. **Abdel-Rahman, M. S., Inman, E. A., and Mohamed, M. A.,** *Hauttuynia streptopelli* [sic] a new cestode from Egyptian doves *"Streptopelia senegalensis,"* Assist. Vet. Med. J., 2, 151—158, 1975.
2. **Abdou, A. H.,** Life cycle of *Davainea proglottina*, Trans. R. Soc. Trop. Med. Hyg., 47, 261—262, 1953.
3. **Abdou, A. H.,** Observations on the life cycle of *Davainea proglottina* in Britain, J. Helminthol., 30, 189—202, 1956.
4. **Abdou, A. H.,** Studies on the development of *Davainea proglottina* in the intermediate host, J. Parasitol., 44, 484—487, 1958.
5. **Abduladze, K. I.,** *Taeniata of Animals and Man and Diseases Caused by Them* (in Russian), Akademiya Nauk SSSR, Moscow, 1964; English transl.: Israel Program for Scientific Translations, Jerusalem, 1970, 548 pp.
6. **Abildgaard, P. C.,** in Mueller, O. F., *Zoologica Danica*, Havniae, 1789.
7. **Abildgaard, P. C.,** *Almindelige Betragtninger over Indvolde-Orme, Skrivter af Naturhistorie Selskabet*, Kjobenhavn, I, pp. 26—64, 1790.
8. **Abildgaard, P. C.,** *Allgemeine Betrachtung der Eingeweidewurmer. Schriften der Naturforsch., Ges. Kopenhagen* I, pp. 24—59, 1793.
9. **Ablasov. N. A., Spasskaya, L. P., and Karabekova, D. N.,** *Rallitaenia porzanae* n. sp., and *R. gallinulae* (Cestoda, Dilepididae) from Rallidae in Kirgizia (in Russian), Izv. Akad. Nauk Kirg. SSSR, 4, 64—70, 1976.
10. **Ackert, J. E.,** On the life-cycle of the fowl cestode, *Davainea cesticillus* (Molin), J. Parasitol., 5, 41—43, 1918.
11. **Ackert, J. E.,** On the life history of *Davainea tetragona* (Molin), a fowl tapeworm, J. Parasitol., 6, 28—34, 1919.
12. **Ackert, J. E.,** On the transmission of the fowl cestode, *Davainea cesticillus* (Molin), Trans. Kansas Acad. Sci., 29, 101—102, 1920.
13. **Ackert, J. E.,** The house fly and tapeworm transmission, Trans. Kansas Acad. Sci., 30, 202—204, 1922.
14. **Ackert, J. E.,** Natural resistance to helminthic infections, J. Parasitol., 28, 1—24, 1942.
15. **Ackert, J. E. and Reid, W. M.,** The cysticercoid of the fowl tapeworm, *Raillietina cesticillus*, Trans. Am. Microsc. Soc., 55, 97—100, 1936.
16. **Ackert, J. E. and Reid, W. M.,** The house fly and fowl tapeworm transmission, J. Parasitol., 22, 526—543, 1936.
17. **Adam, W.,** Sur une larve de cestode de *Mesoplodon bidens* (Sowerby), Bull. Mus. R. Hist. Nat. Belg., 14, 1—17, 1938.
18. **Adam, W.,** Notes sur les céphalopodes. IX. Sur la présence d'une larve de cestode (Tetrarhynchidae) dans la cavité palléale d'un *Octopus* des Iles Andamans, Bull. Mus. R. Hist. Nat. Belg., 14, 1—4, 1938.
19. **Adams, A. R. D. and Webb, L.,** Two further cases of human infestation with *Bertiella studeri* (Blanchard, 1891), Ann. Trop. Med. Parasitol., 27, 471—475, 1933.
20. **Africa, C. M.,** Experimental infection of Philippine *Cyclops* with the coracidium of *Diphyllobothrium mansoni* Cobbold, 1882, Philipp. J. Public Health, 1, 27—31, 1934.
21. **Africa, C. M. and Garcia, E. Y.,** A rat tapeworm (*Raillietina garrisoni* Tubangui, 1931) transmissible to man. With notes on *Davainea madagascariensis* Garrison, 1911, J. Public Health, 1, 44—51, 1934.
22. **Africa, C. M. and Garcia, E. Y.,** The occurrence of *Bertiella* in man, monkey and dog in the Philippines, Philipp. J. Sci., 56, 1—11, 1935.
23. **Ahern, W. B. and Schmidt, G. D.,** Parasitic helminths of the American avocet *Recurvirostra americana*; four new species of the families Hymenolepididae and Acoleidae (Cestoda: Cyclophyllidea), Parasitology, 73, 381—398, 1976.
24. **Akashi, S.,** *Davainea formosana* nov. sp., a new tapeworm from Formosa and Tokyo, J. Parasitol., 3, 182, 1917.
25. **Akmerov, A. K.,** *Amurotaenia percotti* n. gen., n. sp., Vertreter einer neuen Cestoden-Ordnung., C. R. Acad. Sci. U.R.S.S., 30, 191—192, 1941.
26. **Akhmerov, A. K.,** The tapeworms of fishes in the Amur River, Tr. Gel'mintol. Lab. Akad. Nauk SSSR, 10, 15—31, 1960.
27. **Akhmerov, A. K.,** New cestode *Postgangesia orientale* g. et sp. n. New subfamily Postgangesiinae (Cestoda: Proteocephalidae) from silurid fishes in the Amur River (in Russian), Tr. Gel'mintol. Lab. Akad. Nauk SSSR, 20, 3—7, 1969.
28. **Akhumyan, K. S.,** Cestodes of Armenian rodents (in Russian), Tr. Gel'mintol. Lab. Akad. Nauk SSSR, 1, 183—185, 1948.
29. **Akhumyan, K. S.,** Systematics of the cestode genus *Catenotaenia* Janicki, 1904 (in Russian), Gel'mintol. Sborn. 40-Let. Deiatelnost. Skrjabin, Akademiya Nauk SSSR, Moscow, 1946, 37—41.
30. **Alexander, C. G.,** A new species of *Proteocephalus* (Cestoda) from Oregon trout, J. Parasitol., 37, 160—164, 1951.

31. **Alexander, C. G.,** Tetraphyllidean cestodes from the eastern Pacific, *J. Parasitol.,* 19 (Sect. 2, Suppl.), 28—29, 1953.
32. **Alexander, C. G.,** Five new species of *Acanthobothrium* (Cestoda: Tetraphyllidea) from southern California rays, *J. Parasitol.,* 39, 481—486, 1953.
33. **Alexander, C. G.,** Tetraphyllidean and diphyllidean cestodes of New Zealand selachians, *Trans. R. Soc. N.Z. Zool.,* 3, 117—142, 1963.
34. **Alexander, C. G. and Alexander, E. P.,** *Oochoristica crotalicola,* a new anoplocephalid cestode from California rattlesnakes, *J. Parasitol.,* 43, 365, 1957.
35. **Ali, S. M. and Shinde, G. B.,** On a new tapeworm *Sureshia micropusia* gen. et. sp. novo from the house swift *Micropus affinis* in India, *Ind. J. Helminthol.,* Year 1966, 18, 59—64, 1967.
36. **Alicata, J. E.,** The occurrence of *Moniezia benedeni* in a peccary, *J. Parasitol.,* 19, 83, 1932.
37. **Alicata, J. E.,** The amphipod, *Orchestia platensis,* an intermediate host of *Hymenolepis exigua,* a tapeworm of chickens in Hawaii, *J. Parasitol.,* 22, 515—516, 1936.
38. **Alicata, J. E.,** Parasites and parasitic diseases of domestic animals in the Hawaiian Islands, *Pac. Sci.,* 1, 69—84, 1947.
39. **Alicata, J. E. and Chang, E.,** The life-history of *Hymenolepis exigua,* a cestode of poultry in Hawaii, *J. Parasitol.,* 25, 121—127, 1939.
40. **Alicata, J. E. and Jones, M. F.,** The dung beetle, *Ataenius cognatus,* as the intermediate host of *Hymenolepis cantaniana, J. Parasitol.,* 19, 244, 1933.
41. **Allen, G. M. and Lawrence, B.,** Scientific results of an expedition to rain forest regions in Eastern Africa. III. Mammals, *Bull. Mus. Comp. Zool. Harvard Coll.,* 79, 29—126, 1936.
42. **Allen, R. W. and Jackson, P. K.,** The occurrence of the fringed tapeworm, *Thysanosoma actinioides,* in pronghorn antelope, *Proc. Helminthol. Soc. Wash.,* 10, 96—97, 1953.
43. **Alvey, C. H. and Kintner, K. E.,** *Taenia asiota* n. sp. from the screech owl, *J. Parasitol.,* 10, 327—328, 1934.
44. **Ameel, D. J.,** Two larval cestodes from the muskrat, *Trans. Am. Microsc. Soc.,* 61, 267—271, 1942.
45. **Amin, M.,** Two new species of *Avitellina* (Cestoda) from ovines in the Punjab, *Proc. 25th Ind. Sci. Congr. Zool.,* Sect 6, 1939, 158—159.
46. **Amin, M.,** On a new variety of *Avitellina sudanea* (Cestoda) with a note on the anomaly in its genitalia, Proc. 26th Ind. Sci. Congr., Part III, 1940, p. 131.
47. **Amin, M.,** On a new cestode parasite of the genus *Avitellina* from ovines in the Punjab, Proc. 27th Ind. Sci. Congr., Part III, Sect. 7, Madras, 1940; *Zoology,* (Abstr.), 1941, p. 3.
48. **Amin, M.,** A new species of the genus *Avitellina* (Cestoda) from ovines in the Punjab, Proc. 28th Ind. Sci. Congr., Part III, Benares, 1941, (Abstr.) 1942, p. 172.
49. **Amin, M.,** On the genitalia of *Helictometra giardi* (Moniez, 1879), Proc. 30th Ind. Sci. Congr., Part III, 69—70, 1943.
50. **Amin, M.,** On the taxonomy of some species of the genus *Avitellina* (Cestoda), Proc. 29th Ind. Sci. Congr., Part III, p. 152, 1943.
51. **Anantaraman, M.,** The development of *Moniezia,* the large tapeworm of domestic ruminants, *Sci. Cult.,* 17, 155—157, 1951.
52. **Anantaraman, M.,** On the scheme of development in the genus *Mesocestoides* of tapeworms, *Ind. Vet. J.,* 31, 98—101, 1954.
53. **Anderson, M. G.,** The validity of *Taenia confusa* Ward, 1896, *J. Parasitol.,* 20, 207—212, 1934.
54. **Anderson, M. G.,** Some intestinal parasites of *Natrix sepedon* Linn. With notes on the identity of *Ophiotaenia (Taenia) lactea* Leidy with *Ophiotaenia perspicua* La Rue, *Ohio J. Sci.,* 35, 78—80, 1935.
55. **Ando, A. and Ito, S.,** A contribution to the study of *Dibothriocephalus latus, Iji Shinbun Tokyo,* No. 1083, 1921; No. 1094, 1922.
56. **Ando, A. and Matsubara, K.,** Ueber die Mutterwuermer der *Ligula* bei Katzen (Japanese text, German summary), *Hifuka Kiyo Kyoto,* 2, 69—78, 1923.
57. **Andreiko, O. F.,** The hymenolepidid fauna of mammals in south-western USSR, (in Russian), *Vosbuditeli Parazit. Zabolevaniv,* Shtiintsa, Kishinev, U.S.S.R., 1980, 10—23.
58. **Andreiko, O. F., Skvortsov, V. G., and Konovalov, Yu. N.,** Cestodes of bats in Moldavia (in Russian), in *Parasites of Vertebrates,* Spasskii, A. A., Ed., Kartya Moldovenyaska, Kishinev, U.S.S.R., 1969, 31—36.
59. **Andreiko, O. F. and Spasskii, A. A.,** Description of *Triodontolepis Skriabini* and *Coronacanthus integra* and a review of the genus *Triodontolepis* (Cestoda: Hymenolepididae) (in Russian), *Parasit. Zhivotnykh Rastenil,* 7, 27—39, 1971.
60. **Andreiko, O. F. and Yun, T. A.,** *Insinuarotaenia spasskii* n. sp. from carnivores (in Russian), *Izv. Akad. Nauk Moldav SSR Ser. Zool.,* No. 5, 12—19, 1963.
61. **Andrews, J. R. H. and Daniel, M. J.,** A new species of *Hymenolepis* (Cestoda: Hymenolepididae) from the New Zealand long-tailed bat, *Chalinolobus tuberculatus, N.Z. J. Zool.,* 1, 333—336, 1974.

62. **Andry, N.,** *De La Génération des Vers dans le Corps de l'Homme,* Paris, 1701, 470 pp.
63. **Annenkova-Khlopina, N. P.,** Ichthyotaeniidae des poissons d'eau douce de Russie (in Russian), *Izv. Ross. Akad. Nauk,* 12, 2129—2148, 1918.
64. **Annenkova-Khlopina, N. P.,** Deux nouvelles espèces du genre *Caryophyllaeus* (in Russian), *Izv. Ross. Akad. Nauk,* 12, 97—110, 1919.
65. **Annenkova-Khlopina, N. P.,** Notes helminthologiques (I-III) (in Russian), *Ezhegodnik Zool. Muz. Ross. Akad. Nauk,* 24, 32—43, 1923.
66. **Antonow, A.,** Ueber die Art der Kapselbildung bei Hirncysticercose, *Virchows Arch. Pathol. Anat. Physiol.,* 285, 485—493, 1932.
67. **Appy, R. G. and Dailey, M. D.,** Two new species of *Acanthobothrium* (Cestoda: Tetraphyllidea) from elasmobranchs of the eastern Pacific, *J. Parasitol.,* 59, 817—820, 1973.
68. **Appy, R. and Dailey, M. D.,** A new species of *Rhinobothriun* (Cestoda: Tetraphillidea) and redescription of three rhinobothriate species from the round stingray, *Urolophus halleri* Cooper in southern California, *Bull. South. Calif. Acad. Sci.,* 76, 116—127, 1977.
69. **Arenas, Y. R. and Alvarez, G. J. R.,** El *Cisticercus tenuicollis* en nuestros animales de matadero, *Bol. Mens. Col. Vet. Nac. Habana,* 3, 573—574, 1934.
70. **Ariola, V.,** Sulla '*Bothriotaenia plicata*' (Rud.) e sul suo sviluppo, *Atti Soc. Ligust. Sci. Nat. Geogr. Genova,* 7, 117—126, 1896.
71. **Ariola, V.,** Osservazione sopra alcuni Dibrotrii dei pesci, *Atti Soc. Ligust. Sci. Nat. Geogr. Genova,* 10, 60—70, 1896.
72. **Ariola, V.,** Il gen. *Scyphocephalus* Rigg. e proposta di una nuova classificazione dei cestodi, *Atti Soc. Ligust. Sci. Nat. Geogr. Genova,* 10, 160—167, 1899.
73. **Ariola, V.,** Nota sui cestodi parassiti del *Centrolophus pompilius,* Linn. *Boll. Mus. Zool. Ant. Comp.,* (University of Genova), No. 93, 6, 1900.
74. **Ariola, V.,** Rivisione della famiglia Bothriocephalidae, *Arch. Parasitol.,* 3, 369—484, 1900.
75. **Ariola, V.,** Sono i cestodi polizoici, *Atti Soc. Ligust. Sci. Nat. Geogr. Genova,* 13, 236—244, 1902.
76. **Ariola, V.,** La métamèrie et la théorie de la polyzoicité chez les cestodes, *Rev. Gén. Sci.,* 13, 471—476, 1902.
77. **Arnold, J. G., Jr.,** A study of the anoplocephaline cestodes of North American rabbits, *Zoologica,* 23, 31—53, 1938.
78. **Arnold, J. P.,** A study of the anoplocephaline cestodes of North American rabbits, *J. Parasitol.,* 2, 445, 1935.
79. **Arnold, J. P.,** A study of anoplocephaline cestodes of North American rabbits, *Zoologica,* 23, 31—53, 1938.
80. **Artyukh, E. S.,** *Essentials of Cestodology,* Vol. 6, *Davaeineata,* Akademiya Nauk SSSR, Moscow, 1966, 1—511.
81. **Babero, B. B.,** The experimental infection of Alaskan gulls (*Larus glaucescens*) with *Diphyllobothrium* sp., *J. Wash. Acad. Sci.,* 43, 166—168, 1951.
82. **Babero, B. B.,** Diphyllobothriasis in Alaska, Proc. 2nd Alaskan Sci. Conf., Mt. McKinley National Park, Sept. 4th to 8th, 1952, 312—314.
83. **Babero, B. B.,** Studies on the helminth fauna of Alaska. XVI. A survey of the helminth parasites of ptarmigan (*Lagopus* spp.), *J. Parasitol.,* 538—546, 1953.
84. **Babero, B. B. and Rausch, R.,** Some observations on host relationships of *Diphyllobothrium* sp. in cats, *J. Parasitol.,* 39, 226—227, 1953.
85. **Bacigalupo, J.,** *Hymenolepis intermedius* (nueva especie). Su evolución, *Semana Med. (Buenos Aires),* 34, 239—240, 1926.
86. **Bacigalupo, J.,** *Hymenolepis diminuta;* su evolución, *Semana Med. (Buenos Aires),* 33, 67—79, 1927.
87. **Baciaglupo, J.,** Una neuva especie de tenia, *Tenia infantil, Actas Trab. 2, Congr. Nac. Med (1922),* 5, 141—142, 1927.
88. **Bacigalupo, J.,** Estudio sobre la evolucion bilogica de algunos parasitos del genero *Hymenolepis* (Weinland, 1858), *Semana Med. (Buenos Aires),* 35, 1249—1267, 1349—1366, 1428—1444, 1928.
89. **Bacigalupo, J.,** La evolución del *Hymenolepis nana, Semana Med. (Buenos Aires),* 35, 200—201, 1928.
90. **Bacigalupo, J.,** *Hymenolepis microstoma* (Dujardin) chez *Mus musculus* (L.), *C. R. Soc. Biol.,* 99, 2048—2057, 1928.
91. **Bacigalupo, J.,** Difrenciación de los huevos de *Hymenolepis nana* y *Hymenolepis diminuta, Actas Trab. 3. Congr. Nac. Med.,* 7, 615—616, 1928.
92. **Bacigalupo, J.,** *Hymenolepis nana* (von Siebold, 1854) e *Hymenolepis fraterna* (Stiles, 1906), *Rev. Soc. Argent. Biol.,* 5, 599—604, 1929.
93. **Baciagalupo, J.,** El *Ctenocephalus canis* Curtis, en la envolución de la *Hymenolepis fraterna* Stiles, *Semana Med. (Buenos Aires),* 38, 878—880, 1931.
94. **Bacigalupo, J.,** Evolution de l'*Hymenolepis fraterna* Stiles chez *Pulex irritans* L., *Xenopsylla cheopis* Rothschild et *Ctenocephalus canis* Curtis, *Ann. Parasitol.,* 9, 339—343, 1931.

95. **Bacigalupo, J.**, Algunas consideraciónes sobre teniasis par *Hymenolepis nana*, *Arch. Argent. Enfermedades*, 7, 359—364, 1932.
96. **Bacigalupo, J.**, Quel est l'avenir du scolex echinococcique avalé par le lapin?, *C. R. Soc. Biol.*, 114, 89—92, 1933.
97. **Bacigalupo, J.**, El *Anisolabis annulipes* (Lucas) en la transmission de la *Hymenolepis diminuta* y la *Hymenolepis fraterna*, *Rev. Chilena Hist. Nat.*, 39, 127—129, 1936.
98. **Bacigalupo, J.**, Nuevo huesped intermediario de la *Hymenolepis diminuta* (Rudolphi, 1829), *Embia (Rhagadachir) argentina* Navas, *Rev. Med. Trop. Parasitol.*, 4, 45—47, 1938.
99. **Bacigalupo, J.**, Diancyrobothriidae, neuva familia del orden Pseudophyllidea, *Rev. Soc. Argent. Biol.*, 21, 383—392, 1945.
100. **Bacigalupo, J.**, Parasitosis experimental de rata blanca con cepa humana de *Hymenolepis diminuta*, *Rev. Soc. Argent. Biol.*, 2, 138—140, 1951.
101. **Bacigalupo, J. and Fischman, M.**, Botriocefaliasis por un nuevo parasito (*Diancyrobothrium taenioides*). Contribución al estudio clinico y parasitologio de un caso de botriocefaliasis, *Arch. Hosp. Israelita (Buenos Aires)*, 6, 89—94, 1949.
102. **Bacigalupo, J. and Rivero, R.**, Vaginal sphincter, new organ of *Echinococcus granulosus*, *J. Parasitol.*, 34 (Sect. 2, Suppl.), 25, 1948.
103. **Baczynska, H.**, Etudes anatomiques et histologiques sur quelques nouvelles espèces de cestodes d'oiseaux, *Bull. Soc. Sci. Nat. Neuchâtel*, 40, 187—239, 1914.
104. **Badalian, A.**, La localisation des cysticeroides du *Hymenolepis nana* dans l'intestin de la souris blanche (in Russian, French summary), *Med. Parazitol. Parazitar. Bolezni*, 7, 580—583, 1938.
105. **Baer, C. E. V.**, Ueber Linne's im Wasser gefundene Bandwürmer, *Verh. Ges. Nat. Freunde Berlin*, I., 1829.
106. **Baer, J. G.**, Résultats zoologiques de voyage du Dr. P. A. Chappuis au Nil supérieur. III. Helminthes, *Rev. Suisse Zool.*, 30, 337—352, 1923.
107. **Baer, J. G.**, Considérations sur le genre *Anoplocephala*, *Bull. Soc. Sci. Nat. Neuchâtel*, 48, 3—16, 1923.
108. **Baer, J. G.**, On a cestode parasite of East African rock rabbits (*Procavia* sp.), *J. Helminthol.*, 2, 77—80, 1924.
109. **Baer, J. G.**, Contributions à la faune helminthologique sud-africaine. Note préliminaire, *Ann. Parasitol.*, 2, 239—247, 1924.
110. **Baer, J. G.**, Quelques cestodes d'oiseaux nouveaux et peu connus, *Bull. Soc. Sci. Nat. Neuchâtel*, 49, 138—154, 1925.
111. **Baer, J. G.**, Sur quelques cestodes du Congo belge, *Rev. Suisse Zool.*, 32, 239—251, 1925.
112. **Baer, J. G.**, On some cestoda described by Beddard, 1911—1920, *Ann. Trop. Med. Parasitol.*, 19, 1—22, 1925.
113. **Baer, J. G.**, Cestodes nouveaux du Sud-Quest de l'Afrique, *Rev. Suisse Zool.*, 31, 529—548, 1925.
114. **Baer, J. G.**, Une nouvelle phase dans le cycle évolutif de *Diphyllobothrium latum*, *Rev. Suisse Zool.*, 31, 555—561, 1925.
115. **Baer, J. G.**, Contribution to the helminth fauna of South Africa. Mammalian cestodes, *Union S. Afr.*, Dep. Agric. 11th and 12th. Rep. Dir. Vet. Ed. Res. Pretoria, pp. 61—136, 1926.
116. **Baer, J. G.**, Cestodes de mammifères, *Bull. Soc. Sci. Nat. Neuchâtel*, 50, 77—81, 1926.
117. **Baer, J. G.**, Monographie des cestodes de la famille des Anoplocephalides, *Bull. Biol. Fr. Belg.*, 10 (Suppl.), 241, 1927.
118. **Baer, J. G.**, On a new species of *Hymenolepis* from a monkey, *J. Parasitol.*, 14, 48—50, 1927.
119. **Baer, J. G.**, Des cestodes adultes peuvent-ils vivre dans la cavité générale d'oiseaux?, *Ann. Parasitol.*, 5, 337—340, 1927.
120. **Baer, J. G.**, Contributions to the anatomy of some reptilian cestodes, *Parasitology*, 19, 274—283, 1927.
121. **Baer, J. G.**, Die Cestoden der Saugetiere Brasiliens, *Abhandl. Senkenbergischen Naturforsch. Ges.*, 40, 377—386, 1927.
122. **Baer, J. G.**, Notes sur les Ténias des autruches, *Bull. Soc. Sci. Nat. Neuchâtel*, 52, 7—13, 1928.
123. **Baer, J. G.**, A propos d'une nouvelle classification des cestodes du genre *Davainea* R. Bl. s. 1, *Bull. Soc. Zool. Fr.*, 55, 44—57, 1931.
124. **Baer, J. G.**, Sur la position systématique du *Taenia muris-sylvatica* Rudolphi, 1819, *Bull. Soc. Sci. Nat. Neuchâtel*, 55, 35—39, 1931.
125. **Baer, J. G.**, Contribution à faune helminthologique de Suisse, *Rev. Suisse Zool.*, 39, 1—57, 1932.
126. **Baer, J. G.**, Contribution à faune helminthologique de Suisse, *Rev. Suisse Zool.*, 39, 195—228, 1932.
127. **Baer, J. G.**, Contribution à l'étude de la faune helminthologique africaine, *Rev. Suisse Zool.*, 40, 31—84, 1933.
128. **Baer, J. G.**, Etude de quelques helminthes de lémuriens, *Rev. Suisse Zool.*, 42, 275—291, 1935.
129. **Baer, J. G.**, Un cestode nouveau parasite d'un poisson d'ornement, *Bull. Soc. Nat. Acclim. Fr. Année*, 84, 168—173, 1937.

130. **Baer, J. G.**, Un genre nouveau de cestodes d'oiseaux, *Bull. Soc. Sci. Nat. Neuchâtel,* 62, 149—156, 1937.
131. **Baer, J. G.**, Some avian tapeworms from Antigua, *Parasitology,* 32, 174—197, 1940.
132. **Baer, J. G.**, The origin of human tapeworms, *J. Parasitol.,* 26, 174—197, 1940.
133. **Baer, J. G.**, L'existence aventureuse des vers solitaires, Leçon inaugurale prononcée par Jean G. Bear le 26 Novembre 1941 à son installation à la chaire de zoologie et d'anatomie comparée a l'Université de Neuchâtel, 1941.
134. **Baer, J. G.**, La sparganose oculaire, *Acta Trop.,* 2, 155—157, 1945.
135. **Baer, J. G.**, Contribution à l'étude expérimentale des cycles évolutifs des cestodes, *Bull. Soc. Neuchâtel. Sci. Nat.,* 69, 181—183, 1945.
136. **Baer, J. G.**, Le Parasitisme, *Lausanne,* 1946, 232 pp.
137. **Baer, J. G.**, Les helminthes parasites des vertébrés, in *Premier Colloque Francosuisse, Besançon,* 1946.
138. **Baer, J. G.**, La signification des générations larvaires chez les vers plats parasites, *Rev. Sci.,* 84, 262—272, 1946.
139. **Baer, J. G.**, Contributions à l'étude des cestodes de sélaciens. I—IV, *Bull. Soc. Sci. Nat. Neuchâtel,* 71, 63—122, 1948.
140. **Baer, J. G.**, Contributions à la faune helminthologique africaine, *Acta Trop.,* 6, 41—45, 1949.
141. **Baer, J. G.**, Etude critique des helminthes parasites de l'okapi, *Acta Trop.,* 7, 164—186, 1950.
142. **Baer, J. G.**, Phylogénie et cycles évolutifs des cestodes, *Rev. Suisse Zool.,* 57, 553—558, 1950.
143. **Baer, J. G.**, *Ecology of Animal Parasites,* University of Illinois Press, Urbana, Ill., 1951, 224 pp.
144. **Baer, J. G.**, Notes de faunistique éburnéenne. IV. *Bertiella douceti* n. sp. cestode nouveau de l'anomalure de pel, *Acta Trop.,* 10, 79—82, 1953.
145. **Baer, J. G.**, Revision taxinomique et étude biologique des cestodes de la famille des Tetrabothriidae, parasites d'oiseaux de haute mer et de mammifères marins, *Mem. Univ. Neuchâtel Sér. Inquarto,* 1, 1—121, 1954.
146. **Baer, J. G.**, The tapeworm genus *Wyominia* Scott, 1941, *Proc. Helminthol. Soc. Wash.,* 21, 48—52, 1954.
147. **Baer, J. G.**, Un nouveau cas de parasitisme d'un enfant en Afrique orientale par le cestode *Inermicapsifer arvicanthidis* (Kofend, 1917), *Acta Trop.,* 12, 174—176, 1955.
148. **Baer, J. G.**, Cestodes d'un dauphin de l'océan Pacifique, *Bull. Soc. Neuchâtel Sci. Nat.,* 78, 33—36, 1955.
149. **Baer, J. G.**, Revision critique de la sous-famille Idiogeninae Fuhrmann 1932 (Cestodes: Davaineidae) et étude analytique de la distribution des espèces, *Rev. Suisse Zool.,* 62 (fasc. Suppl.), 1—51, 1955.
150. **Baer, J. G.**, Parasitic helminths collected in West Greenland, *Medelelser om Grønland undivne af Kommiss. f. Vidensk. Unders. Grønl.,* 124, 1—55, 1956.
151. **Baer, J. G.**, The taxonomic position of *Taenia madagascariensis* Davaine, 1870, a tapeworm parasite of man and rodents, *Ann. Trop. Med. Parasitol.,* 50, 152—156, 1956.
152. **Baer, J. G.**, Cestoda, *Zoology of Iceland,* Ejnat Munksgaard, Copenhagen, Vol. 2, Part 12, 1962, 1—63.
153. **Baer, J. G.**, *Eurycestus avoceti* Clark, 1954 (Cestode cyclophyllidien) parasite de l'avocette en Carmargue, *Vie Milieu,* 19, 189—198, 1968.
154. **Baer, J. G. and Bona, F.**, Revision des *Cestodes Dilepididae Fuhrm.,* 1907 des ardeiformes. Note preliminaire, *Boll. Instit. Mus. Zool. Univ. Torino,* 6, 1—53, 1960.
155. **Baer, J. G. and Euzet, L.**, *Prosobothrium armigerum* Cohn, 1902 (Cestoda). Histoire, synonymie, description et position systématique, *Recueil Trav. Lab. Bot. Geol. Zool. Univ. Montpellier Ser. Zool.,* 1, 44—45, 1956.
156. **Baer, J. G. and Euzet, L.**, Revision critique des Cestodes tetraphyllides decrits par T. Southwell (lre partie), *Bull. Soc. Neuchâtel. Sci. Nat.,* 85, 143—172, 1962.
157. **Baer, J. G. and Fain, A.**, Les cestodes des pangolins, *Bull. Soc. Neuchâtel. Sci. Nat.,* 3 s., 78, 37—52, 1955.
158. **Baer, J. G. and Fain, A.**, Cestodes, Exploration du Parc National de l'Upemba. Mission G. F. de Witte, Institut des Parcs Nationaux du Congo Belgique, 1955, 38 pp.
159. **Baer, J. G. and Fain, A.**, *Bothriocephalus (Clestobothrium) kivuensis* n. sp., Cestode parasite d'un barbeau du lac Kivu, *Ann. Soc. R. Zool. Belq.,* Years 1957—58, 88, 287—302, 1958.
160. **Baer, J. G. and Joyeux, C. E.**, Les larves cysticercoides de quelques tenias de la musaraigne d'eau *Neomys fodiens* (Schreb.) (Note préliminaire), *Schweiz. Z. Allg. Pathol. Bakteriol.,* 6, 395—399, 1942.
161. **Baer, J. G. and Joyeux, C. E.**, Réalisation expérimentale d'un nouveau cycle évolutif de cestode de la souris blanche (note préliminaire), *Verh. Schweiz. Naturforsch. Ges. (124. Jahresvers.),* pp. 133—134, 1944.
162. **Baer, J. G., Kouri, P., and Sotolongo, F.**, Anatomie, position systématique et epidémiologie de *Inermicapsifer cubenis* (Kouri, 1938) Kouri, 1940, *Acta Trop.,* 6, 120—130, 1949.

163. **Baer, J. G., Kouri, P., and Sotolongo, F.**, Anatomia, posición sistematica y epidemiologia de *Inermicapsifer cubenis* (Kouri, 1938) Kouri, 1940, cestode parasito del hombre en Cuba, *Rev. Kuba Med. Trop. Parasitol.*, 6, 9—13, 1950.
164. **Baer, J. G. and Sandars, D. F.**, The first record of *Raillietina (Raillietina) celebensis* (Janicki, 1902) (Cestoda) in man from Australia, with a critical survey of previous cases, *J. Helminthol.*, 30, 173—182, 1956.
165. **Bahamonde, N. N. and Lopez, M. T.**, *Proboscidosaccus mesodesmatis* n. sp., parasito de *Mesodesma donacium* Lamarck (Cestoda, Tetraphyllidea), *Invest. Zool. Chilenas*, 8, 43—56, 1962.
166. **Bailey, W. S.**, Observation on the rôle of *Tenebrio molitor* as an intermediate host for *Hymenolepis nana* var. *fraterna*, *J. Parasitol.*, 33, 433—434, 1947.
167. **Bailliet, C.**, Recherches sur un cystique polycéphale du lapin, *Mém. Acad. Imp. Toulouse*, 11, 452—482, 1863.
168. **Bailliet, C.**, Histoire Naturelle des Helminthes des Principaux Mammifères Domestiques, *N. Dict. Prat. Méd. Vet.*, 8, 519—687, 1866.
169. **Baird, W.**, *A Catalogue of the Species of Entozoa or Intestinal Worms Contained in the Collection of the British Museum*, London, 1853, 1—132.
170. **Baird, W.**, Descriptions of some new species of intestinal worms, *Proc. R. Soc. London*, 28, 446—448, 1860.
171. **Baird, W.**, Descriptions of some new species of intestinal worms, *Ann. Mag. Nat. Hist.*, 3 s., 7, 228—230, 1861.
172. **Baird, W.**, Descriptions of two new species of cestode worms belonging to the genus *Taenia*, *Proc. Zool. Soc. London*, pp. 18—25, 1862.
173. **Baird, W.**, Description of a new species of entozoon from the intestine of the diamond-snake of Australia *(Morelia spiroletes)*, *Proc. Zool. Soc. London*, pp. 58—59, 1865.
174. **Baker, A. D.**, Records of distribution of internal parasites of poultry in Quebec, *Sci. Agric.*, 13, 127—130, 1932.
175. **Balls, H. H.**, Ueber die Entwicklung der Geschlechtsgänge bei Cestoden, *Z. Wiss. Zool.*, 91, 226—296, 1908.
176. **Bangham, R. V.**, Series of articles on the tapeworms of North American freshwater fishes in Ohio, *Ohio J. Sci.*, 25, 41; *Trans. Am. Fish. Soc.*, 63, 68, 70, 1925—1941.
177. **Bangham, R. V.**, A study of the cestode parasites of the black bass in Ohio, with special reference to their life history and distribution, *Ohio J. Sci.*, 25, 255—270, 1925.
178. **Bangham, R. V.**, A new intermediate host of *Proteocephalus pearsei* La Rue, *J. Parasitol.*, 13, 223, 1927.
179. **Bangham, R. V.**, Life history of bass cestode *Proteocephalus ambloplitis*, *Trans. Am. Fish. Soc. 57th Annu. Meet.*, Hartford, Conn., 1927, 206—209.
180. **Bangham, R. V. and Venard, C. E.**, Studies on the parasites of Reelfoot Lake fishes, *Tenn. Acad. Sci.*, 17, 22—38, 1942.
181. **Barbieri, C.**, Ueber eine neu Species der Gattung *Ichthyotaenia* und ihre Verbreitungsweise, *Centrabl. Bakteriol.*, 49, 334—340, 1909.
182. **Barker, F.**, Some new cases of trihedral *Taenia*, *Science*, 31, 837, 1910.
183. **Barker, F.**, A contribution to the evolution of the cestode rostellum, *Science*, 39, 435, 1914.
184. **Barker, F.**, Parasites of American muskrats *(Fiber zibethicus)*. *J. Parasitol.*, 1, 184—197, 1915.
185. **Barker, F.**, Are polyradiate cestodes mutations?, *Anat. Rec.*, 11, 507, 1916.
186. **Barker, F. D. and Andrews, M.**, *Anomotaenia telescopia* Barker and Andrews, sp. nov., in Barker, F. D., *Parasites of the American musrat*, *Fiber zibethicus*, *J. Parasitol.*, 1, 194—195, 1915.
187. **Barker, F. D. and Andrews, M.**, *Hymenolepis evaginata* Barker and Andrews, sp. nov., *J. Parasitol.*, p. 194, 1915.
188. **Barrois, T.**, Sur quelques Ichthyoténias parasites de serpents, *Bull. Soc. Sci. Agric. Arts Lille*, p. 4, 1898.
189. **Bartel, M. H. and Hansen, M. F.**, *Raillitiena (Raillietina) loeweni* sp. n. (Cestoda: Davaineidae) from the hare in Kansas with notes on *Raillietina* of North American mammals, *J. Parasitol.*, 50, 448—453, 1964.
190. **Bartels, E.**, *Cysticercus fasciolaris*, Beiträge zur Anatomie, Entwicklung und Umwandlung in *Taenia crassicollis*, *Zool. Jahrb. Anat.*, 16, 511—570, 1902.
191. **Bashkirova, E. Y.**, Etude biologique des *Anoplocephala perfoliata* Goeze, 1782, *Vesynik Selskokhozyaustvennol Nauki Vet.*, No. 2, 57—67, 1941.
192. **Basiev, M.**, Contribution to the study of the biology of the tapeworm *Anoplocephala perfoliata* (Goeze, 1782), parasitic in the horse, *Dokl. Akad. Nauk SSSR*, n. s., 30, 576—578, 1941.
193. **Basso, A.**, Richerche morfologiche sulla *Taenia echinococcus* con special riferimento all'apparto sessuale, *Arch. Zool. Ital.*, 9, 165—194, 1920.
194. **Bator, T.**, *Moniezia skrjabini* n. sp. from sheep and goats in the Mongolian Peoples Republic (in Russian), *Parazitologiya*, 5, 73—76, 1971.

195. **Batsch, A. J. G.**, Naturgeschichte der Bandwürmgattung ueberhaupt und ihrer Arten insbesondere, nach den neuern Beobachtungen in einem systematischen Auszuge, Halle, 1786, 298 pp.
196. **Bauer, O.**, Parasite fauna of birds of Alpine part of Borjomsk Region, Uchen (in Russian), *Zapiski Leningrad. Gusdarstv*, 1941.
197. **Baugh, S. C. and Saxena, S. K.**, A discussion on the taxonomy of the genus *Paradilipis* Hsü, 1935, *Anal. Inst. Biol. Univ. Nac. Autón. Mexico Ser. Zool.*, 44, 15—24, 1974. (Dated 1973.)
198. **Baugh, S. C. and Saxena, S. K.**, On cestodes of *Passer domesticus*. I. *Choanotaenia, Raillietina* and *Proparuterina, Angewandte Parasitol.*, 17, 146—160, 1976.
199. **Bavay, M.**, Sur la présence du *Bothriocephalus latus* à Madagascar, *Bull. Soc. Zool. Fr.*, 15, 134—135, 1890.
200. **Baylis, H. A.**, on *Octopetalum*, a new genus of avian Cestodes, *Ann. Mag. Nat. Hist.* (Ser. 8), 14, 414—420, 1914.
201. **Baylis, H. A.**, A new cestode of the genus *Zschokkeela*, *Ann. Mag. Nat. Hist.* (Ser. 8), 16, 40—49, 1915.
202. **Baylis, H. A.**, A collection of Entozoa, chiefly from birds, from the Murman coast, *Ann. Mag. Nat. Hist.* (Ser. 9), 3, 501—513, 1919.
203. **Baylis, H. A.**, On two new species of the cestode genus *Oochoristica* from lizards, *Parasitology*, 11, 405—414, 1919.
204. **Baylis, H. A.**, Notes on some parasitic worms from East Africa, *Ann. Mag. Nat. Hist.* (Ser. 9), 9, 292—295, 1920.
205. **Baylis, H. A.**, A new cestode and other parasitic worms from Spitzbergen, with a note on two leeches. Results of the Oxford University Expedition to Spitzbergen, *Ann. Mag. Nat. Hist.* (Ser. 9), 9, 421—497, 1922.
206. **Baylis, H. A.**, Observations on certain cestodes of rats, with an account of a new species of *Hymenolepis*, *Parasitology*, 14, 1—8, 1922.
207. **Baylis, H. A.**, Some tetrabothriid cestodes from whales of the genus *Balaenoptera*, *J. Linn. Soc. London*, 36, 161—172, 1926.
208. **Baylis, H. A.**, Some parasitic worms from Sarawak, *Sarawak Mus. J. No. 10*, pp. 303—322, 1926.
209. **Baylis, H. A.**, Some parasites of whales, *Nat. Hist. Mag. London*, 1, 55—57, 1927.
210. **Baylis, H. A.**, On two adult cestodes from wild swine, *Ann. Mag. Nat. Hist.* (Ser. 9), 19, 417—425, 1927.
211. **Baylis, H. A.**, The cestode genus *Catenotaenia*, *Ann. Mag. Nat. Hist. (Ser. 9)*, 19, 433—439, 1927.
212. **Baylis, H. A.**, Some parasitic worms from Lake Tanganyika, *Ann. Mag. Nat. Hist.* (Ser. 10), 1, 552—562, 1928.
213. **Baylis, H. A.**, A new species of *Schizotaenia* from the capybara, *Ann. Mag. Nat. Hist.* (Ser. 10), 1, 601—605, 1928.
214. **Baylis, H. A.**, On larval forms of *Acanthotaenia*, *Ann. Mag. Nat. Hist.* (Ser. 10), 4, 224—229, 1929.
215. **Baylis, H. A.**, Some new parasitic nematodes and cestodes from Java, *Parasitology*, 21, 25—65, 1929.
216. **Baylis, H. A.**, List of worms parasitic in Cetacea, *Discovery Reports (Government Dependencies, Falkland Islands)*, Vol. 6, Cambridge University Press, England, 1932, 393—418.
217. **Baylis, H. A.**, On a coenurus from man, *Trans. R. Soc. Trop. Med. Hyg.*, 25, 275—280, 1932.
218. **Baylis, H. A.**, Miscellaneous note on parasitic worms, *Ann. Mag. Nat. Hist.* (Ser. 10), 13, 223—228, 1934.
219. **Baylis, H. A.**, On a collection of cestodes and nematodes from small mammals in Tanganyika Territory, *Ann. Mag. Nat. Hist.* (Ser. 10), 13, 338—353, 1934.
220. **Baylis, H. A.**, Two new species of the genus *Bertiella*, with a note on the presence of uterine pores, *Ann. Mag. Nat. Hist.* (Ser. 10), 14, 412—421, 1934.
221. **Baylis, H. A.**, Notes on four cestodes, *Ann. Mag. Nat. Hist.* (Ser. 10), 13, 587—594, 1934.
222. **Baylis, H. A.**, Some parasitic worms from Australia, *Parasitology*, 26, 129—132, 1934.
223. **Baylis, H. A.**, Some parasitic worms from muskrats in Great Britain, *Ann. Mag. Nat. Hist.* (Ser. 10), 15, 543—549, 1935.
224. **Baylis, H. A.**, Note on the cestode *Moniezia (Fuhrmannella) transvaalensis* (Baer, 1925), *Ann. Mag. Nat. Hist.* (Ser. 10), 15, 673—675, 1935.
225. **Baylis, H. A.**, The plerocercoid larva of *Bothridium*, *Ann. Mag. Nat. Hist.* (Ser. 10), 15, 482—485, 1935.
226. **Baylis, H. A.**, On the probable identity of a cestode of the genus *Diphyllobothrium* occurring in Wales and Eire, *Ann. Trop. Med. Parasitol.*, 39, 41—45, 1945.
227. **Baylis, H. A.**, Some roundworms and flatworms from the West Indies and Surinam. II. Cestodes, *J. Linn. Soc. London Zool.*, 41, 406—414, 1947.
228. **Baylis, H. A.**, On the synonymy of two tetraphyllidean cestodes from rays, *Ann. Mag. Nat. Hist.* (Ser. 12), 1, 293—295, 1948.
229. **Baylis, H. A.**, "*Taenia exigua*" Dujardin, *Ann. Mag. Nat. Hist.* (Ser. 11), 14, 353—358, 1948.
230. **Baylis, H. A.**, A new human cestode infection in Kenya, *Inermicapsifer arvicanthidis*, a parasite of rats, *Trans. R. Soc. Trop. Med. Hyg.*, 42, 531—542, 1949.

231. **Baylis, H. A.,** A review of the species of *Dinobothrium* (Cestoda) with a description of a new species, *Parasitology,* 40, 96—104, 1950.
232. **Baylis, H. A.,** Una nueva infection humana por cestode en Kenya, *Inermicapsifer arvicanthidis,* parasito de la rata, *Rev. Kub. Med. Trop. Parasitol.,* 6, 14—19, 1950.
233. **Baylis, H. A.,** The parasitic worms of British reptiles and amphibia, in *The British Amphibians and Reptiles,* Smith, M., Ed., London, 1951, 267—273.
234. **Bayon, H. P.,** Recent advances in our knowledge of poultry diseases, *Vet. Rec.,* 13, 655—669, 1933.
235. **Bearup, A. J.,** Observations on the life cycle of *Diphyllobothrium (Spirometra) erinacei* in Australia (Cestoda: Diphyllibothriidae), *Aust. J. Sci.,* 10, 183—184, 1948.
236. **Bearup, A. J.,** Life history of a spirometrid tapeworm, causing sparganois in feral pigs, *Aust. Vet.,* 29, 217—224, 1953.
237. **Bearup, A. J.,** Experimental vectors of the first larval stage of *Dibothriocephalus latus* (Cestoda) in Australia, *Aust. J. Exp. Biol. Med. Soc.,* 35, 187—191, 1957.
238. **Bearup, A. J. and Morgan, E. L.,** The occurrence of *Hymenolepis diminuta* (Rud., 1819) and *Dipylidium caninum* (Linn., 1758) as parasites of man in Australia, *Med. J. Aust.,* 1, 104—106, 1939.
239. **Beauchamp, P. M. de,** Etude sur les cestodes des sélaciens, *Arch. Parasitol.,* 9, 463—539, 1905.
240. **Beaver, P. C. and Simer, P. H.,** A re-study of the three existing species of the cestode genus *Marsipometra* Cooper (Amphicotylidae) from the spoonbill, *Polyodon spathula* (Wal.), *Trans. Am. Microsc. Soc.,* 59, 167—182, 1940.
241. **Beck, J. W.,** *Megacirrus megapodii* n. a., n. sp., a cestode from the Malayan brush turkey, *Megapodius laparouse senex* (Dilepididae), *J. Parasitol.,* 37, 405—407, 1951.
242. **Becker, E. R.,** Two observations on helminthes, *Trans. Am. Microsc. Soc.,* 52, 361—362, 1933.
243. **Becker, R.,** Ueber das Vorkommen von Bandwürmern beim Pferde, Berlin, *München Tierärzt. Wochensch.,* 36, 224—225, 1920.
244. **Becker, R.,** Die äussere Gestalt der Pferdebandwürmer, *Centralbl. Bakteriol.,* 87, 110—118, 1921.
245. **Becker, R.,** Weitere Beiträge zur Anatomie der Pferdebandwürmer, *Centralbl. Bakteriol.,* 88, 21—227, 1921.
246. **Becker, R.,** Zur Nomenklatur der Pferdebandwürmer, *Centralbl. Bakteriol.,* 91, 63—67, 1923.
247. **Beddard, F. E.,** Contributions to the anatomy and systematic arrangement of the Cestoidea, *Proc. Zool. Soc. London,* 1911—1916.
    I. On some mammalian Cestoidea, 626—660, 1911.
    II. On two new genera of Cestoides from mammals, 904—1018, 1911.
    III. On a new genus of tapeworms (*Otiditaenia*) from the bustard (*Eupodotes kori*), 194—221, 1912.
    IV. On a new species of *Inermicapsifer* from the hyrax and on the genera *Zschokkeella, Thysanotaenia* and *Hyracotaenia,* 576—607, 1912.
    V. On a new genus (*Dasyurotaenia*) from the Tasmanian devil, *Dasyurus ursinus,* the type of a new family, 667—695, 1912.
    VI. On an asexual tapeworm from the rodent *Fiber zibethicus* showing a new form of asexual reproduction and on the supposed asexual form, 822—850, 1912.
    VII. On six species of tapeworms from reptiles belonging to the genus *Ichthyotaenia* (s.l.), 4—36, 1913.
    VIII. On some species of *Ichthyotaenia* and *Ophidotaenia,* 153—168, 1913.
    IX. On a new genus of Ichthyotaeniids, 243—261, 1913.
    X. On two new species of tapeworms from *Genetta dongolana,* 549—571, 1913.
    XI. On a new tapeworm from *Oedicnemus,* 861—877, 1913.
    XII. Further observations on the genus *Urocystidium,* 1—22, 1914.
    XIII. On two new species belonging to the genera *Oochoristica* and *Linstowia,* 265—269, 1914.
    XIV. On a new species of *Rhabdometra* and on paruterine organs in *Otidotaenia,* 859—887, 1914.
    XV. On a new species and genus of the family Acoleidae, 1039—1055, 1914.
    XVI. On certain points in the anatomy of the genus *Amabilia* and of *Dasyurotaenia,* 175—191, 1915.
    XVII. On *Taenia tauricollis* of Chapman and on the genus *Chapmania,* 429—443, 1915.
    XVIII. On *Taenia struthionis* (Parona) and allied forms, 589—601, 1915.
    XIX. On two new species of cestodes belonging to the genera *Linstowia* and *Cotugnia,* 695—706, 1916.
248. **Beddard, F. E.,** On the scolex of the cestode genus *Duthiersia,* and on the species of that genus, *Proc. Zool. Soc. London,* p. 73, 1917.
249. **Beddard, F. E.,** On a new tentaculate cestode, *Ann. Mag. Nat. Hist.,* (Ser. 9), 5, 203—207, 1920.
250. **Befus, A. D. and Freeman, R. S.,** *Corallobothrium parafimbriatum* sp. n. and *Corallotaenia minutia* (Fritts, 1959) comb. n. (Cestoda: Proteocephaloidea) from Algonquin Park, Ontario, *Can. J. Zool.,* 51, 243—248, 1973.
251. **Belogurov, O. I. and Zueva, L. Z.,** Two new species of cestodes from Charadriiformes in the Far East (in Russian), *Soobshch. Dal'nevost. Fil. V. L. Komarova sib. Otd. Akad. Nauk SSSR,* No. 26, 11—17, 1968.

252. **Belopolskaja, M. M.,** Helminth fauna of tetraonid birds (in Russian), *Rabot. Gel'mintol. 75-Let. Skrjabin,* pp. 47—65, 1954.
253. **Belopolskaya, M. M.,** Parasitic fauna of birds in the Sudzukhinsk Reserve (Primorsk Region). IV. Cestoidea (in Russian), *Tr. Gel'mintol. Lab. Akad. Nauk SSSR,* 13, 144—163, 1963.
254. **Belopolskaya, M. M.,** Cestoda of the genus *Aploparakis* Clerc, 1903 (Hymenolepididae) from Charadriiformes (in Russian), *Parazit. Sb.,* 24, 49—62, 1969.
255. **Belopolskaya, M. M.,** The formation of strobilae of *Wardium amphitricha* n. comb. and transition to dioecism (in Russian), *Parazitologiya,* 4, 201—209, 1970.
256. **Belopolskaya, M. M.,** Cestodes of the family Hymenolepididae from Charadriiformes (in Russian), *Vest. Leningrad. Univ. Ser. Biol.,* 25, 19—33, 1970.
257. **Belopolskaya, M. M.,** *Amoebotaenia oophorae* n. sp. (Dilepididae), the formation and structure of its oophore (in Russian), *Parazitologiya,* 5, 77—82, 1971.
258. **Belopolskaya, M. M.,** Cestodes of the family Progynotaeniidae Burt, 1939 from Charadriiformes of the U.S.S.R. (in Russian), *Parazitologiya,* 7, 44—50, 1973.
259. **Belopolskaya, M. M.,** Cestodes of Charadriiformes from the genera *Anomotaenia* Cohn, 1900 and *Dichoanotaenia* Lopez-Neyra, 1944 (Cestoda: Cyclophyllidea) (in Russian), *Vest. Leningrad. Univ. Ser. Biol.,* 4 (21), 31—44, 1977.
260. **Beneden, P. J. van,** Sur le développement des tétrarhynques, *Bull. Acad. R. Sci. Belg.,* 16, 44—52, 1849.
261. **Beneden, P. J. van,** Notice sur un nouveau genre d'helminthe cestoide, *Bull. Acad. R. Sci. Belg.,* 16, 182—193, 1849.
262. **Beneden, P. J. van,** Les helminthes cestoides considerés sous le rapport de leurs métamorphoses, de leur composition anatomique, et de leur classification, *Bull. Acad. R. Sci. Belg.,* 16, 269—282, 1849; *Bull. Acad. R. Sci. Belg.,* 17, 102—108, 1849—50.
263. **Beneden, P. J. van,** Recherches sur la faune littorale de Belgique; les vers cestoides, *Mém. Acad. R. Sci. Belg.,* 25, 1—4, 1850.
264. **Beneden, P. J. van,** Mémoire sur les vers intestinaux, *Suppl. C. R. Acad. Sci. Paris,* 2, 1—376, 1848 and 1861.
265. **Beneden, P. J. van,** Les poissons des côtes de Belge, leurs parasites et leurs commenseaux, *Mém. Acad. R. Sci. Belg.,* 38, 1—120, 1871.
266. **Beneden, P. J. van,** Recerces sur le développement embryonnaire de quelques Ténias, *Arch. Biol. Liège,* 2, 183—210, 1881.
267. **Beneden, P. J. van,** Deux cestodes nouveaux de *Lamna cornubica, Bull. Acad. R. Sci. Belg.,* 17, 68—74, 1889.
268. **Benedict, H. M.,** On the structure of two fish tapeworms of the genus *Proteocephalus* Weinland, 1858, *J. Morphol.,* 16, 337—368, 1900.
269. **Benham, W. B.,** On the structure of the rostellum in two new species of tapeworm from *Apteryx, Q. J. Microsc. Sci.,* 43, 83—96, 1900.
270. **Beveridge, I.,** On two new davaineid cestodes from Australian marsupials. *J. Helminthol.,* 49, 129—136, 1975.
271. **Beveridge, I.,** A taxonomic revision of the Anoplocephalidae (Cestoda: Cyclophyllidea) of Australian marsupials, *Aust. J. Zool. Suppl. Ser.,* 44, 1—110, 1976.
272. **Beveridge, I.,** On two new linstowiid cestodes from Australian dasyurid marsupials, *J. Helminthol.,* 51, 31—40, 1977.
273. **Beveridge, I.,** *Progamotaenia ruficola* sp. n. (Cestoda: Anoplocephalidae) from the red kangaroo, *Macropus rufus* (Marsupialia), *J. Parasitol.,* 64, 273—276, 1978.
274. **Beveridge, I.,** A taxonomic review of the genera *Cittotaenia* Riehm, 1881, *Ctenotaenia* Railliet, 1893, *Mosgovoyia* Spasskii, 1951, and *Pseudocittotaenia* Tenora, 1976 (Cestoda: Anoplocephalidae), *Mém. Mus. Natl. Hist. Nat. (Nov. Ser.),* 107, 1—64 (plus 24 unnumbered plates), 1978.
275. **Beveridge, I.,** *Progamotaenia* Nybelin (Cestoda: Anoplocephalide): new species, redescriptions and new host records, *Trans. R. Soc. South Aust.,* 104, 67—79, 1980.
276. **Beveridge, I.,** Three new species of *Calostaurus* (Cestoda: Davaineidae) from the New Guinea wallaby *Dorcopsis veterum, Trans. R. Soc. South Aust.,* 105, 139—147, 1981.
277. **Beveridge, I. and Gregory, G. G.,** The identification of *Taenia* species from Australian carnivores, *Aust. Vet. J.,* 52, 369—373, 1976.
278. **Beveridge, I. and Thompson, R. C. A.,** The anoplocephalid cestode parasites of the spectacled hare-wallaby, *Lagorchestes conspicillatus* Gould, 1842 (Marsupialia: Macropodidae), *J. Helminthol.,* 53, 153—160, 1979.
279. **Beverly-Burton, M.,** A new dilepidid cestode, *Mashonolepis dafyddi* n. g., n. sp., from the gray heron, *Ardea cinerea* L., *J. Parasitol.,* 46, 487—490, 1960.
280. **Beverly-Burton, M.,** A new cestode, *Hymenolepis mandabbi* sp. nov., from the tufted duck, *Aythya fuligula* (L.), *Ann. Mag. Nat. Hist.* (Ser. 13, Year 1959), 2, 560—564, 1960.

281. **Beverly-Burton, M.**, Studies on the cestoda of British freshwater birds, *Proc. Zool. Soc. London*, 112, 307—346, 1964.
282. **Beverly-Burton, M. and Thomas, V. G.**, *Rhabdometra lygodaptrion* n. sp. (Cestoda: Paruteriniidae) from willow ptarmigan (*Lagopus lagopus albus*) in northern Ontario, Canada, *Syst. Parasitol.*, 1, 157—165, 1980.
283. **Bhaduri, N. V. and Maplestone, P. A.**, Variations in *Taenia gaigeri* (Hall, 1919), *Rec. Ind. Mus.*, 42, 431—435, 1940.
284. **Bhalerao, G. D.**, On some representatives of the cestode genus *Avitellina* from India, *J. Helminthol.*, 14, 141—162, 1936.
285. **Bienek, G. K. and Grundman, A. W.**, A new tapeworm, *Schizorchodes dipodomi* gen. et sp. n. (Cestoda: Anoplocephalidae), from the Merriam kangaroo rat, *Dipodomys merriami vulcani*, *Proc. Helminthol. Soc. Wash.*, 40, 192—195, 1973.
286. **Bienek, G. K. and Grundman, A. W.**, *Catenotaenia utahensis* sp. n. (Cestoda: Catenotaeniidae) from the Merriam kangaroo rat, *Dipodomys merriami vulcani*, in Utah, *Proc. Helminthol. Soc. Wash.*, 41, 134—139, 1974.
287. **Bilqees, F. M.**, *Catenotaenia karachiensis*, new species (Cestoda) from *Paraechinus micropus* in Karachi, *Pakistan J. Zool.*, 11, 115—117, 1979.
288. **Bilqees, F. M.**, (1980) A new cestode genus *Myrmillorhynchus* (Trypanorhyncha: Gymnorhynchidae) from the fish *Myrmillo manazo* (Bik.), *Pakistan J. Zool.*, 12, 211—215, 1980.
289. **Bilqees, F. M.**, Three new species of *Acanthobothrium* Van Beneden (Cestoda: Tetraphyllidea: Onchobothriidae) in *Myrmillo Manazo* (Bik.) of Karachi coast, *Pakistan J. Zool.*, 12, 239—246, 1980.
290. **Bilquees, F. M. and Malik, N.**, *Hymenolepis mujibi* sp. n. (Cestoda: Hymenolepidide) from *Suncus murinus* L., *Norwegian J. Zool.*, 22, 319—321, 1974.
291. **Bilquees, F. M. and Masood, S.**, Parasites of *Varanus monitor* of Sind (Pakistan), *Sind. Univ. Res. J.*, 7, 41—56, 1973.
292. **Birkeland, I. W.**, "Bothriocephalus anemia", *Diphyllobothrium latum* and pernicious anemia, *Medicine (Baltimore)*, 11, 1—139, 1932.
293. **Birová-Volsinovicová, V.**, Neue Art einer Plathelminthe *Dilepis* (?) *spasskaya* sp. n. (Dilepididae) aus dem Eichalhäher (*Garrulus glandarius* L)., *Biol. Bratislava*, 22, 823—830, 1967.
294. **Bischoff, C. R.**, Cestoden aus *Hyrax*, *Zool. Anz.*, 39, 751—758, 1912.
295. **Bischoff, C. R.**, Cestoden aus *Hyrax*, *Rev. Suisse Zool.*, 21, 225—284, 1913.
296. **Blainville, H.**, Entozoorum synopsis cui accedunt mantissa duplex et indices locupletissimi, auctore Carolo Asmund Rudolphi, Extrait par. M. H. de Blainville, *J. Physique Paris*, 40, 229—234, 1820.
297. **Blair, D.**, *Polyonchobothrium scleropagis* n. sp. (Cestoda: Pseudophyllidea) from the Australian olsteoglossid fish *Scleropages leichardti* (Günther), *J. Helminthol.*, 52, 147—153, 1978.
298. **Blanchard, E.**, Recherches sur l'organisation des vers, *Ann. Sci. Nat. Zool.*, 10, 321—364; *Ann. Sci. Nat. Zool.*, 11, 106—202, 1848—49.
299. **Blanchard, R.**, Sur les helminthes des primates anthropoides, *Mém. Soc. Zool. Fr.*, 4, 186—196, 1891.
300. **Blanchard, R.**, *Historie Zoologique et Médicale des Téniadés du Genre Hymenolepis Weinland*, Paris, 1891, 112 pp.
301. **Blanchard, R.**, Notice sur les parasites de l'homme (3me sér.); sur le *Krabbea grandis* et remarques sur la classification de Bothriocephalines, *C. R. Soc. Biol. Paris*, 46, 699—702, 1894.
302. **Blanchard, R.**, Le genre *Chlamydocephalus* Cohn, 1906 remplacé par *Cephalochlamys*, *Arch. Parasitol.*, 12, 334, 1908.
303. **Blanchard, R.**, *Cotylorhipis furnarii*, nouveau genre de Téniadés, *Arch. Parasitol.*, 13, 477—482, 1909.
304. **Blanchard, R.**, *Bertiella satyri* de l'orang-outang, est aussi parasite de l'homme, *Bull. Acad. Méd.*, 69, 286—296, 1913.
305. **Blanchard, R.**, Tête de *Davainea madagascariensis*, *Bull. Soc. Pathol. Exot.*, 9, 413, 1916.
306. **Blei, R.**, Drei neue Schafszestoden, *Centralbl. Bakteriol.*, 87, 365—387, (1921), 1922.
307. **Bloch, M. E.**, Beitrag zur Naturgeschichte der Würmer, welche in andern Thieren leben, *Beschäft. Berl. Ges. Naturf. Fr.*, 4, 534—561, 1779.
308. **Bloch, M. E.**, Abhandlung von der Erzeugung der Eingeweidewürmer und den Mitteln wider dieselben, Danish Society of Science, Berlin, 1782, 54 pp.
309. **Blochmann, F.**, Ueber freie Nervenendigungen und Sinneszellen bei Bandwürmern, *Biol. Centr.*, 15, 14—25, 1895.
310. **Blochmann, F.**, Zur Epithelfrage bei Cestoden, *Zool. Anz.*, 29, 460—463, 1897.
311. **Boas, J. E. V.**, *Triplotaenia mirabilis*, *Zool. Jahrb. Syst.*, 17, 329—334, 1902.
312. **Bodkin, G. E. and Cleare, L. D.**, Notes on some animal parasites in British Guiana, *Bull. Entomol. Res.*, 7, 179—190, 1916.
313. **Boehm, L. K.**, Beiträge zur Kenntnis tierischer Parasiten, *Centralbl. Bakteriol. I. Abt.*, 87, 407—426, 1921.

314. **Boehm, L. K.,** Ein neuer Bandwurm von Huhn, *Raillietina (Davainea) grobbeni* n. sp., *Z. Wiss. Zool.,* 125, 518—532, 1925.
315. **Boev, S. N., Sokolova, I. B., and Tazieva, Z. K.,** Specificity of species causing cysticerciasis in ruminants (in Russian), in *Parasites of Farm Animals in Kazakhstan,* Boev, S. N., Ed., *Akad. Nauk Kazakh. SSSR,* Alma-Ata, 3, 7—29, 1964.
316. **Bona, F. V.,** Istologia delle capsule uterine di *Choanotaenia marchali* (Mola, 1907) e considerazioni sul genere *Monopylidium* Fuhrm., 1899 (Cestoda, Dilipididae), *Boll. Inst. Mus. Zool. Univ. Torino,* 5, 1—31, 1955—56.
317. **Bona, F. V.,** La formazione dei gusci embrionali e la morfologia dell'utero im *Paricterotaenia porosa* (Rud., 1810) quali elementi di guidizio per la validita del gen. *Paricterotaenia* Fuhrmann, 1932, (Cestoda, Dilepididae), *Riv. Parassitol.,* 18, 155—184, 1957.
318. **Bona, F. V.,** Etude critique et taxonomique des Dilepididae Fuhrm., 1907 (Cestoda) parasites des Ciconiiformes. Considérations sur la spécificité et la spéciation, *Ital. J. Zool. n. s. Monogr.,* 1, 1—750, 1975.
319. **Bona, F. V.,** The genus *Clelandia* Johnston, 1909 and its affinities with *Parvitaenia* and *Neogryporhynchus* (Cestoda, Dilepididae), *Annal. Parasitol.,* 53, 163—180, 1978.
320. **Bondarenko, S. K.,** *Wardium lagopi* n. sp. (Hymenolepididae) from *Lagopis lagopus* (in Russian), *Tr. Gel'mintol. Lab. Akad. Nauk SSSR,* 15, 64—66, 1965.
321. **Bondarenko, S. K.,** New trematodes and cestodes of shorebirds of Northern Asia (in Russian), *Mater. Nauchn. Konf. Vsesoyoznogo Obshchestva Gel'mintol.,* 3, 50—60, 1966.
322. **Bondarenko, S. K.,** Cestodes of the genus *Aploparaksis* Clerc 1903 (Hymenolepididae) from Charadriiformes from the Yenisey lowland and Norilsk lakes (in Russian), *Tr. Gel'mintol. Lab. Akad. Nauk SSSR,* 17, 19—34, 1966.
323. **Bondarenko, S. K.,** A new cestode species — *Aploparaksis parabirulai* n. sp. (Hymenolepididae) from *Macrorhamphus griseus* in Chukotka (in Russian), *Parazitologiya,* 9, 542—544, 1975.
324. **Bondarenko, S. K. and Kontrimavichus, V. L.,** Polymorphism of larvae of the genus *Aploparaksis* Clerc, 1903 (Hymenolepididae), *Folia Parasitol.,* 23, 39—44, 1976.
325. **Bondarenko, S. K. and Kontrimavichus, V. L.,** A new cestode of Charadriiformes—*Wardium chaunense* sp. n. (Hymenolepididae) from *Gallinago gallinago* (L.) from Chukotka (in Russian), *Folia Parasitol.,* 24, 281—284, 1977.
326. **Bondarenko, S. K. and Kontrimavichus, V. L.,** Life cycle of the cestode *Wardium calumnacantha* (Schmidt 1963) comb. n. (Hymenolepididae) from common snipe, *Gallinago gallinago* (L.), *Folia Parasitol.,* 25, 35—39, 1978.
327. **Bondarenko, S. K. and Kontrimavichus, V. L.,** *Dichoanotaenia multifilamenta* sp. n. (Cestoda: Dilepididae) from the black turnstone, *Arenaria melanocephala* (Vigors), in Alaska, *J. Parasitol.,* 66, 825—827, 1980.
328. **Bondarenko, S. K. and Rausch, R. L.,** *Aploparaksii borealis* sp. n. (Cestoda: Hymenolepididae) from passeriform and charadriform birds in Chukotka and Alaska, *J. Parasitol.,* 63, 96—98, 1977.
329. **Bondareva, V. I.,** Role of domestic and wild carnivores in the distribution of larval cestodes (in Russian), *Tr. Inst. Zool. Akad. Nauk Kaz. SSR,* 1, 126—131, 1953.
330. **Bondareva, V. I. and Zverev, M. D.,** Experimental infection of foxes and jackals with *Multiceps multiceps,* (in Russian), *Tr. Inst. Zool. Akad. Nauk Kaz. SSR,* 7, 237—240, 1957.
331. **Bonne, C.,** Over *Bertiella studeri* (Blanchard, 1891), *Bertiella satyri* Stiles & Hassall, 1926, *Bertia satyri* Blanchard, 1891, *Geneesk. Tijdschr. Ned.-Indie,* 80, 2222—2230, 1940.
332. **Borcea, L.,** Note sur *Tetrarhynchobothrium tenuicolle* Diesing, *Ann. Sci. Univ. Jassy,* 17, 565—567, 1933.
333. **Borcea, L.,** Note préliminaire sur les cestodes des Elasmobranches ou Séłaciens de la Mer Noire, *Ann. Sci. Univ. Jassy,* 19, 345—369, 1934.
334. **Borcea, L.,** Nouvelle note sur *Acanthobothrium ponticum, Ann. Sci. Univ. Jassy,* 20, 480—481, 1935.
335. **Borcea, L.,** Sur la présence du cestode *Diphyllobothrium stemmacephalum* Cobbold comme parasite chez le marsouin, *Phocaena phocaena,* de la Mer Noire, *Ann. Sci. Univ. Jassy,* 21, 524—525, 1935.
336. **Borgarenko, L. F.,** *Lateriporus solitariae* n. sp. (Cestoda: Dilepididae) from *Gallinago solitaria* from the Gissarskii mountain range (in Russian), *Izv. Akad. Nauk Tadzh. SSR (Ohboroti Akademija Fanhoi RSS Tocikiston), Otdelenie Biol. Nauki,* 4, 104—107, 1975.
337. **Borgarenko, L. F.,** *Icterotaenia columbi* n. sp. (Cestoda: Dilepididae) in pigeons in Garma (in Russian), *Izv. Akad. Nauk Tadzh. SSR, Otd. Biol. Nauki,* 1, 110—112, 1976.
338. **Borgarenko, L. F.,** A new cestode species, *Capiuterilepis pamirensis* n. sp. (Cestoda: Hymenolepididae) from *Leucosticte brandti pamirensis* in the Pamir (in Russian), *Dokl. Akad. Nauk Tadzh. SSR,* 19, 67—70, 1976.
339. **Borgarenko, L. F.,** Platyhelminths of Alaudidae in Tadzhikstan, *Izv. Akad. Nauk Tadzh. SSR (Ahboroti Akademijai Fanhoi RSS Tocikiston), Biol. Nauki,* 3, 60—65, 1976.
340. **Borgarenko, L. F.,** *Helminths of birds in Tadzhikistan,* Book 1. *Cestodes* (in Russian), Donish, Dushanbe, U.S.S.R., 1981, 1—327.

341. **Borgarenko, L. F., Spasskaja, L. P., and Spasskii, A. A.,** Cestodes of the genus *Tatria* in aquatic birds in Tadzhikistan (in Russian), *Izv. Akad. Nauk Tadzhikskoi SSR Biol. Nauki,* 4, 53—57, 1972.
342. **Borrel, A.,** Tumeurs cancereuses et helminthes, *Bull. Acad. Med.,* 56, 141, 1906.
343. **Bosc, L. A. G.,** *Histoire Naturelle des Vers, Contenant leur Description et leurs Moeurs,* Vol. 1, 324 pp.; Vol. 2, 300 pp.; Vol. 3, 270 pp., Paris, 1802.
344. **Bosc, L. A. G.,** Sur deux nouveaux genres de vers, *Bull. Sci. Soc. Philomat.,* 2, 384—385, 1811.
345. **Bosc, L. A. G.,** Hepatoxylon, *Hepatoxylon, Nouv. Dict. Hist. Nat.,* 14, 334—355, 1817.
346. **Bosc, L. A. G.,** Tentaculaire, *Tentacularia, Nouv. Dict. Hist. Nat.,* 33, 54—55, 1819.
347. **Bosc, L. A. G.,** Tétrarhynque, *Tetrarhynchus, Nouv. Dict. Hist. Nat.,* 33, 429, 1819.
348. **Boughton, I. B.,** Tapeworms in sheep and goats, *Southwest Sheep Goat Raiser,* 8, 26—27, 49—50, 1937.
349. **Boughton, I. B.,** A note on the broad tapeworm of sheep, *Southwest Sheep Goat Raiser,* 8, 11, 1938.
350. **Bounhiol, J.,** Sur l'existence d'une varieté nouvelle de *l'Anoplocephala plicata* Zeder (var. *servei*) chez le cheval barbe (cestodes), *Bull. Soc. Hist. Nat. Afr. Nord.,* 4, 146—147, 1912.
351. **Bourquin, J.,** Cestodes de mammifères; le genre *Bertia, Rev. Suisse Zool.,* 13, 415—503, 1905.
352. **Bourquin, J.,** Une nouveau *Taenia (Davainea)* chez les prosimiens, *Centralbl. Bakteriol. I. Abt.,* 41, 222, 1906.
353. **Bovien, P.,** Caryophyllaeidae from Java, *Medd. Dansk. Naturh. Foren.,* 82, 157—181, 1926.
354. **Brandt, B. B.,** Parasites of certain North Carolina Salientia, *Ecol. Monogr.,* 6, 491—532, 1936.
355. **Brandt, E. K.,** Zwei Fälle von *Taenia cucumerina* Rud. beim Menschen, *Zool. Anz.,* 11, 481—484, 1888.
356. **Braun, M.,** Zur Frage des Zwischenwirts von *Bothriocephalus latus* Brems, *Zool. Anz.,* 5, 39—43, 1882.
357. **Braun, M.,** *Die thierischen Parasiten des Menschen nebst einer Anleitung zur praktischen Beschäftigung mit der Helminthologie,* Würzburg, 1883, 232 pp.
358. **Braun, M.,** Zur Entwicklungsgeschichte des breiten Bandwurms (*Bothriocephalus latus* Brems.,), Würzburg; *Sitzungsber. Nat. Ges. Dorpat,* 6, 528—534, 1883.
359. **Braun, M.,** Ueber parasitische Schnurwürmer, *Centralbl. Bakteriol. I. Abt.,* 3, 16—19, 1888.
360. **Braun, M.,** Die embryonale Entwicklung der Cestoden, *Centralbl. Bakteriol. I. Abt.,* 5, 661—667, 1889.
361. **Braun, M.,** *Gyrocotyle, Amphiptyches* und Verwandte, Zusammenfassender Bericht, *Centralbl. Bakteriol. I. Abt.,* 6, 436—441, 1889.
362. **Braun, M.,** Helminthologische Mittheilungen, *Centralbl. Bakteriol. I. Abt.,* 9, 52—56, 1891.
363. **Braun, M.,** Verzeichniss von Eingeweidewürmern aus Mecklenburg, *Arch. Ver. Fr. Naturg. Mecklenburg,* 45, 97—117, 1892.
364. **Braun, M., Bronn, H. G., Ed.,** *Klassen und Ordnungen des Thierreichs,* Vol. IV, *Vermes*; Part I., *Cestodes,* H. G. Bronn, Leipzig, 1894—1900, 927—1731.
365. **Braun, M.,** Helminthologische Notizen. V. Ein proliferierender Cysticercus und die zugehörige Tänie, *Zool. Anz.,* 19, 417—420, 1896.
366. **Braun, M.,** Zur Entwicklungsgeschichte des *Cysticercus longicollis* Rud., *Zool. Anz.,* 20, 1—2, 1897.
367. **Braun, M.,** Notiz zur Entwicklung der *Taenia tenicollis* Rud., *Centralbl. Bakteriol. I. Abt.,* 39, 54, 1905.
368. **Bray, R. A.,** The cestode *Taenia krepkogorski* (Shulz and Landa, 1934) in the Arabian sand cat (*Felis margarita* Loche, 1858) in Bahrain, *Bull. Br. Mus. Nat. Hist. Zool.,* 24, 183—194, 1972.
369. **Bray, R. A.,** A new genus of dilepidid cestode in *Tilapia nilotica* (L. 1757) and *Phalacrocorax carbo* (L., 1958) in Sudan, *J. Nat. Hist.,* 8, 589—596, 1974.
370. **Bremser, J. G.,** *Nachricht von einer beträchtlichen Sammlung thierischer Eingeweidewürmer,* Vienna, 1811, 31 pp.
371. **Bremser, J. G.,** *Über lebende Würmer in lebenden Menschen,* Vienna, 1819, 284 pp.
372. **Bremser, J. G.,** Icones Helminthum Systema Rudolphii Entozoologicum Illustrantes, Vienna, 1824, 12 pp.
373. **Brendau, V.,** *Pseudoparadilepis ankeli* n. gen. n. sp. (Dilepididae; Dilepidinae) ein neurer Cestode aus der Zwergspitzmaus, *Zool. Anz.,* 182, 83—87, 1969.
374. **Briganti, V.,** Descrizione dele ligule, che abitano nell' addomine de ciprini del lago di Palo in provincia di principato Citra, *Atti R. Accad. Sci. Napoli,* 1, 209—233, 1819.
375. **Briganti, V.,** De novo vermium intestinalium genere, cui nomen Balanoforus descriptio, *Atti R. Accad. Sci. Napoli,* 2, 79—86, 1825.
376. **Brighenti, D.,** Descrizione di una nuova specie di tetrabotride parassita delle foche, *Boll. Zool.,* 2, 137—141, 1931.
377. **Brock, M. E.,** *Hymenolepis stolli,* a new hymenolepidid cestode from the pintail duck, *Wasmann Coll.,* 4, 135—138, 1941.
378. **Brock, M. E.,** A new hymenolepidid tapeworm, *Hymenolepis filumferens* from the blue winged teal, *Trans. Am. Microsc. Soc.,* 61, 181—185, 1942.
379. **Brooks, D. R.,** Six new species of tetraphyllidean cestodes, including a new genus, from a marine stingray *Himantura schmardae* (Werner, 1904) from Colombia, *Proc. Helminthol. Soc. Wash.,* 44, 51—59, 1977.
380. **Brooks, D. R.,** Systematic status of proteocephalid cestodes from reptiles and amphibians in North America with descriptions of three new species, *Proc. Helminthol. Soc. Wash.,* 45, 1—28, 1978.

381. **Brooks, D. R.**, Evolutionary history of the cestode order Proteocephalidea, *Syst. Zool.*, 27, 312—327, 1978.
382. **Brooks, D. A.**, A simulations approach to discerning possible sister-groups of *Dioecotaenia* Schmidt, 1969 (Cestoda: Tetraphyllidea; Dioecotaeniidae), *Proc. Helminthol. Soc. Wash.*, 49, 56—61, 1982.
383. **Brooks, D. R. and Deardorff, T. L.**, Three proteocephalid cestodes from Colombian siluriform fishes, including *Nomimoscolex alovarius* sp. n. (Monticelliidae: Zygobothriinae), *Proc. Helminthol. Soc. Wash.*, 47, 15—21, 1980.
384. **Brooks, D. R. and Mayes, M. A.**, *Hymenolepis asketus* sp. n. (Cestoidea: Hymenolepididae) from the short-tailed shrew *Blarina brevicanda* Say, from Nebraska, *Proc. Helminthol. Soc. Wash.*, 44, 60—62, 1977.
385. **Brooks, D. R. and Mayes, M. A.**, *Acanthobothrium electricolum* sp. n. and *A. lintoni* Goldstein, Henson, and Schlicht, 1969 (Cestoda: Tetraphyllidea) from *Narcine brasiliensis* (Olfer) (Chondrichthyes: Torpedinidae) in Colombia, *J. Parasitol.*, 64, 617—619, 1978.
386. **Brooks, D. R. and Mayes, M. A.**, Cestodes in four species of euryhaline stingrays from Colombia, *Proc. Helminthol. Soc. Wash.*, 47, 22—29, 1980.
387. **Brooks, D. R., Mayes, M. A., and Thorson, T. B.**, Systematic review of cestodes infecting freshwater stingrays (Chondrichthyes: Pomatotrygonidae) including four new species from Venezuela, *Proc. Helminthol. Soc. Wash.*, 48, 43—64, 1981.
388. **Brooks, D. R. and Schmidt, G. D.**, *Acanthotaenia overstreeti* sp. n. (Cestoda: Proteocephalidae) from a Puerto Rican lizard, the first acanthotaeniine in the New World, *Proc. Helminthol. Soc. Wash.*, 45, 193—195, 1978.
389. **Brooks, D. R. and Thorson, T. B.**, Two tetraphyllidean cestodes from the freshwater stingray *Potomatrygon magdalenae* Dumeril, 1852 (Chondrichthyes: Pomatotrygonidae) from Colombia, *J. Parasitol.*, 62, 943—947, 1976.
390. **Brown, F. J.**, Life-history of the fowl tapeworm, *Davainea proglottina, Nature (London)*, 131, 276—277, 1933.
391. **Brumpt, E.**, Evolution de *l'Hymenolepis nana* var. *fraterna, Arch. Zool. Exp. Gen.*, 75, 235—246, 1933.
392. **Brumpt, E.**, Réproduction expérimentale du sarcoma hépatique du rat par le cysticerque du *Taenia taeniaormis* ( = *T. crassicollis*) du chat, *Ann. Parasitol.*, 12, 130—133, 1934.
393. **Brumpt, E. and Joyeux, C.**, Description d'un nouvel echinocoque, *Echinococcus cruzi* n. sp., *Ann. Parasitol.*, 2, 226—231, 1924.
394. **Buchanan, G. D.**, Occurrence of the cestode *Mathevotaenia surinamensis* (Cohn, 1902), Spasskii, 1951 in a North American armadillo, *J. Parasitol.*, 42, 34—38, 1956.
395. **Buckley, J. J. C.**, *Cysticercus bovis* in the liver of a giraffe, *Trans. R. Soc. Trop. Med. Hyg.*, 41, 7, 1950.
396. **Bugge, G.**, Zur Kenntnis des Exkretionsgefäss-systems der Cestoden und Trematoden, *Zool. Jahrb. Anat.*, 16, 177—234, 1902.
397. **Buhler, G. A.**, *Monoecocestus giganticus* sp. n. (Cestoda: Anoplocephalidae) from the porcupine *Erithizon dorsatum* L. (Rodentia), *Proc. Helminthol. Soc. Wash.*, 37, 243—245, 1970.
398. **Burt, D. R. R.**, *Oochoristica lygosomae* sp. n., a cestode from the lizard *Lygosoma punctatum*, *Spolia Zeylanica*, 18, 1—7, 1933.
399. **Burt, D. R. R.**, Two new reptilian cestodes of the genus *Proteocephalus (Ophiotaenia), Ceylon J. Sci.*, 20, 157—179, 1937.
400. **Burt, D. R. R.**, New avian cestodes of the family Dilepididae from *Collocalia unicolor unicolor, Ceylon J. Sci.*, 21, 1—14, 1938.
401. **Burt, D. R. R.**, A new avian cestode, *Pseudochoanotaenia collocaliae* gen. et sp. nov. (Dipylidiinae), *Ceylon J. Sci.*, 21, 15—20, 1938.
402. **Burt, D. R. R.**, A new avian cestodes of the sub-family Dilepidinae from the eastern swallow (*Hirundo rustica gutturalis*) with descriptions of *Vitta magniuncinata* gen. et sp. nov., *Ceylon J. Sci.*, 21, 21—30, 1938.
403. **Burt, D. R. R.**, Some new cestodes of the genus *Paronia, Ceylon J. Sci.*, 21, 209—218, 1939.
404. **Burt, D. R. R.**, Some new species of cestodes from Charadriiformes, Ardeiformes and Pelecaniformes in Ceylon, *Ceylon J. Sci.*, 22, 1—63, 1940.
405. **Burt, D. R. R.**, New avian cestodes of the family Davaineidae from Ceylon, *Ceylon J. Sci.*, 22, 65—77, 1940.
406. **Burt, D. R. R.**, A new avian cestode, *Krimi chrysocolaptis* gen. et sp. nov. from Layard's woodpecker, *Chrysocolaptes guttacristatus stricklandi* (Layard, 1854), *Ceylon J. Sci.*, 22, 162—164, 1944.
407. **Burt, D. R. R.**, New avian species of *Hymenolepis* from Ceylon, *Ceylon J. Sci.*, 22, 165—172, 1944.
408. **Burt, D. R. R.**, On a new anoplocephalid cestode *Biporouterina psittaculae* gen. et sp. nov. from Layard's paroquet *Psittacula calthropae* (Layard, 1849), *Zool. J. Linn. Soc.*, 53, 81—86, 1973.
409. **Burt, D. R. R.**, On two new tetrabothriid cestodes from the brown gannet or booby *Sula leucogaster plotus* (Forster) from the Indian Ocean, *Zool. J. Linn. Soc.*, 58, 309—319, 1976.

410. **Burt, D. R. R.**, On a new species of tetrabothriid cestode from the shag *Phalacrocorax aristotelis aristotelis* (L.), *Zool. J. Linn. Soc.*, 60, 391—395, 1977.
411. **Burt, D. R. R.**, A reappraisal of *"Taenia heterosoma"* including descriptions of two new species of *Tetrabothrius*, *Zool. J. Linn. Soc.*, 62, 365—372, 1978.
412. **Burt, D. R. R.**, A new species of cestode from the small frigate bird *Fregata ariel iredalei* Mathews, taken in Ceylon and an emended account of *Tetrabothriun priestlyi* Leiper and Atkinson, from a frigate bird in South Trinidad, *Zool. J. Linn. Soc.*, 64, 1—7, 1978.
413. **Burt, D. R. R.**, New cestodes of the genus *Eurycestus* Clark 1954 from the avocet *Recurvirostra americana* Gmelin 1788, *Zool. J. Linn. Soc.*, 65, 71—82, 1979.
414. **Burt, M. D. B.**, Cyclophyllidean cestodes from birds in Borneo, *Bull. Br. Mus. Nat. Hist. Zool.*, 17, 281—346, 1969.
415. **Burt, M. D. B. and McLaughlin, J. O.**, On two cestode species, parasites of waterfowl, *Hymenolepis diorchis* Fuhrmann, 1913, and *Diorchis diorchis* Baer, 1962, *Acta Parasitol. Pol.*, 23, 201—205, 1975.
416. **Burt, M. D. B. and Sandeman, I. M.**, Biology of *Bothrimonus* ( = *Diplocotyle*) (Pseudophyllidea: Cestoda). I. History, description, synonomy, and systematics, *J. Fish. Res. Board Can.*, 26, 975—996, 1969.
417. **Buscher, H. N.**, *Raillietina (Raillietina) selfi* sp. n. (Cestoda: Davaineidae) from the desert cottontail in Okalahoma with notes on the distribution of *Raillietina* from North American mammals, *Proc. Okla. Acad. Sci.*, 55, 103—107, 1975.
418. **Bussieras, J. and Aldwin, J. F.**, *Sphyriocephalus dollfusi* n. sp., cestode *Trypanorhyncha* parasite de l'estomac des thon patudo, *Thunnus obesus*, *Ann. Parasitol. Hum. Comp.*, 43, 645—653, 1968.
419. **Butylin, R. Y.**, *Moniezia (Blanchardiezia) kuzetsovi*, a new cestode of sheep and goats (in Russian), *Zool. Zh.*, 53, 626—628, 1974.
420. **Bychowsky, B. E.**, Ontogenetic and phylogenetic relationships of parasitic flatworms, *Bull. Acad. Sci. U.R.S.S. Cl. Sci. Math. Nat. Ser. Biol.*, pp. 1353—1383, 1937.
421. **Bylund, G.**, Studies on the taxonomic status and biology of *Diphyllobothrium vogeli* Kuhlow, 1953, *Commentationes Biol. Soc. Sci. Fennica*, 79, 1—22, 1975.
422. **Byrd, E. E. and Fite, F. W.**, Morphological observations on normal and triradiated *Taenia pisiformis* from the dog, *J. Parasitol.*, 39 (Sect. 2, Suppl.), 24, 1953.
423. **Byrd, E. E. and Fite, F. W.**, Studies on the anatomical features of *Multiceps packi* Christenson, 1929, a cestode parasite of the dog, *J. Parasitol.*, 41, 149—156, 1955.
424. **Byrd, E. E. and Ward, J. W.**, Observations on the segmental anatomy of the tapeworm *Mesocestoides variabilis* Mueller from the opossum, *J. Parasitol.*, 29, 217—226, 1943.
425. **Caballero, C. E. and Vogelsang, E. G.**, Presencia del *Polycephalus serialis* (Gervais, 1847) en la liebre *Lepus californicus* del Norte de Mexico, *Rev. Med. Vet. Parasitol. Caracus*, 6, 79—80, 1947.
426. **Cable, R. M. and Meyers, R. M.**, A dioecious species of *Gyrocoelia* (Cestoda: Acoleidae) from the naped plover, *J. Parasitol.*, 42, 510—515, 1956.
427. **Cable, R. M. and Michaelis, M. B.**, *Plicatobothrium cypseluri* n. gen., n. sp. (Cestoda: Pseudophyllidea) from the Caribbean flying fish, *Cypselurus bahiensis* (Ranzani, 1842), *Proc. Helminthol. Soc. Wash.*, 34, 15—18, 1967.
428. **Cabot, D.**, A new tapeworm, *Hymenolepis arenariae* n. sp. (Cyclophyllidea: Hymenolepididae) from the intestine of the turnstone, *Arenaria interpres* L. (Aves: Charadriiformes) from Galway Bay, Ireland, *Ir. Nat. J.*, 16, 135—138, 1969.
429. **Calentine, R. L.**, *Archigetes iowensis* sp. n. (Cestoda: Caryophyllaeidae) from *Cyprinus carpio* L. and *Limnodrilus hoffmeisteri* Claparède, *J. Parasitol.*, 48, 513—524, 1962.
430. **Calentine, R. L. and Mackiewicz, J. S.**, *Monobothrium ulmeri* n. sp. (Cestoda: Caryophyllaeidae) from North American Catostomidae, *Trans. Am. Microsc. Soc.*, 85, 516—520, 1966.
431. **Calentine, R. L. and Ulmer, M. J.**, *Khawia iowensis* n. sp. (Cestoda: Caryophyllaeidae) from *Cyprinus carpio* L. in Iowa, *J. Parasitol.*, 47, 795—805, 1961.
432. **Callot, J. and Desportes, C.**, Sur le cycle évolutif de *Schistocephalus solidus* (O. F. Mueller), *Ann. Parasitol.*, 12, 35—39, 1934.
433. **Calvo, F. R.**, Contribucion a la estadística de *Inermicapsifer cubensis* (Kouri, 1938) Kouri, 1940, Reporte de seite nuevos casos, *Rev. Kuba. Med. Trop. Parasitol.*, 6, 26—27, 1950.
434. **Cameron, T. W. M.**, The cestode genus *Mesocestoides* Vaillant, *J. Helminthol.*, 3, 33—44, 1925.
435. **Cameron, T. W. M.**, Observations on the genus *Echinococcus* Rudolphi, 1801, *J. Helminthol.*, 4, 13—22, 1926.
436. **Cameron, T. W. M.**, Some modern conceptions on the hydatid, *Proc. R. Soc. Med.*, 20, 272—283, 1927.
437. **Cameron, T. W. M.**, On some parasites of the rusty tiger cat *(Felis planiceps)*, *J. Helminthol.*, 6, 87—98, 1928.
438. **Cameron, T. W. M.**, A new record of the occurrence of tapeworm of the genus *Bertiella* in man, *J. Helminthol.*, 7, 231—234, 1929.

439. **Cameron, T. W. M.**, The internal parasites of land mammals in Scotland, *Proc. R. Phys. Soc.*, 22, 133—154, 1933.
440. **Cameron, T. W. M.**, Studies on the endoparasitic fauna of Trinidad. III. Some parasites of Trinidad carnivora, *Can. J. Res.*, 14, 25—38, 1936.
441. **Cammerlohner, H.**, Über die Larve von *Anthocephalus elongatus*, *Sitzungsber. Akad. Wiss. Wien. Math. Nat. Kl. Abt. I.*, 138, 125—143, 1929.
442. **Campbell, D. M. and Lacroix, J. V.**, *Essentials of Parasitology, Including a Brief Discourse on Zoology*, Hiawatha, Kansas, 1907, 96 pp.
443. **Campbell, R. A.**, New species of *Acanthobothrium* (Cestoda: Tetraphyllidea) from Chesapeake Bay, Virginia, *J. Parasitol.*, 55, 559—570, 1969.
444. **Campbell, R. A.**, Notes on tetraphyllidean cestodes from the Atlantic coast of North America with descriptions of two new species, *J. Parasitol.*, 56, 498—508, 1970.
445. **Campbell, R. A.**, Two new species of *Echeneibothrium* (Cestoda: Tetraphyllidea) from skates in the Western North Atlantic, *J. Parasitol.*, 61, 95—99, 1975.
446. **Campbell, R. A.**, Tetraphyllidean cestodes from Western North Atlantic selachians with description of two new species, *J. Parasitol.*, 61, 265—270, 1975.
447. **Campbell, R. A.**, New tetraphyllidean and tryparorhynch cestodes from deep-sea skates in the Western North Atlantic, *Proc. Helminthol. Soc. Wash.*, 44, 191—197, 1977.
448. **Campbell, R. A.**, A new family of pseudophyllidean cestodes from the deep-sea teleost *Acanthochaenas lutkenii* Gill, 1884, *J. Parasitol.*, 63, 301—305, 1977.
449. **Campbell, R. A.**, Two new genera of pseudophyllidean cestodes from deep-sea fishes, *Proc. Helminthol. Soc. Wash.*, 46, 74—78, 1979.
450. **Campbell, R. A. and Carvajal, J.**, A revision of some typanorhynchs from the Western North Atlantic described by Edwin Linton, *J. Parasitol.*, 61, 1016—1022, 1975.
451. **Campbell, R. A. and Carvajal, J.**, Synonymy of the phyllobothriid genera *Rhodobothrium* Linton, 1889, *Inermiphyllidum* Riser, 1955, and *Sphaerobothrium* Euzet, 1954 (Cestoda: Tetraphyllidea), *Proc. Helminthol. Soc. Wash.*, 46, 88—97, 1979.
452. **Campbell, R. A., Correia, S. J., and Haedrich, R. L.**, A new monogenean and cestode from the deep-sea fish, *Macrourus berglax* Lacépède, 1802, from the Flemish Cape of Newfoundland, *Proc. Helminthol. Soc. Wash.*, 49, 169—175, 1982.
453. **Campbell, R. A. and Gartner, J. V.**, *Pistana eurypharyngis* gen. et sp. n. (Cestoda: Pseudophyllidea) from the bathypelagic gulper eel, *Eurypharynx pelecanoides* Vaillant, 1882, with comments on host and parasite ecology, *Proc. Helminthol. Soc. Wash.*, 49, 218—225, 1982.
454. **Canavan, W. P.**, Notes on the actual appearance of a cestode shell gland, *Parasitology*, 19, 283, 1927.
455. **Canavan, W. P.**, Notes on the occurrence of *Ophiotaenia loennbergi* in Pennsylvania, *J. Helminthol.*, 6, 56, 1928.
456. **Canavan, W. P.**, The spread of broad fish tapeworm of man, *Science*, 75, 270, 1932.
457. **Canavan, W. P.**, A six suckered tapeworm, *Taenia hydatigena* (*Hydatigena* Pallas, 1776), *J. Parasitol.*, 20, 57, 1933.
458. **Cannon, D. G.**, On the parasites of the small intestine of the European starling (*Sturnus vulgaris*) in Quebec, *Can. Field Nat.*, 53, 40—42, 1939.
459. **Capoor, V. N.**, On a new cestode, *Taufikia ghoshi* n. sp., from white gidha, *Neophron percnopterus* linnaeus, from Allahabad (India), *Ind. J. Helminthol. Year 1966*, 18, 172—176, 1967.
460. **Capoor, V. N.**, On a new cestode, *Mogheia bayamegaparuterina* n. sp. from the Indian common baya, *Ploceus philippensis* (Linneaus) from Allahabad, India, with the revision of the diagnosis of the genus *Mogheia* Lopez-Neyra, 1944, *Proc. Natl. Acad. Sci. India Sect. B*, 37, 51—53, 1967.
461. **Capoor, V. N. and Srivastava, V. C.**, On a new cestode, *Mogheia megaparuterina* n. sp., from Allahabad (India), *Proc. Ind. Acad. Sci. Sect. B.*, 64, 293—295, 1966.
462. **Capoor, V. N. and Srivastava, V. C.**, On the synonymy of the genus *Columbia* Srivastava and Capoor, 1966 with *Kelligrewia* Meggitt, 1927 and a new combination *Killigrewia allahabadi* n. comb. and with its redescription, *J. Zool. Soc. India Year 1965*, 17, 123—124, 1967.
463. **Capoor, V. N., Srivastava, V. C., and Chauhan, A. S.**, On a new cestode, *Valipora sultanpurensis* n. sp. of the subfamily Dilepidinae Fuhrmann, 1907, family Dilepididae Railliet et Henry 1909, in Dr. B. S. Chanhan Commemoration Volume, Tiwari, K. K. and Srivastava, C. B., Eds., Orissa, India, *Zool. Soc. India*, pp. 373—376, 1975.
464. **Capoor, V. N., Srivastava, V. C., and Chauhan, A. S.**, On a new cestode of the genus *Oochoristica* Lühe, 1891, from an unidentified snake from Allahabad, India, *J. Zool. Soc. India*, 26, 83—86, 1976.
465. **Carta, A.**, L'identificazione del ciclo evolutivo del *Mesocestoides lineatus* povata sperimentalmente, *Riv. Parassitol.*, 3, 65—81, 1939.
466. **Carta, A.**, Il *Tetrathyridium bailleti* puo riprodursi?, *Profilassi*, 12, 71—73, 1939.
467. **Carta, A.**, Biologia ed asione patogena del *Mesocestoides lineatus* e delle sue forme larvali (Nota risseuntiva), *Profilassi*, 14, 83—93, 1941.

468. **Carter, W. J.**, *Proteocephalus sandgroundi*, a new tetraphyllidean cestode from an East Indian monitor lizard, *Trans. Am. Microsc. Soc.*, 62, 301—305, 1943.
469. **Carvajal, G. J.**, *Grillotia dollfusi* sp. n. (Cestoda: Trypanorhyncha) from the skate, *Raja chilensis*, from Chile, and a note on *G. heptanchi*, *J. Parasitol.*, 57, 1269—1271, 1971.
470. **Carvajal, J.**, Records of cestodes from Chilean sharks, *J. Parasitol.*, 60, 29—34, 1974.
471. **Carvajal, J.**, Description of the adult and larva of *Caulobothrium myliobatidis* sp. n. (Cestoda: Tetraphyllidea) from Chile, *J. Parasitol.*, 63, 99—103, 1977.
472. **Carvajal, J. and Campbell, R. A.**, *Rhinoptericola megacantha* gen. et sp. n., representing a new family of trypanorhynch cestodes from common ray, *Rhinoptera bonasus* (Mitchell, 1815), *J. Parasitol.*, 61, 1023—1030, 1975.
473. **Carvajal, J., Campbell, R. A., and Cornford, E. M.**, Some trypanorhynch cestodes from Hawaiian fishes, with descriptions of four new species, *J. Parasitol.*, 62, 70—77, 1976.
474. **Carvajal, J. and Dailey, M. D.**, Three new species of *Echeneibothrium* (Cestoda: Tetraphyllidea) from the skate, *Raja chilensis* Guichenot, 1848, with comments on mode of attachment and host specificity, *J. Parasitol.*, 61, 89—94, 1975.
475. **Carvajal, G. J. and Goldstein, R. J.**, *Acanthobothrum psammobati* sp. n. (Cestoda: Tetraphyllidea: Onchobothriidae) from the skate, *Psammobatis scobina* (Chondrichthyes: Rajidae) from Chile, *Zool. Anz.*, 182, 432—435, 1969.
476. **Carvajal, J. and Goldstein, R. J.**, *Acanthobothrium annapinkiensis* n. sp. (Cestoda: Tetraphyllidea: Onchobothriidae) from the skate, *Raja chilensis* (Chondrichthyes: Rajiidae) from Chile, *Zool. Anz.*, 186, 158—162, 1971.
477. **Case, A. A. and Ackert, J. E.**, Intermediate hosts of chicken tapeworms found in Kansas, *Trans. Kan. Acad. Sci.*, 42, 437—442, 1939.
478. **Case, A. A. and Ackert, J. E.**, New intermediate hosts of fowl cestodes, *Trans. Kan. Acad. Sci.*, 43, 393—396, 1940.
479. **Cerruti, A.**, Di un tenioide dell' *Alauda arvensis* con riguardo speciale ad un organo parauterino, *Atti R. Accad. Sci. Napoli*, 11, 1—8, 1901.
480. **Cerruti, A.**, Su la *Oochoristica (Taenia) tuberculata* (Rud.), *Boll. Soc. Nat. Napoli*, 16, 311, 1903.
481. **Chaloner, J. W.**, On the cestode parasites of trout with special reference to the plerocercoid disease of trout from Loch Morar, *Rep. Br. Assoc. Sci.*, 1912, 507—509, 1913.
482. **Chandler, A. C.**, A new record of *Taenia confusa* with additional notes on the morphology, *J. Parasitol.*, 7, 34—38, 1920.
483. **Chandler, A. C.**, Species of *Hymenolepis* as human parasites, *J. Am. Vet. Assoc.*, 78, 636—639, 1922.
484. **Chandler, A. C.**, Observations on the life-cycle of *Davainea proglottina* in the United States, *Trans. Am. Microsc. Soc.*, 42, 144—147, 1923.
485. **Chandler, A. C.**, New records of *Bertiella satyri* (Cestoda) in man and apes, *Parasitology*, 17, 421—425, 1925.
486. **Chandler, A. C.**, The helminthic parasites of cats in Calcutta and the relation of cats to human helminthic infections, *Ind. J. Med. Res.*, 13, 213—227, 1925.
487. **Chandler, A. C.**, On a specimen of *Taenia pisiformis* with a completely doubled strobila, *Trans. Am. Microsc. Soc.*, 49, 168—173, 1930.
488. **Chandler, A. C.**, Susceptibility and resistance to helminth infections, *J. Parasitol.*, 18, 135—152, 1932.
489. **Chandler, A. C.**, Notes on the helminth parasites of the opossum (*Didelphys virginiana*) in Southeast Texas, with descriptions of four new species, *Proc. U.S. Natl. Mus.*, 81, 1—15, 1932.
490. **Chandler, A. C.**, Parasites of fishes in Galveston Bay, *Proc. U.S. Natl. Mus.*, 83, 123—157, 1935.
491. **Chandler, A. C.**, A new tetrarhynchid larva from Galveston Bay, *J. Parasitol.*, 21, 214—215, 1935.
492. **Chandler, A. C.**, The genus *Schizotaenia* in porcupines, *J. Parasitol.*, 22, 513, 1936.
493. **Chandler, A. C.**, Helminths of muskrats in Southeast Texas, *J. Parasitol.*, 27, 183—184, 1941.
494. **Chandler, A. C.**, Helminths of tree squirrels in Southeast Texas, *J. Parasitol.*, 28, 135—140, 1942.
495. **Chandler, A. C.**, Some cestodes from Florida sharks, *Proc. U.S. Natl. Mus.*, 92, 25—31, 1942.
496. **Chandler, A. C.**, *Mesocestoides manteri* n. sp. from a lynx, with notes on other North American species of *Mesocestoides*, *J. Parasitol.*, 28, 227—231, 1942.
497. **Chandler, A. C.**, The helminths of raccoons in East Texas, *J. Parasitol.*, 28, 255—268, 1942.
498. **Chandler, A. C.**, First record of a case of human infection with tapeworms of the genus *Mesocestoides*, *Am. J. Trop. Med.*, 22, 493—498, 1942.
499. **Chandler, A. C.**, First case of human infection with *Mesocestoides*, *Science*, 96, 112, 1942.
500. **Chandler, A. C.**, A new species of *Mesocestoides*, *M. Kirbyi*, from *Canis latrans*, *J. Parasitol.*, 30, 273, 1944.
501. **Chandler, A. C.**, Helminths of armadillos, *Dasypus novemcinctus*, in eastern Texas, *J. Parasitol.*, 32, 237—241, 1946.
502. **Chandler, A. C.**, Observations on the anatomy of *Mesocestoides*, *J. Parasitol.*, 32, 242—246, 1946.

503. **Chandler, A. C.**, The anatomy of *Mesocestoides* — corrections, *J. Parasitol.*, 33, 444, 1947.
504. **Chandler, A. C.**, New species of the genus *Schistotaenia* with a key to the known species, *Trans. Am. Microsc. Soc.*, 67, 169—176, 1948.
505. **Chandler, A. C.**, Two new species of *Oochoristica* from Minnesota skunks, *Am. Midl. Nat.*, 48, 69—73, 1952.
506. **Chandler, A. C.**, Cestoda (in Gulf of Mexico, Its Origin, Waters and Marine Life), Fish Bull. (89), U.S. Fish Wildl. Serv., Washington, D.C., 1954, pp. 351—353.
507. **Chandler, A. C. and Melvin, D. M.**, A new cestode, *Oochoristica pennsylvanica,* and some new or rare helminth host records from Pennsylvania mammals, *J. Parasitol.*, 37, 106—109, 1951.
508. **Chandler, A. C. and Suttles, C. L.**, A new rat tapeworm, *Schizotaenia sigmodontis,* from North America, *J. Parasitol.*, 8, 123—128, 1922.
509. **Chapin, E. A.**, *Southwellia ransomi* n. sp., *J. Parasitol.*, 13, 29—33, 1926.
510. **Chapman, H. C.**, Description of a new *Taenia* from *Rhea americana, Proc. Acad. Nat. Sci. Philadelphia,* p. 14, 1876.
511. **Chapman, P. A.**, Some helminth parasites from partridges and other English birds, *J. Helminthol.*, 13, 139—148, 1935.
512. **Chatterji, P. N.**, Two new cestodes of the genera *Idiogenes* Krabbe, 1868 and *Choanotaenia* Railliet, 1896, *J. Parasitol.*, 40, 535—539, 1954.
513. **Chen, H. T.**, A preliminary report on a survey of animal parasites of Canton, China, *Lingnan Sci. J. Canton,* 12, 65—74, 1933.
514. **Cheng, T. C. and James, H. A.**, *Bothriocephalus schilbeodes* n. sp. (Cestodea: Bothriocephalidae), an intestinal parasite of *Schilbeodes insignis, J. Tenn. Acad. Sci.*, 35, 164—168, 1960.
515. **Chernobai, V. F.**, *Satyolepis skryabini* n. sp. (Hymenolepididae), a new cestode from corvid birds (in Russian), *Vestnik Zool.*, 3, 26—28, 1969.
516. **Chernobai, V. F.**, On the validity of *Dilepis brachyarthra* Cholodkovssky, 1906 (in Russian), *Vopr. Parazitol. Zhivotnykh Yugo-Vostoka SSSR Volgograd,* pp. 96—99, 1974.
517. **Chertkova, A. N. and Kosupko, G. A.**, Cestodes of the genus *Mesocestoides* found in domestic and wild animals in USSR and some principles of their systematics (in Russian), *Tr. Vses. Inst. Gel'mintol. im K. I. Skryabina (Teoreticheskie Problemy Vet. Gel'mintol.),* 22, 193—211, 1975.
518. **Chertkova, A. N. and Kosupko, G. A.**, Revision of the suborder Mesocestoidata Skrjabin, 1940 (in Russian), *Tr. Vses. Inst. Gel'mintol. im K. I. Skryabina,* 23, 141—153, 1977.
519. **Chibichanko, N. T.**, Cestodes of Coraciiformes in Kirgizia (in Russian), *Fauna Gel'mintov Zhivotnykl Rastenii Kirgizii,* pp. 59—67, 1974.
520. **Chin, T. G.**, A new species of cestode of the family Anoplocephalidae (Cestoda) from tapira, *Lingnan Sci. J.,* 17, 605—607, 1938.
521. **Chincholikar, L. N. and Shinde, G. B.**, On a new cestode *Yogeshwaria nagabhushani* (Cestoda: appendix to Lecanicephalidea, *insertae sedis*) gen. sp., *Marthwada Univ. J. Sci. Natl. Sci.*, 15, 273—276, 1976.
522. **Chincholikar, L. N. and Shinde, G. B.**, A new species of cestode *Gymnorhynchus cybiumi* (Gymnorhynchidae Dollfus, 1935) from a marine fish at Ratnagiri, India, *Riv. Parasitol.*, 38, 161—164, 1977.
523. **Chincholikar, L. N. and Shinde, G. B.**, On *Cephalobothrium subhapradhi* sp. n. (Cestoda: Lecanicephalidae Braun, 1900) from a marine fish at Ratnagiri, Maharashtra, India, in *All-India Symp. Helminthol.,* Srinigar, Kashmir, 8—11 August, 1977, 1977, pp. 8—9.
524. **Chincholikar, L. N. and Shinde, G. B.**, On a new species of *Circumonchobothrium* Shinde, 1968, (Cestoda: Pseudophyllidea, Carus, 1863), from a fresh water fish in India, *Marathwada Univ. J. Sci. Natl. Sci.*, 16(9), 183—185, 1977.
525. **Chincholikar, L. N. and Shinde, G. B.**, A new cestode *Eniochobothrium trygonis* sp. n. from *Trygon sephen, Folia Parasitol.*, 25, 177—178, 1978.
526. **Chincholikar, L. N. and Shinde, G. B.**, A new cestode *Lecanicephalum maharashtrae* sp. n. (Lecanicephalidae) from a marine fish *Trygon sephen, Folia Parasitol.*, 25, 345—346, 1978.
527. **Chincholikar, L. N. and Shinde, G. B.**, A new cestode *Tylocephalum madhukarii* n. sp. (Lecanicephalidae) from a marine fish, *Trygon sephen, Rev. Parassit.*, 41, 23—26, 1980.
528. **Chincholikar, L. N., Shinde, G. B., and Deshmukh, R. A.**, A new cestode *Ptychobothrium clupeoidesii* (Cestoda: Pseudophyllidea) from a freshwater fish from Aurangabad, India, *Marathwada Univ. J. Sci. Natl. Sci.,* 15, 277—280, 1976.
529. **Chiriai, E.**, Contributii la cunoasterea helmintofaunei păsărilor din R.P.R. (in Rumanian), *Anal. Univ. Buc. Seria Stiint. Nat. Biol.,* 12, 171—180, 1963.
530. **Christi, M. Z.**, On a new species of cestode genus *Choanataenia* Railliet 1896 from *Acridotheres tristis* in Kashmir, *J. Sci. Univ. Kashmir,* 1, 51—54, 1973.
531. **Christi, M. Z.**, On a new species of the cestode genus *Sobolevicanthus* from *Anas platyrhynchos platyrhynchos* in Kashmir, *Ind. J. Parasitol.*, 3 (Suppl.), 13—14, 1980.

532. **Chisti, M. Z. and Khan, A. R.**, *Mayhewia kavini* n. sp. (Hymenolepididae Railliet et Henry, 1909; Cestoda) from *Corvus monedula* in Kashmir, *All-India Symp. Helminthol.*, Spinigar, Kashmir, 8—11, August, 1977, 1977, 38.
533. **Chitwood, B. G., McIntosh, A., and Price, E. W.**, Report of the committee on nomenclature. Authors of combinations of zoological names, *J. Parasitol.*, 32, 519—520, 1946.
534. **Chizhova, T. P.**, On diphyllobothrids of gulls in the Baikal (in Russian), *Zool. Zh.*, 30, 218—223, 1951.
535. **Cholodkovsky, N.**, Helminthologische Notizen, *Centralbl. Bakteriol. Parasitenk.*, 18, 10—12, 1895.
536. **Cholodkovsky, N.**, Contributions à la connaissance des ténias des ruminants, *Arch. Parasitol.*, 6, 145—148, 1902.
537. **Cholodkovsky, N.**, Eine *Idiogenes* species mit wohlentwickeltem Scolex, *Zool. Anz.*, 29, 580—583, 1905.
538. **Cholodkovsky, N.**, Cestodes nouveaux ou peu connus, *Arch. Parasitol.*, 10, 332—347, 1906.
539. **Cholodkovsky, N.**, Ueber eine neue Tänie des Hundes, *Zool. Anz.*, 33, 418—420, 1908; *Arch. Inst. Past. Tunis*, pp. 204—205, 1909.
540. **Cholodkovsky, N.**, Explanatory Catalogue of the Collection of Parasites of the Imperial Military Academy of Medicine, I. Tapeworms (Cyclophyllidea) (in Russian), St. Petersburg, 1912, 1—96.
541. **Cholodkovsky, N.**, Cestodes nouveaux ou peu connus, 2me sér., *Ann. Mus. Acad. Sci. Imp. St. Petersburg*, 18, 221—232 (1913), 1914.
542. **Cholodkovsky, N. A.**, Zur Kenntnis der Bothriocephalen Russlands (in Russian), *Tr. Imp. St. Petersburg Obsh. Estestvois I. Protok. Zasied.*, 45, 59—64, 1914.
543. **Cholodkovsky, N.**, Cestodes nouveaux ou peu connus, 3me sér., *Ann. Mus. Acad. Sci. Imp. St. Petersburg*, 19, 516—523, 1915.
544. **Cholodkovsky, N.**, Notes helminthologiques, *Ann. Mus. Acad. Sci. Imp. St. Petersburg*, 29, 164—166, 1915.
545. **Cholodkovsky, N.**, Contributions to the knowledge of cestodes (Pseudophyllidea) of the Russian fauna (in Russian), *Trav. Soc. Imp. Nat. St. Petersburg*, 46, 59, 1915.
546. **Cholodkovsky, N.**, Sur un nouveau parasite de l'homme (in Russian, French summary), *Zool. Vestnik.*, 1, 231—237, 1916; as abstracted by Leiper in *Trop. Dis. Bull.*, 19 (1922), 669, 1916.
547. **Cholodkovsky, N. A.**, Contribution à la connaissance de cysticerques d'oiseaux, *C. R. Soc. Biol.*, 80, 219—222, 1917.
548. **Cholodkovsky, N. A.**, Notes helminthologiques (3—4) (in Russian) *Izv. Ross. Akad. Nauk.*, 6 s., 12, 2311—2316, 1918.
549. **Christenson, R. O.**, A new cestode reared in the dog, *Multiceps packii* n. sp., *J. Parasitol.*, 16, 49—53, 1929.
550. **Christiansen, M.**, Muskelfinnen beim Reh *(Cervus capreolus)*, *Vidensk. Medd. Naturh. Foren. Kjøb.*, 84, 251—279, 1927.
551. **Christiansen, M.**, Die Muskelfinne des Rehes und deren Bandwurm (Cysticercus et *Taenia cervi* n. sp. and interim), *Z. Parasitenk.*, 4, 75—100, 1931.
552. **Cielecka, D. and Zditowiecki, K.**, The tapeworm *Microsomacanthus shetlandicus* (Hymenolepididae) from the Dominican gull of King George Island (South Shetlands, Antarctic), *Bull. Acad. Pol. Sci.*, 29, 173—180, 1981.
553. **Ciordia, H.**, *Mesocestoides jonesi* n. sp. from the gray fox, with descriptions of the chromosome complement and a dicephalic specimen, *J. Tenn. Acad. Sci.*, 30, 57—63, 1955.
544. **Clapham, P. A.**, On the presence of hooks on the rostellum of *Hymenolepis microps*, *J. Helminthol.*, 17, 21—24, 1939.
555. **Clapham, P. A.**, Some polyradiate specimens of *Taenia pisiformis* and *Dipylidium caninum*, with a bibliography of the abnormalities occurring among cestodes, *J. Helminthol.*, 17, 163—176, 1939.
556. **Clapham, P. A.**, Studies on *Coenurus glomeratus*, *J. Helminthol.*, 18, 45—52, 1940.
557. **Clapham, P. A.**, Further studies on *Coenurus glomeratus*, *J. Helminthol.*, 18, 95—102, 1940.
558. **Clapham, P. A.**, An English case of *Coenurus cerebralis* in the human brain, *J. Helminthol.*, 19, 84—86, 1941.
559. **Clapham, P. A.**, On two new coenuri from Africa and a note on the development of the hooks, *J. Helminthol.*, 20, 25—31, 1942.
560. **Clapham, P. A.**, On identifying *Multiceps* spp. by measurement of the large hook, *J. Helminthol.*, 20, 31—40, 1942.
561. **Clapham, P. A.**, On the occurrence of *Davainea madagascariensis* on the African mainland, *J. Helminthol.*, 22, 47—48, 1947.
562. **Clapham, P. A. and Peters, B. G.**, The differentiation of *Coenurus* species by hook measurements, *J. Helminthol.*, 19, 75—84, 1941.
563. **Clark, D. T.**, Three new dilepidid cestodes, *Dictymetra nummenii* n. gen. n. sp., *Dictymetra paranumenii* n. sp. and *Anomotaenia filovata* n. sp., *Proc. Helminthol. Soc. Wash.*, 19, 18—27, 1952.
564. **Clark, D. T.**, A new cyclophylidean cestode from the avocet, *J. Parasitol.*, 40, 340—346, 1954.

565. **Clark, H. G.,** Cestodes from cormorants from South Australia, *Trans. R. Soc. S. Aust.,* 80, 124—134, 1957.
566. **Clarke, A. S.,** Studies on the life cycle of the pseudophyllidean cestode, *Schistocephalus solidus, Proc. Zool. Soc. London,* 124, 257—302, 1954.
567. **Clarke, M. R.,** *Multiductus physeteris* gen. et sp. nov. — a new diphyllobothriid cestode from a sperm whale, *J. Helminthol.,* 36, 1—10, 1962.
568. **Claus, C.,** Zur morphologischen und polygenetischen Beurtheilung des Bandwurmkörpers, *Weiner Klin. Wochenschr.,* 2, 697—700, 1889.
569. **Clausen, E.,** Recherches Anatomiques et Histologiques sur Quelques Cestodes d'Oiseaux, Thesis, University of Neuchâtel, France, 1915, 111 pp.
570. **Cleland, J. B.,** The parasites of Australian birds, *Trans. R. Soc. S. Aust.,* 46, 85—118, 1922.
571. **Clerc, W.,** Contribution à l'étude de la faune helminthologique de l'Oural. I, *Zool. Anz.,* 25, 569—575, 1902; II, *Zool. Anz.,* 25, 650—664, 1902.
572. **Clerc, W.,** Contribution à l'étude de la faune helminthologique de l'Oural. III, *Rev. Suisse Zool.,* 11, 241—368, 1903.
573. **Clerc, W.,** Quelques remarques à propos d'une critique, *Zool. Anz.,* 28, 243—245, 1904.
574. **Clerc, W.,** Courte notice sur mes excursions zoologiques en 1903 et 1904 (Russes), *Bull. Soc. Ouralienne Sci. Nat.,* 25, 1—11, 1904.
575. **Clerc, W.,** Note sur les cestodes d'oiseaux de l'Oural. *I, Centralbl. Bakteriol. I. Abt.,* 42, 422—426; II, *Centralbl. Bakteriol. I. Abt.,* 42, 713—720, 1906.
576. **Clerc, W.,** Notes sur les cestodes d'oiseaux de l'Oural. III, *Centralbl. Bakteriol. I. Abt.,* 43, 703—708, 1907.
577. **Clerc, W.,** Liste des oiseaux ouverts durant les excursions de 1905 dans la region de l'Oural ente 54 et 57 L. N. afin d'en étudier les parasites, *Bull. Soc. Ouralienne Sci. Nat.,* 30, 91—98, 1910.
578. **Clerc, W.,** Quelques données sur l'origine de l'unisexualite dans le genre *Dioicocestus, Rev. Suisse Zool.,* 37, 147—171, 1930.
579. **Coatney, G. R.,** Some notes on cestodes from Nebraska, *J. Parasitol.,* 22, 409, 1936.
580. **Cobbold, T. S.,** Observations on Entoza with notices of several new species including an account of two experiments in regard to the breeding of *Taenia serrata* and *Taenia cucumerina, Trans. Linn. Soc. London,* 22, 155—172, 1858.
581. **Cobbold, T. S.,** On some new forms of Entozoa, *Trans. Linn. Soc. London,* 22, 363—366, 1859.
582. **Cobbold, T. S.,** Remarks on all the human entozoa, *Proc. Zool. Soc. London,* pp. 288—315, 1862.
583. **Cobbold, T. S.,** *Entozoa: An Introduction to the Study of Helminthology,* London, 1864.
584. **Cobbold, T. S.,** New species of human tapeworm, *Trans. Pathol. Soc. London,* 17, 438—439, 1866.
585. **Cobbold, T. S.,** Parasites: *A Treatise on the Entozoa of Man and Animals,* London, 1879, 1—519.
586. **Cobbold, T. S.,** *Human Parasites,* London, 1882, 88 pp.
587. **Cobbold, T. S.,** Description of *Ligula mansoni,* a new human cestode, *J. Linn. Soc. London,* 17, 78—83, 1883.
588. **Cohn, E.,** Ueber *Diphyllobothrium stemmacephalum* Cobbold, *Inaug. Diss.,* Koenigsberg, 1912, 29 pp.
589. **Cohn, L.,** Untersuchungen ueber das zentrale Nervensystem der Cestoden, *Zool. Jahrb. Anat.,* 12, 89—158, 1898.
590. **Cohn, L.,** Zur Anatomie der *Amabilia lamelligera* (Owen), *Zool. Anz.,* 21, 557—562, 1898.
591. **Cohn, L.,** Zur Systematik der Vogeltaenien. I, *Centralbl. Bakteriol.,* 26, 222—227; II, *Zool. Anz.,* 22, 405—408, 1899; III, *Zool. Anz.,* 22, 415—422, 1899.
592. **Cohn, L.,** Zur Kenntnis einiger Vogeltaenien, *Zool. Anz.,* 23, 91—98, 1900.
593. **Cohn, L.,** Zur Systematik der vogeltaenien. IV, *Centralbl, Bakteriol.,* 27, 325—328, 1900.
594. **Cohn, L.,** Zur Anatomie der Vogelcestoden. I, *Z. Wiss. Zool.,* 67, 255—290, 1900.
595. **Cohn, L.,** Zur Anatomie und Systematik der Vogelcestoden, *Nova Acta Leop. Carol. Akad.,* 79, 269—450, 1901.
596. **Cohn, L.,** Zur Kenntnis des Genus *Wageneria* Monticelli und der anderen Cestoden, *Centralbl. Bakteriol.,* 33, 53—60, 1902.
597. **Cohn, L.,** Helminthologische Mitteilugen, *Arch. Naturg. Jena,* 69, J. 1, 47—68, 1902.
598. **Cohn, L.,** Helminthologische Mitteilugen, *Arch. Naturg. Jena,* 70, 243—248, 1904.
599. **Cohn, L.,** Zur Anatomie der *Amphilina foliacea* (Rud.), *Z. Wiss. Zool.,* 76, 367—387, 1904.
600. **Cohn, L.,** Zur Anatomie Zweier Cestoden, *Centralbl. Bakteriol.,* 40, 362—367, 1906.
601. **Cohn, L.,** Die Orientierung der Cestoden, *Zool. Anz.,* 32, 51—66, 1907.
602. **Cohn, L.,** Die Anatomie eines neuen Fischcestoden, *Centralbl. Bakteriol. Parasitenk.,* 46, 134—139, 1908.
603. **Coil, W. H.,** The genus *Ophiovalipora* Hsü, 1935 (Cestoda: Dilepididae) with the description of *Ophiovalipora minuta* sp. nov. from the green heron (*Butorides virescens* L.), *J. Parasitol.,* 36, 55—61, 1950.
604. **Coil, W. H.,** The genus *Ophiovalipora* Hsü, 1935 (Cestoda: Dilepididae) with the description of *Ophiovalipora minuta* sp. nov. from the green heron (*Butorides virescens* L.), *J. Parasitol.,* 36, 405—407, 1950.

605. **Coil, W. H.**, *Infula macrophallus* sp. nov., a dioecious cestode parasitic in the black-necked stilt, *Himantopus mexicanus*, *J. Parasitol.*, 41, 291—294, 1955.
606. **Coil, W. H.**, *Parvitaenia cochlearii* sp. nov. (Cestoda: Dilepididae) a new tapeworm parasitic in the boat-billed heron, *Cochlearius cochlearius*, *Proc. Helminthol. Soc. Wash.*, 22, 66—68, 1955.
607. **Coil, W. H.**, *Oligorchis cyanocittii* sp. nov., a hymenolepid cestode parasitic in the steller jay, *Cyanocitti stelleri* (Corvidae), *Proc. Helminthol. Soc. Wash.*, 22, 112—114, 1955.
608. **Coil, W. H.**, Two new hymenolepidid cestodes from Mexican birds with observations on *Hymenolepis crocethiae* Webster 1947, *J. Parasitol.*, 42, 584—587, 1956.
609. **Collins, H. M.**, Cestodes from rodents in the Republic of South Africa, *Onderstepoort J. Vet. Res.*, 39, 25—50, 1972.
610. **Condorelli-Francaviglia, M.**, Ricerche sui vermi parassiti del *Gobius avernensis*, Canester, *Boll. Soc. Rom. Stud. Zool.*, 6, 206—210, 1898.
611. **Condorelli-Francaviglia, M.**, Contributo allo studio della tauna elmintologica di taluni pesci della provincia di Roma, *Boll. Soc. Rom. Stud. Zool.*, 7, 110—144, 1898.
612. **Connor, R. S.**, A study of the seasonal cycle of a proteocephalan cestode *Proteocephalus stizostethi* Hunter and Bangham, found in the yellow pikeperch, *Stizostedion vitreum vitreum* (Mitchell), *J. Parasitol.*, 39, 621—624, 1953.
613. **Cooper, A. R.**, On the systematic position of *Haplobothrium globuliforme* Cooper, *Trans. R. Soc. Can.*, 8, 1—5, 1914.
614. **Cooper, A. R.**, A new cestode from *Amia calva*, L., *Trans. R. Can. Inst.*, 10, 81—119, 1914.
615. **Cooper, A. R.**, Contribution to the life-history of *Proteocephalus ambloplitis*, a parasite of the black bass, *Contrib. Can. Biol. Ottawa*, 1911—1914, 2, 177—194, 1915.
616. **Cooper, A. R.**, A morphological study of bothriocephalid cestodes from fishes, *J. Parasitol.*, 4, 33—39, 1917.
617. **Cooper, A. R.**, North American pseudophyllidean cestodes from fishes, *Ill. Biol. Monogr.*, 4, 1—243, 1918.
618. **Cooper, A. R.**, *Glaridacris catostomi* n. g. n. sp., a cestodarian parasite, *Trans. Am. Microsc. Soc.*, 39, 5—24, 1920.
619. **Cooper, A. R.**, Trematoda and Cestoda, in *Reports of Canadian Arctic Expedition*, 1913—1918, Parts G-H, 1921.
620. **Cordero, E. H.**, *Ophiotaenia cohospes* n. sp., de la tortuga fluvial *Hydromedusa tectifera* Cope, una larva plerocercoide en el parenquima de *Temmnocephala brevicornis* Mont., y su probable metamorfosis, *Commun. Zool. Mus. Hist. Nat. Montevideo*, 2, 1—12, 1946.
621. **Cornford, E. M.**, Two tetraphyllidean cestodes from Hawaiian stingrays, *J. Parasitol.*, 60, 942—948, 1974.
622. **Cottelaer, C. and Schyns, P.**, A propos d'une nouvelle espèce de *Nematoparataenia (Nematoparataenia brabantiae* n. sp.) du cygne, décrite pour la prémière fois en Belgique, *Ann. Parasitol. Hum. Comp.* 36, 44—49, 1961.
623. **Cousi, D.**, La cysticercose bovine en Tunisie, *Rev. Vet. Toulouse*, 85, 121—130, 1933.
624. **Coutelen, F.**, Sur l'évolution vesiculaire du scolex échinococciques, *Ann. Parasitol.*, 5, 239—242, 1929.
625. **Coutelen, F.**, Essai de culture in vitro de scolex et hydatides échinococciques *(E. granulosus), Ann. Parasitol.*, 5, 1—19, 1927.
626. **Coutelen, F.**, Contribution à l'étude morphologique des scolex echinococcique, *Ann. Parasitol.*, 5, 243—244, 1927.
627. **Coutelen, F.**, Essai de culture in vitro du cénure serial: vésiculation des scolex, *C. R. Soc. Biol.*, 100, 619—621, 1929.
628. **Coutelen, F.**, Recherches sur le système excréteur des hydatides échinococciques, *Ann. Parasitol.*, 9, 423—455, 1931.
629. **Cram, E. B.**, The presence of *Bertiella delafondi* in the pigeon (*Columba domestica*) in the United States, *J. Parasitol.*, 11, 115, 1924.
630. **Cram, E. B.**, Larval tapeworms in the gizzard of a duck, *J. Parasitol.*, 12, 178—179, 1926.
631. **Cram, E. B.**, Re-description of *Taenia Krabbei* Moniez, *J. Parasitol.*, 19, 34—41, 1926.
632. **Cram, E. B.**, The present status of our knowledge of poultry parasitism, *N. Am. Vet.*, 9, 43—51, 1928.
633. **Cram, E. B.**, A species of the genus *Bertiella* in man and chimpanzee in Cuba, *Am. J. Trop. Med.*, 8, 339-344, 1928.
634. **Cram, E. B. and Jones, M. F.**, Observations on the life-histories of *Raillietina cesticillus* and of *Hymenolepis carioca*, *North Am. Vet.*, 10, 49—51, 1929.
635. **Creplin, F. C. H.**, *Novae Observationes de Entozois*, Berolini, 1829, 134 pp.
636. **Creplin, F. C. H.**, Eingeweidewürmer, Binnenwürmer, Thierwürmer, *Allg. Encykl. Wiss. Kunste (Esch und Gruber)*, 32, 277—302, 1838.
637. **Creplin, F. C. H.**, Helminthologisches aus dem Französischen mitgetheilt, Floricep's neue notizen aus dem Geibiet der Natur und Heilkunde, *N. Notiz. Geb. Nat. Heilk.*, 24, 134—137, 1842.

638. **Creplin, F. C. H.**, Eingeweidewürmer des *Dicholophus critatus*, *Abh. Naturf. Ges. Halle*, 1, 59—68, 1853.
639. **Crety, C.**, Cestodi della *Coturnix communis*, *Boll. Mus. Zool. Anat. Comp. Torino*, 5, 1—16, 1890.
640. **Crety, C.**, Ricerche anatomiche et istologische sul genere *Solenophorus* (Creplin), *Atti R. Accad. Lincei*, 5 s., 6, 384—413, 1890.
641. **Crety, C.**, Sopra alcuni cisticerchi di una foca (*Konachus albiventer* Gray), *Boll. Soc. Nat. Napoli*, 4, 106—108, 1890.
642. **Crowcroft, P. W.**, Note on *Anthobothrium hickmani*, a new cestode from the Tasmanian electric ray (*Narcine tasmaniensis* Richardson), *Pap. Proc. R. Soc. Tasmania* (1946), pp. 1—4, 1947.
643. **Crozier, B. U.**, A new taeniid cestode, *Cladotaenia banghami* from a bald eagle, *Trans. Am. Microsc. Soc.*, 65, 222—227, 1946.
644. **Crusz, H.**, Contributions to the helminthology of Ceylon. I. On *Multiceps serialis*, *Ceylon J. Sci.*, 22, 173—181, 1944.
645. **Crusz, H.**, The early development of the rostellum of *Cysticercus fasciolaris* Rud., and the chemical nature of its hooks, *J. Parasitol.*, 33, 87—98, 1947.
646. **Crusz, H.**, Further studies on the development of *Cysticercus fasciolaris* and *Cysticercus pisiformis*, with special reference to the growth and sclerotization of the rostellar hooks, *J. Helminthol.*, 22, 179—198, 1948.
647. **Crusz, H.**, On the transverse fission of *Cysticercus pisiformis* in experimentally infested rabbits, and the phylogenetic significance of asexual phenomena in cysticerci, *J. Helminthol.*, 22, 164—178, 1948.
648. **Crusz, H. and Sanmugasunderam, V.**, Parasites of the relict fauna of Ceylon. II. New species of cyclophyllidean cestodes from small hill-vertebrates, *Ann. Parasitol. Hum. Comp.*, 46, 575—588, 1971.
649. **Cruz-Reyes, A.**, Céstodes de peces de México. I. Redescripción del subgénero *Otobothrium (Pseudotobothrium)* Dollfus, 1942 y de la especie *Otobothrium (P.) dipsacum* Linton, 1897, *Annal. Instit. Biol., Univ. Nac. Autón. México Zool.*, 44, 25—34, 1974 (dated 1973).
650. **Cruz-Reyes, A.**, Céstodos de peces de México. II. Descripción de una neuva especie del género *Floriceps* Cuvier, 1817, (Trypanorhyncha: Dasyrhynchidae Dollfus, 1935), in *Excerta Parasitologica en Memoria del Doctor Eduardo Caballero y Caballero*, Mexico Universidad Nacional Autonoma de Mexico, 1977, 343—355.
651. **Curtice, C.**, The Animal Parasites of Sheep, Spec. Rep., U.S. Department of Agriculture, Washington, D.C., 1890, 222 pp.
652. **Curtis, W. C.**, *Crossobothrium laciniatum* and developmental stimuli in the Cestoda, *Biol. Bull.*, 5, 125, 1888—1890.
653. **Curtis, W. C.**, The life-history of *Scolex polymorphus* of the Woods Hole region, *J. Morphol.*, 22, 819, 1911.
654. **Cushing, H. B. and Bacal, H. L.**, *Diphyllobothrium latum* (fish tapeworm) infestation in eastern Canada, *Can. Med. J.*, 30, 377—384, 1932.
655. **Cuvier, G.**, *Le Règne Animal Distribué d'après son Organization*, 4 Vols., Paris, 1817.
656. **Cuvillier, E. and Jones, M. F.**, Two new intermediate hosts for the poultry cestode, *Hymenolepis carioca*, *J. Parasitol.*, 19, 245, 1933.
657. **Czaplinski, B.**, *Aploparaksis stefanski* sp. n., nouvelle espèce de la famille Hymenolepididae Fuhrm., 1907 (Cestoda) chez le canard domestique (*Anas platyrhynchos dom.* L.) (in Polish), *Acta Parasitol. Pol.*, 2, 303—318, 1955.
658. **Czaplinski, B.**, Hymenolepididae Fuhrmann, 1907 (Cestoda), Parasites of some domestic and wild Anseriformes in Poland, *Acta Parasitol. Pol.*, 4, 175—357, 1956.
659. **Czaplinski, B.**, Anatomia i cykl rozwojowy tasiemca *Hymenolepis vistulae* sp. n. (Hymenolepididae Fuhrmann, 1907) pasozyta tracza nurogesi — *Mergus merganser* L. (in Czechoslovakian), *Acta Parasitol. Pol.*, 8, 299—314, 1960.
660. **Czaplinski, B.**, *Retinometra guberiana* sp. n. (Cestoda, Hymenolepididae), a new cestode species from *Cygnus olar* (Gm), *Acta Parasitol. Pol.*, 13, 35—39, 1965.
661. **Czaplinski, B.**, Genus *Monosaccanthes* g. n. (Cestoda, Hymenolepididae) and redescription of *M. tenuirostris* (Rud., 1819, p.p.) comb. n. and *M. kazachstanica* (Maksimova, 1963) comb. n., *Acta Parasitol. Pol.*, 14, 327—350, 1967.
662. **Czaplinski, B.**, *Diorchis ovofurcata* sp. n. (Cestoda, Hymenolepididae) from *Aythya fuligena* and *A. nyroca* in Poland, *Acta Parasitol. Pol.*, 20, 63—73, 1972.
663. **Czaplinski, B.**, Redescription of *Sobolevicanthus keneiensis* Schiller (Schiller, 1952) comb. n. (syn. *S. gladium* Spassky and Bobova, 1962) (Cestoda, Hymenolepididae), *Acta Parasitol. Pol.*, 21, 251—261, 1973.
664. **Czaplinski, D. and Czaplinski, B.**, *Retinometra fulicatrae* sp. n. (Cestoda, Hymenolepididae) from *Fulica atra* L., *Acta Parasitol. Pol.*, 20, 229—232, 1972.
665. **Czaplinski, B. and Ryzikov, K. M.**, *Gastrotaenia paracygni* sp. n. (Hymenolepididae), a new cestode of *Cygnus olar* and *L. cygnus*, *Acta Parasitol. Pol.*, 14, 113—119, 1966.

666. **Czaplinski, B. and Szelenbaum, D.**, Morphological and biological differences between *Diorchis ransomi* Johri, 1939 and *Diorchis parvogentialis* Skrjabin and Mathevossian, 1945 (Cestoda, Hymenolepididae), *Acta Parasitol. Pol.*, 22, 113—132, 1974.
667. **Czaplinski, B. and Vaucher, C.**, Révision de *Fuhrmaniella* (sic) *fausti* Tseng Shen, 1932, *Ann. Parasitol.*, 52, 253—258, 1977.
668. **Czaplinski, B. and Vaucher, C.**, Parasitic helminths from Paraguay. I. *Sobolevicanthus dlouhyi* n. sp. (Cestoda, Hymenolepididae) found in *Amazonetta brasiliensis* Gmelin, *Syst. Parasitol.*, 3, 71—76, 1981.
669. **Czaplinski, B. and Vaucher, C.**, Studies on collections of Hymenolepididae Fuhrmann, 1907 (Cestoda) deposited in the Museum of Natural History in Geneva. I. *Retionometra serrata* (Fuhrmann, 1906) Spasskii, 1963, *Acta Parasitol. Pol.*, 28, 141—149, 1981.
670. **Czaplinski, B. and Wilanowicz, H.**, Anatomy and development in the intermediate host of *Monosaccanthes steperae* sp. n. (Cestoda, Hymenolepididae) from the ceca of *Anas stepera* L., *Acta Parasitol. Pol.*, 17, 103—108, 1969.
671. **Daday, J.**, Helminthologische Studien. Einige in Süsswasser-Entomostraken lebende *Cercocystis*-Formen, *Zool. Jahrb. Syst.*, 14, 161—214, 1900.
672. **Dailey, M. D.**, *Litobothrium alopias* and *L. coniformis*, two new cestodes representing a new order from elasmobranch fishes, *Proc. Helminthol. Soc. Wash.*, 36, 218—224, 1969.
673. **Dailey, M. D.**, *Litobothrium gracile* sp. n. (Eucestoda: Litobothridea) from the sand shark *(Odontaspis ferox)*, *J. Parasitol.*, 57, 94—96, 1971.
674. **Dailey, M. D. and Carvajal, J.**, Helminth parasites of *Rhinobatos planiceps* Garman 1880, including two new species of cestodes, with comments on host specificity of the genus *Rhinebothrium* Linton 1890, *J. Parasitol.*, 62, 939—942, 1976.
675. **Dailey, M. D. and Mudry, D. R.**, Two new species of Cestoda from California rays, *J. Parasitol.*, 54, 1141—1143, 1968.
676. **Dailey, M. D. and Overstreet, R. M.**, *Cathetocephalus thatcheri* gen. et sp. n. (Tetraphyllidea: Cathetocephalidae fam. n.) from the bull shark: a species demonstrating multistrobilization, *J. Parasitol.*, 59, 469—473, 1973.
677. **Dailey, M. D. and Vogelbein, W.**, Mixodigmatidae, a new family of cestode (Trypanorhyncha) from a deep sea, planktivorous shark, *J. Parasitol.*, 68, 145—149, 1982.
678. **Daly, E. F.**, A new dilepidid cestode, *Paruterina reynoldsi*, from the southern crow, *Corvus brachyrhynchos paulus* Howell, *Proc. Helminthol. Soc. Wash.*, 25, 34—36, 1958.
679. **Daniels, C. W.**, *Taenia demerariensis*, *Br. Guiana Med. Ann. Hosp. Rep.*, 7, 95—98, 1895.
680. **Daniels, C. W.**, *Taenia demerariensis* (?), *Lancet*, 2, 1455, 1896.
681. **Dansker, V. N. and Pik-Levontin, E. M.**, Comparative data on the infectivity and course of infection of *Hymenolepis fraterna* and *H. nana* (in Russian), *Tr. Leningrad. Inst. Epidemiol. Bakteriol. Pastera*, 3, 50—65, 1937; English summary, 128—129.
682. **Darrah, J. H.**, A new anoplocephalid cestode from the woodchuck, *Marmota flaviventris*, *Trans. Am. Microsc. Soc.*, 49, 252—257, 1930.
683. **Davaine, C.**, *Traité des Entozoaires et des Maladies Vermineuses de l'Homme et des Animaux Domestiques*, Paris, 1860, 838 pp.
684. **Davaine, C.**, Sur une ligule (*Ligula minuta*) de la truite du lac du Genève, *C. R. Soc. Biol. Paris*, 4, 87—88, 1865.
685. **Davidson, W. R., Doster, G. L., and Prestwood, A. K.**, A new cestode, *Imparmargo baileyi* (Dilepididae: Dipylidiinae), from the eastern wild turkey, *J. Parasitol.*, 60, 949—952, 1974.
686. **Davies, T. I.**, The anatomy of *Dilepis undula* (Schrank, 1798), *Proc. Zool. Soc. London*, pp. 717—722, 1935.
687. **Davies, T. I.**, A description of *Hymenolepis neoarctica* n. nom., syn. of *H. fusus* Linton, with a discussion on the synonymity of *Taenia fusus* Krabbe, 1869, *Parasitology*, 30, 339—343, 1938.
688. **Davies, T. I.**, On *Anomotaenia gallinaginis* n. sp. from the intestine of the common snipe, *Gallinago gallinago* (Linn.), *Parasitology*, 30, 344—346, 1938.
689. **Davies, T. I.**, Some factors governing the incidence of helminth parasites in the domestic duck, *Welsh J. Agric.*, 14, 280—287, 1938.
690. **Davies, T. I.**, Four species of *Hymenolepis* Weinland parasitic in the oyster catcher, *Haematopus ostralegus* Linn., *Parasitology*, 31, 401—412, 1939.
691. **Davies, T. I.**, Three closely related species of *Aploparaksis* Clerc, 1903, *Parasitology*, 32, 198—207, 1940.
692. **Davies, T. I. and Rees, G.**, *Andropigynotaenia haematopodis* n.g. n.sp., a new protogynous tapeworm from the oyster catcher *Haematopus ostralegus occidentalis* Neumann, *Parasitology*, 38, 93—100, 1947.
693. **Davis, H. E.**, *Cittotaenia sandgroundi*, a new anoplocephalid cestode from a Javanese tree duck, *J. Parasitol.*, 30, 241—244, 1944.
694. **Davis, H. E.**, A new hymenolepidid cestode, *Hymenolepis javanensis*, from an East Indian tree duck, *Trans. Am. Microsc. Soc.*, 64, 213—219, 1945.

695. **Davis, H. E.,** A new hymenolepidid cestode, *Hymenolepis furcouterina*, from a Celebesian black-bellied snakebird, *Trans. Am. Microsc. Soc.,* 64, 306—310, 1945.
696. **Davis, H. E.,** The tapeworm *"Cittotaenia sandgroundi"* transferred to *Diplogynia, Okla. Acad. Sci.,* 27, 65—66, 1947.
697. **Deardorff, T. L. and Brooks, D. R.,** *Passerilepis schmidti* sp.n. (Cestoidea: Hymenolepididae) from the blue jay, *Cyanocitta cristata* L. in Nebraska, *Proc. Helminthol. Soc. Wash.,* 45, 190—192, 1978.
698. **Deardorff, T. L., Schmidt, G. D., and Kuntz, R. E.,** Tapeworms from Philippine birds, with three new species of *Raillietina (Raillietina), J. Helminthol.,* 50, 133—142, 1976.
699. **Deardorff, T. L., Schmidt, G. D., and Kuntz, R. E.,** *Allohymenolepis palawanensis* sp. n. (Cyclophyllidea: Hymenolepididae) from the Philippine bird, *Nectarinia jugularis* (Tweeddale 1878), *J. Helminthol.,* 52, 211—213, 1978.
700. **Deblock, S.,** Les *Hymenolepis* de Charadriiformes. (Seconde note à propos d'un vingtaine d'autres descriptions dont deux nouvelles.), *Ann. Parasitol. Hum. Comp.,* 39, 695—754, 1964.
701. **Deblock, S.,** Six cestodes d'oiseaux de mer ou de rivage de l'hémisphère austral (Ile Europa). Description de *Tetrabothrius mozambiquus* n. sp. et de *Baerbonaia baerbonae* n. gen., n.sp., *Mém. Mus. Natl. Hist. Nat. Sér. A Zool.,* 41, 103—124, 1966.
702. **Deblock, S., Biquet, J., and Capron, A.,** Contribution à l'étude des cestodes de Lari des côtes de France. I. Le genre *Hymenolepis*. Révision des espèces du genre, à propos de trois descriptions dont une nouvelle, *Ann. Parasitol. Hum. Comp.,* 35, 538—574, 1960.
703. **Deblock, S. and Capron, A.,** *Bertiella lemuriformis* nouveau cestode Anoplocephalidae d'un lémurien de Madagascar *(Lichanotus laniger* Gmel.), *Parasitologia,* 1, 97—111, 1959.
704. **Deblock, S., Capron, A., and Rosé, F.,** *Choanotaenia ridibundum* n.sp. (Dipylidiinae), nouveau cestode des lari des Côtes de France, *Bull. Soc. Zool. Fr.,* 85, 321—330, 1960.
705. **Deblock, S. and Rausch, R. L.,** Les *Hymenolepis* (s. l.) de Charadriiformes, (4 note à propos de deux espèces nouvelles d'Alaska), *Ann. Parasitol. Hum. Comp.,* 42, 303—311, 1967.
706. **Deblock, S. and Rausch, R. L.,** Dix *Aploparaksis* (Cestoda) de Charadriiformes d'Alaska, et quelques autres d'ailleurs, *Ann. Parasitol. Hum. Comp.,* 43, 429—448, 1968.
707. **Deblock, S. and Rosé, F.,** *Liga celermaturus* nouveau dilépidé de Charadriforme des côtes du Nord de la France, *Bull. Soc. Zool. Fr.,* 87, 600—608, 1962.
708. **Deblock, S. and Rosé, F.,** Les *Hymenolepis* (sensu lato) de Charadriiformes (A propos de 23 descriptions.), *Ann. Parasitol. Hum. Comp.,* 37, 767—847, 1962.
709. **Deblock, S. and Rosé, F.,** Hymenolepididae (Cestoda) des Charadriiformes des Côtes de France. Validité du genre *Oligorchis* (Fuhrm. 1906) et description d'*Hymenolepis longocylindrocirrus* n.sp., *Ann. Parasitol.,* 39, 157—178, 1964.
710. **Deblock, S., Rosé, F., Broussart, J., Capron, A., and Brygoo, E. R.,** Miscellanea helminthologica madegascariensia. Cestodes de Madagascar et des îles voisines, *Arch. Inst. Pasteur Madagascar,* 31, 1—87, 1962.
711. **Deblock, S. and Tran-van-ky, O.,** Les *Hymenolepis* (sensu lato) de Charadriiformes. (Troisième note), *Ann. Parasitol. Hum. Comp.,* 40, 131—139, 1965.
712. **Deffke, O.,** Die Entozoen des Hundes, *Arch. Wiss. Prakt. Tierheilk.,* 17, 1—60, 253—289, 1891.
713. **Deiner, E.,** Anatomie der *Anoplocephala latissima* (nom. nov.), *Arb. Zool. Inst. Univ. Wien,* 19, 347—372, 1912.
714. **Della Santa, E.,** Révision du genre *Oochoristica* Lühe (Cestodes), *Rev. Suisse Zool.,* 58, 1—113, 1956.
715. **Delyamure, S. L.,** Helminthofauna of marine mammals (ecology and phylogeny), *Tr. Gel'mintol. Lab. Akad. Nauk SSSR 1955;* Israel Program for Scientific Translations, Jerusalem, 1968, 522 pp.
716. **Delyamure, S. L.,** *Diphyllobothrium ponticum* n. sp., a new diphyllobothriid from *Tursiopus truncatus* (in Russian), *Sbornik Rabot po Gel'mintologii Posvyashchen Go-letiyu so dnya Rozhdeniya Akademika K.I. Skrjabina,* Izdatel'stvo Kolos, Moscow, 1971, 123—125.
717. **Delyamure, S. L., and Krotov, (1955),** see Delyamure, S. L., 1955 (Ref. 715).
718. **Delyamure, S. L. and Parukhin, A. M.,** A new *Diphyllobothrium* from the South African fur seal (in Russian), in Bodyanitski, V. A., Ed., *Biology of Sras. Parasites of Marine Animals* (No. 14), Naukova Dumka, Kiev, 1968, 25—34.
719. **Delyamure, S. L. and Skryabin, A. S.,** *Diphyllobothrium polyrugosum* n.sp. from *Orcinus orca* in the Southern Hemisphere (in Russian), in *Helminth Fauna of Animals in Southern Seas,* Delyamure, S. L., Ed., Naukova Dumka, Kiev, 1966, 3—9.
720. **Delyamure, S. L. and Skryabin, A. S.,** *Diphyllobothrium pterocephalum* n.sp. from *Cystophora cristata* (in Russian), *Helminthologia,* Year 1966, 7, 65—70, 1967.
721. **Dence, W. A.,** The occurrence of "free-living" *Ligula* in Catlin Lake, Central Adirondacks, *Copeia,* p. 140, 1940.
722. **Deshmukh, R. A.,** On a new cestode *Flapocephalus trygonis* gen. et sp. nov. (Cestoda: Lecanicephalidae) from *Trygon sephen* from west coast of India, *Riv. Parassitol.,* 40, 261—265, 1979.

723. **Deshmukh, R. A.,** On a new cestode *Yorkeria southwelli* (Cestoda: Onchobothriidae) from a marine fish, *Curr. Sci.,* 48, 271—272, 1979.
724. **Deshmukh, R. A.,** On a new cestode *Spinocephalum rhinobatii* gen. et sp. nov. (Cestoda: Lecanicephalidae) from a marine fish from west coast of India, *Riv. Parassitol.,* 41, 27—32, 1980.
725. **Deshmukh, R. A. and Shinde, G. B.,** *Spinibiloculus* n. gen. (Cestoda: Onchobothriidae) from a marine fish, *Ginglymostoma concolor* from west coast of India, *Ind. J. Parasitol. (Suppl.); Abstr. 3rd Natl. Congr. Parasitol., Haryana Agric. Univ.,* 1980, p. 15.
726. **Deshmukh, R. A., Shinde, G. B., and Jadav, B. V.,** On a new species of the genus *Platybothrium* Linton 1899 (Cestoda: Onchobothriidae) from a marine fish at Veraval, west coast of India, All-India Symp. Helminthol. Sriniger, Kashmir, 8—11 August, 1977, p. 9.
727. **Dévé, F.,** De l'existence de formes de transition entre l'echinococcose hydatique et l'echinococcose alveolaire chez l'homme, *C. R. Soc. Biol.,* 113, 223—224, 1933.
728. **Dévé, F.,** Intermediate, transitional and pathological forms between hydatid echinococcus and alveolar echinococcus (Bavaro-Tyrolienne) in man, *Aust. N.Z. J. Surg.,* 4, 99—117, 1934.
729. **Dévé, F.,** Formes anatomo-pathologiques intermédiaires et formes de passage entre l'echinococcose hydatique et l'echinococcose alveolaire (Bavaro-Tyrolienne) chez l'homme, *Ann. Anat. Pathol. Med. Chir.,* 10, 1155—1178, 1935.
730. **Devi, P. R.,** *Proteocephalus atretiumi* n.sp. (Cestoda: Proteocephalidea) from the water snake. *Atretium schistosum* (Gunther), *Br. J. Herpetol.,* 5, 346—351, 1971.
731. **Devi, P. R.,** Pseudophyllidean cestodes (Bothriocephalidae) from marine fishes of Waltair Coast, Bay of Bengal, *Riv. Parassitol.,* 36, 279—286, 1975.
732. **Devi, P. R. and Rao, K. H.,** On *Senga visakhapatnamensis* n.sp. (Cestoda: Pseudophyllidea) from the intestine of the fresh water fish *Ophiocephalus punctatus* Bloch, *Riv. Parassitol.,* 34, 281—286, 1973.
733. **Dew, H. R.,** The histogenesis of the hydatid parasite (*Taenia echinococcus*) in the pig, *Med. J. Aust.,* 12, 101—110, 1925.
734. **Dew, H. R.,** Daughter cyst formation in hydatid disease, *Med. J. Aust.,* 12, 497—505, 1925.
735. **Dew, H. R.,** The mechanism of daughter cyst formation in hydatid disease, *Med. J. Aust.,* 13, 451—466, 1926.
736. **Dew, H. R.,** Hydatid disease of the brain, *Surg. Gynecol. Obstet.,* 59, 321—329, 1934.
737. **Dhawan, K. and Capoor, V. N.,** On a new cestode *Davaineia hewetensis* n.sp. from *Gallus gallus, Proc. Natl. Acad. Sci. India,* 42B, 272—274, 1972.
738. **Diamare, V.,** Note sui cestodi, *Boll. Soc. Nat. Napoli,* 7, 9—13, 1893.
739. **Diamare, V.,** Il genere *Dipylidium* Lt., *Atti Accad. Sci. Napoli,* 6, 1—31, 1893.
740. **Diamare, V.,** Le funzioni dell' ovaria nella *Davainea tetragona* (Mol.), *Rend. Accad. Sci. Fis. Mat. Napoli,* 7, 213—217, 1893.
741. **Diamare, V.,** Anatomie der Genitalien des Genus *Amabilia* mihi, *Centralbl. Bakteriol. I Abt.,* 21, 862—872, 1897.
742. **Diamare, V.,** Die genera *Amabilia* und *Diploposthe, Centralbl. Bakt. I Abt.,* 22, 98—99, 1897.
743. **Diamare, V.,** Ueber die weiblichen Geschlechsteile der *Davainea tetragona* (Molin), eine kurze Antwort an Herrn Dr. Holzberg, *Centralbl. Bakt. I Abt.,* 24, 480—483, 1898.
744. **Diamare, V.,** Ueber *Amabilia lamelligera* (Owen), *Centralbl. Bakt. I Abt.,* 25, 357—359, 1899.
745. **Diamare, V.,** Einige Bemerkungen zur Antwort an Herrn Dr. L. Cohn, *Centralbl. Bakt. I Abt.,* 26, 780—782, 1899.
746. **Diamare, V.,** *Paronia carrinoi* n.g. n.sp. di Taenioide a duplici organi genitali, *Boll. Mus. Zool. Anat. Comp. Genova,* 91, 1—8, 1900.
747. **Diamare, V.,** *Paronia carrinoi* n.g. n.sp. von Taenioiden mit doppelten Geschlechtsorganen, *Centralbl. Bakteriol. I. Abt.,* 28, 846—850, 1900.
748. **Diamare, V.,** Zur Kenntnis der Vogelcestoden (Ueber *Paronia carrinoi* mihi), *Centralbl. Bakt. I Abt.,* 39, 359—373, 1901.
749. **Dickey, L. B.,** A new amphibian cestode, *J. Parasitol.,* 7, 129—136, 1921.
750. **Diesing, K. M.,** *Tropisurus* und *Thysanosoma*, zwei neue Gattungen von Binnenwürmern aus Brasiliens, *Med. Jahrb. L.K. Oesterreichischen Staates Wien,* 16, 83—116, 1835.
751. **Diesing, K. M.,** Ueber eine naturgemässe Verteilung der Cephalocotyleen, *Sitzungber. Akad. Wiss. Wien Math. Naturwiss. Kl. Abt. I.,* 13, 555—616, 1854.
752. **Diesing, K. M.,** Sechzehn Gattungen von Binnenwürmern und ihre Arten, *Denkschr. Akad. Wiss. Wien Math. Naturwiss. Kl. Abt. I.,* 9, 171—185, 1855.
753. **Diesing, K. M.,** Zwanzig Arten von Cephalocotyleen, *Denkschr. Akad. Wiss. Wien Math. Naturwiss. Kl. Abt. I.,* 12, 23—38, 1856.
754. **Diesing, K. M.,** Revision der Cephalocotyleen, Abtheilung: Paramecocotyleen, *Sitzungber. Akad. Wiss. Wien Math. Naturwiss. Kl. Abt. I.,* 48, 200—345, 1863.
755. **Diesing, K. M.,** Revision der Cephalocotyleen. Abtheilung: Cyclocotyleen, *Sitzungber. Akad. Wiss. Wien Math. Naturwiss. Kl. Abt. I.,* 49, 357—430, 1863.

756. **Dinnik, I. A.**, *Anomotaenia tarnogradskii* nov. sp. aus *Cinclus cinclus caucasicus*, (in Russian), *Sb. Rabot. Gel'mintol. Posv. K.I. Skrjabinu (Moskva)*, pp. 66—68, 1927.
757. **Dinnik, I. A. and Zvereva, N. S.**, Un cas de parasitisme d'un cestode de genre *Raillietina* Fuhrmann chez un enfant en Caucasie (in Russian), *Med. Parazitol. Parazit. Bolezni*, 9, 458—460, 1940.
758. **Dinnik, J. A. and Sachs, R.**, Zystizerkose der Kreuzbeinwirbel bei Antilopen und *Taenia olngojinei* sp. nov. de Tüpfelhyäne, *Z. Parasitenkd.*, 31, 326—339, 1969.
759. **Dinnik, J. A. and Sachs, R.**, Taeniidae of lions in East Africa, *Z. Tropenmed. Parasitol.*, 23, 197—210, 1972.
760. **Dixit, S., Capoor, V. N., and Rengaragu, V.**, On a new mammalian cestode, *Gopalaia krishnai* n.g., n.sp. from the bat, *Megaderma lyra-lyra*, *Ind. J. Parasitol.*, 4, 101—103, 1980.
761. **Dixit, S. and Capoor, V. N.**, Taxonometric approach in description of a new cestode, *Amoebotaenia madrasiensis* n.sp., *Proc. Ind. Acad. Parasitol.*, 2, 28—30, 1981.
762. **Dixit, S., Capoor, V. N., and Rengaragu, V.**, On a new mammalian cestode, *Gopalaia krishnai* n.g., n.sp. from the bat, *Megaderma lyra-lyra*, *Ind. J. Parasitol.*, 4, 101—103, 1980.
763. **Dobrovolny, C. G. and Dobrovolny, M. F.**, An unusual collection of polyradiate tapeworms from a dog, *Trans. Kan. Acad. Sci.*, 36, 214, 1933.
764. **Dobrovolny, C. G. and Dobrovolny, M. F.**, Polyradiate tapeworms from a dog, *Trans. Am. Microsc. Soc.*, 54, 22—27, 1935.
765. **Dobrovolny, C. G. and Harbaugh, M. J.**, *Cysticercus fasciolaris* from the red squirrel, *Trans. Am. Microsc. Soc.*, 53, 67, 1934.
766. **Dogiel, V. A. and Volkova, M. M.**, Sur le cycle vital du *Diplocotyle* (Cestoda, Pseudophyllidea), *C. R. Acad. Sci. U.R.S.S.*, 53, 385—387, 1946.
767. **Dollfus, R. Ph.**, L'orientation morphologique des *Gyrocotyle* et des cestodes en géneral, *Bull. Soc. Zool. Fr.*, 48, 203—242, 1922.
768. **Dollfus, R. Ph.**, Enumération des cestodes du plancton du des invertébrés marins, *Ann. Parasitol.*, I, 276—300, 363—394, 1923.
769. **Dollfus, R. Ph.**, Le cestode des perles fines des méléagrines de Nossi-Bé, *C. R. Acad. Sci.*, 176, 1265—1267, 1923.
770. **Dollfus, R. Ph.**, Sur *Acanthobothrium crassicolle* K. Wedl, 1855, *Bull. Soc. Zool. Fr.*, 51, 464—470, 1926.
771. **Dollfus, R. Ph.**, Addendum à mon "Enumération des cestodes du plancton et des invertébrés marins", *Ann. Parasitol.*, 7, 325—347, 1929.
772. **Dollfus, R. Ph.**, Sur les Tétrarhynques. I. Définition des genres, *Bull. Soc. Zool. Fr.*, 54, 308—342, 1929.
773. **Dollfus, R. Ph.**, Sur les Tétrarhynques. II, *Dibothriorhynchus Sphyriocephala, Tentacularia, Nybelinia, Mém. Soc. Zool. Fr.*, 29, 139—216, 1930.
774. **Dollfus, R. Ph.**, Nouvel addendum à mon "Enumération des cestodes du plancton et des invertébrés marins", *Ann. Parasitol.*, 9, 552—560, 1931.
775. **Dollfus, R. Ph.**, Identification d'un cestode de la collection du laboratoire de parasitologie de la Faculté de Medicine de Paris, *Bull. Soc. Zool. Fr.*, 57, 246—258, 1932.
776. **Dollfus, R. Ph.**, Mission saharienne Augiéras-Draper, 1927—1928. Cestodes de reptiles, *Bull. Mus. Hist. Nat. (Paris)*, Sér. 2, 4, 539—554, 1932.
777. **Dollfus, R. Ph.**, Sur une larve de Tétrarhynque enkystée chez un *Dentex macrophthalmus* Cuv., *Travaux Publiés par la Station d'Aquiculture et de Pêche de Castiglione: Année 1932, 2me fasc.*, 125—133, 1934.
778. **Dollfus, R. Ph.**, Sur *"Taenia gallinulae"* P. J. van Beneden, 1858, *Ann. Parasitol.*, 12, 267—272, 1934.
779. **Dollfus, R. Ph.**, Sur un cestode pseudophyllide parasite de poisson d'ornement, *Bull. Soc. Zool. Fr.*, 59, 476—490, 1935.
780. **Dollfus, R. Ph.**, Sur quelques Tétrarhynques (notes préliminaires), *Bull. Soc. Zool. Fr.*, 60, 353—357, 1935.
781. **Dollfus, R. Ph.**, Cestodes des invertébrés marins et thalassoides, in *Cestodes. Faune de France*, Vol. 30, Joyeux, C. and Baer, J. G., Eds., Office Central de Faunistique, Paris, 1936, 508—539.
782. **Dollfus, R. Ph.**, Sur un *Cysticercus fasciolaris* Rudolphi teratologique (Polycephale), *Ann. Parasitol.*, 16, 133—141, 1938.
783. **Dollfus, R. Ph.**, Cestodes du genre *Raillietina* récemment observés chez l'homme en Equateur, *Bull. Soc. Pathol. Exot.*, 32, 660—665, 1939.
784. **Dollfus, R. Ph.**, Cestodes du genre *Raillietina* trouvés chez l'homme en Amerique intertropicale, *Ann. Parasitol.*, 17, 542—562, 1940.
785. **Dollfus, R. Ph.**, Etudes critiques sur les Tétrarhynques du Muséum de Paris, *Arch. Mus. Hist. Nat. (Paris)*, 19, 1—466, 1942.
786. **Dollfus, R. Ph.**, Sur les cestodes de *Puma concolor* (L.), *Bull. Mus. Hist. Nat. (Paris)*, 15, 316—320, 1944.

787. **Dollfus, R. Ph.**, Notes diverses sur des tétrarhynques, *Mém. Mus. Hist. Nat. (Paris)*, 22, 179—219, 1946.
788. **Dollfus, R. Ph.**, Coenurose de la cavité abdominale chez un écureuil (*Sciurus vulgaris* L.) à Richelieu (Indre-et-Loire), *Ann. Parasitol.*, 22, 143—147, 1948.
789. **Dollfus, R. Ph.**, Amoenitates helminthologicae. VI. *Raillietina* (R.) *kouridovali* R. Ph. Dollfus et *Inermicapsifer cubensis* (P. Kouri) P. Kouri, *Ann. Parasitol.*, 22, 277—278, 1948.
790. **Dollfus, R. Ph.**, Amoenitates helminthologicae. VI. *Raillietina* (R.) *kouridovali* R. Ph. Dollfus et *Inermicapsifer cubensis* (P. Kouri) P. Kouri, *Rev. Kuba. Med. Trop. Parasitol.*, 6, 13—14, 1950.
791. **Dollfus, R. Ph.**, Un hôte accidentel d'*Hymenolepis diminuta* (Rudolphi, 1819); l'écureuil (*Sciurus vulgaris* L.) en captivité, *Ann. Parasitol.*, 26, 263, 1951.
792. **Dollfus, R. Ph.**, Cystique polycéphale de Taenia chez une gerbille, *Ann. Parasitol.*, 26, 274—278, 1951.
793. **Dollfus, R. Ph.**, Quelques trématodes, cestodes et acanthocéphales, *Miscell. Helminthol. Maroc. Arch. Inst. Pasteur Maroc.*, 4, 104—229, 1951.
794. **Dollfus, R. Ph.**, *Catenotaenia chabaudi* n. sp., de *Xerus (Atlantoxerus) getulus* (Linné, 1758), *Miscell. Helminthol. Maroc. Arch. Inst. Pasteur Maroc.*, 4, 533—540, 1953.
795. **Dollfus, R. Ph.**, Cystique polycéphale chez un *Meriones libycus* K. M. H. Lichtenstein, 1823, *Micell. Helminthol. Maroc. VIII. Arch. Inst. Pasteur Maroc.*, 4, 513—517, 1953.
796. **Dollfus, R. Ph.**, Nouvelles récoltes de cystiques polycéphales chez des Meriones: *M. crassus* Sundevall, 1842; *M. libycus erythrourus* P. Ed. Gray, 1842; *M. persicuas* (Blanford, 1875), *Miscell. Helminthol. Maroc. IX. Arch. Inst. Pasteur Maroc.*, 4, 518—532, 1953.
797. **Dollfus, R. Ph.**, Miscellanea helminthologica maroccana. XVIII. Quelques cestodes du group *Oochoristica* auctorum récoltés du Maroc, *Arch. Inst. Pasteur Maroc.*, 4, 654—714, 1954.
798. **Dollfus, R. Ph.**, Miscellanea helminthologica maroccana. XIX. Nouvelles récoltés d'*Oochoristica* chez des sauriens du Maroc. Famille Linstowiidae (P. Mola, 1929), Genus *Oochoristica* Max Lühe, 1908, emendatum, *Arch. Inst. Pasteur Maroc.*, 5, 272—299, 1957.
799. **Dollfus, R. Ph.**, Miscellanea helminthologica maroccana. XX. Contribution à la connaissance des *Nematotaenia*, *Arch. Inst. Pasteur Maroc.*, 5, 300—328, 1957.
800. **Dollfus, R. Ph.**, Miscellanea helminthologica maroccana. XXI. Quelques cestodes d'Otidiformes, principalement d'Afrique du Nord. Répartition géographique des cestodes d'Otidiformes, *Arch. Inst. Pasteur Maroc.*, 5, 329—402, 1957.
801. **Dollfus, R. Ph.**, Miscellanea helminthologica maroccana. XXXI. Sur deux espèces de Dilepididae de l'intestine d'*Apus pallidus brehmorum* Hartert 1901. (Supplément à Miscellanea helminthologica maroccana XXV.), *Arch. Inst. Pasteur Maroc.*, 5, 587—609, 1958.
802. **Dollfus, R. Ph.**, Sur une collection de tétrarhynques homeacanthes de la famille des Tentaculariidae, récoltés principalement dans la région de Dakar, *Bull. Inst. Français d'Afrique Noire. Sér. A: Sci. Nat.*, 22, 788—852, 1960.
803. **Dollfus, R. Ph.**, Cystique d'un nouveau Taenia, de la cavitée peritoneale d'un *Ctenomys* (Rodentia) de l'Uruguay, *Arch. Soc. Biologia Montevideo*, 25, 47—51, 1960.
804. **Dollfus, R. Ph.**, Mission Yves-J. Golvan et Jean-A. Rioux en Iran. Cestodes d'oiseaux. I. Cestode d'Accipitriforme, *Ann. Parasitol. Hum. Comp.*, 38, 23—27, 1963.
805. **Dollfus, R. Ph.**, Organismes dont la présence dans le plancton marin était, jusqu'à présent, ignorée: larves et postlarves de cestodes tétrarhynques, *C. R. Hebd. Séanc. Acad. Sci. Paris Sér. D*, 262, 2612—2615, 1966.
806. **Dollfus, R. Ph.**, De quelques cestodes tétrarhynques (Hétéracanthes et pécilacanthes) récoltés chez des poissons de la Méditerranée, *Vie Milieu Ser. A.*, 20, 441—542, 1969.
807. **Dollfus, R. Ph.**, Quelques espèces de cestodes Tétrarhynques de la côte Atlantique des Etats Unis, dont l'une n'était pas connue à l'état adulte, *J. Fish. Res. Board Can.*, 26, 1037—1061, 1969.
808. **Dollfus, R. Ph.**, D'un cestode ptychobothrien parasite de cyprinide en Iran. Mission C.N.R.S. (Théodore Monod), Février 1969, *Bull. Mus. Natl. Hist. Nat. 2 Sér.*, 41, 1517—1521, 1970.
809. **Dollfus, R. Ph.**, Énumération des cestodes du plancton et des invertébrés marins. 8 Contribution, *Ann. Parasitol. Hum. Comp.*, 49, 381—410, 1974.
810. **Dollfus, R. Ph.**, Miscellanea helminthologica maroccana. XLII. Cestodes d'oiseaux et de mammifères, *Bull. Mus. Natl. Hist. Nat. 3 Sér. Zool.*, 212(302), 659—684, 1975.
811. **Dollfus, R. Ph. and Carayon, J.**, Larve de cestodes chez un hémiptère hétéroptère, *Ann. Parasitol.*, 22, 276, 1938.
812. **Dollfus, R. Ph. and Chabaud, A. G.**, Miscellanea helminthologica maroccana. II. Cystique polycéphale chez un *Meriones Shawi* (G. L. Duvernoy in C. A. Rozet 1833), *Arch. Inst. Pasteur Maroc.*, 4, 230—235, 1951.
813. **Dönges, J. and Harder, W.**, *Nesolecithus africanus* n.sp. (Cestodaria, Amphilinidea) aus dem Coelom von *Gymnarchus niloticus* Cuvier 1829 (Teleostei), *Z. Parasitol.*, 28, 125—141, 1966.
814. **Donnadieu, A. L.**, Contribution à l'histoire de la ligule, *J. Anat. Physiol.*, 13, 321—370, 1877.
815. **Donnadieu, A. L.**, Contribution sur le développement de la ligule de la tanche, *Lyon Méd.*, 24, 563—567, 1877.

816. **Douglas, L. T.,** The spermatogenesis of two nematotaeniid cestodes, *Va. J. Sci.,* 4, 231—232, 1953.
817. **Douglas, L. T.,** The early embryology of the nematotaeniid cestode, *Baerietta diana* (Helfer, 1948) comb. nov., *J. Parasitol.,* 42(4), 41, 1956.
818. **Douglas, L. T.,** The taxonomy of nematotaeniid cestodes, *J. Parasitol.,* 44, 261—273, 1958.
819. **Douthitt, H.,** Studies on the cestode family Anoplocephalidae, *Ill. Biol. Monogr.,* 1, 1—96, 1915.
820. **Dowell, A. M.,** *Catenotaenia californica* sp. nov., a cestode of the kangaroo rat, *Dipodomys panamintinus mohavensis, Am. Midl. Nat.,* 49, 738—742, 1953.
821. **Drummond, F. H.,** Cestoda, in Lady Julia Percy Island Reports of the Expedition of the McCoy Society for Field Investigations and Research, *Proc. R. Soc. Victoria (Australia),* 49, 401—404, 1937.
822. **Dubinina, M. N.,** New data on the morphology and biology of *Ligula* (in Russian), *Zool. Zh.,* 29, 417—426, 1950.
823. **Dubinina, M. N.,** Destrobilisation in tapeworms and its causes (in Russian), *Zool. Zh.,* 29, 147—151, 1950.
824. **Dubinina, M. N.,** Tapeworms of birds wintering in southern Tadzhikistan (in Russian), *Parazitol. Sb. Zool. Inst. Akad. Nauk SSSR,* pp.351—381, 1950.
825. **Dubinina, M. N.,** On the biology and distribution of *Diphyllobothrium erinaceieuropaei* (Rud., 1819) Iwata, 1933 (in Russian), *Zool. Zh.,* 30, 421—429, 1951.
826. **Dubinina, M. N.,** Cestodes of birds nesting in western Siberia (in Russian), *Parazitol. Sb. Akad. Nauk SSSR,* pp. 117—233, 1954.
827. **Dubinina, M. N.,** Specificity of cestodes in different phases of their life cycles (in Russian), *Parazitol. Sb. Akad. Nauk SSSR,* pp. 234—251, 1954.
828. **Dubinina, M. N.,** Experimental study of the life cycle of *Schistocephalus solidus* (Cestoda: Pseudophyllidea) (in Russian), *Zool. Zh.,* 36, 1647—1658, 1957.
829. **Dubinina, M. N.,** The natural classification of the genus *Schistocephalus* Creplin (Cestoda, Ligulidae), *Zool. Zh.,* 38, 1498—1517, 1959.
830. **Dubinina, M. N.,** Cestodes of the family Ligulidae and their taxonomy, in Parasitic Worms and Aquatic Conditions, Ergens, R. and Ryšavy, B., Eds., Proc. Symp., Czechoslovak Acad. Sci., Prague, Oct. 29 to Nov. 2, 1962, 1964, pp. 173—186.
831. **Dubinina, M. N.,** Ligulidae, Cestoda: Ligulidae, Fauna of SSSR, *Akad. Nauk SSSR Moscow,* p. 261, 1966.
832. **Dubinina, M. N.,** The development of *Amphilina foliacea* (Rud.) at all stages of its life cycle and the position of Amphilinidea in the system of Platyhelminthes, *Parazitol. Sb. Akad. Nauk SSSR,* 26, 9—37, 1974.
833. **Dubinina, M. N.,** Platyhelminths of the class Amphilinoidea (in Russian), in *Problemy Zoologii (Zool. Inst. A.N. SSSR)* Izdatel'-stvo Nauka, Leningrad, 1976, 34—37.
834. **Dubinina, M. N.,** Synonymization of species of the genus *Bothriocephalus* (Cestoda, Bothriocephalidae), parasitic in cyprinid fish of the USSR (in Russian), *Parazitologiya,* 16, 41—45, 1982.
835. **Dubinina, M. N.,** Parasitic Helminths. Class Amphilinida (Platyhelminthes), *Akad. Nauk SSSR, Tr. Zool. Inst., Leningrad,* 100, 1—143, 1982.
836. **Dubnitskii, A. A.,** Data on the biological cycle of the cestode *Multiceps endothoracicus* Kirschenblat, 1947 (in Russian), *Dokl. Akad. Nauk SSSR,* 85, 1193—1195, 1952.
837. **Dubnitskii, A. A.,** A new cestode from the intestine of the Barguzinsk sable, *Karakulevodst. Sverovodst.,* 5(Abstr.), 79, 1952.
838. **Dubnitskii, A. A.,** *Taenia pisiformis,* facultative parasite of the fox (in Russian), *Rabot. Gel'mintol. 75-Let. Skrjabin,* pp. 234—236, 1954.
839. **Duby, (J. P. and Pande, B. P.,** A note on some helminths of the wild duck *(Anas poecilorhyncha), Ind. J. Helminthol.,* 16, 27—32, 1964.
840. **Dujardin, F.,** Sur l'embryon des Entozoaires et sur les mouvements de cet embryon dans l'oeuf, *Ann. Sci. Nat. Zool.,* 8, 303—305, 1837.
841. **Dujardin, F.,** Observations sur les ténias et sur le mouvement de leur embryon dans l'ouef, *Ann. Sci. Nat. Zool.,* 10, 29—34, 1839.
842. **Dujardin, F.,** *Histoire Naturelle des Helminthes ou Vers Intestinaux,* Paris, xvi + 654 + 15 pp., 1845.
843. **Du Noyer, M. and Baer, J. G.,** Etude comparée du *Taenia saginata* et du *Taenia solium, Bull. Sci. Pharmacol.,* 35, 209—234, 1928.
844. **Durie, P. H. and Rick, R. F.,** The role of the dingo and wallaby in the infestation of cattle with hydatids (*Echinococcus granulosus* (Batsch, 1786) Rudolphi, 1805), in Queensland, *Aust. Vet. J.,* 28, 249—254, 1952.
845. **Dutt, S. C. and Mehra, K. N.,** Studies on the life history of *Hymenolepis farciminosa* (Goeze, 1782), a tapeworm of crow and collared myna, *Proc. 42 Ind. Sci. Congr.,* Sect. 7, 283—284, 1955.
846. **Duvernoy, G. L.,** Note sur un nouveau genre de ver intestinal de la famille des Tenioides, le Bothrimone de l'esturgeon *(Bothrimonus sturionis), Ann. Sci. Nat. Zool.,* 18, 123—236, 1842.

847. **Dyer, W. G. and Altig, R.**, *Ophiotaenia olseni* sp. n. (Cestoda: Proteocephalidae) from *Hyla geographica* Spix 1824, in Equador, *J. Parasitol.*, 63, 790—792, 1977.
848. **Dzhalilov, V. D. and Ashurova, M.**, A new species of *Proteocephalus* from fish in waters of the Pamirs (in Russian), *Izv. Akad. Nauk Tadzh. SSR Otd. Biol. Nauk*, 44(3), 110—111, 1971.
849. **Edgar, S. A.**, Artificial evagination of larval tapeworms, *Trans. Kan. Acad. Sci.*, 43, 397—399, 1940.
850. **Edgar, S. A.**, Use of bile salts for the evagination of tapeworm cysts, *Trans. Am. Microsc. Soc.*, 60, 121—128, 1941.
851. **Eguchi, S.**, Studien über *Dibothriocephalus latus*, *Trans. Jpn. Pathol. Soc.*, 15, 254, 1925.
852. **Eguchi, S.**, Studies on *Dibothriocephalus latus*, *Trans. Jpn. Pathol. Soc.*, 16, 102—105, 1926.
853. **Eguchi, S.**, *Dibothriocephalus latus* in *Oncorhynchus gorbuscha* & *O. keta*, *Nihon Kiseichu Gakkai Kiji*, p. 15, 1929.
854. **Eguchi, S.**, Studien über *Dibothriocephalus latus*, besonders über seinen zweiten Zwischenwirt in Japan, *Trans. Jpn. Pathol. Soc.*, 19, 567—572, 1929.
855. **Eguchi, S.**, On the secondary intermediate host of *Diphyllobothrium latum* in Japan, with special reference to fishes of the genus *Oncorhynchus*, *Proc. 5th Pac. Sci. Congr.*, 5, 4145—4149, 1934.
856. **Egizbaeva, Kh. I. and Nasyrova, S. R.**, *Gastrotaenia kazachstanica* n.sp., a new parasite of ducks (in Russian), *Izv. Akad. Nauk Kaz. SSR Biol. Nauki*, 2, 48—52, 1979.
857. **Ehlert, W.**, Zwei neue Cestoden-Arten aus *Branta bernicla* (L.): *Dicranotaenia bernicola* n.sp. und *D. aberrata* n.sp. (Hymenolepididae), *Zool. Anz.*, 182, 423—431, 1969.
858. **Ehrstrom, R.**, Zur Kenntnis der Darmparasiten in Finnland, *Acta Med. Scand.*, 64, 29—68, 1926.
859. **Einarsson, M.**, L'echinococcose en Islande. Sur la mode de contamination humaine, *Ann. Parasitol.*, 4, 172—184, 1926.
860. **Ekbaum, E. K.**, A study of the cestode genus *Eubothrium* of Nybelin in Canadian fishes, *Contrib. Can. Biol. Fish.*, 8, 89—98, 1933.
861. **Ekbaum, E. K.**, *Citharichthys stigmaeus* as a possible intermediate host of *Gilguinia squali* (Fabricius), *Contrib. Can. Biol. Fish.*, 8, 99—100, 1933.
862. **Ekbaum, E. K.**, Notes on the species of *Triaenorphorus* in Canada, *J. Parasitol.*, 21, 260—263, 1935.
863. **Ekbaum, E. K.**, On the maturation and the hatching of the eggs of the cestode *Trianenophorus crassus* Forel from Canadian fish, *J. Parasitol.*, 23, 293—295, 1937.
864. **Elce, J. B.**, On a new cestode, *Choanotaenia larimarina* sp.nov., from the greater black-backed gull, *Larus marinus* L., *J. Helminthol.*, 36, 365—374, 1962.
865. **Elek, S. R. and Finkelstein, L. E.**, *Multiceps serialis* infestation in a baboon, *Zoologica*, 24, 323—328, 1939.
866. **Eltyshev, Yu. A.**, *Molluscotaenia baicalensis* n.sp. (Dilepididae), a new species of cestode from shrews in the Zabaykal region (in Russian), *Parazit. Zhivotnykh Rast. (Kishinev)*, 6, 9—12, 1971.
867. **Eltyshev, Yu. A.**, The helminth fauna of mammals in the Barquzin valley and its geographical analysis. I. Systematic review of helminths (in Russian), *Parazit. Organizmy Severovostoka Azii. Vladivostok USSR Akad. Nauk SSSR Dal' nevostochnyi Nauchnyi Tsentr.*, pp. 135—167, 1975.
868. **Endrigkeit, A.**, Parasitäre Massenerkrankungen als Ursache für Bestandsveränderungen in unserer Vogelwelt, *Dtsch. Vogelwelt*, 65, 70—75, 1940.
869. **Endrigkeit, A.**, Ein durch Parasiten hervorgerufenes Schwanensterben auf dem Nordenburger See, *Berl. Münch. Tierärztl. Wochenschr.*, 13, 148—151, 1940.
870. **Erickson, A. B.**, Parasites of some Minnesota Cricetidae and Zapodidae and a host catalogue of helminth parasites of native American mice, *Am. Midl. Nat.*, 20, 575—589, 1938.
871. **Erlanger, R. S. von**, Der Geschlechtsapparat der *Taenia echinococcus*, *Z. Wiss. Zool.*, 50, 555—559, 1890.
872. **Esch, G. W. and McDaniel, J. S.**, A new cestode from the least sandpiper, *Erolia minutilla*, *Trans. Am. Microsc. Soc.*, 84, 252—254, 1965.
873. **Eschricht, D. F.**, Anatomisch-physiologische Untersuchungen ueber die Bothryocephalen, *Nova Acta Acad. Leopold-Carol. Nat. Curios*, 19(Suppl. 2), 161, 1841.
874. **Essex, H. E.**, The structure and development of *Corallobothrium*, *Ill. Biol. Monogr.*, 11, 257—328, 1927.
875. **Essex, H. E.**, Early development of *Diphyllobothrium latum* in northern Minnesota, *J. Parasitol.*, 14, 106—109, 1927.
876. **Essex, H. E.**, On the life-history of *Bothriocephalus cuspidatus* Cooper, 1917, a tapeworm of the walleyed pike, *Trans. Am. Microsc. Soc.*, 47, 348—355, 1928.
877. **Essex, H. E.**, *Crepidobothrium fragile* n.sp., a tapeworm of the channel catfish, *Parasitology*, 21, 164—167, 1929.
878. **Essex, H. E.**, A report on fishes from the Mississippi River and other waters with respect to infestation by *Diphyllobothrium latum*, *Minn. Med.*, 12, 149—150, 1929.
879. **Essex, H. E.**, The life-cycle of *Haplobothrium globuliforme* Cooper, 1914, *Science*, 69, 677—678, 1929.
880. **Essex, H. E.**, A new larval cestode, probably *Hymenolepis cuneata*, a tape-worm of a wild duck, *J. Parasitol.*, 18, 291—293, 1932.

881. **Essex, H. E.**, The present status of infestation of fishes of Long Lake, Ely, Minnesota, with the larvae of *Diphyllobothrium latum*, *Minn. Med.*, 21, 253—255, 1938.
882. **Essex, H. E. and Hunter, G. W.**, A biological study of fish parasites from the central states, *Trans. Ill. Acad. Sci.*, 19, 151—181, 1926.
883. **Essex, H. E. and Magath, T. B.**, A comparison of the viability of ova of the broad tapeworm, *Diphyllobothrium latum*, from man and dogs: its bearing on the spread of infestation with this parasite, *Am. J. Hyg.*, 14, 698—704, 1931.
884. **Euzet, L.**, *Echinobothrium mathiasi* n.sp. (Cestoda: Diphyllidea), parasite d'une raie: *Leiobatis aquila* L., *Bull. Soc. Zool. Fr.*, 76, 182—188, 1951.
885. **Euzet, L.**, Sur deux cestodes Tetraphyllides, *Bull. Soc. Neuchâtel Sci. Nat.*, 75, 169—178, 1952.
886. **Euzet, L.**, *Dicranobothrium*, un nouveau genre de cestode tétraphyllide, parasite de sélaciens, *Bull. Soc. Neuchâtel Sci. Nat.*, 3. s., 76, 87—91, 1953.
887. **Euzet, L.**, Cestodes tétraphyllides nouveaux ou peu connus de *Dasyatis pastinaca*, *Ann. Parasitol.*, 28, 339—351, 1953.
888. **Euzet, L.**, Une espèce nouvelle d'*Echeneibothrium* van Ben., 1850, *Bull. Soc. Neuchâtel Sci. Nat.* 79, 39—41, 1956.
889. **Euzet, L.**, Cestodes tetraphyllide nouveaux ou peu connus de *Dasyatis pastinaca* (L.), *Ann. Parasitol. Hum. Comp.*, 28, 339—351, 1953.
890. **Euzet, L.**, Recherches sur les Cestodes Tétraphyllides de Sélaciens des Cotes de France. Causse, Graille et Castelneau, D.Sc. Thesis, University of Montpellier, France, 1959, 1—263.
891. **Euzet, L. and Carvajal, J.**, *Rhinebothrium* (Cestoda, Tetraphyllidea) parasites de raies du genre *Psammobatis* au Chile, *Bull. Mus. Natl. Hist. Nat. 3 Sér. Zool.*, 101(137), 779—787, 1973.
892. **Euzet, L. and Jourdane, J.**, Helminthes parasites des micromammiferes des Pyrénées-Orientales. I. Cestodes de *Neomys fodiens* (Schreiber), *Bull. Soc. Neuchâtel Sci. Nat.*, 91, 31—42, 1968.
893. **Evanno, C. H.**, Sur la sparganose oculaire, *Bull. Acad. Vet. Fr.*, 6, 355—356, 1933.
894. **Evans, W. M. R.**, Observations upon some common cestode parasites of the wild rabbit, *Oryctolagus cuniculus*, *Parasitology*, 32, 78—90, 1940.
895. **Eysenhardt, C. G.**, Einiges ueber Eingeweidewürmer. *Ver. Berlin Ges. Naturf. Fr.*, 1, 144—152, 1829.
896. **Ezzat, M. A. E.**, The occurrence of *Multiceps gaigeri* Hall, 1916 in subcutaneous connective tissue of Sudanese sheep and Nubian ibex, 1944, 6 pp.
897. **Ezzat, M. A. E.**, Helminth Parasites of some Ungulates from the Giza Zoological Gardens, Egypt, with an Appendix on some Nematodes from the African Rhinoceros, Thesis, *Bull. 241, Technical and Scientific Service of the Ministry of Agriculture, Vet. Sect.*, Egypt, 1945, 1—104.
898. **Ezzat, M. A. E.**, On the validity of *Avitellina nagatyi* Ezzat, 1945, *J. R. Egypt Med.*, 34, 206—209, 1951.
899. **Ezzat, M. A. E. and Gaafar, S. M.**, *Tetrathyridium* sp. in a Sykes' monkey (*Cercopithecus albigularis*) from Giza zoological gardens, Egypt, *J. Parasitol.*, 37, 392—394, 1951.
900. **Fabricius, O.**, Bidrag til Snylte-Ormenes Historie, *Skrivter af Nat. Hist. Selskabet*, 3, 1—45, 1794.
901. **Fahmy, M. A. M., Mandour, A. M., and El-Naffer, M. K.**, On some cestodes of the freshwater fishes in Assiut Province, Egypt, *Vet. Med. J.*, 24, 253—262, 1978.
902. **Fain, A.**, *Inermicapsifer cubensis* (Kouri, 1938), présence du cestode *I. cubensis*, synonyme de *Inermicapsifer arvicanthidis* (Kofend, 1917), chez un enfant indigène et chez un rat (*Rattus r. rattus* L.) au Ruanda-Urundi (Congo Belge), *Bull. Soc. Pathol. Exot.*, 43, 438—443, 1950.
903. **Fain, A.**, Morphologie et cycle évolutif d *Taenia brauni* Setti, 1897, cestode très commun chez le chien et le chacal en Ituri (Congo belge), *Rev. Suisse Zool.*, 59, 487—501, 1952.
904. **Fain, A.**, Coenurus of *Taenia brauni* Setti parasitic in man and animals from the Belgian Congo and Ruanda-Urundi, *Nature (London)*, 178, 1353, 1956.
905. **Fasbender, M. V.**, A morphological and histological description of a cestode removed from the small intestine of a short-tailed shrew, *Blarina brevicauda*, *Proc. S. Dak. Acad. Sci.*, 35, 131—143, 1957.
906. **Fasten, N.**, The tapeworm infection in Washington trout and its related biological problems, *Am. Nat.*, 16, 439—447, 1922.
907. **Faust, E. C.**, Two new Proteocephalidae, *J. Parasitol.*, 6, 79—83, 1920.
908. **Faust, E. C.**, Preliminary survey of the parasites of vertebrates of North China, *China Med. J.*, 35, 1—15, 1921.
909. **Faust, E. C.**, Infestation experiments in man and other mammalian hosts with the *Sparganum* stage of oriental Diphyllobothriidae, *Proc. Soc. Exp. Biol. Med.*, 26, 252—254, 1928.
910. **Faust, E. C.**, What is *Sparganum mansoni?*, *J. Trop. Med. Hyg.*, 32, 76—77, 1929.
911. **Faust, E. C., Campbell, H. E., and Kellogg, C. R.**, Morphological and biological studies on the species of *Diphyllobothrium* in China, *Am. J. Hyg.*, 9, 560—583, 1929.
912. **Faust, E. C., Neghme, R. A., and Tagie, V. I.**, *Diphyllobothrium latum* indigenous in the Lake district of Chile, *J. Parasitol.*, 37 (Suppl. 2), 24, 1951.

913. **Faust, E. C. and Wassell, C. M.**, Preliminary survey of the intestinal parasites of man in the central Yangtze valley, *China Med. J.*, 35, 532—561, 1931.
914. **Fediushin, A. V.**, On several new forms of cestodes of commercial fowl of northern Kazakhstan and southern Ural (in Russian), *Tr. Inst. Zool. Akad. Nauk Kaz. SSR.*, 1, 182—189, 1953.
915. **Fediushin, A. V.**, *Raillietina (Paroniella) urogalli*. Contribution to the knowledge of *Raillietina (Paroniella) urogalli* parasite of Tetraonidae (in Russian), *Rabot Gelmintol. 75-Let. Skrjabin*, pp. 723—732, 1954.
916. **Feigenbaum, D. L.**, Parasites of the commercial shrimp *Penaeus vannamei* Bovne and *Penaeus brasiliensis* Latreille, *Bull. Mar. Sci.*, 25, 491—514, 1975.
917. **Ferguson, M. S.**, *Diphyllobothrium latum* in the dog, *J. Parasitol.*, 37(Suppl.), 24, 1951.
918. **Fernando, C. H. and Furtado, J. I.**, A study of some helminth parasites of freshwater fishes in Ceylon, *Z. Parasitenk.*, 23, 141—163, 1963.
919. **Fillippi, C.**, Nota preliminare sul systema riproduttore della *Taenia bothrioplitis*, *Bull. Soc. Roma Stud. Zool. An.*, I, 75—79, 1892.
920. **Fillippi, C.**, Ricerche istologiche ed anatomiche sulla *Taenia bothrioplitis* Piana, *Atti R. Accad. Lincei Roma*, 7, 249—294, 1892.
921. **Fillippi, F. de**, Mémoire pour servir à l'histoire génétique des trématodes, *Ann. Sci. Nat.*, 11, 255—284, 1854.
922. **Fischthal, J. H.**, A new genus and species of Caryophyllaeidae (Cestoda) from fishes, *J. Parasitol.*, 36(Sect. 2), 28, 1950.
923. **Fischthal, J. H.**, *Pliovitellaria wisconsinensis* n.g., n.sp. (Cestoda: Caryophyllaeidae) from Wisconsin cyprinid fishes, *J. Parasitol.*, 37, 190—194, 1951.
924. **Fischthal, J. H.**, *Hypocaryophyllaeus gilae* n.sp. (Cestoda: Caryophyllaeidae) from the Utah chub, *Gila straria* in Wyoming, *Proc. Helminthol. Soc. Wash.*, 20, 113—117, 1953.
925. **Fischthal, J. H.**, *Bialovarium necomis* n.g., n.sp. (Cestoda: Caryophyllaeidae) from the hornyhead chub, *Nocomis biguttatus* (Kirtland), *J. Parasitol.*, 39(Sect. 2), 24, 1953.
926. **Fischthal, J. H.**, Parasites of Northwest Wisconsin fishes. IV. Summary and limnological relationships, *Trans. Wisc. Acad. Sci. Arts Lett.*, 42, 84—108, 1953.
927. **Fischthal, J. H.**, *Bialovarium nocomis* Fischthal, 1953 (Cestoda: Caryophyllaeidae) from the Hornyhead Chub, *Nocomis biguttatus*, *Proc. Helminthol. Soc. Wash.*, 21, 117—120, 1954.
928. **Flattley, F. W.**, Considerations on the life-history of tapeworms of the genus *Moniezia*, *Parasitology*, 14, 268—281, 1922.
929. **Flores Barroeta, L.**, Cestodos de vertebrados. II (i), *Rev. Ibér. Parasitol.*, 15, 115—134, 1955.
930. **Flores Barroeta, L.**, Cestodos de vertebrados. III. *Ciencia, Mexico*, 15, 33—38, 1955.
931. **Flores Barroeta, L.**, Helminthos de los perros *Canis familiaris* y gatos *Felis catus* en la Ciudad de México, *Anal. Exc. Nac. Cienc. Biol.*, 8, 159—202, 1955.
932. **Foggie, A.**, On a cestode parasite of the domestic pigeon *(Columba livia)*, *Ann. Mag. Nat. Hist.* (Ser. 10), 12, 168—172, 1933.
933. **Forel, F. A.**, Preparations microscopiques d'une nouvelle espèce de *Triaenophorus*, *Bull. Soc. Vaudoise Sci. Nat. Lausanne*, 9, 696, 1868.
934. **Forti, A.**, L. F. Marsilli e Schistocefalo dello Spinarello, *Arch. Zool. Ital.*, 16, 1437—1439, 1932.
935. **Fortner, H. C.**, The distribution of frog parasites of the Douglas Lake region, Michigan, *Trans. Am. Microsc. Soc.*, 42, 79—90, 1923.
936. **Foster, W. D.**, Two new cases of polyradiate cestodes with a summary of the cases already known, *J. Parasitol.*, 2, 7—19, 1916.
937. **Fotedar, D. N.**, New nematotaeniid cestode from *Bufo viridis* in Kashmir, *Kashmir Sci.*, 3, 17—32, 1966.
938. **Fotedar, D. N. and Bambroo, N.**, On a new species of cestode genus *Thysaniezia* Skrjabin, 1926, from local sheep in Kashmir, *J. Sci. Univ. Kashmir*, 3, 105—108, 1978 (dated 1975).
939. **Fotedar, D. N. and Chishti, M. Z.**, On a new species *Anomotaenia kashmirensis:* (Choanotaeniidae, Mathevossian, 1953) from *Sternus vulgaris* in Kashmir, *J. Sci. Univ. Kashmir*, 1, 48—50, 1973.
940. **Fotedar, D. N. and Chishti, M. Z.**, *Anomotaenia acrocephali* n.sp. and first record of *Anomotaenia galbulae* (Gmelin, 1790) Fuhrmann, 1932 from some birds of Kashmir, *Rev. Parassitol.*, 37, 247—252, 1976 (pub. 1977).
941. **Fotedar, D. N. and Chishti, M. Z.**, Studies on helminth parasites of some passeriform birds of Kashmir, *Abstr. 1st Natl. Congr. Parasitol. Baroda, India, 24—26 February, 1977*, 1977, 47—48.
942. **Fotedar, D. N. and Chishti, M. Z.**, On a new species of the genus *Pseudoschistotaenia* Fotedar and Chishti 1976 from a bird *Podiceps ruficollis* (sic) *capensis* in Kashmir, *All-India Symp. Helminthol.*, Srinigar, Kashmir, 8—11 August, 1977, 1977, 37.
943. **Fotedar, D. N. and Chishti, M. Z.**, *Pseudoschistotaenia indica* gen. et sp. nov. (Amabiliidae Fuhrmann, 1980: Cestoda) from *Podiceps ruficollis capensis* in Kashmir, *Rev. Parassitol.*, 41, 39—43, 1980.
944. **Fotedar, D. N. and Dhar, R. L.**, *Gangesia (Vermaia) jammuensis* n.sp. (Proteocephalidae: Cestoda) from freshwater fish in Jammu, India, *Proc. Ind. Sci. Congr. Assoc.*, 61(Abstr.), 68, 1974.

945. **Fraipont, J.**, Recherches sur l'appareil excreteur des trematodes et des cestoides. Note preliminaire, *Bull. Acad. R. Sci. Belg.*, Sér. 2, 49, 397—402, 1880.
946. **Fraipont, J.**, Recherches sur l'appareil excreteur des trematodes et des cestoides. 2me et 3me communication, *Bull. Acad. R. Sci. Belg.*, 50, 106—107; 265—270, 1880.
947. **Fraipont, J.**, Recherches sur l'appareil excreteur des trematodes et des cestoides. Deuxieme partie. *Arch. Biol.*, 2, 1—40, 1881.
948. **Fredrickson, L. H. and Ulmer, M. J.**, Caryophyllaeid cestodes from two species of redhorse *(Moxostoma)*, *Proc. Iowa Acad. Sci.*, 72, 444—461, 1965.
949. **Freeman, R. S.**, The biology and life history of *Monoecocestus* Beddard, 1914 (Cestoda: Anoplocephalidae) from the porcupine, *J. Parasitol.*, 38, 111—129, 1952.
950. **Freeman, R. S.**, *Paradilepis rugovaginosus* n.sp. (Cestoda: Dilepididae) from the osprey, with notes on the genus *Oligorchis* Fuhrm., 1906, *J. Parasitol.*, 40, 22—28, 1954.
951. **Freeman, R. S.**, Studies on the biology of *Taenia crassiceps* (Zeder, 1800) Rud., 1810, *J. Parasitol.*, 40(Sect. 2), 41, 1954.
952. **Freeman, R. S.**, Life cycle and morphology of *Paruterina rauschi* n.sp. and *P. candelabraria* (Goeze, 1782) (Cestoda) from owls, and significance of plerocercoids in the order Cyclophyllidea, *Can. J. Zool.*, 35, 349—370, 1957.
953. **Freeman, R. S.**, Another hymenolepid with great morphological variation, *Hymenolepis bennetti* n. sp. (Cestoda) from *Napaeozapus insignis algonquinensis* Prince, *Can. J. Zool.*, 38, 737—743, 1960.
954. **Freeman, R. S.**, Do any *Anonchotaenia, Cyathocephalus, Echeneibothrium*, or *Tetragonocephalum* ( = *Tylocephalum*) (Eucestoda) have hookless oncospheres or coracidia?, *J. Parasitol.*, 68, 737—743, 1982.
955. **Freitas, de M. G.**, *Anoplocephala mamillana* (Cestoda, Anoplocephalidae) em equinos no Brasil, *Rev. Brasil. Biol.*, 5, 87—90, 1945.
956. **Freitas, de M. G.**, Sobre um cestoide de pombo domestico em Minas Geras-Brasil (Cestoda-Davaineidae), *An. 3. Congr. Brasil. Vet.*, pp. 256—258, 1946.
957. **Freze, V. I.**, *Principles of Cestodology* (in Russian), Vol. 5, *Proteocephalata*, Skrjabin, K. I., Ed., Izdatel'stvo Nauka, Moscow, 1965; English translation, *Isr. Program Sci. Translations, Jerusalem*, 1965, 1—597.
958. **Freze, V. I.**, Analysis of the classification of Ophiotaeniidae Freze, 1963 (Cestoda; Proteocephalata) (in Russian), *Helminthologia*, 6, 49—59, 1965.
959. **Freze, V. I. and Kazakov, B. E.**, A new species of *Proteocephalus* (Cestoidea: Proteocephalata) from *Coregonus alba* in the north European part of the USSR (in Russian), *Tr. Gel'mintol. Lab.*, 20, 171—175, 1969.
960. **Freze, V. I. and Ryšavý, B.**, Cestodes of the suborder Proteocephalata Spassky, 1957 (Cestoda-Pseudophyllidea) from Cuba and description of a new species *Ophiotaenia habanensis* sp. n., *Folia Parasitol.*, 23, 97—104, 1976.
961. **Freze, V. I. and Sharpilo, V. P.**, Two new cestode species of the genus *Ophiotaenia* LaRue, 1911 (Cestoda: Proteocephalata) from reptiles in the European part of the USSR (in Russian), *Helminthologia, Year 1966*, 7, 71—80, 1967.
962. **Friis, S.**, En hidtil ubeskreven Baendelorm hos Fugle, *Vidensk. Meddel. Naturh. Foren. Kjoebenhavn*, 1, 121—124, 1870.
963. **Frisch, J. L.**, De taeniis in anserum intestinis. *Misc. Berolin.* 3, 42, 1727; *Phys. Med. Abh. K. Acad. Wiss. Berlin*, 1, 155—156, 1727.
964. **Frisch, J. L.**, De taeniis in piscibus, *Misc. Berolin*, 3, 43—44, 1727.
965. **Frisch, J. L.**, De mustelae fluviatilis rapacitate et de taeniis in stomacho hujus piscis, *Misc. Berolin*, 4, 392—393, 1734.
966. **Frisch, J. L.**, De taeniis in pisciculo aculeato qui in Marchina Brandenburgia vocatur "Stecherling", *Misc. Berolin*, 4, 395—396, 1734.
967. **Frisch, J. L.**, De *Taenia capitata, Misc. Berolin*, 6, 129, 1740.
968. **Fritsch, G.**, Die Parasiten des Zitterwelses, *Sitzungber. Kgl. Preuss. Akad. Wiss. Jahrb.*, 1886, pp. 99—108.
969. **Froehlich, J. A.**, Beschreibungen einiger neuen Eingeweidewürmer, *Der Naturforscher, Halle*, 24, 101—162, 1789.
970. **Froehlich, J. A.**, Beiträge zur Naturgeschichte der Eingeweidewürmer, *Der Naturforscher, Halle*, 25, 52—113, 1791.
971. **Froehlich, J. A.**, Beiträge zur Naturgeschichte der Eingeweidewürmer, *Der Naturforscher, Halle*, 29, 5—96, 1802.
972. **Fuhrmann, O.**, Die Taenien der Amphibien, *Vorläufige Mitteilung. Zool. Anz.*, 18, 181—184, 1895.
973. **Fuhrmann, O.**, Die Taenien der Amphibien, *Zool. Jahrb. Anat. Ont.*, 9, 207—236, 1895.
974. **Fuhrmann, O.**, Beitrag zur Kenntnis der Vogeltaenien. I, *Rev. Suisse Zool.*, 3, 433—458, 1895.
975. **Fuhrmann, O.**, Beitrag zur Kenntnis der Vogeltaenien. II, *Rev. Suisse Zool.*, 4, 111—133, 1896.

976. **Fuhrmann, O.**, Beitrag zur Kenntnis der Bothriocephalen. I, *Centralbl. Bakteriol. I Abt.*, 19, 546—550, 1896.
977. **Fuhrmann, O.**, Beitrag zur Kenntnis der Bothriocephalen. II, *Centralbl. Bakteriol. I Abt.*, 19, 605—608, 1896.
978. **Fuhrmann, O.**, Sur un nouveau *Taenia* d'oiseau, *Rev. Suisse Zool.*, 5, 107—117, 1897.
979. **Fuhrmann, O.**, Ist *Bothriocephalus Zschokkei* mihi synonym mit *Schistocephalus nodosus* Rud.?, *Centralbl. Bakteriol.*, 23, 550—551, 1898.
980. **Fuhrmann, O.**, Ueber die Genera *Prosthecocotyle* Monticelli und *Bothriotaenia* Lönnberg, Vorläufige Mitteilung, *Zool. Anz.*, 21, 385—388, 1898.
981. **Fuhrmann, O.**, Das Genus *Prosthecocotyle*, *Zool. Anz.*, 22, 180—183, 1899.
982. **Fuhrmann, O.**, Das Genus *Prosthecocotyle*, *Centralbl. Bakteriol.*, 25, 863—877, 1899.
983. **Fuhrmann, O.**, Mitteilungen ueber Vogeltaenien. I. Ueber *T. depressa* Siebold, *Centralbl. Bakteriol. I Abt.*, 26, 83—86, 1899; II. Zwei eigentümliche Vogeltaenien, *Centralbl. Bakteriol. I Abt.*, 26, 618—622, 1899; III. *T. musculosa* mihi et *T. crateriformis* Goeze (*Monopylidium* nov. gen.), *Centralbl. Bakteriol. I Abt.*, 26, 622—627, 1899.
984. **Fuhrmann, O.**, Deux singuliers Taenias d'oiseaux: *Gyrocoelia perversus* n.g. n.sp. et *Acoleus armatus* n.g. n.sp., *Rev. Suisse Zool.*, 7, 341—451, 1899.
985. **Fuhrmann, O.**, On the anatomy of *Prosthecocotyle torulosa* (Linstow) and *Prosthecocotyle heteroclita* (Dies.), *Proc. R. Soc. Edinburgh*, 22, 641—651, 1899.
986. **Fuhrmann, O.**, Neue eigentümliche Vogelcestoden. Ein getrenntgeschlechtlicher Cestode, *Zool. Anz.*, 23, 48—51, 1900.
987. **Fuhrmann, O.**, Zur Kenntnis der Acoleinae, *Centralbl. Bakteriol.*, 28, 363—376, 1900.
988. **Fuhrmann, O.**, Neue Arten und Genera von Vogeltaenien. Vorläufige Mitteilung, *Zool. Anz.*, 24, 271—273, 1901.
989. **Fuhrmann, O.**, Bemerkungen ueber einige neuere Vogelcestoden, *Centralbl. Bakteriol. I Abt.*, 24, 757—763, 1901.
990. **Fuhrmann, O.**, Sur un nouveau Bothriocephalide d'oiseau *(Ptychobothrium armatum)*, *Arch. Parasitol.*, 5, 440—448, 1902.
991. **Fuhrmann, O.**, Sur deux nouveaux genres de cestodes d'oiseaux. Note préliminaire, *Zool. Anz.*, 25, 357—360, 1902.
992. **Fuhrmann, O.**, Die Anoplocephaliden der Vögel, *Centralbl. Bakteriol. I Abt.*, 32, 122—147, 1902.
993. **Fuhrmann, O.**, L'évolution des Taenias et en particulier de la larve des Ichthyotaenias, *Arch. Sci. Phys. Nat.*, 16, 1—3, 1903.
994. **Fuhrmann, O.**, Ein merkwürdiger getrenntgeschlechtlicher Cestode. Vorläufige Mitteilung, *Zool. Anz.*, 27, 327—331, 1904.
995. **Fuhrmann, O.**, Neue Anoplocephaliden der Vögel. Vorläufige Mitteilung, *Zool. Anz.*, 27, 384—388, 1904.
996. **Fuhrmann, O.**, Ein getrenntgeschlechtlicher Cestode, *Zool. Jahrb. Syst.*, 20, 131—150, 1904.
997. **Fuhrmann, O.**, Die Tetrabothrien der Säugetiere, *Centralbl. Bakteriol. I Abt.*, 35, 744—752, 1904.
998. **Fuhrmann, O.**, Über Öst-Asiatische Vogelcestoden. Reise von Dr. Walther Volz, *Zool. Jahrb. Syst.*, 22, 303—320, 1905.
999. **Fuhrmann, O.**, Das Genus *Diploposthe* Jacobi, *Centralbl. Bakteriol. I Abt.*, 40, 217—224, 1905.
1000. **Fuhrmann, O.**, Die Taenien der Raubvögel, *Centralbl. Bakteriol. I Abt.*, 41, 79—89, 1906.
1001. **Fuhrmann, O.**, Die Hymenolepis Arten der Vögel, *Centralbl. Bakteriol. I Abt.*, 41, 352—358, 440—452, 1906; *Centralbl. Bakteriol.*, 42, 620—628, 730—755, 1906.
1002. **Fuhrmann, O.**, Die Systematik der Ordnung der Cyclophyllidea, *Zool. Anz.*, 32, 289—297, 1907.
1003. **Fuhrmann, O.**, Bekannte und neue Arten und Genera von Vogeltaenien, *Centralbl. Bakteriol. I Abt.*, 45, 512—536, 1907.
1004. **Fuhrmann, O.**, Nouveaux Ténias d'oiseaux, *Rev. Suisse Zool.*, 16, 27—73, 1908.
1005. **Fuhrmann, O.**, Das Genus *Anonchotaenia* und *Biuterina*, *Centralbl. Bakteriol. I Abt.*, 46, 622-631, 1908; *Centralbl. Bakteriol. I Abt.*, 48, 412—428, 1908.
1006. **Fuhrmann, O.**, Cestoden der Vögel, *Zool. Jahrb.*, Suppl. 10, 232, 1908.
1007. **Fuhrmann, O.**, Neue Davaineiden, *Centralbl. Bakteriol. Parasitenk. I Abt.*, 49, 94—124, 1909.
1008. **Fuhrmann, O.**, Cestoden. *Wissenschaftliche Ergebnisse der Schwedischen Expedition nach dem Kilimandjaro, dem Meru und den umgebenden Masai Steppen Deutsch Ost-Afrikas, 1905—1906*, 3(22), Vermes, pp. 11—12, 1909.
1009. **Fuhrmann, O.**, *Triaenophorus robustus* Olsson dans les lacs de Neuchâtel et de Bienne, *Bull. Soc. Sci. Nat. Neuchâtel*, 36, 86—89, 1909.
1010. **Fuhrmann, O.**, La distribution faunistique et géographique des cestodes d'oiseaux, *Bull. Soc. Sci. Nat. Neuchâtel*, 36, 90—101, 1909.
1011. **Fuhrmann, O.**, Die Cestoden der Vögel des Weissen Nils, in *Jägerskiöld: Result of the Swedish Zoological Expedition to Egypt and the White Nile, 1901*, (Part 3), 1910, 55 pp.

1012. **Fuhrmann, O.**, Vogelcestoden der Aru-Inseln, *Abhandl. Senckenberg. Nat. Ges.,* 34, 251—266, 1911.
1013. **Fuhrmann, O.**, Vogelcestoden. Ergebnisse der mit Subvention aus der Erbschaft Treitl unternommenen zoologischen Forschungsreise Dr. F. Werners nach dem aegyptischen Sudan und Nord-Uganda, *Sitzungberg. Akad. Wiss. Wien Math. Naturwiss. Kl. Abt. I,* 121, 181—192, 1912.
1014. **Fuhrmann, O.**, Nordische Vogelcestoden aus dem Museum Göteborg, *Medd. Göteborgs Mus. Zool.,* Afdelning I, 1913, 41 pp.
1015. **Fuhrmann, O.**, Vogelcestoden, Nova Guinea. Résultats de l'expédition scientifique néerlandaise à la Nouvelle Guinée, 1910, *Zoologie,* 9, 467—470, 1913.
1016. **Fuhrmann, O.**, Sur l'origine de *Fimbriaria fasciolaris* Pallas, *9th Congr. Int. Zool.,* Monaco, 1913, 1914, 435—457.
1017. **Fuhrmann, O.**, Ein neuer getrenntgeschlechtlicher Cestode, *Zool. Anz.,* 44, 611—620, 1914.
1018. **Fuhrmann, O.**, Eigentümliche Fischcestoden, *Zool. Anz.,* 46, 385—398, 1916.
1019. **Fuhrmann, O.**, Cestodes des oiseaux de la Nouvelle-Calédonie et des Iles Loyalty, *Nova Caledonia,* 2, 399—449, 1918.
1020. **Fuhrmann, O.**, Considérations générales sur les Davainea, *Festschrift für Zschokke Bale,* 1920, 19 pp.
1021. **Fuhrmann, O.**, Die Cestoden der Deutschen Südpolar Expedition, 1901—1903. Deutsche Südpolar-Expedition, 1901—1903 (1920), 16, *Zoologie,* 8, 469—524, 1920.
1022. **Fuhrmann, O.**, Einige Anoplocephaliden der Vögel, *Centralbl. Bakteriol. I Abt.,* 87, 438—451, 1921.
1023. **Fuhrmann, O.**, Encore le cycle du *Bothriocephalus latus, Rev. Méd. Suisse Romande,* 43me année, pp. 573—575, 1923.
1024. **Fuhrmann, O.**, *Hymenolepis macracanthos* v. Linstow avec considérations sur le genre *Hymenolepis, J. Parasitol.,* 11, 33—43, 1924.
1025. **Fuhrmann, O.**, Questions de nomenclature concernant le genre *Raillietina* Fuhrmann (syn: *Davainea* Bl.), *Ann. Parasitol.,* 2, 312—313, 1924.
1026. **Fuhrmann, O.**, Two new species of reptilian cestodes, *Ann. Trop. Med. Parasitol.,* 18, 505—513, 1924.
1027. **Fuhrmann, O.**, Le phénomène des mutations chez les cestodes, *Rev. Suisse Zool.,* 32, 95—97, 1925.
1028. **Fuhrmann, O.**, Sur le développement et la reproduction asexuée des *Idiogenes otidis* Kr., *Ann. Parasitol.,* 3, 143—150, 1925.
1029. **Fuhrmann, O.**, Cestodes. Catalogue des invertébrés de la Suisse, fasc., 17, *Musée Hist. Nat. Genève,* 1926, 1—150, 1926.
1030. **Fuhrmann, O.**, Brasilianische Cestoden aus Reptilien und Vögeln, *Abh. Senkenb. Naturforsch. Ges.,* 40, 389—401, 1927.
1031. **Fuhrmann, O.**, Dritte Klasse des Cladus Plathelminthes. Cestoidea, in *Kükenthal's Handbuch der Zoologie,* Vol. 2, Kükenthal and Krumbach, Bogen, 1930—1931, 141—416.
1032. **Fuhrmann, O.**, Les Ténias des oiseaux, *Mém. Univ. Neuchâtel,* Vol. 8, 1932, 381 pp.
1033. **Fuhrmann, O.**, Cestodes nouveaux, *Rev. Suisse Zool.,* 40, 169—178, 1933.
1034. **Fuhrmann, O.**, Deux nouveaux cestodes de mammifères d'Angola, *Bull. Soc. Sci. Nat. Neuchâtel,* 58, 97—106, 1933.
1035. **Fuhrmann, O.**, Un cestode aberrant, *Bull. Soc. Sci. Nat. Neuchâtel,* 58, 107—120, 1933.
1036. **Fuhrmann, O.**, Vier Diesing'sche Typen (Cestoda), *Rev. Suisse Zool.,* 41, 545—564, 1934.
1037. **Fuhrmann, O.**, Rectification de nomenclature, *Ann. Parasitol.,* 13, 386, 1935.
1038. **Fuhrmann, O.**, Les Ténias de oiseaux, *Bull. Ornithol. Suisse Romande,* 1, 114—157, 1935.
1039. **Fuhrmann, O.**, *Gynandrotaenia stammeri* n.g. n.sp., *Rev. Suisse Zool.,* 43, 517—518, 1936.
1040. **Fuhrmann, O.**, Un singulier Ténia d'oiseaux, *Gynandrotaenia stammeri* n.g. n.sp., *Ann. Parasitol.,* 14, 261—271, 1936.
1041. **Fuhrmann, O.**, Un cestode extraordinaire, *Nematoparataenia southwelli* Fuhrmann, *C. R. XIIme Congr. Int. Zool. Lisbonne* (1935), pp. 1517—1532, 1937.
1042. **Fuhrmann, O.**, Cestodes d'Angola, *Rev. Suisse Zool.,* 50, 449—471, 1943.
1043. **Fuhrmann, O. and Baer, J. G.**, Zoological Results of the Third Tanganyika Expedition, conducted by W. A. Cunnington, 1904—1905, I, 79—100, 1925.
1044. **Fuhrmann, O. and Baer, J. G.**, Cestodes. Mission biologique Sagan-Omo (Ethiopie méridionale), 1939, dirigée par le professeur Eduardo Zavattari, *Bull. Soc. Sci. Nat. Neuchâtel,* 68, 113—140, 1943.
1045. **Furmaga, S.**, *Spirometra janicki* sp. n. (Diphyllobothriidae) (in Polish, English summary), *Acta Parasitol. Pol.,* 1, 29—59, 1953.
1046. **Furtado, J. I.**, A new caryophyllaeid cestode, *Lytocestus parvulus* sp. nov., from a Malayan catfish, *Ann. Mag. Nat. Hist.,* (Ser. 13), 6, 97—106, 1963.
1047. **Furtado, J. I. and Chau-lan, L.**, Two new helminth species from the fish *Channa micropeltes* Cuvier (Ophiocephalidae) of Malaysia, *Folia Parasitol.,* 18, 365—372, 1971.
1048. **Gabrion, C. and MacDonald, G.**, *Artemia* sp. (Crustacé, anostracé), hôte intermédiare d'*Eurycestus avoceti* Clark, 1954 (cestode cyclophyllide) parasite de l'avocette en Camargue, *Ann. Parasitol. Hum. Comp.,* 55, 327—331, 1980.

1049. **Gagarin, V. G., Chertkova, A. N., and Vshivtsev, V. I.,** The species composition of the genus *Schistocephalus* Creplin, 1829 (Cestoidea: Ligulidae) (in Russian), *Mater. Nauch. Konf. Vses. Obshch. Gel'mintol.*, 3, 71—75, 1966.
1050. **Gaiger, S. H.,** A revised check-list of the animal parasites of domesticated animals in India, *J. Comp. Pathol. Ther. London*, 28, 67—76, 1915.
1051. **Gaikwad, P. M. and Shinde, G. B.,** A new cestode *Parvitaenia yamagutii* (Cestoda: Dilepididae) from a common kite, *Milvus migrans* from Nanded, India, *Proc. Ind. Acad. Parsitol.*, 1, 27—32, 1980.
1052. **Galli Valerio, B.,** Notes de parasitologie, *Centralbl. Bakteriol.*, 27, 305—309, 1900.
1053. **Galli Valerio, B.,** *Bothriocephalus latus* Brems. chez le chat, *Centralbl. Bakteriol. I Abt.*, 32, 285—287, 1902.
1054. **Galli Valerio, B.,** Notes de parasitologie, *Centralbl. Bakteriol. I Abt.*, 35, 85—91, 1904.
1055. **Galli Valerio, B.,** Einige Parasiten von *Arvicola nivialis*, *Zool. Anz.*, 28, 519—522, 1905.
1056. **Galli Valerio, B.,** Recherches helminthologiques. I. Infection expérimentale des Cyclops du lac des quatre cantons avec les embryons du *Dibothriocephalus latus*. II. Les migrations des ascarides dans l'organisme, *Schweiz. Med. Wochenschr.*, 53, 314—316, 1923.
1057. **Galli Valerio, B.,** Notes de parasitologie, *Centralbl. Bakteriol. I Abt. Orig.*, 112, 54—59, 1929.
1058. **Galli Valerio, B.,** Note parasitologiques, *Schweiz. Arch. Tierheilk.*, 77, 643—647, 1935.
1059. **Galliard, H.,** Particularités du cycle évolutif de *Diphyllobothrium mansoni* au Tonkin, *Ann. Parasitol.*, 21, 246—253, 1947.
1060. **Galliard, H.,** Infestation naturelle des batraciens et reptiles par les larves plérocercoides de *Diphyllobothrium mansoni* au Tonkin, *Ann. Parasitol.*, 23, 23—26, 1948.
1061. **Galliard, H.,** Infestation expérimentale par les larves plérocercoides de *Diphyllobothrium mansoni* au Tonkin, *Ann. Parasitol.*, 23, 203—213, 1948.
1062. **Galliard, H. and Dang-van Ngu,** Particularités du cycle évolutif des *Diphyllobothrium mansoni* au Tonkin, *Ann. Parasitol.*, 21, 246—253, 1947.
1063. **Gallien, L.,** *Proboscidosaccus enigmaticus* nov. g. nov. sp. parasite de *Mactra solida* L. (note préliminaire), *Bull. Soc. Zool. Fr.*, 74, 322—326, 1949.
1064. **Gan, K. H.,** Research on the life history of *Diphyllobothrium ranarum* Doc., *Neerland. Indonesia Morb. Trop.*, 1, 90—92, 1949.
1065. **Ganapati, P. N. and Rao, K. H.,** On two species of *Bothriocephalus* Rudolphi (1808) Cestoda from the gut of *Saurida tumbil* (Bloch), *Proc. Ind. Sci. Congr.*, Part III, (Abstr.), 172, 1954.
1066. **Garcia, E. Y. and Africa, C. M.,** *Diphyllobothrium latum* (Linnaeus, 1758) Lühe, 1920 in a native Filipino, *Philipp. J. Sci.*, 57, 451—456, 1935.
1067. **Garkavi, B. L.,** On the question of the biology of the cestode *Fimbriaria fasciolaris* (Pallas, 1781) parasitic in domestic and wild ducks (in Russian), *Tr. Vsesoiuz. Inst. Gel'mintol.*, 4, 5, 1950.
1068. **Garoian, G. S.,** *Schistocephalus thomasi* n. sp., (Cestoda: Diphyllobothriidae) from fish-eating birds, *Proc. Helminthol. Soc. Wash.*, 27, 199—202, 1960.
1069. **Garrison, P. E.,** Preliminary report on the specific identity of the cestode parasites of man in the Philippine Islands with a description of a new species of *Taenia*, *Philipp. J. Sci.*, 2, 537—550, 1907.
1070. **Gasowska, M.,** Les cestodes des oiseaux des environs de Kiew (Ukraine), *Acad. Pol. Sci. Lett. C. R. Mens. Cl. Sci. Math. Nat. Cracovie*, pp. 5—6, 1931.
1071. **Gasowska, M.,** Die Vogelcestoden aus der Umgebung von Kiew (Ukraine), *Bull. Int. Acad. Pol. Sci. Lett.*, (Ser. B.), 11, 599—627, 1932.
1072. **Gässlein, H.,** Die Cestoden der Vertebraten aus der Umgebung von Erlangen, *Z. Parasitenk.*, 16, 443—468, 1954.
1073. **Gassner, F. X. and Thorp, F., Jr.,** Studies on *Thysanosoma actinioides*, *J. Am. Vet. Med. Assoc.*, 96, 410—411, 1940.
1074. **Gastaldi, B.,** Cenni Sopra Alcuni Nuovi Elminti della *Rana esculenta*, con nuove Osservazione sul *Codonocephalus mutabilis* Diesing, Torino, 1854, 14 pp.
1075. **Gauthier, M.,** Développement de l'oeuf et embryon du *Cyathocéphale*, parasite de la truite, *C. R. Acad. Sci.*, 177, 913—916, 1923.
1076. **Gauthier, M.,** Endoparasites de la truite indigène (*T. fario* L.) en Dauphiné, *C. R. Assoc. Fr. Avanc. Sci.*, 49, 442—444, 1926.
1077. **Gedoelst, L.,** *Synopsis de Parasitologie de l'Homme et des Animaux Domestiques*, 1911, 332 pp.
1078. **Gedoelst, L.,** Notes sur la faune parasitaire du Congo Belge, *Rev. Zool. Afr.*, 5, 1—90, 1916.
1079. **Gedoelst, L.,** Un cas de parasitation de l'homme par l'*Hymenolepis diminuta* (Rudolphi), *C. R. Soc. Biol.*, 83, 190—192, 1920.
1080. **Genov, T.,** A new species of cestode of the genus *Hilmylepis* (Hymenolepididae) from shrew in Bulgaria, *Parazitologiya*, 4, 473—475, 1970.
1081. **Genov, T.,** A new cestode from rodents in Bulgaria — *Catenotaenia matovi* sp. nov. (Cestoidea, Catenotaeniidae), *Dokl. Akad. Sel'skokhozyaĭstvennykh Nauk Bolgarii*, 4, 119—122, 1971.

1082. **Genov, T.**, Morphology and taxonomy of the species of genus *Coronacanthus* Spassky, 1954 (Cestoda: Hymenolepididae) in Bulgaria, *Helminthologia,* 17, 245—255, 1980.

1083. **Genov, T. and Tenora, F.**, Re-organization of the system of cestodes of the family Catenotaeniidae Spassky, 1950, *Khelmintologiya,* No. 8, 34—42, 1974.

1084. **Germanos, N. K.**, *Bothriocephalus schistochilos* n. sp. Ein neurer Cestode aus dem Darm von *Phoca barabata, Jen. Z. Med. Naturwiss.,* 30, 1—38, 1895.

1085. **Gervais, H.**, Sur les entozoaires des Dauphins, *C. R. Acad. Sci.,* 71, 779—781, 1870.

1086. **Gervais, P.**, Sur quelques entozoaires taenioïdes et hydatids, *Mém. Acad. Sci. Lett. Montpellier,* 1, 85—103, 1847.

1087. **Ghosh, R. K.**, On a new species of the genus *Aporina* Fuhrmann, 1902 (Cestoda; Anoplocephalidae) along with comments on certain allied genera, in *Dr. B. S. Chauhan Commemoration Volume,* Tiwari, K. K. and Srivastava, C. B., Eds., Zoological Society of India, Orissa, India, 1975, 181—184.

1088. **Gibbs, H. C.**, On the role of rodents in the epidemiology of hydatid disease in the Mackenzie River basin, *Can. J. Comp. Med.,* 21, 287—289, 1957.

1089. **Giebel, C. G. A.**, Chr. L. Nitzsch's helminthologissche Untersuchungen, *Z. Ges. Naturwiss.,* 9, 264—269, 1857.

1090. **Ginezinskaya, T. A.**, Neoteny phenomena in cestodes (in Russian), *Zool. Zh.,* 23, 35—42, 1944.

1091. **Gläser, H.**, Die Entwicklungsgeschichte des *Cysticercus longicollis* Rud., *Z. Wiss. Zool.,* 92, 540—561, 1909.

1092. **Gnedizlov, V. G.**, *Mesocricetus auratus* Waterhouse as the potential definitive host of the tapeworm *Taenia solium* (in Russian), *Zool. Zh.,* 36, 1770—1773, 1957.

1093. **Goette, A.**, Einziges aus der Entwicklungsgeschichte der Cestoden, *Zool. Jahrb. Anat.,* 42, 213—228, 1921.

1094. **Goldberg, O.**, Helminthum Dispositio Systematica, Ph.D. thesis, University of Berlin, 1855, 1—130.

1095. **Goldschmidt, R.**, Zur Entwicklungsgeschichte der Echinococcusköpfchen, *Zool. Jahrb.,* 13, 466—492, 1900.

1096. **Goldstein, R. J.**, A note on the genus *Poecilancistrium* Dollfus, 1929 (Cestoda: Trypanorhyncha), *J. Parasitol.,* 49, 301—304, 1963.

1097. **Goldstein, R. J.**, Species of *Acanthobothrium* (Cestoda: Tetraphyllidea) from the Gulf of Mexico, *J. Parasitol.,* 50, 656—661, 1964.

1098. **Goldstein, R. J.**, The genus *Acanthobothrium* van Beneden, 1849 (Cestoda: Tetraphyllidea), *J. Parasitol.,* 53, 455—483, 1967.

1099. **Goldstein, R. J.**, A tabular synopsis of published data on *Acanthobothrium* van Beneden, 1849, (Cestoda: Tetraphyllidea), *Bull. Ga. Acad. Sci.,* 27, 53—69, 1969.

1100. **Goldstein, R. J., Henson, R. N., and Schlicht, F. G.**, *Acanthobothrium lintoni* (Cestoda: Tetraphyllidea) from the electric ray, *Narcine brasiliensis* (Olfers) in the Gulf of Mexico, *Zool. Anz.,* 181, 435—438, 1968.

1101. **Golovkova, V. I.**, *Diorchis oxyuri* n. sp. (Cestoda: Hymenolepididae), a new cestode from *Oxyuris leucocephala* (in Russian), *Tr. Gel'mintol. Lab. (Ekologiya Taksonomiya Gel'mintov),* 23, 45—48, 1973.

1102. **Golovkova, V. I.**, *Echinocotyle kornyushini* n. sp., a new cestode of Charadriiformes (in Russian), *Tr. Gel'mintol. Lab. (Gel'minty Zhivotnyky Rastenii),* 29, 28—30, 1979.

1103. **Goodsir, J.**, On *Gymnorhynchus horridus*, a new cestoid entozoon, *Edinburgh Philos. J.,* 31, 9—12, 1841.

1104. **Gordadze, G. N., Kamalova, A. N., and Bugianishvili, S. M.**, *Taenia* infections in man in Georgia, *Med. Parasitol. Paras. Dis.* (in Russian), 13, 64—66, 1944.

1105. **Gordon, H. McL.**, A note on the longevity of *Moniezia* spp. in sheep, *Aust. Vet. J.,* 8, 153—154, 1932.

1106. **Gorodilova, L. I.**, Epidemiology and control of *Hymenolepis* infections, *Med. Parasitol. Paras. Dis.* (in Russian), 13, 18—26, 1944.

1107. **Goss, O. M.**, Platyhelminth and acanthocephalan parasites of local shags, *J. R. Soc. West. Aust.,* 26, 1—14, 1941.

1108. **Goto, S. and Ishii, N.**, On a new cestode species, *Amphilina japonica, Jpn. J. Exp. Med.,* 14, 81—83, 1936.

1109. **Gough, L. H.**, A monograph of the tapeworms of the subfamily Avitellinae, being a review of the genus *Stilesia* and an account of the histology of *Avitellina centripunctata* (Riv.), *Q. J. Microsc. Sci.,* 56, 317—383, 1911.

1110. **Gough, L. H.**, The anatomy of *Stilesia globipunctata, Parasitology,* 5, 114—118, 1912.

1111. **Gower, W. C.**, An unusual cestode record from the porcupine in Michigan, *Pap. Mich. Acad. Sci. Arts Lett.,* 24, 149—151, 1938.

1112. **Gower, W. C.**, A host-parasite catalogue of the helminths of ducks, *Am. Midl. Nat.,* 22, 580—628, 1939.

1113. **Graber, M.**, Parasites internes des vértébrés domestiques et sauvages, autres que les primates de la République Populaire du Congo (d'après la collection Cossard-Chambron, 1956—1960), Rôle pathogène-prophylaxie, *Rev. Elevage Méd. Vét. Pays Trop.,* 34, 155—167, 1981.

1114. **Graber, M. and Euzéby, J.**, *Trichocephaloides beauporti* n. sp. Cestode nouveau des Charadriiforms et de certains Passériformes de la Guadeloupe, *Ann. Parasitol.*, 51, 189—198, 1976.
1115. **Graber, M. and Euzéby, J.**, *Hymenolepis guadeloupensis* n. sp. cestode nouveau du canard domestique *Anas boschas* Linné, *Ann. Parasitol. Hum. Comp.*, 51, 199—205, 1976.
1116. **Graber, M. and Euzéby, J.**, Deuxième enquête parasitologique en Guadeloupe. Note 2: les cestodes des oiseaux aquatiques, *Bull. Soc. Sci. Vét. Méd. Comp. Lyon*, 78, 153—171, 1976.
1117. **Graham, R., Torrey, J. P., Mizelle, J. D., and Michael, V. M.**, Internal parasites of poultry, Circular 469, *Ill. Agric. Exp. Stn.*, 1937, 50 pp.
1118. **Gramiccia, G.**, Reperto di *Raillitiena (Raillietina) bothrioplitis* nel Tacchino, *Rend. Ist. Sanita Pubblica Roma*, 4, 708—712, 1941.
1119. **Gramiccia, G.**, Sulla nomenclatura di due specie de tenie dei polli e considerazioni sul genere *Raillietina* Fuhrmann, 1920, *Rend. Ist. Sanita Pubblica Roma*, 4, 539—544, 1941; *Riv. Parasitol. Roma*, 5, 67—71, 1941.
1120. **Grassi, G. B.**, Bestimmung der vier von Dr. E. Parona in einem kleinen Mädchen aus Varese (Lombardei) gefundenen Taenien (*Taenia flavopunctata?* Dr. E. Parona), *Centralbl. Bakteriol. I Abt.*, 1, 257—259, 1867.
1121. **Grassi, G. B.**, Contribuzione allo studio della nostra fauna, *Atti Accad. Gioenia Sci. Nat. Catania*, 18, 241—252, 1885.
1122. **Grassi, G. B.**, Entwicklungsgeschichte der *Taenia nana*, *Centralbl. Bakteriol. I Abt.*, 2, 94—95, 1887.
1123. **Grassi, G. B.**, La pulce del cane (*Pulex serraticeps* Gervais) e l'ordinario, ospite intermedio della *Taenia cucumerina*. Nota preventiva, *Bull. Soc. Entomol. Ital.*, 20, 66, 1888.
1124. **Grassi, G. B.**, Beiträge zur Kenntniss des Entwicklungscyclus von fünf Parasiten des Hundes (*Taenia cucumerina* Goeze; *Ascaris marginata* Rud.; *Spiroptera sanguinolenta* Rud.; *Filaria immitis* Leidy; und *Haematozoon* Lewis), *Centralbl. Bakteriol. I Abt.*, 4, 609—621, 1888.
1125. **Grassi, G. B.**, *Taenia flavopunctata* Weinl., *Taenia leptocephala* Creplin, *Taenia diminuta* Rud., *Atti R. Accad. Sci. Torino*, 23, 492—501, 1888.
1126. **Grassi, G. B. and Rovelli, G.**, Embryologische Forschungen an Cestoden, *Centralbl. Bakteriol. I Abt.*, 5, 370—377, 1889.
1127. **Grassi, G. B. and Rovelli, G.**, Embryologische Forschungen an Cestoden (Schluss), *Zentralbl. Bacteriol. I Abt.*, 5, 401—410, 1889.
1128. **Grassi, G. B. and Rovelli, G.**, Ricerche embriologiche sui Cestodi, *Atti Accad. Gioenia Sci. Nat.*, 4, 1—109, 1892.
1129. **Gresson, R. A. R.**, A new species of *Proteocephalus* from *Coregonus pollan* Thompson, *Ir. Nat. J.*, 10, 308—309, 1952.
1130. **Gresson, R. A. R.**, A morphological study of a fish tapeworm, *Proteocephalus pollanicola*, *Parasitology*, 44, 34—49, 1954.
1131. **Grimes, L. R. and Miller, G. C.**, Caryophyllaeid cestodes in the creek chubsucker, *Erimyzon oblongus* (Mitchill), *J. Parasitol.*, 61, 973—974, 1975.
1132. **Grimm, O.**, Zur Kenntnis einiger wenig bekannten Binnenwürmer, *Nachr. K. Ges. Wiss. Göttingen*, 12, 240—246, 1872.
1133. **Grube, E.**, Bemerkungen ueber einige Helmińthen und Meerwürmer, *Arch. Naturg.*, 21, 137—158, 1855.
1134. **Gruber, A.**, Ein neuer Cestodenwirt, *Zool. Anz.*, 1, 74—75, 1878.
1135. **Gruber, A.**, Zur Kenntnis des *Archigetes Sieboldii*, *Zool. Anz.*, 4, 89—91, 1881.
1136. **Grundmann, A. W.**, A new tapeworm, *Mesocestoides carnivoricolus*, from carnivores of the Great Salt Lake Desert region of Utah, *Proc. Helminthol. Soc. Wash.*, 23, 26—28, 1956.
1137. **Grundmann, A. and Brown, W. C.**, Cestodes of the California gull (*Larus californicus*) in the great Salt Lake Area, *Proc. Utah Acad. Sci.*, 27—28, 74, 1951.
1138. **Gubanov, N. M.**, The Helminthofauna of Commercial Animals of the Sea of Okhotsk and Pacific Ocean, Report and Thesis 1952; cited in Reference 715.
1139. **Guberlet, J. E.**, Morphology of adult and larval cestodes from poultry, *Trans. Am. Microsc. Soc.*, 35, 23—44, 916.
1140. **Guberlet, J. E.**, Studies on the transmission and prevention of cestode infection in chickens, *J. Am. Vet. Med. Assoc.*, 49, 2, 218—237, 1916.
1141. **Guberlet, J. E.**, On the life-history of the chicken cestode, *Hymenolepsis carioca*, *J. Parasitol.*, 6, 35—38, 1919.
1142. **Guevara Pozo, D.**, Cestodes del género *Dinobothrium*, parásitos de grandes selacios, pescados en las costas españolas, *Rev. Ibér. Parasitol.*, pp. 260—270, 1945.
1143. **Guiart, J.**, Classification des Tétrarhynques, *Assoc. Fr. Avanc. Sci.*, 50th Session, Lyon, 1926, pp. 397—401, 1927.
1144. **Guiart, J.**, Considérations historiques sur la nomenclature et sur la classification des Tétrarhynques, *Bull. Inst. Oceanogr. Monaco*, No. 575, 1—27, 1931.

1145. **Guiart, J.,** Contribution à l'étude des cestodes de Calmars, avec description d'une espèce nouvelle, *Diplobothrium pruvoti, Arch. Zool. Exp. Gén.,* 75, 465—473, 1933.
1146. **Guiart, J.,** Le véritable *Floriceps saccatus* de Cuvier n'est pas la larve géant de tétrarhynque vivant dans le foie du môle *(Mola mola), Bull. Inst. Oceanogr. Monaco,* No. 666, 1—15, 1935.
1147. **Guiart, J.,** Cestodes parasites provenant des campagnes scientifiques du Prince Albert Ier de Monaco, *Résultats des Campagnes Scientifiques Accomplies sur son Yacht par Albert Ier Monaco.,* Fasc. 91, 1—115, 1935.
1148. **Guiart, J.,** Etude parasitologique et épidémiologique de quelques poissons de mer, *Bull. Inst. Oceanogr. Monaco,* No. 755, 1—15, 1938.
1149. **Guidal, J. A.,** On the systematic position of *Taenia transfuga* Krabbe 1869, *K. Veterinaer — og Landbohøjsk. Arsskr. Copenhagen,* pp. 72—78, 1960.
1150. **Gulati, A. N.,** Description of a new species of tapeworm, *Dipylidium catus* n. sp., with a note on the genus *Dipylidium* Leuckart, 1863, *Bull. Agric. Res. Inst. Pusa,* 190, 1—14, 1929.
1151. **Gupta, N. K. and Grewal, S. S.,** On a new cestode, *Raillietina (Raillietina) streptopeline* sp. n. from red turtle dove, *Streptopelia tranquebarica tranquebarica, Acta Parasitol. Pol.,* 16, 73—75, 1969.
1152. **Gupta, N. K. and Grewal, S. S.,** On *Raillietina (Raillietina) buckleyi* n. sp. from little brown doves, *Streptopelia senegalensis combayensis* Gmelin, *Zool. Anz.,* 182, 255—258, 1969.
1153. **Gupta, N. K. and Grewal, S. S.,** Studies on two new ophryocotyloid cestodes (family: Davaineidae) from crow, *Corvus splendens* (Vieillot), *Res. Bill. Punjab Univ. Sci.,* Year 1970, 21, 77—86, 1971.
1154. **Gupta, N. K. and Grewal, S. S.,** A new cestode, *Raillietina (Raillietina) inda* n. sp., from Indian spotted dove, *Res. Bull. Panjab Univ.,* 21, 511—513, 1971 (dated 1970).
1155. **Gupta, N. K. and Madhu,** On a new poultry cestode in India, *Proc. Ind. Acad. Sci. Anim. Sci.,* 90, 377—380, 1981.
1156. **Gupta, P. D.,** A new species of cestode of the genus *Schistometra* (Cestoda: Davaineidae: Idiogeninae) from the Great Indian Bustard, *Choriotis nigriceps* (Vigors), *J. Bombay Nat. Hist. Soc.,* 73, 183—186, 1976.
1157. **Gupta, S. P.,** Caryophyllaeids (Cestoda) from freshwater fishes of India, *Proc. Helminthol. Soc. Wash.,* 28, 38—50, 1961.
1158. **Gupta, S. P. and Sinha, N.,** On a new caryophyllaeid *Capingentoides heteropneusti* n. sp. from the intestine of a freshwater fish *Heteropneustes fossilis* (Ham.) from Lucknow, *Ind. J. Helminthol.,* (1979, publ. 1980), 31, 65—68, 1980.
1159. **Gupta, S. P. and Sinha, N.,** On a new cestode *Duthiersia gomatii* n. sp. from a reptilian host *Varanus niloticus* from Lucknow, *Ind. J. Helminthol.,* (1979, publ. 1980), 31, 72—74, 1980.
1160. **Guttowa, A.,** O inwazyjnosci onkosfer *Triaenophorus lucii* (Müll.) i jej zmiennosci, *Acta Parasitol. Pol.,* 3, 447—465, 1956.
1161. **Guyer, M. F.,** On the structure of *Taenia confusa* Ward, *Zool. Jahrb. Syst.,* 11, 469—492, 1898.
1162. **Gvosdev, E. V.,** A new species of cestode of the family Anoplocephalidae from *Ochotona alpina* (in Russian), *Tr. Gel'mintol. Lab. Akad. Nauk SSSR,* 5, 143—145, 1951.
1163. **Gvosdev, E. V.,** Helminth fauna of *Tetraogallus himalayensis* Gray, 1842 (in Russian), *Zool. Zh.,* 33, 39—43, 1954.
1164. **Gvosdev, E. V.,** *Diuterinotaenia spasskyi* n.g., n.sp. from Ochotonidae in the kazakh SSR (in Russian), *Helminthologia,* 3, 139—142, 1961.
1165. **Gvosdev, E. V.,** Cestodes of game birds in southern Kazakhstan (in Russian), *Tr. Inst. Zool. Alma Ata,* 22, 74—109, 1964.
1166. **Gvosdev, E. V. and Maksimova, A. P.,** New cestode species (Cestoda: Hymenolepididae) from *Phoenicopterus roseus* (in Russian), *Izv. Akad. Nauk Kaz. SSR,* Ser. Biol. No. 5, 30—37, 1968.
1167. **Gvosdev, E. V. and Maksimova, A. P.,** To the helminthofauna of the rosy flamingo (*Phoenicopterus roseus* Pall.) in Kazachstan (in Russian), *Tr. Izv. Zool. Acad. Nauk Kaz. SSR,* 31, 41—46, 1971.
1168. **Gvosdev, E. V., Maksimova, A. P., and Kornyushin, A. P.,** *Monotestilepis tadornae* n.g., n.sp. (Hymenolepididae), a new cestode of *Tadorna tadorna* (in Russian), *Tr. Inst. Zool. Alma-Ata,* 31, 47—50, 1971.
1169. **Gwynn, A. M. and Hamilton, A. G.,** Occurrence of larval cestode in the red locust *(Nomadacris septemfasciata), J. Parasitol.,* 27, 551—555, 1935.
1170. **Hall, M. C.,** The flukes and tapeworms of cattle, sheep and swine, with special reference to the infection of meats, *Bur. Anim. Ind. U.S. Dep. Agric.,* 19, 11—136, 1898.
1171. **Hall, M. C.,** A new rabbit cestode, *Cittotaenia mosaica, Proc. U.S. Natl. Mus.,* 34, 691—699, 1908.
1172. **Hall, M. C.,** A new species of cestode parasite (*Taenia balaniceps*) of the dog and of the lynx, with a note on *Proteocephalus punicus, Proc. U.S. Natl. Mus.,* 39, 139—151, 1911.
1173. **Hall, M. C.,** The gid parasite and allied species of the cestode genus *Multiceps.* I. Historical review, *U.S. Bur. Anim. Ind. Bull.,* 125, 1—68, 1911.
1174. **Hall, M. C.,** The gid parasite and allied species of the cestode genus *Multiplex* (sic), *Am. Vet. Rev. N.Y.,* 38 (Abstr.), 591—592, 1911.

1175. **Hall, M. C.**, The coyote as a host of *Multiceps multiceps*, *Science*, 33, 975, 1911.
1176. **Hall, M. C.**, A second case of *Multiceps multiceps* in the coyote, *Science*, 35, 556, 1912.
1177. **Hall, M. C.**, The dog as a carrier of parasites and disease, *U.S. Dept. Agric. Bull.*, 260, 1—27, 1915.
1178. **Hall, M. C.**, A new and economically important tapeworm, *Multiceps gaigeri*, from the dog, *J. Am. Vet. Med. Assoc.*, 50, 214—223, 1916.
1179. **Hall, M. C.**, A synoptical key to the adult taenioid cestodes of the dog, cat and some related carnivores, *J. Am. Vet. Med. Assoc.*, 50, 356—360, 1916.
1180. **Hall, M. C.**, Parasites of the dog in Michigan, *J. Am. Vet. Med. Assoc.*, 51, 383—396, 1917.
1181. **Hall, M. C.**, The adult taenioid cestodes of dogs and cats, and of related carnivores in North America, *Proc. U.S. Natl. Mus.*, 55, 1—94, 1919.
1182. **Hall, M. C.**, Intestinal parasites found in eighteen Alaskan foxes, *North Am. Vet.*, 1, 123—124, 1920.
1183. **Hall, M. C.**, Parasites and parasitic diseases of sheep, *U.S. Dep. Agric. Farmers' Bull.*, 1330, 1—36, 1929.
1184. **Hall, M. C.**, A new cestode reared in the dog, *Multiceps packi* sp. nov., *J. Parasitol.*, 16, 49, 1929.
1185. **Hall, M. C.**, Arthropods as intermediate hosts of helminths, *Smithsonian Misc. Coll.*, 81, 1—77, 1929.
1186. **Hall, M. C.**, Parasites and parasitic diseases of dogs, *U.S. Dep. Agric. Circ.*, p. 338, 1934.
1187. **Hall, M. C.**, The discharge of eggs from segments of *Thysanosoma actinioides*, *Proc. Helminthol. Soc. Wash.*, 1, 6—7, 1934.
1188. **Hall, M. C. and Hoskins, H. P.**, The occurrence of tapeworms, *Anoplocephala* spp., of the horse in the United States, *Cornell Vet.*, 8, 287—292, 1918.
1189. **Hall, M. C. and Wigdor, M.**, A bothriocephalid tapeworm from the dog in North America, with notes on cestode parasites of dogs, *J. Am. Vet. Med. Assoc.*, 53, n.s., v. 6, 355—362, 1918.
1190. **Hamann, O.**, *Taenia lineata* Goeze, eine Taenie mit flächenständigen Geschlechtsöffnungen, *Z. Wiss. Zool.*, 42, 718—744, 1885.
1191. **Hamann, O.**, in *Gammarus pulex* lebende Cysticercoiden mit Schwanzanhängen, *Jena. Z. Naturwiss.*, 17, 1—10, 1890.
1192. **Hamann, O.**, Neue Cysticercoiden mit Schwanzanhängen, *Jena. Z. Naturwiss.*, 18, 553—564, 1891.
1193. **Hamid, A.**, A cestode, *Oochoristica khalili* n.sp., from a snake, *Psammophis schokari* Forskål, *J. Parasitol.*, 24, 238—240, 1932.
1194. **Hamilton, P. C.**, A new species of *Taenia* from a coyote, *Trans. Am. Microsc. Soc.*, 59, 64—69, 1940.
1195. **Hannum, C. A.**, A new species of cestode, *Ophiotaenia magna* n. sp., from the frog, *Trans. Am. Microsc. Soc.*, 44, 148—155, 1925.
1196. **Hansen, M. F.**, Three anoplocephalid cestodes from the prairie meadow vole, with description of *Andrya microti* n.sp., *Trans. Am. Microsc. Soc.*, 66, 279—282, 1947.
1197. **Hansen, M. F.**, *Schizorchis ochotonae* n.g., n.sp. of anoplocephalid cestode, *Am. Midl. Nat.*, 39, 754—757, 1948.
1198. **Hansen, M. F.**, Studies on cestodes of rodents, *Abstr. Doct. Diss. Univ. Nebr.*, pp. 112—119, 1948.
1199. **Hansen, M. F.**, A new dilepidid tapeworm and notes on other tapeworms of rodents, *Am. Midl. Nat.*, 43, 471—479, 1950.
1200. **Hanson, A. J.**, The slug as the intermediate host of the microscopic tapeworm of chickens, *Annu. Rep. West. Wash. Exp. Stn.*, (Bull. 18-W), p. 11, 1930.
1201. **Harding, C. L.**, A new species of Dibothriocephaloidea [sic] from *Oncorhynchus keta*, *Publ. Univ. Wash. Theses Ser. v.*, 2, 313—316, 1937.
1202. **Harkema, R.**, The parasites of some North Carolina rodents, *Ecol. Monogr.*, 2, 151—232, 1936.
1203. **Harkema, R.**, The mourning dove, a new host of the anoplocephalid tapeworm, *Aporina delafondi* (Railliet), *J. Parasitol.*, 28, 495, 1942.
1204. **Harkema, R.**, The cestodes of North Carolina poultry, with remarks on the life of *Raillietina tetragona*, *J. Elisha Mitchell Sci. Soc.*, 59, 127, 1943.
1205. **Harper, W. F.**, On some British larval cestodes from land and freshwater invertebrate hosts, *Parasitology*, 22, 202—213, 1930.
1206. **Harper, W. F.**, A cysticercoid from *Helodrilus (Allolobophora longus* Cede) and *Lumbricus terrestris*, *Parasitology*, 25, 483—484, 1933.
1207. **Harris, J. R. and Hickey, M. D.**, Occurrence of Diphyllobothriidae in Ireland, *Nature (London)*, 156 (correspondence), 447—448, 1945.
1208. **Hart, J. F.**, Cestoda from fishes of Puget Sound. II. Tetrarhynchoidea, *Trans. Am. Microsc. Soc.*, 55, 369—387, 1936.
1209. **Hart, J. F.**, Cestoda from fishes of Puget Sound. III. Phyllobothrioidea, *Trans. Am. Microsc. Soc.*, 55, 488—496, 1936.
1210. **Hart, J. F. and Guberlet, J. E.**, Cestoda from fishes of Puget Sound. I. Spathebothrioidea, a new superfamily, *Trans. Am. Microsc. Soc.*, 55, 199—207, 1936.
1211. **Harwood, D. P. D.**, The helminths parasitic in the amphibia and reptilia of Houston, Texas and vicinity, *Proc. U.S. Natl. Mus.*, 81, 1—71, 1932.

1212. **Harwood, D. P. D.**, The helminths parasitic in water moccasin, with a discussion of the characters of Proteocephalidae, *Parasitology*, 25, 130—142, 1933.
1213. **Harwood, D. P. D.**, Reproductive cycles of *Raillietina cesticillus* of the fowl, *Livro Jubilar Travassos*, Rio de Janeiro, 1938, 213—220.
1214. **Harwood, P. D. and Cooke, V.**, The helminths from a heavily parasitized fox squirrel, *Sciurus niger*, Ohio J. Sci., 49, 146—148, 1949.
1215. **Haskins, P. A.**, *Sigmodon hispidus hispidus*, a new host for the strobilocercus of *Taenia taeniaeformis*, J. Parasitol., 28, 94, 1942.
1216. **Haskins, P. A.**, *Moniezia expansa* infections in sheep, *J. Parasitol.*, 34 (Suppl.), 33, 1948.
1217. **Hassall, A. A.**, Check list of the animal parasites of chickens, *U.S. Bur. Anim. Ind. Circ.*, 9, 1—7, 1896.
1218. **Hassall, A. A.**, Check list of the animal parasites of turkeys, *U.S. Bur. Anim. Ind. Circ.*, 12, 1—3, 1896.
1219. **Hassall, A. A.**, Bibliography of the tapeworms of poultry, *U.S. Bur. Anim. Ind. Bull.*, 12, 81—88, 1896.
1220. **Hassall, A. A.**, Check list of animal parasites of ducks, *U.S. Bur. Anim. Ind. Circ.*, 13, 1—7, 1896.
1221. **Hassall, A. A.**, Check list of animal parasites of geese, *U.S. Bur. Anim. Ind. Circ.*, 14, 1—5, 1896.
1222. **Hassall, A. A.**, Check list of animal parasites of pigeons, *U.S. Bur. Anim. Ind. Circ.*, 15, 1—4, 1896.
1223. **Hassan, S. H.**, *Discobothrium aegyptiacus* n.sp., a cestode from *Raja circularis* in the Mediterranean Sea, Egypt, *J. Egyptian Soc. Parasitol.*, 12, 169—173, 1982.
1224. **Haswell, W. A.**, On a Gyrocotyle from *Chimaera ogilbyi*, and on *Gyrocotyle* in general, *Proc. Linn. Soc. N.S.W.*, 27, 48—54, 1902.
1225. **Haswell, W. A.**, On a cestode of *Cestracion*, *Q. J. Microsc. Sci.*, n. s. (183), 46, 399—415, 1902.
1226. **Haswell, W. A. and Hill, J. P.**, On *Polycercus:* a proliferating cystic parasite of the earthworms, *Proc. Linn. Soc. N.S.W.*, 8, 365—376, 1893.
1227. **Hatch, J. L.**, Parasites in a python, *Trans. Pathol. Soc. Phila.*, 15, 342—343, 1891.
1228. **Hatsushika, R., Shimizu, M., Kawakami, S., and Sawada, I.**, On a new species of *Pseudanoplocephala* Baylis, 1927 (Cestoda: Anoplocephalatidae) from the wild boar in Japan, *Jpn. J. Parasitol.*, 27, 535—542, 1978.
1229. **Hein, W.**, Beiträge zur Kenntnis von *Amphilina foliacea*, *Z. Wiss. Zool.*, 76, 400—438, 1904.
1230. **Heinz, M. L. and Dailey, M. D.**, The Trypanorhyncha (Cestoda) of elasmobranch fishes from southern California and Northern Mexico, *Proc. Helminthol. Soc. Wash.*, 41, 161—169, 1974.
1231. **Heitz, F. A.**, *Salmo salar* Lin., seine Parasitenfauna und seine Ernährung in Meer und in Süsswasser. Eine parasitologisch-biologische Studie, *Arch. Hydrobiol.*, 12, 311—372, 1918.
1232. **Helfer, J. R.**, Two new cestodes from salamanders, *Trans. Am. Microsc. Soc.*, 67, 359—364, 1949.
1233. **Heller, A. F.**, Parasites of cod and other marine fish from the Baie de Chaleur region, *Can. J. Res.*, 27 (Sect. D), 243—264, 1949.
1234. **Hendrickson, G. L., Grieve, R. B., and Kingston, N.**, *Monordotaenia honessi* sp.n. (Cyclophyllidea: Taeniidae), from a dog in Wyoming: a second taeniid from North America with a single circle of hooks, *Proc. Helminthol. Soc. Wash.*, 42, 46—52, 1975.
1235. **Henry, A.**, Tétrathyridium et *Mesocestoides*, *Bull. Soc. Cent. Med. Vet.*, 80, 147—152, 1927.
1236. **Henry, A.**, Les parasites et maladies parasitaires du ragondin, *Bull. Soc. Nat. Acclimat. Fr.*, 78, 421—447, 1931.
1237. **Henson, R. N.**, Cestodes of elasmobranch fishes of Texas, *Texas J. Sci.*, 26, 401—406, 1975.
1238. **Herde, K. E.**, Early development of *Ophiotaenia perspicua* La Rue, *Trans. Am. Microsc. Soc.*, 57, 282—291, 1938.
1239. **Herdman, W. A. and Hornell, J.**, Pearl production, *Rep. Govt. Ceylon Pearl Oyster Fish. Gulf Manaar.*, Part 5, 1—42, 1906.
1240. **Hickey, J. P.**, The diagnosis of the more common helminthic infestation of man, *U.S. Treasury Dep. Public Health Rep.*, 35, 1383—1400, 1920.
1241. **Hickey, M. D. and Harris, J. R.**, Progress of the *Diphyllobothrium* epizootic at Poulaphouca Reservoir, Co. Wicklow, Ireland, *J. Helminthol.*, 22, 13—28, 1947.
1242. **Hickman, J. L.**, Two new cestodes (genus *Oochoristica*) one from the lizard, *Egernia whitii*, the other from the bat, *Nyctophilus geoffroyi*, *Pap. Proc. R. Soc. Tasmania*, 88, 81—104, 1954.
1243. **Hickman, J. L.**, Cestodes of some Tasmanian Anura, *Ann. Mag. Nat. Hist.*, Ser. XIII, 3, 1—23, 1960.
1244. **Hill, J. P.**, A contribution to a further knowledge of the cystic cestodes. II. On a monocercus from *Didymogaster*, *Proc. Linn. Soc. N.S.W.*, 9, 49—84, 1895.
1245. **Hill, W. C.**, *Gryporhynchus tetrorchis*, a new dilepidid cestode from the great blue heron, *J. Parasitol.*, 27, 171—174, 1941.
1246. **Hilmy, J. S.**, *Bothriocephalus scorpii* (Müller, 1776) Cooper, 1917, *Ann. Trop. Med. Parasitol. Liverpool*, 23, 385—396, 1929.
1247. **Hilmy, J. S.**, Parasites from Liberia and French Guinea. III. Cestodes from Liberia, *Publ. Egyptian Univ. Fac. Med.*, 9, 1—72, 1936.
1248. **Hine, P. M.**, New species of *Nippotaenia* and *Amurotaenia* (Cestoda: Nippotaeniidae) from New Zealand freshwater fishes, *J. R. Soc. N.Z.*, 7, 143—155, 1977.

1249. **Hiscook, I. D.**, A new species of *Otobothrium* (Cestoda, Trypanorhyncha) from Australian fishes, *Parasitology*, 44, 65—70, 1954.
1250. **Hjortland, A. L.**, On the structure and life-history of an adult *Triaenophorus robustus*, *J. Parasitol.*, 15, 38—44, 1928.
1251. **Hoberg, E. P.**, *Diorchis pelagicus* sp. nov. (Cestoda: Hymenolepididae) from the whiskered auklet, *Aethia pygmaea*, and the crested auklet, *A. cristatella*, in the western Aleutian Islands, Alaska, *Can. J. Zool.*, 60, 2198—2202, 1982.
1252. **Hockley, A. R.**, On *Skrjabinotaenia cricetomydis* n. sp. (Cestoda: Anoplocephalata) from the Gambian pouched rat, Nigeria, *J. Helminthol.*, 35, 233—254, 1961.
1253. **Hoek, P. P. C.**, Ueber den encystirten Scolex von *Tetrarhynchus*, *Niederl. Arch. Zool.*, 5, 1—18, 1879.
1254. **Hoeppli, R.**, *Mesocestoides corti*, a new species of cestode from the mouse, *J. Parasitol.*, 12, 91—96, 1925.
1255. **Hoff, E. C. and Hoff, H. E.**, *Proteocephalus pugetensis*, a new tapeworm from the stickleback, *Trans. Am. Microsc. Soc.*, 48, 54—61, 1929.
1256. **Hofmann (no initials)**, Einiges ueber die Wanderung von Taenienembryonen, *Thierärztl. Wochenschr. Berlin*, 36, 537—541, 1901.
1257. **Hölldobler, K.**, *Cysticercus multiformis* nov. spez., eine noch nicht beschriebene Finnenform einer Cyclophyllidea, *Z. Parasitenk.*, 9, 523—528, 1937.
1258. **Honda, D.**, On a new cestode, *Raillietina (Raillietina) coreensis* n. sp. from a field mouse, *Apodemus agrarius coreae* in Chosen (in Japanese), *Chosen Igakkai Zasshi*, 29, 229—233, 1939.
1259. **Honess, R. F.**, Un nouveau cestode: *Fossor angertrudae* n.g. n.sp. du blaireau d'Amérique *Taxidea taxus taxus* (Schreber, 1778), *Ann. Parasitol.*, 15, 363—366, 1937.
1260. **Honess, R. F.**, Studies on the Life History of *Thysanosoma actinioides*. Review of Some of the Work at the Wyoming Station, *Rep. Conf. Parasites Parasitic Dis. Ruminants*, Bozeman, Montana, 1954, 18—21.
1261. **Honigberg, B.**, A morphological abnormality in the cestode *Dipylidium caninum*, *Trans. Am. Microsc. Soc.*, 63, 340—341, 1944.
1262. **Hopkins, C. A.**, Notes on the morphology and life history of *Schistocephalus solidus* (Cestoda: Diphyllobothriidae), *Parasitology*, 41, 283—291, 1951.
1263. **Hornell, J.**, Report on the Placuna placenta Pearl Fishery of Lake Tampalakaman, *Rep. Ceylon Mar. Biol. Lab.*, 1, 41—54, 1906.
1264. **Hornell, J.**, New cestodes from Indian fishes, *Rec. Ind. Mus.*, 7, 197—204, 1912.
1265. **Hornell, J. and Nayudu, M. R.**, A contribution to the life history of the Indian sardine with notes on the plankton of the Malabar Coast, *Madras Fish. Bull.*, 17 (5), 129—197, 1924.
1266. **Horsfall, M. W.**, Meal beetles as intermediate hosts of poultry tapeworms, *Poultry Sci.*, 17, 8—11, 1938.
1267. **Horsfall, M. W.**, A new unarmed cysticercoid, *Cysticercus setiferus*, *Parasitology*, 30, 61—64, 1938.
1268. **Horsfall, M. W.**, Observations on the life history of *Raillietina echinobothrida* and of *R. tetragona*, *J. Parasitol.*, 24, 409—421, 1938.
1269. **Horsfall, M. W. and Jones, M. F.**, The life-history of *Choanotaenia infundibulum*, a cestode parasitic in chickens, *J. Parasitol.*, 23, 435—450, 1937.
1270. **Houdemer, E., Dodero (no initials), and Cornet, E.**, Les sparganoses animales et la sparganose oculaire en Indochine, *Bull. Soc. Medicochirug. Indochine*, 11, 425—451, 1933.
1271. **Hovorka, J. and Macko, J. K.**, *Diorchis nitidohamulus* sp. n. (Cestoda) aus dem Wirt *Aythya fuligula*, *Biologia (Bratislava)*, 27, 97—103, 1972.
1272. **Hsü, H. F.**, Contribution à l'étude des cestodes de Chine, *Rev. Suisse Zool.*, 42, 477—570, 1935.
1273. **Hubbard, W. E.**, A remarkable infection of tapeworm larvae in a whipsnake, *Am. Midl. Nat.*, 19, 617—618, 1933.
1274. **Hübscher, H.**, Notes helminthologiques, *Rev. Suisse Zool.*, 42, 459—482, 1937.
1275. **Hudson, J. R.**, Notes on some avian cestodes, *Ann. Mag. Nat. Hist.* (Ser. 19), 14, 314—318, 1934.
1276. **Hudson, J. R.**, A list of cestodes known to occur in East African mammals, birds and reptiles, *J. East Afr. Uganda Nat. Hist. Soc.*, No. 49—50, 205—217, 1934.
1277. **Huey, R. and Dronen, N. O., Jr.**, Nematode and cestode parasites from the roseate spoonbill, *Ajaia ajaja*, including *Paradilepis diminuta* sp. n. (Cestoda: Dilepididae), *J. Parasitol.*, 67, 721—723, 1981.
1278. **Hugerbühler, M.**, Studien an Gyrocotylen und Cestoden. Ergebnisse einer von L. Schultze ausgeführten Zoologischen Forschungsreise in Südafrika, *Jen. Denkschr.*, 16, 495—522, 1910.
1279. **Hughes, R. C.**, The genus *Hymenolepis* Weinland, 1858, *Okla. Agric. Exp. Stn. Tech. Bull.*, 8, 1—2, 1940.
1280. **Hughes, R. C.**, The genus *Oochoristica* Lühe, 1898, *Am. Midl. Nat.*, 23, 368—381, 1940.
1281. **Hughes, R. C.**, The Taeniae of yesterday, *Okla. Agric. Exp. Stn. Tech. Bull.*, 38, 1—83, 1941.
1282. **Hughes, R. C.**, A key to the species of tapeworms in *Hymenolepis*, *Trans. Am. Microsc. Soc.*, 60, 378—414, 1941.
1283. **Hughes, R. C., Baker, J. H., and Dawson, G. B.**, The tapeworms of reptiles. I, *Am. Midl. Nat.*, 25, 454—468, 1941.

1284. **Hughes, R. C., Baker, J. H., and Dawson, G. B.,** The tapeworms of reptiles. II. Host catalogue, *Wasmann Coll.,* 4, 97—104, 1941.
1285. **Hughes, R. C., Baker, J. H., and Dawson, G. B.,** The tapeworms of reptiles. III, *Proc. Okla. Acad. Sci.,* 22, 81—89, 1942.
1286. **Hughes, R. C. and Schultz, R. L.,** genus *Raillietina* Fuhrmann, 1920, *Okla. Agric. Exp. Stn. Tech. Bull.,* 39, 1—53, 1942.
1287. **Humes, A. G.,** Experimental Copeped hosts of the broad tapeworm of man, *Dibothriocephalus latus* (L.), *J. Parasitol.,* 36, 541—547, 1950.
1288. **Hunkeler, P.,** Deux *Pseudohymenolepis* nouveaux (Cestoda, Hymenolepididae) chez les musuraigne de Côte-d'Ivoire, *Zool. Anz.,* 184, 125—129, 1970.
1289. **Hunkeler, P.,** Les cestodes parasites des petit mammifères (ronqeurs et insectivores) de Côte-d'Ivoire et de Haute-Volta. (Note préliminaire), *Bull. Soc. Neuchâteloise Sci. Nat.,* 95, 121—132, 1972.
1290. **Hunkeler, P.,** Les cestodes parasites des petits mammifères (Ronquers et Insectivores) de Côte-d'Ivoire et de Haute-Volta, *Rev. Suisse Zool.,* 80, 809—930, 1974.
1291. **Hunninen, A. V.,** A method of demonstrating cysticercoids of *Hymenolepis fraterna (H. nana* var. *fraterna* Stiles) in the intestinal villi of mice, *J. Parasitol.,* 21, 124—125, 1935.
1292. **Hunninen, A. V.,** Infections of abnormal hosts with the mouse strain of *Hymenolepis fraterna, J. Parasitol.,* 21, 312, 1935.
1293. **Hunninen, A. V.,** Studies on the life-history and host-parasite relations of *Hymenolepis fraterna (H. nana* var. *fraterna* Stiles) in white mice, *Am. J. Hyg.,* 22, 414—443, 1935.
1294. **Hunninen, A. V.,** An experimental study of internal auto-infection with *Hymenolepis fraterna* in white mice, *J. Parasitol.,* 22, 84—87, 1936.
1295. **Hunter, G. W., III,** Notes on the Caryophyllaeidae of North America, *J. Parasitol.,* 14, 16—26, 1927.
1296. **Hunter, G. W., III,** Studies on the Caryophyllaeidae of North America, *Ill. Biol. Monogr.,* 11, 1—186, 1927.
1297. **Hunter, G. W., III,** Contributions to the life-history of *Proteocephalus ambloplitis* (Leidy), *J. Parasitol.,* 14, 229—242, 1928.
1298. **Hunter, G. W., III,** New Caryophyllaeidae from North America, *J. Parasitol.,* 15, 185—192, 1929.
1299. **Hunter, G. W., III,** A case of accidental parasitism, *Science,* n.s. (1799), 69, 645—646, 1929.
1300. **Hunter, G. W., III,** Life-history studies on *Proteocephalus pinguis* La Rue, *Parasitology,* 31, 487—496, 1929.
1301. **Hunter, G. W., III,** Parasites of fishes in the lower Hudson area, *Annu. Rep. (26th) N.Y. State Conservancy Dep. Biol. Surv., Suppl.,* 1937.
1302. **Hunter, G. W., III,** Studies on the parasites of freshwater fishes of Connecticut, *Conn. Geol. Nat. Hist. Surv. Bull.,* 63, 228—288, 1942.
1303. **Hunter, G. W., III and Bangham, R. V.,** Studies on the fish parasites of Lake Erie. II. New Cestoda and Nematoda, *J. Parasitol.,* 19, 304—311, 1933.
1304. **Hunter, G. W., III and Hunninen, A. V.,** A biological survey of the Racquette watershed. X. Studies of the plerocercoid larva of the bass tapeworm, *Proteocephalus ambloplitis* (Leidy) in the small-mouthed bass, *Annu. Rep. (23rd) N.Y. State Conservancy Dep. Biol. Surv., Suppl.,* pp. 255—261, 1934.
1305. **Hunter, G. W., III and Hunter, W. S.,** Further studies on the bass tapeworm, *Proteocephalus ambloplitis* (Leidy), *Annu. Rep. (18th) N.Y. State Conservancy Dep. Biol. Surv., Suppl.,* pp. 198—207, 1929.
1306. **Hunter, G. W., III and Hunter, W. S.,** Studies on the parasites of fishes of the Lake Champlain watershed, *Annu. Rep. (19th) N.Y. State Conservancy Dep. Biol. Surv., Suppl.,* pp. 241—260, 1930.
1307. **Hunter, G. W., III and Hunter, W. S.,** Studies on fish parasites in the St. Lawrence watershed. XX, *Annu. Rep. N.Y. State Conservancy Dep. Biol. Surv., Suppl.,* pp. 197—216, 1931.
1308. **Hunter, G. W., III and Hunter, W. S.,** A biological survey of the Racquette watershed. IX. Studies on fish and bird parasites, *Annu. Rep. (23rd) N.Y. State Conservancy Dep. Biol. Surv., Suppl.,* pp. 245—255, 1934.
1309. **Hunter, G. W., III and Mackenthun, K. M.,** The life cycle of a pseudophyllidean tapeworm from the common sunfish *Lepomis gibbosus, J. Parasitol.,* 26 (Suppl.), 39, 1940.
1310. **Hunter, G. W., III and Rankin, J. S.,** Parasites of northern pike and pickerel, *Trans. Am. Fish. Soc.,* 69, 268—272, 1940.
1311. **Hunter, W. S. and Quay, T. L.,** An ecological study of the helminth fauna of Macgillivray's seaside sparrow, *Ammospiza maritima macgillivraii* (Audubon), *Am. Midl. Nat.,* 50, 407—413, 1953.
1312. **Hussey, K. L.,** *Aporina delafondi* (Railliet), an anoplocephalid cestode from the pigeon, *Am. Midl. Nat.,* 25, 413—417, 1941.
1313. **Hwang, J. C.,** *Cladotaenia (Paracladotaenia) cathartis* n.sp. (Cestoda: Taeniidae) from the intestine of the turkey buzzard, *Cathartes aura septentrionalis* Wied, 1893, *J. Parasitol.,* 47, 205—207, 1966.
1314. **Ibañez Herrera, N.,** Nota sôbre el estudio de una nueve especie del género *Choanotaenia* Railliet, 1876 en la fauna helmintológica peruana (Cestoda, Cyclophyllidea), *Rev. Brasil. Biol.,* 26, 77—79, 1966.

1315. **Ihle, J. E. W.**, Twee Cestoden gevonden in een exemplaar van *Mola mola*, *Tijdschr. Nederl. Dierk.*, Ver. No. 20, 17, 1927.
1316. **Ihle, J. E. W. and Ihle-Landenberg, M. E.**, Ueber einen neuen Cestodarier (*Kosterina kuiperi* n.gen. n.sp.) aus einer Schildkröte, *Zool. Anz.*, 100, 309—316, 1932.
1317. **Ihle, J. E. W. and van Oordt, G.**, Eenige Cestoden van Vogels, *Tijdschr. Diergeneeskd.*, Deel. 52, 949—952, 1926.
1318. **Ijima, I.**, On a new cestode larva parasitic in man *(Plerocercoides prolifer)*, *J. Coll. Sci. Imp. Univ. Tokyo*, 20, 1—21, 1905.
1319. **Inamdar, N. B.**, A new species of avian cestode from India, *Ann. Mag. Nat. Hist.* (Ser. 10), 11, 610—613, 1933.
1320. **Inamdar, N. B.**, Four new species of avian cestodes from India, *Z. Parasitol.*, 7, 198—206, 1934.
1321. **Inamdar, N. B.**, A new species of avian cestode from India, *J. Univ. Bombay*, 11, 77—81, 1942.
1322. **Inamdar, N. B.**, On an undescribed species of avian cestode from Dharvar, India, *Proc. 29 Ind. Sci. Congr.*, Part 3, 152, 1943.
1323. **Inamdar, N. B.**, A new species of avian cestode, *Ophryocotyloides bhaleraoi*, from the purple-rumped sunbird, *Cinnyris zeylonicus* (Linn.), *Proc. 31 Ind. Sci. Congr.*, Part 3, 89, 1944.
1324. **Ingles, L. G.**, Worm parasites of California Amphibia, *Trans. Am. Microsc. Soc.*, 55, 73—92, 1936.
1325. **Iverus, J.**, Sur un cestode du *Rhombus maximus*, *C. R. 6 Congr. Int. Zool. Berne*, pp. 702—703, 1904.
1326. **Iwata, S.**, Some experimental and morphological studies on the postembryonal development of Manson's tapeworm, *Diphyllobothrium erinacei* (Rudolphi), *Jpn. J. Zool.*, 5, 209—247, 1933.
1327. **Iwata, S.**, Some experimental studies on the regeneration of the plerocercoids of Manson's tapeworm, *Diphyllobothrium erinacei* (Rudolphi), with special reference to its relationship with *Sparganum proliferum* Ijima, *Jpn. J. Zool.*, 6, 139—158, 1934.
1328. **Iwata, S.**, *Diphyllobothrium mansonoides* Mueller is the synonym of *D. erinacei* (Rud.) (in Japanese), *Dobuts, Zasshi*, 48, 665—669, 1936.
1329. **Iwata, S.**, On a cestode parasitic in giraffe (in Japanese), *Dobuts. Zasshi*, 52, 48—49, 1940.
1330. **Iwata, S. and Matsuda, S.**, *Ophiotaenia ranarum*, a new amphibian cestode, *Dobuts. Zasshi*, 50, 221—222, 1938.
1331. **Iwata, S. and Tamura, O.**, Some intestinal parasites in the duck from Japan, *Annot. Zool. Jpn.*, 14, 1—6, 1933.
1332. **Jacobi, A.**, *Diploposthe*, eine neue Gattung von Vogeltaenien, *Zool. Anz.*, 19, 268—269, 1896.
1333. **Jacobi, A.**, *Diploposthe laevis*, eine merkwürdige Vogeltaenien, *Zool. Jahrb. Anat.*, 10, 287—306, 1897.
1334. **Jacobi, A.**, *Amabilia* und *Diploposthe*, *Centralbl. Bakteriol.*, 21, 873—874, 1897.
1335. **Jacobi, A.**, Ueber den Bau der *Taenia inflata* Rud., *Zool. Jahrb. Syst.*, 12, 95—104, 1898.
1336. **Jadhav, B. V. and Shinde, G. E.**, A new species of *Oncodiscus* Yamaguti, 1934 (Cestoda — Tetraphyllidea) from India, *Proc. Ind. Acad. Parasitol.*, 2, 26—27, 1981.
1337. **Jadhav, B. V. and Shinde, G. B.**, On a new species of the genus *Mastacembellophyllaeus* Shinde and Chincholikar (1976) (Cestoda: Monoporophyllaeidae Subhapradha, 1957) from a fresh water fish at Aurangabad, India, Abstr. 1st Natl. Congr. Parasitol., Baroda, February 24 to 26, 1977, India, 1977, 9.
1338. **Jadhav, B. V. and Shinde, G. B.**, *Balanobothrium veravalensis* n.sp. (Cestoda: Lecanicephalidae) from a marine fish, *Ind. J. Parasitol.*, 3, 83—85, 1979.
1339. **Jadhav, B. V. and Shinde, G. B.**, A new species of the genus *Tylocephalum* Linton, 1890 (Cestoda: Lecanicephalidea) from an Indian marine fish, *Ind. J. Parasitol.*, 5, 109—111, 1981.
1340. **Jadhav, B. V. and Shinde, G. B.**, *Uncibilocularis veravalensis* n.sp. (Cestoda: Onchobothriidae) from an Indian marine fish, *Ind. J. Parasitol.*, 5, 113—115, 1981.
1341. **Jadhav, B. V., Shinde, G. B., and Deshmukh, R. A.**, On a new cestode *Shindeiobothrium karbharagae* gen.n. sp.n. from a marine fish, *Riv. Parassitol.*, 42, 31—34, 1981.
1342. **Jameson, H. L.**, Studies on pearl oysters and pearls. I. The structure of the shell and pearls of the Ceylon pearl oyster (*Margaritifera vulgaris* Schumacher) with an examination of the cestode theory of pearl formation, *Proc. Zool. Soc. London*, pp. 260—358, 1912.
1343. **Janicki, C.**, Ueber zwei neue Arten des Genus *Davainea* aus celebensischen Säugern., *Arch. Parasitol.*, 6, 257—292, 1902.
1344. **Janicki, C.**, Weitere Angaben ueber *Triplotaenia mirabilis* J. E. V. Boas, *Zool. Anz.*, 27, 243—247, 1904.
1345. **Janicki, C.**, Zur Kenntnis einiger Säugetiercestoden, *Zool. Anz.*, 27, 770—782, 1904.
1346. **Janicki, C.**, Bemerkungen ueber Cestoden ohne Genitalporus, *Centralbl. Bakteriol.*, 36, 222—223, 1904.
1347. **Janicki, C.**, Beutlercestoden der Niederländischen Neu-Guinea Expedition. Zugleich einiges neues aus dem Geschlechtsleben der Cestoden, *Zool. Anz.*, 29, 127—131, 1905.
1348. **Janicki, C.**, Studien an Säugetiercestoden, *Z. Wiss. Zool.*, 81, 505—597, 1906.
1349. **Janicki, C.**, Die Cestoden Neu Guineas. Nova Guinea, *Résultats de l'Expédition Sci. Néerl. à la Nouvelle-Guinée en 1903*. V, 1, 181—200, 1906.

1350. **Janicki, C.**, Zur Embryonalentwicklung von *Taenia serrata* Goeze, *Zool. Anz.*, 30, 763—768, 1906.
1351. **Janicki, C.**, Ueber die Embryonalentwicklung von *Taenia serrata* Goeze, *Z. Wiss. Zool.*, 87, 685—724, 1907.
1352. **Janicki, C.**, Ueber den Bau von *Amphilina liguloidea* Diesing, *Z. Wiss. Zool.*, 89, 568—597, 1908.
1353. **Janicki, C.**, Die Cestoden aus Procavia, *Jen. Denkschr. Med. Naturwiss. Ges.*, 16, 373—396, 1910.
1354. **Janicki, C.**, Neue Studien ueber postembryonale Entwicklung und Wirtswechsel bei Bothriocephalen. I. *Triaenophorus nodulosus* (Pallas), *Correspondenz-Blatt für Schweizer Ärzte*, 48, 1343—1349, 1918.
1355. **Janicki, C.**, Neue Studien ueber die postembryonale Entwicklung und Wirtswechel bei Bothriocephalen. II. Die Gattung *Ligula*, *Correspondenz-Blatt für Schweizer Ärzte*, 49, 915—918, 1919.
1356. **Janicki, C.**, Grundlinien einer "Cercomer Theorie" zur Morphologie der Trematoden und Cestoden, *Festschrift für Zschokke*, No. 30, Basel, 1920, 22 pp.
1357. **Janicki, C.**, Cestodes s. str. aus Fischen und Amphibien, *Results of the Swedish Zool. Expedition Egypt White Nile, 1901*, Part 5, Uppsala, pp. 1—58, 1926.
1358. **Janicki, C.**, Ueber die Lebensgeschichte von *Amphilina foliacea*, dem Parasiten des Wolga-Sterlets, nach Beobachtungen und Experimenten, *Naturwissenschaften*, 16, 820—821, 1928.
1359. **Janicki, C.**, Ueber die jüngsten Zustände von *Amphilina foliacea* in der Fischleibeshöhle, sowie Generelles zur Auffassung des Genus *Amphilina* Wagener, *Zool. Anz.*, 90, 190—205, 1930.
1360. **Janicki, C., and Rosen, F.**, Le cycle évolutif du *Dibothriocephalus latus* L. Recherches expérimentales et observation, *Bull. Soc. Sci. Nat. Neuchâtel.*, 42, 19—53, 1917.
1361. **Janiszewska, J.**, *Paraglaridacris silesiacus* n.g. n.sp. de la famille Caryophillaeidae, *Zool. Pol.*, 5, 67—72, 1950.
1362. **Janiszewska, J.**, Is *Biacetabulum sieboldi* Szidat a mature form of *Archigetes sieboldi* Leuck.? (in Polish), *Proc. 2nd Meet. Polish Parasitol. Soc. Pulawy*, pp. 95—96, 1950; Russian summary p. 121, English summary p. 135.
1363. **Janiszewska, J.**, *Caryophyllaeus brachycollis* n.sp. from cyprinoid fishes, *Zool. Pol.*, 6, 57—68, 1953.
1364. **Janiszewska, J.**, Caryophyllaeidae europjeskie ze Szczególnym uwzględnieniem Polski, *Trav. Soc. Sci. Lett. Wroclaw*, Ser. B, No. 66, 1954, p. 73.
1365. **Jaroń, W.**, *Pseudanomotaenia parachelidonariae* sp.n. (syn. *Anomotaenia chelidonariae* Spasskaja, 1957-sensu Spasskaja, 1959) et *P. chelidonariae* (Spasskaja, 1957) Mathevossian, 1963-sensu Spasskaja, 1957, *Acta Parasitol. Pol.*, 14, 351—355, 1967.
1366. **Jaroń, W.**, The helminth parasites of Hirudinidae of the neighbourhood of Warszawa and Olsztyn, *Acta Parasitol. Pol.*, 16, 137—152, 1969.
1367. **Jarvi, T. H.**, Die kleine Marane, *Coregonus albula* L., als der Zwischenwirt des *Dibothriocephalus latus* L. in den Seen Nord-Tawastlands (Finnland), *Medd. Soc. Fauna Flora Fens. Helsingfors*, 35, 62—67, 1909.
1368. **Jellison, W. L.**, Parasites of porcupines of the genus *Erethizon* (Rodentia), *Trans. Am. Microsc. Soc.*, 52, 42—47, 1933.
1369. **Jellison, W. L.**, The occurrence of the cestode, *Moniezia benedeni* (Anoplocephalidae), in the American moose, *Proc. Helminthol. Soc. Wash.*, 3, 16, 1936.
1370. **Jenkins, J. W. R.**, On a new species of *Moniezia* from the sheep, *Ovis aries*, *Ann. Appl. Biol.*, 10, 267—286, 1923.
1371. **Jensen, L. A.**, *Parabothriocephalus sagitticeps* (Sleggs 1927) comb.n. (Cestoda: Parabothriocephalidae) from *Sebastes paucispinis* of southern and central California, *J. Parasitol.*, 62, 560—562, 1976.
1372. **Jensen, L. A. and Heckmann, R. A.**, *Anantrum histocephalum* sp.n. (Cestoda: Bothriocephalidae) from *Synodus lucioceps* (Synodontidae) of southern California, *J. Parasitol.*, 63, 471—472, 1977.
1373. **Jensen, L. A. and Howell, K. M.**, *Vampirolepis schmidt* sp.n. (Cestoidea: Hymenolepididae) from *Triaenops persicus* (Hipposideridae) of Tanzania, *Proc. Helminthol. Soc. Wash.*, pp. 135—137, 1983.
1374. **Jensen, L. A., Schmidt, G. D., and Kuntz, R. E.**, A survey of cestodes from Borneo, Palawan, and Taiwan, with special reference to three new species, *Proc. Helminthol. Soc. Wash.*, 50, 117—134, 1983.
1375. **Jepps, M. W.**, Note on Apstein's parasites and some very early larval Platyhelminthes, *Parasitology*, 29, 554—558, 1937.
1376. **Jewell, M. E.**, *Cylindrotaenia americana* nov. spec. from the cricket frog, *J. Parasitol.*, 2, 180—192, 1916.
1377. **John, D. D.**, On *Cittotaenia denticulata* (Rud. 1804) with notes as to the occurrence of other helminthic parasites of rabbits found in the Aberystwyth area, *Parasitology*, 18, 436—454, 1926.
1378. **Johnston, T. H.**, Über den Bau von *Amphilina liguloidea* Diesing, *Z. Wiss. Zool.*, 89, 568—597, 1908.
1379. **Johnston, T. H.**, On a cestode from *Dacelo gigas* Bodd., *Rec. Aust. Mus.*, 7, 246—250, 1909.
1380. **Johnston, T. H.**, Notes on Australian Entozoa. I, *Rec. Aust. Mus.*, 7, 329—344, 1909.
1381. **Johnston, T. H.**, Notes on some Australian parasites, *Agric. Gaz. N.S.W.*, 20, 582—584, 1909.
1382. **Johnston, T. H.**, On a new reptilian cestode, *Proc. R. Soc. N.S.W.*, 43, 103—116, 1909.
1383. **Johnston, T. H.**, On a new genus of bird cestodes, *Proc. R. Soc. N.S.W.*, 43, 139—147, 1909.

1384. **Johnston, T. H.,** On the anatomy of *Monopylidium passerinum* Fuhrmann, *Proc. R. Soc. N.S.W.,* 43, 405—411, 1910.
1385. **Johnston, T. H.,** The Entozoa of Monotremata and Australian marsupials, I, *Proc. Linn. Soc. N.S.W.,* 34, 514—523, 1910.
1386. **Johnston, T. H.,** On Australian avian Entozoa, *Proc. R. Soc. N.S.W.,* 44, 84—122, 1910.
1387. **Johnston, T. H.,** The Entozoa of Monotremata and Australian marsupials. II, *Proc. Linn. Soc. N.S.W.,* 36, 45—57, 1911.
1388. **Johnston, T. H.,** New species of avian cestodes, *Proc. Linn. Soc. N.S.W.,* 36, 58—80, 1911.
1389. **Johnston, T. H.,** *Proteocephalus gallardi,* a new cestode from the black snake, *Ann. Queensland Mus.,* No. 10, 175—182, 1911.
1390. **Johnston, T. H.,** On a re-examination of the types of Krefft's species of Cestoda, *Rec. Aust. Mus.,* 9, 1—35, 1912.
1391. **Johnston, T. H.,** New species of cestodes from Australian birds, *Mem. Queensland Mus.,* 1, 211—215, 1912.
1392. **Johnston, T. H.,** Internal parasites recorded from Australian birds, *Emu,* 12, 105—112, 1912.
1393. **Johnston, T. H.,** A census of Australian reptilian Entozoa, *Proc. R. Soc. Queensland,* 23, 233—249, 1912.
1394. **Johnston, T. H.,** Notes on some Entozoa, *Proc. R. Soc. Queensland,* 24, 63—91, 1913.
1395. **Johnston, T. H.,** Cestoda and Acanthocephala, *Rep. Aust. Inst. Trop. Med.,* 1911, 75—96, 1913.
1396. **Johnston, T. H.,** Second report on Cestoda and Acanthocephala in Queensland, *Ann. Trop. Med. Parasitol.,* 8, 105—112, 1914.
1397. **Johnston, T. H.,** Some new Queensland endoparasites, *Proc. R. Soc. Queensland,* 26, 76—84, 1914.
1398. **Johnston, T. H.,** Australian trematodes and cestodes; a study in zoogeography, *Med. J. Aust.,* 1, 243—244, 1914.
1399. **Johnston, T. H.,** Helminthological notes, *Mem. Queensland Mus.,* 5, 186—196, 1916.
1400. **Johnston, T. H.,** A census of the endoparasites recorded as occurring in Queensland, arranged under their hosts, *Proc. R. Soc. Queensland,* 28, 31—79, 1916.
1401. **Johnston, T. H.,** Endoparasites of the dingo, *Canis dingo* Blumb, *Proc. R. Soc. Queensland,* 28, 96—100, 1916.
1402. **Johnston, T. H.,** Notes on certain entozoa of rats and mice, with a catalogue of the internal parasites recorded as occurring in rodents in Australia, *Proc. R. Soc. Queensland,* 30, 53—78, 1918.
1403. **Johnston, T. H.,** The endoparasites of the domestic pigeon in Queensland, *Mem. Queensland Mus.,* 6, 168—174, 1918.
1404. **Johnston, T. H.,** An Australian caryophyllaeid cestode, *Proc. Linn. Soc. N.S.W.,* 49, 339—347, 1924.
1405. **Johnston, T. H.,** An amphilinid from an Australian tortoise, *Aust. J. Exp. Biol. Med. Sci.,* 8, 1—7, 1931.
1406. **Johnston, T. H.,** Remarks on some Australian Cestodaria, *Proc. Linn. Soc. N.S.W.,* 59, 66—70, 1934.
1407. **Johnston, T. H.,** Remarks on the cestode genus *Porotaenia, Trans. Proc. R. S. Aust.,* 59, 164—167, 1935.
1408. **Johnston, T. H.,** Entozoa from the Australian hair seal, *Proc. Linn. Soc. N.S.W.,* 62, 9—16, 1937.
1409. **Johnston, T. H. and Clark, H. G.,** A new cestode, *Raillietina* (R.) *leipoae,* from the mallee hen, *Rec. S. Aust. Mus.,* 9, 87—91, 1948.
1410. **Johnston, T. H. and Muirhead, N. G.,** Some Australian caryophyllaeid cestodes, *Rec. S. Aust. Mus.,* 9, 339—348, 1950.
1411. **Johnstone, J.,** Internal parasites and diseased conditions of fishes. From Herdman's Reports on the Lancashire sea fisheries, *Trans. Biol. Soc. Liverpool,* Vols. 19—26, 1895—1912.
1412. **Johnstone, J.,** *Tetrarhynchus erinaceus* van Beneden. I. Structure of larva and adult worm, *Parasitology,* 4, 364—415, 1912.
1413. **Johnstone, J., Scott, A., and Smith, W. C.,** The parasites and diseases of the cod, *Fish. Invest. G. B.,* 6, 15—27, 1924.
1414. **Johri, G. N.,** On a new cestode from the palm squirrel, *Funambulus palmarum* Linn., *Proc. Nat. Acad. Sci. Allahabad,* 26 (Ser. B, Part 4), 274—277, 1956.
1415. **Johri, G. N.,** A new cestode *Senga lucknowensis* from *Mastacembelus armatus* Lacep., *Curr. Sci.,* 25, 193—195, 1956.
1416. **Johri, G. N.,** Occurrence of two species of the cestode, *Oochoristica* Lühe, 1898, in a South Indian Lizard, *Proc. Natl. Acad. Sci. India,* Sect. B, 28, 242—245, 1958.
1417. **Johri, G. N.,** Descriptions of two amabilid cestodes from the little grebe, *Podiceps ruficollis,* with remarks on the family Amabiliidae Braun, 1900, *Parasitology,* 49, 454—461, 1959.
1418. **Johri, G. N.,** On a remarkable new caryophyllaeid cestode, *Hunteroides mystei* gen. et sp. nov. from a fresh water fish in Delhi State, *Z. Parasitenk.,* 19, 368—374, 1959.
1419. **Johri, G. N.,** Studies on some cestode parasites. III. Variability in the number and position of testes in some unarmed species of *Hymenolepis* from mammals, *Proc. Natl. Acad. Sci. India,* Sect. B, 29, 134—142, 1959.

1420. **Johri, G. N.**, *Vogea vestibularis* n.g., n.sp., a dilepidid cestode from the intestine of the large grey babbler, *Argya malcolmi, J. Parasitol.,* 45, 287—290, 1959.
1421. **Johri, G. N.**, Studies on some cestode parasites. IV. On four new species including a new genus belonging to the family Hymenolepididae, *Proc. Natl. Acad. Sci. India,* Sect. B, 30, 192—202, 1960.
1422. **Johri, G. N.**, Studies on some cestode parasites. V. Two new species of cestodes belonging to the family Hymenolepididae Fuhrmann, 1907, *J. Parasitol.,* 46, 251—255, 1960.
1423. **Johri, G. N.**, A new paruterinid cestode, *Lallum magniparuterina* gen. et sp. nov. from the intestine of a common teal, *Nettion crecca* Linn., *Parasitology,* 50, 269—272, 1960.
1424. **Johri, G. N.**, Studies on some cestode parasites. VII. Some old and new cestodes from Indian reptiles and mammals, *Zool. Anz.,* 167, 296—303, 1961.
1425. **Johri, G. N.**, On a new protogynous cestode with remarks on certain species of the genus *Progynotaenia* Fuhrmann, 1909, *J. Helminthol.,* 37, 39—46, 1963.
1426. **Johri, L. N.**, A new cestode from the grey hornbill in India, *Ann. Mag. Nat. Hist.,* 10. s., 8, 239—242, 1931.
1427. **Johri, L. N.**, On the genus *Houttuynia* Fuhrmann, 1920 (Cestoda), with a description of some species of *Raillietina* from the pigeon, *Zool. Anz.,* 103, 89—92, 1933.
1428. **Johri, L. N.**, Report on a collection of cestodes from Lucknow, *Rec. Ind. Mus.,* 36, 153—177, 1934.
1429. **Johri, L. N.**, On cestodes from Burma, *Parasitology,* 27, 476—479, 1935.
1430. **Johri, L. N.**, On two new species of *Diorchis* (Cestoda) from the Indian Columbiformes, *Rec. Ind. Mus.,* 41, 121—129, 1939.
1431. **Johri, L. N.**, On a collection of cestodes from a peacock (*Pavo cristatus* L., 1758) from the Terai Forest Area, India, *Ann. Trop. Med. Parasitol.,* 33, 211—216, 1939.
1432. **Johri, L. N.**, On two new species of the family Hymenolepididae Fuhrmann, 1907 (Cestoda) from a Burmese cormorant, *Phalacrocorax javanicus* (Horsfield, 1821), *Philipp. J. Sci.,* 74, 83—89, 1941.
1433. **Johri, L. N.**, Report on cestodes collected in India and Burma, *Ind. J. Helminthol.,* 2, 23—24, 1950.
1434. **Johri, L. N.**, On avian cestodes of the family Dilepididae Fuhrm., collected in Burma, *Parasitology,* 41, 11—14, 1951.
1435. **Johri, L. N.**, A new avian cestode, *Thaparea magnivesicula* gen. and sp. nova from the common fantail snipe, *Capella gallinago gallinago* Linn. From Delphi State, *Thapar Comm.,* Vols. 139—142, 1953.
1436. **Johri, L. N.**, On a new cyclophyllidean cestode *Multiceps smythi* n. sp., from dogs in Dublin, Eire, *Parasitology,* 47, 16—20, 1957.
1437. **Johri, L. N.**, On two new avian cestodes belonging to the subfamily Hymenolepidinae Perrier, 1897 from Delhi State, *Proc. Nat. Acad. Sci. India,* Sect. B, 30, 234—240, 1960.
1438. **Johri, L. N.**, On a new avian cestode belonging to the subfamily Hymenolepidinae Perrier, 1897 from Delhi State, *Proc. Natl. Acad. Sci. India,* Sect. B, 32, 200—202, 1962.
1439. **Johri, L. N.**, Report on a new anoplocephalid cestode from Delhi State, *Proc. Natl. Acad. Sci. India,* Sect. B, 32, 351—354, 1962.
1440. **Jones, A.**, *Proteocephalus pentastoma* (Klaptocz, 1906) and *Polyonchobothrium polypteri* (Leydig, 1853) from species of *Polypterus* Geoffroy, 1802 in the Sudan, *J. Helminthol.,* 54, 25—38, 1980.
1441. **Jones, A.**, A redescription of *Triuterina anoplocephaloides* (Fuhrmann, 1902) Cestoda: Anoplocephalidae) from African parrots, *Syst. Parasitol.,* 4, 253—255, 1982.
1442. **Jones, A. and Khalil, L. F.**, The helminth parasites of the lesser flamingo, *Phoeniconais miner* (Geoffroy), from Lake Nakuru, Kenya, including a new cestode, *Phoenicolepis nakurensis* n.g., n.sp., *Syst. Parasitol.,* 2, 61—76, 1980.
1443. **Jones, A. W.**, *Protogynella blarinae* n.g. n.sp., a new cestode from the shrew, *Blarina brevicauda* Say, *Trans. Am. Microsc. Soc.,* 62, 169—173, 1943.
1444. **Jones, A. W.**, *Diorchis reynoldsi* n.sp., a hymenolepidid cestode from the shrew, *Trans. Am. Microsc. Soc.,* 63, 46—49, 1944.
1445. **Jones, A. W.**, *Diorchis ralli* n.sp., a hymenolepidid cestode from the king rail, *Trans. Am. Microsc. Soc.,* 63, 50—53, 1944.
1446. **Jones, A. W.**, Studies in cestode cytology, *J. Parasitol.,* 31, 213—235, 1945.
1447. **Jones, A. W.**, The scolex of *Rhabdometra similis, Trans. Am. Microsc. Soc.,* 65, 357—359, 1946.
1448. **Jones, A. W.**, Speciation in the Cestoda, *J. Parasitol.,* 34 (Suppl.), 16—17, 1948.
1449. **Jones, A. W.**, The chromosomes of *Davainea proglottina, Trans. Am. Microsc. Soc.,* 70, 272—273, 1951.
1450. **Jones, A. W., Clayton, K., and Sneed, K. R.**, New species in the genus *Corallobothrium* Fritsch 1886, *J. Parasitol.,* 40 (Sect. 2, Suppl.), 41, 1954.
1451. **Jones, A. W. and Ward, H. L.**, The application of cytological techniques to cestodes and other helminth material, *J. Parasitol.,* 31 (Suppl.), 16, 1945.
1452. **Jones, M. F.**, *Schistotaenia macrorhyncha* Rud., *J. Parasitol.,* 15, 1—18, 1929.
1453. **Jones, M. F.**, Tapeworms of the genera *Rhabdometra* and *Paruterina* found in the quail and yellow-billed cuckoo, *Proc. U.S. Natl. Mus.,* 75, 1—8, 1929.

1454. **Jones, M. F.**, *Aphodius granarius* (Coleoptera), an intermediate host for *Hymenolepis carioca* (Cestoda), *J. Agric. Res.*, 38, 629—632, 1929.
1455. **Jones, M. F.**, Notes without title on the life-cycle of *Raillietina cesticillus, J. Parasitol.*, 16, 158, 1930; *J. Parasitol.*, 16, 158—159, 1930; *J. Parasitol.*, 16, 164, 1930; *J. Parasitol.*, 17, 57, 1930.
1456. **Jones, M. F.**, On the loss of an experimentally produced infestation of tapeworms in a chicken, *J. Parasitol.*, 17, 234, 1931.
1457. **Jones, M. F.**, On the life histories of species of *Raillietina, J. Parasitol.*, 17, 234, 1931.
1458. **Jones, M. F.**, Additional notes on intermediate hosts of poultry tapeworms, *J. Parasitol.*, 18, 307, 1932.
1459. **Jones, M. F.**, On the systematic position of *Davainea fuhrmanni* Williams, 1931, *J. Parasitol.*, 19, 255, 1933.
1460. **Jones, M. F.**, Notes on cestodes of poultry, *J. Parasitol.*, 20, 66, 1933.
1461. **Jones, M. F.**, Cysticercoids of the crow cestode, *Hymenolepis variabilis* (Mayhew, 1925) Fuhrmann, 1932 (Hymenolepididae), *Proc. Helminthol. Soc. Wash.*, 1, 62—63, 1934.
1462. **Jones, M. F.**, The cestode, *Hymenolepis microps* (Hymenolepididae) in ruffed grouse *(Bonasa umbellias), Proc. Helminthol. Soc. Wash.*, 2, 93, 1935.
1463. **Jones, M. F.**, *Metroliasthes lucida,* a cestode of galliform birds, in arthropod and avian hosts, *Proc. Helminthol. Soc. Wash.*, 3, 26—30, 1936.
1464. **Jones, M. F.**, A new species of cestode, *Davainea meleagridis* (Davaineidae) from the turkey, with a key to the species of *Davainea* from galliform birds, *Proc. Helminthol. Soc. Wash.*, 3, 49—52, 1936.
1465. **Jones, M. F. and Alicata, J. E.**, Development and morphology of the cestode, *Hymenolepis cantaniana,* in coleopteran and avian hosts, *J. Wash. Acad. Sci.*, 25, 237—247, 1935.
1466. **Jones, M. F. and Horsfall, M. W.**, Ants as intermediate hosts for the two species of *Raillietina* parasitic in chickens, *J. Parasitol.*, 21, 442—443, 1935.
1467. **Jones, M. F. and Horsfall, M. W.**, The life-history of a poultry cestode, *Science*, 83, 303—304, 1936.
1468. **Jones, N. V. and Williams, I. C.**, The cestode parasites of the sheathbill, *Chionis alba* (Gmelin), from Signy Island, South Orkney Islands, *J. Helminthol.*, 41, 151—160, 1967.
1469. **Jordano, D.**, Hallazgo en España de *Diplopylidium triseriale* (Lühe) (Cestoda: Dilepididae) y demostración biométrica de la validez de esta especie, *Rev. Ibér. Parasitol.*, 10, 97—126, 1950.
1470. **Jordano, D.**, *Hymenolepis cordobensis* n.sp. (Cestoda: Hymenolepididae) nueva Tenia parasita de la paloma domestica, *Rev. Ibér. Parasitol.*, 12, 59—64, 1952.
1471. **Jordano, D. and Diaz-Ungria, C.**, Cestodes de Venezuela, II. *Dendrometra ginesi* (nov. gen. nov. sp.), (Cestoda: Dilepididae) nueva tenia parasita de la tijereta *(Fregata magnificens), Novedades Cientificas Contrib. Occ. d. Mus. d. Hist. Nat. La Salle Caracas*, Ser. Zool., No. 18, pp. 1—4, 1956.
1472. **Jordano, D., and Diaz-Ungria, C.**, Cestodos de Venezuela. VIII. *Craspedocotyla margaritensis* nov. gen., nov. sp. (Cestoda: Hymenolepididae) neuva tenia parásita del turpial *(Icterus icterus)* (Aves: passeres), *Mem. Soc. Ciencias Nat. La Salle*, 20, 198—210, 1960.
1473. **Joyeux, C.**, Sur le cycle évolutif de quelques cestodes. Note préliminaire, *Bull. Soc. Pathol. Exot.*, 9, 578—583, 1916.
1474. **Joyeux, C.**, *Hymenolepis nana* (v. Siebold, 1852) et *Hymenolepis nana* var. *fraterna* Stiles, 1902, *Bull. Soc. Pathol. Exot.*, 12, 228—231, 1919.
1475. **Joyeux, C.**, Cycle évolutif de quelques cestodes. Recherches expérimentales, *Bull. Biol. Fr. Belg.*, 2 (Suppl.), 1—219, 1920.
1476. **Joyeux, C.**, Développement direct d'un *Hymenolepis* (Téniadés) dans les villosités intestinales du hérisson, *Bull. Soc. Pathol. Exot.*, 14, 386—390, 1921.
1477. **Joyeux, C.**, Recherches sur les Ténias des Ansériformes. Développement larvaire d'*Hymenolepis parvula* chez *Herpobdella octoculata* (L.), *Bull. Soc. Pathol. Exot.*, 15, 45—51, 1922.
1478. **Joyeux, C.**, Recherche sur l'*Urocystis prolifer* Villot. Note préliminaire, *Bull. Soc. Zool. Fr.*, 47, 52—58, 1922.
1479. **Joyeux, C.**, Recherches sur la faune helminthologique africaine, *Arch. Inst. Pasteur Tunis*, 12, 119—167, 1923.
1480. **Joyeux, C.**, Présence de *Dinobothrium plicitum* Linton, 1922, chez *Cetorhinus maximus* (L.), *Ann. Parasitol.*, 1, 344, 1923.
1481. **Joyeux, C.**, Liste de quelques helminthes récoltés dans les colonies portugaises d'Afrique, *Ann. Parasitol.*, 2, 232—235, 1924.
1482. **Joyeux, C.**, Cestodes des poules d'Indochine, *Ann. Parasitol.*, 2, 314—318, 1924.
1483. **Joyeux, C.**, Recherches sur le cycle évolutif des *Cylindrotaenia, Ann. Parasitol.*, 2, 74—81, 1924.
1484. **Joyeux, C.**, Parasites des poules dans la province de Schinchiku (Formosa), *Ann. Parasitol.*, 3, 103, 1925.
1485. **Joyeux, C.**, *Hymenolepis nana* et *Hymenolepis fraterna, Ann. Parasitol.*, 3, 270—280, 1925.
1486. **Joyeux, C.**, Sur quelques cysticercoïdes de *Gammarus pulex, Arch. Schiffs. Trop. Hyg.*, 30, 433—451, 1926.
1487. **Joyeux, C.**, Recherches sur le cycle évolutif d'*Hymenolepis erinacei* (Gmelin, 1789), *Ann. Parasitol.*, 5, 20—26, 1927.

1488. **Joyeux, C.**, Recherches sur la faune helminthologique algérienne (cestodes et trematodes), *Arch. Inst. Pasteur Algérie*, 5, 509—528, 1927.
1489. **Joyeux, C.**, *Diphyllobothrium mansoni* (Cobbold, 1883), Note préliminaire, *Bull. Soc. Pathol. Exot.*, 20, 226—228, 1927.
1490. **Joyeux, C.**, Les ténias extra-intestinaux, *Bull. Inst. Clín. Quir.*, 3, 861—863, 1927.
1491. **Joyeux, C.**, La classification des cestodes d'après quelques travaux récents, *Ann. Parasitol.*, 6, 132—136, 1928.
1492. **Joyeux, C.**, Procédé pour rechercher les cysticercoïdes des petits crustacés, *Ann. Parasitol.*, 7, 112—115, 1929.
1493. **Joyeux, C.**, Sur quelques helminthes récoltés dans la région de Villers-sur-Mer, *Bull. Soc. Linn. Normandie* (Trav. orig), 3, 7—12, 1930.
1494. **Joyeux, C.**, A propos d'une nouvelle classification du genre *Davainea* R. Bl. s. lat., *Bull. Soc. Zool. Fr.*, 55, 44—57, 1931.
1495. **Joyeux, C.**, Note rectificative au sujet des crochets du rostre chez *Raillietina* (R.), *insignis* (Steudner, 1877), *Bull. Soc. Zool. Fr.*, 57, 397, 1932.
1496. **Joyeux, C.**, Les données parasitologiques concernant le kyste hydatique du poumon, *Arch. Med. Gén. Coloniale*, 1, 277—283, 1932.
1497. **Joyeux, C. and Baer, J. G.**, Etude de quelques cestodes provenant des colonies françaises d'Afrique et de Madagascar, *Ann. Parasitol.*, 5, 27-36, 1927.
1498. **Joyeux, C. and Baer, J. G.**, Recherches sur quelques espèces du genre *Bothridium* de Blainville, 1824 (Diphyllobothriidae), *Ann. Parasitol.*, 5, 127—139, 1927.
1499. **Joyeux, C. and Baer, J. G.**, Sur quelques larves de Bothriocephales, *Bull. Soc. Pathol. Exot.*, 20, 921—937, 1927.
1500. **Joyeux, C. and Baer, J. G.**, Sur quelques cestodes de la région d'Entebbé (Uganda), *Ann. Parasitol.*, 6, 179—181, 1928.
1501. **Joyeux, C. and Baer, J. G.**, Note sur quelques helminthes récoltés en Macédoine, *Bull. Soc. Pathol. Exot.*, 21, 214—220, 1928.
1502. **Joyeux, C. and Baer, J. G.**, Recherches sur le cycle évolutif d'*Hymenolepis fraterna*, *C. R. Soc. Biol.*, 99, 1317—1318, 1928.
1503. **Joyeux, C. and Baer, J. G.**, Rectification de nomenclature, *Ann. Parasitol.*, 6, 144, 1928.
1504. **Joyeux, C. and Baer, J. G.**, Note d'Helminthologie tunisienne, *Arch. Inst. Pasteur Tunis*, 17, 347—349, 1928.
1505. **Joyeux, C. and Baer, J. G.**, Les cestodes rares de l'homme, *Bull. Soc. Pathol. Exot.*, 22, 114—136, 1929.
1506. **Joyeux, C. and Baer, J. G.**, *Raillietina* (R.) *celebensis* Janicki, 1902 et *Raillietina* (R.) *baeri* Meggitt and Subramanian, 1927, *Bull. Soc. Pathol. Exot.*, 22, 675—677, 1929.
1507. **Joyeux, C. and Baer, J. G.**, Recherches expérimentales sur le larve plérocercoïde de *Diphyllobothrium ranarum* (Gastaldi, 1854), *C. R. Soc. Biol.*, 101, 294—296, 1929.
1508. **Joyeux, C. and Baer, J. G.**, Etudes sur le ré-encapsulement de *Sparganum ranarum* (Gastaldi, 1854), *C. R. Soc. Biol.*, 102, 305—307, 1929.
1509. **Joyeux, C. and Baer, J. G.**, Cestodes. In Mission Saharienne Augiéras-Draper, 1927—1928, *Bull. Mus. Natl. Hist. Nat.*, (Paris), Sér. 2, II, 217—223, 1930.
1510. **Joyeux, C. and Baer, J. G.**, On a collection of cestodes from Nigeria, *J. Helminthol.*, 8, 59—64, 1930.
1511. **Joyeux, C. and Baer, J. G.**, Evolution des plérocercoïdes de *Diphyllobothrium* (Cestodes, Pseudophyllidiens), *C. R. Soc. Biol.*, 108, 97—99, 1931.
1512. **Joyeux, C. and Baer, J. G.**, Recherches sur les cestodes appartenant au genre *Mesocestoides* Villant, *Bull. Soc. Pathol. Exot.*, 25, 993—1010, 1932.
1513. **Joyeux, C. and Baer, J. G.**, *Hymenolepis fraterna* (Stiles, 1906). II. Le cycle évolutif en Europe, *C. R. Congr. Int. Med. Trop. Hyg.*, 4, 55—56, 1932.
1514. **Joyeux, C. and Baer, J. G.**, Sur le cycle évolutif d'un Ténia de serpent, *C. R. Acad. Sci.*, 196, 1838—1839, 1933.
1515. **Joyeux, C. and Baer, J. G.**, Le ré-encapsulement de quelques larves de cestodes, *C. R. Acad. Sci.*, 197, 493—495, 1933.
1516. **Joyeux, C. and Baer, J. G.**, Sur quelques cestodes de France, *Arch. Mus. Natl. Hist. Nat. Paris*, 11, 157—171, 1934.
1517. **Joyeux, C. and Baer, J. G.**, Les hôtes d'attente dans le cycle évolutif des helminthes, *Biol. Méd. Paris*, 24, 482—506, 1934.
1518. **Joyeux, C. and Baer, J. G.**, Cestodes d'Indochine, *Rev. Suisse Zool.*, 42, 249—273, 1935.
1519. **Joyeux, C. and Baer, J. G.**, Un ténia hyperapolytique chez un mammifère, *C. R. Soc. Biol.*, 120, 334—336, 1935.
1520. **Joyeux, C. and Baer, J. G.**, Recherches sur le cycle évolutif d'*Hymenolepis pistillum* Dujardin, *C. R. Acad. Sci.*, 201, 742—743, 1935.

1521. **Joyeux, C. and Baer, J. G.**, Notices helminthologiques, *Bull. Soc. Zool. Fr.*, 60, 482—501, 1935.
1522. **Joyeux, C. and Baer, J. G.**, Recherches biologiques sur la ligule intestinale; ré-infestation parasitaire, *C. R. Soc. Biol.*, 121, 67—68, 1936.
1523. **Joyeux, C. and Baer, J. G.**, Quelques helminthes nouvelles et peu connus de la musaraigne, *Crocidura russula* Herm., *Rev. Suisse Zool.*, 43, 25—50, 1936.
1524. **Joyeux, C. and Baer, J. G.**, Helminthes des rats de Madagascar. Contribution à l'étude de *Davainea madagascariensis* (Davaine, 1869), *Bull. Soc. Pathol. Exot.*, 29, 611—619, 1936.
1525. **Joyeux, C. and Baer, J. G.**, Cestodes, *Faune de France*, Vol. 30, Paris, 1936, 613 pp.
1526. **Joyeux, C. and Baer, J. G.**, Recherches sur l'évolution des cestodes de gallinacés, *C. R. Acad. Sci.*, 205, 751—753, 1937.
1527. **Joyeux, C. and Baer, J. G.**, Evolution du *Taenia taeniaeformis* Batsch, *C. R. Soc. Biol.*, 126, 359—361, 1937.
1528. **Joyeux, C. and Baer, J. G.**, Sur quelques cestodes de Cochinchine, *Bull. Soc. Pathol. Exot.*, 30, 872—874, 1937.
1529. **Joyeux, C. and Baer, J. G.**, Remarques morphologiques et biologiques sur quelques cestodes de la famille des Taeniidae Ludwig, *Rabot. Gel'mintol.* (Skrjabin), pp. 269—274, 1936.
1530. **Joyeux, C. and Baer, J. G.**, Sur le développement des Pseudophylidea (Cestodes), *C. R. Soc. Biol.*, 127, 1265—1266, 1938.
1531. **Joyeux, C. and Baer, J. G.**, L'évolution des plérocercoïdes de la Ligule intestinale, *C. R. Soc. Biol.*, 129, 314—316, 1938.
1532. **Joyeux, C. and Baer, J. G.**, Recherches sur le début du développement des cestodes chez leur hôte définitif, *Livro Jubilar do Professor Lauro Travassos*, Rio de Janeiro, 1938, pp. 245—250.
1533. **Joyeux, C. and Baer, J. G.**, Sur quelques cestodes de Galliformes, *Trav. Stn. Zool.*, Wimereux, 13, 369—389, 1938.
1534. **Joyeux, C. and Baer, J. G.**, Sur quelques cestodes de Madagascar, *Bull. Soc. Pathol. Exot.*, 32, 39—43, 1939.
1535. **Joyeux, C. and Baer, J. G.**, Sur quelques cestodes de Charadriiformes, *Bull. Soc. Zool. Fr.*, 64, 171—187, 1939.
1536. **Joyeux, C. and Baer, J. G.**, Recherches biologiques sur quelques cestodes Pseudophyllidea, *Volumen Jubilare pro Prof. Sadao Yoshida*, Vol. 2, Osaka Natural History Society, Osaka, 1939, 203—210.
1537. **Joyeux, C. and Baer, J. G.**, Anatomica y posición sistemica de *Raillietina* (R.) *quitensis* Léon, 1935, cestode parásito del hombre, *Rev. Med. Trop. Parasitol. Havana*, 6, 79—88, 1940.
1538. **Joyeux, C. and Baer, J. G.**, Sur quelques cestodes, *Rev. Suisse Zool.*, 47, 381—388, 1940.
1539. **Joyeux, C. and Baer, J. G.**, Un cestode nouveau parasite du plongeon, *Bull. Soc. Neuchâtel. Sci. Nat.*, 65, 21—24, 1941.
1540. **Joyeux, C. and Baer, J. G.**, Morphologie, évolution et position systématique de *Catenotaenia pusilla* (Goeze, 1782), parasite de rongeurs, *Rev. Suisse Zool.*, 52, 13—51, 1945.
1541. **Joyeux, C. and Baer, J. G.**, L'hôte normal de *Railliettina* (R.) *demerariensis* (Daniels, 1895) en Guyane hollandaise, *Acta Trop.*, 6, 141—144, 1949.
1542. **Joyeux, C. and Baer, J. G.**, A propos des Ténias du genre *Inermicapsifer* récemment découverts chez l'homme, *Bull. Soc. Pathol. Exot.*, 42, 581—586, 1949.
1543. **Joyeux, C. and Baer, J. G.**, The status of the cestode genus *Meggittiella* López-Neyra, 1942, *Proc. Helminthol. Soc. Wash.*, 17, 91—95, 1950.
1544. **Joyeux, C. and Baer, J. G.**, Sur quelques espèces nouvelles ou peu connues du genre *Hymenolepis* Weinland, 1858, *Bull. Soc. Neuchâtel. Sci. Nat.*, 73, 51—70, 1950.
1545. **Joyeux, C. and Baer, J. G.**, Le genre *Gyrocotyloides* Fuhrmann, 1931 (Cestodaria), *Bull. Soc. Neuchâtel. Sci. Nat.*, 73, 71—79, 1950.
1546. **Joyeux, C. and Baer, J. G.**, Sobre le posicion sistematica del genero *Inermicapsifer* Janicki, 1910 (Cestoda), *Rev. Kuba Med. Trop. Parasitol.*, 6, 7—9, 1950.
1547. **Joyeux, C. and Baer, J. G.**, Le genre *Gyrocotyle* Diesing, 1850 (Cestodaria), *Rev. Suisse Zool.*, 58, 371—381, 1951.
1548. **Joyeux, C. and Baer, J. G.**, Les cestodes de *Neomys fodiens* (Schreb.) musaraigne d'eau, *Bull. Soc. Neuchâtel. Sci. Nat.*, 75, 87—88, 1952.
1549. **Joyeux, C. and Baer, J. G.**, Cestodes et Acanthocéphales récoltés par M. Patrice Paulian aux Iles Kerguelen et Amsterdam, 1951—52, *Mém. Inst. Sci. Madagascar*, Sér. A, 9, 1—16, 1954; Sér. B, 9, 23—40, 1954.
1550. **Joyeux, C. and Baer, J. G.**, Cestodes d'oiseaux récoltés dans le centre de la France, *Bull. Soc. Zool. Fr.*, 80, 174—196, 1955.
1551. **Joyeux, C., Baer, J. G., and Gaud, J.**, Recherches sur des cestodes d'Indochine et sur quelques Diphyllobothrium (Bothriocephales), *Bull. Soc. Pathol. Exot.*, 43, 482—489, 1950.
1552. **Joyeux, C., Baer, J. G., and Gaud, J.**, Recherches helminthologiques marocaines. Cestodes (deuxième note), *Arch. Inst. Pasteur Maroc.*, 4, 93—102, 1951.

1553. **Joyeux, C., Baer, J. G., and Martin, R.**, Recherches sur les sparganoses, *Bull. Soc. Pathol. Exot.*, 26, 1199—1208, 1933.
1554. **Joyeux, C., Baer, J. G., and Martin, R.**, Sur le cycle évolutif des *Mesocestoides*, *C. R. Soc. Biol.*, 114, 1179—1180, 1933.
1555. **Joyeux, C., Baer, J. G., and Martin, R.**, Sur quelques cestodes de la Somalie-Nord, *Bull. Soc. Pathol. Exot.*, 29, 82—96, 1936.
1556. **Joyeux, C., Baer, J. G., and Martin, R.**, Sur quelques cestodes de la Somalie-Nord (deuxième note), *Bull. Soc. Pathol. Exot.*, 30, 416—423, 1937.
1557. **Joyeux, C. and Dollfus, R. Ph.**, Sur quelques cestodes de la collection du Musée de Munich, *Zool. Jahr. Syst.*, 62, 109—118, 1931.
1558. **Joyeux, C. and Foley, H.**, Recherches épidémiologiques sur l'*Hymenolepis nana* et sur *Hymenolepis fraterna*, *Arch. Inst. Pasteur Algérie*, 7, 31—50, 1929.
1559. **Joyeux, C. and Foley, H.**, Les helminthes de *Meriones shawi* Rozet dans le nord de l'Algérie, *Bull. Soc. Zool. Fr.*, 55, 353—374, 1930.
1560. **Joyeux, C., Gendre, C., and Baer, J. G.**, Recherches sur les helminthes de l'Afrique occidentale française, *Bull. Soc. Pathol. Exot. Monogr.*, 2, 1—120, 1929.
1561. **Joyeux, C. and Houdemer, E.**, Recherches sur la faune helminthologique de l'Indochine (cestodes et trématodes), *Ann. Parasitol.*, 5, 289—309, 1927.
1562. **Joyeux, C. and Houdemer, E.**, Recherches sur la faune helminthologique de l'Indochine (cestodes et trématodes), *Ann. Parasitol.*, 6, 27—58, 1928.
1563. **Joyeux, C., Houdemer, E., and Baer, J. G.**, Etiologie de la sparganose oculaire, *Marseille Méd.*, 69, 405—409, 1932.
1564. **Joyeux, C., Houdemer, E., and Baer, J. G.**, Recherches sur la biologie des Sparganum et l'étiologie de la sparganose oculaire, *Bull. Soc. Pathol. Exot.*, 27, 70—78, 1934.
1565. **Joyeux, C. and Kobozief, N. I.**, Recherches sur l'*Hymenolepis microstoma* (Dujardin, 1845), *C. R. Soc. Biol.*, 97, 12—14, 1927.
1566. **Joyeux, C. and Kobozief, N. I.**, Recherches sur l'*Hymenolepis microstoma* (Dujardin, 1845), *Ann. Parasitol.*, 6, 59—79, 1928.
1567. **Joyeux, C. and Mathias, P.**, Cestodes et trématodes récoltés par le professeur Brumpt au cours de la mission du bourg de Bizas, *Ann. Parasitol.*, 4, 333—336, 1926.
1568. **Joyeux, C. and Millot, J.**, Sur un cysticercoïde nouveau parasite de *Herpobdella atomaria* Carena, 1820, *Travaux Stn. Zool. Wimereux*, 9, 98—101, 1925.
1569. **Joyeux, C., Noyer, R. Du., and Baer, J. G.**, Les Bothriocéphales, *Bull. Sci. Pharmacol.*, 38, 175—435, 1931.
1570. **Joyeux, C., Richet, Ch., and Schulman, E.**, Description d'une cénure trouvé chez la souris blanche de laboratoire, *Bull. Soc. Zool. Fr.*, 47, 181—186, 1922.
1571. **Joyeux, C. and Timon-David, J.**, Sur quelques cestodes d'oiseaux, *Ann. Mus. Hist. Nat. Marseille*, 26, 1—26, 1934.
1572. **Joyeux, C. and Timon-David, J.**, Note sur les cestodes d'oiseaux récoltés dans la région de Marseille, *Ann. Mus. Hist. Nat. Marseille*, 26, 1—8, 1934.
1573. **Joyeux, C. and Timon-David, J.**, Cestodes d'oiseaux de la région marseillaise, *Ann. Fac. Sci. Marseille*, 9, 67—77, 1936.
1574. **Joyeux, C. and Truong-Tan-Ngog,** Les cestodes de quelques oiseaux de bassecour dans la région de Cholon (Viet-Nam), *Rev. Elevage Méd. Vét. Pays Trop.*, n. s., 4, 67—69, 1950.
1575. **Kadam, S. S., Shinde, G. B., and Jadhav, B. V.**, *Gidhaia kolhapurensis* n. sp., (Cestoda: Anoplocephalidae) from *Torgos calvus*, *Curr. Sci.*, 50, 296, 1981.
1576. **Kadenatsii, A. N. and Sulimov, A. D.**, A new cestode from rodents in Tuva (in Russian), *Trudi Omskogo Vet. Inst.*, 22, 89—92, 1964.
1577. **Kahane, Z.**, Anatomie von *Taenia perfoliata* Goeze als Beitrag zur Kenntnis der Cestoden, *Z. Wiss. Zool.*, 34, 175—254, 1880.
1578. **Kalyankar, S. D. and Palladwar, V. D.**, On a new species of avian cestode of the genus *Amoebotaenia* (Dilepididae: Dilepidinae) Cohn, 1900 from India, *An. Fac. Vet. Léon*, 21, 27—31, 1975 (Pub. 1977).
1579. **Kalyankar, S. D. and Palladwar, V. D.**, Study on a new poultry worm *Amoebotaenia kharati* n.sp. (Cestoda: Dilepididae: Dilepidinae) from Aurangabad, *Marathwada Univ. J. Sci.*, (Nat. Sci.) 16, (Sci. no. 9), (9), 233—236, 1977.
1580. **Kalyankar, S. D. and Palladwar, V. D.**, On a new species of the genus *Neyraia* Joyeux and Timon-David, 1934 (Biuterinidae Meggitt, 1927) from Coraciiformes in India, and a key to the species of the genus *Neyraia* Joyeux et David, 1934, All-India Symp. Helminthol., University of Kashmir Grants Commission, Srinigar, Kashmir, August 8 to 11, 1971, 1977, 15—16.
1581. **Kamalova, A. G.**, Comparative characteristics of *Taeniarhynchus saginatus* and *Taenia solium* (in Russian), *Rabot. Gel'mintol., 75-Let. Skrjabin*, pp. 276—283, 1954.

1582. **Kamburov, P.,** *Choanotaenia thraciensis* n.sp. and *Aploparaksis bulgarica* n.sp. in *Scolopax rusticola* (in Bulgarian), *Izv. Tsent. Khelmintol. Lab.*, 13, 185—190, 1969.
1583. **Kamiya, M., Suzuki, H., and Villa-R. B.,** A new anoplocephaline cestode, *Anoplocephaloides romerolagi* sp.n. parasitic in the volcano rabbit, *Romerolagus diazi, Jpn. J. Vet. Res.*, 27, 67—71, 1979.
1584. **Kamo, H. and Miyazaki, I.,** *Diplogonoporus fukuokaensis* sp.nov. (Cestoda: Diphyllobothriidae) from a girl in Japan, *Jpn. J. Parasitol.*, 19, 635—644, 1970.
1585. **Kan, K.,** *Cysticercus pisiformis* and the coenurus of *Polycephalus serialis* in wild rabbits (in Japanese) *Keio Igaku Tokyo,* 14, 505—511, 1934.
1586. **Kataoka, N. and Momma, K.,** A cestode parasitic in *Plecoglossus altivelis, Annot. Zool. Jpn.*, 14, 13—22, 1933.
1587. **Kataoka, N. and Momma, K.,** A preliminary note on the life-history of *Proteocephalus neglectus*, with special reference to its intermediate host, *Bull. Jpn. Soc. Sci. Fish.*, 3, 125—126, 1934.
1588. **Kates, K. C. and Goldberg, A.,** Experimental tapeworm (*Moniezia expansa*) infections in young lambs, *J. Parasitol.*, 35 (Sect. 2), 38, 1949.
1589. **Kates, K. C. and Goldberg, A.,** The pathogenicity of the common sheep tapeworm, *Moniezia expansa, Proc. Helminthol. Soc. Wash.*, 18, 87—101, 1951.
1590. **Kates, K. C. and Runkel, C. E.,** Observations on oribatid mites, vectors of *Moniezia expansa* on pastures, with a report of several new vectors from the U.S., *J. Parasitol.*, 38 (Sect. 2), 15, 1947.
1591. **Kates, K. C. and Runkel, C. E.,** Observations on oribatid mite vectors of *Moniezia expansa* on pastures, with a report of several new vectors from the United States, *Proc. Helminthol. Soc. Wash.*, 15, 19—33, 1948.
1592. **Kawanishi, K.,** Experimental studies on the morphological changes of the blood, and clinical symptoms, in infections with *Taenia solium* of man, *Taiwan Igakkai Zasshi,* 31, 93—94, 1932.
1593. **Kay, M. W.,** A new species of *Phyllobothrium* van Beneden from *Raja binoculata* (Girard), *Trans. Am. Microsc. Soc.*, 61, 261—263, 1942.
1594. **Kennedy, C. R.,** Taxonomic studies on *Archigetes* Leuckart, 1878 (Cestoda: Caryophyllaeidae), *Parasitology,* 55, 439—451, 1965.
1595. **Kennedy, M. J., Killick, L. M., and Beverly-Burton, M.,** *Oochoristica javaensis* n.sp. (Eucestoda: Linstowiidae) from *Gehyra mutilata* and other gekkonid lizards (Lacertilia: Gekkonidae) from Java, Indonesia, *Can. J. Zool.*, 60, 2459—2463, 1982.
1596. **Keppner, E. J.,** *Fossor taxidiensis* (Skinker, 1935) n. comb. with a note on the genus *Fossor* Honess, 1937 (Cestoda: Taeniidae), *Trans. Am. Microsc. Soc.*, 86, 157—158, 1967.
1597. **Kerr, T.,** On *Linstowia echidnae* (Thompson, 1893) Zschokke, 1899: a cestode from the Australian ant eater, *Ann. Mag. Nat. Hist.*, (Ser. 10), 15, 156—160, 1935.
1598. **Kevorkov, N. P. and Vavilova, M. P.,** On intraintestinal autoreinvasion in hymenolepidosis (in Russian), *Med. Parasitol. Parazitar Bolezni,* 13, 31—34, 1944.
1599. **Khalil, L. F.,** On a new genus *Sandonella,* for *Proteocephalus sandoni* Lynsdale, 1960, (Proteocephalidae) and the erection of a new subfamily, Sandonellinae, *J. Helminthol.*, 34, 47—54, 1960.
1600. **Khalil, L. F.,** On a new cestode, *Paradilepis maleki* sp. nov., (Dilepididae), from a sacred ibis in the Sudan, *J. Helminthol.*, 35, 255—258, 1961.
1601. **Khalil, L. F.,** *Bertiella trichosuri* n.sp. from the brush-tail opossum, *Trichosurus vulpeculo* (Kerr) from New Zealand, *Zool. Anz.*, 185, 442—450, 1970.
1602. **Khalil, L. F.,** *Ichthybothrium ichthybori* gen. et sp. nov. (Cestoda: Pseudophyllidea) from the African freshwater fish *Icthyborus besse* (Joannis, 1835), *J. Helminthol.*, 45, 371—379, 1971.
1603. **Khalil, L. F.,** *Gyrometra kunduchi* n.sp., a cestodarian from *Plectorhinchus pictus* (Thunberg, 1792) from the Indian Ocean, *J. Fish. Biol.*, 11, 15—19, 1977.
1604. **Khambata, F. S. and Bal, D. V.,** Five new species of cestodes from marine fishes of Bombay, *Proc. 38. Ind. Sci. Congr. (Bangalore 1951),* Part 3, 211, 1952.
1605. **Khambata, F. S. and Bal, D. V.,** Three new species of the genus *Otobothrium* (Cestoda) from marine fishes of Bombay, *Proc. 40th Ind. Sci. Congr. Lucknow 1953,* Part 3 (Abstr.), Sect. 7, 191, 1954.
1606. **Khambata, F. S. and Bal, D. V.,** Two new species of the genus *Discobothrium* (Cestoda) from the marine fishes of Bombay, *Proc. 40th Ind. Sci. Congr. Lucknow 1953,* Part 3 (Abstr.) Sect. 7, 192, 1954.
1607. **Khan, D. and Habibullah,** Avian cestodes from Lahore, West Pakistan, *Bull. Dep. Zool. Univ. Panjab,* No. 1, 1—34, 1967.
1608. **Khan, D. and Habibullah,** Two new species of hymenolepid cestodes from Lahore, West Pakistan, *Pak. J. Zool.*, 3, 213—216, 1971.
1609. **Kheisin, E. M.,** The structure of the eggs of *Diphyllobothrium latum* and their resistance to various environmental factors (in Russian), *Tr. Leningrad. Inst. Epidemiol. Bakteriol. Pastera,* 3, 40—49, 1937; English summary 127—128, 1937.
1610. **Khlopina, A.,** Two new species of the genus *Caryophyllaeus* parasitic in Cyprinidae (in Russian), *Bull. Acad. Sci. Petrograd,* pp. 97—110, 1919.

1611. **Kiessling, F.,** Ueber den Bau von *Schistocephalus dimorphus* Creplin und *Ligula simplicissima* Rudolphi, *Arch. Naturg.,* 1, 241—280, 1882.
1612. **Kingscote, A. A.,** The occurrence of tapeworms of the genus *Anoplocephala* in the horse, *Rep. Ontario Vet. Coll.,* pp. 61—62, 1931.
1613. **Kintner, K. E.,** Note on the cestodes of English sparrows in Indiana, *Parasitology,* 30, 347—357, 1938.
1614. **Kiribayashi, S.,** Studies on the growth of *Hymenolepis nana* with special reference to the possibility of differentation of *H. nana* var. *fraterna* (Stiles), *Taiwan Igakkai Zasshii,* 32, 117—118, 1933.
1615. **Kirschenblat, J. D.,** Die Gesetzmässigkeiten der Dynamik der Parasitenfauna bei den mäuseähnlichen Nagetieren (Muriden) in Transkaukasien, *Diss. Univ. Leningrad,* 1938, 87 pp.
1616. **Kirschenblat, J. D.,** Intermediate hosts of the cestoids of the family Anoplocephalidae (in Russian), *Priroda,* pp. 83—85, 1940.
1617. **Kirschenblat, J. D.,** On cestodes of the genus *Cittotaenia* Riehm, parasites of ground squirrels (in Russian), *Proc. Acad. Sci. Armenia SSR,* 6, 115—118, 1947.
1618. **Kirschenblat, J. D.,** Life cycle of *Oochoristica ratti* (in Russian), *Priroda,* 39, 49—50, 1950.
1619. **Klaptocz, B.,** Ergebnisse der mit der Subvention aus Erbschaft Treitl unternommenen zoologischen Forschungsreise Dr. Franz Werners in den aegyptischen Sudan und nach Nord-Uganda. Cestoden aus Fischen, aus Varanus und Hyrax, *Sitzungber. Akad. Wiss. Wien Math. Naturwiss. Klasse Abt. I,* 115, 121—144, 1906.
1620. **Klaptocz, B.,** Ergebnisse der mit der Subvention aus Erbschaft Treitl unternommenen zoologischen Forschungsreise Dr. Franz Werners in den aegyptischen Sudan und nach Nord-Uganda. Cestoden aus Fischen, aus Varanus und Hyrax, Cestoden aus *Numida philorhyncha* Licht, *Sitzungber. Akad. Wiss. Wien Math. Naturwiss. Klasse Abt. I,* 115, 963—974, 1906.
1621. **Klaptocz, B.,** *Polyonchobothrium polyteri* (Leydig), *Centralbl. Bakteriol.,* 41, 527—536, 1906.
1622. **Klaptocz, B.,** Neue Phyllobothriden aus *Notidanus (Hexanchus) griseus* Gm., *Arb. Zool. Inst. Wien,* 16, 325—360, 1906.
1623. **Klaptocz, B.,** Vogelcestoden, *Arb. Zool. Inst. Wien,* 17, 1—40, 1908; *Sitzungber. K. Akad. Wiss. Wien Math. Naturwiss Klasse Abt. I,* 117, 259—298, 1908.
1624. **Knoch, J.,** Die Naturgeschichte des Breiten Bandwurms, mit besonderer Berücksichtigung seiner Entwickelungsgeschichte, *Mem. Acad. Imp. Sci. St. Petersbourg,* 5, 1—134, 1862.
1625. **Knoch, J.,** Eie Entwicklungsgeschichte des *Bothriocephalus proboscideus* (B. salmonis Koelliker's) als Beitrag zur Embryologie des *Bothriochephalus latus, Bull. Acad. Imp. Sci. St. Petersbourg,* 9, 290—314, 1866.
1626. **Knoch, J.,** Neue Beiträge zur Embryologie des *Bothriocephalus latus* als Beweis einer direkten Metamorphose des geschlechtsreifen Individuums aus seinem bewimperten Embryo. Zugleich ein Beitrag zur Therapie der Helminthiasis, *Bull. Acad. Imp. Sci. St. Petersbourg,* 14, 176—188, 1870.
1627. **Kobayashi, H.,** Studies on the development of *Diphyllobothrium mansoni* Cobbold, 1882 (Joyeux, 1927). IV. Hatching of the egg, onchosphaera and discarding of the ciliated coat. V. The first intermediate host (in Japanese), *Taiwan Igakkai Zasshi,* 30, 15—16, 23—27, 1931.
1628. **Kobyshev, N. M.,** *Cladotaenia spasskyi* n.sp. (Taeniidae), a new cestode of Falconiformes (in Russian), *Vest. Zool.,* 5, 82—84, 1971.
1629. **Kofend, L.,** Cestoden aus Säugetieren und aus *Agama colonorum.* Vorläufige Mitteilung, *Anz. Akad. Wiss. Wien Math. Naturwiss. Klasse Abt. I.,* 54, 229—321, 1917.
1630. **Kofend, L.,** Cestoden aus Säugetieren und aus *Agama colonorum.* Wissenschaftliche Ergebnisse der mit Unterstützung der Akademie der Wissenschaft in Wien aus der Erbschaft Treitl. von F. Werner unternommenen zoologischen Expedition nach dem Anglo-Aegyptischen Sudan (Kordofan) 1914, *Denkschr. Akad. Wiss. Wien Math. Naturwiss. Klasse Abt. I.,* 98, 1—10, 1921.
1631. **Kofoid, C. A. and Watson, E. E.,** On the orientation of *Gyrototyle* and of the cestode strobila, Advance print from Proc. 7th Int. Zool. Congr. Boston, August, 1907, 1910, 5 pp.
1632. **Kolmakov, D. V.,** Beschreibung einer neuen Gattung *Moniezia rangiferina* beim Renntier (in Russian), *Tr. Omsk. Vet. Inst.,* 11, 101—105, 1938.
1633. **Kolmakov, D. V.,** *Avitellina arctica* n.sp. from wild reindeer, *Tr. Vsesoiuz Inst. Gel'minthol.,* 3, 148—151, 1938.
1634. **Konovalov, Yu. N. and Reznik, V. N.,** A new dilepidid species (Cestoda: Cyclophyllidea) from Charadriiformes (in Russian), *Parasites Anim. Plants,* 4, 74—77, 1968.
1635. **Korneev, K. P.,** A case of multiple *Dithyridium elongatum* in a dog (in Russian), *Sovet. Vet.,* 15, 65—66, 1938.
1636. **Kornyushin, V. V.,** *Vitta alexandri* n.sp. (Cestoda, Choanotaeniidae) from *Glareola nordmanni* in the Ukraine (in Ukranian), *Dopovidi Akad. Nauk Ukrains'koi RSR,* No. 8, 1085—1089, 1966.
1637. **Kornyushin, V. V.,** Cestodes of the genus *Echinocotyle* Blanchard, 1891 (Hymenolepididae) from Charadriformes in the Ukraine, *Parazitologiya,* 3, 542—550, 1969.
1638. **Kornyushin, V. V.,** Cestode fauna of the Black Sea population of *Tadorna tadorna* L. (in Ukranian), *Zbīrnik Prats' Zoologīchnogo Mazeyu,* 33, 36—47, 1969.

1639. **Kornyushin, V. V.**, New species of cestodes, *Wardium squaterolae* n.sp. (Cyclophyllidea: Hymenolepididae), *Dopvidi Akademii Nauk Ukrains'koi RSR*, B, 368—370, 1970.
1640. **Kornyushin, V. V. and Spasskii, A. A.**, *Wardium smogorjevskajae* n.sp. (Cestoda: Cyclophyllidea) from *Tringa totanus* on the Black Sea coast (in Russian), *Vest. Zool.*, 2, 46—50, 1967.
1641. **Koropov, V. M.**, Etude éxpérimentale de l'influence exercée par les produits des helminthes sur le système cardiovasculaire (in Russian), *Med. Parsitol. Paras. Dis.*, 4, 281—287, 1935.
1642. **Korpaczewska, W. and Sulgostowska, T.**, Revision of the genus *Tatria* Kow., 1904 (Cestoda, Amabiliidae), including description of *Tatria iunii* sp.n., *Acta Parasitol. Pol.*, 22, 67—91, 1974.
1643. **Kostylev, N. N.**, La *Taenia crenata* Goeze comme une espèce indépendante, *Ann. Musée Zool. Acad. Sci. Imp. St. Petersbourg*, 20, 127—129, 1916.
1644. **Kotelnikov, G. A.**, Development of the cestode genus *Fimbriaria* Froelich, 1802 (in Russian), *Mater. Nauch. Konf. Vses. Obshch. Gel'mintol.*, Year 1965, Part 3, 136—141, 1965.
1645. **Kotelnikov, G. A.**, The life cycle of *Fimbriaria amurensis* n.sp. parasitic in domestic ducks (in Russian), *Dokl. Akad. Nauk SSSR*, 130, 944—945, 1960.
1646. **Kotlán, A.**, Uj-Guinea Madádar-Cestodák. I. Papagály-Cestodák. (Bird cestodes from New Guinea. I. Parrot cestodes) (in Hungarian), *Ann. Hist. Nat. Musée Nat. Hung.*, 18, 1—27, 1921.
1647. **Kotlán, A.**, Ueber *Sparganum raillieti* Rátz und den zugehörigen geschlectreifen Bandwurm, *Dibothriocephalus raillieti* Rátz, *Centralbl. Bakteriol. Parasitenk.*, 90, 272—285, 1923.
1648. **Kotlán, A.**, Avian cestodes from New Guinea. II. Cestodes from Casuariformes, *Ann. Trop. Med. Parasitol.*, 17, 47—57, 1923.
1649. **Kotlán, A.**, Avian cestodes from New Guinea. III. Cestodes from Galliformes, *Ann. Trop. Med. Parasitol.*, 17, 59—69, 1923.
1650. **Kotlán, A.**, On *Davainea proglottina* and its synonyms, *J. Parasitol.*, 12, 26—32, 1925.
1651. **Kouri, P.**, Tercer informe en relación al *Inermicapsifer cubensis* (Kouri, 1938), *Rev. Med. Trop. Parasitol.*, 10, 107—112, 1944.
1652. **Kouri, P.**, *Inermicapsifer cubensis* (Kouri, 1938) historia, nomenclatura y sinonimia, *Rev. Med. Trop. Parasitol., Habana*, 4, 97—98, 1948.
1653. **Kouri, P. and Doval, J. M.**, Le raillietinosis humaine en Cuba, *Bol. Mens. Clin. Asoc. Damas la Covadonga*, 5, 121—134, 1938.
1654. **Kouri, P. and Doval, J. M.**, Tres casos di parasitismo humano por especies de la familia Davaineidae, *Rev. Med. Trop. Parasitol. Habana*, 4, 207—218, 1938.
1655. **Kouri, P. and Kouri, J.**, Discusiones entorno al *Inermicapsifer cubensis* (Kouri, 1938), *Rev. Kuba Med. Trop. Parasitol.*, 6, 1—7, 1950.
1656. **Kouri, P. and Kouri, J.**, Diskussionen um *Inermicapsifer cubensis* (Kouri 1938), English summary, *Z. Tropenmed. Parasitol.*, 3, 243—253, 1951.
1657. **Kouri, P. and Kouri, J.**, Hallazgo del *Inermicapsifer cubensis* en la rata blanca. Nota previa, *Rev. Med. Trop. Parasitol., Habana*, 8, 27, 1952.
1658. **Kouri, P. and Rappaport, I.**, A new human helminthic infection in Cuba, *J. Parasitol.*, 26, 179—181, 1940.
1659. **Kouri, P., Sotolongo, F., and Baer, J. G.**, Anatomie, position systématique et épidémiologie de *Inermicapsifer cubensis* (Kouri, 1938) Kouri 1940, cestode parasite de l'homme á Cuba. II. Epidémiologie et diagnostic, *Acta Trop.*, 6, 127—130, 1949.
1660. **Kovacs, K. J. and Schmidt, G. D.**, Two new species of cestode (Trypanorhyncha, Eutetrarhynchidae) from the yellow-spotted stingray, *Urolophus jamaicensis*, *Proc. Helminthol. Soc. Wash.*, 47, 10—14, 1980.
1661. **Kowalewski, M.**, Ein Beitrag zur Kenntnis der Excretionsorgane, *Biol. Centralbl.*, 9, 33—47, 1889.
1662. **Kowalewski, M.**, Helminthological studies. (in Polish with French resumé), *Bull. Int. Acad. Sci. Cracovie*, pp. 278—280, 1894.
1663. **Kowalewski, M.**, Sur la tête du *Taenia malleus* Goeze, *Arch. Parasitol.*, 1, 326—329, 1898.
1664. **Kowalewski, M.**, Helminthological studies. VIII. On a new tapeworm, *Tatria biremis* gen. nov. et sp. nov. (in Polish with English summary), *Bull. Int. Acad. Sci. Cracovie*, pp. 367—369, 1904.
1665. **Kowalewski, M.**, Helminthological studies. IX. On two new species of tapeworms of the genus *Hymenolepis* (in Polish with English summary), *Bull. Int. Acad. Sci. Cracovie*, pp. 532—564, 1905.
1666. **Kowalewski, M.**, Mitteilungen ueber eine *Idiogenes*-species, *Zool. Anz.*, 29, 683—686, 1906.
1667. **Kowalewski, M.**, Helminthological studies. X. Contribution à l'étude de deux cestodes d'oiseaux (in Polish with French resumé), *Bull. Int. Acad. Sci. Cracovie Cl. Sci. Math. Nat.*, pp. 774—776, 1907.
1668. **Kozicka, J.**, Some observation on the infection of *Coregonus albula* Linné with *Ichthyotaenia longicollis* Rudin (in Polish with English summary), *Medycyna Wet.*, 5, 438—441, 1949.
1669. **Krabbe, H.**, Helminthologiske undersøgelser i Danmark og paa Island, med saerligt Hensyn til Blaecormlidelserne paa Island (in Danish), *K. Danske Vidensk. Selsk. Skr. Naturvidenskab. Math. Afd.*, 7, 345—408, 1865.
1670. **Krabbe, H.**, Om nogle Baendelormammers udvikling til Baendelorme (in Danish), *Vidensk. Medd. Naturh. Foren. Kjøbenhavn*, pp. 1—10, 1867.

1671. **Krabbe, H.,** Trappens Baendelorme (in Danish), *Vidensk. Medd. Naturh. Foren. Kjøbenhavn,* pp. 122—126; (transl.), *Ann. Mag. Nat. Hist.,* (1869), 4, 47—51, 1868.
1672. **Krabbe, H.,** Bidrag til Kundskab om Fuglenes Baendelorme (in Danish), *K. Danske Vidensk. Selsk. Skr. Naturvidenskab. Math. Afd.,* 8, 249—363, 1868.
1673. **Krabbe, H.,** *Diplocotyle Olrikii,* cestoide non articulé du groupe des bothriocéphales, *J. Zool.,* III, 392—395, 1874.
1674. **Krabbe, H.,** Nye Bidrag til Kundskab om Fuglenes Baendelorme (in Danish), *K. Danske Vidensk. Selsk. Skr. Naturvidenskab. Math. Afd.,* 1, 347—366, 1882.
1675. **Kraemer, H.,** Beiträge zur Anatomie und Histologie der Cestoden der Süsswasserfische, *Z. Wiss. Zool.,* 53, 647—722, 1892.
1676. **Krause, E.,** About *Botriocephalus* [sic] *latus.* (in Hebrew), *Harefuah,* 2, 328—330, 1926; English summary, pp. VII—IX, 1926.
1677. **Krause, K.,** Contribution á l'etude du *Diphyllobothrium latum* (L.) en Palestine. Description d'anomalies chez ce cestode, *Ann. Parasitol.,* 5, 249—251, 1927.
1678. **Krefft, G.,** On Australian Entozoa, *Trans. Entomol. Soc. N.S.W.,* 2, 206—232, 1871.
1679. **Krotov, A. I.,** Hymenolepids of Anserinae of the SSSR (in Russian), *Tr. Gel'mintol. Lab. Akad. Nauk SSSR,* 2, 99—109, 1949.
1680. **Krotov, A. I.,** New cestodes of birds (in Russian), *Tr. Gel'mintol. Lab. Akad. Nauk SSSR,* 5, 130—137, 1951.
1681. **Krotov, A. I.,** New cestodes (Hymenolepididae and Paruterinidae) of birds (in Russian), *Tr. Gel'mintol. Lab. Akad. Nauk SSSR,* 6, 259—272, 1952.
1682. **Krotov, A. I.,** Contribution to the knowledge of the cestode fauna of SSSR (in Russian), *Rabot. Gel'mintol. 75-Lett. Skrjabin,* pp. 326—339, 1954.
1683. **Krull, W. H.,** On the life-history of *Moniezia expansa* and *Cittotaenia* sp. (Cestoda, Anoplocephalidae), *Proc. Helminthol. Soc. Wash.,* 6, 10—11, 1939.
1684. **Krull, W. H.,** Investigations on possible intermediate hosts, other than oribatid mites, for *Moniezia expansa, Proc. Helminthol. Soc. Wash.,* 7, 68—70, 1940.
1685. **Krull, W. H.,** The identification of *Thysanosoma actinioides* infections in sheep by examination of fecal pellets, *Trans. Am. Microsc. Soc.,* 65, 351—353, 1946.
1686. **Kruse, D. N.,** Parasites of the commercial shrimps, *Penaeus aztecus* Ives, *P. duararum* Burkenroad and *P. setiferus* (Linnaeus), *Tulane Stud. Zool.,* 7, 123—144, 1959.
1687. **Küchenmeister, G. F. H.,** Ueber die Umwandlung der Finnen (Cysticerci) in Bandwürmer, *Vierteljahr. Schriften. Prakt. Heilk.,* 9, Jahrg. Prağ, 33, 106—158, 1852.
1688. **Küchenmeister, G. F. H.,** Experimente ueber die Entstehung der Cestoden zweiter Stufe zunächst des *Coenurus cerebralis, Z. Klin. Med. Breslau,* 4, 448—451, 1853.
1689. **Küchenmeister, G. F. H.,** Ueber eine Abart der *Taenia coenurus* d.h. des Bandwurms von der die Quese des Schafes und des Rindes herstammen, *Allg. Dtsch. Naturhist. Ztg.,* 1, 191—194, 1855.
1690. **Küchenmeister, G. F. H.,** Ueber die Umwandlung der Blasenwürmer in Taenien, insbesondere des *Coenurus cerebralis* Gervais, *Wien Med. Wochschr.,* 6, 319—320, 1856.
1691. **Kuczkowski, St.,** Die Entwicklung in Genus *Ichthyotaenia* Lönnberg. Ein Beitrag zur Cercomertheorie auf Grund experimenteller Untersuchungen, *Bull. Int. Acad. Pol. Sci. Lett. Cl. Sci. Math. Nat. Ser. B,* pp. 423—446, 1925.
1692. **Kugi, G. and Sawada, I.,** *Mathevotaenia brasiliensis* n.sp., a tapeworm from the squirrel monkey, *Saimiri sciureus, Jpn. J. Parasitol.,* 19, 467—470, 1970.
1693. **Kugi, G. and Sawada, I.,** A new cestode, *Raillietina (Paroniella) japonica,* from a crow, *Corvus levaillantii,* in Japan., *Jpn. J. Parasitol.,* 21, 135—137, 1972.
1694. **Kugi, G. and Sawada, I.,** *Crepidobothrium macroacetabula* n.sp., a cestode from the anaconda, *Eunectes murinus, Jpn. J. Zool.,* 16, 181—183, 1972.
1695. **Kuhlow, F.,** Beitrag zur Entwicklung und Systematik heimischer *Diphyllobothrium*-Arten., *Z. Tropenmed. Parasitol.,* 4, 203—234, 1953.
1696. **Kuhlow, F.,** Bau und Differentialdiagnose heimischer *Diphyllobothrium* Plerocercoide, *Z. Tropenmed. Parasitol.,* 4, 186—202, 1953.
1697. **Kuhlow, F.,** Über die Entwicklung und Anatomie von *Diphyllobothrium dendriticum* Nitzsch, 1824, *Z. Parasitol.,* 16, 1—35, 1953.
1698. **Kuhlow, F.,** Untersuchungen über die Entwicklung des breiten Bandwurmes *(Diphyllobothrium latum), Z. Tropenmed. Parasitol.,* 6, 213—225, 1955.
1699. **Kulachkova, V. G.,** New species of *Aploparaksis* Clerc, 1903 (Hymenolepididae) from *Numenius arquatus* in the White Sea (in Russian), *Mater. Nauchn. Konf. Vses. Obshch. Gel'mintol.,* Year 1969, Part 1, 138—142, 1969.
1700. **Kulakovskaya, O. P.,** Caryophyllaeidae (Cestoda, Pseudophyllidea) of the USSR (in Russian), *Parazitol. Sbornik,* 20, 339—355, 1961.

1701. **Kulakovskaya, O. P.,** *Breviscolex orientalis* n.g., n.sp. (Caryophyllaeidae, Cestoda) from fish in the Amur basin (in Russian), *Dokl. Akad. Nauk SSSR,* 143, 1001—1004, 1962.
1702. **Kulakovskaya, O. P. and Akmerov, O. K.,** *Markevitschia sagittata* n.g., n.sp. (Cestoda, Lytocestidae) from common carp in the Amur River (in Russian), *Tr. Ukr. Resp. Nauch. Ovo. Parazitol.,* 4, 264—271, 1965.
1703. **Kulasiri, C.,** Some cestodes of the rat, *Rattus rattus* Linnaeus, of Ceylon and their epidemiological significance for man, *Parasitology,* 44, 349—352, 1954.
1704. **Kulikov, M. S. and Chernishova, P. S.,** Cycle of development of the anoplocephalids, *Moniezia expansa* (Rud., 1910), *Moniezia denticulata* (Rud., 1810) and *Anoplocephala magna* (Abildg., 1789) (in Ukrainian, Russian summary), *Nauk. Pr. Ukr. Inst. Eksper. Vet.,* 7, 24—29, 1937.
1705. **Kulmatycki, W. J.,** *Caryophyllaeus niloticus* sp. n., *Result Swedish Zool. Exped. Egypt. White Nile, 1901, Jägerskiöld, Ed., Uppsala,* Part. 5, 1—19, 1928.
1706. **Kunsemüller, F.,** Zur Kenntnis der polycephalen Blasenwürmer, insbesondere des *Coenurus cerebralis* Rudolphi und des *C. serialis* Gervais, *Zool. Jahrb. Abt. Anat.,* 18, 507—538, 1903.
1707. **Kuntz, R. E.,** Cysticercus of *Taenia taeniaeformis* with two strobilae, *J. Parasitol.,* 29, 424—425, 1943.
1708. **Kuperman, B. I.,** New species of the genus *Triaenophorus* Rud. (Cestoda, Pseudophyllidea) (in Russian), *Parazitologiya,* 2, 495—501, 1968.
1709. **Kuperman, B. I.,** Tapeworms of the genus *Triaenophorus* parasites of fishes (English translation by Amerind Publ. Co., New Delhi, 1981), Academy of Sciences of the USSR, Institute of Biology of Inland Waters, Leningrad, 1973, 222 pp.
1710. **Kurashvili, B. E.,** Helminth fauna of game birds of Georgia (in Russian), *Rabot. Gel'mintol. 75-Lett. Skrjabin,* pp. 340—346, 1954.
1711. **Kurimoto, T.,** Ueber eine neue Art *Bothriocephalus, Verhandl.,* 17, *Congr. Inn. Med. Wiesbaden,* pp. 452—456, 1899.
1712. **Kurimoto, T.,** *Diplogonoporus grandis* (E. Blanchard), Beschreibung einer zum ersten Male im menschlichen Darm gefundenen Art *Bothriocephalus, Z. Klin. Med.,* 40, 1—16, 1900.
1713. **Kurimoto, T.,** Ueber einen neuen Bandwurm *Diplogonoporus grandis* (in Japanese), *Mitt. Med. Ges. Tokyo,* pp. 1—10, 1901.
1714. **Kurochkin, Yu. B. and Slenkis, A. Ya.,** New representatives and the composition of the order Litobothridea Dailey, 1969 (Cestoidea) (in Russian), *Parazitologiya,* 7, 502—508, 1973.
1715. **Kuznetsov, M. I.,** *Moniezia (Blanchariezia) autumnalia* n.sp. — a new cestode of sheep and cattle (in Russian), *Parazitologiya,* 1, 431—434, 1967.
1716. **Lampio, T.,** On the occurrence of the bladder worm (*Cysticercus (pisiformis)* in Finland, *Rustatieteellisia Julkaisuja,* pp. 32—39, 1951.
1717. **Larsh, J. E., Jr.,** Life cycle of *Corallobothrium* sp. from *Ameiurus nebulosus, J. Parasitol.,* 25 (Suppl.), 19—20, 1939.
1718. **Larsh, J. E., Jr.,** *Corallobothrium parvum* n.sp., a cestode from the common bullhead, *Ameiurus nebulosus* Le Sueuar, *J. Parasitol.,* 27, 221—227, 1941.
1719. **Larsh, J. E., Jr.,** Increased infectivity of dwarf tapeworm (*Hymenolepis nana* var. *fraterna*) eggs following storage in host feces, *J. Parasitol.,* 29, 417—418, 1943.
1720. **Larsh, J. E., Jr.,** Comparing the percentage development of the dwarf tapeworm, *Hymenolepis nana* var. *fraterna,* obtained from mice of two different localities, *J. Parasitol.,* 29, 423—424, 1943.
1721. **Larsh, J. E., Jr.,** Comparative studies on a mouse strain of *Hymenolepis nana* var. *fraterna,* in different species and varieties of mice, *J. Parasitol.,* 30, 21—25, 1944.
1722. **Larsh, J. E., Jr.,** Studies on the artificial immunization of mice against infection with the dwarf tapeworm, *Hymenolepis nana* var. *fraterna, Am. J. Hyg.,* 39, 129—132, 1944.
1723. **Larsh, J. E., Jr.,** A comparative study of *Hymenolepis* in white mice and golden hamsters, *J. Parasitol.,* 32, 477—479, 1946.
1724. **La Rue, G. R.,** On the morphology and development of a new cestode of the genus *Proteocephalus* Weinland, *Trans. Am. Microsc. Soc.,* 28, 17—49, 1909.
1725. **La Rue, G. R.,** A revision of the cestode family Proteocephalidae, *Zool. Anz.,* 38, 473—482, 1911.
1726. **La Rue, G. R.,** A revision of the cestode family Proteocephalidae, *Ill. Biol. Monogr.,* 1, 1—350, 1914.
1727. **La Rue, G. R.,** A new cestode, *Ophiotaenia cyptobranchi* nov. spec. from *Cryptobranchus alleghaniensis* (Daudin), *16th Rep. Mich. Acad. Sci.,* pp. 11—17, 1914.
1728. **La Rue, G. R.,** A new species of tapeworm of the genus *Proteocephalus* from the perch and rock bass, *Mus. Zool. Univ. Mich. Occ. Pap.,* pp. 1—10, 1919.
1729. **Law, R. G. and Kennedy, A. H.,** *Echinococcus granulosus* in a moose, *North Am. Vet.,* 14, 33—34, 1933.
1730. **Lawler, H. J.,** A new cestode, *Cylindrotaenia quadrijugosa* n.sp. from *Rana pipiens,* with a key to Nematotaeniidae, *Trans. Am. Microsc. Soc.,* 58, 73—77, 1939.
1731. **Lawler, C. H. and Scott, W. B.,** Notes on the geographical distribution and the hosts of the cestode genus *Triaenophorus* in North America, *J. Fish. Res. Bd. Can.,* 11, 884—893, 1954.

1732. **Layman, E. M.**, Uber die parasitischen Würmer der Fische des Baikalsees (in Russian, German summary), *Tr. Baikal. Limnol. Stantsii*, 4, 5—98, 1933.
1733. **Leared, A.**, *Bothriocephalus latus*, *Trans. Pathol. Soc. London*, 25, 263—264, 1874.
1734. **Le Bas, C. Z. L.**, Experimental studies on *Dibothriocephlus latus* in man, *J. Helminthol.*, 2, 151—166, 1924.
1735. **Le Blond, C.**, Quelques observations d'helminthologie, *Ann. Sci. Nat. Zool.*, 6, 289—307, 1836.
1736. **Lee, O. P.**, Some helminths from Malayan wild birds with descriptions of two new species, *Bull. Natl. Mus. St. Singapore*, Part. 11, 77—81, 1966.
1737. **Leidy, J.**, Contributions to helminthology, *Proc. Acad. Nat. Sci. Philadelphia*, 5, 96—98, 239—244, 1850—1851.
1738. **Leidy, J.**, Notices on some tapeworms, *Proc. Acad. Nat. Sci. Philadelphia*, 7, 443—444, 1855.
1739. **Leidy, J.**, A synopsis of Entozoa and some of their congeners observed by the author, *Proc. Acad. Nat. Sci. Philadelphia*, 8, 42—58, 1856.
1740. **Leidy, J.**, Notices of some worms, *Dibothrium cordiceps*, *Proc. Acad. Nat. Sci. Philadelphia*, 23, 305—307, 1871.
1741. **Leidy, J.**, On Ligula in a fish of Susquehanna, *Proc. Acad. Nat. Sci. Philadelphia*, 24, 415—416, 1872.
1742. **Leidy, J.**, Notes on some parasitic worms, *Proc. Acad. Nat. Sci. Philadelphia*, 27, 14—16, 1875.
1743. **Leidy, J.**, On *Amia* and its probable *Taenia*, *Proc. Acad. Nat. Sci. Philadelphia*, 38, 62—63, 1886.
1744. **Leidy, J.**, *Bothriocephalus* in trout, *Proc. Acad. Nat. Sci. Philadelphia*, 38, 122—123, 1886.
1745. **Leidy, J.**, Notices on some parasitic worms, *Proc. Acad. Nat. Sci. Philadelphia*, 39, 20—24, 1887.
1746. **Leidy, J.**, Tapeworms in birds, *J. Comp. Med. Surg.*, 8, 1—11, 1887.
1747. **Leidy, J.**, Parasites of the pickerel, *Proc. Acad. Nat. Sci. Philadelphia*, 40, 169, 1888.
1748. **Leidy, J.**, Parasites of *Mola rotunda*, *Proc. Acad. Nat. Sci. Philadelphia*, 42, 281—282, 1890.
1749. **Leidy, J.**, Notices of Entozoa, *Proc. Acad. Nat. Sci. Philadelphia*, 42, 410—418, 1891.
1750. **Leidy, J.**, Researches in helminthology and parasitology, arranged and edited by J. Leidy, Jr., *Smithsonian Misc. Coll.*, 46, 1—281, 1904.
1751. **Leigh, W. H.**, Variations in a new cestode of the genus *Raillietina (Skrjabinia)* from the prairie chicken, *J. Parasitol.*, 25 (Suppl.), 10, 1939.
1752. **Leigh, W. H.**, Preliminary studies on the parasites of upland game birds and fur-bearing mammals, *Bull. Ill. Nat. Hist. Surv.*, 21, 185—194, 1940.
1753. **Leigh, W. H.**, Variations in a new species of cestode, *Raillietina (Skrjabinia) variabila* from the prairie chicken in Illinois, *J. Parasitol.*, 27, 97—106, 1941.
1754. **Leiper, R. T.**, Report on the helminth parasites of the okapi living in the Society's Gardens, *Proc. Zool. Soc. London*, pp.11—12, 1935.
1755. **Leiper, R. T.**, Crustacea as helminth intermediaries, *Proc. R. Soc. Med.*, 29, 1073-1074, 1936.
1756. **Leiper, R. T.**, Some experiments and observations on the longevity of *Diphyllobothrium* infections, *J. Helminthol.*, 14, 127—130, 1936.
1757. **Leiper, R. T. and Atkinson, E. L.**, Helminths of the British Antarctic Expedition, 1910—1913, *Proc. Zool. Soc. London*, pp. 222—226, 1914.
1758. **Leiper, R. T. and Atkinson, E. L.**, Parasitic worms with a note on a free-living nematode. Report of the British Antarctica "Terra Nova" Expedition, 1910—1913, *Nat. Hist. Rep. Zool.*, 2, 19-60, 1915.
1759. **Léon, L. A.**, Contribución al estudio de la parasitología sudamericana. El género *Raillietina* y su frecuencia en el Ecuador, *Rev. Med. Trop. Parasitol.*, 4, 219—230, 1938.
1760. **Léon, L. A.**, La raillietinosis es una endemia en el Ecuador; neuva aporte al conocimiento de la *Raillietina (Raillietina) equatoriensis*, *Bol. San. Ecuador*, 3, 41—49, 1947.
1761. **Léon, L. A.**, Neuvas consideraciones sobre la raillietinosis humana y nuevos aportes al conocimiento de la *Raillietina* (R.) *quitensis*, *Rev. Kuba Med. Trop. Parasitol.*, 5, 1—4, 1949.
1762. **Leon, N.**, Sur la fenestration du *Bothriocephalus latus*, *Zool. Anz.*, 32, 209—212, 1907.
1763. **Leon, N.**, *Diplogonoporus brauni*, *Zool. Anz.*, 32, 376—379, 1907.
1764. **Leon, N.**, Ein neuer menschlicher Cestode, *Zool. Anz.*, 33, 359—362, 1908.
1765. **Leon, N.**, Deux bothriocéphales monstreux, *Centralbl. Bakteriol. I*, 50, 616—619, 1909.
1766. **Leon, N.**, Ueber eine Missbildung von *Hymenolepis*, *Zool. Anz.*, 34, 609—612, 1909.
1767. **Leon, N.**, Un nouvel cas de *Diplogonoporus brauni*, *Centralbl. Bakteriol. I*, 55, 23—27, 1910.
1768. **Leon, N.**, *Dibothriocephalus taenioides*, *Centralbl. Bakteriol. I*, 78, 503—504, 1916.
1769. **Leon, N.**, Notes sur quelques vers parasites de Roumanie, *Ann. Sci. Univ. Jassy*, 10, 308—313, 1920.
1770. **Leon, N.**, Accouplement et fécondation du *Dibothriocephalus latus*, *Ann. Parasitol.*, 3, 263—266, 1925.
1771. **Leon, N.**, Sur la bifurcation du *Dibothriocephalus latus*, *Ann. Parasitol.*, 4, 236—240, 1926.
1772. **Leonardi, C.**, Un caso di *Taenia mediocanellata* nell' *Himantopus candidus*, *Avicula*, 2, 59, 1898.
1773. **Leonov, V. A. and Belogurov, O. I.**, Cestodes of the superfamily Dilepidoidea Matevosyan, 1962 in marine and fish-eating birds on Wrangel Island (in Russian), *Uch. Zap. Dal'nevost. Gos. Univ.*, 16, 26—36, 1970.

1774. **Le Roux, P.,** Helminths collected from the domestic fowl (*Gallus domesticus*) and the domestic pigeon (*Columba livia*) in Natal, 11th and 12th Rep. Dir. Vet. Educ. Res. Union South Afr. Pretoria, 1927, 209—217.
1775. **Lespés, P. G. C.,** Notes sur une nouvelle espèce du genre *Echinobothrium*, *Ann. Sci. Nat. Zool.*, (Ser. 4), 7, 118—119, 1857.
1776. **Lethbridge, R. C.,** The biology of the oncosphere of cyclophyllidean cestodes, *Helminthol. Abstr. Ser. A, Anim. Hum. Helminthol.*, 49, 59—72, 1980.
1777. **Leuckart, F. S.,** Das Genus *Bothriocephalus* Rud., *Zoologische Bruchstücke I.*, Helmstadt, pp. viii + 70, 1820.
1778. **Leuckart, K. G. F. R.,** Beschreibung zweier neuen Helminthen, *Arch. Naturg.*, XIV Jahrg., 1, 26—29, 1848.
1779. **Leuckart, K. G. F. R.,** Helminthologische Notizen, *Arch. Naturg.*, XVI Jahrg., 1, 9—16, 1850.
1780. **Leuckart, K. G. F. R.,** Parasitismus und Parasiten, *Arch. Physiol. Heilk. Stuttgart* II, pp. 199—259, 1852.
1781. **Leuckart, K. G. F. R.,** Erziehung des *Cysticercus fasciolaris* aus den Eiern von *Taenia crassicollis*, *Z. Wiss. Zool.*, 6, 139, 1855.
1782. **Leuckart, K. G. F. R.,** Die Blasenwürmer und ihre Entwicklung. Zugleich ein Beitrag zur Kenntnis der Cysticercus-Leber, 1856, 162 pp.
1783. **Leuckart, K. G. F. R.,** Die menschlichen Parasiten, *Arch. Sci. Phys. Nat. n. Ser.*, 16, 243—245, 1863.
1784. **Leuckart, K. G. F. R.,** Bericht ueber die wissenschaftlichen Leistungen in der Naturgeschichte der niederen Thiere während der Jahre 1864 und 1865 (Erste Hälfte), *Arch. Naturg., 31 Jahrg.*, Bd. 2, 165—268, 1865.
1785. **Leuckart, K. G. F. R.,** Bericht ueber die wissenschaftlichen Leistungen in der Naturgeschichte der niederen Thiere für 1868 and 1869 (Erste Hälfte), *Arch. Naturg., 35 Jahrg.*, Bd. 2, 207—244, 1869.
1786. **Leuckart, K. G. F. R.,** Archigetes Sieboldi, eine geschlechtsreife Cestodenamme, *Z. Wiss. Zool.*, 30 (Suppl.), 593—606, 1878.
1787. **Leuckart, K. G. F. R.,** Zur Bothriocephalusfrage, *Centralbl. Bakteriol. I Abt.*, 1—6; 33—40, 1887.
1788. **Leuckart, K. G. F. R.,** Ueber *Taenia madagascariensis* Dav., *Verh. Dtsch. Zool. Ges.*, pp. 68—71, 1891.
1789. **Leuckart, K. G. F. R. and Pagenstecher, A.,** Untersuchungen ueber miedere See-Thiere: *Echinobothrium Typus*, *Arch. Anat. Physiol. Wiss. Med.*, 25, 558—613, 1858.
1790. **Levander, K. M.,** Om larver af *Dibothriocephalus latus* L. hos insjölax (Ueber Larven von *Dibothriocephalus latus* L. bei *Salmo lacustris*), *Medd. Soc. Fauna Flora Fenn.*, 32, 93, 1906.
1791. **Le-Van-Hoa,** Cystique d'un nouveaux *Taenia* du singe *Macacus cynamolgus* (L.) du Viet-Nam, *Bull. Soc. Pathol. Exot.*, 57, 23—27, 1964.
1792. **Levine, P. P.,** Observation on the biology of the poultry cestode, *Davainea proglottina*, in the intestine of the host, *J. Parasitol.*, 24, 423—431, 1938.
1793. **Lewis, E. A.,** Helminths of wild birds found in the Aberystwyth area, *J. Helminthol.*, 4, 7—12, 1926.
1794. **Lewis, E. A.,** Helminths collected from horses in the Aberystwyth area, *J. Helminthol.*, 4, 179—182, 1926.
1795. **Lewis, E. A.,** A study of the helminths of dogs and cats of Aberystwyth, Wales, *J. Helminthol.*, 5, 175—182, 1927.
1796. **Lewis, R. C.,** On two new species of tapeworms from the stomach and small intestine of a wallaby, *Lagorchestes conspicillatus*, from Hermite Island, Monte Bello Islands, *Proc. Zool. Soc. London*, pp. 419—433, 1914.
1797. **Leydig, F.,** Ein neuer Bandwurm aus *Polypterus bichir*, *Arch. Naturg.*, 19, 219—222, 1853.
1798. **Li, H. C.,** The life-histories of *Diphyllobothrium decipiens* and *D. erinacei*, *Am. J. Hyg.*, 10, 527—550, 1929.
1799. **Limbourg, J. P. de,** Observationes de ascaridibus et cucurbitinis, et potissimum de Taenia, tam humana quam leporina (in Latin) *Philos. Trans.*, 56, 126—132, 1767.
1800. **Lin, Y.,** Three new species of *Unciunia* from birds in Fujian Province, with notes on the genus *Unciunia* Skrjabin (Cestoda) (in Chinese), *Acta Zool. Sinica*, 22, 89—100, 1976.
1801. **Lincicome, D. R.,** A new tapeworm, *Choanotaenia iola*, from the robin, *J. Parasitol.*, 25, 203—206, 1939.
1802. **Lindner, E.,** Die Bedeutung des Cysticercus-Schwanzes, *Biol. Z.*, 41, 36—41, 1921.
1803. **Linstow, O. F. B., von,** Ueber den *Cysticercus taeniae gracilis*, eine freie Cestodenamme des Barsches, *Arch. Mikr. Anat. Entwicklungsmech*, 8, 535—537, 1872.
1804. **Linstow, O. F. B., von,** Sechs neue Taenien, *Arch. Naturg.*, 38, 55—58, 1872.
1805. **Linstow, O. F. B., von,** Beobachtungen an neuen und bekannten Helminthen, *Arch. Naturg.*, 4, 183—207, 1875.
1806. **Linstow, O. F. B., von,** Helminthologische Beobachtungen, *Arch. Naturg.*, 42, 1—18, 1876.
1807. **Linstow, O. F. B., von,** Helminthologica, *Arch. Naturg.*, 43, 1—18, 1877.
1808. **Linstow, O. F. B., von,** Enthelminthologica, *Arch. Naturg.*, 43, 173—197, 1877.

1809. **Linstow, O. F. B., von,** Neue Beobachtungen an Helminthen, *Arch. Naturg.,* 44, 218—245, 1878.
1810. **Linstow, O. F. B., von,** Helminthologische Untersuchungen, *Jahresber. Ver. Vaterl. Naturkd. Württemberg* (Stuttgart), 35, 313—342, 1879.
1811. **Linstow, O. F. B., von,** Helminthologische Studien, *Arch. Naturg.,* 45, 165—188, 1879.
1812. **Linstow, O. F. B., von,** Helminthologische Untersuchungen, *Arch. Naturg.,* 46, 41—54, 1880.
1813. **Linstow, O. F. B., von,** Helminthologische Studien, *Arch. Naturg.,* 48, 1—25, 1882.
1814. **Linstow, O. F. B., von,** Helminthologisches, *Arch. Naturg.,* 50, 125—145, 1884.
1815. **Linstow, O. F. B., von,** Report on the Entozoa, Reports of the Scientific Results of the Challenger Expedition, *Zoology, Edinburgh,* 23, 1—18, 1888.
1816. **Linstow, O. F. B., von,** Helminthologisches, *Arch. Naturg.,* 55, 235—246, 1889.
1817. **Linstow, O. F. B., von,** Beitrag zur Kenntnis der Vogeltaenien, *Arch. Naturg.,* 56, 171—188, 1890.
1818. **Linstow, O. F. B., von,** Ueber den Bau und die Entwicklung von *Taenia longicollis* Rud. Ein Beitrag zur Kenntnis der Fischtänien, *Jena. Z. Naturwiss.,* 25, 565—576, 1891.
1819. **Linstow, O. F. B., von,** Beobachtungen an Vogeltänien, *Centralbl. Bakteriol.,* 12, 501—504, 1892.
1820. **Linstow, O. F. B., von,** Beobachtungen an Helminthenlarven, *Arch. Mikr. Anat.,* 39, 325—343, 1892.
1821. **Linstow, O. F. B., von,** Helminthen von Süd-Georgien nach der Ausbeute der deutschen Station von 1882—1883, *Jahrb. Hamburg. Wissenschaft. Anstalten,* 9, 59—77, 1892.
1822. **Linstow, O. F. B., von,** Zur Anatomie und Entwicklungsgeschichte der Tänien, *Arch. Mikr. Anat.,* 42, 442—459, 1893.
1823. **Linstow, O. F. B., von,** Helminthologische Studien, *Jena. Z. Naturwiss.,* 28, 328—342, 1893.
1824. **Linstow, O. F. B., von,** Ueber *Taenia (Hymenolepis) nana* v. Siebold und *murina* Dujardin, *Jena. Z. Naturwiss.,* 30, 571—582, 1895.
1825. **Linstow, O. F. B., von,** Helminthologische Mitteilungen, *Arch. Naturg.,* 48, 328—342, 1896.
1826. **Linstow, O. F. B., von,** *Tetrabothrium cylindraceum* Rud. und das Genus *Tetrabothrium, Centralbl. Bakteriol.,* 27, 362—366, 1900.
1827. **Linstow, O. F. B., von,** *Taenia africana* n.sp. eine neue Tänie des Menschen aus Afrika, *Centralbl. Bakteriol.,* 28, 485—490, 1900.
1828. **Linstow, O. F. B., von,** *Taenia horrida, Tetrabothrium macrocephalum* und *Heterakis distans, Arch. Naturg.,* 67, 1—9, 1901.
1829. **Linstow, O. F. B., von,** *Taenia asiatica,* eineneue Tänie aus Menschen, *Centralbl. Bakteriol.,* 29, 982—985, 1901.
1830. **Linstow, O. F. B., von,** Die systematische Stellung von *Ligula intestinalis* Goeze, *Zool. Anz.,* 24, 627—634, 1901.
1831. **Linstow, O. F. B., von,** Beobachtungen an Helminthen des Senckenbergischen Naturhistorischen Museums des Breslauer Zoologischen Instituts und andern, *Arch. Mikr. Anat.,* 58, 182—198, 1901.
1832. **Linstow, O. F. B., von,** Helminthen von den Ufern des Nyassa-Sees. Beitrag zur Helminthenfauna von Süd-Afrika, *Jena. Z. Naturwiss.,* 35, 409-428, 1901.
1833. **Linstow, O. F. B., von,** Entozoa des zoologischen Museums der Kaiserlichen Akademie der Wissenschaft zu St. Petersburg, *Bull. Acad. Imp. Sci. St. Petersburg,* 15, 271—292, 1901.
1834. **Linstow, O. F. B., von,** On *Tetrabothrium torulosum* und *Tetrabothrium auriculatum, Proc. R. Soc. Edinburgh,* 23, 158—160, 1902.
1835. **Linstow, O. F. B., von,** *Taenia trichoglossi, Centralbl. Bakteriol.,* 31, 32, 1902.
1836. **Linstow, O. F. B.,** Zwei neue Parasiten des Menschen, *Centralbl. Bakteriol,* 31, 768—771, 1902.
1837. **Linstow, O. F. B., von,** Eine neue Cysticercus-Form, *Cysticercus Brauni* Setti, *Centralbl. Bakteriol.,* 32, 882—886, 1902.
1838. **Linstow, O. F. B., von,** *Echinococcus alveolaris* und *Plerocercus lachesis, Zool. Anz.,* 26, 162—167, 1902.
1839. **Linstow, O. F. B., von,** Entozoa des zoologischen Museums der kaiserlichen Akademie der Wissenschaft zu St. Petersburg. II, *Ann. Mus. Zool. Acad. Imp. Sci. St. Petersburg,* 8, 265—294, 1903.
1840. **Linstow, O. F. B., von,** Neue Helminthen, *Centralbl. Bakteriol.,* Abt. I, 35, 352—357, 1903.
1841. **Linstow, O. F. B., von,** Drei neue Tänien aus Ceylon, *Centralbl. Bakteriol.,* 33, 532—535, 1903.
1842. **Linstow, O. F. B., von,** Helminthologische Beobachtungen, *Centralbl. Bakteriol. Parasitenk.,* 34, 526—531, 1903.
1843. **Linstow, O. F. B., von,** Parasiten, meistens Helminthen, aus Siam, *Arch. Mikr. Anat.,* 62, 108—121, 1903.
1844. **Linstow, O. F. B., von,** Ueber zwei neue Entozoa aus Acipenseriden, *Ann. Mus. Zool. Acad. Imp. Sci. St. Petersburg,* 9, 17—19, 1904.
1845. **Linstow, O. F. B., von,** Beobachtungen an Nematoden und Cestoden, *Arch. Naturg.,* 70, 299—309, 1904.
1846. **Linstow, O. F. B., von,** Neue Helminthen aus West-Afrika, *Centralbl. Bakteriol.,* 36, 379—383, 1904.
1847. **Linstow, O. F. B., von,** Neue Beobachtungen an Helminthen, *Arch. Mikr. Anat.,* 69, 484—497, 1904.
1848. **Linstow, O. F. B., von,** Neue Helminthen, *Centralbl. Bakteriol.,* 37, 678—683, 1904.
1849. **Linstow, O. F. B., von,** Helminthologische Beobachtungen, *Arch. Mikr. Anat.,* 66, 355—366, 1905.

1850. **Linstow, O. F. B., von,** Helminthen der russischen polar-Expedition 1900—1903, *Mém. Acad. Imp. Sci. St. Petersburg,* 18, 1—17, 1905.
1851. **Linstow, O. F. B., von,** Neue Helminthen, *Arch. Naturg.,* 71, 267—276, 1905.
1852. **Linstow, O. F. B., von,** Helminthen aus Ceylon und aus Arktischen Breiten, *Z. Wiss. Zool.,* 82, 182—193, 1905.
1853. **Linstow, O. F. B., von,** Helminths from the collection of the Colombo Museum, *Ceylon J. Sci.,* 3, 163—188, 1906.
1854. **Linstow, O. F. B., von,** Neue und bekannte Helminthen, *Zool. Jahrb.,* 24, 1—20, 1906.
1855. **Linstow, O. F. B., von,** Neue Helminthen, *Centralbl. Bakteriol.,* 41, 15—17, 1906.
1856. **Linstow, O. F. B., von,** Helminthen von Herrn Eduard Jacobson in Java gesammelt, *Notes Mus. Leiden,* 29, 81—87, 1907.
1857. **Linstow, O. F. B., von,** *Hymenolepis furcifera* und *Tatria biremis,* zwei Tänien aus *Podiceps nigricollis, Centralbl. Bakteriol.,* 56, 38—40, 1908.
1858. **Linstow, O. F. B., von,** Recent additions to the collection of Entozoa in the Indian Museum, *Rec. Ind. Mus.,* 2, 108—109, 1908.
1859. **Linstow, O. F. B., von,** Neue Helminthen aus Deutsch-Südwest-Afrika, *Centralbl. Bakteriol.,* 50, 448—451, 1909.
1860. **Linstow, O. F. B., von,** *Davainea provincialis, Centralbl. Bakteriol.,* 52, 75—77, 1909.
1861. **Linton, E.,** Notes on two forms of cestoid embryos, *Am. Nat.,* 21, 195—210, 1887.
1862. **Linton, E.,** Notes on cestoid Endozoa of marine fishes, *Am. J. Sci. Arts,* 37, 239—240, 1889.
1863. **Linton, E.,** Notes on Entozoa of marine fishes of New England, Annu. Rep. U.S. Comm. Fish Fish. 1886, Vol. 14, Washington, D.C., 1889, 453—511.
1864. **Linton, E.,** Notes on Entozoa of marine fishes of New England. II, Annu. Rep. U.S. Comm. Fish Fish. 1887, Vol. 15, Washington, D.C., 1890, 719—899.
1865. **Linton, E.,** A contribution to the life history of *Dibothrium cordiceps* Leidy, a parasite infesting the trout of Yellowstone Lake, *Bull. U.S. Fish Comm.* (1889), 9, 337—358, 1891.
1866. **Linton, E.,** Notes on Entozoa of marine fishes with descriptions of new species. III, Annu. Rep. U.S. Comm. Fish Fisheries 1888, Vol. 16, Washington, D.C., 1891, 523—542.
1867. **Linton, E.,** On the anatomy of *Thysanocephalum crispum* Linton, a parasite of the tiger shark, Annu. Rep. U.S. Comm. Fish Fish. 1888, Vol. 16, Washington, D.C., 1891, 543—546.
1868. **Linton, E.,** On two species of larval *Dibothria* from the Yellowstone National Park, *Bull. U.S. Comm. Fish Fish.,* 9, 65—79, 1891.
1869. **Linton, E.,** Notes on avian Entozoa, *Proc. U.S. Natl. Mus.,* 15, 87—113, 1892.
1870. **Linton, E.,** On fish Entozoa, Annu. Rep. U.S. Comm. Fish Fish. 1889—1891, Washington, D.C., 1893, 545—564.
1871. **Linton, E.,** Notes on larval cestode parasites, *Proc. U.S. Natl. Mus.,* 19, 787—824, 1897.
1872. **Linton, E.,** Notes on cestode parasites of fishes, *Proc. U.S. Natl. Mus.,* 29, 423—456, 1897.
1873. **Linton, E.,** Fish parasites collected at Woods Hole in 1898, *Bull. U.S. Comm. Fish Fish. 1899,* 19, 267—304, 1900.
1874. **Linton, E.,** Parasites of fishes of the Woods Hole region, *Bull. U.S. Comm. Fish Fish. 1899,* 19, 405—492, 1901.
1875. **Linton, E.,** Parasites of fishes of Beaufort, North Carolina, *Bull. U.S. Bur. Fish 1904,* 24, 321—428, 1905.
1876. **Linton, E.,** Notes on cestode cysts, *Taenia chamissonii,* new species, from a porpoise, *Proc. U.S. Natl. Mus.,* 28, 819—822, 1905.
1877. **Linton, E.,** Notes on *Calyptobothrium,* a cestode genus found in the torpedo, *Proc. U.S. Natl. Mus.,* 32, 275—284, 1907.
1878. **Linton, E.,** Notes on parasites of Bermuda fishes, *Proc. U.S. Natl. Mus.,* 33, 85—126, 1907.
1879. **Linton, E.,** A cestode parasite in the flesh of the butterfish, *Bull. U.S. Bur. Fish. 1906,* 26, 111—132, 1907.
1880. **Linton, E.,** Preliminary report on animal parasites of Tortugas, 5th Yearbook of Carnegie Institution for 1906, Washington, D.C., 1907, 112—117.
1881. **Linton, E.,** Preliminary report on animal parasites of Tortugas, 6th Yearbook of Carnegie Institution for 1907, Washington, D.C., 1908, 114—116.
1882. **Linton, E.,** Helminth fauna of the Dry Tortugas. I. Cestoda, Carnegie Institution, Publ. 102, Washington, D.C., 1909, 157—190.
1883. **Linton, E.,** Cestodes in flesh of marine fishes, *Science,* 29, 715, 1909.
1884. **Linton, E.,** Notes on the flesh parasites of marine food fishes, *Bull. U.S. Bur. Fish. 1908,* 28, 1195—1209, 1910.
1885. **Linton, E.,** Notes on the distribution of Entozoa of North American marine fishes, *Proc. 7th Int. Zool. Congr. Boston, 1907,* pp. 686—696, 1911.

1886. **Linton, E.**, Cestoda. In Summer, Osburn and Cole. Catalogue of marine fauna of Woods Hole and vicinity, *Bull. U.S. Bur. Fish. 1911,* p. 31, 1913.
1887. **Linton, E.**, Cestode cysts in the flesh of marine fishes, *Trans. Am. Fish. Soc. 42nd Annu. Meet.,* pp. 119—127, 1913.
1888. **Linton, E.**, On the seasonal distribution of fish parasites, *Trans. Am. Fish. Soc.,* 44, 48—56, 1914.
1889. **Linton, E.**, Cestode cysts from the muskrat, *J. Parasitol.,* 2, 46—47, 1915.
1890. **Linton, E.**, Notes on two cestodes from the spotted sting ray, *J. Parasitol.,* 3, 34—38, 1916.
1891. **Linton, E.**, *Rhynchobothrium ingens* sp. nov., a parasite of the dusky shark, *Carcharinus obscurus, J. Parasitol.,* 8, 23—32, 1921.
1892. **Linton, E.**, A contribution to the anatomy of *Dinobothrium,* a genus of selachian tapeworms, *Proc. U.S. Natl. Mus.,* 60, 1—13, 1922.
1893. **Linton, E.**, A new cestode from *Liparis liparis, Trans. Am. Microsc. Soc.,* 41, 118—121, 1922.
1894. **Linton, E.**, A new cestode from the maneater and mackerel sharks, *Proc. U.S. Natl. Mus.,* 61, 1—16, 1922.
1895. **Linton, E.**, A new cetacean cestode (*Prosthecocotyle monticellii* sp.n.) with a note on the genus *Tetrabothrius* Rud., *J. Parasitol.,* 10, 51—55, 1923.
1896. **Linton, E.**, A new *Gyrocotyle* from South Africa, *Anat. Rec.,* 26, 355-356, 1923.
1897. **Linton, E.**, Notes on parasites of sharks and rays, *Proc. U.S. Natl. Mus.,* 64, 1—114, 1924.
1898. **Linton, E.**, *Gyrocotyle plana* sp.nov., with notes on South African cestodes of fishes, *Fish. Mar. Biol. Surv. S. Afr.,* Rep. 3, 1—27, 1924.
1899. **Linton, E.**, A new diecian cestode, *J. Parasitol.,* 11, 163—169, 1925.
1900. **Linton, E.**, Notes on cestode parasites of birds, *Proc. U.S. Natl. Mus.,* 70, 1—73, 1927.
1901. **Linton, E.**, Larval cestodes (*Tetrarhynchus elongatus* Rudolphi) from the liver of the pelagic sunfish (*Mola mola*) collected at Woods Hole, Massachusetts, *Trans. Am. Microsc. Soc.,* 47, 464—467, 1928.
1902. **Linton, E.**, A pseudophyllidean cestode from a flying fish, *Trans. Am. Microsc. Soc.,* 53, 66, 1934.
1903. **Linton, E.**, Cestode parasites of teleost fishes of the Woods Hole region, Massachusetts, *Proc. U.S. Natl. Mus.,* 90, 417—442, 1941.
1904. **Little, J. W.**, *Monordotaenia* nom. nov. for the badger taeniid cestodes with one row of hooks, *Proc. Helminthol. Soc. Wash.,* 34, 67—68, 1967.
1905. **Llewellyn, J.**, The evolution of the parasitic platyhelminths, *Evolution of Parasites, 3rd Symp. Br. Soc. Parasitol.,* Vol. 6., Taylor, A. E. R., Ed., Academic Press, New York, 1965, 373—383.
1906. **Locker, B. and Rausch, R.**, Some cestodes from Oregon shrews, with descriptions of four new species of *Hymenolepis* Weinland, 1858, *J. Wash. Acad. Sci.,* 42, 26—31, 1952.
1907. **Loennberg, E.**, Bidrag till kannedomen om i Sverige förekommande Cestoder (in Danish), *Bihang. K. Sven. Vet. Akad. Handl.,* 14, 1—69, 1889.
1908. **Loennberg, E.**, Ueber eine eigentümliche Tetrarhynchidenlarve, *Bihang. K. Sven. Vetenskaps. Akad. Handl.,* 15, 1—48, 1889.
1909. **Loennberg, E.**, Ueber *Amphiptyches* Wagener und *Gyrocotyle urna* (Grube und Wagener) Diesing (eine vorläufige Mitteilung), *Biol. Fören. Förhandl. Stockholm,* 2, 55—61, 1890.
1910. **Loennberg, E.**, Helminthologische Beobachtungen von der Westküste Norvegens, *Bihang. K. Sven. Vetenskaps. Akad. Handl.,* 16, 1—47, 1890.
1911. **Loennberg, E.**, Bemerkungen zum "Elenco degli elminti studiati a Wimereux nella primavera del 1889 dal Dott. F. S. Monticelli," *Biol. Fören. Förhandl. Stockholm,* 3, 4—9, 1890.
1912. **Loennberg, E.**, Mitteilungen ueber einige Helminthen aus dem zoologischen Museum der Universitäut zu Kristiania, *Biol. Fören. Förhandl. Stockholm,* 3, 64—78, 1891.
1913. **Loennberg, E.**, Anatomische Studien ueber Skandinavische Cestoden. I, *Sven. Vetensk. Akad. Handl.,* 24, 1—109, 1891.
1914. **Loennberg, E.**, Anatomische Studien ueber Skandinavische Cestoden. II. Zwei Parasiten aus Walfischen und zwei aus *Lamna cornubica, Sven. Vetensk. Akad. Handl.,* 24, 1—28, 1892.
1915. **Loennberg, E.**, Einige Experimente Cestoden künstlich lebend zu erhalten, *Centralbl. Bakteriol.,* 11, 89—92, 1892.
1916. **Loennberg, E.**, Ueber das Vorkommen des breiten Bandwurms in Schweden, *Centralbl. Bakteriol.,* 11, 189—192, 1892.
1917. **Loennberg, E.**, Bemerkungen ueber einige Cestoden, *Sven. Vetensk. Akad. Handl.,* 18, 1—17, 1893.
1918. **Loennberg, E.**, Ueber eine neue *Tetrabothrium* species und die Verwandtschaftsverhältnisse der Ichthyotaenien, *Centralbl. Bakteriol.,* 15, 801—803, 1894.
1919. **Loennberg, E.**, Beiträge zur Phylogenie der parasitischen Plathelminthen, *Centralbl. Bakteriol.,* 21, 674—684, 1897.
1920. **Loennberg, E.**, Ein neuer Bandworm (*Monorygma chlamydoselachi*) aus *Chlamydoselachus anguineus* Garman, *Arch. Math. Naturv. Kristiania,* 20, 1—11, 1898.
1921. **Loennberg, E.**, Ueber einige Cestoden aus dem Museum zu Bergen, *Bergens Mus. Arb.,* pp. 1—23, 1899.
1922. **Loewen, S. L.**, A new cestode, *Taenia rileyi* n.sp., from a lynx, *Parasitology,* 21, 469—471, 1929.

1923. **Loewen, S. L.,** A new cestode from a bat, *Trans. Kan. Acad. Sci.,* 37, 257-261, 1934.
1924. **Loewen, S. L.,** On some reptilian cestodes of the genus *Oochoristica* (Anoplocephalidae), *Trans. Am. Microsc. Soc.,* 59, 511—518, 1940.
1925. **Loewen, S. L.,** A new host record for the cestode *Bothridium pithonis* de Blainville 1828, *Trans. Kan. Acad. Sci.,* 48, 197—108, 1945.
1926. **Logachev, E. D.,** On the structure and development of calcareous bodies in tapeworm (in Russian), *Dokl. Akad. Nauk SSSR,* 80, 693—695, 1951.
1927. **Logachev, E. D.,** On the nature of the tissue and the physiological importance of the subcuticular cells in tapeworms (in Russian), *Dokl. Akad. Nauk SSSR,* an. 19, n.s., 77, 161—163, 1951.
1928. **Logachev, E. D.,** Investigations on the basic argyrophil substance of tapeworms (in Russian), *Dokl. Akad. Nauk SSSR,* 80, 289—292, 1951.
1929. **Logachev, E. D.,** Development of the vitellarium and formation of yolk sacs in tapeworms (in Russian), *Dokl. Akad. Nauk SSSR,* n.s., 85, 1197—1199, 1952.
1930. **Logachev, E. D.,** Formation and development of calcareous bodies in tapeworms, *Helminthol. Abstr.,* 20, 159, 1952.
1931. **Logachev, E. D.,** The tissue character and physiological importance of the subcuticular cells in tapeworms, *Helminthol. Abstr.,* 20, 90, 1952.
1932. **Logachev, E. D.,** On the development of the egg cell and the importance of the nuclei in the cestode *Raillietina urogalli* Modeer (in Russian), Dokl. Akad. Nauk SSSR, n.s., 88, 181—184, 1953.
1933. **Logachev, E. D.,** On the development of the egg cell and the importance of the nuclei in the cestode *Raillietina urogalli* Modeer, *Helminthol. Abstr.,* 22, 13, 1953.
1934. **Logachev, E. D.,** Structure and histogenesis of the vitelline gland of *Diphyllobothrium latum* (in Russian), *Dokl. Akad. Nauk SSSR,* n.s., 99, 181—184, 1954.
1935. **Long, L. H. and Wiggins, N. E.,** A new species of *Diorchis* (Cestoda) from the canvasback, *J. Parasitol.,* 25, 483—486, 1939.
1936. **Loos-Frank, B.,** *Mesocestoides leptothylacus* n.sp. und das nomenklatorische Problem in der Gattung Mesocestoides Vaillant, 1863 (Cestoda, Mesocestoididae), *Tropenmed. Parasitol.,* 31, 2—14, 1980.
1937. **López-Neyra, C. R.,** Notas helmintologicas, *Bol. Real. Soc. Espan. Hist. Nat. Madrid,* 16, 457—462, 1916.
1938. **López-Neyra, C. R.,** Notas helmintologicas (2. ser.), *Bol. Real. Soc. Espan. Hist. Nat. Madrid,* 18, 145—155, 1918.
1939. **López-Neyra, C. R.,** Notas helmintologicas (3. ser), *Bol. Real. Soc. Espan. Hist. Nat. Madrid,* 20, 75—90, 1920.
1940. **López-Neyra, C. R.,** Estudio critico de las Davaineas parásitas de las gallinas en la región Granadina, *Rev. Real. Acad. Cien. Exact. Fis. Nat. Madrid,* 18, 323—349, 1930.
1941. **López-Neyra, C. R.,** Apuntes para un compendio de helmintologica, Iberica, *Asoc. Exp. Progreso Sci.,* pp. 93—111, 1923.
1942. **López-Neyra, C. R.,** Considérations sur le genre *Dipylidium* Leuckart, *Bull. Soc. Pathol. Exot.,* 20, 434—440, 1927.
1943. **López-Neyra, C. R.,** Sur les cysticercoides de quelques *Dipylidium, Ann. Parasitol.,* 5, 245—248, 1927.
1944. **López-Neyra, C. R.,** Sobre le evolución de la *Joyeuxia chyzeri* v. Ratz, *Bol. Real. Soc. Espan. Hist. Nat. Madrid,* 27, 398—399, 1927.
1945. **López-Neyra, C. R.,** Recherches sur le genre *Dipylidium* avec description de quatre espèces nouvelles, *Bull. Soc. Pathol. Exot.,* 21, 239—253, 1928.
1946. **López-Neyra, C. R.,** Revisión del género *Dipylidium* Leuckart, *Mem. Real. Acad. Cien. Exact. Fis. Nat. Madrid,* 32, 1—112, 1929.
1947. **López-Neyra, C. R.,** Consideraciones sobre el género *Davainea* y description de dos especies neuvas, *Bol. Real. Soc. Espan. Hist. Nat. Madrid,* 29, 345—359, 1929.
1948. **López-Neyra, C. R.,** Davaineidos parasitos humanos y sus relaciones con los de los maníferos, *Bol. Univ. Granada,* pp. 1—28, 1930.
1949. **López-Neyra, C. R.,** Relations du *Davainea madagascariensis* et des espéces parasites des mammiféres. Considérations sur les *Davainea, Ann. Parasitol.,* 9, 162—184, 1931.
1950. **López-Neyra, C. R.,** Revision del género *Davainea* Leuckart, *Mem. Real. Acad. Cien. Exact. Fis. Nat. Madrid,* 1, 1—177, 1931.
1951. **López-Neyra, C. R.,** La *Davainea formosana* y sus relaciones con los davaineidos de los roedores, *Arch. Ital. Zool. Napoli Torino,* 15, 465—472, 1931.
1952. **López-Neyra, C. R.,** Estudios sobre el proceso de fimbriarizacion. Los géneros *Fimbriaria* e *Hymenofimbria* como deformidades de *Hymenolepis* y *Diorchis, Medicina de los Páises Calidos,* 4, 494—501, 1931.
1953. **López-Neyra, C. R.,** Sobre la clasificación que proqusimos del género, *Davainea, Med. Páises Calidos,* 4, 494—501, 1931.
1954. **López-Neyra, C. R.,** La *Fimbriaria fasciolaris* y sus relaciones con el *Diorchis acuminata, Bol. Univ. Granada,* pp. 131—156, 1931.

1955. **López-Neyra, C. R.**, Sur la classification du genre *Davainea* (s. l), *Bull. Soc. Zool. Fr.*, 56, 534—541, 1932.
1956. **López-Neyra, C. R.**, *Hymenolepis pittalugai* n.s. et ses rapports avec les especes similaires *(H. macracanthos), Ann. Parasitol.*, 19, 248—256, 1932.
1957. **López-Neyra, C. R.**, Sobre la clasificación que propusimos de *Davainea* s. l. Respuesta a Fuhrmann, *Bol. Univ. Granada,* 6, 47—55, 1934.
1958. **López-Neyra, C. R.**, Sobre algunos géneros de Dilepididae, *Bol. Acad. Cien. Exact. Nat. Madrid,* 1, 9, 1935.
1959. **López-Neyra, C. R.**, Sobre una tenia críteria del alcaraván, *Bol. Soc. Espan. Hist. Nat. Madrid,* 35, 203—216, 1935.
1960. **López-Neyra, C. R.**, *Fernandezzia* [sic] *goizuetai* nov. gen. nov. sp., parásito intestinal del zorzal y revisión de los "Ophryocotyliinae," *Rev. Acad. Cien. Exact. Fis. Nat. Madrid,* 33, 5—18, 1936.
1961. **López-Neyra, C. R.**, Especies neuvas o insuficientemente conocidas correspondientes al género *Hymenolepis* Weinland (s. l.), *Rev. Iber. Parasitol.*, 1, 133—170, 1941.
1962. **López-Neyra, C. R.**, Division del género *Hymenolepis* Weinland (s. l.) en otros mas naturales, *Rev. Ibér. Parasitol.*, 2, 46—93, 1942; *Rev. Ibér. Parasitol.*, 2, 113—256, 1942.
1963. **López-Neyra, C. R.**, Raillietinosis humanas en la zona tropical, *Med. Colonial Madrid,* 1, 215—242, 1943.
1964. **López-Neyra, C. R.**, Consideraciones sobre el género *Liga* Weinland, 1857, *Rev. Ibér. Parasitol.*, 3, 61—68, 1943.
1965. **López-Neyra, C. R.**, Las raillietinas parasitas humanas, *Rev. Ibér. Parasitol.*, 3, 141—168, 1943; *Rev. Ibér. Parasitol.*, 257—258, 1943.
1966. **López-Neyra, C. R.**, *Paradicranotaenia anormalis* n.g. n.sp. y consideraciones sobre los Nematoparataeniidae, *Rev. Ibér. Parasitol.*, 3, 229—254, 1943.
1967. **López-Neyra, C. R.**, Una neuva especie de *Deltokeras* y su situación entre los ciclofilididos, *Rev. Ibér. Parasitol.*, 3, 351—358, 1943.
1968. **López-Neyra, C. R.**, *Nematotaenia tarentolae* n.sp., parasite intestinal de geckonidos, *Rev. Ibér. Parasitol.*, 4, 123—137, 1944.
1969. **López-Neyra, C. R.**, Los cestodes hispanos, *Rev. Acad. Cien. Exact. Fis Nat. Madrid,* 39, 97—136, 1945.
1970. **López-Neyra, C. R.**, Bibliografia equinoccósica ibérica, *Arch. Int. Hidat.*, 7, 131—174, 1947.
1971. **López-Neyra, C. R.**, Raillietinosis humanas. Estudios de parasitologia comparada sobre Raillietinae parasitas humanas y en especial de las formas Neotropicales, *Rev. Ibér. Parasitol.*, 9, 299—362, 1949.
1972. **López-Neyra, C. R.**, La parasitologia humana en el Marruecos espanol, *Rev. Ibér. Parasitol.*, 9, 373—443, 1949.
1973. **López-Neyra, C. R.**, Revision del género *Cotugnia*, motivada por el estudio de una especie neuva hallada en la tortola de Granada, *Rev. Ibér. Parasitol.*, 10, 57—96, 1950.
1974. **López-Neyra, C. R.**, La *Tenia madagascariensis* Davaine 1869, oriunad cubana, y el *Inermicapsifer cubensis* (Kouri, 1939) son una misma especie, correspondiente al género *Raillietina* (Fuhrmann, 1920) López-Neyra 1934, *Rev. Ibér. Parasitol.*, 10, 187—203, 1950.
1975. **López-Neyra, C. R.**, Raillietinosis humanas. Estudios de parasitologia comparada sobre Raillietininae parasitas humanas y en especial de las formas Neotropicales, *Rev. Kuba Med. Trop. Parasitol.*, 6, 19—26, 1950.
1976. **López-Neyra, C. R.**, Análisis critico de los géneros *Choanotaenia, Anomotaenia* y afines con redescripción de la *Taenia porosa* Rudolphi, 1819 e invalidez del género *Paricterotaenia* (Primera parte), *Rev. Ibér. Parasitol.*, 11, 337—368, 1951.
1977. **López-Neyra, C. R.**, Análsis critico de los géneros *Choanotaenia, Anomotaenia* y afines con redescripción de la *Taenia porosa* Rud. 1810 e invalidez del género *Paricterotaenia* Fuhrm., 1932, *Rev. Ibér. Parasitol.*, 12, 1—58, 1952.
1978. **López-Neyra, C. R.**, *Gyrocoelia albaredai* n.sp. Relaciones con Tetrabothriidae y Dilepididae, *Rev. Ibér. Parasitol.*, 12, 319—344, 1952.
1979. **López-Neyra, C. R.**, Las especies de *Tatria Kowalewsky,* 1904 (Amabiliidae) consideradas teratologias de Hymenolepididae, *Thapar Commemorative Volume,* Lucknow, 1953, 185—192.
1980. **López-Neyra, C. R.**, Consideraciones sobre Acoleidae, Amabiliidae y Nematoparataenidae, *Rev. Ibér. Parasitol.*, 13, 119—184, 1953.
1981. **López-Neyra, C. R.**, Anoplocephalidae, *Rev. Ibér. Parasitol.*, 14, 13—130, 225—290, 303—396, 1954.
1982. **López-Neyra, C. R.**, Anoplocephalidae, *Rev. Ibér. Parasitol.*, 25, 33—84, 1955.
1983. **López-Neyra, C. R.**, *Hymenosphenacanthus* nomen novum par *Sphenacanthus* López-Neyra, 1942 (Cestode-Hymenolepididae) nec Agassiz, 1837 (Pez fósil), *Rev. Ibér. Parasitol.*, (personal communication dated May 20, 1958), 1958.
1984. **López-Neyra, C. R. and Diaz-Ungria, C.**, Cestodes de Venezuela. III. Sobre unos cestodes intestinales de reptiles y mamiferos venezolanos, *Mem. Soc. Cinc. Nat. La Salle,* 17, 28—63, 1957.

1985. **López-Neyra, C. R. and Diaz-Ungria, C.**, Cestodes de Venezuela. V. Cestodes de vertebrados venezolanos (Segunda nota), *Nova Ciencia,* 23, 1—41, 1958.
1986. **López-Neyra, C. R. and Muñoz-Medina, J. M.**, *Dipylidium quinquecoronatum* nov. sp., parasito intestinal del gato doméstico, *Bol. Real. Soc. Espan. Hist. Nat. Madrid,* 21, 421—426, 1921.
1987. **López-Neyra, C. R. and Soler, P. M.**, Revision del género *Echinococcus* Rud. y descripción de una especie neuva parásita intestinal del perro en Almería, *Rev. Ibér. Parasitol.,* 3, 169—210, 1943.
1988. **Lorincz, F., Burghoffer, G., and Bodrogi, G.**, Beitrag zur Echinokokkenkrankheit in Ungarn, *Centralbl. Bakteriol.,* 124, 16—22, 1932.
1989. **Lovekar, C. D., Seth, D., and Deshmukh, P. G.**, *Schizorchis esarsi* n.sp. (Cestoda: Anoplocephalidae) from albino mouse, *Mus musculus, Riv. Parassitol.,* 33, 17—20, 1972.
1990. **Loveland, A. E.**, On the anatomy of *Taenia crassicollis, J. Comp. Med. Vet. Arch. Philadelphia,* 15, 67—89, 1894; Suppl. note by C. W. Stiles, p. 85.
1991. **Loveridge, A.**, Notes on East African birds (chiefly nesting habits and endoparasites) collected 1920—1923, *Proc. Zool. Soc. London,* 1923, 899—921, 1923.
1992. **Lucet, A. and Marotel, E.**, Les cestodes du dindon nature, zoologique et rôle pathogène, *Rev. Med. Vet. Paris,* 81, 162—168, 1904.
1993. **Lühe, M.**, Beiträge zur Kenntnis des Rostellums und der Skolexmuskulatur der *Taenia, Zool. Anz.,* 17, 279—282, 1893.
1994. **Lühe, M.**, Zur Morphologie des Tänienskolex, Inaugural Dissertation, University of Konigsberg, 1895, 1—133.
1995. **Lühe, M.**, Zur Kenntnis der Muskulatur des Taenienkörpers, *Zool. Anz.,* 19, 260—264, 1895.
1996. **Lühe, N.**, Mitteilungen ueber einige wenig bekannte bzw. neue süd-amerikanische Taenien des K.K. naturhistorischen Hof-Museums in Wien, *Arch. Naturg.,* 61, 199—212, 1895.
1997. **Lühe, M.**, Das Nervensystem von *Ligula* in seinen Beziehungen zur Anordnung der Muskulatur, *Zool. Anz.,* 19, 383—384, 1895.
1998. **Lühe, M.**, *Bothriocephalus zschokkei* Fuhrmann, *Centralbl. Bakteriol.* 22, 586, 1897.
1999. **Lühe, M.**, Die Anordnung der Muskulatur bei dem Dibothrien, *Centralbl. Bakteriol.,* 22, 739—747, 1897.
2000. **Lühe, M.**, Beiträge zur Helminthenfauna der Berberei, *Sitzungber. Kgl. Preuss. Akad. Wiss.,* 40, 619—628, 1898.
2001. **Lühe, M.**, Die Gliederung von *Ligula, Centralbl. Bakteriol.,* 23, 280—286, 1898.
2002. **Lühe, M.**, *Oochoristica* nov. gen. Taeniadarum (Vorläufige Mitteilung), *Zool. Anz.,* 21, 650—652, 1898.
2003. **Lühe, M.**, Zur Anatomie und Systematik der Bothriocephaliden, *Verhandl. Dtsch. Zool. Ges.,* 9, 30—55, 1899.
2004. **Lühe, M.**, Beiträge zur Kenntnis der Bothriocephaliden. I, II. Bothriocephaliden mit marginalen Genitalöffnungen, *Centralbl. Bakteriol.,* 26, 702—719, 1899.
2005. **Lühe, M.**, Zur Kenntnis einiger Distomen, *Zool. Anz.,* 22, 524—539, 1899.
2006. **Lühe, M.**, Ueber *Bothrimonus* Duvernoy und verwandte Bothriocephaliden, *Zool. Anz.,* 23, 8—14, 1900.
2007. **Lühe, M.**, Untersuchungen euber Bothriocephaliden mit marginalen Genitalöffnungen, *Z. Wiss. Zool.,* 68, 97—99, 1900.
2008. **Lühe, M.**, Beiträge zur Kenntnis der Bothriocephaliden. III. Bothriocephaliden der landbewohnenden Reptilien, *Centralbl. Bakteriol.,* 27, 29—217, 252—258, 1900.
2009. **Lühe, M.**, Ueber einen eigentümlichen Cestoden aus *Acanthias, Zool. Anz.,* 24, 347—349, 1901.
2010. **Lühe, M.**, Referat ueber v. Ariola's Revisione della famiglia Bothriocephalidae s. str. *(Bothriotaenia longispicula* Stoss. = *Acoleus longispiculus* Stoss.), *Centralbl. Bakteriol.,* 29, 415, 1901.
2011. **Lühe, M.**, Ueber die Fixierung der Helminthen an der Darmwandung ihrer Wirte und die dadurch verursachten pathologisch-anatomischen Veränderungen des Wirtsdarmes, *Verhandl. Int. Zool. Congr. Berlin,* pp. 698—706, 1902.
2012. **Lühe, M.**, *Urogonoporus armatus:* ein eigentümlicher Cestode aus *Acanthias,* mit anschliessenden Bemerkungen ueber die sogenannten Cestodarier, *Arch. Parasitol. Paris,* 5, 209—250, 1902.
2013. **Lühe, M.**, Revision meines Bothriocephaliden-systems, *Centralbl. Bakteriol.,* 31, 318—331, 1902.
2014. **Lühe, M.**, Bemerkungen ueber die Cestoden aus *Centrolophus pompilius.* I. Zur Synonymie der *Centrolophus*-Cestoden, *Centralbl. Bakteriol.,* 31, 629—637, 1902.
2015. **Lühe, M.**, Cestoden, in Brauer, A., *Die Süsswasserfauna Deutschlands,* Vol. 18, Fischer, Jena, 1910, 1—153.
2016. **Lühe, M.**, Cystotänien südamerikanischer Feliden, *Zool. Jahrb.,* Suppl., 13, 687—710, 1910.
2017. **Lungwitz, M.**, *Taenia ovilla* Rivolta, ihr anatomischer Bau und die Entwicklung ihrer Geschlechtsorgane, *Arch. Wiss. Prakt. Thierheilk,* 21, 105—159, 1895.
2018. **Luther, A.**, Ueber *Triaenophorus robustus* Olsson und *Henneguya zschokkei* Gurley als Parasiten von *Coregonus albula* aus dem See Sapsojärvi, *Medd. Soc. Fauna Flora Fenn.,* 35, 58—59, 1909.
2019. **Lyman, R. A.**, Studies on the genus *Cittotaenia, Trans. Am. Microsc. Soc.,* 23, 173—190, 1902.
2020. **Lynch, J. E.**, Re-description of the species of *Gyrocotyle* from the ratfish, *Hydrolagus colliei* (Lay and Bebbet), with notes on the morphology and taxonomy of the genus, *J. Parasitol.,* 31, 418—446, 1945.

2021. **Lynsdale, J. A.**, On a remarkable new cestode, *Meggittina baeri* gen. et sp. nov. (Anoplocephalinae) from rodents in southern Rhodesia, *J. Helminthol.*, 27, 129—142, 1953.
2022. **Lynsdale, J. A.**, On two new species of *Lytocestus* from Burma and the Sudan, respectively, *J. Helminthol.*, 30, 87—96, 1956.
2023. **Lynsdale, J. A.**, On a new species of *Proteocephalus* from Brazil, *J. Helminthol.*, 33, 145—150, 1959.
2024. **Lynsdale, J. A.**, On *Proteocephalus sandoni* n.sp. from the Sudan, *J. Helminthol.*, 34, 43—46, 1960.
2025. **Lyon, M. W., Jr.**, Native cases of infestation by fish tapeworm, *Diphyllobothrium latum*, *JAMA* 86, 264—265, 1926.
2026. **Lyster, L. L.**, Parasites of some Canadian sea mammals, *Can. J. Res., (Sect. D, Zool. Sci.)*, 18, 395—409, 1940.
2027. **MacCallum, G. A.**, Some new forms of parasitic worms, *Zoopathologica*, 1, 43—75, 1917.
2028. **MacCallum, G. A.**, Studies in helminthology, *Zoopathologica*, 1, 229—294, 1921.
2029. **MacCallum, G. A. and MacCallum, W. G.**, On the structure of *Taenia gigantea*, *Zool. Jahrb. Syst.*, 32, 379—388, 1912.
2030. **MacDonough, E. J. M.**, Parasitos de peces comestibles. II. Larvas de un cestode trypanorhynchido de la pescadila, *Semana Med. Buenos Aires*, 34, 373—376, 1927.
2031. **MacDonough, E. J. M.**, Parasitos de peces comestibles. VI. Sobre una *"Ichthyotaenia"* y oncosfera del pejerrey, *Semana Med. Buenos Aires*, 39, 1917—1921, 1932.
2032. **Mace, T. F. and Shepard, C. D.**, Helminths of a Vancouver Island marmot, *Marmota vancouverensis* Swarth, 1911, with a description of *Diandrya vancouverensis* sp. nov. (Cestoda: Anoplocephalidae), *Can. J. Zool.*, 59, 790—792, 1981.
2033. **Mackiewicz, J. S.**, *Monobothrium hunteri* sp.n. (Cestoidea: Caryophyllaeidae) from *Catostomus commersoni* (Lacépède) (Pices: Catostomidae) in North America, *J. Parasitol.*, 49, 723—730, 1963.
2034. **Mackiewicz, J. S.**, *Isoglaridacris bulbocirrus* gen. et sp.n. (Cestoidea: Caryophyllaeidae) from *Catostomus commersoni* in North America, *J. Parasitol.*, 51, 377—381, 1965.
2035. **Mackiewicz, J. S.**, *Isoglaridacris hexacotyle* comb. n. (Cestoidea: Caryophyllidea) from catostomid fishes in southwestern North America, *Proc. Helminthol. Soc. Wash.*, 35, 193—196, 1968.
2036. **Mackiewicz, J. S.**, Two new caryophyllaeid cestodes from the spotted sucker, *Minytrema melanops* (Raf.) (Catostomidae), *J. Parasitol.*, 54, 808—813, 1968.
2037. **Mackiewicz, J. S.**, *Penarchigetes oklensis* gen. et sp.n. and *Biacetabulum carpiodi* sp.n. (Cestoidea: Caryophyllaeidae) from catostomid fish in North America, *Proc. Helminthol. Soc. Wash.*, 36, 119—126, 1969.
2038. **Mackiewicz, J. S.**, *Edlintonia ptychocheila* gen. n., sp.n. (Cestoidea: Capingentidae) and other caryophyllid tapeworms from cyprinid fishes of North America, *Proc. Helminthol. Soc. Wash.*, 37, 110—118, 1970.
2039. **Mackiewicz, J. S.**, Two new species of caryophyllid tapeworms from catostomid fishes in Tennessee, *J. Parasitol.*, 58, 1075—1081, 1972.
2040. **Mackiewicz, J. S.**, Caryophyllidea (Cestoidea): a review, *Exp. Parasitol.*, 31, 417—512, 1972.
2041. **Mackiewicz, J. S.**, *Calentinella etnieri* gen. et sp.n. (Cestoidea: Caryophyllaeidae) from *Erimyzon oblongus* (Mitchill) (Cypriniformes: Catostomidae) in North America, *Proc. Helminthol. Soc. Wash.*, 41, 42—45, 1974.
2042. **Mackiewicz, J. S.**, The genus *Caryophyllaeus* Gmelin (Cestoidea: Caryophyllidea) in the Nearctic, *Proc. Helminthol. Soc. Wash.*, 41, 184—191, 1974.
2043. **Mackiewicz, J. S.**, *Isoglaridacris calentinei* n.sp. (Cestoidea: Caryophyllidea) from catostomid fish in western United States, *Trans. Am. Microsc. Soc.*, 93, 143—147, 1974.
2044. **Mackiewicz, J. S.**, *Glaridacris vogei* n.sp. (Cestoidea: Caryophyllidea) from catostomid fishes in western North America, *Trans. Am. Microsc. Soc.*, 95, 92—97, 1976.
2045. **Mackiewicz, J. S.**, Caryophyllidea (Cestoidea): evolution and classification, *Adv. Parasitol.*, 19, 139—206, 1981.
2046. **Mackiewicz, J. S.**, Caryophyllidea: perspectives, *Parasitology*, 84, 397—417, 1982.
2047. **Mackiewicz, J. S. and Beverly-Burton, M.**, *Monobothrioides woodlandi* sp.nov. (Cestoidea: Caryophyllidea) from *Clarias mellandi* Boulenger (Cypriniformes: Clariidae) in Zambia, Africa, *Proc. Helminthol. Soc. Wash.*, 34, 125—128, 1967.
2048. **Mackiewicz, J. S. and Blair, D.**, Balanotaeniidae fam.n. and *Balanotaenia newguinensis* sp.n. (Cestoidea: Caryophyllidea) from *Tandanus* (Siluriformes: Plotosidae) in New Guinea, *J. Helminthol.*, 52, 199—203, 1978.
2049. **Mackiewicz, J. S. and Deutsch, W. G.**, *Rowardleus* and *Janiszewskella*, new caryophyllid genera (Cestoidea: Caryophyllaeidae) in eastern North America, *Proc. Helminthol. Soc. Wash.*, 43, 9—17, 1976.
2050. **Mackiewicz, J. S. and McCrae, R.**, *Hunterella nodulosa* gen.n., sp.n. (Cestoidea: Caryophyllaeidae) from *Catostomus commersoni* (Lacépède) (Pices: Catostomidae) in North America, *J. Parasitol.*, 48, 798—806, 1962.

2051. **Mackiewicz, J. S. and McCrae, R. C.,** *Biacetabulum biloculoides* n.sp. (Cestoidea: Caryophyllaeidae) from *Catostomus commersoni* (Lacépède) in North America, *Proc. Helminthol. Soc. Wash.,* 32, 225—228, 1965.
2052. **Mackiewicz, J. S. and Murhar, B. M.,** Redescription of *Bovienia serialis* (Bovien, 1926) (Cestoidea: Caryophyllidea) from the catfish *Clarias batrachus* (L.) in India, *J. Helminthol.,* 46, 399—405, 1972.
2053. **Macko, J. K.,** Neue Art eines Bandwurms *Wardium filamentoovatum* sp.n. (Hymenolepididae) aus dem Wirt *Larus minutus* Pall., *Biol. Bratislava,* 17, 355—364, 1962.
2054. **Macko, J. K.,** Neue Art eines Bandwurms aus der Gattung *Echinocotyle* Blanchard 1891 (Hymenolepididae), *Biol. Bratislava,* 17, 503—507, 1962.
2055. **Macko, J. K.,** *Krimi rallida* sp. nova (Choanotaeniidae; Cestoda) aus dem Wirt *Rallus aquaticus* L., *Annot. Zool. Bot. Bratislav.,* 29, 1—7, 1966.
2056. **Macko, J. K.,** *Anomotaenia porata* sp.n. (Choanotaeniidae) from the host *Porzana parva* (Ralliformes), *Folia Parasitol.,* 15, 263—265, 1968.
2057. **Macko, J. K.,** *Wardium zmorayi* n.sp. (Cestoda) aus dem Wirt löffelente — *Anas clypeata, Biol. Bratislava,* 25, 321—329, 1970.
2058. **Macko, J. K. and Ryśavý, B.,** *Diorchis longihamulus* sp.n. (Hymenolepididae) a new cestode from *Gallinula chloropes cerceris* (Ralliformes), *Folia Parasitol.,* 15, 267—270, 1968.
2059. **MacLulich, D. A.,** *Proteocephalus parallacticus,* a new species of tapeworm from lake trout, *Cristivomer namaycush, Can. J. Res., (Sect. D, Zool. Sci.),* 21, 145—149, 1943.
2060. **Macy, R. W.,** A key to the species of *Hymenolepis* found in bats, and the description of a new species, *H. christensoni,* from *Myotis lucifugus, Trans. Am. Microsc. Soc.,* 50, 344—347, 1931.
2061. **Macy, R. W.,** Parasites found in certain Oregon bats with the description of a new cestode, *Hymenolepis gertschi, Am. Midl. Nat.,* 37, 375—378, 1947.
2062. **Macy, R. W.,** *Cinclotaenia filamentosa* gen. et sp.n. (Cestoda: Dilepididae) from the dipper in Oregon, *Proc. Helminthol. Soc. Wash.,* 40, 201—204, 1973.
2063. **Macy, R. W. and Rausch, R. L.,** Morphology of a new species of bat cestode, *Hymenolepis roudabushi,* and a note on *Hymenolepis christensoni* Macy, *Trans. Am. Microsc. Soc.,* 65, 173—175, 1946.
2064. **Magalhães, P. S.,** de, Notes d'helminthologie brésilienne, *Bull. Soc. Zool. Fr.,* 17, 145—146, 1892.
2065. **Magalhães, P. S.,** de, Notes d'helminthologie brésilienne. Deux nouveaux ténias de la poule domestique, *Arch. Parasitol.,* 1, 442—451, 1898.
2066. **Magalhães, P. S. de,** *Davainea oligophora* de Magalhães, 1898 et *T. cantaniana* Polonio, 1860, *Arch. Parasitol.,* 3, 480—482, 1899.
2067. **Magalhães, P. S., de** Notes d'helminthologie brésilienne, *Arch. Parasitol.,* 9, 305—318, 1905.
2068. **Magalhães, P. S., de,** Notes d'helminthologie brésilienne, *Ann. Policlin. Ger. do Rio de Janeiro,* 3, 69—92, 1918.
2069. **Magalhães, P. S.,** de, Notes d'helminthologie brésilienne. Le *Davainea bothrioplitis* (Piana, 1881—1882), *Arch. Parasitol.,* 16, 481—502, 1919.
2070. **Magath, T. B.,** *Ophiotaenia testudo,* a new species from *Amyda spinifera, J. Parasitol.* 11, 44—49, 1924.
2071. **Magath, T. B.,** Experimental studies on *Diphyllobothrium latum, Am. J. Trop. Med.,* 9, 17—48, 1929.
2072. **Magath, T. B.,** The early life-history of *Crepidobothrium testudo* (Magath, 1924), *Ann. Trop. Med. Parasitol.,* 23, 121—128, 1929.
2073. **Magath, T. B.,** *Diphyllobothrium latum, Am. J. Clin. Pathol.,* 1 (Editorial), 187—189, 1931.
2074. **Magath, T. B.,** Hydatid *(Echinococcus)* disease in Canada and the United States, *Am. J. Hyg.,* 25, 107—134, 1936.
2075. **Magath, T. B. and Essex, H. E.,** Concerning the distribution of *Diphyllobothrium latum* in North America, *J. Prev. Med.,* 5, 227—242, 1931.
2076. **Magath, T. B. and Essex, H. E.,** A comparison of the viability of ova of the broad fish tapeworm, *D. latum,* from man and dogs, *Am. J. Hyg.,* 14, 698—704, 1931.
2077. **Mahon, J.,** Observations on the abnormal occurrence of *Hymenolepis nana fraterna* cysticercoids in the liver of a rodent, *Proc. Zool. Soc. London,* 124, 527—529, 1954.
2078. **Mahon, J.,** Occurrence of larvae of *Taenia taeniaeformis* (Batsch, 1786) in the American rabbit, *Lepus americanus, J. Parasitol.,* 40, 698, 1954.
2079. **Mahon, J.,** Contributions to the genus *Paradilepis* Hsü, 1935, *Parasitology,* 45, 63—78, 1955.
2080. **Mahon, J.,** *Dendrouterina pilherodiae* sp.nov. (Dilepididae) from *Pilherodias pileatus* (Bodd), *Can. J. Zool.,* 34, 28—34, 1956.
2081. **Mahon, J.,** On a collection of avian cestodes from Canada, *Can. J. Zool.,* 34, 104—119, 1956.
2082. **Mahon, J.,** *Deltokeras synallaxis* sp.nov. (Dilepididae) from *Synallaxis rutilans* Temm., *Can. J. Zool.,* 35, 441—447, 1957.
2083. **Mahon, J.,** Helminth parasites of reptiles, birds and mammals of Egypt. V. Avian cestodes, *Can. J. Zool.,* 36, 577—605, 1958.

2084. **Makarenko, V. K.,** *Raillietina apivori* n.sp. — a new species of cestode from *Pernis apivorus* (in Russian), in *Helminths of Man, Animals and Plants and their Control: Papers on Helminthology Presented to Academician K. I. Skrjabin on his 85th Birthday, Izdatel'stvo Akademii Nauk SSSR,* Moscow 1963, 160—162.

2085. **Maksimova, A. P.,** Cestodes of wild aquatic birds from the Turgay lakes (in Russian), *Tr. Inst. Zool., Alma-Ata,* 19, 101—116, 1963.

2086. **Maksimova, A. P.,** New species of cestodes from *Cygnus olor* in the Kazakh SSR (in Russian), *Tr. Inst. Zool., Alma-Ata,* 19, 126—132, 1963.

2087. **Maksimova, A. P.,** *Retinometra oxyurae* n.sp. (Cestoda: Hymenolepididae) from the wild goose of Kazakhstan (in Russian), *Helminthologia,* Year 1966, 7, 291—295, 1967.

2088. **Maksimova, A. P.,** A new cestode — *Fimbriarioides tadornae* n.sp. from *Tadorna tadorna* and its development in the intermediate host (in Russian), *Parazitologiya,* 10, 17—24, 1976.

2089. **Malhotra, S. K. and Capoor, V. N.,** Introduction of taxonometric approach to differentiate *Vampirolepis somariensis* n.sp., *Geobios,* 7, 302—308, 1980.

2090. **Malhotra, S. K. and Capoor, V. N.,** Taxonometric evaluation of a new mammalian cestode, *Hydatigera himalayotaenia* sp.n. (Cestoda: Taeniidae) with a note on strobilocercus larva, *Helminthologia,* 19, 121—127, 1982.

2091. **Malhotra, S. K. and Capoor, V. N.,** An avian cestode *Ophryocotyloides srinagarensis* n.sp. from *Corvus macrarhynchos* (Wagler) and *Corvus splendens* (Vieillot) from India with a revised key, *Bioresearch,* 3, 35—38, 1979.

2092. **Malhotra, S. K., Capoor, V. N., and Pundir, K.,** Taxometric evaluation of a new fish cestode genus *Tortocephalus songi* n.g., n.sp., from hillstream fishes in Garhwal Himalayas, *Sci. Environ.,* 2, 197—205, 1980.

2093. **Malhotra, S. K., Capoor, V. N., and Shinde, G. B.,** Introduction of a taxonomic device to evaluate a new proteocephalid cestode *Gangesia sanehensis* n.sp. from freshwater fishes of Garhwal Himalayas with a revised key to species of genus *Gangesia, Marathwada Univ. J. Sci. Nat. Sci.,* 19, 41—52, 1980.

2094. **Malhotra, S. K., Dixit, S., and Capoor, V. N.,** Taxometric evaluation of a new fish cestode *Gangesia mehamdabadensis* n.sp., from *Mystus tengra* from Mehamdabad, Gujarat, *Sci. Environ.,* 3, 7—20, 1981.

2095. **Malkani, P. G.,** A rapid method of evaginating the scolices in parasitic cysts, *Ind. Vet. J.,* 9, 193, 1933; *Ind. Vet. J.,* 10, 122—124, 1933.

2096. **Malviya, H. C.,** Studies on the life history, biology and control of common cestode parasites of the domestic pigeon, *Agra Univ. J. Res. Sci.,* 20, 99—104, 1971.

2097. **Malviya, H. C. and Dutt, S. C.,** A new species of *Cotugnia* (Cestoda: Davaineidae) from the domestic pigeon in India, *Parasitology,* 59, 397—400, 1969.

2098. **Malviya, H. C. and Dutt, S. C.,** Morphology and life history of *Cotugnia sravastavai* n.sp. (Cestoda: Davaineidae) from the domestic pigeon, in Singh, K. S., Tandan, B. K., Eds., "H.D. Srivastava Commemoration Volume", *Ind. Vet. Res. Instit.,* Izatnagar, pp. 103—108, 1970.

2099. **Malviya H. C. and Dutt, S. C.,** Morphology and Life-history of *Raillietina (Raillietina) mehrai* sp.n., *Ind. J. Anim. Sci.,* 41 1003—1007, 1971.

2100. **Malviya, H. C. and Dutt, S. C.,** Morphology and life history of *Raillietina (Raillietina) singhi* n.sp. (Cestoda, Davaineidae), *Ind. J. Helminthol.,* 23, 1—10, 1971.

2101. **Mamaev, Yu. L.,** *Eubothrium vittevitellatus* n.sp. from marine fish of Kamchatka (in Russian), in Skryabin, K. I. and Yu. L. Mamaev, Eds., *Helminths of Animals of the Pacific Ocean, Izdatl'stvo Nauka.,* Moscow, 1968, 28—29.

2102. **Mamaev, Yu. L. and Okhotina, M. V.,** *Prochoanotaenia spasskii* n.sp. from *Mogera robusta* (in Russian), in *Parasites of Animals and Plants, Izdatel'stvo Nauka.,* Moscow, 1968, 116—117.

2103. **Manger, B. R.,** Some cestode parasites of the elasmobranchs *Raja batis* and *Squalus acanthias* from Iceland, *Bull. Br. Mus. Nat. Hist. Zool.,* 24, 161—181, 1972.

2104. **Mankau, S. K.,** Studies on *Echinococcus alveolaris* (Klemm, 1883) from St. Lawrence Isl., Alaska. I. Histogenesis of the alveolar cyst in white mice, *J. Parasitol.,* 43, 153—159, 1957.

2105. **Manter, H. W.,** *Gyrocotyle,* a peculiar parasite of the elephant fish in New Zealand, *Tuatara,* 5, 49—51, 1953.

2106. **Maplestone, P. A.,** Notes on Australian cestodes. I and II, *Ann. Trop. Med. Parasitol.,* 15, 403—412, 1921.

2107. **Maplestone, P. A.,** Notes on Australian cestodes. III. *Cotugnia oligorchis* n.sp., *Ann. Trop. Med. Parasitol.,* 16, 55—60, 1922.

2108. **Maplestone, P. A.,** Notes on Australian cestodes. VI. *Schizotaenia cacatuae* n.sp., *Ann. Trop. Med. Parasitol.,* 16, 305—310, 1922.

2109. **Maplestone, P. A. and Southwell, T.,** Notes on Australian cestodes. IV. *Gyrocoelia australiensis* Johnston, *Ann. Trop. Med. Parasitol.,* 16, 61—68, 1922.

2110. **Maplestone, P. A. and Southwell, T.,** Notes on Australian cestodes. V. Three cestodes from the black swan, *Ann. Trop. Med. Parasitol.,* 16, 189—198, 1922.

2111. **Maplestone, P. A. and Southwell, T.**, Notes on Australian cestodes, *Ann. Trop. Med. Parasitol.*, 17, 317—331, 1923.
2112. **Marchi, P.**, Sopra una *Taenia* della *Loxia curvirostra*, *Atti Soc. Ital. Sci. Nat.*, 12, 534—535, 1869.
2113. **Marchi, P.**, Sopra un nuovo cestode trovato nell' *Ascalobotes mauritanicus*, *Atti Soc. Ital. Sci. Nat. Milano* 15, 305—306, 1873.
2114. **Marchi, P.**, Sur le développement du cysticerque des geckos en cestode parfait chez les *Strix nocuta*, *C. R. Assoc. Fr. Avanc. Sci.*, 7, 757, 1878.
2115. **Markov, G.**, The survival of a broad tapeworm's plerocercoids (*Diphyllobothrium latum L.*) in artificial media, *C. R. Acad. Sci. U.R.S.S.*, 19, 511—512, 1938.
2116. **Markowski, S.**, Evolution de *Cladotaenia cylindracea* (Bloch), *Ann. Parasitol.*, 6, 431—439, 1928.
2117. **Markowski, S.**, Untersuchungen ueber die Helminthfauna der Raben (Corvidae) von Polen (in Polish with German summary), *C. R. Acad. Pol. Sci. Classe Sci. Math. Nat.*, 5, 1—65, 1933.
2118. **Markowski, S.**, Contribution à la connaissance de devéloppement de la larve *Tetrathyridium variabile* (Diesing, 1850) (in Polish with French summary), *C. R. Acad. Pol. Sci. Classe Sci. Math. Nat.*, 5, 5—6, 1933.
2119. **Markowski, S.**, Beitrag zur Kenntnis der Entwicklung der Larve *Tetrathyridium variabile* (Diesing, 1850) (German summary of 1933), *Mem. Acad. Pol. Sci. Classe Sci. Math. Nat.*, pp. 43—51, 1934.
2120. **Markowski, S.**, Ueber den Entwicklungszyklus von *Bothriocephalus scorpii* (Muller, 1776) (German text), *Bull. Int. Acad. Pol. Sci. Lett. Cl. Sci. Math. Nat. B: Sci. Nat.*, II, 1—17, 1935.
2121. **Markowski, S.**, The cestodes of seals from the Antarctic, *Bull. Br. Mus. (Nat. Hist.) Zool.*, 1, 123—150, 1952.
2122. **Markowski, S.**, Cestodes of whales and dolphins from the Discovery collections, *Discovery Rep.*, 27, 377—395, 1955.
2123. **Marotel, G.**, Sur une téniadé du *Bothrops lanceolatus* (note préliminaire), *C. R. Soc. Biol.*, 50, 99—101, 1898.
2124. **Marotel, G.**, Etude zoologique de l' *Ichthyotaenia calmettei* Barrois, *Arch. Parasitol.*, 2, 34—42, 1899.
2125. **Marotel, G.**, Sur une téniadé du Blaireau, *C. R. Soc. Biol.*, 51, 21—23, 1899.
2126. **Marotel, G.**, Sur deux cestodes parasites des oiseaux (note préliminaire), *C. R. Soc. Biol.*, 51, 935—937, 1899.
2127. **Marotel, G.**, Nouveau cestode de mouton, *Bull. Soc. Sci. Vét. Lyon*, 1912.
2128. **Marotel, G.**, Nouveau mode de présentation des cestodes avec application aux parasites des ruminants, *9th Congr. Int. Zool. Monaco*, pp. 662—663, 1913.
2129. **Marshall, W. S. and Gilbert, N. C.**, Notes on the food and parasites of some freshwater fishes from the lakes at Madison, Wisconsin, *Annu. Rep. U.S. Bur. Fish. 1904*, pp. 513—522, 1905.
2130. **Martinez Gomes, F. and Hernandez Rodriguez, S.**, Helmintos parasitos de la oveja (*ovis aries*) en Cordoba. II. Descripción de *Stilesia vittata* Railliet, 1896, primera cita en España, y segunda relación de helmintos, *Riv. Ibér. Parasitol.*, 33, 11—20, 1973.
2131. **Mas-Coma, S.**, Helminthes de micromammifères. Spécificité, évolution et phylogénie des cestodes Arostrilepididae Mas-Coma et Tenora, 1981 (Cyclophyllidea: Hymenolepidoidea), in Second symposium on host specificity among parasites of vertebrates, 13-17 April 1981, *Mém. Mus. Natl. Hist. Nat. Ser. A Zool.*, 123, 185—193, 1981.
2132. **Mas-Coma, S. and Jourdane, J.**, Description de l'adult de *Staphylocystis biliarius* Villot, 1877 (Cestoda: Hymenolepididae), parasite de *Crocidura russula* Hermann, 1780 (Insectivora: Soricidae), *Ann. Parasitol. Hum. Comp.*, 52, 609—614, 1977.
2133. **Mas-Coma, S., Tenora, F., and Gallego, J.**, Consideraciones sobre los hymenolepididos inermes de roedores, con especial referencia a la problemática entorno a *Hymenolepis diminuta*, *Circ. Farm.*, 38, 137—152, 1980.
2134. **Mateo, E. and Bullock, W. L.**, *Neobothriocephalus aspinosus* gen. et sp.n. (Pseudophyllidea: Parabothriocephalidae), from the Peruvian marine fish, *Neptomenus crassus*, *J. Parasitol.*, 52, 1070—1073, 1966.
2135. **Mathevossian, E. M.**, An analysis of the specific components of the genus *Diploposthe*: cestodes from Anatidae, *C. R. Acad. Sci. U.R.S.S.*, 34, 265—268, 1942.
2136. **Mathevossian, E. M.**, New cestodes of birds in Russia, *Collect. Pap. on Helmintol. Ded. Skrjabin*, 1946, 178—188.
2137. **Mathevossian, E. M.**, *Taenia schavarschi* n.sp. from *Larus ichthyaetus*, (in Russian), *Tr. Moskovskogo Zooparka*, 4, 292—296, 1949.
2138. **Mathevossian, E. M.**, Morphological-systematic characteristics of cestodes (Paruterinidae) of birds of chase and an attempt to establish the path of their phylogenetic evolution (in Russian), *Tr. Gel'mintol. Lab. Akad. Nauk SSSR*, 4, 261—263, 1950.
2139. **Mathevossian, E. M.**, Cestode fauna of birds of southern Kirgizia (in Russian), *Tr. Gel'mintol. Lab. Akad. Nauk SSSR*, 4, 84—89, 1950.
2140. **Mathevossian, E. M.**, On the reorganization of the cestodes Dilepididae (in Russian), *Rabot. Gel'mintol. 75-Let. Skrjabin*, pp. 392—397, 1954.

2141. **Mathevossian, E. M.**, *Essentials of Cestodology*, Vol. III. *Dilepidoidea — Cestode Helminths of Domestic and Wild Animals*, Academiya Nauk SSSR, Moscow, 1963, 1—687.
2142. **Mathevossian, E. M.**, Analysis of some paruterinid species (in Russian), *Mater. Nauchn. Konf. Vses. Ova. Gel'mintol.*, Year 1964, Part 1, 252—255, 1964.
2143. **Mathevossian, E. M.**, Revision of the classification of Paruterinoidea (Cestoda) (in Russian), *Mater. Nauchn. Konf. Vses. Ova. Gel'mintol.*, Year 1965, Part II, 150—156, 1965.
2144. **Mathevossian, E. M.**, An analysis of some Biuterininae (Cestoda: Paruterinoidea) (in Russian), *Mater. Nauchn. Konf. Vses. Ova. Gel'mintol.*, Year 1966, 3, 166—170, 1966.
2145. **Mathevossian, E. M.**, The analysis of some hymenolepidids (Cestoda: Hymenolepididae) (in Russian), *Tr. Vses. Inst. Gel'mintol.*, 14, 229—232, 1968.
2146. **Mathevossian, E. M.**, *Essentials of Cestodology*, Vol. VII. *Paruterinoidea, Cestodes of Domestic and Wild Birds*, Akademiya Nauk SSSR, Moscow, 1969, 303 pp.
2147. **Mathevossian, E. M. and Krotov, A. K.**, Two new species of Echinocotyle (Cestoda) from aquatic birds (in Russian), *Tr. Gel'mintol Lab. Akad. Nauk SSSR*, 2, 96—98, 1949.
2148. **Mathevossian, E. M. and Movsesyan, S. O.**, *Pentocoronaria rusannae* n.g., n.sp. (Cestoda: Davaineidae) from *Turtur turtur* (in Russian), *Mater. Nauchn. Konf. Vses. Ova. Gel'mintol.*, 3, 170—175, 1966.
2149. **Mathevossian, E. M. and Movsesyan, S. O.**, Revision of the genera *Ascometra* and *Octopetalum* (Cestoda: Paruterinoidea) (in Russian), *Tr. Vses. Instit. Gel'mintol. K.I. Skrjabina*, 16, 137—146, 1970.
2150. **Mathevossian, E. M. and Sailov, D. I.**, *Tatria azarbaijanica* n.sp. from grebes of the Kyzyl-Agachsk State Reserve (in Russian), *Tr. Vses. Inst. Gel'mintol.*, 10, 8—11, 1963.
2151. **Matoff, K. N. and Jantscheff, J.**, Kann *Echinococcus granulosus* in Darm des Fuchses *(Canis vulpes)* geschlechtsreif entwickeln?, *Acta Vet.*, 4, 411—418, 1954.
2152. **Matz, F.**, Beiträge zur Kenntnis der Bothriocephalen, *Arch. Naturg.*, 58, 97—122, 1892.
2153. **Mayes, M. A.**, *Proteocephalus buplanensis* sp.n. (Cestoda: Proteocephalidae) from the creek chub, *Semotilus atromaculatus* (Mitchill), in Nebraska, *Proc. Helminthol. Soc. Wash.*, 43, 34—37, 1976.
2154. **Mayes, M. A., Brooks, D. R., and Thorson, T. B.**, Two new species of *Acanthobothrium* van Beneden 1849 (Cestoidea: Tetraphyllidea) from freshwater stingrays in South America, *J. Parasitol.*, 64, 838—841, 1978.
2155. **Mayes, M. A., Brooks, D. R., and Thorson, T. B.**, Two new tetraphyllidean cestodes from *Pomatotrygon circularis* Garman (Chondrichthyes: Pomatotrygonidae) in the Itacuaí River, Brazil, *Proc. Helminthol. Soc. Wash.*, 48, 38—42, 1981.
2156. **Mayhew, R. L.**, Studies on the avian species of the cestode family Hymenolepididae, *Ill. Biol. Monogr.*, 10, 1—125, 1925.
2157. **Mayhew, R. L.**, The genus *Diorchis*, with descriptions of four new species from North America, *J. Parasitol.*, 15, 251—258, 1929.
2158. **Mazza, S., Parodi, S., and Fiora, A.**, Cestode anoplocefálido n.sp. de vizcacha de la sierra *(Lagidium tucumanus* Thos.) de la provincia de Jujuy. 7, *Reunión Soc. Argent. Patol. Reg. Norte (Tucuman)*, 2, 1046—1054, 1932.
2159. **McCrae, R. C.**, *Biacetabulum macrocephalum* sp.n. (Cestoda: Caryophyllaeidae) from the white sucker *Catostomus commersoni* (Lacépède) in Northern Colorado, *J. Parasitol.*, 48, 807—811, 1962.
2160. **McDonald, M. E.**, Catalogue of helminths of waterfowl (Anatidae), Special Sci. Rep. U.S. Fish Wildl. Serv., Washington, D.C., 1969, 1—126.
2161. **McEwin, B. W.**, Cestodes from mammals, *B.A.N.Z. Antarct. Res. Exped. 1929-1931, Rep. Ser. B*, 6 (4), 75—90, 1957.
2162. **McIntosh, A.**, New host records for *Diphyllobothrium mansonoides* Mueller, 1935, *J. Parasitol.*, 23, 313—315, 1937.
2163. **McIntosh, A.**, Description of the adult of *Taenia twitchelli* from an Alaskan wolverine, *Proc. Helminthol. Soc. Wash.*, 5, 14—15, 1938.
2164. **McIntosh, A.**, A new taenioid cestode, *Cladotaenia foxi*, from a falcon, *Proc. Helminthol. Soc. Wash.*, 7, 71—74, 1940.
2165. **McIntosh, A.**, A new dilepidid cestode, *Catenotaenia linsdalei*, from a pocket gopher in California, *Proc. Helminthol. Soc. Wash.*, 8, 60—62, 1941.
2166. **McIntosh, W. C.**, Notes on the food and parasites of the *Salmo salar* of the Tay, *Proc. Zool. Soc. London (Zool.)*, 7, 145—154, 1864.
2167. **McLaughlin, J. D.**, A redescription of *Paradilepis longivaginosus* (Mayhew, 1925) (Cestoda: Dilepididae) and a comparison with *Paradilepis simoni* Rausch, 1949 and *Paradilepis rugovaginosus* Freeman, 1954, *Can. J. Zool.*, 52, 1185—1190, 1974.
2168. **McLaughlin, J. D. and Burt, M. D. B.**, A contribution to the systematics of three cestode species of the genus *Diorchis* Clerc, 1903 reported from birds of the genus *Fulica* L., *Acta Parasitol. Pol.*, 23, 213—221, 1975.
2169. **McLaughlin, J. D. and Burt, M. D. B.**, Studies on the hymenolepid cestodes of waterfowl from New Brunswick, Canada, *Can. J. Zool.*, 57, 34—79, 1979.

2170. **McLaughlin, J. D. and Burt, M. D. B.**, On the validity of *Diorchis maroccana* Dollfus, 1975 (Cestoda: Hymenolepididae), *Can. J. Zool.*, 58, 882—885, 1980.
2171. **McLeod, J. A.**, A parasitological survey of the genus *Citellus* in Manitoba, *Can. J. Res., Sect. D., Zool. Sc.*, 9, 108—127, 1933.
2172. **McVicar, A. H.**, *Echinobothrium harfordi* sp.nov. (Cestoda: Diphyllidea) from *Raja naevus* in the North Sea and English Channel, *J. Helminthol.*, 50, 31—38, 1976.
2173. **Meade-Thomas, G. and Riser, N. W.**, Observations on the morphology and systematic position of *Thysanocephalum thysanocephalum* (Linton, 1889), *Proc. Helminthol. Soc. Wash.*, 23, 98—106, 1957.
2174. **Meggitt, F. J.**, The structure and life-history of a tapeworm *(Ichthyotaenia filicollis* Rud.) parasitic in the stickleback, *Proc. Zool. Soc. London*, pp. 113—138, 1914.
2175. **Meggitt, F. J.**, On the anatomy of a fowl tapeworm, *Amoebotaenia sphenoides* v. Linstow, *Parasitology*, 7, 262—277, 1914.
2176. **Meggitt, F. J.**, A new species of tapeworm from a parakeet, *Brotogerys typica*, *Parasitology*, 8, 42—55, 1915.
2177. **Meggitt, F. J.**, A contribution to our knowledge of the tapeworms of fowls and sparrows, *Parasitology*, 8, 390—410, 1916.
2178. **Meggitt, F. J.**, A new species of cestode *(Oochoristica erinacei)* from the hedgehog, *Parasitology*, 12, 310—313, 1920.
2179. **Meggitt, F. J.**, A contribution to our knowledge of the tapeworms of poultry, *Parasitology*, 12, 301—309, 1920.
2180. **Meggitt, F. J.**, On two new tapeworms from the ostrich, with a key to the species of *Davainea*, *Parasitology*, 13, 1—24, 1921.
2181. **Meggitt, F. J.**, On a new cestode from the pouched rat, *Cricetomys gambianum*, *Parasitology*, 13, 195—204, 1921.
2182. **Meggitt, F. J.**, On two new species of Cestoda from a mongoose, *Parasitology*, 16, 48—54, 1924.
2183. **Meggitt, F. J.**, On the collection and examination of tapeworms, *Parasitology*, 16, 266—268, 1924.
2184. **Meggitt, F. J.**, Tapeworms of the Rangoon pigeon, *Parasitology*, 16, 303—312, 1924.
2185. **Meggitt, F. J.**, On the life-history of a reptilian tapeworm *(Sparganum reptans)*, *Ann. Trop. Med. Parasitol.*, 18, 195—204, 1924.
2186. **Meggitt, F. J.**, On the occurrence of *Ligula ranarum* in a frog, *Ann. Mag. Nat. Hist.* (Ser. 9), 13, 216—219, 1924.
2187. **Meggitt, F. J.**, On the life-history of an amphibian tapeworm, *Diphyllobothrium ranarum*, *Ann. Mag. Nat. Hist.* (Ser. 9), 16, 654—655, 1925.
2188. **Meggitt, F. J.**, The tapeworms of the domestic fowl, *J. Burma Res. Soc.*, 15, 222—243, 1926.
2189. **Meggitt, F. J.**, On a collection of Burmese cestodes, *Parasitology*, 18, 232—237, 1926.
2190. **Meggitt, F. J.**, A list of cestodes collected in Rangoon during the years 1923-26, *J. Burma Res. Soc. Rangoon*, 16, 200—201, 1927.
2191. **Meggitt, F. J.**, Remarks on the cestode families Monticelliidae and Ichthyotaeniidae, *Ann. Trop. Med. Parasitol.*, 21, 69—87, 1927.
2192. **Meggitt, F. J.**, On cestodes collected in Burma, *Parasitology*, 19, 141—153, 1927.
2193. **Meggitt, F. J.**, Report on a collection of cestodes mainly from Egypt. I. Families Anoplocephalidae, Davaineidae, *Parasitology*, 19, 314—327, 1927.
2194. **Meggitt, F. J.**, Report on a collection of cestodes mainly from Egypt. II. Cyclophyllidea: family Hymenolepididae, *Parasitology*, 19, 420—448, 1927.
2195. **Meggitt, F. J.**, Report on a collection of cestodes mainly from Egypt. III. Cyclophyllidea (conclusion); Tetraphyllidea, *Parasitology*, 20, 315—328, 1928.
2196. **Meggitt, F. J.**, Report on a collection of cestodes mainly from Egypt. IV. Conclusion, *Parasitology*, 22, 338—345, 1930.
2197. **Meggitt, F. J.**, On cestodes collected in Burma. II, *Parasitology*, 23, 250—263, 1931.
2198. **Meggitt, F. J.**, Cestodes collected from animals dying in the Calcutta Zoological Gardens during 1931, *Rec. Ind. Mus.*, 35, 145—165, 1933.
2199. **Meggitt, F. J.**, The theory of host specificity as applied to cestodes, *Ann. Trop. Med. Parasitol.*, 28, 99—105, 1934.
2200. **Meggitt, F. J.**, On some tapeworms from the bull snake *(Pityopis sayi)* with remarks on the species of the genus *Oochoristica* (Cestoda), *J. Parasitol.*, 20, 182—189, 1934.
2201. **Meggitt, F. J.**, On two tapeworms from a Burmese snake, *Ann. Mag. Nat. Hist.*, (Ser. 11), 5, 255—256, 1940.
2202. **Meggitt, F. J. and Maung Po Saw**, On a new tapeworm from a duck, *Ann. Mag. Nat. Hist.*, 14, 324—326, 1924.
2203. **Meggitt, F. J. and Subramanian, K.**, The tapeworms of rodents of the subfamily Murinae, with special reference to those occurring in Rangoon, *J. Burma Res. Soc.*, 17, 190—237, 1927.

2204. **Mégnin, P.,** De la caducité des crochets et du scolex lui-même chez les ténias, *C. R. Acad. Sci.*, 90, 715—717, 1880.
2205. **Mégnin, P.,** Notes sur les helminthes rapportés des côtes de la Laponie par M. le professeur Pouchet, *Bull. Soc. Zool. Fr.*, 8, 153—156, 1883.
2206. **Mégnin, P.,** Un nouveau ténia du pigeon ou plutôt une espèce douteuse de Rudolphi réhabilitée, *C. R. Soc. Biol.*, 3, 751—753, 1891.
2207. **Mégnin, P.,** Un ténia du pigeon ramier *(Palombus torquatus), Davainea bonini* n.sp., *Vol. Jub. Cinquantenaire Soc. Biol.*, pp. 279—281, 1899.
2208. **Mehlis, E.,** Novae observationes de entozois, *Isis*, 68—99, 166—199, 1831.
2209. **Mehra, H. R.,** On a new species of *Caryophyllaeus* O. F. Müller from Kashmir with remarks on *Lytocestus indicus* (Moghe) 1925, *Proc. 17th Ind. Sci. Congr.*, p. 247, 1930.
2210. **Mehra, K. N.,** Studies on the life history of *Hymenolepis fraterna* (Stiles, 1906), a tapeworm of rat, *Proc. 42nd Ind. Sci. Congr.*, Sect. 7, p. 284, 1955.
2211. **Mehra, K. N.,** Studies on the life history of *Hymenolepis diminuta,* common tapeworm of rat and man, *Proc. 42nd Ind. Sci. Congr.*, Part 3, Sect. 7, 284, 1955.
2212. **Mehra, K. N. and Srivastava, H. D.,** Studies on the life history of *Moniezia expansa* (Rudl, 1810), a broad tapeworm of ruminants, *Proc. 42nd Ind. Sci. Congr.*, Part. 3, Sect. 9, 352, 1955.
2213. **Mehra, K. N. and Srivastava, H. D.,** Studies on the life history of *Moniezia benedeni* (Moniez, 1879), a tapeworm of ruminants, *Proc. 42nd Ind. Sci. Congr.*, Part 3, Sect. 9, 352, 1955.
2214. **Meijer, W. C. P.,** *Cysticercus cellulosae* bij den hond (in Dutch), *Nederl. Ind. Bladen voor Diergeneeskunde en Dierenteelt*, 45, 135—137, 1933.
2215. **Meijer, W. C. P. and Sahar** (no initials), Over een lintworm van den hond, *Diphyllobothrium raillieti* Rátz, en het bijbehorende plerocercoid, *Sparganum* raillieti Rátz, van het varken (in Dutch), *Nederl. Ind. Bladen voor Geneeskunde en Dierenteelt*, 46, 1—12, 1934.
2216. **Meinkoth, N. A.,** Notes on the life-cycle and taxonomic position of *Haplobothrium globuliforme* Cooper, a tapeworm of *Amia calva* L., *Trans. Am. Microsc. Soc.*, 66, 256—261, 1947.
2217. **Meissner, F.,** Zur Entwicklungsgeschichte und Anatomie der Bandwürmer, *Z. Wiss. Zool.*, 5, 380—391, 1854.
2218. **Mello, U.,** *Anoplocephala minima* n.sp. del fagiano, *Monitore Zool. Ital.*, 23, 124—130, 1912.
2219. **Melnikow, N. M.,** Ueber die Jugendzustände der *Taenia cucumerina, Arch. Naturg.*, 35, 62—70, 1869.
2220. **Melvin, D. M.,** Studies on the life cycle and biology of *Monoecocestus sigmodontis* (Cestoda: Anoplocephalidae) from the cotton rat, *Sigmodon hispidus, J. Parasitol.*, 38, 346—355, 1952.
2221. **Mendes, M. V.,** Sobre a larva de *Dibothriorhynchus dinoi* sp.n. parasita dos Rhizostomata, *Arq. Mus. Paranaense*, 4, 47—82, 1945.
2222. **Mendheim, H.,** Bemerkungen zur Biologie des Katzenbandwurms, *Berl. München. Tierärztl. Wochnschr.*, 68, 117, 1955.
2223. **Mendivil-Nerrera, J.,** *Gyrocotyle meandrica* n.sp., del intestino espiral del pez gallo, *Callorhynchus callorhynchus* (L.), *Comun. Zool. Mus. Hist. Nat. Montevideo*, 2, 1—12, 1946.
2224. **Mendonça, M. M., de,** *Mathevotaenia cruzsilvai,* n.sp. (Cestoda, Anoplocephalidae), parasite de *Macaca irus* F. Cuvier, 1818, *Bull. Mus. Natl. Hist. Nat.*, Ser. 3, Sect. A, 4, 1081—1085, 1981.
2225. **Metchnikow, E.,** Entwicklungsgeschichtliche Beiträge, *Bull. Acad. Imp. Sci. St. Petersbourg*, 13, 284—300, 1869.
2226. **Mettrick, D. F.,** A new tapeworm, *Inermicapsifer rhodesiensis* sp.nov. from a scaly ant-eater, *Manis temminkii,* in Southern Rhodesia, *J. Helminthol.*, 33, 273—276, 1959.
2227. **Mettrick, D. F.,** A new cestode, *Anonomotaenia prinopsia* sp.nov. from the straight crested helmet shrike, *Prinops plumata,* in Southern Rhodesia, *J. Helminthol.*, 33, 277—280, 1959.
2228. **Mettrick, D. F.,** A new cestode, *Ophiotaenia ophiodex* n.sp., from a night adder, *Causus rhombeatus* (Licht.), in Southern Rhodesia, *Proc. Helminthol. Soc. Wash.*, 27, 275—278, 1960.
2229. **Mettrick, D. F.,** A new tapeworm, *Choanotaenia angolensis,* n.sp. from the Angola pitta, *Pitta angolensis,* in Southern Rhodesia, *J. Parasitol.*, 46, 398—399, 1960.
2230. **Mettrick, D. F.,** Two new species of the genus *Paruterina* Fuhrmann, 1906, from passeriform birds in Southern Rhodesia, *Proc. Helminthol. Soc. Wash.*, 27, 181—184, 1960.
2231. **Mettrick, D. F.,** *Ethiopotaenia trachyphonoides* gen. n., sp.n. from the crested barbet, *Trachyphonus vaillantii* (Ranzani) (Aves), in Southern Rhodesia, *J. Parasitol.*, 47, 875—877, 1961.
2232. **Mettrick, D. F.,** *Onderstepoortia coronati* sp.nov., a new cestode from a crowned plover, *Stephanibyx coronatus* (Boddaert), *Rev. Zool. Botanique Afr.*, 64, 133—137, 1961.
2233. **Mettrick, D. F.,** Some cestodes of the subfamily Paruterininae Fuhrmann, 1907 from birds in Central Africa, *J. Helminthol.*, 37, 319—328, 1963.
2234. **Mettrick, D. F.,** Some cestodes of reptiles and amphibians from the Rhodesias, *Proc. Zool. Soc. London*, 141, 239—250, 1963.
2235. **Mettrick, D. F.,** Some cestodes of the family Davaineidae from birds in Central Africa, *Proc. Zool. Soc. London*, 140, 469—484, 1963.

2236. **Mettrick, D. F.**, Some cestodes from birds of prey of the family Aquilidae, *Helminthol. Soc.*, 30, 237—244, 1963.
2237. **Mettrick, D. F.**, Some cestodes from Ardeiformes and Charadriiformes in Central Africa, *Rev. Zool. Bot. Afr.*, 75, 333—362, 1967.
2238. **Mettrick, D. F. and Beverly-Burton, M.**, Some cyclophyllidean cestodes from Carnivores in Southern Rhodesia, *Parasitology*, 51, 533—544, 1961.
2239. **Mettrick, D. F. and Beverly-Burton, M.**, A new cestode, *Anomotaenia caenodex* sp. nov. from a mistle thrush, *Turdus viscivorous viscivorous* (L.), *J. Helminthol.*, 36, 157—160, 1962.
2240. **Mettrick, D. F. and Beverly-Burton, M.**, Two new cestodes *Raillietina (Raillietina) bembezi* n.sp. and *Raillietina (Raillietina) bumi* n.sp. from the spotted eagle owl *Bubo africanus* (Temminck), *Rev. Biol. Lisbon*, 3, 87—94, 1962.
2241. **Meyer, M. C.**, Coenuriasis in varying hare in Maine, with remarks on the validity of *Multiceps serialis*, *Trans. Am. Microsc. Soc.*, 74, 163—169, 1955.
2242. **Meyner, R.**, Anatomie und Histologie zweier neuer Tänien-Arten des Subgenus *Bertia*, *Z. Naturwiss.*, 68, 1—106, 1895.
2243. **Michajlow, W.**, *Triaenophorus crassus* Forel *(T. robustus* Olsson) et son développement, *Ann. Parasitol.*, 10, 257—270, 1932.
2244. **Michajlow, W.**, Les adaptations graduelles des copépodes comme premiers hôtes intermédiaires de *Triaenophorus nodulosus* (Pall.), *Ann. Parasitol.*, 10, 334—344, 1932.
2245. **Michajlow, W.**, Les stades larvaires de *Triaenophorus nodulosus* (Pall.), *Ann. Parasitol.*, 11, 339—358, 1933.
2246. **Michajlow, W.**, Sur les stades larvaires de *Triaenophorus nodulosus* (Pall.). Le procercoide, *C. R. Acad. Pol. Sci. Lett. Classe Sci. Math. Nat.*, pp. 53—66, 1933.
2247. **Michajlow, W.**, Ueber die Entwicklung der Eier von *Triaenophorus lucii* (Müll.) in Süss-und Meerwasser, *Zool. Pol.*, 3, 251—259, 1939.
2248. **Mikhail, J. W. and Fahmy, M. A. M.**, Two new species of *Hymenolepis* from insectivores, *Egyptian J. Vet. Sci.*, 13, 69—75, 1976 (published 1977).
2249. **Mikhail, J. W. and Fahmy, M. A. M.**, Two new records of the genus *Mathevotaenia* (Cestodes) with description of a new species and a review of the genus, *Zool. Anz.*, 180, 335—344, 1968.
2250. **Mikhail, J. W. and Fahmy, M. A. M.**, A new species of the genus *Hymenandrya* with a short review of the family Anoplocephalidae, *Zool. Anz.*, 180, 436—441, 1968.
2251. **Mikhail, J. W. and Fahmy, M. A. M.**, Study on some members of genus *Skrjabinotaenia* with a description of a new species and a review of the subfamily Catenotaeniinae Spasskii, 1946, *Zool. Anz.*, 181, 439—450, 1968.
2252. **Millemann, R.**, Studies on the Biology of the Cestode, *Oochoristica deserti* n.sp., Thesis, University of California, Berkeley, 1954.
2253. **Millemann, R.**, Studies on the life history and biology of *Oochoristica deserti* n.sp. (Cestoda: Linstowiidae) from desert rodents, *J. Parasitol.*, 41, 424—440, 1955.
2254. **Miller, H. M., Jr.**, Experiments on immunity of the white rat to *Cysticercus fasciolaris*, *Proc. Soc. Exp. Biol. Med.*, 27, 926—927, 1930.
2255. **Miller, H. M., Jr.**, Superinfection of cats with *Taenia taeniaeformis*, *J. Prev. Med.*, 6, 17—29, 1932.
2256. **Miller, R. B.**, The life history of the pike-whitefish tapeworm, *Triaenophorus crassus*, Rep. Fish. Branch Dep. Lands Mines, province of Alberta, Canada, March 1945 (published 1946).
2257. **Millzner, T. M.**, On the cestode genus *Dipylidium* from cats and dogs, *Univ. Calif. Publ. Zool.*, 28, 317—356, 1926.
2258. **Misra, V. R.**, On a new species of the genus *Oochoristica* from the intestine of *Calotes versicolor*, *Proc. Ind. Acad. Sci. Sect. B*, 22, 1—5, 1945.
2259. **Miyata, I.**, On an encysted larval nematode together with two species of cysticercoids occurring in rat fleas from ships at Kobe, *Vol. Jub. Yoshida*, Vol. I, Osaka Natural History Society, Osaka, 1939, 85—99.
2260. **Mlodzianowska, B.**, Ueber die jüngsten Entwicklungsstadien von *Cysticercus fasciolaris* Rud., der Larve von *Taenia taeniaeformis* Bloch, auf Grund von Experimentaluntersuchungen, *Bull. Int. Acad. Pol. Sci. Lett. Cl. Sci. Math. Nat. S. B. Sci. Nat.*, II, 475—511, 1931.
2261. **Moghe, M. A.**, *Caryophyllaeus indicus* n.sp. (Cestoda) from the catfish *(Clarias batrachus* Bl.), *Parasitology*, 17, 232—235, 1925.
2262. **Moghe, M. A.**, A new species of *Monopylidium*, *M. chandleri*, from the red-nettled lapwing (*Sarcogrammus indicus* Stoliczka), with a key to the species of *Monopylidium*, *Parasitology*, 17, 385—400, 1925.
2263. **Moghe, M. A.**, Two new species of cestodes from Indian Columbidae, *Rec. Ind. Mus.*, 27, 431—437, 1925.
2264. **Moghe, M. A.**, Two new species of cestodes from Indian lizards, *Rec. Ind. Mus.*, 28, 53—60, 1926.
2265. **Moghe, M. A.**, A supplementary note on *Monopylidium chandleri* and other related species, *Parasitology*, 18, 267—268, 1926.

2266. **Moghe, M. A.**, A supplementary description of *Lytocestus indicus* Moghe (syn. *Caryophyllaeus indicus* Moghe, 1925, Cestoda), *Parasitology,* 23, 84—87, 1931.
2267. **Moghe, M. A.**, Four new species of avian cestodes from India, *Parasitology,* 25, 333—341, 1933.
2268. **Moghe, M. A. and Inamdar, N. B.**, Some new species of avian cestodes from India, with a description of *Biuterina intricata* (Krabbe, 1882), *Rec. Ind. Mus.,* 36, 7—16, 1934.
2269. **Mokhtur-Maamouri, F. and Zamali**, *Phyllobothrium pastinacae* n.sp. (Cestoda, Tetraphyllidea, Phyllobothriidae) parasite de *Dasyatis pastinaca* (Linnaeus, 1758), *Ann. Parasitol. Hum. Comp.,* 56, 375—379, 1981.
2270. **Mola, P.**, Su di un cestode del *Carcharodon rondoletti* M. Hle., *Arch. Zool. Napoli,* 1, 345—366, 1903.
2271. **Mola, P.**, Di alcuni species poco studiate o mal noti di cestodi, *Ann. Mus. Zool. R. Univ. Napoli,* n.s., 2, 1—12, 1906.
2272. **Mola, P.**, La ventosa apicale a chi è omologa?, *Zool. Anz.,* 30, 37—44, 1907.
2273. **Mola, P.**, Sopra la *Davainea circumvallata* Krabbe, *Zool. Anz.,* 30, 126—130, 1907.
2274. **Mola, P.**, Les organes génitaux de *Taenia nigropunctata* Crety et en particulier l'organe paruterin, *C. R. Acad. Sci.,* 145, 87—90, 1907.
2275. **Mola, P.**, Di un nuovo cestode del genero *Davainea* Blanch, *Biol. Zentralbl.,* 17, 575—578, 1907.
2276. **Mola, P.**, Un nuovo elminto della *Gallinula chloropus*, *Bull. Acad. R. Belg.,* 64, 886—898, 1907.
2277. **Mola, P.**, Ueber eine neue Cestodenform, *Centralbl. Bakteriol.,* 44, 256—260, 1907.
2278. **Mola, P.**, Nota intorno ad una forma de cestode di pesce fluviatili, *Boll. Soc. Zool. Ital.,* 8, 67—73, 1907.
2279. **Mola, P.**, *Choanotaenia infundibulum* Bloch, *Boll. Soc. Zool. Ital.,* 17, 167—177, 1908.
2280. **Mola, P.**, Due nuove forme di Tetraphyllidae, *Boll. Soc. Adriat. Sci. Nat.,* 24, 1—16, 1908.
2281. **Mola, P.**, *Davainea pluriuncinata* (Crety) e sinonima della *D. circumvallata* Krabbe, *Arch. Parasitol.,* 15, 432—441, 1912.
2282. **Mola, P.**, Die parasiten des *Cottus gobio* Linn. Beitrag zu der helminthologischen Fauna der Teleostei, *Centralbl. Bakteriol. I. Orig.,* 65, 491—504, 1912.
2283. **Mola, P.**, Nuovi ospiti di uccelli contributo al genero *Hymenolepis*, *Biol. Zentralbl.,* 33, 208—222, 1913.
2284. **Mola, P.**, Cestodes avium. Contributo alla fauna elmintologica Sarda, *Arch. Parasitol.,* 22, 577—578, 1919.
2285. **Mola, P.**, Vi e sinonima tra *Davainea bothrioplitis* (Piana) e *Davainea echinobothrida* (Mégnin), *Studi Sassaresi,* 5, 487—491, 1927.
2286. **Mola, P.**, Vermi parassiti dell' ittiofauna italiana. Contributo all patologia ittica, *Boll. di Pesca di Piscic. e Idrobiol.,* Vol. 4, 1928, 48 pp.
2287. **Mola, P.**, Il nuovo genero *Viscoia* Mola, 1929 (Nota), *Studi Sassaresi,* p. 7, 1929.
2288. **Mola, P.**, Descriptio platodorum sine exstis, *Zool. Anz.,* 86, 101—113, 1929.
2289. **Mola, P.**, Due nuove specie de *Onchobothrium* de Blainville (1828), *Biol. Abstr.,* 8, 2082, 1934.
2290. **Molin, R.**, Notzie elmintologiche, *Atti Reale Ist. Veneto Sci. Lett. Arti* 3.s. 2, 146—152, 1857; 216—233, 1857.
2291. **Molin, R.**, Prospectus helminthum, quae in prodromo faunae helminthologicae venetae continentur, *Sitzungber. Akad. Wiss. Wien Math. Naturwiss. Classe Abt. I.,* 30, 127—158, 1858.
2292. **Molin, R.**, Cephalocotylia e nematoidea, *Sitzungber. Akad. Wiss. Wien Math. Naturwiss. Class Abt. I.,* 38, 7—38, 1859.
2293. **Molin, R.**, Prospectus helminthum, quae in parti secunda prodromi faunae helminthologicae venetae continentur, *Sitzungber, Akad. Wiss. Wien Math. Naturwiss. Classe Abt. I.,* 38, 287—302, 1859.
2294. **Molin, R.**, Prodromus faunae helminthologicae venetae, *Denkschr. Akad. Wiss. Wien Math. Naturwiss. Classe Abt. II,* 18, 230—233, 1860.
2295. **Molin, R.**, Prodromus faunae helminthologicae venetae adjestic desquisitionibus anatomicis et criticis, *Denkschr. Akad. Wiss. Wien Math. Naturwiss. Class Abt. II,* 19, 189—338, 1861.
2296. **Moll, A. M.**, Animal parasites of rats at Madison, Wisconsin, *J. Parasitol.,* 4, 89—90, 1917.
2297. **Molnár, K.**, *Bothriocephalus phoxini* sp.n. (Cestoda, Pseudophyllidea) from *Phoxinus phoxinus* L., *Folia Parasitol.,* 15, 83—86, 1968.
2298. **Moniez, R.**, Sur l'embryogénie des cestoides, *C. R. Acad. Sci.,* 85, 974—976, 1877.
2299. **Moniez, R.**, Note sur le *Taenia Krabbei*, espèce nouvelle de Taenia armé, *Bull. Sci. Dép. Nord.,* Scr. 2, 2, 61—163, 1879.
2300. **Moniez, R.**, Note sur deux espèces nouvelles de taenias inermes, *Bull. Sci. Dép. Nord.,* Ser. 2, 2, 163—164, 1879.
2301. **Moniez, R.**, Note sur l'histiologie des tétrarhynques, *Bull. Sci. Dép. Nord.,* Ser. 2, 2, 393—398, 1879.
2302. **Moniez, R.**, Note sur le *Taenia Giardi* et sur quelques espéces du groupe inermes, *C. R. Acad. Sci.,* 88, 1094—1096, 1879.
2303. **Moniez, R.**, Essai monographique sur les cysticerques. Travaux de l'institut zoologique de Lille de la station maritime de Wimereux, Paris, *Trav. Inst. Zool. Lille,* 3, 1—190, 1880.
2304. **Moniez, R.**, Etudes sur les cestodes, *Bull. Sci. Dep. Nord.,* Sér. 2, 3, 240—246, 356—358, 407—409, 1880.

2305. **Moniez, R.**, Notes sur les vaisseaux de l' *Abothrium gadi, Bull. Sci. Dep. Nord.,* Sér. 2, 3, 448, 1880.
2306. **Moniez, R.**, Mémoires sur les cestodes. Travaux de l'institut zoologique de Lille et de la station maritime de Wimereux, Paris, *Trav. Inst. Zool. Lille,* 3, 1—238, 1881.
2307. **Moniez, R.**, Sur quelques types de cestodes, *C. R. Acad. Sci.,* 154, 661—663, 1882.
2308. **Moniez, R.**, Sur le *Taenia nana,* parasite de l'homme, et son cysticerque supposé *(Cysticercus tenebrionis), C. R. Acad. Sci.,* 106, 368—370, 1888.
2309. **Moniez, R.**, Sur la larve du *Taenia Grimaldii* n.sp., parasite du dauphin, *C. R. Acad. Sci.,* 2, 825—827, 1889.
2310. **Moniez, R.**, Le *Gymnorhynchus reptans* Rud. et sa migration, *C. R. Acad. Sci.,* 113, 870—871, 1891.
2311. **Moniez, R.**, Note sur les helminthes, *Rev. Biol. Nord Fr.,* 4, 22—34, 65—79, 108—118, 1891.
2312. **Moniez, R.**, Note sur les helminthes, *Rev. Biol. Nord Fr.,* 150—151, 279, 1892.
2313. **Mönnig, H. O.**, Three new helminths, *Trans. R. Soc. S. Afr.,* 13, 291—298, 1926.
2314. **Mönnig, H. O.**, Helminthological notes. The anatomy and life-history of the fowl tapeworm *(Amoebotaenia sphenoides),* 11th and 12th Rep. Dir. Vet. Educ. Research Union S. Afr. Pretoria, 1926, 199—206.
2315. **Mönnig, H. O.**, Check list of the worm parasites of domesticated animals in South Africa, 13th and 14th Rep. Dir. Vet. Educ. Res. Union S. Afr. Pretoria, 1928, 801—837.
2316. **Monticelli, F. S.**, Contribuzioni allo studio della fauna elmintologica del Golfo di Napoli. I. Ricerche sulle *Scolex polymorphus* Rud., *Mitt. Zool. Stat. Neapel,* 8, 85—152, 1888.
2317. **Monticelli, F. S.**, Intorno allo *Scolex polymorphus* Rud., *Boll. Soc. Nat. Napoli,* 2, 13—16, 1888.
2318. **Monticelli, F. S.**, Notes on some Entozoa in the collection of the British Museum, *Proc. Zool. Soc. London,* pp. 321—325, 1889.
2319. **Monticelli, F. S.**, Elenco degli elminti raccolti dal Capitano G. Chierchia durante il viaggio di circumnavigazione della r. corvetta "Vettor Pisani", *Boll. Soc. Nat. Napoli,* 3, 67—71, 1889.
2320. **Monticelli, F. S.**, *Gyrocotyle* Diesing-*Amiphiptyches* Grube et Wagener. Nota preliminare, *Atti Reale Accad. Lincei Cl. Sci. Fis. Mat. Nat. Rend.,* 5, 228—230, 1889.
2321. **Monticelli, F. S.**, Sul sistema nervoso dell' *Amphiptyches urna* Grube et Wagener, *Zool. Anz.,* 12, 142—144, 1889.
2322. **Monticelli, F. S.**, Alcuni considerazioni biologiche sul genere *Gyrocotyle, Atti Soc. Ital. Sci. Nat. Milano,* 32, 327—329, 1890.
2323. **Monticelli, F. S.**, Note elmintologiche, *Boll. Soc. Nat. Napoli,* 4, 189—208, 1890.
2324. **Monticelli, F. S.**, Elenco degli elminti studiati a Wimereux nella primavera del 1889, *Bull. Sci. Fr. Belg.,* 22, 417—444, 1890.
2325. **Monticelli, F. S.**, Un mot de réponse à Monsieur Loennberg, *Bull. Sci. Fr. Belg.,* 23, 355—357, 1891.
2326. **Monticelli, F. S.**, Notizie su di alcuni specie di *Taenia, Boll. Soc. Nat. Napoli,* 6, 151—174, 1892.
2327. **Monticelli, F. S.**, Appunti sui Cestodaria, *Atti Accad. Sci. Fis. Mat. Nat. Napoli,* 5, 1—11, 1892.
2328. **Monticelli, F. S.**, Sulla cosidetta subcuticola dei Cestodi, *Rend. Accad. Sci. Fis. Mat. Nat. Napoli,* 6, 158—166, 1892.
2329. **Monticelli, F. S.**, Nota intorno a due forme di Cestodi, *Boll. Mus. Zool. Anat. Comp. Univ. Torino,* 7, 1—9, 1892.
2330. **Monticelli, F. S.**, Sul genere *Bothrimonus* Duv. e proposte per una classificazione dei Cestodi, *Monit. Zool. Ital.,* 3, 100—108, 1892.
2331. **Monticelli, F. S.**, Intorno ad alcuni elminti della collezione del museo zoologico della reale università de Palermo, *Nat. Siciliano,* 12, 167—180, 208—216, 1893.
2332. **Monticelli, F. S.**, Sul *Tetrabothrium Gerrardii* Baird, *Atti Soc. Mat. Nat. Modena* (Ser. 4), Ann. 33, 1, 9—26, 1899.
2333. **Monticelli, F. S. and Crety, C.**, Ricerche intorno alla sottofamiglia Solenophorinae Montic. et Crety, *Mem. R. Accad. Sci. Torino,* 31, 381—402, 1891.
2334. **Moore, J. T.**, *Sparganum mansoni,* first reported American case, *Am. J. Trop. Dis.,* 2, 518—529, 1914.
2335. **Morell, A.**, Anatomisch-histologische Studien an Vogeltänien, *Arch. Naturg.,* 61, 81—102, 1895.
2336. **Morenas, L. and Coudert, J.**, Sur un cas d'infestation par le taenia *Hymenolepis diminuta* chez un nourrisson, *Arch. Mal. Appar. Dig.,* 38, 496, 1949.
2337. **Morgan, B. B. and Waller, E. F.**, Severe parasitism in a raccoon *(Procyon lotor lotor* Linnaeus), *Trans. Am. Microsc. Soc.,* 59, 523—547, 1940.
2338. **Motomura, J.**, On *Caryophyllaeus gotoi* n.sp., a new monozoic cestode from Korea, *Sci. Rep. Tohoku Imp. Univ.,* 4 (Ser. 3), 51—53, 1928.
2339. **Motomura, I.**, On the early development of the monozoic cestode *Archigetes appendiculatus,* including the oogenesis and fertilisation, *Annot. Zool. Jpn.,* 12, 109—129, 1929.
2340. **Movsesyan, S. O.**, Two new species of *Raillietina* (Cestoda: Davaineidae) from Columbiformes in South Kirgiz SSR, *Mater. Nauch. Konf. Vses. Ova. Gel'mintol.,* Year 1965, Part. IV., 162—167, 1965.
2341. **Movsesyan, S. O.**, Revision of *Raillietina* Fuhrmann, 1920 (Cestoda: Davaineidae) (in Russian), *Temat. Sb. Rab. Gel'mintol.,* 12, 5—10, 1966.

2342. **Movsesyan, S. O.**, Revision of the genus *Raillietina* Fuhrmann, 1920 (Cestoda: Davaineidae) (in Russian), *Temat. Sb. Rab. Gel'mintol.*, 13, 17—40, 1967.
2343. **Movsesyan, S. O.**, *Raillietina (Raillietina) coturnixi* n.sp. (in Russian), *Temat. Sb. Rab. Gel'mintol.*, 13, 44—77, 1967.
2344. **Movsesyan, S. O.**, Analysis of *Raillietina (Skrjabinia) cesticillus* (Molin, 1858) and a description of *Skrjabinia (Skrjabinia) piiogenesia* n.sp. (Cestoda: Davaineidae) (in Russian). *Papers on Helminthology Presented to Academician K. I. Skrjabin on his 90th Birthday*, Izdat. Akademiya Nauk SSSR, Moscow, 1968, 253—262.
2345. **Movsesyan, S. O.**, *Idiogenes skrjabini* n.sp. and *Raillietina (Raillietina) gvosdevi* n.sp. (Cestoidea: Idiogenidae and Davaineidae) (in Russian), *Parazitologiya*, 2, 454—464, 1968.
2346. **Movsesyan, S. O.**, Revision of the genus *Cotugnia* Diamare, 1893 (Cestoidea: Davaineidae) (in Russian), *Tr. Vses. Inst. Gel'mintol.*, 15, 195—217, 1969.
2347. **Movsesyan, S. O.**, Revision of the genus *Idiogenes* Krabbe, 1867 (Cestoidea: Idiogenidae Mola, 1929) (in Russian), *Sb. Rab. po Gel'mintol. Posvashchen go-letiyu so Dnya Rozhdenniya Akademika K. I. Skrjabina*, Izdatel'stvo KOLOS, Moscow, 1971, 227—247.
2348. **Mrázek, A.**, O Cysticerkoidech našich korýsu sladkovudnich *(in Czech), Věstnik Českoslov. Akad. Zemedelske Bull. Czech. Acad. Agric.*, 1, 226—248, 1890.
2349. **Mrázek, A.**, Recherches sur le développement de quelques Tenias des oiseaux (in Czech with French summary), *Věstnik Českoslov. Akad. Zemedelske*, 1, 97—131, 1891.
2350. **Mrázek, A.**, Ueber die Larve von *Caryophyllaeus mutabilis* Rud., *Centralbl. Bakteriol. I. Orig.*, 29, 485—491, 1891.
2351. **Mrázek, A.** Zur Entwicklungsgeschichte einiger Tänien (German summary), *Sitzungsber. K. Böhm. Ges. Wiss. Prag. Math. Naturwiss. Cl.*, Part 2, 38, 1—16, 1896.
2352. **Mrázek, A.**, *Archigetes appendiculatus* Rátz, *Sitzungsber. K. Böhm. Ges. Wiss. Prag. Math. Naturwiss. Cl.*, 3, 1—47, 1897.
2353. **Mrázek, A.**, Ueber die Larve von *Caryophyllaeus mutabilis* Rud., *Centralbl. Bakteriol.*, 29, 485—491, 1901.
2354. **Mrázek, A.**, Ueber das Verhalten der Längsnerven bei *Abothrium rectangulum* Rud., *Centralbl. Bakteriol.*, 29, 569—571, 1901.
2355. **Mrázek, A.**, Ueber *Taenia acanthorhyncha* Wedl., *Sitzungsber. K. Böhm. Ges. Wiss. Prag. Math. Naturwiss. Cl.*, 7, 1—24, 1905.
2356. **Mrázek, A.**, Cestodenstudien. I. Cysticercoiden aus *Lumbriculus variegatus*, *Zool. Jahrb. Syst.*, 24, 591—624, 1907.
2357. **Mrázek, A.**, Ueber eine neue Art der Gattung *Archigetes. Vorläufige Mitteilung, Centralbl. Bakteriol.*, 46, 719—723, 1908.
2358. **Mrázek, A.**, Ein neues Cysticercoid aus *Tubifex, Centralbl. Bakteriol. I. Abt. Orig.*, 53, 315—317, 1910.
2359. **Mrázek, A.**, Cestodenstudien. II. Die morphologische Bedeutung der Cestodenlarven, *Zool. Jahrb. Anat.*, 39, 515—584, 1916.
2360. **Mrázek, A.**, Organisace a ontogenie larvy druhu *Tatria acanthorhyncha* (Weld.) (Anatomy and ontogeny of the larva of *Tatria acanthorhyncha*, Cestoda.), *Věstník Kralovske České Společnosti Nauk*, 7, 1—12, 1927; French summary pp. 10—12.
2361. **Mudaliar, S. V.**, *Cotugnia brotugerys* Meggitt, 1915, from *Gallus domesticus*, Hosur cattle farm, Madras, *Ind. J. Vet. Sci. Anim. Husb.*, 9, 333, 1939.
2362. **Mudaliar, S. V.**, *Cotugnia bhaleraoi* n.sp., *Ind. J. Vet. Sci. Anim. Husb.*, 13, 166—167, 1943.
2363. **Mudaliar, S. V. and Iyer, K. S. G.**, *Pseudanoplocephala crawfordi* Baylis, 1927, *Ind. J. Vet. Soc. Anim. Husb.*, 8, 235—237, 1938.
2364. **Muehling, P.**, Die Helminthen-Fauna der Wirbeltiere Ostpreussens, *Arch. Naturg.*, 1—118, 1898.
2365. **Mueller, J. F.**, Two new species of the cestode genus *Mesocestoides*, *Trans. Am. Microsc. Soc.*, 46, 294, 1927; The genus *Mesocestoides* in mammals, *Zool. Jahrb. Syst. Oekol.*, 55, 403—418, 1928.
2366. **Mueller, J. F.**, Cestodes of the genus *Mesocestoides* from the opossum and the cat, *Am. Midl. Nat.*, 12, 81—90, 1930.
2367. **Mueller, J. F.**, A *Diphyllobothrium* from cats and dogs in the Syracuse region, *J. Parasitol.*, 21, 114—121, 1935.
2368. **Mueller, J. F.**, Comparative studies on certain species of *Diphyllobothrium*, *J. Parasitol.*, 22, 471—478, 1936.
2369. **Mueller, J. F.**, Spargana in *Natrix*, *Science*, 85, 519—520, 1937.
2370. **Mueller, J. F.**, A repartition of the genus *Diphyllobothrium*, *J. Parasitol.*, 23, 308—310, 1937.
2371. **Mueller, J. F.**, New host records for *Diphyllobothrium mansonoides* Mueller, 1935, *J. Parastiol.*, 23, 313—315, 1937.
2372. **Mueller, J. F.**, The hosts of *Diphyllobothrium mansonoides* (Cestoda: Diphyllobothriidae), *Proc. Helminthol. Soc. Wash.*, 4, 68, 1937.

2373. **Mueller, J. F.**, The life history of *Diphyllobothrium mansonoides* Mueller, 1935, and some considerations with regard to sparganosis in the United States, *Am. J. Trop. Med.*, 18, 41—66, 1938.
2374. **Mueller, J. F.**, Studies on *Sparganum mansonoides* and *Sparganum proliferum*, *Am. J. Trop. Med.*, 18, 303—328, 1938.
2375. **Mueller, J. F.**, An additional species of *Diphyllobothrium* (Subgenus *Spirometra*) from the United States, *Livro Jubil. Prof. Travassos*, Rio de Janeiro, 1938, 337—341.
2376. **Mueller, J. F. and Coulston, F.**, Experimental human infection with the sparganum larva of *Spirometra mansonoides* (Mueller, 1935), *Am. J. Trop. Med.*, 21, 399—425, 1941.
2377. **Mueller, J. F. and Goldstein, F.**, Experimental human infection with *Sparganum mansonides* (Mueller, 1935), *J. Parasitol.*, 25 (Suppl.), 31—32, 1939.
2378. **Mueller, J. F. and Van Cleave, H. J.**, Parasites of Oneida Lake fishes. II. Descriptions of new species and some general taxonomic considerations, especially concerning the trematode family Heterophyidae, *Roosev. Wildlife Ann.*, 3, 79—137, 1932.
2379. **Mukherjee, R. P.**, Two new cestodes from Passeriformes birds, *Ind. J. Helminthol.*, 16, 65—70, 1964.
2380. **Mukherjee, R. P.**, On two species of cestodes from babbler, *J. Zool. Soc. India Year 1965*, 17, 32—36, 1967.
2381. **Mukherjee, R. P.**, Fauna of Rajesthan, India. IX. Cestoda, *Rec. Zool. Surv. India Year 1964*, 62, 191—215, 1970.
2382. **Müller, O. F.**, Von Bandwürmern, *Naturforscher (Halle)*, 14, 129—203, 1780.
2383. **Müller, O. F.**, Vom Bandwürme des Stichlings und vom milchigten Plattwurme, *Naturforscher (Halle)*, 18, 21—37, 1782.
2384. **Müller, O. F.**, Verzeichniss der bisher entdeckten Eingeweidewürmer der Thiere, in welchen sie gefunden wurden, und besten Schriften die dieselben erwähnen, *Naturforscher (Halle)*, 22, 33—86, 1787.
2385. **Muñoz Medina, J. M.**, Cestodes Parásitos Intestinales del Perro y del Gato Domésticos, Tesis, University of Madrid, 1923, 50 + 1 pp.
2386. **Murai, E.**, Cestodes of bats in Hungary, *Parasitol. Hung.*, 9, 41—62, 1976.
2387. **Murai, E. and Tenora, F.**, *Hymenolepis meszarosi* sp.n. (Cestoidea), a parasite of *Alticola roylei* (Rodentia) in Mongolia, *Ann. Hist. Nat. Mus. Natl. Hung.*, 67, 61—63, 1975.
2388. **Murai, E., Tenora, F., and Rocamora, J.-M.**, *Paranoplocephala mascomai* sp.n. (Cestoda: Anoplocephalidae) a parasite of *Microtus cabrerae* (Rodentia) in Spain, *Parasitol. Hung.*, 13, 35—37, 1980.
2389. **Murav'eva, S. I.**, *Tetrabothrius morschtini* n.sp. from the polar gull (in Russian), *Nauchn. Dokl. Vyssh. Shk. Biol. Nauk*, 4, 11—13, 1968.
2390. **Murav'eva, S. I. and Popov, V. N.**, Taxonomic position and some data on the ecology of *Anophryocephalus skrjabini* (Cestoda, Tetrabothriidae), a parasite of pinnipeds (in Russian), *Zool. Zh.*, 55, 1247—1250, 1976.
2391. **Murav'eva, S. I. and Treshchev, V. V.**, *Priapocephalus eschrichtii* n.sp. (Cestoda, Tetrabothriidae), parasitic in whales in Chukchi Sea (in Russian), *Vest. Zool.*, 4, 84—86, 1970.
2392. **Murhar, B. M.**, *Crescentovitus biloculus* gen.nov., sp.nov., a fish cestode (Caryophyllaeidae) from Nagpur, India, *Parasitology*, 53, 413—418, 1963.
2393. **Murie, J.**, On a probably new species of *Taenia* from the rhinoceros, *Proc. Zool. Soc. London*, pp. 608—610, 1870.
2394. **Muro, P., De**, *Choanotaenia littoriae* sp.n., nuovo cestode nell' intestino della civetta *(Athene noctua)*, *Croce Rossa*, 9, 486—490, 1934.
2395. **Nagaty, H. F.**, An account of the anatomy of certain cestodes of the genera *Stilesia* and *Avitellina*, *Ann. Trop. Med. Parasitol.*, 23, 349—380, 1929.
2396. **Nagaty, H. F. and Ezzat, A. E.**, On the identity of *Multiceps multiceps* (Leske, 1780), *M. gaigeri* Hall, 1916 and *M. serialis* (Gervais, 1845), with a review of these and similar forms in man and animals, *Proc. Helminthol. Soc. Wash.*, 13, 33—44, 1946.
2397. **Nagaty, H. F., Hegab, S. M., and Meguid Fahmy, M. A.**, On the identity of *Avitellina woodlandi* and *A. nagatyi*, with further new records of some parasites from Egyptian food mammals, *J. Egypt. Med. Assoc.*, 30, 401—403, 1947.
2398. **Nagoya, T.**, Route of migration of the orally fed *Ligula mansoni* Cobbold in frog and mouse, *Jpn. J. Exp. Med. Tokyo*, 8, 39—54, 1930.
2399. **Nama, H. S.**, *Cylindrotaenia roonwali* sp.n. (Cestoda: Nematotaeniidae) from *Rana cyanophlyctis*, *Proc. Natl. Acad. Sci. India*, 42B, 335—337, 1972.
2400. **Nama, H. S.**, On a new species of *Hymenolepis* from *Funambulus pennanti*, *Proc. Natl. Acad. Sci. India*, 44, 71—74, 1974.
2401. **Nama, H. S.**, A note on some cestodes of goat, *Ind. J. Helminthol.*, 24, 52—55, 1974 (dated 1972).
2402. **Nama, H. S.**, On a new species of *Myotolepis* Spassky, 1954 (Cestoda: Hymenolepididae), *Geobios*, 1, 139—140, 1974.
2403. **Nama, H. S.**, On a new species of *Mathevotaenia* (Cestoda: Anoplocephalidae) from the hedgehog, *Paraechinus micropus micropus* Blyth, *Rev. Brasil. Biol.*, 35, 117—119, 1975.

2404. **Nama, H. S.,** On a new species of *Staphylocystis* Villot, 1877 (Cestoda, Hymenolepididae) from *Suncus murinus sindensis, Acta Parasitol. Pol.,* 24, 19—22, 1976.
2405. **Nama, H. S.,** On a new species of *Oochoristica* (Cestoda, Anoplocephalidae) from the house lizard, *Hemidactylus falviviridis* Ruppell, *Rev. Brasil. Biol.,* 37, 121—123, 1977.
2406. **Nama, H. S.,** On a new species of *Paronia galli* (Cestoda: Anoplocephalidae) from *Gallus domesticus,* in India, *Curr. Sci.,* 47, 352—353, 1978.
2407. **Nama, H. S.,** On a new species of *Mosgovoyia* from the hare, *Lepus* sp. (Cestoda, Anoplocephalidae), *Rev. Brasil. Biol.,* 40, 689—691, 1980.
2408. **Nama, H. S. and Khichi, P. S.,** On a new species of *Oochoristica* Lühe, 1898 from *Varanus monitor, Proc. Natl. Acad. Sci. India,* 42B, 240—244, 1972.
2409. **Nama, H. S. and Khichi, P. S.,** On a new species of *Mathevotaenia* from mongoose, *Herpestes* sp., *Zool. Anz.,* 191, 132—135, 1973.
2410. **Nama, H. S. and Khichi, P. S.,** On a new species of cestode from *Calotes versicolor, Folia Parasitol.,* 21, 373—375, 1974.
2411. **Nama, H. S. and Khichi, P. S.,** A new cestode *Staphylocystis sanchorensis* sp.n. (Hymenolepididae) from the shrew, *Suncus marinus sindensis, Folia Parasitol.,* 22, 93—95, 1975.
2412. **Nama, H. S. and Khichi, P. S.,** Studies on cestodes (Hymenolepididae) from *Columba livia* and *Rattus rattus, Acta Parasitol. Pol.,* 23, 223—228, 1975.
2413. **Narihara, N.,** Studies on the post-embryonal development of *Hymenolepis diminuta.* I. On the hatching of the eggs (in Japanese, with English summary, 730—731), *Taiwan Igakkai Zasshi,* pp. 713—729, 1937.
2414. **Narihara, N.,** Studies on the post-embryonal development of *Hymenolepis diminuta.* II. On the development of the cysticercoid within the definitive host (in Japanese, with English summary, 780—784), *Taiwan Igakkai Zasshi,* 36, 732—784, 1937.
2415. **Neghme, R. A.,** *Diphyllobothrium latum* en Chile. II. Primera encuesta en el Lago Colico (English summary), *Rev. Kuba Med. Trop. Parasitol.,* 6, 134, 1950.
2416. **Neghme, R. A.,** An autochthonous focus of *Diphyllobothrium latum* in the southern hemisphere, *Thapar Commemorative Volume,* 1953, 223—226.
2417. **Neghme, R. A. and Bertin, S. V.,** Estado actual de las investigaciones sobre *Diphyllobothrium latum* en Chile, *Rev. Med. Chile,* 79, 637—640, 1951.
2418. **Neghme, R. A., Donckaster, R. R., and Silva, C. R.,** *Diphyllobothrium latum* en Chile. Primer caso autóctono en el hombre, *Rev. Méd. Chile,* 78, 410—411, 1950.
2419. **Neghme, R. A., et al.,** *Diphyllobothrium latum* en Chile. II. Primera encuesta en el Lago Colico, *Bol. Inf. Parasitar. Chilenas,* 5, 16—17, 1950.
2420. **Neiland, K. A.,** Helminths of Northwestern mammals. II. *Oligorchis nonarmatus* n.sp. (Cestoda: Hymenolepididae) from the yellow-bellied squirrel, *J. Parasitol.,* 38, 341—345, 1952.
2421. **Neiland, K. A.,** A new species of *Proteocephalus* Weinland, 1858, (Cestoda) with notes on its life history, *J. Parasitol.,* 38, 540—545, 1952.
2422. **Neiland, K. A.,** Helminths of Northwestern mammals. V. Observations on cestodes of shrews with the descriptions of new species of *Liga* Weinland, 1857, and *Hymenolepis* Weinland, 1858, *J. Parasitol.,* 39, 487—495, 1953.
2423. **Neiland, K. A.,** The helminth fauna of Nicaragua. I. A new genus and species of cestode (Dilepidinae) from the hummingbird, *Phaecochroa cuvierii roberti, J. Parasitol.,* 41, 495—498, 1955.
2424. **Neiland, K. A. and Senger, C. M.,** Helminths of northwestern mammals. I. Two new species of *Hymenolepis, J. Parasitol.,* 38, 409—414, 1952.
2425. **Neslobinsky, N.,** Zur Kenntnis der Vogeltaenien Mittelrusslands, *Centralbl. Bakteriol.,* 57, 436—442, 1911.
2426. **Neslobinsky, N.,** *Dilepis brachyarthra* Chol. und *Dilepis undulata* Schr., *Centralbl. Bakteriol.,* 59, 416—417, 1911.
2427. **Neumann, L. G.,** Observations sur les Ténias du mouton, *Bull. Soc. Hist. Nat. Toulouse,* 24, 6—9, 1891.
2428. **Neumann, L. G.,** Sur la place de *Taenia ovilla* Riv. dans la classification, *Bull. Soc. Hist. Nat. Toulouse,* 26, 12—14, 1892.
2429. **Neumann, L. G.,** Note sur les téniadés du chien et du chat, *Mém. Soc. Zool. Fr.,* 9, 171—184, 1896.
2430. **Neumüller, O.,** *Echinococcus granulosus (Taenia echinococcus)* beim Fuchs, ein Beitrag zur Entstehung der Hülsenwurmkrankheit, *Z. Fleisch. Milchyg.,* 43, 3—4, 1932.
2431. **Nevenič, Vladislav V. and Markovič, B.,** La chèvre comme un hôte intermédiaire pour *Taenia serialis* (Gervais, 1847) (in Serbo-Croatian), *Acta Vet. Beograd,* 1, 128—131, 1951.
2432. **Newton, M. V. B.,** The biology of *Triaenophorus tricuspidatus* (Bloch, 1889) in western Canada, *Contrib. Can. Biol. Fish.,* n.s. 7, 341—360, 1932.
2433. **Nicholson, D.,** Fish tapeworm; intestinal infection in man, the infestation in Manitoba lakes, *Can. Med. Assoc. J.,* 19, 25—33, 1928.
2434. **Nicholson, D.,** The *Triaenophorus* parasite in the flesh of the tullibee *(Leucichthys), Can. J. Res.,* Sect. D, 6, 162—165, 1932.

2435. **Nicholson, D.**, *Diphyllobothrium* infection in *Esox lucius, Can. J. Res.*, Sect. D, 6, 166—170, 1932.
2436. **Nicherson, W. S.**, The broad tapeworm in Minnesota, with the report of a case of infection acquired in the state, *JAMA*, 46, 711—713, 1906.
2437. **Nicoll, W.**, A contribution toward a knowledge of the Entozoa of British marine fishes. I, *Ann. Mag. Nat. Hist.* (Ser. 7), 19, 66—94, 1907.
2438. **Nicoll, W.**, A contribution towards a knowledge of the Entozoa of British marine fishes. II., *Ann. Mag. Nat. Hist.*, (Ser. 8), 4, 1—25, 1909.
2439. **Nicoll, W.**, On the Entozoa of fishes from the Firth of Clyde, *Parasitology* 3, 322—359, 1910.
2440. **Nicoll, W.**, Recent progress in our knowledge of parasitic worms, *Parasitology*, 6, 141—152, 1913; *Parasitology*, 14, 378—410, 1922.
2441. **Niemiec, J.**, Recherches sur le système nerveux des Ténias, *Rec. Zool. Suisse*, 2, 589—648, 1885.
2442. **Niemiec, J.**, Untersuchungen ueber das Nervensystem der Cestoden, *Arb. Zool. Inst. Wien*, 7, 1—60, 1888.
2443. **Nigrelli, R. F.**, Parasites of the swordfish, *Xiphias gladius* Linnaeus, *Am. Mus. Nov.*, No. 996, 1—16, 1938.
2444. **Nitsche, H.**, Untersuchungen ueber den Bau der Tänien, *Z. Wiss. Zool.*, 23, 181—197, 1873.
2445. **Nitzsch, C. L.**, *Bothriocephalus*, in Ersch, J. S. und Gruber, J. G., *Allgemeine Encyclopaedie der Wissenschaften und Künste*, Ersch and Gruber, Leipzig, 12, 94—99, 1824.
2446. **Noble, G. A. and Tesh, R. B.**, *Monoecocestus diplomys* sp.n. (Cestoda: Anoplocephalidae) from the rat, *Diplomys darlingi, J. Parasitol.*, 60, 605—607, 1974.
2447. **Noda, R.**, Studies on the parasites in the digestive tract of poultry. II. On the tapeworm, *Amoebotaenia sphenoides* (in Japanese), *Jpn. J. Vet. Sci.*, 13, 261, 386, 1951.
2448. **Nuñez, M. O.**, de, Estudios preliminares sobra la fauna parasitaria de algunes elasmobranquios del litoral bon aerense, Mar del Plata, Argentina. I. Cestodes y trematodes de *Psammobatis microps* (Günther) y *Zapteryx brevirostris* (Müller y Henle), *Physis*, 30, 425—446, 1971.
2449. **Nybelin, O.**, Notizen ueber Cestoden. I. Ueber *Progynotaenia odhneri* einen neuen Vogelcestoden aus Schweden, *Zool. Bidrag. Uppsala*, 3, 225—230, 1914.
2450. **Nybelin, O.**, Neue Tetrabothriiden aus Vögeln., *Zool. Anz.*, 47, 297—301, 1916.
2451. **Nybelin, O.**, Zur Frage der Entwicklungsgeschichte einiger Bothriocephaliden, *Göteborgs Kgl. Vetenskaps-Akad. Handl.*, 19, 1—12, 1918.
2452. **Nybelin, O.**, Zur Anatomie und systematischen Stellung von "*Tetrabothrium norvegicum*" Olsson, *Göteborgs Kgl. Vetenskaps-Akad. Handl.*, 20, 1918, 25 pp.
2453. **Nybelin, O.**, Zur Entwicklungsgeschichte von *Schistocephalus solidus, Centrabl. Bakteriol.*, 83, 295—297, 1919.
2454. **Nybelin, O.**, Anatomisch-systematische Studien ueber Pseudophyllidien, *Göteborgs Kgl. Vetenskap-Akad. Handl.*, 26, 1—228, 1922.
2455. **Nybelin, O.**, *Anoplocephala gorillae, Arkiv. Zool.*, 19, 1—3, 1927.
2456. **Nybelin, O.**, Zwei neue Cestoden aus Bartenwalen, *Zool. Anz.*, 78, 309—314, 1928.
2457. **Nybelin, O.**, Säugetier-und Vogelcestoden von Juan Fernandez, *Nat. Hist. Juan Fernandez Easter Island*, Skottsberg, C., Ed., Vol. 3, *Zoology*, 1931, 493—523.
2458. **Nybelin, O.**, Sur le *Tetrarhynchus minutus* P. J. van Beneden, *Göteborgs Kgl. Vetenk. Vetters-Samh. Handl.* (Ser. B), 6, 1—20, 1940.
2459. **Nybelin, O.**, Zur Helminthfauna der Süsswasserfische Schwedens. II. Die Cestoden des Welses, *Göteborgs Kgl. Vetenskaps-Akad. Handl.*, Ser. B, 1, 1—24, 1942.
2460. **Obersteiner, W.**, Ueber eine neue Tetraphyllide *(Bilocularia* n. *hyperapolytica* n.), *Zool. Anz.*, 42, 57—58, 1913.
2461. **Obersteiner, W.**, Ueber eine neue Cestodenform *Bilocularia hyperapolytica* nov. gen. nov. spec. aus *Centrophorus granulosus, Arb. Zool. Inst. Univ. Wien*, 20, 109—124, 1914.
2462. **Obitz, K.**, Recherches sur les oeufs de quelques Anoplocéphalides, *Ann. Parasitol.*, 12, 40—55, 1934.
2463. **Obushenkov, I. N.**, *Monordotaenia alopexi* sp.n. (Cestoda: Taeniidae) a cestode from polar fox of Chukotka, *Parazitologiya*, 17, 61—83, 1983.
2464. **Odening, K.**, Zum systematischen Status und zur Verbreitung der in Europäischen Schlangen schmarotzenden Proteocephaliden (Cestoidea: Proteocephala) nebst Bemerken zur Gattungszugehörigkeit einer madagassischen Proteocephalidae — Art aus Schlangen, *Z. Parasitenkd.*, 23, 226—234, 1963.
2465. **Odhner, T.**, *Urogonoporus armatus* Lühe, 1901, die reifen Proglottiden von *Trilocularia gracilis* Olsson, 1869, *Arch. Parasitol.*, 8, 465—471, 1904.
2466. **Odhner, T.**, Die Homologien der weiblichen Genitalwege bei den Trematoden und Cestoden, *Zool. Anz.*, 39, 337—351, 1912, 1913.
2467. **Ogren, R. E.**, Development and morphology of the oncospheres of *Mesocestoides corti*, a tapeworm of mammals, *J. Parasitol.*, 42, 414—428, 1956.
2468. **Ogren, R. E.**, Embryonic development and morphology of onchospheres of the tapeworm, *Oochoristica symmetrica, J. Parasitol.*, 42, Sect. 2, 30, 1956.

2469. **Ogren, R. E.,** Morphology and development of oncosphere of the cestode *Oochoristica symmetrica* Baylis, 1927, *J. Parasitol.,* 43, 505—520, 1957.
2470. **Okorokov, V. I. and Tkachev, V. A.,** A new cestode, *Tatria jubilaea* n.sp., from *Podiceps auritus* and *P. ruficollis* (in Russian), *Voprosy Zool.,* 3, 75—78, 1973.
2471. **Okumura, T.,** An experimental study of the life history of *Sparganum mansoni* Cobbold (a preliminary report), *Kitasato Arch. Exp. Med.,* 3, 190—197, 1919.
2472. **Oldham, J. N.,** On *Hymenolepis sinensis* n.sp., a cestode from the grey sandhamster *(Cricetulus griseus), J. Helminthol.,* 7, 235—246, 1929.
2473. **Oldham, J. N.,** The helminth parasites of marsupials, *Imp. Bur. Agric. Parasitol. Notes Memo.,* No. 10, 1930, 62 pp.; reprinted from *J. Helminthol.,* 8, 1930.
2474. **Oldham, J. N.,** On the arthropod intermediate hosts of *Hymenolepis diminuta* (Rudolphi, 1819), *J. Helminthol.,* 9, 21—28, 1931.
2475. **Oldham, J. N.,** The helminth parasites of common rats, *J. Helminthol.,* 9, 49—90, 1931.
2476. **Oldham, J. N.,** Hand-list of helminth parasites of the rabbit, *Imp. Bur. Agric. Parasitol. Notes Memo.* No. 2, 12 pp; reprinted from *J. Helminthol.,* 9, 105—116, 1931.
2477. **Oldham, J. N.,** The helminth parasites of deer, *Imp. Bur. Agric. Parasitol. Notes Memo.,* No. 4; reprinted from *J. Helminthol.,* 9, 217—248, 1931.
2478. **Oliger, I. M.,** Parasite fauna of tetraonid birds of the forest zone of the European part of the RSFSR (in Russian), *Tr. Gel'mintol. Lab. Akad. Nauk SSSR,* 6, 411—412, 1952.
2479. **Olsen, O. W.,** A new species of cestode, *Dendrouterina nycticoracis* (Dilepididae) from the black-crowned night heron *(Nycticorax nycticorax hoactli* Gmelin), *Proc. Helminthol. Soc. Wash.,* 4, 30—32, 1937.
2480. **Olsen, O. W.,** A new species of cestode, *Dendrouterina lintoni* (Dilepididae) from the little green heron *(Butorides virescens virescens* Linn.) *Proc. Helminthol. Soc. Wash.,* 4, 72—75, 1937.
2481. **Olsen, O. W.,** Anoplocephaliasis in Minnesota horses, *J. Am. Vet. Med. Assoc.,* 92, 557—559, 1938.
2482. **Olsen, O. W.,** *Deltokeras multilobatus,* a new species of cestode (Paruterinae, Dilepididae) from the twelve-wired bird of paradise *Selucides melanoleucus* Dudin (Passeriformes), *Zoologica,* 24, 341—344, 1939.
2483. **Olsen, O. W.,** The cysticercoid of the tapeworm, *Dendrouterina nycticoracis* Olsen, 1937 (Dilepididae), *Proc. Helminthol. Soc. Wash.,* 6, 20—21, 1939.
2484. **Olsen, O. W.,** Schizotaeniasis in muskrats, *J. Parasitol.,* 25, 279, 1939.
2485. **Olsen, O. W.,** *Tatria duodecacantha,* a new species of cestode (Amabiliidae Braun, 1910) from the piedbilled grebe *(Podilymbus podiceps podiceps* Linn.), *J. Parasitol.,* 25, 495—499, 1939.
2486. **Olsen, O. W.,** *Diplogynia americana,* a new species of cestode (Hymenolepididae) from the eastern little green heron *(Butorides virescens* Linn.), *Trans. Am. Microsc. Soc.,* 59, 183—186, 1940.
2487. **Olsen, O. W.,** *Diplophallus taglei* n.sp. (Cestoda: Cyclophyllidea) from the viccacha, *Lagidium peruanum* Meyer, 1832 (Chinchillidae) from the Chilean Andes, *Proc. Helminthol. Soc. Wash.,* 33, 49—53, 1966.
2488. **Olsen, O. W.,** *Hymenolepis pribilofensis* n.sp. (Cestoda: Hymenolepididae) from the Pribilof shrew *(Sorex pribilofensis* Merriam) from the Pribilof Islands, Alaska, *Can. J. Zool.,* 47, 449—454, 1969.
2489. **Olsen, O. W.,** *Aploparaksis tinamoui* n.sp., cestode (Hymenolepididae) from the Chilean tinamou *(Notoprocta perdicaria* [Kittlitz, 1830] Tinaniformes), *Rev. Iber. Parasitol.,* 30, 701—718, 1970.
2490. **Olsen, O. W.,** *Monoecocestus torresi* n.sp. (Cestoda: Cyclophyllidea: Anoplocephalidae) from the tucotuco *Ctenomys maulinus brunneus* Osgood, 1943 (Hystrichomorpha: Rodentia), *Rev. Parasitol.,* 36, 209—217, 1976.
2491. **Olsen, O. W., Haskins, A. G., and Braun, C. E.,** *Rhabdometra alpinensis* n.sp. (Cestoda: Paruterinidae: Dilepididea) from southern white-tailed ptarmigan *(Lagopus leucurus altipetens* Osgood) in Colorado, U.S.A., with a key to the species of *Rhabdometra* Cholodkowsky, 1906, *Can. J. Zool.,* 56, 446—450, 1978.
2492. **Olsen, O. W. and Kuntz, R. E.,** *Staphylocystis (Staphylocystis) suncensis* sp.n. (Cestoda: Hymenolepididae) from the musk shrew, *Suncus murinus* (Soricidae), from Taiwan, with a key to the known species of *Staphylocystis* Villot, 1877, *Proc. Helminthol. Soc. Wash.,* 45, 182—189, 1978.
2493. **Olsen, O. W. and Kuntz, R. E.,** *Fuhrmannetta (Fuhrmannetta) bandicotensis* sp.n. of cestoda (Eucestoda, Davaineidea, Davaineidae) from the bandicoot *(Bandicota indica nemorivaga* Hodgson, 1836) from Taiwan, *Proc. Helminthol. Soc. Wash.,* 46, 79—83, 1979.
2494. **Olsson. P.,** Entozoa, iakttagna hos Skandinaviska Hafsfiskar. Platyhelminthes. I, *Lunds. Univ. Arsskr.,* 3, 41—59, 1868.
2495. **Olsson, P.,** Entozoa, iakttagna hos Skandinaviska Hafsfiskar. Platyhelminthes. (Forts.) (in Swedish), *Lunds. Univ. Arsskr.,* 4, 1—64, 1868.
2496. **Olsson, P.,** Berattelse om en zoologisk resa till Bohuslan och Skagerack somsaren 1868 (in Swedish), *Ofversigt Kgl. Svenska Vetenskaps-Akad. Handl.,* 25, 471—485, 1868.
2497. **Olsson, P.,** Om entozoernas geografiska utbredning och foerekomst hos olika djur (in Swedish), *Forhandl. Skandinavisk. Naturforsk.,* 31, 481—515, 1869.

2498. **Olsson, P.**, Bidrag till Skandinaviens helminth fauna. I. (in Swedish), *Kgl. Svenska Vetenskaps-Akad. Handl.*, 14, 1—35, 1876.
2499. **Olsson, P.**, Bidrag till Skandinaviens helminth fauna. II. (in Swedish), *Kgl. Sevenskas Vetenskaps-Akad. Handl.*, 25, 1—41, 1893.
2500. **Olsson, P.**, Sur *Chimaera monstrosa* et ses parasites, *Mém. Soc. Zool. Fr.*, 9, 499—512, 1896.
2501. **Omran, L. A. M., El-Naffar, M. K., and Mandour, A. M.**, *Biuterinoides aegypti*, a new cestode from the intestine of *Upupa epops*, *J. Egyptian Soc. Parasitol.*, 11, 171—174, 1981.
2502. **Ortiz, C. I.**, Communicacion preliminar sobre una posible neuva parasitosis intestinal en Venezuela, *Bol. Lab. Clin. "Luis Razetta"*, 6, 287—291, 1945.
2503. **Ortlepp, R. J.**, A new davaineid cestode — *Raillietina (Paroniella) macropa* sp.n. from a wallaby, *Ann. Mag. Nat. Hist.*, 9, 602—612, 1922.
2504. **Ortlepp, R. J.**, On a collection of helminths from a South African farm, *J. Helminthol.*, 4, 127—142, 1926.
2505. **Ortlepp, R. J.**, *Joyeuxia furhmanni* Baer, 1924, a hitherto unrecorded cestode parasite of the domesticated cat in South Africa, *Onderstepoort J. Vet. Sci. Anim. Ind.*, 1, 97—98, 1933.
2506. **Ortlepp, R. J.**, *Echinococcus* in dogs from Pretoria and vicinity, *Onderstepoort J. Vet. Sci. Anim. Ind.*, 3, 97—108, 1934.
2507. **Ortlepp, R. J.**, South African helminths. I, *Onderstepoort J. Vet. Sci. Anim. Ind.*, 9, 311—336, 1937.
2508. **Ortlepp, R. J.**, On two cestodes recovered from a South African kite, *Livro Jub. Travassos*, 1938, 353—358.
2509. **Ortlepp, R. J.**, South African helminths. II. Some taenias from large wild carnivores, *Onderstepoort J. Vet. Sci. Anim. Ind.*, 10, 253—278, 1938.
2510. **Ortlepp, R. J.**, South African helminths. III. Some mammalian and avian cestodes, *Onderstepoort J. Vet. Sci. Anim. Ind.*, 11, 23—50, 1938.
2511. **Ortlepp, R. J.**, South African helminths. IV. Cestodes from Columbiformes, *Onderstepoort J. Vet. Sci. Anim. Ind.*, 11, 51—61, 1938.
2512. **Ortlepp, R. J.**, South African helminths. V. Some avian and mammalian helminths, *Onderstepoort J. Vet. Sci. Anim. Ind.*, 11, 63—104, 1938.
2513. **Ortlepp, R. J.**, South African helminths. VI. Some helminths, chiefly from rodents, *Onderstepoort J. Vet. Sci. Anim. Ind.*, 12, 75—101, 1939.
2514. **Ortlepp, R. J.**, South African helminths. VII. Miscellaneous helminths, chiefly cestodes, *Onderstepoort J. Vet. Sci. Anim. Ind.*, 14, 97—110, 1940.
2515. **Ortlepp, R. J.**, On two new *Catenotaenia* tapeworms from a South African rat with remarks on the species of the genus, *Onderstepoort J. Vet. Res.*, 29, 11—19, 1962.
2516. **Ortlepp, R. J.**, Observations on cestode parasites of Guinea fowl from Southern Africa, *Onderstepoort J. Vet. Res.*, 30, 95—118, 1963.
2517. **Ortlepp, R. J.**, *Pseudandrya mkuzii* sp.nov. (Cestoda: Hymenolepididae) from *Ichneumia albicanda*, *Onderstepoort J. Vet. Res.*, 30, 127—132, 1963.
2518. **Oshmarin, P. G.**, *Skrjabinoparaksis orsenjevi* n.sp. and its position in the Hymenolepididae (in Russian), *Papers on Helminthology Presented to Academician K. I. Skrjabin on his 80th Birthday*, Moscow, Izdatelstvo Academii Nauk SSSR, Moscow, 1958, 257—260.
2519. **Oshmarin, P. G.**, *Polytestilepis chitinocloacis* n.g., n.sp. from ducks, Soobshcheniya Dalnevostochnogo Filiala im. V.L. Komarova Sibirskogo Otdeleniya, Akad. Nauk SSSR, pp. 133—136, 1960.
2520. **Oshmarin, P. G.**, *Helminths of Mammals and Birds in the Primorsk Region*, Izdatelstvo Akademii Nauk SSSR, Moscow, 1963, 323 pp.
2521. **Oshmarin, P. G.**, New species and genus of cestode from *Turnix tanki*, *Parazitologiya*, 6, 558—561, 1972.
2522. **Oshmarin, P. G. and Morozov, F. N.**, Substitution of the fixative function of suckers in the cestode *Aploparaksis sobolevi* nov.sp. (in Russian), *Dokl. Akad. Nauk SSSR*, n.s., 59, 1509—1511, 1948.
2523. **Osler, C. P.**, A new cestode from *Rana clamitans* Latr., *J. Parasitol.*, 17, 183—186, 1931.
2524. **Oswald, V. H.**, Three new hymenolepidid cestodes from the smoky shrew, *Sorex fumeus* Miller, *J. Parasitol.*, 37, 573—576, 1951.
2525. **Oswald, V. H.**, The taxonomics of the genus *Protogynella* Jones, 1943 (Cestoda: Hymenolepididae), with a description of *Protogynella pauciova* n.sp. provis, *Ohio J. Sci.*, 55, 200—208, 1955.
2526. **Oswald, V. H.**, A redescription of *Pseudodiorchis reynoldsi* (Jones, 1944) (Cestoda: Hymenolepididae), a parasite of the short-tailed shrew, *J. Parasitol.* 43, 464—468, 1957.
2527. **Otto, G. F.**, Human infestation with the dwarf tapeworm *(Hymenolepis nana)* in the southern United States, *Am. J. Hyg.*, 23, 25—32, 1936.
2528. **Owen, R.**, Notes on the anatomy of the flamingo, *Phoenicopterus ruber* Linn., *Proc. Zool. Soc. London*, pp. 141—142, 1832.
2529. **Owen, R.**, Notes on the anatomy of *Corythaix porphyreolopha*, *Proc. Zool. Soc. London*, pp. 3—5, 1834.

2530. **Owen, R.,** Description of a new species of tapeworm *Taenia lamelligera, Trans. Zool. Soc. London,* 1, 385—386, 1935.
2531. **Owen, R. L.,** A new species of cestode, *Choanotaenia numenii,* from the long-billed curlew, *Trans. Am. Microsc. Soc.,* 65, 346—350, 1946.
2532. **Pagenstecher, H.,** Beitrag zur Kenntnis der Geschlechtsorgane der Taenien, *Z. Wiss. Zool.,* 9, 523—528, 1858.
2533. **Pagenstecher, H.,** Ueber *Echinococcus* bei *Macropus major, Verhandl. Nat. Med. Ver. Heidelberg,* 5, 181—186, 1871.
2534. **Pagenstecher, H.,** Ueber *Echinococcus* bei *Tapirus bicolor, Verhandl. Nat. Med. Ver. Heidelberg,* 6, 93—95, 1872.
2535. **Pagenstecher, H.,** Zur Naturgeschichte der Cestoden, *Z. Wiss. Zool.,* 30, 171—193, 1877.
2536. **Palacios, N. M.,** Céstodes de Vertebrados, Thesis, Universidad National Antonoma de Mexico, Mexico City, 1963, 81 pp.
2537. **Palacios, N. M. and Barroeta,** Céstodes de vertebrados. XI, *Rev. Ibér. Parsitol.,* 27, 43—62, 1967.
2538. **Palais, M.,** Les anomalies des cestodes. Recherches expérimentales sur *Hymenolepis diminuta* (Rud.), *Ann. Fac. Sci. Marseille,* 6, 111—163, 1933.
2539. **Palais, M.,** Résistance des rats à l'infestation d' *Hymenolepis diminuta, C. R. Soc. Biol.,* 117, 1016—1017, 1934.
2540. **Pallas, P. S.,** Bemerkungen ueber die Bandwürmer in Menschen und Thieren. N., *Nord Beytr. Phys. Geogr. Erd- Völkerbeschr.,* I, 39—112, 1781.
2541. **Palmer, D. G.,** Three cestodes from the black swan: *Hymenolepis chenopsis* sp.n. and redescriptions of *Parabisaccanthes bisucculina* (Szpotańska, 1931) Maksimova, 1963, and *Monosaccanthes kazachstanica* (Maksimova, 1963) Czaplinski, 1967, *Acta Parasitol. Pol.,* 28, 125—139, 1981.
2542. **Pandey, K. C.,** Studies on some cestodes from fishes, birds and mammals, *Ind. J. Zootomy,* 14, 221—226, 1973.
2543. **Pandey, K. C. and Tayal, V.,** On two new cestodes of the genus *Staphylepis* Spassky and Oshmarin, 1954, *Ind. J. Parasitol.,* 5, 43—46, 1981.
2544. **Pandey, K. C. and Tayal, V.,** On a rare cestode, *Malika woodlandi* n.sp., from the intestine of *Burhinus oedicnemus* (Linn.), *Ind. J. Parasitol.,* 5, 47—48, 1981.
2545. **Pandey, K. C. and Tayal, V.,** On a rare cestode, *Spiniglans southwelli,* n.sp., from a grey quail, *Coturnix coturnix* (Linn.), *Ind. J. Parasitol.,* 5, 69—70, 1981.
2546. **Pandey, K. C. and Tayal, V.,** *Malika chauhani* n.sp. from *Burhinus oedicnemus* (Linn.), *Ind. J. Parasitol.,* 5, 71—73, 1981.
2547. **Parihar, A. and Nama, H. S.,** *Catenotaenia indica* sp.n. (Cestoda: Catenotaenidae) from the Indian gerbil, *Tatera indica indica* Hardwicke, *Ind. J. Parasitol.,* 1, 137—139, 1977.
2548. **Parodi, S. E. and Widakowich, V.,** Sobre una nueva especie de *Taenia, Prensa Méd. Argent.,* 2, 337—338, 1916.
2549. **Parodi, S. E. and Widakowich, V.,** Cestodes del genero *Bothriocephalus* parásitos de algunas especies de nuestros felinos salvajes, *Rev. Jard. Zool. Buenos Aires,* 13, 222—227, 1917.
2550. **Parona, C.,** Osservazioni intorno ad un caso di cisticerco nel mufflone de Sardegna, *Ann. Reale Accad. Agric. Torino,* 26, 91—97, 1883.
2551. **Parona, C.,** Di alcuni elminti raccolti nel Sudan orientale da O. Beccari e P. Malgretti, *Ann. Mus. Civ. Stor. Nat. Genova,* 22, 424—445, 1885.
2552. **Parona, C.,** Elmintologia Sarda. Contribuzione allo studio del vermi parassiti in animali di Sardegna, *Ann. Mus. Civ. Stor. Nat. Genova,* 24, 275—384, 1887.
2553. **Parona, C.,** Res ligusticae. II. Vermi parassiti in animalia della Liguria, *Ann. Mus. Civ. Stor. Nat. Genova,* 24, Ser. 2, v. 4, 483—501, 1887.
2554. **Parona, C.,** *Hymenolepis moniezi* n.sp. parassita del *Pteropus medius* et *H. acuta (Taenia acuta* Rud.) dei pipistrelli nostrali, *Atti Soc. Ligust. Sci. Nat. Genova,* 4, 202—206, 1893.
2555. **Parona, C.,** Note intorno agli elminti del museo zoologico Torino, *Boll. Mus. Zool. Anat. Comp. Torino,* 11, 1896, 6 pp.
2556. **Parona, C.,** Elminti raccolti dal Dottore Elio Modigliani alle isole Mentawei, Engano e Sumatra, *Ann. Mus. Civ. Stor. Nat. Genova,* 19, 102—124, 1898.
2557. **Parona, C.,** Catalogo di elminti caccolti in vertebrati dell' isola d'Elba dal Dott. Giacomo Damiani, *Atti Soc. Ligust. Sci. Nat. Genova,* 10, 85—100, 1899.
2558. **Parona, C.,** Helminthum ex Conradi Paronae, Museo Catalogus, Cestodes, Genova, 1900, 6 pp.
2559. **Parona, C.,** Di alcuni elminti del museo nacional di Buenos Aires, *Commun. Mus. Nac. Buenos Aires,* 1, 190—197, 1900.
2560. **Parona, C.,** Catalogus Helminthum ex Conradi Paronae Museo, Geneva, Sect. II, 1900, 6 pp.
2561. **Parona, C.,** Di alcuni cestodi brasiliani raccolti dal Dott. Adolfo Lutz, *Boll. Mus. Zool. Anat. Comp. Genova,* No. 102, 1901, 12 pp; *Atti Soc. Ligust. Sci. Nat. Genova,* 12, 3—14, 1901.

2562. **Parona, C.**, Catalogo di elminti raccolti in vertebrati dell' isola d'Elba, *Boll. Mus. Zool. Anat. Comp. Genova*, 1902, 20 pp.
2563. **Parona, C.**, Due casi rari di *Coenurus serialis* Gerv., *Boll. Mus. Zool. Anat. Comp. Genova*, 1902, 1—6.
2564. **Parona, C.**, Elminti. Osservazioni Scientifiche Eseguite durante la Spedizione Polare de S.A.R. Luigi Amedeo si Savoia, duca degli Abruzzi, 1899—1900, Milano, 1903, 633—635.
2565. **Parona, C.**, Vermi parassiti di vertebrati, *Ruwenzori Parte Sci.*, 1, 415—422, 1909.
2566. **Parra, B. E.**, *Paranoplocephala threlkeldi,* a new species of tapeworm from *Lagidium peruanum, J. Tenn. Acad. Sci.*, 27, 205, 1952.
2567. **Parra, O. B.**, *Perutaenia threlkeldi* n.g., n.sp. (Cestoda: Anoplocephalidae) from *Lagidium peruanum, J. Parasitol.*, 39, 252—255, 1953.
2568. **Parrot, L. and Joyeux, C.**, Les cysticercoides de *Tarentola mauritanica* L. et les ténias du chat, *Bull. Soc. Pathol. Exot.*, 13, 687—695, 1920.
2569. **Paspalewa, A. and Woidowa, S. M.**, *Raillietina macracanthos* n.sp. (Davaineidae) vom specht *Picus viridis* (L.), *Izvest. Zool. Inst. Muzei*, 30, 133—139, 1969.
2570. **Pasquali, A.**, Le tenie dei polli di Massaua. Descrizione de una nouva specie, *Giorn. Int. Sci. Med. Napoli*, 12, 905—910, 1890.
2571. **Patwardhan, S. S.**, On two new species of cestodes from a snipe, *Zool. Jahrb. Syst.*, 66, 541—548, 1935.
2572. **Pearse, A. S.**, Observations on parasitic worms from Wisconsin in fishes, *Trans. Wisc. Acad. Sci. Arts Lett.*, 21, 147—160, 1924.
2573. **Pearse, A. S.**, The parasites of lake fishes, *Trans. Wisc. Acad. Sci.*, 21, 161—194, 1924.
2574. **Peery, H. J.**, A new unarmed tapeworm from a spotted skunk, *J. Parasitol.*, 25, 481—490, 1939.
2575. **Penfold, H. B.**, The life-history of *Cysticercus bovis* in the tissues of the ox, *Med. J. Aust.*, 24th Year, I, 579—583, 1937.
2576. **Penfold, W. J., Penfold, H. B., and Phillips, M.**, A survey of the incidence of *Taenia saginata* infestation in the population of the State of Victoria from January, 1934, to July, 1935, *Med. J. Aust.*, 23rd Year, I, 283—285, 1936.
2577. **Penfold, W. J., Penfold, H. B., and Phillips, M.**, Ridding pasture of *Taenia saginata* ova by grazing with cattle or sheep, *J. Helminthol.*, 14, 135—140, 1936.
2578. **Penfold, W. J., Penfold, H. B., and Phillips, M.**, *Taenia saginata:* its growth and propagation, *J. Helminthol.*, 15, 41—48, 1937.
2579. **Penfold, W. J., Penfold, H. B., and Phillips, M.**, The criteria of life and viability of mature *Taenia saginata* ova, *Med. J. Aust.*, 24th Year, I, 1—5, 1937.
2580. **Penner, L. R.**, A hawk tapeworm which produces proliferating cysticercus in mice, *J. Parasitol.*, 24 (Suppl.), 25, 1938.
2581. **Pérez-Vigueras, I.**, *Ophiotaenia barbouri* n.sp. (Cestoda) parásito de *Tretanorhynchus variabilis* (Reptilia), *Mem. Soc. Cubana Hist. Nat.*, 8, 231—234, 1934.
2582. **Pérez-Vigueras, I.**, *Proteocephalus manjuariphilus* n.sp. (Cestoda) parasito de *Atractosteus tristoechus* (Bloch and Schn.) (Pisces), *Rev. Parasitol. Clin. Lab. Havana*, 2, 17—18, 1936.
2583. **Pérez-Vigueras, I.**, Nota sobre *Hymenolepis chiropterophila* n.sp. y clave para la determinacion de *Hymenolepis* de Chiroptera, *Univ. Habana*, 6, 152—163, 1941.
2584. **Pérez-Vigueras, I.**, Notas sobre algunos cestodes encontrados en Cuba, *Libro Hommaje al Dr. Eduardo Caballero y Caballero, Jubileo, Universidad Nacional Autónoma de Mexico, 1930-1960*, 1960, 377—397.
2585. **Perkins, K. W.**, A new cestode *Raillietina (R.) multitesticulata* n.sp. from the red howler monkey, *J. Parasitol.*, 36, 293—296, 1950.
2586. **Perrenoud, W.**, Recherches anatomiques et histologiques sur quelques cestodes de Sélaciens, *Rev. Suisse Zool.*, 38, 469—555, 1931.
2587. **Perrier, E.**, Description d'un genre nouveau de cestoïdes (Genre *Duthiersia* E.P.), *Arch. Zool. Exp. Gén.*, 2, 349—362, 1873.
2588. **Perrier, E.**, Classification des cestoides, *C. R. Acad. Sci.*, 86, 552—554, 1897.
2589. **Perroncito, E.**, Di un nuova specie de *Taenia (T. alba), Ann. Reale Accad. Agric. Torino*, 21, 127—130, 1879.
2590. **Perroncito, E.**, Ueber eine neue Bandwurmart *(Taenia alba), Arch. Naturg.*, 45, 235—237, 1879.
2591. **Peters, B. G.**, Some recent development in helminthology, *Proc. R. Soc. Med.*, 29, 1074—1084, 1936.
2592. **Peters, W.**, Ueber eine neue durch ihre riesige Grösse ausgezeichnete *Taenia, Monatschrift Kgl. Preuss. Akad. Wiss.*, November 1856, p. 469, 1857.
2593. **Peters, W.**, Note on the *Taenia* from the rhinoceros lately described by R. J. Murie, *Proc. Zool. Soc. London*, pp. 146—147, 1871.
2594. **Petrochenko, V. I. and Kireev, V.**, Life cycle of *Raillietina (Skrjabinia) caucasica* n.sp. from turkeys (in Russian), *Dokl. Akad. Nauk SSSR*, 166, 1491—1493, 1966.

2595. **Petrov, A. M. and Spasskii, A. A.**, Cestodes — Mesocestoidata of domestic and wild animals, *Tr. Gel' mintol., Lab. Akad. Nauk SSSR*, 7, 320—330, 1954.
2596. **Petrov, M. I.**, New diphyllobothriids of man (in Russian with English summary), *Med. Parasitol. Parasit. Dis. (USSR)*, 7, 406—413, 1938.
2597. **Petruschevsky, G. K.**, Ueber die Verbreitung der Plerocercoide von *Diphyllobothrium latum* in den Fischen der Newabucht, *Zool. Anz.*, 94, 139—147, 1931.
2598. **Petruschevsky, G. K.**, Ueber die Infektion der Fische des Onega Sees mit Plerocercoiden von *Diphyllobothrium latum*, *Tr. Berodinks. Biol. Stantsii Karelii*, 6, 71—75, 1933.
2599. **Petrushevsky, G. K. and Boldyr, E. D.**, Ueber die Verbreitung der Plerocerkoide des *Diphyllobothrium latum* in den Fischen des Onega-Sees (In Russian), *Tr. Berodinsk, Biol. Stantsii Karelii*, 8, 89—96, 1935.
2600. **Petruschevsky, G. K. and Boldyr, E. D.**, Propagation du bothricephale *(Diphyllobothrium latum)* et de ses larves plérocercoides dans la région du nordouest de l'U.S.S.R., *Ann. Parasitol.*, 13, 327—337, 1935.
2601. **Petruschevsky, G. K. and Bychowskaja-Pavlovskaja, I.**, Ueber die Verbreitung der Larven von *Diphyllobothrium latum* in Fischen aus Kareliens (in Russian), *Tr. Berodinsk. Biol. Stantsii Karelii*, 6, 4—26, 1933.
2602. **Petruschevsky, G. K. and Tarassow, V. A.**, Versuche ueber die Ansteckung des Menschen mit verschiedenen Fischplerozerkoiden, *Arch. Schiffs Trop. Hyg.*, 37, 370—372, 1933.
2603. **Pfeiffer, H.**, *Hymenosphenacanthus bulbocirrosus* spec. nov. (Hymenolepididae), ein neuer Bandwurm des Schwarzhalsschwanes, *Z. Parasitenkd.*, 20, 345—349, 1960.
2604. **Piana, G. P.**, Di una nuova specie di Taenia dell gallo domestico *(Taenia botrioplotis)* e di un nuova cisticerco della lumachelle terrestri *(Cysticercus botrioplitis)*, *Mem. Accad. Sci. Ist. Bologna*, 1880—1881, 4, s. v. 2, 387—395, 1882.
2605. **Pierantoni, U.**, *Bertia hamadryadis*, n.sp. di cestode anoplocefalo parassita di *Hamadryas hamadryas*, *Ann. Mus. Zool. Reale Univ. Napoli*, 5, 1—3, 1928.
2606. **Pillai, K. M. and Peter, C. T.**, Studies on tapeworms commonly encountered in fowls, *Ind. Vet. J.*, 48, 430—431, 1971.
2607. **Pintner, T.**, Untersuchungen ueber den Bau des Bandwurmkörpers mit besonderer Berucksichtigung der Tetrabothrien und Tetrarhynchen, *Arb. Zool. Inst. Univ. Wien*, 3, 163—242, 1880.
2608. **Pintner, T.**, Neue Untersuchungen ueber den Bau des Bandwurmkörpers. I. Zur Kenntnis der Gattung *Echinobothrium*, *Arb. Zool. Inst. Univ. Wien*, 8, 371—420, 1889.
2609. **Pintner, T.**, Neue Beiträge zur Kenntnis des Bandwurmkörpers, *Arb. Zool. Inst. Univ. Wien*, 9, 57—84, 1890.
2610. **Pintner, T.**, Studien an Tetrarhynchen nebst Beobachtungen an anderen Bandwurmern. I. Mitteilung, *Sitzungsber. Akad. Wiss. Wien Math. Naturwiss. Kl. Abt. I*, 102, 605—650, 1893.
2611. **Pintner, T.**, Versuch einer morphologischen Erklarung des Tetrarhynchenrussels, *Biol. zentralb.*, 16, 258—267, 1896.
2612. **Pintner, T.**, Studien uber Tetrarhynchen nebst Beobachtungen an anderen Bandwürmern. II. Mitteilung. Ueber eine Tetrarhynchenlarve aus dem Magen von *Heptanchus*, nebst Bemerkungen ueber das Exkretionssystem verschiedener Cestoden, *Sitzungsber. Akad. Wiss. Wien Math. Naturwiss. Kl. Abt. I*, 105, 652—682, 1896.
2613. **Pintner, T.**, Die Rhynchodäaldrüsen der Tetrarhynchen, *Arb. Zool. Inst. Univ. Wien Math. Naturwiss. Kl. Abt. I*, 12, 1—24, 1899.
2614. **Pintner, T.**, Studien über Tetrarhynchen nebst Beobachtungen an anderen Bandwürmern. III. Mitteilung. Zwei eigentümliche Drüsensysteme bei *Rhynchobothrius adenoplusius* n. und histologische Notizen ueber *Acanthocephalus, Amphilina* und *Taenia saginata*, *Sitzungsber. Akad. Wiss. Wien Math. Naturwiss. Kl. Abt. I*, 112, 541—597, 1903.
2615. **Pintner, T.**, Das Verhalten des Exkretionssystems in Endgliede von *Rhynchobothrium ruficollis* (Eysenhardt), *Zool. Anz.*, 30, 576—578, 1906.
2616. **Pintner, T.**, Das ursprüngliche Hinterende einiger Rhynchobothriumketten, *Arb. Zool. Inst. Univ. Wien*, 18, 113—132, 1909.
2617. **Pintner, T.**, Eigentümlichkeiten des Sexualapparates der Tetrarhynchen. Verhandl. III. *Int. Zool. Congr. Graz.*, 1910, Jena, 1912, 776—780.
2618. **Pintner, T.**, Vorarbeiten zu einer Monographie der Tetrarhynchoideen, *Sitzungsber. Akad. Wiss. Wien Math. Naturwiss. Kl. Abt. I*, 122, 171—253, 1913.
2619. **Pintner, T.**, Die Entstehung der Russel der Tetrarhynchiden, *Zool. Anz.*, 59, 100—104, 1924.
2620. **Pintner, T.**, Bemerkenswerte Strukturen im Köpfe von Tetrarhynchoideen, *Z. Wiss. Zool.*, 125, 1—34, 1925.
2621. **Pintner, T.**, Topographie des Genitalapparats von *Euterarhynchus ruficollis* (Eysenhardt), *Zool. Jahrb. Anat.*, 47, 212—245, 1925.
2622. **Pintner, T.**, Kritische Beiträge zum System der Tetrarhynchen, *Zool. Jahrb. Syst.*, 53, 559—590, 1927.
2623. **Pintner, T.**, Helminthologische Mitteilungen. I, *Zool. Anz.*, 76, 318—322, 1928.

2624. **Pintner, T.**, Die sogenannte Gamobothriidae Linton, 1899, *Zool. Jahrb. Anat.*, 50, 55—116, 1928.
2625. **Pintner, T.**, Studien ueber Tetrarhynchen nebst Beobachtungen an anderen Bandwürmern. IV. Mitteilung. Ueber einige Diesing'sche Originale und verwandte Formen, *Sitzungsber. Akad. Wiss. Wien Math. Naturwiss. Kl. Abt. I,* 138, 145—166, 1929.
2626. **Pintner, T.**, Tetrarhynchen von den Forschungsreisen des Dr. Sixten Bock, *Göteborgs Kgl. Vetensk. Handl. Ser. B,* 1, 1—46, 1929.
2627. **Pintner, T.**, Helminthologische Mitteilungen. II, *Zool. Anz.,* 84, 1—8, 1929.
2628. **Pintner, T.**, *Tetrarhynchus erinaceus, Anzeiger Akad. Wiss. Wien,* 61, 186—188, 1929.
2629. **Pintner, T.**, Tetrarhynchen aus Pacific Grove, California, U.S.A., *Anz. Akad. Wiss. Wien,* 67, 26—28, 1930.
2630. **Pintner, T.**, Weiteres ueber Anatomie und Systematik der Tetrarhynchen, *Anz. Akad. Wiss. Wien,* 67, 70, 1930.
2631. **Pintner, T.**, Wenigbekanntes und Unbekanntes von Russelbandwürmern. I, *Anz. Akad. Wiss. Wien,* 67, 148, 1930.
2632. **Pintner, T.**, Wenigbekanntes und Unbekanntes von Rüsselbandwürmern. II. *Sitzungsber. Akad. Wiss. Wien Math. Naturwiss. Kl. Abt. I,* 139, 445—537; 140, 777—820, 1931.
2633. **Pintner, T.**, Ueber fortgesetzte Tetrarhynchenuntersuchungen, *Anz. Akad. Wiss. Wien,* 68, 72—75, 1931.
2634. **Pintner, T.**, Ueber fortgesetzte Tetrarhynchenuntersuchungen. II, *Anz. Akad. Wiss. Wien,* 68, 141—142, 1931.
2635. **Pintner, T.**, Weiteres ueber Strukturen im Tetrarhynchenkopfe, *Anz. Akad. Wiss. Wien,* 69, 189—190, 1932.
2636. **Pintner, T.**, Sinnespapillen am Genitalatrium der Tetrarhynchen, *Zool. Anz.,* 98, 295—298, 1932.
2637. **Pintner, T.**, Zur Kenntnis des Exkretionssystems der Cestoden, *Sitzungsber. Akad. Wiss. Wien Math. Naturwiss. Kl. Abt. I,* 142, 205—211, 1933.
2638. **Pintner, T.**, Bruchstücke zur Kenntnis der Rüsselbandwürmer, *Zool. Jahrb. Anat.,* 58, 1—20, 1934.
2639. **Pintner, T.**, Ueber Entwicklungsvorgänge in der Cestodenkette, *Anz. Akad. Wiss. Wien,* 71, 256—258, 1934.
2640. **Pintner, T.**, Ueber die Gewebe des Cestoden, *Anz. Akad. Wiss. Wien,* 72, 6—10, 1935.
2641. **Pintner, T.**, Berichtigung, *Zool. Anz.,* 190, 271—272, 1935.
2642. **Pipkin, A. C., Rizk, E., and Balikian, G. P.**, Echinoccosis in the Near East and its incidence in animal hosts, *Trans. R. Soc. Trop. Med. Hyg.,* 45, 253—260, 1951.
2643. **Plotnikov, N. N.**, On the spread of the plerocercoids of *Diphyllobothrium latum* in fishes of the rivers Irtysh and Tobol, *Med. Parasitol. Parazit. Bolezni,* 4, 330, 1935.
2644. **Plotz, M.**, *Diphyllobothrium latum* infestation in the eastern seaboard. Twenty-one cases from New York, *JAMA,* 98, 313—314, 1932.
2645. **Poche, F.**, Zur Kenntnis der Amphilinidea, *Zool. Anz.,* 54, 276—287, 1922.
2646. **Poche, F.**, Ueber die systematische Stellung des Cestodengenus *Wageneria* Mont., *Zool. Anz.,* 56, 20—27, 1923.
2647. **Poche, F.**, Die Entstehung der Rüssel der Tetrarhynchiden, *Zool. Anz.,* 59, 100—104, 1924.
2648. **Poche, F.**, Zur Kenntnis von *Amphilina foliacea*, *Z. Wiss. Zool.,* 125, 585—619, 1925.
2649. **Poche, F.**, Das System der Platodaria, *Arch. Naturg.,* 91, 241—458, 1926.
2650. **Poche, F.**, On the morphology and systematic position of the cestode, *Gigantolina magna* (Southwell), *Rec. Ind. Mus.,* 28, 1—27, 1926.
2651. **Poche, F.**, Zur Erklärung der Configuration des Exkretionssystems in den freien Proglottiden von *Wageneria* and ueber die Berichtigung der Gattung *Wageneria* (Tetrarhynchoidea), *Livro Jub. Travassos,* 1938, 403—406.
2652. **Podesta, R. B. and Holmes, J. C.**, Hymenolepidid cysticercoids in *Hyalella azteca* of Cooking Lake, Alberta: life cycles and descriptions of four new species, *J. Parasitol.,* 56, 1124—1134, 1970.
2653. **Podiapolskaja, V. P.**, Determining the viability of the oncosphere of *Taeniarhynchus saginatus* (in Russian), *Sb. Rab. Gel'mintol.,* pp. 169—173, 1948.
2654. **Podiapolskaja, V. P.**, Zur Kenntnis der parasitischen Würmer bei Ratten (in Russian), *Rev. Microbiol. Epid. Saratov,* 3, 280—290, 1924.
2655. **Podiapolskaja, V. P. and Gnedina, M. P.**, *Diphyllobothrium tungussicum* n.sp., ein neuer Parasit des Menschen, *Centralbl. Bakteriol.,* 126, 415—419, 1932.
2656. **Polk, S. J.**, *Dilepis hilli,* a new dilepidid cestode from a little blue heron, *Wasmann Coll.,* 4, 131—134, 1941.
2657. **Polk, S. J.**, The genus *Dilepis* Weinland, 1858, *Wasmann Coll.,* 5, 25—32, 1942.
2658. **Polk, S. J.**, *Hymenolepis mastigopraedita,* a new cestode from a pintail duck, *J. Parasitol.,* 28, 141—145, 1942.
2659. **Polk, S. J.**, A new hymenolepidid cestode, *Hymenolepis dafilae,* from a pintail duck, *Trans. Am. Microsc. Soc.,* 61, 186—190, 1942.

2660. **Polonio, A. F.,** Catalogo dei cefalocotilei italiani e alcuni osservazioni sul loro sviluppo, *Atti Soc. Ital. Sci. Nat. Milano,* 2, 217—229, 1860.
2661. **Polonio, A. F.,** Novae helminthum species, *Lotos (Prague),* 10, 21—23, 1860.
2662. **Polonio, A. F.,** Eine neue Art von *Ligula, Lotos (Prague),* 10, 179—180, 1860.
2663. **Pomeroy, M. K. and Burt, M. D. B.,** Cestode of the herring gull, *Larus argentatus* Pontoppidan, 1763, from New Brunswick, Canada, *Can. J. Zool.,* 42, 959—973, 1964.
2664. **Pont, A. M., del,** Sobre una neuva especie de *Taenia (Taenia furnarii), Anales Circulo Med. Argent. (Buenos Aires),* 1906, 15 pp.
2665. **Pont, A. M., del,** Contribución al estudio de los zooparásitos de los animales salvajes, *Semana Med.,* 33, 16—22, 1926.
2666. **Popoff, N. P.,** *Caryophyllaeus skrjabini* nov.sp., eine neue Cestode von *Abramis brama* Russ., *Gidrobiol. Zh. Saratow,* 3, 153—258, 1924.
2667. **Popov, P.,** Sur le développement de *Diplopylidium skrjabini* n.sp., *Ann. Parasitol.,* 13, 322—326, 1935.
2668. **Porta, A.,** Nuovo botriocefalo *(B. andresi)* e appunti elmintologici, *Zool. Anz.,* 38, 373—378, 1911.
2669. **Porter, A.,** A survey of the intestinal Entozoa, both protozoal and helminthic, observed among natives in Johannesburg, *Publ. S. Afr. Inst. Med. Res.,* 11, 1—58, 1918.
2670. **Porter, D. A.,** On the occurrence of tapeworms, *Moniezia expansa* and *Moniezia benedeni* in cattle and sheep, *Proc. Helminthol. Soc. Wash.,* 20, 93—94, 1953.
2671. **Posselt, A.,** Es giebt keine Uebergangs-oder Zwischenformen zwischen beiden Arten des Blasenwurmes *(Echinococcus cysticus* und *Echinococcus alveolaris), Frankfurter Z. Pathol.,* 47, 194—230, 1934.
2672. **Potemkina, V. A.,** Contribution to the biology of *Moniezia expansa* (Rudolphi, 1810), a tapeworm parasitic in sheep and goats (in Russian), *C. R. Acad. Sci. U.S.S.R.,* 30, 474—476, 1941.
2673. **Potemkina, V. A.,** On the decipherment of the biological cycle in *Moniezia benedeni* (Moniez, 1870), a tapeworm parasitic in cattle and other domestic animals (in Russian), *Dokl. Akad. Nauk SSSR,* n.s., 42, 146—148, 1944.
2674. **Potemkina, V. A.,** Contribution to the study of the development of *Thysaniezia ovilla* (Rivolta, 1878), a tapeworm parasitic in ruminants (in Russian), *Dokl. Akad. Nauk SSSR,* 43, 43—44, 1944.
2675. **Potemkina, V. A.,** A study of the biology of *Moniezia expansa* (Rudolphi, 1810) (in Russian), *Sb. Rab. Gel'mintol. (40. Nauchn. Deiat. Skrjabin),* pp. 177—184, 1948.
2676. **Potselueva, V. A.,** Development of *Cysticercus pisiformis* in the organism of the rabbit (in Russian), *Rab. Gel'mintol. 75-Lett. Skrjabin,* pp. 564—566, 1954.
2677. **Potter, C. C.,** A new cestode from a shark *(Hypoprion brevirostris* Poey), *Proc. Helminthol. Soc. Wash.,* 4, 70—72, 1937.
2678. **Pratt, H. S.,** The cuticula and subcuticula of trematodes and cestodes, *Am. Nat.,* 43, 705—729, 1909.
2679. **Premvati, G.,** Studies on *Haplobothrium bistrobilae* sp. nov. (Cestoda: Pseudophyllidea) from *Amia calva* L., *Proc. Helminthol. Soc. Wash.,* 36, 55—60, 1969.
2680. **Prenant, L. A.,** Recherches sur les vers parasites des poissons, *Bull. Soc. Sci. Nancy,* 7, 206—230, 1886.
2681. **Prenant, M.,** Recherches sur le parenchyme de Plathelminthes, *Arch. Morphol. Gen. Exp.,* 5, 1—175, 1922.
2682. **Prestwood, A. K. and Reid, W. M.,** *Drepanidotaenia watsoni* sp.n. (Cestoda, Hymenolepididae) from the wild turkey of Arkansas, *J. Parasitol.,* 52, 432—436, 1966.
2683. **Price, E. W.,** A new host for *Duthiersia fimbriata, J. Parasitol.,* 19, 84, 1932.
2684. **Pritchard, M. H. and Kruse, G. O.,** *The Collection and Preservation of Animal Parasites,* University of Nebraska Press, Lincoln, 1982, 141 pp.
2685. **Prokopič, Ya.,** A new cestode *Triodontolepis kurashvilii* n.sp. from *Neomys fodiens* (in Russian), *Parazitol. Sb. Tbilisi,* 2, 161—164, 1971.
2686. **Prokopič, J.,** *Triodontolepis rysavyi* sp.n. (Hymenolepididae) a new cestode species from *Neomys anomalus, Folia Parasitol.,* 19, 281—284, 1972.
2687. **Prokopič, J. and Matsaberidze, G.,** Cestode species new for the parasite fauna of micromammalians from Georgia, *Věstn. Česk. Společnosti Zool.,* 36, 214—220, 1972.
2688. **Protasova, E. N.,** The systematics of cestodes of the Pseudophyllidea, parasitic in fish (in Russian), *Tr. Gel'mintol. Lab.* 24, 133—144, 1974.
2689. **Protasova, E. N.,** *Paraechinophallus* n.g. (Pseudophyllidea) from the marine fish *Psenopsis anomala* (in Russian), *Tr. Gel'mintol. Lab.,* 25, 109—115, 1975.
2690. **Protasova, E. N.,** *Essentials of Cestodology,* Vol. 8, Bothriocephalata — Tapeworm Helminths of Fish (in Russian), Akademiya Nauk SSSR, Moscow, 1977, 298 pp.
2691. **Prudhoe, S.,** Trematoda, Cestoda and Acanthocephala, *Exploration Hydrobiol. Lac Tanganyika Res. Sci.,* 3, 1—10, 1951.
2692. **Prudhoe, S. and Manger, B. R.,** Three species of hymenolepidid cestodes from a charadriiform bird in North America, *Ann. Mag. Nat. Hist.,* (Ser. 13), 9, 539—549, 1967.
2693. **Prudhoe, S. and Manger, B. R.,** A collection of cestodes from Malayan bats, *J. Nat. Hist.,* 3, 131—143, 1969.

2694. **Pujatti, D.**, Il *Gryllodes sigillatus* W. e ospite intermedio della *Hymenolepis nana* Siebold, 1852, *Ann. Mus. Civ. Storia Nat Genova*, 63, 235—241, 1949.
2695. **Pujatti, D.**, Un nuovo ospite intermedio del *Diplopylidium nölleri*, *Ann. Mus. Civ. Storia Nat. Geneva*, 63, 294—300, 1949.
2696. **Pujatti, D.**, Forme larvali di *Mesocestoides lineatus* Goeze in *Pitymys multiplex* Fatio, *Doriana*, 1, 1—7, 1953.
2697. **Pujatti, D.**, Rara anomalia in un *Cysticercus fasciolaris* (Rudolphi, 1808), *Doriana*, 1, 1—5, 1953.
2698. **Pujatti, D.**, Sulla posizione sistematica della *Sparganum hamadryadis* Teodoro, 1917, *Doriana*, 1, 1—3, 1953.
2699. **Pullar, E. M.**, A survey of Victorian canine and vulpine parasites, *Aust. Vet. J.*, 22, 12—21, 1946.
2700. **Purvis, G. B.**, Cestodes from domestic animals in Malaya, with descriptions of new species, *Vet. Rec.*, 12, 1407—1409, 1932.
2701. **Pushmenkov, E. P.**, A contribution to the knowledge of the developmental cycle of the larvae of cestodes parasitic in the liver of the reindeer, *Dokl. Akad. Nauk SSSR*, 49, 303—304, 1945.
2702. **Quentin, J. C.**, Description de *Cotugnia daynesi* n.sp., cestode parasite de la poule domestique à Madagascar, *Bull. Soc. Pathol. Exot.*, 56, 243—251, 1963.
2703. **Quentin, J. C.**, Etude anatomique et histologique (Cestodes) de trois *Raillietina* s. str. d'oiseaux africaines, *Ann. Parasitol. Hum. Comp.*, 39, 179—200, 1964.
2704. **Quentin, J. C.**, Cestodes de rongeurs de République Centrafricaine, *Cahiers Maboké*, 2, 117—140, 1964.
2705. **Quentin, J. C.**, *Skrajabinotaenia pauciproglottis* n.sp., cestode nouveau parasite de rongeurs de Republique Centre Africaine, *Bull. Mus. Natl. Hist. Nat. Paris* (Ser. 2), 37, 357—362, 1965.
2706. **Quentin, J. C.**, *Atriotaenia (Ershovia) baltazardi* n.sp. (Cestoda, Linstowiidae) parasite d'un rongeur du Bresil: *Galea spixii* (Wagner), *Bull. Mus. Natl. Hist. Nat.* Ser. 2, 39, 595—603, 1967.
2707. **Quentin, J. C.**, Cestodes *Skrjabinotaenia* de rongeurs muridés et dendromuridés de Centrafrique, Hypothèse sur l'evolution des cestodes Catenotaeniinae, *Cahiers Maboké (Paris)*, 9, 57—59, 1971.
2708. **Querner, F.**, Revision zweier von Diesing beschriebenen Rhynchobothriens *(R. tenuicolle* et *R. caryophyllum)*, *Ann. Nat. Mus. Wien*, 38, 107—117, 1925.
2709. **Railliet, A.**, Elements de Zoologie Medicale et Agricole, Fasc. 2., xv + 801—1053, Paris, 1886.
2710. **Railliet, A.**, Sur la classification des téniadés, *Centralbl. Bakteriol. Parasitenk.*, 26, 32—34, 1889.
2711. **Railliet, A.**, Les parasites des animaux domestiques au Japon, *Naturaliste* (Ser. 2), 4, 142—143, 1890.
2712. **Railliet, A.**, Sur un tenia du pigeon domestique représentant une espèce nouvelle *(Taenia delafondi)*, *C. R. Soc. Biol.*, 4, 49—53, 1892.
2713. **Railliet, A.**, Notices parasitologiques: *Taenia tenuirostris* Rud. chez l'oie domestique: remarques sur la classification des cestodes parasites des oiseaux, *Bull. Soc. Zool. Fr.*, 17, 110—117, 1892.
2714. **Railliet, A.**, Quelques rectifications de nomenclature des parasites, *Rec. Med. Vet.*, 3, 157—161, 1896.
2715. **Railliet, A.**, Sur quelques parasites du dromadaire, *C. R. Soc. Biol.*, 48, 489—492, 1896.
2716. **Railliet, A.**, Anomalies des scolex chez le *Coenurus seriale*, *C. R. Soc. Biol.*, 51, 18—21, 1899.
2717. **Railliet, A.**, Sur les cestodes du Blaireau *(Meles taxus)*, *C. R. Soc. Biol.*, 51, 23—25, 1899.
2718. **Railliet, A.**, Sur la classification des téniadés, *Centralbl. Bakteriol.*, 26, 32—34, 1899.
2719. **Railliet, A.**, Sur la synonymie du genre *Tetrarhynchus* Rud., 1809, *Arch. Parasitol.*, 2, 319—320, 1899.
2720. **Railliet, A.**, Encore un mot sur le *Davainea oligophora* Polonio, *Arch. Parasitol.*, 2, 482, 1899.
2721. **Railliet, A.**, The food of slugs and the development of Anoplocephalidae, *Ann. Appl. Biol.*, 3, 52, 1916.
2722. **Railliet, A.**, Les cestodes des oiseaux. (A propos de la communication precedente). Determination du parasite, *Rev. Pathol. Comp.*, 16, 307—308, 1916.
2723. **Railliet, A.**, Les cestodes des oiseaux domestiques. Determination pratique, *Rec. Med. Vet.*, 97, 185—205, 1921.
2724. **Railliet, A. and Henry, A.**, Etude de *Taenia* recuelli au Tonkin par M. le Dr. Lacour, *Ann. Hyg. Med. Colon.*, 8, 288—293, 1905.
2725. **Railliet, A. and Henry, A.**, Les cestodes des oiseaux par O. Fuhrmann, *Rec. Méd. Vét.*, 86, 337—338, 1909.
2726. **Railliet, A. and Henry, A.**, Helminthes du porc recueillis par M. Bauche en Annam, *Bull. Soc. Pathol. Exot.*, 4, 693—699, 1911.
2727. **Railliet, A. and Henry, A.**, Helminthes recueillis par l'expédition antarctique française du Pourquoi-Pas. I. Cestodes d'oiseaux, *Bull. Mus. Natl. Hist. Nat.*, 18, 35—39, 1912.
2728. **Railliet, A. and Henry, A.**, Helminthes recueillis par l'expédition antarctique française du Pourquoi-Pas. II. Cestodes de phoques, *Bull. Mus. Natl. Hist. Nat.*, 18, 153—159, 1912.
2729. **Railliet, A., and Henry, A.**, Sur un cénure de la gerbille à pieds velus, *Bull. Soc. Pathol. Exot.*, 8, 173—177, 1915.
2730. **Railliet, A., Henry, A., and Bauche, J.**, Sur les helminthes de l'éléphant d' Asie. I. Trématodes et cestodes, *Bull. Soc. Pathol. Exot.*, 7, 78—83, 1914.
2731. **Railliet, A. and Lucet, A.**, Développement experimental du *Cysticercus tenuicollis* chez le chevreau, *Bull. Soc. Zool. Fr.*, 16, 157—158, 1891.

2732. **Railliet, A. and Lucet, A.,** Sur le *Davainea proglottina* Davaine, *Bull. Soc. Zool. Fr.*, 17, 105—106, 1892.
2733. **Railliet, A. and Lucet, A.,** Sur l'identite du *Davainea oligophora* Magalhães, 1898 et du *Taenia cantaniana* Polonio, 1860, *Arch. Parasitol.*, 2, 144—146, 1899.
2734. **Railliet, A. and Marullaz, M. M.,** Sur un cénure nouveau du bonnet chinois *(Macacus sinicus), Bull. Soc. Pathol. Exot.*, 12, 223—228, 1919.
2735. **Railliet, A. and Mouquet, A.,** Cénure du coypou, *Bull. Soc. Cent. Med. Vet.*, 72, 204—211, 1919.
2736. **Raina, M. K.,** A monograph on the genus *Avitellina* Gough, 1911 (Avitellinidae: Cestoda), *Zool. Jahrb. Abt. Syst. Ökolo. Geog. Tiere*, 102, 508—552, 1975.
2737. **Ramadevi, P.,** *Echinobothrium reesae* (Cestoda: Diphyllidea) from the sting rays of Waltair Coast, *Ann. Parasitol. Hum. Comp.*, 44, 231—239, 1969.
2738. **Ramadevi, P.,** *Lytocestus longicollis* sp.nov. (Cestoidea: caryophyllidea) from the catfish *Clarias batrachus* (L.) in India, *J. Helminthol.*, 47, 415—420, 1973.
2739. **Ramadevi, P.,** *Proteocephalus hanumanthai* n.sp. (Cestoda: Proteocephalidea) from the intestine of frog, *Rana cyanophlyctis* Schneider, *Ind. J. Helminthol.*, 24, 47—51, 1974 (dated 1972).
2740. **Rankin, J. S.,** An ecological study of parasites of some North Carolina salamanders, *Ecol. Monogr.*, 7, 171—262, 1937.
2741. **Ransom, B. H.,** A new avian cestode, *Metroliasthes lucida* n.g. n.sp., *Trans. Am. Microsc. Soc.*, 21, 213—226, 1900.
2742. **Ransom, B. H.,** On *Hymenolepis carioca* (Magalhães) and *H. megalops* (Nitzsch) with remarks on the classification of the group, *Trans. Am. Microsc. Soc.*, 23, 151—172, 1902.
2743. **Ransom, B. H.,** Notes on the spiny-suckered tapeworms of chickens *(Davainea echinobothrida — Taenia bothrioplites)* and *D. tetragona*, *U.S. Bur. Anim. Ind. Bull.*, 60, 55—72, 1904.
2744. **Ransom, B. H.,** An account of the tapeworms of the genus *Hymenolepis* parasitic in man, with several new cases of the dwarf tapeworm *(H. nana)* in the United States, *U.S. Public Health Service Hyg. Lab. Bull.*, 18, 1—138, 1904.
2745. **Ransom, B. H.,** The tapeworms of American chickens and turkeys, *21st Annu. Rep. U.S. Bur. Anim. Ind.*, 1905, 268—285.
2746. **Ransom, B. H.,** Tapeworm cysts, *Dithyridium cynocephali* n.sp., in the muscles of a marsupial wolf *(Thylacinus cynocephalus), Trans. Am. Microsc. Soc.*, 27, 31—32, 1907.
2747. **Ransom, B. H.,** The taenioid cestodes of North American birds, *U.S. Natl. Mus. Bull.*, 69, 1—141, 1909.
2748. **Ransom, B. H.,** A new cestode from an African bustard, *Proc. U.S. Natl. Mus.*, 40, 637—647, 1911.
2749. **Ransom, B. H.,** The name of the sheep measle worm *(Taenia ovis), Science*, 38, 230, 1913.
2750. **Ransom, B. H.,** *Cysticercus ovis*, the cause of tapeworm cysts in mutton, *J. Agric. Res.*, 1, 15—58, 1913.
2751. **Ransom, B. H.,** List of parasites from the Island of Guam, *J. Parasitol.*, 2, 93—94, 1916.
2752. **Ransom, B. H.,** Yellow-bellied sapsucker infested with tapeworms, *Can. Field Nat.*, 11, 67, 1926.
2753. **Rao, K. H.,** A new bothriocephalid parasite (Cestoda) from the gut of the fish *Saurida tumbil* (Bloch), *Curr. Sci. (Bangalore)*, 23, 333—334, 1954.
2754. **Rao, K. H.,** On *Ptychobothrium cypseluri* n.sp. (Cestoda: Pseudophyllidea) from the flying fish, *Cypselurus poecilopterus* caught off Waltair, *J. Helminthol.*, 33, 267—272, 1959.
2755. **Rao, K. H.,** Studies on *Penetrocephalus ganapati*, new genus (Cestoda: Pseudophyllidea) from the marine teleost *Saurida tumbil* (Bloch), *Parasitology*, 50, 155—163, 1960.
2756. **Rao, M. A. N.,** On a species of *Joyeuxia* Lopez-Neyra, 1927, from a cat *(Felis catus domestica), Ind. J. Vet. Sci. Anim. Husb.*, 9, 377—378, 1927.
2757. **Rao, M. A. N. and Ayyar, L. S. P.,** Triradiate tapeworms from hounds and jackals, *Ind. J. Vet. Sci.*, 2, 397—399, 1932.
2758. **Rao, N. S. K. and Choquette, L. P. E.,** On the finding of an intermediary host for *Moniezia expansa* (Rud., 1810) in eastern Quebec (English and French summaries), *Can. J. Comp. Med.*, 15, 12—14, 1951.
2759. **Rao, S. R., Bhatavdekar, M. Y. and Detha, K. T.,** Morphology and development of *Coenurus gaigeri* Hall in a ewe with particular reference to the taxonomy of the genus *Multiceps*, *Bombay Vet. Coll. Mag.*, 6, 12—18, 1957.
2760. **Rao, V.,** *Acanthobothrium hanumantharaoi* sp.n. (Cestoda: Tetraphyllidea, Oncobothriidae) from the Nieuhof's eagle ray, *Myliobatis nieuhofii* (Bloch and Schneider) of Waltair Coast, Bay of Bengal, *Riv. Parassitol.*, 38, 277—283, 1977.
2761. **Raper, A. B. and Dockeray, G. C.,** Coenurus cysts in man: five cases from East Africa, *Ann. Trop. Med. Parasitol.*, 50, 121—128, 1956.
2762. **Rasīn, K.,** Beiträge zur postembryonalen Entwicklung der *Amphilina foliacea* nebst einer Bemerkung über die Laboratoriumskultur von *Gammarus pulex* (L.), *Z. Wiss. Zool.*, 138, 555—579, 1931.
2763. **Ratz, I. von,** *Dipylidium chyzeri* n.sp. (in Hungarian, with German summary, pp. 259—266), *Termeszet Fuzetek (Budapest)*, 20, 197—203, 1897.
2764. **Ratz, I. von,** Beiträge zur Parasitenfauna der Balatonfische, *Centralbl. Bakteriol.*, 22, 443—453, 1897.

2765. **Ratz, I. von,** Drei neue Cestoden aus Neu-Guinea. Vorläufige Mitteilung, *Centralbl. Bakteriol.,* 28, 657—660, 1900.
2766. **Ratz, I. von,** Trois nouveaux cestodes de reptiles, *C. R. Soc. Biol.,* 52, 980—981, 1900.
2767. **Ratz, I. von,** Parasitoligiai jegyzetek, *Veterinarius (Budapest),* 23, 525—534, 1900.
2768. **Ratz, I. von,** Une larve plérocercoide du porc, *Presse Méd.,* 20, 867—868, 1912.
2769. **Ratz, S. von,** Ein Plerocercoid vom Schwein, *Centralbl. Bakteriol.,* 47, 523—527, 1913.
2770. **Ratzel, F.,** Zur Entwicklungsgeschichte der Cestoden, *Arch. Naturg.,* 34, 138—149, 1868.
2771. **Ratzel, F.,** Beschreibung einiger neuen Parasiten, *Arch. Naturg.,* 34, 150—156, 1868.
2772. **Ratzel, F.,** *Cysticercus lumbriculi, Q. J. Microsc. Sci.,* 9, 315—316, 1869.
2773. **Raum, J.,** Beiträge zur Entwicklungsgeschichte der Cysticerken, *Inaugural Dissertation, University of Dorpat,* 1883.
2774. **Rausch, R.,** *Paranoplocephala troeschi,* a new species of cestode from the meadow vole, *Microtus p. pennsylvanicus* Ord., *Trans. Am. Microsc. Soc.,* 65, 354—356, 1946.
2775. **Rausch, R.,** *Andrya sciuri* n.sp., a cestode from the northern flying squirrel, *J. Parasitol.,* 33, 316—318, 1947.
2776. **Rausch, R.,** *Bakererpes fragilis* n.g. n.sp., a cestode from the nighthawk (Cestoda: Dilepididae), *J. Parasitol.,* 33, 435—438, 1947.
2777. **Rausch, R.,** A redescription of *Taenia taxidiensis* Skinker, 1935, *Proc. Helminthol. Soc. Wash.,* 13, 73—75, 1947.
2778. **Rausch, R.,** Observations on some helminths parasitic in Ohio turtles, *Am. Midl. Nat.,* 38, 434—442, 1947.
2779. **Rausch, R.,** *Dendrouterina botauri* n.sp. a cestode parasitic in bitterns, with remarks on other members of the genus, *Am. Midl. Nat.,* 39, 431—436, 1948.
2780. **Rausch, R.,** Notes on cestodes of the genus *Andrya* Railliet, 1883, with the description of *A. ondatrae* n.sp. (Cestoda: Anoplocephalidae), *Trans. Am. Microsc. Soc.,* 67, 187—191, 1948.
2781. **Rausch, R.,** Observations on cestodes in North American owls with the description of *Choanotaenia speotytonis* (Cestoda: Dipylidinae), *Am. Midl. Nat.,* 40, 462—471, 1948.
2782. **Rausch, R.,** Some additional observations on the morphology of *Dendrouterina botauri* Rausch, 1948 (Cestoda: Dilepididae), *J. Parasitol.,* 35, 76—78, 1949.
2783. **Rausch, R.,** *Paradilepis simoni* n.sp., a cestode parasitic in the osprey, *Zoologica,* 34, 1—3, 1949.
2784. **Rausch, R.,** Observations on the life-cycle and larval development of *Paruterina candelabraria* (Goeze, 1782) (Cestoda: Dilepididae), *Am. Midl. Nat.,* 42, 713—721, 1949.
2785. **Rausch, R.,** Studien an der Helminthenfauna von Alaska. IV. *Haploparaxis galli* n.sp., ein Cestode aus dem Schneehuhn, *Lagopus rupestris* (Gmelin), *Z. Parasitol.,* 15, 1—3, 1951.
2786. **Rausch, R.,** Studies on the helminth fauna of Alaska. VII. On some helminths from arctic marmots with the description of *Catenotaenia reggiae* n.sp. (Cestode: Anoplocephalidae), *J. Parasitol.,* 37, 415—418, 1951.
2787. **Rausch, R.,** Helminths from the round-tailed muskrat, *Neofiber alleni nigrescens* Howell, with descriptions of two new species, *J. Parasitol.,* 38, 151—156, 1952.
2788. **Rausch, R.,** Studies on the helminth fauna of Alaska. XI. Helminth parasites of microtine rodents — taxonomic considerations, *J. Parasitol.,* 38, 415—444, 1952.
2789. **Rausch, R.,** Studies on the helminth fauna of Alaska. XXI. Taxonomy, morphological variation, and ecology of *Diphyllobothrium ursi* n.sp. provis. on Kodiak Island, *J. Parasitol.,* 40, 540—563, 1954.
2790. **Rausch, R.,** Studies on the helminth fauna of Alaska. XXII. *Paranoplocephala wigginsi* n.sp., a cestode from an arctic ground squirrel, *Trans. Am. Microsc. Soc.,* 73, 380—383, 1954.
2791. **Rausch, R.,** *Cyclustra ardeae* n.sp. and the status of *Dendrouterina* Fuhrm., 1912 (Cestoda: Dilepididae), *Proc. Helminthol. Soc. Wash.,* 22, 25—29, 1955.
2792. **Rausch, R.,** Studies on the helminth fauna of Alaska. XXVIII. The description and occurrence of *Diphyllobothrium dalliae* n.sp. (Cestoda), *Trans. Am. Microsc. Soc.,* 75, 180—187, 1956.
2793. **Rausch, R.,** Distribution and specificity of helminths in microtine rodents: evolutionary implications, *Evolution,* 11, 361—368, 1957.
2794. **Rausch, R.,** Studies on the helminth fauna of Alaska. XXXVII. Description of *Schizorchis caballeroi* n.sp. (Cestoda: Anoplocephalidae), with notes on other parasites of *Ochotona, Libro Homenaje al Dr. Eduardo Caballero y Caballero, Jubileo, 1930—1960, Universidad Nacional Autónoma de Mexico,* 1960, 339—405.
2795. **Rausch, R. L.,** *Schizorchis yamashitai* sp.n. (Cestoda: Anoplocephalidae) from the northern pika *Ochotona hyperborea* Pallas in Hokkaido, *J. Parasitol.,* 49, 479—482, 1963.
2796. **Rausch, R. L.,** Taxonomic characters in the genus *Echinococcus* (Cestoda: Taeniidae), *Bull. WHO,* 39, 1—4, 1968.
2797. **Rausch, R. L.,** Diphyllobothriid cestodes from the Hawaiian monk seal, *Monachus schauinslandi* Matschie, from Midway Atoll, *J. Fish. Res. Bd. Can.,* 26, 947—956, 1969.

2798. **Rausch, R. L.,** Studies on the helminth fauna of Alaska. XLV. *Schistotaenia srivastavai* n.sp. (Cestoda: Amabiliidae) from the red-necked grebe, *Podiceps grisegena* (Boddaert), in Singh, K. S. and Tandan, B. K., Eds., *H. D. Srivastava Commemoration Volume,* Indian Veterinary Research Institute, Izatnagar, U. P., 1970, 109—115.

2799. **Rausch, R. L.,** *Davainea lagopodis* sp.nov. (Cestoda: Davaineidae) from grouse (Tetraonidae). Studies on the helminth fauna of Alaska. XLVI, *Helminthologia,* Year 1969, 10, 185—190, 1971.

2800. **Rausch, R. L.,** Cestodes of the genus *Hymenolepis* Weinland, 1858 (Sensu lato) from bats in North America and Hawaii, *Can. J. Zool.,* 53, 1537—1551, 1975.

2801. **Rausch, R. L.,** The genera *Paranoplocephala* Lühe, 1910 and *Anoplocephaloides* Baer, 1923, *Ann. Parasitol. Hum. Comp.,* 51, 513—562, 1976.

2802. **Rausch, R. L.,** Redescription of *Diandrya composita* Darrah, 1930 (Cestoda: Anoplocephalidae) from nearctic marmots (Rodentia: Sciuridae) and the relationships of the genus *Diandrya* emend., *Proc. Helminthol. Soc. Wash.,* 47, 157—164, 1980.

2803. **Rausch, R. L.,** Morphological and biological characteristics of *Taenia rileyi* Loewen, 1929 (Cestoda: Taeniidae), *Can. J. Zool.,* 59, 653—666, 1981.

2804. **Rausch, R. L. and Bernstein, J. J.,** *Echinococcus vogeli* sp.n. (Cestoda: Taenidae) from the bush dog, *Speothos venaticus* (Lund.) *Z. Tropenmed. Parasitol.,* 23, 25—34, 1972.

2805. **Rausch, R. and Kuns, M. L.,** Studies on some North American shrew cestodes, *J. Parasitol.,* 36, 433—438, 1950.

2806. **Rausch, R. L. and Margolis, L.,** *Plicobothrium globicephalae* gen. et sp.nov. (Cestoda: Diphyllobothriidae) from the pilot whale, *Globicephala melaena* Traill, in Newfoundland waters, *Can. J. Zool.,* 47, 745-750, 1969.

2807. **Rausch, R. L. and Maser, C.,** *Monoecocestus thomasi* sp.n. (Cestoda: Anoplocephalidae) from the northern flying squirrel. *Glaucomys sabrinus* (Shaw), in Oregon, *J. Parasitol.,* 63, 793-799, 1977.

2808. **Rausch, R. and Morgan, B. B.,** The genus *Anonchotaenia* (Cestoda: Dilepididae) from North American birds, with the description of a new species, *Trans. Am. Microsc. Soc.,* 66, 203-211, 1947.

2809. **Rausch, R., Morgan, B. B., and Schiller, E. L.,** Studies on species of *Paranoplocephala* parasitic in North American rodents (Cestoda: Anoplocephalidae), *J. Parasitol.,* 34 (Sect. 2, Suppl.), 23, 1948.

2810. **Rausch, R. L. and Nelson, G. S.,** A review of the genus *Echinococcus* Rudolphi, 1800, *Ann. Trop. Med. Parasitol.,* 57, 127-135, 1963.

2811. **Rausch, R. L. and Ohbayashi, M.,** On some anoplocephaline cestodes from pikas, *Ochotona* spp. (Lagomorpha), in Nepal, with the description of *Ectopocephalium abei* gen et sp.n., *J. Parastiol.,* 60, 596-604, 1974.

2812. **Rausch, R. and Schiller, E. L.,** A critical study of North American cestodes of the genus *Andrya,* with special reference to *A. macrocephala* Douthitt, 1915 (Cestoda: Anoplocephalidae), *J. Parasitol.,* 35, 306-314, 1949.

2813. **Rausch, R. and Schiller, E. L.,** A contribution to the study of North American cestodes of the genus *Paruterina* Fuhrmann, 1906, *Zoologica,* 34, 5-8, 1949.

2814. **Rausch, R. and Schiller, E. L.,** Some observations on cestodes of the genus *Paranoplocephala* Lühe, parasitic in North American voles, *Proc. Helminthol. Soc. Wash.,* 16, 23-31, 1949.

2815. **Rausch, R. and Schiller, E. L.,** Studies on the helminth fauna of Alaska. XXIV. *Echinococcus sibiriensis* n.sp. from St. Lawrence Isl., *J. Parasitol.,* 40, 659—662, 1954.

2816. **Rausch, R., Schiller, E. L., and Morgan, B. B.,** Studies on the cestode genus *Paruterina* (Cestoda: Dilepididae), *J. Parasitol.,* 34 (Sect. 2, Suppl.), 23, 1948.

2817. **Rausch, R., Schiller, E. L., and Morgan, B. B.,** Variations in *Andrya macrocephala* Douthitt, 1915 (Cestoda: Anoplocephalidae), *J. Parasitol.,* 34 (Sect. 2, Suppl.), 23, 1948.

2818. **Rausch, R. and Tiner, J. D.,** Studies on the parasitic helminths of the north central states. II. Helminths of voles *(Microtus* spp.), *Am. Midl. Nat.,* 41, 665-694, 1949.

2819. **Rausch, R. and Williamson, S. L.,** Studies on the helminth fauna Alaska. XXXIII. The description and occurrence of *Diphyllobothrium alascense* n.sp. (Cestoda), *Z. Tropenmed. Parasitol.,* 9, 64-72, 1958.

2820. **Rausch, R. and Yamashita, J.,** The occurrence of *Echinococcus multilocularis* Leuckart, 1863, in Japan, *Proc. Helminthol. Soc. Wash.,* 24, 128-133, 1957.

2821. **Rawson, D.,** The anatomy of *Eubothrium crassum* (Block) from the pyloric ceca and small intestine of *Salmo trutta* L. J., *J. Helminthol.,* 31, 103-120, 1957.

2822. **Rayski, C.,** Observations on the life-history of *Moniezia* with special reference to the bionomics of the oribatid mites, *Rep. 14 Int. Vet. Congr.,* 2, 51-55, 1952.

2823. **Read, C. P.,** Preliminary studies on the intermediate metabolism of the cestode *Hymenolepis diminuta, J. Parasitol.,* 35 (Suppl.), 26-27, 1949.

2824. **Rees, F. G.,** Studies on *Cittotaenia pectinata* (Goeze, 1782) from the common rabbit, *Oryctolagus cuniculus.* I. Anatomy and histology. II. Developmental changes in the egg and attempts at direct infestation, *Proc. Zool. Soc. London,* 1933, 239-257, 1933.

2825. **Rees, F. G.**, The musculature and nervous system of the plerocercoid larva of *Dibothriorhynchus grossum* (Rud.), *Parasitology*, 33, 373—389, 1941.
2826. **Rees, F. G.**, The scolex of *Aporhynchus norvegicus* (Olss.), *Parasitology*, 33, 433-438, 1941.
2827. **Rees, F. G.**, The anatomy of *Anthobothrium auriculatum* (Rud.) from *Raja batis* L., *Parasitology*, 35, 1-10, 1943.
2828. **Rees, F. G.**, A new cestode of the genus *Grillotia* from a shark, *Parasitology*, 35, 180-185, 1944.
2829. **Rees, F. G.**, The anatomy of *Phyllobothrium dohrnii* (Oerley) from *Hexanchus griseus* (Gmelin), *Parasitology*, 37, 163-171, 1946.
2830. **Rees, F. G.**, The plerocercoid larva of *Grillotia heptanchi* (Vaullegeard), *Parasitology*, 40, 265-272, 1950.
2831. **Rees, F. G.**, Some parasitic worms from fishes off the coast of Iceland. I. Cestoda, *Parasitology*, 43, 4-14, 1953.
2832. **Rees, G.**, The scolex of *Tetrabothrius affinis* (Lönnberg), a cestode from *Balaenoptera musculus* L., the blue whale, *Parasitology*, 46, 425-442, 1956.
2833. **Rees, G.**, *Echinobothrium acanthinophyllum*, n.sp. from the spiral valve of *Raja montagui* Fowler, *Parasitology*, 51, 407-414, 1961.
2834. **Rees, G.**, Studies on the functional morphology of the scolex and the genitalia in *Echinobothrium brachysoma* Pintner and *E. affine* Diesing from *Raja clavata* L., *Parasitology*, 51, 193-226, 1961.
2835. **Rees, G.**, Cestodes from Bermuda fishes and an account of *Acompsocephalum tortum* (Linton, 1905) gen.nov. from the lizard fish *Synodus intermedius* (Agassiz), *Parasitology*, 59, 519-548, 1969.
2836. **Rees, G.**, Cysticercoids of three species of *Tatria* (Cyclophyllidea: Amabiliidae) including *T. octacantha* sp.nov. from the haemocoele of the damsel-fly nymphs *Pyrrtrosoma nymphula*, Sulz and *Enallagma cyathigerum*, Charp, *Parasitology*, 66, 423-446, 1973.
2837. **Reeves, J. D.**, A new tapeworm of the genus *Bothriocephalus* from Oklahoma salamanders, *J. Parasitol.*, 35, 600—604, 1949.
2838. **Refuerzo, P. G. and Cabrera, D. J.**, Two species of *Raillietina* (Cestoidea: Davaineidae) from Filippino children, *Philipp. J. Anim. Ind.*, 8, 127-133, 1941.
2839. **Regel', K. V. and Bondarenko, S. K.**, The systematic status and life-cycles of *Wardium retracta* (Linstow, 1905) n.comb. and *W. pararetracta* n.sp. (Cyclophyllidea, Hymenolepididae) (in Russian), *Zool. Zh.*, 61, 325-335, 1982.
2840. **Rego, A. A.**, Nota prévia sôbre uma nova espécie do gênero *Monoecocestus* Beddard, 1914 (Cestoda, Cyclophyllidea), *Atas Soc. Biol. Rio de Janeiro*, 4, 67-68, 1960.
2841. **Rego, A. A.**, Nota prévia sôbre um novo *Monoecocestus* parasito de preá (Cestoda, Cyclophylllidea), *Atas Soc. Biol. Rio de Janerio*, 4, 73-74, 1960.
2842. **Rego, A. A.**, Nota prévia sôbre um novo *Vampirolepis* parasito de quirópteros (Cestoda, Hymenolepididae), *Atas Soc. Biol. Rio de Janeiro*, 5, 32—34, 1961.
2843. **Rego, A. A.**, Revisão do genero *Monoecocestus* Beddard, 1914 (Cestoda, Anoplocephalidae), *Mem. Inst. O. Cruz*, 59, 325-354, 1961.
2844. **Rego, A. A.**, Sôbre alguns *Vampirolepis* parasitos de quirópteros (Cestoda, Hymenolepididae), *Rev. Brasil. Biol.*, 22, 129-136, 1962.
2845. **Rego, A. A.**, Nova espécie do gênero *Mathevotaenia* Akhumian, 1946 parasita de quirópteros (Cestoda, Anoplocephalidae), *Rev. Brasil. Biol.*, 23, 31-34, 1963.
2846. **Rego, A. A.**, *Lentiella machadoi* g.n., sp.n. e *Raillietina* (R.) *trinitatae* (Cameron and Reesal, 1951), parasitos de roedor (Cestoda, Cyclophyllidea), *Rev. Brasil. Biol.*, 24, 211—220, 1964.
2847. **Rego, A. A.**, Sôbre alguns cestódeos parasitos de roedores do Brasil (Cestoda, Cyclophyllidae), *Mem. Inst. O. Cruz*, 65, 1—18, 1967.
2848. **Rego, A. A.**, Un nôvo gênero de cestódeo de ave Charadriiformes (Dilepididae, Dipylidiinae), *Atas Soc. Biol. Rio de Janiero*, 11, 43—45, 1967.
2849. **Rego, A.A.**, Sôbre alguns cestódeos parasitos de répteis, *Rev. Brasil. Biol.*, 27, 181—187, 1967.
2850. **Rego, A. A.**, Un novo cestodeo parasito de *Agouti paca* L. (Anoplocephalidae, Linstowiinae), *Atas Soc. Biol. Rio de Janeiro*, 11, 79—80, 1967.
2851. **Rego, A. A.**, Sôbre tres cestódeos de aves Charadriiformes, *Mem. Inst. O. Cruz*, 66, 107—115, 1968.
2852. **Rego, A. A.**, Uma nova espécie de *Rodentolepis* parasita de roedor (Cestoda, Hymenolepididae), in Sing., K. S. and Tandan, B. K., Eds., *H. D. Srivastava Commemmoration Volume*, Indian Veterinary Research Institute, Izatnager, U.P., 1970, 251—254.
2853. **Rego, A. A.**, *Echinocotyle ibanezi* sp.n. parasito de ave do Peru (Cestoda, Hymenolepididae), *Atas Soc. Biol. Rio de Janeiro*, 16, 63—65, 1973.
2854. **Rego, A. A.**, Estudos de cestóides de peixes do Brasil. $2^a$ nota: revisão do genero *Monticellia* LaRue, 1911 (Cestoda, Proteocephalidae), *Rev. Brasil. Biol.*, 35, 567—586, 1975.
2855. **Rego, A. A.**, Contribuiçao ao conhecimento dos helmintos de raias fluviais Paratrygonidae, *Rev. Brasil. Biol.*, 39, 879—890, 1979.
2856. **Rego, A. A. and Dias, A. P. L.**, Estudos de Cestóides de peixes do Brasil. $3^a$ nota: cestóides de raias fluviais Paratrygonidae, *Rev. Brasil. Biol.*, 36, 941—956, 1976.

2857. **Rego, A. A. and Ibanez, H. N.,** Duas novas espécies de *Oochoristica,* parasitas de lagartixas do Peru (Cestoda, Anoplocephalidae), *Mem. Inst. O. Cruz,* 63, 67—73, 1965.
2858. **Rego, A. A. and Oliveira Rodrigues, H.,** Sôbre duas *Oochoristica* parasitas de lacertilios (Cestoda, Cyclophyllidea), *Rev. Brasil. Biol.,* 25, 59—65, 1965.
2859. **Rego, A. A., Santos, J. C., and Silva, P. P.,** Estudos de cestóides de peixes do Brasil, *Mem. Inst. O. Cruz,* 72, 187—204, 1974.
2860. **Rego, A. A. and Vicente, J. J.,** Uma nova *Choanotaenia* parasita de ave do Peru (Cestoda, Dilepididae), *Rev. Brasil. Biol.,* 28, 7—10, 1968.
2861. **Rehana, R. and Bilqees, F. M.,** *Gangesia sindensis,* new species (Cestoda, Proteocephalidae), from the fish *Wallago atu* of Kalri Lake, Sind, West Pakistan, *Pakistan J. Zool.,* 3, 217—219, 1971.
2862. **Reid, W. M.,** Penetration glands in cyclophyllidean onchospheres, *Trans. Am. Microsc. Soc.,* 67, 177—182, 1948.
2863. **Reid, W. M. and Ackert, J. E.,** The cysticercoid of *Choanotaenia infundibulum* (Bloch) and the housefly as its host, *Trans. Am. Microsc. Soc.,* 56, 99—104, 1937.
2864. **Reid, W. M., Ackert, J. E., and Case, A. A.,** Studies on the life-history and biology of the fowl tapeworm *Raillietina cesticillus* (Molin), *Trans. Am. Microsc. Soc.,* 57, 65—76, 1938.
2865. **Reid, W. M. and Nugara, D.,** Description and life cycle of *Raillietina georgiensis* n.sp., a tapeworm from wild and domestic turkeys, *J. Parasitol.,* 47, 885—889, 1961.
2866. **Reimer, L. W.,** Larven de Ordnung Trypanorhyncha (Cestoda) aus Teleostiern des Indischen Ozeans, *Angew. Parasitol.,* 21, 221—231, 1980.
2867. **Rendtorff, R. C.,** Investigations on the life-cycle of *Oochoristica ratti,* A cestode from rats and mice, *J. Parasitol.,* 34, 243—252, 1948.
2868. **Rennie, J. and Reid, A.,** The cestoda of the Scottish Antarctic Expedition (Scotia), *Trans. R. Soc. Edinburgh,* 48, 441—454, 1912.
2869. **Retzius, A. A.,** *Bothriocephalus pythonis,* eine neue Art., *Isis (Oken),* pp. 1347—1350, 1831.
2870. **Richard, J.,** Sur la présence d'un cysticercoid chez un calanide d'eau douce, *Bull. Soc. Zool.,* 17, 17—18, 1892.
2871. **Richardson, L. R.,** Observations on the parasites of the speckled trout in Lake Edward, Quebec, *Trans. Am. Fish. Soc.,* p. 66, 1937.
2872. **Rider, C. L.,** Preliminary survey of the helminth parasites of muskrats in Northwestern Oregon, with description of *Hymenolepis ondatrae* n.sp., *Trans. Am. Microsc. Soc.,* 66, 176—181, 1947.
2873. **Riehm, G.,** Untersuchungen an den Bandwürmern der Hasen und Kaninchen, *Z. Ges. Naturwiss.,* 54, 200, 1881.
2874. **Riehm, G.,** Studien an Cestoden, *Z. Ges. Naturwiss.,* 54, 545—610, 1881.
2875. **Riehm, G.,** Fütterungsversuche mit *Ligula simplicissimus, Z. Ges. Naturwiss.,* 55, 328—330, 1882.
2876. **Rietschel, P. E.,** Ueber eine neue *Hymenolepis* aus einem Kolibri. Zugleich ein Beitrag zum Rechts-Links-Problem bei den Cestoden, *Zool. Anz.,* 105, 113—123, 1934.
2877. **Rietz, J. H.,** Animal parasites of chickens in Ohio and West Virginia, *J. Am. Vet. Med. Assoc.,* 77, 154—156, 1930.
2878. **Riggenbach, E.,** *Taenia dendritica* Goeze, *Centralbl. Bakteriol.,* 17, 710—716, 1895.
2879. **Riggenbach, E.,** Beiträge zur Kenntnis der Tänien der Süsswasserfische. Vorläufige Mitteilung, *Centralbl. Bakteriol.,* 18, 609—613, 1895.
2880. **Riggenbach, E.,** Das Genus *Ichthyotaenia, Rev. Suisse Zool.,* 4, 165—275, 1896.
2881. **Riggenbach, E.,** Bemerkungen ueber das Genus *Bothriotaenia* Railliet, *Centralbl. Bakteriol. Parasitenk.,* 20, 222—231, 1896.
2882. **Riggenbach, E.,** *Bothriotaenia chilensis* nov.spec. (transl. from German by G. Newmann), *Actes Soc. Sci. Chili,* 7, 66—73, 1897.
2883. **Riggenbach, E.,** *Scyphocephalus bisulcatus* n.g. n.sp., ein neuer Cestode aus *Varanus,* Vorläufige Mittheilung., *Zool. Anz.,* 21, 565—566, 1898.
2884. **Riggenbach, E.,** *Cyathocephalus catinatus, Zool. Anz.,* 21, 639, 1898.
2885. **Riggenbach, E.,** *Scyphocephalus bisulcatus* n.g. n.sp., ein neuer Reptiliencestode, *Zool. Jahrb. Syst.,* 12, 145—153, 1899.
2886. **Riggenbach, E.,** *Cyathocephalus carinatus* n.sp., *Zool. Jahrb. Syst.,* 12, 154—160, 1899.
2887. **Rigney, C. C.,** A new davaineid tapeworm, *Raillietina (Paroniella) centuri* from the red-bellied woodpecker, *Trans. Am. Microsc. Soc.,* 62, 398—403, 1943.
2888. **Riley, W. A.,** Some recent work on the development of hymenopterous parasites, *Entomol. News,* 18, 9—11, 1907.
2889. **Riley, W. A.,** The longevity of the fish tapeworm of man, *Dibothriocephalus latus, J. Parasitol.,* 5, 193—194, 1919.
2890. **Riley, W. A.,** An annotated list of the animal parasites of foxes, *Parasitology,* 13, 86—96, 1921.
2891. **Riley, W. A.,** *Diphyllobothrium latum* in Minnesota, *J. Parasitol.,* 10, 188—190; *J. Parasitol.,* 11, 101, 1924.

2892. **Riley, W. A.,** Reservoirs of *Echinococcus* in Minnesota, *Minn. Med.*, 16, 744—745, 1933.
2893. **Riley, W. A.,** The need for data relative to the occurrence of hydatids and of *Echinococcus granulosus* in wildlife, *J. Wildl. Manage.*, 3, 255—257, 1939.
2894. **Riley, W. A.,** Maintenance of *echinococcus* in the United States, *JAMA*, 95, 170—172, 1939.
2895. **Riley, W. A. and Shannon, W. R.,** The rat tapeworm, *Hymenolepis diminuta*, in man, *J. Parasitol.*, 8, 109—117, 1922.
2896. **Rindfleisch, E.,** Zur Histologie der Cestoden, *Arch. Mikrosk. Anat.*, 1, 138—142, 1885.
2897. **Riser, N. W.,** A new proteocephalid from *Amphiuma tridactylum* Cuvier, *Trans. Am. Microsc. Soc.*, 61, 391—397, 1942.
2898. **Riser, N. W.,** Observations on the nervous system of the cestodes, *J. Parasitol.*, 25 (Suppl.), 27, 1949.
2899. **Riser, N. W.,** *Amphilina bipunctata* n.sp., a North American cestodarian, *J. Parasitol.*, 34, 479—485, 1949.
2900. **Riser, N. W.,** Studies on cestode parasites of sharks and skates, *J. Tenn. Acad. Sci.*, 30, 265—311, 1955.
2901. **Riser, N. W.,** Early larval stages of two cestodes from elasmobranch fishes, *Proc. Helminthol. Soc. Wash.*, 23, 120—124, 1956.
2902. **Riser, N. W.,** Observations on the plerocercoid larva of *Pelichnibothrium speciosum* Montic., 1889, *J. Parasitol.*, 42, 31—33, 1956.
2903. **Riser, N. W.,** The hooks of taenioid cestodes from North American felids, *Am. Midl. Nat.*, 56, 133—137, 1956.
2904. **Rivolta, S.,** Di una nuova specie di taenie nella pecora, *T. ovilla*, *Giorn. Anat. Fis. Pat. Anim. Pisa*, 10, 303—308, 1878.
2905. **Rivolta, S.,** Sopra alcune specie di tenie della pecora, *Studi Fatti N. Gab. Anat. Patol. Pisa*, p. 79, 1879.
2906. **Rizhikov, K. M.,** *Microsomacanthus melanittae* n.sp. from *Melanitta deglandi* (in Russian), *Tr. Gel'mintol. Lab.*, 12, 102—105, 1962.
2907. **Rizhikov, K. M.,** Three new cestodes from Anseriformes on Chukotka: *Microsomacanthus minimus* n.sp., *M. borealis* n.sp., *M. somateriae* n.sp. (Cyclophyllidea, Hymenolepididae), *Tr. Gel'mintol. Lab.*, 15, 132—139, 1965.
2908. **Rizhikov, K. M. and Gubanov, N. M.,** *Lateriparus mathevossianae* n.sp. from Anseriformes (in Russian), *Tr. Gel'mintol. Lab.*, 12, 106—108, 1962.
2909. **Rizzo, C.,** La diagnosi biologica di cisticercosi del nevrasse. A proposito di un quarto caso di cisticercosi cerebrale diagnosticato in vita, *Riv. Patol. Nerv.*, 41, 193—216, 1933.
2910. **Roberts, F. H. S.,** A survey of the helminth parasites of the domestic fowl and domestic pigeon in Queensland, *Queensland Agric. J.*, 38, 344—347, 1932.
2911. **Roberts, F. H. S.,** Worm parasites of domesticated animals in Queensland, *Queensland Agric. J.*, 41, 245—252, 1934.
2912. **Roberts, F. H. S.,** *Zygoribatula longiporosa* Hammer (Oribatei: Acarina) an intermediate host of *Moniezia benedeni* (Moniez) (Anoplocephalidae: Cestoda) in Australia, *Aust. J. Zool.*, 1, 239—241, 1953.
2913. **Robinson, E. S.,** Some new cestodes from New Zealand marine fishes, *Trans. R. Soc. N. Z.*, 86, 381—392, 1959.
2914. **Robinson, E. S.,** Cestoda (Tetraphyllidea and Trypanorhyncha) from marine fishes of New South Wales, *Rec. Aust. Mus.*, 26, 341—348, 1965.
2915. **Roboz, Z. von,** Beiträge zur Kenntnis der Cestoden, *Z. Wiss. Zool.*, 37, 263—285, 1882.
2916. **Rodgers, L. O.,** A new dilepid tapeworm from a cardinal, *Trans. Am. Microsc. Soc.*, 60, 273—275, 1941.
2917. **Roitman, V. A.,** *Fissurobothrium unicum* n.g., n.sp. Amphicotylidae, Marsipometrinae), a new pseudophyllidean from fish in the Amur basin (in Russian), *Tr. Gel'mintol. Lab.*, 15, 127—131, 1965.
2918. **Roitman, V. A. and Freze, V. I.,** New species of *Gangesia* (Cestoda: Proteocephalata) from fish from the Amur basin (in Russian), *Tr. Gel'mintol. Lab.*, 14, 170—181, 1964.
2919. **Rolan, R. G. and Leidahl, G.,** *Mayhewia nebraskensis*, sp.n., a cestode from the rock dove, *Columba livia*, *Am. Midl. Nat.*, 82, 598—600, 1969.
2920. **Roman, E.,** Hôtes intermediaires nouveaux d' *Hymenolepis diminuta* (cestodes hyménolépididés), *C. R. Soc. Biol. (Paris)*, 126, 26—28, 1937.
2921. **Romanov, I. V.,** New species of helminths from *Martes zibellina* (in Russian), *Tr. Gel'mintol. Lab. Akad. Nauk SSSR*, 6, 323—330, 1952.
2922. **Romanovitch, M.,** Quelques helminthes du Renne *(Tarandus rangifer)*, *C. R. Soc. Biol.*, 78, 451—453, 1915.
2923. **Ronka, E. K. F.,** Infestation with *Diphyllobothrium latum*, fish tapeworm, *N. Engl. J. Med.*, 210, 582—583, 1934.
2924. **Rosen, F.,** Recherches sur le développement embryonnaire des cestodes. I. Le cycle évolutif des Bothriocephales, *Bull. Soc. Sci. Nat. Neuchâtel.*, 43, 241—300, 1919.
2925. **Rosen, F.,** Recherches sur le développement embryonnaire des cestodes. II. Le cycle évolutif de la Ligule et quelques questions générales sur le développement des Bothriocephales, *Bull. Soc. Sci. Nat. Neuchâtel.*, 44, 259—280, 1920.

2926. **Rosseter, T. B.**, Cysticercoids parasitic in *Cypris cinerea, J. Microsc. Nat. Sci.*, n.s., 3, 241—247, 1890.
2927. **Rosseter, T. B.**, Sur un cysticercoide des Ostracodes capable de se développer dans l'intestine du canard, *Bull. Soc. Zool. Fr.*, 16, 224—229, 1891.
2928. **Rosseter, T. B.**, On a new cysticercus and a new tapeworm, *J. Quekett Microsc. Club*, Ser. 2, 4, 361—366, 1892.
2929. **Rosseter, T. B.**, On the cysticercus of *T. microsoma* and a new cysticercus from *Cyclops agilis, J. Quekett Microsc. Club*, Ser. 2, 5, 179—182, 1893.
2930. **Rosseter, T. B.**, On *Cysticercus quadricurvatus* Ross, *J. Quekett Mirosc. Club*, Ser. 2, 5, 338—343, 1894.
2931. **Rosseter, T. B.**, *Cysticercus venusta* (Rosseter), *J. Quekett Microsc. Club*, Ser. 2, 6, 305—313, 1897.
2932. **Rosseter, T. B.**, Cysticercus of *Taenia liophallus, J. Quekett Microsc. Club*, Ser. 2, 6, 314—317, 1897.
2933. **Rosseter, T. B.**, On experimental infection of ducks with *Cysticercus coronula, Cyst. gracilis, Cyst. tenuirostris, J. Quekett Microsc. Club*, Ser. 2, 6, 397—405, 1897.
2934. **Rosseter, T. B.**, On the generative organs of *Drepanidotaenia venusta, J. Quekett Microsc. Club*, Ser. 2, 7, 10—23, 1898.
2935. **Rosseter, T. B.**, The anatomy of *Dicranotaenia coronula, J. Quekett Microsc. Club*, Ser. 2, 7, 355—370, 1900.
2936. **Rosseter, T. B.**, On the anatomy of *Drepanidotaenia tenuirostris, J. Quekett Microsc. Club*, Ser. 2, 8, 399—406, 1903.
2937. **Rosseter, T. B.**, The genital organs of *Taenia sinuosa, J. Quekett Microsc. Club*, Ser. 2, 9, 81—90, 1904.
2938. **Rosseter, T. B.**, On *Drepanidotaenia undulata* (Krabbe), *J. Quekett Microsc. Club*, Ser. 2, 9, 269—274, 1906.
2939. **Rosseter, T. B.**, On a new tapeworm, *Drepanidotaenia sagitta, J. Quekett Microsc. Club*, Ser. 2, 9, 275—278, 1906.
2940. **Rosseter, T. B.**, On the tapeworm, *Hymenolpeis nitida* Krabbe and *H. nitidulans* Krabbe, *J. Quekett Microsc. Club*, Ser. 2, 10, 31—40, 1907.
2941. **Rosseter, T. B.**, On *Hymenolepis fragilis, J. Quekett Microsc. Club*, Ser. 2, 10, 229—234, 1908.
2942. **Rosseter, T. B.**, *Hymenolepis farciminalis, J. Quekett Microsc. Club*, Ser. 2, 10, 295—310, 1908.
2943. **Rosseter, T. B.**, *Hymenolepis acicula sinuata*, a new species of tapeworm, *J. Quekett Microsc. Club*, Ser. 2, 10, 393—402, 1909.
2944. **Rosseter, T. B.**, *Hymenolepis upsilon*, a new species of avian tapeworm, *J. Quekett Microsc. Club*, Ser. 2, 11, 147—160, 1911.
2945. **Rothman, A. H.**, The larval development of *Hymenolepis diminuta* and *H. citelli, J. Parasitol.*, 43, 643—648, 1957.
2946. **Roudabush, R. L.**, Abnormalities in *Taenia pisiformis, Trans. Am. Microsc. Soc.*, 60, 371—374, 1941.
2947. **Rudin, E.**, Studien an *Fistulicola plicatus* Rud., *Rev. Suisse Zool.*, 22, 321—363, 1914.
2948. **Rudin, E.**, *Oochoristica truncata* Krabbe, *Zool. Anz.*, 47, 75—78, 81—85, 1916.
2949. **Rudin, E.**, Die Ichthyotaenien der Reptilien, *Rev. Suisse Zool.*, 25, 179—381, 1917.
2950. **Rudolphi, C. A.**, Beobachtungen ueber die Eingeweidewürmer, *Arch. Zool. Zoot.*, 2, 1—65, 1801.
2951. **Rudolphi, C. A.**, Fortsetzung der Beobachtungen ueber die Eingeweidewürmer, *Arch. Zool. Zoot.*, 2, 1—67, 1802; *Arch. Zool. Zoot.* 3, 61—125, 1802.
2952. **Rudolphi, C. A.**, Neue Beobachtungen ueber die Eingeweidewürmer, *Arch. Zool. Zoot.*, 3, 1—32, 1803.
2953. **Rudolphi, C. A.**, Bemerkungen aus dem Gebiet der Naturgeschichte, Medicin und Thierarzneikunde, auf einer Reise durch einen Theil von Deutschland, Holland und Frankreich I. Theil., Berlin, 1804, viii + 296 pp.
2954. **Ruether, R.**, *Davainea mutabilis*, Beitrag zur Kenntnis der Bandwürmer des Huhnes. *Dtsch. Thierärztl. Wochenschr.*, 9, 353—357, 362—364, 1901.
2955. **Ruszkowski, J. S.**, Etudes sur le cycle évolutif et sur la structure des cestodes de mer. I. *Echinobothrium benedeni* n.sp., ses larves et son hôte intermédiaire, *Hippolyte varians* (Leach), *Bull. Int. Akad. Pol. Sci. Math. Nat.*, Ser B, pp. 719—738, 1928.
2956. **Ruszkowski, J. S.**, Le cycle évolutif du cestode *Drepanidotaenia lanceolata* (Bloch), *Bull. Int. Akad. Pol. Sci. Math. Nat.*, Ser. B., pp. 1—8, 1932.
2957. **Ruszkowski, J. S.**, Etudes sur le cycle évolutif et sur la structure des cestodes de mer. II. Sur les larves de *Gyrocotyle urna* (Gr. et Wagen.), *Bull. Int. Akad. Pol. Sci. Math. Nat.*, Ser. B, pp. 629—641, 1932.
2958. **Ruszkowski, J. S.**, Etudes sur le cycle évolutif et la structure des cestodes de mer. III. Le cycle évolutif du tétrarhynque *Grillotia erinaceus* (van Ben., 1858), *Acad. Pol. Sci. Lett C. R. Mens. Cl. Sci. Math. Nat.*, p. 6, 1932.
2959. **Ruszkowski, J. S.**, Etudes sur le cycle évolutif et la structure des cestodes de mer. III. Le cycle évolutif du tétrarhynque *Grillotia erinaceus* (van Ben., 1858), *Mem. Int. Acad. Pol. Sci. Lett. Cl. Sci. Math. Nat.*, Ser. B, 1—9, 1934.
2960. **Rutkevich, N. L.**, *Diphyllobothrium giljacicum* nov. sp. and *Diphyllobothrium luxi* n.sp., two new tapeworms of man from Sakhalin (in Russian) *Rab. Gel'mintol. (Skrjabin)*, pp. 574—580, 1937.

2961. **Ryjikov, K. M. and Tolkatcheva, L. M.**, *Essentials of Cestodology. X. Acoleata — Cestode helminths of birds*, Academiya Nauk SSSR, Moscow, 1981, 215 pp.
2962. **Ryšavý, B.**, Dalši poznatky o helmintofauně ptäku v Československu (in Czech), *Česk. Parasitol.*, 4, 299—329, 1957.
2963. **Ryšavý, B.**, Tasemnice vodniho ptactva z rybinčni oblasti jižnich Čech. I. Hymenolepididae Fuhrman 1907 (in Czech), *Česk. Parasitol.* 8, 325—363, 1961.
2964. **Ryšavý, B.**, Neue Bufunde von Bandwürmen (Cestoidea) der Vögel der Ordnung Passeriformes aus dem Bebiete der Tschechoslowakei, *Věstn. Česk. Zool. Společnosti*, 26, 14—24, 1962.
2965. **Ryšavý, B.**, New findings of bird cestodes in Czechoslovakia, *Česk. Parasitol.*, 12, 255—262, 1965.
2966. **Ryšavý, B.**, Nuevas especies de Céstodos (Cestoda: Cyclophyllidea) de aves para Cuba, *Poeyana*, Ser. A (19), 1966, 22 pp.
2967. **Ryšavý, B.**, Nuevas especies de céstodos del género *Nadejdolepis* Spassky et Spasskaja, 1954. (Cestodos: Hymenolepididae) de aves Cubanas del orden Charadriiformes, *Poeyana*, Ser. A (47), 1967, 12 pp.
2968. **Ryšavý, B.**, *Vampirolepis novadomensis* sp.n. (Hymenolepididae), a new cestode species from *Myotis mystacinus* Kuhl, *Folia Parasitol.*, 18, 281—283, 1971.
2969. **Ryšavý, B., Barus, V., and Daniel, M.**, Helminths from *Ithaginis cruentus cruentus* (Phasianidae) from Nepal, in Dr. B. S. Chauhan Commemoration Volume. Tiwari, K. K. and Srivastava, C. B., Eds., Zoolological Society of India, Lucknow, 1975, 151—157.
2970. **Ryšavý, B. and Macko, J. K.**, Bird cestodes of Cuba. I. Cestodes of birds of the orders Podicipediformes, Pelicaniformes and Ciconiiformes, *Anal. Instit. Biol. Univ. Nac. Autón. Mexico Zool.*, 42, 1—28, 1973 (dated 1971).
2971. **Ryšavý, B. and Macko, J. K.**, *Raillietina (R.) cabelleroi* sp.nov., a new cestode species from Cuba, in *Excerta Parasitológica en Memoria del Doctor Eduardo Caballero y Caballero*, Mexico, 1977, 367—370.
2972. **Ryšavý, B. and Moravec, F.**, *Bothriocephalus aegyptiacus* sp.n. (Cestoda: Pseudophyllidea) from *Barbus bynni* and its life cycle, *Věstn. Česk. Společnosti Zool.*, 39, 68—72, 1975.
2973. **Ryšavý, B. and Tenora, F.**, *Raillietina (R.) afghana* sp.n. (Cestoidea: Davaineidae) — a parasite of *Blandfodimys afghana* (Rodentia) in Afghanistan, *Acta Univ. Agric. Fac. Agron. (Brno)*, 22, 111—114, 1974.
2974. **Ryšavý, B. and Tenora, F.**, *Raillietina (Paroniella) kratochvili* sp.n. (Cestoda: Davaineidae), a new bird cestode from *Pica pica bactriana* from Afghanistan, *Folia Parasitol.*, 21, 69—72, 1979.
2975. **Ryzikov, K. M. and Tolkacheva, L. M.**, Taxonomic review of Amabiliidae (Cestoda, Cyclophyllidea) (in Russian), *Zool. Zh.*, 54, 498—502, 1975.
2976. **Ryzhikov, K. M. and Tolkacheva, L. M.**, A taxonomic review of cestodes from the family Dioecocestidae (Acoleata, Cyclophyllididae) (in Russian), *Mater. Nauchn. Konf. Vses. Ova. Gel'mintol.*, (Biologicheski Osnovy bor'by s gel'mintozami cheloveka izhivotnykh), pp. 147—153, 1978.
2977. **Ryzhikov, K. M. and Tolkacheva, L. M.**, *Essentials of Cestodology. X. Acoleata — Cestode Helminths of Birds*, Akademiya Nauk SSSR, Moscow, 1981, 215 pp.
2978. **Saakova**, Two new genera of the family Hymenolepididae from birds of the Danube Delta (in Russian) *Papers on Helminthology Presented to Academecian K. I. Skrjabin on his 80th birthday*, Izdatelstvo Akademii Nauk SSSR, Moscow, 1958, 310—314.
2979. **Sadovskaya, N. P.**, Cestode fauna of insectivores in the Primorsk region. (in Russian), in *Parasitic Worms of Domestic and Wild Animals: Papers on Helminthology Presented to Prof. A. A. Sobolev on the 40th Anniversary of his Scientific and Teaching Activity*, Dalnevostochnii Gosudarstvennii Universitet, Vladivostok, 1965, 290—297.
2980. **Sadykhov, J. A.**, New species of cestode, *Mesocestoides petrowi* n.sp. from the intestine of *Vulpes vulpes*, in *Sbornik Rabot po Gel'mintologii posvyaschchen go-letiyu so dnya rozhdeniye Akademika K. I. Skrjabina*, Izdatel'stvo KOLOS, Moscow, 1971, 351—353.
2981. **Saeki, Y.**, Experimental studies on the development of *Hymenolepis nana* (in Japanese), *Jikwa Zasshi*, pp. 203—244, 1920.
2982. **Sahay, S. N. and Sahay, U.**, On a new caryophylaeid cestode, *Djombangia caballeroi* sp.nov., from freshwater fish *Heteropneustes fossilis* in Chotanagpur with an emendation of the generic character, in *Exerta Parasitológica en Memoria del Doctor Eduardo Caballero y Caballero*, Mexico Universidad Nacional Autonoma de Mexico, Mexico, 1977, 371—376.
2983. **Sailov, D. I.**, New species of cestodes from fish-eating birds in the Kyzyl-Agach State Reserve named after S. M. Kirov (in Russian), *Dokl. Akad. Nauk Az. SSR*, 18, 45—51, 1962.
2984. **Salensky, W.**, Ueber den Bau und die Entwicklungsgeschichte der *Amphilina (Monostomum foliaceum* Rud.), *Z. Wiss. Zool.*, 24, 291—342, 1874.
2985. **Sandars, D. F.**, A study of Diphyllobothriidae (Cestoda) from Australian hosts, *Proc. R. Soc. Queensland*, 63, 65—70, 1953.
2986. **Sandars, D. F.**, *Mirandula parva* gen. et sp.nov. (Cestoda, Dilepididae) from the long-nosed bandicoot *(Perameles nasuta* Geoff.), *J. Helminthol.*, 30, 183—188, 1956.

2987. **Sandars, D. F.,** Cestoda from *Rattus assimilis* (Gould, 1858) from Australia, *J. Helminthol.*, 31, 65—78, 1957.
2988. **Sandars, D. F.,** *Hymenolepis miniopteri* n.sp., (Cestoda), from an Australian bat, *Miniopterus blepotis* (Temm., 1840), *J. Helminthol.*, 31, 79—84, 1957.
2989. **Sandars, D.,** Redescription of some cestodes from marsupials. I. Taeniidae, *Ann. Trop. Med. Parasitol.*, 51, 317—329, 1957.
2990. **Sandars, D.,** Redescription of some cestodes from marsupials. II. Davaineidae, Hymenolepididae and Anoplocephalidae, *Ann. Trop. Med. Parasitol.*, 51, 330—339, 1957.
2991. **Sandeman, I.,** *Capsulata edenensis* gen. et sp.nov. a new cestode with an unusual type of growth, from *Limosa lapponica* (L.); with systematic notes on the genera *Southwellia* Moghe, 1925 and *Malika* Woodland, 1929, *J. Helminthol.*, 33, 171—188, 1959.
2992. **Sandeman, I.,** Une espece nouvelle du genre *Proterogynotaenia* Fuhrmann, *Ann. Parasitol. Hum. Comp.*, 34, 265—270, 1959.
2993. **Sandeman, I.,** A contribution to the revision of the dilepid tapeworms from charadriiformes. Preliminary note, *Zool. Anz.*, 163, 278—288, 1959.
2994. **Sandground, J. H.,** A new mammalian cestode from Brazil, *Contrib. Harvard Inst. Trop. Biol. Med.*, pp. 284—291, 1926.
2995. **Sandground, J. H.,** Some new cestode and nematode parasites from Tanganyika Territory, *Proc. Boston Soc. Nat. Hist.*, 39, 131—150, 1928.
2996. **Sandground, J. H.,** Notes and descriptions of some parasitic helminths collected by the expedition, *Contrib. Dep. Trop. Med. Inst. Trop. Biol. Med.*, 1, 462—486, 1930.
2997. **Sandground, J. H.,** Two new helminths from *Rhinoceros sondaicus*, *J. Parasitol.*, 19, 192—204, 1933.
2998. **Sandground, J. H.,** On species of *Moniezia* (Cestoda, Anoplocephalidae) harboured by the hippopotamus, *Proc. Helminthol. Soc. Wash.*, 3, 52—53, 1936.
2999. **Sandground, J. H.,** On a coenurus from the brain of a monkey, *J. Parasitol.*, 23, 482—490, 1937.
3000. **Santos, J. C. dos and Rolas, F. J. T.,** Sobre alguns cestóides de *Bothrops* e de *Liophis miliaris*, *Atas Soc. Biol. Rio de Janeiro*, 17, 35—40, 1973.
3001. **Saoud, M. F. A.,** On a new cestode, *Anthobothrium taeniuri* n.sp. (Tetraphyllidea) from the Red Sea sting ray and the relationship between *Anthobothrium* van Beneden, 1850, *Rhodobothrium* Linton, 1889 and *Inermiphyllidium* Riser, 1955, *J. Helminthol.*, 37, 135—144, 1963.
3002. **Saoud, M. F. A., Ramadan, M. M., and Hassan, S. I.,** On *Echinobothrium helmymohamedi* n.sp. (Cestoda: Diphyllidea); a parasite of the sting ray *Taeniura lymma* from the Red Sea, *J. Egyptian Soc. Parasitol.*, 12, 199—202, 1982.
3003. **Satpute, L. R. and Agarwal, S. M.,** "Diverticulosis" of the fish duodenum infested with cestodes, *Ind. J. Exp. Biol.*, 12, 373—375, 1974.
3004. **Satpute, L. R. and Agarwal, S. M.,** *Introvertus raipurensis* gen.nov. sp.nov. a fish cestode (Cestoda: Caryophyllidea: Lytocestidae) from Raipur, India, *Proc. Ind. Acad. Parasitol.*, 1, 17—19, 1980.
3005. **Saunders, L. G.,** A survey of helminth and protozoan incidence in man and dogs at Fort Chipewyan, Alberta, *J. Parasitol.*, 35, 31—34, 1949.
3006. **Sauter, K.** Beiträge zur Anatomie, Histologie, Entwicklungsgeschichte und Systematik der Rindertaenien, Dissertation, University of Munich, 1917, 79 pp.
3007. **Savazzini, L. A.,** *Cylindrotaenia americana* en nuestro *Leptodactylus ocellatus*, *Semana Méd. (Buenos Aires)*, 36, 868—870, 1929.
3008. **Sawada, I.,** Observation on the ecology and life history of the poultry cestode, *Raillietina tetragona*, *Rep. Nara Gakugei Univ.*, 1, 211—223, 1952.
3009. **Sawada, I.,** On the life history of chicken cestode *Raillietina cesticillus*, *Rep. Nara Gakugei Univ.*, 1, 235—243, 1952.
3010. **Sawada, I.,** Morphological studies on the fowl cestode, *Raillietina (Paroniella) kashiwarensis* n.sp., *Zool. Mag. Dobuts. Zasshi*, 62, 179—185, 1953.
3011. **Sawada, I.,** Studies on the life history of the poultry cestode, *Raillietina (Paroniella) kashiwarensis* Sawada, *Nara Gakugei Univ. Bull.*, 2, 147—159, 1953.
3012. **Sawada, I.,** Morphological studies on the chicken tapeworm *Raillietina (Raillietina) echinobothrida*, *Dobuts. Zasshi*, 63, 200—203, 1954.
3013. **Sawada, I.,** On the situation of the genital pore as one of the important characters of the chicken tapeworms, belonging to the genus *Raillietina*, *Dobuts. Zasshi*, 63, 384—387, 1954.
3014. **Sawada, I.,** *Raillietina (Raillietina) galli* Yamaguti, is a synonym of *Raillietina (R.) tetragona* (Molin), *Dobuts. Zasshi*, 64, 105—107, 1955.
3015. **Sawada, I.,** Studies on the tapeworms of the domestic fowl found in Japan, *Annot. Zool. Jpn.*, 28, 26—32, 1955.
3016. **Sawada, I.,** Two new species of avian tapeworm belonging to the genus *Choanotaenia* Railliet, 1896, *Annot. Zool. Jpn.*, 35, 47—50, 1962.

3017. **Sawada, I.,** Two new cestodes from the domestic fowl in Sudan, Africa, *Annot. Zool. Jpn.,* 37, 233—237, 1964.
3018. **Sawada, I.,** On the genus *Raillietina* Fuhrmann, 1920 (I), *J. Nara Gakugei Univ.,* 12, 19—36, 1964.
3019. **Sawada, I.,** On the genus *Raillietina* Fuhrmann, 1920 (II), *J. Nara Gakugei Univ.,* 13, 5—38, 1965.
3020. **Sawada, I.,** On a new tapeworm, *Vampirolepis isensis,* found in bats with the table of the morphological features of tapeworms in *Vampirolepis, Jpn. J. Med. Sci. Biol.,* 19, 51—57, 1966.
3021. **Sawada, I.,** Helminth parasites of bats in Japan. I, *Annot. Zool. Jpn.,* 40, 61—66, 1967.
3022. **Sawada, I.,** Helminth fauna of bats in Japan. II, *Jpn. J. Parasitol.,* 16, 103—106, 1967.
3023. **Sawada, I.,** Helminth fauna of bats in Japan. III, *Annot. Zool. Jpn.,* 40, 177—179, 1967.
3024. **Sawada, I.,** Helminth fauna of bats in Japan. IV, *Annot. Zool. Jpn.,* 41, 9—10, 1968.
3025. **Sawada, I.,** Helminth fauna of bats in Japan. V, *Annot. Zool. Jpn.,* 41, 168—171, 1968.
3026. **Sawada, I.,** Helminth fauna of bats in Japan. VI, *Annot. Zool. Jpn.,* 43, 50—52, 1970.
3027. **Sawada, I.,** Helminth fauna of bats in Japan. VII, *Bull. Nara Univ. Educ.,* 19, 73—80, 1970.
3028. **Sawada, I.,** Two new avian cestodes, *Raillietina (Raillietina) somalensis* and *Cotugnia shohoi* from *Acryllium valturinum* in Somalia, *Jpn. J. Zool.,* 16, 131—134, 1971.
3029. **Sawada, I.,** Helminth fauna of bats in Japan. VIII, *Annot. Zool. Jpn.,* 44, 175—178, 1971.
3030. **Sawada, I.,** Helminth fauna of bats in Japan. IX, *Bull. Nara Univ. Educ.,* 20, 1—5, 1971.
3031. **Sawada, I.,** Helminth fauna of bats in Japan. X, *Annot. Zool. Jpn.,* 45, 22—28, 1972.
3032. **Sawada, I.,** Helminth fauna of bats of Japan. XI, *Bull. Nara Univ. Educ.,* 21, 27—30, 1972.
3033. **Sawada, I.,** Helminth fauna of bats. XII, *Annot. Zool. Jpn.,* 45, 245—249, 1972.
3034. **Sawada, I.,** Helminth fauna of bats in Japan. XV, *Annot. Zool. Jpn.,* 47, 103—106, 1974.
3035. **Sawada, I.,** Helminth fauna of bats in Japan. XVI, *Annot. Zool. Jpn.,* 48, 43—48, 1975.
3036. **Sawada, I.,** *Mayhewia shibuei* n.sp., a new cestode from the painted snipe, *Rostratula benghalensis,* in *Dr. B. S. Chauhan Commemoration Volume* Tiwari, K. K. and Srivastava, C. B., Eds., Zoological Society of India, Lucknow, 1975, 134—144.
3037. **Sawada, I.,** Helminth fauna of bats in Japan. XXII, *Annot. Zool. Jpn.,* 53, 194—201, 1980.
3038. **Sawada, I.,** A new species of cestode, *Diorchis siamensis* n.sp., from wild duck, *Anas* sp., in Thailand, *Jpn. J. Parasitol.,* 29, 83—86, 1980.
3039. **Sawada, I.,** Helminth fauna of bats in Japan. XXV, *Annot. Zool. Jpn.,* 55, 26—31, 1982.
3040. **Sawada, I. and Chikada, T.,** *Raillietina (Raillietina) toyohashiensis* n.sp. from a guinea fowl at Toyohashi Zoo, *Jpn. J. Parasitol.,* 21, 408—410, 1972.
3041. **Sawada, I. and Funabashi, F.,** A new avian cestode, *Metroliasthes coturnix* n.sp. from the intestine of a Japanese quail, with an avian cestode from a macaw, *Jpn. J. Parasitol.,* 21, 395—399, 1972.
3042. **Sawada, I. and Iijima, T.,** On cestodes from aquatic birds in Yumanashi Prefecture, *Jpn. J. Med. Sci. Biol.,* 17, 33—37, 1964.
3043. **Sawada, I. and Kifune, T.,** A new species of anoplocephaline cestode from *Macaca iris, Jpn. J. Parasitol.,* 23, 366—368, 1974.
3044. **Sawada, I. and Kifune, T.,** Studies on the helminth fauna of Kyushu. II. Four new cestodes from wild birds in Fukuoka Prefecture, *Bull. Nara Univ. Educ.,* 23, 15—29, 1974.
3045. **Sawada, I. and Kugi, G.,** A new cestode, *Mesocestoides paucitesticulus,* from a badger *Nycterutes procyonoides,* in Japan, *Jpn. J. Parasitol.,* 22, 45—47, 1973.
3046. **Sawada, I. and Kugi, G.,** A new species of *Scyphocephalus* (Cestoda) from Malayan monitor, *Varanus salvator, Jpn. J. Parasitol.,* 22, 126—130, 1973.
3047. **Sawada, I. and Kugi, G.,** On cestodes genus *Bothridium* obtained from reptiles died in the Amazonland at Beppu City during 1970, *Bull. Nara Univ. Educ.,* 22, 43—65, 1973.
3048. **Sawada, I. and Kugi, G.,** Studies on the helminth fauna of Kyushu. I. Three new cestodes from wild birds and rabbit, *Annot. Zool. Jpn.* 47, 261—266, 1974.
3049. **Sawada, I. and Kugi, G.,** Studies on the helminth fauna of Kyushu. III. Three species of Hymenolepididae from anserine birds in Ôita Prefecture, *Bull. Nara Univ. Educ.,* 24, 5—12, 1975.
3050. **Sawada, I. and Kugi, G.,** Studies on the helminth fauna of Kyushu. V. Cestode parasites of wild mammals and birds from Ôita Prefecture, *Annot. Zool. Jpn.,* 52, 133—141, 1979.
3051. **Sawada, I. and Kugi, G.,** Studies on the helminth fauna of Kyushu. VI. Cestode parasites of wild birds from Ôita Prefecture, *Annot. Zool. Jpn.,* 53, 269—279, 1980.
3052. **Sawada, I. and Kugi, G.,** Studies on the helminth fauna of Kyushu. VII. Cestode parasites of wild ducks and mammals from Ôita Prefecture, *Annot. Zool. Jpn.,* 54, 256—258, 1981.
3053. **Sawada, I. and Okada, H.,** Studies on the morphology of successive stages in the development of *Raillietina (Skrjabinia) cesticillus* oncosphere to mature cysticercoid (in Japanese with English summary), *Dobut. Zasshi,* 64, 316—320, 1955.
3054. **Sawada, I. and Papasarathorn, T.,** *Paranoplocephala indicata* n.sp. (Cestoda: Anoplocephalidae) from the Malayan tapir, *Tapirus indicus, Jpn. J. Zool.,* 15, 125—128, 1966.
3055. **Saxena, S. K.,** Studies on cestodes of the common Indian vulture, *Neophron percnopterus* (Linn.). I. On the morphology of a new genus of an anoplocephalid cestode, *Zool. Anz.,* 179, 466—474, 1967.

3056. **Saxena, S. K.**, Studies on cestodes of the common Indian vulture, *Neophron percnopterus* (Linn.). II. On the morphology of *Neophronia luteus* sp. nov., *Zool. Anz.*, 180, 328—335, 1968.
3057. **Saxena, S. K.**, Studies on cestodes of the common Indian vulture, *Neophron percnopterus* (Linn.). III. On *Nephronia irregularis* sp.nov., *Zool. Anz.*, 181, 140—145, 1968.
3058. **Saxena, S. K.**, Studies on cestodes of the common Indian vulture, *Neophron percnopterus* (Linn.). IV. On *Neophronia melanotus* sp. nov., *Zool. Anz.*, 181, 146—152, 1968.
3059. **Saxena, S. K.**, Studies on cestodes of the common vulture, *Neophron percnopterus* (Linn.). V. On the morphology and systematic position of *Neophronia percnopteri* (Singh, 1952) n.comb. with notes on other species of the genus *Neophronia* Saxena, 1967, *Zool. Anz.*, 183, 455—462, 1969.
3060. **Saxena, S. K.**, Studies on the tapeworms of the common Indian kite, *Milvus migrans* (Boddaert), *Ann. Parasitol. Hum. Comp.*, 45, 405—420, 1970.
3061. **Saxena, S. K.**, Studies on cestodes of the common Indian mynahs *Acridotheres tristis* and *A. ginginianus* (Aves), *Neth. J. Zool.*, 22, 307—334, 1972.
3062. **Saxena, S. K. and Baugh, S. C.**, The taxonomy of the cestode genus *Aporina* with a discussion on the systematic positions of the species assigned to it, *Angew. Parasitol.*, 14, 236—245, 1973.
3063. **Saxena, S. K. and Baugh, S. C.**, On cestodes of *Passer domesticus*. II. *Anonchotaenia* and *Mathevotaenia*, *Angew. Parasitol.*, 19, 85—106, 1978.
3064. **Schad, G. A.**, Helminth parasites of mice in Northeastern Quebec and the coast of Labrador, *Can. J. Zool.*, 32, 215—224, 1954.
3065. **Schaefer, G. B. and Self, J. T.**, *Bothriocephalus euryciensis* n.sp. (Cestoidea, Pseudophyllidea) from the cave salamander *Eurycea longicauda*, *Proc. Okla. Acad. Sci.*, 58, 154—155, 1978.
3066. **Schauinsland, H.**, Die embryonale Entwicklung der Bothriocephalen, *Jena. Z. Naturwiss.*, 19, 520—573, 1885.
3067. **Schauinsland, H.**, Ueber die Körperschichten und deren Entwicklung bei den Plattwürmern., *Sitzungsber. Ges. Morphol. Physiol. München*, 2, 7—10, 1886.
3068. **Scheuring, L.**, Beobachtungen zur Biologie des Genus *Triaenophorus* und Betrachtungen ueber die jahreszeitliche Auftreten von Bandwürmern., *Z. Parasitenkd.*, 2, 157—177, 1929.
3069. **Schiefferdecker, P.**, Beiträge zur Kenntnis des feineren Baues des Taenien, *Jena. Z. Naturwiss.*, 8, 459—487, 1874.
3070. **Schiller, E. L.**, Experimental infection with cysticerci of *Taenia taeniaeformis* in laboratory animals, *J. Parasitol.*, 35, 37—38, 1949.
3071. **Schiller, E. L.**, *Hymenolepis rauschi* n.sp., a cestode from the ruddy duck, *J. Parasitol.*, 26, 1—4, 1950.
3072. **Schiller, E. L.**, Studies on the helminth-fauna of Alaska. VI. The parasites of the emperor goose *(Philacte canagica L.)* with the description of *Hymenolepis philactes*, n.sp., *J. Parasitol.*, 37, 217—220, 1951.
3073. **Schiller, E. L.**, Studies on the helminth-fauna of Alaska. I. Two new cestodes from Sabine's gull *(Xema sabini)*, *J. Parasitol.*, 37, 266—272, 1951.
3074. **Schiller, E. L.**, *Hymenolepis hopkinsi*, n.sp., a cestode from the black duck, *Am. Midl. Nat.*, 45, 253—256, 1951.
3075. **Schiller, E. L.**, The cestoda of Anseriformes of the North Central States, *Am. Midl. Nat.*, 46, 444—457, 1951.
3076. **Schiller, E. L.**, Studies on the helminth-fauna of Alaska. VIII. Some cestode parasites of the Pacific Kittiwake *(Rissa tridactyla* Ridgway) with the description of *Haploparaxis rissae* n.sp., *Proc. Helminthol. Soc. Wash.*, 18, 122—125, 1951.
3077. **Schiller, E. L.**, Studies on the helminth-fauna of Alaska. IX. The cestode parasites of the white-fronted goose *(Anser albifrons)* with the description of *Hymenolepis barrowensis* n.sp., *J. Parasitol.*, 38, 32—34, 1952.
3078. **Schiller, E. L.**, Studies on the helminth fauna of Alaska. X. Morphological variation in *Hymenolepis horrida* (von Linstow, 1901), (Cestoda: Hymenolepididae), *J. Parasitol.*, 38, 554—568, 1952.
3079. **Schiller, E. L.**, Studies on the helminth-fauna of Alaska. III. *Hymenolepis kenaiensis* n.sp., a cestode from the greater scaup *(Aythya marila nearctica)* with remarks on endemicity, *Trans. Am. Microsc. Soc.*, 71, 146—149, 1952.
3080. **Schiller, E. L.**, *Hymenolepis johnsoni*, n.sp., a cestode from the vole *Microtus pennsylvanicus drummondii*, *J. Wash. Acad. Sci.*, 42, 53—55, 1952.
3081. **Schiller, E. L.**, Studies on the helminth fauna of Alaska. XIV. Some cestode parasites of the Aleutian teal *(Anas crecca)* with description of *Diorchis longiovum* n.sp., *Proc. Helminthol. Soc. Wash.*, 20, 7—10, 1953.
3082. **Schiller, E. L.**, Studies on the helminth fauna of Alaska. XV. Some notes on the cysticercus of *Taenia polyacantha* Leuckart, 1856, from a vole *(Microtus oeconomus operarius* Nelson), *J. Parasitol.*, 39, 344—346, 1953.
3083. **Schiller, E. L.**, Studies on the helminth fauna of Alaska. XVII. Notes on the intermediate host stages of some helminth parasites of the sea otter, *Biol. Bull.*, 106, 107—121, 1954.

3084. **Schiller, E. L.,** Studies on the helminth fauna of Alaska. XVIII. Cestode parasites in young anseriformes on the Yukon delta nesting grounds, *Trans. Am. Microsc. Soc.,* 73, 194—201, 1954.
3085. **Schiller, E. L.,** Studies on the helminth fauna of Alaska. XIX. An experimental study on blowfly *(Phormia regina)* transmission of hydatid disease, *Exp. Parasitol.,* 3, 161—166, 1954.
3086. **Schiller, E. L.,** Studies on the helminth fauna of Alaska. XXIII. Some cestode parasites of eider ducks, *J. Parasitol.,* 41, 79—88, 1955.
3087. **Schiller, E. L.,** Some cestode parasites of the old-squaw, *Clangula hyemalis* (L.), *Proc. Helminthol. Soc. Wash.,* 22, 41, 1955.
3088. **Schiller, E. L.,** Studies on the helminth fauna of Alaska. XXXII. *Hymenolepis echinorostrae* n.sp., a cestode from the lesser scaup, *Aythya affinis* (Eyton), *J. Parasitol.,* 43, 233—235, 1957.
3089. **Schiller, E. L. and Rausch, R. L.,** A vole *(Microtus)* an important natural intermediate host of *Echinococcus granulosus, J. Parasitol.,* 36, 30, 1950.
3090. **Schmelz, O.,** Quelques cestodes nouveaus d'oiseaux d'Asie, *Rev. Suisse Zool.,* 48, 143—199, 1941.
3091. **Schmidt, F.,** Beiträge zur Kenntnis der Entwicklung der Geschlechtsorgane einiger Cestoden, *Z. Wiss. Zool.,* 46, 155—187, 1888.
3092. **Schmidt, F. L.,** A new cestode, *Cladotaenia oklahomensis,* from a hawk, *Trans. Am. Microsc. Soc.,* 59, 519—522, 1940.
3093. **Schmidt, G. D.,** *Hymenolepis calumnacantha* sp.nov. from Wilson's snipe, *Capella gallinago delicata* (Ord.) in Colorado, *Parasitology,* 53, 409—411, 1963.
3094. **Schmidt, G. D.,** *Dioecotaenia cancellata* (Linton, 1890) gen et comb. n., a dioecious cestode (Tetraphyllidea) from the cow-nosed ray, *Rhinoptera bonasus* (Mitchell) in Chesapeake Bay, with the proposal of a new family, Dioecotaeniidae, *J. Parasitol.,* 55, 271—275, 1969.
3095. **Schmidt, G. D.,** *How to Know the Tapeworms,* Wm. C. Brown, Dubuque, 1970, 1-266.
3096. **Schmidt, G. D.,** Cyclophyllidean cestodes of Australian birds, with three new species, *J. Parasitol.,* 58, 1085—1094, 1972.
3097. **Schmidt, G. D.,** *Acanthobothrium urolophi* sp.n., a tetraphyllidean cestode (Oncobothriidae) from an Australian stingaree, *Proc. Helminthol. Soc. Wash.,* 40, 91—93, 1973.
3098. **Schmidt, G. D.,** The taxonomic status of *Spirometra* Faust, Campbell et Kellogg, 1929 (Cestoidea: Diphyllobothriidae), *J. Helminthol.,* 48, 175—177, 1974.
3099. **Schmidt, G. D.,** New records of helminths from New Guinea, including descriptions of three new cestode species, one in the new genus *Wallabicestus* N. G., *Trans. Am. Microsc. Soc.,* 94, 189—196, 1975.
3100. **Schmidt, G. D.,** *Phyllobothrium kingae* sp.n., a tetraphyllidean cestode from a yellow-spotted stingray in Jamaica, *Proc. Helminthol. Soc. Wash.,* 45, 132—134, 1978.
3101. **Schmidt, G. D.,** *Baerietta allisonae* n.sp. (Cestoda: Nematotaeniidae) from a New Zealand gecko, *Hoplodactylus maculatus, N.Z. J. Zool.,* 7, 7—9, 1980.
3102. **Schmidt, G. D. and Bush, A. D.,** *Parvitaenia ibisae* sp.n. (Cestoidea: Dilepididae), from birds in Florida, *J. Parasitol.,* 58, 1095—1097, 1972.
3103. **Schmidt, G. D. and Courtney, C. H.,** *Parvitaenia heardi* sp.n. (Cestoidea: Dilepididae) from the great blue heron, *Ardea herodias,* in South Carolina, *J. Parasitol.,* 59, 821—823, 1973.
3104. **Schmidt, G. D. and File, S.,** *Tupaiataenia quentini* gen. et sp.n. (Anoplocephalidae: Linstowinae) and other tapeworms from the common tree shrew, *Tupaia glis, J. Parasitol.,* 63, 473—475, 1977.
3105. **Schmidt, G. D. and Kuntz, R. E.,** Tapeworms from Philippine reptiles, with two new species of Proteocephalata, *Proc. Helminthol. Soc. Wash.,* 41, 195—199, 1974.
3106. **Schmidt, G. D. and Martin, R. L.,** Tapeworms of the Chaco Boreal, Paraguay, with two new species, *J. Helminthol.,* 52, 205—209, 1978.
3107. **Schmidt, G. D. and Neiland, K. A.,** *Hymenolepis (Hym.) deblocki* sp.n., and records of other helminths from charadriiform birds, *Can. J. Zool.,* 46, 1037—1040, 1968.
3108. **Schmidt, G. D. and Neiland, K. A.,** Helminth fauna of Nicaragua. IV. *Sacciuterina mathevossiani* sp.nov. (Dilepididae), and other cestodes of birds, *Parasitology,* 62, 145—149, 1971.
3109. **Schmidt, J. E.,** Die Entwicklungsgeschichte und der anatomische Bau der *Taenia anatina* (Krabbe), *Arch. Naturg.,* 60, 65—112, 1894.
3110. **Schneider, G. E.,** Ueber das Vorkommen von Larven des Bandwurms *Bothriotaenia proboscidea* Batsch im Magen und Darm von Ostseeheringen (*Clupea harengus membras* L.), *Sitzungsber. Gesell. Naturforsch (Freunde Berlin),* pp. 28—30, 1902.
3111. **Schneider, G. E.,** Ichthyologische Beiträge. III. Ueber die in Fischen des finnischen Meerbusens vorkommende Endoparasiten, *Acat. Soc. Fauna Flora Fenn.,* 22, 1—87, 1902.
3112. **Schneider, G. E.,** *Bothrimonus nylandicus* n.sp., *Arch. Naturg.,* 68, 72—78, 1902.
3113. **Schneider, G. E.,** *Caryophyllaeus fennicus* n.sp., *Arch. Naturg.,* 68, 65—71, 1902.
3114. **Schneider, G. E.,** Beiträge zur Kenntnis der Helminthfauna des finnischen Meerbusens, *Acta Soc. Fauna Flora Fenn.,* 26, 1—34, 1903.
3115. **Schneider, G. E.,** Ueber zwei Endoparasiten aus Fischen des finnischen Meerbusens, *Medd. Soc. Fauna Flora Fenn.,* 29, 75—76, 1904.

3116. **Schneider, G.,** Die Ichthyotaenien des finnischen Meerbusens, *Festschrift Palmen Helsingfors*, 1, 1—31, 1905.
3117. **Schnur, L. F.,** Observations on cestodes of the genera *Anoplocephala* Blanchard, 1848 and *Inermicapsifer* Janicki, 1910 from hyraxes in East Africa, *J. Helminthol.*, 46, 251—269, 1972.
3118. **Schrank, P. F. von,** Förtekning på nagra hittils obeskrifne intestinalkrak, *Kgl. Svenska Vetenskapsakad. Handl.* (Stockholm), 11, 118—126, 1790.
3119. **Schultz, G.,** The twentieth helminthological expedition in U.S.S.R. in Novotscherkassk, in *Res. of 28th Helminthol. Exped. in U.S.S.R.* (in Russian, English Abstr.), 1927.
3120. **Schultz, R. L.,** *Hymenolepis scalopi*, n. sp., *Am. Midl. Nat.*, 21, 641—644, 1939.
3121. **Schultz, R. L.,** A new tapeworm from Swainson's hawk, *Trans. Am. Microsc. Soc.*, 58, 448—451, 1939.
3122. **Schultz, R. L.,** Some observations on the amabiliid cestode, *Tatria duodecacantha* Olsen, 1939, *J. Parasitol.*, 26, 101—103, 1940.
3123. **Schultz, R. L.,** The genus *Diorchis* Clerc, 1903, *Am. Midl. Nat.*, 23, 382—389, 1940.
3124. **Schultze, T. F. S.,** Ueber die Begattung der Bandwürmer, *Ann. Ges. Heilkunde*, 2, 127—128, 1825.
3125. **Schulz, R. E. S. and Landa, D. M.,** Parasitische Würmer der grossen Rennmaus — *Rhombomys opimus* Licht (in Russian), *Vest. Mikrobiol. Epidemiol. Parazitol. (1934)*, 13, 305—315, 1935.
3126. **Schumacher, G.,** Cestoden aus *Centrolophus pompilius* L., *Zool. Jahrb. Syst.*, 36, 149—198, 1914.
3127. **Schwartz, B.,** A new proliferating larval tapeworm from a porcupine, *Proc. U.S. Natl. Mus.*, 66, 1—4, 1924.
3128. **Schwartz, B.,** The chicken as a host for *Metroliasthes lucida*, *J. Parasitol.*, 12, 112, 1925.
3129. **Schwartz, B.,** A subcutaneous tumour in a primate caused by tapeworm larvae experimentally reared to maturity in dogs, *J. Agric. Res.*, 35, 471—480, 1927.
3130. **Schwartz, B.,** The species of *Dipylidium* parasitic in dogs and cats in the United States, *J. Parasitol.*, 14, 68—69, 1927.
3131. **Schwartz, B.,** The life-history of tapeworms of the genus *Mesocestoides*, *Science*, 66, 17—18, 1927.
3132. **Schwarz, R.,** Die Ichthyotaenien der Reptilien und Beitrage zur Kenntnis der Bothriocephalen, *Inaug. Diss., University of Basel*, 1908, 52 pp.
3133. **Scott, H. H.,** Contribution to the experimental study of the life-history of *Hymenolepis fraterna* and *Hymenolepis longior* in the mouse, *J. Helminthol.*, 1, 193—196, 1923.
3134. **Scott, H. H.,** Stages in the direct development of *Hymenolepis longior* Baylis, *J. Helminthol.*, 2, 173—174, 1924.
3135. **Scott, J. W.,** Experiments with tapeworms. I. Some factors producing evagination of a cysticercus, *Biol. Bull.*, 25, 304—312, 1913.
3136. **Scott, J. W.,** A new genus and species of tapeworm from the bighorn sheep, *Anat. Rec.*, 81, 65—66, 1914.
3137. **Scott, J. W.,** The development of two new *Cladotaenia* in the ferrugineous rough-leg hawk, *J. Parasitol.*, 17, 115, 1930.
3138. **Seddon, H. R.,** On the life of *Moniezia expansa* within the sheep, *Ann. Trop. Med. Parasitol.*, 25, 437—442, 1931.
3139. **Sekutowicz, S.,** Etudes sur le développement et sur la biologie de *Caryophyllaeus laticeps* (Pallas), *C. R. Mens. Cl. Sci. Math. Nat. Acad. Pol.*, p. 4, 1932.
3140. **Self, J. T.,** Parasites of the goldeye, *Hidon alosoides* (Raf.), in Lake Texoma, *J. Parasitol.*, 40, 386—389, 1954.
3141. **Self, J. T. and Esslinger, J. H.,** A new species of bothriocephalid cestode from the fox squirrel, *Sciurus niger*, *J. Parasitol.*, 41, 256—258, 1955.
3142. **Self, J. T. and Janovy, J., Jr.,** *Kowalewskiella totani*, n.sp. (Cestoda: Dilepididae) from *Totanus flavipes*, *Proc. Helminthol. Soc. Wash.*, 32, 169—171, 1965.
3143. **Self, J. T. and McKnight, T. J.,** Platyhelminthes from fur bearers in the Wichita Mountains Wildlife Refuge, with special reference to *Oochoristica* spp., *Am. Midl. Nat.*, 43, 58—61, 1950.
3144. **Semenova, N. S.,** *Eranuides mathevossianae* n.g., n.sp. (Anoplocephalidae), a new cestode from reindeer on Taimyr Peninsula (in Russian), *Tr. Vses. Inst. Gel'mintol. K.I. Skryabina*, 19, 171—175, 1972.
3145. **Senger, C. M.,** Observations on cestodes of the genus *Hymenolepis* in North American shrews, *J. Parasitol.*, 41, 167—170, 1955.
3146. **Serdyukov, A. M.,** *Diphyllobothrium arctomarinum* n.sp. from *Stercorarius parasiticus* (in Russian), *Mater. Nauchn. Konf. Veses. Ova. Gel'mintol.*, Part 1, 254—258, 1969.
3147. **Setti, E.,** Sulle tenie dell' *Hyrax dello* Scioa, *Atti Soc. Ligust. Sci. Nat. Geogr. Genova*, 2, 316—324, 1891.
3148. **Setti, E.,** Elminti dell' Eritrea e delle regioni limitrofe, *Boll. Mus. Zool. Genova*, 1892, 19 pp.
3149. **Setti, E.,** *Dipylidium gervaisi* n.sp., *Atti Soc. Ligust. Sci. Nat. Geogr. Genova*, 6, 99—106, 1895.
3150. **Setti, E.,** Nuovie lminti dell' Eritrea, *Atti Soc. Ligust. Sci. Nat. Geogr. Genova*, 8, 198—247, 1897.
3151. **Setti, E.,** Nuove osservazioni sui cestodi parassiti degli iraci, *Atti Soc. Ligust. Sci. Nat. Geogr. Genova*, 9, 188—202, 1898.

3152. **Setti, E.**, La pretesa *Taenia mediocanellata* dell' *Himantopus candidus* e invece la *T. saginata*, *Boll. Mus. Zool. Genova*, pp. 1—4, 1899.
3153. **Setti, E.**, Une nuova tenia nel cane ( *Taenia brachysoma* n.sp.), *Atti Soc. Ligust. Sci. Nat. Geogr. Genova*, 10, 11—20, 1899.
3154. **Seurat, G.**, Sur un cestode parasite des huîtres perlières determinant la production des perles fines aux îles Gambier, *C. R. Acad. Sci.*, 142, 801—803, 1906.
3155. **Sezen, Y. and Price, C. E.**, The parasites of Turkish fishes. II. Proposal of a new genus to contain the plerocercoids of *Nybelinia* Poche, 1926 (Cestoda: Trypanorhyncha), *Riv. Parassitol.*, 30, 35—38, 1969.
3156. **Shah, M. and Bilqees, F. M.**, *Nybelinia elongata*, new species from the fish *Pellonia elongata* of Kerachi coast, *Pakistan J. Zool.*, 11, 231—233, 1979.
3157. **Shaharom, F. M. and Lester, F. J. G.**, Description of and observations on *Grillotia branchi* n.sp., a larval trypanorhynch from the branchial arches of the Spanish mackerel *Scomberomorus commersioni*, *Syst. Parasitol.*, 4, 1—6, 1982.
3158. **Shaikenov, B.**, New species of helminths from rodents in Kazakhstan (in Russian), *Zhiznennye Tsikly Ekologiya Morfologiya Gel'mintov Zhivotnykh Kazakhstana*, Alma-Ata, pp. 143—150, 1978.
3159. **Shakhtakhtinskaya, Z. M.**, Helminth fauna of game birds in Azerbaidzhan (in Russian), *Tr. Azerb. Pedogogicheskogo Inst. V.I. Lenina*, 1, 29, 34, 1953.
3160. **Shaldibin, L. S.**, A new cestode from *Citellus suslica* in the Gorki district (in Russian), *Uchen. Zap. Gorkov. Gos. Pedagog. Inst.*, pp. 89—92, 1965.
3161. **Shaldybina, E. S.**, Infestation of different species of oribatids and their role in the epizootiology of monieziasis on pastures of the Gorkii oblast (in Russian), *Rab. Gel'mintol. 75-Let. Skrjabin*, pp. 740—746, 1954.
3162. **Sharma, K. N.**, Note on cestodes collected in Nepal, *Ind. Vet. J.*, 26, 53—67, 1943.
3163. **Sharpilo, V. P. and Kornyushin, V. V.**, A new genus of cestodes, *Markewitschitaenia* n.g. (Cestoda, Linstowiidae), *Parazity i Parazitoly Zhivotnykh i Cheloveka*, *Nauk Dumka*, Kiev, 1975, 217—222.
3164. **Sharpilo, V. P., Kornyushin, V. V., and Lisitsina, O. I.**, *Batrachotaenia carpathica* sp.n. (Cestoda: Ophiotaeniidae) a new species of proteocephalid cestodes from amphibians of Europe, *Helminthologia*, 16, 259—264, 1979.
3165. **Shaw, J. N., Simms, B. T., and Muth, O. H.**, Some diseases of Oregon fish and game and identification of parts of game animals, *Stn. Bull. Oregon Agric. Exp. Stat.*, p. 23, 1934.
3166. **Shepard, W.**, A new hymenolepidid cestode, *Hymenolepis parvisaccata*, from a pintail duck, *Trans. Am. Microsc. Soc.*, 62, 174—178, 1943.
3167. **Shimidsu, S., Harimi, T., and Inagami, H.**, Occurrence of *Hymenolepis diminuta* in Japanese mole, *Mogera wogura wogura* (Temm., 1842) (in Japanese), *Eisei Dobutsu*, 2, 62—63, 1951.
3168. **Shinde, G. B.**, On two new species of *Sureshia* Ali and Shinde, 1966, from *Micropus affinis* in India, *Riv. Parassitol.*, 29, 197—202, 1968.
3169. **Shinde, G. B.**, On *Circumonchobothrium ophiocephali* n.gen., n.sp. from a fresh water fish *(Ophiocephalus leucopunctatus)* in India, *Riv. Parassitol.*, 29, 111—114, 1968.
3170. **Shinde, G. B.**, A new species of cestode, *Nematotaenia mabuiae* (Nematotaeniidae Lühe, 1910) from a reptile, *Mabuia carinata* in India, *Riv. Parassitol.*, 29, 115—118, 1968.
3171. **Shinde, G. B.**, On a new genus, *Yogeshwaria malabarica* from a lapwing in India, *Riv. Parassitol.*, 29, 257—260, 1968.
3172. **Shinde, G. B.**, On a new tapeworm *Aliezia indica* gen. et sp. novo from *Ovis bharal* in India, *Zool. Anz.*, 182, 449—452, 1969.
3173. **Shinde, G. B.**, A new species of cestode, *Davainea indica* (Davaineidae) from a fowl, *Gallus domesticus* in India, *Marathwada Univ. J. Sci.*, 8, 85—87, 1969.
3174. **Shinde, G. B.**, A known and two new species of the genus *Cotugnia* Diamare, 1893 from the Columbiformes birds in Maharashtra, India, *Riv. Parassitol.*, 30, 39—44, 1969.
3175. **Shinde, G. B.**, *Mehdiangularia swifti* gen. et sp.nov. from the common house swift, *Apus affinis* in India, *Zool. Anz.*, 182, 453—456, 1969.
3176. **Shinde, G. B.**, A new species and two new varieties of the genus *Lytocestoides* Baylis, 1928 in fresh water fishes in Maharashtra, India, *Marathwada Univ. J. Sci.*, 9, 173—178, 1970.
3177. **Shinde, G. B.**, New avian cestodes of the genus *Amoebotaenia* Cohn, 1900 in India, *Marathwada University J. Sci.*, Sect. B. (Biol. Sci.), 11, 5—15, 1972.
3178. **Shinde, G. B.**, New avian cestodes of the genus *Lapwingia* Singh, 1952, *Marathwada Univ. J. Sci.*, Sect. B. (Biol. Sci.), 11, 21—29, 1972.
3179. **Shinde, G. B.**, On new species of the genus *Neyraia* Joyeux et David, 1934, *Marathwada Univ. J. Sci.*, Sect. B. (Biol. Sci.), 11, 17—20, 1972.
3180. **Shinde, G. B.**, On a new species of *Circumonchobothrium* Shinde, 1968 (Cestoda: Pseudophyllidea, Carus, 1863) from a fresh water fish in India, *Marathwada Univ. J. Sci.*, Nat. Sci. 16, 129—132, 1977.
3181. **Shinde, G. B.**, On a new species of *Pithophorus* Southwell, 1925 (Cestoda: Phyllobothriidae Braun, 1900), from a marine fish at Ratnagiri, *Riv. Parassitol.*, 39, 85—88, 1978.

3182. **Shinde, G. B. and Chincholikar, L. N.,** On a new cestode *Uncibilocularis southwelli* (Cestoda: Onchobothriidae) from a marine fish at Ratnagiri, India, *Marathwada Univ. J. Sci. (Nat. Sci.),* 15, 263—267, 1976.

3183. **Shinde, G. B. and Chincholikar, L. N.,** *Mastacembellophylaeus nandedensis* (Cestoda: Cestodaria Monticelli, 1892) n.g. et n.sp. from a freshwater fish at Nanded, M.S., India, *Riv. Parassitol.,* 38, 171—175, 1977.

3184. **Shinde, G. B. and Chincholikar, L. N.,** On a new species of *Circumonchobothrium* Shinde 1968 (Cestoda: Pseudophyllidea Carus 1863) from a fresh water fish in India, *Marathwada Univ. J. Sci. (Nat. Sci.),* 16 (9), 177—179, 1977.

3185. **Shinde, G. B. and Chincholikar, L. N.,** *Schyphophyllidium arabiansis* (Cestoda: Phyllobothriidae Braun, 1900) n.sp. from a marine fish at Ratnagiri, India, *Riv. Parassitol.,* 38, 177—180, 1977.

3186. **Shinde, G. B. and Chincholikar, L. N.,** *Mixophyllobothrium okamuri* gen.nov. sp.nov. (Cestoda: Tetraphyllidea) from *Trygon sephen* at Ratnagiri, India, *Riv. Parassitol.,* 41, 413—417, 1981 (dated 1980).

3187. **Shinde, G. B. and Deshmukh, R. A.,** On a new species of *Flapocephalus* Deshmukh R.A., 1977 (Cestoda: Lecanicephalidae Braun, 1900) from a marine fish at Veraval, West Coast of India, *Rev. Parassitol.,* 40, 295—298, 1979.

3188. **Shinde, G. B. and Deshmukh, R. A.,** Two new species of *Marsupiobothrium* Yamaguti, 1952 (Cestoda: Phyllobothriidae) from marine fishes, *Curr. Sci.,* 49, 643—644, 1980.

3189. **Shinde, G. B., Deshmukh, R. A., and Jadhav. B. V.,** On a new cestode *Echeneibothrium smitii* sp.n. from marine fish, *Riv. Parassitol.,* 42, 267—270, 1981.

3190. **Shinde, G. B. and Ghare, D. N.,** On a new cestode *Davinea* (sic) Blanchard, 1891, from a fowl, *Gallus domesticus* in India, *Marathwada Univ. J. Sci. (Nat. Sci.),* 16 (9), 191—193, 1977.

3191. **Shinde, G. B. and Ghare, D. N.,** A review of genus *Sureshia* (Cestoda) Ali and Shinde, 1966, 2nd Int. Congr. Syst. Evolutionary Biology, Univ. Br. Columbia, Vancouver, July 17 to 24, 1980, 345.

3192. **Shinde, G. B., Ghare, D. N., and Suryewanshi, V. M.,** *Amoebotaenia domesticus* n.sp. (Cestoda: Dilepididae) from *Gallus domesticus* at Udgir, Maharashtra, *Ind. J. Parasitol.,* 3 (Suppl.), 36, 1980.

3193. **Shinde, G. B. and Jadhav, B. V.,** New species of the genus *Circumonchobothrium* Shinde, 1968 (Cestoda: Pseudophyllidea) from a fresh water fish, from Maharashtra, India, *Marathwada Univ. J. Sci. (Nat. Sci.),* 15 (8), 269—272, 1976.

3194. **Shinde, G. B. and Jadhav, B. V.,** On a new cestode *Mastacembellophyllaeus paithanensis* from a freshwater fish in India, *Riv. Parassitol.,* 39, 9—12, 1978.

3195. **Shinde, G. B. and Jadhav, B. V.,** Four new species of the genus *Polypocephalus* Braun, 1878 (Cestoda: Lecanicephalidae) from the marine fishes of India, *Ind. J. Parasitol.,* 5, 1—7, 1981.

3196. **Shinde, G. B., Jadhav, B. V., and Deshmukh, R. A.,** Two new species of the genus *Pedibothrium* Linton, 1909 (Cestoda: Oncobothriidae), *Proc. Ind. Acad. Parasitol.,* 1, 21—26, 1980.

3197. **Shinde, G. B., Jadhav, B. V., and Kadam, S. S.,** *Neoliga singhi* n.sp. (Cestoda: Dilepididae) from *Micropus affinis* at *Parbhani, Curr. Sci.,* 50, 1083—1084, 1981.

3198. **Shinde, G. B., Jadhav, B. V., and Mohekar, A. D.,** Note on *Anthobothrium veravalensis,* a new cestode from the *Rhynchobatus djeddensis, Ind. J. Parasitol.,* 3 (Suppl.), 36, 1980.

3199. **Shinde, G. B., Jadhav, B. V., and Mohekar, A. D.,** *Anthobothrium veravalensis* sp. nov., (Cestoda: Tetraphyllidea) from *Rhynchobatus djiddensis* (Forsk.) of Veraval, West Coast, *Ind. J. Parasitol.,* 5, 107—108, 1981.

3200. **Shinde, G. B. and Mitra, K. B.,** *Davainea domesticusi* n.sp. from *Gallus domesticus* of Aurangabad, *Ind. J. Parasitol.,* 3 (Suppl.), 35, 1980.

3201. **Shipley, A. E.,** On *Drepanidotaenia hemignathi,* a new species of tapeworm, *Q. J. Microsc. Sci.,* 40, 613—621, 1898.

3202. **Shipley, A. E.,** Entozoa, *Fauna Hawaiiensis,* 2, 427—441, 1900.

3203. **Shipley, A. E.,** A description of Entozoa collected by Dr. Willey during his sojourn in the western Pacific, *Willey Zoological Results,* Part V, Cambridge, 1900, 531—536.

3204. **Shipley, A. E.,** On a new species of *Bothriocephalus, Proc. Cambridge Phil. Soc.,* 11, 209—211, 1901.

3205. **Shipley, A. E.,** On a collection of parasites from the Soudan, *Arch. Parasitol.,* 6, 604—612, 1902.

3206. **Shipley, A. E.,** Some parasites from Ceylon, *Ceylon J. Sci.,* 1, 1—11, 1903.

3207. **Shipley, A. E.,** On the ento-parasites collected by the "Skeat Expedition" to Lower Siam and the Malay Peninsula in the year 1899—1900, *Proc. Zool. Soc. London,* 2, 145—156, 1903.

3208. **Shipley, A. E.,** Notes on a collection of parasites belonging to the museum of University College, Dundee, *Proc. Cambridge Phil. Soc.,* 13, 95—102, 1905.

3209. **Shipley, A. E.,** Note on the occurrence of *Triaenophorus nodulosus* Rud. in the Norfolk Broads, *Parasitology,* 1, 280—281, 1908.

3210. **Shipley, A. E.,** *Anthobothrium crispum, Zool. Anz.,* 34, 641, 1909.

3211. **Shipley, A. E.,** The tapeworms (Cestoda) of the red grouse *(Lagopus scoticus), Proc. Zool. Soc. London,* 2, 351—363, 1909.

3212. **Shipley, A. E. and Hornell, J.**, The parasites of the pearl oyster, in Herdman's Report to the government of Ceylon on the pearl oyster fisheries of the Gulf of Manaar, *Tr. R. Soc. Trop. Med. Hyg.*, Part II, 77—106, 1904.
3213. **Shipley, A. E. and Hornell, J.**, Further report on parasites found in connection with the pearl oyster fisheries in Ceylon, *R. Soc. London*, Part III, 49—56, 1905.
3214. **Shipley, A. E. and Hornell, J.**, Report on cestode and nematode parasites from the marine fishes of Ceylon, in Herdman's Report to the government of Ceylon on the pearl oyster fisheries of the Gulf of Manaar, *R. Soc. London*, Part V, 43—96, 1906.
3215. **Sholl, L. B.**, Marked taeniasis in a dog, *J. Am. Vet. Med. Assoc.*, 84, 805—806, 1934.
3216. **Shoop, W. L. and Corkum, K. C.**, *Proteocephalus micruricola* sp.n. (Cestoda: Proteocephalidae) from *Micrurus diastema affinis* in Oaxaca, Mexico, *Proc. Helminthol. Soc. Wash.*, 49, 62—64, 1982.
3217. **Shorb, D. A.**, Host parasite relations of *Hymenolepis fraterna* in the rat and the mouse, *Am. J. Hyg.*, 18, 74—113, 1933.
3218. **Shubaderov, V. Ya.**, Revision of genus *Moniezia* Blanchard, 1891 (in Russian), *Tr. Vses. Inst. Gel'mintol. K. I. Skryabina*, 20, 219—225, 1973.
3219. **Shuler, R. H.**, Some cestodes of fish from Tortugas, Florida, *J. Parasitol.*, 24, 57—63, 1938.
3220. **Shumilo, R. P. and Spasskaya, L. P.**, Cestodes of Turdidae in Moldavia (in Russian), *Parazit. Zhivotn. Rast.*, 11, 53—73, 1975.
3221. **Siddiqi, A. H.**, On a new unisexual cestode, *Neodioecocestus cablei* n.g., n.sp., from the little grebe, *Podiceps ruficollis*, *Z. Parasitenkd.*, 20, 381—384, 1960.
3222. **Siebold, C. T. von**, Zur Entwicklungsgeschichte der Helminthen, in *Burdach, K. F., Die Physiologie als Erfahrungswissenschaft. II. Aufl.*, Vol. II, Leipzig, 1837, 183—213.
3223. **Siebold, C. T. von**, Ueber den Generationswechsel der Cestoden nebst einer Revision der Gattung *Tetrarhynchus*, *Z. Wiss. Zool.*, 2, 198—253, 1850.
3224. **Siebold, C. T. von**, Ueber die Verwandlung des *Cysticercus pisiformis* in *Taenia serrata*, *Z. Wiss. Zool.*, 5, 400—409, 1853.
3225. **Siebold, C. T. von**, Ueber die Verwandlung der *Echinococcus*-Brut in Taenien, *Z. Wiss. Zool.*, 5, 409—424, 1853.
3226. **Siebold, C. T. von**, Ueber die Band-und Blasenwürmer nebst einer Einleitung ueber die Entstehung der Eingeweidewürmer. Leipzig. (French transl.), in *Ann. Sci. Nat.*, 4, 48—90, 1854.
3227. **Silverman, P. H.**, Studies on the biology of some tapeworms of the genus *Taenia*. II. The morphology and development of taeniid hexacanth embryo and its enclosing membranes, with some notes on the state of development and propagation of gravid segments, *Ann. Trop. Med. Parasitol.*, 48, 356—366, 1954.
3228. **Simer, P. H.**, A preliminary study of the cestodes of the spoonbill, *Polyodon spathula* W(al.), *Trans. Ill. Acad. Sci.*, 22, 139—145, 1930.
3229. **Simić, C. P., Nevenić, V., and Petrovitch, Z.**, *Citellus citellus* animaux de choix pour démontrer l'identité biologique entre *Hymenolepis nana* de l' homme et *H. nana* var. *fraterna* du rat, *Arch. Inst. Pasteur Algérie*, 31, 84—90, 1953.
3230. **Simić, C. P. and Petrovitch, Z.**, La réinfestation de *Citellus citellus* par *Hymenolepis nana* après le sommeil hibernal, est-elle possible?, *Arch. Inst. Pasteur Algérie*, 31, 397—399, 1953.
3231. **Simms, B. T. and Shaw, J. N.**, Studies on the fish-borne tapeworm, *Dibothrium cordiceps*, *JAMA*, 32, 199—205, 1931.
3232. **Simon, F.**, A new cestode, *Raillietina centrocerci*, from the sage grouse, *Centrocercus urophasianus*, *Trans. Am. Microsc. Soc.*, 56, 340—343, 1937.
3233. **Simon, F.**, Parasites of the sage grouse, *Centrocercus urophasianus*, *Univ. Wyoming Publ.*, 7, 77—100, 1940.
3234. **Singal, D. P.**, On a new cestode, *Oochoristica bailea* n.sp. from the common Indian wall lizard, *Hemidactylus flaviviridis* (Ruppell), from Delhi State, *Ind. J. Helminthol.*, 13, 74—78, 1961.
3235. **Singh, K. P.**, *Echinorhynchotaenia lucknowensis* n.sp. (Hymenolepididae: Cestoda) from darter, *Anhinga melanogaster* Pennant, *Sci. (Bangalore)*, 25, 59, 1956.
3236. **Singh, K. P.**, *Choanotaenia aurantia* n.sp. (Dilepididae: Cestoda) from a tern, *Sterna aurantia* Gray, from India, *Ind. J. Helminthol.*, 8, 107—111, 1958.
3237. **Singh, K. P.**, Some avian cestodes from India. I. Species belonging to families Davaineidae and Biuterinidae, *Ind. J. Helminthol.*, Year 1959, 11, 1—24, 1960.
3238. **Singh, K. P.**, Some avian cestodes from India. II. Species belonging to the family Dilepididae, *Ind. J. Helminthol.*, Year 1959, 11, 25—42, 1960.
3239. **Singh, K. P.**, Some avian cestodes from India. III. Species belonging to family Hymenolepididae, *Ind. J. Helminthol.*, Year 1959, 11, 43—62, 1960.
3240. **Singh, K. P.**, Some avian cestodes from India. IV. Species belonging to families Amabiliidae, Diploposthidae and Progynotaeniidae, *Ind. J. Helminthol.*, Year 1959, pp. 63—74, 1960.

3241. **Singh, K. P. and Singh, K. S.,** Two new species of avian cestodes of the genus *Panuwa* Burt, 1940 from India, *Libro Homenaje al Dr. Eduardo Caballero y Caballero, Jubileo, 1930—1960*, Universidad Nacional Autónoma de Mexico, 1960, 407—413.
3242. **Singh, K. S.,** On a new cestode, *Gangesia lucknowia* (Proteocephalidae) from a freshwater fish, *Eutropiichthys vacha* Day, with a revised key to the species of the genus, *Ind. J. Helminthol.*, 1, 41—46, 1948.
3243. **Singh, K. S.,** Cestode parasites of birds, *Ind. J. Helminthol.*, 4, 1—72, 1952.
3244. **Singh, K. S.,** *Hymenolepis vogeae* n.sp. from an Indian field mouse, *Mus buduga* Thomas, 1881, *Trans. Amer. Microsc. Soc.*, 75, 252—255, 1956.
3245. **Singh, K. S.,** *Oochoristica tandani* n.sp. (Cestoda), from a snake, *Lycodon aulis*, from India, *J. Parasitol.*, 43, 377, 1957.
3246. **Singh, K. S.,** *Hymenolepis bahli* n.sp. from grey musk shrew, *Crocidura caerulea* (Kerr, 1792) Paters, 1870 from India, *J. Parasitol.*, 44, 446—448, 1958.
3247. **Singh, K. S.,** Parasitological survey of Kumaun region. IX. *Indotaenia indica* n.g., n.sp. (Cestoda: Anoplocephalidae) from the Himalayan flying squirrel, *Ind. J. Helminthol.*, 14, 86—91, 1962.
3248. **Singh, K. S.,** Parasitological survey of Kumaun region. XIII. *Ophryocotyloides picusi* n.sp. (Davaineidae: Cestoda), from a woodpecker and a key to the species of the genus, *Ind. J. Helminthol.*, 14, 116—121, 1962.
3249. **Singh, K. S.,** Parasitological survey of Kumaun region. XIV. *Ophryocotyloides makundi* n.sp. (Davaineidae: Cestoda) from the little scaly-bellied green woodpecker, *Ind. J. Helminthol.*, 14, 122—126, 1962.
3250. **Singh, K. S.,** Parasitological survey of Kumaun region. XV. *Ivritaenia mukteswarensis* n.g., n.sp. (Cestoda: Dipylidinae, Dilepididae) from a woodpecker, *Ind. J. Helminthol.*, 14, 127—132, 1962.
3251. **Singh, K. S.,** Parasitological survey of Kumaun region. XVII. *Raillietina (Raillietina) thapari* n.sp. (Davaineidae: Cestoda) from a woodpecker, *Ind. J. Helminthol.*, 15, 1—5, 1963.
3252. **Singh, K. S.,** On six new avian cestodes from India, *Parasitology*, 54, 177—194, 1964.
3253. **Singh, K. S. and Tandan, B. K.,** *Dilepis kumaunensis* sp.n. (Cestoda: Dilepididae) from two Himalayan birds, *Zool. Anz.*, 169, 485—488, 1962.
3254. **Singh, K. S. and Tandan, B. K.,** *Mayhewia yamaguti* sp.n. (Cestoda: Hymenolepididae) from *Dicrurus leucophaeus longicaudatus*, the grey drongo, *J. Zool. Soc. India*, Year 1965, 17, 48—51, 1967.
3255. **Singh, S. S.,** On *Lytocestus fossilis* n.sp. (Cestoda: Lytocestidae) from *Heterapneustes fossilis* from Nepal, in *Dr. B. S. Chauhan Commemoration Volume*, Tiwari, K. K. and Srivastava, C. B., Eds., Zoological Society of India, Orissa, 1975, 79—82.
3256. **Sinha, P. K.,** *Raillietina dattai* n.sp. from poultry *(Gallus gallus domesticus)* in India, *J. Parasitol.*, 46, 485—486, 1960.
3257. **Sinitsin, D. F.,** Entoparasitic worms of birds in the vicinity of Warsaw (in Russian with French summary), *Varshavsk. Univ. Izvest.*, pp. 1—22, 1896.
3258. **Sinitsin, D. F.,** A glimpse into the life-history of the tapeworm of the sheep, *Moniezia expansa*, *J. Parasitol.*, 17, 223—227, 1931.
3259. **Skarbilobich, T. S.,** Helminth fauna of bats in Russia (in Russian), *Coll. Pap. Helminthol. Ded. Skrjabin*, pp. 235—244, 1946.
3260. **Skinker, M. S.,** A new species of *Oochoristica* from a skunk, *J. Wash. Acad. Sci.*, 25, 59—65, 1935.
3261. **Skinker, M. S.,** A redescription of *Taenia tenuicollis* Rudolphi 1819, and its larva, *Cysticercus talpae* Rudolphi, 1819, *Parasitology*, 27, 175—185, 1935.
3262. **Skinker, M. S.,** Miscellaneous notes on cestodes, *Proc. Helminthol. Soc. Wash.*, 2, 68, 1935.
3263. **Skinker, M. S.,** Two new species of tapeworms from carnivores and a redescription of *Taenia laticollis* Rudolphi, 1819, *Proc. U.S. Natl. Mus.*, 83, 211—220, 1935.
3264. **Skrjabin, K. I.,** Fischparasiten aus Turkestan. I. Hirudinea et Cestodaria, *Arch. Naturg.*, 79, Abt. A, 1—10, 1913.
3265. **Skrjabin, K. I.,** Vergleichende Charakteristik der Gattungen *Chapmania* Mont. und *Schistometra* Cholodk., *Centralbl. Bakteriol.*, 73, 397—405, 1914.
3266. **Skrjabin, K. I.,** Zwei neue Cestoden der Hausvögel, *Z. Infektionskr. Parasitol. Krankh. Hyg. Haustiere*, 15, 249—260, 1914.
3267. **Skrjabin, K. I.,** Zwei Vogelcestoden mit gleicher Scolexbewaffnung und verschiedener Organisation, *Hymenolepis collaris* Batsch und *Hymenolepis compressa* Linton, *Centralbl. Bakteriol.*, 74, 275—279, 1914.
3268. **Skrjabin, K. I.,** Beitrag zur Kenntnis einiger Vogelcestoden, *Centralbl. Bakteriol. I. Abt. Orig.*, 75, 59—83, 1914.
3269. **Skrjabin, K. I.,** Vogelcestoden aus Russisch Turkestan, *Zool. Jahrb. Syst.*, 37, 411—492, 1914.
3270. **Skrjabin, K. I.,** *Hymenolepis fasciata* Rud. (in Russian), *Vest. Obsh. Vet. St. Petersbourg*, 27, 225—229, 1915.
3271. **Skrjabin, K. I.,** Studien zur Erforschung der parasitischen Wurmer der Raubtiere. I. Ein neuer Hundebandwurm: *Dipylidium rossicum* n.sp., *Arkh. Nauchn. Prakt. Vet.*, 1, 20—27, 1923.

3272. **Skrjabin, K. I.,** Faune des vers parasites dans les steppes du Turkestan. I. Parasites des rongeurs (in Russian), *Tr. Gosudartv. Inst. Iksper. Vet.,* 2, 78—91, 1924.
3273. **Skrjabin, K. I.,** *Progynopylidium noelleri* n.g., n.sp., ein neuer Bandwurm der Katze, Berlin, *Tierarzt. Wochenschr.,* 40, 420—422, 1924.
3274. **Skrjabin, K. I.,** Au sujet d'un nouveau remaniement de la systematique de la famille des Anoplocephalidae Cholodk., 1902, *Bull. Soc. Zool. Fr.,* 58, 84—86, 1933.
3275. **Skrjabin, K. I. and Mathevossian, E. M.,** Stages in the postembryonic development of cestodes of the family Hymenolepididae and an attempt to establish morphological types of their larvicysts (in Russian), *Dokl. Akad. Nauk SSSR,* n.s., an. 10, 35, 83—85, 1942.
3276. **Skrjabin, K. I. and Mathevossian, E. M.,** Typical morphological modifications of the chitinous organs of the scolex in cestodes from the family Hymenolepididae (in Russian), *Dokl. Akad. Nauk SSSR,* n.s., an. 10, 35, 86—88, 1942.
3277. **Skrjabin, K. I. and Mathevossian, E. M.,** Types of topographical correlations of sexual glands in cestodes of the family Hymenolepididae and their taxonomic significance (in Russian), *Dokl. Akad. Nauk SSSR,* n.s., an. 10, 36, 32—35, 1942.
3278. **Skrjabin, K. I. and Matevosyan, E. M.,** Corrections to errors and controversies in the taxonomy of the cestodes of the family Hymenolepididae, *Dokl. Akad. Nauk SSSR,* n.s., an. 10, 36, 188—191, 1942.
3279. **Skrjabin, K. I. and Popoff, N. P.,** Bericht ueber die Tätigkeit der helminthologischen Expedition in Armenien, 1923 (in Russian), *Russk. Zh. Trop. Med.,* 1, 58—63, 1924.
3280. **Skrjabin, K. I. and Schulz, R. E.,** Affinités entre *Dithyridium* des souris et le *Mesocestoides lineatus* (Goeze, 1782) des carnivores, *Ann. Parasitol.,* 4, 68—73, 1926.
3281. **Skrjabin, K. I. and Schulz, R. E.,** Ueber den Umfang der medizinischen Helminthologie (in Russian), *Russk. Zh. Trop. Med.,* 6, 145—152, 1928.
3282. **Skryabin, A. S.,** *Tetragonoporus calyptocephalus* n.g., n.sp. from the sperm whale (in Russian), *Helminthologia,* 3, 311—315, 1961.
3283. **Skryabin, A. S.,** *Polygonoporus giganticus* n.g., n.sp., a parasite of sperm whales (in Russian), *Parazitologiya,* 1, 131—136, 1967.
3284. **Skryabin, A. S. and Murav'eva, S. I.,** The new cestode *Tetrabothrius egregius* n.sp. a parasite of *Balaenoptera physalus* (in Russian), *Nauchn. Dokl. Vyssh. Shk. Biol. Nauki,* 7, 17—21, 1971.
3285. **Skvortsov, A. A.,** Egg structure of *Taeniarhynchus saginatus* and its control (in Russian with English summary), *Zool. Zh.,* 21, 10—18, 1942.
3286. **Skvortsov, A. A. and Talyzin, F. F.,** Cycle of development of the minor tapeworm *(Diphyllobothrium minus* Chol.) (in Russian), *Dokl. Akad. Nauk SSSR,* n.s., 27, 618—620, 1940.
3287. **Sleggs, G. F.,** Notes on cestodes and trematodes of marine fishes of Southern California, *Bull. Scripps Inst. Oceanogr.* Tech. Ser., 1, 63—72, 1927.
3288. **Sluiter, C. P.,** *Taenia plastica* n.sp., eine neue kurzgliedrige *Taenia* aus *Galeopithecus volans, Centralbl. Bakteriol. I. Abt.* 19, 941—946, 1896.
3289. **Smith, A. J.,** Synopsis of studies in metazoan parasitology in the McManus Laboratory of Pathology, University of Pennsylvania, *Univ. Penn. Med. Bull.,* 20, 262—282, 1908.
3290. **Smith, A. J., Fox, H., and White, C. Y.,** Contributions to systematic helminthology, *Univ. Penn. Med. Bull.,* 20, 283—294, 1908.
3291. **Smith, C. F.,** Two anoplocephalid cestodes, *Cittotaenia praecoquis* Stiles and *Cittotaenia megasacca* n.sp., from the western pocket gopher, *Thomomys talpoides* of Wyoming, *J. Parasitol.,* 37, 312—316, 1951.
3292. **Smith, C. F.,** Four new species of cestodes of rodents from the high plains, central and southern rockies and notes on *Catenotaenia dendritica, J. Parasitol.,* 40, 245—254, 1954.
3293. **Smithers, S. R.,** On a new anoplocephalid cestode, *Pulluterina nestoris* gen. et sp. nov. from the kea *(Nestor notabilis), J. Helminthol.,* 28, 1—8, 1954.
3294. **Smyth, J. D.,** Studies on tapeworm physiology. III. Aseptic cultivation of larval Diphyllobothriidae in vitro, *J. Exp. Biol.,* 24, 374—386, 1947.
3295. **Smyth, J. D.,** Development of cestodes in vitro: production of fertile eggs: cultivation of plerocercoid fragments, *Nature (London),* 161, 138, 1948.
3296. **Smyth, J. D.,** Studies on tapeworm physiology. IV. Further observations on the development of *Ligula intestinalis* in vitro, *J. Exp. Biol.,* 26, 1—14, 1949.
3297. **Smyth, J. D.,** Parthenogenetic development of eggs of a cestode cultured aseptically in vitro, *Nature* (Correspondence), 165, 492—493, 1950.
3298. **Smyth, J. D.,** Studies on tapeworm physiology. V. Further observation on the maturation of *Schistocephalus solidus* (Diphyllobothriidae) under sterile conditions in vitro, *J. Parasitol.,* 36, 371—383, 1950.
3299. **Smyth, J. D.,** The biology of cestode life cycles, *Commonw. Agric. Bur. Tech. Commun.,* 34, 1—38, 1963.
3300. **Smyth, J. D. and Heath, D. D.,** Pathogenesis of larval cestodes in mammals, *Helminthol. Abstr.* (Ser. A.), 39, 1—22, 1970.

3301. **Smyth, J. D. and Hopkins, C. A.,** Ester wax as a medium for embedding tissue for the histological demonstration of glycogen, *Q. J. Microsc. Sci.*, 89, 431—435, 1948.
3302. **Sneed, K. E.,** The genus *Corallobothrium* from catfishes in Lake Texoma, Oklahoma, with a description of two new species, *J. Parasitol.*, 36, 43, 1950.
3303. **Soldatova, A. P.,** A contribution to the study of the developmental cycle of the cestode *Mesocestoides lineatus* (Goeze, 1782) parasitic in carnivorous mammals, *Dokl. Akad. Nauk SSSR*, 45, 310—312, 1944.
3304. **Soldatova, A. P.,** A contribution to the study of the biology of oribatid mites, intermediate hosts of cestodes of the family Anoplocephalidae, *Dokl. Akad. Nauk SSSR*, 46, 343—344, 1945.
3305. **Soldatova, A. P.,** On the biology of the oribatid mites as intermediate hosts of anoplocephalid cestodes, parasitic in sheep, cattle and horses (in Russian), *Sb. Rab. Gel'mintol. 40. Nauchn. Deiat, Skrjabin*, pp. 209—213, 1948.
3306. **Soler, P. M.,** de los Angeles, El genero *Nematotaenia* y descripción de una nueva especie, *Rev. Ibér. Parasitol. Tomo Extraordinario*, pp. 67—72, 1945.
3307. **Soliman, K. N.,** Observations on some helminth parasites from ducks in southern England, *J. Helminthol.*, 29, 17—26, 1955.
3308. **Solomon, S. G.,** On the experimental development of *Bothridium* ( = *Solenophorus*) *pythonis* De Blainville, 1824, in *Cyclops viridis* Jurine, 1820, *J. Helminthol.*, 10, 67—74, 1932.
3309. **Solomon, S. G.,** Some points in the early development of *Cysticercus pisiformis* (Bloch, 1780), *J. Helminthol.*, 12, 197—204, 1934.
3310. **Solonitzin, J. A.,** Mehrfacher Tetrathyridios der serösen Höhlen des Hundes, *Z. Infektionskr. Parasitol. Krankh. Hyg. Haustiere*, 45, 144—156, 1933.
3311. **Solowiow, P. F.,** Helminthologische Beobachtungen. Cestodes avium, *Centralbl. Bakteriol., I. Orig.*, 60, 93—132, 1911.
3312. **Soltys, A.,** (The helminths of muridae of the National Parc (sic) of Biaowieza (in Polish with English summary), *Ann. Univ. Mariae Curie-Sklodowska* (Sect. C), 4, 233—259, 1949.
3313. **Sommer, F. B. G.,** Ueber den Bau und die Entwicklung der Geschlechtsorgane von *Taenia mediocanellata* und *Taenia solium, Z. Wiss. Zool.*, 24, 499—563, 1874.
3314. **Sommer, F. B. G. and Landois, L.,** Ueber den Bau der geschlechtsreifen Glieder von *Bothriocephalus latus, Z. Wiss. Zool.*, 22, 40—99, 1872.
3315. **Sondhi, G.,** Tapeworm parasites of dogs in the Punjab, *Parasitology*, 15, 59—66, 1923.
3316. **Sonsino, P.,** Ricerche sugli ematozoi del cane el sul ciclo vitale della *Taenia cucumerina, Atti Soc. Tosc. Sci. Nat. Mem.*, 10, 20—64, 1889.
3317. **Sonsino, P.,** Studie e notzie elmintologiche, *Atti Soc. Tosc. Sci. Nat. Mem.* (Proc. Verb.), 6, 273—285, 1889.
3318. **Sonsino, P.,** Di alcuni entozoi raccolti in Egitto, finora non descritti, *Monitore Zool. Ital. Firenze*, 6, 121—125, 1895.
3319. **Southwell, T.,** On the determination of the adult of the pearl-inducing worm. Report on certain scientific work done on the Ceylon pearl banks during the year 1909, *Ceylon Mar. Biol. Rep.*, Part IV, 169—172, 1910.
3320. **Southwell, T.,** A note on endogenous reproduction discovered in the larvae of *Tetrarhynchus unionifactor* inhabiting the tissues of the pearl oyster, *Ceylon Mar. Biol. Rep.*, Part IV, 173—174, 1910.
3321. **Southwell, T.,** Description of nine new species of cestode parasites including two new genera from marine fishes of Ceylon, *Ceylon Mar. Biol. Rep.*, Part V, 216—225, 1911.
3322. **Southwell, T.,** Some remarks on the occurrence of cestodes in Ceylon, *Spolia Zeylanica*, pp. 194—196, 1911.
3323. **Southwell, T.,** A description of ten new species of cestode parasites from marine fishes of Ceylon, with notes on other cestodes from the same region, *Ceylon Mar. Biol. Rep.*, 1, 259—278, 1912.
3324. **Southwell, T.,** Parasites from fish. Notes from the Bengal Fisheries Laboratory, *Rec. Ind. Mus.*, 9, 79—103, 1913.
3325. **Southwell, T.,** On some Indian Cestoda. I, *Rec. Ind. Mus.*, 9, 279—300, 1913.
3326. **Southwell, T.,** A brief review of the scientific work done on the Ceylon pearl banks since the year 1902, *J. Econ. Biol.*, 8, 22—34, 1913.
3327. **Southwell, T.,** Notes from the Bengal Fisheries Laboratory, Indian Museum. No. 2. On some Indian parasites of fish, with a note on carcinoma in trout, *Rec. Ind. Mus.*, 11, 311—330, 1915.
3328. **Southwell, T.,** On some Indian Cestoda. II, *Rec. Ind. Mus.*, 12, 5—20, 1916.
3329. **Southwell, T.,** A note on the occurrence of certain cestodes in new hosts. A new species of cestode (*Anoplocephala vulgaris*) from an African rhinoceros, *Ann. Trop. Med. Parasitol.*, 14, 295—297, 1921.
3330. **Southwell, T.,** Cestodes from Indian poultry, *Ann. Trop. Med. Parasitol.*, 15, 161—166, 1921.
3331. **Southwell, T.,** Cestodes from African rats, *Ann. Trop. Med. Parasitol.*, 15, 167—168, 1921.
3332. **Southwell, T.,** A new species of Cestoda from a cormorant, *Ann. Trop. Med. Parasitol.*, 15, 169—171, 1921.

3333. **Southwell, T.,** Fauna of the Chilka Lake. On larval cestode from the umbrella of a jelly fish, *Mem. Ind. Mus.,* 5, 561—562, 1921.
3334. **Southwell, T.,** Cestodes in the collection of the Indian Museum, *Ann. Trop. Med. Parasitol.,* 16, 127—152, 1922.
3335. **Southwell, T.,** Cestodes from Indian birds, with a note on *Ligula intestinalis, Ann. Trop. Med. Parasitol.,* 16, 355—382, 1922.
3336. **Southwell, T.,** The pearl-inducing worm in the Ceylon pearl oyster, *Ann. Trop. Med. Parasitol.,* 18, 37—53, 1924.
3337. **Southwell, T.,** Notes on certain cestodes in the School of Tropical Medicine, Liverpool, *Ann. Trop. Med. Parasitol.,* 18, 177—182, 1924.
3338. **Southwell, T.,** Notes on some tetrarhynchid parasites from Ceylon marine fishes, *Ann. Trop. Med. Parasitol.,* 18, 459—491, 1924.
3339. **Southwell, T.,** The genus *Tetracampos* Wedl, 1861, *Ann. Trop. Med. Parasitol.,* 19, 71—91, 1925.
3340. **Southwell, T.,** A monograph on the Tetraphyllidea, *Liverpool School Trop. Med. Mem.,* (n.s.), No. 2, Liverpool University Press, 1925, 1—368.
3341. **Southwell, T.,** On a new cestode from Nigeria, *Ann. Trop. Med. Parasitol.,* 19, 243—246, 1925.
3342. **Southwell, T.,** On the genus *Tetracampos* Wedl, 1861, *Ann. Trop. Med. Parasitol.,* 19, 315—317, 1925.
3343. **Southwell, T.,** Cestodes in the collection of the Liverpool School of Tropical Medicine, *Ann. Trop. Med. Parasitol.,* 20, 221—228, 1926.
3344. **Southwell, T.,** On a collection of cestodes from marine fishes of Ceylon, *Ann. Trop. Med. Parasitol.,* 21, 351—373, 1927.
3345. **Southwell, T.,** Cestodaria from India and Ceylon, *Ann. Trop. Med. Parasitol.,* 22, 319—326, 1928.
3346. **Southwell, T.,** Cestodes of the order Pseudophyllidea recorded from India and Ceylon, *Ann. Trop. Med. Parasitol.,* 22, 419—448, 1928.
3347. **Southwell, T.,** On the classification of the Cestoda, *Ceylon J. Sci.,* 15 (Part I), 49—72, 1929.
3348. **Southwell, T.,** Notes on the anatomy of *Stilesia hepatica* and on the genera of the sub-family Thysanosominae (including Avitellininae), *Ann. Trop. Med. Parasitol.,* 23, 47—66, 1929.
3349. **Southwell, T.,** A monograph on cestodes of the order Trypanorhyncha from Ceylon and India. *I, Ceylon J. Sci.,* 15 (Part III), 169—312, 1929.
3350. **Southwell, T. and Adler, A.,** A note on *Ophiotaenia punica* (Cholodkowsky, 1908) La Rue, 1911, *Ann. Trop. Med. Parasitol.,* 17, 333—335, 1923.
3351. **Southwell, T. and Hilmy, I. S.,** On a new species of *Phyllobothrium (P. microsomum)* from an Indian shark, *Ann. Trop. Med. Parasitol.,* 23, 381—383, 1929.
3352. **Southwell, T. and Hilmy, I. S.,** *Jardugia paradoxa,* a new genus and species of cestode with some notes on the families Acoleidae and Diploposthidae, *Ann. Trop. Med. Parasitol.,* 23, 397—406, 1929.
3353. **Southwell, T. and Kirshner, A.,** Description of a polycephalic cestode larva from *Mastomys erythroleucus* and its probable identity, *Ann. Trop. Med. Parasitol.,* 31, 37—42, 1937.
3354. **Southwell, T. and Kirshner, A.,** Parasitic infections in a swan and in a brown trout, *Ann. Trop. Med. Parasitol.,* 31, 427—433, 1937.
3355. **Southwell, T. and Lake, F.,** On a collection of Cestoda from Belgian Congo, *Ann. Trop. Med. Parasitol.,* 33, 63—90, 107—123, 1939.
3356. **Southwell, T. and Maplestone, P. A.,** A note on the synonymy of the genus *Zschokkeella* Ransom, 1909, and of the species *Z. guineensis* (Graham, 1908), *Ann. Trop. Med. Parasitol.,* 15, 455—456, 1921.
3357. **Southwell, T. and Maplestone, P. A.,** Notes on Australian cestodes, *Ann. Trop. Med. Parasitol.,* 16, 189—198, 1922.
3358. **Southwell, T. and Prashad, B.,** Notes from the Bengal Fisheries Laboratory, No. 4. Cestode parasites of Hilsa, *Rec. Ind. Mus.,* 15, 77—88, 1918.
3359. **Southwell, T. and Prashad, B.,** Notes from the Bengal Fisheries Laboratory, No. 5. Parasites of Indian fishes, with a note on carcinoma in the climbing perch, *Rec. Ind. Mus.,* 15, 341—355, 1918.
3360. **Southwell, T. and Prashad, B.,** Methods of asexual and parthenogenetic reproduction in cestodes, *J. Parasitol.,* 4, 122—129, 1918.
3361. **Southwell, T. and Prashad, B.,** A revision of the Indian species of the genus *Phyllobothrium, Rec. Ind. Mus.,* 19, 1—8, 1920.
3362. **Southwell, T. and Prashad, B.,** A further note on *Ilisha parthenogenetica,* a cestode parasite of the Indian shad, *Rec. Ind. Mus.,* 25, 197—198, 1923.
3363. **Southwell, T. and Walker, A. J.,** Notes on a larval cestode from a fur seal, *Ann. Trop. Med. Parasitol.,* 39, 91—100, 1936.
3364. **Spasskaja, L. P.,** Two new species of cestodes from birds in Tuva (in Russian), *Papers on Helminthology Presented to Academician K. I. Skrjabin on His 80th Birthday,* Izdatelstvo Akademii Nauk SSSR, Moscow, 1958, 349—353.
3365. **Spasskaja, L. P.,** Cestodes of birds of Tuva. IV. Hymenolepididae of aquatic birds (in Russian), *Acta Vet. (Budapest),* 11, 311—337, 1961.

3366. **Spasskaja, L. P.,** The species of the family Idiogenidae (Cestoda) from the Tuva region (in Russian), *Acta Vet. Hung.,* 11, 423—440, 1961.
3367. **Spasskaja, L. P.,** *Biglandatrium* n.g. (Cestoda: Hymenolepididae) (in Russian), *Helminthologia,* 3, 340—345, 1961.
3368. **Spasskaja, L. P.,** Genus *Platyscolex,* gen.nov. (Cestoda: Dilepididae) (in Russian), *Acta Vet. Acad. Sci. Hung.,* 12, 207—211, 1962.
3369. **Spasskaja, L. P.,** *Microsomacanthus skrjabini* n.sp. — a new hymenolepidid from *Histrionicus histrionicus* in northern Kamchatka (in Russian), in *Helminths of Man, Animals and Plants and Their Control. Papers on Helminthology Presented to Academician K. I. Skrjabin on his 85th Birthday,* Izdatel'stvo Academii Nauk SSSR, Moscow, 1963, 163—166.
3370. **Spasskaja, L. P.,** Cestodes of birds of the USSR. Hymenolepididae (in Russian), Izdatel'stvo "Nauka", Moscow, 1966, 698 pp.
3371. **Spasskaja, L. P.,** Revision of the genus *Unciunia* (Cestoda: Dilepididae) (in Russian), *Kishinev RIO Akad. Nauk Moldavskoi SSR,* pp. 36—38, 1970.
3372. **Spasskaja, L. P.,** *Alkataenia* n.q. (Cestoda: Dilepididae) (in Russian), *Parazit. Zhivotn. Rast.,* 6, 12—14, 1971.
3373. **Spasskaja, L. P.,** *Chitonorecta metaskrjabini* n. sp., a new species of dilepid cestode of Limicolae, in Gagarin, V. G. (Ed.), *Probl. Obshchei Prikladnoi Gel'mintol. (Moscow),* 1973, 129—133.
3374. **Spasskaja, L. P.,** *Acanthocirrus retirostris* (Cestoda: Dilepididae) from Passeriformes (in Russian), *Parazit. Zhivotn. Rast.,* 10, 66—71, 1974.
3375. **Spasskaja, L.P. and Kolotilova, E. M.,** *Rissotaenia* n.g. (Cestoda: Dilepididae) parasitic in gulls (in Russian), *Parazit. Zhivotn. Rast.,* 8, 51—58, 1972.
3376. **Spasskaja, L. P. and Makarenko, V. K.,** New genus of cestode of birds — *Sobolevitaenia* n.g. (Cestoda: Dilepididae) (in Russian), In *Parasitic Worms of Domestic and Wild Animals. Papers on Helminthology Presented to Prof. A. A. Sobolev on the 40th Anniversary of his Scientific and Teaching Activity,* Dalnevostochnii Gosudarstvennii Universitet, Vladivostok, 1965, 298—302.
3377. **Spasskaja, L. P. and Shumilo, R. P.,** Cestode fauna of Charadriiformes and Ciconiiformes in Moldavia. I (in Russian), *Parazit. Zhivotn. Rast.,* 7, 3—27, 1971.
3378. **Spasskaja, L. P. and Shumilo, R. P.,** Three new cestode species from Moldavian Charadriiformes, *Izvest. Akad. Nauk Moldavskoi SSR,* 1, 56—64, 1971.
3379. **Spasskaja, L. P. and Shumilo, R. P.,** Obligatory parasites (Cestoda: Dilepididae) of the oriole and starling in Moldavia (in Russian), *Izvest. Akad. Nauk Moldavskoi SSR Biol. Khim. Nauki,* 6, 66—72, 1973.
3380. **Spasskaja, L. P. and Spasskii, A. A.,** Cestodes of birds of Tuva. II. *Microsomacanthus* (Hymenolepididae), *Acta Vet. (Budapest),* 11, 13—53, 1961.
3381. **Spasskaja, L. P. and Spasskii, A. A.,** Species composition of the genus *Polycercus* (Cestoda: Dilepididae). *Parasites Anim. Plants,* (in Russian), *Kishnev RIO Akad. Nauk Moldavskoi SSR,* 5, 38—44, 1970.
3382. **Spasskaja, L. P. and Spasskii, A. A.,** *Cestodes of Birds in Tuva* (in Russian), Izdatel'stvo Shtiintsa, Kishinev, 1971, 252 pp.
3383. **Spasskaja, L. P. and Spasskii, A. A.,** *Dictymetra belopolskajae* n.sp. (Cestoda: Dilepididae) a new cestode from birds (in Russian), *Izvest. Akad. Nauk Mold. SSR Biol. Khim. Nauki,* pp. 63—68, 1973.
3384. **Spasskaja, L. P. and Spasskii, A. A.,** Cestodes of charadrii birds in the Kamchatka region (in Russian), *Parazit. Zhvotnykh Rastenii (Kishinev),* 9, 49—78, 1973.
3385. **Spasskaja, L. P. and Spasskii, A. A.,** *Cestodes of Birds of SSSR. Dilepidids of Lake Birds,* Akademiya Nauk Moldavskoi SSR, Moscow, 1977, 300 pp.
3386. **Spasskaja, L. P., Spasskii, A. A., and Borgarenko, L. F.,** *Diporotaenia colymbi* G.N., Sp.N. — new species new genus of amabiliid cestode from grebes, *Izvest. Akad. Nauk Mold. SSR Biol. Khim. Nauki,* 6, 49—53, 1971.
3387. **Spasskii, A. A.,** A contribution to the knowledge of helminths of birds in Russia (in Russian), *Collected Papers on Helminthology Dedicated to Skrjabin,* 1946, 252—261.
3388. **Spasskii, A. A.,** On the position of the genus *Echinorhynchotaenia* Fuhrmann, 1909, in the system of cestodes (in Russian), *Dokl. Akad. Nauk SSSR,* 58, 513—515, 1947.
3389. **Spasskii, A. A.,** A new cestode family, Skrjabinochoridae n.fam., characterised by complete absence of uteri (in Russian), *Dokl. Akad. Nauk SSSR,* 59, 409—412, 1948.
3390. **Spasskii, A. A.,** *Mathevolepis petroschenkoi* n.g. n.sp., a new cestode genus with uterine canals for the production of the eggs (in Russian), *Dokl. Akad. Nauk SSSR,* 59, 1513—1515, 1948.
3391. **Spasskii, A. A.,** Change of function of the attachment apparatus in the cestode, *Insinuarotaenia schikhovalovi* n.g. n.sp. (in Russian), *Dokl. Akad. Nauk SSSR,* 59, 825—827, 1948.
3392. **Spasskii, A. A.,** *Metadilepis* n.g., a new cestode parasitizing birds (in Russian), *Biul. Moskov. Obsh. Ispyt. Prir. Otdel Biol. an.* 120, 54, 50—54, 1949.
3393. **Spasskii, A. A.,** On the systematic position of the cestode, *Cittotaenia sandgroundi* Davis 1944 (in Russian), *Tr. Gel'mintol. Lab. Akad. Nauk SSSR,* 2, 60—61, 1949.

3394. **Spasskii, A. A.,** A new cestode from *Erinaees* (sic) *(Hemiechinus) auritus, Mathevotaenia skrjabini* n.sp. (in Russian), *Tr. Gel'mintol. Lab. Akad. Nauk SSSR,* 2, 55—59, 1949.

3395. **Spasskii, A. A.,** A new cestode *Vigisolepis barboscolex* n.sp. and its position within the tribe Hymenolepaea Skrjabin et Mathevossian 1941 (in Russian), *Tr. Gel'mintol. Lab. Akad. Nauk SSSR,* 2, 50—54, 1949.

3396. **Spasskii, A. A.,** A new family of tapeworms — Catenotaeniidae fam. nov. and a review of the systematics of the anoplocephalids (Cestoda: Cyclophyllidea) (in Russian), *Dokl. Akad. Nauk SSSR,* 75, 597—599, 1950.

3397. **Spasskii, A. A.,** A new approach to the structure and systematics of the hymenolepids (Cestoda: Hymenolepididae) (in Russian), *Dokl. Akad. Nauk SSSR,* 75, 895—898, 1950.

3398. **Spasskii, A. A.,** An attempt to reconstruct the Anoplocephalata on a philogenetic basis (in Russian), *Tr. Gel'mintol. Lab. Akad. Nauk SSSR,* 3, 80—86, 1950.

3399. **Spasskii, A. A.,** A new species of *Paranoplocephala* from marmots of Tian-Shan (in Russian), *Tr. Gel'mintol. Lab. Akad. Nauk SSSR,* 3, 119—124, 1950.

3400. **Spasskii, A. A.,** On the nomenclature of several representative cestodes of the family Hymenolepididae Fuhrmann, 1907 (in Russian), *Tr. Gel'mintol. Lab. Akad. Nauk SSSR,* 4, 30—31, 1950.

3401. **Spasskii, A. A.,** On the characteristics of the cestode *Sciurus vulgaris, Catenotaenia dendritica* (in Russian), *Tr. Gel'mintol. Lab. Akad. Nauk SSSR,* 4, 25—29, 1950.

3402. **Spasskii, A. A.,** *Essentials of Cestodology,* Vol. 1, *Anoplocephalata,* Akademiya Nauk SSSR, Moscow, 1951; (English transl., Israel Program for Scientific Translations, Jerusalem, 1961).

3403. **Spasskii, A. A.,** Anoplocephalata cestodes of domestic and wild animals, *Osnovy Tsestodologii,* 1, 1—735, 1951.

3404. **Spasskii, A. A.,** The biological and taxonomic importance of the reticulated uterus in anoplocephalids (Cestoda) (in Russian), *Dokl. Akad. Nauk SSSR,* 76, 165—168, 1951.

3405. **Spasskii, A. A.,** Reorganization of the genus *Cittotaenia* Riehm, 1881 together with a new genus *Mosgovoyia* gen.nov. (in Russian), *Tr. Gel'mintol. Lab. Akad. Nauk SSSR,* 5, 28—33, 1951.

3406. **Spasskii, A. A.,** On the nomenclature of the genus *Diorchis* (Cestoda: Hymenolepididae) (in Russian), *Tr. Gel'mintol. Lab. Akad. Nauk SSSR,* 6, 74—75, 1952.

3407. **Spasskii, A. A.,** On the systematic position of hymenolepids with scolices armed with double row of hooks (in Russian), *Tr. Gel'mintol. Lab. Akad. Nauk SSSR,* 6, 76—78, 1952.

3408. **Spasskii, A. A.,** On the question of the alternation of generations in cestodes (in Russian), *Dokl. Akad. Nauk SSSR,* 91, 445—447, 1953.

3409. **Spasskii, A. A.,** Pseudoparasitism of cestodes — hymenolepids of rapacious animals (in Russian), *Dokl. Akad. Nauk SSSR,* 94, 597—599, 1954.

3410. **Spasskii, A. A.,** On the question of the validity of the species *Oligorchis nonarmatus* Neiland, 1952, (Cestoda, Hymenolepididae) (in Russian), *Tr. Gel'mintol. Lab. Akad. Nauk SSSR,* 7, 168—171, 1954.

3411. **Spasskii, A. A.,** On the position of *Meggittiella* Lopez-Neyra and *Skrjabinolepis* Mathevossian in the cestoda (in Russian), *Tr. Gel'mintol. Lab. Akad. Nauk SSSR,* 7, 172—175, 1954.

3412. **Spasskii, A. A.,** On the cycle of development of *Lateriporus* (Cestoda: Dilepididae) (in Russian), *Tr. Gel'mintol. Lab. Akad. Nauk SSSR,* 7, 176—179, 1954.

3413. **Spasskii, A. A.,** On the question of the division of the genus *Moniezia* by subgenera (in Russian), *Tr. Gel'mintol. Lab. Akad. Nauk SSSR,* 7, 180—181, 1954.

3414. **Spasskii, A. A.,** On the presence in *Rajotaenia gerbilli* Wertheim, 1954, of an isolated ovary and on the allocation of this cestode to the family Catenotaeniidae (in Russian), *Dokl. Akad. Nauk,* 103, 945—948, 1955.

3415. **Spasskii, A. A.,** Breve revisione di Hymenolepididae. I, *Parassitologia,* 3, 159—178, 1961.

3416. **Spasskii, A. A.,** Breve revisione di Hymenolepididae. II, *Parassitologia,* 3, 179—198, 1961.

3417. **Spasskii, A. A.,** Key to the species of *Aploparaksis* Clerc (Hymenolepididae) (in Russian), *Helminthologia,* 3, 358—363, 1961.

3418. **Spasskii, A. A.,** *Principles of Cestodology,* Vol. 2, Skrjabin, K. I., Ed., *Hymenolepididae — Tapeworms of Wild and Domestic Birds,* Part 1 (in Russian), Akademii Nauk SSSR, Moscow, 1963, 418 pp.

3419. **Spasskii, A. A.,** Two new hymenolepidid genera from birds, *Ortleppolepis* n.g. and *Satyolepis* n.g. (Cestoda, Cyclophyllidea) (in Russian), *Tr. Gel'mintol. Lab.,* 15, 145—150, 1965.

3420. **Spasskii, A. A.,** Revision of the genus *Dilepis* (Cestoda: Cyclophyllidea) (in Russian), *Parasites Anim. Plants,* 1, 65—83, 1965.

3421. **Spasskii, A. A.,** Cestodes of the genus *Aploparaksis* from birds on Kamchatka (in Russian), in *Parasitic Worms of Domestic and Wild Animals: Papers on Helminthology Presented to Prof. A. A. Sobolev on the 40th Anniversary of His Scientific and Teaching Activity,* Dalnevostochnii Gosudarstvennii Universitet, Vladivostok, 1965, 303—311.

3422. **Spasskii, A. A.,** Heterogenesis of the genus *Anomotaenia* (Cestoda: Dilepididae), *Dokl. Akad. Nauk SSSR,* 169, 1483—1485, 1966.

3423. **Spasskii, A. A.,** Species composition of the genus *Notopentorchis* (Paruterinidae) and its position in the classification of cestodes (in Russian), in Tokobaev, M. M., Ed., *Helminths of Animals in Kirgizia and Adjacent Territories,* Izdatelstvo ILIM, Frunze, 1966, 57—61.

3424. **Spasskii, A. A.,** Phylogenetic analysis of cestodes of the genus *Lateriporus* (Cyclophyllidea) (in Russian), *Parasites Anim. Plants,* 2, 50—63, 1966.

3425. **Spasskii, A. A.,** A comparative morphological, ecological and geographical analysis of dilepidids of the genus *Anomotaenia* (Cestoda: Cyclophyllidea) (in Russian), *Parasites Anim. Plants,* 4, 23—52, 1968.

3426. **Spasskii, A. A.,** A comparative ecological and morphological analysis of cestodes of the genus *Choanotaenia* (in Russian), in *Parasites of Vertebrates,* Spasskii, A. A., Ed., Kishinev, "Kartva Moldovenyaske", Izdat, 1969, 3—30.

3427. **Spasskii, A. A.,** The plural origins of the composite genus *Similuncinus* (Cestoda, Dilepididae) (in Ukrainian), *Paraziti. Parazitozi Shlyakhi ikh Lividatsii vipusk "Nauka Dumka",* 1, 167—170, 1972.

3428. **Spasskii, A. A.,** New genera of cyclophyllidean cestodes, *Parasites Anim. Plants (Kishinev),* 9, 38—48, 1973.

3429. **Spasskii, A. A.,** The species composition of the genera *Oligorchis* and *Wardium* and remarks on the systematics of hymenolepidids from Charadriiformes (in Russian), *Parazit. Zhivotn. Rast. (Kishinev),* 11, 3—26, 1975.

3430. **Spasskii, A. A.,** Birovilepis gen.n. (Cestoda, Dilepididae) — new genus of cestode from land birds of Eurasia, *Izv. Akad. Nauk Mold. SSR Biol. Khim. Nauki,* 2, 88—89, 1975.

3431. **Spasskii, A. A.,** The synonymy of *Hexaparuterina* with *Metroliasthes (Cestoda),* Cyclophyllidea) and some notes of the systematics of paruterinids (in Russian), *Izv. Akad. Nauk Mold. SSR Biol. Khim. Nauki,* 4, 65—70, 1977.

3432. **Spasskii, A. A.,** The tribes of the subfamily Davaineinae (in Russian), *Izv. Akad. Nauk Mold. SSR Biol. Khim. Nauki,* 4, 85, 1977.

3433. **Spasskii, A. A.,** Identification of a series of species of cyclophyllid cestodes (in Russian), *Izv. Akad. Nauk Mold. SSR Biol. Khim. Nauki,* 5, 72—77, 1978.

3434. **Spasskii, A. A.,** On the alien taxa in the family Davaineidae Braun, 1900 (Cestoda, Cyclophyllidea) (in Russian), *Izv. Akad. Nauk Mold. SSR Biol. Khim. Nauki,* 1, 67—70, 1979.

3435. **Spasskii, A. A.,** Hymenolepididae of pigs and boars (family Suidae), *Helminthologia,* 18, 3—9, 1981.

3436. **Spasskii, A. A. and Andreiko, O. F.,** *Triodontolepis skrjabini* n.sp. (Cestoda: Hymenolepididae) from secondary aquatic Micromammalia and its life cycle, *Dokl. Akad. Nauk SSSR,* 178, 1442—1445, 1968.

3437. **Spasskii, A. A. and Andreiko, O. F.,** Cestodes of insectivores in Moldavia (in Russian), *Parasites Anim. Plants,* RIO Akad. Nauk Moldavskoi SSR, Kishinev, 5, 44—59, 1970.

3438. **Spasskii, A. A. and Andreiko, O. F.,** *Molluscotaenia* (Cestoda: Cyclophyllidea), a dilepidid genus from insectivores (in Russian), *Parazit. Zhivotn. Rast. (Kishinev),* 6, 3—9, 1971.

3439. **Spasskii, A. A. and Bobova, L. P.,** Three new species of *Aploparaksis* (Hymenolepididae) (in Russian), *Helminthologia,* 3, 346—357, 1961.

3440. **Spasskii, A. A. and Bobova, L. P.,** Hymenolepididae from aquatic birds in Kamchatka (in Russian), *Tr. Gel'mintol. Lab.,* 12, 172—200, 1962.

3441. **Spasskii, A. A. and Borgarenko, L. F.,** *Stenovaria* n.g. (Cestoda: Dilepididae) — new genus of cestode from the stone curlew (in Russian), *Izv. Akad. Nauk Tadzhikskoi SSR Biol. Nauk,* 2, 87—93, 1973.

3442. **Spasskii, A. A., Borgarenko, L. F., and Spasskaja, L. P.,** *Gruitaenia latissima* n.g., n.sp., a new dilepidid cestode from cranes (in Russian), *Izv. Akad. Nauk Mold. SSR Biol. Khim. Nauki,* 5, 61—65, 1971.

3443. **Spasskii, A. A. and Dao, V. T.,** Two new species of the genus *Wardium* (Hymenolepididae) from birds in North Vietnam, *Izv. Akad. Nauk Mold. SSSR* (Ser. Zool.), 5, 3—11, 1963.

3444. **Spasskii, A. A., Dong Van Ngy, and Yurpalova, N. M.,** Three new species of hymenolepidids from wild and domestic birds of Viet Nam (in Russian), *Parazit. Zhivotn. Rast. Mold.,* 75—83, 1963.

3445. **Spasskii, A. A. and Freze, V. I.,** A review of *Aploparaksis* Clerc, 1903 (Cestoda: Hymenolepididae), *Česk. Parasitol.,* 8, 385—389, 1961.

3446. **Spasskii, A. A. and Gubanov, N. M.,** An unusual form of dioecious cestode (in Russian), *Tr. Inst. Morfologii Zhivotn. A. N. Severtsova (Moscow),* 27, 91—100, 1959.

3447. **Spasskii, A. A. and Kolotilova, Z. M.,** New genus of Cyclophyllidea — *Rissotaenia,* gen.n. (Cestoda: Dilepididae), parasites of gulls, *Parazit. Zhivotn. Rast.,* 8, 51—58, 1972.

3448. **Spasskii, A. A. and Konovalov, Yu. N.,** *Anomotaenia reticulata* n.sp. (Dilepididae) in *Phalaropus fulicarius* (in Russian), *Vest. Zool.,* 1, 43—48, 1967.

3449. **Spasskii, A. A. and Konovalov, Yu. N.,** Two new dilepidid species (Cestoda: Cyclophyllididae) from Charadriiformes of chukotka (in Russian), *Helminthologia,* Year 1966, 7, 343—351, 1967.

3450. **Spasskii, A. A. and Konovalov, Y. N.,** Two new species of *Dichoanotaenia* (Cestoda: Cyclophyllidea) (in Russian), *Parazitologiya,* 1, 207—212, 1967.

3451. **Spasskii, A. A. and Konovalov, Yu. N.,** Two new species of *Anomotaenia* (Cestoda: Dilepididae) from Charadriiformes in the Chukotsk tundra (in Russian), *Helminthologia,* Year 1969, 10, 191—202, 1971.

3452. **Spasskii, A. A. and Konovalov, Y. N.**, Rare species of cestodes from birds in the Amur area (in Russian), *Parazit. Zhivotn. Rast.*, 8, 58—68, 1972.
3453. **Spasskii, A. A. and Kornyushin, V. V.**, Morphological evolution of gonads in hymenolepidid and dilepidid cestodes, *Dokl. Akad. Nauk SSSR*, 198, 1232—1234, 1971.
3454. **Spasskii, A. A. and Kornyushin, V. V.**, The heterogeneity of the genus *Ophryocotyle* Friis, 1870, and of the subfamily Ophryocotylinae Fuhrmann, 1907 (Cestoda, Davaineidae) (in Russian), VIII, *Nauchn. Konf. Parazitol. Ukrainy. Donetsk. Sentyabr' 1975, Kiev, USSR,* (Abstr.), pp. 141—144, 1975.
3455. **Spasskii, A. A. and Kornyushin, V. V.**, Revision of the classification of Ophryocotylidae (Cestoda, Davaineoidea) (in Russian), *Vest. Zool.*, 5, 34—42, 1977.
3456. **Spasskii, A. A. and Morozov, Y. F.**, New hymenolepidids from insectivores, *Věsnik Česk. Zool. Společnosti*, 23, 182—191, 1959.
3457. **Spasskii, A. A. and Oshmarin, P. G.**, Parasitic worms of Corvidae. On the helminthfauna of birds of the Gorkii region (in Russian), *Tr. Gorkovsk. Gusudarst. Pedagoy. Inst.*, 4, 45—70, 1949.
3458. **Spasskii, A. A. and Oshmarin, P. G.**, A new genus of hymenolepids — *Staphylepis* gen. nov. from domestic and wild gallinaceus birds (in Russian), *Tr. Gel'mintol. Lab. Akad. Nauk SSSR*, 7, 182—184, 1954.
3459. **Spasskii, A. A. and Poznakomkin, S.**, *Fuhrmanolepis* (Cestoda: Dilepididae) (in Russian), *Parasites Anim. Plants*, 2, 87—92, 1966.
3460. **Spasskii, A. A. and Reznik, V. N.**, A revision of the genus *Drepanidotaenia* (Cestoda: Hymenolepididae) (in Russian), *Parazit. Zhivotn. Rast. Mold. (Kishinev)*, pp. 84—90, 1963.
3461. **Spasskii, A. A. and Reznik, V. N.**, Revision of the genus *Liga* (Cestoda: Dilepididae), *Parasites Anim. Plants*, 2, 64—74, 1966.
3462. **Spasskii, A. A. and Shumilo, R. P.**, The post-larval development of the scolex and hooks in cestodes of the genus *Triaenorhina* n.gen. (Paruterinidae), *Dokl. Akad. Nauk SSSR*, 164, 1436—1438, 1965.
3463. **Spasskii, A. A. and Shumilo, R. P.**, Characteristics of the genus *Triaenorhina* (Cestoda: Paruterinidae) from insectivorous birds (in Russian), *Parasites Anim. Plants*, 3, 47—61, 1968.
3464. **Spasskii, A. A. and Spasskaja, L. P.**, Systematic structure of the hymenolepids parasitic in birds (in Russian), *Tr. Gel'mintol. Lab. Akad. Nauk SSSR*, 7, 55—119, 1954.
3465. **Spasskii, A. A. and Spasskaja, L. P.**, Cestodes of *Riparia riparia* and *Hirundo rustica* in phylogenetically distant, but ecologically related birds (in Russian), *Helminthologia*, 1, 85—98, 1959.
3466. **Spasskii, A. A. and Spasskaja, L. P.**, *Passerilepis* and *Variolepis* (Cestoda: Hymenolepididae) (in Russian), *Čzlká. Parasitol.*, 11, 247—255, 1964.
3467. **Spasskii, A. A. and Spasskaja, L. P.**, Revision of the genus *Paricterotaenia* (Cestoda: Dilepididae) (in Russian), *Parasites Anim. Plants*, 1, 84—103, 1965.
3468. **Spasskii, A. A. and Spasskaja, L. P.**, Subfamily Echinorhynchotaeniinae (Cestoda, Cyclophyllidea), *Acta Parasitol. Pol.*, 23, 299—304, 1975.
3469. **Spasskii, A. A. and Spasskaja, L. P.**, Morphological and ecological analysis of the genus *Amoebotaenia* (Cestoda: Dilepididae), *Parasites Anim. Plants*, 2, 75—86, 1966.
3470. **Spasskii, A. A. and Spasskaja, L. P.**, Critical analysis of the genera *Vitta* and *Neoliga* (Cestoda: Dilepididae) (in Russian), *Izv. Akad. Nauk Moldav. SSR Biol. Khim. Nauki*, pp. 3—13, 1966.
3471. **Spasskii, A. A. and Spasskaja, L. P.**, On the place of the protandrous cestode *Hymenocoelia chauhani* and the subfamily Hymenocoelinae (Dioecocestidae) in the Hymenolepididae (in Russian), *Dokl. Akad. Nauk SSSR*, 181, 1294—1296, 1968.
3472. **Spasskii, A. A. and Spasskaja, L. P.**, Insufficient morphological criteria of the genus *Himantocestus* (Gyrocoeliinae: Diploposthidae) (in Russian), *Helminthologia*, Year 1967—68, 8—9, 531—536, 1968.
3473. **Spasskii, A. A. and Spasskaja, L. P.**, A new genus of bird cestodes — *Markewitchella* n.g. (Cestoda, Davaineidae) (in Ukrainian), in *Paraziti, Parazitozi ta Shlyakh Ikh Likvidatsii Vipusk, I,* Naukova Dumka, Kiev, 1972, 171—174.
3474. **Spasskii, A. A. and Spasskaja, L. P.**, The genera *Monorcholepis* Oshmarin, 1961 and *Chimaerolepis* n.g., and the subfamily Aploparaksinae (Cestoda: Hymenolepididae), *Parazit. Zhivotn. Rast. (Kishinev),* 8, 69—75, 1972.
3475. **Spasskii, A. A. and Spasskaja, L. P.**, Genus *Schmidneila*, gen.n. (Cestoda: Metadilepididae) (in Russian), *Izv. Akad. Nauk Mold. SSR Ser. Biol. Khim. Nauki,* 1, 58—60, 1973.
3476. **Spasskii, A. A. and Spasskaja, L. P.**, Gryporhynchinae n.subf., (Cestoda: Cyclophyllidea) (in Russian), *Izv. Akad. Nauk Mold. SSR Biol. Khim. Nauki,* 5, 56—58, 1973.
3477. **Spasskii, A. A. and Spasskaja, L. P.**, Tauffikiini n.tribe (Cestoda, Cyclophyllidea) and a review of the genus *Taufikia* Woodland, 1928 (in Russian), *Izv. Akad. Nauk Mold. SSR Biol. Khim. Nauki,* 6, 54—57, 1974.
3478. **Spasskii, A. A. and Spasskaja, L. P.**, The genetic link between Paruterinidae from nocturnal birds and Taeniidae from diurnal birds of prey, *Dokl. Akad. Nauk SSSR,* 220, 254—255, 1975.

3479. **Spasskii, A. A. and Spasskaja, L. P.**, The composition and taxonomic status of the conglomerate genera *Pseudanomotaenia* and *Choanotaenia* (Cestoda, Dilepididae) (in Russian), in *VIII Nauknaya Konferentsiya Parazitol. Ukrainy. Donetsk.*, Sentyabr' 1975, Kiev, 1975, 144—148.
3480. **Spasskii, A. A. and Spasskaja, L. P.**, Characteristics of the genus *Rallitaenia* (Cestoda, Cyclophyllidea), *Izv. Akad. Nauk Mold. Biol. Khim. Nauki*, 2, 80, 1975.
3481. **Spasskii, A. A. and Spasskaja, L. P.**, Subfamily Echinorhynchotaeniinae (Cestoda, Cyclophyllidea), *Acta Parasitol. Pol.*, 23, 299—304, 1975.
3482. **Spasskii, A. A. and Spasskaja, L. P.**, The systematics of amabiliids and Davaineidae (in Russian), *Parazit. Teplokrovnykh Zhivotn. Mold. Izdatel'stvo Shtiintsa, Kishinev*, 3—30, 1976.
3483. **Spasskii, A. A. and Spasskii, Y. A.**, *Bucerolepis* n.g., (Cestoda: Dilepididae) (in Russian), *Acta Parasitol. Lit.*, 7, 107—110, 1968.
3484. **Spasskii, A. A. and Tolkacheva, L. M.**, *Anserilepis* n.g. (Cyclophyllidea, Hymenolepididae) new genus of cestode from anseriform birds (in Russian), *Tr. Gel'mintol. Lab.*, 15, 151—155, 1965.
3485. **Spasskii, A. A. and Yurpalova, N. M.**, The position of the genus *Dilepidoides* (Cestoda: Cyclophyllidea) in the family Dilepididae (in Russian), *Acta Vet. Acad. Sci. Hung.*, 12, 343—350, 1962.
3486. **Spasskii, A. A. and Yurpalova, H. M.**, *Orientolepis* n.g. (Cestoda: Hymenolepididae) from domestic chickens (in Russian), *Tr. Gel'mintol. Lab.*, 14, 197—200, 1964.
3487. **Spasskii, A. A. and Yurpalova, N. M.**, *Echinatrium* n.g., a new hymenolepidid genus from Anseriformes in Chukotsk. (in Russian), *Parasites Anim. Man*, 1, 104—112, 1965.
3488. **Spasskii, A. A. and Yurpalova, N. M.**, Cestodes of the genus *Microsomacanthus* (Hymenolepididae) in Anseriformes in Chukotka, *Parasites Anim. Plants*, 2, 15—49, 1966.
3489. **Spasskii, A. A. and Yurpalova, N. M.**, *Dendrouterina egrettae* n.sp., a new cestode of *Egretta garzetta* (in Russian), *Mater. Nauch. Konf. Vses. Ova. Gel'mintol.*, Year 1966, pp. 301—308, 1967.
3490. **Spasskii, A. A. and Yurpalova, N. M.**, Cestodes of the genus *Nadejdolepis* (Hymenolepididae) from Charadriiformes of Chukotka (in Russian), *Parazitologiya*, 2, 249—257, 1968.
3491. **Spasskii, A. A. and Yurpalova, N. M.**, *Wardium pacificum* n.sp. (Cestoda: Hymenolepididae) — a new tapeworm from gulls in Anadyr (in Russian), *Papers in Helminthology Presented to Academician K. I. Skrjabin on His 90th Birthday*, Izdatelstvo Akademii Nauk SSSR, Moscow, 1968, 313—316.
3492. **Spasskii, A. A. and Yurpalova, N. M.**, *Aploparaksis lateralis* n.sp. (Cestoda: Hymenolepididae), a new species from Charadriiformes and the erection of *Tanureria* n.subg. (in Russian), *Parasites Anim. Plants*, 3, 30—37, 1968.
3493. **Spasskii, A. A. and Yurpalova, N. M.**, *Panuwa metaskrjabini* n.sp., a new cestode from *Lobivanellus indicus indicus* and an analysis of the genus *Panuwa* (Cyclophyllidea), *Parasites Anim. Plants*, 3, 38—46, 1968.
3494. **Spasskii, A. A. and Yurpalova, N. M.**, Cestodes of the genus *Aploparaxis* from Charadriiformes of Chukotsk and a brief review of their zoogeography (in Russian), in *Parasites of Vertebrates*, A. A. Spasskii, Ed., Izdat. "Kartya Moldovenyaske", Kishinev, 1969, 46—73.
3495. **Spasskii, A. A. and Yurpalova, N. M.**, *Bancroftiella sudarikovi* n.sp. and a revision of the genus *Bancroftiella* (Cestoda, Dilepididae) (in Russian), *Izv. Akad. Nauk Mold. SSR Biol. Khim. Nauki*, 1, 46—50, 1970.
3496. **Spasskii, A. A. and Yurpalova, N. M.**, Hymenolepidids of birds in Viet Nam (in Russian), *Helminthologia*, Year 1969, 10, 203—243, 1971.
3497. **Spasskii, A. A. and Yurpalova, N. M.**, *Skrjabinotaurus interruptus* n.g., n.sp. (Cestoda, Davaineidae) in Columbidae in Viet Nam, in Gararin, V. G., Ed., *Problemy obshchei i Prikladnoi Gel'mintol.*, Izdatel'stvo Nauko, Moscow, 1973, 133—136.
3498. **Spasskii, A. A. and Yurpalova, N. M.**, *Daovantienia metacentropi* n.sp. (Cestoda, Davaineidae) from birds in Vietnam (in Russian), *Izv. Akad. Nauk Moldav. SSR Biol. Khim. Nauki*, 3, 55—60, 1976.
3499. **Spasskii, A. A., Yurpalova, N. M., and Kornyushin, V. V.**, New genus of dilepidid — *Anomolepis* gen.n. (Cestoda, Cyclophyllidea), *Vestn. Zool.*, 2, 46—51, 1968.
3500. **Spätlich, W.**, Untersuchungen ueber Tetrabothrien. Ein Beitrag zur Kenntnis des Cestodenkörpers, *Zool. Jahrb. Anat.*, 28, 539—594, 1909.
3501. **Spencer, W. B.**, The anatomy of *Amphiptyches urna* (Grube and Wagner), *Trans. R. Soc. Victoria*, 1, 138—151, 1889.
3502. **Spengel, J. W.**, Die Monozootie der Cestoden, *Z. Wiss. Zool.*, 82, 252—287, 1905.
3503. **Sproston, N. G.**, On the genus *Dinobothrium* von Beneden (Cestoda) with a description of two new species from sharks and a note on *Monorygma* species from the electric ray, *Parasitology*, 39, 73—90, 1948.
3504. **Srámek, A.**, Helminthen der an der zoologischen Station in Podiebrad (Böhmen) untersuchten Fische, *Arch. Naturwiss. Landesdurchforsch. Böhmen*, 11, 94—118, 1901.
3505. **Srivastava, A. K.**, On a new cestode, *Neyraia sultanpurensis* sp.n., of the Paruterininae Fuhrmann, 1907, family Dilepididae Railliet et Henry, 1909 from *Upupa epops* (Linnaeus), *Helminthologia*, 17, 153—158, 1980.

3506. **Srivastava, A. K. and Capoor, V. N.**, On a new cestode *Dicranotaenia alcippina* Tl. sp., *Proc. 62nd Ind. Sci. Congr.*, Part III, Group B, Delhi, 1975, 210.
3507. **Srivastava, A. K. and Capoor, V. N.**, On a new cestode *Valipora amethiensis*, *Proc. 62nd Ind. Sci. Congr.*, Part III, Group B, Delhi, 1975, 210.
3508. **Srivastava, A. K. and Capoor, V. N.**, On a new cestode, *Vampirolepis molus* sp.n., *Helminthologia*, 16, 195—198, 1979.
3509. **Srivastava, A. K. and Capoor, V. N.**, On a cyclophyllidean cestode, *Batrachotaenia junglensis* n.sp. from the intestines of common frog, *Rana tigrina* from suburban area of Amethi, Sultanpur, *Ind. J. Parasitol.*, 3, 39, 1980.
3510. **Srivastava, A. K. and Capoor, V. N.**, On *Acanthobothrium dighaensis* sp.n. (Onchobothriidae Braun, 1900) from *Trygon marginatus*, *Helminthologia*, 17, 165—170, 1980.
3511. **Srivastava, A. K. and Capoor, V. N.**, A new cestode, *Ophryocotylus dinopii* gen.n. et sp.n. (Cyclophyllidea: Davaineidae) from *Dinopium benghalense* (L.), *Helminthologia*, 19, 129—134, 1982.
3512. **Srivastava, A. K. and Tewari, J. P.**, Studies on a new cestode, *Onchobothrium capoori* n.sp. from the intestines of sting ray, *Torpedo* sp. from Sasoon Dock, Bombay, *Ind. J. Parasitol.*, 3, 39, 1980.
3513. **Srivastava, H. D.**, A study of the life-history of a common tapeworm, *Mesocestoides lineatus*, of Indian dogs and cats, *Ind. J. Vet. Sci.*, 9, 187—190, 1939.
3514. **Srivastava, V. C. and Capoor, V. N.**, On a new cestode, *Columbia allahabadi* n.g., n.sp., from the Indian pigeon, *Columbia livia* (Gmelin) from Allahabad (India) with a revision of the key to the various genera of the subfamily, Thysanosomatinae (Skrjabin, 1933), *Proc. Natl. Acad. Sci.* (Section D, Biol. Sci.), 35, 371—374, 1966.
3515. **Srivastava, V. C. and Pandey, G. P.**, A new species, *Echinocotyle singhi* n.sp. (Cestoda, Hymenolepididae) from the blue winged teal, *Querquedula circia* from Allahabad (India), *Proc. Ind. Acad. Parasitol.*, 1, 45—49, 1980.
3516. **Srivastava, V. C. and Sawada, I.**, A new cestode from the grey partridge, *Francolinus pondicerianus*, *Annot. Zool. Jpn.*, 53, 120—123, 1980.
3517. **Stammer, H. J.**, Die Entoparasiten der in Schlesien 1935 beobachteten Flamingos, *Ber. Ver. Schles. Ornithol.*, 21, 15—17, 1936.
3518. **Steelman, G. M.**, *Oochoristica whitentoni*, a new anoplocephalid cestode from a land tortoise, *J. Parasitol.*, 25, 479—482, 1939.
3519. **Steelman, G. M.**, A new cestode from the Texas horned lizard, *Trans. Am. Microsc. Soc.*, 58, 452—455, 1939.
3520. **Steelman, G. M.**, A new cestode, *Diorchis longibursa*, from the coot, *Am. Midl. Nat.*, 22, 637—639, 1939.
3521. **Steenstrup, J. J. S.**, Jagttagelser og Bemaerkninger om Hundesteilens Baendelorm, *Fasciola intestinalis* Linn., *Schistocephalus solidus* (C.F.M. Prod. Z.D.), *Overs. K. Dansk. Vidensk. Selsk. Forh.*, pp. 186—195, 1857.
3522. **Steenstrup, J. J. S. and Lütken, C. F.**, Spolia Atlantica. Bidrag til Kundskab om Klumpeller Maanefiskene (Molidae), *D. Kgl. Danske Videnskab. Selsk. Skr.*, pp. 1—102, 1898.
3523. **Stephens, J. W. W.**, Two new human cestodes and a new linguatulid, *Ann. Trop. Med. Parasitol.*, 1, 549—556, 1908.
3524. **Steudener, F.**, Untersuchungen ueber den feineren Bau der Cestoden, *Abh. Naturf. Ges. Halle*, 13, 277—316, 1877.
3525. **Stewart, M. A.**, The validity of *Dipylidium sexcoronatum* von Ratz, 1900 (Cestoda), *J. Parasitol.*, 25, 185—186, 1939.
3526. **Stieda, L.**, Ein Beitrag zur Kenntnis der Taenien, *Arch. Naturg.*, 28, 208—209, 1862.
3527. **Stieda, L.**, Ein Beitrag zur Anatomie des *Bothriocephalus latus*, *Arch. Anat. Phys.*, pp. 174—212, 1864.
3528. **Stieda, L.**, Beitrage zur Anatomie der Plattwürmer, *Arch. Anat. Physiol. Wiss. Med.*, pp. 52—63, 1867.
3529. **Stiles, C. W.**, Notes sur les parasites. XIII. Sur le *Taenia giardi*, *C. R. Soc. Biol.*, 4, 664—665, 1892.
3530. **Stiles, C. W.**, Notes sur les parasites. XIV. Sur le *Taenia expanasa* Rudolphi, *C. R. Soc. Biol.*, 4, 665—666, 1892.
3531. **Stiles, C. W.**, Ueber die topographische Anatomie des Gefass-systems, in der Familie Taeniidae, *Centralbl. Bakteriol.*, 13, 457—465, 1893.
3532. **Stiles, C. W.**, Notes on parasites. A double-pored cestode with occasional single pores, *Centralbl. Bakteriol.*, 13, 457—459, 1895.
3533. **Stiles, C. W.**, Notes on parasites. XXXVIII. Preliminary note to "A revision of the adult leporine cestodes", *Vet. Mag.*, 2, 341—346, 1895.
3534. **Stiles, C. W.**, A revision of the adult tapeworms of hares and rabbits, *Proc. U.S. Natl. Mus.*, 19, 145—235, 1896.
3535. **Stiles, C. W.**, The type species of the cestode genus *Hymenolepis*, *U.S. Public Health Service Hyg. Lab. Bull.*, 13, 19—21, 1903.

3536. **Stiles, C. W.**, Illustrated key to the cestode parasites of man, *U.S. Public Health Service Hyg. Lab. Bull.*, 25, 1—104, 1906.

3537. **Stiles, C. W.**, The occurrence of a proliferating cestode larva *(Sparganum proliferum)* in man in Florida, *U.S. Public Health Service Hyg. Lab. Bull.*, 40, 5—18, 1908.

3538. **Stiles, C. W. and Baker, C. E.**, Key-catalogue of parasites reported for Carnivora (cats, dogs, bears, etc.) with their possible public health importance, *Natl. Inst. Health Bull.*, 173, 913—1223, 1935.

3539. **Stiles, C. W. and Hassall, A.**, A revision of the adult cestodes of cattle, sheep and allied animals, *U.S. Bur. Anim. Ind. Bull.*, pp. 1—134, 1893.

3540. **Stiles, C. W. and Hassall, A.**, A preliminary catalogue of the parasites contained in the collections of the U.S. Bureau of Animal Industry, U.S. Army Medical Museum, Biological Department of the University of Pennsylvania (Leidy collection), and in the collections of Stiles and of Hassall, *Vet. Mag.*, 1, 245—253, 1894.

3541. **Stiles, C. W. and Hassall, A.**, *Ctenotaenia denticulata* (Rud., 1804) Stiles et Hassal, 1896. Notes on parasites. XLI, *Centralbl. Bakteriol. I Abt.*, 19, 70—72, 1896.

3542. **Stiles, C. W. and Hassall, A.**, Report upon present knowledge of the tapeworms of poultry, *U.S. Bur. Anim. Ind. Bull.*, No. 12, 1896.

3543. **Stiles, C. W. and Hassall, A.**, Notes on parasites. XLVII. Priority of *Cittotaenia* Riehm, 1881 over *Ctenotaenia* Railliet, 1891, *Vet. Mag.*, 3, 407, 1896.

3544. **Stiles, C. W. and Hassall, A.**, Internal parasites of the fur seal. Report of Fur Seal Investigations, Washington. The Fur Seals and Fur Seal Islands of the North Pacific Ocean, Part 3, 1899, 99—177.

3545. **Stiles, C. W. and Hassall, A.**, *Bertiella*, new name for the cestode genus *Bertia* Bl., 1891, *Science*, 16, 434, 1902.

3546. **Stiles, C. W. and Hassall, A.**, Index catalogue of medical and veterinary zoology, Parts 1—36, *U.S. Bur. Anim. Ind. Bull.*, 39, 1—2703, 1902—1912.

3547. **Stiles, C. W. and Hassall, A.**, Index catalogue of medical and veterinary zoology. Subjects: Cestoda and Cestodaria, *U.S. Public Health Serv. Hyg. Lab. Bull.*, 85, 1—467, 1912.

3548. **Stiles, C. W. and Hassall, A.**, Key catalogue of the worms reported for man, *U.S. Public Health Serv. Hyg. Lab. Bull.*, 142, 69—196, 1926.

3549. **Stiles, C. W., Hassall, A., and Nolan, O.**, Key catalogue of parasites reported for primates (monkeys and lemurs) with their possible public importance, and key catalogue of primates for which parasites are reported, *U.S. Public Health Serv. Hyg. Lab. Bull.*, 152, 409—601, 1929.

3550. **Stiles, C. W. and Nolan, M. (Orleman)**, The cestode genus *Hydatigera* Lamarck, 1816, species *reditaenia* Sambon, 1924, *J. Trop. Med. Hyg.*, 28, 249—250, 1925.

3551. **Stiles, C. W. and Nolan, M. (Orleman)**, La nomenclature des genres de cestodes *Raillietina, Ransomia* et *Johnstonia*, *Ann. Parasitol.*, 4, 65—67, 1926.

3552. **Stiles, C. W. and Stevenson, E. C.**, The synonymy of *Taenia (T. crassicollis, T. marginata, T. serialis)* and *Echinococcus*, *U.S. Bur. Anim. Ind. Bull.*, 80, 1—14, 1905.

3553. **Stiles, C. W. and Taylor, J. L.**, An adult cestode *(Diplogonoporus grandis)* which may possibly occur in returning American troops, *U.S. Bur. Anim. Ind. Bull.*, 35, 43—47, 1902.

3554. **Stiles, C. W. and Taylor, J. L.**, A larval cestode *(Sparganum mansoni)* of man which may possibly occur in returning American troops, *U.S. Bur. Anim. Ind. Bull.*, 35, 47—56, 1902.

3555. **Stock, T. M. and Holmes, J. C.**, *Pararetinometra lateralacantha* gen. et sp.nov. (Cestoda: Hymenolepididae) in grebes (Podicipedidae) from Alberta, *Can. J. Zool.*, 59, 2319—2321, 1982.

3556. **Stossich, M.**, Prospetto della fauna del mare Adriatico. IV. Vermes, *Boll. Soc. Adriat. Sci. Nat. Trieste*, 7, 168—242, 1882.

3557. **Stossich, M.**, Elminti veneti raccolti dal Dr. Alessandro Conte di Ninni, *Boll. Soc. Adriat. Sci. Nat. Trieste*, 12, 49—56, 1890.

3558. **Stossich, M.**, Elminto veneti raccolti dal Dr. Alessandro Conte de Ninni, e descritti da Michaele Stossich, *Boll. Soc. Adriat. Sci. Nat. Trieste*, 13, 109—116, 1891.

3559. **Stossich, M.**, Notizie elmintologiche, *Boll. Soc. Adriat. Sci. Nat. Trieste*, 16, 33—46, 1895.

3560. **Stossich, M.**, Ricerche elmintologiche, *Boll. Soc. Adriat. Sci. Nat. Trieste*, 17, 121—136, 1896.

3561. **Stossich, M.**, Elminti trovati in um *Orthagoriscus mola*, *Boll. Soc. Adriat. Sci. Nat. Trieste*, 17, 189—191, 1896.

3562. **Stossich, M.**, Saggio di una fauna elmintologica di Trieste e provincie contermini, *Prog. Civ. Scuola Sup. Trieste*, pp. 1—162, 1898.

3563. **Stossich, M.**, Appunti di elmintologia, *Boll. Soc. Adriat. Sci. Nat. Trieste*, 19, 1—6, 1899.

3564. **Stossich, M.**, Osservazioni elmintologia, *Boll. Soc. Adriat. Sci. Nat. Trieste*, 20, 89—104, 1900.

3565. **Strand, E.**, Miscellanea nomenclatorica zoologica et palaeontologica. I-II, *Arch. Naturg.*, 92, 30—75, 1928.

3566. **Stroh, G.**, *Coenurus cerbralis* bei der Gemse, *Berl. Tierärztl. Wochenschr.*, 48, 465—466, 1932.

3567. **Stunkard, H. W.**, The tapeworms of the rhinoceroses, a study based on material from the Belgian Congo, *Am. Mus. Novitates*, No. 210, 1—17, 1926.

3568. **Stunkard, H. W.**, The resistance of European rabbits and hares to superinfestation by different species of the genus *Cittotaenia, J. Parasitol.*, 19, 156, 1932.
3569. **Stunkard, H. W.**, Studies on the life-history of anoplocephaline cestodes, *Z. Parasitenkd.*, 6, 481—507, 1934.
3570. **Stunkard, H. W.**, The life cycle of *Moniezia expansa, Science*, 86, 312; *Biol. Bull.*, 73, 370, 1937.
3571. **Stunkard, H. W.**, The physiology, life cycles and phylogeny of the parasitic flatworms, *Am. Mus. Novitates*, No. 908, 1—27, 1937.
3572. **Stunkard, H. W.**, The life cycle of anoplocephaline cestodes, *J. Parasitol.*, 23, 569, 1937.
3573. **Stunkard, H. W.**, Parasitic flatworms in Yucatan, *Publ. Carnegie Inst. Wash.*, pp. 33—50, 1938.
3574. **Stunkard, H. W.**, *Oochoristica parvula* n.nom. for *Oochoristica parva* Stunkard, 1938, preoccupied, *J. Parasitol.*, 24, 554, 1938.
3575. **Stunkard, H. W.**, The development of *Moniezia expansa* in the intermediate hosts, *Parasitology*, 30, 491—501, 1939.
3576. **Stunkard, H. W.**, The role of oribatid mites as transmitting agents and intermediate hosts of ovine cestodes, *Verhandl. VII. Int. Kongr. Entom. Berlin*, 3, 1669—1674, 1939.
3577. **Stunkard, H. W.**, The life cycle of the rabbit cestode, *Cittotaenia ctenoides*, *Z. Parasitenkd.*, 10, 753—754, 1939.
3578. **Stunkard, H. W.**, The morphology and life history of the cestode *Bertiella studeri*, *Am. J. Trop. Med.*, 20, 305—333, 1940.
3579. **Stunkard, H. W.**, Observations on the development of the cestode, *Bertiella studeri*, *Proc. 3 Int. Congr. Microbiol.*, New York, 1940, 460—462.
3580. **Stunkard, H. W.**, Tapeworm infections in the West Indies, *Rev. Med. Trop. Parasitol.*, 6, 283—288, 1940.
3581. **Stunkard, H. W.**, Studies on the life history of the anoplocephaline cestodes of hares and rabbits, *J. Parasitol.*, 27, 299—325, 1941.
3582. **Stunkard, H. W.**, The Syrian hamster, *Cricetus auratus*, host of *Hymenolepis nana*, *J. Parasitol.*, 31, 151, 1945.
3583. **Stunkard, H. W.**, On certain pseudophyllidean cestodes from Alaskan pinnipeds, *J. Parasitol.*, 33, 19, 1947.
3584. **Stunkard, H. W.**, Pseudophyllidean cestodes from Alaskan pinnipeds, *J. Parasitol.*, 34, 211—228, 1948.
3585. **Stunkard, H. W.**, Notes on *Diphyllobothrium stemmacephalum* Cobbold, 1858, *J. Parasitol.*, 34, 16, 1948.
3586. **Stunkard, H. W.**, *Diphyllobothrium stemmacephalum* Cobbold, and *D. latum* (Linn., 1758), *J. Parasitol.*, 35, 613—624, 1949.
3587. **Stunkard, H. W.**, *Paratriotaenia oedipomidatis* gen. et sp.n. (Cestoda), from a marmoset, *J. Parasitol.*, 51, 545—551, 1965.
3588. **Stunkard, H. W.**, *Panceriella* nom.n., for *Pancerina* Fuhrmann, 1899, preoccupied by *Pancerina* Chun, 1879, and systematic relations of the genus, *J. Parasitol.*, 55, 1162—1168, 1969.
3589. **Stunkard, H. W.**, Studies on tetraphyllidean and tetrarhynchidean metacestodes from squids taken on the New England coast, *Biol. Bull.*, 153, 387—412, 1977.
3590. **Stunkard, H. W. and Lynch, W. F.**, A new anoplocephaline cestode *Oochoristica anniellae*, from the Californian limbless lizard, *Trans. Am. Microsc. Soc.*, 63, 165—169, 1944.
3591. **Stunkard, H. W. and Milford, J. J.**, Notes on the cestodes of North American sparrows, *Zoologica*, 22, 177—183, 1937.
3592. **Stunkard, H. W. and Schoenborn, H. W.**, Notes on the structure, distribution and synonymy of *Diphyllobothrium lanceolatum*, *Am. Mus. Novitates*, No. 880, 1—9, 1936.
3593. **Subhapradha, C. K.**, On the genus *Polypocephalus* Braun, 1878 (Cestoda) together with descriptions of six new species from Madras, *Proc. Zool. Soc. London*, 121, 205—235, 1951.
3594. **Subhapradha, C. K.**, Cestode parasites of fishes of Madras coast, *Ind. J. Helminthol.*, 7, 41—132, 1957.
3595. **Subramanian, K.**, On a new tapeworm *(Raillietina rangoonica)* from the fowl, *J. Burma Res. Soc.*, 18, 78—79, 1928.
3596. **Subramanian, K.**, Studies on cestode parasites of fishes. I. *Biporophyllaeus madrassensis* gen. et sp.nov., with a note on its systematic position, *Rec. Ind. Mus.*, 41, 131—150, 1939.
3597. **Subramanian, K.**, On a new species of *Echeneibothrium* from *Rhinobatus granulatus* Cuv., *Rec. Ind. Mus.*, 42, 457—464, 1940.
3598. **Subramanian, K.**, Studies on cestode parasites of fishes. II. The nervous system of *Tylocephalum dierama* Shipley and Hornell, *Rec. Ind. Mus.*, 43, 269—280, 1941.
3599. **Subramanian, K.**, Sympathetic innervation of proglottides in *Avitellina lahorea* Woodland, *Curr. Sci.*, 10, 441—443, 1941.
3600. **Sudarikov, V. E.**, New cestodes of birds from the central Volga basin (in Russian), *Tr. Gel'mintol. Lab. Akad. Nauk SSSR*, 3, 142—151, 1950.

3601. **Sugimoto, M.,** Morphological studies on the avian cestodes from Formosa (in Japanese), *Rep. Govt. Res. Inst. Dep. Agric. Formosa,* 64, 1—52, 1934.
3602. **Sulgostowska, T.,** Redescription of the species *Diploposthe laevis* (Bloch, 1782) and *D. bifaria* (Siebold in Creplin, 1846) and revision of the genus (Cestoda, Hymenolepididae), *Acta Parasitol. Pol.,* 24, 231—248, 1977.
3603. **Sulgostowska, T. and Korpaczewska, W.,** *Schistotaenia rufi* sp.n. (Cestoda: Amabiliidae), *Acta Parasitol. Pol.,* 17, 131—138, 1969.
3604. **Sumner, F. B., Osburn, R. C., and Cole, L. J.,** A biological survey of the waters of Woods Hole and vicinity. III. A catalogue of the marine fauna, *Bull. Bur. Fish. Wash.,* 31, 545—794, 1913.
3605. **Sunkes, E. J. and Sellers, T. F.,** Tapeworm infestation in the southern United States, *Am. J. Public Health,* 27, 893—898, 1937.
3606. **Sutton, C. A.,** Un nuevo eucestode parasito de *Myocastor coypus bonariensis* Commerson, *Neotropica,* 19, 38—42, 1973.
3607. **Swales, W. E.,** *Rhabdometra odiosa* (Leidy, 1887) Jones, 1929, a cestode parasite of *Pedioectes phasianellus* in Quebec, *J. Parasitol.,* 20, 313—314, 1934.
3608. **Swingle, L. D.,** The morphology of the sheep tapeworm, *Thysanosoma actinioides, Bull. Agric. Exp. Sta. Wyo.,* No. 102, 103—116, 1914.
3609. **Szidat, L.,** *Archigetes* R. Leuckart, 1878, die progenetische Larve einer fur Europa neuen Caryophyllaeiden-Gattung *Biacetabulum* Hunter, 1927, *Zool. Anz.,* 119, 166—172, 1937.
3610. **Szidat, L.,** Uber einige neue Caryophyllaeiden aus ostpreussischen Fischen, *Z. Parasitol.,* 9, 771—786, 1937.
3611. **Szidat, L.,** *Brachyurus gobii* n.g. n.sp., eine neue Caryophyllaeiden-Art aus dem Grundling, *Gobio fluviatilis* Cuv., *Zool. Anz.,* 124, 249—258, 1938.
3612. **Szidat, L.,** Uber die Caryophyllaeiden-Gattung *Khawia* H.F. Hsü, 1935 und eine neue Art dieser Gattung *Khawia baltica* n.spec., *Z. Parasitenkd.,* 12, 120—132, 1941.
3613. **Szidat, L.,** *Echinococcus patagonicus* (Cestoda), parasito del zorro *Dusicyon culpaeus culpaeus* (Mol.), *Neotropica (Buenos Aires),* 6, 13—16, 1960.
3614. **Szidat, L.,** *Echinococcus pampianus* una nueva especie de la Argentina, parasita de *Felis colocolo pajeros* Desmarest, 1916 (Cestoda), *Neotropica,* 13, 90—96, 1967.
3615. **Szidat, L.,** *Echinococcus cepanzoi* sp.nov. (Cestoda, Taeniidae) del zorro gris *Dusicyon gymnocercus,* de la provincia de Buenos Aires, *Neotropica,* 17, 1—4, 1971.
3616. **Szidat, L. and Nani, A.,** Las remoras del Atlántico Austral con un estudio di su nutrición natural y do sus parásitos (Pisc. Echeneidae), *Rev. Inst. Nac. Invest. Cien. Nat. Buenos Aires Cien. Zool.,* 2, 385—417, 1951.
3617. **Szidat, L. and Soria, M. F.,** Nuevos parasitos de *"Leptodactylus ocellatus"* (L.) de la republica argentina, *Commun. Inst. Nac. Inv. Cien. Nat. Cien. Zool.,* 2, 189—210, 1954.
3618. **Szidat, L. and Soria, M. F.,** Difilobotriasis en nuestro pais. Sobre una nueva especie de *Sparganum,* parásita de Salmones, y d *Diphyllobothrium,* parásita de gaviotas, del lago Naheul Huapi, *Bol. Mus. Argentino Cien. Nat. "Bernardino Rividavia" Instit. Nacional Invest. Cien. Nat.,* 9, 1—22, 1957.
3619. **Szpotanska, I.,** Un nouveau genre, sous-genre et quelques nouvelles espèces de la famille Tetrabothriidae, (in Polish, French summary 918—921), *Sprawog. Posied. Towarzyst. Nauk Warszawsk., Wydz.,* 10, 909—918, 1917.
3620. **Szpotanska, I.,** Etude sur les Tetrabothriidae des Procellariiformes (in Polish with French summary), *Bull. Int. Acad. Pol. Sci. Cl. Sci. Math. Nat. B,* 1925, 673—727, 1926.
3621. **Szpotanska, I.,** Recherches sur quelques Tetrabothriidae d'oiseaux (in Polish with French summary), *Bull. Int. Acad. Pol. Sci. Cl. Sci. Math. Nat. B,* 1928, 129—152, 1929.
3622. **Szpotanska, I.,** Note sur un espèce du genre *Liga* Weinland (in French), *Ann. Mus. Zool. Pol.,* 9, 237—246, 1931.
3623. **Szpotanska, I.,** Quelques especes nouvelles ou peu connues des Hymenolepididae Fuhrmann (Cestodes), *Ann. Mus. Zool. Pol.,* 9, 247—266, 1931.
3624. **Szpotanska, I.,** A propos du genre *Porotaenia, Ann. Parasitol.,* 9, 484, 1931.
3625. **Szpotanska, I.,** Z. Badan nad anatomja *Hymenolepis villosoides* Solowiow, *Ann. Mus. Zool. Pol.,* 10, 327—332, 1934.
3626. **Szymanski, M.,** New bird tapeworms, *Bull. Int. Acad. Sci. Cracov. Cl. Sci. Math. Nat.,* 733—735; *Arch. Naturg.,* 2 (Abstr.), 1907, 46, 1905.
3627. **Tadros, G.,** On the classification of the family Bothriocephalidae Blanchard, 1849 (Cestode), *J. Vet. Sci. United Arab Rep.,* 3, 39—43, 1966.
3628. **Tadros, G.,** On a new cestode *Bothriocephalus prudhoei* sp. nov. from the Nile catfish *Clarias anguillaris* with some remarks on the genus *Clestobothrium* Lühe, 1899, *Bull. Zool. Soc. Egypt Year 1966—67,* 21, 74—88, 1967.

3629. **Tadros, G.**, A re-description of *Polyonchobothrium clarias* (Woodland, 1925) Meggett, 1930 (Bothriocephalidae Cestoda) with a brief review of the genus *Polyonchobothrium* Diesing, 1854 and the identity of the genera *Tetracampos* Wedl, 1861, *Senga* Dollfus, 1935, and *Oncobothriocephalus* Yamaguti, 1959, *J. Vet. Sci. United Arab Rep.*, 5, 53—84, 1968.
3630. **Takeishi, H.**, On the direct development of *Hymenolepis fraterna* in the intestine of the rat, and the influence of the serum of the rat immunized with the substance of *Diphyllobothrium mansoni* upon the infection of *Hymenolepis fraterna, Keio Igaku,* 17, 1521—1535, 1937.
3631. **Talysin, T.**, *Dibothriocephalus minor* Chol., der kleine Bandwurm Transbaikaliens, *Z. Parasitenkd.,* 2, 535—550, 1930.
3632. **Talysin, T.**, *Dibothriocephalus strictus* n. sp., Menschenparasit des Baikalgestades, *Z. Parasitenkd.,* 4, 722—729, 1932.
3633. **Talysin, T.**, Zur Frage der morphologischen Charakteristik der Strobila bei *Diphyllobothrium minus* Chol., *Zool. Anz.,* 106, 209—215, 1934.
3634. **Tamura, O. and Iwata, S.**, Intestinal parasites of ducks from Osaka City and its vicinity, with a possibly new species of *Raillietina, Proc. Jpn. Parasitol. Soc.,* 2, 38—39, 1930.
3635. **Tandan, B. K. and Singh, K. S.**, *Ophryocotyloides dasi* sp. n. (Cestoda: Davaineidae) from a Himalayan barbet, *Zool. Anz.,* 173, 441—443, 1964.
3636. **Tandan, B. K. and Singh, K. S.**, *Mayhewia levinei* sp. n. (Cestoda: Hymenolepididae) from *Tundus merula simillimus*, the blackbird, *Parasitologia,* 5, 217—220, 1963.
3637. **Tarassow, W.**, Ueber die Verbreitung von *Dipyllobothrium latum* und andere Darmparasiten bei der Bevölkerung des Gebiets der Seegruppen Kontschosero, *Tr. Borodin Biol. Stn.*, No. 6, 27—50, 1933.
3638. **Tarassow, W.**, Die Behaftungsgrad der Bevölkerung Karaliens mit Eingeweidewürmern, *Tr. Borodin Biol. Stn.*, No. 6, 57—96, 1933.
3639. **Tarassow, W.**, Das Schwein und der Hund als engültige Träger des *Diphyllobothrium latum, Arch. Schiffs Tropenhyg.,* 38, 156—159, 1934.
3640. **Tarassow, W.**, Beiträge zum Problem des Kampfes gegen *Diphyllobothrium latum* in Nord-Westgebiet., *Arch. Schiffs Tropenhyg.,* 38, 477—486, 1934.
3641. **Taylor, E. L.**, *Moniezia,* a genus of cestode worms and the proposed reduction of its species to three, *Proc. U.S. Natl. Mus.*, 74, 1—9, 1928.
3642. **Taylor, E. L.**, *Davainea proglottina* and disease in fowls, *Vet. J.,* 89, 500—504, 1933.
3643. **Taylor, E. L.**, *Fimbriaria fasciolaris* in the proventriculus of a swan associated with bacterial infection and ulcer formation, *Parasitology* 26, 359—360, 1934.
3644. **Telford, S. R., Jr.**, A new nematotaeniid cestode from California lizards, *Jpn. J. Exp. Med.,* 35, 301—303, 1965.
3645. **Temirova, S. I., and Skriabin, A. S.**, *Essentials of Cestodology,* Vol. 9, *Tetrabothriata (Ariola, 1899) Skrjabin, 1940,* Akademiya Nauk SSSR, Moscow, 1978, 1—154.
3646. **Templeton, R.**, A catalogue of the species of annulose animals, and of rayed ones, found in Ireland, *Mag. Nat. Hist. London,* 9, 233—240, 301—305, 417—421, 466—472, 1836.
3647. **Tendeiro, J.**, Notas de helmintologia guineense. Um novo cestoide, *Ophryocotyle fuhrmanni* n. sp., parasita do neio-macarico, *Numenius phaeopus phaeopus* (L.), *Bol. Cult. Guiné Port.,* 8, 659—666, 1953.
3648. **Tendeiro, J.**, Notas de helmintologia guineense. Sobre a *Cotugnia meleagridis* Joyeux, Baer e Martin 1936 (Cestoda, Davaineidae), *Bol. Cult. Guiné Port.,* 8, 651—657, 1953.
3649. **Tenora, F.**, Revision of the classification of platyhelminths of the family Catenotaeniidae Spassky, 1950 (in Russian), *Zool. Zh.,* 38, 1322—1334, 1959.
3650. **Tenora, F.**, On the systematic situation of tapeworms of the family Catenotaeniidae Spassky, 1950, *Zool. Listy Brno,* 13, 333—352, 1964.
3651. **Tenora, F.**, Studies on parasitic worms in some vertebrates from Afghanistan. Research report of the University of Agriculture, Brno (in Czech), *Vysoká Škola Zemědětská,* 1975, 106 pp.
3652. **Tenora, F.**, Reorganization of the system of cestodes of the genus *Catenotaenia* Janicki, 1904. Evolutionary implications, *Acta Univ. Agric. Brno,* 2, 163—169, 1977.
3653. **Tenora, F. and Barus, V.**, *Armadolepis spasskyi* sp. n., nový druh tasemnice z hlodavců čeledi Myoxidae, *Zool. Listy Brno,* 7, 339—342, 1958.
3654. **Tenora, F. and Barus, V.**, Nové poznatky o tasemnicich netopýro (Microchiroptera) v Čsr (in Czech), *Československí Parasitol.,* 7, 343—349, 1960.
3655. **Tenora, F. and Mas-Coma, S.**, Records of *Galligoides artaai* (Mobedi and Ghadirian, 1977) n. comb. (Cestoda: Anoplocephalidae) in *Apodemus sylvaticus* L. from Western Europe. Proposition of *Gallegoides* nov. gen., *Säugtierkundliche Mitteilungen,* 26, 222—226, 1978.
3656. **Tenora, F., Mas-Coma, S., Murai, E., and Feliu, C.**, The system of cestodes of the suborder Catenotaeniata Spassky, 1963, *Acta Parasitol. Hung.,* 13, 39—57, 1980.
3657. **Tenora, F. and Murai, E.**, Cestodes recovered from rodents (Rodentia) in Mongolia, *Ann. Hist. Nat. Mus. Natl. Hung.,* 67, 65—70, 1975.

3658. **Tenora, F. and Murai, E.,** Anoplocephalidae (Cestoda) parasites of Leporidae and Sciuridae in Europe, *Acta Zool. Acad. Sci. Hung.,* 24, 415—429, 1978.
3659. **Tenora, F. and Murai, E.,** The genera *Anoplocephaloides* and *Paranoplocephala* (Cestoda) parasites of Rodentia in Europe, *Acta Zool. Acad. Sci. Hung.,* 26, 263—284, 1980.
3660. **Tenora, F., Murai, E., and Berg, C.,** *Mosgovoyia pectinata* (Goeze, 1782) (Cestoda, Anoplocephaloidae) — a parasite of *Lepus timidus* L. (Lagomorpha) in Norway, *Parasitol. Hung.,* 12, 53—54, 1979.
3661. **Testa, J. and Dailey, M. D.,** Five new morphotypes of *Phyllobothrium delphini* (Cestoda: Tetraphyllidea), their relationship to existing morphotypes, and their zoogeography, *Bull. Calif. Acad. Sci.* 76, 99—110, 1977.
3662. **Theiler, G.,** On the classification of the cestode genus *Moniezia* (Blanchard, 1891), *Ann. Trop. Med. Parasitol.,* 18, 109—123, 1924.
3663. **Thienemann, J.,** Untersuchungen ueber *Taenia tenuicollis* Rud. mit Berücksichtigung der übrigen Musteliden-Taenien, *Arch. Naturg.,* 72 J., 1, 227—248, 1906.
3664. **Thomas, L. J.,** A new bothriocephalid from *Diemictylus viridescens* with notes on the life history, *J. Parasitol.,* 14, 128, 1927.
3665. **Thomas, L. J.,** Notes on the life history of *Haplobothrium globuliforme* Cooper, a tapeworm of *Amia. calva, Anat. Rec.,* 44, 262; *J. Parasitol.,* 16, 140—145, 1929.
3666. **Thomas, L. J.,** Notes on the hatching of *Diphyllobothrium latum* eggs, *J. Parasitol.,* 16, 244—245, 1930.
3667. **Thomas, L. J.,** Notes on the life history of *Ophiotaenia saphena* from *Rana clamitans* Latr., *J. Parasitol.,* 17, 187—195, 1931.
3668. **Thomas, L. J.,** Further studies on the life cycle of a frog tapeworm, *Ophiotaenia saphena* Osler, *J. Parasitol.,* 20, 291—294, 1934.
3669. **Thomas, L. J.,** Notes on the life cycle of *Ophiotaenia perspicua*, a cestode of snakes, *Anat. Rec.,* 60, 79—80, 1934.
3670. **Thomas, L. J.,** A new source of *Diphyllobothrium* infection, *Science,* 85, 119, 1936.
3671. **Thomas, L. J.,** *Bothriocephalus rarus* n. sp., a cestode from the newt *Triturus viridescens* Raf., *J. Parasitol.,* 23, 119—132, 1937.
3672. **Thomas, L. J.,** Environmental relations and life history of the tapeworm, *Bothriocephalus rarus* Thomas, *J. Parasitol.,* 23, 133—152, 1937.
3673. **Thomas, L. J.,** On the life cycle of a tapeworm, *Diphyllobothrium* sp., from the herring gull, *Larus argentatus* Pont., *J. Parasitol.,* 24 (Suppl.), 28—29, 1938.
3674. **Thomas, L. J.,** Further studies on the life cycle of a cestode from the herring gull, *J. Parasitol.,* 25 (Suppl.), 20, 1939.
3675. **Thomas, L. J.,** The life cycle of *Ophiotaenia perspicua* La Rue, a cestode of snakes, *Rev. Med. Trop. Parasitol. Bact. Clin. Lab.,* 7, 74—78, 1941.
3676. **Thomas, L. J.,** New pseudophyllidean cestodes from the Great Lakes region. I. *Diphyllobothrium oblongatum* n. sp. from gulls, *J. Parasitol.,* 32, 1—6, 1946.
3677. **Thomas, L. J.,** Notes on the life cycle of *Schistocephalus* sp., a tapeworm from gulls, *J. Parasitol.,* 33 (Suppl.), 10, 1947.
3678. **Thomas, L. J.,** Interrelations of *Diphyllobothrium* with fish-eating birds of North Lake Michigan, *J. Parasitol.,* 35, (Sect. 2), 27, 1949.
3679. **Thomas L. J.,** *Bothriocephalus abyssmus*, a cestode from the deep-sea fish *Echistoma tanneri* (Gill) with notes on its development, *J. Parasitol.* 38, (Sect. 2, Suppl.), 23, 1952.
3680. **Thomas, L. J., and Babero, B. B.,** Some helminths of mammals from St. Lawrence Island, Alaska, with a discussion on the nomenclature of *Echinococcus* in voles, *J. Parasitol.,* 42, 500, 1956.
3681. **Thomas, L. J.,** *Bothricephalus abyssmus* n. sp., a tapeworm from the deep-sea fish, *Echiostoma tanneri* (Gill) with notes on its development, *Thapar Commemorative Volume,* 1953, pp. 269—276.
3682. **Thompson, E. W.,** Note on a tapeworm from echidna *(Taenia echidna* n. sp.), *J. R. Micr. Soc.,* 3, 297, 1893.
3683. **Thorson, R. E. and Jordan, E. M.,** A pseudophyllidean tapeworm from a dog in southeastern United States, *Proc. Helminthol. Soc. Wash.,* 21, 123—124, 1954.
3684. **Tinkle, D. P.,** Description and natural intermediate hosts of *Hymenolepis peromysci* n. sp., a new cestode from deer mice *(Peromyscus), Trans. Am. Microsc. Soc.* 91, 66—69, 1972.
3685. **Tkachev, V. A.,** A new species of cestode, *Pseudamphicotyla mamaevi* n. sp. (Pseudophyllidea: Echinophallidae) from *Seriolella* (in Russian), *Izv. Tikhookean. Nauchno-Issled. Inst. Rybn. Khoz. Okeanogr. (TINRO),* 102, 133—135, 1978.
3686. **Tkachev, V. A.,** *Amphicotyle ceratias* n. sp. (Pseudophyllidea: Amphicotylidae), a parasite of the marine fish *Ceratias holboelli* (in Russian), *Parazitologiya,* 13, 549—552, 1979.
3687. **Tkachev, V. A.,** A new cestode species, *Amphicotyle kurochkini* n. sp. (Pseudophyllidea; Amphocotylidae), a parasite of the marine fish *Seriolella* sp. (in Russian), *Mater. Nauchn. Konf. Vses. Ova. Gel'mintol. (Tsestody i tsestodozy),* 31, 142—146, 1979.

3688. **Tkachev, V. A.,** A new cestode species, *Paraechinophallus hyperoglyphe* from the marine fish, *Hyperoglyphe japaniia, Biol. Morya Vladivostok,* No. 5, 81—82, 1979.
3689. **Todd, K. S., Jr.,** *Aploparaksis picae* sp. n. (Cestoda, Hymenolepididae) from the black-billed magpie, *Pica pica hudsonia* (Sabine, 1823), *J. Parasitol.,* 53, 350—351, 1967.
3690. **Tokobaev, M. M.,** Helminth fauna of rodents in the Kirgiz SSR (in Russian), *Tr. Inst. Zool. Parazit. Akad. Nauk Kirgizskoi SSR,* 7, 133—142, 1959.
3691. **Tolkacheva, L. M.,** *Microsomacanthus spasskii* n. sp. (Cyclophyllidea, Hymenolepididae) from Anseriformes (in Russian), *Tr. Gel'mintol. Lab.,* 15, 167—171, 1965.
3692. **Tolkacheva, L. M.,** New cestode species, *Echinatrium clanguli* n. sp. and *Microsomacanthus strictophallus* n. sp. (Hymenolepididae), from Anseriformes, *Sbornik Rabot po Gel'mintologii Posuyeshchev po-letiyu so duya rozhdemiya Akademika K. I. Skrjabina,* Izdatel'stvo KOLOS, Moscow, pp. 406—410, 1971.
3693. **Tolkacheva, L. M.,** *Echinoshipleya semipalmati* nov. gen. et sp. — new dioeceous cestode of shorebird (in Russian), *Helminths Anim. Plants,* 29, 149—154, 1979.
3694. **Tonn, R. J.,** The white sucker, *Catostomus commersonii* (Lacépède) a new host of *Caryophyllaeus terebrans* (Linton, 1893), *J. Parasitol.,* 41, 219, 1955.
3695. **Tosh, J. R.,** Internal parasites of the Tweed salmon, chiefly marine in character, *Ann. Nat. Hist.,* Ser. 7, 16, 115—119, 1905.
3696. **Tower, W. L.,** The nervous system in the cestode, *Moniezia expansa, Zool. Jahrb. Abt. Anat.,* 13, 359—381, 1900.
3697. **Tretiakova, O. N.,** Two new helminths of birds of the Cheliabinsk oblast — *Philophthalmus muraschkinzewi* and *Tatria skrjabini* n. sp. (in Russian), *Sb. Rab. Gel'mintol. (40 Nauchn. Deiat. Skrjabin), pp.* 232—236, 1948.
3698. **Troncy, P. M.,** Nouvelles observations sur les parasites des poissons du bassin tchadien, *Bull. Inst. Fondamental Afr. Noire,* 40, 528—552, 1978.
3699. **Trowbridge, A. H. and Hefley, H. M.,** Preliminary studies on the parasite fauna of Oklahoma anurans, *Proc. Okla. Acad. Sci.,* 14, 16—19, 1934.
3700. **Tschertkova, A. N. and Kosupko, G. A.,** *Essentials of Cestodology,* Vol. 9, *Order Mesocestoidata Skrjabin, 1940,* Akademiya Nauk SSSR, Moscow, 1978, 155—229.
3701. **Tseng, Shen,** Etude sur les cestodes d'oiseaux de Chine, *Ann. Parasitol.,* 10, 105—128, 1932.
3702. **Tseng, Shen,** Studies on avian cestodes from China. I. Cestodes from charadriiform birds, *Parasitology,* 24, 87—106, 1932.
3703. **Tseng, Shen,** Studies on avian cestodes from China. II. Cestodes from charadriiform birds, *Parasitology,* 24, 500—511, 1933.
3704. **Tseng, Shen,** Study on some cestodes from fishes, *J. Sci. Nat. Univ. Shantung Tsingtao, China,* 2, 1—21, 1933.
3705. **Tsimbalyuk, A. K., Andronova, K. Ya., and Kulikov, V. V.,** *Monorcholepis sobolevi* n. sp. (Hymenolepididae) and *Similuncinus leonovi* n. sp. (Choanotaenidae) from birds on islands of the Bering Sea (in Russian), *Soobshsch. Dal'nevost. Fil. V. L. Komarova Sib. Otd. Akad. Nauk SSSR,* No. 26, 126—131, 1968.
3706. **Tsuchiya, H. and Rohlfing, E. H.,** *Hymenolepis nana,* report of additional cases and an experimental transmission from man to rat, *Am. J. Dis. Children,* 43, 865—872, 1932.
3707. **Tsunoo, S., Yokota, S., and Morokuma, M.,** Ein kasuistischer Beitrag zu *Diplogonoporus grandis, Nagasaki Igakkwai Zasshi,* 12, 1200—1202, 1934.
3708. **Tubangui, M. A.,** Metazoan parasites of Philippine domesticated animals, *Philipp. J. Si.,* 28, 11—37, 1925.
3709. **Tubangui, M. A.,** Worm parasites of the brown rat *(Mus norvegicus)* in the Philippine Islands, with special reference to those forms that may be transmitted to human beings, *Philipp. J. Sci.,* 46, 531—591, 1931.
3710. **Tubangui, M. A.,** Pseudophyllidean cestodes occurring in the Philippines, *Livro Jubilar Lauro Travassos, 1938,* Rio de Janeiro, 1938, 489—494.
3711. **Tubangui, M. A. and Masiluñgan, V. A.,** *Oochoristica excelsa,* a new reptilian cestode, *Philipp. J. Sci.,* 61, 75—79, 1936.
3712. **Tubangui, M. A. and Masiluñgan, V. A.,** Tapeworm parasites of Philippine birds, *Philipp. J. Sci.,* 62, 409—438, 1937.
3713. **Turner, M.,** On a coenurus in the rat, *Ann. App. Biol.,* 6, 136—141, 1919.
3714. **Turner, M. and Leiper, R. T.,** On the occurrence of *Ceonurus glomeratus* in man in West Africa, *Trans. R. Soc. Trop. Med. Hyg.,* 13, 23—24, 1919.
3715. **Udekem, J. de,** Notices sur deux nouvelles espèces de scolex, *Bull. Acad. R. Belg.,* 22, 528—533, 1855.
3716. **Ukoli, F. M. A.,** Three cestodes from the families Diploposthidae Poche, 1926, Dioecocestidae Southwell, 1933, and Progynotaeniidae Fuhrmann, 1936 found in the black-winged stilt, *Himantopus himantopus himantopus* (Linn., 1758) in Ghana, *J. Helminthol.,* 39, 383—398, 1965.
3717. **Ukoli, F. M. A.,** Some dilepidid and hymenolepidid cestodes from birds in Ghana, *J. West Afr. Sci. Assoc.,* 12, 65—93, 1967.

3718. **Ukoli, F. M. A.**, Occurrence, morphology and systematics of caryophyllaeid cestodes of the genus *Wenyonia* Woodland, 1923 from fishes in River Niger, Nigeria, *J. West Afr. Sci. Assoc.*, 17, 49—67, 1972.
3719. **Ulmer, M. J. and James, H. A.**, *Nematotaenoides ranae* gen. et sp. n. (Cyclophyllidea: Nematotaeniidae), from the leopard frog *(Rana pipiens)* in Iowa, *Proc. Helminthol. Soc. Wash.*, 43, 185—191, 1976.
3720. **Ulmer, M. J. and James, H. A.**, Studies on the helminth fauna of Iowa. II. Cestodes of amphibians, *Proc. Helminthol. Soc. Wash.*, 43, 191—200, 1976.
3721. **Uzhakhov, D. I.**, *Catenotaenia pigulevski* n. sp. in Muridae in the Dagestan SSR (in Russian), *Uchen. Zap. Checheno-Ingush. Gos. Pedagog. Univ.*, 23, 21—25, 1964.
3722. **Vaillant, L.**, Sur deux helminthes cestoides de la genett, *Inst. Paris*, 31, 87—88, 1863.
3723. **Vaillant, L.**, Note sur les hydatides développées chez un oiseau et des vers cestoides trouvés chez la genette ordinaire, *C. R. Soc. Biol.*, 5, 48, 1864.
3724. **Van Cleave, H. J.**, Ctenophores as the host of a cestode, *Trans. Am. Microsc. Soc.*, 46, 214—215, 1927.
3725. **Van Cleave, H. J.**, An index to the International Rules of Zoological Nomenclature, *Trans. Am. Microsc. Soc.*, 46, 322—325, 1933.
3726. **Van Cleave, H. J.**, Some interesting pre-Linnean names, *Trans. Ill. Acad. Sci.*, 28, 263—265, 1935.
3727. **Van Cleave, H. J., and Mueller, J. F.**, Parasites of the Oneida Lake fishes, I. Descriptions of new genera and new species, *Roosevelt Wild Life Ann.* 3, 1—71, 1932.
3728. **Van Cleave, H. J. and Mueller, J. F.**, Parasites of Oneida Lake fishes, III. A biological and ecological survey of the worm parasites, *Roosevelt Wild Life Ann.*, 3, 161—334, 1934.
3729. **Van Gundy, C. O.**, *Hymenolepis anthocephalus*, a new tapeworm from the mole shrew, *Blarina brevicauda* Say, *Trans. Am. Microsc. Soc.*, 54, 240—244, 1935.
3730. **Vannucci Mendes, M.**, Sobre a larva de *Dibothriorhynchus dinoi* sp. n. parasita dos Rhizostomata (Cest. Tetrarhynchidea), *Arq. Mus. Paranaense*, 4, 47—82, 1954.
3731. **Van Volkenburg, H. L.**, Animal parasitology, *Rep. Puerto Rico Agric. Exp. Stn. 1937*, pp. 103—104, 1938.
3732. **Vasilev, I.**, A new cestode species — *Raillietina (R.) carneostrobilata* n. sp. from turkey and pheasant, *C. R. Acad. Biol. Sci.*, 20, 855—858, 1967.
3733. **Vaucher, C.**, Les cestodes parasites des Soricidae d'Europe. Etude anatomique, révision taxonomique et biologie, *Rev. Swisse Zool.*, 78, 1—113, 1971.
3734. **Vaucher, C.**, Cestodes parasites de chiroptères en Amérique de Sud: révision de *Hymenolepis elongatus* (Rego, 1962) et description de *Hymenolepis phyllostomi* n. sp., *Rev. Swisse Zool.*, 89, 451—459, 1982.
3735. **Vaucher, C. and Tenora, F.**, Sur trois *Hymenolepis* (Cestoda) parasites de *Suncus murinus* L. en Afganistan, *Acta Univ. Agric. Fac. Agron. Brně*, 19, 337—341, 1971.
3736. **Vaullegeard, A.**, Métamorphose et migration de *Tetrarhynchus ruficollis* (Eisenhardt), *Bull. Soc. Linn. Normandie*, 8, 112—143, 1895.
3737. **Vaullegeard, A.**, Recherches sur les Tétrarhynques, *Mem. Soc. Linn. Normandie*, 19, 187—376, 1899.
3738. **Velikova, V. P.**, *Diplopylidium polyacantha* n. sp. (Cestoda, Dipylididae), a new cestode of carnivores (in Russian), *Vestnik Zool.*, 1, 20—24, 1982.
3739. **Venard, C. E.**, Morphology, bionomics and taxonomy of the cestode *Dipylidium caninum*, *Ann. N.Y. Acad. Sci.*, 37, 273—328, 1938.
3740. **Venard, C. E.**, Studies on parasites of Reelfoot Lake fish. I. Parasites of the large-mouthed black bass, *Huro salmoides* (Lacépède), *J. Tenn. Acad. Sci.*, 15, 43—63, 1940.
3741. **Venard, C. E. and Ellis, P. L.**, On cestodes from dogs, *Vet. Alumni Q.* (Ohio State University), 21, 20—23, 1933.
3742. **Vergeer, T.**, *Diphyllobothrium latum* (Linn., 1758), the broad tapeworm of man; experimental studies, *JAMA*, 90, 673—678, 1928.
3743. **Vergeer, T.**, Canadian fish, a source of the broad tapeworm of man in the United States, *JAMA*, 90, 1687—1688, 1928.
3744. **Vergeer, T.**, An important source of broad tapeworm in America, *Science*, 68, 14—15, 1928.
3745. **Vergeer, T.**, New sources of broad tapeworm infestation, *JAMA*, 91, 396—397, 1928.
3746. **Vergeer, T.**, Dissemination of the broad tapeworm by wild carnivora, *Can. Med. Assoc. J.*, 19, 692—694, 1928.
3747. **Vergeer, T.**, The broad tapeworm in America, *J. Infect. Dis.*, 44, 1—12, 1929.
3748. **Vergeer, T.**, The dog, a reservoir of the broad tapeworm, *JAMA*, 92, 607—608, 1929.
3749. **Vergeer, T.**, Causes underlying increased incidence of broad tapeworm in man in North America, *JAMA*, 95, 1579—1581, 1930.
3750. **Vergeer, T.**, *Diphyllobothrium laruei* sp. nov. and *Sparganum pseudosegmentatum* sp. nov., two cestodes from the Great Lakes region, *Anat. Rec.*, 60 (Suppl.), 77, 1934.
3751. **Vergeer, T.**, The origin of the genus *Diplogonoporus* Loennberg, 1892, *J. Parasitol.*, 21, 133—135, 1935.
3752. **Vergeer, T.**, The eggs and coracidia of *Diphyllobothrium latum*, *Pap. Mich. Acad. Sci. Arts Lett.*, 21, 715—726, 1936.

3753. **Vergeer, T.,** Two new pseudophyllidean tapeworms of general distribution in the Great Lakes region, *Trans. Am. Microsc. Soc.,* 61, 373—382, 1942.
3754. **Verma, S. C.,** On a new proteocephalid cestode from an Indian freshwater fish, *Allahabad Univ. Stud.,* 2, 353—362, 1926.
3755. **Verma, S. C.,** Some cestodes from Indian fishes including four new species of Tetraphyllidea and revised keys to the genera *Acanthobothrium* and *Gangesia, Allahabad Univ. Stud.,* 4, 119—176, 1928.
3756. **Verma, S. L.,** Helminth parasites of fresh water fishes. I. On two caryophyllaeids from fresh water fishes of Lucknow, *Ind. J. Helminthol.,* 23, 71—80, 1971.
3757. **Verster, A.,** A taxonomic revision of the genus *Taenia* Linnaeus, 1758, s. str., *Onderstepoort J. Vet. Res.,* 36, 3—58, 1969.
3758. **Vevers, G. M.,** On the cestode parasites from mammalian hosts which died in the Gardens of the Zoological Society of London during the years 1919—1921; with a description of a new species of *Cyclorchida, Proc. Zool. Soc. London (1922),* Part 4, 921—928, 1923.
3759. **Vialli, M.,** *Sparganum lanceolatum* Molin forma larvale di *Diphyllobothrium ranarum* Gastaldi, *Monit. Zool. Ital.,* 40, 90—94, 1929.
3760. **Vialli, M.,** Nota sinonimica sui botriocefali dei ricci, *Monit. Zool. Ital.,* 42, 209—212, 1931.
3761. **Vigener, I.,** Ueber dreikantige Bandwürmer aus der Familie der Taeniiden, *Jahrb. Nassa. Ver. Naturk.,* 56, 113—177, 1903.
3762. **Vigueras, I. P.,** Sobre la presencia en Cuba, de *Diphyllobothrium mansoni* (Cobbold), *Mem. Soc. Cubana Hist. Nat.,* 8, 351—352, 1934.
3763. **Vigueras, I. P.,** *Ophiotaenia barbouri* n. sp. (Cestoda), parasito de *Tretanorhynus variabilis* (Reptilia), *Mem. Soc. Cubana Hist. Nat.,* 8, 231—234, 1934.
3764. **Vigueras, I. P.,** *Proteocephalus manjuariphilus* n. sp. (Cestoda) parasito de *Atractosteus tristoechus* (Bloch et Schn.) (Pisces), *Rev. Parasitol. Clin. Lab.,* 2, 17—18, 1936.
3765. **Vigueras, I. P.,** Note sobre varios vermes encontrados en al "flamenco" *(Phoenicopterus ruber), Mem. Soc. Cubana Hist. Nat.,* 15, 327—336, 1941.
3766. **Vigueras, I. P.,** Nota sobre *Hymenolepis chiropterophila* n. sp., y clave para la determinación de *Hymenolepis* de Chiroptera, *Rev. Univ. Habana,* 6, 152—163, 1941.
3767. **Vigueras, I. P.,** Un genero y cinco especies nuevas de helmintos Cubanos, *Rev. Univ. Habana,* 8, 315—356, 1943.
3768. **Vijayakumaran, N. K. and Nadakal, A. M.,** *Raillietina (Paroniella) nedumanyadensis* from pigeon, *Columba livia domestica, Jpn. J. of Parasitol.,* 30, 241—244, 1981.
3769. **Vik, R.,** Investigations on the pseudophyllidean cestodes of fish, birds, and mammals in the Anya water system in Trøndelag. I. *Cyathocephalus truncatus* and *Schistocephalus solidus, Nytt Mag. Zool.,* 2, 5—51, 1954.
3770. **Vik, R.,** Studies of the helminth fauna of Norway. I. Taxonomy and ecology of *Diphyllobothrium norvegicum* n. sp. and the plerocercoid of *Diphyllobothrium latum* (L.), *Nytt Mag. Zool.,* 5, 25—93, 1957.
3771. **Viljoen, N. F.,** Cysticercosis in swine and bovines, with special reference to South African conditions, *Onderstepoort J. Vet. Sci. Anim. Ind.,* 9, 337—570, 1937.
3772. **Villot, F. C. A.,** Recherches sur les helminthes libres ou parasites des côtes de la Bretagne, *Arch. Zool. Exp. Gen.,* 4, 451—482, 1875.
3773. **Villot, F. C. A.,** *Classification des Cystiques des Ténias Fondée sur les Divers Modes de Formation de la Vésicule Caudale,* Montpellier, 1882, 1—9.
3774. **Villot, F. C. A.,** Mémoire sur les cystiques des ténias, *Ann. Sci. Nat. Zool.,* 15, 1—61, 1883.
3775. **Vlasenko, P. V.,** Zur Parasitenfauna der Schwarzmeerfische (in Russian; German summary), *Tr. Karadag Biol. Stantsii,* 4, 88—136, 1931.
3776. **Voelker, J.,** Zwei neue Cestoden de Gattung *Inermicapsifer* in *Procavia capensis welwitchii* Gray, *Z. Tropenmed. Parasitol.,* 11, 316—324, 1960.
3777. **Voge, M.,** A new anoplocephalid cestode, *Andrya neotomae,* from the wood rat, *Neotoma fuscipes, J. Parasitol.,* 32, 36—39, 1946.
3778. **Voge, M.,** A new anoplocephalid cestode, *Paranoplocephala kirbyi* from *Microtus californicus californicus, Trans. Am. Microsc. Soc.,* 67, 299—303, 1949.
3779. **Voge, M.,** New rodent hosts for *Catenotaenia linsdalei,* with an additional description of this cestode, *Trans. Am. Microsc. Soc.,* 67, 266—267, 1949.
3780. **Voge, M.,** *Mesogyna hepatica* n. g., n. sp. (Cestoda: Cyclophyllida) from kitfox, *Vulpes macrotis, Trans. Am. Microsc. Soc.,* 71, 350—354, 1952.
3781. **Voge, M.,** Variability of *Hymenolepis diminuta* in the laboratory rat and in the ground squirrel, *Citellus leucurus, J. Parasitol.,* 38, 454—456, 1952.
3782. **Voge, M.,** Variation in some unarmed Hymenolepididae (Cestoda) from rodents, *Univ. Calif. Publ. Zool.,* 57, 1—52, 1952.
3783. **Voge, M.,** *Hymenolepis parvissima* n. sp., a minute cestode from the shrew *Sorex bendirei bendirei* (Merriam) in California, *J. Parasitol.,* 39, 599—602, 1953.

3784. **Voge, M.,** New host records for *Mesocestoides* (Cestoda: Cyclophyllidea) in California, *Am. Midl. Nat.,* 49, 249—251, 1953.
3785. **Voge, M.,** Exogenous proliferation in a larval taeniid (Cestoda: Cyclophyllidea) from the body cavity of Peruvian cestodes, *J. Parasitol.,* 39 (Sect. 2, Suppl.), 32, 1953.
3786. **Voge, M.,** *Oochoristica antrozoi* n. sp., a tapeworm from the pallid bat in California, *Trans. Am. Microsc. Soc.,* 73, 404—407, 1954.
3787. **Voge, M.,** Notes on four hymenolepidid cestodes from shrews, *J. Parasitol.,* 41, 74—76, 1955.
3788. **Voge, M.,** *Hymenolepis virilis* n. sp., a cestode from the shrew *Sorex trowbridgei* in California, *J. Parasitol.,* 41, 270—272, 1955.
3789. **Voge, M.,** *Hymenolepis pulchra* n. sp., a cestode from the shrew *Sorex trowbridgei* in California, *Proc. Helminthol. Soc. Wash.,* 22, 90—92, 1955.
3790. **Voge, M.,** Experimental demonstration of the life cycle of *Hymenolepis citelli* (McLeod, 1933) (Cestoda: Cyclophyllidea), *J. Parasitol.,* 41 (Sect. 2), 31, 1955.
3791. **Voge, M.,** Studies on the life history of *Hymenolepis citelli* (McLeod, 1933) (Cestoda: Cyclophyllidea), *J. Parasitol.,* 42, 485—489, 1956.
3792. **Voge, M.,** Systematics of cestodes — present and future, in *Problems in Systematics of Parasites,* Schmidt, G. D., Ed., University Park Press, Baltimore, 1969, 49—72.
3793. **Voge, M. and Fox, W.,** A new anoplocephalid cestode, *Oochoristica scelopori* n. sp., from the Pacific fence lizard, *Sceloporus occidentalis occidentalis, Trans. Am. Microsc. Soc.,* 69, 236—242, 1950.
3794. **Voge, M. and Heyneman, D.,** Development of *Hymenolepis nana* and *Hymenolepis diminuta* (Cestoda: Hymenolepididae) in the intermediate host *Tribolium confusum, Univ. Calif. Publ. Zool.,* 59, 549—580, 1957.
3795. **Voge, M. and Rausch, R.,** Occurrence and distribution of hymenolepidid cestodes in shrews, *J. Parasitol.,* 41, 566—573, 1955.
3796. **Voge, M. and Read, C. P.,** *Diplophallus andinus* n. sp. and *Monoecocestus rheiphilus* n. sp., avian cestodes from the high Andes, *J. Parasitol.,* 39, 558—567, 1953
3797. **Voge, M. and Read, C. P.,** A new record of the cestode *Infula burhini* Burt (Cyclophyllidea: Dioecocestidae) from Australia, *J. Parasitol.* 40, 483, 1954.
3798. **Voge, M. and Read C. P.,** A description of *Parafimbriaria websteri* n. g., n. sp., a cestode from grebes, and notes on three species of *Hymenolepis, J. Parasitol.* 40 (Sect. 1), 564—570, 1954.
3799. **Vogel, H.,** Helminthologische beobachtungen in Ostpreussen, insbesondere ueber *Dibothriocephalus latus* and *Opisthorchis felineus, Dtsch. Med. Wochenschr.,* 55, 1631—1633, 1929.
3800. **Vogel, H.,** Beobachtungen ueber *Dibothriocephalus latus, Arch. Schiffs-u. Tropenhyg.,* 33, 164—168, 1929.
3801. **Vogel, H.,** Studien zur Entwicklung von *Diphyllobothrium.* I. Teil, *Z. Parasitenkd.,* 2, 213—222, 1929.
3802. **Vogel, H.,** Studien zur Entwicklung von *Diphyllobothrium.* II. Teil, *Z. Parasitenkd.,* 2, 629—644, 1930.
3803. **Vogel, R.,** Ein Cysticercus des regen wurmes als Jugendform der Vogeltaenie, *Dilepis undula* (Schrank), *Centralbl. Bakteriol.,* 85, 370—372, 1921.
3804. **Volz, W.,** Die Cestoden der einheimischen Corviden, *Zool. Anz.,* 22, 265—268, 1899.
3805. **Volz, W.,** Beitrag zur Kenntnis einiger Vogelcestoden. Inaug. Diss., Basel, *Arch. Naturg.,* 66, 1, 115—174, 1900.
3806. **Vosgien, Y.,** Le *Cysticercus cellulosae* chez l'Homme et chez les Animaux. Thèse de la Faculté de Medicine, Paris, *Bull. Soc. Centr. Med. Vét.,* 66, 270—275, 1911.
3807. **Vukovic, V. and Varenika, D.,** *Tetrathyridium bailleti* in the pine martin *(Mustela martes), Vet. Sarajevo,* 3, 651—652, 1954.
3808. **Waele, A. de,** Nieuwe bevindingen over der levenscylus der Cestoden, *Natuurwiss. Tijdschr. (Belg.),* 16, 60—69, 1934.
3809. **Waele, A. de,** Le mécanisme physiologique des migrations et de la spécificité chez les cestodes, *C. R. XII Congr. Int. Zool.,* Lisbon, 1935, 1936, 312—328.
3810. **Waele, A. de and Dedeken, L.,** Le phénomène de l'evagination chez *Cysticercus bovis* et la migration du parasite chez l'homme, *Mém. Mus. Roy. Hist. Nat. Belg.,* Ser. 2, 369—373, 1936.
3811. **Wagener, G. R.,** Enthelminthica, Dissertation, Berolini, 1848, 31 pp.
3812. **Wagener, G. R.,** Enthelminthica. I. Ueber Tetrarhynchus, *Arch. Anat. Physiol. Wiss. Med.,* pp. 211—220, 1851.
3813. **Wagener, G. R.,** Enthelminthica. III. Ueber einen neuen in der *Chimaera monstrosa* gefundenen Eingeweideworm, *Amphiptyches urna* Gr. et Wag, *Arch. Anat. Physiol. Wiss.* pp. 543—554, 1852.
3814. **Wagener, G. R.,** Die Entwicklung der Cestoden, *Tagesberichte ueber die Fortschritte der Natur. Heilkunde,* 3 (Abt. Zool.), 65—71, 1852.
3815. **Wagener, G. R.,** Die Entwicklung der Cestoden nach eigenen Untersuchungen, *Verhandl. kaiserl. Leopold Carol. Akad. Naturforscher Nova Acta Akad. Naturwiss. Curios,* 24 (Suppl.), 1—91, 1854.
3816. **Wagener, G. R.,** Beiträge zur Entwicklungsgeschichte der Eingeweidewürmer, *Natuurk. Verhandel. Hollandsche Maatschap Wetenshap. Haarlem II Verzam,* XIII, Deel., 1—112, 1857.

3817. **Wagener, G. R.,** Enthelminthica. V. Ueber *Amphilina foliacea* mihi *(M. foliaceum* Rud.), *Gyrocotyle* Diesing und *Amphiptyches* Gr. W., *Arch. Naturg.,* 24, 244—249, 1858.
3818. **Wagin, W.,** Zur Frage der Helminthfauna der Pinnipedia (in Russian with German summary), *Trans. Arctic Inst. Leningrad,* 3, 51—62, 1933.
3819. **Wagner, E. D.,** A new species of *Proteocephalus* Weinland, 1858 (Cestoda) with notes on its life history, *Trans. Am. Microsc. Soc.,* 72, 364—369, 1953.
3820. **Wagner, E. D.,** The life history of *Proteocephalus tumidocollus* Wagner 1953 (Cestoda) in rainbow trout, *J. Parasitol.,* 40, Sect. 1, 489—498, 1954.
3821. **Wagner, E. D.,** Morphology and Natural History of *Proteocephalus tumidocollus* n. sp. (Cestoda) from the Rainbow Trout, Thesis, University of Southern California, Los Angeles, 1954.
3822. **Wagner, E. D.,** Morphology and natural history of *Proteocephalus tumidocollus* n. sp. (Cestoda) from the rainbow trout, *Abstr. Diss. Univ. South. Calif. (1954),* pp. 82—84, 1955.
3823. **Wagner, O.,** Ueber den Entwicklungsgang einer Fischtaenie, *Zool. Anz.,* 46, 70—75, 1915.
3824. **Wagner, O.,** Uener die Taenien der Süsswasserfische, *Naturwiss. Wochenschr.,* 31, 15, 421—423, 1916.
3825. **Wagner, O.,** Ueber Entwicklungsgang und Bau einer Fischtaenie *(Ichthyotaenia torulosa* Batsch), Jena. *Z. Naturwiss.,* 55, 1—66, 1917.
3826. **Wahid, S.,** Systematic studies on some cestodes of reptiles and birds, *J. Helminthol.,* 35, 169—180, 1961.
3827. **Waitz, J. A. and Mehra, K. N.,** *Baerietta idahoensis* n. sp. a nematotaeniid cestode from the intestine of *Plethodon vandykei idahoensis* from northern Idaho, *J. Parasitol.,* 47, 806—808, 1961.
3828. **Wallace, F. G.,** Parasites collected from the moose, *Alces americanus,* in northern Minnesota, *J. Am. Vet. Med. Assoc.,* 84, 770—775, 1934.
3829. **Wallace, J. C. and Grant, M.,** Infestation by broad tapeworm, *JAMA,* 78, 1050, 1922.
3830. **Walter, H.,** Helminthologische Studien. VII, *Ber. Offenbacher Ver. Naturk. (1865—1866),* pp. 51—79, 1866.
3831. **Walter, H.,** Nachträgliche Mittheilung zu. "Helminthologische Studien". VII, *Ber. Offenbacher Ver. Naturk. (1865—1866),* pp. 133—134, 1866.
3832. **Walton, A. C.,** The Cestoda as parasites of Amphibia, *Contrib. Biol. Lab. Knox College,* No. 64, 1—31, 1939.
3833. **Walton, A. C.,** Notes on some helminths from California Amphibia, *Trans. Am. Microsc. Soc.,* 60, 53—57, 1941.
3834. **Wantland, W. W.,** *Cysticercus fasciolaris* in the wild rat and the development of this strobilocercus in the white rat, *J. Parasitol.,* 39 (Sect. 2), 29; *Cysticercus fasciolaris* in the Syrian hamster, *J. Parasitol.,* p. 29, 1953.
3835. **Ward, H. B.,** A preliminary report on the worms (mostly parasitic) collected in Lake St. Clair, in the summer of 1893, *Mich. Fish. Comm. Bull.,* 4, 49—54, 1894.
3836. **Ward, H. B.,** Some notes on the biological relations of the fish parasites of the Great Lakes, *Proc. Neb. Acad. Sci.,* 4, 8—11, 1894.
3837. **Ward, H. B.,** The parasitic worms of man and the domestic animals, *Annu. Rep. Neb. Bd. Agric. (1894),* pp. 225—348, 1895.
3838. **Ward, H. B.,** A new human tapeworm *(Taenia confusa* n. sp.), *West. Med. Rev.,* 1, 35—36, 1896.
3839. **Ward, H. B.,** The parasitic worms of domesticated birds, *Stud. Zool. Lab. Univ. Neb.,* pp. 1—18, 1898.
3840. **Ward, H. B.,** Internal parasites of Nebraska birds, *Proc. Neb. Ornithol. Union,* 2, 63—70, 1901.
3841. **Ward, H. B.,** A new bothriocephalid parasite of man, *Science,* n. s., 23, 258, 1906.
3842. **Ward, H. B.,** Some points in the migration of Pacific salmon as shown by its parasites, *Trans. Am. Fish. Soc.,* 37, 92—100, 1908.
3843. **Ward, H. B.,** Internal parasites of the Sebago salmon, *Bull. U.S. Bur. Fish. (1908),* 28, 1151—1194, 1910.
3844. **Ward, H. B.,** The discovery of *Archigetes* in America, with a discussion of its structure and affinities, *Science,* n. s., 33, 272—273, 1911.
3845. **Ward, H. B.,** Means for the accurate determination of human internal parasites, *Ill. Med. J.,* 22, 417—434, 1912.
3846. **Ward, H. B.,** The distribution and frequence of animal parasites and parasitic diseases in North American freshwater fishes, *Trans. Am. Fish. Soc. 1911,* pp. 207—241, 1912.
3847. **Ward, H. B.,** Some points in the general anatomy of *Gyrocotyle, Zool. Jahr. Suppl. 15, Festschr. 60, Geburtstag J. W. Spengel,* pp. 717—738, 1912.
3848. **Ward, H. B.,** Cestoda, *Buck's Ref. Handb. Med. Sci.,* 2, 761—780, 1913.
3849. **Ward, H. B.,** Intestinal parasites in children, *Arch. Pediatr.,* 33, 116—123, 1916.
3850. **Ward, H. B.,** On the structure and classification of North American parasitic worms, *J. Parasitol.,* 4, 1—12, 1917.
3851. **Ward, H. B.,** A study on the life history of the broad fish tapeworm in North America, *Science,* n.s. (1904), 66, 197—199, 1927.

3852. **Ward, H. B.,** Studies on the broad fish tapeworm in Minnesota, *JAMA*, 92, 389—390, 1929.
3853. **Ward, H. B.,** The introduction and spread of the fish tapeworm *(Diphyllobothrium latum)* in the United States, *De Lamar Lectures 1929—1930*, Baltimore, 1930, 36 pp.
3854. **Ward, H. B.,** The longevity of *Diphyllobothrium latum*, Recueil des Travaux Dédiés au 25me Anniversaire Scientifique de Professeur Eugene Pavlovsky, 1909—1934, Moscow, 1935, 288—294.
3855. **Wardle, R. A.,** The Cestoda of Canadian fishes. I. The Pacific coast region, *Contrib. Can. Biol. Fish*, n.s., 7, 221—243, 1932.
3856. **Wardle, R. A.,** The Cestoda of Canadian fishes. II. The Hudson Bay drainage system, *Contrib. Can. Biol. Fish.*, n.s., 7, 377—403, 1932.
3857. **Wardle, R. A.,** On the technique of tapeworm study, *Parasitology*, 24, 241—252, 1932.
3858. **Wardle, R. A.,** The limitations of metromorphic characters in the differentiation of Cestoda, *Trans. R. Soc. Canada*, 26, 193—204, 1932.
3859. **Wardle, R. A.,** Significant factors in the plerocercoid environment of *Diphyllobothrium latum*, *J. Helminthol.*, 11, 25—44, 1932.
3860. **Wardle, R. A.,** The parasitic helminths of Canadian animals. I. The Cestodaria and Cestoda, *Can. J. Res.*, 8, 317—333, 1933.
3861. **Wardle, R. A.,** The viability of tapeworm in artificial media, *Physiol. Zool.*, 7, 36—61, 1934.
3862. **Wardle, R. A.,** The Cestoda of Canadian fishes. III. Additions to the Pacific coastal fauna, *Contrib. Can. Biol. Fish.*, 8, 77—87, 1935.
3863. **Wardle, R. A.,** Fish tapeworm, *Biol. Board Can. Bull.*, 45, 1—25, 1935.
3864. **Wardle, R. A.,** The physiology of the sheep tapeworm, *Moniezia expansa*, *Can. J. Res.*, 15, 117—126, 1937.
3865. **Wardle, R. A.,** The physiology of tapeworms, in Manitoba Essays, 60th Anniversary Commemoration Volume, University of Manitoba, Winnipeg, 1937, 8—361.
3866. **Wardle, R. A.,** The distribution of tapeworms in the Pacific area and the conclusions to be drawn therefrom, *Ann. Trop. Med. Parasitol.*, 45, 122—126, 1951.
3867. **Wardle, R. A., Gotschall, M. J., and Horder, L. J.,** The influence of *Diphyllobothrium latum* infestation upon dogs, *Trans. R. Soc. Can.*, 31, 59—69, 1937.
3868. **Wardle, R. A. and Green, N. K.,** The rate of growth of the tapeworm, *Diphyllobothrium latum* (L.), *Can. J. Res.*, 19, 245—251, 1941.
3869. **Wardle, R. A. and McColl, E. L.,** The taxonomy of *Diphyllobothrium latum* (L.) in western Canada, *Can. J. Res.*, 15, 163—175, 1937.
3870. **Wardle, R. A. and McLeod, J. A.,** *The Zoology of Tapeworms*, University of Minnesota Press, Minneapolis, 1952, 780 pp.
3871. **Wardle, R. A., McLeod, J. A., and Radinovsky, S.,** *Advances in the Zoology of Tapeworms, 1950—1970*, University of Minnesota Press, Minneapolis, 1974, 274 pp.
3872. **Wardle, R. A., McLeod, J. A., and Stewart, I. E.,** Lühe's "*Diphyllobothrium*" (Cestoda), *J. Parasitol.*, 33, 319—330, 1947.
3873. **Warthin, A. S.,** Increasing human incidence of broad tapeworm infestation in the Great Lakes region, *JAMA*, 90, 2080—2082, 1928.
3874. **Wason, A. and Johnson, S.,** A new genus of hymenolepid cestodes from the Indian gerbil, *Tatera indica*, *J. Helminthol.*, 51, 309—312, 1977.
3875. **Watanabe, S., Nagayama, F., and Saito, J.,** Observations on the intermediate host of fowl cestode, *Raillietina (Skrjabinia) cesticillus* (Molin, 1858), *Rep. Gov. Exp. Stn. Anim. Hyg.*, 27, 277—287, 1953.
3876. **Watson, D. E. and Thorson, T. B.,** Helminths from elasmobranchs in Central American fresh waters, in *Investigations of the Ichthyofauna of Nicaraguan Lakes*, Thorson, T. B., Ed., University of Nebraska Press, Lincoln, 1976, 629—640.
3877. **Watson, E. E.,** The genus *Gyrocotyle* and its significance for problems of cestode structure and phylogeny, *Univ. Calif. Publ. Zool.*, 6, 353—468, 1911.
3878. **Webster, J. D.,** The type of *Gyrocoelia milligani* Linton, 1927, *J. Parasitol.*, 28, 230, 1942.
3879. **Webster, J. D.,** A revision of the Fimbriariinae (Cestoda, Hymenolepididae), *Trans. Am. Microsc. Soc.*, 62, 390—397, 1943.
3880. **Webster, J. D.,** A new cestode from the bobwhite, *Trans. Am. Microsc. Soc.* 63, 44—45, 1944.
3881. **Webster, J. D.,** Helminths from the bobwhite in Texas, with descriptions of two new cestodes, *Trans. Am. Microsc. Soc.*, 66, 339—343, 1944.
3882. **Webster, J. D.,** Two cestodes from a nighthawk, *J. Parasitol.* 34, 93—95, 1948.
3883. **Webster, J. D.,** Fragmentary studies on the life history of the cestode *Mesocestoides latus*, *J. Parasitol.* 34, 83—90, 1949.
3884. **Webster, J. D.,** Records of *Ophryocotyle* (Cestoda: Davaineidae) from shore birds, *Trans. Am. Microsc. Soc.*, 68, 104—106, 1949.
3885. **Webster, J. D.,** Systematic notes on North American Acoleidae (Cestoda), *J. Parasitol.* 37, 111—118, 1951.

3886. **Webster, J. D.**, Three new forms of *Aploparaksis* (Cestoda: Hymenolepididae), *Trans. Am. Microsc. Soc.*, 74, 45—51, 1955.
3887. **Webster, J. D. and Addis, C. J.**, Helminths from the bob-white quail in Texas, *J. Parasitol.*, 31, 286—287, 1945.
3888. **Wedl, K.**, Helminthologische Notizen, *Sitzungsber. Akad. Wiss. Wein. Math. Naturwiss. Klasse*, Abt. I., 16, 371—395, 1855.
3889. **Wedl, K.**, Charakteristik mehrerer grössenteils neuer Taenien, *Sitzungsber. Akad. Wiss. Wien. Math. Naturwiss. Klasse*, Abt. I, 18, 5—27, 1856.
3890. **Wedl, K.**, Zur Helminthenfauna Aegyptens (2. Abt.), *Sitzungsber. Akad. Wiss. Wien. Math. Naturwiss. Klasse*, Abt. I. 44, 463—482, 1861.
3891. **Weekes, P. J.**, *Pullutrina nestoris*, an anoplocephalid cestode parasite of the kea, *Nestor notabilis*, *N.Z. J. Zool.*, 8, 387—390, 1981.
3892. **Weerekoon, A. C. J.**, A new avian cestode, *Cotugnia platycerci*, from Stanley's rosella parakeet, *Platycercus icterotis*, *Ceylon J. Sci.*, Sect. Zool., 22, 155—159, 1944.
3893. **Weimer, B. R., Hedden, R. S., and Cowdery, K.**, Flatworm and roundworm parasites of wild rabbits of the northern Panhandle, *Proc. West. Va. Acad. Sci.*, 7, 54—55, 1934.
3894. **Weinland, D. F.**, Observations on a new genus of taenioids, *Proc. Boston Soc. Nat. Hist.*, 6, 59—63, 1857.
3895. **Weinland, D. F.**, Beschreibung zweier neuer Taenioiden aus dem Menschen. Versuch einer Systematik der Taenien ueberhaupt, *Nov. Act. Akad. Nat. Curios.*, 28, 1—24, 1861.
3896. **Weithofer, M.**, Vogelcestoden aus Sennar und Kordofan, *Anz. Akad. Wiss. Wien 53 Math. Naturwiss. Kl.*, pp. 312—313, 1916.
3897. **Welch, F. H.**, Observations on the anatomy of *Taenia mediocanellata*, *Q. J. Microsc. Sci.*, 15, 1—23, 1875.
3898. **Welch, F. H.**, The anatomy of two parasitic forms of the family Tetrarhynchidae, *J. Linn. Soc. London*, 12, 329—342, 1876.
3899. **Wenniger, F. J.**, Some notes on *Dibothriocephalus perfoliatus*, *Am. Midl. Nat.*, 11, 97—99, 1928.
3900. **Wenniger, F. J.**, The anatomy of *Stenobothrium macrobothrium* Diesing, *Am. Midl. Nat.*, 11, 503—533, 1929.
3901. **Wenniger, F. J.**, Some structural peculiarities of *Stenobothrium macrobothrium* Dies, *Proc. Indiana Acad. Sci. (1929)*, 39, 307—308, 1930.
3902. **Werner, P. C. F.**, Vermium Intestinalium Praesertim Taeniae Humanae Brevis Expositio, Leipzig, 1782, 144 pp.
3903. **Wertheim, G.**, A new anoplocephalid cestode from the gerbil, *Parasitology*, 44, 446—449, 1954.
3904. **Wertheim, G. and Greenberg, Z.**, Helminths of mammals and birds from Israel. II. *Sinaiotaenia witenbergi* gen. et sp. n. (Cestoda: Anoplocephalidae) from desert rodents, *Proc. Helminthol. Soc. Wash.*, 38, 93—96, 1971.
3905. **Wetzel, R.**, Zur Kenntnis des weniggliedrigen Hühnerbandwurmes, *Davainea proglottina*, *Arch. Wiss. Prakt. Tierheilk*, 65, 595—625, 1932.
3906. **Wetzel, R.**, Zur Kenntnis des Entwicklungskreises des Hühnerbandwurms, *Raillietina cesticillus*, *Dtsch. Tierärztl. Wochenschr.*, 41, 465—467, 1933.
3907. **Wetzel, R.**, Untersuchungen ueber den Entwicklungskreis des Hühnerbandwurmes *Raillietina cesticillus* (Molin, 1858), *Arch. Wiss. Prakt. Tierheilk*, 68, 221—232, 1934.
3908. **Wetzel, R.**, Die Entwicklung der Geflügelbandwürmer und ihre Bekämpfung, *Dtsch. Tierärztl. Wochenschr.*, 43, 188—191, 1935.
3909. **Wetzel, R.**, Neuere Ergebnisse über die Entwicklung von Hühnerbandwürmern, *Verhandl. Dtsch. Zool. Ges.*, 38 Jahrsvers. 195—200, (*Zool. Anz.*, 9 Suppl.-bd.), 1936.
3910. **Wetzel, R.**, Insekten als Zwischenwirte von Bandwürmen der Hühnervögel (Sammelreferat), *Z. Hyg. Zool.*, 30, 84—92, 1938.
3911. **Whittaker, F. H.**, *Raillietina (R.) garciai* sp. n. (Cestoda: Davaineidae) from the Greater Antillean grackle, *Quiscalus niger brachyptrus* Cassin, in Puerto Rico, *Proc. Helminthol. Soc. Wash.*, 40, 50—51, 1973.
3912. **Whittaker, F. H. and Hill, L. G.**, *Proteocephalus chologasteri* sp. n. (Cestoda: Proteocephalidae) from the spring cavefish *Chologaster agassizi* Putman, 1782 (Pices: Amblyopsidae) of Kentucky, *Proc. Helminthol. Soc. Wash.*, 35, 15—18, 1968.
3913. **Whittaker, F. H. and Zobor, S. J.**, *Proteocephalus poulsoni* sp. n. (Cestoda: Proteocephalidae) from the northern cavefish *Amblyopsis spelaea* Dekay, 1842 (Pisces: Amblyopsidae) of Kentucky, *Folia Parasitol.*, 25, 277—280, 1978.
3914. **Will, H.**, Anatomie von *Caryophyllaeus mutabilis* Rud. Ein Beitrag zur Kenntnis der Cestoden, *Z. Wiss. Zool.*, 56, 1—39, 1893.
3915. **Willemoes-Suhm, R.**, Helminthologische Notizen. I. Zur Entwicklung von *Schistocephalus dimorphus* Creplin, *Z. Wiss. Zool.*, 19, 469—475, 1869.

3916. **Williams, A. D. and Campbell, R. A.**, A new tetraphyllidean cestode, *Glyphobothrium zwerneri* gen. et sp. n., from the cownose ray, *Rhinoptera bonasus* (Mitchell 1815), *J. Parasitol.*, 63, 775—779, 1977.
3917. **Williams, A. D. and Campbell, R. A.**, *Duplicibothrium minutum* gen. et sp. n. (Cestoda: Tetraphyllidea) from the cownose ray, *Rhinoptera bonasus* (Mitchell 1815), *J. Parasitol.*, 64, 835—837, 1978.
3918. **Williams, A. D. and Campbell, R. A.**, *Echinobothrium bonasum* sp. n., a new cestode from the cownose ray, *Rhinoptera bonasus* (Mitchell 1815), in the western North Atlantic, *J. Parasitol.*, 66, 1036—1038, 1981.
3919. **Williams, D. D.**, *Biacetabulum oregoni* sp. n. (Cestoda: Caryophyllaeidae) from *Catostomus macrocheilus*, *Iowa State J. Res.*, 52, 397—400, 1978.
3920. **Williams, D. D. and Sutherland, D. R.**, *Khawia sinensis* (Caryophyllidea: Lytocestidae) from *Cyprinus carpio* in North America, *Proc. Helminthol. Soc. Wash.*, 48, 253—255, 1981.
3921. **Williams, E. H., Jr.**, Two new species of *Monobothrium* (Cestoda: Caryophyllaeidae) from catostamid fishes in the southeastern United States, *Trans. Am. Fish. Soc.*, 103, 610—615, 1974.
3922. **Williams, E. H., Jr.**, *Dieffluvium unipapillatum* n.g., n. sp. (Cestoda: Caryophyllaeidae) from the river redhorse *Moxostoma carinatum* (Cope) from the southeastern United States, *Trans. Am. Microsc. Soc.*, 97, 601—605, 1978.
3923. **Williams, E. H., Jr.**, *Penarchigetes fessus* sp. n. from the lake chubsucker, *Erimyzon sucetta* (Lacépède) in the southeastern United States, *Proc. Helminthol. Soc. Wash.*, 46, 84—87, 1979.
3924. **Williams, E. H., Jr.**, *Rogersus rogersi* gen. et sp. n. (Cestoda: Caryophyllaeidae) from the blacktail redhorse, *Moxostoma poecilurum* (Jordan), (Osteichthyes) in the southeastern United States, *J. Parasitol.*, 66, 564—568, 1980.
3925. **Williams, E. H., Jr. and Rogers, O. A.**, *Isoglaridaeris agminis* sp. n. (Cestoda: Caryophyllaeidae) from the lake chubsucker, *Erimyzon sucetta* (Lacépède), *J. Parasitol.*, 58, 1082—1084, 1972.
3926. **Williams, H. H.**, *Acanthobothrium* sp. nov. (Cestoda: Tetraphyllidae) and a comment on the order Biporophyllaeida, *Parasitology*, 52, 67—76, 1962.
3927. **Williams, H. H.**, Some new and little known cestodes from Australian elasmobranchs with a brief discussion on their possible use in problems of host taxonomy, *Parasitology*, 54, 737—748, 1964.
3928. **Williams, H. H.**, The ecology, functional morphology and taxonomy of *Echeneibothrium* Beneden, 1849 (Cestoda: Tetraphyllidea), a revision of the genus and comments on *Discobothrium* Beneden, 1870, *Pseudanthobothrium* Baer, 1956, and *Phormobothrium* Alexander, 1963, *Parasitology*, 56, 227—285, 1966.
3929. **Williams, H. H.**, *Phyllobothrium piriei* sp. nov. (Cestoda: Tetraphyllidea) from *Raja naevus* with a comment on its habitat and mode of attachment, *Parasitology*, 58, 929—937, 1968.
3930. **Williams, H. H.**, *Acanthobothrium quadripartitum* sp. nov. (Cestoda: Tetraphyllidea) from *Raja naevus* in the North Sea and English Channel, *Parasitology*, 58, 105—110, 1968.
3931. **Williams, H. H.**, The taxonomy, ecology and host-specificity of some Phyllobothriidae (Cestoda: Tetraphyllidea), a critical revision of *Phyllobothrium* Beneden, 1849 and comments on some allied genera, *Phil. Trans. R. Soc.*, Ser. B., 253, 231—307, 1968.
3932. **Williams, H. H.**, The genus *Acanthobothrium* Beneden, 1849 (Cestoda: Tetraphyllidea), *Nytt Mag. Zool.*, 17, 1—56, 1969.
3933. **Williams, O. L.**, Cestodes from the eastern wild turkey, *J. Parasitol.*, 18, 14—20, 1931.
3934. **Williams, S. R.**, Variation in *Moniezia expansa* Rudolphi, *Ohio J. Sci.*, 39, 37—42, 1939.
3935. **Williamson, F. S. and Rausch, R. L.**, Studies on the helminth fauna of Alaska. XLII. *Aploparaksis turdi* sp. n., a hymenolepidid cestode from thrushes, *J. Parasitol.*, 51, 249—252, 1965.
3936. **Winfield, G. F. and Chang, C. F.**, *Raillietina (Raillietina) sinensis*, a new tapeworm from the domestic fowl, *Peking Nat. Hist. Bull.*, 11, 35—37, 1936.
3937. **Winfield, G. F. and Chang, C. F.**, *Raillietina (Raillietina) shantungensis*, a new name for *Raillietina (R.) sinensis* Winfield and Chang, 1936, *Peking Nat. Hist. Bull.*, 11, 161, 1936.
3938. **Wisniewski, L. W.**, *Archigetes cryptobothrius* n. sp., nebst Angaben ueber die Entwicklung im Genus *Archigetes* R. Leuck, *Zool. Anz.*, 77, 113—124, 1928.
3939. **Wisniewski, L. W.**, Das Genus *Archigetes* R. Leuck. Eine Studie zur Anatomie, Histogenese, Systematik und Biologie, *Mem. Acad. Pol. Sci. Cl. Sci. Math. Nat. B*, 2, 1—160, 1930.
3940. **Wisniewski, L. W.**, Zur postembryonalen Entwicklung von *Cyathocephalus truncatus* Pallas, *Zool. Anz.*, 98, 213—218, 1932.
3941. **Wisniewski, L. W.**, *Cyathocephalus truncatus* Pallas. I. Die postembryonale Entwicklung und Biologie. II. Allgemeine Morphologie, *Bull. Acad. Pol. Sci. Cl. Sci. Math. Nat. B*, 3, 237—252, 311—327, 1933.
3942. **Wiss, M. A.**, Sur les cestodes de deux genettes capturées en Tunisie, *Bull. Soc. Hist. Nat. Afr. Nord*, 2, 113—115, 1910.
3943. **Wisseman, C. L., Jr.**, Morphology of the cysticercoid of the fowl tapeworm, *Raillietina cesticillus* (Molin), *Trans. Am. Microsc. Soc.*, 64, 145—150, 1945.
3944. **Witenberg, G.**, On the cestode subfamily Dipylidiinae Stiles, *Z. Parasitenkd.*, 4, 542—584, 1932.
3945. **Witenberg, G.**, Studies on the cestode genus *Mesocestoides*, *Arch. Zool. Ital.*, 20, 467—509, 1934.

3946. **Wolf, E.,** Beiträge zur Entwicklungsgeschichte von *Cyathocephalus truncatus* Pallas, *Zool. Anz.,* 30, 37—45, 1906.
3947. **Wolffhügel, K.,** Verläufige Mitteilung ueber die Anatomie von *T. polymorpha* Rud., *Zool. Anz.,* 21, 211—213, 1898.
3948. **Wolffhügel, K.,** *Taenia malleus* Goeze, Repraesentant einer eigenen Cestodenfamilie: Fimbriariidae, *Vorl. Mitteilung. Zool. Anz.,* 21, 388—389, 1898.
3949. **Wolffhügel, K.,** Beitrag zur Kenntnis der Anatomie einiger Vogelcestoden, *Zool. Anz.,* 22, 217—223, 1899.
3950. **Wolffhügel, K.,** Rechtfertigung gegenueber Cohn's Publikation "zur Systematik der Vogeltaenien II," *Centralbl. Bakteriol.,* 26, 632—635, 1899.
3951. **Wolffhügel, K.,** Beitrag zur Kenntnis der Vogelhelminthen, Inaug. Diss., Basel, 1900, 1—204.
3952. **Wolffhügel, K.,** *Drepanidotaenia lanceolata* Bloch, *Centralbl. Bakteriol.,* 28, 49—56, 1900.
3953. **Wolffhügel, K.,** *Stilesia hepatica* nov. spec., ein Bandwurm aus den Gallengängen von Schafen und Ziegen Ostafrikas, *Berlin Tierärztl. Wochenschr.,* 43, 661—665, 1903.
3954. **Wolffhügel, K.,** Ein interessantes Exemplar des Taubenbandwormes *Bertia delafondi* (Railliet), *Berl. Tierärztl. Wochenschr.,* pp. 45—48, 1904.
3955. **Wolffhügel, K.,** Los Zooparásitos de los animales domesticos en la República Argentina, *Rev. Centro Estud. Agron. Vet. (Buenos Aires),* 1911, 108 pp.
3956. **Wolffhügel, K.,** Cestode nuevo, parasito del estómago succenturiado de un cisne *(Cygnus melanocoryphus* Molin), *Rev. Med. Vet.,* I, 226—227, 1916.
3957. **Wolffhügel, K.,** Die Parasiten der Haustiere in Südamerika, besonders in den La Plata Staaten, *Festschr. Zschokke,* No. 29, 1—18, 1920.
3958. **Wolffhügel, K.,** Fimbriariinae (Cestoda), *Z. Infektionskr. Hyg. Haustiere,* 49, 257—291, 1936.
3959. **Wolffhügel, K.,** Nematoparataeniidae, *Z. Infektionskr. Hyg. Haustiere,* 53, 9—42, 1938.
3960. **Wolffhügel, K.,** Ergebnisse von Nematoparataeniidae, *Vol. Jub. Yoshida,* 2, 211—220, 1939.
3961. **Wolffhügel, K.,** *Ophiotaenia noei* n. sp. (Cestoda), *Biológica (Santiago, Chile),* 5, 15—27, 1948.
3962. **Wolffhügel, K. and Vogelsang, W. E. G.,** *Dibothriocephalus decipiens* (Diesing) y su larva *Sparganum reptans* en el Uruguay, *Rev. Med. Vet. Añ.* 9, 2, 433—434, 1926.
3963. **Wolffhügel, K. and Vogelsang, W. E. G.,** *Dibothriocephalus decipiens* (Dies.) y su larva *Sparganum reptans* en el Uruguay, *Actas Trab. 3 Congr. Nac. Med.,* 7, 500—501, 1928.
3964. **Wolfgang, R. W.,** Indian and Eskimo Diphyllobothriasis, *Can. Med. Assoc. J.,* 70, 536—539, 1954.
3965. **Wolfgang, R. W.,** Helminth parasites of reptiles, birds, and mammals in Egypt. II. *Catenotaenia agyptiaca* sp. nov. from myomorph rodents, with additional notes on the genus, *Can. J. Zool.,* 34, 6—20, 1956.
3966. **Woodberry, L. A.,** The development of *Diphyllobothrium cordiceps* ( = *Dibothrium cordiceps*) in *Pelicanus erythrorhynchus, J. Parasitol.,* 18, 304—305, 1932.
3967. **Woodger, J. H.,** Notes on a cestode occurring in the haemocoele of house-flies in Mesopotamia, *Ann. Appl. Biol.,* 7, 345—351, 1921.
3968. **Woodland, W. N. F.,** On *Amphilina paragonopora* sp. n. and a hitherto undescribed phase in the life history of the genus, *Q. J. Microsc. Sci.,* 67, 47—84, 1923.
3969. **Woodland, W. N. F.,** On some remarkable new forms of Caryophyllaeidae from the Anglo-Egyptian Sudan, and a revision of the families of the Cestodaria, *Q. J. Microsc. Sci.,* 67, 435—472, 1923.
3970. **Woodland, W. N. F.,** On *Ilisha parthenogenetica* Southwell and Baini Prashad, 1918, from the pyloric caeca of a fish, *Hilsa ilisha* (Ham. Buch.) and a comparison with other plerocercoid larvae of cestodes, *Parasitology,* 15, 128—136, 1923.
3971. **Woodland, W. N. F.,** On the life-cycle of *Hymenolepis fraterna* (*H. nana* var. *fraterna* Stiles) of the white mouse, *Parasitology,* 16, 69—83, 1924.
3972. **Woodland, W. N. F.,** On the development of the human *Hymenolepis nana* (Siebold, 1852) in the white mouse, with remarks on "*H. fraterna,*" "*H. longior*" and "*H. diminuta,*" *Parasitology,* 16, 424—435, 1924.
3973. **Woodland, W. N. F.,** On a new species of the cestodarian genus *Caryophyllaeus* from an Egyptian siluroid, *Proc. Zool. Soc. London,* pp. 529—532, 1924.
3974. **Woodland, W. N. F.,** On a new *Bothriocephalus* and a new genus of Proteocephalidae from Indian freshwater fishes, *Parasitology,* 16, 441—451, 1924.
3975. **Woodland, W. N. F.,** On some remarkable *Monticellia*-like and other cestodes from Sudanese siluroids, *Q. J. Microsc. Sci.,* 69, 703—729, 1925.
3976. **Woodland, W. N. F.,** *Tetracampos* Wedl, 1861, as a genus of the Bothriocephalidae, *Ann. Trop. Med. Parasitol.,* 19, 185—189, 1925.
3977. **Woodland, W. N. F.,** On *Proteocephalus marenzelleri, P. naiae* and *P. viperis, Ann. Trop. Med. Parasitol.,* 19, 265—279, 1925.
3978. **Woodland, W. N. F.,** On three new proteocephalids (Cestoda) and a revision of the genera of the family, *Parasitology,* 17, 370—394, 1925.

3979. **Woodland, W. N. F.,** On the genera and possible affinities of the Caryophyllaeidae; a reply to Drs. O. Fuhrmann and J. G. Baer, *Proc. Zool. Soc. London,* pp. 46—49, 1926.
3980. **Woodland, W. N. F.,** On three new species of *Avitellina* (Cestoda) from India and the Anglo-Egyptian Sudan, with a redescription of the type species, *A. centripunctata* (Rivolta, 1874), *Ann. Trop. Med. Parasitol.,* 21, 385—414, 1927.
3981. **Woodland, W. N. F.,** A revised classification of the tetraphyllidean Cestoda, with descriptions of some Phyllobothriidae from Plymouth, *Proc. Zool. Soc. London,* pp. 519—548, 1927.
3982. **Woodland, W. N. F.,** On *Dinobothrium septaria* van Beneden, 1889 and *Parabothrium bulbiferum* Nybelin, 1922, *J. Parasitol.,* 13, 231—248, 1927.
3983. **Woodland, W. N. F.,** On a new genus of avitelline tapeworm from ruminants in East Africa, *Parasitology,* 20, 56—65, 1928.
3984. **Woodland, W. N. F.,** On some new avian cestodes from the Sudan, *Parasitology,* 20, 305—314, 1928.
3985. **Woodland, W. N. F.,** On a new species of *Rhabdometra* with a note on the nematodiform embryos of *Anonchotaenia globata, Proc. Zool. Soc. London,* pp. 25—29, 1929.
3986. **Woodland, W. N. F.,** On some new avian cestodes from India, *Parasitology,* 21, 168—179, 1929.
3987. **Woodland, W. N. F.,** On three new cestodes from birds, *Parasitology,* 22, 214—229, 1930.
3988. **Woodland, W. N. F.,** On the genus *Polycephalus* Braun, 1878 (Cestoda), *Proc. Zool. Soc. London,* pp. 347—354, 1930.
3989. **Woodland, W. N. F.,** On the anatomy of some fish cestodes described by Diesing from the Amazon, *Q. J. Microsc. Sci.,* 76, 175—208, 1933.
3990. **Woodland, W. N. F.,** On two new cestodes from the Amazon siluroid fish *Brachyplatystoma vaillanti* Cuv, *Parasitology,* 25, 485—490, 1933.
3991. **Woodland, W. N. F.,** On a new subfamily of proteocephalid cestodes — the Othinoscoloecinae — from the Amazon siluroid fish *Platystomatichthys sturio* (Kner), *Parasitology,* 25, 491—500, 1933.
3992. **Woodland, W. N. F.,** On the Amphilaphorchidinae, a new subfamily of proteocephalid cestodes, and *Myzophorus admonticellia* gen. et sp. n., parasitic in *Pirinampus* spp. from the Amazon, *Parasitology,* 26, 141—149, 1934.
3993. **Woodland, W. N. F.,** On six new cestodes from Amazon fishes, *Proc. Zool. Soc. London,* pp. 33—44, 1934.
3994. **Woodland, W. N. F.,** On some remarkable new cestodes from the Amazon siluroid fish, *Brachyplatystoma filamentosum* (Lichtenstein), *Parasitology,* 26, 268—277, 1934.
3995. **Woodland, W. N. F.,** Some more remarkable cestodes from Amazon siluroid fish, *Parasitology,* 27, 207—225, 1935.
3996. **Woodland, W. N. F.,** Additional cestodes from the Amazon siluroids, Pirarará, Dorád and Sudobim, *Proc. Zool. Soc. London,* pp. 851—862, 1935.
3997. **Woodland, W. N. F.,** Some new proteocephalids and ptychobothriid (Cestoda) from the Amazon, *Proc. Zool. Soc. London,* pp. 619—623, 1935.
3998. **Woodland, W. N. F.,** A new species of avitelline tapeworm, *Avitellina sandgroundi,* from *Hippotragus equinus, Ann. Trop. Med. Parasitol.,* 29, 185—189, 1935.
3999. **Woodland, W. N. F.,** Some cestodes from Sierra Leone. I. On *Wenyonia longicauda* sp. n. and *Proteocephalus bivitellatus* sp. n., *Proc. Zool. Soc. London,* pp. 931—937, 1937.
4000. **Woodland, W. N. F.,** Some cestodes from Sierra Leone. II. A new caryophyllaeid, *Marsypocephalus,* and *Polyonchobothrium, Proc. Zool. Soc. London,* pp. 189—197, 1937.
4001. **Woodland, W. N. F.,** On the species of the genus *Duthiersia* Perrier, 1873 (Cestoda), *Proc. Zool. Soc. London,* pp. 17—36, 1938.
4002. **Woodland, W. N. F.,** A revision of the African and Asiatic forms of *Duthiersia* (Cestoda), *Proc. Zool. Soc. London,* pp. 207—218, 1941.
4003. **Wright, R. R.,** Contributions to American helminthology, *Proc. Can. Inst.,* 1, 54—75, 1879.
4004. **Wu, C.,** The description of two new species of *Schizorchis* (Anoplocephalidae) (in Chinese), *Acta Parasitol. Sinica,* 2, 151—154, 1965.
4005. **Wunder, W.,** Das jahreszeitliche Auftreten des Bandwurmes *(Caryophyllaeus laticeps* Pall.) im Darm des Karpfens (*Cyprinus carpio* L.), *Z. Parasitenkd.,* 10, 704—713, 1939.
4006. **Yakimov, V. L.,** Sur la question des Blastocystis, *Bull. Soc. Pathol. Exot.,* 16, 326—330, 1923.
4007. **Yamada, S., Asada, J., and Miyata, I.,** Studies on the life-history of a common tapeworm, *Hymenolepis diminuta* (Rud.), especially on the relation between this tapeworm and rat fleas (in Japanese), *Zool. Mag.,* 48, 437—457, 1936.
4008. **Yamaguti, S.,** Studies on the helminth fauna of Japan. IV. Cestodes of fishes, *Jpn. J. Zool.,* 6, 1—112, 1934.
4009. **Yamaguti, S.,** Studies on the helminth fauna of Japan. VI. Cestodes of birds, I, *Jpn. J. Zool.,* 6, 183—232, 1935.
4010. **Yamaguti, S.,** Studies on the helminth fauna of Japan. VII. Cestodes of mammals and snakes, *Jpn. J. Zool.,* 6, 233—246, 1935.

4011. **Yamaguti, S.**, Studies on the helminth fauna of Japan. XXII. Two new species of frog cestodes, *Jpn. J. Zool.*, 7, 553—558, 1938.
4012. **Yamaguti, S.**, Studies on the helminth fauna of Japan. XXVIII. *Nippotaenia chaenogobii*, a new cestode representing new order from freshwater fishes, *Jpn. J. Zool.*, 8, 285—289, 1939.
4013. **Yamaguti, S.**, Studies on the helminth fauna of Japan. XXX. Cestodes of birds. II, *Jpn. J. Med. Sci. Sect. VI*, 1, 175—211, 1940.
4014. **Yamaguti, S.**, *Studies on the Helminth Fauna of Japan. XLII. Cestodes of Mammals.* II, Publ. by author, 1945, 18 pp.
4015. **Yamaguti, S.**, On the meaning to be attached to the expression "Le plus anciennement désigné" used in Article 25 of the International Code, with special reference to the case of *Ophiotaenia ranarum* Iwata and Matuda, 1938, and *Ophiotaenia ranae* Yamaguti, 1938 (Class Cestoidea, Order Proteocephalidea), *Bull. Zool. Nomenclature*, 1, 102, 1945.
4016. **Yamaguti, S.**, Studies on the helminth fauna of Japan. XLIX. Cestodes of fishes. II, *Acta Med. Okayama*, 8, 1—78, 1952.
4017. **Yamaguti, S.**, Parasitic worms mainly from Celebes. VI. Cestodes of fishes, *Acta Med. Okayama*, 8, 353—374, 1954.
4018. **Yamaguti, S.**, Parasitic worms mainly from Celebes. VII. Cestodes of reptiles, *Acta med. Okayama*, 8, 375—385, 1954.
4019. **Yamaguti, S.**, Helminth fauna of Mt. Ontake. II. Trematoda and Cestoda, *Acta Med. Okayama*, 8, 393—405, 1954.
4020. **Yamaguti, S.**, Studies on the Helminth Fauna of Japan. L. Cestodes of Birds, III, Publ. by author, 1956, 23 pp.
4021. **Yamaguti, S.**, Parasitic Worms Mainly from Celebes. II. Cestodes of Birds, Publ. by author, 1956, 44 pp.
4022. **Yamaguti, S.**, *Systema Helminthum, The Cestodes of Vertebrates*, Vol. 2, Interscience, New York, 1959, 860 pp.
4023. **Yamaguti, S.**, Studies on the helminth fauna of Japan. LVI. Cestodes of fishes, III, *Publ. Seto Mar. Biol. Lab.*, 8, 41—50, 1960.
4024. **Yamaguti, S.**, Cestode parasites of Hawaiian fishes, *Pac. Sci.*, 22, 21—36, 1968.
4025. **Yamaguti, S. and Miyata, I.**, A new tapeworm *(Oochoristica ratti)* of the family Anoplocephalidae, from *Rattus rattus rattus* and *R. r. alexandrinus*, *Jpn. J. Zool.*, 7, 501—503, 1937.
4026. **Yamaguti, S. and Miyata, I.**, *Nippotaenia mogurndae* n. sp. (Cestoda) from a Japanese freshwater fish, *Mogurnda obscura* (Temm. et Schleg.), *Jpn. J. Med. Sci.*, Sect. VI, 1, 213—214, 1940.
4027. **Yamane, Y., Kamo, H., Tazaki, S., Fukumoto, S., and Maejima, J.**, On a new marine species of the genus *Diphyllobothrium* (Cestoda: Pseudophyllidea) found from a man in Japan, *Jpn. J. Parasitol.*, 30, 101—111, 1981.
4028. **Yamashita, J., Mori, H., and Kobayashi, T.**, Epidemiological survey of parasites of domestic animals in Hokkaido. II. A survey of the horses within the jurisdiction of Kitami City, *Mem. Fac. Agric. Hokkaido Univ.*, 1, 513—521, 1953.
4029. **Yarinsky, A.**, *Hymenolepis pitymi* n. sp. a hymenolepid cestode from the pine mouse, *J. Tenn. Acad. Sci.*, 27, 150—152, 1952.
4030. **Yaroslavsky, W. A. and Solowieff, A. I.**, *Dibothriocephalus dividocapitis* nov. spec., *Berliner Tierärztl. Wochenschr.*, 46, 296—298, 1930.
4031. **Yeh, L. S.**, On a new tapeworm *Bothriocephalus gowkongensis* n. sp. (Cestoda: Bothriocephalidae) from freshwater fish in China, *Acta Zool. Sinica*, 7, 69—74, 1955.
4032. **Yeh, L. S.**, A new tapeworm, *Diphyllobothrium salvelini* sp. nov., from a salmon, *Salvelinus alpinus*, in Greenland, *J. Helminthol.*, 29, 37—43, 1955.
4033. **Yeh, L. S.**, A new species of *Anomotaenia* (Cestoda) from the Gough Island bunting, *Rowettia goughensis*, *J. Helminthol.*, pp. 297—300, 1957.
4034. **Yokogawa, S.**, Report on experiments with *Sparganum mansoni* undertaken in an endeavour to clarify the nature of *Sparganum proliferum*, *Taiwan Igakkai Zasshi*, 32, 114—116, 1933.
4035. **Yokogawa, S. and Kobayashi, H.**, On the species of *Diphyllobothrium mansoni* sensu lato, *Trans. 8th Congr. Far Eastern Assoc. Trop. Med. Siam*, 2, 205—226, 1930.
4036. **Yorke, W. and Southwell, T.**, Lappeted Anoplocephala in horses, *Ann. Trop. Med. Parasitol.*, 15, 249—264, 1921.
4037. **Yoshida, S.**, On three new species of *Hymenolepis* found in Japan, *Anot. Zool. Jpn.*, 7, 235—246, 1910.
4038. **Yoshida, S.**, Cestodes from sharks and rays of Japan (in Japanese), *Dobuts. Zasshi*, 26, 10—19, 1914.
4039. **Yoshida, S.**, On a second and third case of infection with *Plerocercoides prolifer* Ijima found in Japan, *Parasitology*, 7, 219—225, 1914.
4040. **Yoshida, S.**, The occurrence of *Bothriocephalus liguloides* Leuckart, with special reference to its development, *J. Parasitol.*, 3, 171—176, 1917.
4041. **Yoshida, S.**, Some cestodes from Japanese selachians, *Parasitology*, 9, 560—592, 1917.

4042. **Yoshino, K.,** Studies on the post-embryonal development of *Taenia solium, Taiwan Igakkai Zasshi,* 32, 1392—1409, English summary, Suppl. 139—141; 32, 1569—1586, English summary, Suppl. 155—158; 32, 1717—1736, Engl. Summary, Suppl. 166—169, 1933.
4043. **Yoshino, K.,** On the evacuation of eggs from the detached gravid proglottids of *Taenia solium* and on the structure of its eggs, *Taiwan Igakkai Zasshi,* 33, 47—58; English summary, Suppl. 3—4, 1934.
4044. **Young, M. R.,** Helminth parasites of New Zealand, *Publ. Imp. Bur. Agric. Parasitol. Helminthol.,* pp. 1—19, 1938.
4045. **Young, R. T.,** The histogenesis of *Cysticercus pisiformis, Zool. Jahrb. Anat.,* 26, 183—254, 1908.
4046. **Young, R. T.,** The somatic nuclei of certain cestodes, *Arch. Zellforsch.,* 6, 140—163, 1912.
4047. **Young, R. T.,** The histogenesis of the reproductive organs of *Taenia pisiformis, Zool. Jahrb. Anat.,* 35, 355—418, 1913.
4048. **Young, R. T.,** The degeneration of yolk glands and cells in cestodes, *Biol. Bull.,* 36, 309—311, 1919.
4049. **Young, R. T.,** Association of somatic and germ cells in cestodes, *Biol. Bull.,* 36, 312—314, 1919.
4050. **Young, R. T.,** Gametogenesis in cestodes, *Arch. Zellforsch.,* 16, 419—437, 1923.
4051. **Young, R. T.,** Some unsolved problems of cestode structure and development, *Trans. Am. Microsc. Soc.,* 54, 229—239, 1935.
4052. **Young, R. T.,** Cestodes of California gulls, *J. Parasitol.,* 36, 9—12, 1950.
4053. **Young, R. T.,** Is *Hymenolepis californicus* (Young, 1950) a synonym of *Hymenolepis (Wardium) fryi* (Mayhew, 1925)?, *J. Parasitol.,* 38, 367, 1952.
4054. **Young, R. T.,** The larva of *Hymenolepis californicus* in the brine shrimp *(Artemia salina), J. Wash. Acad. Sci.,* 42, 385—388, 1952.
4055. **Young, R. T.,** Cestodes of sharks and rays in southern California, *Proc. Helminthol. Soc. Wash.,* 21, 106—112, 1954.
4056. **Young, R. T.,** A note on the life cycle of *Lacistorhynchus tenuis* (van Beneden, 1858), cestode of the leopard shark, *Proc. Helminthol. Soc. Wash.,* 21, 111, 1954.
4057. **Young, R. T.,** Two new species of *Echeneibothrium* from the stingray *Urobatis halleri, Trans. Am. Microsc. Soc.,* 74, 232—234, 1955.
4058. **Young, S.,** Ueber das Wachstum der *Diphyllobothrium mansoni* (Cobbold, 1883) Joyeux, 1928, im Darme des Endwirtes (Hundes) und die von diesen Bandwurm hervorgerufene Anämie, *J. Shanghai Sci. Inst.,* 3, 51—113, 1934.
4059. **Yun, L.,** A survey on the helminths of birds from Bai-Yang-Dien Lake, Hopeh Province, China. III. Cestodes (in Chinese), *Acta Zool. Sinica,* 19, 257—266, 1973.
4060. **Yun, L.,** A survey on the cestodes of birds from Weishan Lake, Shangdong Province, China, *Acta Zootaxonomica Sinica,* 7, 27—31, 1982.
4061. **Yurakhno, M. V.,** *Diphyllobothrium macroovatum* n. sp. (Cestoda, Diphyllobothriidae), a parasite of the grey whale (in Russian), *Vestn. Zool.,* 6, 25—30, 1973.
4062. **Yurpalova, N. M. and Spasskii, A. A.,** Five species of *Echinocotyle* (Cestoda: Hymenolepididae) from Charadriiformes in Chukotka (in Russian), *Parasites Anim. Plants,* RIO Akad. Nauk Moldavskoi SSR, Kishinev, 5, 60—67, 1970.
4063. **Yurpalova, N. M. and Spasskii, A. A.,** Cestode fauna of birds in Central Asia (in Russian), *Parazit. Zhivotn. Rast.,* 7, 39—56, 1971.
4064. **Yutuc, L. M.,** Observation on Manson's tapeworm, *Diphyllobothrium erinacei* Rudolphi, 1819, in the Philippines, *Philipp. J. Sci.,* 40, 33—51, 1951.
4065. **Zaidi, D. A. and Khan, D.,** Cestodes of fishes from Pakistan, *Biologia,* 22, 157—179, 1976.
4066. **Zdzitowiecki, K. and Rutkowska, M. A.,** The helmintofauna of bats (Chiroptera) from Cuba. II. A review of cestodes with description of four new species and a key to Hymenolepididae of American bats, *Acta Parasitol. Pol.,* 26, 187—200, 1980.
4067. **Zeliff, C. C.,** A new species of cestode, *Crepidobothrium amphiumae,* from *Amphiuma tridactylum, Proc. U.S. Natl. Mus.,* 81, 1—3, 1932.
4068. **Zernecke, E.,** Untersuchungen ueber den feinern Blau der Cestoden, *Zool. Jahrb. Anat.,* 9, 92—161, 1895.
4069. **Ziegler, H. E.,** Das Ektoderm der Plathelminthen, *Verhandl. Dtsch. Zool. Gesell.,* 15, 35—41, 1905.
4070. **Zilluff, H.,** Vergleichende Studien ueber die Muskulatur des Skolex der Cestoden, *Arch. Naturg.,* pp. 1—33, 1912.
4071. **Zimmermann, H. R.,** Life-history studies on cestodes of the genus *Dipylidium* from the dog, *Z. Parasitenkd.,* 9, 717—729, 1937.
4072. **Zmeev, G. Ia.,** Les trématodes et les cestodes des poissons de l'Amur (in Russian, French summary), *Parazit. Sb. Zool. Inst. Akad. Nauk SSSR,* 6, 405—436, 1936.
4073. **Zograf, N. J.,** Structure de la forme cystique de *Gymnorhynchus reptans,* Rud., *Invest. Soc. Imp. Sci. Nat. Moscow,* 50, 259—284, 1886.
4074. **Zograf, N. J.,** Zur Frage ueber die Existenz ectrodermatischer Hüllen bei erwachsenen Cestoden, *Biol. Zentralbl.,* 10, 422, 1890.

4075. **Zschokke, F.**, Recherches sur l'organisation et la distribution zoologique des vers parasites der poissons d'eau douce, *Arch. Biol.*, 5, 153—241, 1884.
4076. **Zschokke, F.**, Ueber den Bau der Geschlechtswerkzeuge von *Taenia litterata*, *Zool. Anz.*, 8, 380—384, 1885.
4077. **Zschokke, F.**, Studien ueber den anatomischen und histologischen Bau der Cestoden, *Centralbl. Bakteriol. Parasitenkd.*, 1, 161—165, 193—199, 1887.
4078. **Zschokke, F.**, Helminthologische Bemerkungen, *Mitt. Zool. Sta. Neapel*, 7, 264—271, 1887.
4079. **Zschokke, F.**, Ein Beitrag zur Kenntnis der Vogeltänien, *Centralbl. Bakteriol.*, 3, 2—6, 41—46, 1888.
4080. **Zschokke, F.**, Ein weiterer Zwischenwirt des *Bothriocephalus latus*, *Centralbl. Bakteriol.*, 3, 417—419, 1888.
4081. **Zschokke, F.**, Recherches sur la structure anatomique et histologique des cestodes des poissons marins, *Mem. Inst. Nat. Genevois*, 17, 1—396, 1888.
4082. **Zschokke, F.**, Ueber Bothriocephalen-larven in *Trutta salar*, *Centralbl. Bakteriol.*, 7, 393—396, 435—439, 1890.
4083. **Zschokke, F.**, Erster Beitrag zur Parasiten-Fauna von *Trutta salar*, *Verhandl. Naturf. Ges. Basel*, 8, 761—793, 1890.
4084. **Zschokke, F.**, Die Parasiten-fauna von *Trutta salar*, *Centralbl. Bakteriol.*, 10, 694—699, 738—745, 792—801, 829—838, 1891.
4085. **Zschokke, F.**, *Davainea contorta* sp. aus *Manis pentadactyla* L., *Centralbl. Bakteriol.*, 17, 634—645, 1895.
4086. **Zschokke, F.**, Zur Faunistik der parasitischen Würmer von Süsswasserfischen, *Centralbl. Bakteriol.*, 19, 772—784, 815—825, 1896.
4087. **Zschokke, F.**, Die Taenien der aplacentalen Säugetiere, *Zool. Anz.*, 19, 481—482, 1896.
4088. **Zschokke, F.**, Weitere Untersuchungen an Taenien der aplacentalen Säugetiere, *Zool. Anz.*, 21, 477—479, 1898.
4089. **Zschokke, F.**, Die Cestoden der Marsupialia und Monotremata, *Denkschr. Med. Naturwiss. Ges. Jena*, 8, 357—380, 1898.
4090. **Zschokke, F.**, Neue Studien an Cestoden aplacentaler Säugetiere, *Z. Wiss. Zool.*, 65, 404—445, 1899.
4091. **Zschokke, F.**, *Hymenolepis (Drepanidotaenia) lanceolata* als Schmarotzer im Menschen, *Centralbl. Bakteriol.*, 31, 331—335, 1902.
4092. **Zschokke, F.**, *Hymenolepis (Drepanidotaenia) lanceolata* aus Ente und Gans als Parasit des Menschen, *Zool. Anz.*, 25, 337—338, 1902.
4093. **Zschokke, F.**, Ein neuer Fall von *Dipylidium caninum* (L.) beim Menschen, *Centralbl. Bakteriol.*, 34, 42—43, 1903.
4094. **Zschokke, F.**, Die Arktischen Cestoden, in *Fauna Arctica*, Römer, F. und Schaudinn, F., Eds., 3, 1-32,—32, 1903.
4095. **Zschokke, F.**, Marine Schmarotzer in Süsswasserfischen, *Verhandl. Naturf. Ges. Basel*, 16, 118—157, 1903.
4096. **Zschokke, F.**, Die Darmcestoden der amerikanischen Beuteltiere, *Centralbl. Bakteriol.*, 36, 51—62, 1904.
4097. **Zschokke, F.**, Die Cestoden der südamerikanischen Beuteltiere, *Zool. Anz.*, 27, 290—293, 1904.
4098. **Zschokke, F.**, Das Genus *Oochoristica*, *Z. Wiss. Zool.*, 83, 51—67, 1905.
4099. **Zschokke, F.**, *Dipylidium caninum* (L.) als Schmarotzer des Menschen, *Centralbl. Bakteriol.*, 38, 534, 1905.
4100. **Zschokke, F.**, *Moniezia diaphana* n. sp. Ein weiterer Beitrag zur Kenntnis der Cestoden aplacentaler Säugetiere, *Centralbl. Bakteriol.*, 44, 261—264, 1907.
4101. **Zschokke, F.**, *Dibothriocephalus parvus* J. J. W. Stephens, *Rev. Suisse Zool.*, 25, 425—440, 1917.
4102. **Zschokke, F. and Heitz, A.**, Entoparasiten aus Salmoniden von Kamschatka, *Rev. Suisse Zool.*, 22, 195—256, 1914.
4103. **Zürn, F. A.**, Die tierischen Parasiten und Krankheiten des Hausgeflügels, Weimer, 1882, 1—237.

# INDEX

## A

*ababili, Neoangularia*, 385
*abassenae, Choanotaenia*, 361
*abassenae, Spreotaenia*, 361
*abberatus, Inermicapsifer*, 465
*abdominalis, Ligula*, 87
*abei, Ectopocephalium*, 446
*aberrans, Proteocephalus*, 186
*aberrata, Bertiella*, 438
*aberrata, Dicranotaenia*, 321
*aberrata, Prototaenia*, 438
*aberratus, Metacapsifer*, 465
*Abortilepis*, 327
*abortiva, Abortilepis*, 327
*abortiva, Hymenolepis*, 327
*abortiva, Microsomacanthus*, 327
*abortiva, Weinlandia*, 327
*Abothrium*, diagnosis, 108
Abothriinae, key to genera, 108
*abruptus, Cephalobothrium*, 123
*abruptus, Hexacanalis*, 123
*abscisus, Choanoscolex*, 178
*abuladze, Diorchis*, 281
*Abuladzugnia*, diagnosis, 245
*abyssinicus, Inermicapsifer*, 465
*abyssmus, Bothriocephalus*, 101
*acanthiae-vulgaris, Phyllobothrideum*, 144
*acanthinophyllum, Echinobothrium*, 166
*Acanthobothrium*, diagnosis, 137
*Acanthobothroides*, diagnosis, 134
*acanthocirrosa, Anoplocephaloides*, 436
*acanthocirrosa kivuensis, Paranoplocephala*, 436
*acanthocirrosa, Paranoplocephala*, 436
*Acanthocirrus*, 378
*acanthocirrus, Aploparaksis*, 275
*acanthodes, Atelemerus*, 105
*Acanthophallus*, 105
*acanthorhyncha, Tatria*, 229
*acanthoscolex, Grillotia*, 64
*Acanthotaenia*, diagnosis, 181
Acanthotaeniinae, key to genera, 180
*acanthotretra, Cysticercus*, 354
*acanthotretra, Diplopylidium*, 354
*Acanthotrias*, 223
*acanthovagina, Nonarmiella*, 253
*acanthovagina, Raillietina (P.)*, 253
*acapillicirrosa, Unciunia*, 386
*accipitris, Anomotaenia*, 410
*accipitris, Cladotaenia*, 413
*accipitris, Pseudanomotaenia*, 410
*acephala, Taenia*, 463
*acheilognathi, Bothriocephalus*, 102
*aciculasinuatus, Diorchis*, 281
*aciculasinuatus, Hymenolepis*, 281
*acinomyxi, Taenia*, 223
*acipenserinum, Abothrium*, 108
*acirrosa, Hymenolepis*, 336

Acoleidae
  diagnosis, 201
  key to genera, 230
*Acoleorhynchus*, 73
*Acoleus*, diagnosis, 231
*acollum, Anomotaenia*, 404
Acompsocephalidae, 82
*Acompsocephalum*, 104
*Acotylolepis*, diagnosis, 273
*acotylus, Dioecocestus*, 213
*acridotheresi, Choanotaenia*, 361
*acridotheridis, Hymenolepis*, 336
*Acrobothrium*, 137
*acrocephala, Anomotaenia*, 404
*actinioides, Thysanosoma*, 416
*aculeata, Taenia*, 420
*aculeata, Wageneria*, 76
*acuminata, Diorchis*, 279
*acuminata, Drepanidotaenia*, 281
*acuminata, Wenyonia*, 26
*acuminatus, Diorchis*, 281
*acus, Microsomacanthus*, 328
*acuta, Staphylocystis*, 301
*acuta, Taenia*, 301
*acutissima, Taenia*, 294
*acutus, Caryophyllaeus*, 33
Acystidea, key to families, 48
*addisi, Bakererpes*, 387
*adelaidae, Lapwingia*, 403
*Adelataenia*, diagnosis, 445
Adelobothriidae, diagnosis, 119, 127
*Adelobothrium*, diagnosis, 127
*adenoplusius, Rhynchobothrius*, 77
*Adenoscolex*, diagnosis, 37
*Adenocephalus*, 93
*adhaerens, Lytocestus*, 25
*adherens, Ichthyotaenia*, 177
*adiposa, Crepidobothrium*, 183
*adiposa, Ophiotaenia*, 183
*admonticellia, Myzophorus*, 194
*aduncihami, Hymenolepis*, 334
*aecophylus, Lateriporus*, 373
*aegypti, Biuterinoides*, 339
*aegypti, Neyraia*, 339
*aegyptiaca, Avitellina*, 419
*aegyptiaca, Catenotaenia*, 236
*aegyptiaca, Choanotaenia*, 395
*aegyptiaca, Icterotaenia*, 395
*aegyptiaca, Meggittina*, 236
*aegyptiaca, Paricterotaenia*, 395
*aegyptiaca, Polycercus*, 395
*aegyptiaca, Taenia*, 304, 395
*aegyptiacus, Bothriocephalus*, 102
*aegyptiacus, Discobothrium*, 126
*aegyptiacus, Lecanicephalum*, 126
*aegyptica, Hymenandrya*, 333
*aegyptica, Joyeuxia*, 353
*aegyptica, Mathevotaenia*, 462

aegyptica, Multitesticulata, 366
aepyprymni, Progamotaenia, 445
aequabilis, Dicranotaenia, 321
aequabilis, Dicranotaenia (D.), 321
aequabilis, Drepanidotaenia, 321
aequabilis, Hymenolepis, 321
aequabilis, Hymenolepis (Drepanidotaenia), 321
aequabilis, Taenia, 321
aethechini, Mathevotaenia, 462
aetiobatidis, Adelobothrium, 127
aetiobatidis, Lecanicephalum, 126
aetobati, Rhynchobothrium, 76
aetiobatidis, Staurobothrium, 123
aetiobatidis, Tetragonocephalum, 126
aetiobatidis, Tylocephalum, 126
aetiobatis, Acanthobothrium, 138
aetobatus, Prochristianella, 59
aetodex, Metabelia, 383
affine, Echineiobothrium, 150
affine, Echinobothrium, 166
affine, Phormobothrium, 150
affine, Sparganum, 94
affine, Tetrarhynchobothrium, 59
affinis, Bothriocephalus, 102
affinis, Neoliga, 391
affinis, Polypocephalus, 121
affinis, Sureshia, 391
affinis, Taenia, 409
affinis, Tetrabothrius, 207
afghana, Catenotaenia, 234
afghana, Raillietina, (R.), 257
africana, Andrya, 431
africana, Aprostatandrya (Sudarikovina), 431
africana, Biuterina, 348
africana, Cittotaenia, 423
africana, Kotlania, 257
africana, Nybelinia, 73
africana, Oochoristica, 455
africana, Paronia, 423
africana, Raillietina (R.), 257
africana, Raillietina (Ransomia), 257
africana, Sudarikovina, 431
africana, Taenia, 222
africana var. ookispensis, Oochoristica, 455
africanus, Nesolecithus, 471
agamai, Oochoristica, 455
agile, Rhynchobothrium, 55
agile, Tetrarhynchobothrium, 77
agkistrondontis, Ophiotaenia, 183
agminis, Isoglaridacris, 37
agnosta, Chitinorecta, 403
agonis, Proteocephalus, 186
agraensis, Gangesia, 176
aichesoni, Tetrabothrius, 208
akhuminae, Cotugnia, 245
akodontis, Rodentolepis, 303
akodontis, Vampirolepis, 303
alagea, Davainea, 257
alagea, Idiogenoides, 257
alagea, Raillietina (R.), 257
alascense, Diphyllobothrium, 93

alaskensis, Hymenolepis, 336
alaskensis, Ophryocotyle, 244
alata, Anomotaenia, 404
alaudae, Mesocestoides, 202
alaudae, Taenia, 341
alba, Aporina, 424
alba, Moniezia, 442
alba, Taenia, 442
alba, var. dubia, Moniezia, 442
alba var. longicollis, Moniezia, 442
alba var. nova, Moniezia, 442
albani, Paricterotaenia, 396
albani, Polycercus, 396
albani, Sacciuterina, 396
albaredae, Gyrocoelia, 215
albaredae, Infula, 215
alberti, Sphyriocephalus, 72
albertinii, Tetrabothrius, 206
albotaenia, Gyrometra, 472
albulae, Proteocephalus, 186
alburni, Ligula, 87
alcae, Taenia, 410
Alcataenia, 404
alcippina, Dicranotaenia, 321
alepocephalus-rostratus, Tetrarhynchus, 77
alessandrinii, Bothriocephalus, 102
alestesi, Lytocestis, 25
Aleurotaenia, diagnosis, 357
aleuti, Hymenolepis, 317
aleuti, Hymenosphenacanthus, 317
aleuti, Retinometra, 317
aleuti, Sphenacanthus, 317
alexandri, Vitta, 393
Aliezia, diagnosis, 417
alii, Neoliga, 391
alii, Polypocephalus, 121
alii, Sureshia, 391
allahabadi, Columbia, 419
allisonae, Baerietta, 212
Allohaploparaxis, diagnosis, 274
Allohymenolepis, diagnosis, 269
alloiotica, Nybelinia, 73
allomyodes, Davainea, 257
allomyodes, Kotlania, 257
allomyodes, Raillietina (R.), 257
Alloptychobothrium, diagnosis, 100
almiquii, Hymenolepis, 336
alopexi, Monordotaenia, 222
alopias, Litobothrium, 169
alopias, Marsupiobothrium, 147
alouattae, Raillietina (R.), 257
alovarius, Nomimoscolex, 192
alpestris, Hymenolepis, 301
alpestris, Staphylocystis, 301
alpinensis, Rhabdometra, 346
altaica, Schizorchis, 440
alternans, Digramma, 87
alternans, Liga, 390
alternans, Ligula, 87
alternans, Ophiotaenia, 183
alternans, Taenia, 390

*alternatus, Neoligorchis*, 334
*aluterae, Ancistrocephalus*, 112
*alvedea, Diorchis*, 281
*Alveococcus*, 222
*alveolaris, Taenia*, 222
*Alyselminthus*, 223, 352
*Amabilia*, diagnosis, 227
Amabiliidae
  diagnosis, 200
  key to genera, 227
*amazonensis, Acanthobothrium*, 138
*amazonensis, Pomatotrygonocestus*, 142
*ambajogaiensis, Davainea*, 247
*ambigua, Moniezia*, 423, 442
*ambigua, Paronia*, 423
*ambigua, Parvitaenia*, 412
*ambiguum, Rhynchobothrium*, 76
*ambiguus, Hymenolepis*, 328
*ambiguus, Mesocestoides*, 202
*ambiguus, Microsomacanthus*, 328
*ambloplitis, Proteocephalus*, 186
*ambloplitis, Taenia*, 186
*ameivae, Linstowia*, 453
*ameivae, Oochoristica*, 453
*americana, Andrya*, 426
*americana, Bertia*, 426
*americana, Bertiella*, 426
*americana, Cylindrotaenia*, 209
*americana, Diplogynia*, 273
*americana, Monoecocestus*, 426
*americana, Oochoristica*, 453
*americana, Progynotaenia*, 219
*americana, Schizotaenia*, 426
*americanum, Acanthobothrium*, 138
*americanus, Diorchis*, 281
*americanus, Diphyllobothrium*, 94
*americanus turkestanicus, Diorchis*, 282
*Amerina*, 340
*amethiensis, Valipora*, 371
*Amirthalingamia*, diagnosis, 374
*Amoebotaenia*, diagnosis, 387
*amphibia, Moniezia*, 442
*amphiboluri, Ophiotaenia*, 183
*Amphicotyle*, diagnosis, 106
Amphicotylidae
  diagnosis, 83
  key to subfamilies, 106
Amphicotylinae, key to genera, 106
Amphilaphorchidinae, 196
*Amphilaphorchis*, 196
*Amphilina*, diagnosis, 468
Amphilinidae
  diagnosis, 467
  key to genera, 468
Amphilinidea
  diagnosis, 467
  key to families, 467
*Amphipetrovia*, diagnosis, 290
*amphisbaenae, Oochoristica*, 453
*amphisbaenae, Semenoviella*, 452
*amphisbaenae, Taenia*, 452

*amphisbeteta, Mathevotaenia*, 463
*amphisbeteta, Oochoristica*, 463
*Amphitretus*, 105
*amphitricha, Dicranotaenia*, 321
*amphitricha, Hymenolepis*, 321
*amphitricha, Hymenolepis (Drepanidotaenia)*, 321
*amphitricha, Limnolepis*, 321
*amphitricha, Taenia*, 321
*amphitricha, Wardium*, 321
*amphitricha, Weinlandia*, 321
*amphiumae, Ophiotaenia*, 183
*amphiumicola, Proteocephalus*, 186
*Amphoterocotyle*, 207
*Amphoteromorphus*, diagnosis, 191
*amplifica, Renyxa*, 169
*amurensis, Fimbriatia, Fimbriaria*, 268
*amurensis, Triaenophorus*, 113
*Amurotaenia*, diagnosis, 171
*anacetabula, Acotylolepis*, 273
*anacetabula, Hymenolepis*, 273
*anacolum, Caulobothrium*, 153
*anadryensis, Echinocotyle*, 285
*anadyrensis, Anomotaenia*, 404
*anantaramanorum, Nybelinia*, 73
*Anantrum*, diagnosis, 104
*anapolytica, Bertiella*, 438
*anatina, Davainea*, 266
*anatina, Dilepis*, 285
*anatina, Drepanidotaenia*, 285
*anatina, Echinocotyle*, 285
*anatina, Hymenolepis*, 285
*anatina, Raillietina*, 266
*anatina, Taenia*, 285
*Anatinella*, diagnosis, 312
*anatis-marilae, Taenia*, 338
*anceps, Dicranotaenia*, 321
*anceps, Hymenolepis*, 321
*Ancistrocephalus*, diagnosis, 112
*ancora, Anomotaenia*, 408
*ancora, Rauschitaenia*, 408
*andersoni, Ophiotaenia*, 183
*andinus, Diplophallus*, 232
*andrei, Aploparaksis*, 275
*andrei, Davainea*, 247
*andrejewoi, Dicranotaenia*, 328
*andrejewoi, Microsomacanthus*, 328
*andrejewoi, Myxolepis*, 328
*andrejewoi, Sphenacanthus*, 328
*Andrepigynotaenia*, diagnosis, 219
*andresi, Bothriocephalus*, 102
*Andrya*, diagnosis, 434
*angeli, Grillotia*, 64
*angertrudae, Fossor*, 222
*angolensis, Anomotaenia*, 404
*angolensis, Choanotaenia*, 361, 404
*angolensis, Inermicapsifer*, 465
*anguillae, Bothriocephalus*, 102
*anguillae, Nybelinia*, 73
*anguillae, Taenia*, 102
*anguillae, Triaenophorus*, 113
*Angularella*, diagnosis, 393

*Angularia*, 393
*angularostris, Hymenolepis*, 336
*angulata, Dilepis*, 380
*angulata, Taenia*, 298, 380
*angusta, Hymenolepis*, 334
*angusta, Raillietina (R.)*, 257
*angustata, Paruterina*, 350
*angustatus, Bothriocephalus*, 102
*angustatus, Mesocestoides*, 202
*angusticeps, Bothriocephalus*, 102
*angusticollis, Tetrarhynchus*, 77
*angustum, Orygmatobothrium*, 161
*anivi, Diorchis*, 281
*ankeli, Pseudoparadilepis*, 284
*annandalei, Dicranotaenia*, 321
*annandalei, Hornelliella*, 63
*annandalei, Hymenolepis*, 321
*annandelei, Limnolepis*, 321
*annandalei, Tetrarhynchus*, 63
*annandalei* var. *longosacco, Dicranotaenia*, 321
*annapinkiensis, Acanthobothrium*, 138
*anniellae, Oochoristica*, 453
*anolis, Oochoristica*, 453
*Anomaloporus*, diagnosis, 341
*Anoncocephalus*, diagnosis, 113
*anomala, Hymenolepis*, 289
*anomalus, Diorchis*, 281
*Anomolepis*, 379
*Anomotaenia*, diagnosis, 404
*Anoncotaenia*, diagnosis, 340
*Anootypus*, 418
*Anophryocephalus*, diagnosis, 205
*anophrys, Anophryocephalus*, 206
*Anoplocephala*, diagnosis, 432
Anoplocephalidae, 415
 diagnosis, 202
 key to subfamilies, 415
Anoplocephalinae, key to genera, 421
*anoplocephaloides, Baerfainia*, 247
*Anoplocephaloides*, diagnosis, 436
*anoplocephaloides, Monoecocestus*, 426
*anoplocephaloides, Raillietina (R.)*, 247
*anoplocephaloides, Schizotaenia*, 426
*anoplocephaloides, Taenia*, 424
*anoplocephaloides, Triuterina*, 424
*Anoplotaenia*, diagnosis, 383
*anormalis, Paradicranotaenia*, 272
*ansa, Nadejdolepis*, 310
*Anserilepis*, 294, 327
*anseris, Hymenolepis*, 330, 336
*anseris, Taenia*, 294
*anserum, Taenia*, 294
*antarctica, Anomotaenia*, 405, 410
*antarctica, Pseudanomotaenia*, 405, 410
*antarctica, Taenia*, 223
*antarcticus, Dibothriocephalus*, 93
*antarcticus, Dibothrium*, 93
*antarcticus, Diplogonoporus*, 93
*antarcticus, Diphyllobothrium*, 93
*antarcticus, Glandicephalus*, 93
*antarcticus, Tetrabothrius*, 209

*anterophallum, Rhabdotobothrium*, 154
*Anteropora*, 164
Anteroporidea, 163
*anteroporus, Gilquinia*, 59
*anteroporus, Tetrarhynchus*, 59
*Anthemobothrium*, 121
*anthicosum, Nybelinia*, 73
*Anthobothrium*, diagnosis, 157
*anthocephala, Taenia*, 92
*Anthocephalum*, 160
*Anthocephalus*, 52, 53
*anthocephalus, Cryptocotylepis*, 288
*anthocephalus, Hymenolepis*, 288
*anthocephalus, Pyramicocephalus*, 92
*anthusi, Anomotaenia*, 404
*anthusi, Choanotaenia*, 361, 404
*anthusi, Sobolevitaenia*, 361
*antipini, Tatria*, 230
*antrozoi, Atriotaenia*, 462
*antrozoi, Mathevotaenia*, 462
*antrozoi, Oochoristica*, 462
*Anurina*, 340
*Aocobothrium*, diagnosis, 147
*Aphanobothrium*, 126, 227
*aphroditae, Tetrarhynchus*, 77
*apicaris, Hymenolepis*, 321
*apivori, Raillietina (R.)*, 257
*Aploparaksis*, diagnosis, 275
*aploparaksioides, Hymenolepis*, 336
*apogonis, Bothriocephalus*, 102
*Apora*, diagnosis, 129
*aporalis, Drepanidotaenia*, 294
*Aporhynchus*, diagnosis, 60
Aporidea, 13, 129
*Aporina*, diagnosis, 424
*Aporodiorchis*, diagnosis, 278
*apospasmation, Inermicapsifer*, 465
*appendiculata, Brumptiella*, 253
*Appendiculata, Davainea*, 253
*appendiculata, Delamurella*, 253
*appendiculata, Raillietina (P.)*, 253
*appendiculata, Tatria*, 229
*appendiculatum, Biacetabulum*, 30
*appendiculatus, Archites*, 30
*appendiculatus, Caryophyllaeus*, 33
*appendiculatus, Ophiotaenia*, 189
*appendiculatus, Proteocephalus*, 189
*appendiculatus, Tejidotaenia*, 189
*appendiculatus, Tetrarhynchus*, 77
*Aprostatandrya*, 434
*apterygis, Choanotaenia*, 396
*apterygis, Drepanidotaenia*, 396
*apterygis, Icterotaenia*, 396
*apterygis, Paricterotaenia*, 396
*apterygis, Polycercus*, 396
*aquilastur, Cladotaenia*, 413
*arabiansis, Scyphophyllidium*, 147
*araii, Insectivorolepis*, 288
*arandasi, Proteocephalus*, 186
*araya, Eutetrarhynchus*, 57
*araya, Tentacularia*, 57

*archeri, Dibothriocephalus*, 95
*Archigetes*, diagnosis, 29
*arciuterus, Passerilepis*, 298
*arctica, Andrya*, 435
*arctica, Avitellina*, 419
*arcticum, Choanotaenia*, 366
*arcticum, Eubothrium*, 107
*arcticum, Monopylidium*, 366
*arcticum, Notobothrium*, 267
*arcticus, Decacanthus*, 324
*arcticus, Hymenolepis*, 324
*arcticus, Proteocephalus*, 186
*arcticus, Tetrabothrius*, 207
*arcticus, Variolepis*, 324
*arcticus, Wardium*, 324
*arctocephalinus, Cordicepalus*, 94
*arctomarinum, Diphyllobothrium*, 88, 93
*Arctotaenia*, diagnosis, 369
*arcuata, Hispaniolepidoides*, 306
*arcuata, Hispaniolepis*, 306
*arcuata, Hymenolepis*, 306
*arcuata, Microsomacanthus*, 306
*arcuata, Weinlandia*, 306
*arcuatum, Bothridium*, 85
*ardeae, Bancroftiella*, 399
*ardeae, Cyclustra*, 380
*ardeae, Dilepis*, 380
*ardeius, Mashonalepis*, 372
*ardeola, Tubanguiella*, 395
*ardeolae, Dilepis*, 380
*ardeolae, Parvitaenia*, 412
*arenaria, Panceria*, 452
*arenariae, Hymenolepis*, 336
*arfaai, Anoplocephaloides*, 436
*arfaai, Galligoides*, 436
*arfaai, Schizorchis*, 436
*argentina, Taenia*, 241
*argentina, Tetrarhynchus*, 77
*argentinum, Tetrabothrius*, 207
*arguata, Choanotaenia*, 396
*arguata, Icterotaenia*, 396
*arguata, Paricterotaenia*, 396
*arguata, Polycercus*, 396
*arguata, Sacciuterina*, 396
*arguei, Hymenolepis*, 336
*arhyncha, Anoncotaenia*, 341
*Arhynchotaenia*, 290, 465
*Arhynchotaeniella*, diagnosis, 290
*arkita, Anomotaenia*, 410
*arkita, Pseudanomotaenia*, 410
*arkteios, Pentorchis*, 331
*Armacetabulum*, 253
*Armadoskrjabinia*, diagnosis, 296
*Armandia*, 55
*armata, Hymenolepis*, 336
*armatus, Acoleus*, 231
*armatus, Oncobothriocephalus*, 95
*armatus, Ptychobothrius*, 95
*armatus, Urogonoporus*, 144
*armeniacus, Caryophyllaeus*, 33
*armigera, Cladotaenia*, 413

*armigerum, Prosobothrium*, 177
*armillaris, Alcataenia*, 410
*armillaris, Anomotaenia*, 410
*armillaris, Pseudanomotaenia*, 410
*armillaris, Taenia*, 410
*Arostellina*, diagnosis, 425
*arsenjevi, Diorchis*, 281
*arsenjevi, Skrjabinoparaxis*, 274
*arsenyevi, Tetrabothrius*, 207
*artibei, Vampirolepis*, 303
*articulata, Acanthotaenia*, 182
*articulata, Crepidobothrium*, 182
*articulatus, Proteocephalus*, 182
*aruensis, Davainea*, 257
*aruensis, Proparuterina*, 411
*aruensis, Raillietina (R.)*, 257
*aruensis, Raillietina (Ransomia)*, 257
*aruensis, Kotlania*, 257
*arvicanthidis, Inermicapsifer*, 465
*arvicolae, Hymenolepis*, 303, 334
*arvicolina, Hymenolepis*, 289
*Ascodilepis*, 379
*Ascometra*, diagnosis, 237
*Ascotaenia*, diagnosis, 421
*asiatica, Catenotaenia*, 234
*asiatica, Diorchis*, 281
*asiatica, Raillietina (R.)*, 257
*asiota, Taenia*, 223
*asketus, Hymenolepis*, 334
*asper, Dioecocestus*, 213
*asper, Taenia*, 213
*Aspidorhynchus*, 73
*aspinosa, Thysaniezia*, 420
*aspinosus, Neobothriocephalus*, 116
*aspirantica, Hymenolepis*, 308
*aspirantica, Sobolevicanthus*, 308
*asymmetrica, Anomotaenia*, 404
*asymmetrica, Choanotaenia*, 396
*asymmetrica, Hymenolepis*, 303, 325
*asymmetrica, Icterotaenia*, 396
*asymmetrica, Paricterotaenia*, 396
*asymmetrica, Polycercus*, 396
*asymmetrica, Rodentolepis*, 303
*asymmetrica, Vampirolepis*, 303
*asymmetrica, Variolepis*, 325
*Atelemerus*, diagnosis, 105
*atlanticum, Diphyllobothrium*, 93
*atlanticus, Philobythos*, 111
*Atractolytocestus*, diagnosis, 22
*atretiumi, Proteocephalus*, 186
*Atriotaenia*, diagnosis, 461
*attenuata, Dilepis*, 380
*attenuata, Taenia*, 380
*attenuatus, Dibothriorhynchus*, 76
*attenuatus, Tetrarhynchus*, 77
*audubonensis, Angularella*, 393
*aulicus, Oochoristica*, 453
*aurangabadensis, Cotugnia*, 245
*aurangabadensis, Lytocestoides*, 19
*aurangabadensis, Sureshia*, 402
*aurantia, Choanotaenia*, 361

*auricula, Phyllobothrium*, 162
*auriculatum, Anthobothrium*, 157, 161
*auriculatum, Monobothrium*, 32
*auriculatum, Platybothrium*, 137
*auriculatus, Tetrabothrius*, 208
*australensis, Capiuterilepis*, 271
*australiensis, Dilepis*, 380
*australiensis, Gyrocoelia*, 215
*australiensis, Oochoristica*, 454
*australiensis, Vampirolepis*, 303
*Australiolepis*, diagnosis, 292
*australis, Acanthobothrium*, 138
*australis, Aploparaksis*, 275
*australis, Davainea*, 257
*australis, Kotlania*, 257
*australis, Lateriporus*, 373
*australis, Raillietina (Ransomia)*, 257
*australis, Proteocephalus*, 186
*australis, Raillietina (R.)*, 257
*australis, Taenia*, 257
*Austramphilina*, diagnosis, 472
Austramphilinidae
  diagnosis, 468
  key to genera, 471
*austrinum, Echeneibothrium*, 151
*autumnalia, Moniezia*, 442
*averini, Anomolepis*, 380
*averini, Dilepis*, 380
*averini, Fuhrmanolepis*, 380
*avicola, Diplopylidium*, 352
*avicola, Dipylidium*, 352
*avicola, Diskrjabiniella*, 352
*avicola, Progynopylidium*, 352
*Avitellina*, diagnosis, 418
*avium, Ligula*, 87
*avoceti, Eurycestus*, 389
*Avocettolepis*, diagnosis, 314
*awogera, Amoebotaenia*, 387
*azerbaijanica, Tatria*, 229

**B**

*bacigalupoi, Hymenolepis*, 305
*bacillaris, Staphylocystis*, 301
*bacillaris, Taenia*, 301
*bacilligera, Anomotaenia*, 404
*bacilligera, Taenia*, 404
*baczynskae, Amoebotaenia*, 396
*baczynskae, Polycercus*, 396
*Baerbonaia*, 371
*Baerfainia*, diagnosis, 247
*baeri, Anomotaenia*, 404
*baeri, Anoplocephaloides*, 436
*baeri, Aploparaksis*, 322
*baeri, Ascometra*, 237
*baeri, Baerietta*, 212
*baeri, Catenotaenia*, 236
*baeri, Davainea*, 247
*baeri, Dichoanotaenia*, 404
*baeri, Eutetrarhynchus*, 57

*baeri, Kotlania*, 257
*baeri, Meggittina*, 236
*baeri, Microsomacanthus*, 328
*baeri, Moniezia*, 442
*baeri, Platybothrium*, 137
*baeri, Raillietina (R.)*, 257
*baeri, Rhinebothrium*, 155
*baeri, Tetrabothrius*, 207
*baeri, Vadifresia*, 257
*baeri, Vampirolepis*, 303
*Baeria*, 419
*baeribonae, Baerbonia*, 371
*baeribonae, Valipora*, 371
*Baeriella*, 446
*Baerietta*, diagnosis, 211
*bagariusi, Circumoncobothrium*, 96
*bahli, Cotugnia*, 245
*bahli, Hymenolepis*, 303
*bahli, Rodentolepis*, 303
*bahli, Vampirolepis*, 303
*baicalensis, Choanotaenia*, 361
*baicalensis, Molluscotaenia*, 361
*bailea, Oochoristica*, 454
*baileyi, Imparmargo*, 365
*bairdi, Tetrabothrius*, 207
*bairdii, Taenia*, 328
*bajaensis, Acanthobothrium*, 138
*bakamovi, Hymenolepis*, 334
*Bakererpes*, diagnosis, 387
*bakeri, Raillietina (R.)*, 257
*balacea, Diorchis*, 281
*balaenopterae, Diplogonoporus*, 90
*balaniceps, Hydatigera*, 224
*balaniceps, Taenia*, 224
Balanobothriidae, diagnosis, 120
*Balanobothrium*, diagnosis, 120
*Balanotaenia*, diagnosis, 17
Balanotaeniidae, diagnosis, 17
*balistes-caprisci, Tetrarhynchus*, 77
*balistidis, Tetrarhynchus*, 77
*balli, Otobothrium*, 68
*balsaci, Vampirolepis*, 303
*baltazardi, Atriotaenia*, 462
*baltica, Khawia*, 20
*bancrofti, Balanotaenia*, 17
*bancrofti, Cittotaenia*, 445
*bancrofti, Dilepis*, 427
*bancrofti, Hemiparonia*, 427
*bancrofti, Progamotaenia*, 445
*Bancroftiella*, diagnosis, 399
*bandicotensis, Fuhrmannetta*, 265
*bandicotensis, Raillietina (F.)*, 265
*banghami, Biacetabulum*, 30
*banghami, Cladotaenia*, 413
*barbara, Choanotaenia*, 396
*barbara, Paricterotaenia*, 396
*barbara, Polycercus*, 396
*barbeti, Ophriocotyloides*, 243
*barboscolex, Vigisolepis*, 293
*barbouri, Ophiotaenia*, 183
*barbus, Bothriocephalus*, 102

*bargetzii, Raillietina (P.)*, 253
*bargusinica, Soricina*, 287
*barmerensis, Corvinella*, 253
*barmerensis, Raillietina (P.)*, 253
*barrosii, Hymenolepis*, 334
*barrowensis, Anserilepis*, 294, 328
*barrowensis, Drepanidotaenia*, 294, 328
*barrowensis, Hymenolepis*, 294, 328
*barrowensis, Microsomacanthus*, 328
*barrowensis, Vigissotaenia*, 328
*baschkiriensis, Hymenolepis*, 336
*basimegacantha, Nybelinia*, 73
*basipunctata, Pseudogrillotia*, 66
*bassani, Tetrabothrius*, 207
*bassarisci, Mesocestoides*, 202
*Bathybothrium*, diagnosis, 108
*bathyphilum, Echeneibothrium*, 151
*batrachii, Capingentoides*, 40
*Batrachotaenia*, 182
*bayamegaparuterina, Mogheia*, 419
*baylisi, Baylisia*, 91
*Baylisia*, diagnosis, 90
*Baylisiella*, diagnosis, 92
*beauchampi, Echeneibothrium*, 151
*beauchampi, Proteocephalus*, 186
*beauforti, Moniezia*, 423
*beauforti, Paronia*, 423
*beauporti, Trichocephaloidis*, 370
*beddardi, Acanthobothrium*, 181
*beddardi, Crepidobothrium*, 181
*beddardi, Davainea*, 250
*beddardi, Proteocephalus*, 181
*beddardi, Rostellotaenia*, 181
*beema, Angularella*, 393
*beema, Angularia*, 393
*bellieri, Hymenolepis*, 334
*belones, Proteocephalus*, 186
*belones, Ptychobothrium*, 99
*belopolskaiae, Hymenolepis*, 310, 336
*belopolskaiae, Nadejdolepis*, 310
*belopolskaja, Anomotaenia*, 405
*belopolskaja, Dictymetra*, 405
*bembezi, Raillietina (R.)*, 257
*benedeni, Acanthobothrium*, 138
*benedeni, Echinobothrium*, 166
*benedeni, Moniezia*, 442
*benedeni, Taenia*, 442
*benedeni, Rhynchobothrium*, 65
*bengalense, Acanthobothrium*, 138
*bengalensis, Bothriocephalus*, 102
*bengalensis, Gangesia*, 176
*bengalensis, Nybelinia*, 73
*bengalensis, Ophriocotyle*, 176
*bennetti, Hymenolepis*, 335
*beppuensis, Raillietina (P.)*, 253
*bergini, Mesocestoides*, 203
*bernicla, Dicranotaenia*, 321
*Bertia*, 438
*Bertiella*, diagnosis, 438
*besnardi, Senga*, 96
*bhaleraoi, Cotugnia*, 245

*bhaleraoi, Ophriocotyloides*, 243
*bhattacharai, Choanotaenia*, 362
*Biacetabulum*, diagnosis, 30
*biacetabulum, Phyllobothrium*, 161
*biaculata, Amphipetrovia*, 290
*biaculeata, Hymenolepis*, 290
*Bialovarium*, diagnosis, 34
*bialowiezensis, Andrya*, 435
*bicolor, Bothriocephalus*, 72
*bicolor, Tentacularia*, 72
*bicoronata, Brasiliolepis*, 380
*bicoronata, Dilepis*, 380
*bidentatus, Vampirolepis*, 303
*bifaria, Taenia*, 232
*bifidum, Anthobothrium*, 157
*bifidum, Echeneibothrium*, 151
*bifurcatus, Bothriocephalus*, 138
*bigemina, Avitellina*, 419
*Biglandatrium*, diagnosis, 314
*biglandatrium, Biglandatrium*, 314
*bilateralis, Choanotaenia*, 394
*bilateralis, Drepanidotaenia*, 294
*bilateralis, Hymenolepis*, 294
*bilateralis, Laterorchites*, 394
*bilharzi, Taenia*, 325
*bilharzi, Variolepis*, 325
*biliaris, Hymenolepis*, 335
*bilobatum, Echeneibothrium*, 151
*Bilocularia*, 160
*biloculoides, Biacetabulum*, 30
*biloculus, Crescentovitus*, 24
*binuncum, Rhynchobothrium*, 76
*binzui, Anomotaenia*, 405
*biorchidum, Rhinebothrium*, 155
*bipapillosa, Moniezia*, 442
*bipapillosa, Taenia*, 442
Biporophyllidea, 163
*Biporouterina*, diagnosis, 427
*bipunctata, Amphilina*, 468
*bipunctatus, Bothriocephalus*, 101
*biremis, Tatria*, 229
*birmanica, Idiogenes*, 265
*birmanica, Raillietina (F.)*, 265
*birmanica, Echinocotyle*, 285
*birmanica, Hymenolepis*, 285
*birmanicus, Lytocestus*, 25
*biroi, Acanthotaenia*, 181
*biroi, Crepidobothrium*, 181
*biroi, Ichthyotaenia*, 181
*biroi, Proteocephalus*, 181
*birostrata, Choanotaenia*, 362
*birostrata, Trichocephaloidis*, 370
*Birovilepis*, 379
*birulai, Aploparaksis*, 275
*Bisaccanthes*, diagnosis, 313
*bisaccata, Drepanidotaenia (D.)*, 313
*bisaccata, Sobolevicanthus*, 313
*bisaccatus, Bisaccanthes*, 313
*bisaccatus, Hymenolepis*, 313
*bisacculina, Drepanidotaenia*, 314
*bisacculina, Parabisaccanthes*, 314

*bistrobilae, Haplobothrium*, 85
*bisulcata, Nybelinia*, 74
*bisulcata, Tetrarhynchus*, 74
*bisulcatum, Tetrarhynchus*, 73
*bisulcatus, Scyphocephalus*, 85
*biuncinata, Echinorhynchotaenia*, 272
*Biuterina*, diagnosis, 347
*biuterina, Paronia*, 423
*Biuterinoides*, 339
*biuterinus, Lateriporus*, 373
*bivitellatus, Proteocephalus*, 186
*bivetellolobata, Oochoristica*, 454
*bivittata, Linstowia (Opossumia)*, 462
*bivittata, Mathevotaenia*, 462
*bivittata, Oochoristica*, 462
*blanchardi, Anoplocephala*, 434, 436
*blanchardi, Anoplocephaloides*, 436
*blanchardi, Brumptiella*, 253
*blanchardi, Davainea*, 253
*blanchardi, Delamurella*, 253
*blanchardi, Paranoplocephala*, 434, 436
*blanchardi, Raillietina (P.)*, 253
*blanchardi, Taenia*, 366, 434, 436
*blanchardi, Viscoia*, 366
*blanksoni, Himantocestus*, 232
*blarinae, Hymenolepis*, 303
*blarinae, Protogynella*, 274
*blarinae, Rodentolepis*, 303
*blarinae, Vampirolepis*, 303
*boae, Tetrabothrium*, 189
*bobica, Anonchotaenia*, 346
*bobica, Orthoskrjabinia*, 346
*bobica, Skrjabinerina*, 346
*bocki, Paronia*, 423
*bodkini, Raillietina (S.)*, 255
*boehmi, Raillietina (S.)*, 255
*boeti, Raillietina (R.)*, 257
*boe-tigridis, Dibothrium*, 85
*boholensis, Bertiella*, 438
*boisii, Echinobothrium*, 166
*bolivari, Meggittia*, 255
*bolivari, Raillietina (S.)*, 255
*Bombycirhynchus*, 75
*bomensis, Raillietina (P.)*, 253
*bonariensis, Ophiotaenia*, 183
*bonasum, Echinobothrium*, 166
*bondarevae, Monopylidium*, 366
*bondarevae, Rodentotaenia*, 366
*bonini, Brumptiella*, 255
*bonini, Davainea*, 255
*bonini, Markewitchella*, 255
*bonini, Raillietina (S.)*, 255
*borealis, Anomotaenia*, 405
*borealis, Aploparaksis*, 275
*borealis, Aporina*, 425
*borealis, Choanotaenia*, 425
*borealis, Icterotaenia*, 425
*borealis, Neoaporina*, 425
*borealis, Paricterotaenia*, 425
*borealis, Taenia*, 405
*Borgarenkolepis*, 404

*boscii, Tentacularia*, 72
*botauri, Dendrouterina*, 378
*bothridiopunctata, Grillotia*, 64
*Bothridiotaenia*, 207
*Bothridium*, diagnosis, 85
Bothrimonidae fam. n.
  diagnosis, 43
  key to genera, 45
*Bothrimonus*, diagnosis, 45
Bothriocephalidae
  diagnosis, 82
  key to genera, 100
*Bothriocephalus*, diagnosis, 101
*Bothriocotyle*, diagnosis, 109
*bothrioplitis, Davainea*, 257
*Bothrioscolex*, 20
*botrioplites, Taenia*, 259
*bouchei, Dilepidoides*, 316
*bouchi, Hymenolepis*, 316
*boueti, Kotlania*, 257
*bovieni, Metaparonia*, 253
*bovieni, Paruterina*, 350
*bovieni, Raillietina (P.)*, 253
*Bovienia*, diagnosis, 24
*brabantiae, Nematoparataenia*, 129
*brachyacantha, Taenia*, 224
*brachyacanthum, Acanthobothrium*, 138
*brachyarthra, Dilepis*, 380
*brachyascum, Inermiphyllidium*, 158
*brachyascum, Rhodobothrium*, 158
*brachycephala, Anatinella*, 312
*brachycephala, Echinocotyle*
*brachycephala, Hymenolepis*, 312
*brachycephala, Hymenolepis (Echinocotyle)*, 312
*brachycephala, Hymenolepis (H.)*, 312
*brachycephala, Monosaccanthes*, 312
*brachycephala, Taenia*, 312
*brachycephala, Tschertkovilepis*, 313
*brachycollis, Caryophyllaeus*, 33
*brachydera, Taenia*, 304
*brachyphallos, Aploparaksis*, 275
*brachyphallos, Hymenolepis*, 275
*brachyphallos, Taenia*, 275
*brachyrhyncha, Chapmania*, 241
*brachyrhyncha, Davainea*, 241
*brachyrhyncha, Taenia*, 241
*brachysoma, Echinobothrium*, 166
*brachysoma, Oochoristica*, 454
*brachysoma, Taenia*, 224
*Brachyurus*, 35
*brachyurus, Archigetes*, 30
*brachyurus, Brachyurus*, 30
*brachyurus, Glaridacris*, 30
*bramae, Bothriocephalus*, 102
*branchii, Grillotia*, 64
*branchiostegi, Bothriocephalus*, 102
*branchiuterina, Proterogynotaenia*, 220
*brandti, Taenia*, 420
*brasiliensis, Anomotaenia*, 405
*brasiliensis, Anoncotaenia*, 341
*brasiliensis, Fuhrmannia*, 390

*brasiliensis, Hymenolepis*, 325
*brasiliensis, Linstowia*, 461
*brasiliensis, Mathevotaenia*, 462
*brasiliensis, Oochoristica*, 453, 461
*brasiliensis, Ophryocotyle*, 244
*brasiliensis, Variolepis*, 325
*Brasiliolepis*, 379
*brassica, Phyllobothrium*, 161
*brauni, Diplogonoporus*, 90
*brauni, Multiceps*, 224
*brauni, Taenia*, 224
*Braunia*, 87
*bremneri, Taenia*, 224
*bresslauei, Diphyllobothrium*, 93
*bresslauei, Spirometra*, 93
*bresslaui, Culcitella*, 349
*bresslaui, Oochoristica*, 454
*breviannulata, Hymenolepis*, 336
*breviceps, Bothriocephalus*, 102
*breviceps, Taenia*, 341
*brevicirrosus, Hymenolepis*, 328
*brevicirrosus, Microsomacanthus*, 328
*brevicollis, Amoebotaenia*, 387, 396
*brevicollis, Davainea*, 266
*brevicollis, Raillietina*, 266
*brevicollis, Taenia*, 224, 227, 266
*brevihamatus, Oligorchis*, 333
*brevis, Amoebotaenia*, 390
*brevis, Anomotaenia*, 405
*brevis, Choanotaenia*, 405
*brevis, Crepidobothrium*, 189
*brevis, Diorchis*, 281
*brevis, Gyrocoelia*, 216
*brevis, Hymenolepis*, 298
*brevis, Liga*, 390
*brevis, Paradilepis*, 375
*brevis, Paranoplocephala*, 436
*brevis, Passerilepis*, 298
*brevis, Porotaenia*, 208
*brevis, Taenia*, 390
*brevis, Tetrabothrius*, 189
*brevis, Tetrarhynchus*, 77
*brevis, Weinlandia*, 298
*Breviscolex*, diagnosis, 38
*brevispine, Mecistobothrium*, 55
*brevispine, Pedibothrium*, 143
*brevispine, Rhynchobothrium*, 55, 76
*brevissime, Acanthobothrium*, 139
*brittanicum, Phyllobothrium*, 161
*Brochocephalus*, 215
*brotogerys, Cotugnia*, 245
*brotulae, Bothriocephalus*, 102
*browni, Cotugnia*, 246
*brumpti, Raillietina, (R.)*, 259
*Brumptiella*, 252
*brustae, Hymenolepis*, 335
*bubalinae, Avitellina*, 419
*bubesei, Taenia*, 224
*bucerotidarum, Brumptiella*, 265
*bucerotidarum, Raillietina (F.)*, 265
*bucerotina, Paruterina*, 350

*buckleyi, Parvitaenia*, 412
*buckleyi, Raillietina, (R.)*, 257
*bucorvi, Idiogenes*, 239
*buecki, Ophryocotyle*, 244
*buencaminoi, Dipylidium*, 353
*bufonis, Distoichometra*, 212
*bufonis, Ophiotaenia*, 184
*bufonis, Taenia*, 213
*bulbifer, Rhynchobothrium*, 65
*bulbiferum, Parabothrium*, 109
*bulbocirrosus, Retinometra*, 317
*bulbocirrus, Aploparaksis*, 275
*bulbocirrus, Isoglaridacris*, 37
*bulbodes, Diorchis*, 281
*bulbularum, Raillietina (P.)*, 253
*bulgarica, Aploparaksis*, 275
*bumi, Raillietina, (R.)*, 257
*buplanensis, Proteocephalus*, 186
*burgeri, Echeneibothrium*, 151
*burhini, Choanotaenia*, 360
*burhini, Infula*, 216
*burmanense, Eugonodaeum*, 355
*burmanensis, Oligorchis*, 375
*burmanensis, Paradilepis*, 375
*bursaria, Cittotaenia*, 451
*burti, Choanotaenia*, 362
*burti, Paricterotaenia*, 396
*burti, Polycercus*, 396
*Burtiella*, 244
*butasteri, Idiogenes*, 239
*buteonis*, Idiogenes, 240
*buzzardia, Kowalewskiella*, 395
*buzzardia, Tubanguiella*, 395
*bybralis, Eugonodaeum*, 355
*bycanistis, Davainea*, 257
*bycanistis, Kotlania*, 257
*bycanistis, Raillietina, (R.)*, 257

# C

*caballeroi, Aploparaksis*, 275
*caballeroi, Djombangia*, 23
*caballeroi, Floriceps*, 53
*caballeroi, Panuwa*, 367
*caballeroi, Paradilepis*, 375
*caballeroi, Raillietina (F.)*, 265
*caballeroi, Schizorchis*, 440
*cablei, Neodioecocestus*, 214
*cacatuae, Hemiparonia*, 427
*cacatuae, Schizotaenia*, 427
*cacatuinae, Davainea*, 257
*cacatuinae, Kotlania*, 257
*cacatuinae, Raillietina, (R.)*, 257
*cacatuinae, Raillietina (Ransomia)*, 257
*cadenati, Nybelinia*, 74
*caenodex, Anomotaenia*, 405
*caestus, Mesocestoides*, 203
*calcaria, Davainea*, 257
*calcaria, Kotlania*, 257
*calcaria, Raillietina (R.)*, 257

*calcaria, Raillietina (Ransomia)*, 257
*calcauterina, Paronia*, 423
*caledonica, Anomotaenia*, 405
*caledonica, Choanotaenia*, 405
*calentinei, Isoglaridacris*, 37
*Calentinella*, diagnosis, 27
*californica, Catenotaenia*, 234
*californicus, Hymenolepis*, 322
*callariae, Bothriocephalus*, 102
*Calliobothrium*, diagnosis, 135
*Callitetrarhynchus*, diagnosis, 53
*calmettei, Crepidobothrium*, 184
*calmettei, Ichthyotaenia*, 184
*calmetti, Ophiotaenia*, 184
*calmetti, Proteocephalus*, 184
*calmettei, Taenia*, 184
*Calostaurus*, diagnosis, 251
*calotes, Oochoristica*, 454
*calumnacantha, Dicranotaenia*, 321
*calumnacantha, Hymenolepis*, 321
*calumnacantha, Wardium*, 321
*calva, Davainea*, 255
*calva, Taenia*, 255
*Calycobothrium*, diagnosis, 122
*calyptocephalus, Tetragonoporus*, 89
*calyptomenae*, Kotlania, 257
*calyptomenae, Raillietina (R.)*, 257
*calyptomenae, Raillietina, (Ransomia)*, 257
*Calyptrobothrium*, 160
*cambrensis, Hymenolepis*, 310
*cambrensis, Nadejdolepis*, 310
*cameroni, Diphyllobothrium*, 93
*cameroni, Echinococcus*, 222
*cameroni, Hymenolepis*, 336
*campanulata, Biuterina*, 348
*campanulata, Choanotaenia*, 396
*campanulata, Davainea*, 266
*campanulata, Daveneolepis*, 266
*campanulata, Icterotaenia*, 396
*campanulata, Parabertiella*, 432
*campanulata, Paricterotaenia*, 396
*campanulata, Polycercus*, 396
*campanulata, Raillietina*, 266
*campanulata, Taenia*, 348
*campanulatum, Crossobothrium*, 161
*campanulatus, Tetrabothrius*, 208
*campbelli, Dioecotaenia*, 198
*campestris, Anoplòcephala*, 434, 436
*campylacantha, Alcataenia*, 410
*campylacantha, Anomotaenia*, 410
*campylacantha, Pseudanomotaenia*, 410
*campylacantha, Taenia*, 410
*campylancristrota, Dilepis*, 380
*campylancristrota, Taenia*, 380
*campylometra, Deltokeras*, 347
*canadensis, Diphyllobothrium*, 88
*cancellata, Dioecotaenia*, 198
*cancellata, Rhinobothrium*, 198
*candelabraria, Paruterina*, 350
*candelabraria, Taenia*, 350
*caninum, Dipylidium*, 353

*caninum solium, Taenia*, 226
*canislagopodis, Mesocestoides*, 203
*cannochaeti, Helictometra*, 420
*cannocaeti, Thysaniezia*, 420
*cantaniana, Davainea*, 291
*cantaniana, Hymenolepis*, 291
*cantaniana, Staphylepis*, 291
*cantaniana, Taenia*, 291
*capellae, Anomolepis*, 380
*capellae, Dicranotaenia*, 321
*capellae, Dilepis*, 380
*capellae, Hymenolepis*, 321
*capensis, Catenotaenia*, 236
*capensis, Hymenolepis*, 335
*capensis, Inermicapsifer*, 465
*capensis, Zschokkeella*, 465
*capetownensis, Debloria*, 312
*capetownensis, Hymenolepis (H.)*, 312
*capillaris, Confluaria*, 325
*capillaris, Davainea*, 258
*capillaris, Dubininolepis*, 325
*capillaris, Hymenolepis*, 325
*capillaris, Kotlania*, 258
*capillaris, Raillietina (R.)*, 258
*capillaris, Raillietina (Ransomia)*, 258
*capillaris, Taenia*, 325
*capillaris, Variolepis*, 325
*capillaroides, Confluaria*, 325
*capillaroides, Dicranotaenia*, 325
*capillaroides, Dubininolepis*, 325
*capillaroides, Hymenolepis*, 325
*capillaroides, Variolepis*, 325
*capillicollis, Bothriocephalus*, 102
*Capingens*, diagnosis, 37
Capingentidae, key to genera, 37
*Capingentoides*, diagnosis, 39
*capitellata, Diplacanthus (Dilepis)*, 297
*capitellata, Hymenolepis*, 297
*capitellata, Taenia*, 297, 330
*capito, Cyclustera*, 389
*capito, Taenia*, 389
*Capiuterilepis*, diagnosis, 270
*capoori, Oncobothrium*, 136
*capoori, Raillietina (P.)*, 253
*caprae, Moniezia*, 442
*caprae, Taenia*, 442
*capreoli, Taenia*, 441
*caprimulgi, Raillietina (S.)*, 255
*caprimulgina, Dilepis*, 381
*caprimulgorum, Dilepis*, 376
*caprimulgorum, Hymenolepis*, 325
*caprimulgorum, Metadilepis*, 376
*caprimulgorum, Variolepis*, 325
*Capsodavainea*, 241
*Capsulata*, diagnosis, 359
*carangis, Bothriocephalus*, 102
*carangis, Rhynchobothrius*, 76
*carchariae, Abothrium*, 76
*carchariae, Pierretia*, 76
*carchariae-rondolettii, Tetrabothrium*, 143
*carcharias, Tetrarhynchus*, 77

*carcharidis, Tetrarhynchus,* 77
*caribbensis, Discobothrium,* 125
*caribbensis, Eutetrarhynchus,* 57
*carioca, Davainea,* 283
*carioca, Dicranotaenia,* 283
*carioca, Echinolepis,* 283
*carioca, Hymenolepis,* 283
*carneostrobilata, Raillietina (R.),* 258
*carnivoricolus, Mesocestoides,* 203
*caroli, Hymenolepis,* 317
*caroli, Hymenosphenacanthus,* 317
*caroli, Retinometra,* 317
*caroli, Taenia,* 317
*caroyoni, Eutetrarhynchus,* 57
*carpatheca, Batrachotaenia,* 184
*carpatheca, Ophiotaenia,* 184
*carpionis, Ligula,* 87
*Carpobothrium,* diagnosis, 156
*carpophagi, Kotlania,* 258
*carpophagi, Raillietina (R.),* 258
*carribaenis, Parvitaenia,* 412
*carrinii, Moniezia,* 423
*carrinii, Paronia,* 423
*carrucci, Aocobothrium,* 147
*cartagenensis, Acanthobothrium,* 138
*Caryoaustralus,* diagnosis, 21
Caryophyllaeidae, key to genera, 26
*Caryophyllaeides,* diagnosis, 19
*Caryophyllaeus,* diagnosis, 32
Caryophyllidea, 11
    key to families, 17
*caryophyllum, Poecilancistrum,* 69
*caryophyllum, Rhynchobothrium,* 69
*caspicus, Caryophyllaeus,* 33
*castellanii, Anoncotaenia,* 341
*casuari, Davainea,* 258
*casuari, Kotlania,* 258
*casuari, Kotlanotaurus,* 258
*casuari, Raillietina (R.),* 258
*casuari, Raillietina (Ransomia),* 258
*cataeniformisvulpes, Taenia,* 203
*catenatum, Aphanobothrium,* 228
*Catenotaenia,* diagnosis, 234
Catenotaeniidae, diagnosis, 201, 232
*cathartis, Cladotaenia,* 413
*catherineae, Tetrabothrius,* 208
Cathetocephalidae, diagnosis, 131
*Cathetocephalus,* diagnosis, 132
*catostomi, Glaridacris,* 36
*caucasica, Raillietina (S.),* 255
*caudatum, Pelichnibothrium,* 149
*Caulobothrium,* diagnosis, 152
*cavoarmata, Hymenolepis,* 328
*cavoarmata, Microsomacanthus,* 328
*cavoarmata, Tschertkovilepis,* 328
*cayennensis, Choanotaenia,* 362
*cayennensis, Monopylidium,* 405
*cayennensis* var. *africana, Choanotaenia,* 362
*cayennensis* var. *scolopacis, Choanotaenia,* 363
*cebidarum, Hymenolepis,* 289
*celebensis, Davainea,* 258

*celebensis, Kotlania,* 258
*celebensis, Meggittia,* 266
*celebensis, Raillietina (R.),* 258
*celebensis, Raillietina (Ransomia),* 258
*celebensis, Raillietina* var. *paucicapsulata,* 258
*celebesensis, Cotugnia,* 246
*celebesensis, Oochoristica,* 454
*centripunctata, Avitellina,* 418, 419
*centripunctata, Stilesia,* 418
*centripunctata, Taenia,* 418
*centrocerci, Raillietina, (S.),* 255
*centropi, Brumptiella,* 255
*centropi, Daovantienia,* 255
*centropi, Davainea,* 255
*centropi, Raillietina (S.),* 255
*centrurum, Phyllobothrium,* 161
*centuri, Choanotaenia,* 362
*centuri, Raillietina (P.),* 253
*cepanzoi, Echinococcus,* 222
Cephalochlamydidae
    diagnosis, 81
    key to genera, 84
*Cephalochlamys,* diagnosis, 84
*cepolae, Bothriocephalus,* 102
*cepolae-rubescentis, Tetrarhynchus,* 77
*ceratias, Amphicotyle,* 106
*Ceratobothrium,* diagnosis, 133
*cercopitheci, Bertiella,* 438
*cercopitheci, Vampirolepis,* 303
*cerebralis leporiscuniculi, Coenurus,* 226
*cernuae, Proteocephalus,* 186
*cervi, Taenia,* 224
*cervinum, Platybothrium,* 137, 143
*cervotestis, Dicranotaenia,* 321
*cervotestis, Hymenolepis,* 321
*cesticillus, Brumptiella,* 255
*cesticillus, Davainea,* 255
*cesticillus, Raillietina (S.),* 255
*cesticillus, Taenia,* 255
Cestodaria, key to orders, 467
Cestoidea, keys to subclasses, 11
*cestracii, Acanthobothrium,* 139
*cestraciontis, Acanthobothrium,* 139
*cestus, Bothriocephalus,* 102
*ceylonica, Davainea,* 258
*ceylonica, Kotlania,* 258
*ceylonica, Raillietina (R.),* 258
*ceylonica, Raillietina (Ransomia),* 258
*ceylonicum, Echeneibothrium,* 151
*ceylonicus, Tetrarhynchus,* 77
*chabaudi, Catenotaenia,* 233
*chabaudi, Hemicatenotaenia,* 233
*chabaudi, Oochoristica,* 454
*chaenogobii, Nippotaenia,* 172
*Chaetophallus,* diagnosis, 206
*chaeturichthydis, Pterobothrium,* 70
*chalcophapsi, Diorchis,* 281
*chalinolobi, Hymenolepis,* 335
*chalmersi, Avitellina,* 419
*chalmersius, Caryophyllaeus,* 33
*chalmersius, Monobothrioides,* 24

*chamisoni, Taenia*, 224
*chandleri, Anomotaenia*, 405
*chandleri, Choanotaenia*, 405
*chandleri, Monopylidium*, 405
*changtuensis, Schizorchis*, 440
*Chapmania*, diagnosis, 241
*charadrii, Anomotaenia*, 405
*charadrii, Hymenolepis*, 310
*charadrii, Mesocestoides*, 203
*charadrii, Paraprogynotaenia*, 221
*charadrii, Proterogynotaenia*, 221
*charadrii, Trichocephaloidis*, 370
*chauhani, Anoncotaenia*, 341
*chauhani, Aporina*, 424
*chauhani, Davainea*, 247
*chauhani, Malika*, 356
*chaunense, Dicranotaenia*, 321
*chaunense, Wardium*, 321
*chavenoni, Oochoristica*, 454
*cheilancristrota, Acanthocirrus*, 371
*cheilancristrota, Taenia*, 371
*cheilancristrotus, Gryporhynchus*, 371
*cheilancristrotus, Neogryporhynchus*, 371
*Chelacanthus*, 280, 320
*chelidonariae, Anomotaenia*, 410
*chelidonariae, Pseudanomotaenia*, 410
*chengi, Acanthobothrium*, 139
*chenopsis, Hymenolepis*, 336
*chetekensis, Isoglaridacris*, 37
*Chettusiana*, diagnosis, 391
*chikugoensis, Aploparaksis*, 276
*childi, Hymenolepis*, 328
*childi, Microsomacanthus*, 328
*chilensis, Anoncocephalus*, 113
*chilensis, Rhinebothrium*, 155
*chilmei, Raillietina (R.)*, 258
*chiloscyllii, Carpobothrium*, 156
*chiloscyllius, Eulacistorhynchus*, 63
*chimaerae, Crobylophorus*, 473
*Chimaerolepis*, 294
*chinata, Davainea*, 266
*chinensis, Oochoristica*, 454
*chinensis, Retinometra*, 317
*chionis, Choanotaenia*, 362
*chionis, Dicranotaenia*, 321
*chionis, Hymenolepis*, 321
*chionis, Icterotaenia*, 362
*chionis, Paricterotaenia*, 362
*chionis, Weinlandia*, 321
*chironemi, Rhynchobothrius*, 76
*chiropterophila, Vampirolepis*, 303
*Chitinolepis*, diagnosis, 332
*Chitinorecta*, diagnosis, 402
*chlamyderae, Choanotaenia*, 396
*chlamyderae, Icterotaenia*, 396
*chlamyderae, Paricterotaenia*, 396
*chlamyderae, Polycercus*, 396
*chlamyderae, Taenia*, 396
*Chlamydocephalus*, 84
*chlamydoselachi, Monorygma*, 163
*chlorurae, Paruterina*, 350

*Choanofuhrmannia*, 361
*Choanoscolex*, diagnosis, 177
*Choanotaenia*, diagnosis, 361
*choatica, Taenia*, 395
*Cholodkovskia*, 367
*cholodkowskii, Deltokeras*, 345
*cholodkowskii, Dictyuterina*, 345
*cholodkowskii, Paruterina*, 345
*cholodkowskyi, Choanotaenia*, 362
*cholodkowskyi, Hymenolepis*, 306
*chologasteri, Proteocephalus*, 186
*christensoni, Vampirolepis*, 303
*Christianella*, diagnosis, 55
*chrysaeti, Taenia*, 204
*chyseri, Dipylidium*, 353
*chyseri, Joyeuxia*, 353
*chrysochloridis, Hymenolepis*, 301
*chrysochloridis, Staphylocystis*, 301
*chrysocolaptis, Krimi*, 394
*ciconia, Hymenolepis*, 336
*ciliata, Anomotaenia*, 384
*ciliata, Platyscolex*, 384
*ciliotheca, Polyoncobothrium*, 97
*ciliotheca, Tetracampos*, 97
*Cinclotaenia*, diagnosis, 367
*cingulata, Anomotaenia*, 405
*cingulata, Dilepis*, 405
*cingulifera, Choanotaenia*, 360
*cingulifera, Kowalewskiella*, 360
*cingulifera, Monopylidium*, 360
*cingulifera, Taenia*, 360
*cingulum, Ligula*, 87
*circi, Cladotaenia*, 413
*circularis, Rhinebothroides*, 154
*circumcincta, Davainea*, 266
*circumcincta, Kotlania*, 266
*circumcincta, Raillietina*, 266
*circumcincta, Taenia*, 266
*Circumoncobothrium*, diagnosis, 95
*circumvallata cadarachensis, Raillietina (S.)*, 256
*circumvallata, Davainea*, 255
*circumvallata, Meggittia*, 256
*circumvallata, Raillietina (R.)*, 255
*circumvallata, Raillietina (Ransomia)*, 255
*circumvallata, Raillietina (S.)*, 255
*circumvallata siberica, Raillietina (S.)*, 256
*circumvallata, Taenia*, 255
*cirroflexa, Raillietina (P.)*, 253
*cirrosa, Aploparaksis*, 322
*cirrosa, Dicranotaenia*, 322
*cirrosa, Haploparaksis*, 322
*cirrosa, Monorchis*, 322
*cirrosa, Taenia*, 322
*cirrospinosa, Choanotaenia*, 362
*cirrospinosa, Paricterotaenia*, 362
*cirrostilifer, Hymenolepis*, 317
*cirrostilifer, Hymenosphenacanthus*, 317
*cirrostilifer, Retinometra*, 317
*cirrovaginata, Soricina*, 287
*citelli, Hymenolepis*, 289
*citrus, Anomotaenia*, 405

citrus, *Choanotaenia*, 405
citrus, *Dichoanotaenia*, 405
citrus, *Taenia*, 405
*Cittotaenia*, diagnosis, 443
*Cladogynia*, diagnosis, 315
*Cladotaenia*, diagnosis, 412
clairae, *Raillietina (R.)*, 258
clandestina, *Dicranotaenia*, 322
clandestina, *Hymenolepis*, 322
clandestina, *Taenia*, 322
clanguli, *Echinatrium*, 320
clariae, *Pseudolytocestus*, 40
clarias, *Polyoncobothrium*, 97
clausa, *Skorikowia*, 280
clausovaginata, *Arhynchotaenia*, 290
clausovaginata, *Arhynchotaeniella*, 290
clava, *Anonchotaenia*, 341
clavata, *Aploparaksis*, 276
clavata, *Taenia*, 341
clavatum, *Diphyllobothrium*, 93
clavatus, *Pseudodiorchis*, 278
clavibothrium, *Bothriocephalus*, 102
claviceps, *Bothriocephalus*, 102
clavicirrosa, *Davainea*, 258
clavicirrosa, *Kotlania*, 258
clavicirrosa, *Raillietina (Ransomia)*, 258
clavicirrosa, *Raillietina (R.)*, 258
clavicirrus, *Dicranotaenia*, 322
clavicirrus, *Hymenolepis*, 322
clavifer, *Coenurus*, 224
clavifer, *Taenia*, 224
claviformis, *Baerietta*, 212
claviger, *Bothriocephalus*, 76
clavigera, *Anomotaenia*, 405
clavigera, *Choanotaenia*, 405
clavigera, *Dichoanotaenia*, 405
clavigera, *Taenia*, 405
clavipera, *Parvitaenia*, 412
clavulus, *Aploparaksis*, 276
clavulus, *Biuterina*, 348
clavulus, *Davainea*, 348
clavulus, *Taenia*, 348
*Cleberia*, diagnosis, 456
clelandi, *Zosteropicola*, 340
*Clelandia*, diagnosis, 382
clerci, *Aploparaksis*, 276
clerci, *Biuterina*, 348
clerci, *Echinocotyle*, 285
clerci, *Kotlania*, 258
clerci, *Lateriporus*, 373
clerci, *Polycercus*, 396
clerci, *Raillietina, (R.)*, 258
clerci, *Raillietina, (Ransomia)*, 258, 264
*Clestobothrium*, diagnosis, 97
*Cloacotaenia*, diagnosis, 337
*Clujia*, 75
clupeoidesii, *Ptychobothrium*, 99
*Clydonobothrium*, diagnosis, 156
coatsi, *Diphyllobothrium*, 95
cobitidis, *Ligula*, 87
cobraeformis, *Discobothrium*, 125

cobraeformis, *Proteocephalus*, 186
cochlearii, *Parvitaenia*, 412
*Coelobothrium*, diagnosis, 97
*Coelodela*, diagnosis, 422
coelorhynchi, *Microbothriorhynchus*, 75
coenoformum, *Echinobothrium*, 166
*Coenomorphus*, 62
coherens, *Diplocotyle*, 45
cohni, *Amoebotaenia*, 387
cohni, *Davainea*, 258
cohni, *Kotlania*, 258
cohni, *Raillietina (R.)*, 258
cohni, *Raillietina (Ransomia)*, 258
cohospes, *Ophiotaenia*, 184
coili, *Diplophallus*, 232
colinia, *Raillietina (R.)*, 258
collaris, *Dicranotaenia*, 328
collaris, *Hymenolepis*, 328
collaris, *Microsomacanthus*, 328
collaris, *Myxolepis*, 328
collaris, *Sobolevicanthus*, 328
collaris, *Taenia*, 328
collaris, *Weinlandia*, 328
collini, *Cotugnia*, 246
collini, *Ershovitugnia*, 246
collocaliae, *Notopentorchis*, 350
collocaliae, *Pseudochoanotaenia*, 355
colluriones, *Biuterina*, 348
columbae, *Bothriocephalus*, 95
columbae, *Davainea*, 265
colombianum, *Acanthobothrium*, 139
coloradensis, *Sobolevicanthus*, 308
columbae, *Cittotaenia*, 422
columbae, *Cotugnia*, 246
columbae, *Davainea*, 255
columbae, *Diplopylidium*, 352
columbae, *Dipylidium*, 352
columbae, *Diskrjabiniella*, 352
columbae, *Hymenolepis*, 308
columbae, *Icterotaenia*, 366
columbae, *Moniezia*, 423
columbae, *Monopylidium*, 366
columbae, *Paronia*, 423
columbae, *Progynopylidium*, 352
columbae, *Raillietina (S.)*, 255
columbae, *Sobolevicanthus*, 308
columbae, *Taenia*, 308
*Columbia*, diagnosis, 419
columbiella, *Raillietina (R.)*, 258
columbina, *Hymenolepis*, 325
columbina, *Variolepis*, 325
colymba, *Schistotaenia*, 230
colymbi, *Diporotaenia*, 227
*Colymbilepis*, 320
comani, *Phascolotaenia*, 448
comitata, *Davainea*, 258
comitata, *Kotlania*, 258
comitata, *Raillietina (R.)*, 258
comitata, *Raillietina (Ransomia)*, 258
communis, *Caryophyllaeus*, 32
commutatus, *Rhynchobothrius*, 76

*compacta, Brumptiella*, 253
*compacta, Catenotaenia*, 234
*compacta, Davainea*, 253
*compacta, Raillietina (P.)*, 253
*compacta, Taenia*, 303
*compacta* var. *polytestis, Raillietina (P.)*, 253
*compactum, Phyllobothrium*, 161
*composita, Dyandrya*, 444
*compressa, Hymenolepis*, 328
*compressa, Microsomacanthus*, 328
*compressa, Taenia*, 328
*conardi, Taenia*, 283
*conferta, Bertiella*, 438
*conferta, Taenia (Bertia)*, 438
*Confluaria*, 324
*confusa, Glaridacris*, 36
*confusa, Marsipometra*, 110
*confusa, Taenia*, 222
*confusum, Acanthobothrium*, 139
*Congeria*, 73
*congolensis, Bertiella*, 438
*congolensis, Inermicapsifer*, 465
*congolensis, Ophiotaenia*, 184
*congolensis, Raillietina (R.)*, 258
*congri, Gongeria*, 74
*congri, Nybelinia*, 74
*conica, Anonchotaenia*, 347
*conica, Orthoskrjabinia*, 347
*conica, Skrjabinerina*, 347
*coniceps, Bothriocephalus*, 93
*coniceps, Diphyllobothrium*, 94
*conicus, Echinorhynchus*, 77
*coniformis, Litobothrium*, 169
*conjugens, Moniezia*, 442
*conocephala, Taenia*, 224
*conoideis, Otiditaenia*, 240
*conoideis, Schistometra*, 240
*conoideis, Taenia*, 240
*conopophilae, Brumptiella*, 253
*conopophilae, Davainea*, 253
*conopophilae, Raillietina (P.)*, 253
*conscripta, Hymenolepis*, 328
*conscripta, Microsomacanthus*, 328
*conscripta, Tschertkovilepis*, 328
*constricta, Anomotaenia*, 409
*constricta, Choanotaenia*, 409
*constricta, Pseudanomotaenia*, 409
*constricta, Taenia*, 409
*continua, Acanthotaenia*, 182
*continua, Crepidobothrium*, 182
*continuus, Proteocephalus*, 182
*contorta, Brumptiella*, 251
*contorta, Davainea*, 251
*contorta, Diorchiraillietina*, 251
*contorta, Raillietina (P.)*, 251
*contorta, Raillietina (R.)*, 251
*contorta, Nippotaenia*, 172
*contortrix, Ligula*, 87
*contracta, Hymenolepis*, 335
*convolutum, Oncobothrium*, 136
*Copesoma*, 413

*coraciae, Hymenolepis*, 336
*coracii, Biuterina*, 348
Corallobothriinae, key to genera, 177
*Corallobothrium*, diagnosis, 180
*Corallotaenia*, diagnosis, 179
*cordatum, Diphyllobothrium*, 93
*Cordicephalus*, 91, 93
*cordiceps, Diphyllobothrium*, 88
*cordobensis, Hymenolepis*, 291
*cordobensis, Staphylepis*, 291
*coreensis, Raillietina (R.)*, 258
*coregoni, Proteocephalus*, 186
*cormoranti, Microsomacanthus*, 328
*cornicis, Taenia*, 298
*cornucopia, Anthobothrium*, 148, 157
*corollatum, Calliobothrium*, 138
*corollatus, Tetrarhynchus*, 57
*coronata, Choanotaenia*, 362
*coronata, Paricterotaenia*, 362
*coronata, Taenia*, 362
*coronati, Onderstepoortia*, 360
*coronatum, Acanthobothrium*, 138
*coronatum, Echinobothrium*, 166
*coronatus, Polypocephalus*, 121
*coronea, Corvinella*, 254
*coronea, Raillietina (P.)*, 254
*coronina, Taenia*, 409
*coronoides, Variolepis*, 325
*coronula, Dicranotaenia*, 321
*coronula, Diplacanthus (Lepidotrias)*, 321
*coronula, Drepanidotaenia*, 321
*coronula, Hymenolepis*, 321
*coronula, Hymenolepis (Drepanidotaenia)*, 321
*coronula micracantha, Dicranotaenia*, 321
*coronula, Taenia*, 321
*coronula, Weinlandia*, 321
*corrariella, Dicranotaenia*, 322
*corti, Mesocestoides*, 203
*corvi, Choanotaenia*, 362, 364
*corvi, Dicranotaenia*, 299
*corvi, Hymenolepis*, 299
*corvi, Weinlandia*, 299
*corvi-cornicis, Taenia*, 298
*corvifrugilegi, Taenia*, 298
*corvina, Brumptiella*, 254
*corvina, Davainea*, 254
*corvina, Raillietina (P.)*, 254
*corvinella*, 253
*corvorum, Ophriocotyloides*, 243
*corvorum, Taenia*, 298
*coryllidis, Paronia*, 423
*corymbum, Rhinebothrium*, 155
*coryphaenae, Tentacularia*, 72
*coryphicephala, Ichthyotaenia*, 194
*coryphicephala, Monticellia*, 194
*coryphicephala, Proteocephalus*, 194
*coryphicephala, Taenia*, 194
*coryphicephala, Tetracotylus*, 194
*Cotugnia*, diagnosis, 245
*coturnix, Metroliasthes*, 344
*coturnixi, Raillietina (R.)*, 258

*Cotylorhipis*, diagnosis, 383
*courdurieri*, *Oochoristica*, 454
*Cracticotaenia*, 404
*crassa*, *Cotugnia*, 246
*crassa*, *Culcitella*, 349
*crassa*, *Duthiersia*, 86
*crassa*, *Gyrocoelia*, 216
*crassa*, *Hymenolepis*, 303
*crassa*, *Raillietina* (*Johnstonia*), 265
*crassa*, *Rodentolepis*, 303
*crassa*, *Taenia*, 107
*crassa*, *Vampirolepis*, 303
*crassiceps*, *Alyselminthus*, 224
*crassiceps*, *Clestobothrium*, 97
*crassiceps*, *Oochoristica*, 454
*crassiceps*, *Rhynchobothrius*, 76
*crassiceps*, *Taenia*, 224
*crassicollis*, *Acanthobothrium*, 139
*crassicollis*, *Diorchis*, 281
*crassicollis*, *Moniezia*, 442
*crassicollis* var. *nova*, *Moniezia*, 442
*crassicollis*, *Pterobothrium*, 70
*crassipenis*, *Aploparaksis*, 276
*crassipora*, *Taenia*, 224
*crassirostrata*, *Dilepis*, 381
*crassirostris*, *Aploparaksis*, 276
*crssirostris*, *Dicranotaenia*, 276
*crassirostris*, *Hymenolepis*, 276
*crassirostris*, *Monorchis*, 276
*crassirostris*, *Taenia*, 276
*crassitestata*, *Choanotaenia*, 396
*crassitestata*, *Icterotaenia*, 396
*crassitestata*, *Paricterotaenia*, 396
*crassitestata*, *Polycercus*, 396
*crassitestata*, *Sacciuterina*, 396
*crassivesicula*, *Cyclorchida*, 378
*crassoides*, *Eubothrium*, 107
*crassula*, *Davainea*, 258, 265
*crassula*, *Kotlania*, 265
*crassula*, *Raillietina* (*F.*), 265
*crassula*, *Taenia*, 265
*crassus*, *Eubothrium*, 107
*crassus*, *Traenophorus*, 113
*crateriformis*, *Choanotaenia*, 366
*crateriformis*, *Monopylidium*, 366
*crateriformis*, *Taenia*, 259, 366
*crawfordi*, *Pseudanoplocephala*, 434
*creani*, *Tetrabothrius*, 208
*crecca*, *Hymenolepis*, 331
*crenacollis*, *Otobothrium*, 68
*crenata*, *Dicranotaenia*, 298
*crenata*, *Hymenolepis*, 298
*crenata*, *Mayhewia*, 298
*crenata*, *Passerilepis*, 298
*crenata*, *Taenia*, 298
*crenata*, *Variolepis*, 298
*crenulatum*, *Orygmatobothrium*, 163
*Crepidobothrium*, diagnosis, 188
*creplini*, *Dicranotaenia*, 322
*creplini*, *Hymenolepis*, 322
*creplini*, *Taenia*, 322

*Crescentovitus*, diagnosis, 24
*criceti*, *Dicranotaenia*, 303
*criceti*, *Hymenolepis*, 303
*criceti*, *Rodentolepis*, 303
*criceti*, *Staphylocystis*, 303
*criceti*, *Vampirolepis*, 303
*cricetomydis*, *Meggittina*, 236
*cricetomydis*, *Skrjabinotaenia*, 236
*cricetorum*, *Catenotaenia*, 234
*crimensis*, *Myotolepis*, 287
*criniae*, *Baerietta*, 212
*crispatissimum*, *Phyllobothrium*, 161
*crispum*, *Thysanocephalum*, 133
*critica*, *Arhynchotaenia*, 465
*criticus*, *Inermicapsifer*, 465
*croaxum*, *Choanotaenia*, 362
*Crobylophorus*, 473
*crocethia*, *Debloria*, 312
*crocethia*, *Dicranotaenia*, 312
*crocethia*, *Echinocotyle*, 312
*crocethia*, *Nadejdolepis*, 312
*crocethia*, *Hymenolepis* (*H.*), 312
*crocethiae*, *Echinocotyle*, 285
*crocethiae*, *Hymenoleois* (*Echinocotyle*), 285
*crociduri*, *Hymenolepis*, 335
*crocutae*, *Taenia*, 224
*Crossobothrium*, 160
*crotalicola*, *Oochoristica*, 454
*crotophopeltis*, *Ophiotaenia*, 184
*cruciata*, *Brumptiella*, 254
*cruciata*, *Davainea*, 254
*cruciata*, *Raillietina* (*P.*), 254
*cruciata*, *Taenia*, 254
*crucigera*, *Moniezia*, 442
*crucigera*, *Taenia*, 442
*cruzi*, *Echinococcus*, 222
*cruzsilvai*, *Mathevotaenia*, 462
*cryptacantha*, *Davainea*, 258
*cryptacantha*, *Hymenolepis*, 336
*cryptacantha*, *Kotlania*, 258
*cryptacantha*, *Raillietina* (*R.*), 258
*cryptacantha*, *Raillietina* (*Ransomia*), 258
*cryptobothrium*, *Ichthyotaenia*, 454
*cryptobothrium*, *Oochoristica*, 454
*cryptobothrius*, *Archigetes*, 30
*cryptobranchi*, *Ophiotaenia*, 184
*cryptocotyle*, *Davainea*, 256
*cryptocotyle*, *Raillietina* (*S.*), 256
*Cryptocotylepis*, diagnosis, 288
*crypturi*, *Davainea*, 258
*crypturi*, *Kotlania*, 258
*crypturi*, *Raillietina* (*R.*), 258
*crypturi*, *Raillietina* (*Ransomia*), 258
*ctenoides*, *Cittotaenia*, 451
*ctenoides*, *Ctenotaenia*, 451
*ctenoides*, *Mosgovoyia*, 451
*ctenoides*, *Taenia*, 451
*Ctenotaenia*, diagnosis, 448
*cubana*, *Mathevotaenia*, 462
*cubanus*, *Sobolevicanthus*, 308
*cubensis*, *Inermicapsifer*, 464

*cubensis, Raillietina,* 464
*cubensis, Thysanotaenia,* 464
*Cucurbilepis,* 293
*cucurbitina canis, Taenia,* 226
*Culcitella,* diagnosis, 349
*culiauana, Raillietina (P.),* 254
*cuneata, Amoebotaenia,* 387
*cuneata, Cotugnia,* 246
*cuneata, Hymenolepis,* 328
*cuneata, Microsomacanthus,* 328
*cuneata, Sphenacanthus,* 328
*cuneata, Taenia,* 240, 387
*cuneata, Tschertkovilepis,* 328
*cuneata* var. *nervosa, Cotugnia,* 246
*cuneata* var. *tenuis, Cotugnia,* 247
*cuniculi, Andrya,* 435
*cuniculi, Anoplocephala,* 435
*cuniculi, Coenurus,* 226
*cuniculi, Moniezia,* 435
*cunningtoni, Monobothrioides,* 24
*cunningtoni, Proteocephalus,* 187
*curilensis, Tetrabothrius,* 208
*curiosa, Anatinella,* 313
*curiosa, Drepanidotaenia,* 313
*curiosa, Monosaccanthes,* 313
*curtum, Otobothrium,* 68
*curtum, Rhynchobothrium,* 68
*cuspidatus, Bothriocephalus,* 102
*cyanocittii, Oligorchis,* 333
*cyathiformis, Anomotaenia,* 410
*cyathiformis, Dichoanotaenia,* 410
*cyathiformis, Pseudanomotaenia,* 410
*cyathiformis, Taenia,* 410
*cyathiformoides, Anomotaenia,* 405, 410
*cyathiformoides, Pseudoanomotaenia,* 410
Cyathocephalidae, diagnosis, 43
*Cyathocephalus,* diagnosis, 44
*Cyatocotyle,* diagnosis, 147
*cybiumi, Gymnorhynchus,* 61
*Cyclobothrium,* 122
Cyclophyllidea, 16
    key to families, 199
*cyclops, Proteocephalus,* 187
*Cyclorchida,* diagnosis, 378
*Cycloskrjabinia,* diagnosis, 456
*Cyclustera,* diagnosis, 388
*cygni, Gastrotaenia,* 130
*cygni, Hymenolepis,* 314
*cygni, Parabisaccanthes,* 314
*cylindracea, Cladotaenia,* 413
*cylindracea, Taenia,* 413
*cylindrica, Biuterina,* 348
*cylindraceum, Polyoncobothrium,* 97
*cylindraceus, Tetrabothrius,* 208
*cylindrica, Rhabdometra,* 346
*cylindrica, Taenia,* 338, 373
*Cylindrophorus,* diagnosis, 143
*Cylindrotaenia,* diagnosis, 209
*cynocephali, Bertiella,* 438
*cyprinorum, Caryophyllaeus,* 33
*cypselina, Dilepis,* 381

*cypseluri, Plicatobothrium,* 100
*cypseluri, Ptychobothrium,* 100
*Cysticercus,* 223
*cyrtoides, Hymenolepis,* 317
*cyrtoides, Hymenosphenacanthus,* 317
*cyrtoides, Retinometra,* 317
*cyrtoides, Sphenacanthus,* 317
*cyrtoides, Weinlandia,* 217
*cyrtus, Davainea,* 259
*cyrtus, Kotlania,* 259
*cyrtus, Raillietina (R.),* 259
*cyrtus, Raillietina (Ransomia),* 259
Cystidea, key to families, 49
*Cystotaenia,* 223

## D

*dacelonis, Similuncinus,* 359
*daetensis, Raillietina (R.),* 259
*dafilae, Hymenolepis,* 308
*dafilae, Sobolevicanthus,* 308
*dafyddi, Mashonalepis,* 372
*dagnallium, Phyllobothrium,* 161
*dahurica, Hymenolepis,* 298
*dahurica, Mayhewia,* 298
*dahurica, Passerilepis,* 298
*dahurica, Taenia,* 298
*daileyi, Acanthotaenia,* 182
*daileyi, Litobothrium,* 169
*dakari, Nybelinia,* 74
*dalli, Tetrabothrius,* 208
*dalliae, Diphyllobothrium,* 88, 94
*dalmatinus, Fistulicola,* 113
*danielae, Oochoristica,* 454
*danutae, Diorchis,* 281
*daouensis, Paruterina,* 350
*Daoventienia,* 252, 253
*darensis, Oochoristica,* 454
*dartevellei, Raillietina (R.),* 259
*dasi, Ophryocotyloides,* 243
*dasybati, Acanthobothrium,* 139
*dasybati, Phyllobothrium,* 161
*dasybati, Pterobothrium,* 70
*dasymidis, Andrya,* 435
*dasyuri, Anoplotaenia,* 383
*dattai, Raillietina (R.),* 259
*Daveneolepis,* 253
*daviesi, Malika,* 356
*Didymobothrium,* 45
*dipsacum, Otobothrium,* 68
Dasyrhynchidae
    diagnosis, 49
    key to genera, 52
*Dasyrhynchus,* diagnosis, 52
*Dasyurotaenia,* diagnosis, 241
*Davainea,* diagnosis, 247
*davainei, Raillietina (R.),* 259
Davaineidae
    diagnosis, 201
    key to subfamilies, 237

key to genera, 245
*Davaineoides*, diagnosis, 248
*daveyi, Marsipocephalus*, 190
*daviesi, Aploparaksis*, 276
*daviesi, Raillietina (S.)*, 256
*dayali, Cotugnia*, 246
*daynesi, Cotugnia*, 246
*debilis, Davainea*, 259
*debilis, Kotlania*, 259
*debilis, Raillietina (R.)*, 259
*debilis, Raillietina, (Ransomia)*, 259
*deblocki, Hymenolepis (H.)*, 317
*deblocki, Retinometra*, 317
*Deblocktaenia*, diagnosis, 189
*Debloria*, diagnosis, 311
*decacantha, Choanotaenia*, 362
*decacantha, Icterotaenia*, 362
*decacantha, Paricterotaenia*, 362
*decacantha, Tatria*, 229
*Decacanthus*, 324
*decidua, Amurotaenia*, 171
*decipiens, Bothriocephalus*, 94
*decipiens, Dibothrium*, 94
*decipiens, Diphyllobothrium*, 94
*decipiens, Spirometra*, 94
*decipiens, Vampirolepis*, 303
*decrescens, Monoecocestus*, 426
*decrescens, Taenia*, 426
*deglandi, Dicranotaenia*, 321
*dehiscens, Anomotaenia*, 405
*dehiscens, Taenia*, 405
*delachauxi, Burhinotaenia*, 396
*delachauxi, Deltokeras*, 347
*delachauxi, Icterotaenia*, 396
*delachauxi, Oligorchis*, 375
*delachauxi, Paradilepis*, 375
*delachauxi, Polycercus*, 396
*delachauxi* var. *mesacantha, Polycercus*, 396
*delafondi, Aporina*, 428
*delafondi, Bertia*, 428
*delafondi, Bertiella*, 428
*delafondi, Killigrewia*, 428
*delafondi, Taenia*, 428
*delalandei, Raillietina (R.)*, 259
*Delamurella*, 253
*Delmuretta*, 253
*delphini, Phyllobothrium*, 161
*delphini, Tetrabothrius*, 208
*Deltokeras*, diagnosis, 347
*demerariensis, Raillietina*, 259
*demerariensis* var. *trinitatae, Raillietina (R.)*, 264
*demerariensis* var. *venezolanensis, Raillietina (R.)*, 259
*demerariensis, Taenia*, 259
*demeriensis, Raillietina (R.)*, 254
*demeusiae, Echeneibothrium*, 151
*Demidovella*, 253
*dendritica, Catenotaenia*, 235
*dendritica, Taenia*, 235
*dendriticum, Diphyllobothrium*, 88
*dendrocitta, Anoncotaenia*, 341

*dendrocitta, Rhabdometra*, 341
*dendrocopina, Raillietina (P.)*, 254
*Dendrometra*, diagnosis, 346
*Dendrouterina*, diagnosis, 377
*dentata, Anoplocephala*, 436
*dentata, Anoplocephaloides*, 436
*dentatum, Phyllobothrium*, 161
*dentatus, Hymenolepis*, 307
*denticulata, Alyselminthus*, 443
*denticulata, Cittotaenia*, 443
*denticulata, Ctenotaenia*, 443
*denticulata, Moniezia*, 442, 443
*denticulata, Taenia*, 441, 442, 443
*denticulatus, Alyselminthus*, 442
*depidocolpos, Paradilepis*, 375
*depressa, Anomotaenia*, 391, 402
*depressa, Liga*, 391
*depressa, Neoliga*, 391
*depressa, Sureshia*, 402
*depressa, Taenia*, 391, 402
*deserti, Mathevotaenia*, 463
*deserti, Oochoristica*, 463
*desmognathi, Baerietta*, 212
*destitutus, Lateriporus*, 373
*destitutus, Taenia*, 373
*deweti, Raillietina (S.)*, 256
*dhuncheta, Raillietina (S.)*, 256
*Diagonaliporus*, 371
*diagonalis, Aploparaksis*, 276
*diana, Baerietta*, 212
*diana, Proteocephalus*, 212
*Diancrobothrium*, 93
*Diandrya*, diagnosis, 444
*diaphana, Cittotaenia*, 445
*diaphana, Ditestolepis*, 272
*diaphana, Hepatotaenia*, 445
*diaphana, Hymenolepis*, 272
*diaphana, Moniezia*, 445
*diaphana, Progamotaenia*, 445
*Dibothriorhynchus*, 62
*Dicranobothrium*, 137
*Dicranotaenia*, diagnosis, 320
*dicruri, Biuterina*, 348
*Dictymetra*, 404
*Dictyuterina*, diagnosis, 344
*didelphidis, Mathevotaenia*, 463
*didelphus, Mesocestoides*, 203
*didelphydis, Diphyllobothrium*, 94
*didelphydis, Oochoristica*, 463
*didelphydis, Taenia*, 463
*Dieffluvium*, diagnosis, 30
*dierama, Lecanicephalum*, 126
*dierama, Tylocephalum*, 126
*Diesingella*, 75
*Diesingia*, 75
*Diesingiella*, 75
*diesingii, Monticellia*, 194
*Diesingium*, 75
*differtus, Pseudolytocestus*, 40
*dighaensis, Acanthobothrium*, 139
*diglobovary, Sinuterilepis*, 270

*digonopora, Cotugnia,* 245
*digonopora, Taenia,* 245
*Digramma,* diagnosis, 87
*dilatatus, Proteocephalus,* 187
Dilepididae
  diagnosis, 201
  genera of uncertain status, 413
  key to subfamilies, 339
Dilepidinae, key to genera, 368
*Dilepidoides,* diagnosis, 316
*Dilepis,* diagnosis, 379
*diminuens, Aploparaksis,* 276
*diminuta, Anoplocephala,* 433
*diminuta, Hymenolepis,* 289
*diminuta, Paradilepis,* 375
*diminuta, Taenia,* 289
*diminutoides, Hymenolepis,* 289
*dimorphus, Schistocephalus,* 88
*Dinobothrium,* diagnosis, 159
*dinoi, Dibothriorhynchus,* 76
*dinopii, Ophryocotylus,* 242
*dinopteri, Proteocephalus,* 187
*Diochetos,* diagnosis, 452
Dioecocestidae
  diagnosis, 200
  key to genera, 213
*Dioecocestus,* diagnosis, 213
*Dioecotaenia,* diagnosis, 197
Dioecotaeniidae, diagnosis, 197
Dioecotaeniidea ord. n., 14, 197
*dioica, Shipleya,* 215
*diomedea, Tetrabothrius,* 208
*Diorchirailllietina,* diagnosis, 251
*Diorchis,* diagnosis, 281
*diorchis, Diorchis,* 281
*diorchis, Hymenolepis,* 328
*diorchis, Microsomacanthus,* 328
*diorchis, Weinlandia,* 328
Diphyllidea, 14, 165
  key to families, 165
Diphyllobothriidae
  diagnosis, 81
  key to genera, 85
  species of uncertain status, 95
*Diphyllobothrium*
  diagnosis, 93
  in birds, 88
  in mammals, 93
  in reptiles, 86
*dipi, Mathevotaenia,* 463
*dipi, Skrjabinia,* 463
*dipi, Taenia,* 463
*diplacantha, Neoliga,* 391
*Diplochetos,* 404
*Diplocotyle,* 45
*Diplogynia,* diagnosis, 222
*Diplomonorchis,* 281
*diplomys, Monoecocestus,* 426
*Diplopylidium,* diagnosis, 354
*diplocoronatus, Skrjabinacanthus,* 299
*Diplogonimus,* 62
*Diplogonoporus,* diagnosis, 89

*Diplootobothrium,* diagnosis, 67
*Diplophallus,* diagnosis, 231
*Diploposthe,* diagnosis, 232
Diploposthidae, 230
*diplosoma, Prosthecocotyle,* 208
*dipodomi, Schizorchoides,* 440
*Diporotaenia,* diagnosis, 227
*dipsadomorphi, Anthobothrium,* 157
Dipylidiinae, key to genera, 352
*Dipylidium,* diagnosis, 352
*Discobothrium,* diagnosis, 125
*Discocephalum,* 120
*discophorus, Tetrarhynchus,* 77
Discocephalidae, 120
*discoidea, Anomotaenia,* 405
*discoidea, Choanotaenia,* 405
*discoidea, Dichoanotaenia,* 405
*discoidea, Taenia,* 405
Disculicepitidae, diagnosis, 119
*Disculiceps,* diagnosis, 120
*Diskrjabiniella,* diagnosis, 352
*dispar, Choanotaenia,* 362, 408
*dispar, Nematotaenia,* 213
*dispar, Taenia,* 213
*dissimilis, Mesocestoides,* 203
*distincta, Biuterina,* 348
*distincta, Dilepis,* 381
*distincta, Taenia,* 381
*Distoichometra,* diagnosis, 212
Ditestolepidinae, 272
*Ditestolepis,* diagnosis, 272
Ditrachybothridiidae, diagnosis, 165
*Ditrachybothridium,* diagnosis, 166
*ditrema, Prodicoelia,* 85
*ditremum, Diphyllobothrium,* 88
*Diuterinotaenia,* diagnosis, 449
*djeirani, Taenia,* 224
*Djombangia,* diagnosis, 22
*dlouhyi, Sobolevicanthus,* 308
*dobrogica, Paricterotaenia,* 396
*dobrogica, Polycercus,* 396
*dodecantha, Hymenolepis,* 302
*dodecacantha, Icterotaenia,* 396
*dodecacantha, Paricterotaenia,* 396
*dodecacantha, Polycercus,* 396
*dodecantha, Staphylocystis,* 302
*dodecacantha, Taenia,* 396
*dogieli, Anomotaenia,* 362
*dogieli, Apora,* 129
*dogieli, Choanotaenia,* 362
*dogieli, Rhabdometra,* 346
*dohrnii, Phyllobothrium,* 161
*dolguschini, Flamingolepis,* 296
*dolichocephala, Grillotia,* 64
*dolichoporum, Echeneibothrium,* 151
*dollfusi, Cotugnia,* 246
*dollfusi, Crepidobothrium,* 189
*dollfusi, Dilepis,* 251, 381
*dollfusi, Dollfusoquenta,* 251
*dollfusi, Echeneibothrium,* 154, 158
*dollfusi, Grillotia,* 64
*dollfusi, Rhabdotobothrium,* 154

*dollfusi, Rhodobothrium,* 158
*dollfusi, Sphyriocephalus,* 72
*Dollfusoquenta,* diagnosis, 250
*dolosa, Echinocotyle,* 285
*domesticus, Amoebotaenia,* 387
*domesticusi, Davainea,* 248
*dominicanus, Anomotaenia,* 405
*dominicanus, Icterotaenia,* 405
*dominicanus, Paricterotaenia,* 405
*dominicanus, Rissotaenia,* 405
*dongolense, Dipylidium,* 353
*donis, Diorchis,* 281
*dorad, Myzophorus,* 194
*dorcopsis, Calostaurus,* 252
*dougi, Proterogynotaenia,* 220
*Drepanidotaenia,* diagnosis, 294
*dromedarii, Cysticercus,* 225
*drygalskii, Tetrabothrius,* 208
*dubia, Anomotaenia,* 405
*dubia, Bothrioscolex,* 20
*dubia, Hymenolepis,* 336
*dubia, Khawia,* 20
*dubininae, Anomotaenia,* 405
*dubininae, Echinocotyle,* 285
*dubininae, Multiuterina,* 343
*dubininae, Paracaryophyllaeus,* 28
*dubininae, Paricterotaenia,* 396
*dubininae, Polycercus,* 396
*dubininae, Tatria,* 229
*dubinini, Liga,* 390
*dubinini, Ophiotaenia,* 184
*Dubiniolepis,* 324
*dubium, Echeneibothrium,* 151
*dubius, Davainea,* 247
*dubius, Proteocephalus,* 187
*duboisi, Paradilepis,* 375
*ductilis, Drepanidotaenia (D.),* 328
*ductilis, Hymenolepis,* 328
*ductilis, Microsomacanthus,* 328
*dujardini, Aploparaksis,* 276
*dujardini, Monorchis,* 276
*dujardini, Monorcholepis,* 276
*dujardini neoarcticus, Aploparaksis,* 276
*dujardini, Taenia,* 276
*dujardinii, Acanthobothrium,* 139
*dujardinii, Prosthecobothrium,* 139
*dunganica, Biuterina,* 348
*duodecantha, Tatria,* 229
*duosyntesticulata, Raillietina (P.),* 254
*Duplicibothrium,* diagnosis, 150
*dusmeti, Drepanidotaenia (D.),* 336
*dusmeti, Hymenolepis,* 336
*Duthiersia,* diagnosis, 86
*dutti, Choanotaenia,* 361
*dysbiotos, Taenia,* 138

## E

*eburnea, Pseudhymenolepis,* 271

*Echeneibothrium,* diagnosis, 150
*echeneis, Bothriocephalus,* 152
*echidnae, Linstowia,* 461
*echidnae, Taenia,* 461
*Echidnotaenia,* diagnosis, 455
*echinata, Raillietina,* 266
*echinatia, Parvitaenia,* 412
*Echinatrium,* diagnosis, 319
*echinobothrida, Kotlania,* 259
*echinobothrida, Raillietina (Fuhrmannetta),* 259
*echinobothrida, Raillietina (Johnstonia),* 259
*echinobothrida, Raillietina (R.),* 259
*echinobothrida, Taenia,* 259
Echinobothriidae, diagnosis, 165
*Echinobothrium,* diagnosis, 165
*Echinococcifer,* 222
*Echinococcus,* diagnosis, 222
*echinococcus, Taenia,* 222
*Echinocotyle,* diagnosis, 284
*echinocotyle, Echinocotyle,* 285
*echinocotyle, Hymenolepis,* 285
*Echinolepis,* diagnosis, 282
Echinophllidae
    diagnosis, 82
    key to genera, 105
*Echinophallus,* diagnosis, 105
*Echinoproboscilepis,* 293
*echinorhynchoides, Joyeuxiella,* 353
*echinorhynchoides, Taenia,* 353
*echinorostrae, Dicranotaenia,* 322
*echinorostrae, Hymenolepis,* 322
*Echinorhynchotaenia,* diagnosis, 272
Echinorhynchotaeniinae, 272
*Echinoshipleya,* diagnosis, 215
*echinovatum, Aploparaksis,* 276
*Ectopocephalium,* diagnosis, 445
*edifontaineus, Anootypus,* 419
*edifontaineus, Avitellina,* 419
*Edlintonia,* diagnosis, 38
*edmondi, Taufikia,* 426
*edonensis, Capsulata,* 359
*edulis, Bertia,* 438
*edulis, Bertiella,* 438
*edulis, Prototaenia,* 438
*edwardsi, Acanthobothrium,* 139
*edwinlintoni, Nybelinia,* 74
*effigia, Wallabicestus,* 447
*egregius, Tetrabothrius,* 208
*egrettae, Dendrouterina,* 378
*eisenbergi, Pseudhymenolepis,* 271
*elaphis, Oochoristica,* 454
*elapsideae, Ophiotaenia,* 184
*electricollum, Acanthobothrium,* 139
*Electrotaenia,* diagnosis, 174
*elegans, Diphyllobothrium,* 94
*elegans, Duthiersia,* 86
*elegans, Monorygma,* 163
*elegans, Onchobothrium,* 135
*elegantissimum, Anthobothrium,* 147, 156
*elegantissimum, Clydonobothrium,* 156
*elegantissimum, Scyphophyllidium,* 157

*elisae, Aploparaksis*, 276, 279
*elisae, Diorchis*, 279
*elisae, Schillerius*, 279
*ellipticum, Sparganum*, 94
*ellipticus, Bothriocephalus*, 102
*ellisi, Hymenolepis*, 336
*ellisoni, Dicranotaenia*, 322
*ellisoni, Hymenolepis*, 322
*elloraii, Sureshia*, 402
*elongata, Austramphilina*, 472
*elongata, Bertia*, 438
*elongata, Bertiella*, 438
*elongata, Catenotaenia*, 235
*elongata, Davainea*, 265
*elongata, Drepanidotaenia*, 294
*elongata, Hymenolepis*, 294
*elongata, Kotlania*, 265
*elongata, Nybelinia*, 74
*elongata, Oochoristica*, 454
*elongata, Ohiotaenia*, 184
*elongata, Parataenia*, 121
*elongata, Prototaenia*, 438
*elongata, Raillietina (F.)*, 265
*elongata, Raillietina (Johnstonia)*, 265
*elongatus, Anthocephalus*, 53
*elongatus, Mesocestoides*, 203
*elongatus, Proteocephalus*, 187
*elongatus, Tetrarhynchus*, 77
*elongatus, Vampirolepis*, 303
*Emberizotaenia*, 385
*embiensis, Schistometra*, 240
*embryo, Choanotaenia*, 396
*embryo, Icterotaenia*, 396
*embryo, Paricterotaenia*, 396
*embryo, Polycercus*, 396
*embryo, Taenia*, 396
*emperus, Davainea*, 259
*emperus, Kotlania*, 259
*emperus, Raillietina (R.)*, 259
*emperus, Raillietina (Ransomia)*, 259
*endacantha, Aploparaksis*, 276
Endorchiinae, key to genera, 193
*Endorchis*, diagnosis, 193
*endothoracica, Taenia*, 224
*endothoracicus, Coenurus*, 224
*enigmaticum, Rhodobothrium*, 158
*enigmaticus, Proboscidosaccus*, 158
*Eniochobothrium*, diagnosis, 124
*enteraneidis, Baerietta*, 212
*enteraneidis, Proteocephalus*, 212
*eperlani, Proteocephalus*, 187
Ephedrocephalinae, 195
*Ephedrocephalus*, diagnosis, 195
*epistocotyle, Tetrarhynque*, 79
*equatoriensis, Raillietina (R.)*, 259
*equi, Taenia*, 433
*equidentata, Nybelinia*, 74
*equidentata, Tetrarhynchus*, 74
*equina perfoliata, Taenia*, 433
*equina, Taenia*, 433
*eranui, Lateriporus*, 373

*Eranuides*, diagnosis, 448
*erethizontis, Monoecocestus*, 426
*ericetorum, Anomotaenia*, 405
*ericetorum, Taenia*, 405
*erinacei, Mathevotaenia*, 463
*erinacei, Oochoristica*, 463
*erinacei, Rodentolepis*, 303
*erinacei, Spirometra*, 94
*erinacei, Taenia*, 303
*erinacei, Vampirolepis*, 303
*erinacei* var. *rodentinum, Oochoristica*, 459
*erinacei* var. *steudeneri, Dicranotaenia*, 303
*erinaceieuropaei, Diphyllobothrium*, 94
*erinacei-europaei, Dubium*, 94
*erinaceus, Grillotia*, 64
*erinaceus, Tetrarhynchus*, 64
*eroliae, Anomotaenia*, 405
*eriocis, Bothriocephalus*, 102
*erostris, Tetrabothrius*, 208
*erostris* var. *minor, Bothridiotaenia*, 209
*erschovi, Mesocestoides*, 203
*erschovi, Raillietina (R.)*, 259
*erschovi, Tatria*, 230
*Ershovilepis*, 379
*Ershovitugnia*, 245
*Ersinogenes*, diagnosis, 238
*erythraea, Nybelinia*, 74
*erythrea, Taenia*, 224
*esarsi, Schizorchis*, 440
*eschrichtii, Calliobothrium*, 135
*eschrichtii, Priapocephalus*, 205
*esocis, Ichthyotaenia*, 187
*esocis, Proteocephalus*, 187
*estavarensis, Choanotaenia*, 362
*estigmana, Nybelinia*, 74
*etaplesensis, Debloria*, 312
*etaplesensis, Hymenolepis (H.)*, 312
*Ethiopotaenia*, diagnosis, 383
*etneri, Calentinella*, 28
*Eubothrioides*, diagnosis, 114
*Eubothrium*, diagnosis, 107
Eucestoda, keys to orders, 11
*eudocimi, Parvitaenia*, 412
*eudromii, Anomotaenia*, 405
*eudypidis, Neotetrabothrius*, 208
*eudyptidis, Tetrabothrius*, 208
*Eugonodaeum*, diagnosis, 355
*Eulacistorhynchus*, diagnosis, 63
*eumecis, Oochoristica*, 454
*eunectes, Taenia*, 224
*eupodotidis, Otiditaenia*, 240
*eupodotidis, Raillietina (R.)*, 259
*eureia, Nybelinia*, 74
*europaea, Ophiotaenia*, 184
*Eurycestus*, diagnosis, 389
*euryciensis, Bothriocephalus*, 102
*eurypharygis, Pistana*, 115
Eutetrarhynchidae
  diagnosis, 51
  key to genera, 55
*Eutetrarhynchus*, diagnosis, 56

*euzeti, Echinobothrium,* 166
*evaginata, Hymenolepis,* 303
*evaginata, Progynotaenia,* 219
*evaginata, Rodentolepis,* 303
*evaginata, Vampirolepis,* 303
*evansi, Staphylocystis,* 302
*ewersi, Progamotaenia,* 447
*ewersi, Wallabicestus,* 447
*excelsa, Oochoristica,* 454
*excentricus, Diorchis,* 281
*exceptum, Phoreiobothrium,* 141
*excisus, Dibothriorhynchus,* 76
*exigua, Choanotaenia,* 366
*exigua, Monopylidium,* 366
*exigua, Taenia,* 366
*exiguum, Anthobothrium,* 157
*exiguus, Hymenolepis,* 317
*exiguus, Hymenosphenacanthus,* 317
*exiguus, Proteocephalus,* 187
*exiguus, Retinometra,* 317
*exiguus, Sphenacanthus,* 317
*exile, Diphyllobothrium,* 88
*exile, Rhynchobothrium,* 76
*exilis, Hymenolepis,* 336
*expansa, Duthiersia,* 86
*expansa, Moniezia,* 441
*expansa, Taenia,* 441
*expansus, Alyselminthus,* 441

# F

*fabulosum, Diplopylidium,* 354
*fabulosum, Progynopylidium,* 354
*facile, Davainea,* 254
*facile, Raillietina (P.),* 254
*facilis, Anomotaenia,* 390
*facilis, Liga,* 390
*fahmi, Schistocephalus,* 88
*falcatus, Hymenolepis,* 329
*falcatus, Microsomacanthus,* 329
*falciformis, Eurycestus,* 389
*falciformis, Fimbriaria,* 269
*falciformis, Fimbriariella,* 269
*falciformis, Fimbriarioides,* 269
*falconis, Unciunia,* 386
*falcoris, Cladotaenia,* 413
*falcoris, Paracladotaenia,* 413
*falculata, Hymenolepis,* 303
*falculata, Rodentolepis,* 303
*falculata, Staphylocystis,* 303
*falculata, Vampirolepis,* 303
*fallax, Bertiella,* 439, 463
*fallax, Bothrimonus,* 45
*fallax, Biuterina,* 348
*fallax, Dicranotaenia,* 322
*fallax, Diplacanthus (Lepidotrias),* 322
*fallax, Discobothrium,* 125
*fallax, Hymenolepis,* 322
*fallax, Proteocephalus,* 187
*fallax, Taenia,* 322

*falsata, Hispaniolepis,* 306
*falsata, Hymenolepis,* 306
*falsificata, Choanotaenia,* 397
*falsificata, Paricterotaenia,* 397
*falsificata, Polycercus,* 397
*falsificata, Stenovaria,* 397
*famosa, Kotlania,* 259
*famosa, Raillietina, (R.),* 259
*fanatica, Hymenolepis,* 336
*fania, Cladotaenia,* 413
*faranciae, Ophiotaenia,* 184
*farciminosa, Dicranotaenia,* 324
*farciminosa, Hymenolepis,* 324
*farciminosa, Taenia,* 324
*farciminosa, Variolepis,* 324
*farmeri, Calliobothrium,* 136
*farmeri, Oncobothrium,* 136
*farrani, Pseudoshipleya,* 216
*farrani, Shipleyia,* 216
*fasciatus, Diplogonoporus,* 90
*fasciolaris, Fimbriaria,* 267
*fasciolaris, Taenia,* 267
*fastigata, Cotugnia,* 246
*fatalis, Brumptiella,* 256
*fatalis, Raillietina (S.),* 256
*fausti, Dicranotaenia,* 329
*fausti, Diphyllobothrium,* 94
*fausti, Fuhrmanniella,* 329
*fausti, Gyrocoelia,* 216
*fausti, Hymenolepis,* 329
*fausti, Microsomacanthus,* 328, 329
*faxanum, Echeneibothrium,* 151
*fecunda, Raillietina (P.),* 254
*fedtschenkoi, Hispaniolepis,* 306
*fedtschoenkoi, Hymenolepis,* 306
*felidis, Echinococcus,* 222
*felis, Bothriocephalus,* 94
*fellicola, Hepatotaenia,* 445
*fennicus, Caryophyllaeides,* 19
*fernandensis, Variolepis,* 325
*Fernandezia,* diagnosis, 243
*fessus, Penarchigetes,* 29
*festiva, Cittotaenia,* 445
*festiva, Hepatotaenia,* 445
*festiva, Moniezia,* 445
*festiva, Progamotaenia,* 445
*festiva, Taenia,* 445
*feuta, Cladotaenia,* 413
*fevita, Dioecocestus,* 213
*fibrata, Oochoristica,* 454
*fibriata, Triplotaenia,* 416
*ficta, Macrobothriotaenia,* 182
*ficticia, Hymenolepis,* 375
*ficticia, Meggittiella,* 375
*ficticia, Paradilepis,* 375
*ficticia, Skrjabinolepis,* 375
*fictum, Crepidobothrium,* 182
*fictus, Proteocephalus,* 182
*fidelis, Hymenolepis,* 330
*fieldingi, Anomotaenia,* 366
*fieldingi, Choanotaenia,* 366

fieldingi, Monopylidium, 366
figurata, Oochoristica, 462
fila, Cotugnia, 246
filamentoovatum, Wardium, 324
filamentosa, Cinclotaenia, 367
filamentosa, Hymenolepis, 366
filamentosa, Taenia, 366
filamentosum, Choanotaenia, 366
filamentosum, Monopylidium, 366
filamentosum, Rodentotaenia, 366
filarioides, Ophiotaenia, 184
fileri, Nototaenia, 401
filicolle, Synbothrium, 70
filicolle, Syndesmobothrium, 70
filicolle var. benedenii, Acanthobothrium, 138
filicollis, Acanthobothrium, 139
filicollis, Proteocephalus, 185
filicollis, Pterobothrium, 70
filiforme, Phyllobothrium, 161
filiforme, Raillietina (P.), 254
filiformis, Lytocestus, 25
filiformis, Tetrabothrius, 208
filiformis, Taenia, 261
filirostris, Hymenolepis, 329
filirostris, Microsomacanthus, 329
filirostris, Taenia, 329
filosomum, Echinatrium, 320
filovata, Anomotaenia, 410
filovata, Pseudanomotaenia, 410
filum, Aploparaksis, 275
filum, Drepanidotaenia, 275
filum, Haploparaxis, 275
filum, Hymenolepis, 275
filum, Monopylidium, 275
filum, Monorchis, 275
filum, Taenia, 275
filum var. brachyphallos, Taenia, 275
filumferens, Hymenolepis, 308
filumferens, Sobolevicanthus, 308
fima, Ophiotaenia, 184
Fimbriaria, diagnosis, 267
Fimbriariella, diagnosis, 269
Fimbriariinae, key to genera, 267
Fimbriarioides, diagnosis, 268
fimbriata, Duthiersia, 86
fimbriata, Echinocotyle, 285
fimbriata, Gyrocotyle, 473
fimbriata, Taenia, 306, 416
fimbriata, Tatria, 229
fimbriatum, Corallobothrium, 180
fimbriatus, Oncodiscus, 101
fimbriceps, Caryophyllaeus, 33
fimula, Hymenolepis, 317
fimula, Hymenosphenacanthus, 317
fimula, Retinometra, 317
fimula, Sphenacanthus, 317
finlayi, Bertiella, 439
Finna, 223
finta, Hymenolepis, 336
Fischiosoma, 223
fissiceps, Diphyllobothrium, 88

Fissurobothrium, diagnosis, 110
fista, Hymenolepis, 317
fista, Hymenosphenacanthus, 318
fista, Retinometra, 317
fista, Sphenacanthus, 317
Fistulicola, diagnosis, 113
fixa, Ophiotaenia, 184
Flabelloskrjabinia, diagnosis, 435
flabralis, Kotlania, 259
flabralis, Raillietina (R.), 259
flaccida, Kotlania, 259
flaccida, Raillietina, (R.), 259
flaccida, Progynotaenia, 220
flaccida, Proterogynotaenia, 220
flagellatus, Hymenolepis, 308
flagellatus, Sobolevicanthus, 308
flagellum, Idiogenes, 240
flagellum, Taenia, 240
flaminata, Hymenolepis, 336
flaminata, Raillietina (R.), 259
flamingo, Flamingolepis, 296
flamingo, Hymenolepis, 296
Flamingolepis, diagnosis, 295
Flapocephalus, diagnosis, 123
flava, Ophiotaenia, 184
flavescens, Diorchis, 281
flavescens, Taenia, 281
flavopunctata, Taenia, 289
fleari, Cotugnia, 246
flexile, Echeneibothrium, 155
flexile, Rhinebothrium, 155
floraformis, Anthobothrium, 147
floreata, Hymenolepis, 330
floresbarroetae, Anoplocephaloides, 436
Floriceps, diagnosis, 52
floridensis, Acanthobothrium, 139
floriforme, Phyllobothrium, 161
fluviatile, Tetrarhynchobothrium, 77
fluviatilis, Bothriocephalus, 102
fluviatilis, Proteocephalus, 187
fluviatilis, Schizocotyle, 46
fluxa, Raillietina, 266
fodientis, Hymenolepis, 335
foederata, Bertiella, 439
foetida, Progynotaenia, 219
fogeli, Acanthobothrium, 139
foinae, Taenia, 224
fola, Hymenolepis, 298
fola, Passerilepis, 298
foliacea, Amphilina, 468
foliaceum, Monostomum, 468
foliatum, Phyllobothrium, 161
folium, Dibothrium, 95
folius, Isoglaridacris, 37
follilisi, Lucknowia, 18
fona, Hymenolepis, 336
forcipata, Bertiella, 439
formosa, Bothriocephalus, 102
formosa, Hymenolepis, 329
formosa, Microsomacanthus, 329
formosana, Davainea, 258

*formosana, Kotlania*, 258
*formosana, Raillietina*, 258
*formosensis, Diorchis*, 279
*formosensis, Schillerius*, 279
*formosoides, Microsomacanthus*, 329
*forna, Bancroftiella*, 399
*forsteri, Prosthecocotyle*, 208
*forsteri, Taenia*, 208
*forsteri, Tetrabothrius*, 208
*forte, Marsupiobothrium*, 147
*forte, Orygmatobothrium*, 147
*fortobothria, Janiszewskella*, 36
*fortunata, Anomotaenia*, 406
*fossae, Monobothrium*, 32
*fossata, Ichthyotaenia*, 187
*fossatus, Proteocephalus*, 187
*fossilis, Lytocestus*, 25
*Fossor*, 222
*fotedori, Choanotaenia*, 362
*foteria, Cyclorchida*, 378
*fovea, Dendrouterina*, 378, 381
*fovea, Dilepis*, 381
*foveata, Hymenolepis*, 336
*foveolatus, Tetrarhynchus*, 77
*foxi, Cladotaenia*, 413
*fragile, Crepidobothridium*, 184
*fragile, Proteocephalus*, 184
*fragile, Syndesmobothrium*, 70
*fragili, Pterobothrium*, 70
*fragili, Synbothrium*, 70
*fragilis, Bakererpes*, 387
*fragilis, Eubothrium*, 107
*fragilis, Hymenolepis*, 309
*fragilis, Nippotaenia*, 172
*fragilis, Ophiotaenia*, 184
*fragilis, Porotaenia*, 208
*fragilis, Prochristianella*, 59
*fragilis, Raillietina, (R.)*, 259
*fragilis, Sobolevicanthus*, 308
*fragilis, Sphenacanthus*, 309
*fragilis, Taenia*, 308
*fragmentata, Drepanidotaenia*, 287
*fragmentata, Gvosdevilepis*, 287
*francolini, Hymenolepis*, 329
*francolini, Microsomacanthus*, 329
*fraterna, Vampirolepis*, 304
*frayi, Kotlania*, 259
*frayi, Raillietina (R.)*, 259
*freani, Cladotaenia*, 413
*fregatae, Tetrabothrius*, 209
*freitasi, Rhinebothroides*, 154
*freitasi, Rhinebothrium*, 154
*frezei, Kapsulotaenia*, 180
*friedbergeri, Davainea*, 259
*friedbergeri, Kotlania*, 259
*friedbergeri, Raillietina (R.)*, 259
*friedbergeri, Raillietina (Ransomia)*, 259
*friedbergeri, Taenia*, 259
*frigida, Amoebotaenia*, 391
*frigida, Liga*, 391
*fringillarum, Aploparaksis*, 297

*fringillarum, Hymenolepis*, 297
*fringillarum, Taenia*, 297
*frivola, Killigrewia*, 428
*frontina, Davainea*, 259
*frontina, Kotlania*, 259
*frontina, Raillietina (R.)*, 259
*frontina, Raillietina (Ransomia)*, 259
*frontina, Taenia*, 259
*fructifera, Hymenolepis*, 336
*fruticosa, Hymenolepis*, 336
*fryei, Dicranotaenia*, 322
*fryei, Hymenolepis*, 322
*fryei, Wardium*, 322
*fuelleborni, Staphylocystis*, 302
*Fuhrmanacanthus*, 373
*Fuhrmannella*, diagnosis, 442
*fuhrmanni, Amoebotaenia*, 397
*fuhrmanni, Aporina*, 428
*fuhrmanni, Biuterina*, 348
*fuhrmanni, Caryophyllaeus*, 32
*fuhrmanni, Choanotaenia*, 361
*fuhrmanni, Cotugnia*, 246
*fuhrmanni, Culcitella*, 349
*fuhrmanni, Cyclorchida*, 378
*fuhrmanni, Cyclustera*, 389
*fuhrmanni, Davainea*, 260
*fuhrmanni, Dilepis*, 381
*fuhrmanni, Dioecocestus*, 213
*fuhrmanni, Diphyllobothrium*, 94
*fuhrmanni, Gyrocoelia*, 216
*fuhrmanni, Icterotaenia*, 397
*fuhrmanni, Joyeuxia*, 353
*fuhrmanni, Killigrewia*, 428
*fuhrmanni, Kotlania*, 260
*fuhrmanni, Lateriporus*, 349
*fuhrmanni, Oochoristica*, 453
*fuhrmanni, Ophryocotyle*, 244
*fuhrmanni, Paricterotaenia*, 397
*fuhrmanni, Paruterina*, 351
*fuhrmanni, Polycercus*, 397
*fuhrmanni, Progynotaenia*, 219
*fuhrmanni, Raillietina (R.)*. 260
*fuhrmanni, Raillietina (R.) var. idiogenoides*, 260
*fuhrmanni, Raillietina (R.) var. intermedia*, 260
*fuhrmanii, Sphaeruterina*, 351
*fuhrmanni, Tatria*, 229
*fuhrmanni, Tetrabothrius*, 208
*Fuhrmannia*, 389
*Fuhrmanniella*, 327
*Fuhrmannodes*, diagnosis, 446
*fuhrmannoides, Polycercus*, 397
*Fuhrmanolepis*, 379
*fujiensis, Vampirolepis*, 303
*fukuokaensis, Diplogonoporus*, 90
*fulgidum, Polyoncobothrium*, 97
*fulicatrae, Retinometra*, 318
*fulicicola, Microsomacanthus*, 329
*fuligulosa, Aploparaksis*, 276
*fuliginosa, Hymenolepis*, 337
*fulvia, Raillietina (P.)*, 254
*fulvida, Metroliasthes*, 344

*funerbis, Raillietina*, 266
*fungosa, Hymenolepis*, 336
*furcata, Staphylocystis*, 302
*furcifera, Confluaria*, 325
*furcifera, Diplacanthus (Dilepis)*, 325
*furcifera, Dubininolepis*, 325
*furcifera, Hymenolepis*, 325
*furcifera, Taenia*, 325
*furcifera, Variolepis*, 325
*furcigera, Aploparaksis*, 276
*furcigera, Dicranotaenia*, 276
*furcigera, Haploparaksis*, 276
*furcigera, Hymenolepis*, 276
*furcigera, Taenia*, 276
*furcouterina, Hymenolepis*, 336
*furnarii, Cotylorhipis*, 383
*furnarii, Taenia*, 383
*fusa, Dicranotaenia*, 322
*fusa, Taenia*, 322
*fusa, Wardium*, 322
*fusca, Oochoristica*, 454
*fuscum, Diphyllobothrium*, 94
*fusus, Haploparaxis*, 322
*fusus, Hymenolepis*, 322
*futilis, Hymenolepis*, 336

# G

*gabonica, Ophiotaenia*, 184
*gadi, Abothrium*, 108
*gadi-aeglefini, Tetrarhynchus*, 77
*gadi-morrhuae, Tetrarhynchus*, 77
*gaigeri, Multiceps*, 224
*gaigeri, Polycephalus*, 224
*gaigeri, Taenia*, 224
*galbulae, Anomotaenia*, 366
*galbulae, Choanotaenia*, 366
*galbulae, Icterotaenia*, 366
*galbulae, Monopylidium*, 366
*galbulae, Parachoanotaenia*, 366
*galbulae, Taenia*, 366
*galeritae, Davainea*, 260
*galeritae, Kotlania*, 260
*galeritae, Raillietina (R.)*, 260
*galeritae, Raillietina (Ransomia)*, 260
*gallardi, Acanthotaenia*, 182
*gallardi, Crepidobothrium*, 182
*gallardi, Ophiotaenia*, 182
*gallardi, Proteocephalus*, 182
*galli, Aploparaksis*, 276
*galli, Paronia*, 423
*galli, Raillietina (R.)*, 257
*gallica, Oochoristica*, 454
*gallica* var. *pleionorcheis, Oochoristica*, 454
*Galligoides*, 436
*gallinae, Hymenolepis*, 336
*gallinagilis, Anomotaenia*, 410
*gallinagilis, Pseudanomotaenia*, 410
*gallinulae, Liga*, 390
*gallinulae, Rallitaenia*, 390

*gallinulae, Taenia*, 390
*gambianum, Thysanosoma*, 466
*gambianum, Thysanotaenia*, 466
*gambianum, Zschokkeella*, 466
*ganapati, Penetrocephalus*, 104
*ganapattii, Bothriocephalus*, 102
*ganfini, Oncobothrium*, 136
*Gangesia*, diagnosis, 175
Gangesiinae, key to genera, 173
*gangeticum, Poecilancistrum*, 69
*ganii, Anomotaenia*, 406
*ganii, Dictymetra*, 406
*garciai, Raillietina (R.)*, 260
*garmi, Raillietina (R.)*, 260
*garrisoni, Raillietina (R.)*, 261
*garrulae, Biuterina*, 348
*garrulae, Paruterina*, 348, 750
*gasterostei, Bothriocephalus*, 102
*gasterostei, Taenia*, 88
*Gastrolecithus*, diagnosis, 158
*Gastrotaenia*, diagnosis, 129
*Gatesius*, 93
*gaugi, Anoncotaenia*, 341
*gendrei, Kotlania*, 260
*gendrei, Raillietina (R.)*, 260
*gendrei, Raillietina (Ransomia)*, 260
*genettae, Anoplocephala*, 457
*genettae, Diplopylidium*, 354
*genettae, Dipylidium*, 353
*genettae, Oschmarenia*, 457
*gennarii, Drepanidotaenia*, 397
*gennarii, Taenia*, 397
*geographicus, Lateriporus*, 373
*geophiloides, Taenia*, 224
*georgiensis, Raillietina, (R.)*, 260
*geosciuri, Catenotaenia*, 233
*geosciuri, Hemicatenotaenia*, 233
*Gephyrolina*, diagnosis, 468
*geraschmidti, Eutetrarhynchus*, 57
*gerbilli, Meggittina*, 236
*gerbilli, Rajotaenia*, 236
*gerbilli, Skrjabinotaenia*, 236
*gerrardii, Tetrabothrium*, 189
*gerrardii, Crepidobothrium*, 189
*gerrardi* var. *minus, Crepidobothrium*, 189
*gerrhonoti, Baerietta*, 212
*gertschi, Vampirolepis*, 303
*gevreyi, Raillietina, (R.)*, 260
*ghoshi, Taufikia*, 426
*giardi, Helictometra*, 420
*giardi, Taenia*, 420
*giardi, Thysanosoma*, 420
*Gidhaia*, diagnosis, 429
*gigantea, Anoplocephala*, 433
*gigantea, Plagiotaenia*, 433
*gigantea, Taenia*, 433
*giganteum, Anthobothrium*, 147
*giganteum, Corallobothrium*, 178
*giganteum, Corallobothrium (Megathylacoides)*, 178
*giganteum, Megathylacoides*, 178
*giganteum, Scyphophyllidium*, 146

*giganteus, Anthocephalus*, 52
*giganteus, Dasyrhynchus*, 52
*giganticus, Monoecocestus*, 426
*giganticus, Polygonoporus*, 89
*gigantium, Bialovarium*, 34
*gigantocirrosa, Diorchis*, 281
*Gigantolina*, diagnosis, 470
*gigas, Gymnorhynchus*, 61
*gigas, Scolex*, 61
*gilae, Hypocaryophyllaeus*, 28
*giljacica, Diphyllobothrium*, 94
*giljacica, Spirometra*, 94
*gilloni, Hymenolepis*, 335
*Gilquinia*, diagnosis, 59
Gilquiniidae
   diagnosis, 51
   key to genera, 59
*ginesi, Dendrometra*, 346
*giranensis, Hymenolepis*, 317
*giranensis, Hymenosphenacanthus*, 317
*giranensis, Retinometra*, 317
*giranensis, Sphenacanthus*, 317
*giranensis, Sphenacanthus (Retinometra)*, 317
*glaber, Eutetrarhynchus*, 57
*glaciale, Diphyllobothrium*, 94
*glaciloides, Tetrabothrius*, 209
*gladium, Sobolevicanthus*, 309
*glandarii, Halysis*, 298
*Glandicephalus*, diagnosis, 92
*glandularis, Anomotaenia*, 399
*glandularis, Bancroftiella*, 399
*glandularis, Pseudocittotaenia*, 450
*glandularis, Rhinebothroides*, 154
*glanduliger, Proteocephalus*, 187
*glareola, Anomolepis*, 381
*glareola, Dilepis*, 381
*glareolae, Choanotaenia*, 360
*glareolae, Echinocotyle*, 285
*glareolae, Kowalewskiella*, 360
*Glaridacris*, diagnosis, 35
*globacantha, Dilepis*, 376
*globacantha, Metadilepis*, 376
*Globarilepis*, diagnosis, 270
*globata, Anoncotaenia*, 341
*globata, Taenia*, 341
*globicephalae, Plicobothrium*, 91
*globicephalae, Trigonocotyle*, 206
*globicephalum, Pedibothrium*, 143
*globiceps, Anoplocephala*, 436
*globiceps, Flabelloskrjabinia*, 436
*globiceps, Paranoplocephala*, 436
*globiceps, Taenia*, 436
*globifera, Cladotaenia*, 413
*globipunctata, Stilesia*, 418
*globipunctata, Taenia*, 418
*globirostris, Davainea*, 260
*globirostris, Hymenolepis*, 304
*globirostris, Kotlania*, 260
*globirostris, Raillietina (R.)*, 260
*globirostris, Raillietina (Ransomia)*, 260
*globirostris, Rodentolepis*, 304

*globirostris, Vampirolepis*, 304
*globocaudata, Davainea*, 265
*globocaudata, Kotlania*, 265
*globocaudata, Raillietina (F.)*, 265
*globocaudata, Raillietina (Johnstonia)*, 265
*globocephala, Davainea*, 266
*globocephala, Hymenolepis*, 325
*globocephala, Raillietina*, 266
*globocephala, Variolepis*, 325
*globosa, Biuterina*, 348
*globosa, Hymenolepis*, 288
*globosa, Insectivorolepis*, 288
*globosoides, Dicranotaenia*, 288
*globosoides, Insectivorolepis*, 288
*globuliforme, Haplobothrium*, 85
*globulosa, Armadoskrjabinia*, 297
*globulosa, Hymenolepis*, 297
*globulus, Anomotaenia*, 406
*globulus, Choanotaenia*, 406
*globulus, Dichoanotaenia*, 406
*globulus, Taenia*, 406
*glomerata, Taenia*, 225
*glomeratus, Coenurus*, 225
*glomeratus, Multiceps*, 225
*glomovaginata, Valipora*, 371
*Glossobothrium*, diagnosis, 115
*Glyphobothrium*, diagnosis, 148
*gnedini, Lateriporus*, 373
*gobii, Brachyurus*, 30, 36
*gobii, Glaridacris*, 30, 36
*gobionis, Ligula*, 87
*gobiorum, Proteocephalus*, 187
*Goezeella*, 194
*goezei, Ctenotaenia*, 443
*goezei, Moniezia*, 443
*goezei, Taenia*, 443
*Goeziana*, 223
*gogonka, Microsomacanthus*, 329
*goizuetai, Fernandezia*, 243
*goldsteini, Acanthobothrium*, 139
*gomatii, Duthiersia*, 86
*gondo, Diphyllobothrium*, 94
*gondokorensis, Inermicapsifer*, 465
*gondwana, Choanotaenia*, 362
*gongyla, Choanotaenia*, 363, 397
*gongyla, Icterotaenia*, 363, 397
*gongyla, Paricterotaenia*, 363
*gongyla, Polycercus*, 397
*Gonoscolex*, diagnosis, 283
*gonyamai, Taenia*, 225
*Gopalaia*, diagnosis, 278
*gordoni, Polyoncobothrium*, 97
*goreensis, Nybelinia*, 74
*gorillae, Anoplocephala*, 433
*gorsakii, Ophiovalipora*, 377
*gorsakii, Parvitaenia*, 377
*gotoi, Caryophyllaeus*, 33
*goughi, Avitellina*, 419
*goura, Davainea*, 260
*goura, Kotlania*, 260
*goura, Raillietina (R.)*, 260

goura, *Raillietina (Ransomia),* 260
govinda, *Cotugnia,* 246
gowkongensis, *Bothriocephalus,* 102
gracewileyae, *Oochoristica,* 454
gracile, *Acanthobothrium,* 139
gracile, *Anthobothrium,* 161
gracile, *Anthocephalum,* 161
gracile, *Echeneibothrium,* 151
gracile, *Eniochobothrium,* 124
gracile, *Litobothrium,* 169
gracile, *Phyllobothrium,* 161
gracile, *Rhynchobothrium,* 65
gracilis, *Acanthotaenia,* 182
gracilis, *Anthocephalus,* 54
gracilis, *Callitetrarhynchus,* 54
gracilis, *Crepidobothrium,* 182
gracilis, *Davainea,* 260
gracilis, *Diphyllobothrium,* 94
gracilis, *Fuhrmanniella,* 307
gracilis *Halysis,* 307
gracilis, *Hymenolepis,* 307
gracilis, *Hymenolepis (Drepanidotaenia),* 307
gracilis, *Hymenolepis (Weinlandia),* 307
gracilis, *Ichthyotaenia (Acanthotaenia),* 182
gracilis, *Ophiotaenia,* 184
gracilis, *Parabothriocephalus,* 118
gracilis, *Paranoplocephala,* 434
gracilis, *Proteocephalus,* 182
gracilis, *Raillietina (R.),* 260
gracilis, *Sobolevicanthus,* 307
gracilis, *Sphenacanthus,* 307
gracilis, *Spirometra,* 94
gracilis, *Taenia,* 307
gracilis, *Tetrabothrius,* 208
gracilis, *Tetrarhynchus,* 77
gracilis, *Trilocularia,* 144
gracilis, *Weinlandia,* 307
gracillimum, *Callitetrarhynchus,* 54
graeca, *Hymenolepis,* 336
graeca, *Kotlania,* 260
graeca, *Raillietina (R.),* 260
granaia, *Diphyllobothrium,* 94
granatensis, *Deltokeras,* 347
grandiceps, *Acanthobothrium,* 139
grandiporus, *Idiogenes,* 240
grandis, *Diplogonoporus,* 90
grandis, *Krabbea,* 90
grandis, *Ophiotaenia,* 184
grandis, *Priapocephalus,* 204
grandis, *Solenophorus,* 85
granularis, *Bothriocephalus,* 102
granulosus, *Echinococcus,* 222
gretellati, *Ophryocotyle,* 244
*Grillotia,* diagnosis, 63
grisea, *Vampirolepis,* 287
grisea, *Myotolepis,* 287
grisea, *Taenia,* 287
grobbeni, *Kotlania,* 260
grobbeni, *Raillietina (R.),* 260
grobbeni, *Raillietina (Ransomia),* 260
groenlandica, *Aploparaksis,* 276

groenlandica, *Hymenolepis,* 276
groenlandica, *Taenia* 276
grossus, *Tetrarhynchus,* 77
*Gruitaenia,* 404
*Gryporhynchus,* diagnosis, 378
guadeloupensis, *Hymenolepis,* 336
guarany, *Vampirolepis,* 304
guberiana, *Retinometra,* 318
guiarti, *Anomotaenia,* 406
guiarti, *Choanotaenia,* 406
guiarti, *Monopylidium,* 406
guineensis, *Paruterina,* 350
guinensis, *Davainea,* 466
guinensis, *Inermicapsifer,* 466
guinensis, *Zschokkeella,* 466
guivillensis, *Davainea,* 259
gundii, *Andrya,* 435
gundlachi, *Monoecocestus,* 426
guschanskoi, *Dicranotaenia,* 310
guschanskoi, *Hymenolepis,* 310, 336
guschanskoi, *Nadejdolepis,* 310
gutterae, *Abuladzugnia,* 245
gutterae, *Cotugnia,* 245
gutterae, *Octopetalum,* 342
*Gvosdevilepis,* diagnosis, 286
*Gvosdevinia,* 253
gwiletica, *Hymenolepis,* 306
Gymnorhynchidae
   diagnosis, 51
   key to genera, 60
gymnorhynchoides, *Rhopalothyrax,* 75
*Gymnorhynchus,* diagnosis, 63
gymnorhynchus, *Tetrarhynchus,* 77
gynandrolinearis, *Progamotaenia,* 445
*Gynandrotaenia,* diagnosis, 217
*Gyrocoelia,* diagnosis, 215
*Gyrocotyle,* diagnosis, 473
Gyrocotylidae
   diagnosis, 472
   key to genera, 473
Gyrocotylidea, 472
Gyrocotylidea, diagnosis, 467
*Gyrocotyloides,* diagnosis, 474
*Gyrometra,* diagnosis, 471

# H

habanensis, *Ophiotaenia,* 184
haemacephala, *Ophryocotyloides,* 243
haemantopodis, *Fimbriarioides,* 268
haematopodis, *Andrepigynotaenia,* 219
hagmanni, *Monoecocestus,* 426
hagmanni, *Schizotaenia,* 426
hainanensis, *Oochoristica,* 454
haldemani, *Aploparaksis,* 276
*Halisis,* 352
halli, *Raillietina (R.),* 261
*Halysiorhynchus,* diagnosis, 69
hamadryadis, *Bertia,* 439
hamadryadis, *Bertiella,* 439

*hamanni, Vampirolepis*, 304
*hamasigi, Dicranotaenia*, 322
*hamasigi, Dilepis*, 370
*hamasigi, Hymenolepis*, 322
*hamasigi, Lateriporus*, 370
*hamasigi, Limnolepis*, 322
*Hamatolepis*, 327
*hamulacanthos, Hymenolepis*, 318
*hamulacanthos, Hymenosphenacanthus*, 318
*hamulacanthos, Retinometra*, 318
*hamulacanthos, Sphenacanthus*, 318
*hanseni, Pseudanthobothrium*, 148
*hanumanthai, Proteocephalus*, 187
*hanumantharaoi, Acanthobothrium*, 139
Haplobothriidae, 84
   diagnosis, 81
*Haplobothrium*, diagnosis, 84
*Haploparaxis*, 275
*hardoiensis, Mathevotaenia*, 463
*harfordi, Echinobothrium*, 166
*harpago, Acanthobothrium*, 139
*harpago, Dicranobothrium*, 139
*harvathi, Dilepis*, 381
*hassalli, Hymenolepis*, 325
*hassalli, Variolepis*, 325
*hastata, Marsipometra*, 110
*hattorii, Rodentolepis*, 304
*hattorii, Vampirolepis*, 304
*hawaiiensis, Pterobothrium*, 70
*hawaiiensis, Rhinebothrium*, 155
*heardi, Parvitaenia*, 412
*heckmanni, Parvitaenia*, 412
*hedleyi, Acoleus*, 231
*heimi, Anomotaenia*, 406
*Helictometra*, 420
*helmymohamedi, Echinobothrium*, 166
*Hemicatenotaenia*, diagnosis, 233
*hemidactyli, Oochoristica*, 454
*hemignathi, Drepanidotaenia*, 298
*hemignathi, Hymenolepis*, 298
*hemignathi, Passerilepis*, 298
*Hemiparonia*, diagnosis, 427
*hemisphericus, Proteocephalus*, 187
*hemuli, Pterobothrium*, 71
*hepatica, Choanotaenia*, 367
*hepatica, Mesogyna*, 204
*hepatica, Monopylidium*, 367
*hepatica, Rodentotaenia*, 367
*hepatica, Stilesia*, 418
*Hepatotaenia*, 444
Hepatoxylidae, diagnosis, 49
*Hepatoxylon*, diagnosis, 62
*heptanchi, Tetrarhynchus*, 65
*herdmani, Acanthobothrium*, 139
*herdmani, Nybelinia*, 74
*herdmani, Tetrarhynchus*, 74
*hermaphroductus, Abothrium*, 108
*herodiae, Dendrouterina*, 378
*herodiae, Ophryocotyle*, 244
*heroniensis, Echinobothrium*, 166
*herpestis, Mathevotaenia*, 463

*herpestis, Oochoristica*, 463
*hertwigi, Davainea*, 265
*hertwigi, Kotlania*, 265
*hertwigi, Raillietina (F.)*, 265
*hertwigi, Raillietina (Johnstonia)*, 265
*hesperiphonae, Anomaloporus*, 342
*heteracantha, Prochristianella*, 59
*heteracanthum, Onchobothrium*, 135
*heteracanthum, Pterobothrium*, 70
*heterobranchus, Marsipocephalus*, 190
*heteroclitus, Tetrabothrius*, 208, 209
*heterocoronata, Anomotaenia*, 406
*heterocoronata, Choanotaenia*, 406
*heterodonti, Acanthobothrium*, 139
*heteromegacanthus, Parachristianella*, 58
*heteromerum, Rhynchobothrius*, 76
*heteropleura, Amphicotyle*, 106
*heteropneusti, Capingentoides*, 40
*heterosoma, Tetrabothrius*, 208
*heterospine, Rhynchobothrium*, 76
*heterospinus, Microsomacanthus*, 329
*Heterotetrarhynchus*, 63
*hewetensis, Davainea*, 248
*Hexacanalis*, diagnosis, 123
*Hexagonoporus*, diagnosis, 88
*Hexaparuterina*, diagnosis, 211
*Hexastichorchis*, 418
*hians, Diphyllobothrium*, 94
*hickmani, Anthobothrium*, 157
*hidaensis, Vampirolepis*, 304
*hieraticos, Oligorchis*, 333
*hilli, Dilepis*, 381
*hilmyi, Hispaniolepis*, 306
*Hilmylepis*, diagnosis, 293
*hilmylum, Anomotaenia*, 406
*himalayai, Thysaniezia*, 420
*himalayana, Choanotaenia*, 362
*himalayotaenia, Hydatigera*, 225
*himalayotaenia, Taenia*, 225
*Himantocestus*, 231
*himantopodis, Davainea*, 248
*himantopodis, Dicranotaenia*, 322
*himantopodis, Dicranotaenia (Dicranolepis)*, 322
*himantopodis, Hymenolepis*, 322
*himantopodis, Malika*, 357
*himantopodis, Taenia*, 322
*himantopodis, Wardium*, 322
*Himantaurus*, 247
*himanturi, Rhinebothrium*, 155
*hipposideri, Hymenolepis*, 335
*hira, Pterobothrium*, 70
*hirondellei, Bothriocephalus*, 102
*hirsuta, Aploparaksis*, 276
*hirsuta, Hymenolepis*, 276
*hirsuta, Monorchis*, 276
*hirsuta, Taenia*, 276
*hirundina, Angularella*, 393
*hirundina, Anomotaenia*, 393
*hirundina, Vitta*, 393
*Hispaniolepidoides*, 306
*Hispaniolepis*, diagnosis, 306

*hispida, Prochristianella*, 59
*hispida, Rhynchobothrium*, 59
*hispidum, Acanthobothrium*, 139
*hispidum, Rhynchobothrium*, 76
*histiophorus, Bothriocephalus*, 103
*histocephalum, Anantrum*, 104
*hlosei, Taenia*, 225
*hoeppli, Anomotaenia*, 406
*hoffmani, Biacetabulum*, 30
*holorhini, Acanthobothrium*, 139
*hominis, Taenia*, 222
*honessi, Monordotaenia*, 222
*hookensis, Neyraia*, 339
*hoplites, Dilepis*, 381
*hoplites, Taenia*, 381
*hoploporus, Hymenolepis, (Weinlandia)*, 325
*hoploporus, Variolepis*, 325
*Hornelliella*, diagnosis, 63
Hornelliellidae, diagnosis, 52
*horrida, Hymenolepis*, 289
*horrida, Taenia*, 289
*horridus, Gymnorhynchus*, 61
*horridus, Idiogenes*, 240
*horridus, Molicola*, 61
*horridus* var. *africanus, Idiogenes*, 240
*houdemeri, Ophiovalipora*, 377
*houghtoni, Spirometra*, 94
*Houttuynia*, diagnosis, 250
*hoyeri, Tetrabothrius*, 208
*Hsuolepis*, 433
*huebscheri, Raillietina (P.)*, 254
*hughesi, Hybridolepis*, 325
*hughesi, Hymenolepis (H.)*, 325
*hughesi, Variolepis*, 325
*hui, Echeneibothrium*, 151
*humile, Dinobothrium*, 159
*Hunterella*, diagnosis, 33
*hunteri, Monobothrium*, 32
*Hunteroides*, diagnosis, 469
*huronensis, Atractolytocestus*, 22
*hutsoni, Pachybothrium*, 142
*hutsoni, Pedibothrium*, 142
*hutsoni, Phyllobothroides*, 142
*Halysis*, 223
*Hybridolepis*, 324
*Hydatigena*, 223, 225
*Hydatigera*, 223
*Hydatis*, 223
*Hydatula*, 223
*hydrochelidonis, Anomotaenia*, 406
*hydrochelidonis, Choanotaenia*, 406
*hydrochelidonis, Laritaenia*, 406
*hydrochoeri, Monoecocestus*, 426
*hydrochoeri, Schizotaenia*, 426
*Hygroma*, 223
*hylae, Nematotaenia*, 213
*hylae, Ophiotaenia*, 184
*hyalina, Ophiotaenia*, 184
*hyanae, Taenia*, 225
*Hymenandrya*, diagnosis, 332
*Hymenofimbria*, diagnosis, 292

Hymenolepididae, 267
  diagnosis, 201
  key to subfamilies, 267
Hymenolepidinae, key to genera, 272
Hymenolepidinae of birds, appendix to, 335
Hymenolepidinae of mammals, appendix to, 334
*Hymenolepis*, diagnosis, 288
*Hymenosphenacanthoides*, 316
*Hymenosphenacanthus*, 316
*hyodori, Dicranotaenia*, 322
*hyperapolytica, Bilocularia*, 162
*hyperborea, Hydatigera*, 225
*hyperbores, Taenia*, 224, 225
*hyperoglyphe, Echinophallus*, 105
*hyperoglyphe, Paraechinophallus*, 105
*Hypocaryophyllaeus*, diagnosis, 28
*hypoleuci, Echinocotyle*, 285
*hypoleucia, Choanotaenia*, 360, 362
*hypoprioni, Platybothrium*, 137
*hypsipetis, Unciunia*, 386
*hyracis, Anoplocephala*, 465
*hyracis, Hyracotaenia*, 465
*hyracis, Inermicapsifer*, 465
*hyracis, Taenia*, 465
*hyracis* var. *hepatica, Anoplocephala*, 465
*hyracis, Zschokeella*, 465
*Hyracotaenia*, 465
*hystrix, Microsomacanthus*, 329

# I

*ibanezi, Choanotaenia*, 362
*ibanezi, Echinocotyle*, 285
*ibidis, Hymenolepis*, 336
*ibisae, Parvitaenia*, 412
*icelandicum, Acanthobothrium*, 139
*ichneumontis, Mathevotaenia*, 463
*ichneumontis, Oochoristica*, 463
*ichthybori, Ichtybothrium*, 98
*Icthybothrium*, diagnosis, 98
*Ichthyotaenia*, 185
*Icterotaenia*, 361
*idahoensis, Baerietta*, 212
*Idiogenes*, diagnosis, 239
Idiogeninae, key to genera, 237
*Idiogenoides*, 253
*idiogenoides, Raillietina (R.)*, 260
*iduncula, Paruterina*, 351
*iheringi, Linstowia*, 461
*iheringi, Linstowia (Paralinstowia)*, 463
*iheringi, Oochoristica*, 461
*ijimai, Acanthobothrium*, 139
*ilisha, Otobothrium*, 68
*ilisha, Poecilancistrum*, 68
*ilisha, Rhynchobothrius*, 68
*ilisha, Tentacularia*, 68
*ilishai, Bovienia*, 25
*ilocana, Cotugnia*, 246
*imbutiformis, Mesocestoides*, 203
*immatura, Mathevotaenia*, 463

*immerina, Taenia,* 207
*Imparmargo,* diagnosis, 365
*imparspine, Rhynchobothrium,* 64
*impudens, Wageneria,* 76
*inaequalis, Cotugnia,* 246
*inchoatum, Phyllobothrium,* 161
*incisa, Atriotaenia,* 462
*incisa, Oochoristica,* 462
*incognita, Oschmarenia,* 457
*incognita, Taenia,* 139
*incognita, Thysanotaenia,* 457
*incruis, Taenia,* 433
*inda, Raillietina (R.),* 260
*indiana, Amoebotaenia,* 387
*indiana, Chettusiana,* 392
*indica, Aliezia,* 417
*indica, Anoncotaenia,* 341
*indica, Anteropora,* 164
*indica, Capingentoides,* 40
*indica, Catenotaenia,* 235
*indica, Davainea,* 248
*indica, Djombangia,* 23
*indica, Gidhaia,* 429
*indica, Indotaenia,* 440
*indica, Mosgovoyia,* 451
*indica, Ophiotaenia,* 184
*indica, Oochoristica,* 454
*indica, Progynotaenia,* 219
*indica, Pseudocapingentoides,* 40
*indica, Pseudocaryophyllaeus,* 41
*indica, Pseudoschistotaenia,* 230
*indica, Raillietina (R.),* 260
*indica, Raillietina (S.),* 256
*indica, Retinometra,* 318
*indica, Schistotaenia,* 229
*indica, Stylolepis,* 318
*indicata, Flabelloskrjabinia,* 436
*indicata, Paranoplocephala,* 436
*indicus, Bothriocephalus,* 102
*indicus, Fernandezia,* 244
*indicus, Lytocestus,* 25
*indicus, Ophryocotyle,* 244
*Indotaenia,* diagnosis, 439
*Inermicapsifer,* diagnosis, 465
Inermicapsiferinae, key to genera, 467
*Inermiphyllidium,* 158
*inermis, Amerina,* 341, 346
*inermis, Hymenolepis,* 291
*inermis, Shipleya,* 215
*inermis, Taenia,* 334
*inermis, Trichocephaloides,* 370
*infirma, Insectivorolepis,* 288
*inflata, Diorchis,* 281
*inflata, Diplacanthus,* 281
*inflata, Hymenolepis,* 281
*inflata, Taenia,* 281
*inflatocirrosa, Amphipetrovia,* 326
*inflatus, Aspidorhynchus,* 73
*infrequens, Anoplocephala,* 436
*infrequens, Anoplocephaloides,* 436
*infrequens, Biacetabulum,* 30

*infrequens, Hymenolepis,* 291
*infrequens, Paranoplocephala,* 436
*infrequens, Staphylepis,* 291
*infrequens, Raillietina (R.),* 260
*Infula,* diagnosis, 216
*infundibuliformis, Abothrium,* 108
*infundibuliformis, Choanotaenia,* 361
*infundibuliformis, Taenia,* 255, 361
*infundibuliformis* var. *polyorchis, Monopylidium,* 386
*infundibulum, Choanotaenia,* 361, 406
*infundibulum, Drepanidotaenia,* 361
*infundibulum, Monopylidium,* 361
*infundibulum, Taenia,* 361
*ingens, Dasyrhynchus,* 52
*ingens, Rhynchobothrium,* 52
*ingwei, Taenia,* 225
*inhamata, Hymenolepis,* 291
*inhamata, Staphylepis,* 291
*innominata, Choanotaenia,* 397
*innominata, Hymenolepis,* 315
*innominata, Microsomacanthus,* 315
*innominata, Paricterotaenia,* 397
*innominata, Polycercus,* 397
*innominatus hypudaci, Cysticercus,* 227
*innominatus, Tetrabothrius,* 208
*Insectivorolepis,* diagnosis, 288
*insigna, Caulobothrium,* 153
*insigna, Rhinobothrium,* 153
*insigne, Dasyrhynchus,* 52
*insigne, Otobothrium,* 68
*insigne, Rhynchobothrium,* 52
*insignis, Davainea,* 260
*insignis, Kotlania,* 260
*insignis, Ophryocotyle,* 244
*insignis, Raillietina (R.),* 260
*insignis, Raillietina (Ransomia),* ;260
*insignis, Taenia,* 260
*insignis, Tentacularia,* 52
*Insinuarotaenia,* diagnosis, 221
*institata, Grillotia,* 64
*institata, Heterotetrarhyncus,* 64
*insulaemargaritae, Oochoristica,* 454
*intermedia, Choanotaenia,* 397
*intermedia, Corallotaenia,* 179
*intermedia, Cotugnia,* 246
*intermedia, Fimbriaria,* 268
*intermedia, Fimbriarioides,* 268
*intermedia, Glaridacris,* 36
*intermedia, Icterotaenia,* 397
*intermedia, Paricterotaenia,* 397
*intermedia, Polycercus,* 397
*intermedia, Sacciuterina,* 397
*intermedia, Taenia,* 225
*intermedium, Acanthobothrium,* 139
*intermedium, Acrobothrium,* 139
*intermedium, Corallobothrium,* 179
*intermedium, Megathylacoides,* 179
*intermedius, Bothrimonus,* 45
*intermedius, Dicranotaenia, (D.),* 298
*intermedius, Hymenolepis,* 298, 305

*intermedius, Mayhewia,* 298
*intermedius, Passerilepis,* 298
*intermedius, Tetrabothrius,* 208
*intermedius, Weinlandia,* 298
*intermedius* var. *exulans, Tetrabothrius,* 209
*interpositus, Inermicapsifer,* 465, 466
*interpositus* var. *sinaitica, Inermicapsifer,* 466
*interrupta, Ligula,* 87
*interrupta, Raillietina (R.),* 260
*interruptum, Pterobothrium,* 70
*interruptus, Skrjabinotaurus,* 260
*intestinalis, Ligula,* 87
*intrepidus, Tetrabothrius,* 208
*intricata, Hymenolepis,* 300
*intricata, Lockerrauschia,* 300
*intricata, Neyraia,* 339
*intricata, Taenia,* 339
*Introvertus,* 18
*introversa, Hymenolepis,* 328
*introversa, Weinlandia,* 321
*inuzensis, Insectivorolepis,* 288
*inversa, Choanotaenia,* 397
*inversa, Icterotaenia,* 397
*inversa, Parachoanotaenia,* 397
*inversa, Paricterotaenia,* 397
*inversa, Polycercus,* 397
*inversa, Taenia,* 397
*ioensis, Khawia,* 21
*iola, Choanotaenia,* 367
*iola, Monopylidium,* 367
*iowensis, Archigetes,* 30
*iranica, Taufikia,* 426
*iriei, Hymenolepis,* 335
*irregularis, Dilepis,* 381
*irregularis, Neophronia,* 429
*isacantha, Anomotaenia,* 410
*isacantha, Pseudanomotaenia,* 410
*ischnorhyncha, Amoebotaenia,* 218
*ischnorhyncha, Leptotaenia,* 218
*ischnorhyncha, Taenia,* 218
*isensis, Vampirolepis,* 304
*Isoglaridacris,* diagnosis, 36
*isometra, Inermicapsifer,* 466
*isomydis, Anoplocephaloides,* 436
*isomydis, Paranoplocephala,* 436
*isomydis, Taenia,* 436
*isoniciphora, Paruterina,* 351
*isuri, Gymnorhynchus,* 62
*isuri, Molicola,* 62
*iunii, Tatria,* 229
*Ivritaenia,* diagnosis, 364
*iwanizki, Anomotaenia,* 406
*iwatensis, Vampirolepis,* 304

# J

*jacobii, Diorchis,* 282
*jacobsoni, Dilepis,* 381
*jacobsoni, Hymenolepis,* 304
*jacobsoni, Rodentotaenia,* 304
*jacobsoni, Staphylocystis,* 304
*jacobsoni, Vampirolepis,* 304
*jacutensis, Skrjabinacanthus,* 299
*jaegerskioeldi, Baerietta,* 212
*jaegerskioeldi, Hymenolepis,* 329
*jaegerskioeldi, Microsomacanthus,* 329
*jaegerskioeldi, Nematotaenia,* 212
*jaegerskioeldi, Weinlandia,* 329
*jaegerskioldi, Tetrabothrius,* 208
*jaenschi, Hymenolepis,* 336
*jägerskiöldi, Progynotaenia,* 219
*jaisalmerensis, Myotolepis,* 288
*jakhalsi, Taenia,* 225
*jammuensis, Gangesia,* 126
*jamunicus, Hymenolepis,* 329
*jamunicus, Microsomacanthus,* 329
*jandia, Megathylacus,* 178
*jandia, Proteocephalus,* 187
*janicki, Baerietta,* 212
*janicki, Nematotaenia,* 212
*janicki, Nesolecithus,* 471
*janickii, Diphyllobothrium,* 94
*Janiszewskella,* diagnosis, 36
*januaria, Cotugnia,* 246
*japonense, Taphrobothrium,* 101
*japonensis, Aploparaksis,* 276
*japonensis, Bothrioscolex,* 21
*japonensis, Caryophyllaeus,* 21
*japonensis, Khawia,* 21
*japonensis, Ophiotaenia,* 184
*japonensis, Passerilepis,* 298
*japonica, Amphilina,* 468
*japonica, Baerietta,* 212
*japonica, Confluaria,* 325
*japonica, Dubininolepis,* 325
*japonica, Hymenolepis,* 325
*japonica, Paraechinophallus,* 105
*japonica, Raillietina (P.),* 254
*japonica, Variolepis,* 325
*japonicum, Discobothrium,* 125
*japonicum, Prosobothrium,* 177
*japonicus, Bothriocephalus,* 102
*jarara, Ophiotaenia,* 184
*Jardugia,* diagnosis, 232
*jassyensis, Braunia,* 87
*jasuta, Hymenolepis,* 336
*javaensis, Oochoristica,* 454
*javanensis, Hymenolepis,* 309
*javanensis, Sobolevicanthus,* 309
*javanica, Dilepis,* 381
*javanica, Paruterina,* 351
*javanicum, Echeneibothrium,* 151
*javanicum, Tiarabothrium,* 151
*javanicus, Caryophyllaeus,* 33
*jayapaulazariahi, Nybelinia,* 74
*jerratta, Hymenolepis,* 336
*jimenezi, Paraprogynotaenia,* 221
*jodhpurensis, Diorchis,* 282
*jodhpurnesis, Oochoristica,* 454
*johnsoni, Hymenolepis,* 304
*johnsoni, Progamotaenia,* 445

*johnsoni, Rodentolepis*, 304
*johnsoni, Vampirolepis*, 304
*johnstonei, Tentacularia*, 77
*johnstoni, Paramoniezia*, 449
*johnstoni, Thalophyllaeus*, 24
*Johnstonia*, 252
*johri, Raillietina (R.)*, 260
*jonesi, Isoglaridacris*, 37
*jonesi, Mesocestoides*, 203
*Jonesius*, diagnosis, 280
*joubini, Tetrabothrius*, 208
*joyeuxi, Choanotaenia*, 362
*joyeuxi, Cotugnia*, 246
*joyeuxi, Hymenolepis*, 309
*Joyeuxia*, 353
*joyeuxibaeri, Choanotaenia*, 363
*joyeuxbaeri, Raillietina (R.)*, 261
*Joyeuxiella*, diagnosis, 353
*joyeuxi, Kotlania*, 261
*joyeuxi, Raillietina (R.)*, 261
*joyeuxii, Coenomorphus*, 57
*jubilaea, Tatria*, 229, 230
*julievansium, Echeneibothrium*, 151
*junceus, Tetrabothrius*, 207
*junglensis, Batrachotaenia*, 184
*junglensis, Ophiotaenia*, 184
*junkea, Oochoristica*, 454
*junlanae, Multiuterina*, 343
*jurii, Anomotaenia*, 406

# K

*kainjii, Wenyonia*, 26
*kaiseris, Hispaniolepis*, 306
*kaiseris, Hymenolepis*, 306
*kakia, Raillietina (S.)*, 256
*kalawewaensis, Malika*, 357
*kamayuta, Aploparaksis*, 276
*kamienae, Zyxibothrium*, 144
*kantipura, Raillietina (R.)*, 261
*kaparari, Nomimoscolex*, 192
*Kapsulotaenia*, diagnosis, 180
*kapul, Bertiella*, 439
*kapurdiensis, Choanotaenia*, 362
*karachiense, Acanthobothrium*, 139
*karachiensis, Catenotaenia*, 235
*karachii, Thysanocephalum*, 133
*karajasicus, Dendrouterina*, 378
*karajasicus, Lateriporus*, 373
*karatchvili, Raillietina (P.)*, 254
*karbharae, Echeneibothrium*, 151
*karbharae, Shindeiobothrium*, 151
*karuatayi, Anthobothrium*, 157
*kashiwarensis, Raillietina (P.)*, 254
*kashmirensis, Anomotaenia*, 406
*kashmirensis, Nematotaenia*, 213
*kashmiriensis, Caryophyllaeus*, 33
*katpurensis, Polypocephalus*, 121
*kazachstanica, Gastrotaenia*, 130
*kazachstanica, Parabisaccanthes*, 314

*kedroviensis, Echinoproboscilepis*, 293
*kempi, Dilepis*, 375
*kempi, Hymenolepis*, 375
*kempi, Meggittella*, 375
*kempi, Paradilepis*, 375
*kenaiensis, Hymenolepis*, 309
*kenaiensis, Hymenosphenicanthus*, 309
*kenaiensis, Sobolevicanthus*, 309
*kenki, Hymenolepis*, 288
*kenki, Insectivorolepis*, 288
*kerivoulae, Mathevotaenia*, 463
*kerivoulae, Oochoristica*, 463
*kerivoulae, Vampirolepis*, 304
*kerkhami, Phyllobothroides*, 143
*ketae, Phyllobothrium*, 161
*khami, Circumoncobothrium*, 96
*Khawia*, diagnosis, 20
*khalili, Oochoristica*, 454
*khalili, Raillietina (R.)*, 261
*khalili, Vampirolepis*, 304
*kharati, Amoebotaenia*, 387
*kiewietti, Gyrocoelia*, 216
*Killigrewia*, diagnosis, 428
*kingae, Phyllobothrium*, 161
*kirbyi, Mesocestoides*, 203
*kirghisica, Paruterina*, 351
*kirgizica, Catenotaenia*, 235
*kirghizica, Raillietina, (R.)*, 261
*kivuensis, Bothriocephalus*, 102
*klebergi, Raillietina, (R.)*, 261
*kodonodes, Diorchis*, 282
*kodrensis, Hilmylepis*, 294
*kolbei, Idiogenes*, 240
*kolhapurensis, Gidhaia*, 429
*konoresniki, Anomotaenia*, 406
*kontrimavichusi, Anoplocephaloides*, 436
*kordofanensis, Raillietina*, 264
*kori, Idiogenes*, 240
*korkhaani, Otiditaenia*, 240
*korkei, Kotlania*, 261
*korkei, Raillietina (Fuhrmannetta)*, 261
*korkei, Raillietina, (R.)* 261
*kornyushini, Echinocotyle*, 285
*Kosterina*, 472
*Kotlania*, 252
*Kotlanotaurus*, 252, 253
*Kowalewskiella*, diagnosis, 359
*kowalewskii, Dicranotaenia*, 322
*kowalewskii, Hymenolepis*, 322
*kowalewskii, Porotaenia*, 208
*kowalewskii, Tetrabothrius*, 208
*kowalewskii, Wardium*, 322
*Kowalewskius*, 327
*kozloffi, Distoichometra*, 212
*Krabbea*, 89
*krabbeella, Hymenolepis*, 309
*krabbeella, Sobolevicanthus*, 309
*krabbei, Neyraia*, 339
*krabbei, Taenia*, 225
*kratochvili, Catenotaenia*, 235
*krepkogorski, Hydatigera*, 225

*krepkogorski, Taenia*, 225
*Krimi*, diagnosis, 394
*krishna, Hymenolepis*, 304
*krishna, Vampirolepidoides*, 304
*krishna, Vampirolepis*, 304
*krishnai, Cittotaenia*, 443
*krishnai, Gopalaia*, 278
*krotovi, Diphyllobothrium*, 94
*kuantanensis, Ophiotaenia*, 184
*kubanica, Fimbriaria*, 268
*kugii, Bothridium*, 86
*kuhhi, Tylocephalum*, 126
*kuiperi, Kosterina*, 472
*kullmanni, Catenotaenia*, 235
*kunduchi, Gyrometra*, 472
*kurashvilii, Triodontolepis*, 301
*kurochkini, Amphicotyle*, 106
*kutassi, Dicranotaenia*, 321
*kuvaria, Cittotaenia*, 422
*kuvaria, Coelodela*, 422
*kuyukuyu, Proteocephalus*, 187
*kuznetsovi, Moniezia*, 442
*kwangensis, Oligorchis*, 333
*Kystocephalus*, 126
*kyushuensis, Microsomacanthus*, 329
*kyushuensis, Weinlandia*, 329

# L

*labiatus, Solenophorus*, 86
*labracis, Bothriocephalus*, 102
*lacazii, Ophryocotyle*, 244
*laccocephalus, Tetrabothrius*, 208
*lachesidis, Crepidobothrium*, 189
*lachesidis, Tetrabothrius*, 189
*laciniatum, Anthobothrium*, 157
*laciniatum, Phyllobothrium*, 161
*laciniatum, Rhynchobothrium*, 75
*laciniatum longicollis, Crossobothrium*, 153
*laciniatus, Bothriocephalus*, 102
*laciniosa, Avitellina*, 418
Lacistorhynchidae
    diagnosis, 50
    key to genera, 63
*Lacistorhynchus*, diagnosis, 65
*lactea, Ophiotaenia*, 184
*lactuca, Phyllobothrium*, 161
*laevicolle, Echinobothrium*, 166
*laevigata, Choanotaenia*, 397
*laevigata, Icterotaenia*, 397
*laevigata, Monopylidium*, 397
*laevigata, Parachoanotaenia*, 397
*laevigata, Paricterotaenia*, 397
*laevigata, Polycercus*, 397
*laevigata, Taenia*, 397
*laevis, Diploposthe*, 232
*laevis, Taenia*, 232
*lagenicollis, Taenia*, 361
*lagopi, Dicranotaenia*, 323
*lagopi, Wardium*, 323

*lagopodis, Davainea*, 248
*lagorchestis, Cittotaenia*, 445
*lagorchestis, Progamotaenia*, 445
*laguri, Catenotaenia*, 235
*lahorea, Avitellina*, 418
*lahorensis, Oligorchis*, 333
*lali, Hymenolepis*, 336
*lali, Proparuterina*, 411
*Lallum*, diagnosis, 343
*lambi, Anomaloporus*, 342
*lamellata, Hymenolepis*, 291
*lamellata, Staphylepis*, 291
*lamellatum, Eubothrioides*, 114
*lamilligera, Amabilia*, 228
*lamilligera, Taenia*, 228
*lamonteae, Nybelinia*, 74
*lamtoensis, Hymenolepis*, 335
*lanceolata, Drepanidotaenia*, 294
*lanceolata, Hymenolepis (Drepanidotaenia)*, 294
*lanceolata, Taenia*, 88, 213, 285, 294
*lanceolata var. lobata, Drepanidotaenia*, 295
*lanceolato-lobatus, Alyselminthus*, 92
*lanceolatum, Diphyllobothrium*, 94
*langrangei, Oochoristica*, 454
*Lapwingia*, diagnosis, 403
*largoproglottis, Proteocephalus*, 187
*lari, Hymenolepis*, 318
*lari, Hymenosphenacanthus*, 318
*lari, Microsomacanthus*, 329
*lari, Retinometra*, 318
*lari, Sphenacanthus*, 318
*lari, Tetrabothrius*, 208
*Laricanthus*, 294
*larimarina, Choanotaenia*, 362
*larina, Anomotaenia*, 404, 406
*larina, Aploparaksis*, 276
*larina, Haploparaksis*, 276
*larina, Pseudanomotaenia*, 406
*larina, Rissotaenia*, 406
*larina, Taenia*, 406
*Laritaenia*, 404
*laruei, Diphyllobothrium*, 94
*laruei, Glaridacris*, 36
*laruei, Proteocephalus*, 187
*laruei, Taenia*, 225
*lashleyi, Diphyllobothrium*, 94
*lasionycteridis, Hymenolepis*, 335
*lasiopeius, Neogryporhynchus*, 371
*lasium, Phoreiobothrium*, 141
*lata, Diploposthe*, 232
*lata, Linstowia*, 249
*lata, Polycoelia*, 249
*lateolabracis, Bothriocephalus*, 102
*lateralacantha, Pararetinometra*, 316
*lateralis, Aploparaksis*, 276
*lateralis, Davainea*, 261
*lateralis, Drepanidotaenia*, 295
*lateralis, Kotlania*, 261
*lateralis, Laricanthus*, 295
*lateralis, Weinlandia*, 295
*lateralis, Raillietina (R.)*, 261

*lateralis, Raillietina (Ransomia),* 261
*Lateriporus,* diagnosis, 373
Lateroporidea, 163
*Laterorchites,* diagnosis, 394
*Laterotaenia,* diagnosis, 400
*laticanalis, Davainea,* 256, 265
*laticanalis, Kotlania,* 265
*laticanalis, Raillietina (F.),* 265
*laticanalis, Raillietina (Johnstonia),* 265
*laticauda, Podicipitilepis,* 325
*laticauda, Variolepis,* 325
*laticeps, Caryophyllaeus,* 32
*laticeps, Solenophorus,* 85
*laticollis, Hydatigera,* 226
*laticollis, Taenia,* 226
*latifrons, Moniezia,* 442
*latissima, Anomotaenia,* 406
*latissima, Anoplocephala,* 433
*latissima, Cittotaenia,* 443
*latissima, Duthiersia,* 86
*latissima, Gruitaenia,* 406
*latissima, Moniezia,* 443
*latissima, Plagiotaenia,* 433
*latissima, Schizotaenia,* 433
*latissima, Taenia,* 443
*latissimum, Dipylidium,* 443
*latissimus, Eurycestus,* 389
*latum, Acanthobothrium,* 139
*latum, Diphyllobothrium,* 94
*latus, Mesocestoides,* 203
*lauriei, Hymenolepis,* 311
*lauriei, Nadejdolepis,* 311
*lavieri, Davainea,* 256
*lavieri, Raillietina (S.),* 256
*lazera, Stocksia,* 24
*lebasquei, Dilepis,* 381
*leblondii, Tetrarhynchus,* 77
Lecanicephalidae
    diagnosis, 119
    key to genera, 121
Lecanicephalidea, 13
    key to families, 119
*Lecanicephalum,* diagnosis, 126
*Lecanicephalum,* unidentifiable species, 127
*leiblei, Rhinebothrium,* 155
*leioformum, Clydonobothrium,* 157
*leiperi, Megacapsula,* 464
*leipoae, Raillietina (R.),* 261
*lemmi, Anoplocephaloides,* 436
*lemmi, Paranoplocephala,* 436
*lemuriformis, Bertiella,* 439
*lemuris, Coenurus,* 225
*lemuris, Multiceps,* 225
*lemuris, Oochoristica,* 462
*lemuris, Taenia,* 225
*lemuris, Thysanotaenia,* 464
*lenha, Monticellia,* 194
*lenha, Nomimoscolex,* 193
*lenha, Othinoscolex,* 195
*lenha, Peltidocotyle,* 195
*lenha, Proteocephalus,* 193

*Lentiella,* diagnosis, 437
*leoni, Raillietina (R.),* 259
*leonovi, Aploparaksis,* 276
*leonovi, Liga,* 390
*leonovi, Similuncinus,* 359
*lepida, Tentacularia,* 54
*lepidocybii, Pseudeubothrioides,* 108
*lepidocybii, Pseudeubothrium,* 108
*lepidoleprus-trachyrhynchus, Tetrarhynchus,* 77
*lepidopteri, Dibothriorhynchus,* 63
*lepidum, Callitetrarhynchus,* 54
*leporina, Taenia,* 451
*leporis, Inermicapsifer,* 466
*leporis, Monostomum,* 226
*leptacantha, Davainea,* 261
*leptacantha, Kotlania,* 261
*leptacantha, Raillietina (R.),* 261
*leptacantha, Raillietina (Ransomia),* 261
*leptaleum, Mixodigma,* 75
*leptocephala, Taenia,* 289
*leptodera, Taenia,* 297
*leptophallus, Dilepis,* 381
*leptoptili, Taenia,* 307
*leptosoma, Davainea,* 261
*leptosoma, Kotlania,* 261
*leptosoma, Raillietina (R.),* 261
*leptosoma, Raillietina (Ransomia),* 261
*leptosoma, Proteocephalus,* 187
*leptosoma, Taenia,* 261
*Leptotaenia,* diagnosis, 218
*leptotrachela, Davainea,* 265
*leptotrachela, Demidovella,* 265
*leptotrachela, Kotlania,* 265
*leptotrachela, Raillietina (F.),* 265
*leptotrachela, Raillietina (Johnstonia),* 265
*leuce, Gyrocoelia,* 216
*leuci, Phyllobothrium,* 161
*leucisci, Ligula,* 87
*leuckarti, Ctenotaenia,* 451
*leuckarti, Dipylidium,* 451
*leuckarti, Moniezia,* 451
*leuckarti, Taenia,* 451
*Leuckartia,* 107
*leuckartii, Calliobothrium,* 135
*leucomelanus, Eutetrarhynchus,* 57
*leucomelanus, Tetrarhynchus,* 57
*leucoranica, Liga,* 390
*leucura, Progynotaenia,* 219
*lichiae, Floriceps,* 53
*lichiae-vadiginis, Tetrarhynchus,* 77
*Liga,* diagnosis, 389
*Ligula,* diagnosis, 87
*liguloidea, Amphilina,* 470
*liguloideum, Monostomum,* 470
*liguloides, Aporodiorchis,* 296
*liguloides, Dicranotaenia,* 296
*liguloides, Flamingolepis,* 296
*liguloides, Halysis,* 296
*liguloides, Hymenolepis (Drepanidotaenia),* 296
*liguloides, Schizochoerus,* 470
*liguloides, Sphenacanthus,* 296

liguloides, Bothriocephalus, 94
lilliiformis, Taenia, 157
limicolum, Dicranotaenia, 323
limicolum, Wardium, 323
limnaei, Cysticercus, 327
limnocrypti, Aploparaksis, 276
limnodrili, Archigetes, 30
limnodrili, Glaridacris, 30
limnodromi, Hymenolepis, 336
Limnolepis, 320
limosa, Dilepis, 381
linderi, Hymenolepis, 291
linderi, Schmelzia, 291
linea, Hymenolepis, 336
lineata, Halisis, 203
lineata, Ptychophysa, 203
lineata, Taenia, 203
lineatum, Acanthobothrium, 139
lineatus, Eutetrarhynchus, 57
lineatus, Mesocestoides, 203
lineatus, Rhynchobothrium, 57
lineola, Hymenolepis, 304
lineola, Rodentolepis, 304
lineola, Vampirolepis, 304
lingualis, Nybelinia, 73
lingualis, Tetrarhynchus, 73
linguatula, Tetrarhynchus, 77
linsdalei, Catenotaenia, 235
linstowi, Aploparaksis, 297
linstowi, Davainea, 250
linstowi, Echinocotyle, 285
linstowi, Hymenolepis, 457
linstowi, Inermicapsifer, 458
linstowi, Linstowia, 457
linstowi, Multicapsiferina, 457
linstowi, Otobothrium, 68
linstowi, Taenia, 457
linstowi, Zschokkea, 457
linstowi, Zschokkeella, 457
Linstowia, 457
Linstowia, diagnosis, 460
Linstowiinae, key to genera, 451
Linstowius, diagnosis, 280
lintonella, Hymenolepis, 336
lintoni, Acanthobothrium, 139
lintoni, Anthobothrium, 157
lintoni, Dendrouterina, 377
lintoni, Diorchis, 282
lintoni, fimbriarioides, 268
lintoni, Hymenosphenacanthus, 318
lintoni, Oncobothrium, 136
lintoni, Ophiovalipora, 377
lintoni, Pedibothrium, 143
lintoni, Polypocephalus, 122
lintoni, Pterobothrium, 70
lintoni, Retinometra, 318
lintoni, Sphenacanthus, 318
lintoni, Spongiobothrium, 157
lintoni, Synbothrium, 71
lintoni, Tetrarhynchus, 77
lintoni, Trigonocotyle, 206

Lintoniella, 53, 176
liphallos, Hymenolepis, 336
literrata, Taenia, 203
literratus, Mesocestoides, 203
Litobothridae
  diagnosis and
  key to genera, 169
Litobothridea, 15, 169
Litobothrium, diagnosis, 169
litocephalus, Eutetrarhynchus, 57
litoralis, Hymenolepis (Echinocotyle), 31
litoralis, Nadejdolepis, 311
littorie, Choanotaenia, 362
lloydi, Meggittiella, 375
lloydi, Paradilepis, 375
lobata, Biuterina, 339
lobata, Catenotaenia, 236
lobata, Drepanidotaenia, 295
lobata, Hymenolepis, 336
lobata, Proorchida, 374
lobata, Skrjabinotaenia, 236
lobata szpotanskaica, Drepanidotaenia, 295
Lobatolepis, diagnosis, 319
lobatus, Alyselminthus, 433
lobatus, Tetrabothrius, 207
lobipluviae, Choanotaenia, 360
lobipluviae, Kowalewskiella, 360
lobipluviae, Onderstepoortia, 360
lobivanelli, Panuwa, 367
lobosa, Corallobothrium, 196
lobosa, Ephedrocephalus, 196
lobosa, Rudolphiella, 196
lobulata, Hymenolepis, 319
lobulata, Lobatolepis, 319
Lockerrauschia, diagnosis, 300
loculatum, Phyllobothrium, 161
loennbergi, Ophiotaenia, 184
loeweni, Vadifresia, 261
loeweni, Raillietina, (R.) 261
loliginis, Phyllobothrium, 161
lomentaceum, Diesingium, 75
lomentaceum, Rhynchobothrium, 75
longa, Plagiotaenia, 433
longae, Diorchis, 282
longi, Staphylocystis, 302
longiannulata, Kowalewskiella, 360
longibursa, Diorchis, 279
longicauda, Wenyonia, 26
longicephalum, Bothridium, 86
longiceps, Biuterina, 351
longiceps, Sphaeruterina, 351
longiceps, Taenia, 351
longicirrata, Oochoristica, 454
longicirrata, Progynotaenia, 219
longicirrosa, Cotugnia, 246
longicirrosa, Davainea (Chapmania), 240
longicirrosa, Diorchis, 282
longicirrosa, Hymenolepis, 318
longicirrosa, Hymenosphenacanthus, 318
longicirrosa, Retinometra, 318
longicirrosum, Octopetalum, 343

longicolle, *Caulobothrium*, 153
longicolle, *Echeneibothrium*, 153
longicolle, *Echinobothrium*, 166
longicolle, *Orygmatobothrium*, 163
longicollis, *Bothriocephalus*, 95, 266
longicollis, *Bothriotaenia*, 266
longicollis, *Cysticercus*, 224
longicollis, *Davainea*, 266
longicollis, *Lytocestus*, 25
longicollis, *Proteocephalus*, 187
longicollis, *Raillietina*, 266
longicollis, *Taenia*, 256
longicollis, *Tetrarhynchus*, 77
longicorne, *Rhynchobothrium*, 76
longihamulus, *Diorchis*, 282
longiorum, *Bothridium*, 86
longiovata, *Amerina*, 341
longiovata, *Anoncotaenia*, 341
longiovata, *Hymenolepis*, 337
longiovum, *Diorchis*, 279
longiovum, *Schillerius*, 279
longirostellata, *Amoebotaenia*, 388
longirostris, *Echinocotyle*, 285
longirostris, *Nadejdolepis*, 285
longirostris, *Taenia*, 285
longisacculus, *Amoebotaenia*, 388
longispiculus, *Acoleus*, 231
longispiculus, *Bothriocephalus*, 231
longispina, *Brumptiella*, 253
longispina, *Davainea*, 253
longispina, *Raillietina (P.)*, 253
longispine, *Pedibothrium*, 143
longispine, *Rhynchobothrium*, 76
longissimum, *Abothrium*, 108
longistylosa, *Hymenolepis*, 318
longistylosa, *Retinometra*, 318
longistylosa, *Sphenacanthus*, 318
longistylosa, *Stylolepis*, 318
longistriatus, *Mesocestoides*, 203
longivaginata, *Hymenolepis*, 318
longivaginata, *Hymenosphenacanthus*, 318
longivaginata, *Retinometra*, 318
longivaginosus, *Oligorchis*, 375
longivaginosus, *Paradilepis*, 375
longmanni, *Ophiotaenia*, 184
longocylindrocirrus, *Hymenolepis*, 337
longus, *Isoglaridacris*, 37
longus, *Scyphocephalus*, 85
*Lönnbergia*, 190
loossi, *Staphylocystis*, 302
lopas, *Inermicapsifer*, 466
lopesi, *Ophiotaenia*, 184
lopezneyrai, *Bertiella*, 439
lopezneyrai, *Nematotaenia*, 213
lophii, *Bothriocephalus*, 102
lophii-piscatorii, *Tetrarhynchus*, 77
lophoceri, *Raillietina (F.)*, 265
*Lophurolepis*, diagnosis, 293
lotae, *Tetrarhynchus*, 77
louiseuzeti, *Grillotia*, 64
lowzowi, *Coenurus*, 226

loxiae, *Halisis*, 341
lubeti, *Rhodobothrium*, 158
lubeti, *Sphaerobothrium*, 158
lucida, *Catenotaenia*, 235
lucida, *Metroliasthes*, 344
lucii, *Triaenophorus*, 112
luciopercae, *Proteocephalus*, 187
lucknowensis, *Echinorhynchotaenia*, 272
lucknowensis, *Neophronia*, 429
lucknowensis, *Senga*, 96
*Lucknowia*, diagnosis, 18
lucknowia, *Gangesia*, 176
ludificans, *Lecanicephalum*, 127
ludificans, *Tylocephalum*, 127
leuckarti, *Anomotaenia*, 406
leuckarti, *Taenia*, 406
*Lueheella*, 93
luehei, *Anomotaenia*, 406
luehei, *Bothriocephalus*, 103
luengoi, *Drepanidotaenia*, *(D.)*, 329
luengoi, *Microsomacanthus*, 329
luengoi, *Tschertkovilepis*, 329
*Lüheella*, 93
luisaleoni, *Raillietina (R.)*, 259
lumbrici, *Amoebotaenia*, 395
lumbrici, *Paricterotaenia*, 395
lumbrici, *Polycercus*, 395
luteus, *Neophronia*, 429
lutzi, *Davainea*, 261
lutzi, *Kotlania*, 261
lutzi, *Raillietina (R.)*, 261
lutzi, *Raillietina (Ransomia)*, 261
lutzi, *Tetrabothrius*, 209
luxi, *Diphyllobothrium*, 94
lycaontis, *Echinococcus*, 222
lycaontis, *Taenia*, 225
lygodaptrion, *Rhabdometra*, 346
lygosomae, *Oochoristica*, 454
lygosomatis, *Oochoristica*, 454
lyncis, *Cysticercus*, 225
lyncis, *Hydatigera*, 225
lyncis, *Taenia*, 225
*Lyruterina*, 345
Lytocestidae, key to genera, 18
*Lytocestoides*, diagnosis, 19
*Lytocestus*, diagnosis, 25

# M

mabuiae, *Nematotaenia*, 213
macassarensis, *Raillietina*, *(P.)*, 254
maccallumi, *Dibothriorhynchus*, 76
maccallumi, *Echeneibothrium*, 155
maccallumi, *Rhinebothrium*, 155
macdonaghi, *Ichthyotaenia*, 187
macdonaghi, *Proteocephalus*, 187
macfiei, *Tentacularia*, 77
machadoi, *Lentiella*, 437
mackiewiczi, *Monobothrium*, 32
mackiewiczi, *Monoecocestus*, 426

*mackoi, Dendrouterina,* 378
*maclaudi, Vampirolepis,* 304
*macquariae, Monorygma,* 161
*macqueeni, Otiditaenia,* 241
*macqueeni, Paraschistometra,* 241
*macqueeni, Schistometra,* 241
*macracantha, Amirthalingamia,* 374
*macracantha, Anomotaenia,* 406
*macracantha, Choanotaenia,* 397, 406
*macracantha, Hamatofuhrmannia,* 397
*macracantha, Icterotaenia,* 397
*macracantha, Monopylidium,* 406
*macracantha, Multiceps,* 225
*macracantha, Parachoanotaenia,* 397
*macracantha, Paradilepis,* 374, 375
*macracantha, Paraicterotaenia,* 397
*macracantha, Polycercus,* 397
*macracantha, Sacciuterina,* 397
*macracantha, Taenia,* 225
*macracanthissima, Hymenolepis,* 311
*macracanthissima, Nadejdolepis,* 311
*macracanthissima, Sphenacanthus,* 311
*macracanthoides, Anomotaenia,* 406
*macracanthoides, Hymenolepis,* 318
*macracanthoides, Sphenacanthus,* 318
*macracanthos, Hymenolepis,* 318
*macracanthos, Hymenosphenacanthus,* 318
*macracanthos, Paspalia,* 261
*macracanthos, Raillietina, (R.),* 261
*macracanthos, Retinometra,* 318
*macracanthos, Sphenacanthus,* 318
*macracanthos, Taenia,* 318
*macracanthum, Acanthobothrium,* 139
*Macracanthus,* 361
*macrarhyncha, Raillietina, (R.),* 261
*macrascum, Echeneibothrium,* 151
*macroacetabula, Crepidobothrium,* 189
*macrobothria, Ophiotaenia,* 184
*Macrobothriotaenia,* diagnosis, 182
*macrobothrium, Stenobothrium,* 72
*macraobursatum, Monoecocestus,* 426
*macrocephala, Anoncotaenia,* 341
*macrocephala, Chapmania,* 241
*macrocephala, Choanotaenia,* 362
*macrocephala, Dilepis,* 381
*macrocephala, Taenia,* 187
*macrocephalum, Biacetabulum,* 30
*macrocephalum, Ditrachybothridium,* 167
*macrocephalum, Pentaloculum,* 145
*macrocephalus, Halysiorhynchus,* 70
*macrocephalus, Proteocephalus,* 187
*macrocephalus, Tetrabothrius,* 207
*macrocephalus, Tetrarhynchus,* 70
*macrocirrosa, Davainea,* 261
*macrocirrosa, Kotlania,* 261
*macrocirrosa, Porotaenia,* 208
*macrocirrosa, Raillietina, (R.),* 261
*macrocirrosa, Raillietina (Ransomia),* 261
*macrocirrus, Schistotatnia,* 230
*macrocotyle, Monticellia,* 194
*macrocystis, Cysticercus,* 225

*macrocystis, Hydatigera,* 225
*macrocystis, Taenia,* 225
*macrones, Gangesia,* 176
*macroovatum, Diphyllobothrium,* 94
*macropa, Raillietina (Paroniella),* 252
*macropeos, Dilepis,* 371, 379
*macropeos, Taenia,* 279
*macrophallica, Parvitaenia,* 412
*macrophallus, Bothriocephalus,* 93
*macrophallus, Infula,* 216
*macrophallus, Proteocephalus,* 187
*macropodis, Progamotaenia,* 445
*macroporus, Tetrarhynchus,* 77
*macropterygis, Anomotaenia,* 402
*macropterygis, Parvitaenia,* 402
*macropterygis, Sureshia,* 402
*macropus, Calostaurus,* 252
*macrorhyncha, Amabilia,* 229
*macrorhyncha, Drepanidotaenia,* 229
*macrorhyncha, Schistotaenia,* 229
*macrorhyncha, Taenia,* 229
*macrorostratum, Acanthocirrus,* 379
*macrorotratum, Gryporhynchus,* 379
*macroscelidarum, Vampirolepis,* 304
*macroscolecina, Davainea,* 261
*macroscolecina, Kotlania,* 261
*macroscolecina, Raillietina, (R.),* 261
*macroscolecina, Raillietina (Ransomia),* 261
*macroscolecina, Tetracisdicotyla,* 413
*macrosphincter, Dilepis,* 381
*macrostrobilodes, Weinlandia,* 321
*macrotesticulatus, Rodentolepis,* 304
*macrotesticulatus, Vampirolepis,* 304
*macroti, Vampirolepis,* 304
*macrotrachelus, Eutetrarhynchus,* 57
*macrouri, Parabothriocephalus,* 118
*macrourum, Pterobothrium,* 70
*maculatum, Acanthobothrium,* 139
*maculatum, Tetrabothrium,* 147
*maculatus, Bothriocephalus,* 94
*macyi, Hymenolepis,* 287
*macyi, Soricina,* 287
*madagascariensis, Davainea,* 261
*madagascariensis, Kotlania,* 261
*madagascariensis, Raillietina (R.),* 261
*madagascariensis, Raillietina (Ransomia),* 261
*madagascariensis, Skrjabinotaenia,* 236
*madagascariensis, Taenia,* 261
*madhukarii, Lecanicephalum,* 126
*madhukarii, Tylocephalum,* 126
*madrasiensis, Amoebotaenia,* 388
*madrassensis, Biporophyllaeus,* 63
*magdalenensis, Pomatotrygonocestus,* 142
*magellanicus, Mesocestoides,* 204
*magna, Amphilina,* 470
*magna, Anoplocephala,* 433
*magna, Cotugnia,* 246
*magna, Gigantolina,* 470
*magna, Nomimoscolex,* 193
*magna, Ophiotaenia,* 184
*magna, Taenia,* 433

*magnicirrosa, Choanotaenia*, 397
*magnicirrosa, Paricterotaenia*, 397
*magnicirrosa, Polycercus*, 397
*magnicirrosus, Diorchis*, 282
*magnicoronata, Davainea*, 256
*magnicoronata, Raillietina (S.)*, 256
*magnihamata, Choanotaenia*, 362
*magninumida, Raillietina (P.)*, 254
*magniovatus, Hymenolepis*, 329
*magniovatus, Microsomacanthus*, 329
*magniparuterina, Lallum*, 344
*magniphallum, Rhinebothrium*, 155
*magnireceptaculatus, Pseudoligorchis*, 333
*magnirostellata, Vampirolepis*, 304
*magnisaccis, Hymenolepis*, 311
*magnisaccis, Nadejdolepis*, 311
*magnisaccis, Sphenacanthus*, 311
*magnisomum, Parvirostrum*, 400
*magnisomum, Taufikia*, 426
*magniuncinata, Armadoskrjabinia*, 297
*magniuncinata, Hymenolepis*, 297
*magniuncinata, Vitta*, 392
*magniuterina, Anoncotaenia*, 341
*magnivesicula, Thaparea*, 391
*magnum, Oncobothrium*, 136
*magnum, Otobothrium*, 68
*magnum, Phyllobothrium*, 161
*magnum, Polyoncobothrium*, 97
*magnum, Tetracampos*, 97
*mahamdabadensis, Gangesia*, 176
*maharashtri, Amoebotaenia*, 388
*maharashtrae, Lecanicephalum*, 126
*maharashtrae, Oncodiscus*, 101
*mahdiaensis, Lateriporus*, 373
*mahonae, Hymenolepis*, 337
*mahonae, Raillietina (R.)*, 262
*major, Notolytocestus*, 23
*makundi, Dicranotaenia*, 321
*makundi, Hymenolepis*, 321
*makundi, Ophryocotyloides*, 243
*malabarica, Lapwingia*, 404
*malabarica, Yogeshwaria*, 400
*malaccensis, Hymenolepis*, 337
*malakartis, Raillietina (F.)*, 265
*malayensis, Hymenolepis*, 304
*malayensis, Vampirolepis*, 304
*Malika*, diagnosis, 356
*maliki, Paradilepis*, 375
*malleum, Pterobothrium*, 70
*malleus, Taenia*, 267
*malopteruri, Electrotaenia*, 174
*mamaevi, Globarilepis*, 270
*mamaevi, Pseudamphicotyla*, 106
*mamillana, Anoplocephala*, 436
*mamillana, Anoplocephaloides*, 436
*mamillana, Paranoplocephala*, 436
*mamillana, Taenia*, 436
*manazo, Nybelinia*, 74
*mancocapaci, Choanotaenia*, 362
*mandabbi, Hymenolepis*, 337
*mandapemensis, Oochoristica*, 454

*mandube, Anthobothrium*, 158
*mandube, Endorchis*, 193
*manidis, Hymenolepis*, 304
*manidis, Rodentolepis*, 304
*manidis, Vampirolepis*, 304
*manipurensis, Anomotaenia*, 367
*manipurensis, Choanotaenia*, 367
*manipurensis, Monopylidium*, 367
*manjuariphilus, Proteocephalus*, 187
*mansoni, Bothriocephalus*, 94
*mansoni, Dibothrium*, 94
*mansoni, Ligula*, 94
*mansoni, Sparganum*, 94
*mansonoides, Diphyllobothrium*, 94
*mansonoides, Spirometra*, 94
*Mantaurus*, 251
*manteri, Mesocestoides*, 204
*manubriata, Anoplocephala*, 433
*manubriatum, Dicranotaenia*, 323
*manubriatum, Wardium*, 323
*manubriformis, Bothriocephalus*, 103
*maplestonei, Raillietina (R.)*, 262
*marchali, Choanotaenia*, 362
*marchali, Taenia*, 362
*marchesettii, Cyatocotyle*, 148
*marchii, Taenia*, 360
*marcuseni, Lytocestus*, 26
*marenzelleri, Ophiotaenia*, 184
*margareta, Cotugnia*, 246
*margaritifera, Taenia*, 204
*margaritiferae, Lecanicephalum*, 127
*margaritiferae, Tylocephalum*, 127
*marginatum, Phyllobothrium*, 161
*marginatus, Bothriocephalus*, 95
*mariae, Paricterotaenia*, 397
*mariae, Polycercus*, 397
*mariae, Sacciuterina*, 397
*Markevitschia*, diagnosis, 18
*Markewitchella*, 253
*markewitschi, Diorchis*, 282
*Markewitchitaenia*, diagnosis, 459
*marmosae, Mathevotaenia*, 463
*marmosae, Oochoristica*, 463
*marmotae, Ctenotaenia*, 448
*marmotae, Moniezia*, 448
*marmotae, Taenia*, 448
*maroteli, Brumptiella*, 256
*maroteli, Choanotaenia*, 256
*maroteli, Raillietina (S.)*, 256
*Marsipocephalus*, diagnosis, 190
*Marsipometra*, diagnosis, 109
Marsipometrinae, key to genera, 109
*Marsupiobothrium*, diagnosis, 147
*marsupium, Lecanicephalum*, 126
*marsupium, Tylocephalum*, 126
*martis, Halysis*, 225
*masaldani, Vampirolepis*, 304
*mascomai, Paranoplocephala*, 434
*Mashonalepis*, diagnosis, 372
*Mastacembellophyllaeus*, 164
*mastigophora, Idiogenes*, 240

*mastigopraeditus, Hymenolepis*, 309
*mastigopraeditus, Sobolevicanthus*, 309
*matheri, Tetrarhynchus*, 77
*Mathevolepis*, diagnosis, 287
*mathevossianae, Dicranotaenia*, 323
*mathevossianae, Diorchis*, 282
*mathevossianae, Eranuides*, 448
*mathevossianae, Hymenolepis*, 323, 335
*mathevossianae, Lateriporus*, 373
*mathevossianae, Raillietina (R.)*, 262
*mathevossianae, Schistotaenia*, 230
*mathevossianae, Tatria*, 230
*mathevossianae, Wardium*, 323
*Mathevossianetta*, 253
*mathevossiani, Polycercus*, 397
*mathevossiani, Sacciuterina*, 397
*mathevossiani, Schmidneila*, 397
*Mathevotaenia*, diagnosis, 462
*mathiasi, Acanthobothrium*, 139
*mathiasi, Echinobothrium*, 166
*matovi, Catenotaenia*, 234
*matovi, Pseudocatenotaenia*, 234
*mavensis, Spiniloculus*, 137
*mawsonae, Bertiella*, 439
*mawsoni, Tetrabothrius*, 209
*maxima, Dilepis*, 358
*maxima, Gyrocotyle*, 473
*maxima, Neodilepis*, 358
*maxima, Paradilepis*, 358
*mayhewi, Microsomacanthus*, 329
*mayhewi, Weinlandia*, 329
*Mayhewia*, 297
*mcconnelli, Wenyonia*, 26
*meandrica, Gyrocotyle*, 473
*Mecistobothrium*, diagnosis, 55
*media, Skrjabinotaenia*, 236
*medici, Armadoskrjabinia*, 297
*medici, Dicranotaenia*, 297
*medici, Echinorhynchotaenia*, 297
*medici, Hymenolepis*, 297
*medici, Taenia*, 297
*medici, Weinlandia*, 297
*medium, Diphyllobothrium*, 95
*medusarum, Gyrocotyle*, 473
*medusia, Parataenia*, 122
*medusia, Polypocephalus*, 122
*megabothridia, Grillotia*, 64
*megabothridia, Tentacularia*, 64
*megabothrius, Tetrarhynchus*, 73, 77
*megacantha, Choanotaenia*, 397
*megacantha, Icterotaenia*, 397
*megacantha, Paricterotaenia*, 397
*megacantha, Passerilepis*, 298
*megacantha, Polycercus*, 397
*megacantha, Rhinoptericola*, 71
*megacantha, Taenia*, 397
*Megacapsula*, diagnosis, 464
*megacephala, Monticellia*, 194
*megacephalus, Solenophorus*, 85
*megacephalus, Tetrarhyncus*, 78
*megacephalus, Zygobothrium*, 192

*megacirrosa, Dilepis*, 381
*Megacirrus*, diagnosis, 382
*megacotyla, Monorygma*, 163
*Megalacanthus*, 361
*megalhystera, Hymenolepis*, 321
*megalocephala, Taenia*, 369, 433
*megalocephala, Trichocephaloidis*, 369
*Megalonchos*, 133
*megaloon, Hymenolepis*, 289
*megalops, Cloacotaenia*, 338
*megalops, Hymenolepis*, 338
*megalops, Orlovilepis*, 338
*megalorchis, Diplacanthus (Dilepis)*, 296
*megalorchis, Flamingolepis*, 296
*megalorchis, Hymenolepis*, 296
*megalorchis, Hymenolepis (Drepanidotaenia)*, 296
*megalorchis, Taenia*, 296
*megalorchis, Weinlandia*, 296
*megalorhyncha, Dilepis*, 381
*megalorhyncha, Drepanidotaenia*, 381
*megalorhyncha, Hymenolepis*, 381
*megalorhyncha, Taenia*, 381
*megalostellis, Hymenolepis*, 328
*megaparuterina, Anoncotaenia*, 341
*megaparuterina, Mogheia*, 341, 419
*megapodii, Megacirrus*, 382
*megascolecina, Anomotaenia*, 406
*megascolecis, Amoebotaenia*, 388
*megastoma, Mathevotaenia*, 463
*megastoma, Oochoristica*, 463
*megastoma, Taenia*, 463
*Megathylacoides*, diagnosis, 178
*Megathylacus*, diagnosis, 178
*meggitteella, Hymenolepis*, 337
*meggitti, Biuterina*, 348
*meggitti, Cotugnia*, 246
*meggitti, Hymenolepis*, 318
*meggitti, Hymenosphenacanthus*, 318
*meggitti, Ophiotaenia*, 184
*meggitti, Ophryocotyloides*, 243
*meggitti, Paruterina*, 348
*meggitti, Retinometra*, 318
*meggitti, Stylolepis*, 318
*Meggittia*, 252
*Meggittiella*, 375
*Meggittina*, diagnosis, 236
*megistacantha, Burhinotaenia*, 398
*megistacantha, Choanotaenia*, 398
*megistacantha, Icterotaenia*, 398
*megistacantha, Paricterotaenia*, 398
*megistacantha, Polycercus*, 398
*meglosoma, Echeneibothrium*, 151
*Megocephalos*, 223
*Mehdiangularia*, 391
*mehrai, Raillietina (R.)*, 262
*meinertzhageni, Alcataenia*, 410
*meinertzhageni, Anomotaenia*, 410
*meinertzhageni, Pseudanomotaenia*, 410
*melanittae, Microsomacanthus*, 329
*melanocephala, Taenia*, 225
*melanotus, Neophronia*, 429

*meleagridis, Cotugnia,* 246
*meleagridis, Davainea,* 248
*meleagris, Dicranotaenia,* 307
*meleagris, Drepanidotaenia,* 307
*meleagris, Hymenolepis,* 307, 337
*meleagris, Weinlandia,* 307
*melesi, Taenia,* 225
*meliphagicola, Capiuterilepis,* 271
*meliphagidarum, Choanotaenia,* 355
*meliphagidarum, Pseudochoanotaenia,* 355
*meliphagidarum, Ptilotolepis,* 355
*membranacei, Proteocephalus,* 187
*menpachi, Metabothriocephalus,* 117
*mephitis, Oochoristica,* 457
*mephitis, Oschmarenia,* 457
*merganseri, Hymenofimbria,* 292
*mergi, Dicranotaenia,* 321
*mergi, Hymenolepis,* 321
*meridianum, Biacetabulum,* 34
*meridianum, Bialovarium,* 34
*meridionalis, Acoleus,* 231
*meridionalis, Bertia,* 231
*meridionalis, Triaenophorus,* 113
*merionidis, Monopylidium,* 367
*merionidis, Rodentotaenia,* 367
*merlangi, Tetrarhyncus,* 78
*merlangi-vulgaris, Tetrabothriorhynchus,* 77
*meropina, Biuterina,* 348
*meropina macrancristrota, Biuterina,* 348
*meropina, Taenia,* 348
*merops, Lateriporus,* 373
*merotomocheta, Hemiparonia,* 427
*mertoni, Biuterina,* 348
*mesacantha, Hymenolepis,* 337
*Mesocestoides,* diagnosis, 202
Mesocestoididae, 202
 diagnosis, 199
 key to subfamilies, 202
Mesocestoidinae, 202
*mesodermatis, Proboscidosaccus,* 158
*mesodermatum, Rhodobothrium,* 158
*Mesogyna,* diagnosis, 204
Mesogyninae, 204
*mesorchis, Mesocestoides,* 204
*mesovitellinica, Catenotaenia,* 233
*mesovitellinica, Quentinotaenia,* 233
*meszarosi, Vampirolepis,* 304
*Metabelia,* diagnosis, 382
*Metabothriocephalus,* diagnosis, 116
*Metacapsifer,* diagnosis, 465
*metacentropi, Daovantienia,* 256
*metacentropi, Raillietina (S.),* 256
*Metadilepis,* diagnosis, 376
*Metaparonia,* 253
*metaskrjabini, Chitinorecta,* 403
*metaskrjabini, Panuwa,* 367
*Metroliasthes,* diagnosis, 344
*mettami, Moniezia,* 442
*mexicana, Anoncotaenia,* 341
*mexicana, Hexaparuterina,* 211
*michaelseni, Kotlania,* 262

*michaelseni, Raillietina (R.),* 262
*michaelseni, Mesocestoides,* 204
*michiae, Tentacularia,* 77
*michiganensis, Taenia,* 225
*micracantha, Acanthobothrium,* 139
*micracantha, Anomotaenia,* 440
*micracantha, Davainea,* 262
*micracantha dominicana, Anomotaenia,* 405
*micracantha, Hymenolepis,* 335
*micracantha, Kotlania,* 262
*micracantha, Ophiovalipora,* 377
*micracantha, Prochristianella,* 59
*micracantha, Pseudanomotaenia,* 410
*micracantha, Raillietina (R.),* 262
*micracantha, Taenia,* 410
*micracanthus, Staphylocystis,* 301
*micrancristrota, Hymenolepis,* 337
*Microbothriorhynchus,* 75
*microcephala, Hymenolepis,* 307
*microcephala, Oshmarinolepis,* 307
*microcephala, Taenia,* 307
*microcephalus, Ancistrocephalus,* 112
*microcephalus, Caryophyllaeus,* 33
*microcephalus, Ephedrocephalus,* 195
*microcephalus, Proteocephalus,* 187
*microcirrosa, Dicranotaenia,* 323
*microcirrosa, Diorchis,* 282
*microcirrosa, Hymenolepis,* 323
*microcirrosa, Weinlandia,* 323
*microcirrus, Globarilepis,* 270
*microcordiceps, Diphyllobothrium,* 88
*microcordiceps, Sparganum,* 88
*microcotyle, Brumptiella,* 256
*microcotyle, Davainea,* 256
*microcotyle, Raillietina (S.),* 256
*microdisciformis, Bothridium,* 86
*micropalamae, Taeniarhynchaena,* 387
*microphallica, Parvitaenia,* 412
*microphallos, Anomotaenia,* 406
*microphallos, Choanotaenia,* 406
*microphallos, Dichoanotaenia,* 406
*microphallos, Taenia,* 406
*microps, Paradicranotaenia,* 272
*microps, Taenia,* 255
*micropteri, Proteocephalus,* 187
*micropus, Notopentorchis,* 350
*micropusia, Sureshia,* 402
*microrhyncha, Anomotaenia,* 404
*microrhyncha, Raillietina (R.),* 262
*microrhyncha, Taenia,* 404
*microscolecina, Davainea,* 262
*microscolecina, Hymenolepis,* 325
*microscolecina, Kotlania,* 262
*microscolecina, Raillietina (R.),* 262
*microscolecina, Raillietina (Ransomia),* 262
*microscolecina, Variolepis,* 325
*microscolex, Oochoristica,* 454
*microscopius, Proteocephalus,* 187
*microskrjabini, Microsomacanthus,* 329
*microsoma, Choanotaenia,* 358
*microsoma, Diplacanthus (Dilepis),* 327

*microsoma, Hymenolepis,* 327
*microsoma, Hymenolepis (Drepanidotaenia),* 327
*microsoma, Microsomacanthus,* 327
*microsoma, Prochoanotaenia,* 358
*microsoma, Spiniglans,* 358
*microsoma, Taenia,* 327
*microsoma, Weinlandia,* 327
*Microsomacanthus,* diagnosis, 327
*microsomum, Phyllobothrium,* 161
*microstoma, Rodentolepis,* 304
*microstoma, Taenia,* 304
*microstoma, Vampirolepis,* 304
*Microtaenia,* 352
*microthrix, Grillotia,* 64
*migratorius, Tetrabothriorhynchus,* 77
*micruricola, Ophiotaenia,* 185
*micruricola, Proteocephalus,* 185
*milliapharyngens, Dibothrium,* 85
*milligani, Gyrocoelia,* 216
*milvi, Paricterotaenia,* 398
*milvi, Polycercus,* 398
*milvi, Sacciuterina,* 398
*minima, Dilepis,* 375
*minima, Moniezia,* 441
*minima, Paradilepis,* 375
*minima, Prochristianella,* 59
*minima, Taenia,* 289
*minimedius, Hymenolepis,* 335
*minimum, Echeneibothrium,* 148, 151
*minimum, Pterobothrium,* 70
*minimus, Tetrarhynchus,* 78
*miniopteri, Hymenolepis,* 301
*miniopteri, Triodontolepis,* 301
*minor, Grillotia,* 64
*minor, Hymenolepis,* 337
*minor, Monoecocestus,* 426
*minor, Notolytocestus,* 23
*minor, Priapocephalus,* 205
*minor, Tetrabothrius,* 209
*minus, Calyptrobothrium,* 162
*minus, Diphyllobothrium,* 95
*minus, Lecanicephalum,* 127
*minus, Tylocephalum,* 127
*minuta, Anomotaenia,* 406
*minuta, Christianella,* 55
*minuta, Davainea,* 248
*minuta, Dichoanotaenia,* 406
*minuta, Drepanidotaenia,* 406
*minuta, Himantaurus,* 248
*minuta, Hymenolepis,* 287
*minuta, Ophiovalipora,* 377
*minuta, Ophryocotyle,* 129
*minuta, Raillietina (P.),* 254
*minuta, Wenyonia,* 26
*minutia, Corallotaenia,* 179
*minutilla, Echinocotyle,* 285
*minutissima, Echinocotyle,* 285
*minutissima, Staphylocystis,* 302
*minutium, Corallobothrium,* 179
*minutiuncinata, Vitta,* 393
*minuto-striatus, Tetrarhynchus,* 78

*minutum, Anthobothrium,* 158
*minutum, Duplicibothrium,* 150
*minutum, Echeneibothrium,* 151
*minutum, Lecanicephalum,* 127
*minutum, Ophryocotyle,* 245
*minutum, Nematoparataenia,* 129
*minutum, Phyllobothrium,* 161, 162
*minutum, Rhynchobothrium,* 56
*minutum, Tylocephalum,* 127
*minutus, Bothriocephalus,* 103
*minutus, Tetrabothrius,* 209
*minutus, Tetrarhynchus,* 78
*minytremi, Promonobothrium,* 35
*mirabilis, Cittotaenia,* 415
*mirabilis, Microsomacanthus,* 329
*mirabilis, Triplotaenia,* 415
*Mirandula,* diagnosis, 368
*misrai, Thysanosoma,* 417
*mitudori, Allohymenolepis,* 270
*Mixodigma,* diagnosis, 74
Mixodigmatidae, 74
  diagnosis, 49
*Mixophyllobothrium,* diagnosis, 145
*mjobergi, Ophiotaenia,* 185
*mjoebergi, Chitinolepis,* 332
*mkuzii, Pseudandrya,* 373
*mobile, Diphyllobothrium,* 95
*mobilis, Dibothriocephalus,* 95
*moculata, Taenia,* 298
*modigliani, Dilepis,* 381
*modigliani, Hymenolepis,* 381
*moennigi, Ophiotaenia,* 185
*moensis, Aploparaksis,* 276
*moghei, Capingentoides,* 40
*moghei, Neyraia,* 339
*Mogheia,* diagnosis, 419
*mogurdnae, Amurotaenia,* 171
*moldavica, Aploparaksis,* 276
*moldavica, Monopylidium,* 367
*Molicola,* diagnosis, 61
*mollis, Anomotaenia,* 406
*mollis, Taenia,* 406
*Molluscotaenia,* 361, 404
*molpastina, Raillietina (P.),* 254
*molus, Vampirolepis,* 304
*monacanthis, Ophryocotyloides,* 243
*monardi, Anootypus,* 419
*monardi, Avitellina,* 419
*monardi, Moniezia,* 442
*monardi, Pseudandrya,* 372
*monedulae, Dilepis,* 381
*mongolica, Idiogenes,* 239
*mongolica, Paraidiogenes,* 239
*moniezi, Hymenolepis,* 335
*moniezi, Vaullegeardia,* 61
*Moniezia,* diagnosis, 441
*Moniezioides,* diagnosis, 421
*Monobothrioides,* diagnosis, 24
*Monobothrium,* diagnosis, 31
*monodi, Andrya,* 431
*monodi, Aprostatandrya (Sudarikovina),* 431

*monodi, Coelobothrium,* 98
*monodi, Sudarikovina,* 431
*Monoecocestus,* diagnosis, 426, 432
*monogramma, Ligula,* 87
*monomegacantha, Parachristianella,* 58
*monoophoroides, Progynopylidium,* 354
*monoophorum, Diplopylidium,* 354
*monoophorum, Dipylidium,* 354
*monophorum, Progynopylidium,* 354
Monoporophyllaeidae, 164
*Monoporophyllaeus,* 164
*monoposthe, Hymenolepis,* 329
*monoposthe, Microsomacanthus,* 329
*monoposthe, Tschertkovilepis,* 329
*Monopylidium,* diagnosis, 366
*Monorchis,* 275
*monorchis, Bothriocephalus,* 103
*Monorcholepis,* 275
*Monordotaenia,* diagnosis, 222
*Monorygma,* diagnosis, 162
*Monosaccanthes,* 312
*monostephanos, Fossor,* 222
*monostephanos, Monordotaenia,* 222
*monostephanos, Taenia,* 222, 225
*Monotestilepis,* 275
*montana, Baerietta,* 212
*montana, Vampirolepis,* 304
*monticelli, Bothriocephalus,* 103
*monticelli, Diesingella,* 75
*monticelli, Diesingium,* 75
*monticelli, Prosthecocotyle,* 206
*monticelli, Tetrabothrius,* 209
*Monticellia,* diagnosis, 194
Monticellidae, key to subfamilies, 190
*monticellii, Trigonocotyle,* 206
Monticelliinae, 194
*mopoyemi, Hymenolepis,* 335
*moralarai, Rhinebothroides,* 154
*moralarai, Rhinebothrium,* 154
*morchitini, Tetrabothrius,* 209
*morenoi, Nadejdolepis,* 311
*morgani, Biuterina,* 348
*morgani, Paruterina,* 348
*morhuae, Tetrarhynchus,* 78
*moroccana, Diorchis,* 279
*morrhuae, Abothrium,* 108
*Mosgovoyia,* diagnosis, 450
*motacillabrasiliensis, Biuterina,* 348
*motacillabrasiliensis, Taenia,* 348
*motacillacayanae, Biuterina,* 348
*motacillacayanae, Taenia,* 348
*motellae, Bothriocephalus,* 103
*moucheti, Echeneibothrium,* 151
*mozambiquus, Tetrabothrius,* 209
*mucronata, Bertia,* 439
*mucronata, Bertiella,* 439
*mucronata, Taenia (Bertia),* 439
*muelleri, Probothriocephalus,* 115
*mugil-auratus, Tetrarhynchus,* 78
*mugilis, Otobothrium,* 68
*mujibi, Acanthobothrium,* 139

*mujibi, Hymenolepis,* 335
*mukteswarensis, Ivritaenia,* 365
*mulli-barbati, Dibothriorhynchus,* 76
*mulli-rubescentis, Tetrarhynchus,* 78
*mulli, Tetrarhynchus,* 78
*multicanalis, Profimbriaria,* 269
*Multicapsiferina,* diagnosis, 457
*multicapsulata, Davainea,* 262
*multicapsulata, Kotlania,* 262
*multicapsulata, Raillietina (R.),* 262
*multicapsulata, Raillietina (Ransomia),* 262
*Multiceps,* 223
*multiceps, Taenia,* 225
*Multicotugnia,* 245
*Multiductus,* diagnosis, 89
*multifilamenta, Anomotaenia,* 407
*multifilamenta, Dichoanotaenia,* 407
*multiformis, Cysticercus,* 224, 225
*multiformis, Taenia,* 224, 225, 307, 405
*multiglandularis, Echinocotyle,* 285
*multiglandularis, Hymenolepis (Echinocotyle),* 285
*multihamata, Hymenolepis,* 376
*multihamata, Meggittiella,* 376
*multihamata, Paradilepis,* 376
*multihamata, Skrjabinolepis,* 376
*multihamatus, Vampirolepis,* 304
*multihami, Hymenolepis,* 335
*multilobatus, Deltokeras,* 347
*multilocularis, Alveococcus,* 222
*multilocularis, Echinococcus,* 222
*multiloculatum, Echeneibothrium,* 151
*multiorchidum, Caulobothrium,* 153
*multistriata, Colymbilepis,* 323
*multistriata, Dicranotaenia,* 323
*multistriata, Dubininolepis,* 323
*multistriata, Hymenolepis,* 323
*multistriata, Hymenolepis (Drepanidotaenia),* 323
*multistriata, Taenia,* 323
*multitesticulata, Raillietina (R.),* 262
*multiuncinata, Hispaniolepis,* 306
*multiuncinata, Ortleppolepis,* 306
*Multiuterina,* diagnosis, 343
*mundayi, Calostaurus,* 252
*muricola, Inermicapsifer,* 466
*muricola, Zschokkeella,* 466
*murina, Hymenolepis,* 304
*murina, Oochoristica,* 463
*murinae, Hymenolepis,* 335
*murisdecumani, Taenia,* 304
*murissylvatici, Staphylocystis,* 302
*murissylvatici, Taenia,* 302
*murisvariegatai, Hymenolepis,* 304
*murisvariegati, Rodentolepis,* 304
*murisvariegati, Vampirolepis,* 304
*murium, Raillietina (R.),* 262
*murmanica, Aploparaksis,* 277
*murudensis, Anomotaenia,* 407
*musasabi, Bertiella,* 439
*musculara, Grillotia,* 64
*musculara, Tentacularia,* 64
*musculicola, Grillotia,* 64

*musculicola, Pintneriella,* 64
*musculosa, Anomotaenia,* 366
*musculosa, Choanotaenia,* 366
*musculosa, Davainea,* 366
*musculosa, Dicranotaenia (D.),* 321
*musculosa, Drepanidotaenia,* 321
*musculosa, Hymenolepis,* 321
*musculosa, Monopylidium,* 366
*musculosa, Nadejdolepis,* 311
*musculosum, Acanthobothrium,* 139
*musculosum, Acrobothrium,* 139
*musculosus, Bothriocephalus,* 103
*musculosus, Chaetophallus,* 207
*mustelae, Taenia,* 225
*mustelae vulgaris, Taenia,* 227
*musteli, Anthobothrium,* 161
*musteli, Echinobothrium,* 166
*musteli, Orygmatobothrium,* 163
*musteli, Phyllobothrium,* 161
*musteli, Platybothrium,* 137
*musteli, Prochristianella,* 59
*Mustelicola,* 57
Mustelicolidae, 51
*mustelis, Cladotaenia,* 413
*mustelis, Paracladotaenia,* 413
*mutabilis, Anomotaenia,* 407
*mutabilis, Caryophyllaeus,* 32
*mutabilis, Raillietina (R.),* 255
*mutabilis, Taenia,* 407
*mutabilis, Valipora,* 371
*mutata, Hymenolepis,* 305
*myliobati, Mecistobothrium,* 55
*myliobatidis, Caulobothrium,* 153
*myliobatidis, Discobothrium,* 125
*myliobatidis, Rhoptrobothrium,* 145
*myoides, Amphilaphorchis,* 196
*myoides, Rudolphiella,* 196
*myopotami, Monoecocestus,* 427
*Myotolepis,* diagnosis, 287
*Myrmillorhynchus,* diagnosis, 60
*mystei, Hunteroides,* 470
*Myxolepis,* 327
*Myzocephalus,* 132
*myzofera, Othinoscolex,* 195
*myzofera, Peltidocotyle,* 195
*myzofera, Woodlandiella,* 195
*myzomelae, Raillietina (P.),* 254
*Myzophorus,* diagnosis, 193
*Myzophyllobothrium,* diagnosis, 145
*myzorhynchum, Echeneibothrium,* 151

# N

*Nadejdolepis,* diagnosis, 310
*nagabhusheni, Yogeshwaria,* 164
*nagatyi, Avitellina,* 419
*nagatyi, Hilmylepis,* 294
*nagatyi, Hymenolepis,* 294
*nagpurensis, Kotlania,* 262
*nagpurensis, Raillietina (R.),* 262

*naja, Capiuterilepis,* 271
*naja, Dicranotaenia,* 271
*naja, Hymenolepis,* 271
*naja, Hymenolepis (Weinlandia),* 271
*naja, Taenia,* 271
*najae, Ophiotaenia,* 185
*nakayamai, Aporina,* 424
*nakurensis, Phoenicolepis,* 310
*namaquensis, Cephalochlamys,* 84
*namaquensis, Chlamydocephalus,* 84
*nana, Davainea,* 248
*nana, Diplacanthus,* 304
*nana, Echinorhyncholtaenia,* 272
*nana fraterna, Hymenolepis,* 305
*nana, Idiogenes,* 240
*nana, Lepidotrias,* 304
*nana, Taenia,* 304
*nana, Vampirolepis,* 304
*nandedensis, Mastacembellophyllaeus,* 164
*nankingensis, Ophiotaenia,* 185
*nannocephala, Gilquinia,* 60
*nannocephala, Tetrarhynchobothrium,* 60
*narinari, Myzocephalus,* 133
*narinari, Nybelinia,* 74
*narinari, Tetrarhynchus,* 74
*nasuta, Dilepis,* 356
*nasuta, Nasutaenia,* 356
*Nasutaenia,* diagnosis, 356
*natricis, Oochoristica,* 454
*nattereri, Laterotaenia,* 400
*nattereri, Ophiotaenia,* 185
*nebraskensis, Choanotaenia,* 362
*nebraskensis, Mayhewia,* 298
*nebraskensis, Passerilepis,* 298
*nebraskensis, Rodentotaenia,* 362
*necomis, Bialovarium,* 34
*nedumangadensis, Raillietina (P.),* 254
*negevi, Rodentolepis,* 305
*negevi, Vampirolepis,* 305
*neglectus, Bothriocephalus,* 103
*neglectus, Proteocephalus,* 187
*nelsoni, Tetrabothrius,* 209
*nemachili, Schistocephalus,* 88
*Nematoparataenia,* diagnosis, 129
Nematoparataeniidae, key to genera, 129
*nematosoma, Proteocephalus,* 187
*Nematotaenia,* diagnosis, 212
Nematotaeniidae
   diagnosis, 199
   key to genera, 209
*Nematotaenoides,* diagnosis, 210
*nenzi, Diphyllobothrium,* 95
*Neoangularia,* diagnosis, 384
*Neoaporina,* diagnosis, 424
*neoarctica, Dicranotaenia* 322
*neoarctica, Hymenolepis,* 322
*neoarctica, Proterogynotaenia,* 220
*neoarctica, Wardium,* 322
*Neobothriocephalus,* diagnosis, 115
*Neoctenotaenia,* diagnosis, 450
*Neodilepis,* diagnosis, 357

*Neodioecocestus*, diagnosis, 213
neofibrinus, *Anoplocephaloides*, 436
neofibrinus, *Paranoplocephala*, 436
*Neogryporhynchus*, diagnosis, 370
*Neoliga*, diagnosis, 391
*Neoligorchis*, diagnosis, 334
neomeggittilis, *Hymenolepis*, 285
neomidis, *Vampirolepis*, 305
*Neophronia*, diagnosis, 429
*Neoskrjabinolepis*, diagnosis, 301
neosouthwelli, *Hymenolepis*, 322
*Neotaenia*, 223
*Neotetrabothrius*, 207
*Neovalipora*, 371
*Nepalesia*, 428
nepalis, *Raillietina (F.)*, 265
neritina, *Amphilina*, 468
*Nesolecithus*, diagnosis, 470
nestoris, *Pulluterina*, 431
neumanni, *Moniezia*, 442
newguinensis, *Balanotaenia*, 17
neyrai, *Raillietina (R.)*, 262
*Neyraia*, diagnosis, 339
ngoci, *Raillietina (P.)*, 254
nicaraguensis, *Phyllobothrium*, 161
nigriceps, *Otiditaenia*, 241
nigricollis, *Ophiotaenia*, 185
nigromaculata, *Rhabdometra*, 346
nigropunctata, *Lyruterina*, 345
nigropunctata, *Rhabdometra*, 345
nigropunctata, *Taenia*, 345
nigropunctatus, *Bothriocephalus*, 103
nigrosetosa, *Gyrocotyle*, 473
niimiensis, *Insectivorolepis*, 288
nilotica, *Acanthotaenia*, 182
nilotica, *Caryophyllaeus*, 26
nilotica, *Choanotaenia*, 362
nilotica, *Crepidobothrium*, 182
nilotica, *Ichthyotaenia (Acanthotaenia)*, 182
nilotica, *Rostellotaenia*, 182
nilotica, *Taenia*, 362
nilotica, *Wenyonia*, 26
niloticus, *Proteocephalus*, 182
nipponensis, *Pseudanoplocephala*, 434
nipponica, *Nybelinia*, 74
nipponicum, *Glossobothrium*, 115
*Nippotaenia*, diagnosis, 171
Nippotaeniidae
  diagnosis, 171
  key to genera, 171
Nippotaeniidea, 16, 171
nishidai, *Hymenolepis*, 335
nitida, *Echinocotyle*, 285
nitida, *Hymenolepis (Echinocotyle)*, 285
nitida, *Hymenolepis (Hymenolepis)*, 285
nitida, *Taenia*, 285
nitidohamulus, *Diorchis*, 282
nitidulans, *Echinocotyle*, 310
nitidulans, *Hymenolepis (Echinocotyle)*, 310
nitidulans, *nadejdolepis*, 310
nitidulans, *Taenia*, 310

niuginii, *Proteocephalus*, 187
noctua, *Cotugnia*, 246
nodosa, *Taenia*, 112, 298
nodosum, *Calliobothrium*, 135
nodosus, *Triaenophorus*, 112
nodulosa, *Hunterella*, 34
nodulosus, *Triaenophorus*, 112
nodulus, *Triaenophorus*, 112
noei, *Ophiotaenia*, 185
noelleri, *Diplopylidium*, 354
nöleri, *Progynopylidium*, 354
*Nomimoscolex*, diagnosis, 192
nonarmatus, *Oligorchis*, 289
*Nonarmiella*, 253
*Nonarmina*, 253
norhalli, *Inermicapsifer*, 466
norhalli, *Pericapsifer*, 466
norvegicum, *Aporhynchus*, 60
norvegicum, *Diphyllobothrium*, 95
norvegicum, *Tetrabothrium*, 60
notabilicirrus, *Platyscolex*, 384
notidanus, *Tetrarhynchus*, 78
*Notolytocestus*, diagnosis, 23
*Notopentorchis*, diagnosis, 349
*Nototaenia*, diagnosis, 401
novadomensis, *Vampirolepis*, 305
novaeguineae, *Dioecocestus*, 213
novaehollandiae, *Dioecocestus,*, 213
novaehollandiae, *Taenia*, 213
novaezealandae, *Oochoristica*, 455
novella, *Taenia*, 226
nripendra, *Raillietina (R.)*, 262
nullicollis, *Moniezia*, 442
nullicollis, *Rhabdometra*, 346
numenii, *Aploparaksis*, 277
numenii, *Choanotaenia*, 362
numida, *Davainea*, 254
numida, *Malika*, 357
numida, *Numidella*, 254
numida, *Octopetalum*, 342
numida, *Raillietina (P.)*, 254
numida, *Rhabdometra*, 342
*Numidella*, 253
numguae, *Thomasitaenia*, 218
nybelini, *Gyrocotyloides*, 474
nybelini, *Ophiotaenia*, 185
*Nybelinia*, diagnosis, 73
nycticoracis, *Anomotaenia*, 412
nycticoracis, *Dendrouterina*, 377
nycticoracis, *Gryporhynchus*, 371, 379
nycticoracis, *Ophiovalipora*, 377
nycticoracis, *Parvitaenia*, 412
nyctophili, *Mathevotaenia*, 413
nylandica, *Diplocotyle*, 45
nylandicus, *Bothrimonus*, 45
nymphaea, *Anomotaenia*, 407
nymphaea, *Choanotaenia*, 407
nymphaea, *Dichoanotaenia*, 407
nymphaea, *Taenia*, 407
nymphoides, *Dilepis*, 381
nyrocae, *Dicranotaenia (D.)*, 295

nyrocae, Diorchis, 282
nyrocae, Drepanidotaenia, 295
nyrocae, Hymenolepis, 295
nyrocae, Wardiodes, 295
nyrocae, Wardium, 295
nyrocoides, Diorchis, 282

# O

obesa, Bertia, 439
obesa, Bertiella, 439
obesa, Prototaenia, 439
obesa, Taenia, 439
obesa, Tentacularia, 77
oblongatum, Diphyllobothrium, 88
oblongiceps, Moniezia, 441
oberatum, Bothridium, 86
obtusa, Taenia, 301
obvelata, Hymenolepis, 327
obvelata, Octacanthus, 327
obvelata, Sphenacanthus, 327
obvelata, Taenia, 327
obvoluta, Halysis, 213
occidentalis, Aploparaksis, 277
occidentalis, Passerilepis, 298
occidentalis, Skrjabinotaenia, 236
occlusa, Diorchis, 296
occlusus, Aporodiorchis, 279
occlusus, Diorchis, 279
occidentale, Calyptrobothrium, 162
occidentalis, Anomotaenia, 407
occidentalis, Bothriocephalus, 103
occidentalis, Dichoanotaenia, 407
ocellatus, Proteocephalus, 187
ochotensis, Anophryocephalus, 206
ochotensis, Dicranotaenia, 323
ochotensis, Wardium, 323
ochotonae, Schizorchis, 440
ochropodis, Anomolepis, 381
ochropodis, Dilepis, 381
ochropodis, Spasskytaenia, 381
octacantha, Aploparaksis, 277
octacantha, Dicranotaenia, 309
octacantha, Drepanidotaenia, 309
octacantha, Hymenolepis, 309
octacantha, Sobolevicanthus, 309
octacantha, Sphenacanthus, 309
octacantha, Taenia, 309
octacantha, Tatria, 229
octacantha, Weinlandia, 309
octacanthoides, Dicranotaenia, 309
octacanthoides, Hymenolepis, 309
octacanthoides, Sobolevicanthus, 309
octacanthoides, Sphenacanthus, 309
Octacanthus, diagnosis, 326
octocoronata, Rodentolepis, 305
octocoronata, Taenia, 305
octocoronata, Vampirolepis, 305
Octopetalum, diagnosis, 342
octopodiae, Tetrabothriorhynchus, 73

octorchis, Echeneibothrium, 151
odaensis, Hymenolepis, 335
odhneri, Dilepis, 381
odhneri, Progynotaenia, 219
odiosa, Rhabdometra, 346
odiosa, Taenia, 346
odontacantha, Pseudonybelinia, 67
oedicnemi, Eugonodaeum, 355
oedicnemus, Malika, 356
oedipomidatis, Paratriotaenia, 461
oena, Hymenolepis, 298
oena, Passerilepis, 298
oenopopeliae, Aporina, 428
oenopopeliae, Killigrewia, 428
ogaensis, Vampirolepis, 305
ognewi, Hymenolepis, 289
oidemiae, Microsomacanthus, 329
oitana, Mosgovoyia, 451
oitensis, Raillietina (R.), 262
okabei, Bertiella, 439
okamuri, Mixophyllobothrium, 145
okapi, Stilesia, 418
oklahomensis, Cladotaenia, 413
oklahomensis, Oochoristica, 457
oklahomensis, Oschmarenia, 457
oklensis, Penarchigetes, 29
okomotoi, Insectivorolepis, 288
okumurai, Spirometra, 94
olgae, Paricterotaenia, 398
olgae, Polycercus, 398
olgae, Sacciuterina, 398
oligacantha, Davainea, 256
oligacantha, Raillietina (S.), 256
oligarthra, Echinococcus, 222
oligarthra, Taenia, 222
oligophora, Davainea, 291
oligorchida, Davainea, 262
oligorchida, Dilepis, 381
oligorchida, Idiogenoides, 262
oligorchida, Raillietina (R.), 262
oligorchida, Raillietina (Ransomia), 262
oligorchidum, Anthobothrium, 158
oligorchis, Amoebotaenia, 388
oligorchis, Cleberia, 457
oligorchis, Cotugnia, 273
oligorchis, Diplogynia, 273
oligorchis, Gangesia, 176
oligorchis, Glaridacris, 36
Oligorchis, diagnosis, 333
oligorhyncha, Anomotaenia, 407
oligotesticulare, Echeneibothrium, 151
oligotoma, Taenia, 397
olivieri, Hymenolepis, 335
olngojenei, Taenia, 225
olor, Batrachotaenia, 185
olor, Crepidobothrium, 185
olor, Ophiotaenia, 185
olrikii, Diplocotyle, 45
olseni, Acanthobothrium, 139
olseni, Ophiotaenia, 185
olsoni, Hymenolepis, 305

*olsoni, Rodentolepis*, 305
*olsoni, Vampirolepis*, 305
*omalancristrota, Cyclorchida*, 378
*omalancristrota, Taenia*, 378
*omissa, Echinococcus*, 225
*omissa, Taenia*, 225
*omphalodes, Anoplocephala*, 434
*omphalodes, Bertiella*, 434
*omphalodes, Halysis*, 434
*omphalodes, Paranoplocephala*, 434
*omphalodes, Taenia*, 434
Oncobothriidae
   diagnosis, 131
   key to genera, 132
*Oncobothriocephalus*, 96
*Oncobothrium*, diagnosis, 136
*Oncodiscus*, diagnosis, 100
*Oncomegas*, diagnosis, 58
*oncorhynchi, Eubothrium*, 107
*ondatrae, Hymenolepis*, 318
*ondatrae, Hymenosphenacanthoides*, 318
*ondatrae, Retinometra*, 318
*Onderstepoortia*, diagnosis, 360
*Oochoristica*, diagnosis, 453
*oodes, Nybelinia*, 74
*oophorae, Amoebotaenia*, 388
*opatula, Anoplocephala*, 433
*ophia, Oochoristica*, 455
*ophiocephali, Circumoncobothrium*, 96
*ophiocephalina, Anchistrocephalus*, 96
*ophiocephalina, Senga*, 96
*ophiodex, Ophiotaenia*, 185
*Ophiotaenia*, diagnosis, 182
*Ophiovalipora*, diagnosis, 376
*Ophryocotyle*, diagnosis, 244
Ophryocotylinae, key to genera, 241
*Ophriocotyloides*, diagnosis, 242
*Ophryocotylus*, diagnosis, 242
*opisthorchis, Caulobothrium*, 153
*opisthorchis, Echeneibothrium*, 153
*opsariichthydis, Bothriocephalus*, 102
*opuntioides, Taenia*, 226
*oranensis, Catenotaenia*, 235
*oranensis, Skrjabinotaenia*, 235
*oratum, Bothridium*, 86
*orbiuterina, Mogheia*, 419
*oregonensis, Hymenolepis*, 305
*oregonensis, Rodentolepis*, 305
*oregonensis, Vampirolepis*, 305
*oregoni, Biacetabulum*, 30
*oreini, Adenoscolex*, 38
*Oriana*, 207
*orientale, Postgangesia*, 193
*orientalis, Anomotaenia*, 407
*orientalis, Aploparaksis*, 277
*orientalis, Bothridium*, 86
*orientalis, Breviscolex*, 38
*orientalis, Choanotaenia*, 362
*orientalis, Dilepis*, 381
*orientalis, Neoliga*, 391
*orientalis, Paratetrabothrius*, 207
*orientalis, Sobolevitaenia*, 362
*orientalis, Triaenophorus*, 113
*orioli, Choanotaenia*, 362
*oriolina, Anoncotaenia*, 341
*orinocoensis, Pomatotrygonocestus*, 142
*Orlovilepis*, 337
*ornatum, Bothridium*, 86
*ornitheios, Deltokeras*, 347
*ornithis, Mathevotaenia*, 463
*orthacantha, Hymenolepis*, 326
*orthacantha, Sphenacanthus*, 326
*orthaeantha, Variolepis*, 326
*Orthoskrjabinia*, diagnosis, 346
*ortleppi, Raillietina (R.)*, 262
*Ortleppolepis*, 306
*Orygmatobothrium*, diagnosis, 163
*osakensis, Raillietina (R.)*, 262
*osburni, Proteocephalus*, 187
*oschmarini, Diorchis*, 282
*oschmarini, Diorchis (Nudorchis)*, 325
*Oschmarinia*, diagnosis, 457
*osculata, Gangesia*, 176
*osculatus, proteocephalus*, 187
*osculatus, Taenia*, 187
*osensis, Insectivorolepis*, 288
*osheroffi, Oochoristica*, 455
*oshimai, Hymenolepis*, 318
*oshimai, Hymenosphenacanthus*, 318
*oshimai, Retinometra*, 318
*oshimai, Sphenacanthus*, 318
*oshmarini, Anonchotaenia*, 347
*oshmarini, Orthoskrjabinia*, 347
*oshmarini, Skrjabinerina*, 347
*Oshmarinolepis*, diagnosis, 307
*osipovi, Raillietina (R.)*, 262
*osmeri, Bothriocephalus*, 103
*osmeri, Diphyllobothrium*, 88
*Othinoscolex*, 195
*otidis, Anomotaenia*, 410
*otidis, Idiogenes*, 239
*otidis, Inermicapsifer*, 458
*otidis, Paruterina*, 351
*otidis, Pseudanomotaenia*, 410
*otidis, Sobolevina*, 458
*otidis, Taenia*, 306
*Otiditaenia*, diagnosis, 240
Otobothriidae
   diagnosis, 51
   key to genera, 67
*otobothrioides, Paranybelinia*, 67
*Otobothrium*, diagnosis, 68
*otocyonis, Dipylidium*, 353
*otomyos, Paranoplocephala*, 436
*otomys, Multiceps*, 226
*otomys, Taenia*, 226
*ovata, Taenia*, 226
*overstreeti, Acanthotaenia*, 182
*ovifusa, Anomotaenia*, 407
*ovifusa, Dichoanotaenia*, 407
*ovilla, Moniezia*, 420
*ovilla, Taenia*, 420

*ovilla, Thysaniezia*, 420
*ovilla, Thysanosoma*, 420
*ovilla* var. *macilenta, Moniezia*, 420
*ovina, Halysis*, 441
*ovina, Taenia*, 441
*ovipariens, Cysticercus*, 226
*oviparus, Cysticercus*, 226
*ovipunctata, Stilesia*, 418
*ovipunctata, Taenia*, 418
*ovis, Cysticercus*, 226
*ovis, Taenia*, 226
*ovofurcata, Diorchis*, 282
*ovolaciniata, Anomotaenia*, 407
*ovolaciniata, Taenia*, 407
*oweni, Calostaurus*, 252
*oweni, Drepanidotaenia*, 286
*oweni, Echinocotyle*, 286
*oxneri, Floriceps*, 53
*oxycephalus, Caryophyllaeus*, 33
*oxyuri, Diorchis*, 282
*oxyuri, Retinometra*, 318
*ozensis, Vampirolepis*, 305

# P

*pachipora, Valipora*, 371
*Pachybothrium*, diagnosis, 142
*pachycephala, Hymenolepis*, 329
*pachycephala, Microsomacanthus*, 329
*pachycephala, Taenia*, 329
*pachysoma, Prosthecocotyle*, 208
*paciferum, Reesium*, 159
*pacificum, Dicranotaenia*, 323
*pacificum, Diphyllobothrium*, 95
*pacificum, Wardium*, 323
*pacificus, Adenocephalus*, 94
*pacificus, Dasyrhynchus*, 52
*packi, Multiceps*, 226
*packi, Polycephalus*, 226
*packi, Taenia*, 226
*pagenstecheri, Anoplocephala*, 466
*pagenstecheri, Inermicapsifer*, 466
*pagenstecheri, Pericapsifer*, 466
*pagollae, Gyrocoelia*, 216
*pahangensis, Polyoncobothrium*, 97
*pahangensis, Senga*, 97
*paithanensis, Mastacembellophyllaeus*, 164
*pakistanensis, Pithophorus*, 146
*palasoorahi, Hornelliella*, 63
*palawanensis, Allohymenolepis*, 270
*palawanensis, Raillietina (R.)*, 262
*paleaceum, Rhynchobothrium*, 73
*palliata, Nybelinia*, 74
*palliata, Tetrarhynchus*, 74
*pallida, Moniezia*, 442
*palmarum, Hymenolepis*, 289
*palombii, Echeneibothrium*, 151
*palumbi, Bothriocephalus*, 103
*pamelae, Killigrewia*, 428
*pamirensis, Capiuterilepis*, 271

*pamirensis, Proteocephalus*, 187
*pammicrum, Phyllobothrium*, 161
*pampeanus, Echinococcus*, 222
*Panceria*, 451
*Panceriella*, diagnosis, 451
*pancerii, Ligula*, 94
*Pancerina*, 451
*panjadi, Anthobothrium*, 158
*pantayi, Retinometra*, 318
*Panuwa*, diagnosis, 367
*papilla, Anomotaenia*, 407
*papilla, Drepanidotaenia*, 407
*papilla, Taenia*, 407
*papillatus, Hymenolepis*, 309
*papillatus, Sobolevicanthus*, 309
*papillifer, Tetrarhynchus*, 78
*papillifera, Dilepis*, 381
*papilligerum, Onchobothrium (Acanthobothrium)*, 139
*papillosa, Pseudhymenolepis*, 271
*papillosum, Copesoma*, 413
*papillosus, Tetrarhynchus*, 78
*Parabertiella*, diagnosis, 432
*Parabiglandatrium*, diagnosis, 313
*parabirulai, Aploparaksis*, 277
*Parabisaccanthes*, diagnosis, 314
Parabothriocephalidae
    diagnosis, 83
    key to genera, 115
*Parabothriocephaloides*, diagnosis, 117
*Parabothriocephalus*, diagnosis, 117
*Parabothrium*, diagnosis, 108
*Paracaryophyllaeus*, diagnosis, 28
*parachelidonariae, Pseudanomotaenia*, 410
*Parachoanotaenia*, 356
*Parachristianella*, diagnosis, 57
*Paracladotaenia*, 412
*paraclavicirrus, Dicranotaenia*, 323
*paraclavicirrus, Wardium*, 323
*paracompressa, Hymenolepis*, 330
*paracompressa, Microsomacanthus*, 330
*paracygni, Gastrotaenia*, 130
*Paradicranotaenia*, diagnosis, 271
*Paradilepis*, diagnosis, 375
*paradisea, Biuterina*, 348
*paradisea, Brumptiella*, 254
*paradisea, Davainea*, 254
*paradisea, Raillietina (P.)*, 254
*paradoxa, Brochocephalus*, 216
*paradoxa, Choanotaenia*, 395
*paradoxa, Drepanidotaenia*, 395
*paradoxa, Gyrocoelia*, 216
*paradoxa, Icterotaenia*, 395
*paradoxa, Jardugia*, 232
*paradoxa, Nematoparataenia*, 129
*paradoxa, Parachoanotaenia*, 395
*paradoxa, Paricterotaenia*, 395
*paradoxa, Polycercus*, 395
*paradoxa, Sacciuterina*, 395
*paradoxa, Taenia*, 381, 395
*paradoxa* var. *gasowskae, Sacciuterina*, 395

*paradoxus (Echinocotyle), Hymenolepis*, 284
*paradoxus, Gonoscolex*, 284
*paradoxuri, Taenia*, 226
*paraechinis, Mathevotaenia*, 463
*Paraechinophallus*, 105
*parafilum, Aploparaksis*, 277
*Parafimbriaria*, diagnosis, 292
*parafimbriatum, Corallobothrium*, 180
*Paraglaridacris*, diagnosis, 35
*paragonopora, Amphilina*, 469
*paragonopora, Gephyrolina*, 469
*Paragrillotia*, 64
*paraguayae, Mathevotaenia*, 463
*paraguayensis, Ophiotaenia*, 185
*Paraidiogenes*, diagnosis, 239
*parallacticus, Proteocephalus*, 187
*parallelepideda, Paruterina*, 351
*parallelepipeda, Taenia*, 351
*paramicrorhyncha, Anomotaenia*, 410
*paramicrorhyncha, Pseudanomotaenia*, 410
*paramicrosoma, Hymenolepis*, 330
*paramicrosoma, Microsomacanthus*, 330
*Paramoniezia*, diagnosis, 449
*paranitidulans, Echinocotyle*, 311
*paranitidulans, Nadejdolepis*, 311
*Paranoplocephala*, diagnosis, 434
*paranumenii, Choanotaenia*, 362
*Paranybelinia*, diagnosis, 67
Paranybeliniidae
    diagnosis, 48
    key to genera, 66
*Paraoligorchis*, diagnosis, 331
*paraporale, Dicranotaenia*, 323
*paraporale, Wardium*, 323
*Paraprogynotaenia*, diagnosis, 220
*Paraproteocephalus*, diagnosis, 179
*Pararetinometra*, diagnosis, 316
*pararetracta, Dicranotaenia*, 323
*Paraschistometra*, 240
*parasiluri, Corallobothrium*, 180
*parasiluri, Gangesia*, 176
*parasiluri, Paraproteocephalus*, 180
*parasiluri, Proteocephalus*, 188
*Parataenia*, 121
*paratarius, Hypocaryophyllaeus*, 28
*Paratetrabothrius*, 207
*Paratriotaenia*, diagnosis, 461
*paratrygoni, Rhinebothrium*, 155
*parbata, Raillietina (P.)*, 254
*parcitesticulatus, Monoecocestus*, 427
*parechinobothrida, Davainea*, 259
*parenchymatosa, Taenia*, 226
*Paricterotaenia*, 395
*parkamoo, Amphoteromorphus*, 192
*parina, Anomotaenia*, 363
*parina, Choanotaenia*, 362
*parina, Dicranotaenia (D.)*, 298
*parina, Drepanidotaenia*, 362
*parina, Hymenolepis*, 298
*parina, Icterotaenia*, 362
*parina, Paricterotaenia*, 363

*parina, Passerilepis*, 298
*parina, Taenia*, 362
*parina, Weinlandia*, 298
*paronae, Inermicapsifer*, 466
*paronai, Dioecocestus*, 213
*paronai, Porogynia*, 249
*paronai, Taenia*, 249
*Paronia*, diagnosis, 422
*Parorchites*, diagnosis, 399
*Paruterina*, diagnosis, 350
Paruterininae, key to genera, 339
*parva, Atriotaenia*, 462
*parva, Clelandia*, 382
*parva, Corallotaenia*, 179
*parva, Cotugnia*, 246
*parva, Davaineia*, 463
*parva, Hydatigera*, 226
*parva, Hymenolepis*, 305
*parva, Marsipometra*, 110
*parva, Mathevotaenia*, 463
*parva, Mirandula*, 368
*parva, Moniezia*, 442
*parva, Neyraia*, 339
*parva, Oochoristica*, 455, 462, 463
*parva, Polyoncobothrium*, 97
*parva, Rodentolepis*, 305
*parva, Senga*, 97
*parva, Taenia*, 226
*parva, Vampirolepis*, 305
*parva, Yorkeria*, 134
*parviceps, Chelacanthus*, 280, 323
*parviceps, Dicranotaenia*, 323
*parviceps, Diorchis*, 280, 323
*parviceps, Hymenolepis*, 280
*parviceps, Taenia*, 323
*parvirostellata, Hymenolepis*, 337
*parvirostris, Choanotaenia*, 398
*parvirostris, Icterotaenia*, 398
*parvirostris, Parachoanotaenia*, 398
*parvirostris, Paricterotaenia*, 398
*parvirostris, Polycercus*, 398
*parvirostris, Sacciuterina*, 398
*parvirostris, Taenia*, 398
*Parvirostrum*, diagnosis, 460
*parvisaccata, Dicranotaenia*, 321
*parvisaccata, Hymenolepis*, 321
*parvispine, Neovalipora*, 371
*parvispine, Platyscolex*, 371
*parvispine, Valipora*, 371
*parvissima, Staphylocystis*, 302
*Parvitaenia*, diagnosis, 411
*parvitaeniunca, Valipora*, 371
*parviuncinata, Kotlania*, 262
*parviuncinata, Raillietina (R.)*, 262
*parviuncinata, Raillietina (Ransomia)*, 262
*parviuncinatum, Acanthobothrium*, 139
*parviuncinatus, Coenurus*, 226
*parviuncinatus, Multiceps*, 226
*parviuncinatus, Taenia*, 226
*parvogenitalis, Diorchis*, 279
*parvogenitalis, Schillerius*, 279

*parvovaria, Diochetos,* 452
*parvovaria, Oochoristica,* 452
*parvula, Dicranotaenia (D.),* 330
*parvula, Hymenolepis,* 330
*microsoma, Hymenolepis (Drepanidotaenia),* 330
*parvula, Microsomacanthus,* 330
*parvula, Oochoristica,* 455
*parvula, Weinlandia,* 330
*parvulus, Inermicapsifer,* 465
*parvulus, Kowalewskius,* 330
*parvulus, Lytocestus,* 26
*parvum, Acanthobothrium,* 139
*parvum, Anthobothrium,* 158
*parvum, Balanobothrium,* 120
*parvum, Bothridium,* 86
*parvum, Diphyllobothrium,* 94
*parvum, Eubothrium,* 107
*parvum, Platybothrium,* 137
*parvus, Bothriocephalus,* 103
*parvus, Calostaurus,* 252
*parvus, Caryophyllaeus,* 33
*parvus, Corallobothrium,* 179
*parvus, Hymenolepis,* 335
*pascualei, Joyeuxiella,* 353
Paspalia, 252, 253
*pasqualaeiformis, Joyeuxia,* 353
*pasqualei, Diplidium,* 353
*passerellae, Choanotaenia,* 398
*passerellae, Icterotaenia,* 398
*passerellae, Paricterotaenia,* 398
*passerellae, Polycercus,* 398
*passeriformicola, Raillietina (R.),* 262
Passerilepis, diagnosis, 297
*passerina, Anomotaenia,* 363
*passerina, Biuterina,* 348
*passerina, Choanotaenia,* 363
*passerina, Hymenolepis,* 299
*passerina, Monopylidium,* 363
*passerina, Passerilepis,* 299
*passerina, Weinlandia,* 299
*passeris, Dicranotaenia,* 297
*passeris, Passerilepis,* 297
*passeris, Taenia,* 297
*passerum, Anomotaenia,* 410
*passerum, Pseudanomotaenia,* 410
*pastinaca, Prochristianella,* 59
*pastinacae, Grillotia,* 64
*pastinacea, Phyllobothrium,* 161
*patagonicus, Echinococcus,* 223
*patersoni, Hymenolepis,* 337
*patriciae, Paradilepis,* 376
*pauciannulata, Abortilepis,* 330
*pauciannulata, Choanotaenia,* 398
*pauciannulata, Hymenolepis,* 330
*pauciannulata, Icterotaenia,* 398
*pauciannulata, Microsomacanthus,* 330
*pauciannulata, Paricterotaenia,* 398
*pauciannulata, Polycerus,* 398
*pauciannulata, Sacciuterina,* 398
*pauciova, Protogynella,* 274
*pauciovata, Abortilepis,* 330

*pauciovata, Hymenolepis,* 330
*pauciovata, Microsomacanthus,,* 330
*pauciovatus, Hymenolepis,* 318
*pauciovatus, Hymenosphenacanthus,* 319
*pauciovatus, Retinometra,* 318
*pauciproglottis, Hymenolepis,* 302
*pauciproglottis, Skrjabinotaenia,* 236
*pauciproglottis, Staphylocystis,* 302
*paucisegmentata, Davainea,* 248
*paucisegmentata* var. *dahomeensis, Davainea,* 248
*paucitesticulata, Anomotaenia,* 407
*paucitesticulata, Davainea,* 262
*paucitesticulata, Kotlania,* 262
*paucitesticulata, Raillietina (R.),* 262
*paucitesticulata, Raillietina (Ransomia),* 262
*paucitesticulata, Progynotaenia,* 219
*paucitesticulatus, Mesocestoides,* 203
*paucitesticulatus, Oligorchis,* 333
*pauliani, Tetrabothrius,* 209
*paulum, Acanthobothrium,* 140
*paulum, Orygmatobothrium,* 163
*paulus, Acanthobothrium,* 139
*pavlovskyi, Similuncinus,* 359
*pearsei, Hymenolepis,* 305
*pearsei, Rodentolepis,* 305
*pearsei, Vampirolepis,* 305
*pearsei, Proteocephalus,* 188
*pearsoni, Acanthobothrium,* 140
*pearsoni, Myrmillorhynchus,* 60
*pearsoni, Tetrarhynchus,* 60
*pectinata, Andrya,* 435
*pectinata, Cittotaenia,* 451
*pectinata, Ctenotaenia,* 451
*pectinata, Halysis,* 451
*pectinata, Moniezia,* 451
*pectinata, Mosgovoyia,* 451
*pectinata, Taenia,* 451
*pectinatum, Dipylidium,* 451
*pectinatum, Phoreiobothrium,* 141
*pectinatus, Alyselminthus,* 435, 451
Pedibothrium, diagnosis, 143
*pediformis, Taenia,* 267
*pedunculata, Oochoristica,* 457
*pedunculata, Oschmarenia,* 457
*peipingensis, Hymenolepis,* 389
*pekinensis, Amoebotaenia,* 388
*pelecani, Tetrabothrius,* 209
*pelegicus, Diorchis,* 282
Pelichnibothrium, diagnosis, 149
*pellucida, Bertiella,* 439
*pellucida, Dicranotaenia,* 299
*pellucida, Hymenolepis,* 299
*pellucida, Moniezia,* 442
*pellucida, Passerilepis,* 299
*pellucida, Prototaenia,* 439
*pellucida, Weinlandia,* 299
*pellucidum, Calliobothrium,* 136
*peltatum, Lecanicephalum,* 126
Peltidocotyle, diagnosis, 195
Peltidocotylinae, 195
*peltocephalus, Diplogonoporus,* 90

*penaei, Prochristianella,* 59
*penaeus, Eutetrarhynchus,* 57
*penaeus, Renibulbus,* 57
*Penarchigetes,* diagnosis, 29
*penelopina, Davainea,* 262
*penelopina, Kotlania,* 262
*penelopina, Raillietina (R.),* 262
*penelopina, Raillietina (Ransomia),* 262
*penetrans, Aploparaksis,* 277
*penetrans, Davainea,* 259, 263
*penetrans, Djombangia,* 23
*penetrans, Monorchis,* 277
*penetrans nova, Raillietina (R.),* 263
*penetrans, Otobothrium,* 68
*penetrans, Raillietina (R.),* 263
*Penetrocephalus,* diagnosis, 103
*penicillata, Anomotaenia,* 410
*penicellata, Pseudanomotaenia,* 410
*peniculus, Amphoteromorphus,* 142
*pennanti, Hymenolepis,* 289
*pennensis, Rowardleus,* 31
*pennsylvania, Mathevotaenia,* 463
*pennsylvanica, Oochoristica,* 463
*Pentaloculum,* diagnosis, 144
*pentamyzos, Biuterina,* 348
*pentamyzos, Paruterina,* 348
*pentastoma, Proteocephalus,* 188
*Pentocoronaria,* diagnosis, 249
*Pentorchis,* diagnosis, 331
*penphrikos, Otobothrium,* 69
*peradenica, Raillietina (R.),* 263
*peramelidarum, Vampirolepis,* 305
*percae, Proteocephalus,* 188
*percae, Taenia,* 188
*percnopteri, Aporina,* 429
*percnopteri, Neophronia,* 429
*percotti, Amurotaenia,* 171
*perelica, Callitetrarhynchus,* 54
*perelica, Tentacularia,* 54
*perfectum, Monorygma,* 163
*perfidus, Tetrabothrius,* 207
*perfoliata, Anoplocephala,* 433
*perfoliata, Halysis,* 433
*perfoliata megnini, Taenia,* 433
*perfoliata, Taenia,* 433
*perfoliata* var. *zebrae, Anoplocephala,* 433
*perfoliatus, Alyselminthus,* 433
*perfoliatus, Glandicephalus,* 93
*Pericapsifer,* diagnosis, 466
*perideraea, Nybelinia,* 74
*perigrinatoris, Tetrabothrius,* 209
*perisorei, Anomotaenia,* 363
*perisorei, Choanotaenia,* 363
*perlata, Halysis,* 204
*perlata, Taenia,* 204
*perlatus, Mesocestoides,* 204
*permista, Raillietina (R.),* 263
*peromysci, Catenotaenia,* 235
*peromysci, Choanotaenia,* 363
*peromysci, Hymenolepis,* 335
*peromysci, Prochoanotaenia,* 363

*perplexa, Raillietina (R.),* 263
*perplexus, Proteocephalus,* 188
*perreti, Railltietina (P.),* 254
*perspicua, Ophiotaenia,* 183
*peruanum, Anthobothrium,* 158
*Perutaenia,* diagnosis, 437
*perversa, Gyrocoelia,* 215
*petaurina, Bertiella,* 439
*petauristae, Aprostatandrya (A.),* 434
*petaruristae, Paranoplocephala,* 434
*petrocinclae, Hispaniolepis,* 299
*petrocinclae, Hymenolepis,* 299
*petrocinclae, Passerilepis,* 299
*petrocinclae, Taenia,* 299
*petrodromi, Hymenolepis,* 305
*petrodromi, Rodentolepis,* 305
*petrodromi, Vampirolepis,* 305
*petrotchenkoi, Mathevolepis,* 287
*petrovi, Raillietina, (S.),* 256
*petrowi, Mesocestoides,* 204
*petteri, Hymenolepis,* 293
*petteri, Lophurolepis,* 293
*phaenicopteri, Parabiglandatrium,* 313
*phalacrocoracis, Paradilepis,* 376
*phalacrocoracis, Tetrabothrius,* 209
*phalacrocorax, Hymenolepis,* 291
*phalacrocorax, Weinlandia,* 291
*phalacrocorax, Woodlandia,* 291
*Phascolotaenia,* diagnosis, 447
*phasianina, Hymenolepis,* 298
*phasianina, Mayhewia,* 298
*pheidolae, Raillietina (R.),* 263
*philactes, Drepanidotaenia,* 295
*philactes, Hymenolepis,* 295
*philactes, Parabisaccanthes,* 314
*philauti, Cylindrotaenia,* 210
*philippinensis, Sparganum,* 94
*phillipsi, Ophiotaenia,* 185
Philobythiidae
    diagnosis, 83
    key to genera, 110
*Philobythoides,* diagnosis, 111
*Philobythos,* diagnosis, 110
*phocae-foetidae, Bothriocephalus,* 92
*Phoenicolepis,* diagnosis, 309
*phoeniconaiadis, Cladogynia,* 316
*phoeniconaiadis, Hymenolepis,* 316
*Phoreiobothrium,* diagnosis, 140
*Phormobothrium,* diagnosis, 149
*phoxini, Bothriocephalus,* 103
*phrynosomatis, Diochetos,* 452
*phrynosomatis, Oochoristica,* 452
*phycis-medeterranei, Tetrarhynchus,* 78
Phyllobothriidae
    diagnosis, 132
    key to genera, 145
*Phyllobothrium,* diagnosis, 160
*phyllostomi, Hymenolepis,* 335
*Physchiosoma,* 223
*physeteris, Hexagonoporus,* 89
*physeteris, Multiductus,* 89

*picae, Aploparaksis,* 277
*pici, Raillietina (R.),* 263
*picusi, Choanotaenia,* 363
*picusi, Ophryocotyloides,* 243
*pifanoi, Dilepis,* 381
*pigmentata, Hymenolepis,* 337
*pigmentata, Ophiotaenia,* 185
*pigmentatum, Echinobothrium,* 166
*pigulevski, Catenotaenia,* 235
*pileatus, Discocephalum,* 121
*pileatus, Disculiceps,* 121
*pilherodiae, Dendrouterina,* 378
*pilidiatus, Rhynchobothrius,* 76
*pillersi, Tentacularia,* 77
*pindchii, Pseudoschistotaenia,* 230
*pingi, Dicranotaenia (D.),* 322
*pingi, Hymenolepis,* 322
*pingi, Wardium,* 322
*Pinguicollum,* diagnosis, 142
*pinguicollum, Onchobothrium* 142
*pinguicollum, Pinguicollum,* 142
*pinguis, Anoplocephala,* 243
*pinguis, Bertia,* 243
*pinguis, Bertiella,* 243
*pinguis, Chapmania,* 243
*pinguis, Lecanicephalum,* 126
*pinguis, Ophryocotyloides,* 243
*pinguis, Otiditaenia,* 243
*pinguis, Proteocephalus,* 188
*pinguis, Tylocephalum,* 126
*pinnae, Tetrarhynchus,* 78
*pinsonae, Raillietina (P.),* 254
*pintneri, Davainea,* 263
*pintneri, Hexastichorchis,* 418
*pintneri, Kotlania,* 263
*pintneri, Nybelinia,* 74
*pintneri, Raillietina (R.),* 263
*pintneri, Raillietina (Ransomia),* 263
*Pintneriella,* 63
*pipistrelli, Vampirolepis,* 305
*piracatinga, Monticellia,* 194
*piracatinga, Nomimoscolex,* 193
*piraeeba, Amphoteromorphus,* 192
*piraeeba, Endorchis,* 193
*piraeeba, Nomimoscolex,* 192
*piramutab, Anthobothrium,* 158
*piramutab, Goezeella,* 194
*piramutab, Monticellia,* 194
*piranabu, Amphilaphorchis,* 196
*piranabu, Rudolphiella,* 196
*pirara, Myzophorus,* 194
*piriei, Phyllobothrium,* 162
*piriformis, Anoncotaenia,* 341
*piscium, Ligula,* 87
*piscium-aliorum, Tetrarhynchus,* 78
*piscium, Tricuspidaria,* 112
*pisiformis, Cystotaenia,* 226
*pisiformis, Hydatigena,* 226
*pisiformis, Taenia,* 226
*pisiformis, Vesicaria,* 226
*Pistana,* diagnosis, 114

*pistillum, Staphylocystis,* 301
*pistillum, Taenia,* 301
*pithonis, Bothridium,* 85
*Pithophorus,* diagnosis, 146
*pittae, Malika,* 357
*pittalugai, Hymenolepis,* 319
*pittalugai, Hymenosphenacanthus,* 319
*pittalugai, Retinometra,* 319
*pittalugai, Sphenacanthus,* 319
*pitymi, Hymenolepis,* 335
*Plagiotaenia,* 432
*plana, Fimbriaria,* 267
*plana, Gyrocotyle,* 473
*planestici, Hymenolepis,* 326
*planestici, Variolepis,* 326
*planestici, Weinlandia,* 326
*planiceps, Lacistorhynchus,* 65
*planicipitis, Aleurotaenia,* 357
*planirostris, Biuterina,* 348
*planirostris, Taenia,* 348
*planissima, Moniezia,* 442
*planissima* var. *lobata, Moniezia,* 442
*planus, Gastrolecithus,* 159
*plastica, Bertia,* 439
*plastica, Bertiella,* 439
*plastica, Prototaenia,* 439
*plastica, Taenia,* 439
*Platybothrium,* diagnosis, 137
*platycephala, Choanotaenia,* 363
*platycephala, Taenia,* 363
*platycephalum, Pterobothrium,* 70
*platycephalus, Bothriocephalus,* 103
*platycephalus, Tetrarhynchus,* 70
*platycerci, Cotugnia,* 246
*platydera, Taenia,* 226
*platyrhyncha, Anomotaenia,* 407
*platyrhyncha, Dichoanotaenia,* 407
*platyrhyncha, Spasskytaenia,* 407
*platyrhyncha, Taenia,* 407
*Platyscolex,* 371
   diagnosis, 384
*platystomi, Proteocephalus,* 188
*plecoglossi, Proteocephalus,* 188
*plegadis, Anomotaenia,* 407
*plegadis, Choanotaenia,* 407
*pleistacantha, Pseudogrillotia,* 66
*Pleronybelinia,* 73
*pleuronectis-limandae, Tetrarhynchus,* 78
*pleuronectis-maximi, Tetrarhynchus,* 78
*plicata, Halysis,* 433
*plicata, Taenia,* 433
*plicata* var. *pedunculata, Anoplocephala,* 433
*plicata* var. *restricta, Anoplocephala,* 433
*plicata* var. *servei, Anoplocephala,* 433
*plicata* var. *strangulata, Anoplocephala,* 433
*Plicatobothrium,* diagnosis, 99
*plicatum, Orygmatobothrium,* 163
*plicatum, Rhynchobothrium,* 76
*plicatus, Alyselminthus,* 433
*plicatus, Bothriocephalus,* 103
*plicatus, Epision,* 267

*plicatus, Fistulicola,* 113
*plicitum, Diplobothrium,* 159
*Plicobothrium,* diagnosis, 91
*Pliovitellaria,* diagnosis, 34
*pluriuncinata, Cotugnia,* 246
*pluriuncinata, Davainea,* 265
*plutiuncinata, Kotlania,* 265
*pluriuncinata, Raillietina (F.),* 265
*pluriuncinata, Raillietina (Johnstonia),* 265
*pluriuncinata, Taenia,* 265
*pocilifera, Hymenolepis,* 337
*podicipedis, Diphyllobothrium,* 88
*podicipina, Confluaria,* 326
*podicipina, Dicranotaenia,* 326
*podicipina, Dubininolepis,* 326
*podicipina, Hymenolepis,* 326
*podicipina, Variolepis,* 326
*Podicipitilepis,* 324
*podifufi, Echinocotyle,* 286
*podocesi, Paruterina,* 351
*Poecilancistrum,* diagnosis, 69
*polaris, Tetrabothrius,* 209
*pollachii, Proteocephalus,* 188
*pollanicola, Proteocephalus,* 188
*pollonae, Echeneibothrium,* 151
*polyacantha, Cotugnia,* 246
*polyacantha, Echinocotyle,* 286
*polyacantha, Taenia,* 226
*polyacantha* var. *oligorchida, Cotugnia,* 246
*polyacantha* var. *paucimusculosa, Cotugnia,* 246
*polycalcarata, Davainea,* 254
*polycalcaria, Taenia,* 226
*polycalceola, Brumptiella,* 249
*polycalceola, Davainea,* 249
*polycalceola, Davaineoides,* 249
*polycalceolus, Dibothriocephalus,* 94
*Polycephalus,* 223
*Polycercus,* diagnosis, 395
*polychalix, Davainea,* 263
*polychalix, Kotlania,* 263
*polychalix, Raillietina (R.),* 260, 263
*Polycoelia,* 249
*Polygonoporus,* 88
*polyhamata, Raillietina (S.),* 256
*polymorpha, Taenia,* 232
*polymorphus, Diplophallus,* 232
*Polyoncobothrium,* diagnosis, 96
*Polipobothrium,* 157
*polyonchis, Gangesia,* 176
*polyorchis, Bertia,* 438
*polyorchis, Choanotaenia,* 361, 386
*polyorchis, Tetrabothrius,* 209
*polyorchis, Unciunia,* 386
*Polypocephalus,* diagnosis, 121
*polyptera, Anchistrocephalus,* 96
*polypteri, Polyoncobothrium,* 97
*polyrugosum, Diphyllobothrium,* 95
*polystictae, Aploparaksis,* 277
*polytelidis, Cotugnia,* 246
*polytestis, Proterogynotaenia,* 220
*polytuberculosus, Coenurus,* 226

*polytuberculosus, Cysticercus,* 226
*polytuberculosus, Multiceps,* 226
*polytuberculosus, Taenia,* 226
*polyuterinea, Brumptiella,* 256
*polyuterina, Davainea,* 256
*polyuterina, Raillietina (S.),* 256
*Pomatotrygonocestus,* diagnosis, 141
*ponticum, Acanthobothrium,* 139
*ponticum, Diphyllobothrium,* 95
*porale, Dicranotaenia,* 323
*porale, Glareolepis,* 323
*porale, Hymenolepis,* 323
*porale, Sphenacanthus,* 323
*porale, Wardium,* 323
*poralis, Hymenolepis,* 337
*porata, Anomotaenia,* 407
*Porogynia,* diagnosis, 249
*porosa, Choanotaenia,* 363
*porosa, Drepanidotaenia,* 363
*porosa, Icterotaenia,* 363
*porosa, Parachoanotaenia,* 363
*porosa, Paricterotaenia,* 363
*porosa, Taenia,* 363
*Porotaenia,* 207
*porrecta, Wageneria,* 55
*porrogenitalis, Oochoristica,* 455
*portei, Diagonaliporus,* 371
*portei, Valipora,* 371
*porzana, Aploparaksis,* 277
*porzana, Hymenolepis,* 277
*porzanae, Liga,* 390
*porzanae, Rallitaenia,* 390
*posteroporus, Cylindrophorus,* 143
*Postgangesia,* diagnosis, 193
*Postgangesiianae, 191*
*poulsoni, Proteocephalus,* 188
*praecoquis, Cittotaenia,* 450
*praecoquis, Ctenotaenia,* 450
*praecoquis, Pseudocittotaenia,* 450
*praecox, Anomotaenia,* 407
*praecox, Taenia,* 407
*praeputialis, Amphoteromorphus,* 192
*praeputialis, Hymenolepis,* 311
*praeputialis, Nadejdolepis,* 311
*praeputialis, Sobolevicanthus,* 311
*pretoriensis, Diphyllobothrium,* 95
*pretoriensis, Spirometra,* 95
*Priapocephalus,* diagnosis, 204
*pribilofensis, Hymenolepis,* 335
*priestlyi, Tetrabothrius,* 209
*primaverus, Proteocephalus,* 188
*primordialis, Andrya,* 435
*primordialis,* var. *grundii, Andrya,* 435
*prionacis, Phyllobothrium,* 162
*prionodes, Inermicapsifer,* 466
*prionodes* var. *intermedia, Inermicapsifer,* 466
*prinopsia, Anomotaenia,* 410
*prinopsia, Pseudanomotaenia,* 410
*pristis, Anthobothrium,* 158
*pristis, Phyllobothrium,* 162
*proboscideus, Bothriocephalus,* 103

*Proboscidosaccus*, 158
*Probothriocephalus*, diagnosis, 115
*procaviae, Hyracotaenia*, 466
*procera, Hymenolepis*, 335
*procerum, Corallobothrium*, 179
*procerum, Corallobothrium (Megathylacoides)*, 179
*procerum, Megathylacoides*, 179
*procerus, Tetrabothrius*, 209
*Prochoanotaenia*, 361
*Prochristianella*, diagnosis, 58
*procirrosa, Anomotaenia*, 407
*procyonis, Atriotaenia*, 462
*procyonis, Oochoristica*, 462
*Prodicoelia*, 85
*producta, Choanotaenia*, 398
*producta, Icterotaenia*, 398
*producta, Paricterotaenia*, 398
*producta, Polycercus*, 398
*producta, Taenia*, 398
*Profimbriaria*, diagnosis, 269
*Progamotaenia*, diagnosis, 444
*progenesia, Raillietina (S.)*, 256
*progenesia, Skrjabinia (S.)*, 256
*proglottina, Davainea*, 242
*proglottina, Taenia*, 247
*proglottis, Wageneria*, 76
*Progrillotia*, 64
*Progynopylidium*, 354
*Progynotaenia*, diagnosis, 218
Progynotaeniidae
  diagnosis, 200
  key to genera, 217
*prokopici, Hilmylepis*, 294
*proliferum, Sparganum*, 94
*Promonobothrium*, diagnosis, 35
*pronosomum, Otobothrium*, 69
*pronosomum, Rhynchobothrium*, 69
*Proorchida*, diagnosis, 374
*Proparuterina*, diagnosis, 411
*propecysticum, Otobothrium*, 69
*propteres, Lateriporus*, 373
*Prosobothrium*, diagnosis, 176
*Prosthecobothrium*, 137
*Prosthecocotyle*, 207
Proteocephalidea, 16, 173
  key to families, 173
Proteocephalidae, key to subfamilies, 173
Proteocephalinae, key to genera, 182
*Proteocephalus*, diagnosis, 185
*proterogyna, Baeriella*, 446
*proterogyna, Fuhrmannodes*, 446
*proterogyna, Progamotaenia*, 446
*Proterogynotaenia*, diagnosis, 219
*proteus, Ophryocotyle*, 244
*Protogynella*, diagnosis, 273
*Prototaenia*, 438
*provincialis, Davainea*, 263
*provincialis, Kotlania*, 263
*provincialis, Raillietina (R.)*, 263
*provincialis, Raillietina (Ransomia)*, 263
*prudhoei, Bothriocephalus*, 103

*prudhoei, Ophryototyle*, 245
*prudhoei, Tetrabothrius*, 209
*prunellae, Choanotaenia*, 363
*prussica, Bothrioscolex*, 21
*prussica, Khawia*, 21
*pruvoti, Scyphophyllidium*, 147
*przewalskii, Drepanidotaenia*, 295
*przewalskii, Hymenolepis*, 295
*psammobati, Acanthobothrium*, 140
*psammonomi, Skrjabinotaenia*, 236
*Pseudamphicotyla*, diagnosis, 106
*Pseudandrya*, diagnosis, 372
*Pseudangularia*, diagnosis, 393
*Pseudanomotaenia*, diagnosis, 409
*Pseudanoplocephala*, diagnosis, 433
*Pseudanthobothrium*, diagnosis, 148
*Pseudeubothrioides*, 107
*Pseudeubothrium*, diagnosis, 107
Pseudhymenolepidinae, key to genera, 269
*Pseudhymenolepis*, diagnosis, 271
*Pseudoaporina*, 428
*Pseudocapingentoides*, 39
*Pseudocaryophyllaeus*, diagnosis, 41
*Pseudocatenotaenia*, diagnosis, 233
*Pseudocephalochlamys*, diagnosis, 84
*pseudochiri, Bertiella*, 439
*pseudochiri, Prototaenia*, 439
*Pseudochoanotaenia*, diagnosis, 355
*Pseudocittotaenia*, diagnosis, 449
*pseudocoronula, Dicranotaenia*, 321
*pseudocotylea, Oochoristica*, 455
*pseudocucumerina, Taenia*, 203
*pseudocyclorchida, Parvitaenia*, 412
*pseudocyrtus, Raillietina (R.)*, 263
*pseudodera, Tentacularia*, 54
*Pseudodiorchis*, diagnosis, 278
*pseudoechinobothrida, Raillietina (F.)*, 265
*pseudoerinaceus, Grillotia*, 64
*pseudofilum, Aploparaksis*, 277
*pseudofilum, Monorchis*, 277
*pseudofurcata, Hymenolepis*, 335
*pseudofurcigera, Aploparaksis*, 277
*pseudofusa, Hymenolepis*, 322
*pseudofusum, Wardium*, 322
*Pseudogrillotia*, diagnosis, 66
Pseudogrillotiidae, 63
*pseudoinflata, Hyemnolepis*, 337
*Pseudoligorchis*, diagnosis, 33
*Pseudolytocestus*, diagnosis, 40
*pseudomicrorhyncha, Anomotaenia*, 407
*Pseudonybelinia*, diagnosis, 66
*Pseudoparadilepis*, diagnosis, 284
Pseudophyllidea, 13, 81
  key to families, 81
*pseudopodis, Oochoristica*, 453
*pseudopodis, Taenia*, 453
*pseudopolypteri, Polyoncobothrium*, 97
*pseudoporus, Tetrabothrius*, 208
*pseudorostellatus, Hymenolepis*, 330
*pseudorostellatus, Microsomacanthus*, 330
*Pseudoschistotaenia*, diagnosis, 230

*pseudosecessivus, Aploparaksis,* 277
*pseudosetigera, Hymenolepis,* 337
*Pseudoshipleya,* diagnosis, 216
Pseudotobothrium, 68
*pseudotropii, Vermaia,* 175
*pseudouncinatum, Oncobothrium,* 136
*psittacea, Davainea,* 263
*psittacea, Kotlania,* 263
*psittacea, Raillietina (R.),* 263
*psittacea, Raillietina (Ransomia),* 263
*psittacea, Stringopotaenia,* 424
*psittulae, Biporouterina,* 428
Pterobothriidae
    diagnosis, 50
    key to genera, 69
*Pterobothrium,* diagnosis, 70
*pterocephalum, Diphyllobothrium,* 95
*pteroclesi, Rhabdometra,* 346
*pterocleti, Gvosdevinia,* 256
*pterocleti, Raillietina (S.),* 256
*pteroplateae, Anthobothrium,* 158
*Ptilotolepis,* 355
Ptychobothriidae
    diagnosis, 82
    key to genera, 95
*Ptychobothrioides,* diagnosis, 98
*Ptychobothrium,* diagnosis, 98
*ptychocheila, Edlintonia,* 39
*ptychocheilus, Proteocephalus,* 188
*pubescens, Taenia,* 276
*pugetensis, Proteocephalus,* 188
*pujehuni, Stocksia,* 24
*pulcher, Anthemobothrium,* 122
*pulcher, Polypocephalus,* 122
*pulchra, Hymenolepis,* 288
*pulchra, Insectivorolepis,* 288
*pullae, Hymenolepis,* 283
*Pulluterina,* diagnosis, 430
*pulvinatum, Rhodobothrium,* 158
*puncta, Anomotaenia,* 409
*puncta, Taenia,* 409
*punctata, Biuterina,* 351
*punctata, Liga,* 390
*punctata, Sphaeruterina,* 351
*punctata, Taenia,* 298, 390
*punctatissima, Nybelinia,* 74
*punctatus, Bothriocephalus,* 101
*pungitii, Schistocephalus,* 88
*pungutchui, Taenia,* 226
*punica, Taenia,* 226
*puriensis, Pleurocercus,* 164
*purpurata, Dilepis,* 351
*purpurata, Paruterina,* 351
*purpurata, Sphaeruterina,* 351
*pusilla, Catenotaenia,* 234
*pusilla, Hymenolepis,* 337
*pusilla, Taenia,* 234
*pusillum, Gryporhynchus,* 379
*pusillus, Gryporhynchus,* 371
*pusillus, Proteocephalus,* 188
*puthurensii, Amoebotaenia,* 388

*pycnomera, Bothriocephalus,* 96
*pycnomera, Senga,* 96
*pycnonoti, Eugonodaeum,* 355
*pycnonoti, Paronia,* 423
*pycnonoti, Raillietina (P.),* 254
*pycnonoti, Variolepis,* 326
*pygargi, Ascotaenia,* 421
*pygargi, Avitellina,* 421
*pygargi, Thysanosoma,* 421
*pygoscelis, Dibothriocephalus,* 95
*Pyramicocephalus,* diagnosis, 91
*pyramidalis, Hymenolepis,* 380
*pyriformis, Anomotaenia,* 411
*pyriformis, Pseudanomotaenia,* 411
*pyriformis, Taenia,* 411

# Q

*quadrata, Anomotaenia,* 407
*quadrata, Taenia,* 407
*quadratum, Diphyllobothrium,* 95
*quadribothria, Anthobothrium,* 158
*quadribothria, Taenia,* 158
*quadrijugosa, Cylindrotaenia,* 210
*quadriloba, Taenia,* 433
*quadrilobata, Taenia,* 433
*quadripapillosus, Tetrarhynchus,* 78
*quadripartitum, Acanthobothrium,* 140
*quadrirostris, Echinorhynchus,* 76, 77
*quadrirostris gadi-horhuae, Echinorhynchus,* 78
*quadrirostrus, Echinorhynchus,* 78
*quadrisurculi, Discobothrium (nomen nudum),* 125
*quadritesticulata, Kotlania,* 263
*quadritesticulata, Raillietina (R.),* 263
*quasioweni, Echinocotyle,* 286
*quasioweni, Hymenolepis,* 286
*quelea, Paruterina,* 351
*quentini, Tupaiataenia,* 459
*Quentinotaenia,* diagnosis, 232
*querquedula, Dicranotaenia,* 323
*querquedula, Hymenolepis,* 323
*querquedula, Weinlandia,* 323
*querquedulae, Hymenolepis,* 323
*quinonesi, Acanthobothrium,* 140
*quinquarii, Amphicotyle,* 106
*quinquarii, Pseudamphicotyla,* 106
*quinquecoronatum, Diplopylidium,* 354
*quinquecoronatum, Dipylidium,* 354
*quiscali, Anoncotaenia,* 341
*quitensis, Raillietina (R.),* 259, 263

# R

*racemosa, Oochoristica,* 184
*racemosa, Ophiotaenia,* 185
*racovitzai, Clujia,* 75
*radians, Multiceps,* 226
*radians, Taenia,* 226
*radiatus, Polypocephalus,* 121

radioductum, Phyllobothrium, 162
ragazzi, Taenia (Anoplocephala), 465
rahmi, Manitaurus, 251
rahmi, Raillietina (R.), 251
raillieti, Ichthyotaenia, 184
raillieti, Spirometra, 94
Raillietina
   diagnosis, 252
   key to subgenera, 252
   species of unknown subgeneric status, 266
   subgenera, see Raillietina (Fuhrmannetta); Raillietina (Paroniella); Raillietina (Raillietina); Raillietina (Skrjabinia)
Raillietina (Fuhrmannetta), species in, 265
Raillietina (Paroniella), species in, 253
Raillietina (Raillietina), species in, 256
Raillietina (Skrjabinia), species in, 255
raipurensis, Introvertus, 18
raipurensis, Lucknowia, 18
rajae, Anthobothrium, 158
rajae-asperae, Tetrarhynchus, 78
rajae-clavatae, Tetrarhynchus, 78
rajae-megarhynchae, Tetrarhynchus, 78
rajasthanensis, Laterorchites, 394
raji, Echinobothrium, 166
Rajotaenia, 236
ralli, Diorchis, 280
ralli, Jonesius, 280
rallida, Krimi, 394
Rallitaenia, 389
ramosus, Multiceps, 226
ramosus, Taenia, 226
ranae, Nematotaenoides, 211
ranae, Ophiotaenia, 185
ranarum, Ligula, 94
ranarum, Ophiotaenia, 185
ranarum, Spirometra, 94
rangdonensis, Microsomacanthus, 330
rangifer, Cysticercus, 225
rangifer, Taenia, 225
rangiferina, Moniezia, 441
rangonica, Raillietina (P.), 255
rangoonica, Raillietina, 266
rangoonicus, Hymenolepis, 319
rangoonicus, Hymenosphenicanthus, 319
rangoonicus, Retinometra, 319
rankini, Echeneibothrium, 151
ransomi, Choanotaenia, 398
ransomi, Davainea, 256
ransomi, Diorchis, 279
ransomi, Paricterotaenia, 398
ransomi, Polycercus, 398
ransomi, Raillietina (S.), 256
ransomi, Schillerius, 279
ransomi, Southwellia, 380
raoi, Plicatobothrium, 100
raoii, Circumoncobothrium, 96
rapacicola, Culcitella, 349
rapida, Drepanidotaenia, 295
rapida, Hymenolepis, 295
rarus, Bothriocephalus, 103

rarus, Hymenolepis, 331
rarus, Microsomacanthus, 331
rashidi, Sobolevicanthus, 309
rashomonensis, Hymenolepis, 335
ratti, Oochoristica, 462
ratticola, Bothriocephalus, 95
ratticola, Choanotaenia, 367
ratticola, Monopylidium, 367
ratticola, Rodentotaenia, 367
ratzi, Hymenolepis, 337
rauschi, Aploparaksis, 277
rauschi, Echinocotyle, 286
rauschi, Hymenolepis, 286
rauschi, Paruterina, 351
Rauschitaenia, diagnosis, 408
raviensis, Oligorchis, 333
recapta, Arctotaenia, 369
recapta, Dilepis, 369
recapta, Laterinorus, 369
rectacantha, Hymenolepis, 330
rectacantha, Microsomacanthus, 330
rectangula, Biuterina, 348
rectangula, Triaenorhina, 348
rectangulum, Bathybothrium, 108
rectangulus, Marsipocephalus, 140
recurvata, Microsomacanthus, 330
recurvirostrae, Dicranotaenia, 323
recurvirostrae, Diorchis, 282
recurvirostrae, Drepanidotaenia, 323
recurvirostrae, Hymenolepis, 323
recurvirostrae, Taenia, 323
recurvirostroides, Dicranotaenia, 323
recurvirostroides, Hymenlepis, 323
recurvispinus, Grillotia, 64
redactum, Discobothrium (nomen nudum), 125
Reditaenia, 223
reditus, Tetrabothrius, 209
redonica, Pseudhymenolepis, 271
reesae, Echinobothrium, 166
Reesium, diagnosis, 159
reggiae, Catenotaenia, 235
regis, Taenia, 226
relicta, Hymenolepis, 289
relicta, Taenia, 289
remotus, Inermicapsifer, 466
Renibulbus, 56
Renyxa, diagnosis, 169
reptans, Gymnorhynchus, 61
reptans, Ligula, 94
reptans, Sparganum, 94
reptans, Spirometra, 94
resimum, Diphyllobothrium, 95
restiformis, Bothriocephalus, 99
restiformis, Dibothrium, 99
restricta, Anoplocephala, 433
reticulata, Anomotaenia, 407
reticulata, Arostellina, 425
reticulatum, Parvirostrum, 400
reticulosa, Lapwingia, 403
Retinometra, diagnosis, 316
retirostris, Acanthocirrus, 379

*retirostris, Dilepis,* 379
*retirostris, Gryporhynchus,* 379
*retirostris, Taenia,* 379
*retracta, Amphipetrovia,* 326
*retracta, Hymenolepis,* 326
*retracta, Taenia,* 226
*retracta, Variolepis,* 326
*retracta, Wardium,* 326
*retractilis, Brumptiella,* 254
*retractilis, Davainea,* 254
*retractilis, Raillietina (P.),* 254
*retroversa, Aploparaksis,* 277
*retusa, Davainea,* 256
*retusa, Meggittia,* 256
*retusa, Raillietina (S.),* 256
*retzii, Tritaphros,* 164
*reutensis, Anomotaenia,* 407
*reynoldsae, Raillietina (P.),* 254
*reynoldsi, Biuterina,* 349
*reynoldsi, Paruterina,* 349
*reynoldsi, Pseudodiorchis,* 278
*Rhabdometra,* diagnosis, 345
*rhabdophidis, Ophiotaenia,* 185
*Rhabdotobothrium,* diagnosis, 154
*rheiphilus, Monoecocestus,* 427
*Rhinebothrium,* diagnosis, 155
*Rhinebothroides,* diagnosis, 153
*rhinobati, Acanthobothrium,* 140
*rhinobati, Echeneibothrium,* 151
*rhinobati, Rhinebothrium,* 156
*rhinobatidis, Polypocephalus,* 122
*rhinobatii, Lecanicephalum,* 126
*rhinobatii, Spinocephalum,* 126
*rhinocheti, Anomotaenia,* 407
*rhinoptera, Echinobothrium,* 166
*Rhinoptericola,* diagnosis, 71
Rhinoptericolidae, 71
 diagnosis, 51
*rhodesiensis, Anoplocephala,* 433
*rhodesiensis, Hymenolepis,* 291
*rhodesiensis, Inermicapsifer,* 466
*rhodesiensis, Schmelzia,* 291
*Rhodobothrium,* diagnosis, 158
*rhombi, Bothriocephalus,* 103
*rhombiodea, Dicranotaenia,* 276
*rhomboidea, Taenia,* 276
*rhombomidis, Catenotaenia,* 235
*rhopalocephala, Andrya,* 435
*rhopalocephala, Anoplocephala,* 435
*rhopalocephala, Taenia,* 435
*Rhopalothyrax,* 75
*Rhoptrobothrium,* 145
*rhynchichthydis, Schistocephalus,* 88
*rhynchobati, Marsupiobothrium,* 147
*rhynchobatidis, Polypocephalus,* 122
*rhynchobatidis, Tetrarhynchus,* 78
*rhynchopis, Choanotaenia,* 398
*rhynchopis, Icterotaenia,* 398
*rhynchopis, Paricterotaenia,* 398
*rhynchopis, Polycercus,* 398
*rhynchota, Brumptiella,* 254

*rhynchota, Davainea,* 254
*rhynchota, Raillietina (P.),* 254
*rhynchota, Soninotaurus,* 254
*rhinobati, Marsupiobothrium,* 147
*ricardi, Anootypus,* 419
*ricardi, Avitellina,* 419
*ricci, Anomotaenia,* 407
*ricci, Choanotaenia,* 407
*ricci, Dictymetra,* 407
*richardi, Sphyriocephalus,* 72
*ridibundum, Choanotaenia,* 363
*riduculum, Thysanocephalum,* 133
*riggenbachi, Dicranotaenia,* 323
*riggenbachi, Hymenolepis,* 323
*riggii, Calyptrobothrium,* 162
*riggii, Phyllobothrium,* 162
*rigida, Bertia,* 439
*rigida, Bertiella,* 439
*rigida, Prototaenia,* 439
*rileyi, Hydatigera,* 226
*rileyi, Taenia,* 226
*rimandoi, Cotugnia,* 246
*riparia, Anomotaenia,* 407
*ripariae, Angularella,* 393
*ris, Catenotaenia,* 235
*riseri, Nybelinia,* 74
*rissae, Aploparaksis,* 277
*Rissotaenia,* 404
*ritae, Proteocephalus,* 188
*robertsi, Hymenolepis,* 337
*robusta, Dasyurotaenia,* 242
*robusta, Duthiersia,* 86
*robusta, Nybelinia,* 74
*robusta, Tetrarhynchus,* 74
*robustum, Acanthobothrium,* 140
*robustum, Poecilancistrum,* 69
*robustus, Chaetophallus,* 207
*robustus, Triaenophorus,* 113
*rodentinum, Markewitchitaenia,* 459
*rodentinum, Mathevotaenia,* 459
*Rodentolepis,* 302
*Rodentotaenia,* 366
*rogersi, Rogersus,* 29
*Rogersus,* diagnosis, 28
*römeri, Dibothriocephalus,* 93
*romerolagi, Anoplocephaloides,* 437
*roonwali, Cylindrotaenia,* 210
*rosaeformis, Taenia,* 157
*rosenthali, Hymenolepis,* 326
*rosenthali, Octacanthus,* 326
*rosenthali, Sphenacanthus,* 326
*rosickyi, Anomotaenia,* 407
*rosittensis, Bothrioscolex,* 21
*rosittensis, Khawia,* 21
*rosseteri, Echinocotyle,* 285
*rosseteri (Echinocotyle), Hymenolepis,* 285
*rossicum, Dipylidium,* 353
*rossicum, Joyeuxiella,* 353
*rossii, Rhynchobothrium,* 76
*rostellata, Anonchotaenia,* 347
*rostellata, Armadoskrjabinia,* 297

rostellata, Choanotaenia, 363
rostellata, Monopylidium, 363
rostellata, Oochoristica, 454, 455
rostellata, Orthoskrjabinia, 347
rostellata, Skrjabinerina, 347
rostellata, Taenia, 297
Rostellotaenia, 181
rostrata, Anomotaenia, 363, 407
rostrata, Choanotaenia, 363, 407
rostrata, Monopylidium, 363
rostratula, Tetrabothrius, 207
rostratulae, Dilepis, 381
rostratum, Monopylidium, 407
rothlisbergeri, Raillietina, 266
rotunda, Choanotaenia, 398
rotunda, Icterotaenia, 398
rotunda, Paricterotaenia, 398
rotunda, Polycercus, 398
rotunda, Sacciuterina, 398
rotundata, Oochoristica, 453
rotundata, Taenia, 453
rotundum, Phyllobothrium, 162
roudabushi, Vampirolepis, 305
rougetcampanae, Nybelinia, 74
rouxi, Moniezioides, 422
rouxi, Proterogynotaenia, 220
Rowardleus, diagnosis, 31
rowei, Grillotia, 64
rowettiae, Anomotaenia, 411
rowettiae, Pseudanomotaenia, 411
Roytmania, 252, 253
rubromaculatum, Pterobothrium, 71
rubromaculatum, Rhynchobothrium, 71
rubrum, Acanthobothrium, 140
rubrum, Myzophyllobothrium, 145
rudicornis, Anthocephalus, 76
rudolphiana, Taenia, 341
rudolphica, Hymenolepis, 337
Rudolphiella, diagnosis, 196
Rudolphiellinae, 196
rudolphii, Didymobothrium, 45
rudolphii, Diplocotyle, 45
Rufferia, 73
rufi, Schistotaenia, 230
ruficola, Progamotaenia, 445
ruficolle, Rhynchobothrium, 57
ruficollis, Eutetrarhynchus, 57
ruficollis, Tetrarhynchus, 57, 70
rufum, Diphyllobothrium, 93
rugata, Monticellia, 194
rugata, Spatulifer, 194
rugosa, Gyrocotyle, 473
rugosa, Moniezia, 442
rugosa, Monticellia, 194
rugosa, Peltidocotyle, 195
rugosa, Taenia, 231, 442
rugosa, Tentacularia, 72
rugosos, Rhynchobothrius, 72
rugosum, Anthobothrium, 147
rugosum, Eubothrium, 107
rugosus, Acoleus, 231

rugosus, Hymenolepis, 309
rugosus, Sobolevicanthus, 309
rugovaginosus, Paradilepis, 376
rundi, Tetrabothrius, 209
rupicaprae, Moniezia, 442
rusannae, Pentocoronaria, 249
russelli, Ophiotaenia, 185
rustica, Anomotaenia, 411
rustica, Dicranotaenia, 283
rustica, Hymenolepis, 283
rustica, Pseudanomotaenia, 411
rustica, Weinlandia, 283
rybickae, Hymenolepis, 321
rybickae, Raillietina (R.), 263
ryjikovi, Anoplocephaloides, 437
ryjikovi, Echinocotyle, 286
ryjikovi, Paranoplocephala, 437
rysavyi, Passerilepis, 299
rysavyi, Triodontolepis, 301
rysavyi, Vampirolepis, 305

# S

saccatus, Floriceps, 53
saccifera, Acanthotaenia, 180
saccifera, Crepidobothrium, 180
saccifera, Ichthyotaenia, 180
saccifera, Kapsulotaenia, 180
saccifera, Proteocephalus, 180
sacciperum, Hymenolepis, 321
Sacciuterina, 395
sachalinensis, Aploparaksis, 277
saginatus, Taeniarhynchus, 222
sagitta, Allohaploparaxis, 275
sagitta, Drepanidotaenia, 275
sagitta, Proteocephalus, 188
sagitta, Taenia, 188
sagittata, Markevitschia, 18
sagitticeps, Parabothriocephalus, 118
saguei, Nadejdolepis, 311
saigoni, Taenia, 226
salensis, Oochoristica, 455
salmoni, Davainea, 266
salmoni, Raillietina (F.), 266
salmoni, Raillietina (Johnstonia), 266
salmonidicola, Proteocephalus, 188
salmonis, Kotlania, 266
salmonis, Phyllobothrium, 149, 162
salmonisomus, Protecephalus, 188
salmonisumblae, Proteocephalus, 188
salvelini, Bothriocephalus, 103
salvelini, Eubothrium, 107
salvelini, Proteocephalus, 188
samfyia, Parvitaenia, 412
sanbernardinensis, Ophiotaenia, 185
sanchorensis, Staphylocystis, 302
sanchovensis, Mathevotaenia, 463
sandgroundi, Atriotaenia, 462
sandgroundi, Avitellina, 414
sandgroundi, Cittotaenia, 273

*sandgroundi, Diplogynia*, 273
*sandgroundi, Kapsulotaenia*, 180
*sandgroundi, Oochoristica*, 462
*sandgroundi, Proteocephalus*, 180
*sandgroundi, Vampirolepis*, 305
*Sandonella*, diagnosis, 177
*sandoni, Proteocephalus*, 177
*sandoni, Sandonella*, 177
*sanehensis, Gangesia*, 176
*sanjuanensis, Aploparaksis*, 277
*saphena, Ophiotaenia*, 185
*sarasini, Tetrabothrius*, 209
*sarasinorum, Bertia*, 439
*sarasinorum, Bertiella*, 439
*sarasinorum, Prototaenia*, 439
*sarawakensis, Duthiersia*, 86
*sardinellae, Platybothrium*, 137
*sarrakowahi, Vermaia*, 175
*Sartica, Davainea*, 263
*sartica, Kotlania*, 263
*sartica, Raillietina (R.)*, 263
*sartica, Raillietina, (Ransomia)*, 263
*sartica* var. *massiliensis, Raillietina, (R.)*, 263
*sartica* var. *mediterranea, Raillietina (R.)*, 263
*Satyolepis*, 297
*satyri, Bertia*, 438
*satyri, Bertiella*, 438
*satyri, Taenia*, 438
*saurashtri, Flapocephalus*, 124
*sauridae, Bothriocephalus*, 103
*sauridae, Oncodiscus*, 101
*sawadai, Bothridium*, 86
*sawadai, Myotolepis*, 288
*sbesteriometra, Anoncotaenia*, 314
*Sbesterium*, 52
*scalaris, Staphylocystis*, 302
*scalopi, Hymenolepis*, 289
*scelophori, Oochoristica*, 455
*schaeferi, Tetrabothrius*, 209
*schaldybini, Neoskrjabinolepis*, 301
*schavarschi, Taenia*, 226
*schikhobalovae, Diagonaliporus*, 372
*schikhobalovae, Valipora*, 372
*schikhobolovi, Insinuarotaenia*, 221
*schilbiodies, Bothriocephalus*, 103
*schilleri, Aploparaksis*, 277
*schilleri, Hymenolepis*, 305
*schilleri, Staphylocystis*, 305
*schilleri, Vampirolepis*, 305
*Schillerius*, diagnosis, 279
*Schistocephalus*, diagnosis, 87
*schistochilos, Bothriocephalus*, 93
*Schistometra*, 240
*Schistotaenia*, diagnosis, 229
*schizacanthium, Oncobothrium*, 136
*Schizochoerus*, diagnosis, 470
*Schizocotyle*, diagnosis, 45
*Schizorchis*, diagnosis, 440
*Schizorchoides*, diagnosis, 440
*Schizotaenia*, 426
*Schmelzia*, diagnosis, 291

*schmidti, Eutetrarhynchus*, 57
*schmidti, Passerilepis*, 299
*schmidti, Vampirolepis*, 305
*schultzei, Batrachotaenia*, 185
*schultzei, Crepidobothrium*, 185
*schultzei, Ichthyotaenia*, 185
*schultzei, Ophiotaenia*, 185
*sciaenae, Bothriocephalus*, 103
*sciaenae-aquilae, Tetrarhynchus*, 78
*sciuricola, Choanotaenia*, 363
*sciuricola, Rodentotaenia*, 363
*scleropagis, Polyoncobothrium*, 97
*Schmidneila*, 395
*scobinae, Rhinebothrium*, 156
*scolecina, Dilepis*, 375
*scolecina, Grillotia*, 65
*scolecina, Paradilepis*, 375
*scolecina, Taenia*, 375
*scolecina, Tetrarhynchus*, 65
*scolopacina, Choanotaenia*, 363
*scolopacis, Aploparaksis*, 277
*scolopendra, Schistotaenia*, 230
*scolopendra, Taenia*, 230
*scolopendra, Tatria*, 230
*scomber-gobius, Tetrarhynchus*, 78
*scomber-pelamys, Tetrarhynchus*, 78
*scomber-rochei, Tetrarhynchus*, 78
*scomber-thynnus, Tetrarhynchus*, 78
*scombri, Tetrabothriorhynchus*, 77
*scorpii, Bothriocephalus*, 101
*scorzai, Rhinebothroides*, 154
*scorzai, Rhinebothrium*, 154
*scoticum, Diphyllobothrium*, 95
*scotti, Dibothriocephalus*, 95
*scutigerum, Monopylidium*, 363
*scyllium-canicula, Tetrarhynchus*, 78
*scymni, Tetrarhynchus*, 78
*scymni-rostrati, Tetrarhynchus*, 78
*scymnus-micaeensis, Tetrarhynchus*, 78
*scymnus-rostratus, Tetrarhynchus*, 78
*Scyphocephalus*, diagnosis, 85
*Scyphophyllidium*, diagnosis, 146
*secessivus, Aploparaksis*, 277
*secunda, Anomotaenia*, 407
*secunda, Choanotaenia*, 407
*secunda, Cladotaenia*, 413
*secunda, Monopylidium*, 407
*secunda, Taenia*, 226
*secunda, Vigisolepis*, 293
*secundus, Scyphocephalus*, 85
*segmentatus, Parabothriocephaloides*, 117
*selfi, Raillietina (R.)*, 263
*Semenoviella*, diagnosis, 452
*semipalmata, Echinoshipleya*, 215
*semivesiculum, Acanthobothrium*, 140
*semoni, Linstowia*, 461
*semoni, Taenia*, 461
*semoni* var. *acanthocirrus, Linstowia*, 461
*senaariensis, Davainea*, 263
*senaariensis, Kotlania*, 263
*senaariensis, Raillietina (R.)*, 263

senaariensis, *Raillietina* (*Ransomia*), 263
senegalensis, *Nybelinia*, 74
*Senga*, diagnosis, 96
sengeri, *Hymenolepis*, 302
sengeri, *Staphylocystis*, 302
seni, *Cotugnia*, 246
sepentata, *Diorchis*, 275
sepiae, *Tetrarhynchus*, 73
septaria, *Dinobothrium*, 159
septaria, *Hymenolepis*, 337
septemsororum, *Hymenolepis*, 299
septemsororum, *Passerilepis*, 299
septentrionale, *Acanthobothrium*, 140
septentrionalis, *Adenocephalus*, 94
septicolle, *Polyoncobothrium*, 97
septotesticulata, *Paruterina*, 351
sequens, *Raillietina* (*R.*), 263
serialis, *Bovienia*, 25
serialis, *Caryophyllaeus*, 25
serialis, *Coenurus*, 226
serialis, *Cystotaenia*, 226
serialis, *Taenia*, 226
serpentatus, *Diorchis*, 280
serpentatus, *Linstowius*, 280
serpentiformis turturis, *Taenia*, 308
serpentis, *Diphyllobothrium*, 86
serpentis, *Spirometra*, 86
serpentulus, *Dicranotaenia* (*D.*), 298
serpentulus, *Diplacanthus*, 298
serpentulus, *Hymenolepis*, 298
serpentulus, *Hymenolepis* (*Drepanidotaenia*), 298
serpentulus, *Mayhewia*, 298
serpentulus sturni, *Hymenolepis*, 298
serpentulus turdi, *Hymenolepis*, 298
serrata, *Retinometra*, 319
serrata, *Taenia*, 226
serratum, *Dibothrium*, 94
serratum, *Phyllobothrium*, 162
serratus birmanicus, *Sobolevicanthus*, 309
serratus, *Hymenolepis*, 319
serratus, *Sobolevicanthus*, 319
serrula, *Staphylocystis*, 302
setigera, *Diplacanthus* (*Dilepis*), 330
setigera, *Drepanidotaenia*, 330
setigera, *Hymenolepis*, 330
setigera, *Microsomacanthus*, 330
setigera, *Taenia*, 330
setigera, *Tschertkovilepis*, 330
setigera, *Vigissotaenia*, 330
setosa, *Amoebotaenia*, 388
setosa, *Rhabdometra*, 346
setti, *Inermicapsifer*, 466
settiense, *Eutetrarhynchus*, 57
settiense, *Tetrarhynchobothrium*, 57
settii, *Echinophallus*, 105
sedowi, *Dilepis*, 381
sexorchidum, *Anthobothrium*, 158
shantungensis, *Raillietina* (*R.*), 263
sharmai, *Ophryocotyloides*, 243
shengi, *Hsuolepis*, 434
shensiensis, *Hsuolepis*, 434

shen-tsengi, *Hymenolepis*, 337
shentsengi, *Liga*, 390
shetlandicus, *Microsomacanthus*, 330
shindei, *Circumoncobothrium*, 96
shindei, *Cotugnia*, 246
shindei, *Sureshia*, 402
*Shindeiobothrium*, 150
*Shipleya*, diagnosis, 214
shipleyanus, *Tetrarhynchus*, 70
shipleyi, *Acanthotaenia*, 181
shipleyi, *Echeneibothrium*, 156
shipleyi, *Rhinebothrium*, 156
shipleyi, *Tetrarhynchus*, 78
shohoi, *Choanotaenia*, 363
shohoi, *Cotugnia*, 246
siamensis, *Diorchis*, 282
siamensis, *Raillietina* (*P.*), 255
sibirica, *Taenia*, 227
sibericensis, *Echinococcus*, 222
sibiricus, *Diorchis*, 282
sieboldi, *Archigetes*, 30
sieboldi, *Biacetabulum*, 30
sigmodontis, *Monoecocestus*, 427
sigmodontis, *Raillietina* (*R.*), 263
sigmodontis, *Schizotaenia*, 427
sigmoides, *Oochoristica*, 454
signachiana, *Drepanidotaenia*, 295
silesiacus, *Paraglaridacris*, 30, 35
siluri, *Goezeella*, 195
siluri, *Monticellia*, 195
siluri, *Silurotaenia*, 174
*Silurotaenia*, diagnosis, 174
silvestris, *Raillietina*, 265
simbae, *Taenia*, 227
simile, *Diplobothrium*, 159
simile, *Lecanicephalum*, 126
simile, *Rhynchobothrium*, 76
simile, *Tylocephalum*, 126
similis, *Bothriocephalus*, 95
similis, *Choanotaenia*, 363
similis, *Paruterina*, 351
similis, *Rhabdometra*, 351
similis, *Sobolevitaenia*, 363
*Similuncinus*, diagnosis, 358
simmonsi, *Grillotia*, 65
simoni, *Paradilepis*, 376
simplex, *Dicranotaenia*, 323
simplex, *Echeneibothrium*, 158
simplex, *Hymenolepis*, 323
simplex, *Spathebothrium*, 44
simplex, *Weinlandia*, 323
simplicissima, *Ligula*, 87
simplicessimus, *Proteocephalus*, 188
simulans, *Hymenolepis*, 330
simulans, *Microsomacanthus*, 330
*Sinaiotaenia*, diagnosis, 459
sinaitica, *Inermicapsifer*, 466
sindensis, *Duthiersia*, 86
sindensis, *Gangesia*, 176
sindensis, *Staphylocystis*, 302
sinensis, *Anomotaenia*, 408

*sinensis, Aploparaksis*, 277
*sinensis, Choanotaenia*, 408
*sinensis, Khawia*, 20
*sinensis, Raillietina (R.)*, 263
*sinensis, Rodentotaenia*, 305
*sinensis, Vampirolepis*, 305
*singapurensis, Raillietina (P.)*, 255
*singhi, Biuterina*, 349
*singhi, Echinocotyle*, 286
*singhi, Neoliga*, 391
*singhi, Raillietina (R.)*, 263
*singhi, Sureshia*, 402
*singhi, Lapwingia*, 404
*singhia, Capingentoides*, 40
*singhii, Lecanicephalum*, 126
*singhii, Polypocephalus*, 122
*singhii, Tylocephalum*, 126
*singularis, Capingens*, 37
*singularis, Hymenolepis*, 301
*singularis, Neoskrjabinolepis*, 301
*singularis, Proteocephalus*, 188
*sinuosa, Drepanidotaenia*, 328
*sinuosa, Hymenolepis*, 328
*sinuosa, Hymenolepis (Drepanidotaenia)*, 328
*sinuosa, Lepidotrias*, 328
*sinuosa, Taenia*, 328
*Sinuterilepis*, diagnosis, 270
*siriragi, Raillietina (R.)*, 261
*sjöstedti, Stilesia*, 418
*skarbilowitschi, Diorchis*, 282
*skoogi, Tetrabothrius*, 209
*skorikowi, Ichthyotaenia*, 188
*skorikowi, Proteocephalus*, 188
*Skorikowia*, diagnosis, 280
*Skrjabinacanthus*, diagnosis, 299
*skrjabinariana, Vampirolepis*, 303
*Skrjabinerina*, 346
*skrjabini, Anomotaenia*, 363, 406, 408
*skrjabini, Anophryocephalus*, 206
*skrjabini, Aploparaksis*, 277
*skrjabini, Caryophyllaeides*, 19
*skrjabini, Choanotaenia*, 363
*skrjabini, Cucurbilepis*, 293
*skrjabini, Diagonaliporus*, 372
*skrjabini, Dibothriocephalus*, 95
*skrjabini, Diorchis*, 274
*skrjabini, Diploposthe*, 232
*skrjabini, Diplopylidium*, 354
*skrjabini, Echinatrium*, 320
*skrjabini, Echinocotyle*, 286
*skrjabini, Hymenolepis*, 319
*skrjabini, Hymenosphenacanthus*, 319
*skrjabini, Idiogenes*, 240
*skrjabini, Lateriporus*, 373
*skrjabini, Leptotaenia*, 218
*skrjabini, Malika*, 357
*skrjabini, Mathevotaenia*, 463
*skrjabini, Microsomacanthus*, 330
*skrjabini, Moniezia*, 442
*skrjabini, Multiceps*, 227
*skrjabini, Multiuterina*, 343
*skrjabini, Raillietina (R.)*, 263
*skrjabini, Retinometra*, 319
*skrjabini, Schillerius*, 279
*skrjabini, Sphenacanthus*, 319
*skrjabini, Taenia*, 227
*skrjabini, Tatria*, 229
*skrjabini, Trigonocotyle*, 206
*skrjabini, Triodontolepis*, 301
*skrjabini, Valipora*, 372
*skrjabini, Vigisolepis*, 293
*skrjabinia, Choanotaenia*, 406
*skrjabiniana, Anomotaenia*, 411
*skrjabiniana, Pseudanomotaenia*, 411
*skrjabinini, Paruterina*, 351
*skrjabinissima, Aploparaksis*, 277
*skrjabinissima, Hymenolepis*, 337
*Skrjabinochora*, 453
*Skrjabinolepis*, 375
*Skrjabinoparaxis*, diagnosis, 274
*Skrjabinotaenia*, diagnosis, 235
*Skrjabionotarurus*, 252, 253
*slesvicensis, Choanotaenia*, 363
*slesvicensis, Icterotaenia*, 363
*slesvicensis, Paricterotaenia*, 363
*slesvicensis, Taenia*, 363
*smaridis-gorae, Tetrarhynchus*, 65
*smaridis-maenae, Tetrarhynchus*, 78
*smaridium, Rhynchobothrius*, 76
*smarisgora, Grillotia*, 65
*smaris-gora, Tetrarhynchus*, 65
*smitii, Echeneibothrium*, 152
*smogorjevskajae, Dicranotaenia*, 323
*smogorjevskajae, Wardium*, 323
*smythi, Hymenolepis*, 337
*smythi, Multiceps*, 227
*smythi, Taenia*, 227
*sobolevi, Aploparaksis*, 277
*sobolevi, Biuterina*, 349
*sobolevi, Choanotaenia*, 363
*sobolevi, Dicranotaenia*, 323
*sobolevi, Dilepis*, 381
*sobolevi, Diorchis*, 282
*sobolevi, Microsomacanthus*, 330
*sobolevi, Monorcholepis*, 277
*sobolevi, Oochoristica*, 455
*sobolevi, Skrjabinchora*, 455
*sobolevi, Sobolevitaenia*, 363
*sobolevi, Wardium*, 323
*Sobolevina*, diagnosis, 458
*Sobolevicanthus*, diagnosis, 307
*Sobolevitaenia*, 361
*sobrinum, Echeneibothrium*, 152
*socialis, Taenia*, 410
*Solenophorus*, 85
*Solenotaenia*, 182
*solidum, Corallobothrium*, 180
*solidus, Schistocephalus*, 88
*solidus, Tetrarhynchus*, 78
*solinosomum, Bothriocotyle*, 109
*solisoricis, Vampirolepis*, 305
*solitaria, Staphylocystis*, 302

*solitaria, Weinlandia,* 302
*solitariae, Lateriporus,* 373
*solium, Taenia,* 223
*solowiowi, Hymenolepis,* 311
*solowiowi, Nadejdolepis,* 311
*somalensis, Raillietina (R.),* 264
*somariensis, Vampirolepis,* 305
*songi, Tortocephalus,* 88
*Soninotaurus,* 253
*sonoti, Choanotaenia,* 363
*soricina, Choanotaenia,* 363
*Soricina,* diagnosis, 287
*soricis, Soricina,* 287
*southwelli, Anomotaenia,* 367
*southwelli, Australiolepis,* 292
*southwelli, Avitellina,* 418
*southwelli, Choanotaenia,* 367
*southwelli, Hymenolelpis,* 292
*southwelli, Monopylidium,* 367
*southwelli, Nematoparataenia,* 129
*southwelli, Paruterina,* 351
*southwelli, Raillietina (P.),* 255
*southwelli, Sphenacanthus,* 292
*southwelli, Spiniglans,* 358
*southwelli, Uncibilocularis,* 134
*southwelli, Yorkeria,* 134
*spalacis, Coenurus,* 227
*spalacis, Taenia,* 227
*Spartoides,* diagnosis, 40
*spasskajae, Anomotaenia,* 408
*spasskajae, Diorchis,* 282
*spasskii, Aploparaksis,* 277
*spasskii, Choanotaenia,* 363
*spasskii, Confluaria,* 326
*spasskii, Cotugnia,* 246
*spasskii, Dicranotaenia,* 299, 324
*spasskii, Insinuarotaenia,* 221
*spasskii, Microsomacanthus,* 330
*spasskii, Passerilepis,* 299
*spasskii, Prochoanotaenia,* 363
*spasskii, Sobolevicanthus,* 309
*spasskii, Vampirolepis,* 305
*spasskii, Variolepis,* 326
*spasskii, Wardium,* 324
*spasskyi, Cladotaenia,* 413
*spasskyi, Diagonaliporus,* 372
*spasskyi, Diuterinotaenia,* 449
*spasskyi, Multiuterina,* 343
*spasskyi, Ophiotaenia,* 185
*spasskyi, Sinuterilepis,* 270
*spasskyi, Trigonocotyle,* 206
*spasskyi, Valipora,* 372
*Spasskytaenia,* 379, 404
Spathebothriidae, diagnosis, 43
Spathebothriidea, 11, 43
  key to families, 43
*Spathebothrium,* diagnosis, 43
*spatula, Anoplocephala,* 433
*spatula, Taenia (Anoplocephala),* 433
*Spatulifer,* 194
*spearei, Progamotaenia,* 445

*speciosa, Lintoniella,* 54
*speciosum, Callitetrarhynchus,* 54
*speciosum, Pelichnibothrium,* 149
*speciosum, Rhynchobothrium,* 54
*speciosus, Bothriocephalus,* 103
*speciosus, Dibothriorhynchus,* 76
*speotytonis, Choanotaenia,* 363
*spermophili, Choanotaenia,* 363
*spermophili, Prochoanotaenia,* 363
*spermophili, Rodentotaenia,* 363
*Sphaerobothrium,* 158
*sphaerocephala, Dilepis,* 381
*sphaeroides, Davainea,* 266
*sphaeroides, Raillietina,* 266
*sphaerophora, Hymenolepis,* 337
*Sphaeruterina,* diagnosis, 351
*sphenocephala, Taenia,* 428
*sphecotheridis, Brumptiella,* 255
*sphecotheridis, Davainea,* 255
*sphecotheridis, Raillietina (P.),* 255
*Sphenacanthus,* 316
*Sphenacanthus (Retinometra),* 316
*sphenocephala, Hymenolepis,* 308
*sphenocephala, Taenia,* 308
*sphenocephala, Weinlandia,* 308
*sphenoides, Amoebotaenia,* 387
*sphenoides, Dicranotaenia,* 387
*sphenoides, Taenia,* 387
*sphenomorphus, Hymenolepis,* 302
*sphenomorphus, Staphylocystis,* 302
*sphyraena-argentei, Tetrarhynchus,* 78
*sphyraenaicum, Bombycirhynchus,* 75
*Sphyriocephala,* 71
Sphyriocephalidae, 71
  diagnosis, 49
*Sphyriocephalus,* diagnosis, 71
*sphyrnae, Nybelinia,* 74
*Sphyroncotaenia,* diagnosis, 237
*spiculigera, Dicranotaenia,* 324
*spiculigera, Taenia,* 324
*spiculigera* var. *varsoviensis, Dicranotaenia,* 324
*spilonotopteri, Alloptychobothrium,* 100
*spinachiae, Bothriocephalus,* 103
*spinata, Diorchis,* 282
*spinatum, Ersinogenes,* 238
*Spinibiloculus,* 164
*spinifer, Eutetrarhynchus,* 57
*Spiniglans,* diagnosis, 358
*Spinilepis,* diagnosis, 409
*Spiniloculus,* diagnosis, 136
*Spinocephalum,* 126
*spinocirrosa, Hymenolepis,* 337
*spinosa, Amoebotaenia,* 388
*spinosocapite, Anomotaenia,* 367
*spinosocapite, Choanotaenia,* 367
*spinosocapite, Monopylidium,* 367
*spinosocapite, Sobolevitaenia,* 367
*spinosima, Davainea,* 244
*spinosissima, Grillotia,* 65
*spinossima, Fernandezia,* 244
*spinosum, Dinobothrium,* 159

*spinosus, Globarilepis*, 270
*spinosus, Kowalewskius*, 331
*spinosus, Lateriporus*, 374
*spinulifera, Monticellia*, 195
*spinulifera, Phylobothroides*, 137
*spinuliferum, Tentacularia*, 75
*spinuliferum, Trigonolobum*, 75
*spinulosa, Anatinella*, 295, 312
*spinulosa, Drepanidotaenia*, 295, 312
*spinulosa, Hymenolepis*, 293
*spinulosa, Vigisolepis*, 293
*spinulosum, Dinobothrium*, 159
*spiracornutus, Rhynchobothrius*, 76
*spiraliceps, Bothriocephalus*, 98
*spiraliceps, Ptychobothrioides*, 98
*spiralicirrata, Microsomacanthus*, 330
*spiralis, Davainea*, 264
*spiralis, Diorchis*, 282
*spiralis, Kotlania*, 264
*spiralis, Raillietina (R.)*, 264
*spiralis, Raillietina (Ransomia)*, 264
*spirillometra, Avitellina*, 419
*Spirometra*, 93
*Spongiobothrium*, 157
*sprenti, Caryoaustralus*, 22
*Spreotaenia*, 361
*springeri, Diplootobothrium*, 68
*squali, Gilquinia*, 59
*squali, Hepatoxylon*, 63
*squali, Phyllobothrium*, 162
*squali, Taenia*, 59
*squaliglauci, Bothriocephalus*, 103
*squalii, Bothriocephalus*, 103
*squatarolae, Dicranotaenia*, 324
*squatarolae, Wardium*, 324
*squatinae, Lecanicephalum*, 127
*squatinae, Tylocephalum*, 127
*srinagarensis, Ophryocotyloides*, 243
*srivastavai, Choanotaenia*, 364
*srivastavai, Cotugnia*, 246
*srivastavai, Rodentolepis*, 305
*srivastavai, Schistotaenia*, 230
*srivastavai, Vampirolepis*, 305
*stagnatilidis, Choanotaenia*, 360
*stagnatilidis, Kowalewskiella*, 360
*stammeri, Gynandrotaenia*, 217
*Staphylepis*, diagnosis, 291
*Staphylocystis*, diagnosis, 301
*Staphylocystoides*, 301
*Staurobothrium*, 122
*steatomidis, Anomotaenia*, 408
*steatomidis, Hymenolepis*, 335
*stefanski, Aploparaksis*, 277
*stefanskii, Diorchis*, 282
*stefanskii, Hymenolepis*, 301
*stefanskii, Neoskrjabinolepis*, 301
*stefanskii, Vampirolepis*, 301
*stefanskii, Zarnowskiella*, 301
*stegostomatis, Balanobothrium*, 120
*steinhardti, Kotlania*, 264
*steinhardti, Raillietina (R.)*, 264

*steinhardti, Raillietina (Ransomia)*, 264
*stellifera, Choanotaenia*, 398
*stellifera, Icterotaenia*, 398
*stellifera, Paricterotaenia*, 398
*stellifera, Polycercus*, 398
*stellifera, Sacciuterina*, 398
*stellifera, Taenia*, 398
*stellorae, Hymenolepis*, 337
*stemmacephalum, Diphyllobothrium*, 93
*Stenobothrium*, 73
*Stenovaria*, 395
*stentorea, Anomotaenia*, 408
*stentorea, Dichoanotaenia*, 408
*stentorea, Taenia*, 408
*stercorarium, Choanotaenia*, 364
*stercorarium, Monopylidium*, 364
*stercorarium, Monopylidium (Macracanthus)*, 364
*sternina, Choanotaenia*, 364
*sternina, Icterotaenia*, 364
*sternina, Paricterotaenia*, 364
*sternina, Taenia*, 364
*steudeneri, Hymenolepis*, 303
*stictus, Diphyllobothrium*, 94
*Stilesia*, diagnosis, 417
*stilesiella, Raillietina (F.)*, 266
*stilesiella, Raillietina (R.)*, 266
*stizostedionis, Triaenophorus*, 113
*stizostethi, Proteocephalus*, 188
*Stocksia*, diagnosis, 24
*stolli, Hymenolepis*, 308, 309
*stolli, Sobolevicanthus*, 309
*straminea, Rodentolepis*, 305
*straminea, Taenia*, 305
*straminea, Vampirolepis*, 305
*strangulatus, Oligorchis*, 333
*strangulatus, Tetrabothrius*, 209
*strangulatus, Tetrarhynchus*, 78
*streptopeli, Cotugnia*, 247
*streptopeliae, Drepanidotaenia*, 299
*streptopeliae, Hymenolepis*, 299
*streptopeliae, Killigrewia*, 428
*streptopeliae, Passerilepis*, 299
*streptopeliae, Raillietina (R.)*, 264
*streptopelii, Houttuynia*, 250
*striata, Ophiotaenia*, 185
*striatus, Tetrarhynchus*, 78
*stricta, Aploparaksis*, 277
*strictophallus, Microsomacanthus*, 330
*strictum, Diphyllobothrium*, 88
*strigium, Choanotaenia*, 364
*Stringopotaenia*, diagnosis, 423
*Strobilocephalus*, diagnosis, 205
*strongyla, Nybelinia*, 74
*strumosus, Tetrarhynchus*, 78
*struthionis, Davainea*, 250
*struthionis, Houttuynia*, 250
*struthionis meogaeae, Houttuynia*, 250
*struthionis, Taenia*, 250
*studeri, Bertia*, 438
*studeri, Bertiella*, 438
*studeri, Taenia*, 438

stunkardi, *Philobythoides*, 111
sturionis, *Bothrimonus*, 45
stylicirrosa, *Panuwa*, 367
styloides, *Hymenolepis*, 330
styloides, *Microsomacanthus*, 330
*Stylopis*, 316
stylosa, *Dicranotaenia*, 299
stylosa, *Diplacanthus*, 299
stylosa, *Hymenolepis*, 299
stylosa, *Passerilepis*, 299
stylosa, *Taenia*, 299
stylosa, *Weinlandia*, 299
subrostellata, *Hymenolepis*, 335
subterranea, *Amoebotaenia*, 363
subterranea, *Anomotaenia*, 408
subtile, *Diphyllobothrium*, 95
sudanea, *Avitellina*, 418
sudanea, *Unciunia*, 343
sudanica, *Raillietina (S.)*, 256
sudarikovi, *Bancroftiella*, 399
*Sudarikovina*, diagnosis, 431
sudobim, *Myzophorus*, 194
sudobim, *Nomimoscolex*, 193
suis, *Paramoniezia*, 449
sujerensis, *Paricterotaenia*, 398
sujerensis, *Polycercus*, 398
sulae, *Tetrabothrius*, 209
sulcatus, *Diphyllobothrium*, 94
sulcatus, *Ichthyotaenia*, 188
sulcatus, *Proteocephalus*, 188
sulcatus, *Tetrabothrius*, 209
sultanpurensis, *Neyraia*, 339
sultanpurensis, *Valipora*, 372
sunci, *Hymenolepis*, 335
suncusensis, *Staphylocystis*, 302
suraishii, *Aploparaksis*, 277
*Sureshia*, diagnosis, 401
suricattae, *Hymenolepis*, 304
surinamensis, *Mathevotaenia*, 463
surinamensis, *Oochoristica*, 463
surinamensis, *Taenia*, 463
surmenicola, *Nybelinia*, 74
surpentulus, *Taenia*, 298
surubim, *Monticellia*, 195
surubim, *Spatulifer*, 195
susanae, *Kowalewskiella*, 360
suslika, *Hymenolepis*, 335
swiderskii, *Dubininolepis*, 326
swiderskii, *Hymenolepis*, 326
swiderskii, *Variolepis*, 326
swifti, *Angularella*, 393
swifti, *Mehdiangularia*, 391
swifti, *Neoliga*, 391
sylvarum, *Choanotaenia*, 364
*Symbothriorhynchus*, 75
symmetrica, *Catenotaenia*, 462
symmetrica, *Mathevotaenia*, 462
symmetrica, *Oochoristica*, 462
symonsii, *Raillietina (P.)*, 255
synallaxis, *Deltokeras*, 347
*Synbothrium*, 70

*Syndesmobothrium*, 70
*Syngenes*, 73
syngenes, *Nybelinia*, 74
synodontis, *Proteocephalus*, 188
synodontis, *Wenyonia*, 27
syrdariensis, *Staphylocystis*, 302
syrdarjensis, *Caryophyllaeus*, 33
*Szidatinus*, 29

# T

taakreei, *Mastacembellophyllaeus*, 164
taborensis, *Cycloskrjabinia*, 456
taborensis, *Oochoristica*, 456
tachyglossi, *Cittotaenia*, 456
tachyglossi, *Echidnotaenia*, 456
tadornae, *Diorchis*, 282
*Taenia*, diagnosis, 223
taeniacrassicipitis, *Cysticercus*, 224
taeniaeformis, *Kowalewskiella*, 360
taeniaformis, *Onderstepoortia*, 360
taeniaeformis, *Taenia*, 227
*Taeniarhynchus*, diagnosis, 221
Taeniidae
 diagnosis, 200
 key to genera, 221
taenioides, *Diancyrobothrium*, 94
taenioides, *Diphyllobothrium*, 94
*Taeniola*, 223
taeniuri, *Anthobothrium*, 158
taenius, *Taenia*, 204
tadornae, *Fimbriarioides*, 268
*Taeniarhynchaena*, diagnosis, 386
taglei, *Diplophallus*, 232
taiwanensis, *Angularella*, 393
taimyrensis, *Aploparaksis*, 277
taiwanensis, *Cotugnia*, 247
taiwanensis, *Hymenolepis*, 299
taiwanensis, *Passerilepis*, 299
taiwanensis, *Raillietina (R.)*, 264
takashii, *Insectivorolepis*, 288
talboti, *Fuhrmannodes*, 447
talicei, *Taenia*, 227
talismani, *Dasyrhynchus*, 52
talpae, *Cysticercus*, 224
tamilnadensis, *Diplootobothrium*, 68
tanakpuris, *Hymenolepis*, 337
tandani, *Aploparaksis*, 277
tandani, *Biacetabulum*, 35
tandani, *Bialovarium*, 35
tandani, *Choanotaenia*, 364
tandani, *Oochoristica*, 455
tandani, *Pleurocercus*, 164
tangalongi, *Spirometra*, 94
tanganyikae, *Inermicapsifer*, 466
tanganyikae, *Pericapsifer*, 466
tanganyikae, *Loennbergia*, 190
tanganyikae, *Lytocestoides*, 19
tanganyikae, *Marsipocephalus*, 190
tangoli, *Pterobothrium*, 71

*tangoli, Rhynchobothrium*, 71
*Taphrobothrium*, diagnosis, 101
*tapika, Chapmania*, 241
*tapika, Idiogenes*, 241
*tapika, Otiditaenia*, 241
*tapirus, Anoplocephala*, 436
*tapirus, Flabelloskrjabinia*, 436
*tarandi, Cysticercus*, 225
*tardae, Taenia*, 306
*tarnogradskii, Anomotaenia*, 411
*tarnogradskii, Pseudanomotaenia*, 411
*taruiensis, Rodentolepis*, 305
*taruiensis, Vampirolepis*, 305
*tasmanica, Linstowia*, 461
*taterae, Paraoligorchis*, 331
*taterae, Sudarikovina*, 432
*tateri, Hymenolepis*, 335
*tatia, Avitellina*, 419
*tatianae, Skryabinoparaxis*, 274
*Tatria*, diagnosis, 228
*Taufikia*, diagnosis, 425
*taunsaensis, Senga*, 96
*tauricollis, Capsodavainea*, 241
*tauricollis, Chapmania*, 241
*tauricollis, Davainea*, 241
*tauricollis, Taenia*, 241
*taxidiensis, Monordotaenia*, 222
*taxidiensis, Taenia*, 222
*taylori, Choanotaenia*, 364
*taylori, Raillietina (R.)*, 264
*tchadensis, Monobothrioides*, 24
*tecta, Baylisiella*, 92
*tectus, Cordicephalus*, 92
*Tejidotaenia*, diagnosis, 189
*telescopica, Anomotaenia*, 408
*telescopica, Rodentotaenia*, 408
*temminckii, Trichocephaloidis*, 370
*tenax, Balanobothrium*, 120
*tenerrimus, Retinometra*, 319
*tenerrimus, Sphenacanthus*, 319
*tenerrimus, Taenia*, 319
*tengizi, Flamingolepis*, 296
*Tentacularia*, diagnosis, 72
Tentaculariidae
    diagnosis, 48
    key to genera, 72
*tenuicaudatus, Tetrarhynchus*, 78
*tenuicirrus, Schistotaenia*, 230
*tenuicollis, Caryophyllaeus*, 33
*tenuicollis, Dicranotaenia*, 324
*tenuicollis, Hymenolepis (Weinlandia)*, 324
*tenuicollis, Taenia*, 227
*tenuicollis, Tetrarhynchus*, 78
*tenuiformis, Raillietina, (P.)*, 255
*tenuirostris, Anatinella*, 312
*tenuirostris, Cysticercus*, 312
*tenuirostris, Dicranotaenia*, 312
*tenuirostris, Drepanidotaenia*, 312
*tenuirostris, Hyemnolepis*, 312
*tenuirostris, Hymenolepis (Drepanidotaenia)*, 312
*tenuirostris, Microsomacanthus*, 312
*tenuirostris, Monosaccanthes*, 312
*tenuirostris, Taenia*, 312
*tenuis, Bancroftiella*, 399
*tenuis, Cotugnia*, 247
*tenuis, Echinocotyle*, 286
*tenuis, Lacistorhynchus*, 65
*tenuis, Mesocestoides*, 203
*tenuis nodus instructa, Taenia*, 298
*tenuis, Nybelinia*, 74
*tenuis, Tetrarhynchus*, 65, 74
*tenuispine, Prochristianella*, 59
*terebrans, Caryophyllaeus*, 36
*terebrans, Glaridacris*, 36
*terebrans, Monobothrium*, 36
*teres, Hymenolepis*, 373
*teres, Lateriporus*, 373
*teres, Taenia*, 373
*teresoides, Drepanidotaenia (D.)*, 331
*teresoides, Hamatolepis*, 331
*teresoides, Hymenolepis*, 331
*teresoides, Microsomacanthus*, 331
*teresoides, Weinlandia*, 331
*tereticolle, Rhynchobothrius*, 77
*terezae, Acanthobothrium*, 140
*tergestinus, Sphyriocephalus*, 72
*terraereginae, Hymenolepis*, 309
*terraereginae, Sobolevicanthus*, 309
*testudo, Ophiotaenia*, 185
*Testudotaenia*, 182
*tetoni, Wyominia*, 417
Tetrabothriidae, 204
    diagnosis, 199
    key to genera, 204
*tetrabothrioides, Arctotaenia*, 369
*tetrabothrioides, Taenia*, 369
*Tetrabothrium*, 207
*tetrabothrium, Gilquinia*, 59
*tetrabothrium, Rhynchobothrium*, 59
*Tetrabothrius*, diagnosis, 207
*Tetracampos*, 96
*tetracis, Hispaniolepis*, 306
*tetracis, Hymenolepis*, 306
*tetracis, Weinlandia*, 307
*Tetracisdicotyla*, 413
*tetraglobus, Orygmatobothrium*, 146
*tetraglobus, Pithophorus*, 146
*tetragona, Davainea*, 256
*tetragona, Kotlania*, 257
*tetragona, Raillietina (R.)*, 256
*tetragona, Raillietina (Ransomia)*, 257
*tetragona var. cohni, Raillietina (R.)*, 264
*tetragona, Taenia*, 256
*tetragonocephala, Mathevotaenia*, 463
*tetragonocephala, Taenia*, 463
*Tetragonicephalum*, diagnosis, 124
*tetragonoides, Kotlania*, 264
*tetragonoides, Raillietina (R.)*, 264
*tetragonoides, Raillietina (Ransomia)*, 264
*Tetragonoporus*, 88
*tetragonus, Bothriocephalus*, 103
*Tetrantaris*, 62

tetraoensis, Davainea, 248
Tetraonetta, 253
tetraonis, Hymenolepis, 272
Tetraphyllidea, 14, 131
  key to families, 131
  species of doubtful or uncertain status, 163
tetrapterus, Diphyllobothrium, 92
tetrapterus, Diplogonoporus, 90
Tetrarhynchobothriidae, 55
Tetrarhynchobothrius, 56
tetrastes, Choanotaenia, 364
tetrorchis, Gryporhynchus, 371, 379
texomensis, Bothriocephalus, 103
thalassius, Eutetrarhynchus, 57
Thalophyllaeus, diagnosis, 23
Thaparea, diagnosis, 391
thapari, Oochoristica, 455
thapari, Polypocephalus, 122
thatcheri, Cathetocephalus, 132
theileri, Diphyllobothrium, 95
theileri, Oochoristica, 455
theileri, Ophiotaenia, 185
thomasi, Monoecocestus, 427
thomasi, Schistocephalus, 88
Thomasitaenia, diagnosis, 217
thomomyis, Hymenandrya, 333
thompsoni, Corallobothrium, 179
thompsoni, Corallobothrium (Megathylacoides), 179
thompsoni, Megathylacoides, 179
thompsoni, Pseudangularia, 394
thorsoni, Acanthobothroides, 135
thraciensis, Choanotaenia, 364
threlkeldi, Paranoplocephala, 437
threlkeldi, Perutaenia, 437
thridax, Phyllobothrium, 162
thylogale, Calostaurus, 252
thylogale, Progamotaenia, 445
thymalli, Ichthyotaenia, 188
thymalli, Proteocephalus, 188
thynni, Tetrarhynchus, 78
thyrsitae, Gymnorhynchus, 62
thyrsitae, Molicola, 62
Thysaniezia, diagnosis, 420
Thysanobothrium, 121
Thysanocephalum, diagnosis, 132
thysanocephalum, Phyllobothrium, 133
thysanocephalum, Thysanocephalum, 133
Thysanosoma, diagnosis, 416
Thysanosomatinae, key to genera, 416
Thysanotaenia, diagnosis, 463
Tiarabothrium, 150
tiara, Staphylocystis, 302
tibetana, Schizorchis, 440
tichodroma, Hymenolepis, 337
tidswelli, Acanthotaenia, 181
tidswelli, Crepidobothrium, 181
tidswelli, Ichthyotaenia (Acanthotaenia), 181
tidswelli, Kapsulotaenia, 181
tidswelli, Proteocephalus, 181
tilori, Diorchis, 282
timuri, Monopylidium, 367

tinamoui, Aploparaksis, 277
tincae, Ligula, 87
tinguiana, Raillietina (P.), 255
tintinnabulus, Bothriocephalus, 103
tobijei, Echeneibothrium, 153
tobijei, Caulobothrium, 153
todari, Dibothriorhynchus, 73
togata, Schistometra, 240
tokyoensis, Raillietina (R.), 264
tomica, Rhabdometra, 346
tonkinensis, Hymenolepis, 292
tonkinensis, Staphylepis, 292
toratugumi, Bancroftiella, 399
tordae, Anomotaenia, 408
tordae, Taenia, 408
torpedinis-ocellatae, Tetrarhynchus, 78
torquata, Houttuynia, 264
torquata, Kotlania, 264
torquata, Raillietina (R.), 264
torquata, var. rajae, Raillietina (R.), 264
torresi, Monoecocestus, 427
Tortocephalus, 87
tortum, Acompsocephalum, 104
tortum, Anatrum, 104
tortum, Anthobothrium, 92
tortum, Dibothrium, 104
torulosus, Proteocephalus, 188
torulosus, Taenia, 188
torulosus, Tetrabothrius, 209
totani, Kowalewskiella, 360
totaniochropodis, Similuncinus, 359
toxometra, Hymenolepis, 302
toxometra, Oligorchis, 333
toxometra, Staphylocystis, 302
toyohashiensis, Raillietina (R.), 264
Trachelocampylus, 223
trachyphonoides, Ethiopotaenia, 384
trachypteri, Bothriocephalus, 103
trachypteriiris, Bothriocephalus, 103
trachypteriliopteri, Bothriocephalus, 103
trachysauri, Oochoristica, 455
trachysauri, Taenia, 455
tragopani, Davainea, 255
tragopani, Raillietina (P.), 255
transcaucasica, Anoncotaenia, 341
transcaucasica, Orthoskrjabinia, 341
transfuga, Ascodilepis, 381
transfuga, Dilepis, 381
transfuga, Taenia, 381
translucens, Kystocephalus, 127
translucens, Lecanicephalum, 127
translucens, Tylocephalum, 127
translucida, Moniezia, 442
transvaalensis, Cotugnia, 247
transvaalensis, Fuhrmannella, 443
transversaria, Anoplocephala, 437
transversaria, Anoplocephaloides, 437
transversaria, Paranoplocephala, 437
transversaria, Taenia, 437
trapezoides, Amoebotaenia, 388
trapezoides, Anomotaenia, 411

*trapezoides, Biuterina*, 349
*trapezoides, Davainea*, 264
*trapezoides, Kotlania*, 264
*trapezoides, Pseudanomotaenia*, 411
*trapezoides, Raillietina (R.)*, 264
*trapezoides, Raillietina (Ransomia)*, 264
*travassosi, Bothriocephalus*, 103
*travassosi, Idogenes*, 240
*travassosi, Oochoristica*, 455
*travassosi, Pomatotrygonocestus*, 142
Triaenophoridae, diagnosis, 83
*triacis, Acanthobothrium*, 140
*triacis, Phyllobothrium*, 162
Triaenophoridae, key to genera, 112
*Triaenophorus*, diagnosis, 112
*Triaenorhina*, 347
*triangula, Biuterina*, 349
*triangulare, Prosthecocotyle*, 205
*triangulare, Tetrabothrius*, 205
*triangularis, Moniezia*, 442
*triangularis, Strobilocephalus*, 205
*triangularis, Tetrabothrius*, 205
*triangulus, Taenia*, 349
*trichiuri, Echinorhynchus*, 63
*Trichocephaloidis*, diagnosis, 369
*trichocephalus, Dilepis*, 382
*trichocirrosa, Unciunia*, 386
*trichoglossi, Moniezia*, 423
*trichoglossi, Paronia*, 423
*trichoglossi, Taenia*, 423
*trichorhynchus, Dicranotaenia*, 331
*trichorhynchus, Hymenolepis*, 331
*trichorhynchus, Microsomacanthus*, 331
*trichosoma, Hymenolepis*, 337
*trichosuri, Bertiella*, 439
*trichuri, Hepatoxylon*, 63
*Tricuspidaria*, 112
*tricuspidata intestinalis, Taenia*, 112
*tricuspidata, Taenia*, 112
*tricuspis, Taenia*, 112
*tridontophora, Hymenolepis*, 300
*tridontophora, Triodontolepis*, 300
*trifidum, Echeneibothrium*, 152
*trifolium, Hymenolepis*, 331
*trifolium, Microsomacanthus*, 331
*trifolium, Weinlandia*, 331
*triganciensis, Choanotaenia*, 363
*triglae, Callitetrarhynchus*, 54
*triglae-hirudinis, Tetrarhynchus*, 78
*triglae-lepidotae, Tetrarhynchus*, 78
*triglae, Tetrarhynchus*, 54
*trigonacantha, Biuterina*, 349
*trigonocephala, Anomotaenia*, 411
*trigonocephala, Pseudanomotaenia*, 411
*trigonocephala, Taenia*, 411
*Trigonocotyle*, diagnosis, 206
*Trigonolobum*, 75
*trigonophora, Moniezia*, 441, 442
*Trilocularia*, diagnosis, 144
Triloculariidae
  diagnosis, 131

  key to genera, 144
*triloculatum, Phoreiobothrium*, 141
*trimeresuri, Ophiotaenia*, 185
*trinchesii, Diplopylidium*, 354
*trinchesii, Dipylidium*, 354
*tringae, Choanotaenia*, 364
*tringae, Dilepis*, 382
*tringae, Diorchis*, 279
*tringae, Schillerius*, 279
*trinitatae, Raillietina (R.)*, 264
*trinitatis, Diphyllobothrium*, 95
*Triodontolepis*, diagnosis, 300
*trionychinum, Proteocephalus*, 188
*Triorchis*, 288
*tripartita, Socicina*, 287
*tripartitum, Acanthobothrium*, 140
*triplacantha, Pseudangularia*, 394
*Triplotaenia*, diagnosis, 415
Triplotaeniinae, 415
*tripunctata, Taenia*, 303
*triseriale, Diplopylidium*, 354
*triseriale, Dipylidium*, 354
*triserrata, Taenia*, 227
*trisignatus, Caryophyllaeus*, 33
*tritesticulata, Echinorhynchotaenia*, 272
*tritesticulata, Hymenolepis*, 312
*tritesticulatus, Microsomacanthus*, 312
*Triuterina*, diagnosis, 424
*trivialis, Anomotaenia*, 390
*trochili, Anoncotaenia*, 341
*troeschi, Anoplocephaloides*, 437
*troeschi, Paranoplocephala*, 437
*Troglodytilepis*, 324
*troglodytis, Troglodytilepis*, 326
*troglodytis, Variolepis*, 326
*trombidacantha, Hymenolepis*, 337
*truncata, Oochoristica*, 455
*truncata, Taenia*, 455
*truncatus, Caryophyllaeus*, 33
*truncatus, Cyathocephalus*, 44
*trychopeus, Sureshia*, 402
*trygonbrucco, Christianella*, 56
*trygon-brucco, Tetrarhynchus*, 56
*trygonis, Echeneibothrium*, 152
*trygonis, Eniochobothrium*, 124
*trygonis, Flapocephalus*, 124
*trygonis, Prosthecobothrium*, 134
*trygonis, Tetragonicephalum*, 125
*trygonis, Tylocephalum*, 125
*trygonis, Uncibilocularis*, 134
*trygonis-brucconis, Tetrarhynchus*, 76
*trygonicola, Prochristianella*, 59
*trygonis, Parachristianella*, 57
*trygonis-brucconis, Tetrarhynchus*, 78
*trygonis-pastinacae, Tetrarhynchus*, 79
*tryonomysi, Raillietina*, 266
Trypanorhyncha, 47
  diagnosis, 11
  genera of uncertain status, 75
  key to suborders, 48
*Tschertkovilepis*, 327

*tsengi, Dicranotaenia*, 324
*tsengi, Monopylidium*, 367
*tsengi, Wardium*, 324
*tshanensis, Diorchis*, 279
*tsingtaoense, Acanthobothrium*, 140
*tsuzurasensis, Hymenolepis*, 335
*tuba, Caryophyllaeus*, 32
*Tubanguiella*, diagnosis, 395
*tuberculata, Oochoristica*, 453
*tuberculata, Taenia*, 232, 453
*tuberculosus, Coenurus*, 226
*tubiceps, Bothriocephalus*, 73
*tubicirrosa, Armadoskrjabinia*, 297
*tubicirrosa, Hymenolepis*, 297
*tugarinovi, Anomotaenia*, 408
*tugarinovi, Choanotaenia*, 408
*tuliensis, Cotugnia*, 247
*tumens, Paradicranotaenia*, 272
*tumidocollus, Proteocephalus*, 188
*tumidula, Callitetrarhynchus*, 59
*tumidula, Prochristianella*, 59
*tumidula, Rhynchobothrium*, 59
*tumidulum, Echeneibothrium*, 152
*tumidum, Phyllobothrium*, 162
*tundra, Anomotaenia*, 408
*tundra, Dichoanotaenia*, 408
*tunetensis, Kotlania*, 264
*tunetensis, Raillietina (R.)*, 264
*tungussicum, Diphyllobothrium*, 95
*Tupaiataenia*, diagnosis, 458
*turaci, Raillietina (R.)*, 264
*turdi, Aploparaksis*, 277
*turdi, Choanotaenia*, 398
*turdi, Dilepis*, 380, 382
*turdi, Paricterotaenia*, 398
*turdi, Polycercus*, 398
*turdina, Ophryocotyle*, 245
*turdorum, Taenia*, 298
*turkestanicus, Diorchis*, 282
*turkmenicus, Multiceps*, 227
*turkmenicus, Taenia*, 227
*turnicis, Spinilepis*, 409
*turturis, Taenia*, 308
*tushigi, Australiolepis*, 292
*tuvensis, Diorchis*, 282
*tuvensis, Idiogenes*, 240
*tuvensis, Microsomacanthus*, 331
*tva, Corallobothrium (Megathylacoides)*, 178
*tva, Megathylacoides*, 178
*twitchelli, Multiceps*, 227
*twitchelli, Taenia*, 227
*Tylocephalum*, 126
*typhlotritonis, Bothriocephalus*, 103
*typica, Amphicotyle*, 106
*typicum, Acanthobothrium*, 140
*typicum, Calycobothrium*, 122
*typicum, Cyclobothrium*, 122
*typicus, Cylindrophorus*, 143, 177
*typus, Echinobothrium*, 165

## U

*uarnakense, Thysanobothrium*, 121
*uelcal, Aploparaksis*, 278
*ugandae, Biuterina*, 349
*ukrainensis, Echinocotyle*, 286
*uliginosa, Hymenolepis*, 337
*ulmeri, Monobothrium*, 32
*umbrella, Chaetophallus*, 207
*Uncibilocularis*, diagnosis, 133
*uncinata, Choanotaenia*, 364
*uncinata, Hymenolepis*, 325
*uncinata, Paricterotaenia*, 364
*uncinatum, Acanthobothrium*, 140
*uncinatus, Floriceps*, 62
*uncinatus, Molicola*, 62
*uncinatus, Rhynchobothrium*, 62
*uncinispinosa, Hymenolepis*, 305
*uncinispinosa, Rodentolepis*, 306
*uncinispinosa, Vampirolepis*, 305
*uncinum, Fissurobothrium*, 110
*Unciunia*, diagnosis, 385
*undosa, Triplotaenia*, 416
*undula, Dilepis*, 356, 380
*undula, Taenia*, 380
*undulata, Bertiella*, 439
*undulata, Davainea*, 264
*undulata, Dilepis*, 380
*undulata, Drepanidotaenia*, 380
*undulata, Hymenolepis*, 380
*undulata, Kotlania*, 264
*undulata, Prototaenia*, 439
*undulata, Raillietina (R.)*, 264
*undulata, Raillietina (Ransomia)*, 264
*undulatoides, Anomotaenia*, 411
*undulatoides, Pseudanomotaenia*, 411
*unicinata, Sphyroncotaenia*, 238
*unicoronata, Anomotaenia*, 364
*unicoronata, Choanofuhrmannia*, 364
*unicoronata, Choanotaenia*, 364
*unicoronata, Icterotaenia*, 364
*unicoronata, Monopylidium*, 364
*unicoronata, Paricterotaenia*, 364
*unilaterale, Acanthobothrium*, 140
*unilaterale, Phyllobothrium*, 162
*unilateralis, Chapmania*, 241, 243
*unilateralis, Dilepis*, 377, 381, 382
*unilateralis, Taenia*, 382
*unionifactor, Lecanicephalum*, 127
*unionifactor, Tylocephalum*, 127
*unipapillatum, Dieffluvium*, 31
*uniserialis, Ligula*, 87
*uniuterina, Davainea*, 242
*uniuterina, Ophriocotyloides*, 242
*upsilon, Hymenolepis*, 327
*upupae, Choanotaenia*, 364
*upupai, Biuterinoides*, 339
*upupai, Neyraia*, 339
*upuparum, Hymenolepis*, 337
*uragahaensis, Dicranotaenia*, 324
*uragahaensis, Hymenolepis*, 324

*uralensis, Echinocotyle*, 286
*uralensis, Fuhrmanniella*, 286
*uralensis, Hymenolepis (Echinocotyle)*, 286
*uranomidis, Hymenolepis*, 335
*uranoscope-scabri, Tetrarhynchus*, 79
*uranoscopi, Symbothriorhynchus*, 75
*urbica, Angularella*, 393
*urceus, Hymenolepis*, 376
*urceus, Paradilepis*, 376
*urceus, Taenia*, 376
*urichi, Diphyllobothrium*, 95
*urichi, Spirometra*, 95
*urna, Amphiptyches*, 473
*urna, Gyrocotyle*, 473
*urna* var. *magnispinosa, Gyrocotyle*, 473
*urna* var. *parvispinosa, Gyrocotyle*, 473
*urobatidium, Echeneibothrium*, 152
*Urocystidium*, 223
*urogalli, Davainea*, 255
*urogalli, Meggittia*, 255
*urogalli, Raillietina (P.)*, 255
*urogalli, Taenia*, 255
*urogalli, Tetraonetta*, 255
Urogonoporidae, 131
*urogymni, Prosthecobothrium*, 138
*urolophi, Acanthobothrium*, 140
*urotrygoni, Acanthobothrium*, 140
*ursi, Diphyllobothrium*, 95
*ursimaritimi, Taenia*, 227
*ursini, Taenia*, 227
*uteriloba, Triuterina*, 424
*urticulifera, Monoderidum*, 203
*utricularis, Taenia*, 226
*utriculenta, Hydatigena*, 226
*uzbekiensis, Biuterina*, 349

# V

*vaccarii, Polipobothrium*, 158
*vacuolata, Oochoristica*, 455
*Vadifresia*, 252, 253
*vaganda, Davainea*, 266
*vaganda, Raillietina*, 266
*vagans, Phyllobothrium*, 162
*vaginata, Avocettolepis*, 315
*vaginata, Hymenolepis*, 315
*vaginata, Hymenosphenacanthus*, 315
*vaginata, Taenia*, 231
*vaginata, Weinlandia*, 315
*vaginatus, Acoleus*, 231
*vaginatus, Sphenacanthus*, 315
*valdiviae, Tetrabothrius*, 208
*Valipora*, diagnosis, 371
*vallei, Bothriocephalus*, 103
*vallei, Hymenolepis*, 311
*vallei, Nadejdolepis*, 311
*vallei, Taenia*, 311
*Vampirolepidoides*, 302
*Vampirolepis*, diagnosis, 302
*vancouverensis, Dyandrya*, 444

*vanderbrandeni, Raillietina (F.)*, 266
*vanelli, Amoebotaenia*, 388
*vanzolinii, Oochoristica*, 455
*varani, Oochoristica*, 455
*varani, Panceriella*, 452
*varani, Taenia*, 452
*varanii, Pancerina*, 452
*varesina, Taenia*, 289
*varia, Acanthotaenia*, 181
*varia, Crepidobothrium*, 181
*varia, Ichthyotaenia*, 181
*varia, Kapsulotaenia*, 181
*variabile, Anthobothrium*, 158
*variabile, Echeneibothrium*, 152
*variabilis angusta, Cittotaenia*, 451
*variabilis, Anomotaenia*, 408
*variabilis, Anoplocephala*, 437
*variabilis, Anoplocephaloides*, 437
*variabilis, Bothriocephalus*, 90, 94
*variabilis, Cephalobothrium*, 123
*variabilis, Choanotaenia*, 408
*variabilis, Cittotaenia*, 451
*variabilis, Dicranotaenia*, 326
*variabilis, Hexacanalis*, 123
*variabilis, Hymenolepis*, 326
*variabilis, Mesocestoides*, 203
*variabilis, Moniezia*, 423
*variabilis, Monoecocestus*, 427
*variabilis, Mosgovoyia*, 451
*variabilis, Paranoplocephala*, 437
*variabilis, Paronia*, 423
*variabilis, Proteocephalus*, 188
*variabilis, Proterogynotaenia*, 220
*variabilis, Raillietina (S.)*, 256
*variabilis, Schizotaenia*, 427
*variabilis, Taenia*, 408
*variabilis, Variolepis*, 326
*variacanthos, Hymenolepis*, 376
*variacanthos, Meggittiella*, 376
*variacanthos, Paradilepis*, 376
*variacanthos, Skrjabinolepis*, 376
*varians, Davainea*, 247
*Variolepis*, diagnosis, 324
*variouncinatus, Dasyrhynchus*, 52
*variouncinnatus, Halsiorhynchus*, 52
*varius, Proteocephalus*, 181
*varsoviensis, Dicranotaenia*, 324
*vasi, Taenia*, 441
*Vaullegeardia*, 61
*veitchi, Aploparaksis*, 278
*velamentum, Orygmatobothrium*, 163
*venezuelensis, Rhinebothroides*, 154
*ventosaloculata, Deblocktaenia*, 190
*ventosaloculata, Ophiotaenia*, 190
*ventropapillatum, Diphyllobothrium*, 95
*venusta, Duthiersia*, 86
*venusta, Drepanidotaenia*, 319
*venusta, Hymenolepis*, 319
*venusta, Retinometra*, 319
*venusta, Sphenacanthus*, 319
*venusta, Taenia*, 319

*venustus, Hymenosphenacanthus*, 319
*veravalensis, Anthobothrium*, 158
*veravalensis, Balanobothrium*, 120
*veravalensis, Pedibothrium*, 143
*veravalensis, Platybothrium*, 137
*veravalensis, Uncibilocularis*, 134
*Vermaia*, diagnosis, 174
*vernetae, Echeneibothrium*, 152
*versatile, Orygmatobothrium*, 163
*verschureni, Echinocotyle*, 286
*verschureni, Hymenolepis*, 286
*verticillatum, Calliobothrium*, 135
*verulamii, Pseudanomotaenia*, 411
*vesicularis, Polypocephalus*, 122
*vesicularis pisiformis, Vermis*, 226
*vesiculigera, Anomotaenia*, 351
*vesiculigera, Paruterina*, 351
*vesiculigera, Taenia*, 351
*vestibularis, Vogea*, 370
*vestita, Ascometra*, 237
*vestita, Chapmania*, 237
*victoriata, Variolepis*, 326
*vietnamense, Raillietina (R.)*, 264
*vigintivasus, Davaineoides*, 249
*vigintivasus, Meggittia*, 249
*vigisi, Diorchis*, 282
*Vigisolepis*, diagnosis, 293
*Vigissotaenia*, 327
*viguerasi, Nadejdolepis*, 311
*villosa, Adelataenia*, 445
*villosa, Cittotaenia*, 445
*villosa, Hispaniolepis*, 306
*villosa, Hymenolepis*, 306
*villosa, Progamotaenia*, 445
*villosa, Taenia*, 306
*villosa, Triplotaenia*, 445
*villosoides, Hispaniolepis*, 307
*villosoides, Hymenolepis*, 307
*vilocirrus, Hymenolepis*, 337
*vinagoi, Raillietina (R.)*, 264
*viperis, Solenotaenia*, 185
*viperus, Ophiotaenia*, 185
*viridis, Sphyriocephalus*, 72
*viridis, Tetrarhynchus*, 72
*virilis, Hymenolepis*, 306
*virilis, Vampirolepis*, 306
*virilis, Wenyonia*, 26
*visakhapatnamensis, Polyoncobothrium*, 97
*visakhapatnamensis, Senga*, 97
*visayana, Diorchis*, 282
*viscaciae, Cittotaenia*, 443
*viscaciae, Mosgovoyia*, 443
*Viscoia*, 366
*vistulae, Hymenolepis*, 337
*vitellaris, Polypocephalus*, 122
*vitellaris, Proteocephalus*, 188
*Vitta*, diagnosis, 392
*vittata, Stilesia*, 418
*vivieni, Raillietina (R.)*, 264
*Vogea*, diagnosis, 370
*vogea, Panuwa*, 367
*vogeae, Hymenolepis*, 289
*vogei, Glaridacris*, 36
*vogeli, Diphyllobothrium*, 88
*vogeli, Echinococcus*, 223
*vogeli, Raillietina (R.)*, 264
*vogti, Anoplocephala*, 442
*vogti, Moniezia*, 442
*vogti, Taenia*, 442
*volvulus, Anomotaenia*, 408
*volvulus, Dictymetra*, 408
*volvulus, Diplochetos*, 408
*volvuta, Hymenolepis*, 327
*volzi, Davainea*, 264
*volzi, Kotlania*, 264
*volzi, Raillietina (R.)*, 263, 264
*volzi, Raillietina (Ransomia)*, 264
*voucheri, Hymenolepis*, 335
*vulgaris, Anoplocephala*, 433
*vulgaris, Plagiotaenia*, 433
*vulpeculae, Pithophorus*, 146
*vulpina, Taenia*, 203
*vulturi, Cladotaenia*, 413

# W

*wageneri, Oochoristica*, 463
*wageneri, Echinophallus*, 105
*wageneri, Monobothrium*, 32
*wageneri, Oncomegas*, 58
*wageneri, Rhynchobothrium*, 58
*Wageneria*, 76
*walga, Echeneibothrium*, 155
*Wallabicestus*, diagnosis, 447
*wallacei, Oochoristica*, 457
*wallacei, Oschmarenia*, 457
*wallago, Gangesia*, 176
*wardi, Spartoides*, 40
*wardii, Tetrarhynchus*, 79
*Wardium*, 320
*wardlei, Hymenolepis*, 337
*watsoni, Chimaerolepis*, 295
*watsoni, Drepanidotaenia*, 295
*websteri, Parafimbriaria*, 292
*wedli, Acanthobothrium*, 140
*Weinlandia*, 320, 327
*weissi, Kotlania*, 264
*weissi, Raillietina (R.)*, 264
*weissi, Raillietina (Ransomia)*, 264
*weissi, Roytmania*, 264
*weissi* var. *valiclusa, Raillietina (R.)*, 264
*Wenyonia*, diagnosis, 26
*werneri, Davainea*, 264
*werneri, Kotlania*, 264
*werneri, Raillietina (R.)*, 264
*werneri, Raillietina (Ransomia)*, 264
*wettsteini, Schistometra*, 240
*whitentoni, Oochoristica*, 455
*wickliffi, Proteocephalus*, 188
*wigginsi, Anoplocephaloides*, 437
*wigginsi, Diorchis*, 282

*wigginsi, Paranoplocephala*, 437
*williamsi, Echeneibothrium*, 152
*williamsi, Phyllobothrium*, 162
*williamsi, Raillietina (R.)*, 265
*wilsoni, Diphyllobothrium*, 95
*wilsoni, Oriana*, 207
*wilsoni, Tetrabothrius*, 207
*wimerosa, Andrya*, 437
*wimerosa, Anoplocephala*, 437
*wimerosa, Anoplocephaloides*, 437
*wimerosa, Paranoplocephala*, 437
*wimerosa, Taenia*, 437
*wisconsinensis, Isoglaridacris*, 37
*wisconsinensis, Pliovitellaria*, 34
*wislockii, Vampirolepis*, 306
*wisniewskii, Sobolevicanthus*, 309
*witenbergi, Sinaiotaenia*, 460
*woodlandi, Acanthotaenia*, 182
*woodlandi, Avitellina*, 418
*woodlandi, Malika*, 357
*woodlandi, Monobothrioides*, 24
*woodlandi, Proteocephalus*, 182
*woodlandi, Rostellotaenia*, 182
*Woodlandia*, diagnosis, 290
*Woodlandiella*, 195
*woodsholei, Acanthobothrium*, 140
*woodsholei, Dubininolepis*, 326
*woodsholei, Hymenolepis*, 326
*woodsholei, Variolepis*, 326
*wrighti, Tetrabothrius*, 209
*wyatti, Orygmatobothrium*, 163
*Wyominia*, diagnosis, 417

# X

*xanthocephalum, Ceratobothrium*, 133
*xenopi, Dibothriocephalus*, 84
*xenopi, Pseudocephalochlamys*, 84
*xiphiados, Pseudeubothrium*, 108
*xiphiae, Dibothriorhynchus*, 76

# Y

*yamagutii, Nybelinia*, 74
*yamagutii, Parvitaenia*, 412
*yamagutii, Pithophorus*, 146
*yamashitai, Schizorchis*, 440
*yamasigi, Amoebotaenia*, 388
*Yogeshwaria*, 164
*Yogeshwaria*, diagnosis, 400
*yogeshwarii, Lapwingia*, 404
*yonagoensis, Diphyllobothrium*, 95
*yorkei, Dilepis*, 376

*yorkei, Lecanicephalum*, 127
*yorkei, Oligorchis*, 376
*yorkei, Paradilepis*, 376
*yorkei, Tylocephalum*, 127
*Yorkeria*, diagnosis, 134
*yoshidai, Insectivorolepis*, 288
*yoshidai, Kowalewskius*, 330
*yoshiyukiae, Vampirolepis*, 306
*youdeoweii, Wenyonia*, 27

# Z

*zacharovae, Mesocestoides*, 204
*zambiensis, Biuterina*, 349
*zambiensis, Paruterina*, 349
*zanthopygiae, Anoncotaenia*, 341
*zanzibarensis, Inermicapsifer*, 466
*zapterycum, Acanthobothrium*, 140
*Zarnowskiella*, 301
*zavattarii, Paronia*, 423
*zebrae, Anoplocephala*, 433
*zebrae, Taenia*, 433
*zederi, Anomotaenia*, 399
*zederi, Parorchites*, 399
*zederi, Taenia*, 399
*zemae, Aploparaksis*, 278
*zeylanica, Burtiella*, 245
*zeylanica, Malika*, 357
*zeylanica, Ophryocotyle*, 245
*zmorayi, Dicranotaenia*, 324
*zmorayi, Wardium*, 324
*zonifera, Paricterotaenia*, 398
*zonifera, Polycercus*, 398
*zonuri, Oochoristica*, 455
*zosteropis, Hymenolepis*, 299
*zosteropis, Passerilepis*, 299
*zosteropis, Weinlandia*, 299
*Zosteropicola*, diagnosis, 340
*Zschokkea*, 457
*Zschokkeella*, 457
*zschokkei, Acanthobothrium*, 140
*zschokkei, Adelataenia*, 445
*zschokkei, Bothriocephalus*, 88
*zschokkei, Cittotaenia*, 445
*zschokkei, Diplopylidium*, 354
*zschokkei, Dipylidium*, 354
*zschokkei, Ophiotaenia*, 185
*zschokkei, Orygmatobothrium*, 163
*zschokkei, Progamotaenia*, 445
*zschokkei, Progynopylidium*, 354
*zwerneri, Glyphobothrium*, 149
Zygobothriinae, key to genera, 191
*Zygobothrium*, diagnosis, 192
*Zyxibothrium*, diagnosis, 144